AF251049

DICTIONARY OF ORGANIC COMPOUNDS

Formula Index

DICTIONARY OF ORGANIC COMPOUNDS

The constitution and physical, chemical and other properties of the principal carbon compounds and their derivatives, together with relevant literature references

FOURTH EDITION

FORMULA INDEX

for the Dictionary and Fifth Cumulative Supplement

NEW YORK
OXFORD UNIVERSITY PRESS
1971

Fourth Edition of the
Dictionary of Organic Compounds
in five volumes published 1965
The first supplement to the Fourth Edition
published simultaneously
The second supplement published 1966
The third supplement published 1967
The fourth supplement published 1968
The fifth (cumulative) supplement published 1969
This Formula Index published 1971

Printed in Great Britain by Richard Clay (The Chaucer Press) Ltd
Bungay, Suffolk
SBN 413 60700 3

PREFACE

This Formula Index covers those compounds which are given molecular formulae in the Main Work and the Fifth Supplement; i.e. the compounds which form the subject of an entry and certain derivatives. The derivatives usually included are ethers and *N*-alkyl derivatives and esters, acid chlorides, anhydrides, amides, and nitriles of carboxylic acids. In general, acyl derivatives, acetates, and benzoates are not given molecular formulae in the Dictionary and, accordingly will not be found in this Index. Compounds of indeterminate composition have been omitted from this Index except where the formula has been limited to two alternatives. In such cases the compound is listed under both formulae.

The elements are listed in alphabetical order after carbon and hydrogen (C and H) in accord with current practice as followed in the Fifth and subsequent Supplements. It should be noted that in the molecular formulae given in the Main Work the elements are given in the order C, H, O, N, Cl, Br, F, S, P with any others following in alphabetical order. In this Index they have been put into alphabetical order after C and H.

Compounds mentioned only in the Fifth Supplement are denoted by a dagger (†) while unannotated entries refer to the Main Work. An asterisk (★) is also used to draw attention to the Main Work. For example, when an entry bears both an asterisk and a dagger it usually means that the molecular formula given in the Main Work has been revised or corrected in the Supplement. These devices refer only to molecular formulae. Other additional information with regard to unannotated compounds may be found in the Fifth Supplement.

The secretarial work involved in compiling this Index has been carried out by Mrs E. J. Stevens, to whom our thanks are due.

The Sixth and subsequent Supplements will carry their own Formula Indices; the Editor is Miss J. B. Thomson, Ph.D., Department of Chemistry, The University, St Andrews, Scotland.

C_1

$CAgNO$
Fulminic Acid, *Silver fulminate*
$CBrClO$
Carbonyl chlorobromide
$CBrCl_3$
Bromotrichloromethane
$CBrN$
Cyanogen bromide
CBr_2Cl_2
Dibromodichloromethane
CBr_2O
Carbonyl bromide
CBr_3NO_2
Bromopicrin
CBr_4
Carbon tetrabromide
$CCdNO$
Fulminic Acid, *Cadmium fulminate*
$CClN$
Cyanogen chloride
$CClNO_3S$
Chlorosulphonyl isocyanate†
CCl_2F_2
Dichlorodifluoromethane
CCl_2I_2
Dichlorodi-iodomethane
$CCl_2N_2O_4$
Dichlorodinitromethane
CCl_2O
Carbonyl chloride
CCl_2S
Thiocarbonyl chloride
CCl_3NO_2
Chloropicrin
CCl_4
Carbon tetrachloride
CCl_4S
Trichloromethanesulphenyl chloride
$CCuNO$
Fulminic Acid, *Copper fulminate*
CF_2O
Carbonyl fluoride

B

CF_3NO_2
Trifluoronitromethane
CF_4
Carbon tetrafluoride
$CHBrCl_2$
Bromodichloromethane
$CHBr_2I$
Dibromoiodomethane
$CHBr_2NO_2$
Dibromonitromethane
$CHBr_3$
Bromoform
$CHClO$
Formyl chloride†
$CHClO_2$
Chloroformic Acid
$CHCl_2I$
Dichloroiodomethane
$CHCl_3$
Chloroform
CHF_3
Fluoroform
CHI_3
Iodoform
CHN
Hydrocyanic Acid
$CHNO$
Cyanic Acid★†
Fulminic Acid
$CHNO_3$
Nitroformaldehyde
$CHNS$
Thiocyanic Acid
CHN_3O_2
Nitrodiazomethane†
CHN_3O_6
Trinitromethane
CH_2BrCl
Bromochloromethane
CH_2BrF
Bromofluoromethane
CH_2BrI
Bromoiodomethane
CH_2BrNO_2
Bromonitromethane
CH_2Br_2
Methylene bromide

CH_2ClNO
Carbamic Acid, *Chloride*
Chloroformaldoxime
CH_2ClNO_2
Chloronitromethane
CH_2Cl_2
Methylene chloride
$CH_2Cl_2O_4S_2$
Methionic Acid, *Dichloride*
CH_2Cl_3PS
Chloromethylphosphonothioic dichloride†
CH_2I_2
Methylene iodide
CH_2N_2
Cyanamide
Diazomethane
Diazirine†
$CH_2N_2O_3$
Methylnitrolic Acid
$CH_2N_2O_4$
Dinitromethane
N-Nitrocarbamic Acid
CH_2N_4
1,2,3,4-Tetrazole
CH_2O
Formaldehyde
CH_2OS
Thioformic Acid
CH_2OS_2
Dithiocarbonic Acid
CH_2O_2
Formic Acid
CH_2O_2S
Thiocarbonic Acid
CH_2O_3
Performic Acid
CH_2O_4S
Methylene sulphate
CH_2S_5
Pentathiane†
CH_3AsCl_2
Dichloromethylarsine
CH_3AsI_2
Methyldi-iodoarsine
CH_3AsO
Methylarsenious oxide
CH_3Br
Methyl bromide
CH_3BrHg
Mercuri-methyl bromide
CH_3Cl
Methyl chloride
CH_3ClHg
Mercuri-methyl chloride
CH_3ClO
Methyl hypochlorite
CH_3ClOS
Methanesulphinic Acid, *Chloride*†
CH_3ClO_2S
Methane-sulphonic Acid, *Chloride*
CH_3Cl_2N
N-Dichloromethylamine
CH_3Cl_2OP
Methylphosphonic Acid, *Dichloride*
CH_3Cl_3S
Methyl sulphur trichloride
CH_3Cu
Methyl copper
CH_3F
Methyl fluoride
CH_3HgI
Mercuri-methyl iodide
CH_3I
Methyl iodide
CH_3NO
Formaldehyde, *Oxime*
Formamide
Formimidic Acid
CH_3NOS
Thiocarbamic Acid
CH_3NO_2
Carbamic Acid
Formhydroxamic Acid
Methyl nitrite
Nitromethane
CH_3NO_3
Methyl nitrate
CH_3NS_2
Dithiocarbamic Acid
CH_3N_3
Methyl azide
$CH_3N_3O_3$
Nitrourea
CH_3Na
Sodium methyl
CH_4
Methane
CH_4BO_2
Methylboric Acid
CH_4ClP
Chloromethylphosphine†
CH_4HgO
Mercuri-methyl hydroxide
CH_4NO_5P
Carbamoyl dihydrogen phosphate
CH_4N_2
Formamidine
CH_4N_2O
Urea
$CH_4N_2O_2$
Hydrazinoformic Acid
Hydroxyurea
Methylnitramine

$CH_4N_2O_3$
Dihydroxyurea†
CH_4N_2S
Thiourea
$CH_4N_2S_2$
Dithiocarbazic Acid
CH_4N_2Se
Selenourea
$CH_4N_3O_3P$
Phosphorazidic Acid, *Mono-Me ester*†
$CH_4N_4O_2$
Nitroguanidine
CH_4O
Methanol
CH_4O_2S
Methanesulphinic Acid†
CH_4O_3
Orthoformic Acid
CH_4O_3S
Methane-sulphonic Acid
CH_4O_4
Orthocarbonic Acid
CH_4O_4S
Methyl hydrogen sulphate
$CH_4O_6S_2$
Methionic Acid
$CH_4O_9S_3$
Methane-trisulphonic Acid
CH_4S
Methanethiol
CH_5As
Methylarsine
CH_5N
Methylamine
CH_5NO
N-Methylhydroxylamine
O-Methylhydroxylamine
CH_5NO_2S
Methane-sulphonic Acid, *Amide*
CH_5NO_3S
Aminomethyl hydrogen sulphite
CH_5N_3
Guanidine
CH_5N_3O
Semicarbazide
CH_5N_3S
Thiosemicarbazide
CH_5O_3P
Methylphosphonic Acid
CH_5P
Methylphosphine
CH_6N_2
Methylhydrazine
$CH_6N_2O_4S_2$
Methionic Acid, *Diamide*
CH_6N_4
Aminoguanidine
CH_6N_4S
Thiocarbazide
CH_7N_5
Diaminoguanidine
CH_8Si_2
Bis-silymethane
1,3-Disilapropane
CIN
Cyanogen iodide
CI_4
Carbon tetraiodide
CKNO
Fulminic Acid, *Potassium fulminate*
CNNaO
Fulminic Acid, *Sodium fulminate*
$(CNO)_x$
Oxycyanogen
CNOTl
Fulminic Acid, *Thallium fulminate*
CN_4
Cyanogen Azide†
CN_4O_8
Tetranitromethane
CO
Carbon monoxide
COS
Carbonyl sulphide
CO_2
Carbon dioxide
CS_2
Carbon disulphide

C_2

C_2BrCl_3O
Trichloroacetic Acid, *Bromide*
C_2BrF_5
Bromopentafluoroethane
C_2Br_2
Dibromoacetylene
C_2Br_3N
Tribromoacetic Acid, *Nitrile*
C_2Br_4
Tetrabromoethylene
C_2Br_4O
Tribromoacetic Acid, *Bromide*
C_2Br_6
Hexabromoethane
C_2ClF_3
Chlorotrifluoroethylene
C_2ClF_5
Chloropentafluoroethane

C_2Cl_2
Dichloroacetylene
$C_2Cl_2F_2$
1,1-Dichloro-2,2-difluoroethylene
C_2Cl_3F
Trichlorofluoroethylene
C_2Cl_3IO
Trichloroacetic Acid, *Iodide*
C_2Cl_3N
Trichloroacetic Acid, *Nitrile*
C_2Cl_4
Tetrachloroethylene
C_2Cl_4O
Trichloroacetic Acid, *Chloride*
$C_2Cl_4O_2$
Trichloromethyl chloroformate
C_2Cl_6
Hexachloroethane
C_2Cu_2
Acetylene, *Cu deriv.*
C_2F_4
Tetrafluoroethylene
$C_2F_4O_2$
Trifluoromethyl fluoroformate†
C_2F_5I
Pentafluoroiodoethane
C_2F_6
Hexafluoroethane
C_2HBr
Bromoacetylene
$C_2HBrClF_3$
2-Bromo-2-chloro-1,1,1-trifluoroethane†
C_2HBrCl_2
Bromo-1,2-dichloroethylene
Bromo-2,2-dichloroethylene
C_2HBrCl_2O
Bromochloroacetic Acid, *Chloride*
$C_2HBrCl_2O_2$
Bromodichloroacetic Acid
C_2HBr_2ClO
Bromochloroacetic Acid, *Bromide*
$C_2HBr_2ClO_2$
Dibromochloroacetic Acid
C_2HBr_2N
Dibromoacetic Acid, *Nitrile*
C_2HBr_3
Tribromoethylene
C_2HBr_3O
Bromal
Dibromoacetic Acid, *Bromide*
$C_2HBr_3O_2$
Tribromoacetic Acid
C_2HBr_5
Pentabromoethane
C_2HCl
Chloroacetylene
C_2HClFN
Chlorofluoroacetic Acid, *Nitrile*†
C_2HClF_2O
Difluoroacetic Acid, *Chloride*
C_2HCl_2FO
Chlorofluoroacetic Acid, *Chloride*†
Dichloroacetic Acid, *Fluoride*
C_2HCl_2N
Dichloroacetic Acid, *Nitrile*
C_2HCl_3
Trichloroethylene
C_2HCl_3O
Chloral
Dichloroacetic Acid, *Chloride*
$C_2HCl_3O_2$
Trichloroacetic Acid
C_2HCl_5
Pentachloroethane
C_2HF_3O
2,2,2-Trifluoroacetaldehyde
C_2HI
Iodoacetylene
$C_2HI_3O_2$
Tri-iodoacetic Acid
C_2HNOS
Thio-oxalic Acid, *Nitrile*
C_2HNa
Acetylene, *Na deriv.*
C_2HNO_2
Cyanoformic Acid
C_2HN_3
Dicyanamide
Diazoacetic Acid, *Nitrile*
$C_2HN_3O_4$
Nitrodiazoacetic Acid†
C_2H_2
Acetylene
$C_2H_2AsCl_3$
2-Chlorovinyldichloroarsine
C_2H_2BrCl
Acetylene bromochloride
C_2H_2BrClO
Bromoacetic Acid, *Chloride*
Chloroacetyl bromide
$C_2H_2BrClO_2$
Bromochloroacetic Acid
$C_2H_2BrCl_2NO$
Bromodichloroacetic Acid, *Amide*
C_2H_2BrF
Acetylene bromofluoride
1-Bromo-1-fluoroethylene
C_2H_2BrN
Bromoacetic Acid, *Nitrile*
$C_2H_2Br_2$
Acetylene dibromide
1,1-Dibromoethylene

$C_2H_2Br_2Cl_2$
1,2-Dibromo-1,1-dichloroethane
1,2-Dibromo-1,2-dichloroethane

$C_2H_2Br_2O$
Bromoacetic Acid, *Bromide*
Dibromoacetaldehyde

$C_2H_2Br_2O_2$
Dibromoacetic Acid

$C_2H_2Br_3NO$
Tribromoacetic Acid, *Amide*

$C_2H_2Br_4$
1,1,1,2-Tetrabromoethane
1,1,2,2-Tetrabromoethane

C_2H_2ClF
Acetylene chlorofluoride

$C_2H_2ClFO_2$
Chlorofluoroacetic Acid†

C_2H_2ClI
Acetylene chloroiodide

C_2H_2ClIO
Iodoacetic Acid, *Chloride*

$C_2H_2ClIO_2$
Chloroiodoacetic Acid

C_2H_2ClN
Chloroacetonitrile

$C_2H_2ClN_3$
5-Chloro-1,2,4-triazole

$C_2H_2ClN_3O$
Azidoacetic Acid, *Chloride*

$C_2H_2Cl_2$
Acetylene dichloride
1,1-Dichloroethylene

$C_2H_2Cl_2O$
Chloroacetyl chloride
Dichloroacetaldehyde

$C_2H_2Cl_2O_2$
Dichloroacetic Acid

$C_2H_2Cl_3NO$
Trichloroacetic Acid, *Amide*

$C_2H_2Cl_4$
1,1,1,2-Tetrachloroethane
1,1,2,2-Tetrachloroethane

$C_2H_2F_2$
1,1-Difluoroethylene

$C_2H_2F_2O_2$
Difluoroacetic Acid

C_2H_2IN
Iodoacetic Acid, *Nitrile*

$C_2H_2I_2$
Acetylene di-iodide

$C_2H_2I_2O_2$
Di-iodoacetic Acid

$C_2H_2N_2$
Aminocyanocarbene†

$C_2H_2N_2O$
Cyanoformic Acid, *Amide*
1,2,4-Oxadiazole★†
1,2,5-Oxadiazole†
1,3,4-Oxadiazole†

$C_2H_2N_2OS_2$
Rhodan Hydrate†

$C_2H_2N_2O_2$
Diazoacetic Acid
Nitroacetic Acid, *Nitrile*

$C_2H_2N_2O_4$
Azoformic Acid

$C_2H_2N_2S$
Cyanothioformamide
1,2,4-Thiadiazole†
1,2,5-Thiadiazole†
1,3,4-Thiadiazole†
Thio-oxamic Acid, *Nitrile*

$C_2H_2N_2S_3$
Xanthan Hydride†

$C_2H_2N_2Se$
1,2,5-Selenadiazole†

$C_2H_2N_4$
1,2,4,5-Tetrazine

$C_2H_2N_4O_8$
1,1,2,2-Tetranitroethane

C_2H_2O
Ketene

$C_2H_2O_2$
Glyoxal

$C_2H_2O_2S_2$
Dithio-oxalic Acid

$C_2H_2O_3$
Glyoxylic Acid

$C_2H_2O_3S$
Thio-oxalic Acid

$C_2H_2O_4$
Oxalic Acid

$C_2H_2O_4S$
Vinylene sulphate†

$C_2H_2O_6$
Di-peroxalic Acid

C_2H_3Br
Bromoethylene

$C_2H_3BrClNO$
Bromochloroacetic Acid, *Amide*

$C_2H_3BrN_2O_3$
Bromonitroacetamide

C_2H_3BrO
Acetyl bromide
Bromoacetaldehyde

$C_2H_3BrO_2$
Bromoacetic Acid

$C_2H_3Br_2NO$
Dibromoacetic Acid, *Amide*

$C_2H_3Br_2NO_2$
1,1-Dibromonitroethane
1,2-Dibromonitroethane
$C_2H_3Br_3$
1,1,2-Tribromoethane
$C_2H_3Br_3O$
2,2,2-Tribromoethanol
$C_2H_3Br_3O_2$
Bromal, *Hydrate*
C_2H_3Cl
Chloroethylene
C_2H_3ClINO
Chloroiodoacetic Acid, *Amide*
$C_2H_3ClN_2O_2$
Chloroglyoxime
C_2H_3ClO
Acetyl chloride
Chloroacetaldehyde
Chloro-oxirane†
$C_2H_3ClO_2$
Chloroacetic Acid
Methyl chloroformate
$C_2H_3ClO_3$
Acetyl hypochlorite
$C_2H_3Cl_2I$
1,1-Dichloro-2-iodoethane
$C_2H_3Cl_2NO$
Dichloroacetic Acid, *Amide*
$C_2H_3Cl_2NO_3$
2,2-Dichloroethanol, *Nitrate*
$C_2H_3Cl_3$
1,1,1-Trichloroethane
1,1,2-Trichloroethane
$C_2H_3Cl_3O$
2,2,2-Trichloroethanol
$C_2H_3Cl_3O_2$
Chloral Hydrate
C_2H_3F
Fluoroethylene
C_2H_3FO
Acetyl fluoride
Fluoroacetaldehyde†
$C_2H_3FO_2$
Fluoroacetic Acid
$C_2H_3F_2NO$
Difluoroacetic Acid, *Amide*
$C_2H_3F_3$
2,2,2-Trifluoroethanol
$C_2H_3F_3O_2$
2,2,2-Trifluoroacetaldehyde, *Hydrate*
C_2H_3HgN
Mercuri-methyl cyanide
C_2H_3I
Iodoethylene
C_2H_3IO
Acetyl iodide
Iodoacetaldehyde
$C_2H_3IO_2$
Iodoacetic Acid
$C_2H_3I_2NO$
Di-iodoacetic Acid, *Amide*
$C_2H_3I_3$
1,1,1-Tri-iodoethane
C_2H_3N
Acetonitrile
Methyl isocyanide
C_2H_3NO
Glycollic Acid, *Nitrile*
Methyl cyanate†
Methyl isocyanate
$C_2H_3NO_2$
Nitroethylene
$C_2H_3NO_2S$
Thio-oxamic Acid
$C_2H_3NO_3$
Hydroxyiminoacetic Acid
Oxamic Acid
$C_2H_3NO_4$
Acetyl nitrate
Iminodicarboxylic Acid
Nitroacetic Acid
C_2H_3NS
Methyl isothiocyanate
Methyl thiocyanate
$C_2H_3N_3$
1,2,3-Triazole
1,2,4-Triazole
$C_2H_3N_3O$
Cyanourea
Diazoacetic Acid, *Amide*
3-Hydroxy-1,2,4-triazole
5-Hydroxy-1,2,3-triazole
$C_2H_3N_3O_2$
Azidoacetic Acid
Tetrahydro-1,2,4-triazole-3,5-dione
$C_2H_3N_3O_6$
1,1,1-Trinitroethane
C_2H_4
Ethylene
C_2H_4BrCl
1-Bromo-1-chloroethane
1-Bromo-2-chloroethane
C_2H_4BrF
1-Bromo-2-fluoroethane
C_2H_4BrI
1-Bromo-1-iodoethane
1-Bromo-2-iodoethane
C_2H_4BrNO
N-Bromoacetamide
Bromoacetic Acid, *Amide*
$C_2H_4BrNO_2$
1-Bromo-1-nitroethane
$C_2H_4Br_2$
1,1-Dibromoethane
1,2-Dibromoethane

$C_2H_4Br_2O$
Di-bromomethyl Ether
1,2-Dibromoethanol
2,2-Dibromoethanol

C_2H_4ClF
1-Chloro-1-fluoroethane
1-Chloro-2-fluoroethane

$C_2H_4ClFO_2S$
1-Chloroethane-1-sulphonic Acid, *Fluoride*

C_2H_4ClI
1-Chloro-1-iodoethane
1-Chloro-2-iodoethane

C_2H_4ClNO
Chloroacetaldehyde, *Oxime*
Chloroformaldoxime, *Me ether*
Chloroacetamide
N-Chloroacetamide
Methylcarbamic Acid, *Chloride*

$C_2H_4ClNO_2$
1-Chloro-1-nitroethane
1-Chloro-2-nitroethane

$C_2H_4ClO_2P$
Ethylene chlorophosphite†

$C_2H_4Cl_2$
1,1-Dichloroethane
1,2-Dichloroethane

$C_2H_4Cl_2O$
2,2-Dichloroethanol
Di-chloromethyl Ether
Dichloromethyl methyl Ether

$C_2H_4Cl_2O_2$
Dichloroacetaldehyde, *Hydrate*

$C_2H_4Cl_2O_2S$
1-Chloroethane-1-sulphonic Acid, *Chloride*
2-Chloroethane-1-sulphonic Acid, *Chloride*
Di-chloromethyl sulphone

$C_2H_4Cl_2O_4S_2$
Ethane-1,2-disulphonic Acid, *Dichloride*

$C_2H_4Cl_2S$
Di-chloromethyl sulphide

C_2H_4FNO
Fluoroacetic Acid, *Amide*

$C_2H_4F_2$
1,1-Difluoroethane
1,2-Difluoroethane

$C_2H_4F_2O$
2,2-Difluoroethanol

C_2H_4INO
Iodoacetamide
N-Iodoacetamide

$C_2H_4I_2$
1,1-Di-iodoethane
1,2-Di-iodoethane

$C_2H_4N_2$
Aminoacetonitrile

$C_2H_4N_2O$
1,3-Diazetidin-2-one

$C_2H_4N_2OS$
Thio-oxamic Acid, *Amide*

$C_2H_4N_2O_2$
Diformylhydrazine
Glyoxime
Oxamide

$C_2H_4N_2O_3$
Allophanic Acid
Ethylnitrolic Acid
Methazonic Acid
Nitroacetic Acid, *Amide*

$C_2H_4N_2O_4$
Azoformic Acid, *Diamide*
1,1-Dinitroethane
1,2-Dinitroethane
Hydrazodiformic Acid
N-Nitrocarbamic Acid, *Me ester*

$C_2H_4N_2S_2$
Dithio-oxalic Acid, *Diamide*

$C_2H_4N_2S_4$
Thiuram disulphide

$C_2H_4N_4$
1-Amino-1,2,3-triazole
3-Amino-1,2,4-triazole
4-Amino-1,2,4-triazole
Dicyandiamide
1-Methyl-1,2,3,4-tetrazole
1-Methyl-1,2,3,5-tetrazole

$C_2H_4N_4O$
Azidoacetic Acid, *Amide*

$C_2H_4N_4O_2$
4-Amino-3,5-dihydroxy-1,2,4-triazole†
4-Aminotetrahydro-3,5-dioxo-1,2,4-triazole
Hexahydro-*s*-tetrazine-3,6-dione

$C_2H_4N_4O_4$
Nitrobiuret

C_2H_4O
Acetaldehyde
Oxiran
Vinyl Alcohol

$(C_2H_4O)_n$
Metaldehyde

C_2H_4OS
Thioacetic Acid

$C_2H_4OS_2$
Methoxydithioformic Acid

$C_2H_4O_2$
Acetic Acid
Glycollaldehyde
Methyl formate

$C_2H_4O_2S$
Thioglycollic Acid

$C_2H_4O_3$
Glycollic Acid
Peracetic Acid

$C_2H_4O_3S$
Ethylene sulphite†
Vinylsulphonic Acid

$C_2H_4O_4S$
Ethylene sulphate†
$C_2H_4O_4S_2$
Mesylsulphene†
$C_2H_4O_5S$
Sulphoacetic Acid
$C_2H_4O_6S_2$
1,3,2,4-Dioxadithian-2,2,4,4-tetroxide
C_2H_4S
Thiiran
$C_2H_4S_2$
Dithioacetic Acid
$C_2H_4S_4$
1,2,4,5-Tetrathiane†
$C_2H_4S_5$
Lenthionine†
C_2H_5
Ethyl (radical)
$C_2H_5AsBr_2$
Dibromoethylarsine
$C_2H_5AsCl_2$
Dichloroethylarsine
C_2H_5AsO
Ethylarsenious oxide
C_2H_5Br
Bromoethane
C_2H_5BrHg
Mercuri-ethyl bromide
C_2H_5BrO
2-Bromoethanol
Bromomethyl methyl Ether
C_2H_5Cl
Chloroethane
C_2H_5ClHg
Mercuri-ethyl chloride
C_2H_5ClO
1-Chloroethanol
2-Chloroethanol
Chloromethyl methyl Ether
Ethyl hypochlorite
$C_2H_5ClO_2S$
Ethanesulphonic Acid, *Chloride*
$C_2H_5ClO_3S$
1-Chloroethane-1-sulphonic Acid
2-Chloroethane-1-sulphonic Acid
Ethyl chlorosulphonate
$C_2H_5ClO_4$
Ethyl perchlorate
$C_2H_5ClO_4S$
Chloromethyl methyl sulphate
$C_2H_5Cl_2N$
N-Dichloroethylamine
$C_2H_5Cl_2OP$
Ethylphosphonic Acid, *Dichloride*
O-Ethylphosphorodichloridite
$C_2H_5Cl_2OPS$
O-Ethylphosphorodichloridothionate
$C_2H_5Cl_2O_2P$
O-Ethylphosphorodichloridate
C_2H_5F
Fluoroethane
C_2H_5FO
2-Fluoroethanol
C_2H_5HgI
Mercuri-ethyl iodide
C_2H_5I
Iodoethane
C_2H_5IO
2-Iodoethanol
Iodomethyl methyl Ether
C_2H_5N
Aziridine
C_2H_5NO
Acetamide
Acetimidic Acid
Aminoacetaldehyde
N-Methylformamide
C_2H_5NOS
Thiocarbamic Acid, S-*Me ester*
Thiocarbamic Acid, O-*Me ester*
Thioglycollic Acid, *Amide*
$C_2H_5NO_2$
Acethydroxamic Acid
Carbamic Acid, *Me ester*
Ethyl nitrite
Formhydroxamic Acid, *Me ester*
Glycine
Glycollic Acid, *Amide*
Methylcarbamic Acid
Nitroethane
$C_2H_5NO_2S$
N-Sulphonylethylamine†
$C_2H_5NO_3$
Ethyl nitrate
Hydroxyaminoacetic Acid
2-Nitroethanol
C_2H_5NS
Thioacetic Acid, *Amide*
$C_2H_5NS_2$
Dithiocarbamic Acid, *Me ester*
Methyldithiocarbamic Acid
$C_2H_5N_3O$
Azidoethanol
$C_2H_5N_3OS$
Thiobiuret
$C_2H_5N_3O_2$
Biuret
$C_2H_5N_3O_3$
N^1-Methyl-N^2-nitrourea
$C_2H_5N_5$
3,5-Diamino-1,2,4-triazole
C_2H_5OS
Methylsulphinyl Carbanion†
$C_2H_5O_3P$
Vinylphosphonic Acid

$C_2H_5O_4P$
Acetyl phosphite
Vinylphosphoric Acid†

$C_2H_5O_5P$
Acetyl phosphate
Phosphonoacetic Acid

C_2H_6
Ethane

C_2H_6AsBr
Cacodyl bromide

C_2H_6AsCl
Cacodyl chloride

$C_2H_6AsCl_3$
Cacodyl trichloride

C_2H_6AsI
Cacodyl iodide

$C_2H_6AsO_4$
Methyl arsenate

C_2H_6Be
Dimethylberyllium

C_2H_6BrN
2-Bromoethylamine

C_2H_6ClN
2-Chloroethylamine

$C_2H_6ClNO_2S$
1-Chloroethane-1-sulphonic Acid, *Amide*

C_2H_6ClOP
Dimethylphosphinic Acid, *Chloride*

$C_2H_6ClO_2PS$
O,O-Dimethylphosphorochloridothionate

C_2H_6Hg
Mercury dimethyl

C_2H_6HgO
Mercuri-ethyl hydroxide

C_2H_6IN
2-Iodoethylamine

$C_2H_6N_2$
Acetamidine
Azomethane

$C_2H_6N_2O$
Aminoacetamide
Dimethylnitrosamine
O-Methylisourea
Methylurea
Urea, O-*Me*

$C_2H_6N_2O_2$
Dimethylnitramine
Ethylnitramine
Hydrazinoacetic Acid
Hydrazinoformic Acid, *Me ester*
Hydroxymethylurea
Methylazoxymethanol†

$C_2H_6N_2S$
S-Methylisothiourea
Methylthiourea

$C_2H_6N_2S_2$
Dithiocarbazic Acid, *Me ester*

$C_2H_6N_3O_3P$
Phosphorazidic Acid, *Di-Me ester*†

$C_2H_6N_4O$
Amidinourea

$C_2H_6N_4O_2$
Biurea
Hydrazodiformic Acid, *Diamide*

$C_2H_6N_4O_4$
N,N′-Dinitroethylenediamine

$C_2H_6N_4S$
Amidinothiourea†

$C_2H_6N_6$
3,4,5-Triamino-1,2,4-triazole

C_2H_6O
Dimethyl Ether
Ethanol

C_2H_6OS
Dimethyl Sulphoxide†
2-Hydroxy-1-ethanethiol

$C_2H_6O_2$
Dimethyl peroxide†
Ethylene Glycol

$C_2H_6O_2S$
Dimethyl sulphone

$C_2H_6O_3$
Orthoacetic Acid

$C_2H_6O_3S$
Dimethyl sulphite
Ethanesulphonic Acid
Ethyl hydrogen sulphite
Methane-sulphonic Acid, *Me ester*

$C_2H_6O_4S$
Dimethyl sulphate
Ethyl hydrogen sulphate
Isethionic Acid

$C_2H_6O_5S_2$
Methane-sulphonic Acid, *Anhydride*

$C_2H_6O_6S_2$
Ethane-1,2-disulphonic Acid

$C_2H_6O_7S_2$
Ethionic Acid

C_2H_6S
Dimethyl sulphide
Ethanethiol

$C_2H_6S_2$
Dimethyl disulphide
1,2-Ethanedithiol

C_2H_6Zn
Zinc dimethyl

C_2H_7As
Dimethylarsine
Ethylarsine

$C_2H_7AsO_2$
Cacodylic Acid

$C_2H_7AsO_3$
Ethylarsinic Acid

$C_2H_7BO_2$
Ethaneboronic Acid

C_2H_7N
Dimethylamine
Ethylamine

C_2H_7NO
1-Aminoethanol
2-Aminoethanol
N-Ethylhydroxylamine
O-Ethylhydroxylamine
N-Methylhydroxylamine, *Me ether*

$C_2H_7NO_2S$
Ethanesulphonic Acid, *Amide*
Hypotaurine
Methylsulphonylmethylamine

$C_2H_7NO_3S$
Taurine

C_2H_7NS
2-Aminoethanethiol

$C_2H_7N_3$
Methylguanidine

$C_2H_7N_3O$
Glycine, *Hydrazide*
1-Methylsemicarbazide
2-Methylsemicarbazide
4-Methylsemicarbazide

$C_2H_7N_3S$
2-Methylthiosemicarbazide
4-Methylthiosemicarbazide

$C_2H_7N_5$
Diguanide

$C_2H_7N_5O_2$
Diaminobiuret

$C_2H_7O_2P$
Dimethylphosphinic Acid

$C_2H_7O_3P$
Ethylphosphonic Acid

C_2H_7P
Ethylphosphine

$C_2H_8NO_3P$
2-Aminoethylphosphonic Acid†

$C_2H_8NO_4P$
2-Aminoethanol, O-*Phosphate*

$C_2H_8N_2$
Ethylenediamine
Ethylhydrazine
1,1-Dimethylhydrazine
1,2-Dimethylhydrazine

$C_2H_8N_2O_2S$
Taurine, *Amide*

$C_2H_8N_2S$
Dimethyl Sulphone Di-imine†

$C_2H_8N_{10}O$
Tetracene

$C_2H_{10}Si_2$
1,4-Disilabutane

$C_2HgN_2O_2$
Fulminic Acid, *Mercury fulminate*

C_2I_2
Di-iodoacetylene

C_2I_4
Tetraiodoethylene

C_2N_2
Cyanogen

C_2N_2S
Cyanogen thiocyanate

$C_2O_2Br_2$
Oxalyl bromide

$C_2O_2Cl_2$
Oxalyl chloride

$C_2O_2F_2$
Oxalyl fluoride

C_3

C_3BrN
Bromocyanoacetylene†

$C_3Br_2Cl_2O_2$
Dibromomalonic Acid, *Dichloride*

$C_3Br_2F_2$
1,2-Dibromo-3,3-difluorocyclopropene†

$C_3Br_2N_2$
Dibromomalonic Acid, *Dinitrile*

C_3Br_4
Tetrabromocyclopropene†

$C_3Br_4O_2$
Dibromomalonic Acid, *Dibromide*

C_3Br_6
Hexabromocyclopropane†

C_3Br_6O
Hexabromoacetone

C_3ClN
Chlorocyanoacetylene†

$C_3Cl_2F_2$
1,2-Dichloro-3,3-difluorocyclopropene†

C_3Cl_3N
Trichloroacrylic Acid, *Nitrile*

$C_3Cl_3N_3$
Cyanuric chloride
3,5,6-Trichloro-1,2,4-triazine†

C_3Cl_4
Tetrachlorocyclopropene†

C_3Cl_4O
Trichloroacrylic Acid, *Chloride*

C_3Cl_6
Hexachloropropene

C_3Cl_6O
Hexachloroacetone

C_3F_4
Perfluoro(methylacetylene)†
Tetrafluoroallene†

C_3F_6
Hexafluoropropene

C_3F_6O
Hexafluoroacetone

C_3F_8
Profluorane

C_3HBrN_2
Bromomalonic Acid, *Dinitrile*

C_3HBr_5O
Pentabromoacetone

C_3HCl_3O
2,3-Dichloroacrylic Acid, *Chloride*
3,3-Dichloroacrylic Acid, *Chloride*

$C_3HCl_3O_2$
Trichloroacrylic Acid

C_3HCl_5
Pentachlorocyclopropane†
1,1,2,3,3-Pentachloropropene

C_3HCl_5O
Pentachloroacetone

C_3HCl_7
1,1,1,2,2,3,3-Heptachloropropane
1,1,1,2,3,3,3-Heptachloropropane

C_3HIO_2
Iodopropriolic Acid

C_3HN
Propiolic Acid, *Nitrile*

C_3H_2BrNO
Bromoacetyl cyanide

$C_3H_2Br_2N_2O$
Dibromomalonic Acid, *Amide-nitrile*

$C_3H_2Br_2O_2$
2,3-Dibromoacrylic Acid
3,3-Dibromoacrylic Acid

$C_3H_2Br_2O_3$
Dibromopyruvic Acid

$C_3H_2Br_2O_4$
Dibromomalonic Acid

$C_3H_2Br_4O$
1,1,1,3-Tetrabromoacetone
1,1,3,3-Tetrabromoacetone

$C_3H_2ClNO_2$
α-Chloroacetyl isocyanate†

C_3H_2ClNS
2-Chlorothiazole

$C_3H_2Cl_2O$
2-Chloroacrylic Acid, *Chloride*

$C_3H_2Cl_2O_2$
2,3-Dichloroacrylic Acid
3,3-Dichloroacrylic Acid
Malonyl chloride

$C_3H_2Cl_2O_4$
Dichloromalonic Acid

$C_3H_2Cl_3NO$
Trichloroacrylic Acid, *Amide*
3,3,3-Trichlorolactic Acid, *Nitrile*

$C_3H_2Cl_4$
1,2,3,3-Tetrachloropropene

$C_3H_2Cl_4O$
1,1,1,3-Tetrachloroacetone
1,1,3,3-Tetrachloroacetone

$C_3H_2Cl_4O_2$
2,2,2-Trichloroethoxycarbonyl Chloride†
3,3,3-Trichlorolactic Acid, *Chloride*

$C_3H_2Cl_6$
1,1,1,2,2,3-Hexachloropropane
1,1,1,2,3,3-Hexachloropropane
1,1,1,3,3,3-Hexachloropropane
1,1,2,2,3,3-Hexachloropropane

C_3H_2INS
2-Iodothiazole†

$C_3H_2I_2O_4$
Di-iodomalonic Acid

$C_3H_2I_4O$
1,1,3,3-Tetraiodoacetone

$C_3H_2N_2$
Malonitrile

$C_3H_2N_2O_3$
Imidazolidine-2,4,5-trione
Isonitrosomalonic Acid, *Nitrile*
4-Nitroisoxazole

$C_3H_2N_2O_4$
Diazomalonic Acid
Nitrocyanoacetic Acid

C_3H_2O
Cyclopropenone†
Propynal

$C_3H_2OS_2$
1,2-Dithiacyclopent-4-en-3-one†
1,3-Dithiacyclopent-4-en-2-one†

$C_3H_2O_2$
Propiolic Acid

$C_3H_2O_3$
Mesoxalic Dialdehyde

$C_3H_2O_4$
Mesoxalaldehydic Acid

$C_3H_2O_5$
Mesoxalic Acid (exists in free state as dihydroxymalonic Acid, $C_3H_4O_6$)

$C_3H_2S_3$
1,3-Dithiacyclopent-4-en-2-thione.†
3*H*-1,2-Dithiole-3-thione†

C_3H_3Br
3-Bromopropyne

$C_3H_3BrO_2$
2-Bromoacrylic Acid
3-Bromoacrylic Acid
Bromomalondialdehyde

$C_3H_3BrO_4$
Bromomalonic Acid

$C_3H_3Br_2ClO$
2,3-Dibromopropionic Acid, *Chloride*

$C_3H_3Br_2N$
2,3-Dibromopropionic Acid, *Nitrile*

$C_3H_3Br_3O$
1,1,1-Tribromoacetone
1,1,3-Tribromoacetone
$C_3H_3Br_3O_2$
2,2,3-Tribromopropionic Acid
2,3,3-Tribromopropionic Acid
$C_3H_3Br_5$
1,1,1,2,2-Pentabromopropane
1,1,2,3,3-Pentabromopropane
C_3H_3Cl
3-Chloropropyne
$C_3H_3ClN_2$
4-Chloropyrazole
C_3H_3ClO
Acrylic Acid, *Chloride*
Chloromethylketene†
$C_3H_3ClO_2$
2-Chloroacrylic Acid
3-Chloroacrylic Acid
Chloromalondialdehyde
$C_3H_3ClO_3$
Malonic Acid, *Monochloride*
Oxalic Acid, *Me ester chloride*
$C_3H_3ClO_4$
Chloromalonic Acid
$C_3H_3Cl_2N$
2,2-Dichloropropionic Acid, *Nitrile*
$C_3H_3Cl_2NO$
2,3-Dichloroacrylic Acid, *Amide*
3,3-Dichloroacrylic Acid, *Amide*
$C_3H_3Cl_3O$
2,2-Dichloropropionic Acid, *Chloride*
2,3-Dichloropropionic Acid, *Chloride*
1,1,1-Trichloroacetone
1,1,3-Trichloroacetone
$C_3H_3Cl_3O_2$
Trichloroacetic Acid, *Me ester*
2,2,3-Trichloropropionic Acid
$C_3H_3Cl_3O_3$
3,3,3-Trichlorolactic Acid
$C_3H_3Cl_5$
1,1,1,2,2-Pentachloropropane
1,1,1,2,3-Pentachloropropane
1,1,2,2,3-Pentachloropropane
1,1,2,3,3-Pentachloropropane
C_3H_3FO
2-Fluoroacraldehyde†
C_3H_3IO
2-Iodoacraldehyde
C_3H_3N
Acrylic Acid, *Nitrile*
Vinyl isocyanide†
C_3H_3NO
Cyanoacetaldehyde
Isoxazole
Oxazole†
Propiolic Acid, *Amide*
Pyruvic Acid, *Nitrile*

C_3H_3NOS
Acetyl isothiocyanate
$C_3H_3NOS_2$
Rhodanine
$C_3H_3NO_2$
Cyanoacetic Acid
Cyanoformic Acid, *Me ester*
$C_3H_3NO_3$
2,4-Oxazolidinedione
2,5-Oxazolidinedione
$C_3H_3NO_4$
Nitromalondialdehyde
$C_3H_3NO_5$
Isonitrosomalonic Acid
$C_3H_3NO_6$
Nitromalonic Acid
C_3H_3NS
Thiazole
$C_3H_3N_2O_2S$
2-Amino-5-nitrothiazole
$C_3H_3N_3$
Aminomalonitrile†
Dicyanamide, N-*Me*
1,2,4-Triazine†
1,3,5-Triazine†
$C_3H_3N_3O_2$
6-Azauracil
Azomycin
Isonitrosomalonic Acid, *Amide*
4(5)-Nitroimidazole*†
1-Nitropyrazole
4-Nitropyrazole
Oxalylguanidine
$C_3H_3N_3O_3$
Cyamelide
Cyanuric Acid
Nitrocyanoacetic Acid, *Amide*
$C_3H_3N_3O_4$
1-Nitrohydantoin
$C_3H_3N_5$
5*H*-*s*-Triazolo[5,1-*c*]-*s*-triazole†
C_3H_4
Allene
Cyclopropene†
Propyne
C_3H_4BrClO
3-Bromo-1-chloroacetone
2-Bromopropionic Acid, *Chloride*
$C_3H_4BrClO_2$
2-Bromo-2-chloropropionic Acid
2-Bromo-3-chloropropionic Acid
3-Bromo-2-chloropropionic Acid
$C_3H_4Br_2$
1,2-Dibromopropene
1,3-Dibromopropene
2,3-Dibromopropene
$C_3H_4Br_2N_2O_2$
Dibromomalonic Acid, *Diamide*

$C_3H_4Br_2O$
2-Bromopropionic Acid, *Bromide*
1,3-Dibromoacetone
2,2-Dibromopropanal
2,3-Dibromopropanal
2,3-Dibromo-2-propen-1-ol

$C_3H_4Br_2O_2$
Dibromoacetic Acid, *Me ester*
2,2-Dibromopropionic Acid
2,3-Dibromopropionic Acid
3,3-Dibromopropionic Acid

$C_3H_4Br_4$
1,1,1,2-Tetrabromopropane
1,1,1,3-Tetrabromopropane
1,1,2,2-Tetrabromopropane
1,1,2,3-Tetrabromopropane
1,1,3,3-Tetrabromopropane
1,2,2,3-Tetrabromopropane

$C_3H_4ClFO_2$
Chlorofluoroacetic Acid, *Me ester*†

C_3H_4ClIO
2-Iodopropionic Acid, *Chloride*
3-Iodopropionic Acid, *Chloride*

$C_3H_4ClIO_2$
2-Iodoethoxycarbonyl Chloride†

C_3H_4ClN
2-Chloropropionic Acid, *Nitrile*
3-Chloropropionic Acid, *Nitrile*

$C_3H_4ClNO_2$
1-Chloro-1-isonitrosoacetone
N-Formylglycine, *Chloride*

$C_3H_4ClNO_3$
3-Nitropropionic Acid, *Chloride*

$C_3H_4Cl_2$
1,1-Dichloropropene
1,2-Dichloropropene
1,3-Dichloropropene
2,3-Dichloropropene
3,3-Dichloropropene

$C_3H_4Cl_2N_2O_2$
Dichloromalonic Acid, *Diamide*

$C_3H_4Cl_2O$
2-Chloropropionic Acid, *Chloride*
3-Chloropropionic Acid, *Chloride*
1,1-Dichloroacetone
1,3-Dichloroacetone
1,1-Dichloro-2,3-epoxypropane
2,3-Dichloropropanal
3,3-Dichloropropanal

$C_3H_4Cl_2O_2$
Chloroformic Acid, 2-*Chloroethyl Ester*
Dichloroacetic Acid, *Me ester*
2,2-Dichloropropionic Acid
2,3-Dichloropropionic Acid
3,3-Dichloropropionic Acid

$C_3H_4Cl_2O_3$
3,3-Dichlorolactic Acid

$C_3H_4Cl_3NO_2$
Carbamic Acid, *Trichloroethyl ester*
Chloralamide
3,3,3-Trichlorolactic Acid, *Amide*

$C_3H_4Cl_4$
1,1,1,2-Tetrachloropropane
1,1,1,3-Tetrachloropropane
1,1,2,2-Tetrachloropropane
1,1,2,3-Tetrachloropropane
1,2,2,3-Tetrachloropropane

$C_3H_4I_2$
1,2-Di-iodopropene

$C_3H_4I_2O$
1,3-Di-iodoacetone

$C_3H_4N_2$
Imidazole
Pyrazole

$C_3H_4N_2O$
Cyanoacetic Acid, *Amide*
4-Hydroxypyrazole
2-Imidazolone
5-Pyrazolone

$C_3H_4N_2OS$
2-Thiohydantoin
4-Thiohydantoin

$C_3H_4N_2O_2$
Diazoacetic Acid, *Me ester*
2-Diazopropionic Acid
Hydantoin

$C_3H_4N_2O_3$
Di-isonitrosoacetone

$C_3H_4N_2O_3S$
Imidazole-2-sulphonic Acid
Imidazole-4-sulphonic Acid

$C_3H_4N_2O_4$
Isonitrosomalonic Acid, *Monoamide*
Oxaluric Acid

$C_3H_4N_2S$
2-Aminothiazole
5-Aminothiazole
5-Methyl-1,2,3-thiadiazole

$C_3H_4N_4O$
6-Azacytosine

$C_3H_4N_4O_2$
Ammelide

C_3H_4O
Acraldehyde
Cyclopropanone†
Ethynyl methyl Ether
Methylketene
2-Propyn-1-ol

$C_3H_4O_2$
Acrylic Acid
Malondialdehyde
Oxetan-2-one†
Pyruvic Aldehyde

$C_3H_4O_3$
2,3-Epoxypropionic Acid
Ethylene carbonate†
Formylacetic Acid
Glycerosone
Glyoxylic Acid, *Me ester*
Pyruvic Acid
Reductone

$C_3H_4O_4$
Hydroxypyruvic Acid
Malonic Acid
Mesoxalic Dialdehyde, *Hydrate*
Oxalic Acid, *Me ester*

$C_3H_4O_5$
Tartronic Acid

$C_3H_4O_6$
Mesoxalic Acid, *Hydrate*

$C_3H_4O_7S$
3-Thietanone

C_3H_4S
Thiet†

C_3H_5Br
Allyl bromide
1-Bromopropene
2-Bromopropene

C_3H_5BrHg
Mercuri-allyl bromide

$C_3H_5BrN_2O_2$
Bromomalonic Acid, *Diamide*

C_3H_5BrO
Bromoacetone
2-Bromoallyl Alcohol
3-Bromoallyl Alcohol
1-Bromo-2,3-epoxypropane
2-Bromopropanal
3-Bromopropanal
Propionic Acid, *Bromide*

$C_3H_5BrO_2$
Bromoacetic Acid, *Me ester*
2-Bromopropionic Acid
3-Bromopropionic Acid

$C_3H_5Br_2NO$
2,3-Dibromopropionic Acid, *Amide*

$C_3H_5Br_3$
1,1,1-Tribromopropane
1,1,2-Tribromopropane
1,2,2-Tribromopropane
1,2,3-Tribromopropane

C_3H_5Cl
Allyl chloride
1-Chloropropene
2-Chloropropene

$C_3H_5ClN_2O_2$
Chloromalonic Acid, *Diamide*

C_3H_5ClO
Chloroacetone
2-Chloro-1,3-epoxypropane
Chloromethyloxiran
2-Chloropropanal
3-Chloropropanal
2-Chloro-2-propen-1-ol
3-Chloro-2-propen-1-ol
2-Chlorovinyl methyl Ether
Propionic Acid, *Chloride*

$C_3H_5ClO_2$
Chloroacetic Acid, *Me ester*
1-Chloro-3-hydroxyacetone
2-Chloropropionic Acid
3-Chloropropionic Acid
Ethyl chloroformate

$C_3H_5ClO_3$
3-Chlorolactic Acid

$C_3H_5Cl_2NO$
2,2-Dichloropropionic Acid, *Amide*
3,3-Dichloropropionic Acid, *Amide*

$C_3H_5Cl_3$
1,1,1-Trichloropropane
1,1,2-Trichloropropane
1,1,3-Trichloropropane
1,2,2-Trichloropropane
1,2,3-Trichloropropane

$C_3H_5Cl_3O$
1,1,1-Trichloro-2-propanol

$C_3H_5Cl_3O_2$
Chloral, *Mono-Me acetal*

C_3H_5F
Allyl fluoride

C_3H_5FO
2-Fluoropropanal†

$C_3H_5FO_2$
Fluoroacetic Acid, *Me ester*

$C_3H_5F_3O$
1,1,1-Trifluoropropan-2-ol†

$C_3H_5F_3O_2$
2,2,2-Trifluoroacetaldehyde, *Me hemiacetal*

C_3H_5HgI
Mercuri-allyl iodide

C_3H_5HgN
Mercuri-ethyl cyanide

C_3H_5I
Allyl iodide
2-Iodopropene

C_3H_5IO
Iodoacetone
2-Iodopropanal
3-Iodopropanal

$C_3H_5IO_2$
Iodoacetic Acid, *Me ester*
2-Iodopropionic Acid
3-Iodopropionic Acid

$C_3H_5IO_3$
2-Hydroxy-3-iodopropionic Acid

C_3H_5N
3-Aminopropyne
Ethyl isocyanide
Propionic Acid, *Nitrile*

C_3H_5NO
Acrylic Acid, *Amide*
2-Azetidinone
Ethyl cyanate†
Ethyl isocyanate
3-Hydroxypropionic Acid, *Nitrile*
Lactic Acid, *Nitrile*
Methoxyacetic Acid, *Nitrile*

$C_3H_5NO_2$
Isonitrosoacetone
1-Nitro-1-propene
2-Nitro-1-propene
3-Nitro-1-propene
Pyruvic Acid, *Amide*

$C_3H_5NO_2S$
Thio-oxamic Acid, *Me ester*

$C_3H_5NO_3$
N-Formylglycine
Hydroxyiminoacetic Acid, *Me ester*
3-Isonitrosopropionic Acid
Malonamic Acid
Methyloxamic Acid
Nitroacetone
Oxamic Acid, *Me ester*

$C_3H_5NO_3S$
(Carbamoylthio)acetic Acid

$C_3H_5NO_4$
Aminomalonic Acid
Hadacidin
Malonhydroxamic Acid
Nitroacetic Acid, *Me ester*
2-Nitropropionic Acid
3-Nitropropionic Acid
Tartronic Acid, *Monoamide*

$C_3H_5NO_5$
Nitrolactic Acid

C_3H_5NS
Ethyl isothiocyanate
Ethyl thiocyanate
2,3-Thiazoline

$C_3H_5N_3$
3-Aminopyrazole
4-Aminopyrazole
4-Methyl-1,2,4-triazole

$C_3H_5N_3O$
2-Imino-4-oxoimidazolidine

$C_3H_5N_3O_2$
2-Azidopropionic Acid
3-Azidopropionic Acid

$C_3H_5N_3O_3$
Isonitrosomalonic Acid, *Diamide*
Oxalan

$C_3H_5N_3O_4$
Methylnitroglyoxime
Nitromalonic Acid, *Diamide*

$C_3H_5N_3O_9$
Nitroglycerin

$C_3H_5N_5$
2,4-Diamino-*s*-triazine

$C_3H_5N_5O$
Ammeline

$C_3H_5O_6P$
Phosphoenolpyruvic Acid†

C_3H_6
Cyclopropane
Propene

C_3H_6AsN
Cacodyl cyanide

C_3H_6BrCl
1-Bromo-1-chloropropane
1-Bromo-2-chloropropane
1-Bromo-3-chloropropane
2-Bromo-1-chloropropane
2-Bromo-2-chloropropane

C_3H_6BrClO
3-Bromo-1-chloro-2-propanol

C_3H_6BrNO
2-Bromopropionic Acid, *Amide*

$C_3H_6Br_2$
1,1-Dibromopropane
1,2-Dibromopropane
1,3-Dibromopropane
2,2-Dibromopropane

$C_3H_6Br_2O$
2,3-Dibromo-1-propanol
1,3-Dibromo-2-propanol

C_3H_6ClI
2-Chloro-1-iodopropane
2-Chloro-2-iodopropane
3-Chloro-1-iodopropane

C_3H_6ClIO
1-Chloro-3-iodo-2-propanol

C_3H_6ClNO
2-Chloropropionic Acid, *Amide*
3-Chloropropionic Acid, *Amide*
Dimethylcarbamic Acid, *Chloride*

$C_3H_6ClNO_2$
2-Amino-3-chloropropionic Acid
1-Chloro-1-nitropropane
1-Chloro-2-nitropropane
1-Chloro-3-nitropropane
2-Chloro-1-nitropropane
2-Chloro-2-nitropropane

$C_3H_6Cl_2$
1,1-Dichloropropane
1,2-Dichloropropane
1,3-Dichloropropane
2,2-Dichloropropane

$C_3H_6Cl_2O$
1-Chloroethyl chloromethyl Ether
2,3-Dichloro-1-propanol
3,3-Dichloro-1-propanol
1,1-Dichloro-2-propanol
1,3-Dichloro-2-propanol

$C_3H_6Cl_2O_4S_2$
Propane-1,2-disulphonic Acid, *Dichloride*
Propane-1,3-disulphonic Acid, *Dichloride*

$C_3H_6Cl_3O$
2-Chloroethyl chloromethyl Ether

$C_3H_6HgO_2$
Mercuri-methyl acetate

C_3H_6INO
2-Iodopropionic Acid, *Amide*
3-Iodopropionic Acid, *Amide*

$C_3H_6I_2$
1,3-Di-iodopropane

$C_3H_6I_2O$
1,3-Di-iodo-2-propanol
2,3-Di-iodo-1-propanol

$C_3H_6N_2$
NN′-Dimethylcarbodi-imide†
Dimethylcyanamide
Ethylcyanamide
1-Pyrazoline†
2-Pyrazoline
Sarcosine, *Nitrile*

$C_3H_6N_2O$
2-Imidazolidone
4-Methyl-1,3-diazetidin-2-one

$C_3H_6N_2O_2$
Malonamide
Methylglyoxime
Methyloxamic Acid, *Amide*
Oxamycin

$C_3H_6N_2O_2S$
Thiohydantoic Acid

$C_3H_6N_2O_3$
Allophanic Acid, *Me ester*
Hydantoic Acid
Ethyl aminoformate, N-*Nitroso*
Malonhydroxamic Acid, *Amide*
2-Methylallophanic Acid
4-Methylallophanic Acid
2-Nitropropionic Acid, *Amide*
Tartronic Acid, *Diamide*

$C_3H_6N_2O_4$
Ethylnitroaminoformate

$C_3H_6N_2O_7$
Glycerol 1,2-Dinitrate★†
Glycerol 1,3-Dinitrate★†

$C_3H_6N_4$
1-Amino-4-methyl-1,2,3-triazole
1-Amino-5-methyl-1,2,3-triazole
3-Amino-5-methyl-1,2,4-triazole
3,4-Diaminopyrazole†

$C_3H_6N_4O$
2-Azidopropionic Acid, *Amide*

$C_3H_6N_4O_3$
Carbonyl-diurea
Mesoxalic Acid, *Di-hydrazide*
Triuret

$C_3H_6N_6$
Melamine

$C_3H_6N_6O_6$
Cyclonite

C_3H_6O
Acetone
Allyl Alcohol
2-Methyloxiran
Methyl vinyl Ether
Oxetane
Propanal
2-Propen-2-ol

C_3H_6OS
Propenylsulphenic Acid
Thioacetic Acid, S-*Me ester*
Thioacetic Acid, O-*Me ester*
Thiopropionic Acid

$C_3H_6OS_2$
Dithiocarbonic Acid, SS-*Di-Me ester*
Methoxydithioformic Acid, *Me ester*
Xanthogenic Acid

$C_3H_6O_2$
Allyl hydroperoxide
1,3-Dioxolan
Ethyl formate
Hydroxyacetone
2-Hydroxymethyloxiran
2-Hydroxypropanal
3-Hydroxypropanal
Methyl acetate
Propionic Acid

$C_3H_6O_2S$
2-Mercaptopropionic Acid
3-Mercaptopropionic Acid
Methylmercaptoacetic Acid
Thietane sulphone
Thiocarbonic Acid, O,S-*Di-Me ester*
Thiocarbonic Acid, O-*Di-Me ester*

$C_3H_6O_3$
1,3-Dihydroxyacetone
Dimethyl carbonate
Ethyl hydrogen carbonate
Glyceraldehyde
Glycollic Acid, *Me ester*
3-Hydroxypropionic Acid
Lactic Acid
Methoxyacetic Acid
Perpropionic Acid
Trioxymethylene

$C_3H_6O_3S$
Vinylsulphonic Acid, *Me ester*

$C_3H_6O_4$
Glyceric Acid

C_3H_6S
2-Methylthiiran†
2-Propene-1-thiol
Thietane

$C_3H_6S_2$
Dithioacetic Acid, *Me ester*
1,2-Dithiolan

$C_3H_6S_3$
s-Trithiane

$C_3H_6Se_3$
Trimethylene triselenide

$C_3H_7AsO_3$
2-Methoxy-1,3,2-dioxa-arsolan

$C_3H_7BaO_6P$
Glycerophosphoric Acid, *Ba salt*

C_3H_7Br
1-Bromopropane
2-Bromopropane

C_3H_7BrHg
Mercuri-isopropyl bromide
Mercuri-propyl bromide

$C_3H_7BrN_2Se$
2-Amino-1,3-selenazol-2-inium bromide†

C_3H_7BrO
2-Bromo-1-propanol
3-Bromo-1-propanol
1-Bromo-2-propanol

$C_3H_7BrO_2$
Glycerol, α-Monobromihydrin

$C_3H_7CaO_6P$
Glycerophosphoric Acid, *Ca salt*

C_3H_7Cl
1-Chloropropane
2-Chloropropane

C_3H_7ClHg
Mercuri-isopropyl chloride
Mercuri-propyl chloride

C_3H_7ClO
1-Chloroethyl methyl Ether
2-Chloroethyl methyl Ether
Chloromethyl ethyl Ether
1-Chloro-2-propanol
2-Chloro-1-propanol
3-Chloro-1-propanol

$C_3H_7ClO_2$
2-Chloropropanal, *Hydrate*
2-Chloro-1,3-propanediol
3-Chloro-1,2-propanediol

$C_3H_7ClO_2S$
Propane-1-sulphonic Acid, *Chloride*
Propane-2-sulphonic Acid, *Chloride*

$C_3H_7ClO_3S$
1-Chloroethane-1-sulphonic Acid, *Me ester*

C_3H_7ClS
1-Chloroethyl methyl sulphide
2-Chloroethyl methyl sulphide

$C_3H_7Cl_2N$
N-Dichloropropylamine

C_3H_7F
1-Fluoropropane

C_3H_7HgI
Mercuri-isopropyl iodide
Mercuri-propyl iodide

C_3H_7I
1-Iodopropane
2-Iodopropane

C_3H_7IO
2-Iodoethanol, *Me ether*
2-Iodo-1-propanol
3-Iodo-1-propanol
1-Iodo-2-propanol

$C_3H_7IO_2$
Glycerol, α-*Monoiodohydrin*
Glycerol, β-*Monoiodohydrin*

$C_3H_7LiO_2P_6$
Glycerophosphoric Acid, *Li salt*

C_3H_7N
Allylamine
Azetidine
Cyclopropylamine
1-Methylaziridine
2-Methylaziridine

C_3H_7NO
O-Allylhydroxylamine
Aminoacetone
Azetidin-3-ol†
Dimethylformamide
Formimidic Acid, *Et ester*
N-Methylacetamide
Methylaminoacetaldehyde
Propanal, *Oxime*
Propionamide

C_3H_7NOS
Ethyl aminothioformate

$C_3H_7NO_2$
Alanine
3-Amino-2-hydroxypropanal
3-Aminopropionic Acid
Dimethylcarbamic Acid
Ethyl aminoformate
Formhydroxamic Acid, *Et ester*
Glycine, *Me ester*
N-Hydroxymethylacetamide
Isopropyl nitrite
Lactamide
Methoxyacetic Acid, *Amide*
1-Nitropropane
2-Nitropropane
Propyl nitrite
Sarcosine

$C_3H_7NO_2S$
3-Amino-2-mercaptopropionic Acid
Cysteine

$C_3H_7NO_3$
Glyceric Acid, *Amide*
Isopropyl nitrate
Isoserine
Lacthydroxamic Acid
2-Nitropropanol
3-Nitropropanol
1-Nitro-2-propanol
Propyl nitrate
Serine

$C_3H_7NO_4S$
Cysteinesulphinic Acid

$C_3H_7NO_5$
Glycerol, 1-*Mononitrate*
Glycerol, 2-*Mononitrate*

$C_3H_7NO_5S$
3-Amino-2-sulphopropionic Acid
Cysteic Acid

$C_3H_7NO_5S_2$
Sulphocysteine†

$C_3H_7NO_6S$
Serine, *Sulphuric ester*

C_3H_7NS
Thiazolidine
Thioformic Acid, *Ethylamide*
Thiopropionic Acid, *Amide*

$C_3H_7NS_2$
Dimethyldithiocarbamic Acid
Ethyl dithiocarbamate
Methyldithiocarbamic Acid, *Me ester*

$C_3H_7N_3OS$
ω-Methylthiobiuret

$C_3H_7N_3O_2$
Aminomalonic Acid, *Diamide*
Biuret, N-*Me* deriv.
Guanidinoacetic Acid
Hydantoic Acid, *Amide*
5-Methylsemioxamazide

$C_3H_7N_3O_3$
Hydrazodiformic Acid, *Me ester-amide*

$C_3H_7N_3O_4$
Alanosine†

$C_3H_7OPS_2$
2-Methoxy-1,3,2-dithiaphospholan

$C_3H_7O_3P$
2-Methoxy-1,3,2-dioxaphospholan

$C_3H_7O_4P$
Allyl phosphate

C_3H_8
Propane

C_3H_8Hg
Mercury ethylmethyl

C_3H_8HgO
Mercuri-propyl hydroxide

$C_3H_8NO_5P$
2-Amino-3-phosphonopropionic Acid†

$C_3H_8NO_6P$
Serine, *Phosphoric ester*

$C_3H_8N_2$
Pyrazolidine†

$C_3H_8N_2O$
Alanine, *Amide*
3-Aminopropionic Acid, *Amide*
1,1-Dimethylurea
1,3-Dimethylurea
Ethylurea
Lactamidine
Urea, O-*Et*

$C_3H_8N_2O_2$
2,3-Diaminopropionic Acid
N-Ethyl-*N'*-hydroxyurea
Formamidine acetate†
Hydrazinoformic Acid, *Et ester*
2-Hydrazinopropionic Acid
N-2-Hydroxyethylurea
Hydroxyurea, *Et ether*
Propylnitramine

$C_3H_8N_2O_3$
1,3-*Bis*hydroxymethylurea

$C_3H_8N_2O_4S$
Taurocarbamic Acid

$C_3H_8N_2S$
1,1-Dimethylthiourea
1,3-Dimethylthiourea

$C_3H_8N_2S_2$
Dithiocarbazic Acid, *Et ester*

C_3H_8O
Ethyl methyl Ether
1-Propanol
2-Propanol

C_3H_8OS
2-Hydroxy-1-ethanethiol, S-*Me*

$C_3H_8OS_2$
2-Hydroxy-1,3-propanedithiol
3-Hydroxy-1,2-propanedithiol

$C_3H_8O_2$
2-Hydroxyethyl methyl Ether
Methylal
1,2-Propanediol
1,3-Propanediol

$C_3H_8O_2S$
Ethyl methyl sulphone

$C_3H_8O_3$
Glycerol
Orthopropionic Acid

$C_3H_8O_3S$
Ethanesulphonic Acid, *Me ester*
Ethyl methyl sulphite
Methane-sulphonic Acid, *Et ester*
Propane-1-sulphonic Acid
Propane-2-sulphonic Acid

$C_3H_8O_4S$
Ethyl methyl sulphate

$C_3H_8O_6S_2$
Methionic Acid, *Di-Me ester*
Propane-1,1-disulphonic Acid
Propane-1,2-disulphonic Acid
Propane-1,3-disulphonic Acid
Propane-2,2-disulphonic Acid

C_3H_8S
Ethyl methyl sulphide
1-Propanethiol
2-Propanethiol

$C_3H_8S_2$
Bismethylthiomethane†
1,2-Propanedithiol
1,3-Propanedithiol

C_3H_8Se
Ethyl methyl selenide

C_3H_9Al
Trimethylaluminium

C_3H_9As
Trimethylarsine

C_3H_9B
Trimethylborine

$C_3H_9BF_4O$
Trimethyloxonium Fluoroborate†

$C_3H_9BO_3$
Trimethyl borate

$C_3H_9B_3S_3$
Trimethylborosulphole

C_3H_9BrS
Trimethylsulphonium bromide

C_3H_9ClS
Trimethylsulphonium chloride

C_3H_9IS
Trimethylsulphonium iodide

C_3H_9In
Trimethylindium

C_3H_9N
Methylethylamine
Propylamine
2-Propylamine
Trimethylamine

C_3H_9NO
2-Amino-1-propanol
3-Amino-1-propanol
1-Amino-2-propanol
2-Methoxyethylamine
2-Methylaminoethanol
N-Methylhydroxylamine, *Et ether*
Propanal, *Propanal ammonia*
Propylhydroxylamine
Trimethylamine oxide

C_3H_9NOS
Cysteinol

$C_3H_9NO_2$
3-Amino-1,2-propanediol

$C_3H_9NO_2S$
Methylsulphonylethylamine
Propane-1-sulphonic Acid, *Amide*
Propane-2-sulphonic Acid, *Amide*

$C_3H_9NO_3S$
N-Methyltaurine
Taurine, N-*Me*

$C_3H_9NO_4S_2$
N-Di-[methylsulphonyl]-methylamine

C_3H_9NS
3-Amino-1-propanethiol
1-Amino-2-propanethiol

$C_3H_9N_3$
1,1-Dimethylguanidine
1,3-Dimethylguanidine
Ethylguanidine

$C_3H_9N_3O$
2-(1-Guanidino)ethanol
2-(2-Guanidino)ethanol

$C_3H_9N_3O_3$
Triformoxime

$C_3H_9O_2PS_2$
O,O,S-Trimethylphosphorothiolothionate

$C_3H_9O_3P$
Methylphosphonic Acid, *Di-Me ester*
Propylphosphonic Acid

$C_3H_9O_4P$
Phosphoric Acid, *Tri-Me ester*

$C_3H_9O_6P$
Glycerophosphoric Acid

C_3H_9P
Trimethylphosphine

C_3H_9Sb
Trimethylstibine

C_3H_9Tl
Thallium trimethyl

$C_3H_{10}N_2$
1,3-Diaminopropane
Isopropylhydrazine
2-Methylaminoethylamine
1,2-Propanediamine
Propylhydrazine

$C_3H_{10}N_2O$
1,3-Diamino-2-propanol

$C_3H_{10}N_2O_4S_2$
Propane-1,1-disulphonic Acid, *Diamide*
Propane-1,3-disulphonic Acid, *Diamide*

$C_3H_{10}OSi$
Trimethylsilanol†

$C_3H_{11}N_3$
1,2,3-Triaminopropane

$C_3H_{12}B_3N_3$
N,N′,N″-Trimethylborazole

$C_3H_{12}O_3Si_3$
2,4,6-Trimethylcyclotrisiloxane

$C_3H_{12}Si_2$
1,5-Disilapentane

$C_3H_{12}Si_3$
1,3,5-Trisilacyclohexane

C_3IN
Cyanoiodoacetylene★†

C_3I_6O
Hexaiodoacetone

C_3N_2
Dicyanocarbene†

C_3N_2O
Carbonyl cyanide

C_3O_2
Carbon suboxide

$C_3O_2S_3$
1,3-Dithiolane-2-thione-4,5-dione†

C_4

C_4BrF_7O
Heptafluorobutyric Acid, *Bromide*

$C_4Br_2N_2$
Dibromofumaronitrile†

$C_4Br_2O_3$
Dibromomaleic Acid, *Anhydride*

C_4Br_4S
Tetrabromothiophene

C_4ClF_7O
Heptafluorobutyric Acid, *Chloride*

$C_4Cl_2I_2O_2$
Di-iodofumaric Acid, *Dichloride*

$C_4Cl_2O_3$
Dichloromaleic Acid, *Anhydride*

$C_4Cl_4O_2$
Dichloromaleic Acid, *Dichloride*

C_4Cl_4S
Tetrachlorothiophene
$C_4Cl_6O_3$
Trichloroacetic Acid, *Anhydride*
$C_4F_4N_2$
2,3,5,6-Tetrafluoropyrazine†
3,4,5,6-Tetrafluoropyridazine†
2,4,5,6-Tetrafluoropyrimidine†
C_4F_4O
Tetrafluorofuran†
C_4F_6
Hexafluorocyclobutene
C_4F_7IO
Heptafluorobutyric Acid, *Iodide*
C_4F_7N
Heptafluorobutyric Acid, *Nitrile*
C_4F_8
Octafluorobut-1-ene†
Octafluorobut-2-ene†
Octafluorocyclobutane
Pentafluoro-2-(trifluoromethyl)propene†
C_4F_{10}
Decafluorobutane
C_4HBr
1-Bromobuta-1,3-diyne†
C_4HBrO_3
Bromomaleic Acid, *Anhydride*
$C_4HBr_3O_2$
Bromomaleic Acid, *Dibromide*
Mucobromic Acid, *Bromide*
C_4HBr_4N
2,3,4,5-Tetrabromopyrrole
C_4HCl
1-Chlorobuta-1,3-diyne†
C_4HClF_6
1,1,1,4,4,4-Hexafluoro-2-chlorobut-2-ene†
C_4HClO_3
Chloromaleic Acid, *Anhydride*
$C_4HCl_2NO_2$
Dichloromaleic Acid, *Imide*
$C_4HCl_3O_2$
Chlorofumaric Acid, *Dichloride*
Mucochloric Acid, *Chloride*
C_4HCl_4N
2,3,4,5-Tetrachloropyrrole
$C_4HCl_6NO_2$
Trichloroacetic Acid, *Imide*
$C_4HF_7O_2$
Heptafluorobutyric Acid
Heptafluoroisobutyric Acid
$C_4HF_7O_5S$
Heptafluorobutyric Acid, *Sulphate*
C_4HI
1-Iodobuta-1,3-diyne†
C_4HI_4N
Tetra-iodopyrrole

C_4HN_3
Cyanoform
C_4H_2
1,3-Butadiyne
$C_4H_2Br_2Cl_2$
Dibromofumaric Acid, *Dichloride*
$C_4H_2Br_2N_2O_3$
5,5-Dibromobarbituric Acid
$C_4H_2Br_2O$
2,3-Dibromofuran
2,5-Dibromofuran
3,4-Dibromofuran
$C_4H_2Br_2O_2$
Fumaric Acid, *Dibromide*
$C_4H_2Br_2O_3$
3,5-Dibromofuran-2-carboxylic Acid
4,5-Dibromofuran-2-carboxylic Acid
Mucobromic Acid
$C_4H_2Br_2O_4$
Dibromofumaric Acid
Dibromomaleic Acid
$C_4H_2Br_2S$
2,3-Dibromothiophene
2,4-Dibromothiophene
2,5-Dibromothiophene
3,4-Dibromothiophene
$C_4H_2Br_6$
1,1,2,2,3,4-Hexabromocyclobutane
C_4H_2ClNOS
Thiazole-4-carboxylic Acid, *Chloride*
$C_4H_2Cl_2N_2O_3$
5,5-Dichlorobarbituric Acid
$C_4H_2Cl_2O_2$
Fumaric Acid, *Dichloride*
$C_4H_2Cl_2O_3$
2,3-Dichlorosuccinic Acid, *Anhydride*
Mucochloric Acid
$C_4H_2Cl_2O_4$
Dichloromaleic Acid
$C_4H_2Cl_2S$
2,3-Dichlorothiophene
2,4-Dichlorothiophene
2,5-Dichlorothiophene
3,4-Dichlorothiophene
$C_4H_2Cl_4O_3$
Dichloroacetic Acid, *Anhydride*
$C_4H_2Cl_5NO_3$
Oxamic Acid, *Pentachloroethyl ester*
$C_4H_2F_4O_4$
Tetrafluorosuccinic Acid
$C_4H_2F_6$
1,1,2,2,3,4-Hexafluorocyclobutane
$C_4H_2F_7NO$
Heptafluorobutyric Acid, *Amide*
$C_4H_2I_2O_4$
Di-iodofumaric Acid
$C_4H_2N_2$
Fumaric Acid, *Dinitrile*

$C_4H_2N_2O$
Isoxazole-5-carboxylic Acid, *Nitrile*
$C_4H_2N_2O_4$
Alloxan
$C_4H_2N_2O_4S$
2,4-Dinitrothiophene
2,5-Dinitrothiophene
$C_4H_2N_4O_2$
5-Diazouracil†
$C_4H_2O_2$
1,3-Cyclobutanedione
$C_4H_2O_2S$
Maleic Thioanhydride†
$C_4H_2O_3$
Maleic Anhydride
$C_4H_2O_4$
Acetylenedicarboxylic Acid
Squaric Acid†
C_4H_2S
Thiophyne†
$C_4H_3BrCl_2O_2$
Bromosuccinic Acid, *Dichloride*
C_4H_3BrO
2-Bromofuran
3-Bromofuran
$C_4H_3BrO_4$
Bromofumaric Acid
Bromomaleic Acid
C_4H_3BrS
2-Bromothiophene
3-Bromothiophene
$C_4H_3Br_2NO_2$
Mucobromic Acid, *Amide*
$C_4H_3Br_3O_4$
Tribromosuccinic Acid
C_4H_3Cl
Chlorobutatriene†
$C_4H_3ClN_2O_3$
5-Chlorobarbituric Acid
C_4H_3ClO
2-Chlorofuran
3-Chlorofuran
Tetrolic Acid, *Chloride*
$C_4H_3ClO_4$
Chlorofumaric Acid
Chloromaleic Acid
C_4H_3ClS
2-Chlorothiophene
$C_4H_3ClS_2$
2-Chloro-5-thiophenthiol†
$C_4H_3Cl_2N$
4,4-Dichlorocrotonic Acid, *Nitrile*
$C_4H_3Cl_2NO_2$
Mucochloric Acid, *Amide*
$C_4H_3Cl_2NO_3$
Dichloromaleic Acid, *Monoamide*
$C_4H_3Cl_3O$
4,4-Dichlorocrotonic Acid, *Chloride*
$C_4H_3Cl_3O_2$
Chlorosuccinic Acid, *Dichloride*
Trichloroacetic Acid, *Vinyl ester*
Trichloroacrylic Acid, *Me ester*
$C_4H_3F_7O$
2,2,3,3,4,4,4-Heptafluoro-1-butanol
C_4H_3IO
2-Iodofuran
3-Iodofuran
$C_4H_3IO_2$
Iodopropiolic Acid, *Me ester*
$C_4H_3IO_4$
Iodofumaric Acid
Iodomaleic Acid
C_4H_3IS
2-Iodothiophene
3-Iodothiophene
$C_4H_3NO_2$
Maleimide
$C_4H_3NO_2S$
2-Nitrothiophene
3-Nitrothiophene
1-Thiauracil†
Thiazole-2-carboxylic Acid
Thiazole-4-carboxylic Acid
Thiazole-5-carboxylic Acid
$C_4H_3NO_3$
Isoxazole-3-carboxylic Acid
Isoxazole-5-carboxylic Acid
2-Nitrofuran
3-Nitrofuran
$C_4H_3N_3O_2$
5-Nitropyrimidine†
$C_4H_3N_3O_4$
5-Isonitrosobarbituric Acid
Isovioluric Acid
5-Nitrouracil
Oxonic Acid
$C_4H_3N_3O_5$
5-Nitrobarbituric Acid
$C_4H_3N_5$
8-Azapurine†
$C_4H_3N_5O$
8-Azahypoxanthine
$C_4H_3N_5O_2$
8-Azaxanthine
C_4H_4
Butatriene
1-Buten-3-yne
Cyclobutadiene†
C_4H_4BrN
4-Bromocrotonic Acid, *Nitrile*
$C_4H_4Br_2Cl_2O$
3,3-Dibromo-1,1-dichlorobutan-2-one†
$C_4H_4Br_2OS$
Di-2-bromovinyl sulphoxide

$C_4H_4Br_2O_2$
2,3-Dibromocrotonic Acid
2,3-Dibromoisocrotonic Acid
2,3-Dibromosuccindialdehyde
Succinyl bromide

$C_4H_4Br_2O_3$
Bromoacetic Acid, *Anhydride*

$C_4H_4Br_2O_4$
2,3-Dibromosuccinic Acid

$C_4H_4Br_4$
1,1,2,2-Tetrabromocyclobutane
1,2,3,4-Tetrabromocyclobutane

C_4H_4ClN
2-Chlorocrotonic Acid, *Nitrile*
4-Chlorocrotonic Acid, *Nitrile*
3-Chloro-2-methylacrylic Acid, *Nitrile*

C_4H_4ClNS
2-Chloroallyl isothiocyanate

$C_4H_4Cl_2$
1,1-Dichloro-1,3-butadiene
1,4-Dichloro-1,3-butadiene
2,3-Dichloro-1,3-butadiene

$C_4H_4Cl_2O$
2-Chlorocrotonic Acid, *Chloride*
3-Chlorocrotonic Acid, *Chloride*
4-Chlorocrotonic Acid, *Chloride*
3-Chloroisocrotonic Acid, *Chloride*

$C_4H_4Cl_2OS$
Di-2-chlorovinyl sulphoxide

$C_4H_4Cl_2O_2$
2,3-Dichloroacrylic Acid, *Me ester*
2,3-Dichlorocrotonic Acid
4,4-Dichlorocrotonic Acid
Methylmalonic Acid, *Dichloride*
Succinyl chloride

$C_4H_4Cl_2O_2S$
Di-2-chlorovinyl sulphone

$C_4H_4Cl_2O_3$
Chloroacetic Acid, *Anhydride*
Diglycollic Acid, *Dichloride*
Malic Acid, *Dichloride*

$C_4H_4Cl_2O_4$
2,3-Dichlorosuccinic Acid

$C_4H_4Cl_2S$
Di-2-chlorovinyl sulphide

$C_4H_4Cl_3N$
4,4,4-Trichlorobutyric Acid, *Nitrile*

$C_4H_4Cl_4O$
2,2,3-Trichlorobutyric Acid, *Chloride*
2,4,4-Trichlorobutyric Acid, *Chloride*

$C_4H_4F_6O$
Di-(2,2,2-trifluoroethyl) Ether

$C_4H_4I_2N_2O_2$
Di-iodofumaric Acid, *Diamide*

$C_4H_4I_2O_3$
Iodoacetic Acid, *Anhydride*

$C_4H_4KO_7Sb$
Tartar emetic

$C_4H_4N_2$
Methylmalonic Acid, *Dinitrile*
Pyrazine
Pyridazine
Pyrimidine
Succinonitrile

$C_4H_4N_2O$
Hydroxy-methylmalonic Acid, *Dinitrile*
2-Hydroxypyrazine†
2-Hydroxypyrimidine
4-Hydroxypyrimidine
Pyridazine★, *Mono*-N-*oxide*†

$C_4H_4N_2OS$
Thiazole-2-carboxylic Acid, *Amide*
Thiazole-4-carboxylic Acid, *Amide*
2-Thiouracil

$C_4H_4N_2O_2$
Acetylenedicarboxylic Acid, *Diamide*
Cellocidin
2,5-Dihydroxypyrimidine†
4,5-Dihydroxypyrimidine†
4,6-Dihydroxypyrimidine†
2 Hydroxypyrazine, 4-N-*Oxide*†
Imidazole-1-carboxylic Acid
Imidazole-4-carboxylic Acid
Isoxazole-3-carboxylic Acid, *Amide*
Isoxazole-5-carboxylic Acid, *Amide*
Maleic Hydrazide
2-Nitropyrrole†
3-Nitropyrrole†
Pyrazole-1-carboxylic Acid
Pyrazole-3-carboxylic Acid
Pyrazole-4-carboxylic Acid
Pyridazine★, *Di*-N-*oxide*†
Tartaric Acid, *Dinitrile*
Uracil

$C_4H_4N_2O_2S$
5-Methyl-1,2,3-thiadiazole-4-carboxylic Acid
Thiobarbituric Acid

$C_4H_4N_2O_3$
Barbituric Acid
5-Hydroxyuracil
1-Methylimidazolidine-2,4,5-trione
2-Oxoimidazoline-4-carboxylic Acid

$C_4H_4N_2O_4$
5-Hydroxybarbituric Acid
Nitrocyanoacetic Acid, *Me ester*

$C_4H_4N_2O_5$
Alloxanic Acid

$C_4H_4N_2S$
Thiodiglycollic Acid, *Nitrile*

$C_4H_4N_2S_2$
2,4-Dithiouracil†

$C_4H_4N_4$
4-Amino-5-cyanoimidazole†
Diaminofumaronitrile†
Diaminomaleonitrile†

$C_4H_4N_4O_4$
Isocyanilic Acid
β-Isocyanilic Acid

$C_4H_4N_6$
8-Aza-adenine

$C_4H_4N_6O$
8-Azaguanine

C_4H_4O
3-Butyn-2-one
Furan
Methylcyclopropenone†
Tetrolaldehyde

C_4H_4OS
2-Thienol

$C_4H_4O_2$
Diketene
1,4-Dioxin
2-Hydroxyfuran
3-Hydroxyfuran
Maleic Dialdehyde
Propiolic Acid, *Me ester*
Tetrolic Acid

$C_4H_4O_3$
3,4-Dihydroxycrotonolactone
2,3-Dioxotetrahydrofuran
Formylacrylic Acid
4-Hydroxy-2-butynoic Acid
Succinic Anhydride

$C_4H_4O_3S$
Thiodiglycollic Acid, *Anhydride*

$C_4H_4O_4$
Diglycollic Acid, *Anhydride*
2,3-Dioxobutyric Acid
2,5-Dioxodioxan
Fumaric Acid
Maleic Acid
Malic Acid, *Anhydride*
Methylenemalonic Acid
Oxalic Acid, *Ethylene ester*

$C_4H_4O_5$
Formylmalonic Acid
Oxalacetic Acid
Oxiran-1,2-dicarboxylic Acid
Tartaric Acid, *Anhydride*

$C_4H_4O_6$
Dihydroxyfumaric Acid
Dihydroxymaleic Acid
Methane-tricarboxylic Acid

C_4H_4S
Thiophene

$C_4H_4S_2$
1,2-Dithiin†
1,4-Dithiin

C_4H_4Se
Selenole

C_4H_5Br
1-Bromo-1,3-butadiene
2-Bromo-1,3-butadiene

$C_4H_5BrCl_2O_2$
Bromodichloroacetic Acid, *Et ester*

C_4H_5BrO
2-Bromocrotonaldehyde

$C_4H_5BrO_2$
2-Bromoacrylic Acid, *Me ester*
2-Bromocrotonic Acid
3-Bromocrotonic Acid
4-Bromocrotonic Acid
2-Bromoisocrotonic Acid
3-Bromoisocrotonic Acid
3-Bromo-2-methylacrylic Acid

$C_4H_5BrO_4$
Bromosuccinic Acid

$C_4H_5BrO_5$
3-Bromomalic Acid

$C_4H_5Br_2ClO$
2,3-Dibromobutyric Acid, *Chloride*

$C_4H_5Br_2N$
3,4-Dibromobutyric Acid, *Nitrile*

$C_4H_5Br_3O_2$
Tribromoacetic Acid, *Et ester*
2,2,3-Tribromopropionic Acid, *Me ester*

C_4H_5Cl
1-Chloro-1,3-butadiene
4-Chloro-1,2-butadiene
Chloroprene

$(C_4H_5Cl)_n$
Chloroprene (μ-Polychloroprene)

C_4H_5ClO
2-Chlorocrotonaldehyde
3-Chlorocrotonaldehyde
Crotonic Acid, *Chloride*
Cyclopropanecarboxylic Acid, *Chloride*
2-Methylacrylic Acid, *Chloride*

$C_4H_5ClO_2$
2-Chloroacrylic Acid, *Me ester*
2-Chlorobutyrolactone
3-Chlorobutyrolactone
4-Chlorobutyrolactone
2-Chlorocrotonic Acid
3-Chlorocrotonic Acid
4-Chlorocrotonic Acid
Chloroformic Acid, *Allyl ester*
2-Chloroisocrotonic Acid
3-Chloroisocrotonic Acid
3-Chloro-2-methylacrylic Acid
2-Propen-2-ol, *Chloroformyl*

$C_4H_5ClO_3$
Malonic Acid, *Monochloride, Me ester*
Oxalic Acid, *Et ester, chloride*

$C_4H_5ClO_4$
Chlorosuccinic Acid

$C_4H_5ClO_5$
3-Chloromalic Acid

$C_4H_5Cl_2N$
3,4-Dichlorobutyric Acid, *Nitrile*

$C_4H_5Cl_2NO$
4,4-Dichlorocrotonic Acid, *Amide*

$C_4H_5Cl_2NO_2$
N-Chloroacetylchloroacetamide

$C_4H_5Cl_3O$
2,2-Dichlorobutyric Acid, *Chloride*
2,3-Dichlorobutyric Acid, *Chloride*

$C_4H_5Cl_3O$ (*continued*)
2,2,3-Trichlorobutanal
2,2,4-Trichlorobutanal

$C_4H_5Cl_3O_2$
Trichloroacetic Acid, *Et ester*
2,2,3-Trichlorobutyric Acid
2,2,4-Trichlorobutyric Acid
2,3,3-Trichlorobutyric Acid
2,4,4-Trichlorobutyric Acid
4,4,4-Trichlorobutyric Acid
2,2,3-Trichloropropionic Acid, *Me ester*

$C_4H_5Cl_3O_3$
3,3,3-Trichlorolactic Acid, *Me ester*

$C_4H_5Cl_5$
1,2,2,3,4-Pentachlorobutane

$C_4H_5FO_4$
Fluorosuccinic Acid

$C_4H_5IO_2$
2-Iodocrotonic Acid
3-Iodocrotonic Acid
4-Iodocrotonic Acid

$C_4H_5IO_3$
2-Iodoacetoacetic Acid

$C_4H_5IO_4$
Iodosuccinic Acid

C_4H_5N
Allyl cyanide
Crotonic Acid, *Nitrile*
Cyclopropanecarboxylic Acid, *Nitrile*
Isocrotonic Acid, *Nitrile*
2-Methylacrylic Acid, *Nitrile*
Pyrrole

C_4H_5NO
Acetoacetic Acid, *Nitrile*
3-Formylpropionic Acid, *Nitrile*
3-Methylisoxazole
5-Methylisoxazole
2-Oxobutyric Acid, *Nitrile*
Tetrolic Acid, *Amide*

C_4H_5NOS
4-Methyl-2-thiazolone

$C_4H_5NO_2$
Cyanoacetic Acid, *Me ester*
Cyanoformic Acid, *Et ester*
2-Cyanopropionic Acid
3-Cyanopropionic Acid
3-Methylisoxazolone
Succinimide

$C_4H_5NO_2S$
Carbethoxyl isothiocyanate

$C_4H_5NO_3$
Diglycollic Acid, *Imide*
Maleamic Acid
Malimide†

$C_4H_5NO_4$
Tartrimide†

$C_4H_5NO_5$
Isonitrosomalonic Acid, *Me ether*

C_4H_5NS
Allyl isothiocyanate
2-Aminothiophene★†
3-Aminothiophene
2-Methylthiazole
4-Methylthiazole
5-Methylthiazole

C_4H_5NSe
2-Methylselenazole

$C_4H_5N_3$
2-Aminopyrazine†
2-Aminopyrimidine
4-Aminopyrimidine
5-Aminopyrimidine
Iminodiacetic Acid, *Dinitrile*

$C_4H_5N_3O$
2-Amino-4-hydroxypyrimidine
4-Amino-5-hydroxypyrimidine†
Cytosine
Imidazole-4-carboxylic Acid, *Amide*
Pyrazole-1-carboxylic Acid, *Amide*

$C_4H_5N_3O_2$
6-Azathymine
Malonuric Acid, *Nitrile*
3-Methyl-4-nitropyrazole
4-Nitroimidazole, N-*Me*
5-Nitroimidazole, N-*Me*

$C_4H_5N_3O_3$
Isouramil
3-Methyl-4-nitropyrazole-5-one
Uramil

$C_4H_5N_3O_4$
Alloxanic Acid, *Amide*

$C_4H_5N_5O$
Azulmic Acid

$C_4H_5N_7$
Diaminoazapurine

C_4H_6
1,2-Butadiene
1,3-Butadiene
1-Butyne
2-Butyne
Cyclobutene
Bicyclo[1,1,0]butane†
Trimethylenemethane†

$C_4H_6As_2O_4$
Arsenoacetic Acid

C_4H_6BrClO
2-Bromobutyric Acid, *Chloride*

$C_4H_6BrClO_2$
Bromochloroacetic Acid, *Et ester*
3-Bromo-2-chloropropionic Acid, *Me ester*

C_4H_6BrN
4-Bromobutyric Acid, *Nitrile*

C_4H_6BrNO
4-Bromocrotonic Acid, *Amide*
4-Bromo-3-hydroxybutyric Acid, *Nitrile*

$C_4H_6Br_2$
1,1-Dibromo-1-butene
1,2-Dibromo-1-butene

$C_4H_6Br_2$ (*continued*)
2,3-Dibromo-1-butene
2,4-Dibromo-1-butene
3,4-Dibromo-1-butene
1,3-Dibromo-2-butene
1,4-Dibromo-2-butene
1,1-Dibromocyclobutane
1,2-Dibromocyclobutane
1,1(or 3,3)-Dibromo-2-methylpropene
1,3-Dibromo-2-methylpropene

$C_4H_6Br_2O$
2-Bromobutyric Acid, *Bromide*
2-Bromoisobutyric Acid, *Bromide*
2,3-Dibromobutanal
1,1-Dibromo-2-butanone†
3,3-Dibromo-2-butanone
3,4-Dibromo-2-butanone

$C_4H_6Br_2O_2$
Dibromoacetic Acid, *Et ester*
2,2-Dibromobutyric Acid
2,3-Dibromobutyric Acid
3,4-Dibromobutyric Acid
2,3-Dibromo-1,4-dioxan
2,3-Dibromo-2-methylpropionic Acid
2,2-Dibromopropionic Acid, *Me ester*
2,3-Dibromopropionic Acid, *Me ester*

$C_4H_6Br_4$
1,1,2,2-Tetrabromobutane
1,1,2,3-Tetrabromobutane
1,1,3,3-Tetrabromobutane
1,2,2,3-Tetrabromobutane
1,2,2,4-Tetrabromobutane
1,2,3,4-Tetrabromobutane
2,2,3,3-Tetrabromobutane

$C_4H_6ClFO_2$
Chlorofluoroacetic Acid, *Et ester*†

C_4H_6ClN
2-Chlorobutyric Acid, *Nitrile*
3-Chlorobutyric Acid, *Nitrile*
4-Chlorobutyric Acid, *Nitrile*
2-Chloro-2-methylpropionic Acid, *Nitrile*

C_4H_6ClNO
2-Chlorocrotonic Acid, *Amide*
3-Chlorocrotonic Acid, *Amide*
4-Chlorocrotonic Acid, *Amide*
4-Chloro-3-hydroxybutyric Acid, *Nitrile*
3-Chloro-2-hydroxy-2-methylpropionic Acid, *Nitrile*
3-Chloroisocrotonic Acid, *Amide*

$C_4H_6ClNO_2$
N-Acetylchloroacetamide
Acetylglycine, *Chloride*

$C_4H_6Cl_2$
3-Chloro-2-chloromethylpropene
1,2-Dichloro-1-butene
3,4-Dichloro-1-butene
1,1-Dichloro-2-butene
1,2-Dichloro-2-butene
1,3-Dichloro-2-butene
1,4-Dichloro-2-butene
2,3-Dichloro-2-butene
1,2-Dichlorocyclobutane
1,1-Dichloro-2-methylpropene
1,3-Dichloro-2-methylpropene
3,3-Dichloro-2-methylpropene

$C_4H_6Cl_2O$
2-Chlorobutyric Acid, *Chloride*
3-Chlorobutyric Acid, *Chloride*
4-Chlorobutyric Acid, *Chloride*
2-Chloro-2-methylpropionic Acid, *Chloride*
2,2-Dichlorobutanal
2,3-Dichlorobutanal
3,3-Dichloro-2-butanone
3,4-Dichloro-2-butanone
1,2-Dichlorovinyl ethyl Ether
2,2-Dichlorovinyl ethyl Ether

$C_4H_6Cl_2O_2$
Dichloroacetic Acid, *Et ester*
2,2-Dichlorobutyric Acid
2,3-Dichlorobutyric Acid
3,4-Dichlorobutyric Acid
4,4-Dichlorobutyric Acid
2,3-Dichloro-1,4-dioxan
2,5-Dichloro-1,4-dioxan
2,2-Dichloropropionic Acid, *Me ester*
2,3-Dichloropropionic Acid, *Me ester*

$C_4H_6Cl_2O_3$
3-Chloro-2-chloromethyl-2-hydroxypropionic Acid
3,3-Dichloro-2-hydroxy-2-methylpropionic Acid

$C_4H_6Cl_3NO$
2,2,3-Trichlorobutyric Acid, *Amide*
4,4,4-Trichlorobutyric Acid, *Amide*

$C_4H_6Cl_3NO_2$
Chloralacetamide
Chloralamide, *Me ether*

$C_4H_6Cl_4$
1,1,1,2-Tetrachlorobutane
1,2,2,3-Tetrachlorobutane
1,2,3,3-Tetrachlorobutane
1,2,3,4-Tetrachlorobutane
2,2,3,3-Tetrachlorobutane

$C_4H_6CrO_6$
Chromyl acetate

$C_4H_6F_2O_2$
Difluoroacetic Acid, *Et ester*

$C_4H_6INO_3$
Iodosuccinic Acid, β-*Monoamide*

$C_4H_6N_2$
3-Iminobutyronitrile
4-Methylene-Δ^1-pyrazoline†
1-Methylimidazole
2-Methylimidazole
4-Methylimidazole
1-Methylpyrazole
3(5)-Methylpyrazole
4-Methylpyrazole

$C_4H_6N_2O$
2-Cyanopropionic Acid, *Amide*
3-Cyanopropionic Acid, *Amide*
α-Diazobutyraldehyde†
4-Imidazolylmethanol
3-Methyl-2-pyrazolin-5-one
4-Methyl-2-pyrazolin-5-one

$C_4H_6N_2OS$
2-Thiohydantoin, 3-N-*Me*

$C_4H_6N_2O_2$
5-Aminomethyl-3-hydroxyisoxazole†
Aspartic Acid, *Imide*
β-Cyano-α-alanine
2-Diazopropionic Acid, *Me ester*
2-Diazobutyric Acid
Diazoacetic Acid, *Et ester*
Dihydrouracil†
2,5-Dioxopiperazine
Fumaric Acid, *Diamide*
Hexahydro-2,4-dioxopyrimidine
Maleamide
1-Methylhydantoin
3-Methylhydantoin
5-Methylhydantoin
2-Methyl-2-nitropropionic Acid, *Nitrile*

$C_4H_6N_2O_3$
Oxalacetic Acid, *Diamide*

$C_4H_6N_2O_4$
Azoformic Acid, *Di-Me ester*
Isonitrosomalonic Acid, *Monoamide*, *Me ether*
Malonuric Acid
Oxaluric Acid, *Me ester*

$C_4H_6N_2O_5$
5,6-Dihydro-2,4,5,5,6-pentahydroxypyrimidine

$C_4H_6N_2S$
Methimazole

$C_4H_6N_4$
2,4-Diaminopyrimidine
2,5-Diaminopyrimidine
4,5-Diaminopyrimidine
4,6-Diaminopyrimidine

$C_4H_6N_4O_2$
Divicine†
Hexahydro-2,6-dioxoimidazole[4,5-*d*]imidazole

$C_4H_6N_4O_3$
Allantoin

$C_4H_6N_4O_3S_2$
Acetazolamide

C_4H_6O
3-Buten-2-one
2-Butyn-1-ol
3-Butyn-1-ol
3-Butyn-2-ol
Crotonaldehyde
Cyclobutanone
Dimethylketene
Divinyl Ether
Ethyl ethynyl Ether
4-Hydroxy-1,2-butadiene
Methylcyclopropanone†
Methyl 2-propynyl Ether

$C_4H_6O_2$
Acetoacetaldehyde
Acrylic Acid, *Me ester*
Biacetyl
3-Butenoic Acid
2-Butyne-1,4-diol
Butyrolactone
Crotonic Acid
Cyclopropanecarboxylic Acid
1,2,3,4-Diepoxybutane
3,6-Dioxabicyclo[3,1,0]hexane
1,4-Dioxene
Formic Acid, *Allyl ester*
3-Hydroxy-2-methylpropionic Acid, *Lactone*
Isocrotonic Acid
2-Methylacrylic Acid
Succindialdehyde

$C_4H_6O_2S$
Diacetyl sulphide
β-Methylthioacrylic Acid†
3-Sulpholene†

$C_4H_6O_2S_2$
Diacetyl disulphide
1,2-Dithiolane-3-carboxylic Acid†
Dithio-oxalic Acid, *Di-Me ester*

$C_4H_6O_3$
Acetic Anhydride
Acetoacetic Acid
3-Formylpropionic Acid
Glyoxylic Acid, *Et ester*
2-Hydroxy-3-butenoic Acid
3-Hydroxy-2-butenoic Acid
4-Hydroxy-2-butenoic Acid
2-Methyloxiran-2-carboxylic Acid
3-Methyloxiran-2-carboxylic Acid
2-Oxobutyric Acid
Pyruvic Acid, *Me ester*
Reductone, *Mono-Me ether*

$C_4H_6O_3S$
Thio-oxalic Acid, *Di-Me ester*

$C_4H_6O_4$
Acetoxyacetic Acid
Acetyl peroxide
Dimethyl oxalate
Glycerol 1,2-Carbonate†
Methylmalonic Acid
Oxalic Acid, *Et ester*
Succinic Acid
Tartraldehyde

$C_4H_6O_4S$
Mercaptosuccinic Acid
Thiodiglycollic Acid

$C_4H_6O_4S_2$
Disulphidoacetic Acid

$C_4H_6O_5$
Diglycollic Acid
Glycollic Acid, *Anhydride*
Hydroxy-methylmalonic Acid
Hydroxymethyl-malonic Acid
Malic Acid

$C_4H_6O_6$
Tartaric Acid

$C_4H_6O_6S$
Sulphonyldiacetic Acid

$C_4H_6O_8$
Dioxosuccinic Acid

C_4H_6S
Divinyl sulphide

$C_4H_6S_3$
Dithioacetic Acid, "*Anhydride*"

C_4H_6Si
Silacyclopentadiene

C_4H_7Br
1-Bromo-1-butene
1-Bromo-2-butene
2-Bromo-1-butene
2-Bromo-2-butene
3-Bromo-1-butene
4-Bromo-1-butene
1-Bromo-2-methylpropene
3-Bromo-2-methylpropene

C_4H_7BrO
2-Bromoallyl Alcohol, *Me ether*
3-Bromoallyl Alcohol, *Me ether*
2-Bromobutanal
3-Bromobutanal
2-Bromocrotyl Alcohol
2-Bromoisobutyraldehyde
Butyric Acid, *Bromide*

$C_4H_7BrO_2$
Bromoacetic Acid, *Et ester*
2-Bromobutyric Acid
3-Bromobutyric Acid
4-Bromobutyric Acid
2-Bromoisobutyric Acid
3-Bromoisobutyric Acid
2-Bromopropionic Acid, *Me ester*
3-Bromopropionic Acid, *Me ester*

$C_4H_7BrO_3$
2-Bromo-3-hydroxybutyric Acid
2-Bromo-4-hydroxybutyric Acid
3-Bromo-2-hydroxybutyric Acid
4-Bromo-3-hydroxybutyric Acid
3-Bromo-2-hydroxy-2-methylpropionic Acid

$C_4H_7Br_3$
1,1,2-Tribromobutane
1,2,2-Tribromobutane
1,2,3-Tribromobutane
1,2,4-Tribromobutane
2,2,3-Tribromobutane
1,1,2-Tribromo-2-methylpropane
1,2,3-Tribromo-2-methylpropane

$C_4H_7Br_3O$
1,3,4-Tribromo-2-butanol
1,1,1-Tribromo-2-methyl-2-propanol

C_4H_7Cl
1-Chloro-1-butene
2-Chloro-1-butene
3-Chloro-1-butene
4-Chloro-1-butene
1-Chloro-2-butene
2-Chloro-2-butene
1-Chloro-2-methyl-1-propene
3-Chloro-2-methyl-1-propene

C_4H_7ClO
Butyric Acid, *Chloride*
1-Chloro-2-butanone
3-Chloro-2-butanone
4-Chloro-2-butanone
2-Chloro-2-buten-1-ol
3-Chloro-2-buten-1-ol
2-Chlorobutyraldehyde
3-Chlorobutyraldehyde
4-Chlorobutyraldehyde
2-Chloro-2-methylpropanal
2-Chlorovinyl ethyl ether
Isobutyric Acid, *Chloride*

$C_4H_7ClO_2$
2-Chlorobutyric Acid
3-Chlorobutyric Acid
4-Chlorobutyric Acid
Chloroformic Acid, *Propyl ester*
Chloroformic Acid, *Isopropyl ester*
2-Chloro-2-methylpropionic Acid
3-Chloro-2-methylpropionic Acid
2-Chloropropionic Acid, *Me ester*
3-Chloropropionic Acid, *Me ester*
Ethoxyacetic Acid, *Chloride*
Ethyl chloroacetate
2-Methoxypropionic Acid, *Chloride*
3-Methoxypropionic Acid, *Chloride*

$C_4H_7ClO_3$
2-Chloro-3-hydroxybutyric Acid
3-Chloro-2-hydroxybutyric Acid
4-Chloro-2-hydroxybutyric Acid
4-Chloro-3-hydroxybutyric Acid
2-Chloro-3-hydroxy-2-methylpropionic Acid
3-Chloro-2-hydroxy-2-methylpropionic Acid
3-Chlorolactic Acid, *Me ester*

$C_4H_7Cl_2NO$
2,2-Dichlorobutyric Acid, *Amide*
3,4-Dichlorobutyric Acid, *Amide*

$C_4H_7Cl_2NO_2$
1,3-Dichloro-2-propyl carbamate

$C_4H_7Cl_2O_4P$
Dichlorvos†

$C_4H_7Cl_3$
1,1,3-Trichlorobutane
1,2,3-Trichlorobutane
1,2,4-Trichlorobutane
2,2,3-Trichlorobutane

$C_4H_7Cl_3O$
Chloretone
2,2,3-Trichloro-1-butanol
1,3,4-Trichloro-2-butanol

$C_4H_7Cl_3O_2$
Chloral, *Di-Me acetal*
Chloral, *Mono-Et acetal*
2,2,3-Trichloro-1,1-butanediol

C_4H_7FO
2-Fluorobutanal†

$C_4H_7FO_2$
Fluoroacetic Acid, *Et ester*

$C_4H_7F_3O_2$
2,2,2-Trifluoroacetaldehyde, *Et hemiacetal*

C_4H_7HgN
Mercuri-propyl cyanide

C_4H_7I
1-Iodo-1-butene
1-Iodo-2-butene

C_4H_7IO
Butyric Acid, *Iodide*

$C_4H_7IO_2$
Iodoacetic Acid, *Et ester*
2-Iodobutyric Acid
4-Iodobutyric Acid
2-Iodo-2-methylpropionic Acid
3-Iodo-2-methylpropionic Acid
2-Iodopropionic Acid, *Me ester*
3-Iodopropionic Acid, *Me ester*

C_4H_7N
3-Aminopropyne, N-*Me*
Butyric Acid, *Nitrile*
Isobutyric Acid, *Nitrile*
Isopropyl isocyanide
3-Methylamino-1-propyne
Propyl isocyanide
3-Pyrroline

C_4H_7NO
Acrylic Acid, *N*-Methylamide
3-Butenoic Acid, *Amide*
Crotonic Acid, *Amide*
Cyclopropanecarboxylic Acid, *Amide*
Ethoxyacetic Acid, *Nitrile*
Formamide, N-*Allyl*
2-Hydroxybutyric Acid, *Nitrile*
3-Hydroxybutyric Acid, *Nitrile*
4-Hydroxybutyric Acid, *Nitrile*
2-Hydroxy-2-methylpropionic Acid, *Nitrile*
Isocrotonic Acid, *Amide*
Isopropyl cyanate†
Isopropyl isocyanate
2-Methoxypropionic Acid, *Nitrile*
3-Methoxypropionic Acid, *Nitrile*
2-Methylacrylic Acid, *Amide*
Propyl cyanate†
2-Pyrrolidone
3-Pyrrolidone

C_4H_7NOS
4-Methyl-2-oxazolidinethione

$C_4H_7NO_2$
Acetoacetic Acid, *Amide*
3-Aminocrotonic Acid
4-Aminocrotonic Acid
Azetidine-2-carboxylic Acid
Carbamic Acid, *Allyl ester*
Diacetamide
Homoserine, *Lactone*
2-Hydroxy-3-butenoic Acid, *Amide*
Isonitrosoacetone, *Me ether*
2-Methyl-1-nitro-1-propene
2-Oxobutyric Acid, *Amide*

$C_4H_7NO_2S$
Thio-oxamic Acid, *Et ester*

$C_4H_7NO_3$
Acetylglycine
Dimethyloxamic Acid
Ethyloxamic Acid
N-Formyloxymethyl-*N*-methylformamide†
Hydroxyiminoacetic Acid, *Et ester*
Methyloxamic Acid, *Me ester*
Oxamic Acid, *Et ester*
Succinamic Acid

$C_4H_7NO_3S$
(Carbamoylthio)acetic Acid, *Me ester*
Mercaptosuccinic Acid, 4-*Monoamide*
Thiodiglycollic Acid, *Monoamide*

$C_4H_7NO_4$
Aminomethylmalonic Acid
Aspartic Acid
Carbomethoxyglycine
Diglycollic Acid, *Monoamide*
Hadacidin, *Me ester*
Iminodiacetic Acid
α-Malamic Acid
β-Malamic Acid
Methylaminomalonic Acid
2-Methyl-2-nitropropionic Acid
Nitroacetic Acid, *Et ester*
2-Nitrobutyric Acid

$C_4H_7NO_5$
Tartramidic Acid

C_4H_7NS
Isopropyl isothiocyanate
Isopropyl thiocyanate
Propyl isothiocyanate
Thiocyanic Acid, *Propyl ester*

$C_4H_7NS_2$
Dithiocarbamic Acid, *Allyl ester*

$C_4H_7N_3$
1,5-Dimethyl-1,2,3-triazole
4,5-Dimethyl-1,2,3-triazole
3,5-Dimethyl-1,2,4-triazole
4-Imidazolylmethylamine

$C_4H_7N_3O$
Alacreatinine
Creatinine

$C_4H_7N_3O_2$
Azidoacetic Acid, *Et ester*
2-Azidobutyric Acid
4-Azidobutyric Acid
2-Azido-2-methylpropionic Acid
Trigonic Acid

$C_4H_7N_3O_3$
Isonitrosomalonic Acid, *Diamide, Me ether*
Malonuric Acid, *Amide*

C_4H_8
1-Butene
2-Butene
Cyclobutane
Methylcyclopropane
2-Methylpropene

C_4H_8BCl
1-Chloroborolane

C_4H_8BrNO
2-Bromobutyric Acid, *Amide*
3-Bromobutyric Acid, *Amide*
2-Bromoisobutyric Acid, *Amide*

$C_4H_8Br_2$
1,1-Dibromobutane
1,2-Dibromobutane
1,3-Dibromobutane
1,4-Dibromobutane
2,2-Dibromobutane

$C_4H_8Br_2$ (*continued*)
2,3-Dibromobutane
1,2-Dibromo-2-methylpropane
1,3-Dibromo-2-methylpropane

$C_4H_8Br_2O$
2,3-Dibromo-1-butanol
3,4-Dibromo-1-butanol
1,3-Dibromo-2-butanol
1,4-Dibromo-2-butanol
2,3-Dibromo-2-methyl-1-propanol
1,3-Dibromo-2-propanol, *Me ether*

C_4H_8ClNO
Chloroacetamide, N-*Et*
Chloroacetamide, N-*Di-Me*
2-Chlorobutyric Acid, *Amide*

$C_4H_8ClO_2P$
2-Chloro-1,3,2-dioxaphosphepan†

$C_4H_8Cl_2$
1,1-Dichlorobutane
1,2-Dichlorobutane
1,3-Dichlorobutane
1,4-Dichlorobutane
2,2-Dichlorobutane
2,3-Dichlorobutane
1,1-Dichloro-2-methylpropane
1,2-Dichloro-2-methylpropane
1,3-Dichloro-2-methylpropane

$C_4H_8Cl_2O$
Di-1-chloroethyl Ether
Di-2-chloroethyl Ether
1,2-Dichloroethyl ethyl Ether

$C_4H_8Cl_2OS$
Di-1-chloroethyl sulphoxide
Di-2-chloroethyl sulphoxide

$C_4H_8Cl_2O_2$
Dichloroacetaldehyde, *Di-Me-acetal*
Dichloroacetaldehyde, *Et acetal*
1,4-Dichloro-2,3-butanediol

$C_4H_8Cl_2O_2S$
Di-1-chloroethyl sulphone
Di-2-chloroethyl sulphone

$C_4H_8Cl_2S$
Di-1-chloroethyl sulphide
Di-2-chloroethyl sulphide

$C_4H_8HgO_2$
Mercuri-ethyl acetate

$C_4H_8I_2$
1,3-Di-iodobutane
1,4-Di-iodobutane

$C_4H_8NO_7P$
β-Aspartyl phosphate

$C_4H_8N_2$
2-Amino-2-methylpropionic Acid, *Nitrile*
1,1′-Biaziridinyl
2-Butyne-1,4-diamine
N-Dimethylglycine, *Nitrile*
N-Ethylglycine, *Nitrile*
3-Methyl-1-pyrazoline†
4-Methyl-1-pyrazoline†
3-Methyl-2-pyrazoline
5-Methyl-2-pyrazoline
Lysidine

$C_4H_8N_2O$
Allylurea
Gyromitrin†

$C_4H_8N_2O_2$
Acetylglycine, *Amide*
N^1-Acetyl-N^2-methylurea
1,1-Diacetylhydrazine
1,2-Diacetylhydrazine
Dimethylglyoxime
Dimethyloxamic Acid, *Amide*
N,N-Dimethyloxamide
Ethyloxamic Acid, *Amide*
Malonamide, N-*Me*
Methylglyoxime, *Mono-Me ether*
Methylmalonic Acid, *Diamide*
Succinamide

$C_4H_8N_2O_2S$
Thiodiglycollic Acid, *Diamide*

$C_4H_8N_2O_2S_2$
Disulphidoacetic Acid, *Diamide*

$C_4H_8N_2O_3$
Allophanic Acid, *Et ester*
Aminomethylmalonic Acid, *Monoamide*
α-Asparagine★ (Isoasparagine†)
β-Asparagine
Diglycollic Acid, *Diamide*
Glycylglycine
Hydroxy-methylmalonic Acid, *Diamide*
Iminodiacetic Acid, *Monoamide*
Isoasparagine† (α-Asparagine★)
Malamide
2-Methyl-2-nitropropionic Acid, *Amide*
2-Ureidopropionic Acid
3-Ureidopropionic Acid

$C_4H_8N_2O_4$
Allophanic Acid, *Mono-glycol ester*
O-Carbamoyl-D-serine
2,3-Diaminosuccinic Acid
Ethylenedicarbamic Acid
Hydrazodiformic Acid, *Di-Me ester*
β-Hydroxyasparagine†
Tartramidic Acid, *Amide*

$C_4H_8N_2S$
Allylthiourea

$C_4H_8N_2S_4$
Dimethylthiuram disulphide
Ethylene*bis*dithiocarbamic Acid

$C_4H_8N_4$
l-Amino-4,5-dimethyl-1,2,3-triazole
l-Amino-2,5-dimethyl-1,3,4-triazole

$C_4H_8N_4O$
2-Azidobutyric Acid, *Amide*
2-Azido-2-methylpropionic Acid, *Amide*

$C_4H_8N_4O_4$
Allantoic Acid

C_4H_8O
Allyl methyl Ether
Butanal
2-Butanone
2-Buten-1-ol
3-Buten-1-ol

C_4H_8O (*continued*)
3-Buten-2-ol
Cyclobutanol
2,2-Dimethyloxiran
2,3-Epoxybutane
Epoxybutane
Ethoxyethylene
Isobutanal
2-Methyl-2-propen-1-ol
2-Propen-2-ol, *Me ether*
Tetrahydrofuran

C_4H_8OS
1,4-Oxathiane
Thioacetic Acid, S-*Et ester*
Thioacetic Acid, O-*Et ester*
Thiobutyric Acid
Thiopropionic Acid, S-*Me ester*

$C_4H_8OS_2$
Isopropylxanthic Acid
Methoxydithioformic Acid, *Et ester*
Xanthogenic Acid, *Me ester*

$C_4H_8O_2$
Aldol
3-Butene-1,2-diol
Butyric Acid
1,3-Dioxan
1,4-Dioxan
Ethyl acetate
Formic Acid, *Propyl ester*
Glycollaldehyde, *Et ether*
1-Hydroxy-2-butanone
3-Hydroxy-2-butanone
4-Hydroxy-2-butanone
2-Hydroxymethyloxiran, *Me ether*
2-Hydroxy-2-methylpropanal
Isobutyric Acid
Methoxyacetone
Methyl propionate
1,2,3,4-Tetrahydro-3-hydroxyfuran†

$C_4H_8O_2S$
2-Mercaptobutyric Acid
3-Mercaptobutyric Acid
4-Mercaptobutyric Acid
2-Mercapto-2-methylpropionic Acid
3-Mercapto-2-methylpropionic Acid
3-Mercaptopropionic Acid, *Me ester*
Methylmercaptoacetic Acid, *Me ester*
2-Methylmercaptopropionic Acid
3-Methylmercaptopropionic Acid
2-Methylthietane sulphone
Sulpholane†
Thioglycollic Acid, *Et ester*
Thioglycollic Acid, S-*Et*

$C_4H_8O_2S_2$
4,5-Dihydroxy-1,2-dithiane†
1α,2β-Dimethylethylenethionosulphite†
1β,2β-Dimethylethylenethionosulphite†

$C_4H_8O_3$
2-Hydroxybutyric Acid
3-Hydroxybutyric Acid
4-Hydroxybutyric Acid
2-Hydroxy-2-methylpropionic Acid
3-Hydroxy-2-methylpropionic Acid
3-Hydroxypropionic Acid, *Me ester*
Ethoxyacetic Acid
Glycollic Acid, *Et ester*
Glyoxal★, *Mono-Et hemiacetal*†
Lactic Acid, *Me ester*
Methoxyacetic Acid, *Me ester*
2-Methoxypropionic Acid
3-Methoxypropionic Acid
1,2-Methyleneglycerol
1,3-Methyleneglycerol

$C_4H_8O_3S$
4-Hydroxy-1-butanesulphonic Acid Sultone†
Vinylsulphonic Acid, *Et ester*

$C_4H_8O_4$
2,3-Dihydroxybutyric Acid
2,3-Dihydroxy-1,4-dioxan
2,5-Dihydroxy-1,4-dioxan
2,3-Dihydroxy-2-methylpropionic Acid
Erythrose
Erythrulose
Glyceric Acid, *Me ester*
α-Monoformin
Threose

$C_4H_8O_5$
Erythronic Acid
Threonic Acid

$C_4H_8O_5S$
Sulphoacetic Acid★, C-*Et ester*†

$C_4H_8O_8$
Octahydroxycyclobutane†

C_4H_8S
Allyl methyl sulphide
2-Methylthietane
Tetrahydrothiophene

$C_4H_8S_2$
1,2-Dithian
1,3-Dithian
1,4-Dithian
Dithioacetic Acid, *Et ester*

$C_4H_8S_3$
3,5-Dimethyl-1,2,4-trithiolane†

$C_4H_9AsO_3$
2-Ethoxy-1,3,2-dioxa-arsolan

C_4H_9Br
1-Bromobutane
2-Bromobutane
1-Bromo-2-methylpropane
2-Bromo-2-methylpropane

C_4H_9BrHg
Mercuri-butyl bromide
Mercuri-*tert*-butyl bromide
Mercuri-1-methylpropyl bromide
Mercuri-2-methylpropyl bromide

C_4H_9BrO
2-Bromo-1-butanol†
3-Bromo-1-butanol†
4-Bromo-1-butanol†
1-Bromo-2-butanol
3-Bromo-2-butanol
4-Bromo-2-butanol†

C_4H_9BrO (*continued*)
1-Bromoethyl ethyl Ether
2-Bromoethyl ethyl Ether
3-Bromo-1-propanol, *Me ether*

C_4H_9Cl
1-Chlorobutane
2-Chlorobutane
1-Chloro-2-methylpropane
2-Chloro-2-methylpropane

C_4H_9ClHg
Mercuri-butyl chloride
Mercuri-*tert*-butyl chloride
Mercuri-1-methylpropyl chloride

C_4H_9ClO
tert-Butyl hypochlorite
2-Chloro-1-butanol
3-Chloro-1-butanol
4-Chloro-1-butanol
1-Chloro-2-butanol
3-Chloro-2-butanol
4-Chloro-2-butanol
1-Chloroethyl ethyl Ether
2-Chloroethyl ethyl Ether
1-Chloro-2-methyl-2-propanol
2-Chloro-2-methyl-1-propanol
1-Chloro-2-propanol, *Me ether*
1-Chloro-1-propyl methyl Ether
1-Chloro-2-propyl methyl Ether
2-Chloro-1-propyl methyl Ether
3-Chloro-1-propyl methyl Ether

$C_4H_9ClO_2$
Chloroacetaldehyde, *Di-Me acetal*
2-Chloro-1,3-propanediol, *Me ether*

C_4H_9ClS
2-Chloroethyl ethyl sulphide
2-Chloro-1-propyl methyl sulphide
3-Chloro-1-propyl methyl sulphide

$C_4H_9Cl_3OP$
Butylphosphonic Acid, *Dichloride*†

C_4H_9F
1-Fluorobutane
2-Fluorobutane

C_4H_9HgI
Mercuri-butyl iodide
Mercuri-2-methylpropyl iodide

C_4H_9I
1-Iodobutane
2-Iodobutane
1-Iodo-2-methylpropane
2-Iodo-2-methylpropane

C_4H_9IO
Ethyl-2-iodoethyl Ether
3-Iodo-1-propanol, *Me ether*

C_4H_9N
1-Amino-2-butene
Cyclobutylamine
N-Methylallylamine
Pyrrolidine

C_4H_9NO
Acetimidic Acid, *Et ester*
Butyric Acid, *Amide*
N-Dimethylacetamide
N-Ethylacetamide
Formimidic Acid, *Isopropyl ester*
Isobutyric Acid, *Amide*
Morpholine

C_4H_9NOS
Thiocarbamic Acid, O-*Propyl ester*
Thiocarbamic Acid, S-*Propyl ester*
Thiocarbamic Acid, S-*Isopropyl ester*
Thioglycollic Acid, S-*Et*, *Amide*

$C_4H_9NO_2$
Acethydroxamic Acid, *Et ether*
Alanine, *Me ester*
2-Aminobutyric Acid
3-Aminobutyric Acid
4-Aminobutyric Acid
2-Amino-2-methylpropionic Acid
3-Amino-2-methylpropionic Acid
3-Aminopropionic Acid, *Me ester*
Butyl nitrite
2-Butyl nitrite
tert-Butyl nitrite
Carbamic Acid, *Propyl ester*
Dimethylcarbamic Acid, *Me ester*
N-Dimethylglycine
Ethoxyacetic Acid, *Amide*
N-Ethylglycine
Ethyl methylaminoformate
Glycine, *Et ester*
3-Hydroxybutyric Acid, *Amide*
2-Hydroxy-2-methylpropionic Acid, *Amide*
Isobutyl nitrite
2-Methoxypropionic Acid, *Amide*
3-Methoxypropionic Acid, *Amide*
2-Methylaminopropionic Acid
3-Methylaminopropionic Acid
2-Methyl-2-nitropropane
2-Methyl-3-nitropropane
1-Nitrobutane
2-Nitrobutane

$C_4H_9NO_2S$
Cysteine, *Me ester*
Cysteine, S-*Me*†
Homocysteine
1,4-Thiazan, 1,1-*Di-amide*

$C_4H_9NO_3$
3-Amino-2-hydroxybutyric Acid
3-Amino-4-hydroxybutyric Acid
4-Amino-3-hydroxybutyric Acid
3-Amino-2-hydroxy-2-methylpropionic Acid
Butyl nitrate
2-Butyl nitrate
tert-Butyl nitrate
Homoserine
Isobutyl nitrate
2-Methyl-2-nitropropanol
2-Nitro-1-butanol
1-Nitro-2-butanol
3-Nitro-2-butanol
2-Nitroethanol, *Et ether*
Serine, *Me ester*
Threonine

$C_4H_9NO_3S$
S-Methyl-L-cysteine sulphoxide

$C_4H_9NO_4$
Erythronic Acid, *Amide*
Threonic Acid, *Amide*

C_4H_9NS
1,4-Thiazan

$C_4H_9NS_2$
Dimethyldithiocarbamic Acid, *Me ester*
Dithiocarbamic Acid, *Propyl ester*
Dithiocarbamic Acid, *Isopropyl ester*
Methyldithiocarbamic Acid, *Et ester*

$C_4H_9N_3O_2$
Aminomethylmalonic Acid, *Diamide*
L-Aspartic Acid, *Diamide*
Biuret, N-*Et*
Creatine
DL-2-Guanidinopropionic Acid
3-Guanidinopropionic Acid
Iminodiacetic Acid, *Diamide*

$C_4H_9OPS_2$
2-Ethoxy-1,3,2-dithiaphospholan

$C_4H_9O_3P$
2-Ethoxy-1,3,2-dioxaphospholan
2-Methoxy-1,3,2-dioxaphosphane

$C_4H_9O_3PS$
2-Ethoxy-1,3,2-dioxaphospholan-2-thione

C_4H_{10}
Butane
Isobutane

$C_4H_{10}ClO_2P$
O,O-Diethylphosphorochloridite

$C_4H_{10}ClO_2PS$
O,O-Diethylphosphorochloridothionate

$C_4H_{10}ClO_3P$
O,O-Diethylphosphorochloridate

$C_4H_{10}Hg$
Mercury diethyl

$C_4H_{10}HgO$
Mercuri-butyl hydroxide

$C_4H_{10}N_2$
tert-Butyldi-imide†
Butyramidine
Piperazine

$C_4H_{10}N_2O$
2-Amino-2-methylpropionic Acid, *Amide*
2-Aminobutyric Acid, *Amide*
Butyrhydrazide
Diethylnitrosamine
1-Ethyl-3-methylurea
Isopropylurea
Propylurea

$C_4H_{10}N_2O_2$
2-Amino-3-methylaminopropionic Acid†
Butylnitramine
2-Butylnitramine
2,3-Diaminobutyric Acid
2,4-Diaminobutyric Acid
2-Hydrazinobutyric Acid
2-Hydrazine-2-methylpropionic Acid

$C_4H_{10}N_2O_3$
Canaline

$C_4H_{10}N_2S$
1-Ethyl-3-methylthiourea
Isopropylthiourea
Propylthiourea

$C_4H_{10}N_3O_3P$
Phosphorazidic Acid, *Di-Et ester*†

$C_4H_{10}O$
1-Butanol
2-Butanol
Diethyl Ether
Isobutanol
Isopropyl methyl Ether
2-Methyl-2-propanol
Methyl propyl Ether

$C_4H_{10}OS$
tert-Butylsulphenic Acid†
Diethyl sulphoxide
2-Hydroxy-1-ethanethiol, S-*Et*
3-Methylmercaptopropanol

$C_4H_{10}OS_2$
3-Hydroxy-1,2-propanedithiol, 3-*Me ether*

$C_4H_{10}O_2$
Acetaldehyde, *Di-Me acetal*
1,2-Butanediol
1,3-Butanediol
1,4-Butanediol
2,3-Butanediol
Diethyl peroxide
2-Ethoxyethanol
Ethylene Glycol, *Di-Me ether*
2-Methylpropanediol
1,2-Propanediol, 1-*Me ether*
1,3-Propanediol, *Me ether*

$C_4H_{10}O_2S$
Diethyl sulphone
2,2′-Thiodiglycol

$C_4H_{10}O_2S_2$
Diethyl disulphoxide
2,3-Dihydroxybutane-1,4-dithiol†

$C_4H_{10}O_3$
1,2,4-Butanetriol
1,3,3-Butanetriol
Di-(2-hydroxyethyl) Ether
Glycerol, 1-*Me ether*
Glycerol, 2-*Me ether*
Glycollaldehyde, *Di-Me acetal*
1-Methylglycerol
Orthoformic Acid, *Tri-Me ester*

$C_4H_{10}O_3S$
Diethyl sulphite
Ethanesulphonic Acid, *Et ester*

$C_4H_{10}O_4$
Erythritol
Threitol

$C_4H_{10}O_4S$
Diethyl sulphate
Isethionic Acid, *Et ether*

$C_4H_{10}S$
Butanethiol
2-Butanethiol
Diethyl sulphide

$C_4H_{10}S$ (*continued*)
Isobutanethiol
Isopropyl methyl sulphide
2-Methyl-2-propanethiol

$C_4H_{10}S_2$
Diethyl disulphide
1,2-Ethanedithiol, *Di-Me ether*
1,2-Ethanedithiol, *Mono-Et ether*

$C_4H_{10}Se$
Diethyl selenide
Methyl propyl selenide

$C_4H_{10}Se_2$
Diethyl diselenide

$C_4H_{10}Sn$
Tin diethyl

$C_4H_{10}Zn$
Zinc diethyl

$C_4H_{11}As$
Diethylarsine

$C_4H_{11}BO_2$
Butaneboronic Acid
Ethaneboronic Acid, *Di-Et ester*

$C_4H_{11}N$
Butylamine
2-Butylamine
tert-Butylamine
Diethylamine
N-Ethyldimethylamine
Isobutylamine
Isopropylmethylamine
Methylpropylamine

$C_4H_{11}NO$
2-Amino-1-butanol
3-Amino-1-butanol
4-Amino-1-butanol
1-Amino-2-butanol
3-Amino-2-butanol
4-Amino-2-butanol
2-Aminoethyl ethyl Ether
1-Amino-2-methyl-2-propanol
Butoxyamine
N-Diethylhydroxylamine
O,N-Diethylhydroxylamine
2-Dimethylaminoethanol
2-Ethylaminoethanol
3-Methylaminopropanol

$C_4H_{11}NOS$
3-Aminopropyl methyl sulphoxide

$C_4H_{11}NO_2$
3-Amino-1,2-propanediol, N-*Me*
Di-(2-hydroxyethyl)amine
3-Methylamino-1,2-propanediol

$C_4H_{11}NO_2S$
2-Amino-1-methylethyl methylsulphone
3-Aminopropyl methyl sulphone

$C_4H_{11}NO_3S$
Taurine, N-*Di-Me*
Taurine, N-*Et*

$C_4H_{11}NO_4S_2$
N-Di-[methylsulphonyl]ethylamine

C

$C_4H_{11}NS$
2-Amino-1-methylethyl methyl sulphide
3-Aminopropyl methyl sulphide

$C_4H_{11}N_3$
1-Propylguanidine

$C_4H_{11}N_3O$
Creatinol

$C_4H_{11}N_5$
Ethyldiguanide
Metformin

$C_4H_{11}O_2P$
Dimethylphosphinic Acid, *Et ester*

$C_4H_{11}O_2PS_2$
O,O-Diethyl hydrogenphosphorodithioate

$C_4H_{11}O_3P$
Butylphosphonic Acid†

$C_4H_{11}O_3PS$
O,O-Diethyl-*O*-hydrogenphosphorothionate

$C_4H_{12}As_2$
Cacodyl

$C_4H_{12}As_2O$
Cacodyl oxide

$C_4H_{12}As_2S$
Cacodyl sulphide

$C_4H_{12}BrN$
Tetramethylammonium bromide

$C_4H_{12}ClN$
Tetramethylammonium chloride

$C_4H_{12}FN_2OP$
N,N,N',N'-Tetramethylphosphorodiamidic fluoride★†

$C_4H_{12}IN$
Tetramethylammonium iodide

$C_4H_{12}NO_2P$
Ethylphosphoramidite

$C_4H_{12}NO_2PS$
Ethylphosphoramidothionate

$C_4H_{12}NO_3P$
Ethylphosphoramidate

$C_4H_{12}N_2$
1,3-Diaminobutane
2,3-Diaminobutane
1,2-Diamino-2-methylpropane
1,3-Diamino-2-methylpropane†
1,1-Diethylhydrazine
1,2-Diethylhydrazine
N-Dimethylethylenediamine
NN'-Dimethylethylenediamine
N^1-Isopropyl-N^2-methylhydrazine
Putrescine

$C_4H_{12}N_2O$
2-(2-Aminoethyl)aminoethanol
Di-(2-aminoethyl) Ether

$C_4H_{12}N_2OS_2$
Cystamine *S*-monoxide†

$C_4H_{12}N_2O_2S_2$
2,2'-Disulphinylbisethylamine

$C_4H_{12}N_2S$
Di-(2-aminoethyl)sulphide
$C_4H_{12}N_2S_2$
Dithiodimethylamine
$C_4H_{12}O_3P_2$
Dimethylphosphinic Acid, *Anhydride*
$C_4H_{12}O_4Si$
Methyl orthosilicate†
$C_4H_{12}Pb$
Lead tetramethyl
$C_4H_{12}S_2Si_2$
Tetramethylcyclodisilthiane
$C_4H_{12}Si$
Tetramethylsilane
$C_4H_{12}Sn$
Tin tetramethyl
$C_4H_{13}NO$
Tetramethylammonium hydroxide
$C_4H_{13}NO_2$
Hydroxymethyltrimethylammonium Hydroxide
$C_4H_{13}N_3$
Di-2-aminoethylamine
$C_4H_{14}Cl_2N_2O_2S_2$
2-Aminoethyl 2-aminoethanethiolsulphonate dihydrochloride†
$C_4H_{14}NO_2PS_2$
Ammonium diethyl thiophosphate
$C_4H_{14}NO_3PS$
Di-*O*,*O*-ethyl ammonium phosphorothionate
$C_4H_{16}O_4Si_4$
2,4,6,8-Tetramethylcyclotetrasiloxane
$C_4I_2N_2$
Di-iodofumaronitrile†
C_4N_2
Acetylenedicarboxylic Acid, *Dinitrile*
C_4NiO_4
Nickel carbonyl

C_5

C_5Br_5N
Pentabromopyridine
C_5Cl_5N
Pentachloropyridine
C_5Cl_6
Hexachlorocyclopentadiene
C_5Cl_6O
Hexachlorocyclopent-2-en-1-one
Hexachlorocyclopent-3-en-1-one
C_5Cl_8
Octachlorocyclopentene
C_5F_5N
Pentafluoropyridine†
C_5F_8
Octafluoropenta-1,2-diene†
Octafluoropenta-2,3-diene†
C_5F_{12}
Dodecafluoropentane
C_5HBr_4N
2,3,4,5-Tetrabromopyridine
2,3,4,6-Tetrabromopyridine
2,3,5,6-Tetrabromopyridine
C_5HCl_4N
2,3,4,5-Tetrachloropyridine
2,3,4,6-Tetrachloropyridine
2,3,5,6-Tetrachloropyridine
$C_5H_2Br_2O_2$
3,5-Dibromo-4-pyrone
$C_5H_2Br_3N$
2,3,4-Tribromopyridine
2,3,5-Tribromopyridine
2,3,6-Tribromopyridine
2,4,5-Tribromopyridine
2,4,6-Tribromopyridine
3,4,5-Tribromopyridine
$C_5H_2Br_6O_2$
1,1,1,5,5,5-Hexabromo-2,4-pentanedione
$C_5H_2Br_6O_3$
Bromalide
$C_5H_2ClNO_4$
5-Nitrofuran-2-carboxylic Acid, *Chloride*
$C_5H_2Cl_3N$
2,3,4-Trichloropyridine
2,3,5-Trichloropyridine
2,3,6-Trichloropyridine
2,4,5-Trichloropyridine
2,4,6-Trichloropyridine
3,4,5-Trichloropyridine
$C_5H_2Cl_6O_2$
1,1,1,5,5,5-Hexachloro-2,4-pentanedione
$C_5H_2Cl_6O_3$
Chloralide
$C_5H_2O_5$
Croconic Acid
$C_5H_3BrO_2S$
3-Bromothiophene-2-carboxylic Acid
4-Bromothiophene-2-carboxylic Acid
5-Bromothiophene-2-carboxylic Acid
2-Bromothiophene-3-carboxylic Acid
4-Bromothiophene-3-carboxylic Acid
5-Bromothiophene-3-carboxylic Acid
$C_5H_3BrO_3$
2-Bromofuran-3-carboxylic Acid
4-Bromofuran-2-carboxylic Acid
5-Bromofuran-2-carboxylic Acid
5-Bromofuran-3-carboxylic Acid
$C_5H_3Br_2ClO_3$
Dibromofumaric Acid, *Me ester chloride*
$C_5H_3Br_2N$
2,3-Dibromopyridine
2,5-Dibromopyridine
2,6-Dibromopyridine

$C_5H_3Br_2N$ (*continued*)
3,4-Dibromopyridine
3,5-Dibromopyridine

$C_5H_3Br_2NO$
3,5-Dibromo-2-hydroxypyridine
3,5-Dibromo-4-hydroxypyridine

$C_5H_3Br_4N$
2,3,4,5-Tetrabromopyrrole, N-*Me*

$C_5H_3ClN_2O_2$
2-Chloro-3-nitropyridine
2-Chloro-5-nitropyridine
4-Chloro-3-nitropyridine
5-Chloro-2-nitropyridine

$C_5H_3ClN_2O_3$
Pyrazole-3,5-dicarboxylic Acid, *Chloride*

C_5H_3ClOS
Thiophene-2-carboxylic Acid, *Chloride*
Thiophene-3-carboxylic Acid, *Chloride*

$C_5H_3ClO_2$
Furan-2-carboxylic Acid, *Chloride*
Furan-3-carboxylic Acid, *Chloride*
Glutaconic Acid, *Chloro-anhydride*

$C_5H_3ClO_2S$
5-Chlorothiophene-2-carboxylic Acid

$C_5H_3ClO_3$
Aconic Acid, *Chloride*
3-Chlorofuran-2-carboxylic Acid
4-Chlorofuran-2-carboxylic Acid
5-Chlorofuran-2-carboxylic Acid
3-Chloroglutaconic Acid, *Anhydride*

$C_5H_3Cl_2N$
2,3-Dichloropyridine
2,4-Dichloropyridine
2,5-Dichloropyridine
2,6-Dichloropyridine
3,4-Dichloropyridine
3,5-Dichloropyridine

$C_5H_3Cl_2NO$
2,4-Dichloro-6-hydroxypyridine
3,4-Dichloro-2-hydroxypyridine
3,5-Dichloro-2-hydroxypyridine

$C_5H_3Cl_4N$
2,3,4,5-Tetrachloropyrrole, N-*Me*

$C_5H_3F_7O_2$
Heptafluorobutyric Acid, *Me ester*

C_5H_3NO
Furan-2-carboxylic Acid, *Nitrile*

$C_5H_3NO_4$
5-Nitrofuran-2-aldehyde
5-Nitrofuran-3-aldehyde

$C_5H_3NO_5$
3-Nitrofuran-2-carboxylic Acid
5-Nitrofuran-2-carboxylic Acid
5-Nitrofuran-3-carboxylic Acid

C_5H_3NS
Thiophene-2-carboxylic Acid, *Nitrile*

$C_5H_3N_3O_5$
Caffolide

C_5H_4
1,3-Pentadi-yne

C_5H_4BrHgN
3-Mercuri-pyridyl bromide

C_5H_4BrN
2-Bromopyridine
3-Bromopyridine
4-Bromopyridine

$C_5H_4BrNO_2$
4-Bromofuran-2-carboxylic Acid, *Amide*

$C_5H_4Br_2N_2$
2-Amino-3,5-dibromopyridine
2-Amino-4,6-dibromopyridine
3-Amino-2,5-dibromopyridine
3-Amino-2,6-dibromopyridine
4-Amino-2,6-dibromopyridine
4-Amino-3,5-dibromopyridine

$C_5H_4Br_2O_3$
Dibromomaleic Acid, *Me ester*
Mucobromic Acid, *Me ester*
Mucobromic Acid, *ψ-Me ester*

C_5H_4ClHgN
3-Mercuri-pyridyl chloride
4-Mercuri-pyridyl chloride

C_5H_4ClN
2-Chloropyridine
3-Chloropyridine
4-Chloropyridine

C_5H_4ClNO
Pyrrole-2-carboxylic Acid, *Chloride*

$C_5H_4ClNO_2$
5-Chlorofuran-2-carboxylic Acid, *Amide*

$C_5H_4Cl_2N_2$
2-Amino-3,4-dichloropyridine
2-Amino-3,5-dichloropyridine
2-Amino-4,6-dichloropyridine
3-Amino-2,5-dichloropyridine
3-Amino-2,6-dichloropyridine
4-Amino-2,5-dichloropyridine
4-Amino-2,6-dichloropyridine
4-Amino-3,5-dichloropyridine

$C_5H_4Cl_2O_2$
Citraconic Acid, *Dichloride*
Mesaconic Acid, *Dichloride*
2-Propene-1,2-dicarboxylic Acid, *Dichloride*

C_5H_4HgIN
3-Mercuri-pyridyl iodide

C_5H_4IN
2-Iodopyridine
3-Iodopyridine
4-Iodopyridine

$C_5H_4N_2$
Citraconic Acid, *Dinitrile*
Diazocyclopentadiene†
Glutaconic Acid, *Dinitrile*
Mesaconic Acid, *Dinitrile*

$C_5H_4N_2O_2$
2-Nitropyridine
3-Nitropyridine
4-Nitropyridine
Pyrazinecarboxylic Acid

$C_5H_4N_2O_3$
2-Hydroxy-3-nitropyridine
2-Hydroxy-5-nitropyridine
3-Hydroxy-2-nitropyridine
4-Hydroxy-3-nitropyridine
5-Hydroxy-2-nitropyridine

$C_5H_4N_2O_4$
2,4-Dihydroxy-3-nitropyridine
2,6-Dihydroxy-3-nitropyridine
3,5-Dihydroxy-2-nitropyridine
Imidazole-4,5-dicarboxylic Acid
Methylalloxan
5-Nitrofuran-2-carboxylic Acid, *Amide*
Orotic Acid
Pyrazole-3,5-dicarboxylic Acid
Pyrazole-4,5-dicarboxylic Acid

$C_5H_4N_2S$
Thiazole-4-acetic Acid, *Nitrile*

$C_5H_4N_4$
Purine
Pyrazolo[3,4-*b*]pyrazine†

$C_5H_4N_4O$
Allopurinol†
Hypoxanthine

$C_5H_4N_4O_2$
Isoxanthine
Oxypurinol†
Xanthine

$C_5H_4N_4O_3$
Uric Acid

$C_5H_4N_4O_4$
2-Amino-3-nitropyridine, 2-*Nitramine*

$C_5H_4N_4S$
6-Mercaptopurine

$C_5H_4N_6O_2$
2-Propene-1,2-dicarboxylic Acid, *Di-azide*

C_5H_4O
Cyclopentadienone†

C_5H_4OS
2*H*-Pyran-2-thione
4*H*-Pyran-4-thione
2*H*-Thiin-2-one
4*H*-Thiin-4-one
2-Thiophenealdehyde
3-Thiophenealdehyde
1-Thio-4-pyrone†

$C_5H_4O_2$
4-Cyclopentene-1,3-dione
β-Furaldehyde
Furfural
Protoanemonin
2*H*-Pyran-2-one
4*H*-Pyran-4-one

$C_5H_4O_2S$
Thiophene-2-carboxylic Acid
Thiophene-3-carboxylic Acid

$C_5H_4O_3$
Citraconic Acid, *Anhydride*
Furan-2-carboxylic Acid
Furan-3-carboxylic Acid
Glutaconic Acid, *Hydroxy-anhydride*
3-Hydroxy-2*H*-pyran-2-one
3-Hydroxy-4*H*-pyran-4-one
2-Propene-1,2-dicarboxylic Acid, *Anhydride*

$C_5H_4O_4$
Aconic Acid
2,3-Pentadienedioic Acid
Rubiginol
Squaric Acid, *Mono-Me ester*†

$C_5H_4O_7$
Oxalomalonic Acid

$C_5H_4O_8$
Methane-tetracarboxylic Acid

$C_5H_4S_2$
2*H*-Thiine-2-thione
4*H*-Thiine-4-thione

$C_5H_5BrN_2$
2-Amino-3-bromopyridine
2-Amino-4-bromopyridine
2-Amino-5-bromopyridine
2-Amino-6-bromopyridine
3-Amino-5-bromopyridine
4-Amino-2-bromopyridine
4-Amino-3-bromopyridine
5-Amino-2-bromopyridine

C_5H_5BrO
2-Bromomethylfuran

$C_5H_5ClN_2$
2-Amino-3-chloropyridine
2-Amino-4-chloropyridine
2-Amino-5-chloropyridine
3-Amino-2-chloropyridine
3-Amino-4-chloropyridine
4-Amino-2-chloropyridine
5-Amino-2-chloropyridine

$C_5H_5ClN_2O$
4-Imidazolylacetic Acid, *Chloride*

$C_5H_5ClN_2O_3$
5-Chlorobarbituric Acid, 1-N-*Me*

C_5H_5ClO
2-Chloromethylfuran
3-Chloromethylfuran

$C_5H_5ClO_2$
Chloroglutaconic Dialdehyde

$C_5H_5ClO_3$
Fumaric Acid, *Mono-Me ester monochloride*

$C_5H_5ClO_4$
3-Chloroglutaconic Acid

$C_5H_5Cl_3O_2$
Trichloroacetic Acid, *Allyl ester*
Trichloroacrylic Acid, *Et ester*

C_5H_5Fe
Iron carbonyl

$C_5H_5IO_2$
Iodopropiolic Acid, *Et ester*

C_5H_5N
2,4-Pentadienonitrile
Pyridine

C_5H_5NO
2-Hydroxypyridine
3-Hydroxypyridine
4-Hydroxypyridine
Pyridine-*N*-oxide†
Pyrrole-2-aldehyde

$C_5H_5NO_2$
Cyclopropane-1,1-dicarboxylic Acid, *Mononitrile*
2,3-Dihydroxypyridine
2,4-Dihydroxypyridine
2,5-Dihydroxypyridine
2,6-Dihydroxypyridine
3,4-Dihydroxypyridine
3,5-Dihydroxypyridine
Furan-2-carboxylic Acid, *Amide*
Furan-3-carboxylic Acid, *Amide*
Maleimide, N-*Me*
Pyrrole-*N*-carboxylic Acid
Pyrrole-2-carboxylic Acid
Pyrrole-3-carboxylic Acid

$C_5H_5NO_2S$
4-Methylthiazole-2-carboxylic Acid
2-Methylthiazole-4-carboxylic Acid
2-Methylthiazole-5-carboxylic Acid
4-Methylthiazole-5-carboxylic Acid
1-Thiathymine†
1-Thiauracil, N-*Me*†
Thiazole-2-acetic Acid
Thiazole-4-acetic Acid
Thiazole-5-acetic Acid
Thiazole-5-carboxylic Acid, *Me ester*

$C_5H_5NO_3$
Acetylmalonic Acid, *Nitrile*
Isoxazole-5-carboxylic Acid, *Me ester*
5-Methylisoxazole-3-carboxylic Acid
3-Methylisoxazole-5-carboxylic Acid
2-Methyl-3-nitrofuran
2-Methyl-5-nitrofuran
3-Methyl-2-nitrofuran
3-Methyl-5-nitrofuran
2-Oxalopropionic Acid, *Mononitrile*
2-Oxoglutaric Acid, 1-*Nitrile*

$C_5H_5NO_3S$
Pyridine-2 sulphonic Acid
Pyridine-3-sulphonic Acid
Pyridine-4-sulphonic Acid

$C_5H_5NO_4$
Cyanosuccinic Acid

C_5H_5NS
2-Pyridinethiol
3-Pyridinethiol
4-Pyridinethiol

$C_5H_5N_3$
4-Imidazolylacetic Acid, *Nitrile*

$C_5H_5N_3O$
Pyrazinecarboxylic Acid, *Amide*

$C_5H_5N_3O_2$
2-Amino-3-nitropyridine
2-Amino-5-nitropyridine
3-Amino-2-nitropyridine
4-Amino-3-nitropyridine
5-Amino-2-nitropyridine
5-Amino-3-nitropyridine
5-Aminopyrazine-2-carboxylic Acid†
2-Methyl-5-nitropyrimidine†
2-Nitroaminopyridine
3-Nitroaminopyridine

$C_5H_5N_3O_4$
5-Isonitrosobarbituric Acid, 5-O-*Me ether*
5-Isonitroso-1-methylbarbituric Acid
5-Nitrouracil, 3-N-*Me*

$C_5H_5N_5$
Adenine

$C_5H_5N_5O$
6-Amino-2-hydroxypurine
Guanine

C_5H_6
Cyclopentadiene
2-Methylbuten-3-yne
1-Penten-3-yne
1-Penten-4-yne
3-Penten-1-yne

$C_5H_6BrNO_2$
Bromomalonic Acid, *Et ester*, *Nitrile*

$C_5H_6Br_2O_2$
3,3-Dibromoacrylic Acid, *Et ester*
2,3-Dibromocrotonic Acid, *Me ester*
2,3-Dibromoisocrotonic Acid, *Me ester*

$C_5H_6Br_2O_3$
2,3-Dibromolaevulinic Acid
3,5-Dibromolaevulinic Acid

$C_5H_6Br_2O_4$
Dibromomalonic Acid, *Di-Me ester*

$C_5H_6ClNO_2$
Chloromalonic Acid, *Mono-Et ester*, *Nitrile*

$C_5H_6Cl_2O_2$
Dichloroacetic Acid, *Allyl ester*
2,3-Dichloroacrylic Acid, *Et ester*
3,3-Dichloroacrylic Acid, *Et ester*
4,4-Dichlorocrotonic Acid, *Me ester*
Dimethylmalonic Acid, *Dichloride*
Ethylmalonic Acid, *Dichloride*
Glutaric Acid, *Dichloride*

$C_5H_6Cl_2O_3$
Methoxysuccinic Acid, *Dichloride*

$C_5H_6Cl_3NO$
3,3,4-Trichloro-2-hydroxyvaleric Acid, *Nitrile*

$C_5H_6I_2O_4$
Di-iodomalonic Acid, *Di-Me ester*

$C_5H_6N_2$
2-Aminopyridine
3-Aminopyridine
4-Aminopyridine
Dimethylmalonic Acid, *Dinitrile*
Ethylmalonic Acid, *Dinitrile*
Glutaric Acid, *Di-nitrile*
2-Methylpyrazine
3-Methylpyridazine
2-Methylpyrimidine
4-Methylpyrimidine
5-Methylpyrimidine
Methylsuccinic Acid, *Dinitrile*

$C_5H_6N_2O$
N-Acetylimidazole†
Cyclopropane-1,1-dicarboxylic Acid, *Mononitrile, Amide*
3-Hydroxyglutaronitrile†
2-Hydroxypyrimidine, *Me ether*
4-Hydroxypyrimidine, *Me ether*
Pyrrole-*N*-carboxylic Acid, *Amide*
Pyrrole-2-carboxylic Acid, *Amide*

$C_5H_6N_2OS$
4-Methylthiazole-2-carboxylic Acid, *Amide*
Thiazole-2-acetic Acid, *Amide*
Thiazole-4-acetic Acid, *Amide*
Thiazole-5-acetic Acid, *Amide*

$C_5H_6N_2O_2$
4-Amino-2,6-dihydroxypyridine
4,6-Dihydroxypyrimidine, 6-O-*Me ether*†
2-Hydroxypyrazine, 4-N-*Oxide, Me ether*†
1-Imidazolylacetic Acid
4-Imidazolylacetic Acid
2-Methylimidazole-4-carboxylic Acid
3-Methylimidazole-4-carboxylic Acid
5-Methylimidazole-4-carboxylic Acid
1-Methyluracil
3-Methyluracil
4-Methyluracil
6-Methyluracil
Pyrazole-3-carboxylic Acid, *Me ester*
Thymine

$C_5H_6N_2O_2S$
5-Mercaptomethyluracil†

$C_5H_6N_2O_3$
4,6-Dihydroxy-2-methoxypyrimidine†
Dimethylparabanic Acid
Isonitrosomalonic Acid, *Et ester-nitrile*
1-Methylbarbituric Acid
5-Methylbarbituric Acid

$C_5H_6N_2O_4$
Diazomalonic Acid, *Et ester*
2,4-Dioxo-1-imidazolidylacetic Acid
2,4-Dioxo-3-imidazolidylacetic Acid
2,4-Dioxo-5-imidazolidylacetic Acid
Ibotenic Acid†
Maleuric Acid
Muscazone†
Nitrocyanoacetic Acid, *Et ester*

$C_5H_6N_2O_5$
Alloxanic Acid, *Me ester*

$C_5H_6N_4O$
5-Aminopyrazine-2-carboxylic Acid, *Amide*†

$C_5H_6N_4O_2$
Imidazole-4,5-dicarboxylic Acid, *Diamide*

$C_5H_6N_4O_4$
ψ-Uric Acid

C_5H_6O
Cyclopent-2-en-1-one
Dimethylcyclopropenone†
2-Methylfuran
3-Methylfuran
2,4-Pentadienal
1,4-Pentadien-3-one
2-Penten-4-yn-1-ol
4-Penten-2-yn-1-ol
1-Penten-4-yn-3-ol
γ-Pyran

C_5H_6OS
2-Furfuryl Mercaptan†
2-Hydroxymethylthiophene
2-Thienol, *Me ether*

$C_5H_6O_2$
Angelica lactone
1,2-Cyclopentanedione
1,3-Cyclopentanedione
Furfuryl Alcohol
4-Hydroxy-2-methylenebutyric Acid, *Lactone*†
2-Methyl-1,3-cyclobutanedione†
2,4-Pentadienoic Acid
Pentenedial
2-Pentynoic Acid
Propiolic Acid, *Et ester*

$C_5H_6O_3$
3-Acetylacrylic Acid
2,5-Dihydro-2-furoic Acid†
3,4-Dihydroxy-2-methylenebutyric Acid, *Lactone*†
Formylacrylic Acid, *Me ester*
β-Formylcrotonic Acid†
Glutaric Acid, *Anhydride*
4-Hydroxy-2-butynoic Acid, *Me ether*
Methylsuccinic Anhydride
γ-Methyltetronic Acid
2,3,4-Pentanetrione
Reductic Acid

$C_5H_6O_4$
Acetylpyruvic Acid
Citraconic Acid
Cyclopropane-1,1-dicarboxylic Acid
Cyclopropane-1,2-dicarboxylic Acid
2,3-Dioxobutyric Acid, *Me ester*
Fumaric Acid, *Mono-Me ester*
Glutaconic Acid
Mesaconic Acid
2-Propene-1,2-dicarboxylic Acid
Tetrahydro-5-oxofuran-3-carboxylic Acid

$C_5H_6O_5$
Acetonedicarboxylic Acid
Acetylmalonic Acid
Formylsuccinic Acid
Mesoxalic Acid, *Di-Me ester*
2-Oxalopropionic Acid
2-Oxoglutaric Acid

$C_5H_6O_6$
1,3-Dioxolan-4,5-dicarboxylic Acid

$C_5H_6O_8$
1,2-Dihydroxyethane-1,1,2-tricarboxylic Acid

C_5H_6S
2-Methylthiophene
3-Methylthiophene
γ-Thiopyran

C_5H_7Br
4-Bromocyclopentene†

$C_5H_7BrO_2$
Bromoacetic Acid, *Allyl ester*
2-Bromoacrylic Acid, *Et ester*
2-Bromocrotonic Acid, *Me ester*
4-Bromocrotonic Acid, *Me ester*

$C_5H_7BrO_3$
2-Bromolevulinic Acid
3-Bromolevulinic Acid

$C_5H_7BrO_4$
3-Bromoglutaric Acid
Bromomalonic Acid, *Di-Me ester*

$C_5H_7Br_2ClO$
2,3-Dibromo-3-methylbutyric Acid, *Chloride*
2,5-Dibromovaleric Acid, *Chloride*

$C_5H_7Br_2N$
2,3-Dibromovaleric Acid, *Nitrile*

$C_5H_7Br_3O_2$
2,2,3-Tribromopropionic Acid, *Et ester*

C_5H_7Cl
1-Chloro-2-methyl-1,3-butadiene

C_5H_7ClO
Cyclobutane-carboxylic Acid, *Chloride*
3,3-Dimethylacrylic Acid, *Chloride*
2-Methylcrotonic Acid, *Chloride*
2-Methylcyclopropane-1-carboxylic Acid, *Chloride*
2-Methylenebutyric Acid, *Chloride*
2-Pentenoic Acid, *Chloride*
3-Pentenoic Acid, *Chloride*
4-Pentenoic Acid, *Chloride*

$C_5H_7ClO_2$
Chloroacetic Acid, *Allyl ester*
2-Chloroacrylic Acid, *Et ester*
3-Chloroacrylic Acid, *Et ester*
2-Chlorocrotonic Acid, *Me ester*
3-Chlorocrotonic Acid, *Me ester*
4-Chlorocrotonic Acid, *Me ester*
3-Chloroisocrotonic Acid, *Me ester*
3-Chloro-2-methylacrylic Acid, *Me ester*
4 Chloro-4-valerolactone

$C_5H_7ClO_3$
2Acetoxypropionic Acid, *Chloride*
Dimethylmalonic Acid, *Monochloride*
Oxalic Acid, *Propyl ester chloride*

$C_5H_7ClO_4$
2-Chloro-2-ethylmalonic Acid
2-Chloroglutaric Acid
Chloromalonic Acid, *Di-Me ester*
2-Chloro-2-methylsuccinic Acid

$C_5H_7ClO_5$
2-Chloro-3-hydroxy-2-methylsuccinic Acid
2-Chloro-3-hydroxy-3-methylsuccinic Acid

$C_5H_7Cl_3O_2$
Chloralacetone
2,3-Dichloropropionic Acid, β-*Chloroethyl ester*
Trichloroacetic Acid, *Propyl ester*
2,4,4-Trichlorobutyric Acid, *Me ester*
4,4,4-Trichlorobutyric Acid, *Me ester*
2,2,3-Trichloropropionic Acid, *Et ester*

$C_5H_7Cl_3O_3$
3,3,4-Trichloro-2-hydroxyvaleric Acid
3,3,3-Trichlorolactic Acid, *Et ester*

$C_5H_7FO_2$
Fluoroacetic Acid, *Allyl ester*

C_5H_7N
Angelic Acid, *Nitrile*
Cyclobutane-carboxylic Acid, *Nitrile*
1,4-Dihydropyridine†
3,3-Dimethylacrylic Acid, *Nitrile*
1-Methylcyclopropane-1-carboxylic Acid, *Nitrile*
2-Methylcyclopropane-1-carboxylic Acid, *Nitrile*
N-Methylpyrrole
2-Methylpyrrole
3-Methylpyrrole
2-Pentenoic Acid, *Nitrile*
3-Pentenoic Acid, *Nitrile*
4-Pentenoic Acid, *Nitrile*

C_5H_7NO
1-Cyano-2-hydroxybut-3-ene†
2,4-Dimethyloxazole
2,5-Dimethyloxazole
4,5-Dimethyloxazole
2-Furylmethylamine
2-Methylacetoacetic Acid, *Nitrile*
3-Methyl-2-oxobutyric Acid, *Nitrile*
2-Oxovaleric Acid, *Nitrile*
3-Oxovaleric Acid, *Nitrile*
Tetrahydrofuran-2-carboxylic Acid, *Nitrile*

C_5H_7NOS
1-Cyano-2-hydroxy-3,4-epithiobutane†
Goitrin
Thiophene-2-carboxylic Acid, *Amide*
Thiophene-3-carboxylic Acid, *Amide*
5-Vinyloxazolidine-2-thione†

$C_5H_7NO_2$
2-Acetoxypropionic Acid, *Nitrile*
2-Cyanobutyric Acid
4-Cyanobutyric Acid
3-Cyanopropionic Acid, *Me ester*
Ethyl cyanoacetate
Glutaric Acid, *Imide*
3-Methylisoxazolone, N-*Me*
N-Methylsuccinimide

$C_5H_7NO_3$
Acetylpyruvic Acid, *Amide*
Glutamic Acid★, *Anhydride*†
Mesaconic Acid, α-*Amide*
Mesaconic Acid, β-*Amide*
5-Oxopyrrolidine-2-carboxylic Acid

$C_5H_7NO_4$
2-Oxalopropionic Acid, *Monoamide*

$C_5H_7NO_5$
Acetamidomalonic Acid†
Isonitrosomalonic Acid, *Di-Me ester*

$C_5H_7NO_6$
Nitromalonic Acid, *Di-Me ester*

C_5H_7NS
2-Aminomethylthiophene
2-Aminothiophene, N-*Me*
3-Butenyl isothiocyanate
2,4-Dimethylthiazole
2,5-Dimethylthiazole
4,5-Dimethylthiazole

$C_5H_7N_3$
2-Amino-4-methylpyrimidine
2-Amino-5-methylpyrimidine
4-Amino-2-methylpyrimidine
4-Amino-5-methylpyrimidine
4-Amino-6-methylpyrimidine
5-Amino-4-methylpyrimidine
2-Aminopyrimidine, N-*Me*
4-Aminopyrimidine, N-*Me*
2,3-Diaminopyridine
2,4-Diaminopyridine
2,5-Diaminopyridine
2,6-Diaminopyridine
3,4-Diaminopyridine
3,5-Diaminopyridine
2-Methylaminopyrimidine
4-Methylaminopyrimidine
Methyliminodiacetic Acid, *Dinitrile*

$C_5H_7N_3O$
2-Amino-4 hydroxypyrimidine, *Me ether*
4-Amino-6-methyl-2-pyrimidone
2-Amino-5-methyl-4-pyrimidone
2-Amino-6-methyl-4-pyrimidone
6-Amino-2-methyl-4-pyrimidone
3-Methylcytosine
4-Methylcytosine
5-Methylcytosine
6-Methylcytosine

$C_5H_7N_3O_2$
1,2-Dimethyl-4-nitroimidazole†
1,2-Dimethyl-5-nitroimidazole†
5-Hydroxymethylcytosine

$C_5H_7N_3O_3$
Diazomalonic Acid, *Et ester*, *Amide*
2,4-Dioxo-1-imidazolidylacetic Acid, *Amide*
Isouramil★, *5-Me ether*†

$C_5H_7N_3O_4$
Azaserine

C_5H_7P
1-Methylphosphole†

C_5H_8
Bicyclo[1,1,1]pentane†
Bicyclo[2,1,0]pentane†
Cyclopentene
2-Methyl-1,3-butadiene
2-Methyl-2,3-butadiene
3-Methyl-1-butyne
1-Methylcyclobutene
3-Methylcyclobutene†
Methylenecyclobutane
1,2-Pentadiene
1,3-Pentadiene
1,4-Pentadiene
2,3-Pentadiene
1-Pentyne
2-Pentyne
Vinylcyclopropane†

C_5H_8BrClO
2-Bromo-3-methylbutyric Acid, *Chloride*
3-Bromo-3-methylbutyric Acid, *Chloride*
4-Bromovaleric Acid, *Chloride*
5-Bromovaleric Acid, *Chloride*

$C_5H_8BrClO_2$
3-Bromo-2-chloropropionic Acid, *Et ester*

C_5H_8BrN
2-Bromo-3-methylbutyric Acid, *Nitrile*

$C_5H_8Br_2O$
2-Bromo-3-methylbutyric Acid, *Bromide*
1,1-Dibromo-2-pentanone†
2,4-Dibromo-3-pentanone

$C_5H_8Br_2O_2$
2,3-Dibromobutyric Acid, *Me ester*
3,4-Dibromobutyric Acid, *Me ester*
2,3-Dibromo-3-methylbutyric Acid
2,3-Dibromo-2-methylpropionic Acid, *Me ester*
2,3-Dibromo-1-propanol, *Ac*
2,2-Dibromopropionic Acid, *Et ester*
2,3-Dibromopropionic Acid, *Et ester*
2,2-Dibromovaleric Acid
2,3-Dibromovaleric Acid
2,5-Dibromovaleric Acid
3,4-Dibromovaleric Acid
4,4-Dibromovaleric Acid
4,5-Dibromovaleric Acid

$C_5H_8Br_4$
Pentaerythrityl tetrabromide

$C_5H_8ClFO_2$
Chlorofluoroacetic Acid, *Propyl ester*†

C_5H_8ClN
2-Chloro-2-methylbutyric Acid, *Nitrile*
2-Chloro-3-methylbutyric Acid, *Nitrile*
3-Chloro-2-methylbutyric Acid, *Nitrile*
2-Chlorovaleric Acid, *Nitrile*
3-Chlorovaleric Acid, *Nitrile*

$C_5H_8Cl_2$
1,4-Dichloro-2-methyl-2-butene
3,3-Dichloro-2-methyl-1-butene

$C_5H_8Cl_2O$
2-Chloro-2-methylbutyric Acid, *Chloride*
2-Chloro-3-methylbutyric Acid, *Chloride*
2-Chlorovaleric Acid, *Chloride*
3,3-Dichloro-2-pentanone
3,5-Dichloro-2-pentanone

$C_5H_8Cl_2O_2$
Caldariomycin★†
Dichloroacetic Acid, *Propyl ester*
2,3-Dichlorobutyric Acid, *Me ester*
3,4-Dichlorobutyric Acid, *Me ester*
4,4-Dichlorobutyric Acid, *Me ester*
2,2-Dichloropropionic Acid, *Et ester*
2,3-Dichloropropionic Acid, *Et ester*
3,3-Dichloropropionic Acid, *Et ester*

$C_5H_8Cl_2O_3$
Di-2-chloroethyl carbonate
1,2-Dichloroethyl ethyl carbonate

$C_5H_8Cl_3NO_2$
Chloralacetamide, *Me ether*
3,3,4-Trichloro-2-hydroxyvaleric Acid, *Amide*

$C_5H_8Cl_3NO_3$
Chloralurethane

$C_5H_8Cl_4$
Pentaerythrityl tetrachloride

$C_5H_8I_4$
Pentaerythrityl tetraiodide
$C_5H_8NS_2$(ion)
3,5-Dimethylthiazole[2,3-*b*]thiazolium†
$C_5H_8N_2$
2,3-Diazabicyclo[2,2,1]-2-heptene
1,2-Dimethylimidazole
1,4-Dimethylimidazole
1,5-Dimethylimidazole
2,4-Dimethylimidazole
4,5-Dimethylimidazole
1,3-Dimethylpyrazole
1,4-Dimethylpyrazole
1,5-Dimethylpyrazole
3,4-Dimethylpyrazole
3,5-Dimethylpyrazole
$C_5H_8N_2O$
Acetylalanine, *Nitrile*
2-Cyanobutyric Acid, *Amide*
4-Cyanobutyric Acid, *Amide*
1,5-Dimethyl-3-pyrazolone
2,5-Dimethyl-3-pyrazolone
4,5-Dimethyl-3-pyrazolone
2,1′-Imidazolylethanol
2,2′-Imidazolylethanol
2,4′-Imidazolylethanol
$C_5H_8N_2OS$
2-Thiohydantoin, 1,3-N-*Di-Me*
$C_5H_8N_2O_2$
Carbethoxyglycine, *Nitrile*
Citraconic Acid, *Diamide*
2-Diazobutyric Acid, *Me ester*
2-Diazopropionic Acid, *Et Ester*
Dihydrothymine†
1,3-Dimethylhydantoin
1,5-Dimethylhydantoin
5,5-Dimethylhydantoin
Mesaconic Acid, *Diamide*
3-Methyl-2,5-dioxopiperazine
4-Methyl-2,6-dioxopiperazine
5-Oxopyrrolidine-2-carboxylic Acid, *Amide*
2-Propene-1,2-dicarboxylic Acid, *Diamide*
$C_5H_8N_2O_3$
Allophanic Acid, *Allyl ester*
$C_5H_8N_2O_4$
Oxaluric Acid, *Et ester*
$C_5H_8N_2O_5$
5,6-Dihydro-2,4,5,5,6-pentahydroxypyrimidine, 5-*Me ether*
β-*N*-Oxalyl-L-α,β-diaminopropionic Acid†
$C_5H_8N_4O_3$
1-Methylallantoin
3-Methylallantoin
Pyruvil
$C_5H_8N_4O_3S_2$
Methazolamide
$C_5H_8N_4O_6$
Uroxanic Acid
C_5H_8O
Acetylcyclopropane
Allyl vinyl Ether
2-Butyn-1-ol, *Me ether*
Cyclopentanone
Dihydropyran
2,2-Dimethylcyclopropanone†
3-Ethoxypropyne
2-Methyl-2-butenal
3-Methyl-2-butenal
2-Methyl-3-butyn-2-ol
3-Methylbut-3-en-2-one†
1,2-Pentadien-1-ol†
2,3-Pentadien-1-ol
2,4-Pentadien-1-ol
1,4-Pentadien-3-ol
2-Pentenal
3-Pentenal
4-Pentenal
3-Penten-2-one
4-Penten-2-one
1-Penten-3-one
2-Pentyn-1-ol
3-Pentyn-1-ol
4-Pentyn-1-ol
4-Pentyn-2-ol
1-Pentyn-3-ol
2-Propyn-1-ol, *Et ether*
$C_5H_8O_2$
Acetylacetone
Acrylic Acid, *Et ester*
Allyl acetate
Angelic Acid
2-Butyne-1,4-diol, *Mono-Me ether*
Crotonic Acid, *Me ester*
Cyclobutane-carboxylic Acid
Cyclopropanecarboxylic Acid, *Me ester*
Cyclopropylacetic Acid
3,3-Dimethylacrylic Acid
2,6-Dioxaspiro[3,3]heptane
Isocrotonic Acid, *Me ester*
Levulinic Aldehyde
2-Methylcrotonic Acid
1-Methylcyclopropane-1-carboxylic Acid
2-Methylcyclopropane-1-carboxylic Acid
2-Methylenebutyric Acid
Methyl 2-methylacrylate
Methylsuccindialdehyde
2-Oxopentanal
1,3-Pentanedial
2,3-Pentadione
2-Pentenoic Acid
3-Pentenoic Acid
4-Pentenoic Acid
Propynal, *Di-Me acetal*
Tetrahydrofurfural
Tetrahydro-5-methyl-2-furanone
Tetrahydro-2-methylfuran-3-one†
Tetrahydro-2*H*-2-pyranone
Tetrahydro-4*H*-4-pyranone
$C_5H_8O_2S$
β-Methylthioacrylic Acid, *Me ester*†
Tetrahydrothiophene-2-carboxylic Acid
$C_5H_8O_3$
Acetoacetic Acid, *Me ester*
Acetoxyacetone
2-Acetoxypropionaldehyde

$C_5H_8O_3$ (*continued*)
Arabinal
2,3-Epoxy-2-methylbutyric Acid
2,3-Epoxy-3-methylbutyric Acid
2,3-Epoxypropionic Acid, *Et ester*
2-Formylisobutyric Acid
3-Formyl-2-methylpropionic Acid
3-Formylpropionic Acid, *Me ester*
1-Hydroxy-2-butanone, *Formyl*
4-Hydroxy-2-methylenebutyric Acid†
Levulinic Acid
3-Methoxy-2-oxobutyraldehyde†
2-Methylacetoacetic Acid
3-Methyl-2-oxobutyric Acid
2-Oxovaleric Acid
3-Oxovaleric Acid
Pyruvic Acid, *Et ester*
Tetrahydrofuran-2-carboxylic Acid
Xylal

$C_5H_8O_4$
Acetolactic Acid†
2-Acetoxypropionic Acid
3,4-Dihydroxy-2-methylenebutyric Acid†
Dimethyl malonate
Dimethylmalonic Acid
Ethylmalonic Acid
Glutaric Acid
Malonic Acid, *Mono-Et ester*
Methylene diacetate
Methylmalonic Acid, *Mono-Me ester*
Methylsuccinic Acid
Oxalic Acid, *Me-Et ester*
Oxalic Acid, *Propyl ester*
Oxalic Acid, *Isopropyl Ester*
*Rhamno*tetronic Acid, *γ-Lactone*
Succinic Acid, *Me ester*

$(C_5H_8O_4)_n$
Araban
Xylan

$C_5H_8O_5$
Arabonic Acid, *Lactone*
Citramalic Acid
1,2-Diformin
1,3-Diformin
Diglycollic Acid, *Mono-Me ester*
2-Hydroxyglutaric Acid
3-Hydroxyglutaric Acid
Hydroxy-methylmalonic Acid, *Me ether*
Hydroxymethyl-malonic Acid, *Me ether*
Hydroxymethyl-succinic Acid
2-Hydroxy-3-methylsuccinic Acid
Methoxysuccinic Acid
Ribonic Acid, *γ-Lactone*
Tartronic Acid, *Di-Me ester*
Tartronic Acid, *Et ether*
Xylonic Acid, *γ-Lactone*

$C_5H_8O_6$
*Arabo*keturonic Acid
2,3-Dihydroxyglutaric Acid
2,4-Dihydroxyglutaric Acid
D-Lyxuronic Acid
Tartaric Acid, *Me ester*
Tartaric Acid, *Mono-Me ether*

$C_5H_8O_7$
Trihydroxyglutaric Acid

C_5H_8S
Cyclopentanethione†
6-Thiabicyclo[3,1,0]hexane

C_5H_9Br
Bromocyclopentane
1-Bromo-3-methylbut-2-ene†

C_5H_9BrO
2-Bromoallyl Alcohol, *Et ether*
3-Bromoallyl Alcohol, *Et ether*
1-Bromo-3-pentanone
2-Bromo-3-pentanone
4-Bromo-2-pentanone
5-Bromo-2-pentanone
Valeric Acid, *Bromide*

$C_5H_9BrO_2$
Bromoacetic Acid, *Propyl ester*
Bromoacetic Acid, *Isopropyl ester*
2-Bromobutyric Acid, *Me ester*
4-Bromobutyric Acid, *Me ester*
2-Bromoisobutyric Acid, *Me ester*
3-Bromoisobutyric Acid, *Me ester*
2-Bromo-3-methylbutyric Acid
3-Bromo-3-methylbutyric Acid
2-Bromopropionic Acid, *Et ester*
3-Bromopropionic Acid, *Et ester*
2-Bromovaleric Acid
3-Bromovaleric Acid
4-Bromovaleric-Acid
5-Bromovaleric Acid

$C_5H_9BrO_3$
2-Bromo-3-hydroxybutyric Acid, *Me ether*

$C_5H_9Br_2NO$
2,3-Dibromovaleric Acid, *Amide*

C_5H_9Cl
Chlorocyclopentane
1-Chloro-3-methylbut-3-ene†

C_5H_9ClHg
Mercuri-cyclopentyl chloride

C_5H_9ClO
2-Chlorocyclopentanol
2-Chloro-2-propen-1-ol, *Et ether*
3-Chloro-2-propen-1-ol, *Et ether*
2-Chlorovinyl isopropyl Ether
2,2-Dimethylpropionic Acid, *Chloride*
Isovaleric Acid, *Chloride*
2-Methylbutyric Acid, *Chloride*
Valeric Acid, *Chloride*

$C_5H_9ClO_2$
Chloroacetic Acid, *Propyl ester*
Chloroacetic Acid, *Isopropyl ester*
2-Chlorobutyric Acid, *Me ester*
3-Chlorobutyric Acid, *Me ester*
4-Chlorobutyric Acid, *Me ester*
Chloroformic Acid, *Butyl ester*
Chloroformic Acid, *Isobutyl ester*
2-Chloro-2-methylbutyric Acid
2-Chloro-3-methylbutyric Acid
3-Chloro-2-methylbutyric Acid
3-Chloro-3-methylbutyric Acid
2-Chloro-2-methylpropionic Acid, *Me ester*

$C_5H_9ClO_2$ (*continued*)
1-Chloro-2-propanol, *Ac*
2-Chloropropionic Acid, *Et ester*
3-Chloropropionic Acid, *Et ester*
2-Chlorovaleric Acid
3-Chlorovaleric Acid
4-Chlorovaleric Acid
5-Chlorovaleric Acid

$C_5H_9ClO_3$
4-Chloro-3-hydroxybutyric Acid, *Me ester*
2-Chloro-3-hydroxy-2-methylbutyric Acid
3-Chloro-2-hydroxy-2-methylbutyric Acid
3-Chloro-2-hydroxy-2-methylpropionic Acid, *Me ester*
3-Chlorolactic Acid, *Et ester*

$C_5H_9Cl_3O_2$
Chloral, *Propyl acetal*

$C_5H_9FO_2$
tert-Butyl fluoroformate†

C_5H_9HgN
Mercuri-butyl cyanide

C_5H_9I
Iodocyclopentane

$C_5H_9IO_2$
Iodoacetic Acid, *Propyl ester*
4-Iodobutyric Acid, *Me ester*
2-Iodo-3-methylbutyric Acid
3-Iodo-3-methylbutyric Acid
2-Iodo-2-methylpropionic Acid, *Me ester*
2-Iodopropionic Acid, *Et ester*
3-Iodopropionic Acid, *Et ester*
4-Iodovaleric Acid
5-Iodovaleric Acid

C_5H_9N
1-Azabicyclo[3,1,0]hexane†
tert-Butyl isocyanide
2,2-Dimethylpropionic Acid, *Nitrile*
Isobutyl isocyanide
Isovaleric Acid, *Nitrile*
2-Methylbutyric Acid, *Nitrile*
2-Methyl-2-pyrroline
4-Methyl-2-pyrroline
1-Methyl-3-pyrroline
1,2,3,4-Tetrahydropyridine
1,2,3,6-Tetrahydropyridine
Valeric Acid, *Nitrile*

C_5H_9NO
Acetylacetone Imine
Acrylic Acid, N-*Ethylamide*
Butyl cyanate†
2-Butyl cyanate†
Cyclobutane-carboxylic Acid, *Amide*
3,3-Dimethylacrylic Acid, *Amide*
3-Hydroxy-2,2-dimethylpropionic Acid, *Nitrile*
2-Hydroxy-2-methylbutyric Acid, *Nitrile*
2-Hydroxy-3-methylbutyric Acid, *Nitrile*
3-Hydroxy-3-methylbutyric Acid, *Nitrile*
3-Hydroxypropionic Acid, *Et ether*, *Nitrile*
2-Hydroxyvaleric Acid, *Nitrile*
3-Hydroxyvaleric Acid, *Nitrile*
4-Hydroxyvaleric Acid, *Nitrile*
Isobutyl cyanate†
Isobutyl isocyanate
β-Isovalerolactam†
Lactic Acid, *Nitrile*, *Et ether*
1-Methylcyclopropane-1-carboxylic Acid, *Amide*
2-Methylcyclopropane-1-carboxylic Acid, *Amide*
2-Methylenebutyric Acid, *Amide*
N-Methyl-2-pyrrolidone
4-Methyl-2-pyrrolidone
5-Methyl-2-pyrrolidone
N-Methyl-3-pyrrolidone
2-Pentenoic Acid, *Amide*
3-Pentenoic Acid, *Amide*
4-Pentenoic Acid, *Amide*
α-Piperidone
β-Piperidone
γ-Piperidone

C_5H_9NOS
5,5-Dimethyl-2-oxazolidinethione†
4-Ethyloxazolidine-2-thione†
5-Ethyloxazolidine-2-thione†
3-Hydroxypent-4-enethionamide†

$C_5H_9NOS_2$
3-Methylsulphinylpropyl isothiocyanate†

$C_5H_9NO_2$
2-Amino-4-hydroxyvaleric Acid, *Lactone*
2-Amino-4-pentenoic Acid
Glycine, *Allyl ester*
Isonitrosoacetone, *Et ether*
2-Methylacetoacetic Acid, *Amide*
3-Methylaminocrotonic Acid
1-Methyl-1-nitrocyclobutane
3-Methyl-2-oxobutyric Acid, *Amide*
Nitrocyclopentane
2-Oxovaleric Acid, *Amide*
Proline
Tetrahydrofuran-2-carboxylic Acid, *Amide*

$C_5H_9NO_2S$
1,4-Thiazane-3-carboxylic Acid†

$C_5H_9NO_2S_2$
Cheiroline

$C_5H_9NO_3$
Acetyl-DL-alanine
Acetylglycine, *Me ester*
5-Aminolevulinic Acid†
Dimethyloxamic Acid, *Me ester*
3-Hydroxyproline
4-Hydroxyproline
5-Hydroxyproline
Malonamic Acid, *Et ester*
Methyloxamic Acid, *Et ester*
Oxamic Acid, *Propyl ester*
Oxamic Acid, *Isopropyl ester*
Succinamic Acid, *Me ester*

$C_5H_9NO_4$
Aminoethylmalonic Acid
Aminoglutaric Acid
Aminomalonic Acid, *Di-Me ester*
Aminomethylmalonic Acid, *Mono-Me ester*
2-Amino-2-methylsuccinic Acid
Arabonic Acid, *Nitrile*
L-Aspartic Acid, β-*Me ester*
Carbomethoxyurethane

$C_5H_9NO_4$ (*continued*)
Carbethoxyglycine
Citramalic Acid, 2-*Monoamide*
Glutamic Acid
3-Hydroxyglutaric Acid, *Monoamide*
2-Hydroxy-3-methylsuccinic Acid, *Monoamide*
α-Malamic Acid, *Me ester*
β-Malamic Acid, *Me ester*
N-Methylaspartic Acid
Methyliminodiacetic Acid
2-Methyl-2-nitropropionic Acid, *Me ester*
Nitroacetic Acid, *Propyl ester*
Nitroacetic Acid, *Isopropyl ester*
2-Nitropropionic Acid, *Et ester*
3-Nitropropionic Acid, *Et ester*

$C_5H_9NO_4S$
S-Carboxymethyl-L-cysteine†

$C_5H_9NO_5$
3-Hydroxyglutamic Acid
4-Hydroxyglutamic Acid
Serine, O-*Glycollyl*

C_5H_9NS
Butyl isothiocyanate
2-Butyl isothiocyanate
tert-Butyl isothiocyanate
Isobutyl isothiocyanate
Isobutyl thiocyanate
Thiocyanic Acid, *Butyl ester*

$C_5H_9NS_2$
3-Methylthiopropyl isothiocyanate

$C_5H_9N_3$
Histamine
Isohistamine†

$C_5H_9N_3O$
2-Imino-1,3-dimethyl-4-imidazolidone

$C_5H_9N_3O_2$
2-Azido-3-methylbutyric Acid
2-Azidopropionic Acid, *Et ester*
3-Azidopropionic Acid, *Et ester*

$C_5H_9N_3O_3$
Isonitrosomalonic Acid, *Diamide*, *Et ester*

$C_5H_9N_3O_{10}$
Pentaerythritol Trinitrate†

$C_5H_9O_2P$
1-Methylphospholan-3-one-1-oxide†

$C_5H_9O_3P$
Allylethylene phosphite

C_5H_{10}
Cyclopentane
1,1-Dimethylcyclopropane
1,2-Dimethylcyclopropane
Ethylcyclopropane
2-Methyl-1-butene
3-Methyl-1-butene
2-Methyl-2-butene
Methylcyclobutane
1-Pentene
2-Pentene

$C_5H_{10}BrNO$
2-Bromo-3-methylbutyric Acid, *Amide*
5-Bromovaleric Acid, *Amide*

$C_5H_{10}Br_2$
1,1-Dibromo-2,2-dimethylpropane
1,3-Dibromo-2,2-dimethylpropane
1,2-Dibromo-2-methylbutane
1,4-Dibromo-2-methylbutane
2,3-Dibromo-2-methylbutane
2,4-Dibromo-2-methylbutane
1,2-Dibromopentane
1,3-Dibromopentane
1,4-Dibromopentane
1,5-Dibromopentane
2,2-Dibromopentane
2,3-Dibromopentane
2,4-Dibromopentane

$C_5H_{10}Br_2O$
2,3-Dibromo-3-methyl-1-butanol
1,3-Dibromo-2-propanol, *Et ether*

$C_5H_{10}ClN$
Chloroacetamide, N-*Propyl*

$C_5H_{10}ClNO$
4-Chlorovaleric Acid, *Amide*
Diethylcarbamic Acid, *Chloride*

$C_5H_{10}ClNS$
Diethyldithiocarbamic Acid, *Chloride*

$C_5H_{10}Cl_2$
1,2-Dichloropentane
1,3-Dichloropentane
1,4-Dichloropentane
1,5-Dichloropentane
2,2-Dichloropentane
2,3-Dichloropentane
2,4-Dichloropentane

$C_5H_{10}Cl_2O_2$
Dichloroacetaldehyde, *Me-Et acetal*

$C_5H_{10}HgO_2$
Mercuri-isopropyl acetate
Mercuri-propyl acetate

$C_5H_{10}I_2$
1,4-Di-iodopentane
1,5-Di-iodopentane
2,4-Di-iodopentane

$C_5H_{10}NO_7P$
γ-L-Glutamyl phosphate

$C_5H_{10}N_2$
2-Amino-2-methylbutyric Acid, *Nitrile*
Diethylcarbamic Acid, *Nitrile*
2-Methyl-2-methylaminopropionic Acid, *Nitrile*
Tetrahydro-2-methylpyrimidine
Valine, *Nitrile*

$C_5H_{10}N_2O$
Proline, *Amide*

$C_5H_{10}N_2O_2$
Acetyl-alanine, *Amide*
Butyrylurea
Cucurbitine†
NN'-Diacetylmethylenediamine
Dimethylmalonic Acid, *Diamide*
Ethylmalonic Acid, *Diamide*
Glutaric Acid, *Diamide*
Malonamide, N-*Et*
Methylglyoxime, *Di-Me ether*

$C_5H_{10}N_2O_2$ (*continued*)
Methylsuccinic Acid, *Diamide*
N-Nitropiperidine
Piperazine-*N*-carboxylic Acid

$C_5H_{10}N_2O_2S$
Thiohydantoic Acid, *Et ester*

$C_5H_{10}N_2O_3$
D-Alanylglycine
DL-Alanylglycine
L-Alanylglycine
Allophanic Acid, *Propyl ester*
2-Amino-2-methylsuccinamic Acid
Carbethoxyglycine, *Amide*
Glutamic Acid, 5-*Monoamide*
Glutamine†
Glycyl-alanine
Glycylsarcosine
Hydantoic Acid, *Et ester*
Malamide, *Me ether*
N-Methylaspartic Acid, *Monoamide*
Methyliminodiacetic Acid, *Monoamide*

$C_5H_{10}N_2O_3S$
Glycylcysteine

$C_5H_{10}N_2O_4$
Glycyl-DL-serine
γ-Hydroxyglutamine†

$C_5H_{10}N_2O_5$
Allophanic Acid, *Mono-glyceryl ester*
Trihydroxyglutaric Acid, *Diamide*

$C_5H_{10}N_2O_8$
Pentaerythritol Dinitrate†

$C_5H_{10}N_2S_2$
Tetrahydro-3,5-dimethyl-2*H*-1,3,5-thiadiazine-2-thione

$C_5H_{10}N_4O$
2-Azido-3-methylbutyric Acid, *Amide*

$C_5H_{10}N_4O_4$
2,2-Diureidopropionic Acid

$C_5H_{10}O$
Allyl ethyl Ether
Cyclopentanol
2,2-Dimethylpropanal
1,2-Epoxy-2-methylbutane
2,3-Epoxypentane
α-Hydroxyethylcyclopropane
Isovaleraldehyde
2-Methylbutanal
3-Methyl-2-butanone
3-Methyl-3-buten-2-ol
2-Pentanone
3-Pentanone
2-Penten-1-ol
4-Penten-1-ol
3-Penten-2-ol
4-Penten-2-ol
1-Penten-3-ol
2-Propen-2-ol, *Et ether*
Tetrahydro-2-methylfuran
Tetrahydro-3-methylfuran
Tetrahydropyran
n-Valeraldehyde

$C_5H_{10}OS$
Thiane sulphoxide
Thioacetic Acid, S-*Propyl ester*
Thioacetic Acid, O-*Propyl ester*
Thioacetic Acid, S-*Isopropyl ester*
Thioacetic Acid, O-*Isopropyl ester*
Thiopropionic Acid, O-*Et ester*

$C_5H_{10}OS_2$
Dithiocarbonic Acid, SS-*Di-Et ester*
Methoxydithioformic Acid, *Propyl ester*
Xanthogenic Acid, *Et ester*

$C_5H_{10}O_2$
1,2-Cyclopentanediol
1,3-Cyclopentanediol
2,2-Dimethylpropionic Acid
Ethoxyacetone
Ethylene Glycol, *Mono-allyl ether*
Formic Acid, *Butyl ester*
1-Hydroxy-2-butanone, *Me ether*
3-Hydroxy-2-butanone, *Me ether*
3-Hydroxy-2,2-dimethylpropanal
4-Hydroxy-3-methyl-2-butanone
2-Hydroxymethyloxiran, *Et ether*
2-Hydroxy-2-methylpropionic Acid, *Me ether*
5-Hydroxypentanal
1-Hydroxy-2-pentanone
3-Hydroxy-2-pentanone
4-Hydroxy-2-pentanone
5-Hydroxy-2-pentanone
1-Hydroxy-3-pentanone
2-Hydroxy-3-pentanone
Isobutyl formate
Isopropyl acetate
Isovaleric Acid
Methyl butyrate
2-Methylbutyric Acid
Methyl isobutyrate
2-Methyl-3-oxo-2-butanol
Propionic Acid, *Et ester*
Tetrahydrofurfural Alcohol
Valeric Acid

$C_5H_{10}O_2S$
Dimethylsulphonium Methoxycarbonylmethylide†
2-Mercaptopropionic Acid, *Et ester*
3-Mercaptopropionic Acid, *Et ester*
3-Methylmercaptopropionic Acid, *Me ester*
2-Methyltetramethylene sulphone
Thiane sulphone
Thiocarbonic Acid, O,S-*Di-Et ester*
Thiocarbonic Acid, O-*Di-Et ester*
Thioglycollic Acid, S-*Propyl*

$C_5H_{10}O_3$
Diethyl carbonate
Ethoxyacetic Acid, *Me ester*
1,2-*O*-Ethylideneglycerol
1,3-*O*-Ethylideneglycerol
2-Hydroxybutyric Acid, *Me ether*
3-Hydroxybutyric Acid, *Me ester*
4-Hydroxybutyric Acid, *Me ether*
3-Hydroxy-2,2-dimethylpropionic Acid
2-Hydroxy-2-methylbutyric Acid
2-Hydroxy-3-methylbutyric Acid
3-Hydroxy-3-methylbutyric Acid

$C_5H_{10}O_3$ (*continued*)
2-Hydroxy-2-methylpropionic Acid, *Me ester*
3-Hydroxypropionic Acid, *Et ether*
3-Hydroxypropionic Acid, *Et ester*
2-Hydroxyvaleric Acid
3-Hydroxyvaleric Acid
4-Hydroxyvaleric Acid
5-Hydroxyvaleric Acid
Lactic Acid, *Et ester*
Lactic Acid, *Et ether*
Methoxyacetic Acid, *Et ester*
2-Methoxypropionic Acid, *Me ester*
3-Methoxypropionic Acid, *Me ester*
1,2-Methylene-glycerol, *Me ether*
1,3-Methylene-glycerol, *Me ether*

$C_5H_{10}O_3S$
Vinylsulphonic Acid, *Propyl ester*
Vinylsulphonic Acid, *Isopropyl ester*

$C_5H_{10}O_4$
Cordycepose (3-Deoxyxylose)
2-Deoxyribose
2-Deoxyxylose
3-Deoxyxylose
2,3-Dihydroxybutyric Acid, *Me ester*
2,3-Dihydroxy-2-methylbutyric Acid
2,3-Dihydroxy-3-methylbutyric Acid
2,3-Dihydroxy-2-methylpropionic Acid, *Me ester*
2,3-Dihydroxyvaleric Acid
Glyceric Acid, *Et ester*
α-Mono-acetin
β-Mono-acetin
Pentaerythrose†
*Rhamno*tetrose
Tigliceric Acid

$C_5H_{10}O_4S$
5-Deoxy-5-mercapto-α-D-xylopyranose†
L-Ribothiafuranose†
D-Ribothiapyranose†
D-Xylothiapyranose

$C_5H_{10}O_5$
Adonose
Apiose
Arabinose
Cyclopentanepentol†
2,4-Dihydroxy-3-hydroxymethylbutyric Acid
Lyxose
*Rhamno*tetronic Acid
Ribose
Threonic Acid, *Me ester*
Xylose
Xylulose

$C_5H_{10}O_6$
Apionic Acid
Arabonic Acid
Lyxonic Acid
Ribonic Acid
Xylonic Acid

$C_5H_{10}S$
3-Methylbut-2-enethiol†
Tetrahydro-2-methylthiophene
Tetrahydro-3-methylthiophene
Thiane

$C_5H_{10}S_2$
1,2-Dithiacycloheptane

$C_5H_{11}AsO_3$
2-Propoxy-1,3,2-dioxa-arsolan

$C_5H_{11}Br$
1-Bromo-2-methylbutane
1-Bromo-3-methylbutane
2-Bromo-2-methylbutane
2-Bromo-3-methylbutane
1-Bromopentane
2-Bromopentane
3-Bromopentane

$C_5H_{11}BrHg$
Mercuri-1,1-dimethylpropyl bromide
Mercuri-2-methylbutyl bromide
Mercuri-3-methylbutyl bromide
Mercuri-pentyl bromide

$C_5H_{11}BrO$
3-Bromo-1-propanol, *Et ether*

$C_5H_{11}BrO_2$
2-Bromopropanal, *Di-Me acetal*
3-Bromopropanal, *Di-Me acetal*

$C_5H_{11}Cl$
1-Chloro-2-methylbutane
1-Chloro-3-methylbutane
2-Chloro-2-methylbutane
2-Chloro-3-methylbutane
1-Chloropentane
2-Chloropentane
3-Chloropentane

$C_5H_{11}ClHg$
Mercuri-3-methylbutyl chloride
Mercuri-pentyl chloride

$C_5H_{11}ClHgN_2O_2$
Chlormerodrin

$C_5H_{11}ClO$
3-Chloro-2-butanol, *Me ether*
1-Chloro-2-methyl-2-butanol
3-Chloro-2-methyl-2-butanol
4-Chloro-2-methyl-2-butanol
1-Chloroethanol, *Propyl ether*
1-Chloro-2-propanol, *Et ether*
2-Methyl-2-butylhypochlorite

$C_5H_{11}ClO_2$
3-Chloropropanal, *Di-Me acetal*
2-Chloro-1,3-propanediol, *Et ether*
3-Chloro-1,2-propanediol, *Di-Me ether*
3-Chloro-1,2-propanediol,1-*Et ether*

$C_5H_{11}Cl_2N$
N-Methyl-di-(2-chloroethyl)amine
Mustine

$C_5H_{11}F$
1-Fluoro-2-methylbutane
1-Fluoro-3-methylbutane
2-Fluoro-2-methylbutane
1-Fluoropentane
2-Fluoropentane

$C_5H_{11}HgI$
Mercuri-2-methylbutyl iodide
Mercuri-3-methylbutyl iodide
Mercuri-pentyl iodide

$C_5H_{11}I$
1-Iodo-2-methylbutane
1-Iodo-3-methylbutane
2-Iodo-2-methylbutane
1-Iodopentane
2-Iodopentane
3-Iodopentane

$C_5H_{11}IO$
3-Iodo-1-propanol, *Et ether*

$C_5H_{11}IO_2$
3-Iodopropanal, *Di-Me acetal*

$C_5H_{11}N$
Allylamine, N-*Di-Me*
Allylethylamine
4-Amino-1-pentene
5-Amino-1-pentene
Cyclopentylamine
N-Methylpyrrolidine
2-Methylpyrrolidine
3-Methylpyrrolidine
Piperidine

$C_5H_{11}NO$
O-Allylhydroxylamine, N,N-*Di-Me*
N-Diethylformamide
Dimethylaminoacetone
2,2-Dimethylpropionic Acid, *Amide*
1-Hydroxypiperidine
2-Hydroxypiperidine
3-Hydroxypiperidine
4-Hydroxypiperidine
Isovaleric Acid, *Amide*
2-Methylbutyric Acid, *Amide*
Propionamide, N-*Di-Me*
Propylamine, N-*Ac*
Valeric Acid, *Amide*

$C_5H_{11}NOS$
Ethyl aminothioformate, N-*Di-Me*
Thiocarbamic Acid, S-*Butyl ester*
Thiocarbamic Acid, S-*Isobutyl ester*
Thiocarbamic Acid, O-*Isobutyl ester*

$C_5H_{11}NO_2$
Alanine, *Et ester*
Alanine, *N-Et*
2-Aminobutyric Acid, *Me ester*
3-Aminobutyric Acid, *Me ester*
2-Amino-2-methylbutyric Acid
3-Amino-3-methylbutyric Acid
2-Amino-2-methylpropionic Acid, *Me ester*
3-Aminopropionic Acid, *Et ester*
2-Aminovaleric Acid
3-Aminovaleric Acid
4-Aminovaleric Acid
5-Aminovaleric Acid
Betaine
Carbamic Acid, *Butyl ester*
Carbamic Acid, *Isobutyl ester*
Diethylcarbamic Acid
N-Dimethyl-DL-alanine
Dimethylcarbamic Acid, *Et ester*
N-Dimethylglycine, *Me ester*
Ethyl ethylaminoformate
N-Ethyl-*N*-methylglycine
2-Hydroxy-2-methylbutyric Acid, *Amide*
2-Hydroxy-3-methylbutyric Acid, *Amide*
3-Hydroxypropionic Acid, *Et ether*, *Amide*
4-Hydroxyvaleric Acid, *Amide*
5-Hydroxyvaleric Acid, *Amide*
N-Isopropylglycine
Lactamide, *Et ether*
2-Methylaminobutyric Acid
3-Methylaminobutyric Acid
4-Methylaminobutyric Acid
3-Methylaminopropionic Acid, *Me ester*
3-Methyl-1-butyl nitrite
2-Methyl-2-butyl nitrite
2-Methyl-2-methylaminopropionic Acid
2-Methyl-3-methylaminopropionic Acid
1-Nitropentane
2-Nitropentane
3-Nitropentane
Pentyl nitrite
Sarcosine, *Et ester*
Valine

$C_5H_{11}NO_2S$
Cysteine, S-*Et*
Methionine
Penicillamine

$C_5H_{11}NO_3$
2-Amino-4-hydroxyvaleric Acid†
2-Amino-5-hydroxyvaleric Acid
5-Amino-2-hydroxyvaleric Acid
3-Hydroxyvaline
Isoserine, *Et ester*
3-Methyl-1-butyl nitrate
3-Methyl-2-nitrobutanol
2-Nitro-1-pentanol
4-Nitro-1-pentanol
1-Nitro-2-pentanol
5-Nitro-2-pentanol
Serine, *Et ester*
Serine, *Et ether*
Threonine, *Me ester*
Threonine, *Me ether*

$C_5H_{11}NO_3S$
Methionine Sulphoxide†
Taurine, N-*Allyl*

$C_5H_{11}NO_4$
2-Amino-2-deoxy-α-ribose†
2-Amino-2-deoxy-α-D-xylose†
Arabinosylamine
Lyxosimine
*Rhamno*tetronic Acid, *Amide*
L-Ribosamine
Xylosimine

$C_5H_{11}NO_5$
Arabonic Acid, *Amide*
Ribonic Acid, *Amide*
Xylonic Acid, *Amide*

$C_5H_{11}NO_6$
Pentaerythritol Mononitrate†

$C_5H_{11}NS$
1,4-Thiazan, N-*Me*
Thioformic Acid, *Diethylamide*

$C_5H_{11}NS_2$
Diethyldithiocarbamic Acid
Dimethyldithiocarbamic Acid, *Et ester*

$C_5H_{11}NS_2$ (*continued*)
Dithiocarbamic Acid, *Butyl ester*
Nereistoxin

$C_5H_{11}N_3O_2$
2-Amino-2-methylsuccinamic Acid, *Amide*
4-[1-Guanidino]-butyric Acid
Methylguanidinopropionic Acid
Methyliminodiacetic Acid, *Diamide*

$C_5H_{11}N_3O_3$
4-(1-Guanidino)-3-hydroxybutyric Acid

$C_5H_{11}N_3O_4$
O-Ureidohomoserine†

$C_5H_{11}Na_2O_6P$
Glycerophosphoric Acid, *Di-Me ether, Na salt*

$C_5H_{11}O_3P$
2-Ethoxy-1,3,2-dioxaphosphane
2-Isopropoxy-1,3,2-dioxaphospholan
2-Methoxy-1,3,2-dioxaphosphepan
2-Propoxy-1,3,2-dioxaphospholan

$C_5H_{11}O_8P$
D-Xylose-1-phosphate
D-Xylose-3-phosphate
D-Xylose-5-phosphate

C_5H_{12}
2,2-Dimethylpropane
2-Methylbutane
Pentane

$C_5H_{12}HgO$
Mercuri-pentyl hydroxide

$C_5H_{12}N_2$
Isovaleric Acid, *Amidine*
1-Methylpiperazine
2-Methylpiperazine
Piperazine, N-*Me*

$C_5H_{12}N_2O$
Butylurea
(2-Butyl)urea
tert-Butylurea
Diethylcarbamic Acid, *Amide*
1,3-Diethylurea
Isobutylurea
2-Methylaminopropionic Acid, *Methylamide*
N^1-Methyl-N^1-propylurea
Valine, *Amide*

$C_5H_{12}N_2OS$
Aletheine

$C_5H_{12}N_2O_2$
2-Amino-3-dimethylaminopropionic Acid†
2,3-Diaminopropionic Acid, *Et ester*
2-Hydrazino-3-methylbutyric Acid
N-2-Hydroxyethylurea, *Et ether*
Ornithine

$C_5H_{12}N_2O_3$
γ-Hydroxyornithine†

$C_5H_{12}N_2S$
Butylthiourea
2-Butylthiourea
tert-Butylthiourea
1,1-Diethylthiourea
1,3-Diethylthiourea
Isobutylthiourea
N^1-Methyl-N^2-propylthiourea

$C_5H_{12}N_4O_2$
2-Propene-1,2-dicarboxylic Acid, *Dihydrazide*

$C_5H_{12}N_4O_3$
Canavanine

$C_5H_{12}O$
Butyl methyl Ether
2-Butyl methyl Ether
2,2-Dimethyl-1-propanol
Ethyl isopropyl Ether
Ethyl propyl Ether
Isobutyl methyl Ether
2-Methyl-1-butanol
3-Methyl-1-butanol
2-Methyl-2-butanol
4-Methyl-2-butanol
(2-Methyl-2-propyl) methyl Ether
1-Pentanol
2-Pentanol
3-Pentanol

$C_5H_{12}O_2$
Acetone, *Di-Me acetal*
2,2-Dimethylpropane-1,3-diol†
Formaldehyde, *Diethyl acetal*
2-Hydroxyethyl isopropyl Ether
2-Hydroxyethyl propyl Ether
2-Methyl-2-butylhydroperoxide
1,2-Pentanediol
1,3-Pentanediol
1,4-Pentanediol
1,5-Pentanediol
2,3-Pentanediol
2,4-Pentanediol
Propanal, *Di-Me acetal*
1,2-Propanediol, 1-*Et ether*
1,2-Propanediol, 2-*Et ether*
1,3-Propanediol, *Et ether*

$C_5H_{12}O_3$
1,2,4-Butanetriol, 4-*Me ether*
Di-(2-hydroxyethyl) Ether, *Me ether*
Glycerol, 1,2-*Di-Me ether*
Glycerol, 1,3-*Di-Me ether*
Glycerol, 1-*Et ether*
Orthoacetic Acid, *Tri-Me ester*
1,2,3-Pentanetriol
1,2,4-Pentanetriol
1,2,5-Pentanetriol
1,3,4-Pentanetriol
1,3,5-Pentanetriol
2,3,4-Pentanetriol

$C_5H_{12}O_3S$
Methane-sulphonic Acid, 2-*Ethoxyethyl ester*

$C_5H_{12}O_4$
Glyceraldehyde, *Di-Me acetal*
Methyl orthocarbonate†
Orthocarbonic Acid, *Tetra-Me ester*
Pentaerythritol

$C_5H_{12}O_5$
Adonitol
Arabitol
Xylitol

$C_5H_{12}O_6S_2$
Methionic Acid, *Di-Et ester*

$C_5H_{12}S$
2-Methyl-1-butanethiol
2-Methyl-2-butanethiol
3-Methylbutanethiol
2-Methyl-2-propanethiol, S-*Me*
1-Pentanethiol

$C_5H_{12}Se$
Butyl methyl selenide

$C_5H_{13}ClN_2O_2$
2-Hydrazinopropionic Acid, *Et ester hydrochloride*

$C_5H_{13}N$
1-Amino-2,2-dimethylpropane
1-Amino-2-methylbutane
1-Amino-3-methylbutane
2-Amino-2-methylbutane
2-Amino-3-methylbutane
1-Aminopentane
2-Aminopentane
3-Aminopentane
N-Ethylisopropylamine
N-Ethylpropylamine
Isobutylmethylamine
N-Methyl-1-butylamine
N-Methyl-2-butylamine
N-Methyl-*tert*-butylamine
2-Methyl-1-butylamine
N-Methyldiethylamine

$C_5H_{13}NO$
4-Amino-1-butanol, *Me ether*
2-Amino-3-methyl-1-butanol
5-Amino-1-pentanol
3-Amino-2-pentanol
5-Amino-2-pentanol
2-Amino-3-pentanol
3-Amino-1-propanol, N-*Di-Me*
Ethyl 2-methylaminoethyl Ether
2-Hydroxy-2-methylbutylamine
2-Isopropylaminoethanol
2-Methoxyethylamine, N-*Di-Me*
3-Methylaminobutanol
Neurine
2-Propylaminoethanol

$C_5H_{13}NOS$
4-Aminobutyl methyl sulphoxide

$C_5H_{13}NO_2$
2-Amino-1,5-pentanediol
3-Amino-1,2-propanediol, N-*Di-Me*
3-Amino-1,2-propanediol, N-*Et*
Methylaminoacetaldehyde, *Di-Me acetal*
N-Methyl-di-(2-hydroxyethyl)amine

$C_5H_{13}NO_2S$
4-Aminobutyl methyl sulphone

$C_5H_{13}NO_3S$
Taurobetaine

$C_5H_{13}NO_4$
Arabinamine
Ribitylamine

$C_5H_{13}NO_4S$
Choline Sulphuric Acid (Internal Salt)†

$C_5H_{13}NS$
4-Aminobutyl methyl sulphide

$C_5H_{13}N_3O_2$
2-Amino-3-(2-amino-ethylamino)propionic Acid†

$C_5H_{13}N_3O_3S$
Asterubin

$C_5H_{13}O_3P$
Methylphosphonic Acid, *Di-Et ester*

$C_5H_{13}O_3PS$
O,O-Diethyl-*S*-methylphosphorothiolate
O,O-Diethyl-*O*-methylphosphorothionate

$C_5H_{14}INO$
Hydroxymethyltrimethylammonium Hydroxide, *Me ether Iodide*

$C_5H_{14}NO_4P$
Choline Phosphate†

$C_5H_{14}N_2$
1-Butyl-1-methylhydrazine
Cadaverine
1,4-Diamino-2-methylbutane
2,3-Diaminopentane
2,4-Diaminopentane
1-Ethylamino-2-methylaminoethane
N-Methylputrescine

$C_5H_{14}N_4$
Agmatine

$C_5H_{15}NO_2$
Choline

$C_5H_{15}N_3$
Di-2-aminoethylmethylamine†

$C_5H_{20}O_5Si_5$
2,4,6,8,10-Pentamethylcyclopentasiloxane

$C_5O_2Br_8$
Octabromoacetylacetone

$C_5O_2Cl_8$
Octachloroacetylacetone

C_5O_5
Leuconic Acid

C_6

$C_6Br_4ClNO_4S$
2,3,4,5-Tetrabromo-6-nitrobenzenesulphonic Acid, *Chloride*
2,3,4,6-Tetrabromo-5-nitrobenzenesulphonic Acid, *Chloride*

$C_6Br_4O_2$
Bromanil★†

C_6Br_6
Hexabromobenzene

$C_6Br_6O_3$
Hexabromocyclohexane-1,3,5-trione

$C_6Cl_4O_2$
Chloranil

$C_6Cl_5NO_2$
Pentachloronitrobenzene†

C_6Cl_6
Hexachlorobenzene

$C_6Cl_6O_2$
Hexachlorocyclohex-4-ene-1,3-dione

$C_6Cl_6O_3$
Hexachlorocyclohexane-1,3,5-trione
Trichloroacrylic Acid, *Anhydride*

C_6D_6
Hexadeuterobenzene

$C_6HBrCl_2N_2O_5$
2-Bromo-4,6-dichloro-3,5-dinitrophenol
4-Bromo-2,6-dichloro-3,5-dinitrophenol

$C_6HBr_2N_3O_7$
3,5-Dibromo-2,4,6-trinitrophenol

$C_6HBr_3N_2O_4$
1,2,3-Tribromo-4,5-dinitrobenzene
1,2,3-Tribromo-4,6-dinitrobenzene
1,2,4-Tribromo-3,5-dinitrobenzene
1,3,5-Tribromo-2,4-dinitrobenzene

$C_6HBr_3O_2$
Tribromo-*p*-benzoquinone

$C_6HBr_4ClO_2S$
2,3,4,5-Tetrabromobenzenesulphonic Acid, *Chloride*
2,3,4,6-Tetrabromobenzenesulphonic Acid, *Chloride*

$C_6HBr_4NO_2$
1,2,3,4-Tetrabromo-5-nitrobenzene
1,2,3,5-Tetrabromo-6-nitrobenzene
1,2,4,5-Tetrabromo-3-nitrobenzene

$C_6HBr_4NO_5S$
2,3,4,5-Tetrabromo-6-nitrobenzenesulphonic Acid
2,3,4,6-Tetrabromo-5-nitrobenzenesulphonic Acid

C_6HBr_5
Pentabromobenzene

C_6HBr_5O
Pentabromophenol

$C_6HCl_2N_3O_6$
2,4-Dichloro-1,3,5-trinitrobenzene

$C_6HCl_2N_3O_7$
3,5-Dichloro-2,4,6-trinitrophenol

$C_6HCl_3N_2O_4$
1,2,3-Trichloro-4,5-dinitrobenzene
1,2,3-Trichloro-4,6-dinitrobenzene
1,2,4-Trichloro-3,5-dinitrobenzene
1,2,4-Trichloro-5,6-dinitrobenzene
1,3,5-Trichloro-2,4-dinitrobenzene
2,3,5-Trichloro-1,4-dinitrobenzene

$C_6HCl_3O_2$
Trichloro-*p*-benzoquinone

C_6HCl_5
Pentachlorobenzene

C_6HCl_5O
Pentachlorophenol

C_6H_2
1,3,5-Hexatri-yne

$C_6H_2BrCl_2NO_2$
1-Bromo-2,3-dichloro-5-nitrobenzene
1-Bromo-2,5-dichloro-4-nitrobenzene
1-Bromo-2,6-dichloro-4-nitrobenzene

$C_6H_2BrCl_2NO_3$
2-Bromo-4,6-dichloro-3-nitrophenol
2-Bromo-4,6-dichloro-5-nitrophenol

$C_6H_2BrN_3O_6$
1-Bromo-2,4,6-trinitrobenzene

$C_6H_2Br_2ClNO$
2,6-Dibromopyridine-4-carboxylic Acid, *Chloride*

$C_6H_2Br_2N_2$
2,6-Dibromopyridine-4-carboxylic Acid, *Nitrile*

$C_6H_2Br_2N_2O_4$
1,2-Dibromo-3,4-dinitrobenzene
1,2-Dibromo-3,5-dinitrobenzene
1,2-Dibromo-3,6-dinitrobenzene
1,2-Dibromo-4,5-dinitrobenzene
1,3-Dibromo-2,4-dinitrobenzene
1,3-Dibromo-2,5-dinitrobenzene
1,3-Dibromo-4,5-dinitrobenzene
1,3-Dibromo-4,6-dinitrobenzene
1,4-Dibromo-2,3-dinitrobenzene
1,4-Dibromo-2,5-dinitrobenzene
1,4-Dibromo-2,6-dinitrobenzene

$C_6H_2Br_2N_2O_5$
2,4-Dibromo-3,6-dinitrophenol
2,6-Dibromo-3,4-dinitrophenol
3,5-Dibromo-2,4-dinitrophenol
3,5-Dibromo-2,6-dinitrophenol

$C_6H_2Br_2O_2$
2,5-Dibromo-1,4-benzoquinone
2,6-Dibromo-1,4-benzoquinone

$C_6H_2Br_2O_4$
Bromanilic Acid

$C_6H_2Br_3NO_2$
1,2,3-Tribromo-4-nitrobenzene
1,2,3-Tribromo-5-nitrobenzene
1,2,4-Tribromo-3-nitrobenzene
1,2,4-Tribromo-6-nitrobenzene
1,3,4-Tribromo-6-nitrobenzene
1,3,5-Tribromo-2-nitrobenzene

$C_6H_2Br_3NO_3$
2,3,4-Tribromo-6-nitrophenol
2,3,6-Tribromo-4-nitrophenol
2,4,5-Tribromo-6-nitrophenol
2,4,6-Tribromo-3-nitrophenol

$C_6H_2Br_4$
1,2,3,5-Tetrabromobenzene
1,2,4,5-Tetrabromobenzene

$C_6H_2Br_4N_2O_4S$
2,3,4,5-Tetrabromo-6-nitrobenzenesulphonic Acid, *Amide*

$C_6H_2Br_4O$
2,3,4,5-Tetrabromophenol
2,3,4,6-Tetrabromophenol

$C_6H_2Br_4O_2$
Tetrabromocatechol
Tetrabromoquinol
Tetrabromoresorcinol

$C_6H_2Br_4O_3S$
2,3,4,5-Tetrabromobenzenesulphonic Acid
2,3,4,6-Tetrabromobenzenesulphonic Acid

$C_6H_2Br_5N$
Pentabromoaniline

$C_6H_2ClHgN_3O_6$
Mercuri-2,4,6-trinitrophenyl chloride

$C_6H_2ClN_3O_6$
1-Chloro-2,3,4-trinitrobenzene
1-Chloro-2,3,5-trinitrobenzene
1-Chloro-2,4,5-trinitrobenzene
1-Chloro-2,4,6-trinitrobenzene
1-Chloro-3,4,5-trinitrobenzene

$C_6H_2ClN_3O_7$
3-Chloro-2,4,6-trinitrophenol
3-Chloro-2,5,6-trinitrophenol

$C_6H_2Cl_2N_2$
2,5-Dichloropyridine-3-carboxylic Acid, *Nitrile*
5,6-Dichloropyridine-3-carboxylic Acid, *Nitrile*
2,6-Dichloropyridine-4-carboxylic Acid, *Nitrile*

$C_6H_2Cl_2N_2O_4$
1,2-Dichloro-3,4-dinitrobenzene
1,2-Dichloro-3,5-dinitrobenzene
1,2-Dichloro-4,5-dinitrobenzene
1,3-Dichloro-2,4-dinitrobenzene
1,3-Dichloro-2,5-dinitrobenzene
1,4-Dichloro-2,3-dinitrobenzene
1,4-Dichloro-2,5-dinitrobenzene
1,5-Dichloro-2,4-dinitrobenzene
2,3-Dichloro-1,4-dinitrobenzene
1,3-Dichloro-4,5-dinitrobenzene
2,5-Dichloro-1,3-dinitrobenzene

$C_6H_2Cl_2N_2O_5$
3,5-Dichloro-2,4-dinitrophenol
3,6-Dichloro-2,4-dinitrophenol

$C_6H_2Cl_2O_2$
4,5-Dichloro-1,2-benzoquinone
2,3-Dichloro-1,4-benzoquinone
2,5-Dichloro-1,4-benzoquinone
2,6-Dichloro-1,4-benzoquinone

$C_6H_2Cl_2O_3$
Furan-2,5-dicarboxylic Acid, *Dichloride*

$C_6H_2Cl_2O_4$
Chloranilic Acid

$C_6H_2Cl_3NO_2$
1,2,3-Trichloro-4-nitrobenzene
1,2,3-Trichloro-5-nitrobenzene
1,2,4-Trichloro-3-nitrobenzene
1,2,4-Trichloro-5-nitrobenzene
1,2,4-Trichloro-6-nitrobenzene
1,3,5-Trichloro-2-nitrobenzene

$C_6H_2Cl_4$
1,2,3,4-Tetrachlorobenzene
1,2,3,5-Tetrachlorobenzene
1,2,4,5-Tetrachlorobenzene

$C_6H_2Cl_4O$
2,3,4,5-Tetrachlorophenol
2,3,4,6-Tetrachlorophenol
2,3,5,6-Tetrachlorophenol

$C_6H_2Cl_4O_2$
Tetrachlorocatechol
Tetrachloroquinol
Tetrachlororesorcinol

$C_6H_2Cl_5N$
Pentachloroaniline

$C_6H_2FN_3O_7$
3-Fluoro-2,4,6-trinitrophenol

$C_6H_2IN_3O_6$
2-Iodo-1,3,5-trinitrobenzene

$C_6H_2I_2O_2$
2,6-Di-iodo-1,4-benzoquinone

$C_6H_2I_4$
1,2,3,4-Tetraiodobenzene
1,2,3,5-Tetraiodobenzene
1,2,4,5-Tetraiodobenzene

$C_6H_2N_2O_8$
Nitranilic Acid

$C_6H_2N_2S$
Thiophene-2,3-dicarboxylic Acid, *Dinitrile*
Thiophene-2,5-dicarboxylic Acid, *Dinitrile*

$C_6H_2N_4$
Pyrazine-2,3-dicarboxylic Acid, *Di-nitrile*†

$C_6H_2N_4O_8$
1,2,3,5-Tetranitrobenzene
1,2,4,5-Tetranitrobenzene

$C_6H_2N_4O_9$
2,3,4,6-Tetranitrophenol
2,3,5,6-Tetranitrophenol

$C_6H_2N_4O_{10}$
Tetranitroresorcinol

$C_6H_2O_3$
Propiolic Acid, *Anhydride*

$C_6H_3BrClNO_2$
1-Bromo-2-chloro-3-nitrobenzene
1-Bromo-2-chloro-4-nitrobenzene
1-Bromo-2-chloro-5-nitrobenzene
1-Bromo-3-chloro-2-nitrobenzene
1-Bromo-3-chloro-4-nitrobenzene
1-Bromo-3-chloro-5-nitrobenzene
1-Bromo-4-chloro-2-nitrobenzene
1-Bromo-4-chloro-3-nitrobenzene
1-Bromo-5-chloro-2-nitrobenzene

$C_6H_3BrClNO_3$
2-Bromo-3-chloro-4-nitrophenol
2-Bromo-3-chloro-6-nitrophenol
2-Bromo-4-chloro-6-nitrophenol
2-Bromo-6-chloro-4-nitrophenol
4-Bromo-2-chloro-5-nitrophenol
4-Bromo-2-chloro-6-nitrophenol
4-Bromo-5-chloro-2-nitrophenol
5-Bromo-4-chloro-2-nitrophenol

$C_6H_3BrCl_2$
1-Bromo-2,3-dichlorobenzene
1-Bromo-2,4-dichlorobenzene
1-Bromo-2,5-dichlorobenzene
1-Bromo-2,6-dichlorobenzene
1-Bromo-3,4-dichlorobenzene
1-Bromo-3,5-dichlorobenzene

$C_6H_3BrCl_2O$
2-Bromo-4,6-dichlorophenol
3-Bromo-2,6-dichlorophenol
4-Bromo-2,5-dichlorophenol
4-Bromo-2,6-dichlorophenol

$C_6H_3BrN_2O_4$
1-Bromo-2,3-dinitrobenzene
1-Bromo-2,4-dinitrobenzene
1-Bromo-2,5-dinitrobenzene
1-Bromo-2,6-dinitrobenzene
1-Bromo-3,4-dinitrobenzene
1-Bromo-3,5-dinitrobenzene

$C_6H_3BrN_2O_5$
2-Bromo-4,6-dinitrophenol
3-Bromo-2,4-dinitrophenol
3-Bromo-2,6-dinitrophenol
4-Bromo-2,6-dinitrophenol
5-Bromo-2,4-dinitrophenol

$C_6H_3BrO_2$
Bromo-*p*-benzoquinone

$C_6H_3Br_2Cl$
1,2-Dibromo-3-chlorobenzene
1,2-Dibromo-4-chlorobenzene
1,3-Dibromo-2-chlorobenzene
1,3-Dibromo-5-chlorobenzene
1,4-Dibromo-2-chlorobenzene
2,4-Dibromo-1-chlorobenzene

$C_6H_3Br_2ClO_2S$
2,4-Dibromobenzenesulphonic Acid, *Chloride*
2,5-Dibromobenzenesulphonic Acid, *Chloride*
3,4-Dibromobenzenesulphonic Acid, *Chloride*
3,5-Dibromobenzenesulphonic Acid, *Chloride*

$C_6H_3Br_2NO_2$
1,2-Dibromo-3-nitrobenzene
1,2-Dibromo-4-nitrobenzene
1,3-Dibromo-2-nitrobenzene
1,3-Dibromo-5-nitrobenzene
1,4-Dibromo-2-nitrobenzene
2,4-Dibromo-1-nitrobenzene
2,6-Dibromopyridine-4-carboxylic Acid

$C_6H_3Br_2NO_3$
2,3-Dibromo-5-nitrophenol
2,3-Dibromo-6-nitrophenol
2,4-Dibromo-5-nitrophenol
2,4-Dibromo-6-nitrophenol
2,5-Dibromo-4-nitrophenol
2,6-Dibromo-4-nitrophenol
3,6-Dibromo-2-nitrophenol

$C_6H_3Br_2N_3O_4$
2,3-Dibromo-4,6-dinitroaniline
2,4-Dibromo-3,6-dinitroaniline
2,6-Dibromo-3,4-dinitroaniline
3,4-Dibromo-2,6-dinitroaniline
3,6-Dibromo-2,4-dinitroaniline
4,6-Dibromo-2,3-dinitroaniline

$C_6H_3Br_3$
1,2,3-Tribromobenzene
1,2,4-Tribromobenzene
1,3,5-Tribromobenzene

$C_6H_3Br_3N_2O_2$
2,3,4-Tribromo-6-nitroaniline
2,3,6-Tribromo-4-nitroaniline
2,4,6-Tribromo-3-nitroaniline
3,4,5-Tribromo-2-nitroaniline

$C_6H_3Br_3O$
2,3,4-Tribromophenol
2,3,5-Tribromophenol
2,4,5-Tribromophenol
2,4,6-Tribromophenol

$C_6H_3Br_3O_2$
3,4,5-Tribromocatechol
3,4,6-Tribromocatechol
2,4,6-Tribromoresorcinol

$C_6H_3Br_3O_2S$
2,5-Dibromobenzenesulphonic Acid, *Bromide*

$C_6H_3Br_3O_3$
Tribromophloroglucinol
Tribromopyrogallol

$C_6H_3Br_4N$
2,3,4,6-Tetrabromoaniline
2,3,5,6-Tetrabromoaniline

$C_6H_3Br_4NO_2S$
2,3,4,5-Tetrabromobenzenesulphonic Acid, *Amide*
2,3,4,6-Tetrabromobenzenesulphonic Acid, *Amide*

$C_6H_3ClFNO_2$
1-Chloro-2-fluoro-4-nitrobenzene
1-Chloro-3-fluoro-5-nitrobenzene
1-Chloro-4-fluoro-2-nitrobenzene
2-Chloro-1-fluoro-4-nitrobenzene
4-Chloro-1-fluoro-2-nitrobenzene

$C_6H_3ClI_2O_2S$
2,3-Di-iodobenzenesulphonic Acid, *Chloride*
2,4-Di-iodobenzenesulphonic Acid, *Chloride*
2,5-Di-iodobenzenesulphonic Acid, *Chloride*
3,4-Di-iodobenzenesulphonic Acid, *Chloride*
3,5-Di-iodobenzenesulphonic Acid, *Chloride*

$C_6H_3ClN_2O_4$
1-Chloro-2,3-dinitrobenzene (incorrectly given as 1-Chloro-2,3-dinitrotoluene)
1-Chloro-2,4-dinitrobenzene
1-Chloro-3,5-dinitrobenzene
2-Chloro-1,3-dinitrobenzene
2-Chloro-1,4-dinitrobenzene
4-Chloro-1,2-dinitrobenzene

$C_6H_3ClN_2O_5$
2-Chloro-3,5-dinitrophenol
2-Chloro-4,6-dinitrophenol
3-Chloro-2,4-dinitrophenol
5-Chloro-2,4-dinitrophenol
3-Chloro-2,6-dinitrophenol
4-Chloro-2,3-dinitrophenol
4-Chloro-2,5-dinitrophenol
4-Chloro-2,6-dinitrophenol

$C_6H_3ClN_2O_6S$
2,4-Dinitrobenzenesulphonic Acid, *Chloride*
3,5-Dinitrobenzenesulphonic Acid, *Chloride*

$C_6H_3ClO_2$
Chloro-*p*-benzoquinone
3-Chloro-*o*-benzoquinone
4-Chloro-*o*-benzoquinone

$C_6H_3ClO_3$
2-Furylglyoxylic Acid, *Chloride*

$C_6H_3Cl_2I$
1,2-Dichloro-4-iodobenzene
1,3-Dichloro-2-iodobenzene
1,3-Dichloro-5-iodobenzene
1,4-Dichloro-2-iodobenzene
1,5-Dichloro-2-iodobenzene

$C_6H_3Cl_2NO$
4-Chloropyridine-2-carboxylic Acid, *Chloride*
2-Chloropyridine-3-carboxylic Acid, *Chloride*
5-Chloropyridine-3-carboxylic Acid, *Chloride*
6-Chloropyridine-3-carboxylic Acid, *Chloride*
2-Chloropyridine-4-carboxylic Acid, *Chloride*
4,6-Dichloropyridine-2-aldehyde
5,6-Dichloropyridine-3-aldehyde
2,6-Dichloropyridine-4-aldehyde

$C_6H_3Cl_2NO_2$
1,2-Dichloro-3-nitrobenzene
1,2-Dichloro-4-nitrobenzene
1,3-Dichloro-2-nitrobenzene
1,3-Dichloro-5-nitrobenzene
1,4-Dichloro-2-nitrobenzene
2,4-Dichloro-1-nitrobenzene
3,5-Dichloropyridine-2-carboxylic Acid
4,5-Dichloropyridine-2-carboxylic Acid
4,6-Dichloropyridine-2-carboxylic Acid
2,5-Dichloropyridine-3-carboxylic Acid
2,6-Dichloropyridine-3-carboxylic Acid
5,6-Dichloropyridine-3-carboxylic Acid
2,6-Dichloropyridine-4-carboxylic Acid
3,5-Dichloropyridine-4-carboxylic Acid

$C_6H_3Cl_2NO_3$
4,5-Dichloro-6-hydroxypyridine-2-carboxylic Acid
2,3-Dichloro-6-nitrophenol
2,4-Dichloro-3-nitrophenol
2,4-Dichloro-5-nitrophenol
2,4-Dichloro-6-nitrophenol
2,5-Dichloro-4-nitrophenol
2,6-Dichloro-4-nitrophenol
3,4-Dichloro-2-nitrophenol
3,5-Dichloro-2-nitrophenol
3,5-Dichloro-4-nitrophenol
3,6-Dichloro-2-nitrophenol
4,5-Dichloro-2-nitrophenol

$C_6H_3Cl_2NO_4S$
4-Chloro-2-nitrobenzenesulphonic Acid, *Chloride*
4-Chloro-3-nitrobenzenesulphonic Acid, *Chloride*

$C_6H_3Cl_2NO_6S_2$
5-Nitrobenzene-1,3-disulphonic Acid, *Dichloride*

$C_6H_3Cl_2N_3O_4$
2,3-Dichloro-4,6-dinitroaniline
3,4-Dichloro-2,6(?)-dinitroaniline
3,6-Dichloro-2,4-dinitroaniline

$C_6H_3Cl_3$
1,2,3-Trichlorobenzene
1,2,4-Trichlorobenzene
1,3,5-Trichlorobenzene

$C_6H_3Cl_3Hg$
Mercuri-2,5-dichlorophenyl chloride
Mercuri-3,5-dichlorophenyl chloride

$C_6H_3Cl_3O$
2,3,4-Trichlorophenol
2,3,5-Trichlorophenol
2,3,6-Trichlorophenol
2,4,5-Trichlorophenol
2,4,6-Trichlorophenol
3,4,5-Trichlorophenol

$C_6H_3Cl_3O_2$
3,4,5-Trichlorocatechol
Trichloroquinol
2,4,6-Trichlororesorcinol

$C_6H_3Cl_3O_2S$
2,4-Dichlorobenzenesulphonic Acid, *Chloride*
2,5-Dichlorobenzenesulphonic Acid, *Chloride*
3,4-Dichlorobenzenesulphonic Acid, *Chloride*

$C_6H_3Cl_3O_3$
Aconitic Acid, *Trichloride*
Trichlorophloroglucinol
Trichloropyrogallol

$C_6H_3Cl_3O_6S_3$
Benzene-1,3,5-trisulphonic Acid, *Trichloride*

$C_6H_3Cl_4N$
2,3,4,5-Tetracholoraniline
2,3,5,6-Tetrachloroaniline

$C_6H_3Cl_9$
1,1,2,2,3,4,4,5,6-Nonachlorocyclohexane

$C_6H_3FN_2O_4$
1-Fluoro-2,4-dinitrobenzene†
1-Fluoro-2,6-dinitrobenzene

$C_6H_3FN_2O_5$
2-Fluoro-4,6-dinitrophenol
3-Fluoro-2,4-dinitrophenol
3-Fluoro-2,6-dinitrophenol
4-Fluoro-2,6-dinitrophenol
5-Fluoro-2,4-dinitrophenol

$C_6H_3F_2NO_7S_2$
4-Hydroxy-5-nitrobenzene-1,3-disulphonic Acid, *Difluoride*

$C_6H_3F_7O_2$
Heptafluorobutyric Acid, *Vinyl ester*

$C_6H_3IN_2O_4$
1-Iodo-2,3-dinitrobenzene
1-Iodo-2,4-dinitrobenzene
1-Iodo-3,5-dinitrobenzene
2-Iodo-1,3-dinitrobenzene
2-Iodo-1,4-dinitrobenzene
4-Iodo-1,2-dinitrobenzene

$C_6H_3IN_2O_5$
2-Iodo-3,5-dinitrophenol
2-Iodo-4,5-dinitrophenol
2-Iodo-4,6-dinitrophenol
3-Iodo-2,6-dinitrophenol
4-Iodo-2,3-dinitrophenol
4-Iodo-2,5-dinitrophenol
4-Iodo-2,6-dinitrophenol
4-Iodo-3,5-dinitrophenol
5-Iodo-2,4-dinitrophenol

$C_6H_3IO_2$
Iodobenzoquinone

$C_6H_3I_3$
1,2,3-Tri-iodobenzene
1,2,4-Tri-iodobenzene
1,3,5-Tri-iodobenzene

$C_6H_3I_3O$
2,3,5-Tri-iodophenol
2,4,6-Tri-iodophenol

$C_6H_3I_3O_2$
2,4,6-Tri-iodoresorcinol

$C_6H_3N_3$
Cyclopropane-1,2,3-tricarboxylic Acid, *Trinitrile*

$C_6H_3N_3O_5$
Caffolide, 1-N-*Me*
Caffolide, 3-N-*Me*

$C_6H_3N_3O_6$
Triazine-tricarboxylic Acid
1,2,3-Trinitrobenzene
1,2,4-Trinitrobenzene
1,3,5-Trinitrobenzene

$C_6H_3N_3O_7$
Picric Acid
2,3,6-Trinitrophenol
2,4,5-Trinitrophenol

$C_6H_3N_3O_8$
Styphnic Acid
3,4,5-Trinitrocatechol

$C_6H_3N_3O_9$
1,3,5-Trihydroxy-2,4,6-trinitrobenzene

$C_6H_3N_3O_9S$
2,4,6-Trinitrobenzenesulphonic Acid†

$C_6H_3N_5O_8$
N,2,4,6-Tetranitroaniline
2,3,4,6-Tetranitroaniline

$C_6H_3N_7O_3$
Cyameluric Acid

$C_6H_3N_9$
Mellon

C_6H_4
Benzyne†
1,3-Dehydrobenzene†
1,4-Dehydrobenzene†

C_6H_4BrCl
o-Bromochlorobenzene
m-Bromochlorobenzene
p-Bromochlorobenzene

$C_6H_4BrClHg$
Mercuri-*o*-bromophenyl chloride
Mercuri-*m*-bromophenyl chloride
Mercuri-*p*-bromophenyl chloride

$C_6H_4BrClO_2S$
Bromobenzene-*o*-sulphonic Acid, *Chloride*
Bromobenzene-*p*-sulphonic Acid, *Chloride*
p-Chlorobenzenesulphonic Acid, *Bromide*

C_6H_4BrF
o-Bromofluorobenzene
m-Bromofluorobenzene
p-Bromofluorobenzene

C_6H_4BrI
o-Bromoiodobenzene
m-Bromoiodobenzene
p-Bromoiodobenzene

C_6H_4BrNO
o-Bromonitrosobenzene
m-Bromonitrosobenzene
p-Bromonitrosobenzene

$C_6H_4BrNO_2$
o-Bromonitrobenzene
m-Bromonitrobenzene
p-Bromonitrobenzene
5-Bromopyridine-2-carboxylic Acid
6-Bromopyridine-2-carboxylic Acid
5-Bromopyridine-3-carboxylic Acid

$C_6H_4BrNO_3$
2-Bromo-3-nitrophenol
2-Bromo-4-nitrophenol
2-Bromo-5-nitrophenol
2-Bromo-6-nitrophenol
3-Bromo-2-nitrophenol
3-Bromo-4-nitrophenol
3-Bromo-5-nitrophenol
4-Bromo-2-nitrophenol
4-Bromo-3-nitrophenol
5-Bromo-2-nitrophenol

$C_6H_4BrNO_3S$
3-Nitrobenzenesulphinic Acid, *Bromide*

$C_6H_4BrNO_4S$
2-Nitrobenzenesulphonic Acid, *Bromide*
3-Nitrobenzenesulphonic Acid, *Bromide*

$C_6H_4BrN_3$
p-Bromophenyl azide

$C_6H_4BrN_3O_4$
2-Bromo-3,5-dinitroaniline
2-Bromo-4,5-dinitroaniline
2-Bromo-4,6-dinitroaniline
2-Bromo-5,6-dinitroaniline
3-Bromo-2,6-dinitroaniline
4-Bromo-2,3-dinitroaniline
4-Bromo-2,5-dinitroaniline
4-Bromo-2,6-dinitroaniline
4-Bromo-3,5-dinitroaniline
5-Bromo-2,4-dinitroaniline

$C_6H_4Br_2$
1,2-Dibromobenzene
1,3-Dibromobenzene
1,4-Dibromobenzene

$C_6H_4Br_2N_2O$
2,6-Dibromopyridine-4-carboxylic Acid, *Amide*

$C_6H_4Br_2N_2O_2$
2,3-Dibromo-6-nitroaniline
2,4-Dibromo-3-nitroaniline
2,4-Dibromo-5 nitroaniline
2,4-Dibromo-6-nitroaniline
2,5-Dibromo-4-nitroaniline
2,6-Dibromo-4-nitroaniline
3,5-Dibromo-2-nitroaniline
3,5-Dibromo-4-nitroaniline
3,6-Dibromo-2-nitroaniline
4,5-Dibromo-2-nitroaniline
2,4-Dibromophenylnitramine
2,6-Dibromophenylnitramine

$C_6H_4Br_2O$
2,3-Dibromophenol
2,4-Dibromophenol
2,5-Dibromophenol
2,6-Dibromophenol
3,4-Dibromophenol
3,5-Dibromophenol

$C_6H_4Br_2O_2$
3,4-Dibromocatechol
3,5-Dibromocatechol
4,5-Dibromocatechol
2,5-Dibromoquinol
2,6-Dibromoquinol
2,4-Dibromoresorcinol
4,6-Dibromoresorcinol

$C_6H_4Br_2O_2S$
Bromobenzene-*p*-sulphonic Acid, *Bromide*

$C_6H_4Br_2O_3S$
2,4-Dibromobenzenesulphonic Acid
2,5-Dibromobenzenesulphonic Acid
3,4-Dibromobenzenesulphonic Acid
3,5-Dibromobenzenesulphonic Acid

$C_6H_4Br_3N$
2,4,5-Tribromoaniline
2,4,6-Tribromoaniline
3,4,5-Tribromoaniline

$C_6H_4Br_3NO$
3-Amino-2,4,6-tribromophenol

C_6H_4ClF
o-Chlorofluorobenzene
m-Chlorofluorobenzene
p-Chlorofluorobenzene

$C_6H_4ClFO_2S$
p-Fluorobenzenesulphonic Acid, *Chloride*

$C_6H_4ClHgNO_2$
Mercuri-*o*-nitrophenyl chloride
Mercuri-*m*-nitrophenyl chloride
Mercuri-*p*-nitrophenyl chloride

C_6H_4ClI
o-Chloroiodobenzene
m-Chloroiodobenzene
p-Chloroiodobenzene

C_6H_4ClNO
Pyridine-2-carboxylic Acid, *Chloride*
Pyridine-3-carboxylic Acid, *Chloride*
Pyridine-4-carboxylic Acid, *Chloride*

$C_6H_4ClNO_2$
Chloro-*p*-benzoquinone, 1-*Oxime*
3-Chloropyridine-2-carboxylic Acid
4-Chloropyridine-2-carboxylic Acid
6-Chloropyridine-2-carboxylic Acid
2-Chloropyridine-3-carboxylic Acid
5-Chloropyridine-3-carboxylic Acid
6-Chloropyridine-3-carboxylic Acid
2-Chloropyridine-4-carboxylic Acid
3-Chloropyridine-4-carboxylic Acid
o-Chloronitrobenzene
m-Chloronitrobenzene
p-Chloronitrobenzene

$C_6H_4ClNO_3$
2-Chloro-3-nitrophenol
2-Chloro-4-nitrophenol
2-Chloro-5-nitrophenol
2-Chloro-6-nitrophenol
3-Chloro-2-nitrophenol
3-Chloro-4-nitrophenol
3-Chloro-5-nitrophenol
4-Chloro-2-nitrophenol
4-Chloro-3-nitrophenol
5-Chloro-2-nitrophenol

$C_6H_4ClNO_4$
4-Chloro-2-nitroresorcinol
6-Chloro-4-nitroresorcinol

$C_6H_4ClNO_4S$
2-Nitrobenzenesulphonic Acid, *Chloride*
3-Nitrobenzenesulphonic Acid, *Chloride*
4-Nitrobenzenesulphonic Acid, *Chloride*

$C_6H_4ClNO_5S$
2-Chloro-5-nitrobenzenesulphonic Acid
2-Chloro-6-nitrobenzenesulphonic Acid
3-Chloro-5-nitrobenzenesulphonic Acid
4-Chloro-2-nitrobenzenesulphonic Acid
4-Chloro-3-nitrobenzenesulphonic Acid
5-Chloro-2-nitrobenzenesulphonic Acid

$C_6H_4ClN_3$
1-Chlorobenzotriazole†
5-Chlorobenzotriazole

$C_6H_4ClN_3O_4$
2-Chloro-3,5-dinitroaniline
2-Chloro-4,6-dinitroaniline
3-Chloro-2,6-dinitroaniline
4-Chloro-2,6-dinitroaniline
4-Chloro-3,5-dinitroaniline
5-Chloro-2,4-dinitroaniline

$C_6H_4ClO_3P$
o-Phenylene phosphorochloridate†

$C_6H_4Cl_2$
o-Dichlorobenzene
m-Dichlorobenzene
p-Dichlorobenzene

$C_6H_4Cl_2N_2O$
3,5-Dichloropyridine-2-carboxylic Acid, *Amide*
2,6-Dichloropyridine-4-carboxylic Acid, *Amide*

$C_6H_4Cl_2N_2O_2$
2,3-Dichloro-6-nitroaniline
2,4-Dichloro-3-nitroaniline
2,4-Dichloro-5-nitroaniline
2,4-Dichloro-6-nitroaniline
2,5-Dichloro-3-nitroaniline
2,5-Dichloro-4-nitroaniline
2,6-Dichloro-3-nitroaniline
2,6-Dichloro-4-nitroaniline
3,5-Dichloro-2-nitroaniline
3,5-Dichloro-4-nitroaniline
3,6-Dichloro-2-nitroaniline
4,5-Dichloro-2-nitroaniline

$C_6H_4Cl_2N_2O_4S$
3-Nitrobenzenesulphonic Acid, N-*Dichloroamide*

$C_6H_4Cl_2O$
2,3-Dichlorophenol
2,4-Dichlorophenol
2,5-Dichlorophenol
2,6-Dichlorophenol
3,4-Dichlorophenol
3,5-Dichlorophenol

$C_6H_4Cl_2O_2$
3,4-Dichlorocatechol
3,5-Dichlorocatechol
4,5-Dichlorocatechol

$C_6H_4Cl_2O_2$ (*continued*)
2,3-Dichloroquinol
2,5-Dichloroquinol
2,6-Dichloroquinol
4,6-Dichlororesorcinol

$C_6H_4Cl_2O_2S$
o-Chlorobenzenesulphonic Acid, *Chloride*
p-Chlorobenzenesulphonic Acid, *Chloride*

$C_6H_4Cl_2O_3S$
2,4-Dichlorobenzenesulphonic Acid
2,5-Dichlorobenzenesulphonic Acid
3,4-Dichlorobenzenesulphonic Acid

$C_6H_4Cl_2O_4S_2$
Benzene-*o*-disulphonic Acid, *Dichloride*
Benzene-*m*-disulphonic Acid, *Dichloride*
Benzene-*p*-disulphonic Acid, *Dichloride*

$C_6H_4Cl_2O_5S_2$
Phenol-2,4-disulphonic Acid, *Dichloride*

$C_6H_4Cl_2O_6S_2$
4,6-Dihydroxybenzene-1,3-disulphonic Acid, *Dichloride*

$C_6H_4Cl_3HgN$
Mercuri-2-amino-3,5-dichlorophenyl chloride

$C_6H_4Cl_3N$
2,3,4-Trichloroaniline
2,3,5-Trichloroaniline
2,3,6-Trichloroaniline
2,4,5-Trichloroaniline
2,4,6-Trichloroaniline
3,4,5-Trichloroaniline

$C_6H_4Cl_3NO$
2-Amino-3,4,6-trichlorophenol
3-Amino-2,4,6-trichlorophenol
4-Amino-2,3,5-trichlorophenol
4-Amino-2,3,6-trichlorophenol

$C_6H_4Cl_8$
1,1,2,2,3,4,5,6-Octachlorocyclohexane
1,1,2,3,3,4,5,6-Octachlorocyclohexane
1,1,2,3,4,4,5,6-Octachlorocyclohexane

C_6H_4FI
o-Fluoroiodobenzene
p-Fluoroiodobenzene

$C_6H_4FNO_2$
o-Fluoronitrobenzene
m-Fluoronitrobenzene
p-Fluoronitrobenzene

$C_6H_4FNO_3$
2-Fluoro-4-nitrophenol
2-Fluoro-6-nitrophenol
3-Fluoro-2-nitrophenol
3-Fluoro-4-nitrophenol
3-Fluoro-5-nitrophenol
4-Fluoro-2-nitrophenol
5-Fluoro-2-nitrophenol

$C_6H_4FNO_5S$
4-Hydroxy-3-nitrobenzenesulphonic Acid, *Fluoride*

$C_6H_4F_2$
o-Difluorobenzene
m-Difluorobenzene
p-Difluorobenzene

$C_6H_4F_2O_4S_2$
Benzene-*m*-disulphonic Acid, *Difluoride*

$C_6H_4INO_2$
o-Iodonitrobenzene
m-Iodonitrobenzene
p-Iodonitrobenzene

$C_6H_4INO_3$
2-Iodo-3-nitrophenol
2-Iodo-4-nitrophenol
2-Iodo-5-nitrophenol
2-Iodo-6-nitrophenol
3-Iodo-2-nitrophenol
3-Iodo-4-nitrophenol
4-Iodo-2-nitrophenol
4-Iodo-3-nitrophenol
5-Iodo-2-nitrophenol
5-Iodo-3-nitrophenol

$C_6H_4I_2$
o-Di-iodobenzene
m-Di-iodobenzene
p-Di-iodobenzene

$C_6H_4I_2O$
2,4-Di-iodophenol
2,5-Di-iodophenol
2,6-Di-iodophenol
3,4-Di-iodophenol
3,5-Di-iodophenol

$C_6H_4I_2O_2$
2,6-Di-iodoquinol

$C_6H_4I_2O_3S$
2,3-Di-iodobenzenesulphonic Acid
2,4-Di-iodobenzenesulpnonic Acid
2,5-Di-iodobenzenesulphonic Acid
3,4-Di-iodobenzenesulphonic Acid
3,5-Di-iodobenzenesulphonic Acid

$C_6H_4I_2O_4S$
Sozoiodolic Acid

$C_6H_4I_3N$
2,3,5-Tri-iodoaniline
2,3,6-Tri-iodoaniline
2,4,5-Tri-iodoaniline
2,4,6-Tri-iodoaniline
3,4,5-Tri-iodoaniline

$C_6H_4NO_3S$
2-Acetyl-4-nitrothiophene
2-Acetyl-5-nitrothiophene

$C_6H_4N_2$
2-Cyanopyridine
3-Cyanopyridine
4-Cyanopyridine
Muconic Acid★, *Di-nitrile*†

$C_6H_4N_2O$
Benzofurazan
3-Hydroxypyridine-2-carboxylic Acid, *Nitrile*†

$C_6H_4N_2O_2$
Benzofuroxan
m-Dinitrosobenzene

$C_6H_4N_2O_2S$
1,2,3-Benzothiadiazole 1,1-Dioxide†

$C_6H_4N_2O_3$
o-Nitronitrosobenzene
m-Nitronitrosobenzene
p-Nitronitrosobenzene

$C_6H_4N_2O_4$
o-Dinitrobenzene
m-Dinitrobenzene
p-Dinitrobenzene
2,4-Dinitrosoresorcinol
Pyrazine-2,3-dicarboxylic Acid†
Pyrazine-2,5-dicarboxylic Acid†
Pyrazine-2,6-dicarboxylic Acid†

$C_6H_4N_2O_4S$
2,4-Dinitrobenzenethiol

$C_6H_4N_2O_5$
2,3-Dinitrophenol
2,4-Dinitrophenol
2,5-Dinitrophenol
2,6-Dinitrophenol
3,4-Dinitrophenol
3,5-Dinitrophenol

$C_6H_4N_2O_6$
3,4-Dinitrocatechol
3,5-Dinitrocatechol
4,5-Dinitrocatechol
2,3-Dinitroquinol
2,5-Dinitroquinol
2,6-Dinitroquinol
2,4-Dinitroresorcinol
4,5-Dinitroresorcinol
4,6-Dinitroresorcinol

$C_6H_4N_2O_7$
1,2,3-Trihydroxy-4,5-dinitrobenzene
1,2,3-Trihydroxy-4,6-dinitrobenzene
1,3,5-Trihydroxy-2,4-dinitrobenzene

$C_6H_4N_2O_7S$
2,4-Dinitrobenzenesulphonic Acid
3,5-Dinitrobenzenesulphonic Acid

$C_6H_4N_2S$
1,2,3-Benzothiadiazole★†
2,1,3-Benzothiadiazole

$C_6H_4N_2Se$
2,1,3-Benzoselenadiazole

$C_6H_4N_4$
Pteridine
Pyridazo[4,5-*d*]pyridazine†

$C_6H_4N_4O_2$
Lumazin
4-Nitro-1,2,3-benzotriazole
5-Nitro-1,2,3-benzotriazole

$C_6H_4N_4O_4$
2,4,6,8-Tetrahydroxypyrimido[5,4-*d*]pyrimidine†

$C_6H_4N_4O_6$
2,3,4-Trinitroaniline
2,4,5-Trinitroaniline
2,4,6-Trinitroaniline

$C_6H_4N_4O_7$
3-Amino-2,4,6-trinitrophenol
4-Amino-2,3,5-trinitrophenol
4-Amino-2,3,6-trinitrophenol

C_6H_4O
2-Ethynylfuran

$C_6H_4O_2$
o-Benzoquinone
p-Benzoquinone

$C_6H_4O_3$
3-Hydroxy-1,2-benzoquinone
4-Hydroxy-1,2-benzoquinone
Hydroxy-1,4-benzoquinone

$C_6H_4O_3S$
o-Phenylene Sulphite†
2-Thienylglyoxylic Acid

$C_6H_4O_4$
2,5-Dihydroxy-1,4-benzoquinone
5-Formylfuran-2-carboxylic Acid
2-Furylglyoxylic Acid
2-Oxo-2*H*-pyran-5-carboxylic Acid
2-Oxo-2*H*-pyran-6-carboxylic Acid
4-Oxo-4*H*-pyran-2-carboxylic Acid

$C_6H_4O_4S$
Thiophene-2,3-dicarboxylic Acid
Thiophene-2,4-dicarboxylic Acid
Thiophene-2,5-dicarboxylic Acid
Thiophene-3,4-dicarboxylic Acid

$C_6H_4O_5$
Furan-2,3-dicarboxylic Acid
Furan-2,4-dicarboxylic Acid
Furan-2,5-dicarboxylic Acid
Furan-3,4-dicarboxylic Acid
5-Hydroxy-4-oxopyran-2-carboxylic Acid

$C_6H_4O_5S_2$
Benzene-*o*-disulphonic Acid, *Anhydride*

$C_6H_4O_6$
5,6-Dihydroxy-4-oxopyran-2-carboxylic Acid
Rubiginic Acid
Sarsapic Acid
Tetrahydroxy-*p*-benzoquinone

$C_6H_4S_2$
Thieno[2,3-*b*]thiophene
Thieno[3,2-*b*]thiophene
Thieno[3,4-*b*]thiophen†

$C_6H_5AsBr_2$
Dibromophenylarsine

$C_6H_5AsCl_2$
Dichlorophenylarsine

$C_6H_5AsCl_2O$
Benzenearsonic Acid, *Dichloride*

C_6H_5AsO
Arsenosobenzene

C_6H_5Br
Bromobenzene

C_6H_5BrClN
N-Bromo-2-chloroaniline
N-Bromo-4-chloroaniline
2-Bromo-*N*-chloroaniline
2-Bromo-4-chloroaniline
3-Bromo-4-chloroaniline
4-Bromo-*N*-chloroaniline
4-Bromo-2-chloroaniline
4-Bromo-3-chloroaniline

C_6H_5BrClN (*continued*)
5-Bromo-2-chloroaniline
5-Bromo-3-chloroaniline

C_6H_5BrHg
Mercuri-phenyl bromide

C_6H_5BrHgO
Mercuri-*o*-hydroxyphenyl bromide
Mercuri-*p*-hydroxyphenyl bromide

C_6H_5BrIN
2-Bromo-4-iodoaniline
3-Bromo-4-iodoaniline
4-Bromo-2-iodoaniline
4-Bromo-3-iodoaniline

$C_6H_5BrN_2O_2$
2-Bromo-4-nitroaniline
2-Bromo-5-nitroaniline
2-Bromo-6-nitroaniline
3-Bromo-4-nitroaniline
4-Bromo-2-nitroaniline
4-Bromo-3-nitroaniline
5-Bromo-2-nitroaniline

C_6H_5BrO
o-Bromophenol
m-Bromophenol
p-Bromophenol

$C_6H_5BrO_2$
3-Bromocatechol
4-Bromocatechol
5-Bromomethylfurfural
Bromoquinol
2-Bromoresorcinol
4-Bromoresorcinol
5-Bromoresorcinol

$C_6H_5BrO_3S$
Bromobenzene-*o*-sulphonic Acid
Bromobenzene-*m*-sulphonic Acid
Bromobenzene-*p*-sulphonic Acid

C_6H_5BrS
o-Bromobenzenethiol
p-Bromobenzenethiol

$C_6H_5Br_2HgN$
Mercuri-2-amino-5-bromophenyl bromide
Mercuri-4-amino-2-bromophenyl bromide
Mercuri-4-amino-3-bromophenyl bromide

$C_6H_5Br_2N$
2,3-Dibromoaniline
2,4-Dibromoaniline
2,5-Dibromoaniline
2,6-Dibromoaniline
3,4-Dibromoaniline
3,5-Dibromoaniline

$C_6H_5Br_2NO$
2-Amino-3,5-dibromophenol
2-Amino-4,6-dibromophenol
4-Amino-2,5-dibromophenol
4-Amino-2,6-dibromophenol
5-Amino-2,3-dibromophenol

$C_6H_5Br_2NO_2S$
2,4-Dibromobenzenesulphonic Acid, *Amide*
2,5-Dibromobenzenesulphonic Acid, *Amide*
3,4-Dibromobenzenesulphonic Acid, *Amide*
3,5-Dibromobenzenesulphonic Acid, *Amide*

$C_6H_5Br_3N_2$
3,4,5-Tribromo-*o*-phenylenediamine
2,4,6-Tribromo-*m*-phenylenediamine
2,4,6-Tribromophenylhydrazine

$C_6H_5Br_4N$
2,3,4,5-Tetrabromopyrrole, N-*Et*

C_6H_5Cl
Chlorobenzene

C_6H_5ClFN
2-Chloro-6-fluoroaniline
3-Chloro-4 fluoroaniline
4-Chloro-3-fluoroaniline

C_6H_5ClHg
Mercuri-phenyl chloride

C_6H_5ClHgO
Mercuri-*o*-hydroxyphenyl chloride
Mercuri-*m*-hydroxyphenyl chloride
Mercuri-*p*-hydroxyphenyl chloride

C_6H_5ClIN
2-Chloro-4-iodoaniline
2-Chloro-5-iodoaniline
3-Chloro-4-iodoaniline
4-Chloro-2-iodoaniline
4-Chloro-3-iodoaniline
5-Chloro-3-iodoaniline

$C_6H_5ClN_2$
Benzenediazonium chloride

$C_6H_5ClN_2O$
3-Chloropyridine-2-carboxylic Acid, *Amide*
4-Chloropyridine-2-carboxylic Acid, *Amide*
5-Chloropyridine-3-carboxylic Acid, *Amide*
6-Chloropyridine-3-carboxylic Acid, *Amide*
o-Hydroxybenzenediazonium chloride
p-Hydroxybenzenediazonium chloride

$C_6H_5ClN_2O_2$
2-Chloro-3-nitroaniline
2-Chloro-4-nitroaniline
2-Chloro-5-nitroaniline
2-Chloro-6-nitroaniline
3-Chloro-4-nitroaniline
4-Chloro-2-nitroaniline
4-Chloro-3-nitroaniline
5-Chloro-2-nitroaniline
5-Chloro-3-nitroaniline

$C_6H_5ClN_2O_3$
2-Amino-4-chloro-5-nitrophenol
2-Amino-4-chloro-6-nitrophenol
2-Amino-6-chloro-4-nitrophenol
4-Amino-2-chloro-3-nitrophenol
4-Amino-2-chloro-5-nitrophenol
2-Methylimidazole-4,5-dicarboxylic Acid, 5-*Chloride*

$C_6H_5ClN_2O_4S$
2-Chloro-5-nitrobenzenesulphonic Acid, *Amide*
2-Chloro-6-nitrobenzenesulphonic Acid, *Amide*
3-Chloro-5-nitrobenzenesulphonic Acid, *Amide*
4-Chloro-2-nitrobenzenesulphonic Acid, *Amide*
4-Chloro-3-nitrobenzenesulphonic Acid, *Amide*
5-Chloro-2-nitrobenzenesulphonic Acid, *Amide*
2-Nitroaniline-4-sulphonic Acid, *Chloride*

C_6H_5ClO
o-Chlorophenol
m-Chlorophenol
p-Chlorophenol

C_6H_5ClOS
Benzenesulphinic Acid, *Chloride*
3-Methylthiophene-2-carboxylic Acid, *Chloride*

$C_6H_5ClO_2$
3-Chlorocatechol
4-Chlorocatechol
5-Chloromethylfurfuraldehyde
Chloroquinol
2-Chlororesorcinol
4-Chlororesorcinol
2-Furylacetic Acid, *Chloride*
3-Methylfuran-2-carboxylic Acid, *Chloride*
5-Methylfuran-2-carboxylic Acid, *Chloride*

$C_6H_5ClO_2S$
Benzenesulphonic Acid, *Chloride*

$C_6H_5ClO_3$
4-Chloropyrogallol
5-Chloropyrogallol

$C_6H_5ClO_3S$
o-Chlorobenzenesulphonic Acid
m-Chlorobenzenesulphonic Acid
p-Chlorobenzenesulphonic Acid
Phenol-*m*-sulphonic Acid, *Chloride*
Phenol-*p*-sulphonic Acid, *Chloride*

C_6H_5ClS
o-Chlorobenzenethiol
m-Chlorobenzenethiol
p-Chlorobenzenethiol

$C_6H_5Cl_2HgN$
Mercuri-2-amino-5-chlorophenyl chloride
Mercuri-4-amino-2-chlorophenyl chloride
Mercuri-4-amino-3-chlorophenyl chloride

$C_6H_5Cl_2N$
2,3-Dichloroaniline
2,4-Dichloroaniline
2,5-Dichloroaniline
2,6-Dichloroaniline
3,4-Dichloroaniline
3,5-Dichloroaniline

$C_6H_5Cl_2NO$
2-Amino-3,5-dichlorophenol
2-Amino-4,6-dichlorophenol
3-Amino-4,6-dichlorophenol
4-Amino-2,5-dichlorophenol
4-Amino-2,6-dichlorophenol
4-Amino-3,5-dichlorophenol

$C_6H_5Cl_2NO_2S$
2,4-Dichlorobenzenesulphonic Acid, *Amide*
2,5-Dichlorobenzenesulphonic Acid, *Amide*
3,4-Dichlorobenzenesulphonic Acid, *Amide*

$C_6H_5Cl_2OP$
Phenylphosphonic Acid, *Dichloride*

$C_6H_5Cl_3O_3$
Tricarballylic Acid, *Trichloride*

$C_6H_5Cl_3Te$
Phenyltellurium trichloride

$C_6H_5Cl_7$
1,1,2,3,4,5,6-Heptachlorocyclohexane

C_6H_5Cu
Phenyl copper

C_6H_5F
Fluorobenzene

C_6H_5FHg
Mercuri-phenyl fluoride

$C_6H_5FN_2O_2$
2-Fluoro-4-nitroaniline
2-Fluoro-5-nitroaniline
3-Fluoro-4-nitroaniline
4-Fluoro-2-nitroaniline
4-Fluoro-3-nitroaniline
5-Fluoro-2-nitroaniline
5-Fluoro-3-nitroaniline

C_6H_5FO
o-Fluorophenol
m-Fluorophenol
p-Fluorophenol

$C_6H_5FO_3S$
p-Fluorobenzenesulphonic Acid
Phenol-*p*-sulphonic Acid, *Fluoride*

$C_6H_5F_7O_2$
Heptafluorobutyric Acid, *Et ester*

C_6H_5HgI
Mercuri-phenyl iodide

C_6H_5HgIO
Mercuri-*o*-hydroxyphenyl iodide
Mercuri-*p*-hydroxyphenyl iodide

$C_6H_5HgNO_3$
Mercuri-phenyl nitrate

C_6H_5I
Iodobenzene

$C_6H_5IN_2O_2$
2-Iodo-4-nitroaniline
2-Iodo-5-nitroaniline
4-Iodo-2-nitroaniline
4-Iodo-3-nitroaniline
5-Iodo-2-nitroaniline

C_6H_5IO
o-Iodophenol
m-Iodophenol
p-Iodophenol
Iodosobenzene

$C_6H_5IO_2$
1,2-Dihydroxy-4-iodobenzene
1,3-Dihydroxy-2-iodobenzene
1,3-Dihydroxy-4-iodobenzene
1,3-Dihydroxy-5-iodobenzene
Iodoquinol
Iodoxybenzene

C_6H_5IS
o-Iodobenzenethiol
m-Iodobenzenethiol
p-Iodobenzenethiol

$C_6H_5I_2N$
2,4-Di-iodoaniline
2,5-Di-iodoaniline
2,6-Di-iodoaniline

$C_6H_5I_2N$ (*continued*)
3,4-Di-iodoaniline
3,5-Di-iodoaniline

$C_6H_5I_2NO_2S$
2,4-Di-iodobenzenesulphonic Acid, *Amide*
3,4-Di-iodobenzenesulphonic Acid, *Amide*

C_6H_5NO
2-Furylacetic Acid, *Nitrile*
4-Methylfuran-2-carboxylic Acid, *Nitrile*
Nitrosobenzene
Pyridine-2-aldehyde
Pyridine-3-aldehyde
Pyridine-4-aldehyde

C_6H_5NOS
Thionylaniline

$C_6H_5NO_2$
Amino-*p*-benzoquinone
Nitrobenzene
o-Nitrosophenol
p-Nitrosophenol
Pyridine-2-carboxylic Acid
Pyridine-3-carboxylic Acid
Pyridine-4-carboxylic Acid

$C_6H_5NO_2S$
3-Mercaptopyridine-2-carboxylic Acid
2-Mercaptopyridine-3-carboxylic Acid
6-Mercaptopyridine-3-carboxylic Acid
3-Mercaptopyridine-4-carboxylic Acid
o-Nitrobenzenethiol
p-Nitrobenzenethiol
2-Thienylglyoxylic Acid, *Amide*

$C_6H_5NO_3$
1,2-Dihydroxy-4-nitrosobenzene
1,3-Dihydroxy-2-nitrosobenzene
1,3-Dihydroxy-4-nitrosobenzene
3-Hydroxypyridine-2-carboxylic Acid†
2-Hydroxypyridine-3-carboxylic Acid
4-Hydroxypyridine-3-carboxylic Acid
6-Hydroxypyridine-3-carboxylic Acid
3-Hydroxypyridine-4-carboxylic Acid
o-Nitrophenol
m-Nitrophenol
p-Nitrophenol
2-Oxo-2*H*-pyran-5-carboxylic Acid, *Amide*

$C_6H_5NO_4$
1,3-Dihydroxy-2-nitrobenzene
1,3-Dihydroxy-4-nitrobenzene
1,3-Dihydroxy-5-nitrobenzene
3,4-Dihydroxypyridine-2-carboxylic Acid
4,5-Dihydroxypyridine-2-carboxylic Acid
4,6-Dihydroxypyridine-2-carboxylic Acid
2,4-Dihydroxypyridine-3-carboxylic Acid
2,6-Dihydroxypyridine-3-carboxylic Acid
4,6-Dihydroxypyridine-3-carboxylic Acid
2,6-Dihydroxypyridine-4-carboxylic Acid
Furan-2,5-dicarboxylic Acid, *Mono-amide*
3-Nitrocatechol
4-Nitrocatechol
Pyrrole-2,3-dicarboxylic Acid
Pyrrole-2,4-dicarboxylic Acid
Pyrrole-2,5-dicarboxylic Acid
Pyrrole-3,4-dicarboxylic Acid
2-Nitroquinol
1,3,5-Trihydroxy-2-nitrosobenzene

$C_6H_5NO_4S$
2-Nitrobenzenesulphinic Acid
3-Nitrobenzenesulphinic Acid
4-Nitrobenzenesulphinic Acid

$C_6H_5NO_4S_2$
Benzene-*o*-disulphonic Acid, *Imide*

$C_6H_5NO_5$
3-Methyl-5-nitrofuran-2-carboxylic Acid
5-Methyl-4-nitrofuran-2-carboxylic Acid
2-Methyl-5-nitrofuran-3-carboxylic Acid
5-Nitrofuran-2-carboxylic Acid, *Me ester*
1,2,3-Trihydroxy-4-nitrobenzene
1,2,3-Trihydroxy-5-nitrobenzene
1,2,4-Trihydroxy-3-nitrobenzene
1,2,4-Trihydroxy-5-nitrobenzene
1,3,5-Trihydroxy-2-nitrobenzene

$C_6H_5NO_5S$
2-Nitrobenzenesulphonic Acid
3-Nitrobenzenesulphonic Acid
4-Nitrobenzenesulphonic Acid

$C_6H_5NO_6S$
2-Hydroxy-5-nitrobenzenesulphonic Acid
4-Hydroxy-2-nitrobenzenesulphonic Acid
4-Hydroxy-3-nitrobenzenesulphonic Acid
5-Hydroxy-2-nitrobenzenesulphonic Acid

$C_6H_5NO_8S_2$
4-Nitrobenzene-1,3-disulphonic Acid
5-Nitrobenzene-1,3-disulphonic Acid

$C_6H_5NO_9S_2$
4-Hydroxy-5-nitrobenzene-1,3-disulphonic Acid

$C_6H_5N_2O_8P$
2,4-Dinitrophenyl dihydrogen phosphate†
2,5-Dinitrophenyl dihydrogen phosphate†
2,6-Dinitrophenyl dihydrogen phosphate†

$C_6H_5N_3$
6-Aminopyridine-3-carboxylic Acid, *Nitrile*
Benzotriazole
Imidazo[1,2-*a*]pyrimidine†
Phenyl azide
1*H*-Pyrazole[3,4-*b*]pyridine†
1,3,5-Triaza-indene†
Tricarballylic Acid, *Trinitrile*

$C_6H_5N_3O$
3,7-Dihydroimidazo[1,2-*a*]pyrazin-3-one†
1-Hydroxybenzotriazole

$C_6H_5N_3O_3$
Benzenediazonium nitrate

$C_6H_5N_3O_4$
2-Amino-5-nitropyridine-3-carboxylic Acid
6-Amino-5-nitropyridine-3-carboxylic Acid
2,3-Dinitroaniline
2,4-Dinitroaniline
2,5-Dinitroaniline
2,6-Dinitroaniline
3,4-Dinitroaniline
3,5-Dinitroaniline

$C_6H_5N_3O_5$
2-Amino-3,5-dinitrophenol
2-Amino-4,5-dinitrophenol
2-Amino-4,6-dinitrophenol

$C_6H_5N_3O_5$ (*continued*)
3-Amino-2,4-dinitrophenol
3-Amino-2,6-dinitrophenol
4-Amino-2,3-dinitrophenol
4-Amino-2,5-dinitrophenol
4-Amino-2,6-dinitrophenol
4-Amino-3,5-dinitrophenol
5-Amino-2,4-dinitrophenol
N-(2,4-Dinitrophenyl)hydroxylamine
N-(2,6-Dinitrophenyl)hydroxylamine
N-(3,5-Dinitrophenyl)hydroxylamine

$C_6H_5N_3O_8S$
2,4-Dinitrobenzenesulphonic Acid, *Amide*
3,5-Dinitrobenzenesulphonic Acid, *Amide*

$C_6H_5N_5O_2$
1,5,6,7-(5,6,7,8)-Tetrahydro-6-methyl-5,7-dioxopyrimido[5,4-*e*]-*as*-triazine†
Xanthopterin

$C_6H_5N_5O_3$
Leucopterin

$C_6H_5N_5O_6$
2,4,6-Trinitrophenylhydrazine

C_6H_5Na
Sodium phenyl

C_6H_5NS
N-Thioaniline†
2-Thiopheneacetic Acid, *Nitrile*

$C_6H_5O_2P$
Phenylphosphonic Acid, *Anhydride*

C_6H_6
Benzene
3,4-Dimethylenecyclobutene†
Fulvene†
1,5-Hexadien-3-yne
1,3-Hexadi-yne
1,4-Hexadi-yne
1,5-Hexadi-yne
2,4-Hexadi-yne
Prismane†
Tricyclo[3,1,0,$0^{2,6}$]hex-3-ene†
Trimethylenecyclopropane†

$C_6H_6AsCl_2NO$
(3-Amino-4-hydroxyphenyl)dichloroarsine

$C_6H_6AsNO_6$
2-Hydroxy-5-nitrobenzenearsonic Acid
3-Hydroxy-2-nitrobenzenearsonic Acid
3-Hydroxy-4-nitrobenzenearsonic Acid
4-Hydroxy-3-nitrobenzenearsonic Acid

$C_6H_6BrHgNO$
Mercuri-2-amino-5-bromophenyl hydroxide
Mercuri-4-amino-2-bromophenyl hydroxide
Mercuri-4-amino-3-bromophenyl hydroxide

C_6H_6BrN
o-Bromoaniline
m-Bromoaniline
p-Bromoaniline

C_6H_6BrNO
2-Amino-3-bromophenol
2-Amino-4-bromophenol
2-Amino-5-bromophenol
4-Amino-2-bromophenol
4-Amino-3-bromophenol
5-Amino-2-bromophenol
N-*m*-Bromophenylhydroxylamine
N-*p*-Bromophenylhydroxylamine

$C_6H_6BrNO_2S$
Bromobenzene-*o*-sulphonic Acid, *Amide*
Bromobenzene-*m*-sulphonic Acid, *Amide*
Bromobenzene-*p*-sulphonic Acid, *Amide*

$C_6H_6Br_2N_2$
3,5-Dibromo-*o*-phenylenediamine
3,6-Dibromo-*o*-phenylenediamine
4,5-Dibromo-*o*-phenylenediamine
4,6-Dibromo-*m*-phenylenediamine
2,5-Dibromo-*p*-phenylenediamine
2,6-Dibromo-*p*-phenylenediamine
2,3-Dibromophenylhydrazine
2,4-Dibromophenylhydrazine
2,5-Dibromophenylhydrazine
2,6-Dibromophenylhydrazine
3,4-Dibromophenylhydrazine
3,5-Dibromophenylhydrazine

$C_6H_6Br_2O_4$
Dibromofumaric Acid, *Di-Me ester*
Dibromomaleic Acid, *Di-Me ester*
Dibromomaleic Acid, *Et ester*

$C_6H_6Br_6$
Benzene hexabromide

C_6H_6ClN
o-Chloroaniline
m-Chloroaniline
p-Chloroaniline
2-Chloromethylpyridine
2-Chloro-3-methylpyridine
2-Chloro-4-methylpyridine
2-Chloro-6-methylpyridine
3-Chloromethylpyridine
3-Chloro-4-methylpyridine
4-Chloro-2-methylpyridine
4-Chloro-3-methylpyridine
5-Chloro-2-methylpyridine

C_6H_6ClNO
2-Amino-3-chlorophenol
2-Amino-4-chlorophenol
2-Amino-5-chlorophenol
2-Amino-6-chlorophenol
3-Amino-2-chlorophenol
3-Amino-4-chlorophenol
4-Amino-2-chlorophenol
4-Amino-3-chlorophenol
m-Chlorophenylhydroxylamine
p-Chlorophenylhydroxylamine

$C_6H_6ClNO_2S$
o-Chlorobenzenesulphonic Acid, *Amide*
m-Chlorobenzenesulphonic Acid, *Amide*
p-Chlorobenzenesulphonic Acid, *Amide*

$C_6H_6ClNO_3S$
3-Chloroaniline-2-sulphonic Acid
4-Chloroaniline-2-sulphonic Acid
5-Chloroaniline-2-sulphonic Acid
4-Chloroaniline-3-sulphonic Acid
5-Chloroaniline-3-sulphonic Acid
2-Chloroaniline-4-sulphonic Acid
2-Chloroaniline-5-sulphonic Acid

$C_6H_6Cl_2N_2$
3,5-Dichloro-*o*-phenylenediamine
3,6-Dichloro-*o*-phenylenediamine
4,5-Dichloro-*o*-phenylenediamine
2,5-Dichloro-*m*-phenylenediamine
4,6-Dichloro-*m*-phenylenediamine
2,3-Dichloro-*p*-phenylenediamine
2,5-Dichloro-*p*-phenylenediamine
2,6-Dichloro-*p*-phenylenediamine
2,4-Dichlorophenylhydrazine
2,5-Dichlorophenylhydrazine
3,5-Dichlorophenylhydrazine

$C_6H_6Cl_2N_2O_4S_2$
Dichlorphenamide

$C_6H_6Cl_2O_2$
Dimethylfumaric Acid, *Dichloride*

$C_6H_6Cl_2O_4$
Dichloromaleic Acid, *Di-Me ester*

$C_6H_6Cl_4O_3$
2,2-Dichloropropionic Acid, *Anhydride*

$C_6H_6Cl_6$
Benzene hexachloride

$C_6H_6Cl_9N_3$
Chloralimide

C_6H_6FN
o-Fluoroaniline
m-Fluoroaniline
p-Fluoroaniline

$C_6H_6FNO_2S$
p-Fluorobenzenesulphonic Acid, *Amide*

C_6H_6HgO
Mercuri-phenyl hydroxide

C_6H_6IN
o-Iodoaniline
m-Iodoaniline
p-Iodoaniline

C_6H_6INO
2-Amino-3-iodophenol
2-Amino-4-iodophenol
2-Amino-5-iodophenol
2-Amino-6-iodophenol
3-Amino-5-iodophenol
4-Amino-2-iodophenol
4-Amino-3-iodophenol

$C_6H_6I_2N_2$
4,6-Di-iodo-*m*-phenylenediamine
2,6-Di-iodo-*p*-phenylenediamine
2,4-Di-iodophenylhydrazine

$C_6H_6I_2O_2$
2,6-Di-iodo-1,4-benzoquinone

$C_6H_6I_2O_4$
Di-iodofumaric Acid, *Di-Me ester*

$C_6H_6N_2$
Phenyldi-imide†
Pyrazolo[1,2-*a*]pyrazole†

$C_6H_6N_2O$
Nicotinamide
p-Nitrosoaniline
Pyridine-2-carboxylic Acid, *Amide*
Pyridine-4-carboxylic Acid, *Amide*

$C_6H_6N_2OS_2$
Pyrrothine

$C_6H_6N_2O_2$
3-Aminopyridine-2-carboxylic Acid
5-Aminopyridine-2-carboxylic Acid
2-Aminopyridine-3-carboxylic Acid
4-Aminopyridine-3-carboxylic Acid
5-Aminopyridine-3-carboxylic Acid
6-Aminopyridine-3-carboxylic Acid
2-Aminopyridine-4-carboxylic Acid
3-Aminopyridine-4-carboxylic Acid
2,5-Diamino-1,4-benzoquinone
2,6-Diamino-1,4-benzoquinone
3-Hydroxypyridine-2-carboxylic Acid, *Amide*†
2-Methyl-3-nitropyridine
2-Methyl-4-nitropyridine
2-Methyl-5-nitropyridine
2-Methyl-6-nitropyridine
3-Methyl-2-nitropyridine
3-Methyl-4-nitropyridine
3-Methyl-5-nitropyridine
4-Methyl-2-nitropyridine
4-Methyl-3-nitropyridine
5-Methyl-2-nitropyridine
o-Nitroaniline
m-Nitroaniline
p-Nitroaniline
N-Nitroaniline
Pyrazinecarboxylic Acid, *Me ester*
Urocanic Acid

$C_6H_6N_2O_2S$
2-Amino-5-nitrobenzene-thiol
Thiophene-2,3-dicarboxylic Acid, *Diamide*

$C_6H_6N_2O_3$
2-Amino-3-nitrophenol
2-Amino-4-nitrophenol
2-Amino-5-nitrophenol
2-Amino-6-nitrophenol
3-Amino-4-nitrophenol
4-Amino-2-nitrophenol
4-Amino-3-nitrophenol
5-Amino-2-nitrophenol
5-Amino-3-nitrophenol
Furan-2,5-dicarboxylic Acid, *Diamide*
2-Hydroxy-3-nitropyridine, *Me ether*
2-Hydroxy-5-nitropyridine, N-*Me-pyridone*
2-Hydroxy-5-nitropyridine, *Me ether*
3-Hydroxy-2-nitropyridine, *Me ether*
4-Hydroxy-3-nitropyridine, *Me ether*
4-Hydroxy-3-nitropyridine, N-*Me-pyridone*
3,4′-Imidazolylpyruvic Acid
N-*o*-Nitrophenylhydroxylamine
N-*m*-Nitrophenylhydroxylamine
N-*p*-Nitrophenylhydroxylamine

$C_6H_6N_2O_4$
3-Amino-5-nitrocatechol
4-Amino-3-nitrocatechol
4-Amino-5-nitrocatechol
4-Amino-6-nitrocatechol
2-Amino-5-nitroquinol
2-Amino-4-nitroresorcinol
4-Amino-6-nitroresorcinol
Dimethylalloxan
1-Methylimidazole-4,5-dicarboxylic Acid

$C_6H_6N_2O_4$ (*continued*)
- 2-Methylimidazole-4,5-dicarboxylic Acid
- 1-Methylorotic Acid
- 3-Methylorotic Acid
- Orotic Acid, *Me ester*

$C_6H_6N_2O_4S$
- Benzenediazonium sulphate
- 2-Nitrobenzenesulphonic Acid, *Amide*
- 3-Nitrobenzenesulphonic Acid, *Amide*
- 4-Nitrobenzenesulphonic Acid, *Amide*

$C_6H_6N_2O_5S$
- 2-Nitroaniline-4-sulphonic Acid
- 3-Nitroaniline-4-sulphonic Acid
- 3-Nitroaniline-6-sulphonic Acid
- 4-Nitroaniline-2-sulphonic Acid
- 4-Nitroaniline-3-sulphonic Acid

$C_6H_6N_2O_6S$
- 6-Amino-4-nitrophenol-2-sulphonic Acid
- 4-Amino-6-nitrophenol-2-sulphonic Acid
- 2-Amino-5-nitrophenol-4-sulphonic Acid
- 2-Amino-6-nitrophenol-4-sulphonic Acid

$C_6H_6N_3P$
- Hexamethylphosphorous Triamide†

$C_6H_6N_4$
- 1-Aminobenzotriazole†
- 2-Aminobenzotriazole†
- 4-Aminopyrrolo[2,3-*d*]pyrimidine†
- 2,2′-Bi-imidazole
- 1-Methylpurine†
- 3-Methylpurine†
- 7-Methylpurine†
- 9-Methylpurine†
- Purine, 7-*Me*
- Purine, 9-*Me*
- Tri-(carboxymethyl)amine, *Trinitrile*

$C_6H_6N_4O$
- 2-Methylhypoxanthine
- 3-Methylhypoxanthine
- 7-Methylhypoxanthine
- 9-Methylhypoxanthine

$C_6H_6N_4O_2$
- 1-Methylxanthine
- 3-Methylxanthine
- 7-Methylxanthine
- 8-Methylxanthine
- 9-Methylxanthine
- Pyrazine-2,3-dicarboxylic Acid, *Di-Amide*†
- Toxoflavin

$C_6H_6N_4O_3$
- 1-Methyluric Acid
- 3-Methyluric Acid
- 7-Methyluric Acid
- 9-Methyluric Acid

$C_6H_6N_4O_3S$
- Niridazole†
- 3,5-Dinitro-*o*-phenylenediamine
- 2,4-Dinitro-*m*-phenylenediamine
- 4,6-Dinitro-*m*-phenylenediamine
- 2,3-Dinitro-*p*-phenylenediamine
- 2,5-Dinitro-*p*-phenylenediamine
- 2,6-Dinitro-*p*-phenylenediamine
- 2,4-Dinitrophenylhydrazine
- 2,6-Dinitrophenylhydrazine

C_6H_6O
- 2-Furylethylene
- 5-Hexen-3-yn-2-one
- 3-Hexen-5-yn-2-one
- 4-Hexen-1-yn-3-one
- Phenol

C_6H_6OS
- 2-Acetylthiophene
- 3-Acetylthiophene
- *o*-Hydroxybenzenethiol
- *m*-Hydroxybenzenethiol
- *p*-Hydroxybenzenethiol

$C_6H_6O_2$
- 2-Acetylfuran
- 3-Acetylfuran
- Catechol
- Cyclohex-2-ene-1,4-dione†
- 2-(2-Furyl)acetaldehyde
- 3-Methyl-3-cyclopentene-1,2-dione
- 3-Methylfuran-2-aldehyde
- 5-Methylfuran-2-aldehyde
- 4-Methylfuran-3-aldehyde
- Muconic Aldehyde
- Quinol
- Resorcinol

$C_6H_6O_2Pb$
- Benzeneplumbonic Acid

$C_6H_6O_2S$
- 2-Acetyl-3-hydroxythiophene†
- Benzenesulphinic Acid
- 3-Methylthiophene-2-carboxylic Acid
- 4-Methylthiophene-2-carboxylic Acid
- 5-Methylthiophene-2-carboxylic Acid
- 4-Methylthiophene-3-carboxylic Acid
- 5-Methylthiophene-3-carboxylic Acid
- Thiepin-1,1-dioxide†
- 2-Thiopheneacetic Acid
- 3-Thiopheneacetic Acid
- Thiophene-2-carboxylic Acid, *Me ester*

$C_6H_6O_2S_2$
- Benzenethiolsulphonic Acid
- Thiolbenzenesulphonic Acid

$C_6H_6O_2Se$
- Benzeneselenic Acid

$C_6H_6O_3$
- Acrylic Acid, *Anhydride*
- 1-Butene-2,3-dicarboxylic Acid, *Anhydride*
- 3-Carboxy-2,4-pentadienal Lactol
- Cyclobutane-1,2-dicarboxylic Acid, *Anhydride*
- Cyclobutane-1,3-dicarboxylic Acid, *Anhydride*
- Dimethylmaleic Acid, *Anhydride*
- Ethylmaleic Acid, *Anhydride*
- 2-Formylcyclopentane-1,3-dione†
- Furan-2-carboxylic Acid, *Me ester*
- Furan-3-carboxylic Acid, *Me ester*
- 2-Furylacetic Acid
- 5-Hydroxymethylfurfural
- 3-Hydroxy-2*H*-pyran-2-one, *Me ether*
- 3-Hydroxy-4*H*-pyran-4-one, *Me ether*
- 5-Hydroxy-3-vinylfuran-2(5*H*)-one

$C_6H_6O_3$ (*continued*)
Maltol
3-Methylcyclopentane-1,2,4-trione
1-Methylcyclopropane-1,2-dicarboxylic Acid, *Anhydride*
3-Methylcyclopropane-1,2-dicarboxylic Acid, *Anhydride*
3-Methylfuran-2-carboxylic Acid
4-Methylfuran-2-carboxylic Acid
5-Methylfuran-2-carboxylic Acid
2-Methylfuran-3-carboxylic Acid
4-Methylfuran-3-carboxylic Acid
5-Methylfuran-3-carboxylic Acid
Phloroglucinol
Pyrogallol
1,2,4-Trihydroxybenzene

$C_6H_6O_3S$
Benzenesulphonic Acid
2-Thienylglycollic Acid
3-Thienylglycollic Acid

$C_6H_6O_3Se$
Benzeneselenonic Acid

$C_6H_6O_4$
Acetylenedicarboxylic Acid, *Di-Me ester*
Aconic Acid, *Me ester*
5-Hydroxymaltol
5-Hydroxymethylfuran-2-carboxylic Acid
Isokojic Acid
Kojic Acid
3-Methylenecyclopropane-1,2-dicarboxylic Acid
Muconic Acid
Squaric Acid, *Di-Me ester*†
1,2,3,4-Tetrahydroxybenzene
1,2,3,5-Tetrahydroxybenzene
1,2,4,5-Tetrahydroxybenzene

$C_6H_6O_4S$
Phenol-*o*-sulphonic Acid
Phenol-*m*-sulphonic Acid
Phenol-*p*-sulphonic Acid
Phenyl hydrogen sulphate

$C_6H_6O_5$
4-Formylglutaconic Acid
3-Hydroxykojic Acid
Oxalocrotonic Acid
Pentahydroxybenzene
Tricarballylic Acid, αβ-*Anhydride*
Zymonic Acid

$C_6H_6O_5S$
2,4-Dihydroxybenzenesulphonic Acid

$C_6H_6O_6$
Aconitic Acid
Cyclopropane-1,1,2-tricarboxylic Acid
Cyclopropane-1,2,3-tricarboxylic Acid
Dehydroascorbic Acid
2,5-Dioxoadipic Acid
3,4-Dioxoadipic Acid
Hexahydroxybenzene
Isaconitic Acid
Isocitric Acid, *Lactone*

$C_6H_6O_6S_2$
Benzene-*o*-disulphonic Acid
Benzene-*m*-disulphonic Acid
Benzene-*p*-disulphonic Acid

$C_6H_6O_7$
Oxalosuccinic Acid

$C_6H_6O_7S_2$
Phenol-2,4-disulphonic Acid
Phenol-2,5-disulphonic Acid

$C_6H_6O_8$
Ethane-tetracarboxylic Acid

$C_6H_6O_8S_2$
4,6-Dihydroxybenzene-1,3-disulphonic Acid

$C_6H_6O_9S_3$
Benzene-1,3,5-trisulphonic Acid

C_6H_6S
Benzenethiol

$C_6H_6S_2$
Benzene-1,2-dithiol
Benzene-1,3-dithiol
Benzene-1,4-dithiol

$C_6H_6S_3$
Benzene-1,3,5-trithiol

C_6H_6Se
Selenylbenzene

C_6H_7As
Phenylarsine

$C_6H_7AsO_2$
Benzenearsonous Acid

$C_6H_7AsO_3$
Benzenearsonic Acid

$C_6H_7BO_2$
Benzeneboronic Acid

$C_6H_7BO_3$
3-Hydroxybenzeneboronic Acid
4-Hydroxybenzeneboronic Acid

$C_6H_7BrN_2$
Bromo-*p*-phenylenediamine
4-Bromo-*o*-phenylenediamine
4-Bromo-*m*-phenylenediamine
5-Bromo-*m*-phenylenediamine
o-Bromophenylhydrazine
p-Bromophenylhydrazine

$C_6H_7BrN_4S_2$
4-Bromo-3-methylisothiazole-5-carboxaldehyde thiosemicarbazone†

$C_6H_7BrO_4$
Bromofumaric Acid, *Di-Me ester*
Bromomaleic Acid, *Di-Me ester*

$C_6H_7ClN_2$
3-Chloro-*o*-phenylenediamine
4-Chloro-*o*-phenylenediamine
2-Chloro-*m*-phenylenediamine
4-Chloro-*m*-phenylenediamine
5-Chloro-*m*-phenylenediamine
2-Chloro-*p*-phenylenediamine
o-Chlorophenylhydrazine
m-Chlorophenylhydrazine
p-Chlorophenylhydrazine

$C_6H_7ClN_2O_3$
5-Chlorobarbituric Acid, 1,3-N-*Di-Me*

$C_6H_7ClO_3$
Fumaric Acid, *Mono-Et ester*, *Monochloride*
Mesaconic Acid, α-*Me ester*, β-*Chloride*

$C_6H_7ClO_4$
Chlorofumaric Acid, *Di-Me ester*
Chloromaleic Acid, *Di-Me ester*

$C_6H_7Cl_3$
Cyameluric Acid, *Trichloride*

$C_6H_7Cl_3O_4$
Chloral Hydrate, *Di-Ac*

$C_6H_7FN_2$
o-Fluorophenylhydrazine
m-Fluorophenylhydrazine
p-Fluorophenylhydrazine

$C_6H_7FO_7$
Fluorocitric Acid

$C_6H_7IN_2$
4-Iodo-*o*-phenylenediamine
2-Iodo-*p*-phenylenediamine
4-Iodo-*m*-phenylenediamine
5-Iodo-*m*-phenylenediamine
o-Iodophenylhydrazine
m-Iodophenylhydrazine
p-Iodophenylhydrazine

$C_6H_7IO_4$
Iodofumaric Acid, *Di-Me ester*

C_6H_7N
Aniline
1*H*-Azepine†
1-Cyclopentene-1-carboxylic Acid, *Nitrile*
2-Methylpyridine
3-Methylpyridine
4-Methylpyridine

C_6H_7NO
1-Acetylpyrrole†
2-Acetylpyrrole
o-Aminophenol
m-Aminophenol
p-Aminophenol
2-Hydroxy-3-methylpyridine
2-Hydroxy-4-methylpyridine
2-Hydroxy-6-methylpyridine
3-Hydroxy-2-methylpyridine
3-Hydroxy-4-methylpyridine
3-Hydroxy-5-methylpyridine
4-Hydroxy-2-methylpyridine
5-Hydroxy-2-methylpyridine
2-Hydroxypyridine, *Me-ether*
2-Hydroxypyridine, N-*Me*
3-Hydroxypyridine, *Me ether*
4-Hydroxypyridine, *Me ether*
4-Hydroxypyridine, N-*Me*
N-Methyl-α-pyridone
N-Methyl-γ-pyridone
N-Methylpyrrole-2-aldehyde
3-Methylpyrrole-2-aldehyde
5-Methylpyrrole-2-aldehyde
β-Phenylhydroxylamine
O-Phenylhydroxylamine
2-Pyridinemethanol
3-Pyridinemethanol
4-Pyridinemethanol

C_6H_7NOS
Benzenesulphinic Acid, *Amide*
3-Methylthiophene-2-carboxylic Acid, *Amide*
2-Thiopheneacetic Acid, *Amide*

D

$C_6H_7NO_2$
Aminoquinol
2-Aminoresorcinol
4-Aminoresorcinol
5-Aminoresorcinol
Cyclobutane-1,1-dicarboxylic Acid, *Mononitrile*
Diacetylacetic Acid, *Nitrile*
2,4-Dihydroxy-6-methylpyridine
2,6-Dihydroxy-3-methylpyridine
2,6-Dihydroxy-4-methylpyridine
3,4-Dihydroxypyridine, 3-*Me ether*
Dimethylmaleic Acid, *Imide*
Furan-2-carboxylic Acid, *Methylamide*
Maleimide, N-*Et*
3-Methylfuran-2-carboxylic Acid, *Amide*
5-Methylfuran-2-carboxylic Acid, *Amide*
Pyrrole-2-acetic Acid†
Pyrrole-2-carboxylic Acid, N-*Me*
Pyrrole-2-carboxylic Acid, *Me ester*
Pyrrole-3-carboxylic Acid, *Me ester*

$C_6H_7NO_2S$
Benzenesulphonic Acid, *Amide*
4-Methylthiazole-2-carboxylic Acid, *Me ester*
Thiazole-2-acetic Acid, *Me ester*
Thiazole-2-carboxylic Acid, *Et ester*
Thiazole-4-carboxylic Acid, *Et ester*
Thiazole-5-carboxylic Acid, *Et ester*
1-Thiathymine, N-*Me*†

$C_6H_7NO_3$
4-Acetamido-4-hydroxybut-2-enoic Acid γ-Lactone†
2,5-Dimethyl-3-nitrofuran
Isoxazole-5-carboxylic Acid, *Et ester*
Methyl-2-cyanoacetoacetate
5-Methylisoxazole-3-carboxylic Acid, *Me ester*
3-Oxohexanedioic Acid, 1-*Nitrile*

$C_6H_7NO_3S$
Aniline-*o*-sulphonic Acid
Aniline-*m*-sulphonic Acid
Aniline-*p*-sulphonic Acid
Phenol-*p*-sulphonic Acid, *Amide*
Phenylsulphamic Acid
Pyridine-3-sulphonic Acid, *Betaine*

$C_6H_7NO_4$
Tricarballylic Acid, αβ-*Imide*

$C_6H_7NO_4S$
2-Aminophenol-4-sulphonic Acid
2-Aminophenol-5-sulphonic Acid
3-Aminophenol-4-sulphonic Acid
4-Aminophenol-2-sulphonic Acid
4-Aminophenol-3-sulphonic Acid
5-Aminophenol-2-sulphonic Acid
5-Aminophenol-3-sulphonic Acid

$C_6H_7NO_6S_2$
Aniline-2,4-disulphonic Acid

C_6H_7NS
o-Aminobenzenethiol
m-Aminobenzenethiol
p-Aminobenzenethiol

$C_6H_7N_3O$
2-Aminopyridine-3-carboxylic Acid, *Amide*
4-Aminopyridine-3-carboxylic Acid, *Amide*

$C_6H_7N_3O$ (*continued*)
6-Aminopyridine-3-carboxylic Acid, *Amide*
Pyridine-4-carboxylic Acid, *Hydrazide*

$C_6H_7N_3O_2$
2-Amino-3-methyl-5-nitropyridine
2-Amino-4-methyl-3-nitropyridine
2-Amino-4-methyl-5-nitropyridine
2-Amino-5-methyl-3-nitropyridine
2-Amino-6-methyl-3-nitropyridine
2-Amino-6-methyl-5-nitropyridine
2-Amino-3-nitropyridine, N-*Me*
2-Amino-5-nitropyridine, N-*Me*
5-Aminopyrazine-2-carboxylic Acid, *Me ester*†
5,6-Diaminopyridine-3-carboxylic Acid
o-Nitrophenylhydrazine
m-Nitrophenylhydrazine
p-Nitrophenylhydrazine
3-Nitro-*o*-phenylenediamine
4-Nitro-*o*-phenylenediamine
4-Nitro-*m*-phenylenediamine
5-Nitro-*m*-phenylenediamine
2-Nitro-*p*-phenylenediamine

$C_6H_7N_3O_3$
2,3-Diamino-5-nitrophenol
2,5-Diamino-4-nitrophenol
2,6-Diamino-4-nitrophenol

$C_6H_7N_3O_4$
5-Isonitroso-1,3-dimethylbarbituric Acid
5-Nitrouracil, 1,3-N-*Di-Me*
5-Nitrouracil, 3-N-*Et*

$C_6H_7N_3O_4S$
2-Nitroaniline-4-sulphonic Acid, *Amide*
4-Nitroaniline-2-sulphonic Acid, *Amide*
2-Nitrobenzenesulphonic Acid, *Hydrazide*
3-Nitrobenzenesulphonic Acid, *Hydrazide*
4-Nitrobenzenesulphonic Acid, *Hydrazide*

$C_6H_7N_3O_6S_2$
5-Nitrobenzene-1,3-disulphonic Acid, *Diamide*

$C_6H_7N_5$
2-Amino-6-methylpurine
2-Amino-7-methylpurine
2-Amino-9-methylpurine
8-Amino-9-methylpurine
2-Methyladenine
3-Methyladenine†
7-Methyladenine
9-Methyladenine

$C_6H_7N_5O$
N^2-Methylguanine
1-Methylguanine
7-Methylguanine
8-Methylguanine

$C_6H_7O_2P$
Benzenephosphonous Acid

$C_6H_7O_3P$
Benzenephosphonic Acid
Phenylphosphonic Acid

$C_6H_7O_3Sb$
Benzenestibonic Acid

$C_6H_7O_4P$
Phenol, *Dihydrogen phosphate*

C_6H_7P
Phenylphosphine

C_6H_8
Bicyclo[2,1,1]hex-2-ene†
Bicyclo[2,1,0]pent-2-ene
1,2-Cyclohexadiene†
1,3-Cyclohexadiene
1,4-Cyclohexadiene
1,2-Dimethylenecyclobutane-†
1,3-Dimethylenecyclobutane†
1,3,5-Hexatriene
1-Hexen-3-yne
1-Hexen-4-yne
2-Hexen-4-yne
3-Methylenecyclopentene†
Methylenespiropentane†
3-Methyl-3-penten-1-yne
Tricyclo[2,2,0,0^{2,6}]hexane†
Tricyclo[3,1,0,0^{2,6}]hexane†
Vinylmethylenecyclopropane†

$C_6H_8AsNO_3$
o-Aminobenzenearsonic Acid
m-Aminobenzenearsonic Acid
Arsanilic Acid

$C_6H_8AsNO_4$
2-Aminophenol-4-arsonic Acid

$C_6H_8BrNO_2$
Succinimide, N-2-*Bromoethyl*

$C_6H_8Br_2O_3$
2-Bromopropionic Acid, *Anhydride*

$C_6H_8Br_2O_4$
2,3-Dibromoadipic Acid
2,5-Dibromoadipic Acid
3,4-Dibromoadipic Acid
2,3-Dibromosuccinic Acid, *Di-Me ester*

$C_6H_8Br_4$
1,1,2,2-Tetrabromocyclohexane
1,2,3,4-Tetrabromocyclohexane
1,2,4,5-Tetrabromocyclohexane

$C_6H_8Br_6$
1,2,3,4,5,6-Hexabromohexane

$C_6H_8ClNO_3$
Chlorofumaric Acid, *Et ester amide*

$C_6H_8Cl_2O_2$
Adipic Acid, *Dichloride*
4,4-Dichlorocrotonic Acid, *Et ester*
2,2-Dimethylsuccinic Acid, *Dichloride*

$C_6H_8Cl_2O_4$
2,3-Dichlorosuccinic Acid, *Di-Me ester*
Dimethoxysuccinic Acid, *Dichloride*

$C_6H_8Cl_6$
1,1,1,6,6,6-Hexachlorohexane
1,2,3,4,5,6-Hexachlorohexane
1,3,3,4,4,6-Hexachlorohexane

$C_6H_8F_3O_6S_3$
Benzene-1,3,5-trisulphonic Acid, *Trifluoride*

$C_6H_8N_2$
Adipic Acid, *Dinitrile*
2-Aminomethylpyridine
3-Aminomethylpyridine

$C_6H_8N_2$ (*continued*)
4-Aminomethylpyridine
2-Amino-3-methylpyridine
2-Amino-4-methylpyridine
2-Amino-6-methylpyridine
3-Amino-2-methylpyridine
3-Amino-4-methylpyridine
4-Amino-2-methylpyridine
4-Amino-3-methylpyridine
5-Amino-2-methylpyridine
5-Amino-3-methylpyridine
2-Aminopyridine, N-*Me*
2,3-Dimethylpyrazine
2,5-Dimethylpyrazine
2-6-Dimethylpyrazine
3,4-Dimethylpyridazine
3,6-Dimethylpyridazine
4,5-Dimethylpyridazine
2,4-Dimethylpyrimidine
2,5-Dimethylpyrimidine
4,5-Dimethylpyrimidine
4,6-Dimethylpyrimidine
2,2-Dimethylsuccinic Acid, *Dinitrile*
2,3-Dimethylsuccinic Acid, *Dinitrile*
Isopropylmalonic Acid, *Dinitrile*
2-Methylaminopyridine
4-Methylaminopyridine
2-Methylglutaric Acid, *Dinitrile*
3-Methylglutaric Acid, *Dinitrile*
o-Phenylenediamine
m-Phenylenediamine
p-Phenylenediamine
Phenylhydrazine
Propylmalonic Acid, *Dinitrile*

$C_6H_8N_2O$
3-Acetyl-4-methylpyrazole
4-Acetyl-3-methylpyrazole
2,3-Diaminophenol
2,4-Diaminophenol
2,5-Diaminophenol
2,6-Diaminophenol
3,4-Diaminophenol
3,5-Diaminophenol
4,6-Dimethyl-2-pyrimidone
2,6-Dimethyl-4-pyrimidone
3,6-Dimethyl-4-pyrimidone
5,6-Dimethyl-4-pyrimidone
4-Hydroxyphenylhydrazine

$C_6H_8N_2OS$
2-Thiohydantoin, 3-N-*Allyl*

$C_6H_8N_2O_2$
4,5-Dihydroxy-*o*-phenylenediamine
2,4-Dihydroxy-*m*-phenylenediamine
2,5-Dihydroxy-*m*-phenylenediamine
4,6-Dihydroxy-*m*-phenylenediamine
2,5-Dihydroxy-*p*-phenylenediamine
4,6-Dihydroxypyrimidine, *Di-O-Me ether*†
1,3-Dimethyluracil
1,4-Dimethyluracil
1,6-Dimethyluracil
3,4-Dimethyluracil
3,6-Dimethyluracil
4,5-Dimethyluracil
Imidazole-1-carboxylic Acid, *Et ester*
3,4′-Imidazolylpropionic Acid
1-Methyl-4-imidazoleacetic Acid
3-Methyl-4-imidazoleacetic Acid
3-Methylimidazole-4-carboxylic Acid, *Me ester*
Muconic Acid★, *Di-amide*†
Pyrazole-1-carboxylic Acid, *Et ester*
Thymine, 1-N-*Me*
Thymine, 3-N-*Me*
Uracil, 1-N-*Et*

$C_6H_8N_2O_2S$
Aniline-*o*-sulphonic Acid, *Amide*
Aniline-*m*-sulphonic Acid, *Amide*
5-Mercaptomethyluracil, S-*Me*†
5-Methyl-1,2,3-thiadiazole-4-carboxylic Acid, *Et ester*
Phenylsulphamic Acid, *Amide*
Sulphanilamide

$C_6H_8N_2O_3$
1,3-Dimethylbarbituric Acid
5,5-Dimethylbarbituric Acid
1-Ethylbarbituric Acid
5-Ethylbarbituric Acid
Hexahydro-2,4-dioxopyrimidine, *Ac*
3,4′-Imidazolyl-lactic Acid
Isonitrosomalonic Acid, *Et ester-nitrile, Me ether*
Isonitrosomalonic Acid, *Propyl ester-nitrile*
2-Oxoimidazoline-4-carboxylic Acid, *Et ester*

$C_6H_8N_2O_3S$
m-Hydrazinobenzenesulphonic Acid
p-Hydrazinobenzenesulphonic Acid
o-Phenylenediamine-4-sulphonic Acid
m-Phenylenediamine-4-sulphonic Acid
p-Phenylenediamine-2-sulphonic Acid
Phenylhydrazine-β-sulphonic Acid

$C_6H_8N_2O_4$
2,4-Dioxo-1-imidazolidylacetic Acid, *Me ester*

$C_6H_8N_2O_4S_2$
Benzene-*o*-disulphonic Acid, *Diamide*
Benzene-*m*-disulphonic Acid, *Diamide*
Benzene-*p*-disulphonic Acid, *Diamide*

$C_6H_8N_2O_5$
Alloxanic Acid, *Et ester*
Enteromycin†
1-Methyl-4-imidazoleacetic Acid, *Nitrile*
3-Methyl-4-imidazoleacetic Acid, *Nitrile*

$C_6H_8N_2O_5S_2$
Phenol-2,4-disulphonic Acid, *Diamide*

$C_6H_8N_2S$
Thiodipropionic Acid, *Nitrile*

$C_6H_8N_4$
Piperazine-*N,N′*-dicarboxylic Acid, *Dinitrile*

$C_6H_8N_4O$
2-Aminopyridine-3-carboxylic Acid, *Hydrazide*
4-Aminopyridine-3-carboxylic Acid, *Hydrazide*
6-Aminopyridine-3-carboxylic Acid, *Hydrazide*

$C_6H_8N_4O_2$
1,2,3-Triamino-5-nitrobenzene

$C_6H_8N_5O_4$
1-Methyl-ψ-uric Acid
7-Methyl-ψ-uric Acid
9-Methyl-ψ-uric Acid

C_6H_8O
Cyclohex-2-en-1-one
1-Cyclopentene-1-aldehyde
2-Cyclopentene-1-aldehyde
3-Cyclopentene-1-aldehyde
2,3-Dimethylfuran
2,4-Dimethylfuran
2,5-Dimethylfuran
3,4-Dimethylfuran
2,4-Hexadienal
5-Hexen-3-yn-1-ol
2-Hexen-4-yn-1-ol
5-Hexen-3-yn-2-ol
3-Hexen-5-yn-2-ol
4-Hexen-1-yn-3-ol
1-Hexen-4-yn-3-ol
3-Hexyn-1-al
3-Hexyn-2-one
5-Hexyn-2-one
1-Hexyn-3-one
2-Methyl-2-cyclopenten-1-one
3-Methyl-2-cyclopenten-1-one
4-Methyl-2-cyclopenten-1-one
5-Methyl-2-cyclopenten-1-one
2-Methylenecyclopentanone

C_6H_8OS
2-Furfuryl methyl sulphide†

$C_6H_8O_2$
Acrylic Acid, *Allyl ester*
1,2-Cyclohexanedione
1,3-Cyclohexanedione
1,4-Cyclohexanedione
1-Cyclopentene-1-carboxylic Acid
2-Cyclopentene-1-carboxylic Acid
3-Cyclopentene-1-carboxylic Acid
2,3-Dioxabicyclo[2,2,2]oct-2-ene
Furfuryl methyl Ether
2,4-Hexadienoic Acid
3-Hexene-2,5-dione
Hex-2-en-5-olide†
2-Hexynoic Acid
2-1′-Hydroxyethylfuran
3-Methyl-1,2-cyclopentanedione
4-Methyl-1,2-cyclopentanedione
2-Methyl-1,3-cyclopentanedione
5-Methylfurfuryl Alcohol
4-Methyl-2-pentynoic Acid
2-Methyl-4-pentynoic Acid
4-Oxohex-2-enal†
2,4-Pentadienoic Acid, *Me ester*
2-Pentynoic Acid, *Me ester*
Tetrolic Acid, *Et ester*

$C_6H_8O_3$
3-Acetylacrylic Acid, *Me ester*
Adipic Acid, *Anhydride*
2,5-Dimethyl-4-hydroxy-2,3-dihydrofuran-3-one†
2,2-Dimethylsuccinic Acid, *Anhydride*
2,3-Dimethylsuccinic Acid, *Anhydride*
Ethylsuccinic Acid, *Anhydride*
Formylacrylic Acid, *Et ester*
β-Formylcrotonic Acid, *Me ester*†
2,3,4-Hexanetrione
2,3,5-Hexanetrione
4-Hydroxy-2-butynoic Acid, *Et ester*
2-Oxocyclopentane-1-carboxylic Acid
3-Oxocyclopentane-1-carboxylic Acid
Pyruvic Acid, *Allyl ester*
Reductic Acid, *Me ether*

$C_6H_8O_4$
Acetylpyruvic Acid, *Me ester*
Allylmalonic Acid
1-Butene-1,3-dicarboxylic Acid
1-Butene-1,4-dicarboxylic Acid
1-Butene-2,3-dicarboxylic Acid
1-Butene-2,4-dicarboxylic Acid
2-Butene-1,2-dicarboxylic Acid
2-Butene-1,4-dicarboxylic Acid
Cyclobutane-1,1-dicarboxylic Acid
Cyclobutane-1,2-dicarboxylic Acid
Cyclobutane-1,3-dicarboxylic Acid
Diacetylacetic Acid
1,4:3,6-Dianhydro-D-glucopyranose†
1,4:3,6-Dianhydro-D-mannopyranose†
Dimethylfumaric Acid
Dimethylmaleic Acid
2,3-Dioxobutyric Acid, *Et ester*
3,5-Dioxohexanoic Acid
Ethylfumaric Acid
Ethylmaleic Acid
Fumaric Acid, *Di-Me ester*
Fumaric Acid, *Mono-Et ester*
Lactide
Maleic Acid, *Di-Me ester*
Maleic Acid, *Mono-Et ester*
Mesaconic Acid, α-*Me-ester*
Mesaconic Acid, β-*Me ester*
2-Methylcyclopropane-1,1-dicarboxylic Acid
1-Methylcyclopropane-1,2-dicarboxylic Acid
3-Methylcyclopropane-1,2-dicarboxylic Acid
Methylenemalonic Acid, *Di-Me ester*
2-Methylpropene-1,3-dicarboxylic Acid
2-Propene-1,2-dicarboxylic Acid, *Mono-Me ester*

$C_6H_8O_4S$
Tetrahydrothiophene-2,5-dicarboxylic Acid
Tetrahydrothiophene-3,4-dicarboxylic Acid

$C_6H_8O_5$
Acetonylmalonic Acid
Acetylmalonic Acid, *Mono-Me ester*
Acetylsuccinic Acid
Dimethoxysuccinic Acid, *Anhydride*
Glyoxal, *Di-Ac*
Oxalacetic Acid, *Di-Me ester*
Oxalacetic Acid, 4-*Et ester*
3-Oxalobutyric Acid
2-Oxohexanedioic Acid
3-Oxohexanedioic Acid
Tetrahydrofuran-2,5-dicarboxylic Acid

$C_6H_8O_6$
Acetoxysuccinic Acid
*Arabo*ascorbic Acid
Ascorbic Acid
D-Glucurone
Mannuronic Acid, *Lactone*
Tricarballylic Acid
Triformin

$[C_6H_8O_6]_n$
Alginic Acid

$C_6H_8O_7$
Citric Acid
Isocitric Acid
Saccharic Acid, 1,4-*Lactone*
Saccharic Acid, 3,6-*Lactone*

C_6H_8S
2,3-Dimethylthiophene
2,4-Dimethylthiophene
2,5-Dimethylthiophene
3,4-Dimethylthiophene

$C_6H_9AsO_6$
Arsenic triacetate

C_6H_9Br
1-Bromocyclohexene
3-Bromocyclohexene
4-Bromocyclohexene

C_6H_9BrO
2-Bromocyclohexanone

$C_6H_9BrO_2$
2-Bromocrotonic Acid, *Et ester*
4-Bromocrotonic Acid, *Et ester*
3-Bromo-2-methylacrylic Acid, *Et ester*
2-Bromopropionic Acid, *Allyl ester*

$C_6H_9BrO_3$
2-Bromolevulinic Acid, *Me ester*

$C_6H_9BrO_4$
Bromosuccinic Acid, *Di-Me ester*
Bromosuccinic Acid, *Mono-Et ester*

C_6H_9Cl
1-Chlorocyclohexene
3-Chlorocyclohexene
2-Chloro-3,3-dimethylmethylenecyclopropane

C_6H_9ClO
2-Chlorocyclohexanone
3-Chlorocyclohexanone
4-Chlorocyclohexanone
Cyclopentanecarboxylic Acid, *Chloride*
2,2-Dimethyl-3-butenoic Acid, *Chloride*
2-Ethylcrotonic Acid, *Chloride*
2-Hexenoic Acid, *Chloride*
3-Hexenoic Acid, *Chloride*
4-Hexenoic Acid, *Chloride*
2-Methyl-2-pentenoic Acid, *Chloride*
3-Methyl-2-pentenoic Acid, *Chloride*
4-Methyl-2-pentenoic Acid, *Chloride*

$C_6H_9ClO_2$
2-Chlorocrotonic Acid, *Et ester*
3-Chlorocrotonic Acid, *Et ester*
4-Chlorocrotonic Acid, *Et ester*
3-Chloro-2-ethylcrotonic Acid
2-Chloroisocrotonic Acid, *Et ester*
3-Chloroisocrotonic Acid, *Et ester*
3-Chloro-2-methylacrylic Acid, *Et ester*
Tetrahydropyran-4-carboxylic Acid, *Chloride*

$C_6H_9ClO_3$
Ethyl-2-chloroacetoacetate
Ethyl-4-chloroacetoacetate
Methylmalonic Acid, *Mono-Et ester*, *Chloride*
Oxalic Acid, *Isobutyl ester*, *Chloride*
Succinic Acid, *Et ester*, *Chloride*

$C_6H_9ClO_4$
Chlorosuccinic Acid, *Di-Me ester*

$C_6H_9ClO_5$
3-Chloromalic Acid, *Di-Me ester*

$C_6H_9Cl_2NO_3$
1,3-Dichloro-2-propyl carbamate, *Ac*

$C_6H_9Cl_3O_2$
Trichloroacetic Acid, *Isobutyl ester*
2,2,3-Trichlorobutyric Acid, *Et ester*
2,2,4-Trichlorobutyric Acid, *Et ester*

$C_6H_9Cl_3O_3$
3,3,3-Trichlorolactic Acid, *Propyl ester*

C_6H_9F
1-Fluorocyclohexene

$C_6H_9IO_2$
4-Iodocrotonic Acid, *Et ester*

$C_6H_9IO_3$
2-Iodoacetoacetic Acid, *Et ester*

C_6H_9N
Cyclopentanecarboxylic Acid, *Nitrile*
2,3-Dimethyl-2-butenoic Acid, *Nitrile*
1,2-Dimethylpyrrole
2,3-Dimethylpyrrole
2,4-Dimethylpyrrole
2,5-Dimethylpyrrole
3,4-Dimethylpyrrole
1-Ethylpyrrole
2-Ethylpyrrole
3-Ethylpyrrole
3-Hexenoic Acid, *Nitrile*
3-Methyl-2-pentenoic Acid, *Nitrile*
4-Methyl-2-pentenoic Acid, *Nitrile*
4-Methyl-3-pentenoic Acid, *Nitrile*
Pyrrole, N-*Et*

C_6H_9NO
2,2-Dimethylacetoacetic Acid, *Nitrile*
2-(2-Furyl)ethylamine
N-Methylfurfurylamine
2-Methyl-3-oxovaleric Acid, *Nitrile*
4-Methyl-2-oxovaleric Acid, *Nitrile*
2-Oxopentane-3-carboxylic Acid, *Nitrile*
Tetrahydropyran-3-carboxylic Acid, *Nitrile*
Tetrahydropyran-4-carboxylic Acid, *Nitrile*

C_6H_9NOS
5-(2-Hydroxyethyl)-4-methylthiazole

$C_6H_9NOS_2$
(−)-4-Methylsulphinyl-3-butenyl isothiocyanate

$C_6H_9NO_2$
3-Aminocrotonic Acid, *Me ester*
2-Aminohexa-4,5-dien-1-oic Acid†
2-Cyanopropionic Acid, *Et ester*
3-Cyanopropionic Acid, *Et ester*
2,3-Dimethylsuccinic Acid, *Imide*
Ethylmethylmalonic Acid, *Nitrile*
Isopropylmalonic Acid, *Mononitrile*
α-(Methylenecyclopropyl)glycine
4-Methyleneproline*†
2-Methylglutaric Acid, 1-*Nitrile*
3-Methylisoxazolone, N-*Et*
Propylmalonic Acid, *Mononitrile*
Succinimide, N-*Et*
1,2,3,6-Tetrahydropyridine-2-carboxylic Acid
1,2,5,6-Tetrahydropyridine-3-carboxylic Acid

$C_6H_9NO_3$
1-Butene-1,3-dicarboxylic Acid, *Monoamide*
Cyclotriglycyl
Mesaconic Acid, α-*Me ester*, β-*Amide*
4-Methyl-1-nitro-3-penten-2-one
5-Oxopyrrolidine-2-carboxylic Acid, *Me ester*
4-Oxopipecolic Acid†
3,5,5-Trimethyloxazolidine-2,4-dione

$C_6H_9NO_4$
2,3-Dioxobutyric Acid, α-*Oxime*
γ-Methyleneglutamic Acid

$C_6H_9NO_5$
Acetyliminodiacetic Acid

$C_6H_9NO_6$
Tri-(carboxymethyl)amine

C_6H_9NS
2-Ethyl-4-methylthiazole
4-Ethyl-2-methylthiazole
5-Ethyl-4-methylthiazole
3-Methyl-3-butenyl Isothiocyanate†
4-Pentenylisothiocyanate

$C_6H_9NS_2$
4-Methylthio-3-butenyl isothiocyanate†

$C_6H_9N_2OP$
Phenylphosphonic Acid, *Diamide*

$C_6H_9N_3$
2-Amino-4,5-dimethylpyrimidine
2-Amino-4,6-dimethylpyrimidine
4-Amino-2,5-dimethylpyrimidine
4-Amino-2,6-dimethylpyrimidine
6-Amino-4,5-dimethylpyrimidine
2-Amino-5-methylpyrimidine, 2-N-*Me*
2,2′-Iminodipropionic Acid, *Dinitrile*
3,3′-Iminodipropionic Acid, *Dinitrile*
Spinaceamine*†
1,2,3-Triaminobenzene
1,2,4-Triaminobenzene
1,3,5-Triaminobenzene

$C_6H_9N_3O$
2-Amino-3-(4-imidazolyl) propanal
4-Amino-6-methyl-2-pyrimidone, 6-N-*Me*
2-Amino-5-methyl-4-pyrimidone, 2-N-*Me*
2-Hydroxyhistamine†

$C_6H_9N_3O_2$
Bacimethrin†
Histidine
β-Pyrazol-1-ylalanine

$C_6H_9N_3O_3$
Aconitic Acid, *Triamide*
Cyanuric Acid, O-*Tri-Me ester*
Cyanuric Acid, N-*Tri-Me ester*
Fulminic Acid, *Trimolecular Me ester*
Metronidazole†

$C_6H_9N_3O_4$
Alloxanic Acid, *Et amide*
Caffuric Acid
Enteromycin Carboxamide†
Isocaffuric Acid
Nitrodiazoacetic Acid, tert-*Bu ester*†

$C_6H_9N_3O_4S_2$
Aniline-2,4-disulphonic Acid, *Diamide*

$C_6H_9N_3O_6S_3$
Benzene-1,3,5-trisulphonic Acid, *Triamide*

C_6H_{10}
Bicyclo[2,1,1]hexane†
Cyclohexene
1,3-Dimethylbicyclo[1,1,0]butane†
2,3-Dimethyl-1,3-butadiene
3,3-Dimethyl-1-butyne
3,4-Dimethylcyclobutene†
1,2-Hexadiene
1,3-Hexadiene
1,4-Hexadiene
1,5-Hexadiene
2,4-Hexadiene
1-Hexyne
2-Hexyne
3-Hexyne
1-Methylcyclopentene
3-Methylcyclopentene
4-Methylcyclopentene
4-Methyl-1,2-pentadiene
2-Methyl-1,3-pentadiene
3-Methyl-1,3-pentadiene
4-Methyl-1,3-pentadiene
2-Methyl-1,4-pentadiene
3-Methyl-1-pentyne
4-Methyl-1-pentyne
4-Methyl-2-pentyne
Spiro[3,2]hexane†

$C_6H_{10}BaO_8$
Glyceric Acid, *Ba salt*

$C_6H_{10}Br_2$
1,1-Dibromocyclohexane
1,2-Dibromocyclohexane
1,4-Dibromocyclohexane

$C_6H_{10}Br_2N_2O_2$
2,5-Dibromoadipic Acid, *Diamide*

$C_6H_{10}Br_2O$
1,1-Dibromo-3,3-dimethyl-2-butanone

$C_6H_{10}Br_2O_2$
2,3-Dibromobutyric Acid, *Et ester*
3,4-Dibromobutyric Acid, *Et ester*
2,3-Dibromo-3-methylbutyric Acid, *Me ester*
2,3-Dibromopropionic Acid, n-*Propyl ester*
2,3-Dibromopropionic Acid, *Isopropyl ester*

$C_6H_{10}Br_4$
1,2,2,3-Tetrabromohexane
1,2,3,4-Tetrabromohexane
1,2,4,5-Tetrabromohexane
1,2,5,6-Tetrabromohexane
2,3,4,5-Tetrabromohexane

$C_6H_{10}CaO_8$
Glyceric Acid, *Ca salt*

$C_6H_{10}ClF$
1-Chloro-1-fluorocyclohexane†

$C_6H_{10}ClFO_2$
Chlorofluoroacetic Acid, *Butyl ester*†

$C_6H_{10}ClN$
N-Chlorocyclohexylideneimine†

$C_6H_{10}ClNO$
Piperidine-*N*-carboxylic Acid, *Chloride*

$C_6H_{10}Cl_2$
1,1-Dichlorocyclohexane
1,2-Dichlorocyclohexane
1,4-Dichlorocyclohexane

$C_6H_{10}Cl_2O$
1,1-Dichloro-3,3-dimethyl-2-butanone
3,3-Dichloro-2-hexanone
3,4-Dichloro-4-methyl-2-pentanone

$C_6H_{10}Cl_2O_2$
Dichloroacetic Acid, *Butyl ester*
Dichloroacetic Acid, *Isobutyl ester*
2,2-Dichlorobutyric Acid, *Et ester*
2,3-Dichlorobutyric Acid, *Et ester*
3,4-Dichlorobutyric Acid, *Et ester*
2,5(or 2,6)-Di-chloromethyldioxan

$C_6H_{10}Cl_2O_3$
3-Chloro-2-chloromethyl-2-hydroxypropionic Acid, *Et ester*
3,3-Dichloro-2-hydroxy-2-methylpropionic Acid, *Et ester*

$C_6H_{10}Cl_3NO_3$
Chloralurethane, *Me ether*

$C_6H_{10}F_3NO_2$
5′,5′,5′-Trifluoroleucine

$C_6H_{10}N_2$
2,3-Diazabicyclo[2,2,2]-2-octene
1-Ethyl-2-methylimidazole
2-Ethyl-4-methylimidazole
4-Ethyl-2-methylimidazole
4-Ethyl-5-methylimidazole
1-Ethyl-3-methylpyrazole
1-Ethyl-5-methylpyrazole
3-Ethyl-4-methylpyrazole
3-Imino-2-methylvaleronitrile
3,3-Pentamethylenediaziridine†
Piperidine-*N*-carboxylic Acid, *Nitrile*
1,2,5-Trimethylimidazole
1,4,5-Trimethylimidazole
2,4,5-Trimethylimidazole
1,3,4-Trimethylpyrazole
1,3,5-Trimethylpyrazole
1,4,5-Trimethylpyrazole
3,4,5-Trimethylpyrazole

$C_6H_{10}N_2O$
Ethylmethylmalonic Acid, *Nitrile, Amide*
4-Ethyl-5-methyl-3-pyrazolone
5-Ethyl-4-methyl-3-pyrazolone
4-Imidazolylmethanol, *Et ether*
Isopropylmalonic Acid, *Monoamide, Nitrile*
Propylmalonic Acid, *Mononitrile, Amide*
1,2,3-Trimethylpyrazol-5-one
1,3,4-Trimethylpyrazol-5-one
3,4,4-Trimethylpyrazol-5-one

$C_6H_{10}N_2O_2$
2-Butene-1,4-dicarboxylic Acid, *Diamide*
Cyclobutane-1,1-dicarboxylic Acid, *Diamide*
Cyclobutane-1,2-dicarboxylic Acid, *Diamide*
2,6-Diamino-4-hexynoic Acid†
Diazoacetic Acid, *Butyl ether*
2-Diazobutyric Acid, *Et ester*
4(5)-(2,3-Dihydroxypropyl)imidazole†
1,4-Dimethyl-2,5-dioxopiperazine
3,6-Dimethyl-2,5-dioxopiperazine
Ethylfumaric Acid, *Diamide*
1-Ethyl-3-methylhydantoin
5-Ethyl-5-methylhydantoin
2,2′-Iminodipropionic Acid, *Imide*

$C_6H_{10}N_2O_3$
γ-Methyleneglutamic Acid, *γ-Amide*
Tetrahydrofuran-2,5-dicarboxylic Acid, *Diamide*

$C_6H_{10}N_2O_4$
1-Amino-2-nitrocyclopentane-1-carboxylic Acid†
Azoformic Acid, *Di-Et ester*
Malonuric Acid, *Et-ester*
Piperazine-*N,N′*-dicarboxylic Acid

$C_6H_{10}N_2O_5$
2-Amino-4-oxalylaminobutyric Acid†
4-Amino-2-oxalylaminobutyric Acid†
5,6-Dihydro-2,4,5,6-pentahydroxypyrimidine, *Di-Me ether*
5,6-Dihydro-2,4,5,6-pentahydroxypyrimidine, *5-Et ether*

$C_6H_{10}N_3O_3$
6-Diazo-5-oxonorleucine

$C_6H_{10}N_4$
Cardiazole
Metrazole
1,2,3,4-Tetra-aminobenzene
1,2,3,5-Tetra-aminobenzene
1,2,4,5-Tetra-aminobenzene

$C_6H_{10}N_4O_2$
Viomycidine

$C_6H_{10}N_4O_3$
Allantoin, 3-*Et*
Allantoin, 1,3-*Di-Me*

$C_6H_{10}O$
Acetylcyclobutane
Butyl ethynyl Ether
Cyclohexanone
Cyclohexen-1-ol
Cyclohexen-3-ol
Cyclohexen-4-ol
Diallyl Ether
Diethylketene
Ethyl 1-methyleneallyl Ether
1,4-Hexadien-3-ol
1,5-Hexadien-3-ol
2,4-Hexadien-1-ol
3,4-Hexadien-1-ol†
2-Hexen-1-al
3-Hexen-1-al
3-Hexen-2-one
4-Hexen-2-one
5-Hexen-2-one
1-Hexen-3-one
5-Hexen-3-one
2-Hexen-4-one
2-Hexyn-1-ol
3-Hexyn-1-ol
5-Hexyn-1-ol
5-Hexyn-2-ol
1-Hexyn-3-ol
4-Hexyn-3-ol
2-Methyl-3-butyn-2-ol, *Me ether*

$C_6H_{10}O$ (*continued*)
2-Methylcyclopentanone
3-Methylcyclopentanone
2-Methyl-2-pentenal
3-Methyl-3-penten-2-one
4-Methyl-3-penten-2-one
3-Methyl-4-penten-2-one
4-Methyl-4-penten-2-one
2-Methyl-1-penten-3-one
3-Methyl-1-pentyn-3-ol
4-Methyl-1-pentyn-3-ol

$C_6H_{10}OS$
Diallyl sulphoxide

$C_6H_{10}OS_2$
Allicin

$C_6H_{10}O_2$
Acetoacetaldehyde, *Di-Me acetal*
Acetonylacetone
Acrylic Acid, *Propyl ester*
Acrylic Acid, *Isopropyl ester*
Angelic Acid, *Me ester*
3-Butenoic Acid, *Et ester*
2-Butyne-1,4-diol, *Di-Me ether*
Crotonic Acid, *Et ester*
Cyclobutane-carboxylic Acid, *Me ester*
Cyclohexene hydroperoxide
Cyclopentanecarboxylic Acid
Cyclopropanecarboxylic Acid, *Et ester*
3,3-Dimethylacrylic Acid, *Me ester*
2,3-Dimethyl-2-butenoic Acid
2,2-Dimethyl-3-butenoic Acid
4,4-Dimethyl-4-butyrolactone†
3,7-Dioxabicyclo[3,3,0]octane†
6,8-Dioxabicyclo[3,2,1]octane
7,8-Dioxabicylco[4,2,0]octane
2-Ethylcrotonic Acid
1,5-Hexadiene-3,4-diol
1,6-Hexanedial
2,3-Hexanedione
2,4-Hexanedione
3,4-Hexanedione
4-Hexanolactone
5-Hexanolactone
6-Hexanolactone
2-Hexenoic Acid
3-Hexenoic Acid
4-Hexenoic Acid
5-Hexenoic Acid
2-Hydroxycyclohexanone
4-Hydroxycyclohexanone
3-Hydroxy-2,2-dimethylbutyric Acid, *Lactone*
4-Hydroxy-3,3-dimethylbutyric Acid, *Lactone*
4-Hydroxy-4-methylvaleric Acid, *Lactone*
Isocrotonic Acid, *Et ester*
2-Methylacrylic Acid, *Et ester*
2-Methylcrotonic Acid, *Me ester*
1-Methylcyclopropane-1-carboxylic Acid, *Me ester*
2-Methylenevaleric Acid
3-Methyl-2-methylenebutyric Acid
1-Methyl-2-oxo-1-cyclopentanol
3-Methyl-2-oxo-1-cyclopentanol
2-Methyl-3-oxo-1-cyclopentanol
4-Methyl-2-oxo-1-cyclopentanol
5-Methyl-2-oxo-1-cyclopentanol
3-Methylpentanedial
4-Methyl-2,3-pentanedione
3-Methyl-2,4-pentanedione
2-Methyl-2-pentenoic Acid
3-Methyl-2-pentenoic Acid
4-Methyl-2-pentenoic Acid
2-Methyl-3-pentenoic Acid
3-Methyl-3-pentenoic Acid
4-Methyl-3-pentenoic Acid
3-Pentenoic Acid, *Me ester*
Tetrahydro-3,5-dimethylfuran-2-one
Tetrahydro-4,5-dimethylfuran-2-one

$(C_6H_{10}O_2)_n$
Elaiophylin

$C_6H_{10}O_2S$
Diallyl sulphone

$C_6H_{10}O_2S_2$
Dithio-oxalic Acid, *Di-Et ester*

$C_6H_{10}O_3$
Adipaldehydic Acid
2,5-Dihydroxy-3-methylvaleric Acid, *Lactone*†
2,2-Dimethylacetoacetic Acid
3-Ethoxy-2-oxobutyraldehyde†
Ethyl acetoacetate
3-Formyl-2-methylpropionic Acid, *Me ester*
3-Formylpropionic Acid, *Et ester*
Fucal
Glyceraldehyde, *Monoisopropylidene*
Glyoxylic Acid, n-*Butyl ester*†
2-Hydroxy-3-butenoic Acid, *Et ester*
3-Hydroxy-2-butenoic Acid, *Et ether*
4-Hydroxy-2-butenoic Acid, *Et ester*
1-Hydroxycyclopentane-1-carboxylic Acid
2-Hydroxycyclopentane-1-carboxylic Acid
Levulinic Acid, *Me ester*
2-Methylacetoacetic Acid, *Me ester*
2-Methyl-levulinic Acid
2-Methyloxiran-2-carboxylic Acid, *Et ester*
3-Methyloxiran-2-carboxylic Acid, *Et ester*
2-Methyl-3-oxovaleric Acid
3-Methyl-4-oxovaleric Acid
4-Methyl-2-oxovaleric Acid
4-Methyl-3-oxovaleric Acid
Mevalonic Acid, *Lactone*
2-Oxobutyric Acid, *Et ester*
2-Oxohexanoic Acid
3-Oxohexanoic Acid
4-Oxohexanoic Acid
5-Oxohexanoic Acid
2-Oxopentane-3-carboxylic Acid
3-Oxovaleric Acid, *Me ester*
Propionic Acid, *Anhydride*
Pyruvic Acid, *Propyl ester*
Tetrahydropyran-3-carboxylic Acid
Tetrahydropyran-4-carboxylic Acid
Trimethylpyruvic Acid

$C_6H_{10}O_3S$
Thio-oxalic Acid, *Di-Et ester*

$C_6H_{10}O_4$
Acetoxyacetic Acid, *Et ester*
Adipic Acid
Conduritol†

$C_6H_{10}O_4$ (*continued*)
1,1-Diacetoxyethane
1,4,3,6-Dianhydro-L-iditol
1,4,3,6-Dianhydro-D-mannitol
Diethyl oxalate
Dimethyl succinate
2,2-Dimethylsuccinic Acid
2,3-Dimethylsuccinic Acid
Ethylmethylmalonic Acid
Ethylsuccinic Acid
D-Galactal
Glucal
Glutaric Acid, *Mono-Me ester*
Guaiacol, *Mono-guaiacol carbonate*, *Me ester*
Isopropylmalonic Acid
Malonic Acid, *Mono-propyl ester*
2-Methylglutaric Acid
3-Methylglutaric Acid
Methylmalonic Acid, *Di-Me ester*
Methylmalonic Acid, *Mono-Et ester*
Methylsuccinic Acid, *Mono-Me ester*
Neomannide
Propylmalonic Acid
Succinic Acid, *Et ester*

$C_6H_{10}O_4S$
Thiodiglycollic Acid, *Di-Me ester*
Thiodipropionic Acid

$C_6H_{10}O_4S_2$
Disulphidoacetic Acid, *Di-Me ester*

$C_6H_{10}O_5$
1,2-Anhydrofructose
3,6-Anhydrogalactose
3,6-Anhydro-D-glucose
2,5-Anhydro-D-talose†
Chitose
Citramalic Acid, *Me ether*
6-Deoxy-D-*arabino*-hexofuranos-5-ulose†
Dicrotalic Acid
Diethyl pyrocarbonate†
Diglycollic Acid, *Di-Me ester*
2-Ethylmalic Acid
3-Ethylmalic Acid
Glucodesonic Acid, *Lactone*
Glucoisosaccharinic Acid, 1,4-*Lactone*†
2-Hydroxyadipic Acid
Hydroxy-methylmalonic Acid, *Et ether*
Hydroxymethyl-malonic Acid, *Et ether*
1-Hydroxy-2-methylpropane-1,1-dicarboxylic Acid
Hydroxymethyl-succinic Acid, *Me ether*
Isorhamnonose
Isosaccharolactone
Laevoglucosan
Malic Acid, *Di-Me ester*
Malic Acid, *Et ether*
D-Mannosan
Methoxyacetic Acid, *Anhydride*
Methoxysuccinic Acid, *Mono-Me ester*
Rhamnonic Acid, *γ-Lactone*
Rhamnonic Acid, *δ-Lactone*
Streptose

$(C_6H_{10}O_5)_n$
Arctose
Cellulose
Dextran
Dextrin
Glycogen
Graminin
Irisin
Laminarin
Lichenin
Mannan A
Mannan B
Pustulan†
Starch
Varianose

$C_6H_{10}O_6$
α-Acrosone
2,5-Dihydroxyadipic Acid
Dimethoxysuccinic Acid
Dimethyltartrate
2,5-Dioxohexose†
Galactonic Acid, *γ-Lactone*
Galactosone
Gluconic Acid, *γ-Lactone*
Gluconic Acid, *δ-Lactone*
Glucosone
Gulonic Acid, *γ-Lactone*
D-*ribo*-Hexos-3-ulose
Inosose
Mannonic Acid, *γ-Lactone*
Tartaric Acid, *Et ester*

$C_6H_{10}O_7$
Fructuronic Acid
Galacturonic Acid
Glucuronic Acid
D-*arabino*-Hexulosonic Acid†
D-*xylo*-5-Hexulosonic Acid†
L-Iduronic Acid†
Mannuronic Acid

$C_6H_{10}O_8$
Allomucic Acid
Idosaccharic Acid
Mannosaccharic Acid
Mucic Acid
Saccharic Acid
Talomucic Acid

$C_6H_{10}S$
Cyclohexanethione†
Diallyl sulphide
7-Thiabicyclo[4,1,0]heptane

$C_6H_{10}S_2$
Diallyl disulphide

$C_6H_{11}Br$
Bromocyclohexane

$C_6H_{11}BrHg$
Mercuri-cyclohexyl bromide

$C_6H_{11}BrN_2O_2$
2-Bromo-3-methylbutyrylurea

$C_6H_{11}BrO$
2-Bromocyclohexanol
4-Bromocyclohexanol
3-Bromo-2-hexanone
6-Bromo-2-hexanone
3-Bromo-3-methyl-2-pentanone
4-Bromo-4-methyl-2-pentanone
Hexanoic Acid, *Bromide*

$C_6H_{11}BrO_2$
Bromoacetic Acid, *Butyl ester*
Bromoacetic Acid, *Isobutyl ester*
Bromoacetic Acid, tert-*Butyl ester*
2-Bromobutyric Acid, *Et ester*
3-Bromobutyric Acid, *Et ester*
4-Bromobutyric Acid, *Et ester*
2-Bromohexanoic Acid
3-Bromohexanoic Acid
5-Bromohexanoic Acid
6-Bromohexanoic Acid
2-Bromoisobutyric Acid, *Et ester*
2-Bromo-3-methylbutyric Acid, *Me ester*
2-Bromopropionic Acid, *Isopropyl ester*
3-Bromopropionic Acid, *Propyl ester*

$C_6H_{11}BrO_3$
2-Bromo-3-hydroxybutyric Acid, *Et ether*
4-Bromo-3-hydroxybutyric Acid, *Et ester*

$C_6H_{11}Cl$
Chlorocyclohexane

$C_6H_{11}ClHg$
Mercuri-cyclohexyl chloride

$C_6H_{11}ClO$
Butyl-2-chlorovinyl Ether
2,2-Dimethylbutyric Acid, *Chloride*
2,3-Dimethylbutyric Acid, *Chloride*
2-Ethylbutyric Acid, *Chloride*
Hexanoic Acid, *Chloride*
2-Methylvaleric Acid, *Chloride*
3-Methylvaleric Acid, *Chloride*
4-Methylvaleric Acid, *Chloride*

$C_6H_{11}ClO_2$
Chloroacetic Acid, *Butyl ester*
2-Chlorobutyric Acid, *Et ester*
3-Chlorobutyric Acid, *Et ester*
4-Chlorobutyric Acid, *Et ester*
Chloroformic Acid, 3-*Methylbutyl ester*
3-Chloro-2-methylbutyric Acid, *Me ester*
3-Chloro-3-methylbutyric Acid, *Me ester*
2-Chloro-2-methylpropionic Acid, *Et ester*
3-Chloro-2-methylpropionic Acid, *Et ester*
2-Chloropropionic Acid, *Propyl ester*
3-Chloropropionic Acid, *Propyl ester*
2-Chlorovaleric Acid, *Me ester*
4-Chlorovaleric Acid, *Me ester*
5-Chlorovaleric Acid, *Me ester*

$C_6H_{11}ClO_3$
2-Chloro-3-hydroxybutyric Acid, *Et ester*
4-Chloro-3-hydroxybutyric Acid, *Et ester*
2-Chloro-3-hydroxy-2-methylpropionic Acid, *Et ester*
3-Chloro-2-hydroxy-2-methylpropionic Acid, *Et ester*

$C_6H_{11}Cl_3O_2$
Chloral, *Di-Et acetal*
Chloral, *Butyl acetal*
Chloral, tert-*Butyl acetal*

$C_6H_{11}FO_2$
Fluoroacetic Acid, tert-*Butyl ester*

$C_6H_{11}HgI$
Mercuri-cyclohexyl iodide

$C_6H_{11}HgN$
Mercuri-pentyl cyanide

$C_6H_{11}I$
Iodocyclohexane

$C_6H_{11}IO$
2-Iodobutyric Acid, *Et ester*
4-Iodobutyric Acid, *Et ester*
2-Iodocyclohexanol

$C_6H_{11}IO_2$
2-Iodo-2-methylpropionic Acid, *Et ester*

$C_6H_{11}N$
Azabicyclo[2,2,1]heptane
7-Azabicyclo[4,1,0]heptane
Diallylamine
2,2-Dimethylbutyric Acid, *Nitrile*
2,5-Dimethyl-1-pyrroline
2,4-Dimethyl-2(3)-pyrroline
2,5-Dimethyl-2(3)-pyrroline
2-Ethylbutyric Acid, *Nitrile*
Hexanoic Acid, *Nitrile*
3-Methylbutyl isocyanide
2-Methylvaleric Acid, *Nitrile*
3-Methylvaleric Acid, *Nitrile*
4-Methylvaleric Acid, *Nitrile*
1,2,3,4-Tetrahydro-6-methylpyridine
1,2,5,6-Tetrahydro-1-methylpyridine

$C_6H_{11}NO$
Acetylacetone Imine, N-*Me*
Cyclopentanecarboxylic Acid, *Amide*
2,2-Dimethyl-3-butenoic Acid, *Amide*
2-Ethylcrotonic Acid, *Amide*
Hexanolactam
3-Hexenoic Acid, *Amide*
2-Hydroxy-2,3-dimethylbutyric Acid, *Nitrile*
2-Hydroxy-4-methylvaleric Acid, *Nitrile*
4-Imino-3-methyl-2-pentanone
Lactic Acid, *Nitrile*, *Propyl ether*
2-Methyl-2-pentenoic Acid, *Amide*
3-Methyl-2-pentenoic Acid, *Amide*
4-Methyl-2-pentenoic Acid, *Amide*
3-Methyl-3-pentenoic Acid, *Amide*
N-Methyl-2-piperidone
3-Methyl-2-piperidone
4-Methyl-2-piperidone
5-Methyl-2-piperidone
6-Methyl-2-piperidone
N-Methyl-3-piperidone
N-Methyl-4-piperidone
3-Pentanone, *Cyanhydrin*
Piperidine, N-*Formyl*
Piperidine-2-aldehyde
Piperidine-3-aldehyde
2,4,5-Trimethyl-3-oxazoline†

$C_6H_{11}NOS$
Cleomin
1,4-Thiazan, N-*Ac*

$C_6H_{11}NOS_2$
4-Methylsulphinylbutyl isothiocyanate†

$C_6H_{11}NO_2$
3-Acetamidobutan-2-one†
3-Aminocrotonic Acid, *Et ester*
2,2-Dimethylacetoacetic Acid, *Amide*

$C_6H_{11}NO_2$ *(continued)*
Dipropionamide
3-Methylaminocrotonic Acid, *Me ester*
1-Methyl-1-nitrocyclopentane
1-Methyl-2-nitrocyclopentane
2-Methyl-3-oxovaleric Acid, *Amide*
4-Methyl-2-oxovaleric Acid, *Amide*
3-Methylproline†
4-Methylproline
1-Methylpyrrolidine-2-carboxylic Acid
Nitrocyclohexane
Nitromethylcyclopentane
5-Oxohexanoic Acid, *Amide*
2-Oxopentane-3-carboxylic Acid, *Amide*
Piperidine-*N*-carboxylic Acid
Piperidine-2-carboxylic Acid
Piperidine-3-carboxylic Acid
Piperidine-4-carboxylic Acid
Proline, *Me ester*
Pyrrolidine-2-acetic Acid†
Tetrahydropyran-4-carboxylic Acid, *Amide*

$C_6H_{11}NO_2S$
S-(1-Propenyl)-L-cysteine
Thio-oxamic Acid, *Isobutyl ester*

$C_6H_{11}NO_2S_2$
Erysoline

$C_6H_{11}NO_3$
Acetylglycine, *Et ester*
Adipic Acid, *Amide*
2-Amino-2-methylbutyric Acid, N-*Formyl*
Dimethyloxamic Acid, *Et ester*
2,3-Dimethylsuccinic Acid, *Monoamide*
Ethyl-2-aminoacetoacetate
Ethyloxamic Acid, *Et ester*
4-Hydroxymethylproline
4-Hydroxy-*N*-methyl-L-proline†
4-Hydroxypipecolic Acid†
5-Hydroxypipecolic Acid†
3-Hydroxyproline, O-*Me*
Isopropylmalonic Acid, *Monoamide*
Methylmalonic Acid, *Mono-Et ester, Amide*
Oxamic Acid, *Butyl ester*
Oxamic Acid, *Isobutyl ester*
4-Oxo-norleucine†
Succinamic Acid, *Et ester*

$C_6H_{11}NO_3S$
Alliin
Cycloalliin
S-(1-Propenyl)cysteine sulphoxide

$C_6H_{11}NO_4$
2-Aminoadipic Acid
3-Aminoadipic Acid
L-Aspartic Acid, *Di-Me ester*
L-Aspartic Acid, β-*Et ester*
DL-Aspartic Acid,α-*Et ester*
DL-Aspartic Acid, *Di-Me ester*
Carbethoxyalanine
Carbethoxyurethane
Carbomethoxyglycine, *Et ester*
3-Ethylmalic Acid, *Monamide*
Iminodiacetic Acid, *Di-Me ester*
Iminodiacetic Acid, *Mono-Et ester*
2,2′-Iminodipropionic Acid
2,3′-Iminodipropionic Acid
3,3′-Iminodipropionic Acid
α-Malamic Acid, *Et ester*
Mannonic Acid, *Nitrile*
N-Methylglutamic Acid
2-Methylglutamic Acid
3-Methylglutamic Acid
4-Methylglutamic Acid
Nitroacetic Acid, *Isobutyl ester*
2-Nitrobutyric Acid, *Et ester*

$C_6H_{11}NO_4S$
S-[2-Carboxyethyl]-L-cysteine

$C_6H_{11}NO_5$
Dimethoxysuccinic Acid, *Monoamide*
Gluconic Acid, *Nitrile*

$C_6H_{11}NO_7$
Allomucic Acid, *Amide*
Mucic Acid, *Mono-amide*
Saccharic Acid, *Amide*

$C_6H_{11}NS$
2-Methylbutyl isothiocyanate
3-Methylbutyl isothiocyanate
Thiocyanic Acid, *Pentyl ester*

$C_6H_{11}NS_2$
4-Methylthiobutyl isothiocyanate
Piperidine-*N*-carbodithioic Acid

$C_6H_{11}N_2O_4PS_3$
O,O-Dimethyl-*S*-[2-methoxy-1,3,4-thiadiazol-5(4*H*)-onyl-(4)-methyl]dithiophosphate†

$C_6H_{11}N_3$
3,4′-Imidazolylpropylamine
4-2′-Methylaminoethylimidazole

$C_6H_{11}N_3O$
2-Amino-3-(4-imidazolyl)-1-propanol

$C_6H_{11}N_3O_2$
2-Azidobutyric Acid, *Et ester*
4-Azidobutyric Acid, *Et ester*
2-Azido-2-methylpropionic Acid, *Et ester*

$C_6H_{11}N_3O_3$
Acetyliminodiacetic Acid, *Diamide*
5-Guanidino-2-oxovaleric Acid
Tricarballylic Acid, *Triamide*

$C_6H_{11}N_3O_4$
Citramide
Diglycylglycine
Glycyl-L-asparagine
Isocitric Acid, *Triamide*

$C_6H_{11}O_2P$
1-Methylphospholan-3-one 1-oxide, *Methyl (enol) ether*†

$C_6H_{11}O_3P$
Diallyl phosphite

C_6H_{12}
Cyclohexane
2,3-Dimethyl-1-butene
3,3-Dimethyl-1-butene
2,3-Dimethyl-2-butene
1-Ethyl-2-methylcyclopropane
1-Hexene
2-Hexene
3-Hexene

C_6H_{12} (*continued*)
Methylcyclopentane
2-Methyl-1-pentene
3-Methyl-1-pentene
4-Methyl-1-pentene
2-Methyl-2-pentene
3-Methyl-2-pentene
4-Methyl-2-pentene
1,1,2-Trimethylcyclopropane
1,2,3-Trimethylcyclopropane

$C_6H_{12}As_2$
Acetylene*bis*dimethylarsine

$C_6H_{12}BrNO$
Neuronal

$C_6H_{12}Br_2$
1,2-Dibromo-2,3-dimethylbutane
1,3-Dibromo-2,3-dimethylbutane
1,4-Dibromo-2,2-dimethylbutane
1,4-Dibromo-2,3-dimethylbutane
2,3-Dibromo-2,3-dimethylbutane
3,3-Dibromo-2,2-dimethylbutane
3,4-Dibromo-2,2-dimethylbutane
1,2-Dibromohexane
1,4-Dibromohexane
1,5-Dibromohexane
1,6-Dibromohexane
2,2-Dibromohexane
2,3-Dibromohexane
2,4-Dibromohexane
2,5-Dibromohexane
3,4-Dibromohexane

$C_6H_{12}Br_2O$
Di-3-bromopropyl Ether

$C_6H_{12}ClNO$
Chloroacetamide, N-*Di-Et*

$C_6H_{12}Cl_2$
1,2-Dichlorohexane
1,5-Dichlorohexane
1,6-Dichlorohexane
2,3-Dichlorohexane
2,5-Dichlorohexane
3,4-Dichlorohexane

$C_6H_{12}Cl_2O$
Di-2-chloropropyl Ether
Di-3-chloropropyl Ether
1,2-Dichloropropyl propyl Ether
1,3-Dichloropropyl propyl Ether

$C_6H_{12}Cl_2O_2$
Dichloroacetaldehyde, *Di-Et acetal*

$C_6H_{12}Cl_3N$
Tri-(2-chloroethyl)amine

$C_6H_{12}HgO_2$
Mercuri-butyl acetate

$C_6H_{12}I_2$
1,5-Di-iodohexane
1,6-Di-iodohexane
2,5-Di-iodohexane

$C_6H_{12}N_2$
3-Aminopentane-3-carboxylic Acid, *Nitrile*
1,4-Diazabicyclo[2,2,2]octane
N-Diethylglycine, *Nitrile*
Dimethylketazine
5-Ethyl-4-methyl-2-pyrazoline

$C_6H_{12}N_2O$
1-Methylpyrrolidine-2-carboxylic Acid, *Amide*
Piperidine-*N*-carboxylic Acid, *Amide*

$C_6H_{12}N_2O_2$
Adipic Acid, *Diamide*
4-Aminopiperidine-2-carboxylic Acid
2,3-Dimethylsuccinic Acid, *Diamide*
Ethylmethylmalonic Acid, *Diamide*
Ethylsuccinic Acid, *Diamide*
3,4-Hexanedioxime
Malonamide, N-*Isopropyl*
Piperazine-*N*-carboxylic Acid, *Me ester*
Propylmalonic Acid, *Diamide*

$C_6H_{12}N_2O_2S$
Thiodipropionic Acid, *Diamide*

$C_6H_{12}N_2O_3$
D-Alanyl-L-alanine
L-Alanyl-D-alanine
L-Alanyl-L-alanine
DL-Alanylalanine
Carbethoxy-DL-alanine, *Amide*
Glycylglycine, *Et ester*
2,2′-Iminodipropionic Acid, *Monoamide*
N-Methylaspartic Acid, *Monomethylamide*
Propylmalonic Acid, *Monohydrazide*
2-Ureidopropionic Acid, *Et ester*

$C_6H_{12}N_2O_3S$
S-Acetamidomethyl-L-cysteine†

$C_6H_{12}N_2O_4$
2,5-Diaminoadipic Acid
3,4-Diaminoadipic Acid
2,5-Dihydroxyadipic Acid, *Diamide*
Dimethoxysuccinic Acid, *Diamide*
Ethylenedicarbamic Acid, *Di-Me ester*
Hydrazodiformic Acid, *Di-Et ester*
2,2′-Hydrazodipropionic Acid
N^4-(2-Hydroxyethyl)-L-asparagine

$C_6H_{12}N_2O_4S$
Lanthionine

$C_6H_{12}N_2O_4S_2$
Cystine
Di-(2-amino-1-carboxy-1-ethyl)disulphide

$C_6H_{12}N_2O_4Se_2$
Selenocystine†

$C_6H_{12}N_2O_5S_2$
Cystine *S*-monoxide†

$C_6H_{12}N_2O_6$
Allomucic Acid, *Diamide*
Mannosaccharic Acid, *Diamide*
Saccharic Acid, *Diamide*

$C_6H_{12}N_2S_3$
Tetramethylthiuram sulphide

$C_6H_{12}N_2S_4$
Arasan

$C_6H_{12}N_3O_2S$
Methyl *bis*-(1-aziridinyl)thiophosphinyl carbamate

$C_6H_{12}N_3PS$
Tris-(1-aziridinyl)-phosphine sulphide

$C_6H_{12}N_4$
Hexamethylenetetramine

$C_6H_{12}N_4O_2$
Capreomycidine†

$C_6H_{12}N_4O_3$
Gongrine†
Streptolidine
Tri-(carboxymethyl)amine, *Triamide*

$C_6H_{12}N_4O_4$
Allantoic Acid, *Et ester*

$C_6H_{12}N_5O_2PS_2$
Menazon†

$C_6H_{12}N_6S_3$
Methylenethiourea

$C_6H_{12}O$
Allyl isopropyl Ether
Allyl propyl Ether
Butyl vinyl Ether
Cyclohexanol
Cyclopentylmethanol
2,3-Dimethylbutanal
2-Ethylbutanal
Hexanal
2-Hexanone
3-Hexanone
2-Hexen-1-ol
3-Hexen-1-ol
4-Hexen-1-ol
5-Hexen-1-ol
4-Hexen-2-ol
5-Hexen-2-ol
1-Hexen-3-ol
5-Hexen-3-ol
2-Hexen-4-ol
α-Hydroxyethylcyclobutane
1-Methylcyclopentanol
2-Methylcyclopentanol
3-Methylcyclopentanol
3-Methyl-2-pentanone
4-Methyl-2-pentanone
2-Methyl-3-pentanone
2-Methyl-2-penten-1-ol
3-Methyl-2-penten-1-ol
4-Methyl-2-penten-1-ol
4-Methyl-3-penten-1-ol
2-Methyl-3-penten-2-ol
3-Methyl-3-penten-2-ol
4-Methyl-3-penten-2-ol
2-Methyl-4-penten-2-ol
3-Methyl-4-penten-2-ol
4-Methyl-4-penten-2-ol
3-Methyl-1-penten-3-ol
2-Methylvaleraldehyde
3-Methylvaleraldehyde
4-Methylvaleraldehyde
4-Penten-1-ol, *Me ether*
Tetrahydro-2-methylpyran
1,1,1-Trimethylacetone

$C_6H_{12}OS$
4-Mercapto-4-methylpentan-2-one†
Thioacetic Acid, O-*Butyl ester*
Thioacetic Acid, O-*Isobutyl ester*

$C_6H_{12}O_2$
Butyl acetate
2-Butyl acetate
tert-Butyl acetate
1,2-Cyclohexanediol
1,3-Cyclohexanediol
1,4-Cyclohexanediol
2,2-Dimethylbutyric Acid
2,3-Dimethylbutyric Acid
3,3-Dimethylbutyric Acid
2,2-Dimethylpropionic Acid, *Me ester*
Ethoxyacetic Acid, *Et ester*
Ethyl butyrate
2-Ethylbutyric Acid
Ethyl isobutyrate
Formic Acid, *Pentyl ester*
Hexanoic Acid
Hydroxyacetone, *Propyl ether*
1-Hydroxy-2-butanone, *Et ether*
3-Hydroxy-2-butanone, *Et ether*
4-Hydroxyhexanal
5-Hydroxyhexanal
6-Hydroxyhexanal
1-Hydroxy-2-hexanone
3-Hydroxy-2-hexanone
4-Hydroxy-2-hexanone
5-Hydroxy-2-hexanone
6-Hydroxy-2-hexanone
4-Hydroxy-3-hexanone
5-Hydroxy-3-hexanone
6-Hydroxy-3-hexanone
4-Hydroxy-4-methyl-2-pentanone
2-Hydroxy-2-methylpropionic Acid, *Me ether, Me ester*
1-Hydroxy-2-pentanone, *Me ether*
3-Hydroxy-2-pentanone, *Me ether*
1-Hydroxy-3-pentanone, *Me ether*
2-Hydroxy-3-pentanone, *Me ether*
Isobutyl acetate
Isovaleric Acid, *Me ester*
Ketene, *Di-Et acetal*
3-Methylbutyl formate
2-Methylbutyric Acid, *Me ester*
1-Methyl-1,2-cyclopentanediol
3-Methyl-1,2-cyclopentanediol
4-Methyl-1,2-cyclopentanediol
2-Methylvaleric Acid
3-Methylvaleric Acid
4-Methylvaleric Acid
Propionic Acid, *Propyl ester*
Propionic Acid, *Isopropyl ester*
Valeric Acid, *Me ester*

$C_6H_{12}O_2S$
3-Mercaptobutyric Acid, *Et ester*
4-Mercaptobutyric Acid, *Et ester*
Thioglycollic Acid, S-*Et*, *Et ester*
Thioglycollic Acid, S-*Butyl*

$C_6H_{12}O_3$
Amicetose†
1,2,3-Cyclohexanetriol
1,2,4-Cyclohexanetriol
1,3,5-Cyclohexanetriol
1,2-*O*-Ethylideneglycerol, *Me ether*
1,3-*O*-Ethylideneglycerol, *Me ether*
Glycollic Acid, *Butyl ether*

$C_6H_{12}O_3$ (*continued*)
Glycollic Acid, *Isobutyl ether*
2-Hydroxybutyric Acid, *Et ester*
2-Hydroxybutyric Acid, *Me ether*, *Me ester*
2-Hydroxybutyric Acid, *Et ether*
3-Hydroxybutyric Acid, *Et ester*
4-Hydroxybutyric Acid, *Et ether*
2-Hydroxy-2,3-dimethylbutyric Acid
2-Hydroxy-3,3-dimethylbutyric Acid
3-Hydroxy-2,2-dimethylbutyric Acid
3-Hydroxy-2,3-dimethylbutyric Acid
4-Hydroxy-3,3-dimethylbutyric Acid
3-Hydroxy-2,2-dimethylpropionic Acid, *Me ester*
2-Hydroxyhexanoic Acid
3-Hydroxyhexanoic Acid
4-Hydroxyhexanoic Acid
5-Hydroxyhexanoic Acid
6-Hydroxyhexanoic Acid
2-Hydroxy-2-methylbutyric Acid, *Me ester*
3-Hydroxy-3-methylbutyric Acid, *Me ester*
2-Hydroxymethyl-2-methylbutyric Acid
2-Hydroxy-2-methylpropionic Acid, *Et ester*
2-Hydroxy-2-methylpropionic Acid, *Et ether*
3-Hydroxy-2-methylpropionic Acid, *Et ester*
2-Hydroxy-4-methylvaleric Acid
3-Hydroxy-2-methylvaleric Acid
3-Hydroxy-4-methylvaleric Acid
4-Hydroxy-4-methylvaleric Acid
3-Hydroxypropionic Acid, *Propyl ester*
3-Hydroxypropionic Acid, *Isopropyl ester*
4-Hydroxyvaleric Acid, *Me ether*
5-Hydroxyvaleric Acid, *Me ether*
1,2-Isopropylidene-glycerol
Lactic Acid, *Propyl ester*
Lactic Acid, *Et ether*, *Me ester*
Methoxyacetic Acid, *Propyl ester*
2-Methoxypropionic Acid, *Et ester*
Paraldehyde
Peracetic Acid, tert-*Butyl ester*
Rhodinose★†

$C_6H_{12}O_3S$
Vinylsulphonic Acid, *Butyl ester*

$C_6H_{12}O_4$
Abequose
Ascarylose
Boivinose
Colitose
Corchsularose
Chromose C†
1,2,3,4-Cyclohexanetetrol†
1,2,3,5-Cyclohexanetetrol†
Digitoxose
2,3-Dihydroxybutyric Acid, *Et ester*
2,3-Dihydroxy-2-methylpropionic Acid, *Et ester*
2,3-Dihydroxy-3-methylvaleric Acid★†
2,5-Dihydroxy-3-methylvaleric Acid†
Glyceric Acid, *Di-Me ether*, *Me ester*
Glyceric Acid, *Propyl ester*
Glyoxylic Acid, *Di-Et acetal*
Mevalonic Acid
α-Mono-acetin, *Mono-Me ether*
Oliose†
Pantoic Acid
Propionin
2-Propylglyceric Acid
3-Propylglyceric Acid
Tyvelose

$C_6H_{12}O_4S$
6-Deoxy-5-thio-L-talopyranose†
L-Ribothiafuranose, *Methylglycoside*†
D-Ribothiapyranose, *Me glycoside*†

$C_6H_{12}O_4S_2$
1,6-Dithio-D-glucose†

$C_6H_{12}O_5$
1,5-Anhydro-D-sorbitol
2,5-Anhydro-D-talitol†
Arabinose, 2-*Me ether*
2-Deoxy-D-galactose
3-Deoxygalactose
2-Deoxy-D-glucose
3-Deoxy-α-D-*ribo*-hexofuranose†
5-Deoxy-D-*xylo*-hexofuranose†
3-Deoxy-α-D-*ribo*-hexopyranose†
3-Deoxy-β-D-*ribo*-hexopyranose†
4-Deoxy-D-*xylo*-hexopyranose†
6-Deoxy-L-talose
Fucose
epi-Fucose
D-Gulomethylose
Homoribose†
L(+)-Idomethylose
Isorhodeose
Isosaccharinose
Mannitan
Quercitol
Rhamnose
L-*epi*-Rhamnose
Sorbitan
Styracitol
Tagatomethylose
Talomethylose
3,4,5-Trihydroxyhexanoic Acid
Xylose, *Methyl xyloside*

$C_6H_{12}O_5S$
Glucothiose
L-Idothiapyranose

$C_6H_{12}O_5Se$
1-Deoxy-1-seleno-D-glucose

$C_6H_{12}O_6$
α-Acrose
Allose
Altrose
Arabonic Acid, *Me ester*
Dendroketose
Fructose
Fuconic Acid
Galactose
Galtose
Glucodesonic Acid
Glucoisosaccharinic Acid†
Glucose
Gulose
Hamamelose
D-(+)-Idose
Inositol

$C_6H_{12}O_6$ (*continued*)
Isosaccharic Acid
Lyxose★, 4-O-*Me ether*†
Lyxose★, *Me* α-D-*Lyxofuranoside*†
Mannose
Psicose
Rhamnonic Acid
Sorbose
Tagatose
Talose

$C_6H_{12}O_7$
Galactonic Acid
Gluconic Acid
Gulonic Acid
L-(+)-Idonic Acid
Mannonic Acid
Talonic Acid

$C_6H_{12}S$
Cyclohexanethiol
Thiepan

$C_6H_{13}As$
1-Methylarsane

$C_6H_{13}BO_2$
Butaneboronic Acid, *Cyclic ethylene ester*

$C_6H_{13}Br$
1-Bromohexane
2-Bromohexane
3-Bromohexane

$C_6H_{13}BrHg$
Mercuri-hexyl bromide

$C_6H_{13}BrO$
2-Bromo-1-butanol, *Et ether*†

$C_6H_{13}Cl$
1-Chloro-2,3-dimethylbutane
2-Chloro-2,3-dimethylbutane
3-Chloro-2,2-dimethylbutane
4-Chloro-2,2-dimethylbutane
1-Chlorohexane
2-Chlorohexane
3-Chlorohexane

$C_6H_{13}ClHg$
Mercuri-hexyl-chloride

$C_6H_{13}ClN_2O_3$
Glycyl-alanine, *Me ester hydrochloride*

$C_6H_{13}ClO$
3-Chloro-2-butanol, *Et ether*
1-Chloro-2,3-dimethyl-2-butanol
1-Chloro-3,3-dimethyl-2-butanol
3-Chloro-2,3-dimethyl-2-butanol
1-Chloroethanol, *Isobutyl ether*
1-Chloro-2-hexanol
1-Chloro-3-hexanol
2-Chloro-3-hexanol
5-Chloro-3-hexanol

$C_6H_{13}ClO_2$
Chloroacetaldehyde, *Di-Et acetal*
3-Chlorobutyraldehyde, *Di-Me acetal*

$C_6H_{13}F$
1-Fluorohexane
2-Fluorohexane

$C_6H_{13}HgI$
Mercuri-hexyl iodide

$C_6H_{13}I$
1-Iodohexane
2-Iodohexane

$C_6H_{13}IO_2$
Iodoacetaldehyde, *Di-Et acetal*

$C_6H_{13}N$
Allylpropylamine
5-Amino-1-hexene
4-Amino-2-hexene
1-Amino-1-methylcyclopentane
2-Amino-1-methylcyclopentane
3-Amino-1-methylcyclopentane
Cyclohexylamine
1,2-Dimethylpyrrolidine
1,3-Dimethylpyrrolidine
2,2-Dimethylpyrrolidine
2,4-Dimethylpyrrolidine
2,5-Dimethylpyrrolidine
3,3-Dimethylpyrrolidine
3,4-Dimethylpyrrolidine
N-Ethyl-*N*-methylallylamine
N-Methylpiperidine
2-Methylpiperidine
3-Methylpiperidine
4-Methylpiperidine
Perhydroazepine
Pyrrolidine, N-*Et*

$C_6H_{13}NO$
2-Aminocyclohexanol
4-Amino-4-methyl-2-pentanone
N-Butylacetamide
N-*tert*-Butylacetamide
2,2-Dimethylbutyric Acid, *Amide*
2,3-Dimethylbutyric Acid, *Amide*
3,3-Dimethylbutyric Acid, *Amide*
2-Ethylbutyric Acid, *Amide*
Hexanoic Acid, *Amide*
3-Hydroxypiperidine, N-*Me*
4-Hydroxypiperidine, N-*Me*
N-Isobutylacetamide
N-Isopropyl-*N*-Methylacetamide
2-Methylvaleric Acid, *Amide*
3-Methylvaleric Acid, *Amide*
4-Methylvaleric Acid, *Amide*
2-Piperidinemethanol
3-Piperidinemethanol
4-Piperidinemethanol

$C_6H_{13}NOS$
Thiocarbamic Acid, S-*Pentyl ester*

$C_6H_{13}NO_2$
2-Aminobutyric Acid, *Et ester*
3-Aminobutyric Acid, *Et ester*
4-Aminobutyric Acid, *Et ester*
2-Aminohexanoic Acid
3-Aminohexanoic Acid
4-Aminohexanoic Acid
5-Aminohexanoic Acid
6-Aminohexanoic Acid
2-Amino-2-methylpropionic Acid, *Et ester*
3-Amino-2-methylpropionic Acid, *Et ester*
2-Amino-2-methylvaleric Acid

$C_6H_{13}NO_2$ (*continued*)
5-Amino-2-methylvaleric Acid
3-Aminopentane-3-carboxylic Acid
2-Aminovaleric Acid, N-*Me*
5-Aminovaleric Acid, N-*Me*
Butylcarbamic Acid, *Me ester*
2-Butylcarbamic Acid, *Me ester*
tert-Butylcarbamic Acid, *Me ester*
Carbamic Acid, 3-*Methylbutyl ester*
Carbamic Acid, *Methyl-n-propylcarbinol ester*
Diethylcarbamic Acid, *Me ester*
N-Diethylglycine
N-Dimethyl-DL-alanine *Me ester*
2-Dimethylaminobutyric Acid
3-Dimethylaminobutyric Acid
4-Dimethylaminobutyric Acid
N-Dimethylglycine, *Et ester*
N-Ethylglycine, *Et ester*
Ethyl isopropylaminoformate
N-Ethyl-*N*-methylglycine, *Me ester*
Ethyl propylaminoformate
Glycine★, tert-*Bu ester*†
2-Hydroxyhexanoic Acid, *Amide*
4-Hydroxyhexanoic Acid, *Amide*
2-Hydroxy-4-methylvaleric Acid, *Amide*
4-Hydroxy-4-methylvaleric Acid, *Amide*
Isoleucine
Leucine
2-Methylaminobutyric Acid, *Me ester*
3-Methylaminobutyric Acid, *Me ester*
2-Methylaminopropionic Acid, *Et ester*
3-Methylaminopropionic Acid, *Et ester*
2-Methylaminovaleric Acid
5-Methylaminovaleric Acid
3-Methyl-3-methylaminobutyric Acid
2-Methyl-3-methylaminopropionic Acid, *Me ester*
1-Nitrohexane
2-Nitrohexane
α-Propiobetaine
Valine, *Me ester*
Valine, N-*Me*

$C_6H_{13}NO_2S$
Homocysteine, S-*Et*
Penicillamine, *Me ester*

$C_6H_{13}NO_3$
6-Amino-2-hydroxyhexanoic Acid
6-Amino-4-hydroxyhexanoic Acid
3-Amino-2-hydroxy-2-methylpropionic Acid, *Et ester*
Daunosamine†
3-Dimethylamino-2-hydroxy-2-methylpropionic Acid
3-Hydroxyvaline, *Me ether*
Pantamide

$C_6H_{13}NO_3S$
Cyclamic Acid†
S-Propylcysteine sulphoxide

$C_6H_{13}NO_4$
2-Amino-2,6-dideoxygalactose
2-Amino-2,6-dideoxytalose
4-Amino-4,6-dideoxy-D-galactose†
2-Amino-2,6-dideoxyglucose†
3-Amino-3,6-dideoxyglucose†
4-Amino-4,6-dideoxy-D-glucose†
2-Amino-2,6-dideoxymannose†
3-Amino-3,6-dideoxytalose†
Mycosamine
Perosamine†

$C_6H_{13}NO_5$
6-Amino-6-deoxy-D-galactose
6-Amino-6-deoxyglucose
2-Amino-2-deoxy-α-D-allose†
3-Amino-3-deoxy-D-glucose†
2-Amino-2-deoxy-α-D-gulose†
2-Amino-2-deoxymannose†
2-Amino-2-deoxy-α-D-talose†
Chondrosamine
Galactosaminitol
Galactosylamine
Glucosamine
Glucosylamine
Inosamine
Isosaccharic Acid, *Amide*
Kanosamine
Rhamnonic Acid, *Amide*

$C_6H_{13}NO_6$
Galactonic Acid, *Amide*
Gluconic Acid, *Amide*
Glucosaminic Acid
Gulonic Acid, *Amide*
Mannonic Acid, *Amide*
Talonic Acid, *Amide*

$C_6H_{13}NS$
1,4-Thiazan, N-*Et*

$C_6H_{13}NS_2$
Diethyldithiocarbamic Acid, *Me ester*
Dithiocarbamic Acid, 3-*Methylbutyl ester*
Thialdine

$C_6H_{13}N_3$
Galegine

$C_6H_{13}N_3O$
4-Hydroxygalegine

$C_6H_{13}N_3O_2$
2,2′-Iminodipropionic Acid, *Diamide*
Methylguanidinobutyric Acid

$C_6H_{13}N_3O_3$
Citrulline

$C_6H_{13}N_3O_4$
O-Ureidohomoserine, *Me ester*†

$C_6H_{13}O_3P$
Butyl ethylene phosphite
2-Butyl ethylene phosphite
tert-Butyl ethylene phosphite
2-Isobutoxy-1,3,2-dioxaphospholan
2-Propoxy-1,3,2-dioxaphosphane
Vinylphosphonic Acid, *Di-Et ester*

$C_6H_{13}O_5P$
Phosphonoacetic Acid, 3,3-*Di-Me*-1-*Et-ester*

$C_6H_{13}O_9P$
Fructose-1-phosphate†
Fructose-2-phosphate†
Fructose-6-phosphate†

$C_6H_{13}O_9P$ (*continued*)
D-Galactose-1-phosphate
D-Galactose-3-phosphate
D-Galactose-6-phosphate
Glucose-1-phosphate†
Glucose-2-phosphate†
Glucose-3-phosphate†
Glucose-4-phosphate†
Glucose-5-phosphate†
Glucose-6-phosphate†

C_6H_{14}
2,2-Dimethylbutane
2,3-Dimethylbutane
Hexane
Isohexane
3-Methylpentane

$C_6H_{14}ClO_2P$
O,O-Dipropylphosphorochloridite

$C_6H_{14}ClO_2PS$
O,O-Dipropylphosphorochloridothionate

$C_6H_{14}ClO_3P$
O,O-Dipropylphosphorochloridate

$C_6H_{14}Hg$
Mercury di-isopropyl
Mercury dipropyl

$C_6H_{14}HgO$
Mercuri-hexyl hydroxide

$C_6H_{14}NO_8P$
α-D-Galactosamine-1-phosphate†
Glucosamine*, 1-*Phosphate*†
Glucosamine, 3-*Phosphate*
Glucosamine, 6-*Phosphate*

$C_6H_{14}N_2$
1,2-Diaminocyclohexane
1,3-Diaminocyclohexane
1,4-Diaminocyclohexane
1,2-Dimethylpiperazine
2,5-Dimethylpiperazine
2,6-Dimethylpiperazine
Piperazine, N,N-*Di-Me*

$C_6H_{14}N_2O$
N-Diethylglycine, *Amide*
N-Dimethylglycine, *Dimethylamide*
N^1-Isobutyl-N^1-methylurea
Leucine, *Amide*
N-3-Methylbutylurea

$C_6H_{14}N_2O_2$
3,6-Diaminohexanoic Acid
3-Dimethylamino-2-hydroxy-2-methylpropionic Acid, *Amide*
2-Hydrazino-3-methylbutyric Acid, *Me ester*
2-Hydrazino-2-methylpropionic Acid, *Et ester*
Lysine
Ornithine, 2-N-*Me*

$C_6H_{14}N_2O_3$
Actinorubin (or $C_9H_{22}N_5O_4$)
2-Amino-6-hydroxyaminohexanoic Acid
2-Deoxystreptamine†
5-Hydroxylysine

$C_6H_{14}N_2O_4$
2,3-Diamino-2,3-dideoxy-D-allose†
2,6-Diamino-2,6-dideoxy-α-D-allose†
2,6-Diamino-2,6-dideoxy-α-D-galactose
2,3-Diamino-2,3-dideoxy-α-D-glucose†
2,6-Diamino-2,6-dideoxy-β-D-glucose
2,6-Diamino-2,6-dideoxy-L-idose†
2,6-Diamino-2,6-dideoxy-D-mannose†
Streptamine

$C_6H_{14}N_2O_4Pb$
Lead dipropyldinitrate

$C_6H_{14}N_2O_6$
Mannonic Acid, *Hydrazide*

$C_6H_{14}N_2S$
N^1-2-Butyl-N^2-methylthiourea
N-2-Methylbutylthiourea
N-3-Methylbutylthiourea

$C_6H_{14}N_4O_2$
Arginine
Propylmalonic Acid, *Dihydrazide*

$C_6H_{14}N_4O_3$
γ-Hydroxyarginine

$C_6H_{14}N_4O_5P$
Phosphoarginine†

$C_6H_{14}N_4O_6$
Mucic Acid, *Hydrazide*

$C_6H_{14}O$
Butyl ethyl Ether
2-Butyl ethyl Ether
Di-isopropyl Ether
2,2-Dimethyl-1-butanol
2,3-Dimethyl-2-butanol
3,3-Dimethyl-2-butanol
Dipropyl Ether
Ethyl isobutyl Ether
Ethyl 2-methyl-2-propyl Ether
1-Hexanol
2-Hexanol
3-Hexanol
Isopropyl propyl Ether
4-Methyl-2-butanol, *Me ether*
Methyl 2-methylbutyl Ether
Methyl 3-methylbutyl Ether
2-Methyl-1-pentanol
3-Methyl-1-pentanol
4-Methyl-1-pentanol
2-Methyl-2-pentanol
3-Methyl-2-pentanol
4-Methyl-2-pentanol
2-Methyl-3-pentanol
3-Methyl-3-pentanol
Methyl pentyl Ether
1-Pentanol, *Me ether*

$C_6H_{14}OS$
Di-isopropyl Sulphoxide†
Dipropyl sulphoxide
2-Hydroxy-1-ethanethiol, S-*Butyl*

$C_6H_{14}O_2$
Acetal
1,4-Butanediol, *Di-Me ether*
Butyl 1-hydroxyethyl Ether
Butyl 2-hydroxyethyl Ether

$C_6H_{14}O_2$ (*continued*)
- 2,3-Dimethyl-2,3-butanediol
- Ethylene Glycol, *Di-Et ether*
- 1,2-Hexanediol
- 1,3-Hexanediol
- 1,4-Hexanediol
- 1,5-Hexanediol
- 1,6-Hexanediol
- 2,3-Hexanediol
- 2,4-Hexanediol
- 2,5-Hexanediol
- 3,4-Hexanediol
- 2-Hydroxyethyl 1,1-dimethylethyl Ether
- 2-Hydroxyethyl isobutyl Ether
- 2-Methylpropanediol, 1-*Et ether*
- 1,5-Pentanediol, *Mono-Me ether*
- 1,3-Propanediol, *Propyl ether*

$C_6H_{14}O_2S$
- Butyl ethyl sulphate
- Di-isopropyl sulphone
- Dipropyl sulphone

$C_6H_{14}O_3$
- 1,2,4-Butanetriol, 1-*Et ether*
- Di-(2-hydroxyethyl) Ether, *Di-Me ether*
- Di-(2-hydroxyethyl) Ether, *Et ether*
- 2-(1-Ethoxyethoxy)ethanol
- Glycerol, *Tri-Me ether*
- Glycerol, 1-*Propyl ether*
- Glycollaldehyde, *Di-Et acetal*
- 1,2,3-Hexanetriol
- 1,2,4-Hexanetriol
- 1,2,5-Hexanetriol
- 1,2,6-Hexanetriol
- 1,3,5-Hexanetriol
- 2,3,4-Hexanetriol
- 2,3,5-Hexanetriol
- Orthoacetic Acid, *Di-Me, Et ester*

$C_6H_{14}O_4$
- 1,2,3,4-Hexanetetrol
- 1,2,5,6-Hexanetetrol
- 1,3,4,6-Hexanetetrol
- 2,3,4,5-Hexanetetrol
- Triethylene Glycol

$C_6H_{14}O_5$
- Di-2,3-dihydroxypropyl Ether
- Fucitol
- *epi*-Fucitol
- L-Gulomethylitol
- Idomethylitol
- Isorhodeitol
- Rhamnitol
- *epi*-Rhamnitol

$C_6H_{14}O_6$
- α-Acritol
- Allitol
- Dulcitol
- Iditol
- Mannitol
- Sorbitol
- Talitol

$C_6H_{14}O_6S_2$
- Busulphan
- Ethane-1,2-disulphonic Acid, *Di-Et ester*

$C_6H_{14}O_{10}P_2$
- Mevalonic Acid 5-pyrophosphate†

$C_6H_{14}O_{12}P_2$
- Fructose-1,4-diphosphate†

$C_6H_{14}S$
- Di-isopropyl sulphide
- Dipropyl sulphide
- Methyl 2-methylbutyl sulphide
- Methyl 3-methylbutyl sulphide
- 2-Methyl-2-propanethiol, S-*Et*

$C_6H_{14}S_2$
- Di-isopropyl Disulphide†
- Dithioacetal
- 1,2-Ethanedithiol, *Di-Et ether*

$C_6H_{14}Zn$
- Zinc di-isopropyl
- Zinc dipropyl

$C_6H_{15}Al$
- Triethylaluminium

$C_6H_{15}AlO_3$
- Aluminium ethoxide

$C_6H_{15}As$
- Triethylarsine

$C_6H_{15}AsO_4$
- Ethyl arsenate

$C_6H_{15}B$
- Triethylborine

$C_6H_{15}BF_4O$
- Triethyloxonium, Fluoroborate†

$C_6H_{15}BO_3$
- Triethyl borate

$C_6H_{15}ClN_2O_2$
- Doryl

$C_6H_{15}N$
- 1-Amino-2,2-dimethylbutane
- 2-Amino-2,3-dimethylbutane
- 3-Amino-2,2-dimethylbutane
- 1-Amino-2-ethylbutane
- 1-Aminohexane
- 2-Aminohexane
- 3-Aminohexane
- 1-Amino-2-methylpentane
- 1-Amino-3-methylpentane
- 1-Amino-4-methylpentane
- 4-Amino-2-methylpentane
- Di-isopropylamine
- Dipropylamine
- *N*-Ethylbutylamine
- *N*-Ethyl-2-butylamine
- *N*-Ethylisobutylamine
- *N*-Ethyl-*N*-methylpropylamine
- Isopentylmethylamine
- *N*-Methyl-1-pentylamine
- *N*-Methyl-2-pentylamine
- *N*-Methyl-3-pentylamine
- Triethylamine

$C_6H_{15}NO$
- 4-Amino-1-butanol, *Et ether*
- 1-Amino-2-butanol, *Et ether*
- 4-Amino-2-butanol, *Et ether*

$C_6H_{15}NO$ (*continued*)
2-Amino-4-methyl-1-pentanol
4-Amino-2-methyl-2-pentanol
4-Amino-4-methyl-2-pentanol
1-Amino-2-methyl-2-propanol, N-*Di-Me*
3-Amino-1-propanol, N-*Di-Me, Me ether*
2-Butylaminoethanol
2-*tert*-Butylaminoethanol
2-Diethylaminoethanol
4-Dimethylaminobutanol
2-Dimethylaminoethyl Ether
N-Dipropylhydroxylamine
O,N-Dipropylhydroxylamine
Isobutylaminoethanol

$C_6H_{15}NO_2$
Aminoacetal
Dipropylmalonic Acid, *Mononitrile*

$C_6H_{15}NO_3$
Tri-(2-hydroxyethyl)amine

$C_6H_{15}NO_3S$
Taurine, N-*Di-Et*

$C_6H_{15}NO_5$
Glucamine
Glucosaminol
Mannamine

$C_6H_{15}NS$
4-Dimethylamino-1-butanethiol

$C_6H_{15}N_2O_6P$
L-Threonine ethanolamine phosphate

$C_6H_{15}N_3$
Pentamethylguanidine

$C_6H_{15}O_2PS_2$
O,O,S-Triethylphosphorothiolothionate

$C_6H_{15}O_3P$
Ethylphosphonic Acid, *Di-Et ester*
Triethyl phosphite

$C_6H_{15}O_3PS$
O,O-Diethyl-*S*-ethylphosphorothiolate
Hydrogen *O,O*-dipropylphosphorothionate

$C_6H_{15}O_3PS_2$
2-(Ethylthio)ethyl dimethyl phosphorothionate★†

$C_6H_{15}O_3PSe$
O,O-Diethyl-*Se*-ethylphosphoroselenolate

$C_6H_{15}O_3S_2$
S-(2-[Ethylthio]ethyl)dimethylphosphorothiolate

$C_6H_{15}O_4P$
Phosphoric Acid, *Tri-Et ester*

$C_6H_{15}P$
Triethylphosphine

$C_6H_{15}Sb$
Triethylstibine

$C_6H_{15}Tl$
Thallium triethyl

$C_6H_{16}FN_2OP$
NN'-Di-isopropylphosphorodiamidic fluoride†

$C_6H_{16}INO$
Hydroxymethyltrimethylammonium Hydroxide, *Et ether, Iodide*

$C_6H_{16}N_2$
1,6-Diaminohexane
1,2-Di-ethylaminoethane
2-Diethylaminoethylamine
NN-Diethylethylenediamine†
N-Methylcadaverine
Putrescine, N,N'-*Di-Me*

$C_6H_{16}N_2O_4$
1,2-Diamino-1,2-dideoxymannitol†
1,2-Diamino-1,2-dideoxysorbitol†

$C_6H_{16}N_6$
Arcaine

$C_6H_{16}O_6P_2S$
O,O-Diethylphosphorothiolic *O,O*-dimethylphosphorothiolic anhydrosulphide

$C_6H_{16}S_2$
Dipropyl Disulphide†

$C_6H_{17}NO_2$
α-Methylcholine
β-Methylcholine

$C_6H_{18}B_3N_3$
N,N',N''-Triethylborazole

$C_6H_{18}N_3OP$
Hexamethylphosphoramide†

$C_6H_{18}N_4$
Triethylenetetramine
Tris-(2-aminoethyl)ammonia†

$C_6H_{18}O_3Si_3$
1,1,3,3,5,5-Hexamethylcyclotrisiloxane

$C_6H_{24}O_6Si_6$
1,3,5,7,9,11-Hexamethylcyclohexasiloxane

$C_6I_4O_2$
Tetraiodobenzoquinone

C_6I_6
Hexaiodobenzene

C_6N_4
Tetracyanoethylene★†

C_6N_4O
Tetracyanoethylene Oxide†

C_6N_6
Cyanuric cyanide

$C_6N_6O_3$
Benzotrifurazan†

$C_6N_6O_6$
Benzotrifuroxan†

C_7

C_7Br_5N
Pentabromobenzoic Acid, *Nitrile*

C_7Cl_6O
Pentachlorobenzoic Acid, *Chloride*

C_7HBr_4ClO
2,3,4,6-Tetrabromobenzoic Acid, *Chloride*

C_7HBr_4N
2,3,4,5-Tetrabromobenzoic Acid, *Nitrile*
2,3,4,6-Tetrabromobenzoic Acid, *Nitrile*

$C_7HBr_5O_2$
Pentabromobenzoic Acid

C_7HCl_4N
2,3,4,5-Tetrachlorobenzoic Acid, *Nitrile*
2,3,4,6-Tetrachlorobenzoic Acid, *Nitrile*
2,3,5,6-Tetrachlorobenzoic Acid, *Nitrile*

C_7HCl_5O
2,3,4,5-Tetrachlorobenzoic Acid, *Chloride*

$C_7HCl_5O_2$
Pentachlorobenzoic Acid

C_7HF_{11}
1*H*-Undecafluorobicyclo[2,2,1]heptane†

$C_7H_2BrClN_2O_5$
2-Bromo-3,5-dinitrobenzoic Acid, *Chloride*
4-Bromo-3,5-dinitrobenzoic Acid, *Chloride*

$C_7H_2BrNO_3$
5-Bromopyridine-2,3-dicarboxylic Acid, *Anhydride*

$C_7H_2Br_3ClO$
2,4,6-Tribromobenzoic Acid, *Chloride*
3,4,5-Tribromobenzoic Acid, *Chloride*

$C_7H_2Br_3N$
2,4,6-Tribromobenzoic Acid, *Nitrile*

$C_7H_2Br_3NO$
2,4,6-Tribromo-3-hydroxybenzoic Acid, *Nitrile*

$C_7H_2Br_4O_2$
2,3,4,5-Tetrabromobenzoic Acid
2,3,4,6-Tetrabromobenzoic Acid
2,3,5,6-Tetrabromobenzoic Acid
Tetrabromocatechol, *Methylene ether*

$C_7H_2ClN_3O_4$
2-Chloro-3,5-dinitrobenzoic Acid, *Nitrile*
5-Chloro-2,4-dinitrobenzoic Acid, *Nitrile*

$C_7H_2ClN_3O_7$
2,4,6-Trinitrobenzoic Acid, *Chloride*

$C_7H_2Cl_2N_2O_2$
2,6-Dichloro-3-nitrobenzoic Acid, *Nitrile*

$C_7H_2Cl_2N_2O_5$
5-Chloro-2,4-dinitrobenzoic Acid, *Chloride*

$C_7H_2Cl_3N$
2,3,5-Trichlorobenzoic Acid, *Nitrile*
2,4,5-Trichlorobenzoic Acid, *Nitrile*
2,4,6-Trichlorobenzoic Acid, *Nitrile*

$C_7H_2Cl_4O$
2,3,5-Trichlorobenzoic Acid, *Chloride*
2,4,5-Trichlorobenzoic Acid, *Chloride*
2,4,6-Trichlorobenzoic Acid, *Chloride*
3,4,5-Trichlorobenzoic Acid, *Chloride*

$C_7H_2Cl_4O_2$
2,3,4,5-Tetrachlorobenzoic Acid
2,3,4,6-Tetrachlorobenzoic Acid
2,3,5,6-Tetrachlorobenzoic Acid

$C_7H_2Cl_4O_3$
3,4,5,6-Tetrachlorosalicylic Acid†

$C_7H_2F_{10}$
1-*H*,4-*H*-Decafluorobicyclo[2,2,1]heptane†

$C_7H_2N_4O_7$
3-Hydroxy-2,4,6-trinitrobenzoic Acid, *Nitrile*

C_7H_3BrClN
4-Bromo-2-chlorobenzoic Acid, *Nitrile*

$C_7H_3BrClNO_3$
2-Bromo-5-nitrobenzoic Acid, *Chloride*
4-Bromo-3-nitrobenzoic Acid, *Chloride*

$C_7H_3BrCl_2N_2O_5$
2-Bromo-4,6-dichloro-3,5-dinitrophenol, *Me ether*
4-Bromo-2,6-dichloro-3,5-dinitrophenol, *Me ether*

$C_7H_3BrCl_2O$
2-Bromo-3-chlorobenzoic Acid, *Chloride*
2-Bromo-4-chlorobenzoic Acid, *Chloride*
2-Bromo-6-chlorobenzoic Acid, *Chloride*
3-Bromo-2-chlorobenzoic Acid, *Chloride*
3-Bromo-4-chlorobenzoic Acid, *Chloride*
4-Bromo-2-chlorobenzoic Acid, *Chloride*
4-Bromo-3-chlorobenzoic Acid, *Chloride*
5-Bromo-2-chlorobenzoic Acid, *Chloride*
5-Bromo-3-chlorobenzoic Acid, *Chloride*
6-Bromo-3-chlorobenzoic Acid, *Chloride*

$C_7H_3BrN_2O_2$
2-Bromo-5-nitrobenzoic Acid, *Nitrile*
3-Bromo-4-nitrobenzoic Acid, *Nitrile*
4-Bromo-2-nitrobenzoic Acid, *Nitrile*
4-Bromo-3-nitrobenzoic Acid, *Nitrile*

$C_7H_3BrN_2O_6$
2-Bromo-3,5-dinitrobenzoic Acid
2-Bromo-4,5-dinitrobenzoic Acid
4-Bromo-3,5-dinitrobenzoic Acid
5-Bromo-2,4-dinitrobenzoic Acid

$C_7H_3Br_2ClO$
2,3-Dibromobenzoic Acid, *Chloride*
2,4-Dibromobenzoic Acid, *Chloride*
2,5-Dibromobenzoic Acid, *Chloride*
2,6-Dibromobenzoic Acid, *Chloride*
3,4-Dibromobenzoic Acid, *Chloride*
3,5-Dibromobenzoic Acid, *Chloride*

$C_7H_3Br_2ClO_2$
3,5-Dibromo-2-hydroxybenzoic Acid, *Chloride*
3,5-Dibromo-4-hydroxybenzoic Acid, *Chloride*

$C_7H_3Br_2N$
2,4-Dibromobenzoic Acid, *Nitrile*
2,5-Dibromobenzoic Acid, *Nitrile*
2,6-Dibromobenzoic Acid, *Nitrile*
3,5-Dibromobenzoic Acid, *Nitrile*

$C_7H_3Br_2NO$
3,5-Dibromo-2-hydroxybenzoic Acid, *Nitrile*
3,5-Dibromo-4-hydroxybenzoic Acid, *Nitrile*

$C_7H_3Br_2NO_4$
2,4-Dibromo-5-nitrobenzoic Acid
3,5-Dibromo-2-nitrobenzoic Acid
4,5-Dibromo-2-nitrobenzoic Acid

$C_7H_3Br_2N_2$
3-Amino-2,4,6-tribromobenzoic Acid, *Nitrile*

$C_7H_3Br_2N_3O_7$
3,5-Dibromo-2,4,6-trinitrophenol, *Me ether*

$C_7H_3Br_3O$
2,3,5-Tribromobenzaldehyde
2,4,6-Tribromobenzaldehyde
3,4,5-Tribromobenzaldehyde

$C_7H_3Br_3O_2$
2,3,4-Tribromobenzoic Acid
2,3,5-Tribromobenzoic Acid
2,4,5-Tribromobenzoic Acid
2,4,6-Tribromobenzoic Acid
3,4,5-Tribromobenzoic Acid
2,3,5-Tribromo-6-methyl-1,4-benzoquinone

$C_7H_3Br_3O_3$
2,4,6-Tribromo-3-hydroxybenzoic Acid

$C_7H_3Br_4NO_2$
2-Amino-3,4,5,6-tetrabromobenzoic Acid

$C_7H_3Br_5$
2,3,4,5,6-Pentabromotoluene

$C_7H_3Br_5O$
Pentabromophenol, *Me ether*

$C_7H_3ClFNO_3$
2-Fluoro-4-nitrobenzoic Acid, *Chloride*
4-Fluoro-3-nitrobenzoic Acid, *Chloride*

$C_7H_3ClINO_3$
2-Iodo-4-nitrobenzoic Acid, *Chloride*
2-Iodo-5-nitrobenzoic Acid, *Chloride*

$C_7H_3ClI_2O_2$
2-Hydroxy-3,5-di-iodobenzoic Acid, *Chloride*

$C_7H_3ClN_2O_2$
3-Chloro-4-nitrobenzoic Acid, *Nitrile*
4-Chloro-2-nitrobenzoic Acid, *Nitrile*
4-Chloro-3-nitrobenzoic Acid, *Nitrile*

$C_7H_3ClN_2O_4S$
4-Nitro-2-sulphobenzoic Acid, *Chloride*

$C_7H_3ClN_2O_5$
2,4-Dinitrobenzoic Acid, *Chloride*
2,6-Dinitrobenzoic Acid, *Chloride*
3,4-Dinitrobenzoic Acid, *Chloride*
3,5-Dinitrobenzoic Acid, *Chloride*

$C_7H_3ClN_2O_6$
2-Chloro-3,5-dinitrobenzoic Acid
2-Chloro-4,5-dinitrobenzoic Acid
4-Chloro-3,5-dinitrobenzoic Acid
5-Chloro-2,4-dinitrobenzoic Acid
2-Hydroxy-3,5-dinitrobenzoic Acid, *Chloride*

$C_7H_3Cl_2N$
2,4-Dichlorobenzoic Acid, *Nitrile*
2,5-Dichlorobenzoic Acid, *Nitrile*
2,6-Dichlorobenzoic Acid, *Nitrile*
3,4-Dichlorobenzoic Acid, *Nitrile*
3,5-Dichlorobenzoic Acid, *Nitrile*

$C_7H_3Cl_2NO$
3,5-Dichloro-2-hydroxybenzoic Acid, *Nitrile*
3,5-Dichloro-4-hydroxybenzoic Acid, *Nitrile*

$C_7H_3Cl_2NO_2$
Pyridine-2,3-dicarboxylic Acid, *Dichloride*
Pyridine-2,4-dicarboxylic Acid, *Dichloride*
Pyridine-2,6-dicarboxylic Acid, *Dichloride*
Pyridine-3,5-dicarboxylic Acid, *Dichloride*

$C_7H_3Cl_2NO_3$
2-Chloro-4-nitrobenzoic Acid, *Chloride*
2-Chloro-5-nitrobenzoic Acid, *Chloride*
2-Chloro-6-nitrobenzoic Acid, *Chloride*
3-Chloro-4-nitrobenzoic Acid, *Chloride*
4-Chloro-2-nitrobenzoic Acid, *Chloride*
4-Chloro-3-nitrobenzoic Acid, *Chloride*
5-Chloro-2-nitrobenzoic Acid, *Chloride*
2,4-Dichloro-5-nitrobenzaldehyde
2,5-Dichloro-3-nitrobenzaldehyde
2,6-Dichloro-3-nitrobenzaldehyde
3,5-Dichloro-2-nitrobenzaldehyde
3,6-Dichloro-2-nitrobenzaldehyde
4,5-Dichloro-2-nitrobenzaldehyde

$C_7H_3Cl_2NO_4$
2,3-Dichloro-4-nitrobenzoic Acid
2,4-Dichloro-6-nitrobenzoic Acid
2,5-Dichloro-3-nitrobenzoic Acid
2,5-Dichloro-4-nitrobenzoic Acid
2,6-Dichloro-3-nitrobenzoic Acid
2,6-Dichloro-4-nitrobenzoic Acid
3,5-Dichloro-2-nitrobenzoic Acid
3,6-Dichloro-2-nitrobenzoic Acid
4,5-Dichloro-2-nitrobenzoic Acid

$C_7H_3Cl_2NO_5S$
2-Nitro-4-sulphobenzoic Acid, *Dichloride*
4-Nitro-2-sulphobenzoic Acid, *Dichloride*
5-Nitro-3-sulphobenzoic Acid, *Dichloride*

$C_7H_3Cl_3O$
2,3-Dichlorobenzoic Acid, *Chloride*
2,4-Dichlorobenzoic Acid, *Chloride*
2,5-Dichlorobenzoic Acid, *Chloride*
2,6-Dichlorobenzoic Acid, *Chloride*
3,4-Dichlorobenzoic Acid, *Chloride*
3,5-Dichlorobenzoic Acid, *Chloride*
2,3,4-Trichlorobenzaldehyde
2,3,5-Trichlorobenzaldehyde
2,3,6-Trichlorobenzaldehyde
2,4,5-Trichlorobenzaldehyde
2,4,6-Trichlorobenzaldehyde
3,4,5-Trichlorobenzaldehyde

$C_7H_3Cl_3O_2$
3,5-Dichloro-2-hydroxybenzoic Acid, *Chloride*
2,3,4-Trichlorobenzoic Acid
2,3,5-Trichlorobenzoic Acid
2,3,6-Trichlorobenzoic Acid
2,4,5-Trichlorobenzoic Acid
2,4,6-Trichlorobenzoic Acid
3,4,5-Trichlorobenzoic Acid

$C_7H_3Cl_3O_5S_2$
3,5-Disulphobenzoic Acid, *Trichloride*

$C_7H_3Cl_4NO$
2,3,4,5-Tetrachlorobenzoic Acid, *Amide*

$C_7H_3Cl_4NO_2$
2-Amino-3,4,5,6-tetrachlorobenzoic Acid

$C_7H_3Cl_5$
2,3,4,5,6-Pentachlorotoluene

$C_7H_3Cl_5O$
Pentachlorophenol, *Me ether*

$C_7H_3FN_2O_2$
2-Fluoro-5-nitrobenzonitrile†

$C_7H_3IN_2O_3$
4-Cyano-2-iodo-6-nitrophenol†

$C_7H_3I_2NO$
4-Hydroxy-3,5-di-iodobenzoic Acid, *Nitrile*

$C_7H_3I_3N_2$
3-Amino-2,4,6-tri-iodobenzoic Acid, *Nitrile*

$C_7H_3I_3O_2$
2,3,5-Tri-iodobenzoic Acid
2,4,5-Tri-iodobenzoic Acid
3,4,5-Tri-iodobenzoic Acid

$C_7H_3NO_3$
Pyridine-2,3-dicarboxylic Acid, *Anhydride*

$C_7H_3N_3$
Pyridine-2,6-dicarboxylic Acid, *Dinitrile*

$C_7H_3N_3O_2$
2,6-Dihydroxypyridine-3,5-dicarboxylic Acid, *Dinitrile*

$C_7H_3N_3O_4$
2,4-Dinitrobenzoic Acid, *Nitrile*
2,6-Dinitrobenzoic Acid, *Nitrile*

$C_7H_3N_3O_5$
2-Hydroxy-3,5-dinitrobenzoic Acid, *Nitrile*
2-Hydroxy-5,6-dinitrobenzoic Acid, *Nitrile*

$C_7H_3N_3O_7$
2,4,6-Trinitrobenzaldehyde

$C_7H_3N_3O_8$
2,3,4-Trinitrobenzoic Acid
2,3,5-Trinitrobenzoic Acid
2,3,6-Trinitrobenzoic Acid
2,4,5-Trinitrobenzoic Acid
2,4,6-Trinitrobenzoic Acid
3,4,5-Trinitrobenzoic Acid

$C_7H_3N_3O_9$
3-Hydroxy-2,4,6-trinitrobenzoic Acid

$C_7H_3N_7O_2$
Pyridine-2,4-dicarboxylic Acid, *Diazide*
Pyridine-2,6-dicarboxylic Acid, *Diazide*
Pyridine-3,5-dicarboxylic Acid, *Diazide*

C_7H_4BrClO
o-Bromobenzoyl chloride
m-Bromobenzoyl chloride
p-Bromobenzoyl chloride

$C_7H_4BrClO_2$
2-Bromo-3-chlorobenzoic Acid
2-Bromo-4-chlorobenzoic Acid
2-Bromo-6-chlorobenzoic Acid
3-Bromo-2-chlorobenzoic Acid
3-Bromo-4-chlorobenzoic Acid
4-Bromo-2-chlorobenzoic Acid
4-Bromo-3-chlorobenzoic Acid
5-Bromo-2-chlorobenzoic Acid
5-Bromo-3-chlorobenzoic Acid
6-Bromo-3-chlorobenzoic Acid

C_7H_4BrN
o-Bromobenzonitrile
m-Bromobenzonitrile
p-Bromobenzonitrile

C_7H_4BrNO
3-Bromo-2-hydroxybenzoic Acid, *Nitrile*
3-Bromo-4-hydroxybenzoic Acid, *Nitrile*
5-Bromo-2-hydroxybenzoic Acid, *Nitrile*

$C_7H_4BrNO_3$
2-Bromo-6-nitrobenzaldehyde
3-Bromo-5-nitrobenzaldehyde
4-Bromo-2-nitrobenzaldehyde
4-Bromo-3-nitrobenzaldehyde
5-Bromo-2-nitrobenzaldehyde

$C_7H_4BrNO_4$
2-Bromo-3-nitrobenzoic Acid
2-Bromo-4-nitrobenzoic Acid
2-Bromo-5-nitrobenzoic Acid
2-Bromo-6-nitrobenzoic Acid
3-Bromo-2-nitrobenzoic Acid
3-Bromo-4-nitrobenzoic Acid
3-Bromo-5-nitrobenzoic Acid
4-Bromo-2-nitrobenzoic Acid
4-Bromo-3-nitrobenzoic Acid
5-Bromo-2-nitrobenzoic Acid
5-Bromopyridine-2,3-dicarboxylic Acid
5-Bromopyridine-3,4-dicarboxylic Acid

$C_7H_4BrN_3O_5$
2-Bromo-3,5-dinitrobenzoic Acid, *Amide*
4-Bromo-3,5-dinitrobenzoic Acid, *Amide*

$C_7H_4BrN_3O_6$
3-Bromo-2,4,6-trinitrotoluene

$C_7H_4Br_2N_2$
2-Amino-3,5-dibromobenzoic Acid, *Nitrile*

$C_7H_4Br_2N_2O_4$
2,3-Dibromo-4,6-dinitrotoluene
2,4-Dibromo-3,5-dinitrotoluene
2,6-Dibromo-3,5-dinitrotoluene
3,4-Dibromo-2,6-dinitrotoluene
3,5-Dibromo-2,4-dinitrotoluene
3,5-Dibromo-2,6-dinitrotoluene
3,6-Dibromo-2,4-dinitrotoluene

$C_7H_4Br_2N_2O_5$
3,5-Dibromo-4,6-dinitro-*o*-cresol
4,5-Dibromo-3,6-dinitro-*o*-cresol
4,6-Dibromo-3,5-dinitro-*o*-cresol
2,6-Dibromo-3,5-dinitro-*p*-cresol
3,5-Dibromo-2,6-dinitro-*p*-cresol

$C_7H_4Br_2O_2$
2,3-Dibromobenzoic Acid
2,4-Dibromobenzoic Acid
2,5-Dibromobenzoic Acid
2,6-Dibromobenzoic Acid
3,4-Dibromobenzoic Acid
3,5-Dibromobenzoic Acid
2,3-Dibromo-4-hydroxybenzaldehyde
2,4-Dibromo-5-hydroxybenzaldehyde
3,4-Dibromo-2-hydroxybenzaldehyde
3,5-Dibromo-2-hydroxybenzaldehyde
3,5-Dibromo-4-hydroxybenzaldehyde
4,5-Dibromo-2-hydroxybenzaldehyde
3,5-Dibromo-*p*-toluquinone

$C_7H_4Br_2O_3$
2,4-Dibromo-5-hydroxybenzoic Acid
3,5-Dibromo-2-hydroxybenzoic Acid
3,5-Dibromo-4-hydroxybenzoic Acid
4,5-Dibromo-2-hydroxybenzoic Acid
5,6-Dibromoprotocatechualdehyde†

$C_7H_4Br_2O_4$
2,4-Dibromo-3,5-dihydroxybenzoic Acid

$C_7H_4Br_2O_4$ (*continued*)
3,5-Dibromo-2,4-dihydroxybenzoic Acid
3,5-Dibromo-2,6-dihydroxybenzoic Acid
4,5-Dibromo-2,3-dihydroxybenzoic Acid

$C_7H_4Br_2O_5$
2,6-Dibromo-3,4,5-trihydroxybenzoic Acid

$C_7H_4Br_3NO$
2,4,6-Tribromobenzoic Acid, *Amide*
3,4,5-Tribromobenzoic Acid, *Amide*
2,3,4-Tribromo-6-nitrophenol, *Me ether*
2,4,5-Tribromo-6-nitrophenol, *Me ether*
2,4,6-Tribromo-6-nitrophenol, *Me ether*

$C_7H_4Br_3NO_2$
2-Amino-3,4,5-tribromobenzoic Acid
3-Amino-2,4,6-tribromobenzoic Acid
2,4,6-Tribromo-3-hydroxybenzoic Acid, *Amide*
2,3,5-Tribromo-4-nitrotoluene
2,3,5-Tribromo-6-nitrotoluene
2,3,6-Tribromo-4-nitrotoluene
2,3,6-Tribromo-5-nitrotoluene
2,4,6-Tribromo-3-nitrotoluene
3,4,5-Tribromo-2-nitrotoluene

$C_7H_4Br_4$
α,α,α-4-Tetrabromotoluene
α,2,4,6-Tetrabromotoluene
2,3,4,5-Tetrabromotoluene
2,3,4,6-Tetrabromotoluene
2,3,5,6-Tetrabromotoluene

$C_7H_4Br_4O$
α,α,4,6-Tetrabromo-*o*-cresol
3,4,5,6-Tetrabromo-*o*-cresol
α,2,4,6-Tetrabromo-*m*-cresol
2,4,5,6-Tetrabromo-*m*-cresol
α,α,2,6-Tetrabromo-*p*-cresol
α,2,3,6-Tetrabromo-*p*-cresol
2,3,5,6-Tetrabromo-*p*-cresol
2,3,4,6-Tetrabromophenol, *Me ether*

$C_7H_4Br_4O_2$
Tetrabromocatechol, *Mono-Me ether*

C_7H_4ClFO
2-Chloro-6-fluorobenzaldehyde
o-Fluorobenzoic Acid, *Chloride*
m-Fluorobenzoic Acid, *Chloride*
p-Fluorobenzoic Acid, *Chloride*

C_7H_4ClIO
o-Iodobenzoic Acid, *Chloride*
m-Iodobenzoic Acid, *Chloride*
p-Iodobenzoic Acid, *Chloride*

C_7H_4ClN
o-Chlorobenzonitrile
m-Chlorobenzonitrile
p-Chlorobenzonitrile

C_7H_4ClNO
3-Chloro-4-hydroxybenzoic Acid, *Nitrile*
5-Chloro-2-hydroxybenzoic Acid, *Nitrile*
o-Chlorophenyl isocyanate
m-Chlorophenyl isocyanate
p-Chlorophenyl isocyanate

$C_7H_4ClNO_2$
Chlorzoxazone

$C_7H_4ClNO_2S$
o-Sulphobenzoic Acid, 2-*Chloride*
p-Sulphobenzoic Acid, 4-*Chloride*

$C_7H_4ClNO_3$
2-Chloro-4-nitrobenzaldehyde
2-Chloro-5-nitrobenzaldehyde
2-Chloro-6-nitrobenzaldehyde
4-Chloro-2-nitrobenzaldehyde
4-Chloro-3-nitrobenzaldehyde
5-Chloro-2-nitrobenzaldehyde
o-Nitrobenzoyl chloride
m-Nitrobenzoyl chloride
p-Nitrobenzoyl chloride

$C_7H_4ClNO_4$
2-Chloro-3-nitrobenzoic Acid
2-Chloro-4-nitrobenzoic Acid
2-Chloro-5-nitrobenzoic Acid
2-Chloro-6-nitrobenzoic Acid
3-Chloro-2-nitrobenzoic Acid
3-Chloro-4-nitrobenzoic Acid
4-Chloro-2-nitrobenzoic Acid
4-Chloro-3-nitrobenzoic Acid
5-Chloro-2-nitrobenzoic Acid
5-Chloro-3-nitrobenzoic Acid
2-Hydroxy-3-nitrobenzoic Acid, *Chloride*

$C_7H_4ClNO_5$
3-Chloro-2-hydroxy-5-nitrobenzoic Acid
3-Chloro-6-hydroxy-5-nitrobenzoic Acid

$C_7H_4ClNO_6S$
5-Nitro-3-sulphobenzoic Acid, 3-*Chloride*

C_7H_4ClNS
p-Chlorophenyl isocyanate†

$C_7H_4ClN_3O_6$
3-Chloro-2,4,6-trinitrotoluene

$C_7H_4ClN_3O_7$
3-Chloro-2,4,6-trinitrophenol, *Me ether*

$C_7H_4Cl_2N_2O_4$
2,3-Dichloro-4,6-dinitrotoluene
2,4-Dichloro-3,5-dinitrotoluene
2,6-Dichloro-3,4-dinitrotoluene
2,6-Dichloro-3,5-dinitrotoluene
3,4-Dichloro-2,6-dinitrotoluene
3,5-Dichloro-2,4-dinitrotoluene
3,5-Dichloro-2,6-dinitrotoluene
3,6-Dichloro-2,4-dinitrotoluene

$C_7H_4Cl_2O$
2,3-Dichlorobenzaldehyde
2,4-Dichlorobenzaldehyde
2,5-Dichlorobenzaldehyde
2,6-Dichlorobenzaldehyde
3,4-Dichlorobenzaldehyde
3,5-Dichlorobenzaldehyde

$C_7H_4Cl_2O_2$
3-Chloro-2-hydroxybenzoic Acid, *Chloride*
2,3-Dichlorobenzoic Acid
2,4-Dichlorobenzoic Acid
2,5-Dichlorobenzoic Acid
2,6-Dichlorobenzoic Acid
3,4-Dichlorobenzoic Acid
3,5-Dichlorobenzoic Acid
2,4-Dichloro-3-hydroxybenzaldehyde
2,4-Dichloro-5-hydroxybenzaldehyde

$C_7H_4Cl_2O_2$ (*continued*)
2,6-Dichloro-3-hydroxybenzaldehyde
3,5-Dichloro-2-hydroxybenzaldehyde
3,5-Dichloro-4-hydroxybenzaldehyde
5,6-Dichloro-*o*-toluquinone
5,6-Dichloro-*p*-toluquinone

$C_7H_4Cl_2O_3$
2,4-Dichloro-3-hydroxybenzoic Acid
2,6-Dichloro-3-hydroxybenzoic Acid
3,5-Dichloro-2-hydroxybenzoic Acid
3,5-Dichloro-4-hydroxybenzoic Acid
3,6-Dichloro-2-hydroxybenzoic Acid

$C_7H_4Cl_2O_3S$
o-Sulphobenzoic Acid, *Dichloride*
m-Sulphobenzoic Acid, *Dichloride*
p-Sulphobenzoic Acid, *Dichloride*

$C_7H_4Cl_2O_4$
o-Chlorobenzoic Acid, *Chloride*
m-Chlorobenzoic Acid, *Chloride*
p-Chlorobenzoic Acid, *Chloride*

$C_7H_4Cl_2O_5$
2,6-Dichloro-3,4,5-trihydroxybenzoic Acid

$C_7H_4Cl_2O_6S_2$
3,5-Disulphobenzoic Acid, *Dichloride*

$C_7H_4Cl_3NO$
2,3,5-Trichlorobenzoic Acid, *Amide*
2,4,5-Trichlorobenzoic Acid, *Amide*
2,4,6-Trichlorobenzoic Acid, *Amide*
3,4,5-Trichlorobenzoic Acid, *Amide*

$C_7H_4Cl_4$
α,α,α,2-Tetrachlorotoluene
α,α,α,3-Tetrachlorotoluene
α,α,α,4-Tetrachlorotoluene
α,α,2,5-Tetrachlorotoluene
α,α,2,6-Tetrachlorotoluene
α,α,3,4-Tetrachlorotoluene
α,α,3,5-Tetrachlorotoluene
2,3,4,5-Tetrachlorotoluene
2,3,4,6-Tetrachlorotoluene
2,3,5,6-Tetrachlorotoluene

$C_7H_4Cl_4O$
Tetrachloro-*o*-cresol
Tetrachloro-*m*-cresol
Tetrachloro-*p*-cresol
2,3,4,6-Tetrachlorophenol, *Me ether*

$C_7H_4Cl_4O_2$
Drosophilin A
Tetrachlorocatechol, *Mono-Me ether*

C_7H_4FN
p-Fluorobenzoic Acid, *Nitrile*

$C_7H_4FNO_4$
2-Fluoro-3-nitrobenzoic Acid
2-Fluoro-4-nitrobenzoic Acid
2-Fluoro-5-nitrobenzoic Acid
2-Fluoro-6-nitrobenzoic Acid
4-Fluoro-2-nitrobenzoic Acid
4-Fluoro-3-nitrobenzoic Acid
5-Fluoro-2-nitrobenzoic Acid

$C_7H_4FeO_3$
Cyclobutadieneiron Tricarbonyl†

C_7H_4IN
o-Iodobenzoic Acid, *Nitrile*

$C_7H_4INO_3$
4-Iodo-2-nitrobenzaldehyde
4-Iodo-3-nitrobenzaldehyde

$C_7H_4INO_4$
2-Iodo-3-nitrobenzoic Acid
2-Iodo-4-nitrobenzoic Acid
2-Iodo-5-nitrobenzoic Acid
2-Iodo-6-nitrobenzoic Acid
3-Iodo-2-nitrobenzoic Acid
3-Iodo-4-nitrobenzoic Acid
3-Iodo-5-nitrobenzoic Acid
4-Iodo-2-nitrobenzoic Acid
4-Iodo-3-nitrobenzoic Acid
5-Iodo-2-nitrobenzoic Acid

$C_7H_4I_2O_2$
2,3-Di-iodobenzoic Acid
2,4-Di-iodobenzoic Acid
2,5-Di-iodobenzoic Acid
3,4-Di-iodobenzoic Acid
3,5-Di-iodobenzoic Acid

$C_7H_4I_2O_3$
2-Hydroxy-3,5-di-iodobenzoic Acid
4-Hydroxy-3,5-di-iodobenzoic Acid

$C_7H_4I_3NO_2$
3-Amino-2,4,6-tri-iodobenzoic Acid

$C_7H_4I_4$
2,3,4,5-Tetraiodotoluene
2,3,4,6-Tetraiodotoluene
2,3,5,6-Tetraiodotoluene

$C_7H_4N_2O_2$
o-Nitrobenzonitrile
m-Nitrobenzonitrile
p-Nitrobenzonitrile
Pyridine-2,3-dicarboxylic Acid, *Imide*
Pyridine-2,3-dicarboxylic Acid, 3-*Nitrile*

$C_7H_4N_2O_2S$
5-Nitrobenzothiazole
6-Nitrobenzothiazole
7-Nitrobenzothiazole
o-Nitrophenyl isothiocyanate
m-Nitrophenyl isothiocyanate
p-Nitrophenyl isothiocyanate

$C_7H_4N_2O_2S_2$
2-Mercapto-6-nitrobenzothiazole

$C_7H_4N_2O_3$
2-Hydroxy-3-nitrobenzoic Acid, *Nitrile*
2-Hydroxy-4-nitrobenzoic Acid, *Nitrile*
2-Hydroxy-5-nitrobenzoic Acid, *Nitrile*
2-Hydroxy-6-nitrobenzoic Acid, *Nitrile*
o-Nitrophenyl isocyanate
m-Nitrophenyl isocyanate
p-Nitrophenyl isocyanate

$C_7H_4N_2O_4$
2,4-Dihydroxy-5-nitrobenzoic Acid, *Nitrile*

$C_7H_4N_2O_5$
2,4-Dinitrobenzaldehyde
2,6-Dinitrobenzaldehyde
3,4-Dinitrobenzaldehyde
3,5-Dinitrobenzaldehyde

$C_7H_4N_2O_5S$
4-Nitro-2-sulphobenzoic Acid, *Nitrile*

$C_7H_4N_2O_6$
2,3-Dinitrobenzoic Acid
2,4-Dinitrobenzoic Acid
2,5-Dinitrobenzoic Acid
2,6-Dinitrobenzoic Acid
3,4-Dinitrobenzoic Acid
3,5-Dinitrobenzoic Acid
2-Hydroxy-3,5-dinitrobenzaldehyde
3-Hydroxy-2,6-dinitrobenzaldehyde
3-Hydroxy-4,6-dinitrobenzaldehyde
4-Hydroxy-3,5-dinitrobenzaldehyde

$C_7H_4N_2O_7$
2-Hydroxy-3,5-dinitrobenzoic Acid
2-Hydroxy-4,5-dinitrobenzoic Acid
2-Hydroxy-5,6-dinitrobenzoic Acid
3-Hydroxy-2,4-dinitrobenzoic Acid
4-Hydroxy-3,5-dinitrobenzoic Acid
5-Hydroxy-2,4-dinitrobenzoic Acid

$C_7H_4N_3O_5$
5-Chloro-2,4-dinitrobenzoic Acid, *Amide*

$C_7H_4N_4O_3$
o-Nitrobenzoic Acid, *Azide*
m-Nitrobenzoic Acid, *Azide*
p-Nitrobenzoic Acid, *Azide*

$C_7H_4N_4O_4$
2-Amino-3,5-dinitrobenzoic Acid, *Nitrile*
4,6-Dinitroindazole
5,7-Dinitroindazole

$C_7H_4N_4O_7$
2,4,6-Trinitrobenzoic Acid, *Amide*

$C_7H_4N_4O_8$
2,3,4,6-Tetranitrotoluene

$C_7H_4N_4O_9$
2,4,5,6-Tetranitro-*m*-cresol
2,3,4,6-Tetranitrophenol, *Me ether*
2,3,5,6-Tetranitrophenol, *Me ether*

$C_7H_4N_4O_{10}$
Tetranitroresorcinol, *Mono-Me ether*

$C_7H_4OS_2$
Benzo[*c*]dithiol-3-one

$C_7H_4O_3$
2-Furylpropiolic Acid
o-Phenylene carbonate†

$C_7H_4O_4S$
o-Sulphobenzoic Acid, *Anhydride*

$C_7H_4O_6$
Chelidonic Acid

$C_7H_4O_7$
Furan-2,3,4-tricarboxylic Acid
Furan-2,3,5-tricarboxylic Acid
3-Hydroxy-4-oxo-4*H*-pyran-2,6-dicarboxylic Acid

$C_7H_4S_3$
1,2-Benzodithiol-3-thione

$C_7H_5BrN_2$
3-Bromoindazole
5-Bromoindazole

$C_7H_5BrN_2O_3$
2-Bromo-5-nitrobenzoic Acid, *Amide*
4-Bromo-3-nitrobenzoic Acid, *Amide*

$C_7H_5BrN_2O_4$
α-Bromo-2,4-dinitrotoluene
α-Bromo-2,6-dinitrotoluene
α-Bromo-3,5-dinitrotoluene
2-Bromo-3,5-dinitrotoluene
2-Bromo-4,5-dinitrotoluene
3-Bromo-2,6-dinitrotoluene
4-Bromo-2,3-dinitrotoluene
4-Bromo-2,6-dinitrotoluene
4-Bromo-3,5-dinitrotoluene
5-Bromo-2,4-dinitrotoluene

$C_7H_5BrN_2O_5$
2-Bromo-4,6-dinitrophenol, *Me ether*
4-Bromo-2,6-dinitrophenol, *Me ether*
5-Bromo-2,4-dinitrophenol, *Me ether*

C_7H_5BrO
Benzoyl bromide
o-Bromobenzaldehyde
m-Bromobenzaldhehyde
p-Bromobenzaldehyde

$C_7H_5BrO_2$
o-Bromobenzoic Acid
m-Bromobenzoic Acid
p-Bromobenzoic Acid
2-Bromo-3-hydroxybenzaldehyde
2-Bromo-4-hydroxybenzaldehyde
2-Bromo-5-hydroxybenzaldehyde
3-Bromo-2-hydroxybenzaldehyde
3-Bromo-4-hydroxybenzaldehyde
4-Bromo-2-hydroxybenzaldehyde
4-Bromo-3-hydroxybenzaldehyde
5-Bromo-2-hydroxybenzaldehyde
2-Bromo-5-methyl-*p*-benzoquinone
2-Bromo-6-methyl-*p*-benzoquinone

$C_7H_5BrO_3$
2-Bromo-3,4-dihydroxybenzaldehyde
2-Bromo-4,5-dihydroxybenzaldehyde
3-Bromo-4,5-dihydroxybenzaldehyde
5-Bromo-2,3-dihydroxybenzaldehyde
2-Bromo-3-hydroxybenzoic Acid
2-Bromo-4-hydroxybenzoic Acid
2-Bromo-5-hydroxybenzoic Acid
3-Bromo-2-hydroxybenzoic Acid
3-Bromo-4-hydroxybenzoic Acid
3-Bromo-5-hydroxybenzoic Acid
4-Bromo-2-hydroxybenzoic Acid
4-Bromo-3-hydroxybenzoic Acid
5-Bromo-2-hydroxybenzoic Acid

$C_7H_5BrO_4$
3-Bromo-2,4-dihydroxybenzoic Acid
3-Bromo-4,5-dihydroxybenzoic Acid
4-Bromo-3,5-dihydroxybenzoic Acid
5-Bromo-2,3-dihydroxybenzoic Acid
5-Bromo-2,4-dihydroxybenzoic Acid

$C_7H_5Br_2NO$
2-Amino-3,5-dibromobenzaldehyde
4-Amino-3,5-dibromobenzaldehyde

$C_7H_5Br_2NO_2$
2-Amino-3,5-dibromobenzoic Acid
2-Amino-4,5-dibromobenzoic Acid

$C_7H_5Br_2NO_2$ *(continued)*
4-Amino-3,5-dibromobenzoic Acid
3,5-Dibromo-2-hydroxybenzoic Acid, *Amide*
α,α-Dibromo-2-nitrotoluene
α,α-Dibromo-3-nitrotoluene
α,α-Dibromo-4-nitrotoluene
2,3-Dibromo-5-nitrotoluene
2,4-Dibromo-5-nitrotoluene
2,5-Dibromo-3-nitrotoluene
2,5-Dibromo-4-nitrotoluene
2,6-Dibromo-3-nitrotoluene
2,6-Dibromo-4-nitrotoluene
3,4-Dibromo-5-nitrotoluene
3,5-Dibromo-2-nitrotoluene
3,5-Dibromo-4-nitrotoluene
4,5-Dibromo-2-nitrotoluene

$C_7H_5Br_2NO_3$
3,4-Dibromo-6-nitro-*o*-cresol
5,6-Dibromo-4-nitro-*o*-cresol
2,4-Dibromo-6-nitro-*m*-cresol
2,6-Dibromo-4-nitro-*m*-cresol
2,5-Dibromo-6-nitro-*p*-cresol
2,4-Dibromo-6-nitrophenol, *Me ether*
2,6-Dibromo-4-nitrophenol, *Me ether*

$C_7H_5Br_2NO_4$
2,6-Dibromo-3,4,5-trihydroxybenzoic Acid, *Amide*

$C_7H_5Br_3$
α,α,4-Tribromotoluene
2,3,4-Tribromotoluene
2,3,5-Tribromotoluene
2,3,6-Tribromotoluene
2,4,5-Tribromotoluene
2,4,6-Tribromotoluene
3,4,5-Tribromotoluene

$C_7H_5Br_3O$
3,4,5-Tribromo-*o*-cresol
3,4,6-Tribromo-*o*-cresol
4,5,6-Tribromo-*o*-cresol
2,4,6-Tribromo-*m*-cresol
2,3,5-Tribromo-*p*-cresol
2,3,6-Tribromo-*p*-cresol
2,3,4-Tribromophenol, *Me ether*
2,3,5-Tribromophenol, *Me ether*
2,4,5-Tribromophenol, *Me ether*
2,4,6-Tribromophenol, *Me ether*

$C_7H_5Br_3O_2$
2,4,6-Tribromoresorcinol, *Me ether*

$C_7H_5Br_3O_3$
Tribromophloroglucinol, *Me ether*

$C_7H_5Br_4N$
2,4,5,6-Tetrabromo-*m*-toluidine
2,3,5,6-Tetrabromo-*p*-toluidine

$C_7H_5ClHgO_2$
Mercuri-*o*-carboxyphenyl chloride
Mercuri-*m*-carboxyphenyl chloride
Mercuri-*p*-carboxyphenyl chloride

C_7H_5ClINO
1-Chloro-1,3-dihydro-3-oxo-1,2-benziodazole†

$C_7H_5ClN_2$
3-Chloroindazole
4-Chloroindazole
5-Chloroindazole

$C_7H_5ClN_2O$
2-Amino-5-chlorobenzoxazole†

$C_7H_5ClN_2O_3$
2-Chloro-4-nitrobenzoic Acid, *Amide*
2-Chloro-5-nitrobenzoic Acid. *Amide*
2-Chloro-6-nitrobenzoic Acid, *Amide*
4-Chloro-2-nitrobenzoic Acid, *Amide*
4-Chloro-3-nitrobenzoic Acid, *Amide*
5-Chloro-2-nitrobenzoic Acid, *Amide*
o-Nitrophenylcarbamic Acid, *Chloride*
m-Nitrophenylcarbamic Acid, *Chloride*
p-Nitrophenylcarbamic Acid, *Chloride*

$C_7H_5ClN_2O_4$
α-Chloro-2,4-dinitrotoluene
2-Chloro-α,6-dinitrotoluene
2 Chloro-3,4-dinitrotoluene
2-Chloro-3,5-dinitrotoluene
2-Chloro-3,6-dinitrotoluene
2-Chloro-4,5-dinitrotoluene
2-Chloro-4,6-dinitrotoluene
3-Chloro-2,4-dinitrotoluene
3-Chloro-2,6-dinitrotoluene
4-Chloro-α,2-dinitrotoluene
4-Chloro-2,3-dinitrotoluene
4-Chloro-2,5-dinitrotoluene
4-Chloro-2,6-dinitrotoluene
4-Chloro-2,6-dinitrotoluene
4-Chloro-3,5-dinitrotoluene
5-Chloro-2,4-dinitrotoluene
6-Chloro-2,3-dinitrotoluene
3-Chloro-6-hydroxy-5-nitrobenzoic Acid, *Amide*

$C_7H_5ClN_2O_5$
2-Chloro-3,5-dinitrophenol, *Me ether*
5-Chloro-2,4-dinitrophenol, *Me ether*
4-Chloro-2,6-dinitrophenol, *Me ether*

C_7H_5ClO
Benzoyl chloride
o-Chlorobenzaldehyde
m-Chlorobenzaldehyde
p-Chlorobenzaldehyde
2-Chlorotropone†

$C_7H_5ClO_2$
o-Chlorobenzoic Acid
m-Chlorobenzoic Acid
p-Chlorobenzoic Acid
Chloroformic Acid, *Phenyl ester*
2-Chloro-3-hydroxybenzaldehyde
2-Chloro-4-hydroxybenzaldehyde
2-Chloro-5-hydroxybenzaldehyde
3-Chloro-2-hydroxybenzaldehyde
3-Chloro-4-hydroxybenzaldehyde
4-Chloro-2-hydroxybenzaldehyde
4-Chloro-3-hydroxybenzaldehyde
5-Chloro-2-hydroxybenzaldehyde
2-Chloro-3-methyl-1,4-benzoquinone
2-Chloro-5-methyl-1,4-benzoquinone
2-Chloro-6-methyl-1,4-benzoquinone
Salicylic Acid, *Chloride*

$C_7H_5ClO_3$
5-Chloro-2,4-dihydroxybenzaldehyde
2-Chloro-3-hydroxybenzoic Acid
2-Chloro-4-hydroxybenzoic Acid
2-Chloro-5-hydroxybenzoic Acid

$C_7H_5ClO_3$ (*continued*)
2-Chloro-6-hydroxybenzoic Acid
3-Chloro-2-hydroxybenzoic Acid
3-Chloro-4-hydroxybenzoic Acid
4-Chloro-2-hydroxybenzoic Acid
4-Chloro-3-hydroxybenzoic Acid
5-Chloro-2-hydroxybenzoic Acid
2,4-Dihydroxybenzoic Acid, *Chloride*

$C_7H_5ClO_3S$
o-Formylbenzenesulphonic Acid, *Chloride*
p-Formylbenzenesulphonic Acid, *Chloride*

$C_7H_5ClO_4$
3-Chloro-2,4-dihydroxybenzoic Acid
5-Chloro-2,4-dihydroxybenzoic Acid
Furan-2,4-dicarboxylic Acid, 4-*Me Ester, Chloride*
Gallic Acid, *Chloride*

$C_7H_5ClO_4S$
m-Sulphobenzoic Acid, 3-*Chloride*

$C_7H_5ClO_5S$
2-Hydroxy-5-sulphobenzoic Acid, *Sulphonchloride*

$C_7H_5Cl_2N$
Phenyliminophosgene

$C_7H_5Cl_2NO$
2-Amino-3,5-dichlorobenzaldehyde
2-Amino-3,6-dichlorobenzaldehyde
3-Amino-2,5-dichlorobenzaldehyde
3-Amino-2,6-dichlorobenzaldehyde
4-Amino-2,6-dichlorobenzaldehyde
4-Amino-3,5-dichlorobenzaldehyde
5-Amino-2,4-dichlorobenzaldehyde
2,5-Dichlorobenzoic Acid, *Amide*
2,6-Dichlorobenzoic Acid, *Amide*
3,4-Dichlorobenzoic Acid, *Amide*

$C_7H_5Cl_2NO_2$
2-Amino-3,4-dichlorobenzoic Acid
2-Amino-3,5-dichlorobenzoic Acid
2-Amino-3,6-dichlorobenzoic Acid
2-Amino-4,5-dichlorobenzoic Acid
4-Amino-3,5-dichlorobenzoic Acid
6-Amino-2,3-dichlorobenzoic Acid
3,5-Dichloro-2-hydroxybenzoic Acid, *Amide*
α,α-Dichloro-2-nitrotoluene
α,α-Dichloro-3-nitrotoluene
α,α-Dichloro-4-nitrotoluene
2,3-Dichloro-4-nitrotoluene
2,3-Dichloro-5-nitrotoluene
2,4-Dichloro-5-nitrotoluene
2,4-Dichloro-6-nitrotoluene
2,5-Dichloro-3-nitrotoluene
2,5-Dichloro-4-nitrotoluene
2,6-Dichloro-3-nitrotoluene
2,6-Dichloro-4-nitrotoluene
3,4-Dichloro-5-nitrotoluene
3,5-Dichloro-2-nitrotoluene
4,5-Dichloro-2-nitrotoluene
3,5-Dichloropyridine-2-carboxylic Acid, *Me ester*
4,6-Dichloropyridine-2-carboxylic Acid, *Me ester*
2,6-Dichloropyridine-4-carboxylic Acid, *Me ester*

$C_7H_5Cl_2NO_3$
3,6-Dichloro-4-nitro-*o*-cresol
4,5-Dichloro-6-nitro-*o*-cresol
2,4-Dichloro-6-nitro-*m*-cresol
2,6-Dichloro-4-nitro-*m*-cresol
2,3-Dichloro-6-nitrophenol, *Me ether*
2,4-Dichloro-6-nitrophenol, *Me ether*
2,5-Dichloro-4-nitrophenol, *Me ether*
2,6-Dichloro-4-nitrophenol, *Me ether*
3,4-Dichloro-2-nitrophenol, *Me ether*
3,5-Dichloro-2-nitrophenol, *Me ether*
3,5-Dichloro-4-nitrophenol, *Me ether*
3,6-Dichloro-2-nitrophenol, *Me ether*
4,5-Dichloro-2-nitrophenol, *Me ether*

$C_7H_5Cl_2NO_4S$
N-Dichloro-*p*-sulphonamidobenzoic Acid

$C_7H_5Cl_2NS$
2,6-Dichlorothiobenzamide†

$C_7H_5Cl_3$
α,α,α-Trichlorotoluene
α,α,2-Trichlorotoluene
α,α,4-Trichlorotoluene
2,3,4-Trichlorotoluene
2,3,5-Trichlorotoluene
2,3,6-Trichlorotoluene
2,4,5-Trichlorotoluene
2,4,6-Trichlorotoluene
3,4,5-Trichlorotoluene

$C_7H_5Cl_3Hg$
Phenyltrichloromethylmercury†

$C_7H_5Cl_3O$
3,4,6-Trichloro-*o*-cresol
4,5,6-Trichloro-*o*-cresol
2,4,6-Trichloro-*m*-cresol
2,3,5-Trichloro-*p*-cresol
2,3,6-Trichloro-*p*-cresol
2,3,4-Trichlorophenol, *Me ether*
2,3,5-Trichlorophenol, *Me ether*
2,4,5-Trichlorophenol, *Me ether*
2,4,6-Trichlorophenol, *Me ether*
3,4,5-Trichlorophenol, *Me ether*

$C_7H_5Cl_3O_2$
3,4,5-Trichlorocatechol, 1-*Me ether*
3,4,5-Trichlorocatechol, 2-*Me ether*

$C_7H_5Cl_3O_6S_3$
Toluene-2,4,6-trisulphonic Acid, *Trichloride*

$C_7H_5FN_2O_3$
2-Fluoro-4-nitrobenzoic Acid, *Amide*
4-Fluoro-3-nitrobenzoic Acid, *Amide*

$C_7H_5FN_2O_5$
2-Fluoro-4,6-dinitrophenol, *Me ether*
4-Fluoro-2,6-dinitrophenol, *Me ether*

C_7H_5FO
Benzoyl fluoride
o-Fluorobenzaldehyde
m-Fluorobenzaldehyde
p-Fluorobenzaldehyde

$C_7H_5FO_2$
o-Fluorobenzoic Acid
m-Fluorobenzoic Acid
p-Fluorobenzoic Acid

$C_7H_5F_3$
α,α,α-Trifluorotoluene

C_7H_5HgN
Mercuri-phenyl cyanide

C_7H_5HgNS
Mercuri-phenyl thiocyanate

$C_7H_5IN_2O_3$
2-Iodo-4-nitrobenzoic Acid, *Amide*
2-Iodo-5-nitrobenzoic Acid, *Amide*

$C_7H_5IN_2O_4$
α-Iodo-2,4-dinitrotoluene
α-Iodo-2,6-dinitrotoluene
2-Iodo-3,5-dinitrotoluene
3-Iodo-2,4-dinitrotoluene
3-Iodo-2,6-dinitrotoluene
4-Iodo-3,5-dinitrotoluene
5-Iodo-2,4-dinitrotoluene

$C_7H_5IN_2O_5$
2-Iodo-3,5-dinitrophenol, *Me ether*
2-Iodo-4,5-dinitrophenol, *Me ether*
4-Iodo-3,5-dinitrophenol, *Me ether*
5-Iodo-2,4-dinitrophenol, *Me ether*

C_7H_5IO
Benzoyl iodide
o-Iodobenzaldehyde
m-Iodobenzaldehyde
p-Iodobenzaldehyde

$C_7H_5IO_2$
2-Hydroxy-4-iodobenzaldehyde
2-Hydroxy-5-iodobenzaldehyde
3-Hydroxy-2-iodobenzaldehyde
3-Hydroxy-6-iodobenzaldehyde
4-Hydroxy-2-iodobenzaldehyde
4-Hydroxy-3-iodobenzaldehyde
o-Iodobenzoic Acid
m-Iodobenzoic Acid
p-Iodobenzoic Acid
2-Iodo-6-methylbenzoquinone

$C_7H_5IO_3$
1,3-Dihydro-1-hydroxy-3-oxo-1,2-benziodoxole†
2-Hydroxy-3-iodobenzoic Acid
2-Hydroxy-4-iodobenzoic Acid
2-Hydroxy-5-iodobenzoic Acid
3-Hydroxy-2-iodobenzoic Acid
3-Hydroxy-4-iodobenzoic Acid
4-Hydroxy-2-iodobenzoic Acid
4-Hydroxy-3-iodobenzoic Acid
5-Hydroxy-2-iodobenzoic Acid

$C_7H_5I_2NO_2$
2-Amino-3,5-di-iodobenzoic Acid
2-Amino-4,5-di-iodobenzoic Acid
4-Amino-3,5-di-iodobenzoic Acid

$C_7H_5I_2NO_3$
1,4-Dihydro-3,5-di-iodo-4-oxo-1-pyridylacetic Acid

$C_7H_5I_3$
2,3,4-Tri-iodotoluene
2,3,5-Tri-iodotoluene
2,3,6-Tri-iodotoluene
2,4,5-Tri-iodotoluene
2,4,6-Tri-iodotoluene
3,4,5-Tri-iodotoluene

$C_7H_5I_3N_2O$
3-Amino-2,4,6-tri-iodobenzoic Acid, *Amide*

$C_7H_5I_3O$
2,4,6-Tri-iodophenol, *Me ether*

C_7H_5N
Benzonitrile
Phenyl isocyanide

C_7H_5NO
Anthranil
1,2-Benzisoxazole
Benzoxazole
3-Hydroxybenzoic Acid, *Nitrile*
4-Hydroxybenzoic Acid, *Nitrile*
Phenylisocyanate
Salicylic Acid, *Nitrile*

C_7H_5NOS
Benzothiazolone
2-Hydroxybenzothiazole
5-Hydroxybenzothiazole
2-Mercaptobenzoxazole

$C_7H_5NO_2$
2,1-Benzisoxazolone
Benzoxazolin-2-one
2,4-Dihydroxybenzoic Acid, *Nitrile*
2,5-Dihydroxybenzoic Acid, *Nitrile*
3,4-Dihydroxybenzoic Acid, *Nitrile*
o-Nitrosobenzaldehyde
m-Nitrosobenzaldehyde
p-Nitrosobenzaldehyde

$C_7H_5NO_3$
Gallic Acid, *Nitrile*
4-Methylfuran-2,3-dicarboxylic Acid, 2-*Nitrile*
o-Nitrobenzaldehyde
m-Nitrobenzaldehyde
p-Nitrobenzaldehyde
o-Nitrosobenzoic Acid
m-Nitrosobenzoic Acid
p-Nitrosobenzoic Acid

$C_7H_5NO_3S$
Saccharin
o-Sulphobenzoic Acid, *Nitrile*

$C_7H_5NO_3S_3$
2-Mercaptobenzothiazole-5-sulphonic Acid

$C_7H_5NO_4$
2-Hydroxy-3-nitrobenzaldehyde
2-Hydroxy-4-nitrobenzaldehyde
2-Hydroxy-5-nitrobenzaldehyde
2-Hydroxy-6-nitrobenzaldehyde
3-Hydroxy-2-nitrobenzaldehyde
3-Hydroxy-4-nitrobenzaldehyde
4-Hydroxy-2-nitrobenzaldehyde
4-Hydroxy-3-nitrobenzaldehyde
5-Hydroxy-2-nitrobenzaldehyde
2-Hydroxy-5-nitrosobenzoic Acid
2-Methyl-3-nitrobenzoquinone
o-Nitrobenzoic Acid
m-Nitrobenzoic Acid
p-Nitrobenzoic Acid
Pyridine-2,3-dicarboxylic Acid
Pyridine-2,4-dicarboxylic Acid
Pyridine-2,5-dicarboxylic Acid
Pyridine-2,6-dicarboxylic Acid

$C_7H_5NO_4$ (*continued*)
Pyridine-3,4-dicarboxylic Acid
Pyridine-3,5-dicarboxylic Acid

$C_7H_5NO_5$
2,3-Dihydroxy-5-nitrobenzaldehyde
2,3-Dihydroxy-6-nitrobenzaldehyde
2,4-Dihydroxy-3-nitrobenzaldehyde
2,4-Dihydroxy-5-nitrobenzaldehyde
2,5-Dihydroxy-3-nitrobenzaldehyde
2,6-Dihydroxy-3-nitrobenzaldehyde
3,4-Dihydroxy-2-nitrobenzaldehyde
3,4-Dihydroxy-5-nitrobenzaldehyde
4,5-Dihydroxy-2-nitrobenzaldehyde
2-Hydroxy-3-nitrobenzoic Acid
2-Hydroxy-4-nitrobenzoic Acid
2-Hydroxy-5-nitrobenzoic Acid
2-Hydroxy-6-nitrobenzoic Acid
3-Hydroxy-2-nitrobenzoic Acid
3-Hydroxy-4-nitrobenzoic Acid
3-Hydroxy-5-nitrobenzoic Acid
4-Hydroxy-2-nitrobenzoic Acid
4-Hydroxy-3-nitrobenzoic Acid
5-Hydroxy-2-nitrobenzoic Acid
4-Hydroxypyridine-2,6-dicarboxylic Acid
5-Hydroxypyridine-3,4-dicarboxylic Acid
6-Hydroxypyridine-3,4-dicarboxylic Acid

$C_7H_5NO_6$
2,5-Dihydroxy-3-methyl-6-nitro-1,4-benzoquinone
2,3-Dihydroxy-4-nitrobenzoic Acid
2,3-Dihydroxy-5-nitrobenzoic Acid
2,3-Dihydroxy-6-nitrobenzoic Acid
2,4-Dihydroxy-5-nitrobenzoic Acid
2,5-Dihydroxy-3-nitrobenzoic Acid
2,5-Dihydroxy-4-nitrobenzoic Acid
2,5-Dihydroxy-6-nitrobenzoic Acid
2,6-Dihydroxy-3-nitrobenzoic Acid
3,5-Dihydroxy-4-nitrobenzoic Acid
4,5-Dihydroxy-2-nitrobenzoic Acid
4,6-Dihydroxypyridine-2,3-dicarboxylic Acid
2,6-Dihydroxypyridine-3,4-dicarboxylic Acid
2,6-Dihydroxypyridine-3,5-dicarboxylic Acid
3-Hydroxy-4-oxo-4*H*-pyran-2,6-dicarboxylic Acid, *Monoamide*

$C_7H_5NO_7$
2-Nitrogallic Acid

$C_7H_5NO_7S$
2-Nitro-4-sulphobenzoic Acid
3-Nitro-4-sulphobenzoic Acid
4-Nitro-2-sulphobenzoic Acid
5-Nitro-2-sulphobenzoic Acid
5-Nitro-3-sulphobenzoic Acid

C_7H_5NS
Benzisothiazole
2,1-Benzisothiazole†
Benzothiazole
Phenyl isothiocyanate
Thieno[2,3-*c*]pyridine
Thieno[3,2-*c*]pyridine
Thiocyanic Acid, *Phenyl ester*

$C_7H_5NS_2$
2-Mercaptobenzothiazole

C_7H_5NSe
Benzoselenazole

$C_7H_5N_3$
1,2,4-Benzotriazine

$C_7H_5N_3O$
o-Azidobenzaldehyde
p-Azidobenzaldehyde
Benzoyl azide

$C_7H_5N_3O_2$
4-Amino-3-nitrobenzoic Acid, *Nitrile*
2-Nitrobenzimidazole†
4-Nitrobenzimidazole
5-Nitrobenzimidazole
4-Nitroindazole
5-Nitroindazole
6-Nitroindazole
7-Nitroindazole
o-Nitrophenylcyanamide
m-Nitrophenylcyanamide
p-Nitrophenylcyamide

$C_7H_5N_3O_2S$
2-Amino-4-nitrobenzothiazole
2-Amino-6-nitrobenzothiazole

$C_7H_5N_3O_5$
2,4-Dinitrobenzoic Acid, *Amide*
3,4-Dinitrobenzoic Acid, *Amide*
3,5-Dinitrobenzoic Acid, *Amide*

$C_7H_5N_3O_6$
2-Amino-3,5-dinitrobenzoic Acid
2-Amino-4,5-dinitrobenzoic Acid
3-Amino-2,4-dinitrobenzoic Acid
4-Amino-3,5-dinitrobenzoic Acid
5-Amino-2,4-dinitrobenzoic Acid
2-Hydroxy-3,5-dinitrobenzoic Acid, *Amide*
2,3,4-Trinitrotoluene
2,3,5-Trinitrotoluene
2,3,6-Trinitrotoluene
2,4,5-Trinitrotoluene
2,4,6-Trinitrotoluene
3,4,5-Trinitrotoluene

$C_7H_5N_3O_7$
Picric Acid, *Me ether*
2,3,4-Trinitroanisole
2,3,5-Trinitroanisole
2,4,5-Trinitroanisole
2,4,6-Trinitroanisole
4,5,6-Trinitro-*o*-cresol
2,4,6-Trinitro-*m*-cresol

$C_7H_5N_3O_8$
2-Methoxy-3,5,6-trinitrophenol

$C_7H_5N_4S$
Tetrazole [5,1-*b*]benzothiazole

$C_7H_5N_5$
4-Amino-5-cyanopyrrolo[2,3-*d*]pyrimidine†

$C_7H_5N_5O_3$
2-Amino-4-hydroxypteridine-6-carboxylic Acid

$C_7H_5N_5O_8$
Tetryl

C_7H_6
Benzocyclopropene†

C_7H_6BrCl
α-Bromo-2-chlorotoluene
α-Bromo-3-chlorotoluene
α-Bromo-4-chlorotoluene
2-Bromo-α-chlorotoluene
2-Bromo-3-chlorotoluene
2-Bromo-4-chlorotoluene
2-Bromo-5-chlorotoluene
2-Bromo-6-chlorotoluene
3-Bromo-α-chlorotoluene
3-Bromo-2-chlorotoluene
3-Bromo-4-chlorotoluene
3-Bromo-5-chlorotoluene
4-Bromo-α-chlorotoluene
4-Bromo-2-chlorotoluene
4-Bromo-3-chlorotoluene
5-Bromo-2-chlorotoluene

$C_7H_6BrClO_2S$
6-Chlorotoluene-3-sulphonic Acid, *Bromide*

C_7H_6BrI
α-Bromo-2-iodotoluene
α-Bromo-3-iodotoluene
α-Bromo-4-iodotoluene
2-Bromo-α-iodotoluene
3-Bromo-α-iodotoluene
4-Bromo-α-iodotoluene
2-Bromo-4-iodotoluene
2-Bromo-6-iodotoluene
3-Bromo-4-iodotoluene
3-Bromo-5-iodotoluene
4-Bromo-2-iodotoluene
4-Bromo-3-iodotoluene
5-Bromo-2-iodotoluene

C_7H_6BrNO
2-Amino-4-bromobenzaldehyde
2-Amino-5-bromobenzaldehyde
4-Amino-2-bromobenzaldehyde
o-Bromoaniline, N-*Formyl*
p-Bromoaniline, N-*Formyl*
N-Bromobenzamide
o-Bromobenzamide
m-Bromobenzamide
p-Bromobenzamide

$C_7H_6BrNO_2$
2-Amino-3-bromobenzoic Acid
2-Amino-4-bromobenzoic Acid
2-Amino-5-bromobenzoic Acid
2-Amino-6-bromobenzoic Acid
3-Amino-4-bromobenzoic Acid
3-Amino-5-bromobenzoic Acid
4-Amino-2-bromobenzoic Acid
4-Amino-3-bromobenzoic Acid
5-Amino-2-bromobenzoic Acid
3-Bromo-2-hydroxybenzoic Acid, *Amide*
5-Bromo-2-hydroxybenzoic Acid, *Amide*
α-Bromo-2-nitrotoluene
α-Bromo-3-nitrotoluene
α-Bromo-4-nitrotoluene
2-Bromo-3-nitrotoluene
2-Bromo-4-nitrotoluene
2-Bromo-5-nitrotoluene
2-Bromo-6-nitrotoluene
3-Bromo-2-nitrotoluene
3-Bromo-4-nitrotoluene
3-Bromo-5-nitrotoluene
4-Bromo-2-nitrotoluene
4-Bromo-3-nitrotoluene
5-Bromo-2-nitrotoluene
5-Bromopyridine-3-carboxylic Acid, *Me ester*

$C_7H_6BrNO_3$
α-Bromo-4-nitro-*o*-cresol
α-Bromo-5-nitro-*o*-cresol
α-Bromo-2-nitro-*p*-cresol
3-Bromo-4-nitro-*o*-cresol
3-Bromo-6-nitro-*o*-cresol
4-Bromo-6-nitro-*o*-cresol
6-Bromo-4-nitro-*o*-cresol
4-Bromo-2-nitro-*m*-cresol
4-Bromo-6-nitro-*m*-cresol
6-Bromo-4-nitro-*m*-cresol
2-Bromo-3-nitro-*p*-cresol
2-Bromo-5-nitro-*p*-cresol
2-Bromo-6-nitro-*p*-cresol
5-Bromo-2-nitro-*p*-cresol
2-Bromo-3-nitrophenol, *Me ether*
2-Bromo-4-nitrophenol, *Me ether*
2-Bromo-5-nitrophenol, *Me ether*
2-Bromo-6-nitrophenol, *Me ether*
3-Bromo-2-nitrophenol, *Me ether*
3-Bromo-5-nitrophenol, *Me ether*
4-Bromo-2-nitrophenol, *Me ether*
4-Bromo-3-nitrophenol, *Me ether*

$C_7H_6Br_2$
α,α-Dibromotoluene
α,2-Dibromotoluene
α,3-Dibromotoluene
α,4-Dibromotoluene
2,3-Dibromotoluene
2,4-Dibromotoluene
2,5-Dibromotoluene
2,6-Dibromotoluene
3,4-Dibromotoluene
3,5-Dibromotoluene

$C_7H_6Br_2N_2O$
2-Amino-3,5-dibromobenzoic Acid, *Amide*

$C_7H_6Br_2N_2O_2$
4,6-Dibromo-3-nitro-*o*-toluidine
4,6-Dibromo-5-nitro-*o*-toluidine
2,4-Dibromo-6-nitro-*m*-toluidine
2,6-Dibromo-3-nitro-*p*-toluidine

$C_7H_6Br_2O$
3,5-Dibromo-*o*-cresol
3,6-Dibromo-*o*-cresol
4,6-Dibromo-*o*-cresol
5,6-Dibromo-*o*-cresol
2,4-Dibromo-*m*-cresol
2,6-Dibromo-*m*-cresol
4,6-Dibromo-*m*-cresol
2,3-Dibromo-*p*-cresol
2,5-Dibromo-*p*-cresol
2,6-Dibromo-*p*-cresol
3,5-Dibromo-*p*-cresol
2,4-Dibromophenol, *Me ether*
2,6-Dibromophenol, *Me ether*
3,5-Dibromophenol, *Me ether*

$C_7H_6Br_2O_2$
2,4-Dibromo-5-hydroxybenzyl Alcohol

$C_7H_6Br_2O_2$ (*continued*)
2,6-Dibromo-4-hydroxybenzyl Alcohol
3,5-Dibromo-2-hydroxybenzyl Alcohol
3,5-Dibromo-4-hydroxybenzyl Alcohol

$C_7H_6Br_2O_3S$
2,5-Dibromobenzenesulphonic Acid, *Me ester*

$C_7H_6Br_2O_4S_2$
Toluene-2,4-disulphonic Acid, *Dibromide*

$C_7H_6Br_3N$
3,4,6-Tribromo-*o*-toluidine
2,4,5-Tribromo-*m*-toluidine
2,4,6-Tribromo-*m*-toluidine
4,5,6-Tribromo-*m*-toluidine
2,3,5-Tribromo-*p*-toluidine
2,3,6-Tribromo-*p*-toluidine

C_7H_6ClF
a-Chloro-4-fluorotoluene
2-Chloro-4-fluorotoluene
2-Chloro-6-fluorotoluene
4-Chloro-2-fluorotoluene

$C_7H_6ClFO_2S$
4-Fluorotoluene-2-sulphonic Acid, *Chloride*

$C_7H_6ClHgNO_2$
Mercuri-5-nitro-*o*-tolyl chloride
Mercuri-4-nitro-*m*-tolyl chloride
Mercuri-5-nitro-*m*-tolyl chloride
Mercuri-6-nitro-*m*-tolyl chloride
Mercuri-2-nitro-*p*-tolyl chloride
Mercuri-3-nitro-*p*-tolyl chloride

C_7H_6ClI
α-Chloro-2-iodotoluene
α-Chloro-3-iodotoluene
α-Chloro-4-iodotoluene
2-Chloro-3-iodotoluene
2-Chloro-4-iodotoluene
2-Chloro-5-iodotoluene
2-Chloro-6-iodotoluene
3-Chloro-2-iodotoluene
3-Chloro-4-iodotoluene
3-Chloro-5-iodotoluene
4-Chloro-α-iodotoluene
4-Chloro-2-iodotoluene
4-Chloro-3-iodotoluene
5-Chloro-2-iodotoluene

C_7H_6ClNO
2-Amino-4-chlorobenzaldehyde
4-Amino-2-chlorobenzaldehyde
5-Amino-2-chlorobenzaldehyde
3-Chloroacetylpyridine
4-Chloroacetylpyridine
N-Chlorobenzamide
o-Chlorobenzamide
m-Chlorobenzamide
p-Chlorobenzamide
N-Chloroformanilide
o-Chloroformanilide
m-Chloroformanilide
p-Chloroformanilide
2-Methyl-1,4-benzoquinone, 1-*Chloroimide*

$C_7H_6ClNO_2$
2-Amino-3-chlorobenzoic Acid
2-Amino-4-chlorobenzoic Acid
2-Amino-5-chlorobenzoic Acid
2-Amino-6-chlorobenzoic Acid
3-Amino-2-chlorobenzoic Acid
3-Amino-4-chlorobenzoic Acid
3-Amino-5-chlorobenzoic Acid
4-Amino-2-chlorobenzoic Acid
5-Amino-2-chlorobenzoic Acid
o-Chlorocarbanilic Acid
m-Chlorocarbanilic Acid
p-Chlorocarbanilic Acid
3-Chloro-4-hydroxybenzoic Acid, *Amide*
5-Chloro-2-hydroxybenzoic Acid, *Amide*
α-Chloro-2-nitrotoluene
α-Chloro-3-nitrotoluene
α-Chloro-4-nitrotoluene
2-Chloro-3-nitrotoluene
2-Chloro-4-nitrotoluene
2-Chloro-5-nitrotoluene
2-Chloro-6-nitrotoluene
3-Chloro-2-nitrotoluene
3-Chloro-4-nitrotoluene
3-Chloro-5-nitrotoluene
4-Chloro-2-nitrotoluene
4-Chloro-3-nitrotoluene
5-Chloro-2-nitrotoluene
m-Chlorophenylcarbamic Acid†
4-Chloropyridine-2-carboxylic Acid, *Me ester*
5-Chloropyridine-3-carboxylic Acid, *Me ester*
6-Chloropyridine-3-carboxylic Acid, *Me ester*
3-Chloropyridine-4-carboxylic Acid, *Me ester*

$C_7H_6ClNO_3$
3-Amino-5-chloro-2-hydroxybenzoic Acid
5-Amino-3-chloro-2-hydroxybenzoic Acid
2-Chloro-3-nitroanisole
2-Chloro-4-nitroanisole
2-Chloro-5-nitroanisole
2-Chloro-6-nitroanisole
3-Chloro-2-nitroanisole
3-Chloro-4-nitroanisole
4-Chloro-2-nitroanisole
4-Chloro-3-nitroanisole
5-Chloro-2-nitroanisole
5-Chloro-3-nitroanisole
α-Chloro-3-nitro-*o*-cresol
α-Chloro-4-nitro-*o*-cresol
3-Chloro-4-nitro-*o*-cresol
3-Chloro-6-nitro-*o*-cresol
4-Chloro-6-nitro-*o*-cresol
5-Chloro-4-nitro-*o*-cresol
5-Chloro-6-nitro-*o*-cresol
6-Chloro-3-nitro-*o*-cresol
6-Chloro-4-nitro-*o*-cresol
6-Chloro-5-nitro-*o*-cresol
2-Chloro-4-nitro-*m*-cresol
4-Chloro-6-nitro-*m*-cresol
6-Chloro-4-nitro-*m*-cresol
α-Chloro-2-nitro-*p*-cresol
5-Chloro-2-nitro-*p*-cresol
6-Chloro-2-nitro-*p*-cresol

$C_7H_6ClNO_4$
4-Chloro-2-nitroresorcinol, 1-*Me ether*

$C_7H_6ClNO_4S$
3-Nitrotoluene-α-sulphonic Acid, *Chloride*
4-Nitrotoluene-α-sulphonic Acid, *Chloride*

$C_7H_6ClNO_4S$ (*continued*)
4-Nitrotoluene-2-sulphonic Acid, *Chloride*
5-Nitrotoluene-2-sulphonic Acid, *Chloride*
2-Nitrotoluene-3-sulphonic Acid, *Chloride*
4-Nitrotoluene-3-sulphonic Acid, *Chloride*
6-Nitrotoluene-3-sulphonic Acid, *Chloride*
2-Nitrotoluene-4-sulphonic Acid, *Chloride*
3-Nitrotoluene-4-sulphonic Acid, *Chloride*
Thiazolo[3,2-*a*]pyridinium, *Perchlorate*†

$C_7H_6ClNO_5S$
4-Methoxy-3-nitrobenzenesulphonic Acid, *Chloride*

$C_7H_6ClN_3O_4S_2$
Chlorothiazide

$C_7H_6Cl_2$
α,α-Dichlorotoluene
α,2-Dichlorotoluene
α,3-Dichlorotoluene
α,4-Dichlorotoluene
2,3-Dichlorotoluene
2,4-Dichlorotoluene
2,5-Dichlorotoluene
2,6-Dichlorotoluene
3,4-Dichlorotoluene
3,5-Dichlorotoluene

$C_7H_6Cl_2NO_2$
2-Amino-3,5-dichlorobenzoic Acid, *Amide*

$C_7H_6Cl_2N_2$
2-Methyl-1,4-benzoquinone *Dichloroimide*

$C_7H_6Cl_2N_2O_2$
2,4-Dichloro-6-nitro-*m*-toluidine
3,5-Dichloro-2-nitro-*p*-toluidine

$C_7H_6Cl_2O$
2,4-Dichlorobenzyl Alcohol
2,5-Dichlorobenzyl Alcohol
3,4-Dichlorobenzyl Alcohol
3,5-Dichlorobenzyl Alcohol
4,5-Dichloro-*o*-cresol
4,6-Dichloro-*o*-cresol
2,4-Dichloro-*m*-cresol
2,6-Dichloro-*m*-cresol
4,6-Dichloro-*m*-cresol
2,6-Dichloro-*p*-cresol
2,3-Dichlorophenol, *Me ether*
2,4-Dichlorophenol, *Me ether*
2,5-Dichlorophenol, *Me ether*
2,6-Dichlorophenol, *Me ether*
3,4-Dichlorophenol, *Me ether*
3,5-Dichlorophenol, *Me ether*

$C_7H_6Cl_2O_2$
4,5-Dichlorocatechol, 1-*Me ether*
4,5-Dichloro-2,3-dihydroxytoluene
3,5-Dichloro-2-hydroxybenzyl Alcohol
3,5-Dichlorotoluquinol
3,6-Dichlorotoluquinol
5,6-Dichlorotoluquinol

$C_7H_6Cl_2O_2S$
4-Chlorotoluene-2-sulphonic Acid, *Chloride*
5-Chlorotoluene-2-sulphonic Acid, *Chloride*
6-Chlorotoluene-2-sulphonic Acid, *Chloride*
4-Chlorotoluene-3-sulphonic Acid, *Chloride*
6-Chlorotoluene-3-sulphonic Acid, *Chloride*
2-Chlorotoluene-4-sulphonic Acid, *Chloride*
3-Chlorotoluene-4-sulphonic Acid, *Chloride*

$C_7H_6Cl_2O_4S_2$
Toluene-2,4-disulphonic Acid, *Dichloride*
Toluene-2,5-disulphonic Acid, *Dichloride*
Toluene-2,6-disulphonic Acid, *Dichloride*
Toluene-3,4-disulphonic Acid, *Dichloride*
Toluene-3,5-disulphonic Acid, *Dichloride*

$C_7H_6Cl_3N$
2,4,6-Trichloroaniline, N-*Me*
3,5,6-Trichloro-*o*-toluidine
2,4,5-Trichloro-*m*-toluidine
2,4,6-Trichloro-*m*-toluidine
2,5,6-Trichloro-*m*-toluidine

C_7H_6FI
2-Fluoro-5-iodotoluene
2-Fluoro-6-iodotoluene
3-Fluoro-4-iodotoluene
4-Fluoro-α-iodotoluene
4-Fluoro-3-iodotoluene
5-Fluoro-2-iodotoluene

C_7H_6FNO
o-Fluorobenzoic Acid, *Amide*
m-Fluorobenzoic Acid, *Amide*
p-Fluorobenzoic Acid, *Amide*

$C_7H_6FNO_2$
2-Fluoro-3-nitrotoluene
2-Fluoro-4-nitrotoluene
2-Fluoro-5-nitrotoluene
2-Fluoro-6-nitrotoluene
3-Fluoro-2-nitrotoluene
3-Fluoro-4-nitrotoluene
4-Fluoro-2-nitrotoluene
4-Fluoro-3-nitrotoluene
5-Fluoro-2-nitrotoluene

$C_7H_6FNO_3$
2-Fluoro-4-nitrophenol, *Me ether*
3-Fluoro-2-nitrophenol, *Me ether*
3-Fluoro-4-nitrophenol, *Me ether*
4-Fluoro-2-nitrophenol, *Me ether*
5-Fluoro-2-nitrophenol, *Me ether*

$C_7H_6FNO_4S$
4-Nitrotoluene-2-sulphonic Acid, *Fluoride*
2-Nitrotoluene-4-sulphonic Acid, *Fluoride*

$C_7H_6F_2$
α,α-Difluorotoluene

$C_7H_6F_2O_4S_2$
Toluene-2,4-disulphonic Acid, *Difluoride*

C_7H_6INO
p-Iodoaniline, N-*Formyl*
o-Iodobenzamide
m-Iodobenzamide
p-Iodobenzamide

$C_7H_6INO_2$
2-Amino-3-iodobenzoic Acid
2-Amino-4-iodobenzoic Acid
2-Amino-5-iodobenzoic Acid
2-Amino-6-iodobenzoic Acid
3-Amino-2-iodobenzoic Acid
3-Amino-5-iodobenzoic Acid
4-Amino-2-iodobenzoic Acid
4-Amino-3-iodobenzoic Acid

$C_7H_6INO_2$ (*continued*)
α-Iodo-2-nitrotoluene
α-Iodo-3-nitrotoluene
α-Iodo-4-nitrotoluene
2-Iodo-3-nitrotoluene
2-Iodo-4-nitrotoluene
3-Iodo-2-nitrotoluene
3-Iodo-4-nitrotoluene
4-Iodo-2-nitrotoluene
4-Iodo-3-nitrotoluene
5-Iodo-2-nitrotoluene
6-Iodo-2-nitrotoluene
6-Iodo-3-nitrotoluene

$C_7H_6INO_3$
2-Hydroxy-α-iodo-5-nitrotoluene
4-Hydroxy-α-iodo-3-nitrotoluene
6-Iodo-4-nitro-*o*-cresol
2-Iodo-6-nitro-*p*-cresol
6-Iodo-3-nitro-*p*-cresol
2-Iodo-3-nitrophenol, *Me ether*
2-Iodo-4-nitrophenol, *Me ether*
2-Iodo-5-nitrophenol, *Me ether*
2-Iodo-6-nitrophenol, *Me ether*
3-Iodo-2-nitrophenol, *Me ether*
3-Iodo-4-nitrophenol, *Me ether*
4-Iodo-2-nitrophenol, *Me ether*
4-Iodo-3-nitrophenol, *Me ether*
5-Iodo-2-nitrophenol, *Me ether*
5-Iodo-3-nitrophenol, *Me ether*

C_7H_6INS
2-Iodobenzothiazole†

$C_7H_6I_2$
2,3-Di-iodotoluene
2,4-Di-iodotoluene
2,5-Di-iodotoluene
2,6-Di-iodotoluene
3,4-Di-iodotoluene
3,5-Di-iodotoluene

$C_7H_6I_2NO_2$
2-Amino-3,5-di-iodobenzoic Acid, *Amide*

$C_7H_6I_2O$
2,6-Di-iodo-*p*-cresol
2,4-Di-iodophenol, *Me ether*
2,6-Di-iodophenol, *Me ether*
3,5-Di-iodophenol, *Me ether*

$C_7H_6I_2O_3S$
2,3-Di-iodobenzenesulphonic Acid, *Me ester*
2,4-Di-iodobenzenesulphonic Acid, *Me ester*
2,5-Di-iodobenzenesulphonic Acid, *Me ester*
3,4-Di-iodobenzenesulphonic Acid, *Me ester*
3,5-Di-iodobenzenesulphonic Acid, *Me ester*

$C_7H_6N_2$
o-Aminobenzonitrile
m-Aminobenzonitrile
p-Aminobenzonitrile
Benzimidazole
Imidazo[1,2-*a*]pyridine†
Indazole
5-Methylpyridine-3-carboxylic Acid, *Nitrile*
6-Methylpyridine-3-carboxylic Acid, *Nitrile*
Phenylcyanamide
Pyrrolo[2,3-*b*]pyridine
Pyrrolo[3,2-*b*]pyridine

$C_7H_6N_2O$
3-Amino-2-hydroxybenzoic Acid, *Nitrile*
4-Amino-2-hydroxybenzoic Acid, *Nitrile*
5-Amino-2-hydroxybenzoic Acid, *Nitrile*
Benzimidazolone
2-Hydroxyindazole
3-Hydroxyindazole
4-Hydroxyindazole
5-Hydroxyindazole
6-Hydroxyindazole
7-Hydroxyindazole
3,4-Methylenedioxy-6-nitroaniline
Nudiflorine†
Ricinidine

$C_7H_6N_2O_2S$
m-Sulphobenzoic Acid, *Nitrile*
p-Sulphobenzoic Acid, *Nitrile*

$C_7H_6N_2O_2S_2$
Holomycin

$C_7H_6N_2O_3$
2-Amino-4-nitrobenzaldehyde
2-Amino-5-nitrobenzaldehyde
4-Amino-2-nitrobenzaldehyde
4-Amino-3-nitrobenzaldehyde
5-Amino-2-nitrobenzaldehyde
2-Amino-5-nitrosobenzoic Acid
Benzonitrolic Acid
o-Nitrobenzamide
m-Nitrobenzamide
p-Nitrobenzamide
2-Nitro-3-nitrosotoluene
2-Nitro-4-nitrosotoluene
2-Nitro-5-nitrosotoluene
2-Nitro-6-nitrosotoluene
3-Nitro-2-nitrosotoluene
3-Nitro-4-nitrosotoluene
3-Nitro-6-nitrosotoluene
4-Nitro-2-nitrosotoluene
4-Nitro-4-nitrosotoluene
Pyridine-2,3-dicarboxylic Acid, 1-*Amide*
Pyridine-2,3-dicarboxylic Acid, 2-*Amide*

$C_7H_6N_2O_4$
2-Amino-3-nitrobenzoic Acid
2-Amino-4-nitrobenzoic Acid
2-Amino-5-nitrobenzoic Acid
2-Amino-6-nitrobenzoic Acid
3-Amino-2-nitrobenzoic Acid
3-Amino-4-nitrobenzoic Acid
4-Amino-2-nitrobenzoic Acid
4-Amino-3-nitrobenzoic Acid
5-Amino-2-nitrobenzoic Acid
5-Amino-3-nitrobenzoic Acid
α,α-Dinitrotoluene
2,3-Dinitrotoluene
2,4-Dinitrotoluene
2,5-Dinitrotoluene
2,6-Dinitrotoluene
3,4-Dinitrotoluene
3,5-Dinitrotoluene
2-Hydroxy-3-nitrobenzoic Acid, *Amide*
2-Hydroxy-4-nitrobenzoic Acid, *Amide*
m-Nitrobenzhydroxamic Acid
p-Nitrobenzhydroxamic Acid
o-Nitrophenylcarbamic Acid

$C_7H_6N_2O_4$ (*continued*)
m-Nitrophenylcarbamic Acid
p-Nitrophenylcarbamic Acid

$C_7H_6N_2O_4S$
2,4-Dinitrobenzenethiol, *Me ether*

$C_7H_6N_2O_5$
2-Amino-5-hydroxy-4-nitrobenzoic Acid
3-Amino-2-hydroxy-5-nitrobenzoic Acid
3-Amino-4-hydroxy-5-nitrobenzoic Acid
4-Amino-2-hydroxy-5-nitrobenzoic Acid
4-Amino-3-hydroxy-6-nitrobenzoic Acid
5-Amino-2-hydroxy-3-nitrobenzoic Acid
5-Amino-4-hydroxy-2-nitrobenzoic Acid
2,6-Dihydroxypyridine-3,5-dicarboxylic Acid, *Amide*
2,3-Dinitroanisole
2,4-Dinitroanisole
2,5-Dinitroanisole
2,6-Dinitroanisole
3,4-Dinitroanisole
3,5-Dinitroanisole
2,4-Dinitrobenzyl Alcohol
2,6-Dinitrobenzyl Alcohol
3,5-Dinitro-*o*-cresol
4,5-Dinitro-*o*-cresol
4,6-Dinitro-*o*-cresol
5,6-Dinitro-*o*-cresol
2,4-Dinitro-*m*-cresol
2,6-Dinitro-*m*-cresol
4,6-Dinitro-*m*-cresol
2,5-Dinitro-*p*-cresol
2,3-Dinitro-*p*-cresol
2,6-Dinitro-*p*-cresol
3,5-Dinitro-*p*-cresol
3-Hydroxy-4-oxo-4*H*-pyran-2,6-dicarboxylic Acid, *Diamide*

$C_7H_6N_2O_6$
3,4-Dinitroguaiacol
3,5-Dinitroguaiacol
3,6-Dinitroguaiacol
4,5-Dinitroguaiacol
4,6-Dinitroguaiacol
5,6-Dinitroguaiacol
2,3-Dinitroquinol, *Me ether*
2,4-Dinitroresorcinol, 1-*Me ether*

$C_7H_6N_2O_6S$
2-Nitro-4-sulphobenzoic Acid, 4-*Amide*
5-Nitro-3-sulphobenzoic Acid, 3-*Amide*
p-Nitrobenzenesulphonylcarbamic Acid†

$C_7H_6N_2S$
2-Aminobenzothiazole
4-Aminobenzothiazole
5-Aminobenzothiazole
6-Aminobenzothiazole
2-Mercaptobenzimidazole

$C_7H_6N_2S_2$
2-Benzothiazolesulphenamide

$C_7H_6N_4$
1-Phenyl-1,2,3,4-tetrazole
5-Phenyl-1,2,3,4-tetrazole

$C_7H_6N_4O_2$
1-Methyl-4-nitro-1,2,3-benzotriazole
1-Methyl-5-nitro-1,2,3-benzotriazole
1-Methyl-6-nitro-1,2,3-benzotriazole
1-Methyl-7-nitro-1,2,3-benzotriazole
5-Methyl-7-nitro-1,2,3-benzotriazole
5-Nitro-1,2,3-benzotriazole, N-*Me*

$C_7H_6N_4O_3$
Luciopterin†

$C_7H_6N_4O_6$
2,4,6-Trinitro-*m*-toluidine

$C_7H_6N_4S$
N,N′-Thiocarbonyl-di-imidazole†

C_7H_6O
Benzaldehyde
2,4,6-Cycloheptatrien-1-one
Tetracyclo[3,2,0,$0^{2,7}$,$0^{4,3}$]heptane-3-one†

C_7H_6OS
Thiobenzoic Acid

$C_7H_6OS_2$
2-Hydroxybenzenethionothiolic Acid

C_7H_6OSe
Selenobenzoic Acid

$C_7H_6O_2$
Benzoic Acid
Catechol, *Methylene ether*
2-Furylacraldehyde
m-Hydroxybenzaldehyde
p-Hydroxybenzaldehyde
3-Methyl-1,2-benzoquinone
2-Methyl-1,4-benzoquinone
Phenyl formate
Salicylaldehyde
Tropolone

$C_7H_6O_2S$
o-Mercaptobenzoic Acid
m-Mercaptobenzoic Acid
p-Mercaptobenzoic Acid

$C_7H_6O_3$
2-Acetylcyclopent-4-ene-1,3-dione†
2,3-Dihydroxybenzaldehyde
2,4-Dihydroxybenzaldehyde
2,5-Dihydroxybenzaldehyde
2,6-Dihydroxybenzaldehyde
3,4-Dihydroxybenzaldehyde
3,5-Dihydroxybenzaldehyde
3-α-Furylacrylic Acid
1-(2′-Furyl)propane-1,2-dione†
Gentisylquinone
3-Hydroxybenzoic Acid
4-Hydroxybenzoic Acid
3-Hydroxy-1,2-benzoquinone, *Me ether*
4-Hydroxy-1,2-benzoquinone, *Me ether*
Hydroxy-1,4-benzoquinone, *Me ether*
3-Hydroxytoluquinone
5-Hydroxytoluquinone
6-Hydroxytoluquinone
3,4-Methylenedioxyphenol
Perbenzoic Acid
Salicylic Acid

$C_7H_6O_3S$
2-Hydroxy-4-mercaptobenzoic Acid
2-Hydroxy-5-mercaptobenzoic Acid
2-Thienylglyoxylic Acid, *Me ester*

$C_7H_6O_4$
2,3-Dihydroxybenzoic Acid
2,4-Dihydroxybenzoic Acid
2,5-Dihydroxybenzoic Acid
2,6-Dihydroxybenzoic Acid
3,4-Dihydroxybenzoic Acid
3,5-Dihydroxybenzoic Acid
5-Formylfuran-2-carboxylic Acid. *Me ester*
α-Furoylacetic Acid
Isoclavacin
2-Oxo-2*H*-pyran-5-carboxylic Acid, *Me ester*
Patulin
Terreic Acid
2,3,4-Trihydroxybenzaldehyde
2,3,5-Trihydroxybenzaldehyde
2,4,5-Trihydroxybenzaldehyde
2,4,6-Trihydroxybenzaldehyde
3,4,5-Trihydroxybenzaldehyde

$C_7H_6O_4S$
o-Formylbenzenesulphonic Acid
m-Formylbenzenesulphonic Acid
p-Formylbenzenesulphonic Acid
Thiophene-2,3-dicarboxylic Acid, 2-*Me ester*

$C_7H_6O_5$
Furan-2,4-dicarboxylic Acid, 4-*Me ester*
Gallic Acid
5-Hydroxy-4-oxopyran-2-carboxylic Acid, *Me ether*
2-Methylfuran-3,4-dicarboxylic Acid
4-Methylfuran-2,3-dicarboxylic Acid
5-Methylfuran-2,3-dicarboxylic Acid
5-Methylfuran-2,4-dicarboxylic Acid
2,3,4-Trihydroxybenzoic Acid
2,3,5-Trihydroxybenzoic Acid
2,3,6-Trihydroxybenzoic Acid
2,4,5-Trihydroxybenzoic Acid
2,4,6-Trihydroxybenzoic Acid

$C_7H_6O_5S$
o-Sulphobenzoic Acid
m-Sulphobenzoic Acid
p-Sulphobenzoic Acid

$C_7H_6O_6$
5,6-Dihydroxy-4-oxopyran-2-carboxylic Acid, *Me ester*
2,3,4,5-Tetrahydroxybenzoic Acid
2,3,4,6-Tetrahydroxybenzoic Acid

$C_7H_6O_6S$
2-Hydroxy-4-sulphobenzoic Acid
2-Hydroxy-5-sulphobenzoic Acid
3-Hydroxy-4-sulphobenzoic Acid
3-Hydroxy-5-sulphobenzoic Acid
4-Hydroxy-2-sulphobenzoic Acid
4-Hydroxy-3-sulphobenzoic Acid

$C_7H_6O_7$
2,4,6-Trioxopimelic Acid

$C_7H_6O_8S_2$
2,4-Disulphobenzoic Acid
3,5-Disulphobenzoic Acid

$C_7H_6S_2$
3*H*-1,2-Benzodithiole
Dithiobenzoic Acid

$C_7H_7AsCl_2$
Benzyldichloroarsine

$C_7H_7BiO_7$
Dermatol

C_7H_7Br
α-Bromotoluene
o-Bromotoluene
m-Bromotoluene
p-Bromotoluene

C_7H_7BrHg
Mercuri-benzyl bromide
Mercuri-*o*-tolyl bromide
Mercuri-*m*-tolyl bromide
Mercuri-*p*-tolyl bromide

C_7H_7BrHgO
Mercuri-*o*-hydroxyphenyl bromide, *Me ether*
Mercuri-*p*-hydroxyphenyl bromide, *Me ether*

$C_7H_7BrNO_2$
2-Amino-5-bromobenzoic Acid, *Amide*

$C_7H_7BrN_2O$
o-Bromobenzhydrazide
m-Bromobenzhydrazide
p-Bromobenzhydrazide
o-Bromophenylurea
m-Bromophenylurea
p-Bromophenylurea

$C_7H_7BrN_2O_2$
2-Bromo-4-nitroaniline, N-*Me*
4-Bromo-2-nitroaniline, N-*Me*
5-Bromo-2-nitroaniline, N-*Me*
3-Bromo-6-nitro-*o*-toluidine
4-Bromo-6-nitro-*o*-toluidine
6-Bromo-4-nitro-*o*-toluidine
4-Bromo-2-nitro-*m*-toluidine
4-Bromo-6-nitro-*m*-toluidine
5-Bromo-4-nitro-*m*-toluidine
6-Bromo-2-nitro-*m*-toluidine
2-Bromo-5-nitro-*p*-toluidine
2-Bromo-6-nitro-*p*-toluidine
5-Bromo-2-nitro-*p*-toluidine

$C_7H_7BrN_2S$
o-Bromophenylthiourea
p-Bromophenylthiourea

C_7H_7BrO
o-Bromoanisole
m-Bromoanisole
p-Bromoanisole
o-Bromobenzyl Alcohol
m-Bromobenzyl Alcohol
p-Bromobenzyl Alcohol
3-Bromo-*o*-cresol
4-Bromo-*o*-cresol
5-Bromo-*o*-cresol
6-Bromo-*o*-cresol
2-Bromo-*m*-cresol
4-Bromo-*m*-cresol
5-Bromo-*m*-cresol
6-Bromo-*m*-cresol
2-Bromo-*p*-cresol
3-Bromo-*p*-cresol

$C_7H_7BrO_2$
3-Bromocatechol, 1-*Me ether*

$C_7H_7BrO_2$ (*continued*)
4-Bromocatechol, 1-*Me ether*
4-Bromocatechol, 2-*Me ether*
3-Bromo-4,5-dihydroxytoluene
2-Bromo-4-hydroxybenzyl Alcohol
2-Bromo-5-hydroxybenzyl Alcohol
3-Bromo-4-hydroxybenzyl Alcohol
5-Bromo-2-hydroxybenzyl Alcohol
2-Bromo-5-methylquinol
2-Bromo-6-methylquinol
4-Bromo-5-methylresorcinol
4-Bromoresorcinol, 1-*Me ether*
4-Bromoresorcinol, 3-*Me ether*

$C_7H_7BrO_2S$
Toluene-*o*-sulphonic Acid, *Bromide*
Toluene-*p*-sulphonic Acid, *Bromide*

$C_7H_7BrO_3S$
Bromobenzene-*p*-sulphonic Acid, *Me ester*

C_7H_7BrS
o-Bromobenzenethiol, *Me ether*
p-Bromobenzenethiol, *Me ether*

$C_7H_7Br_2N$
2,4-Dibromoaniline, N-*Me*
3,4-Dibromo-*o*-toluidine
3,6-Dibromo-*o*-toluidine
4,5-Dibromo-*o*-toluidine
4,6-Dibromo-*o*-toluidine
2,4-Dibromo-*m*-toluidine
2,5-Dibromo-*m*-toluidine
4,5-Dibromo-*m*-toluidine
4,6-Dibromo-*m*-toluidine
5,6-Dibromo-*m*-toluidine
2,5-Dibromo-*p*-toluidine
2,6-Dibromo-*p*-toluidine
3,5-Dibromo-*p*-toluidine

$C_7H_7Br_2NO$
2-Amino-3,5-dibromophenol, *Me ether*
4-Amino-2,6-dibromophenol, *Me ether*
2,4-Dibromobenzoic Acid, *Amide*
2,6-Dibromobenzoic Acid, *Amide*
3,4-Dibromobenzoic Acid, *Amide*
3,5-Dibromobenzoic Acid, *Amide*

$C_7H_7Br_2NO_2S$
Toluene-*o*-sulphonamide, N-*Dibromo*
Toluene-*p*-sulphonamide, N-*Dibromo*

C_7H_7Cl
α-Chlorotoluene
o-Chlorotoluene
m-Chlorotoluene
p-Chlorotoluene

C_7H_7ClHg
Mercuri-benzyl chloride
Mercuri-*o*-tolyl chloride
Mercuri-*m*-tolyl chloride
Mercuri-*p*-tolyl chloride

C_7H_7ClHgO
Mercuri-*o*-hydroxyphenyl chloride, *Me ether*
Mercuri-*m*-hydroxyphenyl chloride, *Me ether*
Mercuri-*p*-hydroxyphenyl chloride, *Me ether*
Mercuri-2-hydroxy-*m*-tolyl chloride
Mercuri-4-hydroxy-*m*-tolyl chloride
Mercuri-6-hydroxy-*m*-tolyl chloride

$C_7H_7ClNNaO_2S$
Chloramine-T

$C_7H_7ClN_2O$
2-Amino-4-chlorobenzoic Acid, *Amide*
2-Amino-5-chlorobenzoic Acid, *Amide*
o-Chlorobenzhydrazide
m-Chlorobenzhydrazide
p-Chlorobenzhydrazide
o-Chlorophenylurea
m-Chlorophenylurea
p-Chlorophenylurea

$C_7H_7ClN_2O_2$
2-Chloro-*N*-methyl-4-nitroaniline
2-Chloro-*N*-methyl-5-nitroaniline
2-Chloro-*N*-methyl-6-nitroaniline
4-Chloro-*N*-methyl-2-nitroaniline
5-Chloro-*N*-methyl-2-nitroaniline
3-Chloro-6-nitro-*o*-toluidine
4-Chloro-3-nitro-*o*-toluidine
4-Chloro-5-nitro-*o*-toluidine
4-Chloro-6-nitro-*o*-toluidine
5-Chloro-3-nitro-*o*-toluidine
5-Chloro-4-nitro-*o*-toluidine
5-Chloro-6-nitro-*o*-toluidine
2-Chloro-4-nitro-*o*-toluidine
2-Chloro-6-nitro-*m*-toluidine
4-Chloro-6-nitro-*m*-toluidine
6-Chloro-2-nitro-*m*-toluidine
6-Chloro-4-nitro-*m*-toluidine
2-Chloro-3-nitro-*p*-toluidine
2-Chloro-5-nitro-*p*-toluidine
2-Chloro-6-nitro-*p*-toluidine
5-Chloro-2-nitro-*p*-toluidine

$C_7H_7ClN_2O_3$
2-Amino-4-chloro-5-nitrophenol, *Me ether*

C_7H_7ClO
α-Chloroanisole†
o-Chloroanisole
m-Chloroanisole
p-Chloroanisole
o-Chlorobenzyl Alcohol
m-Chlorobenzyl Alcohol
p-Chlorobenzyl Alcohol
3-Chloro-*o*-cresol
4-Chloro-*o*-cresol
5-Chloro-*o*-cresol
6-Chloro-*o*-cresol
2-Chloro-*m*-cresol
4-Chloro-*m*-cresol
6-Chloro-*m*-cresol
2-Chloro-*p*-cresol
3-Chloro-*p*-cresol

C_7H_7ClOS
Toluene-*p*-sulphinic Acid, *Chloride*

$C_7H_7ClO_2$
3-Chloro-2-hydroxybenzyl Alcohol
5-Chloro-2-hydroxybenzyl Alcohol
3-Chloro-2-methoxyphenol
4-Chloro-2-methoxyphenol
5-Chloro-2-methoxyphenol
2-Chloro-3-methylquinol
2-Chloro-6-methylquinol
5-Chloro-2-methylquinol
4-Chlororesorcinol, 3-*Me ether*

$C_7H_7ClO_2S$
Chloromethyl phenyl sulphone
Toluene-α-sulphonic Acid, *Chloride*
Toluene-*o*-sulphonic Acid, *Chloride*
Toluene-*m*-sulphonic Acid, *Chloride*
Toluene-*p*-sulphonic Acid, *Chloride*

$C_7H_7ClO_3$
3-Chlorofuran-2-carboxylic Acid, *Et ester*
4-Chlorofuran-2-carboxylic Acid, *Et ester*
5-Chlorofuran-2-carboxylic Acid, *Et ester*

$C_7H_7ClO_3S$
p-Chlorobenzenesulphonic Acid, *Me ester*
4-Chlorotoluene-2-sulphonic Acid
5-Chlorotoluene-2-sulphonic Acid
6-Chlorotoluene-2-sulphonic Acid
4-Chlorotoluene-3-sulphonic Acid
6-Chlorotoluene-3-sulphonic Acid
2-Chlorotoluene-4-sulphonic Acid
3-Chlorotoluene-4-sulphonic Acid

$C_7H_7Cl_2N$
2,4-Dichloro-*N*-methylaniline
2,5-Dichloro-*N*-methylaniline
3,4-Dichloro-*o*-toluidine
4,5-Dichloro-*o*-toluidine
4,6-Dichloro-*o*-toluidine
2,4-Dichloro-*m*-toluidine
2,5-Dichloro-*m*-toluidine
2,6-Dichloro-*m*-toluidine
4,5-Dichloro-*m*-toluidine
4,6-Dichloro-*m*-toluidine
2,3-Dichloro-*p*-toluidine
2,5-Dichloro-*p*-toluidine
2,6-Dichloro-*p*-toluidine
3,5-Dichloro-*p*-toluidine

$C_7H_7Cl_2NO$
3-Amino-4,6-dichlorophenol, *Me ether*
4-Amino-3,5-dichlorophenol, *Me ether*
Clopidol†

$C_7H_7Cl_2NO_2S$
Dichloramine-T
Toluene-*o*-sulphonamide, N-*Dichloro deriv.*

$C_7H_7Cl_2NO_4S_2$
o-Toluidine-4,6-disulphonic Acid, *Dichloride*
p-Toluidine-2,5-disulphonic Acid, *Dichloride*

$C_7H_7Cl_2OP$
Benzyl phosphorodichloridite

C_7H_7F
α-Fluorotoluene
o-Fluorotoluene
m-Fluorotoluene
p-Fluorotoluene

C_7H_7FO
o-Fluorophenol, *Me ether*
m-Fluorophenol, *Me ether*
p-Fluorophenol, *Me ether*

$C_7H_7FO_2S$
Toluene-α-sulphonic Acid, *Fluoride*
Toluene-*o*-sulphonic Acid, *Fluoride*
Toluene-*p*-sulphonic Acid, *Fluoride*

$C_7H_7FO_3S$
4-Fluorotoluene-2-sulphonic Acid

$C_7H_7F_7O_2$
Heptafluorobutyric Acid, *Isopropyl ester*

C_7H_7HgI
Mercuri-benzyl iodide
Mercuri-*o*-tolyl iodide
Mercuri-*m*-tolyl iodide
Mercuri-*p*-tolyl iodide

C_7H_7HgIO
Mercuri-*o*-hydroxyphenyl iodide, *Me ether*
Mercuri-*p*-hydroxyphenyl iodide, *Me ether*

$C_7H_7HgNO_3$
Mercuri-benzyl nitrate

C_7H_7I
α-Iodotoluene
o-Iodotoluene
m-Iodotoluene
p-Iodotoluene

$C_7H_7IN_2O$
p-Iodophenylurea

C_7H_7IO
α-Hydroxy-2-iodotoluene
α-Hydroxy-3-iodotoluene
α-Hydroxy-4-iodotoluene
o-Iodoanisole
m-Iodoanisole
p-Iodoanisole
3-Iodo-*o*-cresol
4-Iodo-*o*-cresol
5-Iodo-*o*-cresol
6-Iodo-*o*-cresol
2-Iodo-*m*-cresol
6-Iodo-*m*-cresol
2-Iodo-*p*-cresol
3-Iodo-*p*-cresol

$C_7H_7IO_2$
1,3-Dihydroxy-5-iodobenzene, 1-*Me ether*
4-Iodo-2-methoxyphenol
5-Iodo-2-methoxyphenol

$C_7H_7IO_2S$
Toluene-*p*-sulphonic Acid, *Iodide*

$C_7H_7IO_3$
p-Iodoxyanisole

C_7H_7IS
o-Iodobenzenethiol, *Me ether*
m-Iodobenzenethiol, *Me ether*
p-Iodobenzenethiol, *Me ether*

$C_7H_7I_2N$
3,6-Di-iodo-*o*-toluidine
4,5-Di-iodo-*o*-toluidine
2,4-Di-iodo-*m*-toluidine
2,5-Di-iodo-*m*-toluidine
4,5-Di-iodo-*m*-toluidine
4,6-Di-iodo-*m*-toluidine
5,6-Di-iodo-*m*-toluidine
2,5-Di-iodo-*p*-toluidine
2,6-Di-iodo-*p*-toluidine

$C_7H_7I_3N_2$
3-Amino-2,4,6-tri-iodobenzoic Acid, *Nitrile*

C_7H_7N
3*H*-Pyrrolizine†
2-Vinylpyridine

C_7H_7N (*continued*)
3-Vinylpyridine
4-Vinylpyridine

C_7H_7NO
2-Acetylpyridine
3-Acetylpyridine
4-Acetylpyridine
o-Aminobenzaldehyde
m-Aminobenzaldehyde
p-Aminobenzaldehyde
2-Aminocycloheptatrienone†
3-Aminocycloheptatrienone†
4-Aminocycloheptatrienone†
Benzaldoxime
Benzamide
Formanilide
o-Nitrosotoluene
m-Nitrosotoluene
p-Nitrosotoluene

C_7H_7NOS
Phenylthiocarbamic Acid
Thiocarbamic Acid S-*Phenyl ester*
Thiocarbamic Acid, O-*Phenyl ester*
Thionyl-*o*-toluidine
Thionyl-*m*-toluidine
Thionyl-*p*-toluidine

$C_7H_7NO_2$
m-Aminobenzoic Acid
p-Aminobenzoic Acid
Anthranilic Acid
Benzhydroxamic Acid
Benzyl nitrite
Carbamic Acid, *Phenyl ester*
3-α-Furylacrylic Acid, *Amide*
Homarine
3-Hydroxybenzoic Acid, *Amide*
4-Hydroxybenzoic Acid, *Amide*
3,4-Methylenedioxyaniline
3-Methylpyridine-2-carboxylic Acid
4-Methylpyridine-2-carboxylic Acid
5-Methylpyridine-2-carboxylic Acid
6-Methylpyridine-2-carboxylic Acid
2-Methylpyridine-3-carboxylic Acid
4-Methylpyridine-3-carboxylic Acid
5-Methylpyridine-3-carboxylic Acid
6-Methylpyridine-3-carboxylic Acid
4-Nitroso-*o*-cresol
6-Nitroso-*o*-cresol
4-Nitroso-*m*-cresol
o-Nitrosophenol, *Me ether*
p-Nitrosophenol, *Me ether*
α-Nitrotoluene
o-Nitrotoluene
m-Nitrotoluene
p-Nitrotoluene
Phenylcarbamic Acid
2-Pyridineacetic Acid
3-Pyridineacetic Acid
4-Pyridineacetic Acid
Pyridine-betaine
Pyridine-2-carboxylic Acid, *Me ester*
Pyridine-3-carboxylic Acid, *Me ester*
Pyridine-4-carboxylic Acid, *Me ester*
Salicylic Acid, *Amide*
Trigonelline

$C_7H_7NO_2S$
Methyl-*o*-nitrophenyl sulphide
Methyl-*p*-nitrophenyl sulphide
2-Nitro-α-toluenethiol
3-Nitro-α-toluenethiol
4-Nitro-α-toluenethiol

$C_7H_7NO_3$
2-Amino-3-hydroxybenzoic Acid
2-Amino-4-hydroxybenzoic Acid
2-Amino-5-hydroxybenzoic Acid
2-Amino-6-hydroxybenzoic Acid
3-Amino-2-hydroxybenzoic Acid
3-Amino-4-hydroxybenzoic Acid
4-Amino-2-hydroxybenzoic Acid
4-Amino-3-hydroxybenzoic Acid
5-Amino-2-hydroxybenzoic Acid
Benzyl nitrate
2,4-Dihydroxybenzoic Acid, *Amide*
3,4-Dihydroxybenzoic Acid, *Amide*
1,3-Dihydroxy-4-nitrosobenzene, 1-*Me ether*
4-Hydroxypyridine-3-carboxylic Acid, *Me ether*
6-Hydroxypyridine-3-carboxylic Acid, *Me ester*
6-Hydroxypyridine-3-carboxylic Acid, *Me ether*
2-Methoxy-4-nitrosophenol
o-Nitroansiole
m-Nitroanisole
p-Nitroanisole
o-Nitrobenzyl Alcohol
m-Nitrobenzyl Alcohol
p-Nitrobenzyl Alcohol
3-Nitro-*o*-cresol
4-Nitro-*o*-cresol
5-Nitro-*o*-cresol
6-Nitro-*o*-cresol
2-Nitro-*m*-cresol
4-Nitro-*m*-cresol
5-Nitro-*m*-cresol
6-Nitro-*m*-cresol
2-Nitro-*p*-cresol
3-Nitro-*p*-cresol

$C_7H_7NO_3S$
p-Formylbenzenesulphonic Acid, *Amide*

$C_7H_7NO_4$
2-Amino-3,4-dihydroxybenzoic Acid
2-Amino-3,5-dihydroxybenzoic Acid
2-Amino-4,5-dihydroxybenzoic Acid
2-Amino-5,6-dihydroxybenzoic Acid
3-Amino-2,5-dihydroxybenzoic Acid
3-Amino-4,5-dihydroxybenzoic Acid
3-Amino-4,6-dihydroxybenzoic Acid
3-Amino-5,6-dihydroxybenzoic Acid
4-Amino-3,5-dihydroxybenzoic Acid
1,3-Dihydroxy-2-nitrobenzene, 1-*Me ether*
1,3-Dihydroxy-4-nitrobenzene, 1-*Me ether*
1,3-Dihydroxy-4-nitrobenzene, 3-*Me ether*
α,2-Dihydroxy-3-nitrotoluene
α,2-Dihydroxy-4-nitrotoluene
α,2-Dihydroxy-5-nitrotoluene
α,3-Dihydroxy-4-nitrotoluene
α,4-Dihydroxy-3-nitrotoluene
α,5-Dihydroxy-2-nitrotoluene
2,4-Dihydroxy-3-nitrotoluene

$C_7H_7NO_4$ (*continued*)
2,4-Dihydroxy-5-nitrotoluene
3,4-Dihydroxy-5-nitrotoluene
3,5-Dihydroxy-2-nitrotoluene
3,5-Dihydroxy-4-nitrotoluene
4,5-Dihydroxy-2-nitrotoluene
4,5-Dihydroxypyridine-2-carboxylic Acid, 5-*Me ether*
2,6-Dihydroxypyridine-4-carboxylic Acid, *Me ester*
Gallic Acid, *Amide*
4-Methylfuran-2,3-dicarboxylic Acid, 2-*Amide*
2-Methyl-3-nitroquinol
3-Nitrocatechol, 1-*Me ether*
3-Nitrocatechol, 2-*Me ether*
4-Nitrocatechol, 1-*Me ether*
4-Nitrocatechol, 2-*Me ether*
2-Nitroquinol, 1-*Me ether*
2-Nitroquinol, 4-*Me ether*
Pyrrole-2,3-dicarboxylic Acid, 3-*Me ester*
Pyrrole-2,5-dicarboxylic Acid, *Me ester*
Pyrrole-2,5-dicarboxylic Acid, N-*Me*
1,3,5-Trihydroxy-2-nitrosobenzene, 1-*Me ether*

$C_7H_7NO_4S$
2-Nitrobenzenesulphinic Acid, *Me ester*
4-Nitrobenzenesulphinic Acid, *Me ester*
o-Sulphobenzoic Acid, 1-*Amide*
o-Sulphobenzoic Acid, 2-*Amide*
m-Sulphobenzoic Acid, 3-*Amide*
p-Sulphobenzoic Acid, 4-*Amide*

$C_7H_7NO_5$
2,3-Dihydroxy-5-nitrobenzyl Alcohol
2,4-Dimethyl-5-nitrofuran-3-carboxylic Acid
2,5-Dimethyl-4-nitrofuran-3-carboxylic Acid
5-Methyl-4-nitrofuran-2-carboxylic Acid, *Me ester*
5-Nitrofuran-2-carboxylic Acid, *Et ester*
5-Nitrofuran-3-carboxylic Acid, *Et ester*
1,2,4-Trihydroxy-3-nitrobenzene,2-*Me ether*

$C_7H_7NO_5S$
2-Hydroxy-4-sulphobenzoic Acid, *Sulphonamide*
2-Hydroxy-5-sulphobenzoic Acid, *Sulphonamide*
4-Hydroxy-3-sulphobenzoic Acid, *Sulphonamide*
2-Methoxy-4-nitrobenzenesulphinic Acid
2-Methoxy-5-nitrobenzenesulphinic Acid
2-Nitrotoluene-α-sulphonic Acid
3-Nitrotoluene-α-sulphonic Acid
4-Nitrotoluene-α-sulphonic Acid
4-Nitrotoluene-2-sulphonic Acid
5-Nitrotoluene-2-sulphonic Acid
6-Nitrotoluene-2-sulphonic Acid
2-Nitrotoluene-3-sulphonic Acid
4-Nitrotoluene-3-sulphonic Acid
5-Nitrotoluene-3-sulphonic Acid
6-Nitrotoluene-3-sulphonic Acid
2-Nitrotoluene-4-sulphonic Acid
3-Nitrotoluene-4-sulphonic Acid

$C_7H_7NO_6S$
4-Methoxy-3-nitrobenzenesulphonic Acid

$C_7H_7NO_7S_2$
2,4-Disulphobenzoic Acid, 4-*Amide*

C_7H_7NS
Benzothiazoline
Thiobenzoic Acid, *Amide*

$C_7H_7NS_2$
Phenyldithiocarbamic Acid

C_7H_7NSe
Selenobenzoic Acid, *Amide*

$C_7H_7N_2O$
p-Hydroxymethylbenzenediazonium ion

$C_7H_7N_3$
3-Aminoindazole
4-Aminoindazole
5-Aminoindazole
6-Aminoindazole
7-Aminoindazole
Benzyl azide
o-Hydrazinobenzoic Acid, *Nitrile*
1-Methylbenzotriazole
2-Methylbenzotriazole
4-Methylbenzotriazole
5-Methylbenzotriazole
6-Methylbenzotriazole
p-Tolyl azide

$C_7H_7N_3O_2$
o-Nitrobenzylidenehydrazine
m-Nitrobenzylidenehydrazine
p-Nitrobenzylidenehydrazine
Phenylurea, 1-N-*Nitroso*
Pyridine-2,3-dicarboxylic Acid, *Diamide*
Pyridine-2,4-dicarboxylic Acid, *Diamide*
Pyridine-2,5-dicarboxylic Acid, *Diamide*
Pyridine-2,6-dicarboxylic Acid, *Diamide*
Pyridine-3,4-dicarboxylic Acid, *Diamide*
Pyridine-3,5-dicarboxylic Acid, *Diamide*

$C_7H_7N_3O_2S$
o-Nitrophenylthiourea
m-Nitrophenylthiourea
p-Nitrophenylthiourea

$C_7H_7N_3O_3$
2-Amino-3-nitrobenzoic Acid, *Amide*
2-Amino-5-nitrobenzoic Acid, *Amide*
3-Amino-4-nitrobenzoic Acid, *Amide*
4-Amino-3-nitrobenzoic Acid, *Amide*
2,6-Diamino-3-nitrobenzaldehyde
o-Nitrobenzhydrazide
m-Nitrobenzhydrazide
p-Nitrobenzhydrazide
o-Nitrophenylurea
m-Nitrophenylurea
p-Nitrophenylurea

$C_7H_7N_3O_4$
2,6-Dihydroxypyridine-3,5-dicarboxylic Acid, *Diamide*
3,4-Dinitro-*o*-toluidine
3,5-Dinitro-*o*-toluidine
3,6-Dinitro-*o*-toluidine
4,5-Dinitro-*o*-toluidine
4,6-Dinitro-*o*-toluidine
5,6-Dinitro-*o*-toluidine
2,4-Dinitro-*m*-toluidine
2,6-Dinitro-*m*-toluidine
4,5-Dinitro-*m*-toluidine
4,6-Dinitro-*m*-toluidine
5,6-Dinitro-*m*-toluidine
2,3-Dinitro-*p*-toluidine
2,6-Dinitro-*p*-toluidine

$C_7H_7N_3O_4$ (*continued*)
3,5-Dinitro-*p*-toluidine
3,6-Dinitro-*p*-toluidine
2-Hydroxy-4-nitrophenylurea
N-Methyl-2,4-dinitroaniline
N-Methyl-2,5-dinitroaniline
N-Methyl-2,6-dinitroaniline
N-Methyl-3,4-dinitroaniline

$C_7H_7N_3O_5$
4-Amino-2,6-dinitro-*m*-cresol
6-Amino-2,4-dinitro-*m*-cresol
3-Amino-2,6-dinitro-*p*-cresol
2-Amino-3,5-dinitrophenol, *Me ether*
2-Amino-4,5-dinitrophenol, *Me ether*
2-Amino-4,6-dinitrophenol, *Me ether*
2-Amino-4,6-dinitropnenol, N-*Me*
3-Amino-2,4-dinitrophenol, *Me ether*
4-Amino-2,3-dinitrophenol, *Me ether*
4-Amino-2,5-dinitrophenol, *Me ether*
4-Amino-2,6-dinitrophenol, *Me ether*
4-Amino-2,6-dinitrophenol, N-*Me*
4-Amino-3,5-dinitrophenol, *Me ether*
5-Amino-2,4-dinitrophenol, *Me ether*
Apocaffeine
Caffolide,1,3-N-*Di-Me*
N-(2,4-Dinitrophenyl)hydroxylamine, *Me ether*

$C_7H_7N_3O_5S$
2-Nitro-4-sulphobenzoic Acid, *Diamide*
5-Nitro-3-sulphobenzoic Acid, *Diamide*

$C_7H_7N_5O_2$
Chrysopterin
Fervenulin

C_7H_8
Cycloheptatriene
1,3-Heptadi-yne
1,5-Heptadi-yne
1,6-Heptadi-yne
6-Methylfulvene†
Quadricyclo[2,2,1,$0^{2,6}$,$0^{3,5}$]heptane
Tetracyclo[3,2,0,$0^{2,7}$,$0^{4,6}$]heptane†
Toluene

C_7H_8BrN
α-Amino-2-bromotoluene
α-Amino-3-bromotoluene
α-Amino-4-bromotoluene
3-Bromo-*o*-toluidine
4-Bromo-*o*-toluidine
5-Bromo-*o*-toluidine
6-Bromo-*o*-toluidine
4-Bromo-*m*-toluidine
5-Bromo-*m*-toluidine
6-Bromo-*m*-toluidine
2-Bromo-*p*-toluidine
3-Bromo-*p*-toluidine

C_7H_8BrNO
4-Amino-5-bromo-*o*-cresol
4-Amino-6-bromo-*o*-cresol
6-Amino-4-bromo-*o*-cresol
4-Amino-6-bromo-*m*-cresol
6-Amino-4-bromo-*m*-cresol
6-Amino-5-bromo-*m*-cresol
2-Amino-6-bromo-*p*-cresol
3-Bromo-*o*-anisidine
4-Bromo-*o*-anisidine
5-Bromo-*o*-anisidine
6-Bromo-*o*-anisidine
4-Bromo-*m*-anisidine
5-Bromo-*m*-anisidine
3-Bromo-*p*-anisidine

$C_7H_8BrNO_2S$
Toluene-*o*-sulphonamide, N-*Bromo deriv.*

C_7H_8ClHgN
Mercuri-4-amino-*m*-tolyl chloride
Mercuri-*p*-methylaminophenyl chloride

C_7H_8ClN
α-Amino-2-chlorotoluene
α-Amino-3-chlorotoluene
α-Amino-4-chlorotoluene
2-Amino-α-chlorotoluene
3-Amino-α-chlorotoluene
4-Amino-α-chlorotoluene
2-Chloro-4,6-dimethylpyridine
2-Chloro-5,6-dimethylpyridine
3-Chloro-2,6-dimethylpyridine
4-Chloro-2,6-dimethylpyridine
o-Chloro-*N*-methylaniline
m-Chloro-*N*-methylaniline
p-Chloro-*N*-methylaniline
3-Chloro-*o*-toluidine
4-Chloro-*o*-toluidine
5-Chloro-*o*-toluidine
6-Chloro-*o*-toluidine
2-Chloro-*m*-toluidine
4-Chloro-*m*-toluidine
5-Chloro-*m*-toluidine
6-Chloro-*m*-toluidine
2-Chloro-*p*-toluidine
3-Chloro-*p*-toluidine
Pyridine-neurine, *Chloride*

C_7H_8ClNO
4-Amino-5-chloro-*o*-cresol
4-Amino-6-chloro-*o*-cresol
6-Amino-3-chloro-*o*-cresol
6-Amino-4-chloro-*o*-cresol
6-Amino-5-chloro-*o*-cresol
4-Amino-2-chloro-*m*-cresol
4-Amino-6-chloro-*m*-cresol
6-Amino-4-chloro-*m*-cresol
2-Amino-5-chloro-*p*-cresol
2-Amino-6-chloro-*p*-cresol
o-Chlorobenzylhydroxylamine
p-Chlorobenzylhydroxylamine
3-Chloro-*o*-anisidine
4-Chloro-*o*-anisidine
5-Chloro-*o*-anisidine
6-Chloro-*o*-anisidine
4-Chloro-*m*-anisidine
5-Chloro-*m*-anisidine
6-Chloro-*m*-anisidine
2-Chloro-*p*-anisidine
3-Chloro-*p*-anisidine

$C_7H_8ClNO_2S$
4-Chlorotoluene-2-sulphonic Acid, *Amide*
5-Chlorotoluene-2-sulphonic Acid, *Amide*
6-Chlorotoluene-2-sulphonic Acid, *Amide*
4-Chlorotoluene-3-sulphonic Acid, *Amide*
6-Chlorotoluene-3-sulphonic Acid, *Amide*

$C_7H_8ClNO_2S$ (*continued*)
2-Chlorotoluene-4-sulphonic Acid, *Amide*
3-Chlorotoluene-4-sulphonic Acid, *Amide*
Toluene-*o*-sulphonamide, N-*Chloro deriv.*
o-Toluidine-5-sulphonic Acid, *Chloride*
p-Toluidine-3-sulphonic Acid, *Chloride*

$C_7H_8ClN_3O$
1-*o*-Chlorophenylsemicarbazide
1-*m*-Chlorophenylsemicarbazide
1-*p*-Chlorophenylsemicarbazide

$C_7H_8ClN_3O_4S_2$
Hydrochlorothiazide

$C_7H_8Cl_2N_2$
2,4-Dichloro-3,5-toluenediamine
2,6-Dichloro-3,5-toluenediamine

C_7H_8FN
3-Fluoro-*o*-toluidine
4-Fluoro-*o*-toluidine
6-Fluoro-*o*-toluidine
2-Fluoro-*m*-toluidine
4-Fluoro-*m*-toluidine
6-Fluoro-*m*-toluidine
2-Fluoro-*p*-toluidine

C_7H_8FNO
3-Fluoro-*o*-anisidine
5-Fluoro-*o*-anisidine
2-Fluoro-*p*-anisidine
3-Fluoro-*p*-anisidine

$C_7H_8FNO_2S$
4-Fluorotoluene-2-sulphonic Acid, *Amide*
o-Toluidine-5-sulphonic Acid, *Fluoride*
p-Toluidine-3-sulphonic Acid, *Fluoride*

C_7H_8IN
α-Amino-2-iodotoluene
α-Amino-3-iodotoluene
α-Amino-4-iodotoluene
3-Iodo-*o*-toluidine
4-Iodo-*o*-toluidine
2-Iodo-*m*-toluidine
4-Iodo-*m*-toluidine
5-Iodo-*m*-toluidine
6-Iodo-*m*-toluidine
2-Iodo-*p*-toluidine
3-Iodo-*p*-toluidine

C_7H_8INO
2-Amino-4-iodophenol, *Me ether*
2-Amino-5-iodophenol, *Me ether*
2-Amino-6-iodophenol, *Me ether*
3-Amino-5-iodophenol, *Me ether*
4-Amino-2-iodophenol, *Me ether*

$C_7H_8NO_2$
β-Formylphenylhydrazine

$C_7H_8NO_5P$
p-Nitrobenzylphosphonic Acid†

$C_7H_8N_2$
Benzamidine
Benzeneazomethane
Benzylidenehydrazine
2,5-Dimethylpyrrole-3-carboxylic Acid, *Nitrile*

$C_7H_8N_2O$
o-Aminobenzamide
m-Aminobenzamide
p-Aminobenzamide
Benzamidoxime
Benzhydrazide
2,4-Diaminobenzaldehyde
N-Methyl-4-nitrosoaniline
6-Methylpyridine-2-carboxylic Acid, *Amide*
2-Methylpyridine-3-carboxylic Acid, *Amide*
6-Methylpyridine-3-carboxylic Acid, *Amide*
4-Nitroso-*o*-toluidine
4-Nitroso-*m*-toluidine
Phenylurea
2-Pyridineacetic Acid, *Amide*
3-Pyridineacetic Acid, *Amide*

$C_7H_8N_2OS$
2-Hydroxyphenylthiourea
3-Hydroxyphenylthiourea
4-Hydroxyphenylthiourea

$C_7H_8N_2O_2$
5-Amino-2-hydroxybenzoic Acid, *Amide*
2-Aminopyridine-3-carboxylic Acid, *Me ester*
4-Aminopyridine-3-carboxylic Acid, *Me ester*
5-Aminopyridine-3-carboxylic Acid, *Me ester*
3-Aminopyridine-4-carboxylic Acid, *Me ester*
2,3-Diaminobenzoic Acid
2,4-Diaminobenzoic Acid
2,5-Diaminobenzoic Acid
2,6-Diaminobenzoic Acid
3,4-Diaminobenzoic Acid
3,5-Diaminobenzoic Acid
o-Hydrazinobenzoic Acid
m-Hydrazinobenzoic Acid
p-Hydrazinobenzoic Acid
2-Hydroxyphenylurea
3-Hydroxyphenylurea
4-Hydroxyphenylurea
N-Methyl-*o*-nitroaniline
N-Methyl-*m*-nitroaniline
N-Methyl-*p*-nitroaniline
Methylphenylnitramine
N-Nitroaniline, aci-*Me ether*
o-Nitrobenzylamine
m-Nitrobenzylamine
p-Nitrobenzylamine
4-Nitroso-*o*-anisidine
4-Nitroso-*m*-anisidine
3-Nitro-*o*-toluidine
4-Nitro-*o*-toluidine
5-Nitro-*o*-toluidine
6-Nitro-*o*-toluidine
2-Nitro-*m*-toluidine
4-Nitro-*m*-toluidine
5-Nitro-*m*-toluidine
6-Nitro-*m*-toluidine
2-Nitro-*p*-toluidine
3-Nitro-*p*-toluidine
Phenylhydrazine-α-carboxylic Acid
Phenylhydrazine-β-carboxylic Acid

$C_7H_8N_2O_3$
1-Allylbarbituric Acid
5-Allybarbituric Acid
4-Amino-6-nitro-*o*-cresol
6-Amino-3-nitro-*o*-cresol
6-Amino-4-nitro-*o*-cresol
6-Amino-5-nitro-*o*-cresol

$C_7H_8N_2O_3$ (*continued*)
4-Amino-2-nitro-*m*-cresol
4-Amino-5-nitro-*m*-cresol
4-Amino-6-nitro-*m*-cresol
2-Amino-5-nitro-*p*-cresol
2-Amino-6-nitro-*p*-cresol
4-Amino-2-nitrophenol, N-*Me*
3,5-Diamino-2-hydroxybenzoic Acid
3,5-Diamino-4-hydroxybenzoic Acid
4,5-Diamino-2-hydroxybenzoic Acid
2-Hydroxy-4,6-dimethyl-3-nitropyridine
2-Hydroxy-4,6-dimethyl-5-nitropyridine
4-Hydroxy-2,6-dimethyl-3-nitropyridine
2-Hydroxy-5-nitrobenzylamine
4-Hydroxy-3-nitrobenzylamine
2-Hydroxy-5-nitropyridine, *Et ether*
4-Hydroxy-3-nitropyridine, *Et ether*
5-Hydroxy-2-nitropyridine, *Et ether*
4-Hydroxypyridine-3-carboxylic Acid, *Me ether*, *Amide*
2-Methoxy-3-nitroaniline
2-Methoxy-4-nitroaniline
2-Methoxy-5-nitroaniline
2-Methoxy-6-nitroaniline
3-Methoxy-2-nitroaniline
3-Methoxy-4-nitroaniline
3-Methoxy-5-nitroaniline
4-Methoxy-2-nitroaniline
4-Methoxy-3-nitroaniline
5-Methoxy-2-nitroaniline
N-2-Nitrobenzylhydroxylamine
N-3-Nitrobenzylhydroxylamine
N-4-Nitrobenzylhydroxylamine
O-4-Nitrobenzylhydroxylamine

$C_7H_8N_2O_3S$
m-Sulphobenzoic Acid, *Diamide*

$C_7H_8N_2O_4$
3-Amino-5-nitrocatechol, 1-*Me ether*
4-Amino-3-nitrocatechol, 2-*Me ether*
3,5-Dihydroxy-2-nitropyridine, *Di-Me ether*
Imidazole-4,5-dicarboxylic Acid, *Di-Me ester*
1-Methylorotic Acid, *Me ester*
Orotic Acid, *Et ester*
Pyrazole-3,5-dicarboxylic Acid, *Di-Me ester*
Pyrazole-4,5-dicarboxylic Acid, *Di-Me ester*

$C_7H_8N_2O_4S$
p-Aminobenzenesulphonylcarbamic Acid†
2-Nitrobenzenesulphonic Acid, *Me-amide*
3-Nitrobenzenesulphonic Acid, *Me-amide*
4-Nitrobenzenesulphonic Acid, *Me-amide*
2-Nitrotoluene-α-sulphonic Acid, *Amide*
3-Nitrotoluene-α-sulphonic Acid, *Amide*
4-Nitrotoluene-α-sulphonic Acid, *Amide*
4-Nitrotoluene-2-sulphonic Acid, *Amide*
5-Nitrotoluene-2-sulphonic Acid, *Amide*
6-Nitrotoluene-2-sulphonic Acid, *Amide*
2-Nitrotoluene-3-sulphonic Acid, *Amide*
6-Nitrotoluene-3-sulphonic Acid, *Amide*
2-Nitrotoluene-4-sulphonic Acid, *Amide*
3-Nitrotoluene-4-sulphonic Acid, *Amide*

$C_7H_8N_2O_5S$
4-Methoxy-3-nitrobenzenesulphonic Acid, *Amide*
3-Nitro-*o*-toluidine-5-sulphonic Acid
3-Nitro-*p*-toluidine-6-sulphonic Acid
6-Nitro-*o*-toluidine-4-sulphonic Acid

$C_7H_8N_2O_6S_2$
2,4-Disulphobenzoic Acid, 2,4-*Diamide*

$C_7H_8N_2S$
Phenylthiourea

$C_7H_8N_4O_2$
3,8-Dimethylxanthine
Paraxanthine
Theobromine
Theophylline

$C_7H_8N_4O_3$
1-*o*-Nitrophenylsemicarbazide
1-*m*-Nitrophenylsemicarbazide
1-*p*-Nitrophenylsemicarbazide
4-*m*-Nitrophenylsemicarbazide
4-*p*-Nitrophenylsemicarbazide

C_7H_8O
Anisole
Benzyl Alcohol
o-Cresol
m-Cresol
p-Cresol
2,4-Cycloheptadienone
3,5-Cycloheptadienone†
1,3-Cyclohexadiene-1-aldehyde†
1,5-Dimethyltricyclo[2,1,0,$0^{2,5}$]pentan-3-one†
2,5-Heptadi-yn-4-ol
4,6-Heptadi-yn-2-ol
2-Oxabicyclo[3,3,0]octa-3,7-diene†
Quadricyclo[2,2,1,$0^{2,6}$,$0^{3,5}$]heptane-7-o1
Tricyclo[2,2,1,$0^{2,6}$]heptan-3-one†

C_7H_8OS
2-Acetyl-5-methylthiophene
o-Hydroxybenzenethiol, O-*Me ether*
m-Hydroxybenzenethiol, O-*Me ether*
m-Hydroxybenzenethiol, S-*Me ether*
p-Hydroxybenzenethiol, O-*Me ether*
Methyl phenyl sulphoxide†
2-Propionylthiophene

$C_7H_8O_2$
Cyclohept-4-ene-1,2-dione
1,4-Cyclohexadiene-1-carboxylic Acid★†
α,2-Dihydroxytoluene
α-3-Dihydroxytoluene
α,4-Dihydroxytoluene
2,3-Dihydroxytoluene
2,4-Dihydroxytoluene
2,6-Dihydroxytoluene
3,4-Dihydroxytoluene
3,5-Dihydroxytoluene
3,5-Dimethyl-3-cyclopentene-1,2-dione
3,4-Dimethylfurfuraldehyde
3,5-Dimethylfurfuraldehyde
4,5-Dimethylfurfuraldehyde
2,6-Dimethyl-γ-pyrone
Furfuryl methyl Ketone
3-α-Furylallyl Alcohol
Guaiacol
Methylquinol
Resorcinol, *Mono-Me ether*

$C_7H_8O_2S$
Benzenesulphinic Acid, *Me ester*†

$C_7H_8O_2S$ (*continued*)
Methyl phenyl sulphone
3-Methylthiophene-2-carboxylic Acid, *Me ester*
5-Methylthiophene-2-carboxylic Acid, *Me ester*
2-Thiopheneacetic Acid, *Me ester*
Thiophene-2-carboxylic Acid, *Et ester*
Thiophene-3-carboxylic Acid, *Et ester*
Toluene-*o*-sulphinic Acid
Toluene-*m*-sulphinic Acid
Toluene-*p*-sulphinic Acid

$C_7H_8O_2S_2$
Benzenethiolsulphonic Acid, *Me ester*

$C_7H_8O_3$
2-Acetylcylopentane-1,3-dione†
Cyclopentane-1,3-dicarboxylic Acid, *Anhydride*
3,3-Dimethylcyclopropane-1,2-dicarboxylic Acid, *Anhydride*
3,5-Dimethylfuran-2-carboxylic Acid
2,4-Dimethylfuran-3-carboxylic Acid
2,5-Dimethylfuran-3-carboxylic Acid
4,5-Dimethylfuran-3-carboxylic Acid
2-Ethylideneglutaric Acid, *Anhydride*
2-Ethylidene-3-methylsuccinic Acid, *Anhydride*
2-Ethyl-3-methylmaleic Acid, *Anhydride*
Furan-2-carboxylic Acid, *Et ester*
Furan-3-carboxylic Acid, *Et ester*
2-Furylacetic Acid, *Me ester*
3-(2-Furyl) propionic Acid
4-Hydroxy-3,6-dimethyl-2-pyrone†
3-Hydroxy-2*H*-pyran-2-one, *Et ether*
3-Hydroxy-4*H*-pyran-4-one, *Et ether*
Maltol, *Me ether*
cis-3-Methyl-1-butene-1,2-dicarboxylic Acid, *Anhydride*
3-Methyl-2-butene-1,2-dicarboxylic Acid, *Anhydride*
3-Methylcyclopentane-1,2,4-trione, *Mono-Me ether*
3-Methylfuran-2-carboxylic Acid, *Me ester*
5-Methylfuran-2-carboxylic Acid, *Me ester*
5-Methyl-2-furanacetic Acid
Orthobenzoic Acid
4-Oxocyclohex-2-ene-1-carboxylic Acid
6-Oxocyclohex-1-ene-1-carboxylic Acid
Pentenedial, *Ac*
Phloroglucinol, *Mono-Me ether*
Propylmaleic Acid, *Anhydride*
Pyrogallol, 1-*Me ether*
Pyrogallol, 2-*Me ether*
Sarkomycin
1,2,4-Trihydroxybenzene, 1-*Me ether*
1,2,4-Trihydroxybenzene, 2-*Me ether*
α,2,5-Trihydroxytoluene
α,3,4-Trihydroxytoluene
2,3,4-Trihydroxytoluene
2,3,5-Trihydroxytoluene
2,3,6-Trihydroxytoluene
2,4,5-Trihydroxytoluene
2,4,6 Trihydroxytoluene
3,4,5-Trihydroxytoluene

$C_7H_8O_3S$
Benzenesulphonic Acid, *Me ester*
Methane-sulphonic Acid, *Phenyl ester*
Toluene-α-sulphonic Acid
Toluene-*o*-sulphonic Acid
Toluene-*m*-sulphonic Acid
Toluene-*p*-sulphonic Acid

$C_7H_8O_4$
1-Cyclopentene-1,2-dicarboxylic Acid
1-Cyclopentene-1,3-dicarboxylic Acid
2-Cyclopentene-1,2-dicarboxylic Acid
4-Cyclopentene-1,3-dicarboxylic Acid
Iretol
3-Methylmuconic Acid
Muconic Acid, *Mono-Me* cis-trans *ester*
Opuntiol†
4-Oxo-1,7-heptanedioic Acid, *Anhydride*
2,3-Pentadienedioic Acid, *Di-Me ester*
Rubiginol, *Di-Me ether*
Terremutin†
α,2,3,5-Tetrahydroxytoluene
α,2,4,6-Tetrahydroxytoluene
α,3,4,5-Tetrahydroxytoluene
2,3,4,5-Tetrahydroxytoluene
2,3,4,6-Tetrahydroxytoluene
2,3,5,6-Tetrahydroxytoluene

$C_7H_8O_4S$
4-Hydroxytoluene-2-sulphonic Acid
5-Hydroxytoluene-2-sulphonic Acid
4-Hydroxytoluene-3-sulphonic Acid
3-Hydroxytoluene-4-sulphonic Acid

$C_7H_8O_5$
3,4-Dihydroxy-5-oxo-cyclohex-1-enecarboxylic Acid
3-Methyltricarballylic Acid, 1,2-*Anhydride*
Zymonic Acid, *Me ether*

$C_7H_8O_6$
Aconitic Acid, *Me ester*
2,4-Dioxopimelic Acid
2,6-Dioxopimelic Acid
Homocitric Acid, *Lactone*†
3-Methylcyclopropane-1,1,2-tricarboxylic Acid
1-Methylcyclopropane-1,2,3-tricarboxylic Acid

$C_7H_8O_6S_2$
Toluene-2,4-disulphonic Acid
Toluene-2,5-disulphonic Acid
Toluene-2,6-disulphonic Acid
Toluene-3,4-disulphonic Acid
Toluene-3,5-disulphonic Acid

$C_7H_8O_9S_3$
Toluene-2,4,6-trisulphonic Acid

C_7H_8S
Methyl phenyl sulphide
α-Toluenethiol
o-Toluenethiol
m-Toluenethiol
p-Toluenethiol

C_7H_8Se
Methyl phenyl selenide

$C_7H_9AsN_2O_4$
Carbasone

$C_7H_9AsO_3$
α-Toluenearsonic Acid
o-Toluenearsonic Acid
m-Toluenearsonic Acid
p-Toluenearsonic Acid

$C_7H_9BO_3$
4-Hydroxybenzeneboronic Acid, *Me ether*

$C_7H_9BrN_2$
5-Bromotoluene-2,3-diamine
5-Bromotoluene-2,4-diamine
5-Bromotoluene-3,4-diamine
2-Bromo-*p*-tolylhydrazine
4-Bromo-*o*-tolylhydrazine

$C_7H_9ClN_2$
6-Chloro-2,3-toluenediamine
5-Chloro-2,4-toluenediamine
6-Chloro-2,4-toluenediamine
4-Chloro-2,5-toluenediamine
6-Chloro-2,5-toluenediamine
2-Chloro-3,4-toluenediamine
6-Chloro-3,4-toluenediamine
2-Chloro-3,5-toluenediamine

C_7H_9ClO
2-Chloro-1-cyclohexenealdehyde†
1-Cyclohexene-1-carboxylic Acid, *Chloride*
4,4-Dimethyl-2-pentynoic Acid, *Chloride*
Ethchlorvynol
2-Heptynoic Acid, *Chloride*
3-Methyl-2,4-hexadienoic Acid, *Chloride*

$C_7H_9ClO_3$
α-Acetyl-δ-chloro-γ-valerolactone†
2,2-Dimethyl-5-oxo-oxolan-3-carboxylic Acid, *Chloride*
4-Ethyltetrahydro-5-oxo-3-furoic Acid, *Chloride*
Mesaconic Acid, α-*Et ester*, *Chloride*

$C_7H_9IN_2$
4-Iodo-2-methylphenylhydrazine

C_7H_9N
Benzylamine
1-Cyclohexene-1-carboxylic Acid, *Nitrile*
2,3-Dimethylpyridine
2,4-Dimethylpyridine
2,6-Dimethylpyridine
3,4-Dimethylpyridine
3,5-Dimethylpyridine
2-Ethylpyridine
3-Ethylpyridine
4-Ethylpyridine
2-Heptynoic Acid, *Nitrile*
N-Methylaniline
2-Methyl-1-cyclopentene-1-carboxylic Acid, *Nitrile*
Pyrrole, N-*Allyl*
o-Toluidine
m-Toluidine
p-Toluidine

C_7H_9NO
o-Aminobenzyl Alcohol
m-Aminobenzyl Alcohol
p-Aminobenzyl Alcohol
3-Amino-*o*-cresol
4-Amino-*o*-cresol
5-Amino-*o*-cresol
6-Amino-*o*-cresol
2-Amino-*m*-cresol
4-Amino-*m*-cresol
5-Amino-*m* cresol
6-Amino-*m*-cresol
2-Amino-*p*-cresol
3-Amino-*p*-cresol
o-Anisidine
m-Anisidine
p-Anisidine
N-Benzylhydroxylamine
O-Benzylhydroxylamine
2-Ethylpyrrole-5-aldehyde
2-Hydroxybenzylamine
3-Hydroxybenzylamine
4-Hydroxybenzylamine
2-Hydroxy-3,6-dimethylpyridine
2-Hydroxy-4,6-dimethylpyridine
3-Hydroxy-2,6-dimethylpyridine
4-Hydroxy-2,6-dimethylpyridine
2-Hydroxy-6-methylpyridine, *Me ether*
3-Hydroxy-2-methylpyridine, *Me ether*
4-Hydroxy-2-methylpyridine, *Me ether*
5-Hydroxy-2-methylpyridine, *Me ether*
2-Hydroxypyridine, *Et ether*
2-Hydroxypyridine, N-*Et*
3-Hydroxypyridine *Et ether*
4-Hydroxypyridine, *Et ether*
o-Methylaminophenol
m-Methylaminophenol
p-Methylaminophenol
1-Methyl-2-oxocyclopentane-1-carboxylic Acid, *Nitrile*
Pyridine-neurine
N-*o*-Tolylhydroxylamine
N-*m*-Tolylhydroxylamine
N-*p*-Tolylhydroxylamine

C_7H_9NOS
Toluene-*p*-sulphinic Acid, *Amide*

$C_7H_9NO_2$
Acetonylmalonic Acid, *Me ester*, *Nitrile*
3-Aminoguaiacol
4-Aminoguaiacol
5-Aminoguaiacol
6-Aminoguaiacol
5-Amino-2-hydroxybenzyl Alcohol
4-Aminoresorcinol, 1-*Me ether*
4-Aminoresorcinol, 3-*Me ether*
Cyclopropane-1,1-dicarboxylic Acid, *Mononitrile*, *Et ester*
2,3-Dihydroxybenzylamine
3,4-Dihydroxybenzylamine
3,4-Dihydroxypyridine, 3-*Et ether*
3,5-Dihydroxypyridine, 5-*Et ether*
3,3-Dimethylcyclopropane-1,2-dicarboxylic Acid, *Imide*
2,4-Dimethylfuran-3-carboxylic Acid, *Amide*
3,4-Dimethylglutaconic Acid, *Mononitrile*
3,4-Dimethylpyrrole-2-carboxylic Acid
3,5-Dimethylpyrrole-2-carboxylic Acid
4,5-Dimethylpyrrole-2-carboxylic Acid
2,4-Dimethylpyrrole-3-carboxylic Acid
2,5-Dimethylpyrrole-3-carboxylic Acid
4,5-Dimethylpyrrole-3-carboxylic Acid
2-Ethylidene-3-methylsuccinimide†
2-Ethyl-3-methylmaleic Acid, *Imide*
Furan-2-carboxylic Acid, *Ethylamide*
3-(2-Furyl)propionic Acid, *Amide*
Pyrrole-*N*-carboxylic Acid, *Et ester*

$C_7H_9NO_2$ (*continued*)
Pyrrole-2-carboxylic Acid, *Et ester*
Pyrrole-2-carboxylic Acid, N-*Et*
Succinimide, N-*Allyl*

$C_7H_9NO_2S$
4-Methylthiazole-2-carboxylic Acid, *Et ester*
2-Methylthiazole-4-carboxylic Acid, *Et ester*
2-Methylthiazole-5-carboxylic Acid, *Et ester*
Thiazole-2-acetic Acid, *Et ester*
Thiazole-4-acetic Acid, *Et ester*
Thiazole-5-acetic Acid, *Et ester*
β-2-Thienylalanine†
Toluene-α-sulphonamide
Toluene-*o*-sulphonamide
Toluene-*m*-sulphonamide
Toluene-*p*-sulphonamide

$C_7H_9NO_2S_2$
4-Amino-3-ethoxycarbonylmethylene-1,2-dithiole†

$C_7H_9NO_3$
Aloxidone
α-Amino-3,4,5-trihydroxytoluene
2,3-Dihydro-3-hydroxyanthranilic Acid
Ethyl 2-cyanoacetoacetate
β-3-Furylalanine†
5-Methylisoxazole-3-carboxylic Acid, *Et ester*
Tetrahydropyran-4,4-dicarboxylic Acid, *Mono-nitrile*

$C_7H_9NO_3S$
α-Aminotoluene-4-sulphinic Acid
Anisole-*o*-sulphonamide
Anisole-*m*-sulphonamide
Anisole-*p*-sulphonamide
N-Methylaniline-*o*-sulphonic Acid
N-Methylaniline-*m*-su phonic Acid
N-Methylaniline-*p*-sulphonic Acid
o-Toluidine-3-sulphonic Acid
o-Toluidine-4-sulphonic Acid
o-Toluidine-5-sulphonic Acid
o-Toluidine-6-sulphonic Acid
m-Toluidine-4-sulphonic Acid
m-Toluidine-6-sulphonic Acid
p-Toluidine-2-sulphonic Acid
p-Toluidine-3-sulphonic Acid

$C_7H_9NO_4$
2-Amino-3-hydroxy-3,2′-furylpropionic Acid
Cyanosuccinic Acid, *Di-Me ester*
Zymonic Acid, *Me ether*, *Amide*

$C_7H_9NO_5$
Fumarylalanine★†

$C_7H_9NO_6S_2$
o-Toluidine-4,5-disulphonic Acid
o-Toluidine-4,6-disulphonic Acid
p-Toluidine-2 5-disulphonic Acid
p-Toluidine-2,6-disulphonic Acid
p-Toluidine-3,5-disulphonic Acid

C_7H_9NS
o-Aminobenzenethiol, S-*Me*
m-Aminobenzenethiol, S-*Me*
p-Aminobenzenethiol, S-*Me*
2-Aminothiophene, N-*Et*
5-Amino-*o*-toluenethiol
6-Amino-*m*-toluenethiol
3-Amino-*p*-toluenethiol

$C_7H_9N_3$
Phenylguanidine

$C_7H_9N_3O$
Butane-1,2,4-tricarboxylic Acid, 1,2-*Dinitrile*-4-*amide*
1-Phenylsemicarbazide
2-Phenylsemicarbazide
4-Phenylsemicarbazide

$C_7H_9N_3O_2$
5-Aminopyrazine-2-carboxylic Acid, *Et ester*†
2,3-Diamino-5-nitrotoluene
2,4-Diamino-5-nitrotoluene
2,4-Diamino-6-nitrotoluene
2,5-Diamino-4-nitrotoluene
3,4-Diamino-5-nitrotoluene
4,5-Diamino-2-nitrotoluene
4-Nitro-*o*-phenylenediamine, 1-N-*Me*
2-Nitro-*p*-phenylenediamine, 1-N-*Me*
Spinacine†
2,3,5-Triaminobenzoic Acid
2,4,6-Triaminobenzoic Acid
3,4,5-Triaminobenzoic Acid

$C_7H_9N_3O_2S_2$
Sulphathiourea

$C_7H_9N_3O_3$
2,3-Diamino-5-nitrophenol, *Me ether*
2,5-Diamino-4-nitrophenol, *Me ether*

$C_7H_9N_3O_3S$
Sulphanilylurea

$C_7H_9N_3O_4$
5-Nitrouracil, 1-N-*Me*-3-N-*Et*
5-Nitrouracil, 3-N-*Me*-1-N-*Et*
Willardiine

$C_7H_9N_3O_5S_2$
3,5-Disulphobenzoic Acid, *Triamide*

$C_7H_9N_3S$
1-Phenylthiosemicarbazide
2-Phenylthiosemicarbazide
4-Phenylthiosemicarbazide
Zapotidine

$C_7H_9N_5O_2$
Pyridine-2,4-dicarboxylic Acid, *Dihydrazide*
Pyridine-2,5-dicarboxylic Acid, *Dihydrazide*
Pyridine-2,6-dicarboxylic Acid, *Dihydrazide*

$C_7H_9O_3P$
Benzyl dihydrogen phosphite

C_7H_{10}
1,3-Cycloheptadiene
1,3,5-Heptatriene
1,3,6-Heptatriene
1-Hepten-3-yne
1-Methyl-1,3-cyclohexadiene
2-Methyl-1,3-cyclohexadiene
5-Methyl-1,3-cyclohexadiene
1-Methyl-1,4-cyclohexadiene
2-Methylene[2,1,1]bicyclohexane†
2-Norbornene
Tricyclo[2,2,1,$0^{2,6}$]heptane†

$C_7H_{10}AsNO_4$
2-Aminophenol-4-arsonic Acid, N-*Me*

$C_7H_{10}BrNO_2$
Succinimide, N-3-*Bromopropyl*

$C_7H_{10}Br_2O_2$
2,3-Dibromobutyric Acid, *Allyl ester*
1,2-Dibromocyclohexanecarboxylic Acid
2,3-Dibromocyclohexanecarboxylic Acid
3,4-Dibromocyclohexanecarboxylic Acid

$C_7H_{10}Br_2O_4$
Dibromomalonic Acid, *Di-Et ester*

$C_7H_{10}ClNO$
Pyridine-choline, *Chloride*

$C_7H_{10}Cl_2O$
1-Chlorocyclohexane-1-carboxylic Acid, *Chloride*

$C_7H_{10}Cl_2O_2$
Butylmalonic Acid, *Dichloride*
2,3-Dichloroacrylic Acid, *Bu ester*
Diethylmalonic Acid, *Dichloride*
2,2-Dimethylglutaric Acid, *Dichloride*
3,3-Dimethylglutaric Acid, *Dichloride*
Isobutylmalonic Acid, *Dichloride*
Isopropylsuccinic Acid, *Dichloride*
3-Methyladipic Acid, *Dichloride*
Pimelic Acid, *Dichloride*

$C_7H_{10}Cl_2O_4$
Dichloromalonic Acid, *Di-Et ester*

$C_7H_{10}N_2$
1-Allyl-3-methylpyrazole
1-Allyl-5-methylpyrazole
o-Aminobenzylamine
m-Aminobenzylamine
p-Aminobenzylamine
2-Amino-4,6-dimethylpyridine
3-Amino-2,6-dimethylpyridine
4-Amino-2,6-dimethylpyridine
2-Aminopyridine, N-*Di-Me*
2-Aminopyridine, N-*Et*
Benzylhydrazine
Diallylcyanamide
2,3-Diaminotoluene
2,4-Diaminotoluene
2,5-Diaminotoluene
2,6-Diaminotoluene
3,4-Diaminotoluene
3,5-Diaminotoluene
Diethylmalonic Acid, *Dinitrile*
5-Ethyl-4-methylpyrimidine
Isobutylmalonic Acid, *Dinitrile*
o-Methylaminoaniline
m-Methylaminoaniline
p-Methylaminoaniline
N^1-Methyl-N^1-phenylhydrazine
N^1-Methyl-N^2-phenylhydrazine
Pimelic Acid, *Dinitrile*
4,5,6,7-Tetrahydroindazole
o-Tolylhydrazine
m-Tolylhydrazine
p-Tolylhydrazine

$C_7H_{10}N_2O$
4,6-Diamino-*m*-cresol
2,6-Diamino-*p*-cresol
2,4-Diaminophenol, *Me ether*
2,6-Diaminophenol, *Me ether*
3,5-Dimethylpyrrole-2-carboxylic Acid, *Amide*
2,5-Dimethylpyrrole-3-carboxylic Acid, *Amide*
4-Hydroxyphenylhydrazine, *Me ether*

$C_7H_{10}N_2O_2$
1-Ethyl-5-methylpyrazole-3-carboxylic Acid
2-Ethyl-5-methylpyrazole-3-carboxylic Acid
1-Ethyl-3-methyluracil
3-Ethyl-5-methyluracil
3-Ethyl-6-methyluracil
5-Ethyl-6-methyluracil
1-Imidazolylacetic Acid, *Et ester*
4-Imidazolylacetic Acid, *Et ester*
2-Methylimidazole-4-carboxylic Acid, *Et ester*
5-Methylimidazole-4-carboxylic Acid, *Et ester*
3-Methylmuconic Acid, *Diamide*
Tetrahydropyran-4,4-dicarboxylic Acid, *Mononitrile, Amide*
Thymine, 1,3-N-*Di-Me*

$C_7H_{10}N_2O_2S$
o-Aminomethylbenzenesulphonamide
m-Aminomethylbenzenesulphonamide
Carbimazole★†
Maphenide
5-Mercaptomethyluracil, S-*Et*†
N-Methylaniline-*o*-sulphonic Acid, *Amide*
N-Methylaniline-*p*-sulphonic Acid, *Amide*
o-Toluidine-5-sulphonic Acid, *Amide*
p-Toluidine-3-sulphonic Acid, *Amide*

$C_7H_{10}N_2O_3$
Isonitrosomalonic Acid, *Et ester-nitrile*, *Et ether*

$C_7H_{10}N_2O_4$
2,4-Dioxo-1-imidazolidylacetic Acid, *Et ester*
2,4-Dioxo-3-imidazolidylacetic Acid, *Et ester*

$C_7H_{10}N_2O_4S_2$
Toluene-2,4-disulphonic Acid, *Diamide*
Toluene-2,5-disulphonic Acid, *Diamide*
Toluene-2,6-disulphonic Acid, *Diamide*
Toluene-3,4-disulphonic Acid, *Diamide*
Toluene-3,5-disulphonic Acid, *Diamide*

$C_7H_{10}N_4$
Anilinoguanidine

$C_7H_{10}N_4O_2$
Lathyrine

$C_7H_{10}N_4O_2S$
Sulphanilylguanidine

$C_7H_{10}N_4S$
Thiocarbazide, 1-*Phenyl*

$C_7H_{10}O$
2-Cyclohepten-1-one
1-Cyclohexene-1-aldehyde
3-Cyclohexene-1-aldehyde
2,4-Dimethyl-2-cyclopenten-1-one
2-Ethyl-5-methylfuran
3,5-Heptadien-2-one
2,5-Heptadien-4-one
5-Heptyn-2-one
2-Methyl-2-cyclohexen-1-one
3-Methyl-2-cyclohexen-1-one
4-Methyl-2-cyclohexen-1-one
5-Methyl-2-cyclohexen-1-one

$C_7H_{10}O$ (*continued*)
6-Methyl-2-cyclohexen-1-one
3-Methyl-3-cyclohexen-1-one
4-Methyl-3-cyclohexen-1-one
Norcamphor
Tricyclo[2,2,1,$0^{2,6}$]heptan-3-ol †
2,3,4-Trimethylfuran
2,3,5-Trimethylfuran

$C_7H_{10}O_2$
Crotonic Acid, *Isopropenyl Ester*
1,2-Cycloheptanedione
Cyclohepta-1,3-dione †
1-Cyclohexene-1-carboxylic Acid
2-Cyclohexene-1-carboxylic Acid
3-Cyclohexene-1-carboxylic Acid
3,4-Dimethylcyclopentane-1,2-dione †
3,5-Dimethylcyclopentane-1,2-dione †
4,4-Dimethyl-2-pentynoic Acid
3-Ethylcyclopentane-1,2-dione †
2-Heptynoic Acid
2-Hexynoic Acid, *Me ester*
2-Methylacrylic Acid, *Allyl ester*
2-Methyl-1,3-cyclobutanedione, *Et enol ether* †
3-Methyl-1,2-cyclohexanedione
4-Methyl-1,2-cyclohexanedione
2-Methyl-1,3-cyclohexanedione
4-Methyl-1,3-cyclohexanedione
5-Methyl-1,3-cyclohexanedione
Methyl-1,4-cyclohexanedione
2-Methyl-1-cyclopentene-1-carboxylic Acid
1-Methyl-2-cyclopentene-1-carboxylic Acid
2-Methyl-2,4-hexadienoic Acid
3-Methyl-2,4-hexadienoic Acid
4-Methyl-2,4-hexadienoic Acid
5-Methyl-2,4-hexadienoic Acid
4-Methyl-2-pentynoic Acid, *Me ester*
2,4-Pentadienoic Acid, *Et ester*
2-Pentynoic Acid, *Et ester*

$C_7H_{10}O_3$
Acetoacetic Acid, *Allyl ester*
3-Acetylacrylic Acid, *Et ester*
2,2-Dimethylglutaric Acid, *Anhydride*
2,4-Dimethylglutaric Acid, *Anhydride*
3,3-Dimethylglutaric Acid, *Anhydride*
2-Ethylglutaric Acid, *Anhydride*
3-Ethylglutaric Acid, *Anhydride*
2-Ethyl-2-methylsuccinic Acid, *Anhydride*
2-Ethyl-3-methylsuccinic Acid, *Anhydride*
4-Ethyltetrahydro-5-oxofuran-3-aldehyde
β-Formylcrotonic Acid, *Et ester* †
2,4,6-Heptanetrione
Isopropylsuccinic Acid, *Anhydride*
1-Methyl-2-oxocyclopentane-1-carboxylic Acid
1-Methyl-3-oxocyclopentane-1-carboxylic Acid
2-Methyl-3-oxocyclopentane-1-carboxylic Acid
3-Methyl-2-oxocyclopentane-1-carboxylic Acid
4-Methyl-2-oxocyclopentane-1-carboxylic Acid
4-Methyl-3-oxocyclopentane-1-carboxylic Adid
5-Methyl-3-oxocyclopentane-1-carboxylic Acid
2-Oxocyclohexanecarboxylic Acid
3-Oxocyclohexanecarboxylic Acid
4-Oxocyclohexanecarboxylic Acid
2-Oxocyclopentane-1-carboxylic Acid, *Me ester*
3-Oxocyclopentane-1-carboxylic Acid, *Me ester*
Propyl succinic Acid, *Anhydride*
Reductic Acid, *Di-Me ether*

$C_7H_{10}O_4$
3-Acetyl-4-oxovaleric Acid
Acetylpyruvic Acid, *Et ester*
Allylsuccinic Acid
Allylmethylmalonic Acid
1-Butene-1,4-dicarboxylic Acid, β-*Me ester*
Citraconic Acid, *Di-Me ester*
Cyclopentane-1,1-dicarboxylic Acid
Cyclopentane-1,2-dicarboxylic Acid
Cyclopentane-1,3-dicarboxylic Acid
Cyclopropane-1,1-dicarboxylic Acid, *Di-Me ester*
Cyclopropane-1,2-dicarboxylic Acid, *Di-Me ester*
Diacetylacetic Acid, *Me ester*
2,2-Diacetylpropionic Acid
3,3-Dimethylcyclopropane-1,2-dicarboxylic Acid
Dimethylfumaric Acid, *Me ester*
2,4-Dimethylglutaconic Acid
3,4-Dimethylglutaconic Acid
4,4-Dimethylglutaconic Acid
2,2-Dimethyl-5-oxo-oxolan-3-carboxylic Acid
2,5-Dioxohexane-3-carboxylic Acid
4-Ethylglutaconic Acid
2-Ethylideneglutaric Acid
2-Ethylidene-3-methylsuccinic Acid
2-Ethyl-3-methylmaleic Acid
4-Ethyltetrahydro-5-oxo-3-furoic Acid
4-Ethyltetrahydro-5-oxofuran-3-carboxylic Acid
3-Hydroxy-2,4-dimethylglutaric Acid, *Anhydride*
Mesaconic Acid, *Di-Me ester*
Mesaconic Acid, α-*Et ester*
Mesaconic Acid, β-*Et ester*
2-Methyl-1-butene-1,1-dicarboxylic Acid
3-Methyl-1-butene-1,1-dicarboxylic Acid
cis-3-Methyl-1-butene-1,2-dicarboxylic Acid
trans-3-Methyl-1-butene-1,2-dicarboxylic Acid
3-Methyl-2-butene-1,2-dicarboxylic Acid
2-Methyl-1-butene-1,4-dicarboxylic Acid
3-Methyl-1-butene-1,4-dicarboxylic Acid
2-Methyl-2-butene-1,4-dicarboxylic Acid
3-Methyl-1-butene-2,3-dicarboxylic Acid
2-Methyl-1-butene-3,4-dicarboxylic Acid
2-Propene-1,2-dicarboxylic Acid, *Di-Me ester*
2-Propene-1,2-dicarboxylic Acid, *Mono-Et ester*
Propylfumaric Acid
Propylidenesuccinic Acid
Propylmaleic Acid

$C_7H_{10}O_5$
Acetonedicarboxylic Acid, *Di-Me ester*
Acetonedicarboxylic Acid, *Et ester*
Acetylethylmalonic Acid
2-Acetylglutaric Acid
3-Acetylglutaric Acid
Acetylmalonic Acid, *Mono-Et ester*
2-Acetyl-2-methylsuccinic Acid
2-Acetyl-3-methylsuccinic Acid
Butyrylmalonic Acid
Formylsuccinic Acid, *Di-Me ester*
Glyceraldehyde, 1,2-*Di-Ac*
Mesoxalic Acid, *Di-Et ester*
Oxalacetic Acid, 4-*Me*, 1-*Et ester*
Oxalacetic Acid, 1-*Me*, 4-*Et ester*

$C_7H_{10}O_5$ (*continued*)
2-Oxo-1,7-heptanedioic Acid
4-Oxo-1,7-heptanedioic Acid
Quinide
Shikimic Acid
Tetrahydropyran-2,6-dicarboxylic Acid
Tetrahydropyran-4,4-dicarboxylic Acid
3,4,5-Trihydroxycyclohexanecarboxylic Acid, *Lactone*

$C_7H_{10}O_6$
Butane-1,1,2-tricarboxylic Acid
Butane-1,1,3-tricarboxylic Acid
Butane-1,1,4-tricarboxylic Acid
Butane-1,2,2-tricarboxylic Acid
Butane-1,2,4-tricarboxylic Acid
Butane-1,3,3-tricarboxylic Acid
Butane-2,2,3-tricarboxylic Acid
1,3-Dioxolan-4,5-dicarboxylic Acid, *Di-Me ester*
Methane-triacetic Acid
Methane-tricarboxylic Acid, *Tri-Me ester*
2-Methyltricarballylic Acid
3-Methyltricarballylic Acid
Tricarballylic Acid, α-*Me ester*
Tricarballylic Acid, β-*Me ester*
1,3,5-Trihydroxy-4-oxocyclohexanecarboxylic Acid†
1,3,4-Trihydroxy-5-oxocyclohexanecarboxylic Acid★†

$C_7H_{10}O_7$
Gluco-ascorbic Acid
Homocitric Acid†
Homoisocitric Acid†

$C_7H_{10}S$
2-Ethyl-5-methylthiophen
3-Ethyl-4-methylthiophen
3-Ethyl-5-methylthiophen
2-Propylthiophene
3-Propylthiophene
2,3,4-Trimethylthiophene
2,3,5-Trimethylthiophene

$C_7H_{11}BrO_2$
2-Bromoacrylic Acid, sec-*Butyl ester*
2-Bromoacrylic Acid, tert-*Butyl ester*
2-Bromobutyric Acid, *Allyl ester*
4-Bromocrotonic Acid, *Propyl ester*
1-Bromocyclohexane-carboxylic Acid
2-Bromocyclohexane-carboxylic Acid
3-Bromocyclohexane-carboxylic Acid
4-Bromocyclohexane-carboxylic Acid

$C_7H_{11}BrO_3$
3-Bromolevulinic Acid, *Et ester*

$C_7H_{11}BrO_4$
Bromomalonic Acid, *Di-Et ester*

$C_7H_{11}ClN_2O_4$
Glutamic Acid, 5-*Monoamide*, N-*Chloroacetyl*

$C_7H_{11}ClO$
Cyclohexanecarboxylic Acid, *Chloride*
1-Methylcyclopentane-1-carboxylic Acid, *Chloride*
2-Methylcyclopentane-1-carboxylic Acid, *Chloride*
3-Methylcyclopentane-1-carboxylic Acid, *Chloride*
4-Methyl-2-hexenoic Acid, *Chloride*
5-Methyl-2-hexenoic Acid, *Chloride*
4-Methyl-3-hexenoic Acid, *Chloride*

$C_7H_{11}ClO_2$
2-Chloroacrylic Acid, tert-*Butyl ester*
1-Chlorocyclohexane-1-carboxylic Acid
3-Chloro-2-ethylcrotonic Acid, *Me ester*
3-Chloroisocrotonic Acid, *Propyl ester*

$C_7H_{11}ClO_3$
Oxalic Acid, 3-*Methylbutyl ester*, *Chloride*

$C_7H_{11}ClO_4$
Chloromalonic Acid, *Di-Et ester*

$C_7H_{11}Cl_3O_2$
Trichloroacetic Acid, 2-*Methylbutyl ester*
2,2,3-Trichloropropionic Acid, *Isobutyl ester*

$C_7H_{11}Cl_3O_3$
3,3,4-Trichloro-2-hydroxyvaleric Acid, *Et ester*
3,3,3-Trichlorolactic Acid, *Et ester*, *Et ether*
3,3,3-Trichlorolactic Acid, *Isobutyl ester*

$C_7H_{11}HgN$
Mercuri-cyclohexyl cyanide

$C_7H_{11}N$
Cyclohexanecarboxylic Acid, *Nitrile*
2-Ethyl-3-methylpyrrole
2-Ethyl-4-methylpyrrole
3-Ethyl-4-methylpyrrole
3-Ethyl-5-methylpyrrole
3-Methyl-2-hexenoic Acid, *Nitrile*
5-Methyl-2-hexenoic Acid, *Nitrile*
5-Methyl-3-hexenoic Acid, *Nitrile*
2-Nortropene
Pyrrole, N-*Propyl*
1,2,5-Trimethylpyrrole
2,3,4-Trimethylpyrrole
2,3,5-Trimethylpyrrole

$C_7H_{11}NO$
2-Azabicyclo[2,2,2]octan-3-one
1-Cyclohexene-1-carboxylic Acid, *Amide*
2-Cyclohexene-1-carboxylic Acid, *Amide*
2-Heptynoic Acid, *Amide*
1-Hydroxycyclohexanecarboxylic Acid, *Nitrile*
2β-Hydroxypropylpyrrole
3-Methyl-2,4-hexadienoic Acid, *Amide*
3-Methyl-5-oxohexanoic Acid, *Nitrile*
4-Methyl-5-oxohexanoic Acid, *Nitrile*
Nortropinone

$C_7H_{11}NO_2$
Arecaidine
Butylmalonic Acid, *Mononitrile*
Cyanoacetic Acid, *Butyl ester*
Cyanoacetic Acid, *Isobutyl ester*
2-Cyanobutyric Acid, *Et ester*
4-Cyanobutyric Acid, *Et ester*
Diethylmalonic Acid, *Mononitrile*
2,2-Dimethylglutaric Acid, *Imide*
2,3-Dimethylglutaric Acid, *Imide*
2,4-Dimethylglutaric Acid, *Imide*
3,3-Dimethylglutaric Acid, *Imide*
2-Ethyl-3-methylsuccinic Acid, *Imide*
Hypoglycin A†

$C_7H_{11}NO_2$ (*continued*)
Methylsuccinic Acid, β-*Et ester*, α-*Nitrile*
Nor-ψ-scopine
Norscopoline
Pyridine-choline
Succinimide, N-*Propyl*
1,2,3,6-Tetrahydropyridine-2-carboxylic Acid, *Me ester*
1,2,5,6-Tetrahydropyridine-3-carboxylic Acid, *Me ester*
Trimethylsuccinic Acid, β-*Nitrile*

$C_7H_{11}NO_3$
2-Amino-4-pentenoic Acid, N-*Ac*
Citramalic Acid, 1-*Nitrile*, 2-*Ethyl ester*
Mesaconic Acid, α-*Et ester*, β-*Amide*
1-Methyl-2-oxopyrrolidine-5-acetic Acid
5-Oxopyrrolidine-2-carboxylic Acid, *Et ester*
Paramethadione

$C_7H_{11}NO_4$
3-Acetylglutaric Acid, *Mono-Amide*
L-Aspartic Acid, β-*Allyl ester*
γ-Ethylideneglutamic Acid†
4-Oxo-1,7-heptanedioic Acid, *Monoamide*
Piperidine-2,3-dicarboxylic Acid
Piperidine-3,4-dicarboxylic Acid
Piperidine-2,6-dicarboxylic Acid
Quinic Acid, *Nitrile*

$C_7H_{11}NO_5$
Glutamic Acid, N-*Ac*
Isonitrosomalonic Acid, *Di-Et ester*

$C_7H_{11}NO_6$
Nitromalonic Acid, *Di-Et ester*
4-Nitropimelic Acid

$C_7H_{11}NO_6S$
S-Cysteinosuccinic Acid
S-(αβ-Dicarboxyethyl)cysteine

$C_7H_{11}N_3$
3,5-Diamino-2,6-dimethylpyridine
Spinaceamine, 6-*Me deriv.*†

$C_7H_{11}N_3O$
4-Amino-6-methyl-2-pyrimidone, 6-N-*Et*
2-Hydroxyhistamine, O-*Me ether*†

$C_7H_{11}N_3O_2$
Histidine, *Me ester*
N^{α}-Methylhistidine
1-Methylhistidine†
2-Methylhistidine†
3-Methylhistidine†

$C_7H_{11}N_3O_4S_2$
p-Toluidine-2,5-disulphonic Acid, *Diamide*

$C_7H_{11}N_3O_6S_3$
Toluene-2,4,6-trisulphonic Acid, *Triamide*

C_7H_{12}
Bicyclo[3,1,1]heptane†
Bicyclo[3,2,0]heptane
Bicyclo[4,1,0]heptane
Cycloheptene
3,4-Dimethyl-1,2-pentadiene
4,4-Dimethyl-1,2-pentadiene
2,4-Dimethyl-1,3-pentadiene
2,4-Dimethyl-2,3-pentadiene
1,2-Heptadiene
1,4-Heptadiene
1,5-Heptadiene
1,6-Heptadiene
2,4-Heptadiene
1-Heptyne
2-Heptyne
3-Heptyne
1-Methylcyclohexene
3-Methylcyclohexene
5-Methyl-1,2-hexadiene
3-Methyl-1,3-hexadiene
4-Methyl-1,3-hexadiene
5-Methyl-1,4-hexadiene
2-Methyl-1,5-hexadiene
3-Methyl-1,5-hexadiene
2-Methyl-2,4-hexadiene
3-Methyl-2,4-hexadiene
5-Methyl-1-hexyne
Norbornane
Spiro[3,3]heptane†

$C_7H_{12}Br_2O_2$
2,3-Dibromo-3-methylbutyric Acid, *Et ester*
2,3-Dibromopropionic Acid, n-*Butyl ester*
2,3-Dibromopropionic Acid, *Isobutyl ester*
2,3-Dibromopropionic Acid, tert-*Butyl ester*
2,3-Dibromovaleric Acid, *Et ester*
2,5-Dibromovaleric Acid, *Et ester*

$C_7H_{12}ClNO$
1-Chlorocyclohexane-1-carboxylic Acid, *Amide*

$C_7H_{12}ClN_5$
Simazine†

$C_7H_{12}Cl_2O_2$
Dichloroacetic Acid, *Pentyl ester*
Dichloroacetic Acid, 2-*Pentyl ester*
2,3-Dichloropropionic Acid, tert-*Butyl ester*

$C_7H_{12}Cl_3NO_3$
Chloralurethane, *Et ether*

$C_7H_{12}N_2$
1-Aminocyclohexanecarboxylic Acid, *Nitrile*
1,5-Diazabicyclo[4,3,0]non-5-ene†
1,2-Diethylimidazole
4,5-Diethylimidazole
5-Isopropyl-3-methylpyrazole
3-Piperidineacetic Acid, *Nitrile*
Piperidinoacetic Acid, *Nitrile*
1,3,4,5-Tetramethylpyrazole

$C_7H_{12}N_2O$
1,3-Diaza-6-oxa-adamantane†
Diethylmalonic Acid, *Amide-nitrile*

$C_7H_{12}N_2O_2$
Cyclopentane-1,3-dicarboxylic Acid, *Diamide*
Ectylurea
trans-3-Methyl-1-butene-1,2-dicarboxylic Acid, *Diamide*
Propylfumaric Acid, *Diamide*

$C_7H_{12}N_2O_3$
Glycyl-L-proline

$C_7H_{12}N_2O_5$
Diethoxycarbonylurea

$C_7H_{12}N_4O$
Caffeidine

$C_7H_{12}N_4O_3$
Allantoin, 1,3,8-*Tri-Me*
1,3,6-Trimethylallantoin

$C_7H_{12}O$
Acetylcyclopentane
Cycloheptanone
Cyclohexanealdehyde
Cyclohexen-3-ol, *Me ether*
Cyclohexen-4-ol, *Me ether*
1,5-Heptadien-4-ol
1,6-Heptadien-4-ol
2-Heptenal
3-Heptenal
4-Heptenal †
3-Hepten-2-one
4-Hepten-2-one
5-Hepten-2-one
6-Hepten-2-one
1-Hepten-3-one
6-Hepten-3-one
1-Hepten-4-one
2-Hepten-4-one
3-Heptyn-1-ol
2-Methylcyclohexanone
3-Methylcyclohexanone
4-Methylcyclohexanone
1-Methylcyclopentane Aldehyde
2-Methylcyclopentane Aldehyde
3-Methylcyclopentane Aldehyde
2-Methylenecyclohexanol
3-Methyl-3-hexen-2-one
4-Methyl-3-hexen-2-one
5-Methyl-3-hexen-2-one
5-Methyl-4-hexen-2-one
3-Methyl-5-hexen-2-one
4-Methyl-5-hexen-2-one
5-Methyl-5-hexen-2-one
5-Methyl-1-hexen-3-one
2-Methyl-4-hexen-3-one
4-Methyl-4-hexen-3-one
5-Methyl-4-hexen-3-one
Norborneol
4-Pentyn-1-ol, *Et ether*
Tetramethylcyclopropanone †

$C_7H_{12}O_2$
Acrylic Acid, n-*Butyl ester*
Acrylic Acid, *Isobutyl ester*
Angelic Acid, *Et ester*
Butyric Acid, *Allyl ester*
Crotonic Acid, *Isopropyl ester*
Cyclobutanecarboxylic Acid, *Et ester*
Cyclohexanecarboxylic Acid
Cyclopentanecarboxylic Acid, *Me ester* ★ †
Cyclopentylacetic Acid
3,3-Dimethylacrylic Acid, *Et ester*
3,4-Dimethyl-3-pentenoic Acid
1,6-Dioxaspiro[4,4]nonane
1,7-Heptanedial
2,3-Heptanedione
2,4-Heptanedione
2,5-Heptanedione
2,6-Heptanedione
3,4-Heptanedione
3,5-Heptanedione
2-Heptenoic Acid
3-Heptenoic Acid
4-Heptenoic Acid
5-Heptenoic Acid
6-Heptenoic Acid
2-Hexenoic Acid, *Me ester*
3-Hexenoic Acid, *Me ester*
5-Hexenoic Acid, *Me ester*
Isobutyric Acid, *Allyl ester*
2-Methylcrotonic Acid, *Et ester*
1-Methylcyclopentane-1-carboxylic Acid
2-Methylcyclopentane-1-carboxylic Acid
3-Methylcyclopentane-1-carboxylic Acid
2-Methylenebutyric Acid, *Et ester*
5-Methyl-2,3-hexanedione
3-Methyl-2,4-hexanedione
5-Methyl-2,4-hexanedione
3-Methyl-2,5-hexanedione
2-Methyl-2-hexenoic Acid
3-Methyl-2-hexenoic Acid
4-Methyl-2-hexenoic Acid
5-Methyl-2-hexenoic Acid
2-Methyl-3-hexenoic Acid
3-Methyl-3-hexenoic Acid
4-Methyl-3-hexenoic Acid
5-Methyl-3-hexenoic Acid
3-Methyl-4-hexenoic Acid
5-Methyl-4-hexenoic Acid
4-Methyl-5-hydroxyhexanoic Acid Lactone †
4-Methyl-2-oxo-1-cyclopentanol, *Me ether*
2-Methyl-2-pentenoic Acid, *Me ester*
3-Methyl-2-pentenoic Acid, *Me ester*
3-Methyl-3-pentenoic Acid, *Me ester*
4-Methyl-3-pentenoic Acid, *Me ester*
Mutomycin
2-Pentenoic Acid, *Et ester*
3-Pentenoic Acid, *Et ester*
4-Pentenoic Acid, *Et ester*
Propynal, *Di-Et acetal*
Pyruvic Acid, *Butyl ester*

$C_7H_{12}O_2S$
S-Vinyl-*O*-*tert*-butyl thiolcarbonate †

$C_7H_{12}O_3$
Acetoacetic Acid, *Isopropyl ester*
2,2-Dimethylacetoacetic Acid, *Me ester*
2,3-Epoxy-2-methylbutyric Acid, *Et ester*
2,3-Epoxy-3-methylbutyric Acid, *Et ester*
2-Ethyl-2-methylacetoacetic Acid
2-Formylisobutyric Acid, *Et ester*
3-Formyl-2-methylpropionic Acid, *Et ester*
2-Hexanone-3-carboxylic Acid
1-Hydroxycyclohexanecarboxylic Acid
2-Hydroxycyclohexanecarboxylic Acid
3-Hydroxycyclohexanecarboxylic Acid
4-Hydroxycyclohexanecarboxylic Acid
1-Hydroxycyclopentane-1-carboxylic Acid, *Me ester*
2-Hydroxycyclopentane-1-carboxylic Acid, *Me ester*
1-Hydroxy-2-methylcyclopentane-1-carboxylic Acid
2-Hydroxy-1-methylcyclopentane-1-carboxylic Acid
2-Hydroxy-5-methylcyclopentane-1-carboxylic Acid

$C_7H_{12}O_3$ (*continued*)
2-Hydroxy-3-methylcyclopentane-1-carboxylic Acid
3-Hydroxy-2-methylcyclopentane-1-carboxylic Acid
3-Hydroxy-5-methylcyclopentane-1-carboxylic Acid
Levulinic Acid, *Et ester*
Mesitonic Acid
2-Methylacetoacetic Acid, *Et ester*
3-Methyl-2-oxobutyric Acid, *Et ester*
2-Methyl-5-oxohexanoic Acid
2-Methyl-3-oxohexanoic Acid
3-Methyl-5-oxohexanoic Acid
4-Methyl-5-oxohexanoic Acid
2-Methyl-3-oxovaleric Acid, *Me ester*
3-Methyl-4-oxovaleric Acid, *Me ester*
2-Oxoheptanoic Acid
3-Oxoheptanoic Acid
4-Oxoheptanoic Acid
5-Oxoheptanoic Acid
6-Oxoheptanoic Acid
3-Oxohexanoic Acid, *Me ester*
4-Oxohexanoic Acid, *Me ester*
2-Oxopentane-3-carboxylic Acid, *Me ester*
2-Oxovaleric Acid, *Et ester*
3-Oxovaleric Acid, *Et ester*
Tetrahydrofuran-2-carboxylic Acid, *Et ester*
Tetrahydropyran-4-carboxylic Acid, *Me ester*
Trimethylpyruvic Acid, *Me ester*

$C_7H_{12}O_4$
Acetolactic Acid, *Et ester*†
2-Acetoxypropionic Acid, *Et ester*
Adipic Acid, *Me ester*
Butylmalonic Acid
(2-Butyl)malonic Acid
tert-Butylmalonic Acid
Diethyl malonate
Diethylmalonic Acid
2,2-Dimethylglutaric Acid
2,3-Dimethylglutaric Acid
2,4-Dimethylglutaric Acid
3,3-Dimethylglutaric Acid
Dimethylmalonic Acid, *Di-Me ester*
Dimethylmalonic Acid, *Et ester*
2,2-Dimethylsuccinic Acid, 1-*Me ester*
2,2-Dimethylsuccinic Acid, 4-*Me ester*
2-Ethylglutaric Acid
3-Ethylglutaric Acid
Ethylmalonic Acid, *Di-Me ester*
2-Ethyl-2-methylsuccinic Acid
2-Ethyl-3-methylsuccinic Acid
Glutaric Acid, *Di-Me ester*
Glutaric Acid, *Mono-Et ester*
Isobutylmalonic Acid
Isopropylmalonic Acid, *Mono-Me ester*
Isopropylsuccinic Acid
Malonic Acid, *Monobutyl ester*
2-Methyladipic Acid
3-Methyladipic Acid
3-Methylbutane-2,2-dicarboxylic Acid
Methylpropylmalonic Acid
Methylsuccinic Acid, *Di-Me ester*
Methylsuccinic Acid, *Mono-Et ester*
Pimelic Acid
Propylsuccinic Acid
Succinic Acid, *Me-Et ester*
Succinic Acid, *Propyl ester*
2,4,8,10-Tetroxaspiro[5,6]undecane
Trimethylsuccinic Acid

$C_7H_{12}O_5$
3-Carboxy-3-hydroxy-4-methylpentanoic Acid
Citramalic Acid, *Di-Me ester*
Citramalic Acid, *Et ether*
Diacetin
3-Hydroxy-2,2-dimethylglutaric Acid
3-Hydroxy-2,4-dimethylglutaric Acid
3-Hydroxyglutaric Acid, *Di-Me ester*
1-Hydroxy-3-methylbutane-1,1-dicarboxylic Acid
2-Hydroxy-3-methylsuccinic Acid, *Mono-Et ester*
Hydroxytrimethylsuccinic Acid
Laevoglucosan, 3-*Me ether*
Malic Acid, *Propyl ether*
Malic Acid, *Isopropyl ether*
Methoxysuccinic Acid, *Di-Me ester*
Tartronic Acid, *Di-Et ester*
3,4,5-Trihydroxycyclohexanecarboxylic Acid

$C_7H_{12}O_6$
Cordycepic Acid
L-*Gulo*heptulosan
Mesoxalic Acid, *Di-Et acetal*
Mesoxalic Acid, *Di-Et ester, hydrate*
Quinic Acid
α-*Rhamno*hexonic Acid, *γ-Lactone*
β-*Rhamno*hexonic Acid, *γ-Lactone*
*Rhodeo*hexonic Acid, *γ-Lactone*

$C_7H_{12}O_7$
Fructuronic Acid, *Me ester*
Galacturonic Acid, *Me ester*
Galacturonic Acid, *Methyl glycoside*
*Gala*heptonic Acid, *γ-Lactone*
*Gluco*heptonic Acid, *γ-Lactone*
*Gluco*heptonic Acid, *δ-Lactone*
Glucuronic Acid, *Me ester*
Glucuronic Acid, *Me glycoside*
Glucuronic Acid, 4-*Me ether*
Mannoheptonic Acid, *Lactone*
Trihydroxyglutaric Acid, *Di-Me ester*

$C_7H_{12}S$
3-Thiabicyclo[3,2,1]octane★†

$C_7H_{13}As$
1-Arsabicyclo[3,3,0]octane

$C_7H_{13}BrN_2O_2$
Adalin

$C_7H_{13}BrO$
2-Bromocyclohexanol, *Me ether*

$C_7H_{13}BrO_2$
2-Bromobutyric Acid, *Propyl ester*
2-Bromobutyric Acid, *Isopropyl ester*
2-Bromo-3-methylbutyric Acid, *Et ester*
3-Bromo-3-methylbutyric Acid, *Et ester*
2-Bromopropionic Acid, *Butyl ester*
2-Bromopropionic Acid *Isobutyl ester*
2-Bromovaleric Acid, *Et ester*
3-Bromovaleric Acid, *Et ester*
4-Bromovaleric Acid, *Et ester*
5-Bromovaleric Acid, *Et ester*

$C_7H_{13}ClO$
2-Ethylvaleric Acid, *Chloride*
Heptanoic Acid, *Chloride*
2-Methylhexanoic Acid, *Chloride*
3-Methylhexanoic Acid, *Chloride*
4-Methylhexanoic Acid, *Chloride*

$C_7H_{13}ClO_2$
Chloroacetic Acid, 2-*Methylbutyl ester*
2-Chlorobutyric Acid, *Propyl ester*
3-Chlorobutyric Acid, *Propyl ester*
4-Chlorobutyric Acid, *Propyl ester*
2-Chloro-2-methylbutyric Acid, *Et ester*
2-Chloro-3-methylbutyric Acid, *Et ester*
3-Chloro-3-methylbutyric Acid, *Et ester*
2-Chlorovaleric Acid, *Et ester*
3-Chlorovaleric Acid, *Et ester*
4-Chlorovaleric Acid, *Et ester*
5-Chlorovaleric Acid, *Et ester*
2-Hydroxyvaleric Acid, *Et ether*, *Chloride*
5-Hydroxyvaleric Acid, *Et ether*, *Chloride*

$C_7H_{13}ClO_3$
4-Chloro-3-hydroxybutyric Acid, *Propyl ester*
3-Chloro-2-hydroxy-2-methylpropionic Acid, *Propyl ester*

$C_7H_{13}ClO_7$
*Gala*heptonic Acid, *Chloride*

$C_7H_{13}Cl_3O_2$
Chloral, 3-*Methylbutyl acetal*

$C_7H_{13}FO_2$
2-Fluoroheptanoic Acid †

$C_7H_{13}HgN$
Mercuri-hexyl cyanide

$C_7H_{13}IO$
2-Iodocyclohexanol, *Me ether*

$C_7H_{13}IO_2$
4-Iodovaleric Acid, *Et ester*
5-Iodovaleric Acid, *Et Ester*

$C_7H_{13}N$
1-Azabicyclo[3,2,1]octane
2-Azabicyclo[2,2,2]octane †
6-Azabicyclo[3,2,1]octane
1-Azaspiro[2,5]octane
Heptanoic Acid, *Nitrile*
2-Methylhexanoic Acid, *Nitrile*
3-Methylhexanoic Acid, *Nitrile*
4-Methylhexanoic Acid, *Nitrile*
5-Methylhexanoic Acid, *Nitrile*
Norbornylamine
Nortropane
Pyrrolidine, N-*Allyl*
Pyrrolizidine †
Quinuclidine
3-Vinylpiperidine

$C_7H_{13}NO$
Acrylic Acid, N-*Diethylamide*
Cyclohexanecarboxylic Acid, *Amide*
1-Methylcyclopentane-1-carboxylic Acid, *Amide*
2-Methylcyclopentane-1-carboxylic Acid, *Amide*
3-Methylcyclopentane-1-carboxylic Acid, *Amide*
5-Methyl-2-hexenoic Acid, *Amide*
5-Methyl-4-hexenoic Acid, *Amide*
Nortropine
Piperidinoacetaldehyde

$C_7H_{13}NOS_2$
L(−)-5-Methylsulphinylpentyl isothiocyanate

$C_7H_{13}NO_2$
1-Aminocyclohexanecarboxylic Acid
2-Aminocyclohexanecarboxylic Acid
3-Aminocyclohexanecarboxylic Acid
4-Aminocyclohexanecarboxylic Acid
2-Amino-4-methylhex-4-enoic Acid †
Cyanoacetaldehyde, *Di-Et acetal*
2-Ethyl-2-methylacetoacetic Acid, *Amide*
2-Ethyl-3-methylsuccinic Acid, *Amide*
2-Hexanone-3-carboxylic Acid, *Amide*
1-Hydroxycyclohexanecarboxylic Acid, *Amide*
2-Hydroxycyclohexanecarboxylic Acid, *Amide*
3-Hydroxycyclohexanecarboxylic Acid, *Amide*
3-Methylaminocrotinic Acid, *Et ester*
1-Methyl-1-nitrocyclohexane
1-Methyl-2-nitrocyclohexane
1-Methyl-3-nitrocyclohexane
1-Methyl-4-nitrocyclohexane
1-Methylpyrrolidine-2-carboxylic Acid, *Me ester*
Nitromethylcyclohexane
2-Piperidineacetic Acid
3-Piperidineacetic Acid
4-Piperidineacetic Acid
Piperidine-*N*-carboxylic Acid, *Me ester*
Piperidine-2-carboxylic Acid, *Me ester*
Piperidine-2-carboxylic Acid, N-*Me*
Piperidine-3-carboxylic Acid, *Me ester*
Piperidine-3-carboxylic Acid, N-*Me*
Piperidine-4-carboxylic Acid, *Me ester*
Piperidinoacetic Acid
Proline, *Et ester*
Stachydrine

$C_7H_{13}NO_3$
Acetyl-DL-alanine, *Et ester*
2-Amino-6-hydroxy-4-methylhex-4-enoic Acid †
Betonicine
Diethylmalonic Acid, *Monoamide*
3,3-Dimethylglutaric Acid, *Monoamide*
N-Formyl-isoleucine
N-Formyl-leucine
3-Formyl-2-methylpropionic Acid, *Et ester*, *Oxime*
3-Hydroxyproline, *Et ester*
4-Hydroxyproline, *Et ester*
Isopropylmalonic Acid, *Monoamide*, *Me ester*
Oxamic Acid, 3-*Methylbutyl ester*
Turicine

$C_7H_{13}NO_3S$
S-(But-1-enyl)cysteine sulphoxide †

$C_7H_{13}NO_4$
Aminomalonic Acid, *Di-Et ester*
3-Aminopimelic Acid
L-Aspartic Acid, β-*Propyl ester*
L-Aspartic Acid, β-*Isopropyl ester*
Carbethoxyglycine, *Et ester*
Glutamic Acid, *Mono-Et ester*
Methyl 4-acetamido-4-deoxy-L-erythrofuranoside †
N-Methylaspartic Acid, *Mono-Et ester*
Methyliminodiacetic Acid, *Di-Me ester*
Nitroacetic Acid, *Isopentyl ester*

$C_7H_{13}NO_4S$
S-[β-Carboxy-isopropyl]cysteine
S-[2-Carboxy-1-propyl] cysteine

$C_7H_{13}NO_5$
1-Acetyl-D-*gulo*-2,3,4,5-tetrahydroxypiperidine
3-Hydroxyglutamic Acid, *Di-Et ester*
Quinic Acid, *Amide*
α-*Rhamno*hexonic Acid, *Nitrile*

$C_7H_{13}NO_6$
Mannoheptonic Acid, *Nitrile*

$C_7H_{13}NS_2$
5-Methylthiopentyl isothiocyanate
Piperidine-*N*-carbodithioic Acid, *Me ester*

$C_7H_{13}N_3$
4,4′-Imidazolylbutylamine
1,3,5-Triaza-adamantane

$C_7H_{13}N_3O$
1-Carbamoyl-3,5,5-trimethyl-Δ^2-pyrazoline†

$C_7H_{13}N_3O_2$
2-Azido-3-methylbutyric Acid, *Et ester*
Piperidine-2,6-dicarboxylic Acid, *Diamide*

$C_7H_{13}N_3O_4$
Diglycylalanine
Diglycylglycine, *Me ester*
Glycyl-alanyl-glycine
Glycyl-D-glutamine

$C_7H_{13}O_5P$
Phosphonoacetic Acid, 3,3-*Di-Me*-1-*allyl ester*

C_7H_{14}
Cycloheptane
1,1-Dimethylcyclopentane
1,2-Dimethylcyclopentane
1,3-Dimethylcyclopentane
2,3-Dimethyl-1-pentene
2,4-Dimethyl-1-pentene
3,3-Dimethyl-1-pentene
4,4-Dimethyl-1-pentene
2,3-Dimethyl-2-pentene
2,4-Dimethyl-2-pentene
3,4-Dimethyl-2-pentene
4,4-Dimethyl-2-pentene
3-Ethylpent-2-ene
1-Heptene
2-Heptene
3-Heptene
Methylcyclohexane
2-Methyl-1-hexene
3-Methyl-1-hexene
4-Methyl-1-hexene
5-Methyl-1-hexene
2-Methyl-2-hexene
3-Methyl-2-hexene
4-Methyl-2-hexene
5-Methyl-2-hexene
2-Methyl-3-hexene
3-Methyl-3-hexene

$C_7H_{14}BrF$
1-Bromo-2-fluoroheptane†

$C_7H_{14}BrNO$
Neodorme

$C_7H_{14}Br_2$
1,2-Dibromoheptane
1,5-Dibromoheptane
1,7-Dibromoheptane
2,3-Dibromoheptane
2,4-Dibromoheptane
2,6-Dibromoheptane
3,4-Dibromoheptane

$C_7H_{14}Cl_2$
1,1-Dichloroheptane
1,2-Dichloroheptane
1,7-Dichloroheptane
2,2-Dichloroheptane
3,4-Dichloroheptane

$C_7H_{14}HgO_2$
Mercuri-pentyl acetate

$C_7H_{14}N_2$
2-Aminovaleric Acid, N-*Di-Me*, *Nitrile*
Bispidine
1,5-Diazabicyclo[3,2,2]nonane
2,7-Diazaspiro[4,4]nonane†

$C_7H_{14}N_2O$
2-Aminocyclohexanecarboxylic Acid, *Amide*
1-Methylpyrrolidine-2-carboxylic Acid, *Methylamide*

$C_7H_{14}N_2O_2$
Butylmalonic Acid, *Diamide*
(2-Butyl)malonic Acid, *Diamide*
Diethylmalonic Acid, *Diamide*
2,2-Dimethylglutaric Acid, *Diamide*
Isobutylmalonic Acid, *Diamide*
Malonamide, N-*Isobutyl*
Malonamide, N,N′-*Di-Et*
2-Methyladipic Acid, *Diamide*
3-Methyladipic Acid, *Diamide*
Methylpropylmalonic Acid, *Diamide*
Piperazine-*N*-carboxylic Acid, *Et ester*
Propylsuccinic Acid, *Diamide*

$C_7H_{14}N_2O_3$
Allophanic Acid, *Isopentyl ester*
Glycylvaline
Hydantoic Acid, n-*Butyl ester*
Theanine†

$C_7H_{14}N_2O_4$
2,6-Diaminopimelic Acid
Methylenediurethane

$C_7H_{14}N_2O_4S$
Allocystathionine
Cystathionine

$C_7H_{14}N_2O_4S_2$
Djenkolic Acid

$C_7H_{14}N_2O_4Se$
Selenocystathionine†

$C_7H_{14}N_2O_5$
2,6-Diamino-3-hydroxypimelic Acid†

$C_7H_{14}N_4O_4$
2,2-Diureidopropionic Acid, *Et ester*

$C_7H_{14}O$
Allyl isobutyl Ether
Cycloheptanol
Cyclohexylmethanol

$C_7H_{14}O$ (*continued*)

2,2-Dimethyl-3-pentanone
2,4-Dimethyl-3-pentanone
3-Ethylpentan-2-one
Heptanal
2-Heptanone
3-Heptanone
4-Heptanone
1-Hepten-1-ol
2-Hepten-1-ol
3-Hepten-1-ol
1-Hepten-2-ol
2-Hepten-2-ol
3-Hepten-2-ol
5-Hepten-2-ol
6-Hepten-2-ol
1-Hepten-3-ol
6-Hepten-3-ol
1-Hepten-4-ol
2-Hepten-4-ol
5-Hexen-1-ol, *Me ether*
2-Hexen-4-ol, *Me ether*
1-Methylcyclohexanol
2-Methylcyclohexanol
3-Methylcyclohexanol
4-Methylcyclohexanol
3-Methylhexanal
4-Methylhexanal
3-Methyl-2-hexanone
4-Methyl-2-hexanone
5-Methyl-2-hexanone
2-Methyl-3-hexanone
4-Methyl-3-hexanone
5-Methyl-3-hexanone
5-Methyl-2-hexen-1-ol
3-Methyl-3-hexen-1-ol
5-Methyl-3-hexen-1-ol
2-Methyl-3-hexen-2-ol
3-Methyl-3-hexen-2-ol
2-Methyl-5-hexen-2-ol
5-Methyl-5-hexen-2-ol
5-Methyl-1-hexen-3-ol
2-Methyl-4-hexen-3-ol
3-Methyl-4-hexen-3-ol
4-Methyl-4-hexen-3-ol
2-Methyl-5-hexen-3-ol
3-Methyl-5-hexen-3-ol
4-Methyl-5-hexen-3-ol

$C_7H_{14}O_2$

Butyl propionate
2-Butyl propionate
Butyric Acid, *Propyl ester*
1,2-Cycloheptanediol
1,4-Cyclohexanediol, *Me ether*
2,2-Dimethylbutyric Acid, *Me ester*
3,3-Dimethylbutyric Acid, *Me ester*
2,2-Dimethylpropionic Acid, *Et ester*
2,3-Dimethylvaleric Acid †
2-Ethylbutyric Acid, *Me ester*
2-Ethylvaleric Acid
Heptanoic Acid
Hexanoic Acid, *Me ester*
Hydroxyacetone, *Isobutyl ether*
3-Hydroxy-2-butanone, *Propyl ether*
1-Hydroxy-2-heptanone
2-Hydroxy-4-heptanone
3-Hydroxy-4-heptanone
4-Hydroxy-2-heptanone
4-Hydroxy-3-heptanone
6-Hydroxy-2-heptanone
6-Hydroxy-3-heptanone
7-Hydroxy-2-heptanone
4-Hydroxyhexanal, *Me ether*
5-Hydroxyhexanal, *Me ether*
6-Hydroxyhexanal, *Me ether*
1-Hydroxy-2-pentanone, *Et ether*
3-Hydroxy-2-pentanone, *Et ether*
1-Hydroxy-3-pentanone, *Et ether*
2-Hydroxy-3-pentanone, *Et ether*
4-Hydroxy-2,2,4-trimethyltetrahydrofuran †
Isobutyl propionate
Isobutyric Acid, *Propyl ester*
Isobutyric Acid, *Isopropyl ester*
Isovaleric Acid, *Et ester*
2-Methyl-1-butyl acetate
2-Methyl-2-butyl acetate
3-Methyl-1-butyl acetate
2-Methylbutyric Acid, *Et ester*
4-Methyl-1,1-cyclohexanediol
1-Methyl-1,2-cyclohexanediol
3-Methyl-1,2-cyclohexanediol
4-Methyl-1,2-cyclohexanediol
5-Methyl-1,3-cyclohexanediol
2-Methylhexanoic Acid
3-Methylhexanoic Acid
4-Methylhexanoic Acid
5-Methylhexanoic Acid
Methylketene, *Di-Et acetal*
4-Methyl-5-oxo-1-hexanol
5-Methyl-2-oxo-1-hexanol
2-Methyl-3-oxo-2-hexanol
3-Methyl-4-oxo-2-hexanol
5-Methyl-4-oxo-2-hexanol
2-Methyl-4-oxo-3-hexanol
2-Methyl-5-oxo-3-hexanol
3-Methyl-2-oxo-3-hexanol
4-Methyl-5-oxo-3-hexanol
3-Methylvaleric Acid, *Me ester*
Pentyl acetate
Tetrahydrofurfuryl Alcohol, *Et ether*
Valeric Acid, *Et ester*

$C_7H_{14}O_3$

Amicetose, *Me glycoside*†
Dipropyl carbonate
3-Ethyl-3-hydroxyvaleric Acid
2-Hydroxybutyric Acid, *Me ether*, *Et ester*
2-Hydroxybutyric Acid, *Et ether*, *Me ester*
3-Hydroxy-2,3-dimethylbutyric Acid, *Me ester*
3-Hydroxy-2,2-dimethylpropionic Acid, *Et ester*
3-Hydroxy-2,2-dimethylpropionic Acid, *Et ether*
2-Hydroxyheptanoic Acid †
3-Hydroxyhexanoic Acid, *Me ester*
4-Hydroxyhexanoic Acid, *Me ester*
2-Hydroxy-2-methylbutyric Acid, *Et ester*
2-Hydroxy-3-methylbutyric Acid, *Et ester*
3-Hydroxy-3-methylbutyric Acid, *Et ester*
5-Hydroxypentanal, *Et ether*
3-Hydroxypropionic Acid, *Et ether*, *Et ester*
3-Hydroxy-2,2,3-trimethylbutyric Acid
2-Hydroxyvaleric Acid, *Et ester*

$C_7H_{14}O_3$ *(continued)*
- 2-Hydroxyvaleric Acid, *Et ether*
- 3-Hydroxyvaleric Acid, *Et ester*
- 4-Hydroxyvaleric Acid, *Et ester*
- 5-Hydroxyvaleric Acid, *Et ether*
- 5-Hydroxyvaleric Acid, *Me ether*, *Me ester*
- 1,2-Isopropylidene-glycerol, *Me ether*
- Lactic Acid, *Butyl ester*
- Lactic Acid, *Isobutyl ester*
- Levulinic Aldehyde, *Di-Me acetal*
- 2-Methoxypropionic Acid, *Propyl ester*
- 2-Oxopentanal, *Di-Me acetal*

$C_7H_{14}O_3S$
- 3-Methylcyclohexane-sulphonic Acid
- Vinylsulphonic Acid, *Pentyl ester*

$C_7H_{14}O_4$
- Boivinose, *Me glycoside*
- Butyric Acid, 1-*Glycerol ester*
- Chalcose
- Chromose A†
- Cymarose
- Diginose
- 2,3-Dihydroxybutyric Acid, *Propyl ester*
- 2,3-Dihydroxy-2-isopropylbutyric Acid
- 2,3-Dihydroxy-2-methylpropionic Acid, *Propyl ester*
- Glyceric Acid, *Di-Me ether*, *Et ester*
- α-Mono-acetin, *Di-Me ether*
- α-Mono-isobutyrin
- Mycarose
- Oleandrose
- Olivomose†
- Olivomycose†
- Sarmentose

$C_7H_{14}O_5$
- L-Acofriose
- Acovenose
- Arabinose, 2,3-*Di-Me ether*
- Arabinose, 2,4-*Di-Me ether*
- Arabinose, *Ethyl-glycoside*
- D-Cymaronic Acid
- 3-Deoxy-α-D-*ribo*-hexopyranose, *Me glycoside*†
- 4-Deoxy-D-*xylo*-hexopyranose, *Me glycoside*†
- 6-Deoxy-2-*O*-methyl-D-allose†
- 6-Deoxy-3-*O*-methyl-D-allose†
- Digitalose
- Echimidinic Acid†
- Fucose, 2-*Me ether*
- Thevetose
- Xylose, 2,3-*Di-Me ether*
- Xylose, 2,4-*Di-Me ether*
- Xylose, 2,5-*Di-Me ether*
- Xylose, 3,4-*Di-Me ether*
- Xylose, 3,5-Di-*Me ether*

$C_7H_{14}O_6$
- Arabonic Acid, *Et ester*
- Asteritol†
- Fructose, α-*Methyl glycoside*
- Galactose, *Methyl galactoside*
- Glucose, *Methyl glucoside*
- Hamamelose★, *Methyl β-pyranoside*†
- Laminitol
- Lyxose★, 4-O-*Me ether*, α-*Me-lyxoside*†
- Mannose, 1-*Me ether*
- Mannose, 4-*Me ether*
- 2-*O*-Methylgalactose
- 3-*O*-Methylgalactose
- 4-*O*-Methylgalactose
- 6-*O*-Methylgalactose
- 2-*O*-Methylglucose
- 3-*O*-Methylglucose
- 4-*O*-Methylglucose
- 5-*O*-Methylglucose
- 6-*O*-Methylglucose
- Mytilitol
- *Rhamno*hexose
- *Rhodeo*hexose
- Xylonic Acid, 2,3-*Di-Me ether*
- Xylonic Acid, 3,5-*Di-Me ether*

$C_7H_{14}O_7$
- L-*Galacto*heptulose
- Galactonic Acid, 6-*Me ether*
- *Gala*heptose
- *Gluco*heptose
- *Gluco*heptulose
- α-D-*Gulo*heptose
- L-*Gulo*heptulose
- D-*altro*-3-Heptulose†
- Mannoheptose
- Mannoheptulose
- Mannonic Acid, *Me ester*
- α-*Rhamno*hexonic Acid
- β-*Rhamno*hexonic Acid
- *Rhodeo*hexonic Acid
- Sedoheptulose

$C_7H_{14}O_8$
- *Gala*heptonic Acid
- *Gluco*heptonic Acid
- *Gulo*heptonic Acid
- Mannoheptonic Acid

$C_7H_{14}S$
- 3-Methyl-1-cyclohexanethiol

$C_7H_{14}S_2$
- 1,2-Dithiacyclononane

$C_7H_{15}Br$
- 1-Bromoheptane
- 2-Bromoheptane
- 3-Bromoheptane
- 4-Bromoheptane

$C_7H_{15}BrHg$
- Mercuri-heptyl bromide

$C_7H_{15}BrO$
- 6-Bromohexylmethyl Ether

$C_7H_{15}BrO_2$
- 2-Bromopropanol, *Di-Et acetal*
- 3-Bromopropanol, *Di-Et acetal*

$C_7H_{15}Cl$
- 1-Chloroheptane
- 2-Chloroheptane
- 3-Chloroheptane
- 4-Chloroheptane

$C_7H_{15}ClHg$
- Mercuri-heptyl chloride

$C_7H_{15}ClO$
- 3-Chloro-2-butanol, *Propyl ether*
- 3-Chloro-2-butanol, *Isopropyl ether*

$C_7H_{15}ClO$ (*continued*)
1-Chloroethanol, *Isoamyl ether*
7-Chloro-1-heptanol
1-Chloro-2-heptanol

$C_7H_{15}ClO_2$
Chloroacetone, *Di-Et acetal*
3-Chloropropanal, *Di-Et acetal*
2-Chloro-1,3-propanediol, *Di-Et ether*

$C_7H_{15}Cl_2N_2O_2P$
Cyclophosphamide

$C_7H_{15}F$
1-Fluoroheptane

$C_7H_{15}HgI$
Mercuri-heptyl iodide

$C_7H_{15}I$
1-Iodoheptane
2-Iodoheptane
3-Iodoheptane
4-Iodoheptane

$C_7H_{15}N$
N-Allyldiethylamine
N-Allylisobutylamine
5-Amino-4-methyl-1-hexene
6-Amino-4-methyl-1-hexene (or 6-Amino-4-methyl-2-hexene)
Cycloheptylamine
Cyclohexylmethylamine
2,3-Dimethylpiperidine
2,4-Dimethylpiperidine
2,5-Dimethylpiperidine
2,6-Dimethylpiperidine
3,3-Dimethylpiperidine
4,4-Dimethylpiperidine
1-Ethylpiperidine
2-Ethylpiperidine
3-Ethylpiperidine
4-Ethylpiperidine
N-Methylcyclohexylamine
1-Methylcyclohexylamine
2-Methylcyclohexylamine
3-Methylcyclohexylamine
4-Methylcyclohexylamine
2-Methylperhydroazepine
2-Methylpiperidine, N-*Me*
3-Methylpiperidine, N-*Me*
Perhydroazocine
Perhydro-1-methylazepine
Perhydro-4-methylazepine
N-Propylpyrrolidine
2-Propylpyrrolidine
3-Propylpyrrolidine
Pyrrolidine, N-*Propyl*
1,2,2-Trimethylpyrrolidine
1,2,4-Trimethylpyrrolidine
1,2,5-Trimethylpyrrolidine
1,3,3-Trimethylpyrrolidine
2,2,3-Trimethylpyrrolidine
2,2,4-Trimethylpyrrolidine
2,2,5-Trimethylpyrrolidine
2,3,5-Trimethylpyrrolidine

$C_7H_{15}NO$
2,3-Dimethylvaleric Acid, *Amide*†
2-Ethylvaleric Acid, *Amide*
Heptanoic Acid, *Amide*
1-(2-Hydroxyethyl)piperidine
2-(2-Hydroxyethyl)piperidine
3-Hydroxypiperidine, N-*Et*
Isovaleric Acid, *Iminoether*
2-Methylhexanoic Acid, *Amide*
3-Methylhexanoic Acid, *Amide*
4-Methylhexanoic Acid, *Amide*
5-Methylhexanoic Acid, *Amide*
Propionamide, N-*Di-Et*

$C_7H_{15}NOS$
Ethyl aminothioformate, N-*Di-Et*

$C_7H_{15}NO_2$
Actinine
5-Amino-2-ethylvaleric Acid
2-Aminoheptanoic Acid★†
7-Aminoheptanoic Acid
2-Aminohexanoic Acid, N-*Me*
2-Amino-2-methylbutyric Acid, *Et ester*
3-Amino-3-methylbutyric Acid, *Et ester*
2-Amino-4-methylhexanoic Acid†
2-Aminovaleric Acid, *Et ester*
2-Aminovaleric Acid, *N-Di-Me*
3-Aminovaleric Acid, *Et ester*
4-Aminovaleric Acid, *Et ester*
5-Aminovaleric Acid, N-*Di-Me*
5-Aminovaleric Acid, N-*Et*
Butylurethane
(2-Butyl)urethane
tert-Butylurethane
Diethylcarbamic Acid, *Et ester*
N-Diethylglycine, *Me ester*
N-Dimethyl-DL-alanine, *Et ester*
4-Dimethylaminobutyric Acid, *Me ester*
Emylcamate
Ethyl isobutylaminoformate
N-Ethyl-*N*-methylglycine, *Et ester*
2-Hydroxyvaleric Acid, *Et ether*, *Amide*
Leucine, *Me ester*
2-Methylaminobutyric Acid, *Et ester*
3-Methylaminobutyric Acid, *Et ester*
6-Methylaminohexanoic Acid
N-Methyl-leucine
1-Nitroheptane
2-Nitroheptane
Valine, *Et ester*

$C_7H_{15}NO_2S$
Methionine, *Et ester*
Penicillamine, *Et ester*

$C_7H_{15}NO_3$
Carnitine
3-Dimethylamino-2-hydroxy-2-methylpropionic Acid, *Me ester*
3-Hydroxyvaline, *Me ether*, *Me ester*

$C_7H_{15}NO_4$
3-Amino-3,6-dideoxyglucose, α-*Me glycoside*†
4-Amino-4,6-dideoxy-D-glucose, *Methyl α-glucopyranoside*†
3-Amino-3,6-dideoxytalose, *Me glycoside*†
Garosamine†
Perosamine, *Me perosaminide*†

$C_7H_{15}NO_5$
Glucosamine, 3-*Me ether*

$C_7H_{15}NO_6$
Gluconic Acid, *Methylamide*
Gluconic Acid, 2-*Me ether*, *Amide*
Glucosaminic Acid, 3-*Me ether*
α-*Rhamno*hexonic Acid, *Amide*
*Rhodeo*hexonic Acid, *Amide*

$C_7H_{15}NO_7$
*Gala*heptonic Acid, *Amide*
*Gluco*heptonic Acid, *Amide*
Mannoheptonic Acid, *Amide*

$C_7H_{15}NS_2$
Thialdine, N-*Me*

$C_7H_{15}N_2O_8P$
Aminoacetamide, *Ribotide*

$C_7H_{15}N_3O_2$
Indospicine†

$C_7H_{15}N_3O_3$
2-Amino-6-ureidohexanoic Acid

$C_7H_{15}N_5O_3$
Gigartinine†

$C_7H_{15}O_4P$
Allyl diethyl phosphate

$C_7H_{15}O_{10}P$
Sedoheptulose 7-Phosphate†

C_7H_{16}
2,2-Dimethylpentane
2,3-Dimethylpentane
2,4-Dimethylpentane
3,3-Dimethylpentane
3-Ethylpentane
Heptane
Isoheptane
3-Methylhexane

$C_7H_{16}BrNO$
N-(2-Methoxy-2-propenyl)trimethylammonium bromide

$C_7H_{16}HgO$
Mercuri-heptyl hydroxide

$C_7H_{16}N_2$
4-Amino-2,6-dimethylpiperidine

$C_7H_{16}N_2O$
1,1-Di-isopropylurea
1,3-Di-isopropylurea
1,1-Dipropylurea
1,3-Dipropylurea

$C_7H_{16}N_2O$
3-Methyl-3-methylaminobutyric Acid, *Methylamide*

$C_7H_{16}N_2O_2$
2-Hydrazino-3-methylbutyric Acid, *Et ester*
Lysine, *Me ester*

$C_7H_{16}N_2S$
1,1-Dipropylthiourea
1,3-Dipropylthiourea

$C_7H_{16}N_4O_2$
Arginine, *Me ester*
Homoarginine
2-*N*-Methylarginine
5-*N*-Methylarginine

$C_7H_{16}N_4O_3$
γ-Hydroxyhomoarginine

$C_7H_{16}O$
Butyl isopropyl Ether
tert-Butyl isopropyl Ether
Butyl propyl Ether
tert-Butyl propyl Ether
2,4-Dimethyl-2-pentanol
2,2-Dimethyl-3-pentanol
2,3-Dimethyl-3-pentanol
2,4-Dimethyl-3-pentanol
Ethyl 2-methylbutyl Ether
Ethyl 2-methyl-2-butyl Ether
Ethyl 3-methylbutyl Ether
3-Ethyl-2-pentanol
3-Ethyl-3-pentanol
Ethyl pentyl Ether
1-Heptanol
2-Heptanol
3-Heptanol
4-Heptanol
Isobutyl propyl Ether
2-Methyl-1-hexanol
3-Methyl-1-hexanol
4-Methyl-1-hexanol
5-Methyl-1-hexanol
2-Methyl-2-hexanol
3-Methyl-2-hexanol
5-Methyl-2-hexanol
2-Methyl-3-hexanol
3-Methyl-3-hexanol
4-Methyl-3-hexanol
5-Methyl-3-hexanol
1-Pentanol, *Et ether*

$C_7H_{16}OS$
2-Hydroxy-1-ethanethiol, S-3-*Methylbutyl*

$C_7H_{16}O_2$
Acetone, *Di Et acetal*
1,4-Butanediol, *Me-Et ether*
3-Ethyl-2,3-pentanediol
1,2-Heptanediol
1,4-Heptanediol
1,5-Heptanediol
1,6-Heptanediol
1,7-Heptanediol
2,4-Heptanediol
2,5-Heptanediol
2,6-Heptanediol
3,4-Heptanediol
1,6-Hexanediol, *Mono-Me ether*
2-Hydroxyethyl 1,1-dimethylpropyl Ether
2-Hydroxyethyl pentyl Ether
4-Methyl-1,4-hexanediol
5-Methyl-1,5-hexanediol
3-Methyl-1,6-hexanediol
2-Methyl-2,4-hexanediol
3-Methyl-2,4-hexanediol
5-Methyl-2,4-hexanediol
2-Methyl-2,5-hexanediol
1,5-Pentanediol, *Di-Me ether*
1,5-Pentanediol, *Mono-Et ether*
Propanal, *Di-Et acetal*
1,3-Propanediol, *Di-Et ether*
n-Valeraldehyde, *Di-Me acetal*

$C_7H_{16}O_3$
1,3,3-Butanetriol, *Tri-Me ether*
Glycerol, 1,3-*Di-Et ether*
Glycerol, 1-*Butyl ether*
1,2,3-Heptanetriol
1,3,7-Heptanetriol
1,4,5-Heptanetriol
1,4,7-Heptanetriol
2,3,4-Heptanetriol
Hydroxyacetone, *Di-Et acetal*
5-Hydroxypentanal, *Di-Me acetal*
2-Hydroxypropanal, *Di-Et acetal*
3-Hydroxypropanal, *Di-Et acetal*
Orthoacetic Acid, *Me-di-Et ester*
Orthoacetic Acid, *Me-di-Isobutyl ester*
Orthoformic Acid, *Tri-Et ester*
1,2,5-Pentanetriol, 1,5-*Di-Me ether*

$C_7H_{16}O_4$
Glyceraldehyde, *Di-Et acetal*

$C_7H_{16}O_4S_2$
Sulphonal

$C_7H_{16}O_6$
Mannitol★, 1-O-*Me ether*†
Mannitol★, 2-O-*Me ether*†
Mannitol★, 3-O-*Me ether*†
α-*Rhamno*hexitol
Siphulitol

$C_7H_{16}O_7$
*Gala*heptitol
*Gluco*heptitol
*Gulo*heptitol
Perseitol
Sedoheptitol

$C_7H_{17}N$
2-Amino-2,4-dimethylpentane
3-Amino-2,4-dimethylpentane
1-Aminoheptane
2-Aminoheptane
3-Aminoheptane
4-Aminoheptane
1-Amino-2,4,4-trimethylpentane
3,4-Dimethyl-2-pentylamine
N-Ethyl-3-methylbutylamine
N-Ethyl-*N*-methylisobutylamine
Isobutylpropylamine
Methyldipropylamine
2-Methylhexylamine
3-Methylhexylamine
4-Methylhexylamine

$C_7H_{17}NO$
3-Amino-1-propanol, N-*Di-Et*
2-Diethylamino-1-propanol
3-Diethylamino-1-propanol
1-Diethylamino-2-propanol
5-Dimethylamino-1-pentanol
2-3′-Methylbutylaminoethanol

$C_7H_{17}NO_2$
3-Amino-1,2-propanediol, N-*Di-Et*
Methylaminoacetaldehyde, *Di-Et acetal*

$C_7H_{17}NO_3$
3-Amino-2-hydroxypropanal, *Di-Et acetal*
Acetylcholine

$C_7H_{17}O_3PSe$
O,O-Diethyl *Se*-propylphosphoroselenolate

$C_7H_{17}O_6P$
Glycerolphosphoric Acid, *Di-Me ether*, *Di-Me ester*

$C_7H_{18}INO$
Hydroxymethyltrimethylammonium Hydroxide, *Propyl ether*, *Iodide*

$C_7H_{18}NO_6P$
L-α-Glycerylphosphoryl-2-amino-2-methyl-propanol†

$C_7H_{18}N_2$
1,7-Diaminoheptane
1,3-Diaminopropane, 1,3-N-*Tetra-Me*
1,3-Diaminopropane, 1-N-*Di-Et*
1,2-Propanediamine, 1-N-*Di-Et*

$C_7H_{18}N_2O$
1,3-Di-dimethylamino-2-propanol

$C_7H_{19}NO$
Triethyl-methylammonium hydroxide

$C_7H_{19}NO_2$
Diethyl-(2-hydroxyethyl)methylammonium hydroxide

$C_7H_{19}N_3$
Spermidine

C_8

$C_8Br_4O_3$
Tetrabromophthalic Acid, *Anhydride*

C_8Cl_6S
Hexachlorobenzo[*b*]thiophen

$C_8F_{14}O_3$
Heptafluorobutyric Acid, *Anhydride*

$C_8HBr_4NO_2$
Tetrabromophthalic Acid, *Imide*

$C_8HCl_3O_3$
3,4,5-Trichlorophthalic Acid, *Anhydride*
3,4,6-Trichlorophthalic Acid, *Anhydride*

$C_8HCl_4NO_2$
Tetrachlorophthalic Acid, *Imide*

$C_8H_2Br_2Cl_2O_2$
2,5-Dibromoterephthalic Acid, *Dichloride*

$C_8H_2Br_2O_3$
3,4-Dibromophthalic Acid, *Anhydride*
3,5-Dibromophthalic Acid, *Anhydride*
3,6-Dibromophthalic Acid, *Anhydride*
4,5-Dibromophthalic Acid, *Anhydride*

$C_8H_2Br_4O_4$
Tetrabromoisophthalic Acid
Tetrabromophthalic Acid
Tetrabromoterephthalic Acid

$C_8H_2Cl_2I_4$
Tetraiodoterephthalic Acid, *Dichloride*

$C_8H_2Cl_2N_2O_6$
4,6-Dinitroisophthalic Acid, *Dichloride*

$C_8H_2Cl_2O_3$
3,4-Dichlorophthalic Acid, *Anhydride*
3,5-Dichlorophthalic Acid, *Anhydride*
3,6-Dichlorophthalic Acid, *Anhydride*
4,5-Dichlorophthalic Acid, *Anhydride*

$C_8H_2Cl_3NO_2$
3,4,6-Trichlorophthalic Acid, *Imide*

$C_8H_2Cl_4O_2$
2,5-Dichloroterephthalic Acid, *Dichloride*

$C_8H_2Cl_4O_4$
Tetrachloroisophthalic Acid
Tetrachlorophthalic Acid
Tetrachloroterephthalic Acid

$C_8H_2I_2O_3$
3,4-Di-iodophthalic Acid, *Anhydride*
3,5-Di-iodophthalic Acid, *Anhydride*
3,6-Di-iodophthalic Acid, *Anhydride*
4,5-Di-iodophthalic Acid, *Anhydride*

$C_8H_2I_4O_4$
Tetraiodoisophthalic Acid
Tetraiodophthalic Acid
Tetraiodoterephthalic Acid

$C_8H_2N_2O_7$
3,5-Dinitrophthalic Acid, *Anhydride*
3,6-Dinitrophthalic Acid, *Anhydride*
4,5-Dinitrophthalic Acid, *Anhydride*

$C_8H_3BrCl_2O_2$
4-Bromophthalic Acid, *Dichloride*
Bromoterephthalic Acid, *Dichloride*

$C_8H_3BrO_3$
3-Bromophthalic Acid, *Anhydride*
4-Bromophthalic Acid, *Anhydride*

$C_8H_3Br_2NO_2$
4,6-Dibromoisatin
5,6-Dibromoisatin
5,7-Dibromoisatin
4,5-Dibromophthalic Acid, *Imide*

$C_8H_3ClO_3$
3-Chlorophthalic Acid, *Anhydride*
4-Chlorophthalic Acid, *Anhydride*

$C_8H_3Cl_2F_3N_2$
4,5-Dichloro-2-trifluoromethylbenzimidazole†

$C_8H_3Cl_2IO_2$
2-Iodoisophthalic Acid, *Dichloride*†

$C_8H_3Cl_2NO_2$
1,5-Dichloroisatin
4,5-Dichloroisatin
4,6-Dichloroisatin
4,7-Dichloroisatin
5,6-Dichloroisatin
5,7-Dichloroisatin
3,5-Dichlorophthalic Acid, *Imide*
3,6-Dichlorophthalic Acid, *Imide*

$C_8H_3Cl_2NO_4$
3-Nitrophthalic Acid, *Dichloride*
2-Nitroterephthalic Acid, *Dichloride*

$C_8H_3Cl_2NO_6$
3,5-Dichloro-4-nitrophthalic Acid
2,5-Dichloro-6-nitroterephthalic Acid

$C_8H_3Cl_3O_2$
4-Chlorophthalic Acid, *Dichloride*

$C_8H_3Cl_3O_4$
3,4,5-Trichlorophthalic Acid
3,4,6-Trichlorophthalic Acid

$C_8H_3Cl_5O_2$
Pentachlorobenzoic Acid, *Me ester*

$C_8H_3F_4N$
4,5,6,7-Tetrafluoroindole†

$C_8H_3IO_3$
3-Iodophthalic Acid, *Anhydride*
4-Iodophthalic Acid, *Anhydride*

$C_8H_3I_3NO_3$
3-Amino-2,4,5-tri-iodobenzoic Acid, N-*Formyl*

$C_8H_3NO_2$
Diatetryne II

$C_8H_3NO_5$
3-Nitrophthalic Anhydride
4-Nitrophthalic Anhydride

$C_8H_3N_3O_3$
3-Hydroxy-6-nitrophthalic Acid, *Dinitrile*

C_8H_4BrNO
o-Bromobenzoylformic, *Nitrile*
p-Bromobenzoylformic, *Nitrile*

$C_8H_4BrNO_2$
4-Bromoisatin
5-Bromoisatin
6-Bromoisatin
7-Bromoisatin
4-Bromoisophthalic Acid, 3-*Nitrile*
Phthalimide, N-*Bromo*

$C_8H_4Br_2O_2$
Terephthalic Acid, *Dibromide*

$C_8H_4Br_2O_4$
4,6-Dibromoisophthalic Acid
3,4-Dibromophthalic Acid
3,5-Dibromophthalic Acid
3,6-Dibromophthalic Acid
4,5-Dibromophthalic Acid
2,5-Dibromoterephthalic Acid

$C_8H_4Br_3ClN_2O_3$
2,4,6-Tribromo-3-nitroaniline, N-*Chloro*

$C_8H_4Br_3N$
2,4,6-Tribromophenylacetic Acid, *Nitrile*
2,4,6-Tribromo-*m*-toluic Acid, *Nitrile*

$C_8H_4Br_4O_2$
2,3,4,6-Tetrabromobenzoic Acid, *Me ester*

C_8H_4ClNO
p-Cyanobenzoic Acid, *Chloride*
Isatin α-chloride

C_8H_4ClNOS
Benzothiazole-2-carboxylic Acid, *Chloride*
Benzothiazole-6-carboxylic Acid, *Chloride*

$C_8H_4ClNO_2$
4-Chloroisatin
5-Chloroisatin
6-Chloroisatin
7-Chloroisatin
3-Chlorophthalic Acid, *Imide*
4-Chlorophthalic Acid, *Imide*
Phthalimide, N-*Chloro*

$C_8H_4ClN_3O_3$
6-Nitroindazole-1-carboxylic Acid, *Chloride*
$C_8H_4Cl_2O_2$
4,6-Dichloroisophthalaldehyde
2,5-Dichloroterephthalaldehyde
Isophthalic Acid, *Dichloride*
Phthaloyl chloride
Terephthalic Acid, *Dichloride*
$C_8H_4Cl_2O_4$
4,6-Dichloroisophthalic Acid
3,4-Dichlorophthalic Acid
3,5-Dichlorophthalic Acid
3,6-Dichlorophthalic Acid
4,5-Dichlorophthalic Acid
2,5-Dichloroterephthalic Acid
$C_8H_4Cl_3N$
α,α,α-Trichloro-*o*-toluic Acid, *Nitrile*
$C_8H_4Cl_4O$
α,α,α-Trichloro-*o*-toluic Acid, *Chloride*
α,α,α-Trichloro-*m*-toluic Acid, *Chloride*
α,α,α-Trichloro-*p*-toluic Acid, *Chloride*
$C_8H_4Cl_4O_2$
2,3,5,6-Tetrachlorophenylacetic Acid†
$C_8H_4F_2O_2$
Phthaloyl fluoride
$C_8H_4INO_2$
5-Iodoisatin
7-Iodoisatin
3-Iodophthalic Acid, *Imide*
4-Iodophthalic Acid, *Imide*
$C_8H_4I_2O_4$
3,4-Di-iodophthalic Acid
3,5-Di-iodophthalic Acid
3,6-Di-iodophthalic Acid
4,5-Di-iodophthalic Acid
$C_8H_4N_2$
Isophthalic Acid, *Dinitrile*
Phthalonitrile
Terephthalic Acid, *Dinitrile*
$C_8H_4N_2O_2$
3,6-Dihydroxyphthalic Acid, *Dinitrile*
$C_8H_4N_2O_3$
o-Nitrobenzoylformic Acid, *Nitrile*
$C_8H_4N_2O_4$
5-Nitroisatin
6-Nitroisatin
7-Nitroisatin
3-Nitrophthalic Acid, 1-*Nitrile*
3-Nitrophthalimide
4-Nitrophthalimide
N-Nitrophthalimide
$C_8H_4N_2O_5$
2-Hydroxy-5-nitroisophthalic Acid, *Mononitrile*
$C_8H_4N_2O_6$
4,6-Dinitroisophthalaldehyde
$C_8H_4N_2O_8$
2,4-Dinitroisophthalic Acid
4,6-Dinitroisophthalic Acid
3,4-Dinitrophthalic Acid
3,5-Dinitrophthalic Acid
3,6-Dinitrophthalic Acid
4,5-Dinitrophthalic Acid
2,3-Dinitroterephthalic Acid
2,5-Dinitroterephthalic Acid
2,6-Dinitroterephthalic Acid
$C_8H_4O_2$
Benzocyclobutenedione
$C_8H_4O_2S$
2,3-Dihydrobenzo[b]thiophene-2,3-dione
$C_8H_4O_3$
Dihydrobenzofuran-2,3-dione
Phthalic Anhydride
$C_8H_4O_4$
3-Hydroxyphthalic Acid, *Anhydride*
4-Hydroxyphthalic Acid, *Anhydride*
Phthaloyl peroxide
$C_8H_4O_5$
3,4-Dihydroxyphthalic Acid, *Anhydride*
$C_8H_4O_8S$
Thiophene-tetracarboxylic Acid
$C_8H_4O_9$
Furan-tetracarboxylic Acid
$C_8H_4S_2$
Tetracyano-1,4-dithiin
C_8H_5Br
1-Bromo-2-phenylacetylene
o-Bromophenylacetylene
m-Bromophenylacetylene
p-Bromophenylacetylene
$C_8H_5BrClNO_3$
2-Bromo-5-nitro-*p*-toluic Acid, *Chloride*
C_8H_5BrN
3-Bromoindole†
$C_8H_5BrN_2O_2$
4-Bromo-6-nitro-*o*-toluic Acid, *Nitrile*
2-Bromo-5-nitro-*p*-toluic Acid, *Nitrile*
2-Bromo-6-nitro-*p*-toluic Acid, *Nitrile*
$C_8H_5BrN_2O_6$
2-Bromo-3,5-dinitrobenzoic Acid, *Me ester*
4-Bromo-3,5-dinitrobenzoic Acid, *Me ester*
5-Bromo-2,4-dinitrobenzoic Acid, *Me ester*
$C_8H_5BrO_3$
o-Bromobenzoylformic Acid
p-Bromobenzoylformic Acid
2-Bromo-4,5-methylenedioxybenzaldehyde
3-Bromo-4,5-methylenedioxybenzaldehyde
$C_8H_5BrO_4$
2-Bromoisophthalic Acid
4-Bromoisophthalic Acid
3-Bromophthalic Acid
4-Bromophthalic Acid
Bromoterephthalic Acid
$C_8H_5Br_2ClO$
3,6-Dibromo-*p*-toluic Acid, *Chloride*
$C_8H_5Br_2N$
4,6-Dibromo-*o*-toluic Acid, *Nitrile*
2,6-Dibromo-*p*-toluic Acid, *Nitrile*
3,5-Dibromo-*p*-toluic Acid, *Nitrile*
α,α-Dibromo-*o*-tolunitrile
$C_8H_5Br_3O$
α,α,α-Tribromoacetophenone
α,2,4-Tribromoacetophenone

$C_8H_5Br_3O$ (*continued*)
α,3,5-Tribromoacetophenone
2,4,6-Tribromoacetophenone
3,4,5-Tribromoacetophenone

$C_8H_5Br_3O_2$
2,3,5-Tribromobenzoic Acid, *Me ester*
2,4,6-Tribromobenzoic Acid, *Me ester*
3,4,5-Tribromobenzoic Acid, *Me ester*
2,4,6-Tribromophenylacetic Acid
3,4,5-Tribromophenylacetic Acid
2,4,6-Tribromo-*m*-toluic Acid

$C_8H_5Br_3O_3$
2,4,6-Tribromo-3-hydroxybenzoic Acid, *Me ether*
2,4,6-Tribromo-3-hydroxybenzoic Acid, *Me ester*

$C_8H_5Br_5O$
Pentabromophenol, O-*Et ether*

$C_8H_5ClN_2O$
Indazole-2-carboxylic Acid, *Chloride*

$C_8H_5ClN_2O_2$
4-Chloro-5-nitro-*o*-toluic Acid, *Nitrile*
4-Chloro-6-nitro-*o*-toluic Acid, *Nitrile*
2-Chloro-5-nitro-*p*-toluic Acid, *Nitrile*
5-Chloro-2-nitro-*p*-toluic Acid, *Nitrile*

$C_8H_5ClN_2O_5$
2,4-Dinitrophenylacetic Acid, *Chloride*
4,5-Dinitro-*o*-toluic Acid, *Chloride*
3,5-Dinitro-*p*-toluic Acid, *Chloride*

$C_8H_5ClN_2O_6$
2-Chloro-3,5-dinitrobenzoic Acid, *Me ester*
4-Chloro-3,5-dinitrobenzoic Acid, *Me ester*
5-Chloro-2,4-dinitrobenzoic Acid, *Me ester*

C_8H_5ClO
2-Chlorobenzofuran
3-Chlorobenzofuran
5-Chlorobenzofuran
7-Chlorobenzofuran

$C_8H_5ClO_2$
Benzoylformic Acid, *Chloride*
3-Formylbenzoic Acid, *Chloride*
4-Formylbenzoic Acid, *Chloride*

$C_8H_5ClO_3$
3,4-Methylenedioxybenzoic Acid, *Chloride*
Terephthalic Acid, *Monochloride*

$C_8H_5ClO_4$
4-Chloroisophthalic Acid
5-Chloroisophthalic Acid
3-Chlorophthalic Acid
4-Chlorophthalic Acid
Chloroterephthalic Acid

$C_8H_5ClO_6S$
4-Sulphophthalic Acid, 4-*Chloride*

$C_8H_5Cl_2N$
2,3-Dichloroindole
α,α-Dichlorophenylacetic Acid, *Nitrile*
4,6-Dichloro-*o*-toluic Acid, *Nitrile*
α,α-Dichloro-*o*-tolunitrile
α,α-Dichloro-*m*-tolunitrile
α,α-Dichloro-*p*-tolunitrile

$C_8H_5Cl_2NO_4$
4,6-Dichloro-3-nitro-*o*-toluic Acid

$C_8H_5Cl_2NO_5S$
4-Nitro-6-sulpho-*m*-toluic Acid, *Dichloride*
6-Nitro-4-sulpho-*m*-toluic Acid, *Dichloride*

$C_8H_5Cl_3O$
α,α-Dichloro-*o*-toluic Acid, *Chloride*
α,α-Dichloro-*m*-toluic Acid, *Chloride*
α,α-Dichloro-*p*-toluic Acid, *Chloride*

$C_8H_5Cl_3O_2$
Trichloroacetic Acid, *Phenyl ester*
α,α,α-Trichloro-*o*-toluic Acid
α,α,α-Trichloro-*m*-toluic Acid
α,α,α-Trichloro-*p*-toluic Acid

$C_8H_5Cl_4NO$
2,3,4,5-Tetrachloroaniline, N-*Ac*
2,3,5,6-Tetrachloroaniline, N-*Ac*

$C_8H_5Cl_4NO_2$
2-Amino-3,4,5,6-tetrachlorobenzoic Acid, *Me ester*

$C_8H_5Cl_5O$
Pentachlorophenol, *Et ether*

$C_8H_5F_3N_2$
2-Trifluoromethylbenzimidazole†

$C_8H_5F_3O_2S$
4,4,4-Trifluoro-1-(2-thienyl)-1,3-butanedione

C_8H_5I
1-Iodo-2-phenylacetylene

$C_8H_5IN_2O_7$
4-Iodo-2,5-dinitrophenoxyacetic Acid

$C_8H_5IO_4$
2-Iodoisophthalic Acid★†
4-Iodoisophthalic Acid
5-Iodoisophthalic Acid
3-Iodophthalic Acid
4-Iodophthalic Acid
Iodoterephthalic Acid

$C_8H_5I_2NO_5$
3,5-Di-iodo-1-methyl-4-pyridone-2,6-dicarboxylic Acid

C_8H_5NO
Benzoylformic Acid, *Nitrile*
o-Cyanobenzaldehyde†
m-Cyanobenzaldehyde†
p-Cyanobenzaldehyde†
3-Formylbenzoic Acid, *Nitrile*
4-Formylbenzoic Acid, *Nitrile*

C_8H_5NOS
Benzoyl isothiocyanate

$C_8H_5NO_2$
Agrocybin
Anthroxanaldehyde
Benzoyl isocyanate
o-Cyanobenzoic Acid
m-Cyanobenzoic Acid
p-Cyanobenzoic Acid
Isatin
3,4-Methylenedioxybenzoic Acid, *Nitrile*
o-Nitrophenylacetylene
m-Nitrophenylacetylene
p-Nitrophenylacetylene
Phthalimide

$C_8H_5NO_2S$
- Benzothiazole-2-carboxylic Acid
- Benzothiazole-6-carboxylic Acid

$C_8H_5NO_3$
- Anthroxanic Acid
- Diatetryne I
- 1-Hydroxyisatin
- 5-Hydroxyisatin
- 4-Hydroxyphthalic Acid, *Imide*
- Isatoic Acid, *Anhydride*
- 2-Methylpyridine-3,4-dicarboxylic Acid, *Anhydride*
- 2-Nitrobenzofuran
- 5-Nitrobenzofuran
- 6-Nitrobenzofuran
- 7-Nitrobenzofuran
- Phthaloxime

$C_8H_5NO_4$
- 5,6-Dihydroxyisatin
- 6,7-Dihydroxyisatin
- 4-Nitrophthalide
- 5-Nitrophthalide
- 6-Nitrophthalide

$C_8H_5NO_5$
- 2-Formyl-5-nitrobenzoic Acid
- 4,5-Methylenedioxy-2-nitrobenzaldehyde
- *o*-Nitrobenzoylformic Acid
- *m*-Nitrobenzoylformic Acid
- *p*-Nitrobenzoylformic Acid

$C_8H_5NO_6$
- 4,5-Methylenedioxy-2-nitrobenzoic Acid
- 2-Nitroisophthalic Acid
- 4-Nitroisophthalic Acid
- 5-Nitroisophthalic Acid
- 3-Nitrophthalic Acid
- 4-Nitrophthalic Acid
- 2-Nitroterephthalic Acid
- Pyridine-2,3,4-tricarboxylic Acid
- Pyridine-2,3,5-tricarboxylic Acid
- Pyridine-2,3,6-tricarboxylic Acid
- Pyridine-2,4,5-tricarboxylic Acid
- Pyridine-3,4,5-tricarboxylic Acid

$C_8H_5NO_7$
- 2-Hydroxy-5-nitroisophthalic Acid
- 4-Hydroxy-5-nitroisophthalic Acid
- 3-Hydroxy-6-nitrophthalic Acid
- 4-Hydroxy-3-nitrophthalic Acid
- 4-Hydroxy-5-nitrophthalic Acid

$C_8H_5NO_8$
- 3,4-Dihydroxy-6-nitrophthalic Acid

$C_8H_5N_3O_4$
- 2,6-Dinitrophenylacetic Acid, *Nitrile*
- 3,5-Dinitro-*o*-toluic Acid★, *Nitrile*†
- 2,6-Dinitro-*p*-toluic Acid, *Nitrile*
- 6-Nitroindazole-1-carboxylic Acid
- 6-Nitroindazole-3-carboxylic Acid

$C_8H_5N_3O_5$
- 2-Hydroxy-3,5-dinitrobenzoic Acid, *Me ether, Nitrile*
- 3-Hydroxy-2,6-dinitro-*p*-toluic Acid, *Nitrile*

$C_8H_5N_3O_8$
- 2,4,5-Trinitrobenzoic Acid, *Me ester*
- 2,4,6-Trinitrobenzoic Acid, *Me ester*
- 2,4,6-Trinitrophenylacetic Acid
- 2,3,6-Trinitro-*p*-toluic Acid

$C_8H_5N_5O_6$
- Purpuric Acid

C_8H_6
- Cyclopropenylidenecyclopentadiene†
- Phenylacetylene

C_8H_6BrClO
- α-Bromophenylacetic Acid, *Chloride*
- α-Bromo-*p*-toluic Acid, *Chloride*
- 2-Bromo-*p*-toluic Acid, *Chloride*

$C_8H_6BrClO_2$
- 5-Bromo-2-hydroxy-*m*-toluic Acid, *Chloride*
- *p*-Bromophenoxyacetic Acid, *Chloride*

C_8H_6BrN
- *o*-Bromophenylacetic Acid, *Nitrile*
- *p*-Bromophenylacetic Acid, *Nitrile*
- α-Bromophenylacetonitrile
- 4-Bromo-*o*-toluic Acid, *Nitrile*
- 2-Bromo-*p*-toluic Acid, *Nitrile*
- 3-Bromo-*p*-toluic Acid, *Nitrile*
- α-Bromo-*o*-tolunitrile
- α-Bromo-*m*-tolunitrile
- α-Bromo-*p*-tolunitrile

$C_8H_6BrNO_2$
- *o*-Bromobenzoylformic Acid, *Amide*
- *p*-Bromobenzoylformic Acid, *Amide*
- 2-Bromo-4-hydroxy-3-methoxybenzoic Acid, *Nitrile*
- 2-Bromo-4-hydroxy-5-methoxybenzoic Acid, *Nitrile*
- 3-Bromo-4-hydroxy-5-methoxybenzoic Acid, *Nitrile*

$C_8H_6BrNO_3$
- α-Bromo-α-nitroacetophenone
- α-Bromo-2-nitroacetophenone
- α-Bromo-3-nitroacetophenone
- α-Bromo-4-nitroacetophenone
- 2-Bromo-5-nitroacetophenone
- 3-Bromo-2-nitroacetophenone
- 4-Bromo-3-nitroacetophenone
- 5-Bromo-2-nitroacetophenone

$C_8H_6BrNO_4$
- 2-Bromo-3-nitrobenzoic Acid, *Me ester*
- 2-Bromo-5-nitrobenzoic Acid, *Me ester*
- 3-Bromo-5-nitrobenzoic Acid, *Me ester*
- 4-Bromo-3-nitrobenzoic Acid, *Me ester*
- 5-Bromo-2-nitrobenzoic Acid, *Me ester*
- 4-Bromo-3-nitro-*o*-toluic Acid
- 4-Bromo-5-nitro-*o*-toluic Acid
- 4-Bromo-6-nitro-*o*-toluic Acid
- 2-Bromo-3-nitro-*p*-toluic Acid
- 2-Bromo-5-nitro-*p*-toluic Acid
- 2-Bromo-6-nitro-*p*-toluic Acid
- 5-Bromo-2-nitro-*p*-toluic Acid
- 5-Bromo-3-nitro-*p*-toluic Acid

$C_8H_6Br_2$
- α,β-Dibromostyrene
- β,β-Dibromostyrene
- 2,5-Dibromostyrene
- 3,4-Dibromostyrene

$C_8H_6Br_2N_2O_2$
2,5-Dibromoterephthalic Acid, *Diamide*

$C_8H_6Br_2N_2O_5$
3,5-Dibromo-4,6-dinitro-*o*-cresol, *Me ether*
4,6-Dibromo-3,5-dinitro-*o*-cresol, *Me ether*
2,6-Dibromo-3,5-dinitro-*p*-cresol, *Me ether*

$C_8H_6Br_2O$
α-Bromophenylacetic Acid, *Bromide*
α-Bromo-*p*-toluic Acid, *Bromide*
αα-Dibromoacetophenone
α,3-Dibromoacetophenone
α,4-Dibromoacetophenone
2,4-Dibromoacetophenone
2,5-Dibromoacetophenone
3,5-Dibromoacetophenone

$C_8H_6Br_2O_2$
2,4-Dibromobenzoic Acid, *Me ester*
2,5-Dibromobenzoic Acid, *Me ester*
2,6-Dibromobenzoic Acid, *Me ester*
3,5-Dibromobenzoic Acid, *Me ester*
2,4-Dibromo-5-hydroxybenzaldehyde, *Me ether*
4,5-Dibromo-*o*-toluic Acid
4,6-Dibromo-*o*-toluic Acid
4,6-Dibromo-*m*-toluic Acid
2,6-Dibromo-*p*-toluic Acid
3,5-Dibromo-*p*-toluic Acid
3,6-Dibromo-*p*-toluic Acid

$C_8H_6Br_2O_3$
3,5-Dibromo-2,4-dihydroxyacetophenone
2,4-Dibromo-5-hydroxybenzoic Acid, *Me ester*
2,4-Dibromo-5-hydroxybenzoic Acid, *Me ether*
3,5-Dibromo-2-hydroxybenzoic Acid, *Me ester*
3,5-Dibromo-2-hydroxybenzoic Acid, *Me ether*
3,5-Dibromo-4-hydroxybenzoic Acid, *Me ester*
3,5-Dibromo-4-hydroxy-*o*-toluic Acid
4,6-Dibromo-5-hydroxy-*o*-toluic Acid
3,5-Dibromo-2-hydroxy-*p*-toluic Acid
3,5-Dibromo-4-methoxybenzoic Acid

$C_8H_6Br_2O_5$
2,6-Dibromo-3,4,5-trihydroxybenzoic Acid, *Me ester*

$C_8H_6Br_3NO$
2,4,6-Tribromophenylacetic Acid, *Amide*
2,4,6-Tribromo-*m*-toluic Acid, *Amide*

$C_8H_6Br_3NO_2$
3-Amino-2,4,6-tribromobenzoic Acid, *Me ester*

$C_8H_6Br_3NO_3$
2,3,4-Tribromo-6-nitrophenol, *Et ether*
2,4,6-Tribromo-3-nitrophenol, *Et ether*

$C_8H_6Br_4$
α,α,α′,α′-Tetrabromo-*o*-xylene
3,4,5,6-Tetrabromo-*o*-xylene
α,α,α′,α′-Tetrabromo-*m*-xylene
2,4,5,6-Tetrabromo-*m*-xylene
α,α,α′,α′-Tetrabromo-*p*-xylene
2,3,5,6-Tetrabromo-*p*-xylene

$C_8H_6Br_4O$
3,4,5,6-Tetrabromo-*o*-cresol, *Me ether*
2,4,5,6-Tetrabromo-*m*-cresol, *Me ether*
α,α,2,6-Tetrabromo-*p*-cresol, *Me ether*

$C_8H_6Br_4O_2$
Tetrabromocatechol, *Di-Me ether*

$C_8H_6ClIO_2$
Chloroiodoacetic Acid, *Phenyl ester*

C_8H_6ClN
3-Chloroindole
4-Chloroindole
5-Chloroindole
6-Chloroindole
7-Chloroindole
α-Chlorophenylacetic Acid, *Nitrile*
o-Chlorophenylacetic Acid, *Nitrile*
m-Chlorophenylacetic Acid, *Nitrile*
p-Chlorophenylacetic Acid, *Nitrile*
3-Chloro-*o*-toluic Acid, *Nitrile*
4-Chloro-*o*-toluic Acid, *Nitrile*
5-Chloro-*o*-toluic Acid, *Nitrile*
6-Chloro-*o*-toluic Acid, *Nitrile*
α-Chloro-*m*-toluic Acid, *Nitrile*
α-Chloro-*p*-toluic Acid, *Nitrile*
2-Chloro-*p*-toluic Acid, *Nitrile*
3-Chloro-*p*-toluic Acid, *Nitrile*
α-Chloro-*o*-tolunitrile
α-Chloro-*m*-tolunitrile
α-Chloro-*p*-tolunitrile

C_8H_6ClNO
o-Chloromandelic Acid, *Nitrile*
p-Chloromandelic Acid, *Nitrile*

$C_8H_6ClNO_2$
α-Chloro-4-nitrostyrene
β-Chloro-2-nitrostyrene
β-Chloro-3-nitrostyrene
β-Chloro-4-nitrostyrene
4-Chloro-β-nitrostyrene
Oxanilic Acid, *Chloride*

$C_8H_6ClNO_2S$
4-Sulpho-*o*-toluic Acid, *Chloride-nitrile*
2-Sulpho-*p*-toluic Acid, *Chloride-nitrile*

$C_8N_6ClNO_3$
α-Chloro-2-nitroacetophenone
α-Chloro-3-nitroacetophenone
α-Chloro-4-nitroacetophenone
2-Chloro-5-nitroacetophenone
3-Chloro-2-nitroacetophenone
4-Chloro-2-nitroacetophenone
4-Chloro-3-nitroacetophenone
5-Chloro-2-nitroacetophenone
p-Nitrophenylacetic Acid, *Chloride*
3-Nitro-*o*-toluic Acid, *Chloride*
5-Nitro-*o*-toluic Acid, *Chloride*
6-Nitro-*o*-toluic Acid, *Chloride*
2-Nitro-*m*-toluic Acid, *Chloride*
5-Nitro-*m*-toluic Acid, *Chloride*
2-Nitro-*p*-toluic Acid, *Chloride*
Pyridine-2,3-dicarboxylic Acid, 1-*Me ester*, 2-*Chloride*

$C_8H_6ClNO_4$
2-Chloro-3-nitrobenzoic Acid, *Me ester*
2-Chloro-4-nitrobenzoic Acid, *Me ester*
2-Chloro-5-nitrobenzoic Acid, *Me ester*
2-Chloro-6-nitrobenzoic Acid, *Me ester*
4-Chloro-2-nitrobenzoic Acid, *Me ester*
4-Chloro-3-nitrobenzoic Acid, *Me ester*
5-Chloro-2-nitrobenzoic Acid, *Me ester*
5-Chloro-3-nitrobenzoic Acid, *Me ester*

$C_8H_6ClNO_4$ (*continued*)
4-Chloro-3-nitro-*o*-toluic Acid
4-Chloro-5-nitro-*o*-toluic Acid
4-Chloro-6-nitro-*o*-toluic Acid
2-Chloro-3-nitro-*p*-toluic Acid
2-Chloro-5-nitro-*p*-toluic Acid
3-Chloro-2-nitro-*p*-toluic Acid
3-Chloro-5-nitro-*p*-toluic Acid
5-Chloro-2-nitro-*p*-toluic Acid
2-Hydroxy-5-nitro-*m*-toluic Acid, *Chloride*
4-Methoxy-3-nitrobenzoic Acid, *Chloride*
o-Nitrophenoxyacetic Acid, *Chloride*
m-Nitrophenoxyacetic Acid, *Chloride*
p-Nitrophenoxyacetic Acid, *Chloride*

$C_8H_6ClNO_5$
3-Chloro-2-hydroxy-5-nitrobenzoic Acid, *Me ether*
4-Hydroxy-3-methoxy-5-nitrobenzoic Acid, *Chloride*

$C_8H_6ClNO_6S$
4-Nitro-2-sulphobenzoic Acid, 1-*Me ester*, 2-*Chloride*
4-Nitro-6-sulpho-*m*-toluic Acid, 6-*Chloride*

$C_8H_6ClN_3O_7$
3-Chloro-2,4,6-trinitrophenol, *Et ether*

$C_8H_6Cl_2$
α,β-Dichlorostyrene
β,β-Dichlorostyrene
2,3-Dichlorostyrene
2,4-Dichlorostyrene
2,5-Dichlorostyrene
2,6-Dichlorostyrene
3,4-Dichlorostyrene
3,5-Dichlorostyrene

$C_8H_6Cl_2N_2O_3$
2,4-Dichloro-3-nitroaniline, N-*Ac*
2,4-Dichloro-6-nitroaniline, N-*Ac*
2,5-Dichloro-4-nitroaniline, N-*Ac*
2,6-Dichloro-3-nitroaniline, N-*Ac*
2,6-Dichloro-4-nitroaniline, N-*Ac*
3,5-Dichloro-2-nitroaniline, N-*Ac*
3,5-Dichloro-4-nitroaniline, N-*Ac*
3,6-Dichloro-2-nitroaniline, N-*Ac*
4,5-Dichloro-2-nitroaniline, N-*Ac*

$C_8H_6Cl_2O$
α-Chlorophenylacetic Acid, *Chloride*
p-Chlorophenylacetic Acid, *Chloride*
α,α-Dichloroacetophenone
2,4-Dichloroacetophenone
2,5-Dichloroacetophenone
2,6-Dichloroacetophenone
3,4-Dichloroacetophenone
3,5-Dichloroacetophenone

$C_8H_6Cl_2O_2$
3-Chloro-6-hydroxy-*p*-toluic Acid, *Chloride*
o-Chlorophenoxyacetic Acid, *Chloride*
p-Chlorophenoxyacetic Acid, *Chloride*
Dichloroacetic Acid, *Phenyl ester*
2,4-Dichlorobenzoic Acid, *Me ester*
3,4-Dichlorobenzoic Acid, *Me ester*
3,5-Dichlorobenzoic Acid, *Me ester*
2,4-Dichloro-3-hydroxybenzaldehyde, *Me ether*
2,4-Dichloro-5-hydroxybenzaldehyde, *Me ether*
2,6-Dichloro-3-hydroxybenzaldehyde, *Me ether*
α,α-Dichlorophenylacetic Acid
α,4-Dichlorophenylacetic Acid
2,4-Dichlorophenylacetic Acid
2,5-Dichlorophenylacetic Acid
2,6-Dichlorophenylacetic Acid
3,4-Dichlorophenylacetic Acid
α,α-Dichloro-*o*-toluic Acid
α,α-Dichloro-*m*-toluic Acid
α,α-Dichloro-*p*-toluic Acid
4,6-Dichloro-*o*-toluic Acid
4,6-Dichloro-*m*-toluic Acid
2,5-Dichloro-*p*-toluic Acid
3,5-Dichloro-*p*-toluic Acid

$C_8H_6Cl_2O_3$
2,4-Dichloro-3-hydroxybenzoic Acid, *Me ether*
2,6-Dichloro-3-hydroxybenzoic Acid, *Me ether*
3,5-Dichloro-2-hydroxybenzoic Acid, *Me ester*
3,5-Dichloro-2-hydroxybenzoic Acid, *Me ether*
3,5-Dichloro-4-hydroxybenzoic Acid, *Me ester*
3,5-Dichloro-4-methoxybenzoic Acid,
2,4-Dichlorophenoxyacetic Acid
2,6-Dichlorophenoxyacetic Acid
3,5-Dichlorophenoxyacetic Acid

$C_8H_6Cl_2O_3S$
2-Sulpho-*p*-toluic Acid, *Dichloride*

$C_8H_6Cl_2O_4$
2,4-Dichloro-5-hydroxyphenoxyacetic Acid†
2,5-Dichloro-4-hydroxyphenoxyacetic Acid†

$C_8H_6Cl_2O_4S$
3-Hydroxy-4-sulphobenzoic Acid, *Me ether*, *Dichloride*

$C_8H_6Cl_2O_5$
2,6-Dichloro-3,4,5-trihydroxybenzoic Acid, *Me ester*

$C_8H_6Cl_3NO$
α,α,α-Trichloro-*p*-toluic Acid, *Amide*

$C_8H_6Cl_4$
α,α,α′,α′-Tetrachloro-*o*-xylene
3,4,5,6-Tetrachloro-*o*-xylene
α,α,α′,α′-Tetrachloro-*m*-xylene
2,4,5,6-Tetrachloro-*m*-xylene
α,α,α′,α′-Tetrachloro-*p*-xylene
2,3,5,6-Tetrachloro-*p*-xylene

$C_8H_6Cl_4O$
Tetrachloro-*o*-cresol, *Me ether*
Tetrachlorophenol, *Et ether*

$C_8H_6Cl_4O_2$
Drosophilin A, *Me ether*★†
Tetrachlorocatechol, *Di-Me ether*
Tetrachloroquinol, *Di-Me ether*

$C_8H_6FNO_4$
2-Fluoro-4-nitrobenzoic Acid, *Me ester*

$C_8H_6F_2$
β,β-Difluorostyrene†

C_8H_6IN
p-Iodomethylbenzoic Acid, *Nitrile*
3-Iodo-4-methylbenzoic Acid, *Nitrile*
o-Iodomethylbenzonitrile
p-Iodomethylbenzonitrile
p-Iodophenylacetic Acid, *Nitrile*

$C_8H_6INO_3$
α-Iodo-3-nitroacetophenone
α-Iodo-4-nitroacetophenone

$C_8H_6INO_4$
2-Iodo-4-nitrobenzoic Acid, *Me ester*
2-Iodo-5-nitrobenzoic Acid, *Me ester*
2-Iodo-6-nitrobenzoic Acid, *Me ester*

$C_8H_6I_2O$
α-Di-iodoacetophenone

$C_8H_6I_2O_3$
2-Hydroxy-3,5-di-iodobenzoic Acid, *Me ester*
4-Hydroxy-3,5-di-iodobenzoic Acid, *Me ester*
4-Hydroxy-3,5-di-iodobenzoic Acid, *Me ether*

$C_8H_6N_2$
Cinnoline
Cycloheptimidazole
1,5-Naphthyridine
1,6-Naphthyridine
1,7-Naphthyridine†
1,8-Naphthyridine
2,6-Naphthyridine†
2,7-Naphthyridine
Phthalazine
Quinazoline
Quinoxaline

$C_8H_6N_2O$
Cinnoline 1-Oxide†
Cinnoline 2-Oxide†
o-Cyanobenzoic Acid, *Amide*
m-Cyanobenzoic Acid, *Amide*
p-Cyanobenzoic Acid, *Amide*
3-Hydroxycinnoline
4-Hydroxycinnoline
5-Hydroxycinnoline
6-Hydroxycinnoline
7-Hydroxycinnoline
8-Hydroxycinnoline
2-Hydroxyquinazoline
4-Hydroxyquinazoline
Imesatin
Oxanilic Acid, *Nitrile*
Phthalazone

$C_8H_6N_2OS$
Benzothiazole-6-carboxylic Acid, *Amide*

$C_8H_6N_2O_2$
Anthroxanic Acid, *Amide*
Cinnoline 1,2-Dioxide†
Indazole-1-carboxylic Acid
Indazole-2-carboxylic Acid
Indazole-3-carboxylic Acid
o-Nitrobenzyl cyanide
m-Nitrobenzyl cyanide
p-Nitrobenzyl cyanide
3-Nitroindole
6-Nitroindole
7-Nitroindole
3-Nitro-*o*-toluic Acid, *Nitrile*
4-Nitro-*o*-toluic Acid, *Nitrile*
5-Nitro-*o*-toluic Acid, *Nitrile*
6-Nitro-*o*-toluic Acid, *Nitrile*
2-Nitro-*m*-toluic Acid, *Nitrile*
4-Nitro-*m*-toluic Acid, *Nitrile*
5-Nitro-*m*-toluic Acid, *Nitrile*

F

6-Nitro-*m*-toluic Acid, *Nitrile*
2-Nitro-*p*-toluic Acid, *Nitrile*
3-Nitro-*p*-toluic Acid, *Nitrile*
3-Phenylsyndone†
1,4-Phthalazinedione

$C_8H_6N_2O_2S$
2-Methyl-5-nitrobenzothiazole
2-Methyl-6-nitrobenzothiazole
2-Methyl-7-nitrobenzothiazole
4-Methyl-2-nitrobenzothiazole
6-Methyl-2-nitrobenzothiazole

$C_8H_6N_2O_3$
2-Hydroxy-5-nitrobenzoic Acid, *Me ether*, *Nitrile*
2-Hydroxy-6-nitrobenzoic Acid, *Me ether*, *Nitrile*
Hydroxy-*o*-nitrophenylacetic Acid, *Nitrile*
Hydroxy-*p*-nitrophenylacetic Acid, *Nitrile*
4-Methoxy-3-nitrobenzoic Acid, *Nitrile*
5-Nitro-oxindole

$C_8H_6N_2O_4$
2,6-Dihydroxy-3-nitrobenzoic Acid, *Nitrile*, 6-*Me ether*
α,β-Dinitrostyrene
β,2-Dinitrostyrene
β,3-Dinitrostyrene
β,4-Dinitrostyrene
2,4-Dinitrostyrene
4-Hydroxy-3-methoxy-5-nitrobenzoic Acid, *Nitrile*
o-Nitrobenzoylformic Acid, *Amide*

$C_8H_6N_2O_5$
2,4-Dinitroacetophenone
3,5-Dinitroacetophenone
2-Nitroisophthalic Acid, *Monoamide*
2-Nitro-oxanilic Acid
3-Nitro-oxanilic Acid
4-Nitro-oxanilic Acid
3-Nitrophthalamic Acid

$C_8H_6N_2O_6$
4-Amino-6-nitroisophthalic Acid
2-Amino-5-nitroterephthalic Acid
2,3-Dinitrobenzoic Acid, *Me ester*
2,4-Dinitrobenzoic Acid, *Me ester*
2,5-Dinitrobenzoic Acid, *Me ester*
2,6-Dinitrobenzoic Acid, *Me ester*
3,4-Dinitrobenzoic Acid, *Me ester*
3,5-Dinitrobenzoic Acid, *Me ester*
2,4-Dinitrophenylacetic Acid
2,6-Dinitrophenylacetic Acid
3,5-Dinitro-*o*-toluic Acid
4,5-Dinitro-*o*-toluic Acid
5,6-Dinitro-*o*-toluic Acid
2,6-Dinitro-*m*-toluic Acid
4,6-Dinitro-*m*-toluic Acid
2,3-Dinitro-*p*-toluic Acid
2,6-Dinitro-*p*-toluic Acid
3,5-Dinitro-*p*-toluic Acid
3,6-Dinitro-*p*-toluic Acid
2-Hydroxy-3,5-dinitrobenzaldehyde, *Me ether*
3-Hydroxy-2,6-dinitrobenzaldehyde, *Me ether*
3-Hydroxy-4,6-dinitrobenzaldehyde, *Me ether*
4-Hydroxy-3,5-dinitrobenzaldehyde, *Me ether*
4-Methoxy-3,5-dinitrobenzaldehyde

$C_8H_6N_2O_7$
2,4-Dinitrophenoxyacetic Acid
3,5-Dinitrophenoxyacetic Acid
2-Hydroxy-3,5-dinitrobenzoic Acid, *Me ether*
2-Hydroxy-3,5-dinitrobenzoic Acid, *Me ester*
2-Hydroxy-5,6-dinitrobenzoic Acid, *Me ether*
4-Hydroxy-3,5-dinitrobenzoic Acid, *Me ester*
5-Hydroxy-2,4-dinitrobenzoic Acid, *Me ether*
2-Hydroxy-3,5-dinitro-*p*-toluic Acid
3-Hydroxy-2,6-dinitro-*p*-toluic Acid
4-Methoxy-2,3-dinitrobenzoic Acid
4-Methoxy-3,5-dinitrobenzoic Acid

$C_8H_6N_2S$
5-(2′-Pyridyl)thiazole †

$C_8H_6N_3OS$
Benzothiazole-2-carboxylic Acid, *Hydrazide*

$C_8H_6N_4O_3$
6-Nitroindazole-1-carboxylic Acid, *Amide*

$C_8H_6N_4O_5$
Nitrofurantoin

$C_8H_6N_4O_6$
5,5′-Bibarbituric Acid
5,5′-Dimethyl-4,4′-dinitro-3,3′-bi-isoxazole

$C_8H_6N_4O_7$
2,4,6-Trinitrobenzoic Acid, *Methylamide*

$C_8H_6N_4O_9$
2,3,5,6-Tetranitrophenol, *Et ether*

$C_8H_6N_4O_{10}$
Tetranitroresorcinol, *Mono-Et ether*

$C_8H_6N_6O_{11}$
Pentryl

$C_8H_6N_8O_4$
Alloxantin

C_8H_6O
Benzofuran
Isobenzofuran †
Octa-2,3-diene-5,7-diyn-1-ol †

C_8H_6OS
1,4-Benzoxathiin
3-Hydroxybenzo[*b*]thiophene
Junipal

$C_8H_6OS_2$
Thieno[3,4-*d*]thiepin Sulphoxide †

$C_8H_6O_2$
1,4-Benzodioxin
2,2′-Bifuryl †
Dihydrobenzofuran-3-one
Dihydro-2-oxobenzofuran
4-Hydroxybenzofuran
5-Hydroxybenzofuran
6-Hydroxybenzofuran
Isophthalaldehyde
Phenylglyoxal
o-Phthalaldehyde
Phthalide
Terephthalaldehyde

$C_8H_6O_2S$
Benzo[*b*]thiophene, S-*Oxide*

$C_8H_6O_2S_2$
Thieno[3,4-*d*]thiepin Sulphone †

$C_8H_6O_3$
Benzoylformic Acid
1,4-Cyclohexadiene-1,2-dicarboxylic Acid, *Anhydride*
2,4-Cyclohexadiene-1,2-dicarboxylic Acid, *Anhydride*
2,5-Cyclohexadiene-1,2-dicarboxylic Acid, *Anhydride*
2,6-Cyclohexadiene-1,2-dicarboxylic Acid, *Anhydride*
3,5-Cyclohexadiene-1,2-dicarboxylic Acid, *Anhydride*
2,4-Diformylphenol
2,6-Diformylphenol
2-Formylbenzoic Acid
3-Formylbenzoic Acid
4-Formylbenzoic Acid
4-Hydroxyphthalide †
5-Hydroxyphthalide †
6-Hydroxyphthalide †
7-Hydroxyphthalide †
3,4-Methylenedioxybenzaldehyde

$C_8H_6O_4$
2,4-Dihydroxyisophthalaldehyde
5,7-Dihydroxyphthalide †
3-Formyl-2-hydroxybenzoic Acid
3-Formyl-4-hydroxybenzoic Acid
4-Formyl-3-hydroxybenzoic Acid
5-Formyl-2-hydroxybenzoic Acid
4-(2-Furyl)-2-oxo-3-butenoic Acid
2-Hydroxybenzoylformic Acid
3-Hydroxybenzoylformic Acid
4-Hydroxybenzoylformic Acid
Isophthalic Acid
3,4-Methylenedioxybenzoic Acid
2,6-Octadien-4-ynedioic Acid
Phthalic Acid
Terephthalic Acid

$C_8H_6O_4S$
4-Mercaptophthalic Acid

$C_8H_6O_5$
2,4-Dihydroxybenzoylformic Acid
2,5-Dihydroxybenzoylformic Acid
2,6-Dihydroxybenzoylformic Acid
3,4-Dihydroxybenzoylformic Acid
3,4-Dihydroxycrotonolactone, *Anhydride*★ †
2,5-Dihydroxyphenylglyoxylic Acid
Fomecin B †
2-Formyl-5,6-dihydroxybenzoic Acid
5-Formyl-3,4-dihydroxybenzoic Acid
2-Hydroxyisophthalic Acid
4-Hydroxyisophthalic Acid
5-Hydroxyisophthalic Acid
3-Hydroxyphthalic Acid
4-Hydroxyphthalic Acid
Hydroxyterephthalic Acid
Stipitatic Acid

$C_8H_6O_6$
Butane-1,2,3,4-tetracarboxylic Acid, *Dianhydride*
2,4-Dihydroxyisophthalic Acid
4,6-Dihydroxyisophthalic Acid
3,4-Dihydroxyphthalic Acid

$C_8H_6O_6$ (*continued*)
3,5-Dihydroxyphthalic Acid
3,6-Dihydroxyphthalic Acid
4,5-Dihydroxyphthalic Acid
2,3-Dihydroxyterephthalic Acid
2,5-Dihydroxyterephthalic Acid
2,6-Dihydroxyterephthalic Acid
Puberulic Acid
2,4,5-Trihydroxyphenylglyoxylic Acid

$C_8H_6O_6S$
3-Sulphophthalic Acid
4-Sulphophthalic Acid

$C_8H_6O_7$
3-Hydroxy-4-oxo-4*H*-pyran-2,6-dicarboxylic Acid, *Mono-Me ester*
2,4,6-Trihydroxyisophthalic Acid
4,5,6-Trihydroxyisophthalic Acid
3,4,5-Trihydroxyphthalic Acid
3,4,6-Trihydroxyphthalic Acid
2,3,5-Trihydroxyterephthalic Acid

C_8H_6S
Benzo[*b*[thiophene
Benzo[*c*]thiophene†
Cyclopenta[*c*]thiin†

$C_8H_6S_2$
1,4-Benzodithiin
2,2′-Bithienyl
2,3′-Bithienyl
3,3′-Bithienyl
Thieno[3,4-*d*]thiepin†

$C_8H_6S_3$
2,2′-Dithienyl Sulphide†
2,3′-Dithienyl Sulphide†
3,3′-Dithienyl Sulphide†

C_8H_6Se
Selenonaphthene

C_8H_7Br
Bromocyclo-octatetraene
1-Bromo-1-phenylethylene
1-Bromo-2-phenylethylene
o-Bromophenylethylene
m-Bromophenylethylene
p-Bromophenylethylene

$C_8H_7BrClNO$
2-Bromo-4-chloroaniline, N-*Ac*
3-Bromo-4-chloroaniline, N-*Ac*
4-Bromo-*N*-chloroaniline, N-*Ac*
4-Bromo-2-chloroaniline, N-*Ac*
4-Bromo-3-chloroaniline, N-*Ac*
5-Bromo-2-chloroaniline, N-*Ac*

$C_8H_7BrHgO_2$
Mercuri-*o*-bromophenyl acetate
Mercuri-*m*-bromophenyl acetate
Mercuri-*p*-bromophenyl acetate

$C_8H_7BrN_2$
3-Bromoindazole, 2-N-*Me*
5-Bromoindazole, 1-N-*Me*
5-Bromoindazole, 2-N-*Me*

$C_8H_7BrN_2O_2$
Bromoterephthalic Acid, *Diamide*

$C_8H_7BrN_2O_3$
2-Bromo-4-nitroaniline, N-*Ac*
2-Bromo-5-nitroaniline, N-*Ac*
2-Bromo-6-nitroaniline, N-*Ac*
4-Bromo-2-nitroaniline, N-*Ac*
4-Bromo-3-nitroaniline, N-*Ac*
5-Bromo-2-nitroaniline, N-*Ac*
4-Bromo-6-nitro-*o*-toluic Acid, *Amide*
2-Bromo-5-nitro-*p*-toluic Acid, *Amide*
2-Bromo-6-nitro-*p*-toluic Acid, *Amide*

$C_8H_7BrN_2O_5$
4-Bromo-2,6-dinitrophenol, *Et ether*

C_8H_7BrO
α-Bromoacetophenone
o-Bromoacetophenone
m-Bromoacetophenone
p-Bromoacetophenone
2-Bromo-*p*-tolualdehyde†
Phenylacetic Acid, *Bromide*
o-Toluic Acid, *Bromide*
m-Toluic Acid, *Bromide*
p-Toluic Acid, *Bromide*

$C_8H_7BrO_2$
Bromoacetic Acid, *Phenyl ester*
o-Bromobenzoic Acid, *Me ester*
m-Bromobenzoic Acid, *Me ester*
p-Bromobenzoic Acid, *Me ester*
α-Bromo-4-hydroxyacetophenone
3-Bromo-4-hydroxyacetophenone
5-Bromo-2-hydroxyacetophenone
2-Bromo-3-hydroxybenzaldehyde, *Me ether*
2-Bromo-4-hydroxybenzaldehyde, *Me ether*
2-Bromo-5-hydroxybenzaldehyde, *Me ether*
3-Bromo-4-hydroxybenzaldehyde, *Me ether*
4-Bromo-2-hydroxybenzaldehyde, *Me ether*
4-Bromo-3-hydroxybenzaldehyde, *Me ether*
5-Bromo-2-hydroxybenzaldehyde, *Me ether*
α-Bromophenylacetic Acid
o-Bromophenylacetic Acid
m-Bromophenylacetic Acid
p-Bromophenylacetic Acid
α-Bromo-*o*-toluic Acid
α-Bromo-*m*-toluic Acid
α-Bromo-*p*-toluic Acid
3-Bromo-*o*-toluic Acid
4-Bromo-*o*-toluic Acid
5-Bromo-*o*-toluic Acid
6-Bromo-*o*-toluic Acid
2-Bromo-*m*-toluic Acid†
4-Bromo-*m*-toluic Acid
5-Bromo-*m*-toluic Acid
6-Bromo-*m*-toluic Acid
2-Bromo-*p*-toluic Acid
3-Bromo-*p*-toluic Acid

$C_8H_7BrO_3$
α-Bromo-2,4-dihydroxyacetophenone
α-Bromo-2,6-dihydroxyacetophenone
α-Bromo-3,4-dihydroxyacetophenone
2-Bromo-3,4-dihydroxybenzaldehyde, 4-*Me ether*
2-Bromo-4,5-dihydroxybenzaldehyde, 4-*Me ether*
2-Bromo-3-hydroxybenzoic Acid, *Me ether*
2-Bromo-4-hydroxybenzoic Acid, *Me ether*

$C_8H_7BrO_3$ (*continued*)
2-Bromo-5-hydroxybenzoic Acid, *Me ether*
3-Bromo-2-hydroxybenzoic Acid, *Me ester*
3-Bromo-4-hydroxybenzoic Acid, *Me ester*
3-Bromo-5-hydroxybenzoic Acid, *Me ether*
4-Bromo-2-hydroxybenzoic Acid, *Me ether*
4-Bromo-3-hydroxybenzoic Acid, *Me ester*
4-Bromo-3-hydroxybenzoic Acid, *Me ether*
5-Bromo-2-hydroxybenzoic Acid, *Me ether*
5-Bromo-2-hydroxybenzoic Acid, *Me ester*
2-Bromo-4-hydroxy-3-methoxybenzaldehyde
2-Bromo-4-hydroxy-5-methoxybenzaldehyde
3-Bromo-4-hydroxy-5-methoxybenzaldehyde
3-Bromo-4-hydroxyphenylacetic Acid
5-Bromo-2-hydroxy-*m*-toluic Acid
5-Bromo-4-hydroxy-*m*-toluic Acid
5-Bromo-6-hydroxy-*m*-toluic Acid
3-Bromo-2-hydroxy-*p*-toluic Acid
5-Bromo-2-hydroxy-*p*-toluic Acid
p-Bromomandelic Acid
2-Bromo-4-methoxybenzoic Acid
3-Bromo-4-methoxybenzoic Acid
5-Bromomethyl-2-hydroxybenzoic Acid
o-Bromophenoxyacetic Acid
m-Bromophenoxyacetic Acid
p-Bromophenoxyacetic Acid

$C_8H_7BrO_4$
3-Bromo-4,5-dihydroxybenzoic Acid, *Me ester*
5-Bromo-2,4-dihydroxybenzoic Acid, *Me ester*
2-Bromo-4-hydroxy-3-methoxybenzoic Acid
2-Bromo-4-hydroxy-5-methoxybenzoic Acid
3-Bromo-4-hydroxy-5-methoxybenzoic Acid

$C_8H_7Br_2NO$
2,3-Dibromoacetanilide
2,4-Dibromoacetanilide
2,5-Dibromoacetanilide
2,6-Dibromoacetanilide
3,4-Dibromoacetanilide
3,5-Dibromoacetanilide
4,6-Dibromo-*o*-toluic Acid, *Amide*
4,6-Dibromo-*m*-toluic Acid, *Amide*
2,6-Dibromo-*p*-toluic Acid, *Amide*
3,5-Dibromo-*p*-toluic Acid, *Amide*

$C_8H_7Br_2NO_2$
2-Amino-3,5-dibromobenzoic Acid, *Me ester*
4-Amino-3,5-dibromobenzoic Acid, *Me ester*
4,5-Dibromo-3-nitro-*o*-xylene
4,6-Dibromo-2-nitro-*m*-xylene
2,3-Dibromo-5-nitro-*p*-xylene
2,5-Dibromo-3-nitro-*p*-xylene
3,5-Dibromo-2-nitro-*p*-xylene

$C_8H_7Br_2NO_3$
2,6-Dibromo-4-carbamoylmethyl-4-hydroxy-cyclohexa-2,5-dien-1-one†
2,4-Dibromo-6-nitrophenol, *Et ether*
2,6-Dibromo-4-nitrophenol, *Et ether*
3,6-Dibromo-2-nitrophenol, *Et ether*

$C_8H_7Br_3$
3,4,5-Tribromo-*o*-xylene
3,4,6-Tribromo-*o*-xylene
2,4,5-Tribromo-*m*-xylene
2,4,6-Tribromo-*m*-xylene
4,5,6-Tribromo-*m*-xylene
2,3,5-Tribromo-*p*-xylene

$C_8H_7Br_3O$
3,4,5-Tribromo-*o*-cresol, *Me ether*
3,4,6-Tribromo-*o*-cresol, *Me ether*
2,4,6-Tribromophenol, *Et ether*

$C_8H_7Br_3O_2$
3,4,5-Tribromocatechol, *Di-Me ether*
3,4,6-Tribromocatechol, *Di-Me ether*
2,4,6-Tribromoresorcinol, *Di-Me ether*

C_8H_7Cl
Chlorocyclo-octatetraene
1-Chloro-1-phenylethylene
1-Chloro-2-phenylethylene
o-Chlorophenylethylene
m-Chlorophenylethylene
p-Chlorophenylethylene

C_8H_7ClHgO
Mercuri-phenacyl chloride

$C_8H_7ClHgO_2$
Mercuri-*o*-carboxyphenyl chloride, *Me ester*
Mercuri-*m*-carboxyphenyl chloride, *Me ester*
Mercuri-*p*-carboxyphenyl chloride, *Me ester*

$C_8H_7ClN_2$
3-Chloroindazole, 1-*Me*
3-Chloroindazole, 2-*Me*
o-Chlorophenylglycine, *Amide*
p-Chlorophenylglycine, *Amide*

$C_8H_7ClN_2O_2$
Chloroterephthalic Acid, *Diamide*

$C_8H_7ClN_2O_3$
2-Chloro-4-nitroaniline, N-*Ac*
2-Chloro-5-nitroaniline, N-*Ac*
2-Chloro-6-nitroaniline, N-*Ac*
3-Chloro-4-nitroaniline, N-*Ac*
4-Chloro-2-nitroaniline, N-*Ac*
4-Chloro-3-nitroaniline, N-*Ac*
5-Chloro-2-nitroaniline, N-*Ac*
4-Chloro-3-nitrobenzoic Acid, *Me-amide*
5-Chloro-2-nitrobenzoic Acid, *Me-amide*

$C_8H_7ClN_2O_5$
5-Chloro-2,4-dinitrophenol, *Et ether*
4-Chloro-2,6-dinitrophenol, *Et ether*

C_8H_7ClO
α-Chloroacetophenone
o-Chloroacetophenone
m-Chloroacetophenone
p-Chloroacetophenone
4-Chloro-*o*-tolualdehyde (p. 393—out of alphabetical order★)
2-Chloro-*p*-tolualdehyde (p. 693)
3-Chloro-*p*-tolualdehyde (p. 693)
Phenylacetic Acid, *Chloride*
o-Toluic Acid, *Chloride*
m-Toluic Acid, *Chloride*
p-Toluic Acid, *Chloride*

C_8H_7ClOS
Phenylmercaptoacetic Acid, *Chloride*

$C_8H_7ClO_2$
Anisic Acid, *Chloride*
Chloroacetic Acid, *Phenyl ester*

$C_8H_7ClO_2$ (*continued*)
- *o*-Chlorobenzoic Acid, *Me ester*
- *m*-Chlorobenzoic Acid, *Me ester*
- *p*-Chlorobenzoic Acid, *Me ester*
- Chloroformic Acid, p-*Tolyl ester*
- Chloroformic Acid, *Benzyl ester*
- α-Chloro-2-hydroxyacetophenone
- α-Chloro-4-hydroxyacetophenone
- 4-Chloro-α-hydroxyacetophenone
- 3-Chloro-4-hydroxyacetophenone
- 5-Chloro-2-hydroxyacetophenone
- 2-Chloro-3-hydroxybenzaldehyde, *Me ether*
- 2-Chloro-4-hydroxybenzaldehyde, *Me ether*
- 2-Chloro-5-hydroxybenzaldehyde, *Me ether*
- 3-Chloro-4-hydroxybenzaldehyde, *Me ether*
- 4-Chloro-2-hydroxybenzaldehyde, *Me ether*
- 4-Chloro-3-hydroxybenzaldehyde, *Me ether*
- 5-Chloromethyl-2-hydroxybenzaldehyde
- α-Chlorophenylacetic Acid
- *o*-Chlorophenylacetic Acid
- *m*-Chlorophenylacetic Acid
- *p*-Chlorophenylacetic Acid
- α-Chloro-*o*-toluic Acid
- 3-Chloro-*o*-toluic Acid
- 4-Chloro-*o*-toluic Acid
- 5-Chloro-*o*-toluic Acid
- 6-Chloro-*o*-toluic Acid
- α-Chloro-*m*-toluic Acid
- 2-Chloro-*m*-toluic Acid
- 4-Chloro-*m*-toluic Acid
- 5-Chloro-*m*-toluic Acid
- 6-Chloro-*m*-toluic Acid
- α-Chloro-*p*-toluic Acid
- 2-Chloro-*p*-toluic Acid
- 3-Chloro-*p*-toluic Acid
- 4-Hydroxyphenylacetic Acid, *Chloride*
- 2-Hydroxy-*m*-toluic Acid, *Chloride*
- 6-Hydroxy-*m*-toluic Acid, *Chloride*
- 2-Hydroxy-*p*-toluic Acid, *Chloride*
- *o*-Methoxybenzoic Acid, *Chloride*
- *m*-Methyoxybenzoic Acid, *Chloride*
- Phenoxyacetic Acid, *Chloride*

$C_8H_7ClO_3$
- α-Chloro-2,4-dihydroxyacetophenone
- α-Chloro-3,4-dihydroxyacetophenone
- 2-Chloro-3-hydroxybenzoic Acid, *Me ester*
- 2-Chloro-5-hydroxybenzoic Acid, *Me ester*
- 3-Chloro-2-hydroxybenzoic Acid, *Me ester*
- 3-Chloro-4-hydroxybenzoic Acid, *Me ester*
- 5-Chloro-2-hydroxybenzoic Acid, *Me ester*
- 2-Chloro-3-hydroxybenzoic Acid, *Me ether*
- 2-Chloro-5-hydroxybenzoic Acid, *Me ether*
- 2-Chloro-6-hydroxybenzoic Acid, *Me ether*
- 4-Chloro-3-hydroxybenzoic Acid, *Me ether*
- 5-Chloro-2-hydroxybenzoic Acid, *Me ether*
- α-Chloro-4-hydroxy-*m*-toluic Acid
- 3-Chloro-6-hydroxy-*p*-toluic Acid
- 4-Chloro-6-hydroxy-*m*-toluic Acid
- 6-Chloro-2-hydroxy-*m*-toluic Acid
- *o*-Chloromandelic Acid
- *p*-Chloromandelic Acid
- 2-Chloro-4-methoxybenzoic Acid
- 3-Chloro-4-methoxybenzoic Acid
- 5-Chloromethyl-2-hydroxybenzoic Acid
- *o*-Chlorophenoxyacetic Acid
- *m*-Chlorophenoxyacetic Acid
- *p*-Chlorophenoxyacetic Acid
- Vanillic Acid, *Chloride*

$C_8H_7ClO_3S$
- Phenylsulphonylacetic Acid, *Chloride*

$C_8H_7ClO_4$
- Gallic Acid, 3-*Me ether*, *Chloride*

$C_8H_7ClO_4S$
- *o*-Sulphobenzoic Acid, 1-*Me ester*

$C_8H_7ClO_5S$
- 2-Hydroxy-5-sulphobenzoic Acid, *Sulphonchloride*, *Me ester*
- 3-Hydroxy-4-sulphobenzoic Acid, *Me ether*, *Sulphonchloride*
- 2-Hydroxy-5-sulpho-*m*-toluic Acid, *Sulphonchloride*
- 2-Hydroxy-6-sulpho-*p*-toluic Acid *Sulphonchloride*
- 6-Hydroxy-5-sulpho-*m*-toluic Acid, *Sulphonchloride*

$C_8H_7Cl_2HgNO_2$
- Mercuri-2-amino-3,5-dichlorophenyl acetate

$C_8H_7Cl_2NO$
- 2,3-Dichloroacetanilide
- 2,4-Dichloroacetanilide
- 2,5-Dichloroacetanilide
- 2,6-Dichloroacetanilide
- 3,4-Dichloroacetanilide
- 3,5-Dichloroacetanilide
- α,α-Dichlorophenylacetic Acid, *Amide*
- 2,4-Dichlorophenylacetic Acid, *Amide*
- 2,5-Dichlorophenylacetic Acid, *Amide*
- α,α-Dichloro-*o*-toluic Acid, *Amide*

$C_8H_7Cl_2NO_2$
- 2-Amino-3,5-dichlorobenzoic Acid, *Me ester*
- 2,4-Dichlorophenoxyacetic Acid, *Amide*
- 4,6-Dichloropyridine-2-carboxylic Acid, *Et ester*
- 2,6-Dichloropyridine-4-carboxylic Acid, *Et ester*

$C_8H_7Cl_2NO_3$
- 2,4-Dichloro-6-nitrophenol, *Et ether*
- 2,6-Dichloro-4-nitrophenol, *Et ether*

$C_8H_7Cl_3$
- 3,4,5-Trichloro-*o*-xylene
- 3,5,6-Trichloro-*o*-xylene
- 2,4,5-Trichloro-m-xylene
- 2,3,5-Trichloro-*p*-xylene

$C_8H_7Cl_3O$
- 4,5,6-Trichloro-*o*-cresol, *Me ether*
- 2,4,6-Trichloro-*m*-cresol, *Me ether*
- 2,4,6-Trichlorophenol, *Et ether*

$C_8H_7Cl_3O_2$
- 3,4,5-Trichlorocatechol, *Di-Me ether*
- 2-(2,4,5-Trichlorophenoxy)ethanol
- 2-(2,4,6-Trichlorophenoxy)ethanol

$C_8H_7Cl_3O_3$
- Trichlorophloroglucinol, *Di-Me ether*

$C_8H_7FN_2O_5$
- 2-Fluoro-4,6-dinitrophenol, *Et ether*

C_8H_7FO
α-Fluoroacetophenone
2-Fluoroacetophenone
3-Fluoroacetophenone
4-Fluoroacetophenone
Phenylacetic Acid, *Fluoride*

$C_8H_7FO_2$
o-Fluorobenzoic Acid, *Me ester*
m-Fluorobenzoic Acid, *Me ester*
p-Fluorobenzoic Acid, *Me ester*
p-Fluorophenylacetic Acid
4-Fluoro-*m*-toluic Acid
3-Fluoro-*p*-toluic Acid

C_8H_7HgN
Mercuri-benzyl cyanide
Mercuri-*p*-tolyl cyanide

$C_8H_7HgNO_4$
Mercuri-*p*-nitrophenyl acetate

C_8H_7IO
α-Iodoacetophenone
o-Iodoacetophenone
m-Iodoacetophenone
p-Iodoacetophenone

$C_8H_7IO_2$
2-Hydroxy-4-iodobenzaldehyde, *Me ether*
3-Hydroxy-2-iodobenzaldehyde, *Me ether*
3-Hydroxy-6-iodobenzaldehyde, *Me ether*
o-Iodobenzoic Acid, *Me ester*
m-Iodobenzoic Acid, *Me ester*
p-Iodobenzoic Acid, *Me ester*
2-Iodo-4-methoxybenzaldehyde
3-Iodo-4-methoxybenzaldehyde
o-Iodomethylbenzoic Acid
p-Iodomethylbenzoic Acid
2-Iodo-3-methylbenzoic Acid
2-Iodo-4-methylbenzoic Acid
2-Iodo-5-methylbenzoic Acid
3-Iodo-4-methylbenzoic Acid
o-Iodophenylacetic Acid
m-Iodophenylacetic Acid
p-Iodophenylacetic Acid

$C_8H_7IO_3$
1,3-Dihydro-1-hydroxy-3-oxo-1,2-benziodoxole, 1-*Me ether* †
2-Hydroxy-4-iodobenzoic Acid, *Me ether*
2-Hydroxy-5-iodobenzoic Acid, *Me ether*
3-Hydroxy-2-iodobenzoic Acid, *Me ether*
4-Hydroxy-3-iodobenzoic Acid, *Me ester*
4-Hydroxy-2-iodo-3-methoxybenzaldehyde
4-Hydroxy-5-iodo-3-methoxybenzaldehyde
2-Iodo-4-methoxybenzoic Acid
3-Iodo-4-methoxybenzoic Acid

$C_8H_7I_2NO$
2,4-Di-iodoaniline, N-*Ac*
3,5-Di-iodoaniline, N-*Ac*

$C_8H_7I_3O$
2,3,5-Tri-iodophenol, *Et ether*
2,4,6-Tri-iodophenol, *Et ether*

C_8H_7N
Benzyl cyanide
Indole
Indolizine
1*H*-1-Pyrindine †
o-Tolunitrile
m-Tolunitrile
p-Tolunitrile

C_8H_7NO
Anisic Acid, *Nitrile*
1-Hydroxyindole
4-Hydroxyindole
5-Hydroxyindole
6-Hydroxyindole
7-Hydroxyindole
m-Hydroxymethyl-benzoic Acid, *Nitrile*
p-Hydroxymethyl-benzoic Acid, *Nitrile*
2-Hydroxyphenylacetic Acid, *Nitrile*
3-Hydroxyphenylacetic Acid, *Nitrile*
3-Hydroxy-*o*-toluic Acid, *Nitrile*
2-Hydroxy-*m*-toluic Acid, *Nitrile*
4-Hydroxy-*m*-toluic Acid, *Nitrile*
5-Hydroxy-*m*-toluic Acid, *Nitrile*
6-Hydroxy-*m*-toluic Acid, *Nitrile*
3-Hydroxy-*p*-toluic Acid, *Nitrile*
Indoxyl
Mandelic Acid, *Nitrile*
o-Methoxybenzoic Acid, *Nitrile*
2-Methylbenzoxazole
5-Methylbenzoxazole
6-Methylbenzoxazole
Oxindole
Phenoxyacetic Acid, *Nitrile*
Phthalimidine
o-Tolyl isocyanate
m-Tolyl isocyanate
p-Tolyl isocyanate

C_8H_7NOS
2-Hydroxybenzothiazole, *Me ether*
p-Hydroxybenzyl isothiocyanate †

$C_8H_7NO_2$
Benzomorpholone
Benzoxazolin-2-one, N-*Me*
Benzoylformic Acid, *Amide*
2,3-Dihydro-3-hydroxy-2-oxoindole
1,2-Dihydroxyindole
3,6-Dihydroxyindole
5,6-Dihydroxyindole
α,2-Dihydroxyphenylacetic Acid, *Nitrile*
α,4-Dihydroxyphenylacetic Acid, *Nitrile*
3,4-Dihydroxyphenylacetic Acid, *Nitrile*
3-Formylbenzoic Acid, *Amide*
3-Hydroxy-4-methoxybenzoic Acid, *Nitrile*
Isonitrosoacetophenone
3,4-Methylenedioxybenzaldehyde, *Imide*
α-Nitrostyrene
o-Nitrostyrene
m-Nitrostyrene
p-Nitrostyrene
3-Phenyl-1,3-oxazetidin-2-one †
2-Pyridineacrylic Acid
3-Pyridineacrylic Acid
4-Pyridineacrylic Acid
Vanillic Acid, *Nitrile*

$C_8H_7NO_2S$
4*H*-1,4-Benzothiazine-1,1-dioxide †
Phenylsulphonylacetic Acid, *Nitrile*
p-Toluenesulphonyl cyanide †

$C_8H_7NO_3$
Benzoylnitromethane
Blepharigenin†
3,4-Dihydroxymandelic Acid, *Nitrile*
2-Hydroxy-β-nitrostyrene
3-Hydroxy-β-nitrostyrene
4-Hydroxy-β-nitrostyrene
4-Hydroxy-3-nitrostyrene
Isatic Acid
Isophthalamic Acid
6-Methoxybenzoxazolin-2-one
3,4-Methylenedioxybenzoic Acid, *Amide*
2-Methyl-5-nitrobenzaldehyde†
o-Nitroacetophenone
m-Nitroacetophenone
p-Nitroacetophenone
o-Nitrophenylacetaldehyde
m-Nitrophenylacetaldehyde
p-Nitrophenylacetaldehyde
o-Nitrophenyloxiran
m-Nitrophenyloxiran
p-Nitrophenyloxiran
o-Nitrosobenzoic Acid, *Me ester*
m-Nitrosobenzoic Acid, *Me ester*
p-Nitrosobenzoic Acid, *Me ester*
2-Nitro-*m*-toluic Aldehyde
4-Nitro-*m*-toluic Aldehyde
6-Nitro-*m*-toluic Aldehyde
2-Nitro-*p*-toluic Aldehyde
3-Nitro-*p*-toluic Aldehyde
Oxanilic Acid
Phthalamic Acid
Terephthalamic Acid

$C_8H_7NO_3S$
Saccharin, N-*Me*
Sulphazone

$C_8H_7NO_4$
2-Aminoisophthalic Acid
4-Aminoisophthalic Acid
5-Aminoisophthalic Acid
3-Aminophthalic Acid
4-Aminophthalic Acid
Aminoterephthalic Acid
Apophyllenic Acid
p-Carboxycarbanilic Acid
2,4-Dihydroxy-1,4-benzoxazolin-3-one
2-Hydroxyisophthalic Acid, *Mono-amide*
α-Hydroxy-3-nitroacetophenone
α-Hydroxy-4-nitroacetophenone
2-Hydroxy-3-nitroacetophenone
2-Hydroxy-4-nitroacetophenone
2-Hydroxy-5-nitroacetophenone
3-Hydroxy-2-nitroacetophenone
4-Hydroxy-3-nitroacetophenone
2-Hydroxy-3-nitrobenzaldehyde, *Me ether*
2-Hydroxy-5-nitrobenzaldehyde, *Me ether*
2-Hydroxy-6-nitrobenzaldehyde, *Me ether*
3-Hydroxy-2-nitrobenzaldehyde, *Me ether*
3-Hydroxy-4-nitrobenzaldehyde, *Me ether*
5-Hydroxy-2-nitrobenzaldehyde, *Me ether*
2-Hydroxy-5-nitrosobenzoic Acid, *Me ester*
2-Hydroxy-5-nitro-*m*-tolualdehyde
4-Hydroxy-5-nitro-*m*-tolualdehyde
6-Hydroxy-5-nitro-*m*-tolualdehyde
2-Hydroxy-3-nitro-*p*-tolualdehyde
2-Hydroxy-5-nitro-*p*-tolualdehyde
Isatoic Acid
4-Methoxy-3-nitrobenzaldehyde
4-Methylpyridine-2,3-dicarboxylic Acid
6-Methylpyridine-2,4-dicarboxylic Acid
4-Methylpyridine-2,6-dicarboxylic Acid
2-Methylpyridine-3,4-dicarboxylic Acid
6-Methylpyridine-3,4-dicarboxylic Acid
2-Methylpyridine-3,5-dicarboxylic Acid
4-Methylpyridine-3,5-dicarboxylic Acid
o-Nitrobenzoic Acid, *Me ester*
m-Nitrobenzoic Acid, *Me ester*
p-Nitrobenzoic Acid, *Me ester*
o-Nitrophenylacetic Acid
m-Nitrophenylacetic Acid
p-Nitrophenylacetic Acid
3-Nitro-*o*-toluic Acid
4-Nitro-*o*-toluic Acid
5-Nitro-*o*-toluic Acid
6-Nitro-*o*-toluic Acid
2-Nitro-*m*-toluic Acid
4-Nitro-*m*-toluic Acid
5-Nitro-*m*-toluic Acid
6-Nitro-*m*-toluic Acid
2-Nitro-*p*-toluic Acid
3-Nitro-*p*-toluic Acid
Pyridine-2,3-dicarboxylic Acid, 1-*Me ester*
Pyridine-2,3-dicarboxylic Acid, 2-*Me ester*

$C_8H_7NO_5$
2,4-Dihydroxy-5-nitroacetophenone
2,5-Dihydroxy-3-nitroacetophenone
2,6-Dihydroxy-3-nitroacetophenone
3,4-Dihydroxy-α-nitroacetophenone
2,3-Dihydroxy-5-nitrobenzaldehyde, 3-*Me ether*
2,3-Dihydroxy-6-nitrobenzaldehyde, 3-*Me ether*
2,4-Dihydroxy-3-nitrobenzaldehyde, 4-*Me ether*
2,4-Dihydroxy-5-nitrobenzaldehyde, 4-*Me ether*
2,5-Dihydroxy-3-nitrobenzaldehyde, 5-*Me ether*
3-Hydroxy-4-methoxy-2-nitrobenzaldehyde
3-Hydroxy-4-methoxy-5-nitrobenzaldehyde
4-Hydroxy-3-methoxy-2-nitrobenzaldehyde
4-Hydroxy-3-methoxy-5-nitrobenzaldehyde
4-Hydroxy-5-methoxy-2-nitrobenzaldehyde
5-Hydroxy-4-methoxy-2-nitrobenzaldehyde
2-Hydroxy-3-nitrobenzoic Acid, *Me ester*
2-Hydroxy-4-nitrobenzoic Acid, *Me ether*
2-Hydroxy-4-nitrobenzoic Acid, *Me ester*
2-Hydroxy-5-nitrobenzoic Acid, *Me ether*
2-Hydroxy-5-nitrobenzoic Acid, *Me ester*
2-Hydroxy-6-nitrobenzoic Acid, *Me ether*
3-Hydroxy-2-nitrobenzoic Acid, *Me ether*
3-Hydroxy-4-nitrobenzoic Acid, *Me ether*
3-Hydroxy-4-nitrobenzoic Acid, *Me ester*
4-Hydroxy-2-nitrobenzoic Acid, *Me ester*
4-Hydroxy-3-nitrobenzoic Acid, *Me ester*
5-Hydroxy-2-nitrobenzoic Acid, *Me ether*
Hydroxy-*o*-nitrophenylacetic Acid
Hydroxy-*m*-nitrophenylacetic Acid
Hydroxy-*p*-nitrophenylacetic Acid
2-Hydroxy-5-nitro-*m*-toluic Acid
2-Hydroxy-5-nitro-*p*-toluic Acid
4-Hydroxy-5-nitro-*m*-toluic Acid
6-Hydroxy-5-nitro-*m*-toluic Acid
2-Hydroxy-3-nitro-*p*-toluic Acid
4-Methoxy-2-nitrobenzoic Acid

$C_8H_7NO_5$ (*continued*)
4-Methoxy-3-nitrobenzoic Acid
o-Nitrophenoxyacetic Acid
m-Nitrophenoxyacetic Acid
p-Nitrophenoxyacetic Acid

$C_8H_7NO_6$
2,3-Dihydroxy-4-nitrobenzoic Acid, 2-*Me ether*
2,3-Dihydroxy-5-nitrobenzoic Acid, 3-*Me ether*
2,4-Dihydroxy-5-nitrobenzoic Acid, 4-*Me ether*
2,4-Dihydroxy-5-nitrobenzoic Acid, *Me ester*
2,5-Dihydroxy-3-nitrobenzoic Acid, 5-*Me ether*
2,5-Dihydroxy-3-nitrobenzoic Acid, *Me ester*
2,5-Dihydroxy-4-nitrobenzoic Acid, 5-*Me ether*
2,5-Dihydroxy-6-nitrobenzoic Acid, 5-*Me ether*
3,4-Dihydroxy-2-nitrophenylacetic acid
4,5-Dihydroxy-2-nitrophenylacetic Acid
3-Hydroxy-4-methoxy-5-nitrobenzoic Acid
4-Hydroxy-3-methoxy-2-nitrobenzoic Acid
4-Hydroxy-3-methoxy-5-nitrobenzoic Acid
5-Hydroxy-4-methoxy-2-nitrobenzoic Acid

$C_8H_7NO_6S$
3-Sulphophthalic Acid, 3-*Amide*
4-Sulphophthalic Acid, 4-*Amide*

$C_8H_7NO_7S$
2-Nitro-4-sulphobenzoic Acid, 1-*Me ester*
2-Nitro-4-sulphobenzoic Acid, 4-*Me ester*
4-Nitro-2-sulphobenzoic Acid, 1-*Me ester*
4-Nitro-6-sulpho-*m*-toluic Acid
6-Nitro-4-sulpho-*m*-toluic Acid

C_8H_7NS
3-Aminobenzo[*b*]thiophene
5-Aminobenzo[*b*]thiophene
Benzyl isothiocyanate
Benzyl thiocyanate
6-Mercapto-*o*-toluic Acid, *Nitrile*
2-Methylbenzothiazole
4-Methylbenzothiazole
6-Methylbenzothiazole
o-Methylmercaptobenzoic Acid, *Nitrile*

$C_8H_7NS_2$
2-Mercaptobenzothiazole, *Me ether*

$C_8H_7N_3$
2-Aminocycloheptimidazole
2-Aminoquinoxaline
6-Aminoquinoxaline
1-Phenyl-1,2,3-triazole
2-Phenyl-1,2,3-triazole
4-Phenyl-1,2,3-triazole
1-Phenyl-1,2,4-triazole
3-Phenyl-1,2,4-triazole
4-Phenyl-1,2,4-triazole

$C_8H_7N_3O$
Indazole-2-carboxylic Acid, *Amide*

$C_8H_7N_3O_2$
Luminol†
1-Methyl-5-nitrobenzimidazole
2-Methyl-4-nitrobenzimidazole
2-Methyl-5-nitrobenzimidazole
5-Nitrobenzimidazole, N-*Me*
1-Methyl-4-nitroindazole
1-Methyl-5-nitroindazole
1-Methyl-6-nitroindazole
2-Methyl-4-nitroindazole
2-Methyl-5-nitroindazole
2-Methyl-6-nitroindazole
2-Methyl-7-nitroindazole
4-Methyl-5-nitroindazole
4-Methyl-6-nitroindazole
4-Methyl-7-nitroindazole
5-Methyl-4-nitroindazole
5-Methyl-6-nitroindazole
5-Methyl-7-nitroindazole
6-Methyl-4-nitroindazole
6-Methyl-5-nitroindazole
6-Methyl-7-nitroindazole
7-Methyl-4-nitroindazole
7-Methyl-6-nitroindazole
1-Phenylurazole
4-Phenylurazole

$C_8H_7N_3O_4$
3-Nitro-oxanilic Acid, *Amide*
4-Nitro-oxanilic Acid, *Amide*
3-Nitrophthalic Acid, *Diamide*
4-Nitrophthalic Acid, *Diamide*

$C_8H_7N_3O_5$
2,3-Dinitroacetanilide
2,4-Dinitroacetanilide
2,5-Dinitroacetanilide
2,6-Dinitroacetanilide
3,4-Dinitroacetanilide
3,5-Dinitroacetanilide
2,4-Dinitrophenylacetic Acid, *Amide*
4,5-Dinitro-*o*-toluic Acid, *Amide*
2,6-Dinitro-*p*-toluic Acid, *Amide*
Furazolidone

$C_8H_7N_3O_6$
2-Amino-3,5-dinitrobenzoic Acid, *Me ester*
2-Amino-3,5-dinitrobenzoic Acid, N-*Me*
4-Amino-3,5-dinitrobenzoic Acid, *Me ester*
4-Amino-3,5-dinitrobenzoic Acid, N-*Me*
3-Hydroxy-2,6-dinitro-*p*-toluic Acid, *Amide*
3,4,5-Trinitro-*o*-xylene
3,4,6-Trinitro-*o*-xylene
2,4,5-Trinitro-*m*-xylene
2,4,6-Trinitro-*m*-xylene
4,5,6-Trinitro-*m*-xylene
2,3,5-Trinitro-*p*-xylene

$C_8H_7N_3O_7$
3,5-Dimethyl-2,4,5-trinitrophenol
Picric Acid, *Et ether*
4,5,6-Trinitro-*o*-cresol, *Me ether*
2,4,6-Trinitro-*m*-cresol, *Me ether*
2,3,4-Trinitrophenetole
2,3,5-Trinitrophenetole

$C_8H_7N_3O_8$
1,2-Dimethoxy-3,4,5-trinitrobenzene
1,2-Dimethoxy-3,4,6-trinitrobenzene
Styphnic Acid, *Di-Me ether*

$C_8H_7N_4O_6$
3-Hydroxy-2,6-dinitro-*p*-toluic Acid, *Amide*

$C_8H_7N_5O_8$
N,2,4,6-Tetranitroaniline, N-*Et*

C_8H_8
Barrelene
Benzocyclobutene
Cubane†
Cyclo-octatetraene
Heptafulvene†
Styrene
Tricyclo[5,1,0,$0^{4,8}$]octa-2,5-diene†
6-Vinylfulvene

$(C_8H_8)_x$
Styrene, *Metastyrene*

$C_8H_8AsNNa_2O_4$
Arsacetin

$C_8H_8BrHgNO_2$
Mercuri-2-amino-5-bromophenylacetate
Mercuri-4-amino-2-bromophenylacetate
Mercuri-4-amino-3-bromophenylacetate

C_8H_8BrNO
o-Bromoaniline, N-*Ac*
m-Bromoaniline, N-*Ac*
p-Bromoaniline, N-*Ac*
α-Bromophenylacetic Acid, *Amide*
o-Bromophenylacetic Acid, *Amide*
p-Bromophenylacetic Acid, *Amide*
4-Bromo-*o*-toluic Acid, *Amide*
2-Bromo-*p*-toluic Acid, *Amide*

$C_8H_8BrNO_2$
2-Amino-4-bromobenzoic Acid, *Me ester*
2-Amino-4-bromobenzoic Acid, N-*Me*
2-Amino-5-bromobenzoic Acid *Me ester*
2-Amino-4-bromophenylacetic Acid
3-Amino-4-bromophenylacetic Acid
4-Amino-3-bromophenylacetic Acid
5-Amino-2-bromo-*p*-toluic Acid
3-Bromo-4-hydroxyacetanilide
4-Bromo-3-hydroxyacetanilide
5-Bromo-2-hydroxyacetanilide
5-Bromo-2-hydroxy-*m*-toluic Acid, *Amide*
3-Bromo-4-methoxybenzoic Acid, *Amide*
3-Bromo-5-nitro-*o*-xylene
2-Bromo-4-nitro-*m*-xylene
4-Bromo-2-nitro-*m*-xylene
4-Bromo-5-nitro-*m*-xylene
4-Bromo-6-nitro-*m*-xylene
5-Bromo-4-nitro-*m*-xylene
2-Bromo-5-nitro-*p*-xylene
2-Bromo-6-nitro-*p*-xylene
o-Bromophenoxyacetic Acid, *Amide*
N-*o*-Bromophenylglycine
N-*p*-Bromophenylglycine
o-Bromophenylglycine
p-Bromophenylglycine

$C_8H_8BrNO_3$
α-Bromo-2-nitro-*p*-cresol, *Me ether*
6-Bromo-4-nitro-*m*-cresol, *Me ether*
2-Bromo-3-nitro-*p*-cresol, *Me ether*
2-Bromo-5-nitro-*p*-cresol, *Me ether*
2-Bromo-3-nitrophenol, *Et ether*
2-Bromo-4-nitrophenol, *Et ether*
4-Bromo-2-nitrophenol, *Et ether*
5-Bromo-2-nitrophenol, *Et ether*

$C_8H_8BrN_3O_4$
4-Bromo-2,6-dinitroaniline, N-*Di-Me*

$C_8H_8Br_2$
α,β-Dibromoethylbenzene
α,α′-Dibromo-*o*-xylene
α,α′-Dibromo-*m*-xylene
α,α′-Dibromo-*p*-xylene
3,4-Dibromo-*o*-xylene
4,5-Dibromo-*o*-xylene
2,4-Dibromo-*m*-xylene
2,5-Dibromo-*m*-xylene
4,5-Dibromo-*m*-xylene
4,6-Dibromo-*m*-xylene
2,5-Dibromo-*p*-xylene
2,6-Dibromo-*p*-xylene

$C_8H_8Br_2O$
2,4-Dibromophenol, *Et ether*
2,6-Dibromophenol, *Et ether*
3,5-Dibromophenol, *Et ether*
5,6-Dibromo-2,3-xylenol
3,6-Dibromo-2,5-xylenol
4,6-Dibromo-2,5-xylenol
3,5-Dibromo-2,6-xylenol
4,5-Dibromo-2,6-xylenol
2,5-Dibromo-3,4-xylenol
2,6-Dibromo-3,4-xylenol
2,4-Dibromo-3,5-xylenol

$C_8H_8Br_2O_2$
2,4-Dibromo-5-hydroxybenzyl Alcohol, α-*Me ether*
3,5-Dibromo-4-hydroxybenzyl Alcohol, α-*Me ether*
2,5-Dibromoquinol, *Di-Me ether*
2,6-Dibromoquinol, *Di-Me ether*
4,6-Dibromoresorcinol, *Di-Me ether*

$C_8H_8Br_2O_3$
2,3-Dibromo-4,5-dihydroxybenzyl methyl ether†

$C_8H_8Br_2O_3S$
2,5-Dibromobenzenesulphonic Acid, *Et ester*

$C_8H_8Br_6$
2,3,4,5,7,8-Hexabromobicyclo[4,2,0]octane

$C_8H_8ClHgNO_2$
Mercuri-2-amino-5-chlorophenyl acetate
Mercuri-4-amino-3-chlorophenyl acetate

C_8H_8ClN
o-Amino-β-chlorostyrene

C_8H_8ClNO
N-Chloroacetanilide
o-Chloroaniline, N-*Ac*
m-Chloroaniline, N-*Ac*
p-Chloroaniline, N-*Ac*
α-Chlorophenylacetic Acid, *Amide*
o-Chlorophenylacetic Acid, *Amide*
4-Chloro-*o*-toluic Acid, *Amide*
6-Chloro-*o*-toluic Acid, *Amide*
α-Chloro-*m*-toluic Acid, *Amide*
α-Chloro-*p*-toluic Acid, *Amide*
2-Chloro-*p*-toluic Acid, *Amide*
N-Methylphenylcarbamic Acid, *Chloride*

$C_8H_8ClNO_2$
2-Amino-4-chlorobenzoic Acid, *Me ester*
2-Amino-5-chlorobenzoic Acid, *Me ester*
2-Amino-5-chlorobenzoic Acid, N-*Me*
2-Amino-6-chlorobenzoic Acid, *Me ester*

$C_8H_8ClNO_2$ (*continued*)

2-Amino-6-chlorobenzoic Acid, N-*Me*
3-Amino-4-chlorobenzoic Acid, *Me ester*
4-Amino-2-chlorobenzoic Acid, *Me ester*
2-Amino-4-chlorophenol, *N-Ac*
4-Amino-2-chlorophenol, *N-Ac*
4-Amino-3-chlorophenol, *N-Ac*
2-Amino-5-chloro-*p*-toluic Acid
5-Amino-2-chloro-*p*-toluic Acid
3-Chloro-6-hydroxy-*p*-toluic Acid, *Amide*
p-Chloromandelic Acid, *Amide*
3-Chloro-4-methoxybenzoic Acid, *Amide*
α-Chloro-2-methyl-5-nitrotoluene
α-Chloro-4-methyl-3-nitrotoluene
3-Chloro-4-nitro-*o*-xylene
3-Chloro-5-nitro-*o*-xylene
3-Chloro-6-nitro-*o*-xylene
4-Chloro-3-nitro-*o*-xylene
5-Chloro-4-nitro-*o*-xylene
2-Chloro-4-nitro-*m*-xylene
4-Chloro-5-nitro-*m*-xylene
4-Chloro-6-nitro-*m*-xylene
5-Chloro-(4 or 2)-nitro-*m*-xylene
2-Chloro-3-nitro-*p*-xylene
2-Chloro-5-nitro-*p*-xylene
2-Chloro-6-nitro-*p*-xylene
o-Chlorophenoxyacetic Acid, *Amide*
p-Chlorophenoxyacetic Acid, *Amide*
o-Chlorophenylglycine
m-Chlorophenylglycine
p-Chlorophenylglycine

$C_8H_8ClNO_3$

α-Chloro-3-nitro-*o*-cresol, *Me ether*
α-Chloro-2-nitro-*p*-cresol, *Me ether*
6-Chloro-2-nitro-*p*-cresol, *Me ether*
2-Chloro-3-nitrophenetole
2-Chloro-4-nitrophenetole
2-Chloro-5-nitrophenetole
3-Chloro-2-nitrophenetole
3-Chloro-4-nitrophenetole
3-Chloro-5-nitrophenetole
4-Chloro-2-nitrophenetole
4-Chloro-3-nitrophenetole
5-Chloro-2-nitrophenetole

$C_8H_8ClNO_4$

4-Chloro-2-nitroresorcinol, *Di-Me ether*
6-Chloro-4-nitroresorcinol, *Di-Me ether*

$C_8H_8ClNO_4S$

2,4-Dimethyl-3-nitrobenzenesulphonic Acid, *Chloride*
2,4-Dimethyl-5-nitrobenzenesulphonic Acid, *Chloride*
2,4-Dimethyl-6-nitrobenzenesulphonic Acid, *Chloride*
2,5-Dimethyl-3-nitrobenzenesulphonic Acid, *Chloride*
2,5-Dimethyl-4-nitrobenzenesulphonic Acid, *Chloride*
2,5-Dimethyl-6-nitrobenzenesulphonic Acid, *Chloride*
3,4-Dimethyl-5-nitrobenzenesulphonic Acid, *Chloride*

$C_8H_8ClN_3O_4$

2-Chloro-*N*-dimethyl-4,6-dinitroaniline
4-Chloro-*N*-dimethyl-2,3-dinitroaniline
4-Chloro-*N*-dimethyl-2,6-dinitroaniline
5-Chloro-*N*-dimethyl-2,4-dinitroaniline

$C_8H_8Cl_2$

α,β-Dichloroethylbenzene
α,α-Dichloro-*o*-xylene
α,α-Dichloro-*m*-xylene
α,α-Dichloro-*p*-xylene
α,α′-Dichloro-*o*-xylene
α,α′-Dichloro-*m*-xylene
α,α′-Dichloro-*p*-xylene
3,4-Dichloro-*o*-xylene
3,5-Dichloro-*o*-xylene
3,6-Dichloro-*o*-xylene
4,5-Dichloro-*o*-xylene
2,4-Dichloro-*m*-xylene
4,6-Dichloro-*m*-xylene
2,3-Dichloro-*p*-xylene
2,5-Dichloro-*p*-xylene
2,6-Dichloro-*p*-xylene

$C_8H_8Cl_2O$

4,6-Dichloro-*o*-cresol, *Me ether*
2,6-Dichloro-*p*-cresol, *Me ether*
α,α-Dichloro-4-methoxytoluene
2,4-Dichlorophenol, *Et ether*
5,6-Dichloro-2,3-xylenol
2,5-Dichloro-3,4-xylenol
2,6-Dichloro-3,4-xylenol
5,6-Dichloro-3,4-xylenol
2,4-Dichloro-3,5-xylenol
Filicinic Acid, *Dichloride*

$C_8H_8Cl_2O_2$

4,5-Dichlorocatechol, *Di-Me ether*
2-(2,4-Dichlorophenoxy)ethanol
2,5-Dichloroquinol, *Di-Me ether*
2,6-Dichloroquinol, *Di-Me ether*
4,6-Dichlororesorcinol, *Di-Me ether*

$C_8H_8Cl_2O_4S_2$

o-Xylene-3,5-disulphonic Acid, *Dichloride*
m-Xylene-2,4-disulphonic Acid, *Dichloride*
m-Xylene-4,6-disulphonic Acid, *Dichloride*
p-Xylene-2,5-disulphonic Acid, *Dichloride*
p-Xylene-2,6-disulphonic Acid, *Dichloride*

$C_8H_8Cl_2O_5S_2$

2-Hydroxy-4,6-dimethylbenzene-1,3-disulphonic Acid, *Dichloride*
4-Hydroxy-2,6-dimethylbenzene-1,3-disulphonic Acid, *Dichloride*
Phenetole-2,5-disulphonic Acid, *Dichloride*
Phenol-2,5-disulphonic Acid, *Et ether*, *Dichloride*

$C_8H_8Cl_2O_6S_2$

4,6-Dihydroxybenzene-1,3-disulphonic Acid, *Di-Me ether*

$C_8H_8Cl_3N$

2,4,6-Trichloroaniline, N-*Di-Me*
2,4,6-Trichloroaniline, N-*Et*

$C_8H_8Cl_6O_3$

4,4,4-Trichlorobutyric Acid, *Anhydride*

$C_8H_8FNO_3$

2-Fluoro-4-nitrophenol, *Et ether*
4-Fluoro-2-nitrophenol, *Et ether*

$C_8H_8FNO_4S$
- 2,4-Dimethyl-5-nitrobenzenesulphonic Acid, *Fluoride*
- 2,5-Dimethyl-3-nitrobenzenesulphonic Acid, *Fluoride*

$C_8H_8F_2N_3O_4S_2$
- Hydroflumethiazide

$C_8H_8F_2O_4S_2$
- *m*-Xylene-4,6-disulphonic Acid, *Difluoride*

$C_8H_8F_3N_3O_4S_2$
- 7-Sulphamoyl-6-trifluoromethyl-4*H*-1,4,2-benzothiadiazine-1,1-dioxide

$C_8H_8HgO_2$
- Mercuri-phenyl acetate

$C_8H_8HgO_3$
- Mercuri-*o*-hydroxyphenyl acetate
- Mercuri-*p*-hydroxyphenyl acetate

C_8H_8INO
- *o*-Iodoaniline, N-*Ac*
- *m*-Iodoaniline, N-*Ac*
- *p*-Iodoaniline, N-*Ac*
- 3-Iodo-4-methylbenzoic Acid, *Amide*

$C_8H_8INO_2$
- 2-Amino-4-iodobenzoic Acid, N-*Me*
- 4-Amino-2-iodobenzoic Acid, *Me ester*

$C_8H_8INO_3$
- 6-Iodo-4-nitro-*o*-cresol, *Me ether*
- 6-Iodol-3-nitro-*p*-cresol, *Me ether*
- 2-Iodo-4-nitrophenol, *Et ether*
- 4-Iodo-2-nitrophenol, *Et ether*
- 4-Iodo-3-nitrophenol, *Et ether*
- 5-Iodo-2-nitrophenol, *Et ether*

$C_8H_8I_2$
- α,α′-Di-iodo-*o*-xylene
- α,α′-Di-iodo-*m*-xylene
- α,α′-Di-iodo-*p*-xylene
- 4,5-Di-iodo-*o*-xylene
- 4,6-Di-iodo-*m*-xylene
- 2,5-Di-iodo-*p*-xylene

$C_8H_8I_2O$
- 2,4-Di-iodophenol, *Et ether*
- 2,6-Di-iodophenol, *Et ether*
- 3,5-Di-iodophenol, *Et ether*

$C_8H_8I_2O_2$
- 2,6-Di-iodoquinol, *Di-Me ether*

$C_8H_8I_2O_3S$
- 2,3-Di-iodobenzenesulphonic Acid, *Et ester*
- 2,4-Di-iodobenzenesulphonic Acid, *Et ester*
- 2,5-Di-iodobenzenesulphonic Acid, *Et ester*
- 3,4-Di-iodobenzenesulphonic Acid, *Et ester*
- 3,5-Di-iodobenzenesulphonic Acid, *Et ester*

$C_8H_8I_2O_4S$
- Soziodolic Acid, *Et ether*

$C_8H_8N_2$
- 2-Aminoindole
- 3-Aminoindole
- 5-Aminoindole
- 6-Aminoindole
- α-Aminophenylacetonitrile
- *o*-Aminophenylacetonitrile
- *m*-Aminophenylacetonitrile
- *p*-Aminophenylacetonitrile
- 3-Amino-*o*-toluic Acid, *Nitrile*
- 4-Amino-*o*-toluic Acid, *Nitrile*
- 5-Amino-*o*-toluic Acid, *Nitrile*
- 6-Amino-*o*-toluic Acid, *Nitrile*
- 2-Amino-*m*-toluic Acid, *Nitrile*
- 4-Amino-*m*-toluic Acid, *Nitrile*
- 5-Amino-*m*-toluic Acid, *Nitrile*
- 6-Amino-*m*-toluic Acid, *Nitrile*
- 2-Amino-*p*-toluic Acid, *Nitrile*
- 3-Amino-*p*-toluic Acid, *Nitrile*
- Apoharmine
- Benzylamine-*m*-carboxylic Acid, *Nitrile*
- Benzylamine-*p*-carboxylic Acid, *Nitrile*
- 2,2′-Bipyrrole
- 2,4-Dimethylpyridine-3-carboxylic Acid, *Nitrile*
- 1-Methylbenzimidazole
- 2-Methylbenzimidazole
- 4-Methylbenzimidazole
- 5-Methylbenzimidazole
- 1-Methylindazole
- 2-Methylindazole
- 3-Methylindazole
- 5-Methylindazole
- 6-Methylindazole
- 7-Methylindazole
- *N*-Phenylglycine, *Nitrile*

$C_8H_8N_2O$
- 2-Amino-6-hydroxybenzoic Acid, *Me ether*, *Nitrile*
- 1-Amino-oxindole
- 3-Amino-oxindole
- 5-Amino-oxindole
- 6-Amino-oxindole
- 2-Hydroxyphenylglycine, *Nitrile*

$C_8H_8N_2OS$
- Benzoylthiourea

$C_8H_8N_2O_2$
- Benzoylurea
- 1,3-Diamino-4,6-diformylbenzene
- 4,5-Dimethylpyrrole-2,3-dicarboxylic Acid, 2-*Nitrile*
- 2,5-Dimethylpyrrole-3,4-dicarboxylic Acid, 4-*Nitrile*
- Isophthalic Acid, *Diamide*
- *N*-Nitrosoacetanilide
- Oxanilic Acid, *Amide*
- Phenylglyoxime
- Phthalamide
- Ricinine
- Terephthalaldehyde, *Dioxime*
- Terephthalic Acid, *Diamide*

$C_8H_8N_2O_2S$
- 4-Sulpho-*o*-toluic Acid, *Amide-nitrile*
- 2-Sulpho-*p*-toluic Acid, *Amide-nitrile*

$C_8H_8N_2O_2S_2$
- Thiolutin

$C_8H_8N_2O_3$
- Allophanic Acid, *Phenyl ester*
- α-Amino-3-nitroacetophenone
- α-Amino-4-nitroacetophenone
- 2-Amino-3-nitroacetophenone
- 2-Amino-4-nitroacetophenone
- 2-Amino-5-nitroacetophenone

$C_8H_8N_2O_3$ *(continued)*
3-Amino-2-nitroacetophenone
3-Amino-4-nitroacetophenone
4-Amino-3-nitroacetophenone
5-Amino-2-nitroacetophenone
5-Amino-3-nitroacetophenone
6-Amino-2-nitroacetophenone
2-Amino-5-nitrosobenzoic Acid, *Me ester*
m-Amino-oxanilic Acid
p-Amino-oxanilic Acid
4-Hydroxyisophthalic Acid, *Diamide*
2-*N*-Methylamino-5-nitrosobenzoic Acid
2-(*N*-Methyl-*N*-nitrosoamino)benzoic Acid
Nicotinuric Acid
o-Nitroacetanilide
m-Nitroacetanilide
p-Nitroacetanilide
ω-Nitroacetanilide
m-Nitrobenzamide, N-*Me*
p-Nitrobenzamide, N-*Me*
o-Nitrophenylacetic Acid, *Amide*
m-Nitrophenylacetic Acid, *Amide*
p-Nitrophenylacetic Acid, *Amide*
3-Nitro-*o*-toluic Acid, *Amide*
5-Nitro-*o*-toluic Acid, *Amide*
6-Nitro-*o*-toluic Acid, *Amide*
2-Nitro-*m*-toluic Acid, *Amide*
4-Nitro-*m*-toluic Acid, *Amide*
5-Nitro-*m*-toluic Acid, *Amide*
6-Nitro-*m*-toluic Acid, *Amide*
2-Nitro-*p*-toluic Acid, *Amide*
3-Nitro-*p*-toluic Acid, *Amide*

$C_8H_8N_2O_4$
2-Amino-3-nitrobenzoic Acid, N-*Me*
2-Amino-4-nitrobenzoic Acid, *Me ester*
2-Amino-5-nitrobenzoic Acid, *Me ester*
2-Amino-5-nitrobenzoic Acid, N-*Me*
2-Amino-6-nitrobenzoic Acid, *Me ester*
3-Amino-4-nitrobenzoic Acid, N-*Me*
4-Amino-2-nitrobenzoic Acid, *Me ester*
4-Amino-3-nitrobenzoic Acid, *Me ester*
5-Amino-3-nitrobenzoic Acid, *Me ester*
α-Amino-3-nitrophenylacetic Acid
4-Amino-2-nitrophenylacetic Acid
4-Amino-3-nitrophenylacetic Acid
4-Amino-5-nitro-*o*-toluic Acid
6-Amino-4-nitro-*m*-toluic Acid
4-Amino-6-nitro-*m*-toluic Acid
2-Amino-5-nitro-*p*-toluic Acid
5-Amino-2-nitro-*p*-toluic Acid
5-Amino-3-nitro-*p*-toluic Acid
o-Benzenedicarbamic Acid
m-Benzenedicarbamic Acid
p-Benzenedicarbamic Acid
NN'-Bisuccinimidyl†
Butane-1,2,3,4-tetracarboxylic Acid, *Di-Imide*
4,6-Diaminoisophthalic Acid
3,6-Diaminophthalic Acid
2,5-Diaminoterephthalic Acid
2,6-Diaminoterephthalic Acid
3,4-Dinitro-*o*-xylene
3,5-Dinitro-*o*-xylene
3,6-Dinitro-*o*-xylene
4,5-Dinitro-*o*-xylene
2,4-Dinitro-*m*-xylene
2,5-Dinitro-*m*-xylene
4,5-Dinitro-*m*-xylene
4,6-Dinitro-*m*-xylene
2,3-Dinitro-*p*-xylene
2,5-Dinitro-*p*-xylene
2,6-Dinitro-*p*-xylene
2-Hydroxy-6-nitrobenzoic Acid, *Me ether*, *Amide*
Hydroxy-*o*-nitrophenylacetic Acid, *Amide*
2-Hydroxy-5-nitro-*m*-toluic Acid, *Amide*
4-Methylamino-3-nitrobenzoic Acid
3,4-Methylenedioxy-6-nitroaniline, N-*Me*
o-Nitrophenoxyacetic Acid, *Amide*
m-Nitrophenoxyacetic Acid, *Amide*
p-Nitrophenoxyacetic Acid, *Amide*
o-Nitrophenylcarbamic Acid, *Me ester*
m-Nitrophenylcarbamic Acid, *Me ester*
p-Nitrophenylcarbamic Acid, *Me ester*
N-*o*-Nitrophenylglycine
N-*m*-Nitrophenylglycine
N-*p*-Nitrophenylglycine
Pyrazine-2,3-dicarboxylic Acid, *Di-Me ester*†

$C_8H_8N_2O_4S$
2,4-Dinitrobenzenethiol, *Et ether*

$C_8H_8N_2O_5$
2-Amino-5-hydroxy-4-nitrobenzoic Acid, *Me ether*
3-Amino-4-hydroxy-5-nitrobenzoic Acid, *Me ester*
4-Amino-3-hydroxy-6-nitrobenzoic Acid, *Me ether*
5-Amino-4-hydroxy-2-nitrobenzoic Acid, *Me ether*
4-Amino-2-nitrophenoxyacetic Acid
4-Amino-3-nitrophenoxyacetic Acid
4,5-Dinitro-*o*-cresol, *Me ether*
4,6-Dinitro-*o*-cresol, *Me ether*
2,4-Dinitro-*m*-cresol, *Me ether*
2,6-Dinitro-*m*-cresol, *Me ether*
4,6-Dinitro-*m*-cresol, *Me ether*
2,3-Dinitro-*p*-cresol, *Me ether*
2,5-Dinitro-*p*-cresol, *Me ether*
2,6-Dinitro-*p*-cresol, *Me ether*
3,5-Dinitro-*p*-cresol, *Me ether*
2,3-Dinitrophenol, *Et ether*
2,4-Dinitrophenol, *Et ether*
2,5-Dinitrophenol, *Et ether*
2,6-Dinitrophenol, *Et ether*
3,4-Dinitrophenol, *Et ether*
3,5-Dinitrophenol, *Et ether*
4,6-Dinitro-2,3-xylenol
3,5-Dinitro-2,4-xylenol
5,6-Dinitro-2,4-xylenol
4,6-Dinitro-2,5-xylenol
2,6-Dinitro-3,4-xylenol
2,4-Dinitro-3,5-xylenol
2,6-Dinitro-3,5-xylenol

$C_8H_8N_2O_5S$
2-Hydroxy-1-ethanethiol, S-2,4-*Dinitrophenyl*

$C_8H_8N_2O_6$
1,2-Dimethoxy-3,4-dinitrobenzene
1,2-Dimethoxy-3,5-dinitrobenzene
1,2-Dimethoxy-3,6-dinitrobenzene
1,2-Dimethoxy-4,5-dinitrobenzene
3,5-Dinitrocatechol, 1-*Et ether*

$C_8H_8N_2O_6$ *(continued)*
2,3-Dinitroquinol, *Di-Me ether*
2,5-Dinitroquinol, *Di-Me ether*
2,6-Dinitroquinol, *Di-Me ether*
2,4-Dinitroresorcinol, *Di-Me ether*
2,4-Dinotroresorcinol, 1-*Et ether*
4,5-Dinitroresorcinol, *Di-Me ether*
4,6-Dinitroresorcinol, *Di-Me ether*

$C_8H_8N_2O_6S$
p-Nitrobenzenesulphonylcarbamic Acid, *Me ester* †

$C_8H_8N_2O_6S_2$
Cystine-di-*N*-carboxyanhydride

$C_8H_8N_2O_7$
1,2,3-Trihydroxy-4,5-dinitrobenzene, 1,3-*Di-Me ether*
1,2,3-Trihydroxy-4,6-dinitrobenzene, 1,3-*Di-Me ether*
1,3,5-Trihydroxy-2,4-dinitrobenzene, 1,5-*Di-Me ether*

$C_8H_8N_2S$
2-Mercaptobenzimidazole, 1-N-*Me*
2-Methylaminobenzothiazole

$C_8H_8N_4$
1-*p*-Aminophenyl-1,2,3-triazole
4-Amino-1-phenyl-1,2,3-triazole
5-Amino-1-phenyl-1,2,3-triazole
4-Amino-2-phenyl-1,2,3-triazole
3-Amino-1-phenyl-1,2,4-triazole
5-Amino-1-phenyl-1,2,4-triazole
1-Hydrazinophthalazine

$C_8H_8N_4O_4$
ω-*o*-Nitrophenylbiuret
ω-*m*-Nitrophenylbiuret
ω-*p*-Nitrophenylbiuret

$C_8H_8N_4O_4S$
Thiophene-tetracarboxylic Acid, *Tetra-amide*

$C_8H_8N_4O_6$
3,5-Dimethyl-2,4,6-trinitroaniline

$C_8H_8N_6O_6$
Murexide

C_8H_8O
Acetophenone
Bicyclo[3,2,1]octa-3,6-dien-2-one †
Dihydrobenzofuran
Homotropone †
2-Hydroxystyrene
3-Hydroxystyrene
4-Hydroxystyrene
Phenylacetaldehyde
Phenyloxiran
1-Phenylvinyl Alcohol
2-Phenylvinyl Alcohol
Phenyl vinyl Ether
Phthalan
Tetracyclo[3,3,0,$0^{2,8}$,$0^{4,6}$]octan-3-one †
o-Toluic Aldehyde
m-Toluic Aldehyde
p-Toluic Aldehyde
Vinyl Alcohol, *Phenyl ether*

C_8H_8OS
o-Acetylbenzenethiol
α-Mercaptoacetophenone
p-Methylmercaptobenzaldehyde
Phenylthiolacetic Acid
Thioacetic Acid, S-*Phenyl ester*
Thioacetic Acid, O-*Phenyl ester*
Thiobenzoic Acid, O-*Me ester*
Thiobenzoic Acid, S-*Me ester*
Thio-*o*-toluic Acid
Thio-*p*-toluic Acid

$C_8H_8OS_2$
2-Hydroxybenzenethionothiolic Acid, *Me ester*

$C_8H_8O_2$
Anisaldehyde
1,3-Benzodioxan
1,4-Benzodioxan
3,5-Dimethyl-1,2-benzoquinone
4,5-Dimethyl-1,2-benzoquinone
2,3-Dimethyl-1,4-benzoquinone
2,5-Dimethyl-1,4-benzoquinone
2,6-Dimethyl-1,4-benzoquinone
Formic Acid, *Benzyl ester*
4-(2-Furyl)-3-buten-2-one
2-Hydroxyacetophenone
3-Hydroxyacetophenone
4-Hydroxyacetophenone
α-Hydroxyacetophenone
p-Hydroxymethyl-benzaldehyde
2-Hydroxyphenylacetaldehyde
3-Hydroxyphenylacetaldehyde
4-Hydroxyphenylacetaldehyde
2-Hydroxy-2-phenylacetaldehyde
4-Hydroxy-*o*-tolualdehyde
6-Hydroxy-*o*-tolualdehyde
2-Hydroxy-*m*-tolualdehyde
4-Hydroxy-*m*-tolualdehyde
6-Hydroxy-*m*-tolualdehyde
2-Hydroxy-*p*-tolualdehyde
3-Hydroxy-*p*-tolualdehyde
o-Methoxybenzaldehyde
m-Methoxybenzaldehyde
Methyl benzoate
Phenoxyacetaldehyde
Phenyl acetate
Phenylacetic Acid
o-Toluic Acid
m-Toluic Acid
p-Toluic Acid
Tropolone, *Me ether*

$C_8H_8O_2S$
o-Mercaptobenzoic Acid, *Me ester*
m-Mercaptobenzoic Acid, *Me ester*
m-Mercaptobenzoic Acid, S-*Me*
p-Mercaptobenzoic Acid, S-*Me*
p-Mercaptobenzoic Acid, *Me ester*
o-Mercaptomethylbenzoic Acid
o-Mercaptophenylacetic Acid
6-Mercapto-*o*-toluic Acid
6-Mercapto-*m*-toluic Acid
α-Mercaptophenylacetic Acid †
o-Methylmercaptobenzoic Acid
Phenylmercaptoacetic Acid
Thioglycollic Acid, S-*Phenyl*

$C_8H_8O_3$

Acetylquinol
Anisic Acid
Bicyclo[2,2,2]octane-2,6,7-trione †
Coprinin
1-Cyclohexene-1,2-dicarboxylic Acid, *Anhydride*
1-Cyclohexene-1,3-dicarboxylic Acid, *Anhydride*
2-Cyclohexene-1,2-dicarboxylic Acid, *Anhydride*
3-Cyclohexene-1,2-dicarboxylic Acid, *Anhydride*
4-Cyclohexene-1,2-dicarboxylic Acid, *Anhydride*
α,4-Dihydroxyacetophenone
2,3-Dihydroxyacetophenone
2,4-Dihydroxyacetophenone
2,6-Dihydroxyacetophenone
3,4-Dihydroxyacetophenone
3,5-Dihydroxyacetophenone
2,3-Dihydroxybenzaldehyde, 3-*Me ether*
3,4-Dihydroxy-*o*-tolualdehyde
4,5-Dihydroxy-*o*-tolualdehyde
4,6-Dihydroxy-*o*-tolualdehyde
4,6-Dihydroxy-*m*-tolualdehyde
2,6-Dihydroxy-*p*-tolualdehyde
3-α-Furylacrylic Acid, *Me ester*
1-(2′-Furyl)butane-1,2-dione †
3-Hydroxybenzoic Acid, *Me ester*
4-Hydroxybenzoic Acid, *Me ester*
Hydroxy-1,4-benzoquinone, *Et ether*
3-Hydroxy-4-methoxybenzaldehyde
o-Hydroxymethyl-benzoic Acid
m-Hydroxymethyl-benzoic Acid
p-Hydroxymethyl-benzoic Acid
2-Hydroxyphenylacetic Acid
3-Hydroxyphenylacetic Acid
4-Hydroxyphenylacetic Acid
3-Hydroxy-*o*-toluic Acid
4-Hydroxy-*o*-toluic Acid
5-Hydroxy-*o*-toluic Acid
6-Hydroxy-*o*-toluic Acid
2-Hydroxy-*m*-toluic Acid
4-Hydroxy-*m*-toluic Acid
5-Hydroxy-*m*-toluic Acid
6-Hydroxy-*m*-toluic Acid
2-Hydroxy-*p*-toluic Acid
3-Hydroxy-*p*-toluic Acid
3-Hydroxytoluquinone, *Me ether*
5-Hydroxytoluquinone, *Me ether*
6-Hydroxytoluquinone, *Me ether*
Mandelic Acid
o-Methoxybenzoic Acid
m-Methoxybenzoic Acid
1-Methyl-2-cyclopentene-1,2-dicarboxylic Acid, *Anhydride*
3,4-Methylenedioxybenzyl Alcohol
1-[(5′-Methyl)-2′-furyl]-propane-1,2-dione †
Methyl salicylate
Phenoxyacetic Acid
Vanillin

$C_8H_8O_3S$

2-Hydroxy-5-mercaptobenzoic Acid, S-*Me*
p-Methylsulphonylbenzaldehyde
2-Thienylglyoxylic Acid, *Et ester*
p-Vinylbenzenesulphonic Acid

$C_8H_8O_4$

3-Acetyl-2-hydroxy-6-methylpyran-4-one
Barbatol
1,3-Cyclohexadiene-1,4-dicarboxylic Acid
1,4-Cyclohexadiene-1,2-dicarboxylic Acid
1,4-Cyclohexadiene-1,4-dicarboxylic Acid
1,5-Cyclohexadiene-1,3-dicarboxylic Acid
1,5-Cyclohexadiene-1,4-dicarboxylic Acid
2,4-Cyclohexadiene-1,2-dicarboxylic Acid
2,5-Cyclohexadiene-1,2-dicarboxylic Acid
2,5-Cyclohexadiene-1,4-dicarboxylic Acid
2,6-Cyclohexadiene-1,2-dicarboxylic Acid
3,5-Cyclohexadiene-1,2-dicarboxylic Acid
2,3-Dihydroxybenzoic Acid, 2-*Me ether*
2,3-Dihydroxybenzoic Acid, 3-*Me ether*
2,4-Dihydroxybenzoic Acid, 2-*Me ether*
2,4-Dihydroxybenzoic Acid, 4-*Me ether*
2,4-Dihydroxybenzoic Acid, *Me ester*
2,5-Dihydroxybenzoic Acid, *Me ester*
2,5-Dihydroxybenzoic Acid, 2-*Me ether*
2,5-Dihydroxybenzoic Acid, 5-*Me ether*
2,6-Dihydroxybenzoic Acid, *Me ester*
2,6-Dihydroxybenzoic Acid, *Me ether*
3,4-Dihydroxybenzoic Acid, *Me ester*
3,5-Dihydroxybenzoic Acid, *Me ester*
2,5-Dihydroxy-1,4-benzoquinone, *Di-Me ether*
α,2-Dihydroxyphenylacetic Acid
α,3-Dihydroxyphenylacetic Acid
α,4-Dihydroxyphenylacetic Acid
2,3-Dihydroxyphenylacetic Acid
2,4-Dihydroxyphenylacetic Acid
2,5-Dihydroxyphenylacetic Acid
3,4-Dihydroxyphenylacetic Acid
3,4-Dihydroxy-*o*-toluic Acid
3,5-Dihydroxy-*o*-toluic Acid
4,6-Dihydroxy-*o*-toluic Acid
5,6-Dihydroxy-*o*-toluic Acid
2,4-Dihydroxy-*m*-toluic Acid
2,5-Dihydroxy-*m*-toluic Acid
5,6-Dihydroxy-*m*-toluic Acid
2,5-Dihydroxy-*p*-toluic Acid
2,6-Dihydroxy-*p*-toluic Acid
3,5-Dihydroxy-*p*-toluic Acid
4,6-Dimethyl-2-oxo-2*H*-pyran-5-carboxylic Acid
Fumigatin
α-Furoylacetic Acid, *Me ester*
3-Hydroxy-4-methoxybenzoic Acid
3-Hydroxy-5-methoxy-*p*-toluquinone †
4-Hydroxy-6-(2-oxopropyl)-2*H*-pyran-2-one †
2,4,6-Octatrienedioic Acid
2-Oxo-2*H*-pyran-5-carboxylic Acid, *Et ester*
2-Oxo-2*H*-pyran-6-carboxylic Acid, *Et ester*
4-Oxo-4*H*-pyran-2-carboxylic Acid, *Et ester*
Terreic Acid★, *Me ether* †
α,2,4-Trihydroxyacetophenone
α,3,4-Trihydroxyacetophenone
2,3,4-Trihydroxyacetophenone
2,3,5-Trihydroxyacetophenone
2,3,6-Trihydroxyacetophenone
2,4,5-Trihydroxyacetophenone
2,4,6-Trihydroxyacetophenone
3,4,5-Trihydroxyacetophenone
2,4,5-Trihydroxybenzaldehyde, 4-*Me ether*
2,4,6-Trihydroxybenzaldehyde, 2-*Me ether*
2,4,6-Trihydroxybenzaldehyde, 4-*Me ether*
3,4,5-Trihydroxybenzaldehyde, 3-*Me ether*
3,4,5-Trihydrozybenzaldehyde, 4-*Me ether*
2,4,6-Trihydroxy-*m*-toluic Aldehyde

$C_8H_8O_4$ (*continued*)
2,3,6-Trihydroxy-*p*-toluic Aldehyde
Vanillic Acid

$C_8H_8O_4S$
o-Formylbenzenesulphonic Acid, *Me ester*
Phenylsulphonylacetic Acid
Thiophene-2,3-dicarboxylic Acid, *Di-Me ester*
Thiophene-2,4-dicarboxylic Acid, *Di-Me ester*
Thiophene-2,5-dicarboxylic Acid, *Di-Me ester*
Thiophene-3,4-dicarboxylic Acid, *Di-Me ester*

$C_8H_8O_5$
2,3-Dihydroxymandelic Acid
2,4-Dihydroxymandelic Acid
2,5-Dihydroxymandelic Acid
3,4-Dihydroxymandelic Acid
2,5-Dimethylfuran-3,4-dicarboxylic Acid
Fomecin A★†
Furan-2,3-dicarboxylic Acid, *Di-Me ester*
Furan-2,4-dicarboxylic Acid, *Di-Me ester*
Furan-2,5-dicarboxylic Acid, *Di-Me ester*
Furan-2,5-dicarboxylic Acid, *Mono-Et ester*
Furan-3,4-dicarboxylic Acid, *Di-Me ester*
Gallic Acid, *Me ester*
Gallic Acid, 3-*Me ether*
Gallic Acid, 4-*Me ether*
5-Hydroxy-4-oxopyran-2-carboxylic Acid, *Et ester*
5-Hydroxy-4-oxopyran-2-carboxylic Acid, *Et ether*
Methronic Acid
Oxalacetoacetic Acid Ethyl Ester
cis-2-Pentene-2,3,5-tricarboxylic Acid, *Anhydride*
Spinulosin
α,2,3,5-Tetrahydroxyacetophenone
α,2,4,6-Tetrahydroxyacetophenone
α,3,4,5-Tetrahydroxyacetophenone
2,3,4,5-Tetrahydroxyacetophenone
2,3,4,6-Tetrahydroxyacetophenone
2,3,4-Trihydroxybenzoic Acid, *Me ester*
2,3,4-Trihydroxybenzoic Acid, 4-*Me ether*
2,3,6-Trihydroxybenzoic Acid, *Me ester*
2,4,6-Trihydroxybenzoic Acid, *Me ester*
2,4,6-Trihydroxybenzoic Acid, 4-*Me ether*
4,5,6-Trihydroxy-*o*-toluic Acid
2,4,6-Trihydroxy-*m*-toluic Acid

$C_8H_8O_5S$
m-Sulphobenzoic Acid, 1-*Me ester*
m-Sulphobenzoic Acid, 3-*Me ester*
p-Sulphobenzoic Acid, 1-*Me ester*
p-Sulphobenzoic Acid, 4-*Me ester*
4-Sulpho-*o*-toluic Acid
5-Sulpho-*o*-toluic Acid
4-Sulpho-*m*-toluic Acid
5-Sulpho-*m*-toluic Acid
6-Sulpho-*m*-toluic Acid
2-Sulpho-*p*-toluic Acid
3-Sulpho-*p*-toluic Acid

$C_8H_8O_5S_2$
p-Xylene-2,3-disulphonic Acid, *Anhydride*

$C_8H_8O_6$
5,6-Dihydroxy-4-oxopyran-2-carboxylic Acid, *Di-Me ether*
5,6-Dihydroxy-4-oxopyran-2-carboxylic Acid, *Et ester*
2,5-Dioxocyclohexane-1,4-dicarboxylic Acid
α,2,3,4,6-Pentahydroxyacetophenone
Phorbic Acid, *Dilactone*†
Sarsapic Acid, *Di-Me ester*

$C_8H_8O_6S$
3-Hydroxy-4-sulphobenzoic Acid, *Me ether*
4-Hydroxy-2-sulphobenzoic Acid, *Me ether*
4-Hydroxy-3-sulphobenzoic Acid, *Me ether*
2-Hydroxy-5-sulpho-*m*-toluic Acid
2-Hydroxy-6-sulpho-*p*-toluic Acid
6-Hydroxy-5-sulpho-*m*-toluic Acid

$C_8H_8O_7$
Butane-1,2,3,4-tetracarboxylic Acid, *Mono-anhydride*

$C_8H_8O_8$
1-Butene-1,2,3,4-tetracarboxylic Acid

$C_8H_8S_2$
1,6-Dithiacyclodeca-3,8-diyne†
Dithiobenzoic Acid, *Me ester*

$C_8H_9AsN_2Na_2O_4$
Tryparsamide

$C_8H_9AsO_2$
Benzenearsonous Acid, *Di-Me ester*

$C_8H_9AsO_3$
2-Phenoxy-1,3,2-dioxa-arsolan

$C_8H_9AsO_5$
4-Acetyl-3-hydroxybenzenearsonic Acid
5-Acetyl-2-hydroxybenzenearsonic Acid

C_8H_9Br
1-Bromoethylbenzene
β-Bromoethylbenzene
m-Bromoethylbenzene
p-Bromoethylbenzene
α-Bromo-*o*-xylene
α-Bromo-*m*-xylene
α-Bromo-*p*-xylene
3-Bromo-*o*-xylene
4-Bromo-*o*-xylene
2-Bromo-*m*-xylene
4-Bromo-*m*-xylene
5-Bromo-*m*-xylene
2-Bromo-*p*-xylene

C_8H_9BrHg
Mercuri-*p*-ethylphenyl bromide

C_8H_9BrHgO
Mercuri-*o*-hydroxyphenyl bromide, *Et ether*
Mercuri-*p*-hydroxyphenyl bromide, *Et ether*

$C_8H_9BrN_2O_2$
2-Bromo-4-nitroaniline, N-*Et*
4-Bromo-2-nitroaniline, N-*Et*
5-Bromo-2-nitroaniline, N-*Et*

C_8H_9BrO
4-Bromo-*m*-cresol, *Me ether*
2-Bromo-*p*-cresol, *Me ether*
3-Bromo-*p*-cresol, *Me ether*
2-Bromoethyl phenyl Ether
o-Bromophenetole
p-Bromophenetole
2-Bromo-1-phenylethanol
2-Bromo-2-phenylethanol
3-Bromo-2,4-xylenol

C_8H_9BrO (*continued*)
4-Bromo-2,5-xylenol
4-Bromo-2,6-xylenol
4-Bromo-3,5-xylenol
5-Bromo-2,3-xylenol
5-Bromo-2,4-xylenol
5-Bromo-2,6-xylenol
5-Bromo-3,4-xylenol
6-Bromo-2,4-xylenol
6-Bromo-3,4-xylenol
6-Bromo-3,5-xylenol

$C_8H_9BrO_2$
3-Bromocatechol, *Di-Me ether*
4-Bromocatechol, *Di-Me ether*
5-Bromo-2-hydroxybenzyl Alcohol, *Me ether*
Bromoquinol, *Di-Me ether*
4-Bromoresorcinol, *Di-Me ether*
5-Bromoresorcinol, *Di-Me ether*

$C_8H_9BrO_2S$
m-Xylene-5-sulphonic Acid, *Bromide*

$C_8H_9BrO_3S$
Bromobenzene-*p*-sulphonic Acid, *Et ester*

$C_8H_9Br_2N$
2,4-Dibromoaniline, N-*Di-Me*
2,5-Dibromoaniline, N-*Di-Me*
3,5-Dibromoaniline, N-*Di-Me*
4,6-Dibromo-2,3-xylidine
5,6-Dibromo-2,3-xylidine
3,5-Dibromo-2,4-xylidine
5,6-Dibromo-2,4-xylidine
3,6-Dibromo-2,5-xylidine
4,6-Dibromo-2,5-xylidine
3,5-Dibromo-2,6-xylidine
4,5-Dibromo-2,6-xylidine
2,5-Dibromo-3,4-xylidine
2,6-Dibromo-3,4-xylidine
2,4-Dibromo-3,5-xylidine

$C_8H_9Br_2NO$
2-Amino-3,5-dibromophenol, *Et ether*
2-Amino-4,6-dibromophenol, *Et ether*
4-Amino-2,6-dibromophenol, *Et ether*

C_8H_9Cl
1-Chloroethylbenzene
2-Chloroethylbenzene
o-Chloroethylbenzene
p-Chloroethylbenzene
α-Chloro-*o*-xylene
3-Chloro-*o*-xylene
4-Chloro-*o*-xylene
α-Chloro-*m*-xylene
2-Chloro-*m*-xylene
4-Chloro-*m*-xylene
5-Chloro-*m*-xylene
α-Chloro-*p*-xylene
2-Chloro-*p*-xylene

C_8H_9ClHg
Mercuri-*p*-ethylphenyl chloride

C_8H_9ClHgO
Mercuri-*p*-hydroxyphenyl chloride, *Et ether*

$C_8H_9ClNO_5PS$
Chlorthion

$C_8H_9ClN_2O$
4,5,6,7-Tetrahydroindazole-1-carboxylic Acid, *Chloride*

$C_8H_9ClN_2O_2$
2-Chloro-*N*-dimethyl-3-nitroaniline
2-Chloro-*N*-dimethyl-4-nitroaniline
2-Chloro-*N*-dimethyl-5-nitroaniline
3-Chloro-*N*-dimethyl-2-nitroaniline
3-Chloro-*N*-dimethyl-4-nitroaniline
4-Chloro-*N*-dimethyl-2-nitroaniline
4-Chloro-*N*-dimethyl-3-nitroaniline
5-Chloro-*N*-dimethyl-2-nitroaniline
4-Chloro-*N*-ethyl-2-nitroaniline
5-Chloro-*N*-ethyl-2-nitroaniline

$C_8H_9ClN_2O_3$
4-Amino-2-chloro-3-nitrophenol, *Et ether*
4-Amino-2-chloro-5-nitrophenol, *Et ether*

C_8H_9ClO
Bicyclo[2,2,1]hept-5-ene-2-carboxylic Acid, *Chloride*
4-Chloro-*o*-cresol, *Me ether*
2-Chloro-*m*-cresol, *Me ether*
4-Chloro-*m*-cresol, *Me ether*
6-Chloro-*m*-cresol, *Me ether*
2-Chloro-*p*-cresol, *Me ether*
3-Chloro-*p*-cresol, *Me ether*
2-Chloroethyl phenyl Ether
o-Chlorophenetole
m-Chlorophenetole
p-Chlorophenetole
2-Chloro-1-phenylethanol
4-Chloro-2,3-xylenol
5-Chloro-2,3-xylenol
5-Chloro-2,4-xylenol
6-Chloro-2,4-xylenol
4-Chloro-2,5-xylenol
2-Chloro-3,4-xylenol
5-Chloro-3,4-xylenol
6-Chloro-3,4-xylenol
2-Chloro-3,5-xylenol
4-Chloro-3,5-xylenol

$C_8H_9ClO_2$
o-Chlorophenyl 2-hydroxyethyl Ether
m-Chlorophenyl 2-hydroxyethyl Ether
p-Chlorophenyl 2-hydroxyethyl Ether
4-Chlororesorcinol, *Di-Me ether*
3-Chloroveratrol
4-Chloroveratrol

$C_8H_9ClO_2S$
o-Xylene-3-sulphonic Acid, *Chloride*
o-Xylene-4-sulphonic Acid, *Chloride*
m-Xylene-2-sulphonic Acid, *Chloride*
m-Xylene-4-sulphonic Acid, *Chloride*
m-Xylene-5-sulphonic Acid, *Chloride*
p-Xylenesulphonic Acid, *Chloride*

$C_8H_9ClO_3S$
p-Chlorobenzenesulphonic Acid, *Et ester*
2-Hydroxy-3,5-dimethylbenzenesulphonic Acid, *Chloride*
4-Hydroxytoluene-2-sulphonic Acid, *Me ether, Chloride*
Phenetole-*o*-sulphonic Acid, *Chloride*
Phenetole-*m*-sulphonic Acid, *Chloride*
Phenetole-*p*-sulphonic Acid, *Chloride*

$C_8H_9Cl_2N$
2,4-Dichloro-*N*-dimethylaniline
3,4-Dichloro-*N*-dimethylaniline
4,6-Dichloro-2,3-xylidine
5,6-Dichloro-2,3-xylidine
3,5-Dichloro-2,4-xylidine
3,4-Dichloro-2,5-xylidine
3,6-Dichloro-2,5-xylidine
4,6-Dichloro-2,5-xylidine
2,5-Dichloro-3,4-xylidine
2,6-Dichloro-3,4-xylidine
5,6-Dichloro-3,4-xylidine
2,4-Dichloro-3,5-xylidine
2,6-Dichloro-3,5-xylidine

$C_8H_9Cl_2NO$
4-Amino-2,5-dichlorophenol, *Et ether*

C_8H_9F
o-Ethylfluorobenzene
p-Ethylfluorobenzene
4-Fluoro-*m*-xylene
5-Fluoro-*m*-xylene

C_8H_9FO
o-Fluorophenol, *Et ether*
m-Fluorophenol, *Et ether*
p-Fluorophenol, *Et ether*

$C_8H_9FO_3S$
2-Hydroxy-3,5-dimethylbenzenesulphonic Acid, *Fluoride*

$C_8H_9F_7O_2$
Heptafluorobutyric Acid, tert-*Butyl ester*

C_8H_9HgI
Mercuri-*p*-ethylphenyl iodide

C_8H_9HgIO
Mercuri-*p*-hydroxyphenyl iodide, *Et ether*

$C_8H_9HgNO_2$
Mercuri-*o*-aminophenyl acetate
Mercuri-*p*-aminophenyl acetate

C_8H_9I
α-Iodo-*o*-xylene
α-Iodo-*p*-xylene
3-Iodo-*o*-xylene
4-Iodo-*o*-xylene
2-Iodo-*m*-xylene
4-Iodo-*m*-xylene
5-Iodo-*m*-xylene
Iodo-*p*-xylene

C_8H_9IO
4-Iodo-*o*-cresol, *Me ether*
6-Iodo-*o*-cresol, *Me ether*
2-Iodo-*m*-cresol, *Me ether*
6-Iodo-*m*-cresol, *Me ether*
2-Iodo-*p*-cresol, *Me ether*
2-Iodo-4,5-dimethylphenol
2-Iodo-4,6-dimethylphenol
3-Iodo-2,6-dimethylphenol
4-Iodo-2,3-dimethylphenol
4-Iodo-2,5-dimethylphenol
4-Iodo-2,6-dimethylphenol
4-Iodo-3,5-dimethylphenol
2-Iodoethyl phenyl Ether
o-Iodophenetole
m-Iodophenetole
p-Iodophenetole

$C_8H_9IO_2$
1,3-Dihydroxy-2-iodobenzene, *Di-Me ether*
1,3-Dihydroxy-4-iodobenzene, *Di-Me ether*
2,6-Dihydroxy-3-iodo-*p*-xylene
1-Iodo-2,3-dimethoxybenzene
4-Iodo-1,2-dimethoxybenzene
Iodoquinol, *Di-Me ether*

C_8H_9IS
p-Iodobenzenethiol, *Et ether*

C_8H_9N
o-Aminostyrene
m-Aminostyrene
p-Aminostyrene
2,3-Dihydroindole
Ethylideneaniline
2-Propenylpyridine
4-Propenylpyridine

C_8H_9NO
Acetanilide
α-Aminoacetophenone
o-Aminoacetophenone
m-Aminoacetophenone
p-Aminoacetophenone
2,3-Dihydro-7-methyl-1*H*-pyrrolizin-1-one†
Formo-*o*-toluidide
Formo-*m*-toluidide
Formo-*p*-toluidide
o-Methylaminobenzaldehyde
p-Methylaminobenzaldehyde
N-Methylbenzamide
6-Methylpyridine-2-carboxylic Acid, *Me ester*
6-Methylpyridine-3-carboxylic Acid, *Me ester*
3-Nitroso-*o*-xylene
4-Nitroso-*o*-xylene
2-Nitroso-*m*-xylene
4-Nitroso-*m*-xylene
2-Nitroso-*p*-xylene
Phenylacetic Acid, *Amide*
o-Toluamide
m-Toluamide
p-Toluamide

C_8H_9NOS
Phenylmercaptoacetic Acid, *Amide*
Phenylthiocarbamic Acid, O-*Me ester*
Phenylthiocarbamic Acid, S-*Me ester*
Thiocarbamic Acid, S-*Benzyl ester*

$C_8H_9NO_2$
2-Acetamidophenol
3-Acetamidophenol
4-Acetamidophenol
m-Aminobenzoic Acid, *Me ester*
p-Aminobenzoic Acid, *Me ester*
α-Amino-3-hydroxyacetophenone
α-Amino-4-hydroxyacetophenone
2-Amino-α-hydroxyacetophenone
2-Amino-3-hydroxyacetophenone
3-Amino-4-hydroxyacetophenone
4-Amino-α-hydroxyacetophenone
4-Amino-2-hydroxyacetophenone
5-Amino-2-hydroxyacetophenone
α-Aminophenylacetic Acid
o-Aminophenylacetic Acid
m-Aminophenylacetic Acid
p-Aminophenylacetic Acid

$C_8H_9NO_2$ (*continued*)

3-Amino-*o*-toluic Acid
4-Amino-*o*-toluic Acid
5-Amino-*o*-toluic Acid
6-Amino-*o*-toluic Acid
2-Amino-*m*-toluic Acid
4-Amino-*m*-toluic Acid
5-Amino-*m*-toluic Acid
6-Amino-*m*-toluic Acid
2-Amino-*p*-toluic Acid
3-Amino-*p*-toluic Acid
Anisic Acid, *Amide*
Benzylamine-*o*-carboxylic Acid
Benzylamine-*m*-carboxylic Acid
Benzylamine-*p*-carboxylic Acid
Carbamic Acid, *Benzyl ester*
1-Cyclohexene-1,2-dicarboximide
2-Cyclohexene-1,2-dicarboximide
3-Cyclohexene-1,2-dicarboximide
4-Cyclohexene-1,2-dicarboximide
5-Deoxypyridoxal†
5,6-Dihydroxyindoline†
3,5-Dimethylpyridine-2-carboxylic Acid
4,6-Dimethylpyridine-2-carboxylic Acid
2,4-Dimethylpyridine-3-carboxylic Acid
2,5-Dimethylpyridine-3-carboxylic Acid
2,6-Dimethylpyridine-3-carboxylic Acid
1-Ethyl-1-nitrobenzene
1-Ethyl-3-nitrobenzene
1-Ethyl-4-nitrobenzene
2-Hydroxyphenylacetic Acid, *Amide*
2-Hydroxy-*m*-toluic Acid, *Amide*
6-Hydroxy-*m*-toluic Acid, *Amide*
Mandelamide
o-Methoxybenzoic Acid, *Amide*
o-Methylaminobenzoic Acid
m-Methylaminobenzoic Acid
p-Methylaminobenzoic Acid
Methyl anthranilate
N-Methylphenylcarbamic Acid
4-Nitroso-*o*-cresol, *Me ether*
4-Nitroso-*m*-cresol, *Me ether*
p-Nitrosophenol, *Et ether*
α-Nitro-*o*-xylene
3-Nitro-*o*-xylene
4-Nitro-*o*-xylene
α-Nitro-*m*-xylene
2-Nitro-*m*-xylene
4-Nitro-*m*-xylene
5-Nitro-*m*-xylene
α-Nitro-*p*-xylene
2-Nitro-*p*-xylene
Phenoxyacetic Acid, *Amide*
N-Phenylglycine
α-Picoline-betaine
β-Picoline-betaine
2-Pyridineacetic Acid, *Me ester*
3-Pyridineacetic Acid, *Me ester*
Pyridine-2-carboxylic Acid, *Et ester*
Pyridine-3-carboxylic Acid, *Et ester*
Pyridine-4-carboxylic Acid, *Et ester*
o-Tolylcarbamic Acid
m-Tolylcarbamic Acid
p-Tolylcarbamic Acid

$C_8H_9NO_2S$

Ethyl-*o*-nitrophenyl sulphide
Ethyl-*p*-nitrophenyl sulphide
p-Vinylbenzenesulphonic Acid, *Amide*

$C_8H_9NO_3$

3-Acetyl-2-hydroxy-6-methylpyran-4-one, *Amide*
2-Amino-3-hydroxybenzoic Acid, *Me ether*
2-Amino-4-hydroxybenzoic Acid, *Me ether*
2-Amino-5-hydroxybenzoic Acid, *Me ester*
2-Amino-5-hydroxybenzoic Acid, *Me ether*
2-Amino-6-hydroxybenzoic Acid, *Me ether*
3-Amino-2-hydroxybenzoic Acid, *Me ester*
3-Amino-2-hydroxybenzoic Acid, *Me ether*
3-Amino-4-hydroxybenzoic Acid, *Me ester*
3-Amino-4-hydroxybenzoic Acid, *Me ether*
4-Amino-2-hydroxybenzoic Acid, *Me ester*
4-Amino-2-hydroxybenzoic Acid, *Me ether*
4-Amino-2-hydroxybenzoic Acid, N-*Me*
4-Amino-3-hydroxybenzoic Acid, *Me ester*
5-Amino-2-hydroxybenzoic Acid, *Me ester*
5-Amino-2-hydroxybenzoic Acid, *Me ether*
2-Amino-4-hydroxy-3-methoxybenzaldehyde
2-Aminophenoxyacetic Acid
3-Aminophenoxyacetic Acid
4-Aminophenoxyacetic Acid
1,2-Dihydroxy-4-nitrosobenzene, 1-*Et ether*
1,3-Dihydroxy-2-nitrosobenzene, 3-*Et ether*
1,3-Dihydroxy-4-nitrosobenzene, *Di-Me ether*
1,3-Dihydroxy-4-nitrosobenzene, 1-*Et ether*
1,3-Dihydroxy-4-nitrosobenzene, 3-*Et ether*
α,4-Dihydroxyphenylacetic Acid, *Amide*
2,5-Dihydroxyphenylacetic Acid, *Amide*
2,3-Dimethyl-5-nitrophenol
2,4-Dimethyl-5-nitrophenol
2,4-Dimethyl-6-nitrophenol
2,5-Dimethyl-3-nitrophenol
2,5-Dimethyl-4-nitrophenol
2,5-Dimethyl-6-nitrophenol
2,6-Dimethyl-3-nitrophenol
2,6-Dimethyl-4-nitrophenol
3,4-Dimethyl-2-nitrophenol
3,5-Dimethyl-2-nitrophenol
3,5-Dimethyl-4-nitrophenol
4,5-Dimethyl-2-nitrophenol
2-Hydroxyphenylglycine
4-Hydroxyphenylglycine
3-Hydroxypyridine-2-carboxylic Acid, *Et ester*†
2-Hydroxypyridine-3-carboxylic Acid, *Et ester*
6-Hydroxypyridine-3-carboxylic Acid, *Et ester*
6-Hydroxypyridine-3-carboxylic Acid, *Et ether*
2-Hydroxypyridine-3-carboxylic Acid, *Me ester, Me ether*
Mandelhydroxamic Acid
2-Methoxy-4-nitrosophenol, *Me ether*
3-Nitro-*o*-cresol, *Me ether*
4-Nitro-*o*-cresol, *Me ether*
5-Nitro-*o*-cresol, *Me ether*
6-Nitro-*o*-cresol, *Me ether*
2-Nitro-*m*-cresol, *Me ether*
4-Nitro-*m*-cresol, *Me ether*
5-Nitro-*m*-cresol, *Me ether*
6-Nitro-*m*-cresol, *Me ether*
2-Nitro-*p*-cresol, *Me ether*
3-Nitro-*p*-cresol, *Me ether*
o-Nitrophenetole
m-Nitrophenetole
p-Nitrophenetole

$C_8H_9NO_3$ *(continued)*
1-*m*-Nitrophenylethanol
1-*p*-Nitrophenylethanol
2-*o*-Nitrophenylethanol
2-*p*-Nitrophenylethanol
2-Nitro-1-phenylethanol
Pyridoxal

$C_8H_9NO_3S$
p-Formylbenzenesulphonic Acid, *Methylamide*
2-Hydroxy-1-ethanethiol, S-o-*Nitrophenyl*
2-Hydroxy-1-ethanethiol, S-m-*Nitrophenyl*
2-Hydroxy-1-ethanethiol, S-p-*Nitrophenyl*
Phenylsulphonylacetic Acid, *Amide*

$C_8H_9NO_4$
3-Amino-4,5-dihydroxybenzoic Acid, 5-*Me ether*
3-Amino-4,5-dihydroxybenzoic Acid, *Me ester*
3-Amino-5,6-dihydroxybenzoic Acid, 5-*Me ether*
1,3-Dihydroxy-2-nitrobenzene, *Di-Me ether*
1,3-Dihydroxy-4-nitrobenzene, *Di-Me ether*
1,3-Dihydroxy-5-nitrobenzene, *Di-Me ether*
1,3-Dihydroxy-4-nitrobenzene, 1-*Et ether*
α,2-Dihydroxy-3-nitrotoluene, *Me ether*
α,2-Dihydroxy-5-nitrotoluene, 2-*Me ether*
α,4-Dihydroxy-3-nitrotoluene, 4-*Me ether*
3,4-Dihydroxy-5-nitrotoluene, 4-*Me ether*
3,5-Dihydroxy-2-nitrotoluene, 3-*Me ether*
3,5-Dihydroxy-2-nitrotoluene, 5-*Me ether*
4,5-Dihydroxy-2-nitrotoluene, 3-*Me ether*
3,4-Dihydroxypyridine-2-carboxylic Acid, *Et ester*
4,5-Dihydroxypyridine-2-carboxylic Acid, *Et ester*
4,5-Dihydroxypyridine-2-carboxylic Acid, 5-*Et ether*
4,6-Dihydroxypyridine-2-carboxylic Acid, *Et ester*
2,6-Dihydroxypyridine-3-carboxylic Acid, *Et ester*
4,6-Dihydroxypyridine-3-carboxylic Acid, *Et ester*
4,5-Dimethylpyrrole-2,3-dicarboxylic Acid
3,5-Dimethylpyrrole-2,4-dicarboxylic Acid
2,5-Dimethylpyrrole-3,4-dicarboxylic Acid
Furan-2,3-dicarboxylic Acid, 2-*Et ester-3-amide*
4-Hydroxypyridine-3-carboxylic Acid, *Me ether, Me ester*
3-Nitrocatechol, *Di-Me ether*
4-Nitrocatechol, *Di-Me ether*
4-Nitrocatechol, 1-*Et ether*
2-Nitroquinol, *Di-Me ether*
2-Nitroquinol, *Et ether*
cis-2-Pentene-2,3,5-tricarboxylic Acid, *Imide*
Pyrrole-2,5-diacetic Acid†
Pyrrole-2,5-dicarboxylic Acid, *Di-Me ester*
Pyrrole-2,5-dicarboxylic Acid, N-*Et*
1,3,5-Trihydroxy-2-nitrosobenzene, 1,3-*Di-Me ether*
1,3,5-Trihydroxy-2-nitrosobenzene, 1,5-*Di-Me ether*

$C_8H_9NO_4S$
2-Nitrobenzenesulphinic Acid, *Et ester*
3-Nitrobenzenesulphinic Acid, *Et ester*
4-Nitrobenzenesulphinic Acid, *Et ester*
o-Sulphobenzoic Acid, 1-*Me ester, Amide*
o-Sulphobenzoic Acid, 1-*Methylamide*
4-Sulpho-*o*-toluic Acid, 1-*Amide*
4-Sulpho-*o*-toluic Acid, 4-*Amide*
5-Sulpho-*o*-toluic Acid, 5-*Amide*
4-Sulpho-*m*-toluic Acid, 4-*Amide*
2-Sulpho-*p*-toluic Acid, 2-*Amide*
3-Sulpho-*p*-toluic Acid, 3-*Amide*

$C_8H_9NO_5$
3-Methyl-5-nitrofuran-2-carboxylic Acid, *Et ester*
2-Methyl-5-nitrofuran-3-carboxylic Acid, *Et ester*
1,2,3-Trihydroxy-4-nitrobenzene, 1,3-*Di-Me ether*

$C_8H_9NO_5S$
2,4-Dimethyl-3-nitrobenzenesulphonic Acid
2,4-Dimethyl-5-nitrobenzenesulphonic Acid
2,4-Dimethyl-6-nitrobenzenesulphonic Acid
2,5-Dimethyl-3-nitrobenzenesulphonic Acid
2,5-Dimethyl-4-nitrobenzenesulphonic Acid
2,5-Dimethyl-6-nitrobenzenesulphonic Acid
3,4-Dimethyl-2-nitrobenzenesulphonic Acid
3,4-Dimethyl-5-nitrobenzenesulphonic Acid
4,5-Dimethyl-2-nitrobenzenesulphonic Acid
2-Hydroxy-4-sulphobenzoic Acid, *Sulphonamide, Me ether*
3-Hydroxy-4-sulphobenzoic Acid, S-*Me ether, Sulphonamide*
4-Hydroxy-3-sulphobenzoic Acid, *Me ether, Sulphonamide*
3-Nitrotoluene-α-sulphonic Acid, *Me ester*
4-Nitrotoluene-α-sulphonic Acid, *Me ester*

$C_8H_9NO_6S$
4-Methoxy-3-nitrobenzenesulphonic Acid, *Me ester*

C_8H_9NS
Phenylthiolacetic Acid, *Amide*
Thiobenzoic Acid, *Methylamide*
Thio-*o*-toluic Acid, *Amide*
Thio-*p*-toluic Acid, *Amide*

$C_8H_9NS_2$
Phenyldithiocarbamic Acid, *Me ester*

$C_8H_9N_3$
N-*p*-Aminophenylglycine, *Nitrile*

$C_8H_9N_3O_2$
m-Amino-oxanilic Acid, *Amide*
p-Amino-oxanilic Acid, *Amide*
2-(*N*-Methyl-*N*-nitrosoamino)benzoic Acid, *Amide*
6-Methylpyridine-3,4-dicarboxylic Acid, *Diamide*
ω-Phenylbiuret
ms-Phenylbiuret

$C_8H_9N_3O_3$
o-Nitrobenzylurea
p-Nitrobenzylurea
p-Nitrophenylurea, N′-*Me*

$C_8H_9N_3O_4$
2,4-Dinitro-*N*-dimethylaniline
2,5-Dinitrodimethylaniline
2,6-Dinitrodimethylaniline
3,4-Dinitrodimethylaniline
4,5-Dinitro-2,3-xylidine
4,6-Dinitro-2,3-xylidine
5,6-Dinitro-2,3-xylidine

$C_8H_9N_3O_4$ (*continued*)
3,5-Dinitro-2,4-xylidine
3,6-Dinitro-2,4-xylidine
5,6-Dinitro-2,4-xylidine
3,4-Dinitro-2,5-xylidine
3,6-Dinitro-2,5-xylidine
4,6-Dinitro-2,5-xylidine
3,4-Dinitro-2,6-xylidine
3,5-Dinitro-2,6-xylidine
2,6-Dinitro-3,4-xylidine
5,6-Dinitro-3,4-xylidine
2,4-Dinitro-3,5-xylidine
2,6-Dinitro-3,5-xylidine
N-Ethyl-2,4-dinitroaniline
N-Ethyl-2,5-dinitroaniline
N-Methyl-4,6-dinitro-*o*-toluidine
N-Methyl-2,6-dinitro-*m*-toluidine
N-Methyl-4,6-dinitro-*m*-toluidine
N-Methyl-2,3-dinitro-*p*-toluidine
N-Methyl-2,6-dinitro-*p*-toluidine
N-Methyl-3,6-dinitro-*p*-toluidine

$C_8H_9N_3O_5$
2-Amino-3,5-dinitrophenol, *Et ether*
2-Amino-4,5-dinitrophenol, *Et ether*
2-Amino-4,6-dinitrophenol, N-*Di-Me*
3-Amino-2,4-dinitrophenol, *Et ether*
3-Amino-2,6-dinitrophenol, N-*Di-Me*
4-Amino-2,3-dinitrophenol, *Et ether*
4-Amino-2,5-dinitrophenol, *Et ether*
4-Amino-2,6-dinitrophenol, *Et ether*
4-Amino-2,6-dinitrophenol, N-*Di-Me*

$C_8H_9N_3O_5S$
6-Nitro-4-sulpho-*m*-toluic Acid, *Diamide*

$C_8H_9N_5$
Phenylguanazole

$C_8H_9O_3PS$
Salithion

$C_8H_9O_4P$
Salioxon

C_8H_{10}
Bicyclo[3,2,1]octa-2,6-diene †
Bicyclo[4,2,0]octa-2,4-diene
1,3,5-Cyclo-octatriene
1,3,6-Cyclo-octatriene
Dispiro[2,0,2,2]oct-7-ene †
Ethylbenzene
Tetracyclo[3,2,0,$0^{2,4}$,$0^{3,7}$]octane †
Tetracyclo[3,3,0,$0^{2,4}$,$0^{3,6}$]octane †
o-Xylene
m-Xylene
p-Xylene

$C_8H_{10}AsNO_5$
2-Aminophenol-4-arsonic Acid, N-*Ac*

$C_8H_{10}BrN$
2-Bromo-*N*-dimethylaniline
3-Bromo-*N*-dimethylaniline
4-Bromo-*N*-dimethylaniline
3-Bromo-2,4-xylidine
3-Bromo-2,6-xylidine
4-Bromo-2,5-xylidine
4-Bromo-2,6-xylidine
4-Bromo-3,5-xylidine
5-Bromo-2,4-xylidine
5-Bromo-3,4-xylidine
6-Bromo-2,4-xylidine
6-Bromo-3,4-xylidine
6-Bromo-3,5-xylidine

$C_8H_{10}BrNO$
4-Bromo-*o*-phenetidine
5-Bromo-*o*-phenetidine
2-Bromo-*p*-phenetidine
3-Bromo-*p*-phenetidine

$C_8H_{10}Br_2O_4$
3,5-Dibromocyclohexane-1,2-dicarboxylic Acid
3,6-Dibromocyclohexane-1,2-dicarboxylic Acid
4,5-Dibromocyclohexane-1,2-dicarboxylic Acid
1,2-Dibromocyclohexane-1,3-dicarboxylic Acid
1,3-Dibromocyclohexane-1,3-dicarboxylic Acid
1,6-Dibromocyclohexane-1,3-dicarboxylic Acid
4,5-Dibromocyclohexane-1,3-dicarboxylic Acid
1,2-Dibromocyclohexane-1,4-dicarboxylic Acid
1,4-Dibromocyclohexane-1,4-dicarboxylic Acid
2,3-Dibromocyclohexane-1,4-dicarboxylic Acid
2,5-Dibromocyclohexane-1,4-dicarboxylic Acid
Dibromofumaric Acid, *Di-Et ester*
Dibromomaleic Acid, *Di-Et ester*

$C_8H_{10}ClHgN$
Mercuri-*p*-dimethylaminophenyl chloride
Mercuri-*p*-ethylaminophenyl chloride

$C_8H_{10}ClN$
α-Amino-2-chlorotoluene, N-*Me*
α-Amino-3-chlorotoluene, N-*Me*
o-Chloro-*N*-dimethylaniline
m-Chloro-*N*-dimethylaniline
p-Chloro-*N*-dimethylaniline
N-2-Chloroethylaniline
o-β-Chloroethylaniline
p-β-Chloroethylaniline
o-Chloro-*N*-ethylaniline
m-Chloro-*N*-ethylaniline
p-Chloro-*N*-ethylaniline
4-Chloro-2,3-xylidine
6-Chloro-2,3-xylidine
3-Chloro-2,4-xylidine
5-Chloro-2,4-xylidine
6-Chloro-2,4-xylidine
3-Chloro-2,5-xylidine
4-Chloro-2,5-xylidine
6-Chloro-2,5-xylidine
2-Chloro-3,4-xylidine
5-Chloro-3,4-xylidine
6-Chloro-3,4-xylidine
2-Chloro-3,5-xylidine
3-Chloro-2,6-xylidine
4-Chloro-2,6-xylidine

$C_8H_{10}ClNO$
2-Amino-5-chloro-*p*-cresol, *Me ether*
4-Chloro-*o*-phenetidine
5-Chloro-*o*-phenetidine
6-Chloro-*m*-phenetidine
2-Chloro-*p*-phenetidine
3-Chloro-*p*-phenetidine

$C_8H_{10}FN$
m-Fluorophenethylamine
p-Fluorophenethylamine

$C_8H_{10}IN$
o-Iodo-*N*-dimethylaniline
m-Iodo-*N*-dimethylaniline
p-Iodo-*N*-dimethylaniline
2-Iodo-4,6-dimethylaniline

$C_8H_{10}INO$
4-Amino-2-iodophenol, *Et ether*

$C_8H_{10}IO_2$
3-Iodopropionic Acid, *Isopentyl ester*

$C_8H_{10}I_2O_4$
Di-iodofumaric Acid, *Di-Et ester*

$C_8H_{10}NO_5PS$
O,*O*-Dimethyl-*O*-*p*-nitrophenylthiophosphate†

$C_8H_{10}NO_6P$
Codecarboxylase

$C_8H_{10}N_2$
Benzeneazoethane
Benzylidenemethylhydrazine
Cyclohexane-1,2-dicarboxylic Acid, *Dinitrile*
1,2,3,4-Tetrahydroquinazoline
1,2,3,4-Tetrahydroquinoxaline
m-Tolamidine
p-Tolamidine
Phenylacetamidine

$C_8H_{10}N_2O$
o-Aminoacetanilide
m-Aminoacetanilide
p-Aminoacetanilide
α-Aminophenylacetic Acid, *Amide*
o-Aminophenylacetic Acid, *Amide*
m-Aminophenylacetic Acid, *Amide*
p-Aminophenylacetic Acid, *Amide*
2-Amino-*m*-toluic Acid, *Amide*
4-Amino-*m*-toluic Acid, *Amide*
6-Amino-*m*-toluic Acid, *Amide*
2-Amino-*p*-toluic Acid, *Amide*
Benzylurea
2,3-Diaminoacetophenone
2,4-Diamonoacetophenone
3,4-Diaminoacetophenone
3,5-Diaminoacetophenone
N-Dimethyl-4-nitrosoaniline
4,6-Dimethylpyridine-2-carboxylic Acid, *Amide*
2,4-Dimethylpyridine-3-carboxylic Acid, *Amide*
N-Ethyl-*p*-nitrosoaniline
α-Hydrazinoacetophenone
DL-Mandelamidine
o-Methylaminobenzoic Acid, *Amide*
N^1-Methyl-N^1-phenylurea
N^1-Methyl-N^2-phenylurea
4-Nitroso-*o*-toluidine, N-*Me*
N-Phenylglycine, *Amide*
N-*o*-Tolylurea
N-*m*-Tolylurea
N-*p*-Tolylurea

$C_8H_{10}N_2OS$
2-Hydroxyphenylthiourea, *Me ether*
4-Hydroxyphenylthiourea, *Me ether*

$C_8H_{10}N_2O_2$
2-Amino-6-hydroxybenzoic Acid, *Me ether*, *Amide*
3-Aminophenoxyacetic Acid, *Amide*
4-Aminophenoxyacetic Acid, *Amide*
N-*m*-Aminophenylglycine
N-*p*-Aminophenylglycine
3-Aminopyridine-2-carboxylic Acid, *Et ester*
5-Aminopyridine-2-carboxylic Acid, *Et ester*
2-Aminopyridine-3-carboxylic Acid, *Et ester*
4-Aminopyridine-3-carboxylic Acid, *Et ester*
6-Aminopyridine-3-carboxylic Acid, *Et ester*
Cephalosporidine
2,5-Diaminobenzoic Acid, *Me ester*
3,4-Diaminobenzoic Acid, *Me ester*
2,4-Diaminophenylacetic Acid
4,6-Diamino-*m*-toluic Acid
2,3-Diamino-*p*-toluic Acid
2,5-Diamino-*p*-toluic Acid
3,5-Diamino-*p*-toluic Acid
N-Dimethyl-2-nitroaniline
N-Dimethyl-3-nitroaniline
N-Dimethyl-4-nitroaniline
2,3-Dimethyl-4-nitroaniline
2,3-Dimethyl-5-nitroaniline
2,3-Dimethyl-6-nitroaniline
2,4-Dimethyl-3-nitroaniline
2,4-Dimethyl-5-nitroaniline
2,4-Dimethyl-6-nitroaniline
2,5-Dimethyl-3-nitroaniline
2,5-Dimethyl-4-nitroaniline
2,5-Dimethyl-6-nitroaniline
2,6-Dimethyl-3-nitroaniline
2,6-Dimethyl-4-nitroaniline
3,4-Dimethyl-2-nitroaniline
3,4-Dimethyl-5-nitroaniline
3,5-Dimethyl-2-nitroaniline
3,5-Dimethyl-4-nitroaniline
4,5-Dimethyl-2-nitroaniline
N-Ethyl-3-nitroaniline
N-Ethyl-4-nitroaniline
o-Hydrazinobenzoic Acid, α-N-*Me*
4-Hydroxyphenylglycine, *Amide*
2-Hydroxyphenylurea, *Me ether*
4-Hydroxyphenylurea, *Me ether*
N-Methyl-4-nitro-*o*-toluidine
N-Methyl-5-nitro-*o*-toluidine
N-Methyl-6-nitro-*o*-toluidine
N-Methyl-4-nitro-*m*-toluidine
N-Methyl-6-nitro-*m*-toluidine
N-Methyl-2-nitro-*p*-toluidine
N-Methyl-3-nitro-*p*-toluidine
o-Nitrobenzylamine, N-*Me*
m-Nitrobenzylamine, N-*Me*
p-Nitrobenzylamine, N-*Me*
N-β-Nitroethylaniline
4-Nitroso-*o*-anisidine, N-*Me*
Phenylhydrazine-α-carboxylic Acid, *Me ester*
Phenylhydrazine-β-carboxylic Acid, *Me ester*
2-α-Phenylhydrazinoacetic Acid
2-β-Phenylhydrazinoacetic Acid
4,5,6,7-Tetrahydroindazole-1-carboxylic Acid
4,5,6,7-Tetrahydroindazole-2-carboxylic Acid
4,5,6,7-Tetrahydroindazole-3-carboxylic Acid

$C_8H_{10}N_2O_3$
6-Amino-3-nitro-*o*-cresol, *Me ether*
6-Amino-4-nitro-*o*-cresol, *Me ether*
6-Amino-5-nitro-*o*-cresol, *Me ether*
2-Amino-5-nitro-*p*-cresol, *Me ether*

$C_8H_{10}N_2O_3$ (*continued*)
2-Methoxy-6-nitroaniline, N-*Me*
4-Methoxy-2-nitroaniline, N-*Me*
4-Methoxy-3-nitroaniline, N-*Me*
3-Nitro-*o*-phenetidine
4-Nitro-*o*-phenetidine
5-Nitro-*o*-phenetidine
6-Nitro-*o*-phenetidine
4-Nitro-*m*-phenetidine
5-Nitro-*m*-phenetidine
6-Nitro-*m*-phenetidine
2-Nitro-*p*-phenetidine
3-Nitro-*p*-phenetidine
o-Nitrophenol, β-*Aminoethyl ether*
p-Nitrophenol, β-*Aminoethyl ether*

$C_8H_{10}N_2O_3S$
3-Sulpho-*p*-toluic Acid, *Diamide*

$C_8H_{10}N_2O_4$
3-Amino-5-nitrocatechol, 1,2-*Di-Me ether*
4-Amino-3-nitrocatechol, 1,2-*Di-Me ether*
4-Amino-5-nitrocatechol, 1,2-*Di-Me ether*
2-Amino-5-nitroquinol, *Di-Me ether*
4-Amino-6-nitroresorcinol, *Di-Me ether*
Leucenine
1-Methylimidazole-4,5-dicarboxylic Acid, *Di-Me ester*
1-Methylorotic Acid, *Et ester*

$C_8H_{10}N_2O_4S$
p-Aminobenzenesulphonylcarbamic Acid, *Me ester*†
2,4-Dimethyl-3-nitrobenzenesulphonic Acid, *Amide*
2,4-Dimethyl-5-nitrobenzenesulphonic Acid, *Amide*
2,4-Dimethyl-6-nitrobenzenesulphonic Acid, *Amide*
2,5-Dimethyl-3-nitrobenzenesulphonic Acid, *Amide*
2,5-Dimethyl-4-nitrobenzenesulphonic Acid, *Amide*
2,5-Dimethyl-6-nitrobenzenesulphonic Acid, *Amide*
3,4-Dimethyl-2-nitrobenzenesulphonic Acid, *Amide*
3,4-Dimethyl-5-nitrobenzenesulphonic Acid, *Amide*
4,5-Dimethyl-2-nitrobenzenesulphonic Acid, *Amide*
3-Hydroxy-4-sulphobenzoic Acid, *Me ether, Diamide*
2-Nitrobenzenesulphonic Acid, *Et-amide*
3-Nitrobenzenesulphonic Acid, *Et-amide*
4-Nitrobenzenesulphonic Acid, *Et-amide*
3-Nitrotoluene-α-sulphonic Acid, *Methylamide*
4-Nitrotoluene-α-sulphonic Acid, *Methylamide*
2-Nitrotoluene-4-sulphonic Acid, *Methylamide*

$C_8H_{10}N_2S$
Benzylthiourea
Ethionamide
N[1]-Methyl-*N*[1]-phenylthiourea
N[1]-Methyl-*N*[2]-phenylthiourea
Phenylthiourea, 3-N-*Me*
N-o-Tolylthiourea
N-m-Tolylthiourea
N-p-Tolylthiourea

$C_8H_{10}N_4$
Benzylideneaminoguanidine

$C_8H_{10}N_4O$
Salicylideneaminoguanidine†

$C_8H_{10}N_4O_2$
Caffeine
7-Ethyl-8-methylxanthine
Isocaffeine
Kryogenin

$C_8H_{10}N_4O_3$
1,3,7-Trimethyluric Acid
1,3,9-Trimethyluric Acid
1,7,9-Trimethyluric Acid
3,7,9-Trimethyluric Acid

$C_8H_{10}N_6$
Dihydrallazine

$C_8H_{10}O$
1-Acetylcyclohexa-1,3-diene†
1-Acetylcyclohexa-1,4-diene†
Benzyl methyl Ether
Bicyclo[3,3,0]-1(5)-octen-2-one
o-Cresol, *Me ether*
m-Cresol, *Me ether*
p-Cresol, *Me ether*
2-Cyclobutylidenecyclobutanone†
2,3-Dimethylphenol
2,4-Dimethylphenol
2,5-Dimethylphenol
2,6-Dimethylphenol
3,4-Dimethylphenol
3,5-Dimethylphenol
o-Ethylphenol
m-Ethylphenol
p-Ethylphenol
α-Hydroxy-*o*-xylene
α-Hydroxy-*m*-xylene
α-Hydroxy-*p*-xylene
2,4,6-Octatrienal
Phenetole
1-Phenylethanol
2-Phenylethanol

$C_8H_{10}OS$
2-Butyrylthiophene
o-Hydroxybenzenethiol, O-*Et ether*
o-Hydroxybenzenethiol, *Di-Me ether*
m-Hydroxybenzenethiol, S-*Et ether*
p-Hydroxybenzenethiol, S-*Me, Me ether*
p-Hydroxybenzenethiol, O-*Et ether*
2-Hydroxy-1-ethanethiol, S-*Phenyl*
2-Isobutyrylthiophen
Methyl *p*-tolyl sulphoxide

$C_8H_{10}O_2$
Bicyclo[2,2,1]hept-5-ene-2-carboxylic Acid
Catechol, *Et ether*
Creosol
α,β-Dihydroxyethylbenzene
α,2-Dihydroxytoluene, α-*Me ether*
α,2-Dihydroxytoluene, 2-*Me ether*
α,3-Dihydroxytoluene, 3-*Me ether*
2,3-Dihydroxytoluene, 2-*Me ether*
3,4-Dihydroxytoluene, 4-*Me ether*
3,5-Dihydroxy-*o*-xylene
3,6-Dihydroxy-*o*-xylene

$C_8H_{10}O_2$ *(continued)*

4,5-Dihydroxy-*o*-xylene
2,4-Dihydroxy-*m*-xylene
2,5-Dihydroxy-*m*-xylene
4,5-Dihydroxy-*m*-xylene
4,6-Dihydroxy-*m*-xylene
2,5-Dihydroxy-*p*-xylene
2,6-Dihydroxy-*p*-xylene
1,2-Dimethoxybenzene
4-Ethylresorcinol
5-Ethylresorcinol†
1-(2-Furyl)-2-butanone
3-(2-Furyl)-2-butanone
4-(2-Furyl)-2-butanone
1-Hydroxyethyl phenyl Ether
2-Hydroxyethyl phenyl Ether
4-Hydroxymethyl-*o*-cresol
2-Hydroxymethyl-*m*-cresol
4-Hydroxymethyl-*m*-cresol
6-Hydroxymethyl-*m*-cresol
2-Hydroxymethyl-*p*-cresol
2-(2-Hydroxyphenyl)ethanol
2-(4-Hydroxyphenyl)ethanol
Isophthalyl Alcohol
p-Methoxybenzyl Alcohol
Methylquinol, 2-*Me ether*
Methylquinol, 5-*Me ether*
Norbornene-1-carboxylic Acid
Norbornene-2-carboxylic Acid
Norbornene-5-carboxylic Acid
Norbornene-7-carboxylic Acid
3,5-Octadiene-2,7-dione
Phthalyl Alcohol
Quinol, *Di-Me ether*
Quinol, *Et ether*
Resorcinol, *Di-Me ether*
Resorcinol, *Mono-Et ether*
Terephthalyl Alcohol
3,4,5-Trimethylfurfural

$C_8H_{10}O_2S$

Benzyl methyl sulphone
Ethyl phenyl sulphone
Methyl *p*-tolyl sulphone
2-Thiopheneacetic Acid, *Et ester*
3-Thiopheneacetic Acid, *Et ester*
Thiophene-2-carboxylic Acid, *Propyl ester*
Thiophene-2-carboxylic Acid, *Isopropyl ester*
o-Xylene-3-sulphinic Acid
o-Xylene-4-sulphinic Acid
m-Xylene-4-sulphinic Acid
m-Xylene-5-sulphinic Acid
p-Xylenesulphinic Acid

$C_8H_{10}O_2S_2$

Thiolbenzenesulphonic Acid, *Et ester*

$C_8H_{10}O_3$

2-Acetylcyclohexane-1,3-dione†
Crotonic Acid, *Anhydride*
Cyclohexane-1,2-dicarboxylic Acid, *Anhydride*
Cyclohexane-1,3-dicarboxylic Acid, *Anhydride*
Dibutolactone
Diethylmaleic Acid, *Anhydride*
α,2-Dihydroxy-3-methoxytoluene
α,3-Dihydroxy-4-methoxytoluene
α,4-Dihydroxy-3-methoxytoluene
2,2-Dimethylcyclobutane-1,3-dicarboxylic Acid, *Anhydride*
2,5-Dimethylfuran-3-carboxylic Acid, *Me ester*
4-Ethyl-2-methylfuran-3-carboxylic Acid
5-Ethyl-2-methylfuran-3-carboxylic Acid
4-Ethyl-3-methylglutaconic Acid, *Anhydride*
Filicinic Acid
Furan-2-carboxylic Acid, *Propyl ester*
Furan-2-carboxylic Acid, *Isopropyl ester*
2-Furylacetic Acid, *Et ester*
3-(2-Furyl)propionic Acid, *Me ester*
4-Hydroxy-3,6-dimethyl-2-pyrone, *Me ether*†
1-Methylcyclopentane-1,2-dicarboxylic Acid, *Anhydride*
3-Methylcyclopentane-1,2-dicarboxylic Acid, *Anhydride*
3-Methylfuran-2-carboxylic Acid, *Et ester*
5-Methylfuran-2-carboxylic Acid, *Et ester*
2-Methylfuran-3-carboxylic Acid, *Et ester*
5-Methylfuran-3-carboxylic Acid, *Et ester*
4-Methyl-6-oxo-1-cyclohexene-1-carboxylic Acid
2-Methyl-4-oxo-2-cyclohexene-1-carboxylic Acid
1-Methyl-2-oxo-3-cyclohexene-1-carboxylic Acid
Norcamphor-3-carboxylic Acid
Phloroglucinol, *Di-Me ether*
Phloroglucinol, *Mono-Et ether*
Pyrogallol, 1,2-*Di-Me ether*
Pyrogallol, 1,3-*Di-Me ether*
Pyrogallol, *Mono-Et ether*
Terrein
1,2,4-Trihydroxybenzene, 2-*Et ether*
2,3,5-Trihydroxytoluene, 3-*Me ether*
2,3,6-Trihydroxytoluene, 2-*Me ether*
2,4,5-Trihydroxytoluene, 4-*Me ether*
2,4,6-Trihydroxytoluene, 2-*Me ether*
2,4,6-Trihydroxytoluene, 4-*Me ether*
3,4,5-Trimethylfuran-2-carboxylic Acid
2,4,5-Trimethylfuran-3-carboxylic Acid

$C_8H_{10}O_3S$

Benzenesulphonic Acid, *Et ester*
Toluene-*p*-sulphonic Acid, *Me ester*
o-Xylene-3-sulphonic Acid
o-Xylene-4-sulphonic Acid
m-Xylene-2-sulphonic Acid
m-Xylene-4-sulphonic Acid
m-Xylene-5-sulphonic Acid
p-Xylenesulphonic Acid

$C_8H_{10}O_4$

Acetylenedicarboxylic Acid, *Di-Et ester*
1-Cyclohexane-1,2-dicarboxylic Acid
1-Cyclohexene-1,3-dicarboxylic Acid
1-Cyclohexene-1,4-dicarboxylic Acid
2-Cyclohexene-1,2-dicarboxylic Acid
2-Cyclohexene-1,4-dicarboxylic Acid
3-Cyclohexene-1,2-dicarboxylic Acid
3-Cyclohexene-1,3-dicarboxylic Acid
4-Cyclohexene-1,2-dicarboxylic Acid
4-Cyclohexene-1,3-dicarboxylic Acid
Fumigatol†
Kojic Acid, *Di-Me ether*
1-Methyl-2-cyclopentene-1,2-dicarboxylic Acid
3-Methylenecyclopropane-1,2-dicarboxylic Acid, *Di-Me ester*
3-Methylmuconic Acid, *Mono-Me ester*
Muconic Acid, *Di-Me ester*

$C_8H_{10}O_4$ *(continued)*
Oxalic Acid, *Diallyl ester*
Penicillic Acid
Squaric Acid, *Di-Et ester* †
1,2,3,4-Tetrahydroxybenzene, 1,2-*Di-Me ether*
1,2,3,4-Tetrahydroxybenzene, 1,4-*Di-Me ether*
1,2,3,5-Tetrahydroxybenzene, 1,3-*Di-Me ether*
1,2,3,5-Tetrahydroxybenzene, 2-*Et ether*
1,2,4,5-Tetrahydroxybenzene, 1,4-*Di-Me ether*
2,3,4,5-Tetrahydroxytoluene, 4-*Me ether*
2,3,4,6-Tetrahydroxytoluene, 4-*Me ether*
2,3,5,6-Tetrahydroxy-*p*-xylene
2,4,5,6-Tetrahydroxy-*m*-xylene

$C_8H_{10}O_4S$
2-Hydroxy-3,5-dimethylbenzenesulphonic Acid
2-Hydroxy-3,6-dimethylbenzenesulphonic Acid
2-Hydroxy-4,5-dimethylbenzenesulphonic Acid
3-Hydroxy-2,6-dimethylbenzenesulphonic Acid
5-Hydroxy-2,4-dimethylbenzenesulphonic Acid
4-Hydroxytoluene-3-sulphonic Acid, *Me ester*
4-Hydroxytoluene-3-sulphonic Acid, *Me ether*
Phenetole-*o*-sulphonic Acid
Phenetole-*m*-sulphonic Acid
Phenetole-*p*-sulphonic Acid

$C_8H_{10}O_5$
3,4-Dihydroxy-5-oxo-cyclohex-1-enecarboxylic Acid, *Me ester*
4-Formylglutaconic Acid, *Di-Me ester*
Norcantharidin
Oospolide †
Oxalacetic Acid, *Di-Et ester*
3,5,7-Trioxo-octanoic Acid †
Zymonic Acid, *Me ether*, *Me ester*

$C_8H_{10}O_6$
2,3-Diacetylsuccinic Acid
2,5-Dioxoadipic Acid, *Di-Me ester*
Dioxosuccinic Acid, *Di-Et ester*
cis-2-Pentene-2,3,5-tricarboxylic Acid
2,3,4,5-Tetrahydroxybenzoic Acid, *Me ester*

$C_8H_{10}O_6S_2$
o-Xylene-3,5-disulphonic Acid
o-Xylene-3,6-disulphonic Acid
m-Xylene-2,4-disulphonic Acid
m-Xylene-4,6-disulphonic Acid
p-Xylene-2,3-disulphonic Acid
p-Xylene-2,5-disulphonic Acid
p-Xylene-2,6-disulphonic Acid

$C_8H_{10}O_7$
Oxalomalonic Acid, *Tri-Me ester*

$C_8H_{10}O_7S_2$
2-Hydroxy-4,6-dimethylbenzene-1,3-disulphonic Acid
4-Hydroxy-2,6-dimethylbenzene-1,3-disulphonic Acid
Phenetole-2,5-disulphonic Acid

$C_8H_{10}O_8$
Butane-1,1,2,2-tetracarboxylic Acid
Butane-1,1,2,3-tetracarboxylic Acid
Butane,1,1,2,4-tetracarboxylic Acid
Butane-1,1,3,3-tetracarboxylic Acid
Butane-1,1,3,4-tetracarboxylic Acid
Butane-1,1,4,4-tetracarboxylic Acid
Butane-1,2,2,4-tetracarboxylic Acid
Butane-1,2,3,3-tetracarboxylic Acid
Butane-1,2,3,4-tetracarboxylic Acid
Butane-2,2,3,3-tetracarboxylic Acid
Ethane-tetracarboxylic Acid, sym.-*Di-Me ester*
Succinperoxide

$C_8H_{10}O_9$
Malomalic Acid

$C_8H_{10}S$
Ethyl phenyl sulphide
Methyl *p*-tolyl sulphide
α-Toluenethiol, *Me ether*
m-Xylene-2-thiol
m-Xylene-4-thiol
p-Xylene-2-thiol

$C_8H_{10}S_2$
Benzene-1,4-dithiol, *Di-Me ether*
o-Xylene-α,α′-dithiol
m-Xylene-α,α′-dithiol
p-Xylene-α,α′-dithiol

$C_8H_{11}As$
Phenylarsine, As-*Di-Me*

$C_8H_{11}AsO_3$
Benzenearsonic Acid, *Di-Me ester*

$C_8H_{11}BrN_2$
4-Bromo-*o*-phenylenediamine, 1-N-*Di-Me*

$C_8H_{11}BrN_2O_2$
5-Bromo-3-isopropyl-6-methyluracil

$C_8H_{11}BrO_4$
Bromofumaric Acid, *Di-Et ester*
Bromomaleic Acid, *Di-Et ester*

$C_8H_{11}ClN_2$
2-Amino-4-chloro-*N*-dimethylaniline
2-Amino-5-chloro-*N*-dimethylaniline
3-Amino-4-chloro-*N*-dimethylaniline
4-Amino-3-chloro-*N*-dimethylaniline
5-Amino-2-chloro-*N*-dimethylaniline

$C_8H_{11}ClO$
2-1′-Cyclopentenylpropionic Acid, *Chloride*
2-2′-Cyclopentenylpropionic Acid, *Chloride*
4-Methyl-3-cyclohexene-1-carboxylic Acid, *Chloride*
3-Methyl-1-cyclopentene-1-acetic Acid, *Chloride*
2-Octynoic Acid, *Chloride*

$C_8H_{11}ClO_3$
Cyclopentane-1,3-dicarboxylic Acid, *Chloride*

$C_8H_{11}ClO_4$
Chlorofumaric Acid, *Di-Et ester*
Chloromaleic Acid, *Di-Et ester*

$C_8H_{11}Cl_3O_6$
α-Glucochloralose
β-Glucochloralose

$C_8H_{11}N$
2-Allyl-4-pentenoic Acid, *Nitrile*
α-Aminoethylbenzene
β-Aminoethylbenzene
6-Dimethylaminofulvene †
N-Dimethylaniline
2,3-Dimethylaniline
2,4-Dimethylaniline

$C_8H_{11}N$ (*continued*)
2,5-Dimethylaniline
2,6-Dimethylaniline
3,4-Dimethylaniline
3,5-Dimethylaniline
N-Ethylaniline
o-Ethylaniline
m-Ethylaniline
p-Ethylaniline
2-Ethyl-4-methylpyridine
2-Ethyl-6-methylpyridine
3-Ethyl-4-methylpyridine
3-Ethyl-6-methylpyridine
4-Ethyl-2-methylpyridine
2-Isopropylpyridine
3-Isopropylpyridine
4-Isopropylpyridine
N-Methylbenzylamine
o-Methylbenzylamine
m-Methylbenzylamine
p-Methylbenzylamine
N-Methyl-*o*-toluidine
N-Methyl-*m*-toluidine
N-Methyl-*p*-toluidine
2-Octynoic Acid, *Nitrile*
2-Propylpyridine
3-Propylpyridine
4-Propylpyridine
2,3,4-Trimethylpyridine
2,3,5-Trimethylpyridine
2,3,6-Trimethylpyridine
2,4,5-Trimethylpyridine
2,4,6-Trimethylpyridine
3,4,5-Trimethylpyridine

$C_8H_{11}NO$
o-Aminobenzyl Alcohol, *Me ether*
4-Amino-*o*-cresol, *Me ether*
5-Amino-*o*-cresol, *Me ether*
6-Amino-*o*-cresol, *Me ether*
4-Amino-*m*-cresol, *Me ether*
6-Amino-*m*-cresol, *Me ether*
2-Amino-*p*-cresol, *Me ether*
3-Amino-*p*-cresol, *Me ether*
2-Amino-1-phenylethanol
2-Amino-2-phenylethanol
2-*o*-Aminophenylethanol
2-*p*-Aminophenylethanol
4-Amino-2,3-xylenol
5-Amino-2,3-xylenol
5-Amino-2,4-xylenol
6-Amino-2,4-xylenol
3-Amino-2,5-xylenol
4-Amino-2,5-xylenol
6-Amino-2,5-xylenol
4-Amino-2,6-xylenol
2-Amino-3,4-xylenol
5-Amino-3,4-xylenol
2-Amino-3,5-xylenol
4-Amino-3,5-xylenol
2-Anilinoethanol
Bicyclo[2,2,1]hept-5-ene-2-carboxylic Acid, *Amide*
o-Dimethylaminophenol
m-Dimethylaminophenol
p-Dimethylaminophenol
o-Ethylaminophenol
m-Ethylaminophenol
p-Ethylaminophenol
2-Hydroxybenzylamine, *Me ether*
3-Hydroxybenzylamine, *Me ether*
4-Hydroxybenzylamine, *Me ether*
4-Hydroxy-2,6-dimethylpyridine, *Me ether*
2-Hydroxy-3-methylpyridine, *Et ether*
2-Hydroxy-4-methylpyridine, *Et ether*
2-Hydroxy-6-methylpyridine, *Et ether*
4-Hydroxy-2-methylpyridine, *Et ether*
p-Methoxybenzylamine
α-Methylaminobenzyl Alcohol
3-Methylamino-*p*-cresol
N-Methyl-*o*-anisidine
N-Methyl-*m*-anisidine
N-Methyl-*p*-anisidine
Methyridine†
o-Phenetidine
m-Phenetidine
p-Phenetidine
1-(2-Pyrryl)-1-butanone
Tyramine

$C_8H_{11}NO_2$
Acetonylmalonic Acid, *Et ester nitrile*
5-Amino-2-hydroxybenzyl Alcohol, α-*Me ether*
α-Aminomethyl-*p*-hydroxybenzyl Alcohol
Aminoquinol, *Di-Me ether*
2-Aminoresorcinol, *Di-Me ether*
4-Aminoresorcinol, *Di-Me ether*
4-Aminoresorcinol, 1-*Et ether*
4-Aminoresorcinol, 3-*Et ether*
5-Aminoresorcinol, *Di-Me ether*
3-Aminoveratrol
4-Aminoveratrol
Cyclobutane-1,1-dicarboxylic Acid, *Mononitrile, Et ester*
1,2-Cyclohexanedicarboximide
1-Cyclohexene-1,2-dicarboxylic Acid, *Monoamide*
2-(3,4-Dihydroxyphenyl)ethylamine†
4-Ethyl-3-methylglutaconic Acid, *Mononitrile*
3-Ethyl-4-methylpyrrole-2-carboxylic Acid
3-Ethyl-5-methylpyrrole-2-carboxylic Acid
4-Ethyl-5-methylpyrrole-2-carboxylic Acid
5-Ethyl-3-methylpyrrole-2-carboxylic Acid
5-Ethyl-4-methylpyrrole-2-carboxylic Acid
4-Hydroxy-3-methoxybenzylamine
Pyrrole-2-acetic Acid, *Et ester*†
Pyrrole-2-carboxylic Acid, *Propyl ester*

$C_8H_{11}NO_2S$
S-Allylcysteine
o-Xylene-3-sulphonic Acid, *Amide*
o-Xylene-4-sulphonic Acid, *Amide*
m-Xylene-2-sulphonic Acid, *Amide*
m-Xylene-4-sulphonic Acid, *Amide*
m-Xylene-5-sulphonic Acid, *Amide*
p-Xylenesulphonic Acid, *Amide*

$C_8H_{11}NO_3$
Adermin
2-Amino-1-(3,4-dihydroxyphenyl)ethanol
Butyrylmalonic Acid, *Me ester nitrile*
2-Oxohexanedioic Acid, 6-*Et ester*-1-*nitrile*

$C_8H_{11}NO_3S$
N-Dimethylaniline-2-sulphonic Acid
N-Dimethylaniline-4-sulphonic Acid
N-Ethylaniline-*m*-sulphonic Acid
N-Ethylaniline-*p*-sulphonic Acid
Phenetole-*o*-sulphonic Acid, *Amide*
Phenetole-*m*-sulphonic Acid, *Amide*
Phenetole-*p*-sulphonic Acid, *Amide*

$C_8H_{11}NO_4$
Diethyl cyanomalonate
Methane-tricarboxylic Acid, *Mononitrile-di-Et ester*
3-Methylbutane-1,1,2-tricarboxylic Acid, 4-*Mono-nitrile*
2-Methylbutane-1,1,3-tricarboxylic Acid, 1-*Mono-nitrile*
2-Methylbutane-1,1,4-tricarboxylic Acid, 1-*Mono-nitrile*
2-Methylbutane-1,2,4-tricarboxylic Acid, 2-*Mono-nitrile*
2-Methylbutane-1,3,3-tricarboxylic Acid, 3-*Mono-nitrile*
3-Methylbutane-2,2,3-tricarboxylic Acid, 3-*Mono-nitrile*
Zymonic Acid, *Et ether*

$C_8H_{11}NO_5$
Cyclopropane-1,2,3-tricarboxylic Acid, *Di-Me ester-amide*

$C_8H_{11}NO_5S_2$
Benzene-*o*-disulphonic Acid, *Monoamide*, *Et ester*

$C_8H_{11}NS$
p-Aminobenzenethiol, S-*Et*
5-Amino-*o*-toluenethiol, S-*Me*
3-Amino-*p*-toluenethiol, S-*Me*

$C_8H_{11}N_3$
o-Tolylguanidine
m-Tolylguanidine
p-Tolylguanidine

$C_8H_{11}N_3O$
N-*m*-Aminophenylglycine, *Amide*
N-*p*-Aminophenylglycine, *Amide*
1-Benzylsemicarbazide
2-Benzylsemicarbazide
4-Benzylsemicarbazide
2-Methyl-4-phenylsemicarbazide
4-Methyl-1-phenylsemicarbazide
2-α-Phenylhydrazinoacetic Acid, *Amide*
4-Phenylsemicarbazide, 2-*Me*
4,5,6,7-Tetrahydroindazole-1-carboxylic Acid, *Amide*
4,5,6,7-Tetrahydroindazole-2-carboxylic Acid, *Amide*
1-*o*-Tolylsemicarbazide
1-*m*-Tolylsemicarbazide
1-*p*-Tolylsemicarbazide
2-*m*-Tolylsemicarbazide
4-*o*-Tolylsemicarbazide
4-*m*-Tolylsemicarbazide
4-*p*-Tolylsemicarbazide

$C_8H_{11}N_3O_2$
2,4-Diamino-5-nitrotoluene, 4-N-*Me*
4,5-Diamino-2-nitrotoluene, 4-N-*Me*
2,4-Diamino-6-nitro-*m*-xylene
4,6-Diamino-2-nitro-*m*-xylene
2,3-Diamino-5-nitro-*p*-xylene
3-Nitro-*o*-phenylenediamine, 2-N-*Di-Me*
4-Nitro-*o*-phenylenediamine, 1,2-N-*Di-Me*
4-Nitro-*o*-phenylenediamine, 1-N-*Di-Me*
4-Nitro-*m*-phenylenediamine, 1-N-*Di-Me*
2-Nitro-*p*-phenylenediamine, 4-N-*Di-Me*

$C_8H_{11}N_3O_6$
6-Azauridine
5-(D-Ribofuranosyl)-6-azauracil †

$C_8H_{11}O_3P$
Phenylphosphonic Acid, *Di-Me ester*
Phenylphosphonic Acid, *Et ester*

C_8H_{12}
Bicyclo[2,2,2]oct-2-ene
Bicyclo[4,2,0]oct-2-ene
Bicyclo[4,2,0]oct-3-ene
Bicyclo[4,2,0]oct-7-ene
1,2-Cyclo-octadiene †
1,3-Cyclo-octadiene
1,4-Cyclo-octadiene
1,5-Cyclo-octadiene★ †
Cyclo-octyne
1,3-Dimethylcyclohexa-1,3-diene
1,4-Dimethylcyclohexa-1,3-diene
1,5-Dimethylcyclohexa-1,3-diene
2,3-Dimethylcyclohexa-1,3-diene
2,5-Dimethylcyclohexa-1,3-diene
3,5-Dimethylcyclohexa-1,3-diene
3,6-Dimethylcyclohexa-1,3-diene
5,5-Dimethylcyclohexa-1,3-diene
2,5-Dimethyl-2,3,4-hexatriene †
1,2-Divinylcyclobutane †
2-Methylenenorbornane
Tricyclo[3,2,1,$0^{2,7}$]octane †
Tricyclo[3,2,1,$0^{3,6}$]octane †
Tricyclo[3,3,0,$0^{2,7}$]octane †
Tricyclo[3,3,0,$0^{3,7}$]octane †
Tricyclo[5,1,0,$0^{3,5}$]octane †
Tricyclo[5,1,0,$0^{4,8}$]octane †
1-Vinylcyclohexene
4-Vinylcyclohexene

$C_8H_{12}AsNO_4$
2-Aminophenol-4-arsonic Acid, N-*Di-Me*

$C_8H_{12}Br_2O_3$
2-Bromobutyric Acid, *Anhydride*
2-Bromoisobutyric Acid, *Anhydride*

$C_8H_{12}Br_2O_4$
2,5-Dibromoadipic Acid, *Di-Me ester*
3,4-Dibromoadipic Acid, *Di-Me ester*
2,3-Dibromosuccinic Acid, *Di-Et ester*

$C_8H_{12}Br_4$
1,2,5,6-Tetrabromocyclo-octane †

$C_8H_{12}ClNO$
Chloroacetamide, N-*Isobutyl*

$C_8H_{12}Cl_2$
3,4-Dichloro-1,2,3,4-tetramethylcyclobutene †

$C_8H_{12}Cl_2O_2$
2,4-Dimethyladipic Acid, *Dichloride*
Octanedioic Acid, *Dichloride*

$C_8H_{12}Cl_2O_4$
2,3-Dichlorosuccinic Acid, *Di-Et ester*
Succinic Acid, *Di-2-chloroethyl ester*

$C_8H_{12}N_2$
o-Amino-*N*-dimethylaniline
m-Amino-*N*-dimethylaniline
p-Amino-*N*-dimethylaniline
αβ-Diaminoethylbenzene
2,4-Diaminotoluene, 2-N-*Me*
2,5-Diaminotoluene, 2-N-*Me*
3,4-Diaminotoluene, 4-N-*Me*
α,α′-Diamino-*o*-xylene
α,α′-Diamino-*m*-xylene
α,α′-Diamino-*p*-xylene
3,4-Diamino-*o*-xylene
3,5-Diamino-*o*-xylene
3,6-Diamino-*o*-xylene
4,5-Diamino-*o*-xylene
2,4-Diamino-*m*-xylene
2,5-Diamino-*m*-xylene
4,5-Diamino-*m*-xylene
4,6-Diamino-*m*-xylene
2,3-Diamino-*p*-xylene
2,5-Diamino-*p*-xylene
2,6-Diamino-*p*-xylene
N,N′-Dimethyl-*o*-phenylenediamine
N,N′-Dimethyl-*m*-phenylenediamine
N,N′-Dimethyl-*p*-phenylenediamine
N,N′-Dimethylphenylhydrazine
N-Ethyl-*o*-phenylenediamine
N-Ethyl-*m*-phenylenediamine
N-Ethyl-*p*-phenylenediamine
1-Ethyl-1-phenylhydrazine
1-Ethyl-2-phenylhydrazine
p-Ethylphenylhydrazine
4-Methylpentane-1,1-dicarboxylic Acid, *Dinitrile*
Octanedioic Acid, *Dinitrile*
Phenethylhydrazine
1,4,6,9-Tetrahydropyridazino[1,2-*a*]pyridazine †
Tetramethylpyrazine
Tetramethylsuccinic Acid, *Dinitrile*

$C_8H_{12}N_2O$
2,4-Diaminophenol, *Et ether*
2,6-Diaminophenol, *Et ether*
3,4-Diaminophenol, *Et ether*
4-Hydroxyphenylhydrazine, *Et ether*
4-Hydroxy-2,2,4-trimethylglutaric Acid, *Dinitrile*

$C_8H_{12}N_2O_2$
4,5-Dihydroxy-*o*-phenylenediamine, *Di-Me ether*
2,4-Dihydroxy-*m*-phenylenediamine, 2,4-*Di-Me ether*
Hexamethylene di-isocyanate †
3,4′-Imidazolylpropionic Acid, *Et ester*
1-Methyl-4-imidazoleacetic Acid, *Et ester*
Pyridoxamine

$C_8H_{12}N_2O_2S$
N-Dimethylaniline-2-sulphonic Acid, *Amide*

$C_8H_{12}N_2O_3$
1,3-Diethylbarbituric Acid
3,4′-Imidazolyl-lactic Acid, *Et ester*
Primocarcin †
Veronal

$C_8H_{12}N_2O_3S$
6-Aminopenicillanic Acid

$C_8H_{12}N_2O_4$
2,4-Dioxo-3-imidazolidylacetic Acid, *Propyl ester*

$C_8H_{12}N_2O_4S_2$
o-Xylene-3,5-disulphonic Acid, *Diamide*
o-Xylene-3,6-disulphonic Acid, *Diamide*
m-Xylene-2,4-disulphonic Acid, *Diamide*
m-Xylene-4,6-disulphonic Acid, *Diamide*
p-Xylene-2,5-disulphonic Acid, *Diamide*
p-Xylene-2,6-disulphonic Acid, *Diamide*

$C_8H_{12}N_2O_5S_2$
4-Hydroxy-2,6-dimethylbenzene-1,3-disulphonic Acid, *Diamide*
Phenetole-2,5-disulphonic Acid, *Diamide*
Phenol-2,5-disulphonic Acid, *Et ether*, *Diamide*

$C_8H_{12}N_2O_6$
Butane-1,2,3,4-tetracarboxylic Acid, *Diamide*

$C_8H_{12}N_3O_4$
6-Diazo-5-oxonorleucine★, N-*Ac* †

$C_8H_{12}N_4$
2,2′-Azodi-3-methylpropionic Acid, *Dinitrile*

$C_8H_{12}N_4O_3$
Glycylhistidine †

$C_8H_{12}N_4O_5$
5-Azacytidine †

$C_8H_{12}N_4S$
Thiocarbazide, 1-*Me*-1-*phenyl*

$C_8H_{12}O$
1-Acetylcyclohexene
cis-Bicyclo[3,3,0]octan-2-one
1-Cyclohexenylacetaldehyde
Cyclohexylideneacetaldehyde
2-Cyclo-octenone †
3-Cyclo-octenone
2,3-Dimethyl-2-cyclohexen-1-one
2,5-Dimethyl-2-cyclohexen-1-one
2,6-Dimethyl-2-cyclohexen-1-one
3,5-Dimethyl-2-cyclohexen-1-one
3,6-Dimethyl-2-cyclohexen-1-one
4,6-Dimethyl-2-cyclohexen-1-one
5,5-Dimethyl-2-cyclohexen-1-one
4,6-Dimethyl-3-cyclohexen-1-one
1-Ethynylcyclohexanol
3-Isopropyl-2-cyclopenten-1-one
Laurenone
2-Methyl-1-cyclohexene-1-aldehyde
3-Methyl-1-cyclohexene-1-aldehyde
4-Methyl-1-cyclohexene-1-aldehyde
4-Methyl-3-cyclohexene-1-aldehyde
6-Methyl-3-cyclohexene-1-aldehyde
3-Methyl-3,5-heptadien-2-one
5-Methyl-3,5-heptadien-2-one
6-Methyl-3,5-heptadien-2-one
3-Methyl-2,5-heptadien-4-one
2,4,6-Octatrien-1-ol
3-Octyn-2-one
4-Octyn-3-one
2-Octynal
Tricyclo[2,2,2,$0^{2,6}$]octan-3-ol

$C_8H_{12}O_2$

3-Acetyl-5-hexen-2-one
2-Allyl-4-pentenoic Acid
Cyclohexane-1,2-dialdehyde
1-Cyclohexene-1-carboxylic Acid, *Me ester*
2-Cyclohexene-1-carboxylic Acid, *Me ester*
Cyclohexylideneacetic Acid
1,2-Cyclo-octanedione
1-Cyclopentene-1-carboxylic Acid, *Et ester*
3-Cyclopentene-1-carboxylic Acid, *Et ester*
2,1′-Cyclopentenylpropionic Acid
2,2′-Cyclopentenylpropionic Acid
5,5-Dimethylcyclohexane-1,3-dione
4,4-Dimethyl-2-pentynoic Acid, *Me ester*
2-Heptynoic Acid, *Me ester*
2-Hexynoic Acid, *Et ester*
3-Hydroxy-3-methylcyclohexane-1-carboxylic Acid, *Lactone*
2-Hydroxy-1-methyl-1-cyclopentaneacetic Acid, *Lactone*
2-Methyl-1-cyclohexene-1-carboxylic Acid
3-Methyl-1-cyclohexene-1-carboxylic Acid
4-Methyl-1-cyclohexene-1-carboxylic Acid
5-Methyl-1-cyclohexene-1-carboxylic Acid
6-Methyl-1-cyclohexene-1-carboxylic Acid
2-Methyl-2-cyclohexene-1-carboxylic Acid
3-Methyl-2-cyclohexene-1-carboxylic Acid
5-Methyl-2-cyclohexene-1-carboxylic Acid
6-Methyl-2-cyclohexene-1-carboxylic Acid
2-Methyl-3-cyclohexane-1-carboxylic Acid
3-Methyl-3-cyclohexene-1-carboxylic Acid
4-Methyl-3-cyclohexene-1-carboxylic Acid
5-Methyl-3-cyclohexene-1-carboxylic Acid
6-Methyl-3-cyclohexene-1-carboxylic Acid
2-Methyl-1-cyclopentene-1-acetic Acid
3-Methyl-1-cyclopentene-1-acetic Acid
6-Methyl-5-heptene-2,4-dione
3-Methyl-2,4-hexadienoic Acid, *Me ester*
4-Methyl-2-pentynoic Acid, *Et ester*
2-Methyl-4-pentynoic Acid, *Et ester*
Norbornane-2-carboxylic Acid
4-Octene-2,7-dione
2-Octynoic Acid
3-Octynoic Acid
4-Octynoic Acid
5-Octynoic Acid
6-Octynoic Acid
7-Octynoic Acid
9-Oxa-bicyclo-[3,3,1]-nonan-3-one†

$C_8H_{12}O_3$

3-Acetylacrylic Acid, *Propyl ester*
2,2-Diethylsuccinic Acid, *Anhydride*
2,3-Diethylsuccinic Acid, *Anhydride*
3-Ethyl-3-methylglutaric Acid, *Anhydride*
β-Formylcrotonic Acid, *Propyl ester*†
2-Isopropylpropane-1,3-dicarboxylic Acid, *Anhydride*
Levulinic Acid, *Allyl ester*
1-Methyl-2-oxo-1-cyclohexanecarboxylic Acid
2-Methyl-3-oxo-1-cyclohexanecarboxylic Acid
2-Methyl-4-oxo-1-cyclohexanecarboxylic Acid
3-Methyl-2-oxo-1-cyclohexanecarboxylic Acid
3-Methyl-4-oxo-1-cyclohexanecarboxylic Acid
4-Methyl-2-oxo-1-cyclohexanecarboxylic Acid
4-Methyl-3-oxo-1-cyclohexanecarboxylic Acid
5-Methyl-2-oxo-1-cyclohexanecarboxylic Acid
5-Methyl-3-oxo-1-cyclohexanecarboxylic Acid
1-Methyl-4-oxo-1-cyclohexanecarboxylic Acid
1-Methyl-2-oxocyclopentane-1-carboxylic Acid, *Me ester*
1-Methyl-3-oxocyclopentane-1-carboxylic Acid, *Me ester*
3-Methyl-2-oxocyclopentane-1-carboxylic Acid, *Me ester*
4-Methyl-2-oxocyclopentane-1-carboxylic Acid, *Me ester*
4-Methylpentane-1,3-dicarboxylic Acid, *Anhydride*
Octanedioic Acid, *Anhydride*
4-Oxocyclohexanecarboxylic Acid, *Me ester*
2-Oxocyclopentane-1-carboxylic Acid, *Et ester*
3-Oxocyclopentane-1-carboxylic Acid, *Et ester*
3-Propylglutaric Acid, *Anhydride*
Tetramethylsuccinic Acid, *Anhydride*

$C_8H_{12}O_4$

3-Acetyl-2-methyl-4-oxovaleric Acid
4-Acetyl-5-oxohexanoic Acid
3-Acetyl-4-oxovaleric Acid, *Me ester*
Biacetyl, *Dimer*
1-Butene-1,3-dicarboxylic Acid, *Di-Me ester*
1-Butene-1,4-dicarboxylic Acid, *Di-Me ester*
2-Butene-1,4-dicarboxylic Acid, *Di-Me ester*
3-Carboxy-1-cyclopentylacetic Acid
2-Carboxy-3,3-dimethylcyclopropylacetic Acid
Cyclobutane-1,2-dicarboxylic Acid, *Di-Me ester*
Cyclobutane-1,3-dicarboxylic Acid, *Di-Me ester*
Cyclohexane-1,2-dicarboxylic Acid
Cyclohexane-1,3-dicarboxylic Acid
Cyclohexane-1,4-dicarboxylic Acid
Cyclopentane-1,2-dicarboxylic Acid, *Me ester*
Cyclopentane-1,3-dicarboxylic Acid, *Me ester*
Diacetylacetic Acid, *Et ester*
2,2-Diacetylbutyric Acid
2,3-Diacetylbutyric Acid
Diethylmaleic Acid
3,3-Dimethylbutene-1,4-dicarboxylic Acid
3,3-Dimethylcyclobutane-1,2-dicarboxylic Acid
2,2-Dimethylcyclobutane-1,3-dicarboxylic Acid
3,3-Dimethylcyclopropane-1,2-dicarboxylic Acid, *Me ester*
Dimethylfumaric Acid, *Di-Me ester*
Dimethylfumaric Acid, *Et ester*
Dimethylmaleic Acid, *Di-Me ester*
2,2-Dimethyl-5-oxo-oxolan-3-acetic Acid
2,2-Dimethyl-5-oxo-oxolan-3-carboxylic Acid, *Me ester*
3,5-Dioxohexanoic Acid, *Et ether*
Ethylfumaric Acid, 1-*Et ester*
4-Ethyl-3-methylglutaconic Acid
Fumaric Acid, *Di-Et ester*
2-Hydroxy-3,3,4-trimethylglutaric Acid, *Lactone*
4-Hydroxy-2,2,3-trimethylglutaric Acid, *Lactone*
Isobutylfumaric Acid
Maleic Acid, *Di-Et ester*
Mesaconic Acid, α-*Me*-β-*Et ester*
Mesaconic Acid, β-*Me*-α-*Et ester*
2-Methylcyclopentane-1,1-dicarboxylic Acid
3-Methylcyclopentane-1,1-dicarboxylic Acid

$C_8H_{12}O_4$ (*continued*)
1-Methylcyclopentane-1,2-dicarboxylic Acid
3-Methylcyclopentane-1,2-dicarboxylic Acid
1-Methylcyclopentane-1,3-dicarboxylic Acid
4-Methylcyclopentane-1,3-dicarboxylic Acid
1-Methylcyclopropane-1,2-dicarboxylic Acid, *Di-Me ester*
3-Methylcyclopropane-1,2-dicarboxylic Acid, *Di-Me ester*
3-Methyl-2,5-dioxohexane-3-carboxylic Acid
Methylenemalonic Acid, *Di-Et ester*
4-Methyl-1-pentene-1,2-dicarboxylic Acid
2-Methyl-1-pentene-1,3-dicarboxylic Acid
2-Methylpropene-1,3-dicarboxylic Acid, *Mono-Et ester*
Umbellularic Acid

$C_8H_{12}O_5$
3-Acetonylglutaric Acid†
3-Acetylglutaric Acid, *Me ester*
Acetylmalonic Acid, *Mono-propyl ester*
Acetylsuccinic Acid, *Di-Me ester*
Formylmalonic Acid, *Di-Et ester*
Homoquinaic Acid, *δ-Lactone*
Monocrotalic Acid
Shikimic Acid, *Me ester*
Tetrahydrofuran-2,5-dicarboxylic Acid, *Di-Me ester*

$C_8H_{12}O_6$
Acetoxysuccinic Acid, *Di-Me ester*
Methane-tricarboxylic Acid, *Di-Et ester*
Methane-tricarboxylic Acid, *Di-Me mono-Et ester*
2-Methylbutane-1,1,2-tricarboxylic Acid
3-Methylbutane-1,1,2-tricarboxylic Acid
2-Methylbutane-1,1,3-tricarboxylic Acid
2-Methylbutane-1,1,4-tricarboxylic Acid
3-Methylbutane-1,2,2-tricarboxylic Acid
2-Methylbutane-1,2,3-tricarboxylic Acid
3-Methylbutane-1,2,3-tricarboxylic Acid
2-Methylbutane-1,2,4-tricarboxylic Acid
3-Methylbutane-1,2,4-tricarboxylic Acid
2-Methylbutane-1,3,3-tricarboxylic Acid
3-Methylbutane-2,2,3-tricarboxylic Acid
Pentane-1,3,5-tricarboxylic Acid†

$C_8H_{12}O_8$
Phorbic Acid†

$C_8H_{12}S_2$
1,6-Dithiacyclodeca-3-*cis*,8-*cis*-diene†

$C_8H_{13}BrO_2$
1-Bromocyclohexane-carboxylic Acid, *Me ester*

$C_8H_{13}BrO_4$
Bromosuccinic Acid, *Di-Et ester*

$C_8H_{13}BrO_5$
3-Bromomalic Acid, *Di-Et ester*

$C_8H_{13}ClO$
Cyclohexylacetic Acid, *Chloride*
2-Cyclopentylpropionic Acid, *Chloride*
2-Methylcyclohexane-1-carboxylic Acid, *Chloride*
3-Methylcyclohexane-1-carboxylic Acid, *Chloride*
2-Methylcyclopentane-1-acetic Acid, *Chloride*
3-Methylcyclopentane-1-acetic Acid, *Chloride*

$C_8H_{13}ClO_2$
3-Chloro-2-ethylcrotonic Acid, *Et ester*

$C_8H_{13}ClO_3$
3,3-Dimethylglutaric Acid, *Mono-Me ester Chloride*
Pimelic Acid, *Et ester*, *Chloride*

$C_8H_{13}ClO_4$
Chlorosuccinic Acid, *Di-Et ester*

$C_8H_{13}ClO_5$
3-Chloromalic Acid, *Di-Et ester*

$C_8H_{13}IO_4$
Iodosuccinic Acid, *Di-Et ester*

$C_8H_{13}N$
9-Azabicyclo[3,3,1]non-2-ene
Cyclohexylacetic Acid, *Nitrile*
1,2-Diethylpyrrole
2,5-Diethylpyrrole
3,4-Diethylpyrrole
3,5-Diethylpyrrole
1-Ethyl-2,3-dimethylpyrrole
1-Ethyl-2,5-dimethylpyrrole
2-Ethyl-1,5-dimethylpyrrole
2-Ethyl-3,4-dimethylpyrrole
2-Ethyl-3,5-dimethylpyrrole
2-Ethyl-4,5-dimethylpyrrole
3-Ethyl-1,2-dimethylpyrrole
3-Ethyl-2,4-dimethylpyrrole
3-Ethyl-2,5-dimethylpyrrole
4-Ethyl-2,3-dimethylpyrrole
Heliotridene
2-Isopropyl-5-methylpyrrole
3-Isopropyl-2-(or 4)-methylpyrrole
2-Methylcyclohexane-1-carboxylic Acid, *Nitrile*
3-Methylcyclohexane-1-carboxylic Acid, *Nitrile*
4-Methylcyclohexane-1-carboxylic Acid, *Nitrile*
1-Methylenepyrrolizidine
2-Octenoic Acid, *Nitrile*
3-Octenoic Acid, *Nitrile*
Pyrrole, N-*Butyl*
2,3,4,5-Tetramethylpyrrole
Tropidine

$C_8H_{13}NO$
2-Allyl-4-pentenoic Acid, *Amide*
Arecolone
1-Hydroxy-3-methylcyclohexane-1-carboxylic Acid, *Nitrile*
1-Hydroxy-4-methylcyclohexane-1-carboxylic Acid, *Nitrile*
7β-Hydroxy-1-methylene-8β-pyrrolizidine
7β-Hydroxy-1-methylene-8α-pyrrolizidine
2β-Hydroxypropylpyrrole, N-*Me*
4-Methyl-1-cyclohexene-1-carboxylic Acid, *Amide*
4-Methyl-3-cyclohexene-1-carboxylic Acid, *Amide*
N-Methylfurfurylamine, N-*Et*
2-Octynoic Acid, *Amide*
3-Octynoic Acid, *Amide*
4-Octynoic Acid, *Amide*
5-Octynoic Acid, *Amide*
6-Octynoic Acid, *Amide*
7-Octynoic Acid, *Amide*
3-Oxo-octanoic Acid, *Nitrile*
2-Pentyloxazole

$C_8H_{13}NO$ (*continued*)
Supinidine
Tropidine, N-*Oxide*
Tropinone
Tropan-2-one†

$C_8H_{13}NO_2$
Arecoline
3-Ethyl-3-methylglutaric Acid, *Imide*
Ethylmethylmalonic Acid, *Nitrile*, *Et ester*
1β,2β-Epoxy-1α-hydroxymethyl-8α-pyrrolizidine†
Heliotridine
Isopropylmalonic Acid, *Mono-Et ester*, *Nitrile*
2-Isopropylpropane-1,3-dicarboxylic Acid, *Imide*
β-(Methylenecyclopropyl)-β-methylalanine†
2-Methylglutaric Acid, 1-*Nitrile*, *Et ester*
3-Methylglutaric Acid, *Et ester*, *Nitrile*
4-Methylpentane-1,1-dicarboxylic Acid, *Mononitrile*
Norecgonidine
2-Pentenylpenilloaldehyde
Propylmalonic Acid, *Mononitrile*, *Et ester*
Pyrrolizidine-1-carboxylic Acid†
8*H*α-Pyrrolizidine-1α-carboxylic Acid†
8*H*α-Pyrrolizidine-1β-carboxylic Acid†
8*H*β-Pyrrolizidine-1α-carboxylic Acid†
8*H*β-Pyrrolizidine-1β-carboxylic Acid†
Retronecine
Scopine
Scopoline
Succinimide, N-2-*Butyl*
Succinimide, N-*Isobutyl*
Tetramethylsuccinic Acid, *Imide*
Tropinone, N-*Oxide*

$C_8H_{13}NO_3$
Crotanecine†
Cyclohexane-1,2-dicarboxylic Acid, *Monoamide*
Diethylmaleic Acid, *Monoamide*
1-Methyl-2-oxopyrrolidine-5-acetic Acid, *Me ester*
Norecgonine
Nor-ψ-ecgonine

$C_8H_{13}NO_3S$
2-Isopropylthiamorpholin-3-one-5-carboxylic Acid†

$C_8H_{13}NO_4$
3-Carboxy-4-piperidineacetic Acid
Homoquinaic Acid, *Nitrile*
N-Methylpiperidine-2,6-dicarboxylic Acid
Tropinic Acid

$C_8H_{13}NO_5$
Acetyliminodiacetic Acid, *Di-Me ester*
Oryzacidin

$(C_8H_{13}NO_5)_n$
Chitin

$C_8H_{13}N_2O_5P$
Pyridoxamine phosphate

$C_8H_{13}N_3$
1,2,4-Triaminobenzene, 1-N-*Di-Me*

$C_8H_{13}N_3O_4$
Caffuric Acid, *Et ether*

C_8H_{14}
Bicyclo[2,2,2]octane
cis-Bicyclo[3,3,0]octane
Bicyclo[4,2,0]octane
Bicyclo[3,2,1]octane†
Cyclo-octene
1,2-Dimethylcyclohexene
1,3-Dimethylcyclohexene
1,4-Dimethylcyclohexene
1,5-Dimethylcyclohexene
3,5-Dimethylcyclohexene
3,6-Dimethylcyclohexene
4,4-Dimethylcyclohexene
3,5-Dimethyl-1,2-hexadiene
2,5-Dimethyl-1,3-hexadiene
5,5-Dimethyl-1,3-hexadiene
2,5-Dimethyl-1,5-hexadiene
3,3-Dimethyl-1,5-hexadiene
3,4-Dimethyl-1,5-hexadiene
2,5-Dimethyl-2,3-hexadiene
2,4-Dimethyl-2,4-hexadiene
2,5-Dimethyl-2,4-hexadiene
3,4-Dimethyl-2,4-hexadiene
1-Methylcycloheptene
5-Methylcycloheptene
2-Methyl-1,3-heptadiene
6-Methyl-1,3-heptadiene
3-Methyl-1,5-heptadiene
3-Methyl-2,4-heptadiene
4-Methyl-2,4-heptadiene
6-Methyl-2,4-heptadiene
2,4-Octadiene
2,6-Octadiene
3,5-Octadiene
1-Octyne
2-Octyne
3-Octyne
4-Octyne
1,2,3-Trimethylcyclopentene
1,5,5-Trimethylcyclopentene
Vinylcyclohexane

$C_8H_{14}BrN$
2-Bromo-2-propylvaleric Acid, *Nitrile*

$C_8H_{14}Cl_3NO_2$
N-Dimethylglycine, *Trichloro*-tert-*butyl ester*

$C_8H_{14}FO_3P$
Di-isopropylfluorophosphonate

$C_8H_{14}KNO_9S_2$
Glucocapparin

$C_8H_{14}N_2$
1-Isobutyl-2-methylimidazole

$C_8H_{14}N_2O$
Festucine†
Loline†
4-Methylpentane-1,1-dicarboxylic Acid, *Amide-nitrile*

$C_8H_{14}N_2O_2$
Cyclohexane-1,2-dicarboxylic Acid, *Diamide*
1,4-Diethyl-2,5-dioxopiperazine
3,3-Diethyl-2,5-dioxopiperazine
Isobutylfumaric Acid, *Diamide*
2,2,6-Trimethyl-4-piperidone, N-*Nitroso*

$C_8H_{14}N_2O_4$
2,2′-Azodi-2-methylpropionic Acid
Piperazine-*N,N′*-dicarboxylic Acid, *Di-Me ester*

$C_8H_{14}N_2O_5$
5,6-Dihydro-2,4,5,5,6-pentahydroxypyrimidine, *Di-Et ether*
α-L-Glutamyl-L-alanine
γ-L-Glutamyl-L-alanine†
γ-L-Glutamyl-β-alanine

$C_8H_{14}N_4$
2,2′-Hydrazodi-(2-methylpropionic) Acid, *Dinitrile*

$C_8H_{14}N_4O$
2-Amino-4-hydroxy-6-pentyl-*s*-triazine
Caffeidine, 8-N-*Me*

$C_8H_{14}N_4O_4$
Butane-1,1,2,3-tetracarboxylic Acid, *Tetra-amide*
Butane-1,1,3,4-tetracarboxylic Acid, *Tetra-amide*
Butane-1,2,3,4-tetracarboxylic Acid, *Tetra-amide*

$C_8H_{14}N_4O_5$
Glycylglycylglycylglycine

$C_8H_{14}O$
Acetylcyclohexane
1-Acetyl-1-methylcyclopentane
1-Acetyl-2-methylcyclopentane
1-Acetyl-3-methylcyclopentane
Cycloheptane-aldehyde
Cyclohexen-3-ol, *Et ether*
Cyclo-octanone
Cyclo-oct-2-en-1-ol†
2,2-Dimethylcyclohexanone
2,3-Dimethylcyclohexanone
2,4-Dimethylcyclohexanone
2,5-Dimethylcyclohexanone
2,6-Dimethylcyclohexanone
3,3-Dimethylcyclohexanone
3,4-Dimethylcyclohexanone
3,5-Dimethylcyclohexanone
4,4-Dimethylcyclohexanone
3,4-Dimethyl-3-hexen-2-one
3,4-Dimethyl-4-hexen-2-one
2-Ethyl-5-methylcyclopentanone
3-Ethyl-2-methylcyclopentanone
3-Ethyl-3-methylcyclopentanone
3-Ethyl-4-methylcyclopentanone
1,6-Heptadien-4-ol, *Me ether*
2-Isopropylcyclopentanone
2-Methylcycloheptanone
3-Methylcycloheptanone
4-Methylcycloheptanone
1-Methylcyclohexane-1-aldehyde
2-Methylcyclohexane-1-aldehyde
3-Methylcyclohexane-1-aldehyde
4-Methylcyclohexanealdehyde
5-Methyl-1,5-heptadien-4-ol
4-Methyl-1,6-heptadien-4-ol
3-Methyl-3-hepten-2-one
6-Methyl-3-hepten-2-one
6-Methyl-4-hepten-2-one
3-Methyl-5-hepten-2-one
5-Methyl-5-hepten-2-one
6-Methyl-5-hepten-2-one
5-Methyl-1-hepten-3-one
4-Methyl-4-hepten-3-one
5-Methyl-4-hepten-3-one
6-Methyl-4-hepten-3-one
2-Methyl-5-hepten-3-one
5-Methyl-5-hepten-3-one
4-Methyl-6-hepten-3-one
5-Methyl-2-hepten-4-one
6-Methyl-2-hepten-4-one
Octahydrocyclopenta[*c*]pyran†
2-Octen-4-one
5-Octen-4-one
2-Octyn-1-ol
3-Octyn-2-ol
2-Propylcyclopentanone
3-Propylcyclopentanone
2-Propyn-1-ol, 3-*Methylbutyl ether*
2,2,3-Trimethylcyclopentanone
2,2,4-Trimethylcyclopentanone
2,2,5-Trimethylcyclopentanone
2,3,3-Trimethylcyclopentanone
2,3,5-Trimethylcyclopentanone
2,4,4-Trimethylcyclopentanone
3,3,4-Trimethylcyclopentanone
1-Vinylcyclohexanol

$C_8H_{14}O_2$
Acetoacetaldehyde, *Di-Et acetal*
1-Acetylcyclohexanol†
2-Butyne-1,4-diol, *Di-Et ether*
Cycloheptane-carboxylic Acid
Cyclohexanecarboxylic Acid, *Me ester*
Cyclohexylacetic Acid
1,4-Cyclo-octanedione
2-Cyclopentylpropionic Acid
Cyclopropanecarboxylic Acid, *Isobutyl ester*
2,3-Dimethyl-2-butenoic Acid, *Et ester*
2,2-Dimethyl-3-butenoic Acid, *Et ester*
7,7-Dimethyl-6,8-dioxabicyclo[2,3,1]octane†
3,4-Dimethyl-2,5-hexanedione
2,5-Dimethyl-3,4-hexanedione
1,4-Dioxaspiro[4,5]decane
2-Ethylcrotonic Acid, *Et ester*
4-Heptenoic Acid, *Me ester*
3-Hexenoic Acid, *Et ester*
4-Hydroxyoctanoic Acid, *Lactone*
6-Hydroxyoctanoic Acid, *Lactone*
3-Hydroxy-2,3,4-trimethylvaleric Acid, *Lactone*
3-Hydroxy-3,4,4-trimethylvaleric Acid, *Lactone*
4-Hydroxy-2,2,3-trimethylvaleric Acid, *Lactone*
4-Hydroxy-3,3,4-trimethylvaleric Acid, *Lactone*
Isovaleric Acid, *Allyl ester*
2-Methylacrylic Acid, sec-*Butyl ester*
1-Methylcyclohexane-1-carboxylic Acid
2-Methylcyclohexane-1-carboxylic Acid
3-Methylcyclohexane-1-carboxylic Acid
4-Methylcyclohexane-1-carboxylic Acid
2-Methylcyclopentane-1-acetic Acid
3-Methylcyclopentane-1-acetic Acid
1-Methylcyclopentane-1-carboxylic Acid, *Me ester*
2-Methylenevaleric Acid, *Et ester*
3-Methyl-2,6-heptanedione
6-Methyl-2,3-heptanedione

$C_8H_{14}O_2$ (*continued*)
2-Methyl-2-heptenoic Acid
6-Methyl-2-heptenoic Acid
6-Methyl-3-heptenoic Acid
6-Methyl-4-heptenoic Acid
3-Methyl-6-heptenoic Acid
3-Methyl-2-methylenebutyric Acid, *Et ester*
4-Methyl-2-oxo-1-cyclopentanol, *Et ether*
2-Methyl-2-pentenoic Acid, *Et ester*
3-Methyl-2-pentenoic Acid, *Et ester*
4-Methyl-2-pentenoic Acid, *Et ester*
2-Methyl-3-pentenoic Acid, *Et ester*
3-Methyl-3-pentenoic Acid, *Et ester*
4-Methyl-3-pentenoic Acid, *Et ester*
2,3-Octanedione
2,4-Octanedione
2,7-Octanedione
3,4-Octanedione
3,5-Octanedione
3,6-Octanedione
4,5-Octanedione
2-Octenoic Acid
3-Octenoic Acid
7-Octenoic Acid
2-Pentylimidazole

$C_8H_{14}O_2S_2$
Dithio-oxalic Acid, *Dipropyl ester*
α-Lipoic Acid
5,8-Thioctic Acid

$C_8H_{14}O_3$
Acetoacetic Acid, *Isobutyl ester*
Butyric Acid, *Anhydride*
Cyclohexylglycollic Acid
2,2-Dimethylacetoacetic Acid, *Et ester*
2,2-Dimethyl-3-oxohexanoic Acid
2,2-Dimethyl-5-oxohexanoic Acid
3,3-Dimethyl-5-oxohexanoic Acid
4,4-Dimethyl-5-oxohexanoic Acid
2-Ethyl-3-oxo-hexanoic Acid
2-Ethyl-5-oxo-hexenoic Acid
1-Hydroxycyclohexanecarboxylic Acid, *Me ester*
2-Hydroxycyclohexanecarboxylic Acid, *Me ester*
3-Hydroxycyclohexanecarboxylic Acid, *Me ester*
4-Hydroxycyclohexanecarboxylic Acid, *Me ester*
1-Hydroxycyclopentane-1-carboxylic Acid, *Et ester*
2-Hydroxycyclopentane-1-carboxylic Acid, *Et ester*
2-Hydroxycycloheptanecarboxylic Acid†
1-Hydroxy-2-methylcyclohexane-1-carboxylic Acid
1-Hydroxy-3-methylcyclohexane-1-carboxylic Acid
1-Hydroxy-4-methylcyclohexane-1-carboxylic Acid
2-Hydroxy-1-methylcyclohexane-1-carboxylic Acid
2-Hydroxy-4-methylcyclohexane-1-carboxylic Acid
2-Hydroxy-5-methylcyclohexane-1-carboxylic Acid
3-Hydroxy-2-methylcyclohexane-1-carboxylic Acid
3-Hydroxy-3-methylcyclohexane-1-carboxylic Acid
3-Hydroxy-4-methylcyclohexane-1-carboxylic Acid
3-Hydroxy-5-methylcyclohexane-1-carboxylic Acid
3-Hydroxy-6-methylcyclohexane-1-carboxylic Acid
4-Hydroxy-3-methylcyclohexane-1-carboxylic Acid
4-Hydroxy-4-methylcyclohexane-1-carboxylic Acid
1-Hydroxy-3-methyl-1-cyclopentaneacetic Acid
2-Hydroxy-1-methyl-1-cyclopentaneacetic Acid
Isobutyric Acid, *Anhydride*
Levulinic Acid, *Propyl ester*
2-Methyl-levulinic Acid, *Et ester*
2-Methyl-3-oxohexanoic Acid, *Me ester*
3-Methyl-5-oxohexanoic Acid, *Me ester*
4-Methyl-5-oxohexanoic Acid, *Me ester*
2-Methyl-3-oxovaleric Acid, *Et ester*
3-Methyl-4-oxovaleric Acid, *Et ester*
4-Methyl-2-oxovaleric Acid, *Et ester*
4-Methyl-3-oxovaleric Acid, *Et ester*
5-Oxoheptanoic Acid, *Me ester*
2-Oxohexanoic Acid, *Et ester*
3-Oxohexanoic Acid, *Et ester*
4-Oxohexanoic Acid, *Et ester*
5-Oxohexanoic Acid, *Et ester*
2-Oxo-octanoic Acid
3-Oxo-octanoic Acid
4-Oxo-octanoic Acid
5-Oxo-octanoic Acid
6-Oxo-octanoic Acid
7-Oxo-octanoic Acid
2-Oxopentane-3-carboxylic Acid, *Et ester*
Pyruvic Acid, *Pentyl ester*
Tetrahydropyran-3-carboxylic Acid, *Et ester*
Tetrahydropyran-4-carboxylic Acid, *Et ester*
Trimethylpyruvic Acid, *Et ester*

$C_8H_{14}O_4$
Adipic Acid, *Di Me-ester*
Adipic Acid, *Et ester*
Butylmethylmalonic Acid
tert-Butylmethylmalonic Acid
Butylsuccinic Acid
(2-Butyl)-succinic Acid
tert-Butylsuccinic Acid
Conduritol A, *Di-Me ether*†
Diethyl succinate
2,2-Diethylsuccinic Acid
2,3-Diethylsuccinic Acid
2,2-Dimethyladipic Acid
2,4-Dimethyladipic Acid
2,5-Dimethyladipic Acid
3,3-Dimethyladipic Acid
3,4-Dimethylapidic Acid
3,3-Dimethylglutaric Acid, *Mono-Me ester*
2,2-Dimethylsuccinic Acid, *Di-Me ester*
2,2-Dimethylsuccinic Acid, 1-*Et ester*
2,3-Dimethylsuccinic Acid, *Di-Me ester*
2-Ethyladipic Acid
3-Ethyladipic Acid

$C_8H_{14}O_4$ (*continued*)
2-Ethylbutane-1,1-dicarboxylic Acid
2-Ethyl-3-methylglutaric Acid
2-Ethyl-4-methylglutaric Acid
3-Ethyl-3-methylglutaric Acid
Ethylmethylmalonic Acid, *Di-Me ester*
Ethylmethylmalonic Acid, *Mono-Et ester*
2-Ethyl-2-methylsuccinic Acid, α-*Me ester*
Ethylsuccinic Acid, *Di-Me ester*
Glyceric Acid, *Et ester*, *Mono-acetone*
Hexane-3,3-dicarboxylic Acid
Isopropylmalonic Acid, *Di-Me ester*
Isopropylmalonic Acid, *Mono-Et ester*
2-Isopropyl-2-methylsuccinic Acid †
2-Isopropylpropane-1,3-dicarboxylic Acid
Malonic Acid, *Mono-Et ester*, *Propyl ester*
(2-Methyl-1-butyl)malonic Acid
(2-Methyl-2-butyl)malonic Acid
2-Methylglutaric Acid, *Di-Et ester*
3-Methylglutaric Acid, *Di-Et ester*
Methylmalonic Acid, *Di-Et ester*
4-Methylpentane-1,1-dicarboxylic Acid
4-Methylpentane-1,3-dicarboxylic Acid
4-Methylpentane-2,2-dicarboxylic Acid
2-Methylpimelic Acid
3-Methylpimelic Acid
4-Methylpimelic Acid
2-Methylpropanediol, 1,2-*Di-Ac*
2-Methyl-3-propylsuccinic Acid
Methylsuccinic Acid, *Me-Et ester*
Octanedioic Acid
Oxalic Acid, *Dipropyl ester*
Oxalic Acid, *Di-isopropyl ester*
2-Pentylmalonic Acid
Pimelic Acid, *Me ester*
2-Propylglutaric Acid
3-Propyglutaric Acid
Propylmalonic Acid, *Di-Me ester*
Succinic Acid, *Butyl ester*
Tetramethylsuccinic Acid
2,2,3-Trimethylglutaric Acid
2,2,4-Trimethylglutaric Acid
2,3,3-Trimethylglutaric Acid
2,3,4-Trimethylglutaric Acid

$C_8H_{14}O_4S_2$
Disulphidoacetic Acid, *Di-Et ester*
Mercaptosuccinic Acid, *Di-Et ester*
Thiodiglycollic Acid, *Di-Et ester*
Thiodipropionic Acid, *Di-Me ester*

$C_8H_{14}O_5$
Chromose D †
Diglycollic Acid, *Di-Et ester*
Ethoxyacetic Acid, *Anhydride*
2-Ethyl-3-hydroxymethylglutaric Acid
3-Ethylmalic Acid, *Et ester*
2-Hydroxyadipic Acid, *Di-Me ester*
2-Hydroxy-3,3,4-trimethylglutaric Acid
3-Hydroxy-2,2,3-trimethylglutaric Acid
3-Hydroxy-2,2,4-trimethylglutaric Acid
3-Hydroxy-2,3,4-trimethylglutaric Acid
4-Hydroxy-2,2,3-trimethylglutaric Acid
4-Hydroxy-2,2,4-trimethylglutaric Acid
Malic Acid, *Et ether*, *Di-Me ester*
Malic Acid, *Di-Et ester*
Malic Acid, *Isobutyl ether*
3,4,5-Trihydroxycyclohexanecarboxylic Acid, *Me ester*

$C_8H_{14}O_6$
1,2-Diethoxysuccinic Acid
Diethyl tartrate
2,5-Dihydroxyadipic Acid, *Di-Me ester*
Dimethoxysuccinic Acid, *Di-Me ester*
Homoquinaic Acid
Quinic Acid, *Me ester*
Succinic Acid, *Di-2-hydroxyethyl ester*

$C_8H_{14}O_7$
6-*O*-Acetylglucose
Fructuronic Acid, *Et ester*
*Rhamno*heptonic Acid, γ-*Lactone*

$C_8H_{14}O_8$
Gluco-octonic Acid, γ-*Lactone*
Mucic Acid, *Di-Me ester*
Mucic Acid, *Mono-Et ester*
Saccharic Acid, *Di-Me ester*

$C_8H_{14}S$
Octahydrocyclopenta[*c*]thiin †

$C_8H_{14}SSe$
2-Thia-3-selenaspiro[4,5]decane †

$C_8H_{14}S_2$
2,3-Dithiaspiro[4,5]decane †

$C_8H_{15}B$
9-Borabicyclo[3,3,1]nonane †

$C_8H_{15}BrO_2$
2-Bromobutyric Acid, *Isobutyl ester*
2-Bromohexanoic Acid, *Et ester*
5-Bromohexanoic Acid, *Et ester*
6-Bromohexanoic Acid, *Et ester*
2-Bromoisobutyric Acid, *Isobutyl ester*
2-Bromo-3-methylbutyric Acid, *Propyl ester*
2-Bromo-octanoic Acid
7-Bromo-octanoic Acid
8-Bromo-octanoic Acid
2-Bromopropionic Acid, *Isopentyl ester*
3-Bromopropionic Acid, *Isopentyl ester*
2-Bromo-2-propylvaleric Acid

$C_8H_{15}ClO$
2,5-Dimethylhexanoic Acid, *Chloride*
2-Methylheptanoic Acid, *Chloride*
4-Methylhexane-3-carboxylic Acid, *Chloride*
5-Methylhexane-3-carboxylic Acid, *Chloride*
Octanoic Acid, *Chloride*

$C_8H_{15}ClO_2$
2-Chlorobutyric Acid, *Isobutyl ester*
2-Hydroxyhexanoic Acid, *Et ether*, *Chloride*

$C_8H_{15}ClO_3$
4-Chloro-3-hydroxybutyric Acid, *Et ester*, *Et ether*
4-Chloro-3-hydroxybutyric Acid, n-*Butyl ester*

$C_8H_{15}Cl_3O_3$
Chlorhexadol

$C_8H_{15}HgN$
Mercuri-heptyl cyanide

$C_8H_{15}IO$
2-Iodocyclohexanol, *Et ether*

$C_8H_{15}N$
9-Azabicyclo[3,3,1]nonane
α-Coniceine
β-Coniceine
γ-Coniceine
ε-Coniceine
3-Ethyl-2,4-dimethyl-3-pyrroline
2-Ethylhexanoic Acid★, *Nitrile*†
Heliotridane
D-(−)-Indolizidine
6-Methylheptanoic Acid, *Nitrile*
1-Methyl-2-vinylpiperidine
1-Methyl-3-vinylpiperidine
2-Methyl-6-vinylpiperidine
Octahydroindole
2*H*-Octahydro-2-pyridine†
Octanoic Acid, *Nitrile*
Piperidine, N-*Allyl*
2-Propylvaleric Acid, *Nitrile*
Tropane
3-Vinylpiperidine, N-*Me*

$C_8H_{15}NO$
9-Azabicyclo[3,3,1]nonan-3-ol
1-*tert*-Butyl-3,3-dimethylaziridinone†
Cycloheptane-carboxylic Acid, *Amide*
Cyclohexylacetic Acid, *Amide*
2-Cyclopentylpropionic Acid, *Amide*
1α-Hydroxymethyl-8*H*β-pyrrolizidine†
1β-Hydroxymethyl-8*H*α-pyrrolizidine†
1β-Hydroxymethyl-8*H*β-pyrrolizidine†
2-Hydroxyoctanoic Acid, *Nitrile*
Hygrine
Isopelletierine
1-Methylcyclohexane-1-carboxylic Acid, *Amide*
2-Methylcyclohexane-1-carboxylic Acid, *Amide*
3-Methylcyclohexane-1-carboxylic Acid, *Amide*
4-Methylcyclohexane-1-carboxylic Acid, *Amide*
2-Methylcyclopentane-1-acetic Acid, *Amide*
3-Methylcyclopentane-1-acetic Acid, *Amide*
6-Methyl-2-heptenoic Acid, *Amide*
2-Octenoic Acid, *Amide*
Piperidine-3-aldehyde, N-*Et*
Piperidinoacetone
3-2′-Piperidylpropanal
2-Propionylpiperidine
Retronecanol
Trachelantamidine
2,2,6-Trimethyl-4-piperidone
Tropane, N-*Oxide*
Tropine
ψ-Tropine

$C_8H_{15}NOS_2$
6-Methylsulphinylhexyl isothiocyanate

$C_8H_{15}NO_2$
1-Hydroxy-3-methylcyclohexane-1-carboxylic Acid, *Amide*
1-Hydroxy-4-methylcyclohexane-1-carboxylic Acid, *Amide*
8-Methyl-8-azabicyclo[3,2,1]octane-6,7-diol
1-Methylpyrrolidine-2-carboxylic Acid, *Et ester*
2-Oxobutyric Acid, *Di-Et amide*
2-Oxo-octanoic Acid, *Amide*
3-Oxo-octanoic Acid, *Amide*
Pentylpenilloaldehyde
Piperidine-*N*-carboxylic Acid, *Et ester*
Piperidine-2-carboxylic Acid, *Et ester*
Piperidine-3-carboxylic Acid, *Et ester*
Piperidine-3-carboxylic Acid, N-*Me*, *Me ester*
Piperidine-4-carboxylic Acid, *Et ester*
Piperidinoacetic Acid, *Me ester*
3α,6β-Tropanediol
Tropine, N-*Oxide*
ψ-Tropine, N-*Oxide*
Turneforcidine

$C_8H_{15}NO_3$
4-Methylpentane-1,1-dicarboxylic Acid, *Monoamide*
Octanedioic Acid, *Amide*
Otonecine†
Teloidine

$C_8H_{15}NO_4$
3-Aminosuberic Acid
L-Aspartic Acid, *Di-Et ester*
L-Aspartic Acid, β-*Butyl ester*
L-Aspartic Acid, β-*Isobutyl ester*
DL-Aspartic Acid, *Di-Et ester*
Carbethoxy-DL-alanine, *Et ester*
Iminodiacetic Acid, *Di-Et ester*
3,3′-Iminodibutyric Acid
2,2′-Iminodipropionic Acid, *Di-Me ester*

$C_8H_{15}NO_4S$
Isovalthine

$C_8H_{15}NO_6$
Galacturonic Acid, 2-*Me ether*
Glucosylamine, N-*Ac*
Glucuronic Acid, 4-*Me ether*, *Methylglycoside*, *Amide*

$C_8H_{15}N_3O_2$
2,2′-Hydrazodi-(2-methylpropionic) Acid, *Mononitrile*

$C_8H_{15}N_3O_4$
Glycyl-DL-alanyl-DL-alanine

$C_8H_{15}N_3O_7$
Streptozotocin†

$C_8H_{15}N_5O$
Noformicin

$C_8H_{15}O_4P$
Allyl ethyl phosphate
Diallyl ethyl phosphate

$C_8H_{15}O_5P$
Phosphonoacetic Acid, 3,3-*Di-Et*-1-*vinyl ester*

C_8H_{16}
Cyclo-octane
1,1-Dimethylcyclohexane
1,2-Dimethylcyclohexane
1,3-Dimethylcyclohexane
1,4-Dimethylcyclohexane
1-Ethyl-2-methylcyclopentane
1-Ethyl-3-methylcyclopentane
3-Ethyl-2-methyl-2-pentene
Methylcycloheptane
2-Methyl-1-heptene
6-Methyl-1-heptene
2-Methyl-2-heptene
3-Methyl-2-heptene
4-Methyl-2-heptene

C_8H_{16} (*continued*)
4-Methyl-3-heptene
6-Methyl-3-heptene
1-Octene
2-Octene
3-Octene
4-Octene
Propylcyclopentane
1,1,2-Trimethylcyclopentane
1,1,3-Trimethylcyclopentane
1,2,3-Trimethylcyclopentane
1,2,4-Trimethylcyclopentane
2,3,3-Trimethyl-1-pentene
2,3,4-Trimethyl-1-pentene
2,4,4-Trimethyl-1-pentene
2,3,4-Trimethyl-2-pentene
2,4,4-Trimethyl-2-pentene
3,4,4-Trimethyl-2-pentene

$C_8H_{16}BrN$
5-Azoniaspiro[4,4]nonane, *Bromide*

$C_8H_{16}BrNO$
2-Bromo-2-propylvaleric Acid, *Amide*

$C_8H_{16}Br_2$
1,2-Dibromo-octane
1,4-Dibromo-octane
1,5-Dibromo-octane
1,6-Dibromo-octane
1,7-Dibromo-octane
1,8-Dibromo-octane
3,6-Dibromo-octane
4,5-Dibromo-octane

$C_8H_{16}Br_2O_4$
2,3-Dibromosuccindialdehyde, *Di-dimethylacetal*

$C_8H_{16}ClN$
5-Azoniaspiro[4,4]nonane, *Chloride*

$C_8H_{16}Cl_2$
1,5-Dichloro-octane
1,8-Dichloro-octane
2,3-Dichloro-octane
4,5-Dichloro-octane

$C_8H_{16}HgO_2$
Mercuri-hexyl acetate

$C_8H_{16}N_2$
3-Amino-8-azabicyclo[3,3,1]nonane
2-Amino-8-aza-8-methylbicyclo[3,2,1]octane
2-Amino-octanoic Acid, *Nitrile*
1,4-Diazaspiro[4,5]decane
Ethylmethylketazine
Tropylamine
ψ-Tropylamine

$C_8H_{16}N_2O_2$
Adipic Acid, *Di*-N-*methylamide*
2,4-Dimethyladipic Acid, *Diamide*
4-Hydroxy-2,2,6-trimethylpiperidine, N-*Nitroso*
6-*N*-Methyl-lysine †
4-Methylpentane-1,1-dicarboxylic Acid, *Diamide*
Nocardamin
Octanedioic Acid, *Diamide*
2-Pentylmalonic Acid, *Diamide*
3-Propylglutaric Acid, *Diamide*
Tetramethylsuccinic Acid, *Diamide*

$C_8H_{16}N_2O_3$
L-Alanyl-D-valine
2-Ethyl-3-hydroxymethylglutaric Acid, *Diamide*
Glycyl-leucine
N^5-Isopropyl-L-glutamine †
Leucylglycine

$C_8H_{16}N_2O_4$
2,3-Diaminosuccinic Acid, *Di-Et ester*
Ethylidene-diurethane
1,6-Hexanedicarbamic Acid
2,2′-Hydrazodi-(2-methylpropionic) Acid
2,2′-Hydrazodipropionic Acid, *Di-Me ester*

$C_8H_{16}N_2O_4S$
Homolanthionine

$C_8H_{16}N_2O_4S_2$
Cystine, *Di-Me ester*
Homocystine
NN′-Dimethyl-L-cystine †

$C_8H_{16}N_2O_5$
Trihydroxyglutaric Acid, *Diamide*, *Tri-Me ether*

$C_8H_{16}N_2O_7$
Cycasin †

$C_8H_{16}N_4$
1,3,6,8-Tetrazatricyclo[6,2,1,$1^{5,6}$]dodecane

$C_8H_{16}N_4O_2$
2,2′-Azodi-2-methylpropionic Acid, *Diamide*

$C_8H_{16}N_4O_3$
Gongrine, *Et ester* †

$C_8H_{16}N_6OS_2$
3-Ethoxy-2-oxobutyraldehyde bis(thiosemicarbazone) †

$C_8H_{16}O$
Allyl 3-methylbutyl Ether
Cyclo-octanol
1,2-Dimethylcyclohexanol
1,3-Dimethylcyclohexanol
1,4-Dimethylcyclohexanol
2,2-Dimethylcyclohexanol
2,3-Dimethylcyclohexanol
2,4-Dimethylcyclohexanol
2,5-Dimethylcyclohexanol
2,6-Dimethylcyclohexanol
3,3-Dimethylcyclohexanol
3,4-Dimethylcyclohexanol
3,5-Dimethylcyclohexanol
4,4-Dimethylcyclohexanol
2,5-Dimethyl-3-hexanone
4,4-Dimethyl-3-hexanone
5,5-Dimethyl-3-hexen-1-ol
2,5-Dimethyl-4-hexen-2-ol
3,5-Dimethyl-1-hexen-3-ol
2,2-Dimethyl-4-hexen-3-ol
2,4-Dimethyl-4-hexen-3-ol
2,5-Dimethyl-4-hexen-3-ol
3,5-Dimethyl-4-hexen-3-ol
2,3-Dimethyl-3-hexen-3-ol
2-Ethylhexanal †
1-Ethyl-3-methylcyclopentanol
2-Ethyl-1-methylcyclopentanol
3-Ethyl-3-methyl-2-pentanone
3-Ethyl-4-methyl-2-pentanone
1-Hepten-2-ol, *Me ether*

$C_8H_{16}O$ (*continued*)

2-Hepten-4-ol, *Me ether*
5-Hexen-1-ol, *Et ether*
α-Hydroxyethylcyclohexane
1-Methylcycloheptanol
2-Methylcycloheptanol
4-Methylcycloheptanol
2-Methylcyclohexane-1-methanol
3-Methylcyclohexane-1-methanol
4-Methylcyclohexane-1-methanol
2-Methylheptanal
5-Methylheptanal
4-Methyl-2-heptanone
5-Methyl-2-heptanone
6-Methyl-2-heptanone
2-Methyl-3-heptanone
6-Methyl-3-heptanone
2-Methyl-4-heptanone
3-Methyl-4-heptanone
3-Methyl-5-heptanone
3-Methyl-6-hepten-1-ol
2-Methyl-3-hepten-2-ol
2-Methyl-4-hepten-2-ol
2-Methyl-5-hepten-2-ol
6-Methyl-5-hepten-2-ol
2-Methyl-6-hepten-2-ol
6-Methyl-1-hepten-3-ol
3-Methyl-4-hepten-3-ol
4-Methyl-4-hepten-3-ol
3-Methyl-6-hepten-3-ol
4-Methyl-1-hepten-4-ol
6-Methyl-1-hepten-4-ol
3-Methyl-2-hepten-4-ol
5-Methyl-2-hepten-4-ol
6-Methyl-2-hepten-4-ol
4-Methyl-3-penten-1-ol, *Et ether*
4-Methyl-3-penten-2-ol, *Et ether*
Octanal
2-Octanone
3-Octanone
4-Octanone
2-Octen-1-ol
3-Octen-1-ol
4-Octen-1-ol
5-Octen-1-ol
6-Octen-1-ol
1-Octen-2-ol
1-Octen-3-ol
1-Octen-4-ol
2-Octen-4-ol
1-Propylcyclopentanol
2-Propylcyclopentanol
2-Propylpentanal
1,2,2-Trimethylcyclopentanol
1,2,4-Trimethylcyclopentanol
1,2,5-Trimethylcyclopentanol
2,3,4-Trimethylcyclopentanol
2,2,4-Trimethyl-3-pentanone

$C_8H_{16}O_2$

Butyric Acid, n-*Butyl ester*
Butyric Acid, *Isobutyl ester*
1,4-Cyclohexanediol, *Di-Me ether*
1,2-Cyclo-octanediol
1,4-Cyclo-octanediol
2-Cyclohexyloxyethanol†
2,2-Dimethylbutyric Acid, *Et ester*
2,3-Dimethylbutyric Acid, *Et ester*
3,3-Dimethylbutyric Acid, *Et ester*
2,5-Dimethylhexanoic Acid
2,3-Dimethylvaleric Acid, *Me ester*†
2-Ethylbutyric Acid, *Et ester*
2-Ethylhexanoic Acid
3-Ethylhexanoic Acid
2-Ethylvaleric Acid, *Me ester*
Heptanoic Acid, *Me ester*
Hexanoic Acid, *Et ester*
2-Hexanol, *Ac*
Hydroxyacetone, 3-*Methylbutyl ether*
1-Hydroxy-2-butanone, *Isobutyl ether*
4-Hydroxy-2,5-dimethyl-3-hexanone
1-Hydroxy-2-hexanone, *Et ether*
5-Hydroxy-4-octanone
Isobutyric Acid, *Isobutyl ester*
Isovaleric Acid, *Propyl ester*
Isovaleric Acid, *Isopropyl ester*
Ketene, *Di-propyl acetal*
2-Methyl-2-butyl propionate
3-Methyl-1-butyl propionate
2-Methylbutyric Acid, *Propyl ester*
2-Methylbutyric Acid, *Isopropyl ester*
2-Methylheptanoic Acid
4-Methylheptanoic Acid
5-Methylheptanoic Acid
6-Methylheptanoic Acid
2-Methylhexane-3-carboxylic Acid
3-Methylhexane-3-carboxylic Acid
4-Methylhexane-3-carboxylic Acid
5-Methylhexane-3-carboxylic Acid
2-Methylhexanoic Acid, *Me ester*
3-Methylhexanoic Acid, *Me ester*
4-Methylhexanoic Acid, *Me ester*
5-Methyl-6-oxo-1-heptanol
6-Methyl-2-oxo-1-heptanol
2-Methyl-6-oxo-2-heptanol
4-Methyl-6-oxo-2-heptanol
2-Methyl-6-oxo-3-heptanol
3-Methyl-2-oxo-3-heptanol
3-Methyl-5-oxo-3-heptanol
3-Methyl-2-oxo-4-heptanol
2-Methylvaleric Acid, *Et ester*
3-Methylvaleric Acid, *Et ester*
4-Methylvaleric Acid, *Et ester*
Octanoic Acid
Pentyl propionate
Propionic Acid, *Pentyl ester*
2-Propylvaleric Acid
Valeric Acid, *Propyl ester*

$C_8H_{16}O_3$

2-Hydroxybutyric Acid, *Isobutyl ester*
2-Hydroxybutyric Acid, *Et ether*, *Et ester*
3-Hydroxy-2,2-dimethylbutyric Acid, *Et ester*
3-Hydroxy-2,3-dimethylbutyric Acid, *Et ester*
2-Hydroxyhexanoic Acid, *Et ether*
3-Hydroxyhexanoic Acid, *Et ester*
5-Hydroxyhexanoic Acid, *Et ester*
6-Hydroxyhexanoic Acid, *Et ester*
2-Hydroxymethyl-2-methylbutyric Acid, *Et ester*
2-Hydroxy-4-methylvaleric Acid, *Et ester*
3-Hydroxy-4-methylvaleric Acid, *Et ether*

$C_8H_{16}O_3$ *(continued)*
2-Hydroxyoctanoic Acid
3-Hydroxyoctanoic Acid
4-Hydroxyoctanoic Acid
6-Hydroxyoctanoic Acid
8-Hydroxyoctanoic Acid
2-Hydroxy-2-propylvaleric Acid
2-Hydroxy-2,4,4-trimethylvaleric Acid
3-Hydroxy-2,2,3-trimethylvaleric Acid
3-Hydroxy-2,2,4-trimethylvaleric Acid
3-Hydroxy-2,3,4-trimethylvaleric Acid
3-Hydroxy-3,4,4-trimethylvaleric Acid
4-Hydroxy-2,2,3-trimethylvaleric Acid
4-Hydroxy-3,3,4-trimethylvaleric Acid
2-Hydroxyvaleric Acid, *Et ether*, *Me ester*
Lactic Acid, *Pentyl ester*

$C_8H_{16}O_3S$
Vinylsulphonic Acid, *Hexyl ester*

$C_8H_{16}O_4$
Chalcose, *Methyl chalcoside*
Cladinose
Diginose, *Me-glycoside*
2,3-Dihydroxybutyric Acid, *Butyl ester*
2,3-Dihydroxy-2-methylpropionic Acid, *Butyl ester*
Formylacetic Acid, *Me ester*
Glyoxylic Acid, *Di-Et acetal*, *Et ester*
Isovaleric Acid, *Mono-glycerol ester*
Ketene bis(2-methoxyethyl) acetal†
Metaldehyde, *Tetrameric*
Mycarose, *Methyl glycoside*
Olivomose, *Methyl glycoside*†
Olivomycose, α-*Methyl glycoside*†
Paraldol

$C_8H_{16}O_4Si$
Diallyl dimethyl orthosilicate

$C_8H_{16}O_5$
Apiose, 2,3,4-*Tri-O-Me ether*★†
2,6-Dideoxy-4-*C*-(1′-hydroxyethyl)hexose
Fucose, 2,3-*Di-Me ether*
Fucose, 3,4-*Di-Me ether*
Erythronic Acid, *Butyl ester*
Mycinose
Noviose
Rhamnose, 3,4-*Di-Me ether*
Ribose, 2,3,4-*Tri-Me ether*
Ribose, 2,3,5-*Tri-Me ether*
Thevetose, α-*Methyl thevetoside*
Xylose, 2,3,4-*Tri-Me ether*
Xylose, 2,3,5-*Tri-Me ether*

$C_8H_{16}O_6$
Dambonitol
2,3-Di-*O*-methylgalactose
2,4-Di-*O*-methylgalactose
2,6-Di-*O*-methylgalactose
3,4-Di-*O*-methylgalactose
4,6-Di-*O*-methylgalactose
2,3-Di-*O*-methylglucose
2,4-Di-*O*-methylglucose
2,6-Di-*O*-methylglucose
3,4-Di-*O*-methylglucose
3,5-Di-*O*-methylglucose
3,6-Di-*O*-methylglucose
4,5-Di-*O*-methylglucose
4,6-Di-*O*-methylglucose
5,6-Di-*O*-methylglucose
Fructose, *Ethyl glycoside*
Galactose, *Ethyl galactoside*
Glucose, *Ethyl glucoside*
Glyceraldehyde, 2-*Me ether dimer*
Mannose, 3,4-*Di-Me ether*
Xylonic Acid, 2,3,4-*Tri-Me ether*
Xylonic Acid, 2,3,5-*Tri-Me ether*

$C_8H_{16}O_7$
*Gluco*heptose, *Methyl glycoside*
*Gluco*heptulose, α-*Methyl glycoside*
Gluconic Acid, *Et ester*
Mannonic Acid, *Et ester*
*Rhamno*heptose

$C_8H_{16}O_8$
Gala-octose
Gluco-octose
D-*Gluco*-L-*tagato*-octose
D-*Gluco*-L-*talo*-octose
*Rhamno*heptonic Acid

$C_8H_{16}O_9$
Gluco-octonic Acid

$C_8H_{17}Br$
1-Bromo-octane
2-Bromo-octane

$C_8H_{17}BrHg$
Mercuri-octyl bromide

$C_8H_{17}BrO_2$
Bromoacetaldehyde, *Dipropyl acetal*

$C_8H_{17}Cl$
1-Chloro-octane
2-Chloro-octane

$C_8H_{17}ClHg$
Mercuri-octyl chloride

$C_8H_{17}ClO$
3-Chloro-2-butanol, 3-*Methylbutyl ether*

$C_8H_{17}ClO_2$
Chloroacetaldehyde, *Dipropyl acetal*
Chloroacetaldehyde, *Di-isopropyl acetal*
3-Chlorobutyraldehyde, *Di-Et acetal*

$C_8H_{17}F$
2-Fluoro-octane

$C_8H_{17}I$
1-Iodo-octane
2-Iodo-octane

$C_8H_{17}N$
N-Allyl-3-methylbutylamine
Coniine
Cyclo-octylamine
N-Ethylcyclohexylamine
1-Ethylcyclohexylamine
2-Ethylcyclohexylamine
3-Ethyl-2,4-dimethylpyrrolidine
1-Ethyl-2-methylpiperidine
2-Ethyl-1-methylpiperidine
2-Ethyl-6-methylpiperidine
3-Ethyl-1-methylpiperidine
3-Ethyl-4-methylpiperidine
4-Ethyl-1-methylpiperidine
4-Ethyl-2-methylpiperidine

$C_8H_{17}N$ (*continued*)
5-Ethyl-2-methylpiperidine
1-Isopropyl-2-methylpyrrolidine
1-Isopropyl-3-methylpyrrolidine
2-Isopropyl-5-methylpyrrolidine
2-Isopropylpiperidine
4-Isopropylpiperidine
4-Methyl-1-cyclohexanemethylamine
2-Methylpiperidine, N-*Et*
3-Methylpiperidine, N-*Et*
Perhydroazonine
Piperidine, N-*Propyl*
Piperidine, N-*Isopropyl*
N-Propylpiperidine
3-Propylpiperidine
4-Propylpiperidine
2-Propylpyrrolidine, N-*Me*
Pyrrolidine, N-*Butyl*
1,2,2,5-Tetramethylpyrrolidine
2,2,5,5-Tetramethylpyrrolidine
2,3,4,5-Tetramethylpyrrolidine
1,2,2-Trimethylpiperidine
1,2,6-Trimethylpiperidine
2,2,4-Trimethylpiperidine
2,2,6-Trimethylpiperidine
2,3,6-Trimethylpiperidine
2,4,6-Trimethylpiperidine
3,3,5-Trimethylpiperidine

$C_8H_{17}NO$
5-Azoniaspiro[4,4]nonane(hydroxide)
Conhydrine
ψ-Conhydrine
2,5-Dimethylhexanoic Acid, *Amide*
2-Ethylhexanoic Acid, *Amide*
2-(4-Hydroxycyclohexyl)ethylamine
4-Hydroxy-2,2,6-trimethylpiperidine
Hygroline
ψ-Hygroline
6-Methylheptanoic Acid, *Amide*
2-Methylhexane-3-carboxylic Acid, *Amide*
3-Methylhexane-3-carboxylic Acid, *Amide*
4-Methylhexane-3-carboxylic Acid, *Amide*
5-Methylhexane-3-carboxylic Acid, *Amide*
Octanoic Acid, *Amide*
2-Propylvaleric Acid, *Amide*

$C_8H_{17}NO_2$
2-Aminohexanoic Acid, *Et ester*
2-Aminohexanoic Acid, N-*Di-Me*
2-Aminohexanoic Acid, N-*Et*
3-Aminohexanoic Acid, *Et ester*
6-Aminohexanoic Acid, *Et ester*
2-Amino-octanoic Acid
3-Amino-octanoic Acid
4-Amino-octanoic Acid
6-Amino-octanoic Acid
8-Amino-octanoic Acid
3-Aminopentane-3-carboxylic Acid, *Et ester*
N-Diethylglycine, *Et ester*
3-Dimethylaminobutyric Acid, *Et ester*
4-Dimethylaminobutyric Acid, *Et ester*
6-Dimethylaminohexanoic Acid
N,γ-Dimethyl-L-*allo*isoleucine†
Ethyl 3-methylbutylaminoformate
Forosamine†
2-Hydroxyhexanoic Acid, *Et ether*, *Amide*
2-Hydroxyoctanoic Acid, *Amide*
Leucine, *Et ester*
5-Methylaminovaleric Acid, *Et ester*
3-Methyl-3-methylaminobutyric Acid, *Et ester*
1-Nitro-octane
2-Nitro-octane
3-Nitro-octane

$C_8H_{17}NO_3$
Carnitine, *Me ether*
Desosamine
3-Dimethylamino-2-hydroxy-2-methylpropionic Acid, *Et ester*
Rhodosamine

$C_8H_{17}NO_3S$
Felinine

$C_8H_{17}NO_4$
Amosamine†
Garosamine, *Me glycoside*†
Mycaminose

$C_8H_{17}NO_5$
Tri-(2-hydroxyethyl)ammonium-acetic Acid Betaine†

$C_8H_{17}NO_6$
Galactonic Acid, 2,3-*Di-Me ether*, *Amide*
Galactonic Acid, 2,4-*Di-Me ether*, *Amide*
Galactonic Acid, 3,4-*Di-Me ether*, *Amide*
Galactonic Acid, 4,6-*Di-Me ether*, *Amide*
Gluconic Acid, 3,5-*Di-Me ether*, *Amide*

$C_8H_{17}NO_6S$
Pantoyltaurine

$C_8H_{17}NO_8$
Gluco-octonic Acid, *Amide*

$C_8H_{17}N_3O_2$
Glycyl-leucine, *Amide*
Methylguanidinohexanoic Acid

$C_8H_{17}N_3O_4$
N^{δ}-(2-Amino-2-carboxyethyl)ornithine†

$C_8H_{17}O_5P$
Phosphonoacetic Acid, *Tri-Et ester*

C_8H_{18}
2,2-Dimethylhexane
2,3-Dimethylhexane
2,4-Dimethylhexane
2,5-Dimethylhexane
3,3-Dimethylhexane
3,4-Dimethylhexane
3-Ethylhexane
2-Methylheptane
3-Methylheptane
4-Methylheptane
n-Octane
2,2,3,3-Tetramethylbutane
2,2,3-Trimethylpentane
2,2,4-Trimethylpentane
2,3,3-Trimethylpentane
2,3,4-Trimethylpentane

$C_8H_{18}ClNO_2$
Methacholine chloride

$C_8H_{18}ClO_2P$
O,*O*-Dibutylphosphorochloxidite

$C_8H_{18}ClO_2PS$
O,O-Dibutylphosphorochloridothionate
$C_8H_{18}ClO_3P$
Di-*O*-butylphosphorochloridate
$C_8H_{18}CrO_4$
Di-*tert*-butyl chromate
$C_8H_{18}Hg$
Mercury di-butyl
Mercury di-*tert*-butyl
Mercury di-1-methylpropyl
Mercury di-2-methylpropyl
$C_8H_{18}HgO$
Mercuri-octyl hydroxide
$C_8H_{18}INO_2$
Methacholine iodide†
$C_8H_{18}INO_3$
Lactoylcholine, *Iodide*†
$C_8H_{18}N_2$
Piperazine, N,N-*Di-Et*
1,2,4,5-Tetramethylpiperazine
1,2,4,6-Tetramethylpiperazine
2,2,5,5-Tetramethylpiperazine
2,3,5,6-Tetramethylpiperazine
$C_8H_{18}N_2O$
2-Amino-octanoic Acid, *Amide*
Di-*tert*-butyloxadiaziridine†
$C_8H_{18}N_2O_2$
Di-*tert*-butyl hyponitrite†
Lysine★, 6-N-*Et*†
$C_8H_{18}N_2O_4$
Actinamine
$C_8H_{18}N_2O_4Pb$
Lead dibutyldinitrate
$C_8H_{18}N_4O_2$
2-Pentylmalonic Acid, *Dihydrazide*
$C_8H_{18}N_6O_4$
Streptidine
$C_8H_{18}O$
Dibutyl Ether
Di-2-butyl Ether
Di-isobutyl Ether
2,2-Dimethyl-1-hexanol
2,5-Dimethyl-1-hexanol
3,5-Dimethyl-1-hexanol
2,3-Dimethyl-2-hexanol
2,4-Dimethyl-2-hexanol
2,5-Dimethyl-2-hexanol
3,4-Dimethyl-2-hexanol
3,5-Dimethyl-2-hexanol
5,5-Dimethyl-2-hexanol
2,2-Dimethyl-3-hexanol
2,3-Dimethyl-3-hexanol
2,4-Dimethyl-3-hexanol
2,5-Dimethyl-3-hexanol
3,5-Dimethyl-3-hexanol
4,4-Dimethyl-3-hexanol
5,5-Dimethyl-3-hexanol
2-Ethyl-1-hexanol
3-Ethyl-3-hexanol
Ethyl hexyl Ether
2-Ethyl-2-methyl-1-pentanol
2-Ethyl-3-methyl-1-pentanol
3-Ethyl-2-methyl-2-pentanol
3-Ethyl-2-methyl-3-pentanol
3-Ethyl-4-methyl-2-pentanol
Heptyl methyl Ether
Isobutanol, *Butyl ether*
3-Methyl-1-heptanol
4-Methyl-1-heptanol
5-Methyl-1-heptanol
6-Methyl-1-heptanol
2-Methyl-2-heptanol
3-Methyl-2-heptanol
4-Methyl-2-heptanol
5-Methyl-2-heptanol
6-Methyl-2-heptanol
2-Methyl-3-heptanol
3-Methyl-3-heptanol
4-Methyl-3-heptanol
5-Methyl-3-heptanol
6-Methyl-3-heptanol
2-Methyl-4-heptanol
3-Methyl-4-heptanol
4-Methyl-4-heptanol
4-Methyl-2-pentanol, *Et ether*
1-Octanol
2-Octanol
3-Octanol
4-Octanol
2,2,3-Trimethyl-3-pentanol
2,2,4-Trimethyl-3-pentanol
$C_8H_{18}OS$
Dibutyl sulphoxide
Di-isobutyl sulphoxide
$C_8H_{18}O_2$
Acetaldehyde, *Dipropyl acetal*
Acetaldehyde, *Di-isopropyl acetal*
1,4-Butanediol, *Di-Et ether*
Di-*tert*-butyl peroxide
1,6-Hexanediol, *Di-Me ether*
Isobutanal, *Di-Et acetal*
Isobutyl peroxide
1,2-Octanediol
1,4-Octanediol
1,7-Octanediol
1,8-Octanediol
2,4-Octanediol
2,7-Octanediol
4,5-Octanediol
$C_8H_{18}O_2S$
Dibutyl sulphone
Di-isobutyl sulphone
$C_8H_{18}O_3$
2-(2-Butoxyethoxy)ethanol
2-(*tert*-Butoxethoxy)ethanol
Di-(2-hydroxyethyl) Ether, *Di-Et ether*
Di-(2-hydroxyethyl) Ether, *Butyl ether*
Glycerol, 1-3'-*Methylbutyl ether*
Orthoacetic Acid, *Tri-Et ester*
Orthoformic Acid, *Di-Et*, *Propyl ester*
1,2,5-Pentanetriol, 5-*Me-Et ether*
$C_8H_{18}O_3S$
Dibutyl sulphite
$C_8H_{18}O_3Ti$
Dibutyl metatitanate

$C_8H_{18}O_4$
Erythritol, 1,4-*Di-Et ether*
$C_8H_{18}O_4S$
Dibutyl sulphate
Di-isobutyl sulphate
$C_8H_{18}O_4S_2$
Methylsulphonal
$C_8H_{18}O_5$
Rhamnitol, 3,4-*Di-Me ether*
Tetraethylene Glycol
$C_8H_{18}O_6$
Mannitol, 1,2-*Di-Me ether*
$C_8H_{18}S$
Dibutyl sulphide
Di-isobutyl sulphide
Di-*tert*-butyl Sulphide†
$C_8H_{18}S_2$
Dibutyl disulphide
Di-*tert*-butyl disulphide
$C_8H_{18}S_3$
Dibutyl trisulphide
Di-*tert*-butyl trisulphide
$C_8H_{18}S_4$
Di-*tert*-butyl tetrasulphide
$C_8H_{18}Zn$
Zinc dibutyl
Zinc di-*sec*-butyl
Zinc di-isobutyl
$C_8H_{19}BO$
Dibutylborinic Acid
$C_8H_{19}N$
2-Amino-2,5-dimethylhexane
1-Amino-octane
2-Amino-octane
4-Amino-octane
Dibutylamine
Di-2-butylamine
Di-isobutylamine
N-Ethyldipropylamine
3-Methylbutylpropylamine
N-Methylheptylamine
2-Methylheptylamine
3-Methylheptylamine
4-Methylheptylamine
$C_8H_{19}NO$
1-Diethylamino-2-propanol, *Me ether*
2-Di-isopropylaminoethanol
2-Dipropylaminoethanol
2-Hexylaminoethanol
$C_8H_{19}OPS$
S-Butyl *OO*-diethyl phosphorothiolate†
$C_8H_{19}OPSe$
Se-Butyl *OO*-diethyl phosphoroselenolate†
$C_8H_{19}O_3P$
Dibutyl phosphonate
Di-*tert*-butyl phosphonate
Dibutyl phosphite
Di-*tert*-butyl phosphite
Di-isobutyl phosphite
Di-isobutyl phosphonate

$C_8H_{19}O_3PS$
O,*O*-Dibutyl-*O*-hydrogenphosphorothionate
$C_8H_{19}O_3PS_2$
Diethyl-*S*-(2-[ethylthio]ethyl)phosphorothiolate
Diethyl 2-(ethylthio)ethyl phosphorothionate
$C_8H_{19}O_4P$
Ethyl dipropyl phosphate
$C_8H_{19}PS$
S-*tert*-Butyl *OO*-diethyl phosphorothiolothionate†
$C_8H_{20}As_2$
Arsenic diethyl
$C_8H_{20}As_2O$
Ethyl cacodyl oxide
$C_8H_{20}BrN$
Tetraethylammonium bromide
$C_8H_{20}ClN$
Tetraethylammonium chloride
$C_8H_{20}IN$
Tetraethylammonium iodide
$C_8H_{20}INO$
Hydroxymethyltrimethylammonium Hydroxide, *Butyl ether*, *Iodide*
$C_8H_{20}N_2$
1,8-Diamino-octane
Putrescine, N,N′-*Tetra-Me*
$C_8H_{20}N_2S_2$
Dithiodiethylamine
$C_8H_{20}O_2P_6S$
O,*O*-Diethylphosphorothiolic anhydrosulphide
$C_8H_{20}O_5P_2S_2$
Tetraethyldithiopyrophosphate†
$C_8H_{20}O_6P_2S$
O,*O*-Diethylphosphoric *O*,*O*-diethylphosphorothionic anhydride
$C_8H_{20}O_7P_2$
Tetraethyl pyrophosphate†
$C_8H_{20}Pb$
Lead tetra-ethyl
$C_8H_{20}Si$
Tetraethylsilane
$C_8H_{20}Sn$
Tin tetraethyl
$C_8H_{21}NO$
Tetraethylammonium hydroxide
$C_8H_{21}NOSi_2$
N,*O*-Bis(trimethylsilyl)acetamide†
$C_8H_{24}N_4O_3P_2$
Octamethylpyrophosphoramide†
$C_8H_{24}O_4Si_4$
2,2,4,4,6,6,8,8-Octamethylcyclotetrasiloxane
$C_8H_{26}N_2Si_4$
1,4-Diaza-2,2,3,3,5,5,6,6-octamethyl-2,3,5,6-tetrasilacyclohexane†
C_8N_4S
Tetracyanothiophene

$C_8N_4S_3$
Tricyano-1,4-dithiino[*c*]isothiazole
$C_8O_2Cl_6$
Tetrachlorophthalic Acid, sym.-*Dichloride*
Tetrachlorophthalic Acid, unsym.-*Dichloride*
$C_8O_3Cl_4$
Tetrachlorophthalic Acid, *Anhydride*
$C_8O_3I_4$
Tetraiodophthalic Acid, *Anhydride*

C_9

C_9Br_6O
Hexabromo-1-indanone
C_9Cl_6O
Hexachloro-1-indanone
$C_9F_{12}O$
Tetrakis(trifluoromethyl)cyclopentadienone†
$C_9H_4BrN_3O_4$
3-Bromo-5,7-dinitroquinoline
3-Bromo-6,8-dinitroquinoline
$C_9H_4Br_2N_2O_2$
3,6-Dibromo-8-nitroquinoline
3,8-Dibromo-5-nitroquinoline
5,6-Dibromo-8-nitroquinoline
5,7-Dibromo-8-nitroquinoline
5,8-Dibromo-6-nitroquinoline
6,8-Dibromo-5-nitroquinoline
$C_9H_4Br_4O_4$
Tetrabromophthalic Acid, *Mono-Me ester*
$C_9H_4ClN_3O_4$
5-Chloro-6,8-dinitroquinoline
8-Chloro-5,7-dinitroquinoline
$C_9H_4Cl_2N_2O_2$
2,4-Dichloro-3-nitroquinoline
2,4-Dichloro-6-nitroquinoline
3,6-Dichloro-8-nitroquinoline
4,5-Dichloro-3-nitroquinoline
4,7-Dichloro-3-nitroquinoline
4,7-Dichloro-8-nitroquinoline
5,6-Dichloro-8-nitroquinoline
5,7-Dichloro-8-nitroquinoline
5,8-Dichloro-6-nitroquinoline
$C_9H_4Cl_2O_3$
Phthalonic Acid, *ψ-Dichloride*
$C_9H_4Cl_4O_4$
Tetrachlorophthalic Acid, *Mono-Me ester*
$C_9H_4Cl_8O$
Isobenzan†
$C_9H_4I_4O_4$
Tetraiodophthalic Acid, *Mono-Me ether*
$C_9H_4N_2O_2$
2-Diazaindane-1,3-dione†
C_9H_4O
Non-2-ene-4,6,8-triyn-1-al
$C_9H_4O_2$
Non-2-ene-4,6,8-triynoic Acid

$C_9H_4O_3$
Indane-1,2,3-trione†
$C_9H_4O_4$
Phthalonic Acid, *Anhydride*
$C_9H_4O_5$
Benzene-1,2,3-tricarboxylic Acid, *Anhydride*
Benzene-1,2,4-tricarboxylic Acid, *Anhydride*
4,5-Methylenedioxyphthalic Acid, *Anhydride*
$C_9H_4O_6$
Stipitatonic Acid★†
$C_9H_4O_7$
Puberulonic Acid
C_9H_5BrClN
3-Bromo-4-chloroquinoline
3-Bromo-6-chloroquinoline
6-Bromo-2-chloroquinoline
7-Bromo-4-chloroquinoline
$C_9H_5BrN_2O_2$
2-Bromo-5-nitroquinoline
2-Bromo-8-nitroquinoline
3-Bromo-4-nitroquinoline
3-Bromo-5-nitroquinoline
3-Bromo-6-nitroquinoline
3-Bromo-8-nitroquinoline
5-Bromo-6-nitroquinoline
5-Bromo-8-nitroquinoline
6-Bromo-5-nitroquinoline
6-Bromo-8-nitroquinoline
7-Bromo-8-nitroquinoline
8-Bromo-5-nitroquinoline
8-Bromo-6-nitroquinoline
$C_9H_5BrO_2$
2-Bromo-1,3-indanedione
$C_9H_5BrO_6$
6-Bromobenzene-1,2,4-tricarboxylic Acid
$C_9H_5Br_2N$
2,3-Dibromoquinoline
2,4-Dibromoquinoline
2,5-Dibromoquinoline
2,6-Dibromoquinoline
2,7-Dibromoquinoline
3,4-Dibromoquinoline
3,5-Dibromoquinoline
3,6-Dibromoquinoline
3,7-Dibromoquinoline
3,8-Dibromoquinoline
4,5-Dibromoquinoline
5,6-Dibromoquinoline
5,7-Dibromoquinoline
5,8-Dibromoquinoline
6,7-Dibromoquinoline
6,8-Dibromoquinoline
7,8-Dibromoquinoline
$C_9H_5Br_2NO$
3,6-Dibromo-4-hydroxyquinoline
5,7-Dibromo-8-hydroxyquinoline
6,8-Dibromo-2-hydroxyquinoline
6,8-Dibromo-5-hydroxyquinoline
$C_9H_5Br_5O$
Pentabromophenol, *Allyl ether*
C_9H_5ClINO
Clioquinol† (Vioform★)

$C_9H_5ClI_3NO_2$
3-Amino-2,4,6-tri-iodobenzoic Acid, N-*Ac*, *Chloride*

$C_9H_5ClN_2O_2$
2-Chloro-5-nitroquinoline
2-Chloro-6-nitroquinoline
2-Chloro-8-nitroquinoline
3-Chloro-5-nitroquinoline
3-Chloro-8-nitroquinoline
4-Chloro-3-nitroquinoline
4-Chloro-5-nitroquinoline
4-Chloro-6-nitroquinoline
4-Chloro-8-nitroquinoline
5-Chloro-6-nitroquinoline
5-Chloro-8-nitroquinoline
6-Chloro-5-nitroquinoline
6-Chloro-7-nitroquinoline
6-Chloro-8-nitroquinoline
7-Chloro-6-nitroquinoline
7-Chloro-8-nitroquinoline
8-Chloro-5-nitroquinoline

C_9H_5ClO
3-Phenyl-2-propynoic Acid, *Chloride*

C_9H_5ClOS
Benzo[*b*]thiophene-2-carboxylic Acid, *Chloride*
Benzo[*b*]thiophene-3-carboxylic Acid, *Chloride*

$C_9H_5ClO_2$
Benzofuran-2-carboxylic Acid, *Chloride*
3-Chlorocoumarin
4-Chlorocoumarin
6-Chlorocoumarin
7-Chlorocoumarin

$C_9H_5ClO_4$
Phthalonic Acid, ψ-*Chloride*

$C_9H_5ClO_6$
2-Chlorobenzene-1,3,5-tricarboxylic Acid

$C_9H_5Cl_2N$
1,3-Dichloroisoquinoline
1,4-Dichloroisoquinoline
2,3-Dichloroquinoline
2,4-Dichloroquinoline
2,6-Dichloroquinoline
2,7-Dichloroquinoline
3,4-Dichloroquinoline
4,5-Dichloroquinoline
4,6-Dichloroquinoline
4,7-Dichloroquinoline
4,8-Dichloroquinoline
5,6-Dichloroquinoline
5,7-Dichloroquinoline
5,8-Dichloroquinoline
6,8-Dichloroquinoline
7,8-Dichloroquinoline

$C_9H_5Cl_2NO$
3,7-Dichloro-4-hydroxyquinoline
4,7-Dichloro-8-hydroxyquinoline
5,7-Dichloro-6-hydroxyquinoline
5,7-Dichloro-8-hydroxyquinoline

$C_9H_5Cl_3O$
α,β-Dichlorocinnamic Acid, *Chloride*

$C_9H_5IO_2$
6-Iodocoumarin

$C_9H_5I_2N$
5,6-Di-iodoquinoline
5,7-Di-iodoquinoline
5,8-Di-iodoquinoline

$C_9H_5I_2NO$
8-Hydroxy-5,7-di-iodoquinoline

C_9H_5N
3-Phenyl-2-propynoic Acid, *Nitrile*

C_9H_5NO
Benzofuran-2-carboxylic Acid, *Nitrile*

$C_9H_5NO_2$
Quinoline-5,6-quinone
Quinoline-5,8-quinone

$C_9H_5NO_3$
3-Hydroxy-6-nitrophthalic Acid, *Dinitrile*, *Me ether*
Quinisatin

$C_9H_5NO_4$
Benzene-1,2,3-tricarboxylic Acid, *Imide*
Benzene-1,2,4-tricarboxylic Acid, 2-*Nitrile*
Isatin-1-carboxylic Acid
Isatin-4-carboxylic Acid
Isatin-5-carboxylic Acid
Isatin-6-carboxylic Acid
Isatin-7-carboxylic Acid
Isatogenic Acid
5,6-Methylenedioxyisatin
4,5-Methylenedioxyphthalic Acid, *Imide*
6-Nitrocoumarin
7-Nitrocoumarin
8-Nitrocoumarin
6-Nitro-1,2-indanedione
o-Nitrophenylpropiolic Acid
m-Nitrophenylpropiolic Acid
p-Nitrophenylpropiolic Acid

$C_9H_5N_3O_4$
5,6-Dinitroquinoline
5,7-Dinitroquinoline
6,8-Dinitroquinoline
7,8-Dinitroquinoline

$C_9H_5N_3O_5$
2-Hydroxy-6,8-dinitroquinoline
5-Hydroxy-6,8-dinitroquinoline
8-Hydroxy-5,7-dinitroquinoline

C_9H_6BrClO
α-Bromocinnamic Acid, *Chloride*
p-Bromocinnamic Acid, *Chloride*

C_9H_6BrN
1-Bromoisoquinoline
3-Bromoisoquinoline
4-Bromoisoquinoline
8-Bromoisoquinoline
2-Bromoquinoline
3-Bromoquinoline
4-Bromoquinoline
5-Bromoquinoline
6-Bromoquinoline
7-Bromoquinoline
8-Bromoquinoline

C_9H_6BrNO
2-Bromo-5-hydroxyquinoline
3-Bromo-2-hydroxyquinoline

C_9H_6BrNO (*continued*)
3-Bromo-4-hydroxyquinoline
3-Bromo-6-hydroxyquinoline
4-Bromo-2-hydroxyquinoline
4-Bromo-6-hydroxyquinoline
5-Bromo-2-hydroxyquinoline
5-Bromo-6-hydroxyquinoline
5-Bromo-8-hydroxyquinoline
6-Bromo-2-hydroxyquinoline
6-Bromo-5-hydroxyquinoline
7-Bromo-2-hydroxyquinoline
7-Bromo-4-hydroxyquinoline
7-Bromo-8-hydroxyquinoline
8-Bromo-5-hydroxyquinoline
8-Bromo-6-hydroxyquinoline
8-Bromo-7-hydroxyquinoline

$C_9H_6Br_2N_2$
5-Amino-6,8-dibromoquinoline
6-Amino-3,5-dibromoquinoline
6-Amino-5,8-dibromoquinoline
8-Amino-3,6-dibromoquinoline
8-Amino-5,6-dibromoquinoline
8-Amino-5,7-dibromoquinoline

$C_9H_6Br_2O$
α-Bromocinnamic Acid, *Bromide*

$C_9H_6Br_2O_2$
αβ-Dibromocinnamic Acid

$C_9H_6Br_4O_2$
2,3,4,6-Tetrabromobenzoic Acid, *Et ester*

$C_9H_6ClFO_2$
2-Chloro-6-fluorocinnamic Acid

C_9H_6ClIO
3-*o*-Iodophenylacrylic Acid, *Chloride*
3-*m*-Iodophenylacrylic Acid, *Chloride*

C_9H_6ClN
o-Chlorocinnamic Acid, *Nitrile*
1-Chloroisoquinoline
3-Chloroisoquinoline
8-Chloroisoquinoline
2-Chloroquinoline
3-Chloroquinoline
4-Chloroquinoline
5-Chloroquinoline
6-Chloroquinoline
7-Chloroquinoline
8-Chloroquinoline

C_9H_6ClNO
1-Chloro-3-hydroxyisoquinoline
1-Chloro-4-hydroxyisoquinoline
3-Chloro-1-hydroxyisoquinoline
2-Chloro-4-hydroxyquinoline
2-Chloro-6-hydroxyquinoline
2-Chloro-8-hydroxyquinoline
3-Chloro-2-hydroxyquinoline
3-Chloro-6-hydroxyquinoline
3-Chloro-7-hydroxyquinoline
3-Chloro-8-hydroxyquinoline
4-Chloro-2-hydroxyquinoline
4-Chloro-6-hydroxyquinoline
5-Chloro-2-hydroxyquinoline
5-Chloro-6-hydroxyquinoline
5-Chloro-8-hydroxyquinoline
6-Chloro-2-hydroxyquinoline
6-Chloro-3-hydroxyquinoline
6-Chloro-4-hydroxyquinoline
6-Chloro-5-hydroxyquinoline
7-Chloro-2-hydroxyquinoline
7-Chloro-8-hydroxyquinoline

C_9H_6ClNOS
6-Methylbenzothiazole-2-carboxylic Acid, *Chloride*
2-Methylbenzothiazole-6-carboxylic Acid, *Chloride*

$C_9H_6ClNO_2S$
Quinoline-6-sulphonic Acid, *Chloride*
Quinoline-7-sulphonic Acid, *Chloride*
Quinoline-8-sulphonic Acid, *Chloride*

$C_9H_6ClNO_3$
α-Chloro-2-nitrocinnamaldehyde
α-Chloro-3-nitrocinnamaldehyde
α-Chloro-4-nitrocinnamaldehyde
3-*o*-Nitrophenylacrylic Acid, *Chloride*
3-*p*-Nitrophenylacrylic Acid, *Chloride*

$C_9H_6ClNO_4$
α-Chloro-2-nitrocinnamic Acid
α-Chloro-3-nitrocinnamic Acid
α-Chloro-4-nitrocinnamic Acid
β-Chloro-3-nitrocinnamic Acid
4-Chloro-2-nitrocinnamic Acid
4-Chloro-3-nitrocinnamic Acid
5-Chloro-2-nitrocinnamic Acid

$C_9H_6ClNO_5$
2-Nitroisophthalic Acid, *Mono-Me ester, Chloride*
3-Nitrophthalic Acid, 1-*Me ester*, *Chloride*
3-Nitrophthalic Acid, 2-*Me ester*, *Chloride*

$C_9H_6Cl_2N_2$
3-Amino-4,5-dichloroquinoline
3-Amino-4,7-dichloroquinoline
5-Amino-6,8-dichloroquinoline
6-Amino-2,4-dichloroquinoline
6-Amino-5,8-dichloroquinoline
8-Amino-4,7-dichloroquinoline
8-Amino-5,6-dichloroquinoline
8-Amino-5,7-dichloroquinoline

$C_9H_6Cl_2O$
α-Chlorocinnamic Acid, *Chloride*
o-Chlorocinnamic Acid, *Chloride*

$C_9H_6Cl_2O_2$
2,4-Dichlorocinnamic Acid
3,4-Dichlorocinnamic Acid
α,β-Dichlorocinnamic Acid
Phenylmalonic Acid, *Dichloride*

$C_9H_6Cl_2O_3$
4-Hydroxy-5-methylisophthalic Acid, *Dichloride*

$C_9H_6Cl_4O_2$
2,3,4,5-Tetrachlorobenzoic Acid, *Et ester*

C_9H_6IN
2-Iodoquinoline
3-Iodoquinoline
4-Iodoquinoline
5-Iodoquinoline
6-Iodoquinoline
8-Iodoquinoline

C_9H_6INO
2-Hydroxy-3-iodoquinoline
2-Hydroxy-4-iodoquinoline
3-Hydroxy-6-iodoquinoline
4-Hydroxy-3-iodoquinoline
4-Hydroxy-7-iodoquinoline
5-Hydroxy-8-iodoquinoline
6-Hydroxy-4-iodoquinoline
6-Hydroxy-5-iodoquinoline
6-Hydroxy-8-iodoquinoline
8-Hydroxy-5-iodoquinoline

$C_9H_6INO_4S$
8-Hydroxy-7-iodoquinoline-5-sulphonic Acid

$C_9H_6I_2O_2$
α,β-Di-iodocinnamic Acid

$C_9H_6I_3NO_3$
3-Amino-2,4,6-tri-iodobenzoic Acid, N-*Ac*

$C_9H_6N_2$
o-Cyanophenylacetonitrile
m-Cyanophenylacetonitrile
p-Cyanophenylacetonitrile
Indole-3-carboxylic Acid, *Nitrile*
Indole-6-carboxylic Acid, *Nitrile*
4-Methylisophthalic Acid, *Dinitrile*
3-Methylphthalic Acid, *Dinitrile*
4-Methylphthalic Acid, *Dinitrile*
Phenylmalonic Acid, *Dinitrile*

$C_9H_6N_2O_2$
3-*o*-Nitrophenylacrylic Acid, *Nitrile*
3-Nitroquinoline
4-Nitroquinoline
5-Nitroquinoline
6-Nitroquinoline
7-Nitroquinoline
8-Nitroquinoline
Quinoxaline-2-carboxylic Acid†

$C_9H_6N_2O_3$
2-Hydroxy-3-nitroquinoline
2-Hydroxy-4-nitroquinoline
2-Hydroxy-5-nitroquinoline
2-Hydroxy-6-nitroquinoline
2-Hydroxy-7-nitroquinoline
2-Hydroxy-8-nitroquinoline
3-Hydroxy-4(?)-nitroquinoline
4-Hydroxy-3-nitroquinoline
4-Hydroxy-5-nitroquinoline
4-Hydroxy-6-nitroquinoline
4-Hydroxy-8-nitroquinoline
5-Hydroxy-8-nitroquinoline
6-Hydroxy-5-nitroquinoline
6-Hydroxy-7-nitroquinoline
6-Hydroxy-8-nitroquinoline
7-Hydroxy-8-nitroquinoline
8-Hydroxy-5-nitroquinoline
8-Hydroxy-6-nitroquinoline
Kynuric Acid, β-*Nitrile*
o-Nitrophenylpropiolic Acid, *Amide*

$C_9H_6N_2O_4$
2,4-Dihydroxy-3-nitroquinoline
4,6-Dihydroxy-3-nitroquinoline
5,6-Dihydroxy-8-nitroquinoline
1-Methyl-5-nitroisatin
5-Methyl-7-nitroisatin
1-Nitroindole-2-carboxylic Acid
3-Nitroindole-2-carboxylic Acid
7-Nitroindole-2-carboxylic Acid
6-Nitroindole-3-carboxylic Acid
2-Nitrotoluene-α,4-dicarboxylic Acid, 4-*Nitrile*

$C_9H_6N_2O_6$
2,4-Dinitrocinnamic Acid
2,6-Dinitrocinnamic Acid

$C_9H_6N_2O_8$
4,5-Dinitrophthalic Acid, *Mono-Me ester*

$C_9H_6N_2S$
Benzothiazole-2-acetic Acid, *Nitrile*
2-Methylbenzothiazole-6-carboxylic Acid, *Nitrile*

$C_9H_6N_4$
s-Triazolo[3,4-*a*]phthalazine

$C_9H_6N_4O_4$
4-Amino-6-nitroquinoline, N-*NO*

C_9H_6O
Non-2-ene-4,6,8-triyn-1-ol
3-Phenyl-2-propyn-1-al

C_9H_6OS
2*H*-Benzo[*b*]thiin-2-one
Benzo[*b*]thiin-4-one
2*H*-Chromen-2-thione
1-Thioisocoumarin†

$C_9H_6OS_2$
Di-2-thienyl Ketone
Di-3-thienyl Ketone
2-Thienyl 3-thienyl Ketone

$C_9H_6O_2$
Chromone
Coumarin
Cyclohepta[*c*]furan-6-one†
2,3-Epoxynona-4,6,8-triyn-1-ol†
1,2-Indanedione
1,3-Indanedione
Isocoumarin
Nona-3,4-diene-6,8-diynoic Acid
3-Phenyl-2-propynoic Acid

$C_9H_6O_2S$
Benzo[*b*]thiophene-2-carboxylic Acid
Benzo[*b*]thiophene-3-carboxylic Acid
5-Methylthianaphthenequinone

$C_9H_6O_3$
Benzene-1,3,5-tricarboxaldehyde†
Benzofuran-2-carboxylic Acid
2-Carboxyphenylacetic Acid, *Anhydride*
3-Hydroxycoumarin
4-Hydroxycoumarin
5-Hydroxycoumarin
6-Hydroxycoumarin
8-Hydroxycoumarin
2-Hydroxyphenylpropiolic Acid
3-Hydroxyphenylpropiolic Acid
4-Hydroxyphenylpropiolic Acid
3-Methylphthalic Acid, *Anhydride*
4-Methylphthalic Acid, *Anhydride*
Umbelliferone

$C_9H_6O_3S$
3-Hydroxybenzo[*b*]thiophene-2-carboxylic Acid

$C_9H_6O_4$
Aesculetin
Daphnetin
3,7-Dihydroxychromone
5,7-Dihydroxychromone
7,8-Dihydroxychromone
3,6-Dihydroxycoumarin
4,7-Dihydroxycoumarin
5,7-Dihydroxycoumarin
5-Hydroxy-3-methylphthalic Acid, *Anhydride*
3-Hydroxy-5-methylphthalic Acid, *Anhydride*
3-Hydroxyphthalic Acid, *Me ether*, *Anhydride*
4-Hydroxyphthalic Acid, *Me ether*, *Anhydride*
Ninhydrin
Phthalide-3-carboxylic Acid

$C_9H_6O_5$
2-Formylisophthalic Acid
Phthalonic Acid

$C_9H_6O_6$
Benzene-1,2,3-tricarboxylic Acid
Benzene-1,2,4-tricarboxylic Acid
Benzene-1,3,5-tricarboxylic Acid
4,5-Methylenedioxyphthalic Acid

$C_9H_6O_7$
4-Hydroxybenzene-1,2,3-tricarboxylic Acid
5-Hydroxybenzene-1,2,3-tricarboxylic Acid
3-Hydroxybenzene-1,2,4-tricarboxylic Acid
5-Hydroxybenzene-1,2,4-tricarboxylic Acid
6-Hydroxybenzene-1,2,4-tricarboxylic Acid
3-Hydroxybenzene-1,2,5-tricarboxylic Acid
2-Hydroxybenzene-1,3,5-tricarboxylic Acid

$C_9H_6S_2$
7*H*-Cyclopenta[1,2-*b*:3,4-*b*′]dithiophene†
7*H*-Cyclopenta[1,2-*b*:3,4-*c*′]dithiophene†
7*H*-Cyclopenta[1,2-*b*:4,3-*b*′]dithiophene†
7*H*-Cyclopenta[1,2-*c*:3,4-*c*′]dithiophene†
4*H*-Cyclopenta[2,2-*b*:3,4-*b*′]dithiophene†
7*H*-Cyclopenta[2,1-*b*:3,4-*c*′]dithiophene†

$C_9H_7BrNO_4S_2$ (ion)
"Tyrindoxyl" sulphate†

$C_9H_7BrN_2$
2-Amino-7-bromoquinoline
4-Amino-3-bromoquinoline
5-Amino-3-bromoquinoline
5-Amino-6-bromoquinoline
5-Amino-8-bromoquinoline
6-Amino-3-bromoquinoline
6-Amino-5-bromoquinoline
6-Amino-7-bromoquinoline
6-Amino-8-bromoquinoline
8-Amino-3-bromoquinoline
8-Amino-4-bromoquinoline
8-Amino-5-bromoquinoline
8-Amino-6-bromoquinoline
8-Amino-7-bromoquinoline

$C_9H_7BrN_2O$
3-(*p*-Bromophenyl)-4-methyl-1,2,5-oxadiazole†

$C_9H_7BrN_2O_2$
3-(*p*-Bromophenyl)-4-methyl-1,2,5-oxadiazole, 2-*N-Oxide*†

$C_9H_7BrN_2O_6$
2-Bromo-3,5-dinitrobenzoic Acid, *Et ester*
4-Bromo-3,5-dinitrobenzoic Acid, *Et ester*

C_9H_7BrO
α-Bromocinnamaldehyde
β-Bromocinnamaldehyde
o-Bromocinnamaldehyde
p-Bromocinnamaldehyde

$C_9H_7BrO_2$
α-Bromocinnamic Acid
β-Bromocinnamic Acid
o-Bromocinnamic Acid
m-Bromocinnamic Acid
p-Bromocinnamic Acid

$C_9H_7BrO_4$
Bromoterephthalic Acid, 1-*Me ester*
Bromoterephthalic Acid, 4-*Me ester*

$C_9H_7Br_3O$
2,4,6-Tribromophenol, *Allyl ether*

$C_9H_7Br_3O_2$
2,4,6-Tribromobenzoic Acid, *Et ester*
3,4,5-Tribromobenzoic Acid, *Et ester*
3,4,5-Tribromophenylacetic Acid, *Me ester*

$C_9H_7Br_5O$
Pentabromophenol, *Propyl ether*
Pentabromophenol, *Isopropyl ester*

$C_9H_7ClN_2$
2-Amino-6-chloroquinoline
3-Amino-2-chloroquinoline
3-Amino-4-chloroquinoline
4-Amino-2-chloroquinoline
4-Amino-7-chloroquinoline
5-Amino-6-chloroquinoline
5-Amino-8-chloroquinoline
6-Amino-5-chloroquinoline
6-Amino-8-chloroquinoline
8-Amino-4-chloroquinoline
8-Amino-5-chloroquinoline
8-Amino-6-chloroquinoline
8-Amino-7-chloroquinoline

$C_9H_7ClN_2O_3$
3-*o*-Nitrophenylacrylic Acid, *Chloroamide*
3-*m*-Nitrophenylacrylic Acid, *Chloroamide*
3-*p*-Nitrophenylacrylic Acid, *Chloroamide*

$C_9H_7ClN_2O_6$
2-Chloro-3,5-dinitrobenzoic Acid, *Et ester*
4-Chloro-3,5-dinitrobenzoic Acid, *Et ester*
5-Chloro-2,4-dinitrobenzoic Acid, *Et ester*

C_9H_7ClO
Atropic Acid, *Chloride*
α-Chlorocinnamaldehyde
β-Chlorocinnamaldehyde
o-Chlorocinnamaldehyde
p-Chlorocinnamaldehyde
Cinnamic Acid, *Chloride*

$C_9H_7ClO_2$
Acetophenone-*p*-carboxylic Acid, *Chloride*
α-Chlorocinnamic Acid
β-Chlorocinnamic Acid
o-Chlorocinnamic Acid
m-Chlorocinnamic Acid
p-Chlorocinnamic Acid

$C_9H_7ClO_4$
Myristicic Acid, *Chloride*

$C_9H_7Cl_3O$
2,3-Dichloro-3-phenylpropionic Acid, *Chloride*
3-(2,3-Dichlorophenyl)propionic Acid, *Chloride*
3-(2,4-Dichlorophenyl)propionic Acid, *Chloride*
3-(2,5-Dichlorophenyl)propionic Acid, *Chloride*

$C_9H_7Cl_3O_2$
2,4,5-Trichlorobenzoic Acid, *Et ester*
3,4,5-Trichlorobenzoic Acid, *Et ester*
α,α,α-Trichloro-*o*-toluic Acid, *Me ester*
α,α,α-Trichloro-*m*-toluic Acid, *Me ester*
α,α,α-Trichloro-*p*-toluic Acid, *Me ester*

$C_9H_7Cl_5O$
Pentachlorophenol, *Propyl ether*

$C_9H_7FO_2$
α-Fluorocinnamic Acid
2-Fluorocinnamic Acid
3-Fluorocinnamic Acid

$C_9H_7F_4NO_3$
Tetrafluorotyrosine†

$C_9H_7IN_2$
4-Amino-3-iodoquinoline
5-Amino-6-iodoquinoline
5-Amino-8-iodoquinoline
8-Amino-5-iodoquinoline

$C_9H_7IN_2O$
o-Iodobenzamidoacetic Acid, *Nitrile*
p-Iodobenzamidoacetic Acid, *Nitrile*

$C_9H_7IO_2$
2-Iodo-3-phenylacrylic Acid
3-Iodo-3-phenylacrylic Acid
3-*o*-Iodophenylacrylic Acid
3-*m*-Iodophenylacrylic Acid
3-*p*-Iodophenylacrylic Acid

$C_9H_7IO_4$
Iodoterephthalic Acid, 4-*Me ester*

$C_9H_7I_2NO_3$
N-2,4-Di-iodobenzoylglycine
N-2,5-Di-iodobenzoylglycine
N-3,4-Di-iodobenzoylglycine
N-3,5-Di-iodobenzoylglycine

$C_9H_7I_3O$
2,4,6-Tri-iodophenol, *Allyl ether*

C_9H_7N
5-Aza-azulene
Cinnamic Acid, *Nitrile*
Cyclohepta[*b*]pyrrole
Isoquinoline
Quinoline

C_9H_7NO
Acetophenone-*m*-carboxylic Acid, *Nitrile*
Acetophenone-*p*-carboxylic Acid, *Nitrile*
Carbostyril
m-Coumaric Acid, *Nitrile*
Formylphenylacetic Acid, *Nitrile*
1-Hydroxyisoquinoline
4-Hydroxyisoquinoline
5-Hydroxyisoquinoline
7-Hydroxyisoquinoline
8-Hydroxyisoquinoline
3-Hydroxyquinoline
4-Hydroxyquinoline
5-Hydroxyquinoline
6-Hydroxyquinoline
7-Hydroxyquinoline
8-Hydroxyquinoline
Indole, N-*Formyl*
Indole-2-aldehyde
Indole-3-aldehyde
4-Phenyloxazole
5-Phenyloxazole
3-Phenyl-2-propynoic Acid, *Amide*
γ-Quinolone
p-Toluylformic Acid, *Nitrile*

C_9H_7NOS
Benzo[*b*]thiophene-2-carboxylic Acid, *Amide*
Benzo[*b*]thiophene-3-carboxylic Acid, *Amide*
2-Phenyl-5-thiazolone
4-Phenyl-2-thiazolone

$C_9H_7NO_2$
Acetylanthranil
3-Aminocoumarin
6-Aminocoumarin
7-Aminocoumarin
8-Aminocoumarin
o-Aminophenylpropiolic Acid
m-Aminophenylpropiolic Acid
Benzofuran-2-carboxylic Acid, *Amide*
2-Carboxyphenylacetic Acid, α-*Nitrile*
2-Carboxyphenylacetic Acid, *Imide*
4-Carboxyphenylacetic Acid, 4-*Nitrile*
o-Cyanobenzoic Acid, *Me ester*
m-Cyanobenzoic Acid, *Me ester*
p-Cyanobenzoic Acid, *Me ester*
1,4-Dihydroxyisoquinoline
3,4-Dihydroxyisoquinoline
5,8-Dihydroxyisoquinoline
6,7-Dihydroxyisoquinoline
2,3-Dihydroxyquinoline
2,4-Dihydroxyquinoline
2,6-Dihydroxyquinoline
2,8-Dihydroxyquinoline
3,4-Dihydroxyquinoline
4,6-Dihydroxyquinoline
5,8-Dihydroxyquinoline
3-Hydroxybenzoylformic Acid, *Me ether-nitrile*
4-Hydroxybenzoylformic Acid, *Me ether-nitrile*
Indole-1-carboxylic Acid
Indole-2-carboxylic Acid
Indole-3-carboxylic Acid
Indole-4-carboxylic Acid
Indole-5-carboxylic Acid
Indole-6-carboxylic Acid
Indole-7-carboxylic Acid
Isatin, 2-*Me ether*
Malonanil
3,4-Methylenedioxyphenylacetic Acid, *Nitrile*
N-Methylisatin
4-Methylisatin
5-Methylisatin
6-Methylisatin
7-Methylisatin
3-Methylphthalic Acid, *Imide*
4-Methylphthalic Acid, *Imide*
N-Methylphthalimide
2-Nitroindene

$C_9H_7NO_2$ *(continued)*
5-Nitroindene
Phenylcyanoacetic Acid
2-Phenyl-5(4)-oxazolone

$C_9H_7NO_2S$
Benzothiazole-2-acetic Acid
Benzothiazole-2-carboxylic Acid, *Me ester*
Benzothiazole-6-carboxylic Acid, *Me ester*
S-o-Carboxyphenylthioglycollic Acid, 2-*Nitrile*
4-Methylbenzothiazole-2-carboxylic Acid
6-Methylbenzothiazole-2-carboxylic Acid
2-Methylbenzothiazole-6-carboxylic Acid

$C_9H_7NO_3$
Anthroxanic Acid, *Me ester*
2,6-Dimethylpyridine-3,4-dicarboxylic Acid, *Anhydride*
1-Hydroxyindole-2-carboxylic Acid
3-Hydroxyindole-2-carboxylic Acid
4-Hydroxyindole-2-carboxylic Acid
5-Hydroxyindole-2-carboxylic Acid
6-Hydroxyindole-2-carboxylic Acid
7-Hydroxyindole-2-carboxylic Acid
5-Hydroxyisatin, *Me ether*
Isatoic Acid, N-*Me*
3,4-Methylenedioxymandelic Acid, *Nitrile*
N-Methylisatoic Anhydride
2-Methyl-5-nitrobenzofuran
2-Nitro-1-indanone
4-Nitro-1-indanone
6-Nitro-1-indanone
5-Nitro-2-indanone
3-*o*-Nitrophenylacrolein
3-*m*-Nitrophenylacrolein
3-*p*-Nitrophenylacrolein
3-*m*-Nitrosophenylacrylic Acid
3-*p*-Nitrosophenylacrylic Acid
Oxindole-4-carboxylic Acid
Oxindole-6-carboxylic Acid
Phthalide-3-carboxylic Acid, *Amide*
Phthaloxime, O-*Me*

$C_9H_7NO_3S$
Quinoline-2-sulphonic Acid
Quinoline-3-sulphonic Acid
Quinoline-4-sulphonic Acid
Quinoline-5-sulphonic Acid
Quinoline-6-sulphonic Acid
Quinoline-7-sulphonic Acid
Quinoline-8-sulphonic Acid

$C_9H_7NO_4$
5,6-Dihydroxyindole-2-carboxylic Acid
6,7-Dihydroxyisatin, 7-*Me ether*
3-(2-Hydroxy-3-nitrophenyl)acrolein
3-(2-Hydroxy-5-nitrophenyl)acrolein
3-*o*-Nitrophenylacrylic Acid
3-*m*-Nitrophenylacrylic Acid
3-*p*-Nitrophenylacrylic Acid
Phthalonic Acid, α-*Amide*

$C_9H_7NO_5$
2-Acetyl-3-nitrobenzoic Acid
2-Acetyl-4-nitrobenzoic Acid
2-Acetyl-5-nitrobenzoic Acid
2-Acetyl-6-nitrobenzoic Acid
4-Acetyl-2-nitrobenzoic Acid
Benzene-1,2,4-tricarboxylic Acid, 1-*Amide*
Benzene-1,2,4-tricarboxylic Acid, 2-*Amide*
Benzene-1,2,4-tricarboxylic Acid, 4-*Amide*
α-Hydroxy-2-nitrocinnamic Acid
α-Hydroxy-4-nitrocinnamic Acid
3-(2-Hydroxy-3-nitrophenyl)acrylic Acid
3-(2-Hydroxy-4-nitrophenyl)acrylic Acid
3-(2-Hydroxy-5-nitrophenyl)acrylic Acid
3-(3-Hydroxy-2-nitrophenyl)acrylic Acid
3-(3-Hydroxy-4-nitrophenyl)acrylic Acid
3-(3-Hydroxy-5-nitrophenyl)acrylic Acid
3-(3-Hydroxy-6-nitrophenyl)acrylic Acid
3-(4-Hydroxy-3-nitrophenyl)acrylic Acid
Kynuric Acid
2-(Nitroacetyl)benzoic Acid
o-Nitrobenzoylacetic Acid
m-Nitrobenzoylacetic Acid
p-Nitrobenzoylacetic Acid
3-*o*-Nitrophenyloxiran-2-carboxylic Acid
3-*m*-Nitrophenyloxiran-2-carboxylic Acid
3-*p*-Nitrophenyloxiran-2-carboxylic Acid
3-*o*-Nitrophenylpyruvic Acid
3-*m*-Nitrophenylpyruvic Acid
3-*p*-Nitrophenylpyruvic Acid

$C_9H_7NO_6$
α,3-Dihydroxy-2-nitrocinnamic Acid
2,3-Dihydroxy-5-nitrocinnamic Acid
2,3-Dihydroxy-6-nitrocinnamic Acid
3,4-Dihydroxy-2-nitrocinnamic Acid
3,4-Dihydroxy-5-nitrocinnamic Acid
4,5-Dihydroxy-2-nitrocinnamic Acid
4,5-Methylenedioxy-2-nitrobenzoic Acid, *Me ester*
6-Methylpyridine-2,3,4-tricarboxylic Acid
6-Methylpyridine-2,3,5-tricarboxylic Acid
3-Methylpyridine-2,4,5-tricarboxylic Acid
6-Methylpyridine-2,4,5-tricarboxylic Acid
2-Nitroisophthalic Acid, *Mono-Me ester*
2-Nitromyristicinaldehyde
2-Nitroterephthalic Acid, 1-*Me ester*
2-Nitroterephthalic Acid, 4-*Me ester*
3-Nitrophthalic Acid, 1-*Me ester*
3-Nitrophthalic Acid, 2-*Me ester*
4-Nitrophthalic Acid, 1-*Me ester*
4-Nitrophthalic Acid, 2-*Me ester*
4-Nitrotoluene-α,2-dicarboxylic Acid
5-Nitrotoluene-α,2-dicarboxylic Acid
2-Nitrotoluene-α,4-dicarboxylic Acid

$C_9H_7NO_7$
3-Hydroxy-6-nitrophthalic Acid, *Me ether*
4-Hydroxy-3-nitrophthalic Acid, *Me ether*
4-Hydroxy-5-nitrophthalic Acid, *Me ether*
4-Hydroxy-5-nitrophthalic Acid, *Me ester*
2-Nitromyristicic Acid
o-Nitrophenoxymalonic Acid
m-Nitrophenoxymalonic Acid
p-Nitrophenoxymalonic Acid

$C_9H_7NO_8$
3,4-Dihydroxy-6-nitrophthalic Acid, 4-*Me ether*

C_9H_7NS
2-Phenylthiazole
4-Phenylthiazole
5-Phenylthiazole
2-Quinolinethiol
6-Quinolinethiol
8-Quinolinethiol
2-(2′-Thienyl)pyridine†

C_9H_7NSe
Quinoline-8-selenol

$C_9H_7N_3$
6-Methylindazole-3-carboxylic Acid, *Nitrile*

$C_9H_7N_3O$
Quinoxaline-2-carboxylic Acid, *Amide*†

$C_9H_7N_3O_2$
2-Amino-5-nitroquinoline
2-Amino-6-nitroquinoline
2-Amino-8-nitroquinoline
3-Amino-5-nitroquinoline
3-Amino-6-nitroquinoline
4-Amino-3-nitroquinoline
4-Amino-6-nitroquinoline
4-Amino-8-nitroquinoline
4-Amino-(?)-nitroquinoline
5-Amino-8-nitroquinoline
6-Amino-5-nitroquinoline
7-Amino-8-nitroquinoline
8-Amino-5-nitroquinoline
8-Amino-6-nitroquinoline
2-Nitroaminoquinoline
4-Nitroaminoquinoline

$C_9H_7N_3O_3$
m-Nitrobenzamidoacetic Acid, *Nitrile*
p-Nitrobenzamidoacetic Acid, *Nitrile*

$C_9H_7N_3O_4$
6-Nitroindazole-1-carboxylic Acid, *Me ester*

$C_9H_7N_3O_8$
2,3,4-Trinitrobenzoic Acid, *Et ester*
2,4,5-Trinitrobenzoic Acid, *Et ester*
2,4,6-Trinitrobenzoic Acid, *Et ester*
2,3,6-Trinitro-*p*-toluic Acid, *Me ester*

$C_9H_7N_5O_3$
2-Hydroxy-3,5-dinitrobenzoic Acid, *Et ether*, *Nitrile*

$C_9H_7N_7O_2S$
Azathioprine†

$C_9H_7S_2$ (ion)
3-Phenyl-1,2-dithiolium Cation†
4-Phenyl-1,2-dithiolium Cation†

C_9H_8
Indene
1-Phenylpropyne
p-Tolylacetylene

C_9H_8BrClO
2-Bromo-4,6-dimethylbenzoic Acid, *Chloride*
2-Bromo-3-phenylpropionic Acid, *Chloride*
3-*o*-Bromophenylpropionic Acid, *Chloride*

C_9H_8BrN
2-Bromo-4,6-dimethylbenzoic Acid, *Nitrile*
4-Bromo-2,5-dimethylbenzoic Acid, *Nitrile*
5-Bromo-2,4-dimethylbenzoic Acid, *Nitrile*

C_9H_8BrNO
α-Bromocinnamic Acid, *Amide*
6-Bromo-3,4-dihydro-2-quinoline
7-Bromo-3,4-dihydro-2-quinoline
Bromomalondialdehyde, *Anilino deriv.*

$C_9H_8BrNO_2$
2-Bromo-4,5-dimethoxybenzoic Acid, *Nitrile*

$C_9H_8BrNO_3$
2-Amino-4-bromobenzoic Acid, N-*Ac*
2-Amino-5-bromobenzoic Acid, N-*Ac*
2-Amino-6-bromobenzoic Acid, N-*Ac*
3-Amino-5-bromobenzoic Acid, N-*Ac*
4-Amino-2-bromobenzoic Acid, N-*Ac*
4-Amino-3-bromobenzoic Acid, N-*Ac*

$C_9H_8BrNO_4$
2-Bromo-5-nitrobenzoic Acid, *Et ester*
3-Bromo-2-nitrobenzoic Acid, *Et ester*
3-Bromo-5-nitrobenzoic Acid, *Et ester*
4-Bromo-3-nitrobenzoic Acid, *Et ester*
5-Bromo-2-nitrobenzoic Acid, *Et ester*

$C_9H_8Br_2$
1,2-Dibromoindane
4,6-Dibromoindane
5,6-Dibromoindane

$C_9H_8Br_2O_2$
3-Bromo-3-*p*-bromophenylpropionic Acid
3,4-Dibromobenzoic Acid, *Et ester*
3,5-Dibromobenzoic Acid, *Et ester*
2,3-Dibromo-2-phenylpropionic Acid
2,3-Dibromo-3-phenylpropionic Acid
2,3-Dibromopropionic Acid, *Phenyl ester*
4,6-Dibromo-*m*-toluic Acid, *Me ester*
3,5-Dibromo-*p*-toluic Acid, *Me ester*

$C_9H_8Br_2O_3$
3,5-Dibromo-2-hydroxybenzoic Acid, *Et ester*
3,5-Dibromo-2-hydroxybenzoic Acid, *Me ether*, *Me ester*
3,5-Dibromo-2-hydroxybenzoic Acid, *Et ether*
3,5-Dibromo-4-hydroxybenzoic Acid, *Et ester*
4,6-Dibromo-5-hydroxy-*o*-toluic Acid, *Me ester*
3,5-Dibromo-2-hydroxy-*p*-toluic Acid, *Me ether*
3,5-Dibromo-4-methoxybenzoic Acid, *Me ester*

$C_9H_8Br_2O_4$
2,3-Dibromo-4,5-dimethoxybenzoic Acid
2,5-Dibromo-3,4-dimethoxybenzoic Acid
2,6-Dibromo-3,4-dimethoxybenzoic Acid

$C_9H_8Br_2O_5$
2,6-Dibromo-3,4,5-trihydroxybenzoic Acid, *Et ester*

$C_9H_8Br_3NO$
2,4,6-Tribromobenzoic Acid, *Dimethylamide*

$C_9H_8Br_3NO_2$
3-Amino-2,4,6-tribromobenzoic Acid, *Et ester*

$C_9H_8Br_4O$
2,4,5,6-Tetrabromo-*m*-cresol, *Et ether*

C_9H_8ClN
2-Chloro-4,6-dimethylbenzoic Acid, *Nitrile*
3-*o*-Chlorophenylpropionic Acid, *Nitrile*

C_9H_8ClNO
α-Chlorocinnamic Acid, *Amide*
β-Chlorocinnamic Acid, *Amide*
o-Chlorocinnamic Acid, *Amide*

$C_9H_8ClNO_2$
Hippuric Acid, *Chloride*
N-Methyloxanilic Acid, *Chloride*

$C_9H_8ClNO_3$
o-Chlorohippuric Acid
m-Chlorohippuric Acid

$C_9H_8ClNO_3$ (*continued*)
- *p*-Chlorohippuric Acid
- 3-*o*-Nitrophenylpropionic Acid, *Chloride*
- Pyridine-2,3-dicarboxylic Acid, 1-*Et ester*, 2-*chloride*

$C_9H_8ClNO_4$
- 2-Chloro-3-nitrobenzoic Acid, *Et ester*
- 2-Chloro-5-nitrobenzoic Acid, *Et ester*
- 2-Chloro-6-nitrobenzoic Acid, *Et ester*
- 4-Chloro-3-nitrobenzoic Acid, *Et ester*
- 5-Chloro-3-nitrobenzoic Acid, *Et ester*
- 4-Hydroxy-3-nitrobenzoic Acid, *Et ether*, *Chloride*

$C_9H_8ClNO_5$
- 3-Chloro-6-hydroxy-5-nitrobenzoic Acid, *Et ester*
- 3,4-Dimethoxy-2-nitrobenzoic Acid, *Chloride*

$C_9H_8ClNO_6S$
- 4-Nitro-2-sulphobenzoic Acid, 1-*Et ester*, 2-*chloride*
- 4-Nitro-6-sulpho-*m*-toluic Acid, 6-*Chloride*, *Me ester*

$C_9H_8Cl_2$
- 3,3-Dichloro-1-phenylpropene

$C_9H_8Cl_2O$
- 2-Chloro-2-phenylpropionic Acid, *Chloride*
- 3-*o*-Chloro-phenylpropionic Acid, *Chloride*

$C_9H_8Cl_2O_2$
- Dichloroacetic Acid, *Benzyl ester*
- 2,5-Dichlorobenzoic Acid, *Et ester*
- 3,4-Dichlorobenzoic Acid, *Et ester*
- 2,3-Dichloro-3-phenylpropionic Acid
- 3-(2,3-Dichlorophenyl)propionic Acid
- 3-(2,4-Dichlorophenyl)propionic Acid
- 3-(2,5-Dichlorophenyl)propionic Acid
- 3,3-Dichloro-2-phenylpropionic Acid
- 2,3-Dichloropropionic Acid, *Phenyl ester*

$C_9H_8Cl_2O_3$
- 2,6-Dichloro-3-hydroxybenzoic Acid, *Me ether*, *Me ester*
- 3,5-Dichloro-2-hydroxybenzoic Acid, *Et ester*
- 3,5-Dichloro-4-hydroxybenzoic Acid, *Et ester*
- 2,4-Dichlorophenoxyacetic Acid, *Me ester*

$C_9H_8Cl_2O_4$
- 2,4-Dichloro-5-hydroxyphenoxyacetic Acid, *Me ether*†
- 2,5-Dichloro-4-hydroxyphenoxyacetic Acid, *Me ether*†

$C_9H_8Cl_2O_5$
- 2,6-Dichloro-3,4,5-trihydroxybenzoic Acid, *Et ester*

$C_9H_8Cl_3NO_2S$
- Captan

$C_9H_8Cl_4O_2$
- Tetrachloroquinol, *Me-Et ether*

$C_9H_8Cl_4O_4$
- Methane-tetracetic Acid, *Tetrachloride*

C_9H_8FN
- 5-Fluoro-2-methylindole

$C_9H_8FNO_3$
- *o*-Fluorobenzamidoacetic Acid
- *m*-Fluorobenzamidoacetic Acid
- *p*-Fluorobenzamidoacetic Acid

$C_9H_8FNO_4$
- 2-Fluoro-4-nitrobenzoic Acid, *Et ester*
- 2-Fluoro-5-nitrobenzoic Acid, *Et ester*
- 4-Fluoro-3-nitrobenzoic Acid, *Et ester*

C_9H_8IN
- 2-Iodo-4,6-dimethylbenzoic Acid, *Nitrile*

C_9H_8INO
- 3-*o*-Iodophenylacrylic Acid, *Amide*

$C_9H_8INO_3$
- 2-Acetyl-1,3-dihydro-1-hydroxy-3-oxobenziodazole†
- *o*-Iodobenzamidoacetic Acid
- *m*-Iodobenzamidoacetic Acid
- *p*-Iodobenzamidoacetic Acid

$C_9H_8INO_4$
- 2-Iodo-4-nitrobenzoic Acid, *Et ester*
- 2-Iodo-5-nitrobenzoic Acid, *Et ester*
- 3-Iodo-2-nitrobenzoic Acid, *Et ester*
- 3-Iodo-5-nitrobenzoic Acid, *Et ester*
- 4-Iodo-3-nitrobenzoic Acid, *Et ester*
- 5-Iodo-2-nitrobenzoic Acid, *Et ester*

$C_9H_8I_2O$
- 2,4-Di-iodophenol, *Allyl ether*
- 2,6-Di-iodophenol, *Allyl ether*

$C_9H_8I_2O_2$
- 2,5-Di-iodobenzoic Acid, *Et ester*

$C_9H_8I_2O_3$
- 2-Hydroxy-3,5-di-iodobenzoic Acid, *Et ester*
- 3-(4-Hydroxy-3,5-di-iodophenyl)propionic Acid

$C_9H_8N_2$
- *o*-Aminocinnamic Acid, *Nitrile*
- 1-Aminoisoquinoline
- 3-Aminoisoquinoline
- 4-Aminoisoquinoline
- 5-Aminoisoquinoline
- 8-Aminoisoquinoline
- 2-Aminoquinoline
- 3-Aminoquinoline
- 4-Aminoquinoline
- 5-Aminoquinoline
- 6-Aminoquinoline
- 7-Aminoquinoline
- 8-Aminoquinoline
- Bicyclo[2,2,1]hept-5-ene-2,3-dicarboxylic Acid, *Dinitrile*
- Diazocyclononatetraene†
- Indoline-2-carboxylic Acid, *Nitrile*†
- 2-Methylquinazoline
- 4-Methylquinazoline
- 2-Methylquinoxaline
- 6-Methylquinoxaline
- 1-Phenylimidazole
- 2-Phenylimidazole
- 4-Phenylimidazole
- 1-Phenylpyrazole
- 3(5)-Phenylpyrazole
- 4-Phenylpyrazole
- 1-2′-Pyridyl-pyrrole

$C_9H_8N_2$ (*continued*)
1-4′-Pyridyl-pyrrole
2-2′-Pyridyl-pyrrole
2-3′-Pyridyl-pyrrole
2-4′-Pyridyl-pyrrole
3-2′-Pyridyl-pyrrole
3-3′-Pyridyl-pyrrole
3-4′-Pyridyl-pyrrole

$C_9H_8N_2O$
N-Acetylanthranilic Acid, *Nitrile*
2-Acetylbenzimidazole†
3-Acetylindazole
2-Amino-4-hydroxyquinoline
2-Amino-8-hydroxyquinoline
3-Amino-2-hydroxyquinoline
3-Amino-4-hydroxyquinoline
4-Amino-2-hydroxyquinoline
4-Amino-6-hydroxyquinoline
5-Amino-2-hydroxyquinoline
5-Amino-4-hydroxyquinoline
5-Amino-6-hydroxyquinoline
5-Amino-8-hydroxyquinoline
6-Amino-2-hydroxyquinoline
6-Amino-4-hydroxyquinoline
6-Amino-8-hydroxyquinoline
7-Amino-2-hydroxyquinoline
7-Amino-8-hydroxyquinoline
8-Amino-4-hydroxyquinoline
8-Amino-5-hydroxyquinoline
8-Amino-6-hydroxyquinoline
8-Amino-7-hydroxyquinoline
4-Carboxyphenylacetic Acid, α-*Amide*-4-*nitrile*
Di-2-pyrryl Ketone
Glycorine
Hippuric Acid, *Nitrile*
4-Hydroxyquinazoline, *Me ether*
Indole-1-carboxylic Acid, *Amide*
Phenylcyanoacetic Acid, *Amide*
1-Phenylpyrazol-3-one
1-Phenylpyrazol-5-one
2-Phenylpyrazol-5-one
3-Phenylpyrazol-5-one
Phthalazone, N-*Me*

$C_9H_8N_2OS$
Benzothiazole-2-acetic Acid, *Amide*
4-Methylbenzothiazole-2-carboxylic Acid, *Amide*
6-Methylbenzothiazole-2-carboxylic Acid, *Amide*
2-Methylbenzothiazole-6-carboxylic Acid, *Amide*
2-Thiohydantoin, 3-N-*Phenyl*

$C_9H_8N_2O_2$
m-Carboxyphenylglycine, m-*Nitrile*†
2,4-Dimethyl-3-nitrobenzoic Acid, *Nitrile*
2,4-Dimethyl-5-nitrobenzoic Acid, *Nitrile*
2,4-Dimethyl-6-nitrobenzoic Acid, *Nitrile*
2,5-Dimethyl-4-nitrobenzoic Acid, *Nitrile*
2,6-Dimethylpyridine-3,4-dicarboxylic Acid, *Imide*
Glycosmicine
Indazole-1-carboxylic Acid, *Me ester*
Indazole-2-carboxylic Acid, *Me ester*
Indazole-3-carboxylic Acid, 1-N-*Me*
Indazole-3-carboxylic Acid, 2-N-*Me*
1-Indazolylacetic Acid
2-Indazolylacetic Acid
7-Methylindazole-1-carboxylic Acid
5-Methylindazole-2-carboxylic Acid
7-Methylindazole-2-carboxylic Acid
1-Methylindazole-3-carboxylic Acid
2-Methylindazole-3-carboxylic Acid
5-Methylindazole-3-carboxylic Acid
6-Methylindazole-3-carboxylic Acid
2-Methyl-3-nitroindole
3-Methyl-4-nitrophenylacetic Acid, *Nitrile*
2-*p*-Nitrophenylpropionic Acid, *Nitrile*
N-Phenylglycine-*o*-carboxylic Acid, α-*Nitrile*
N-Phenylglycine-*o*-carboxylic Acid, β-*Nitrile*
1-Phenylhydantoin
3-Phenylhydantoin
5-Phenylhydantoin

$C_9H_8N_2O_2S$
2-Methylaminobenzothiazole-6-carboxylic Acid
Quinoline-6-sulphonic Acid, *Amide*
Quinoline-7-sulphonic Acid, *Amide*
Quinoline-8-sulphonic Acid, *Amide*

$C_9H_8N_2O_3$
2-Hydroxy-5-nitrobenzoic Acid, *Et ether*, *Nitrile*
2-Hydroxy-6-nitrobenzoic Acid, *Et ether*, *Nitrile*
2-Hydroxy-3-*o*-nitrophenylpropionic Acid, *Nitrile*
3-*o*-Nitrophenylacrylic Acid, *Amide*
3-*m*-Nitrophenylacrylic Acid, *Amide*
3-*p*-Nitrophenylacrylic Acid, *Amide*

$C_9H_8N_2O_4$
2-Acetyl-3-nitrobenzoic Acid, *Amide*
α-Amino-2-nitrocinnamic Acid
2-Amino-3-nitrocinnamic Acid
2-Amino-4-nitrocinnamic Acid
4-Amino-3-nitrocinnamic Acid
2,6-Dihydroxy-3-nitrobenzoic Acid, *Nitrile*, 6-*Et ether*
2,3-Dimethoxy-5-nitrobenzoic Acid, *Nitrile*
2,6-Dimethoxy-3-nitrobenzoic Acid, *Nitrile*
4,5-Dimethoxy-2-nitrobenzoic Acid, *Nitrile*
2-Hydroxy-6-nitrobenzoic Acid, *Et ether*, *Amide*
3-*o*-Nitrophenylpyruvic Acid, *Amide*

$C_9H_8N_2O_5$
o-Nitrobenzamidoacetic Acid
m-Nitrobenzamidoacetic Acid
p-Nitrobenzamidoacetic Acid
2-Nitroisophthalic Acid, *Monoamide*, *Me ester*
4-Nitro-oxanilic Acid, *Me ester*

$C_9H_8N_2O_6$
2,3-Dinitrobenzoic Acid, *Et ester*
2,4-Dinitrobenzoic Acid, *Et ester*
2,5-Dinitrobenzoic Acid, *Et ester*
2,6-Dinitrobenzoic Acid, *Et ester*
3,4-Dinitrobenzoic Acid, *Et ester*
3,5-Dinitrobenzoic Acid, *Et ester*
2,4-Dinitrophenylacetic Acid, *Me ester*
2,6-Dinitrophenylacetic Acid, *Me ester*
3,5-Dinitro-*o*-toluic Acid, *Me ester*
4,5-Dinitro-*o*-toluic Acid, *Me ester*
2,6-Dinitro-*m*-toluic Acid, *Me ester*
4,6-Dinitro-*m*-toluic Acid, *Me ester*
3,5-Dinitro-*p*-toluic Acid, *Me ester*

$C_9H_8N_2O_7$
2-Hydroxy-3,5-dinitrobenzoic Acid, *Me ether*, *Me ester*
2-Hydroxy-3,5-dinitrobenzoic Acid, *Et ether*

$C_9H_8N_2O_7$ *(continued)*
2-Hydroxy-3,5-dinitrobenzoic Acid, *Et ester*
2-Hydroxy-5,6-dinitrobenzoic Acid, *Et ether*
4-Hydroxy-3,5-dinitrobenzoic Adid, *Et ester*
4-Hydroxy-3,5-dinitrobenzoic Acid, *Et ether*
5-Hydroxy-2,4-dinitrobenzoic Acid, *Et ether*
4-Methoxy-2,3-dinitrobenzoic Acid, *Me ester*
4-Methoxy-3,5-dinitrobenzoic Acid, *Me ester*

$C_9H_8N_2O_8$
3,4-Dimethoxy-2,6-dinitrobenzoic Acid
4,5-Dimethoxy-2,3-dinitrobenzoic Acid

$C_9H_8N_4O_7$
2,4,6-Trinitrobenzoic Acid, *Dimethylamide*

$C_9H_8N_6O_4$
Lepidopterin

C_9H_8O
Acrylophenone
Cinnamaldehyde
2-Chromene†
3-Chromene†
1-Indanone
2-Indanone
4-Indenol†
7-Indenol†
Isochromene†
2-Methylbenzofuran
3-Methylbenzofuran
5-Methylbenzofuran
6-Methylbenzofuran
7-Methylbenzofuran
Nona-3,4-diene-6,8-diyn-1-ol
1-Phenyl-2-propyn-1-ol
3-Phenyl-2-propyn-1-ol
2-Propyn-1-ol, *Phenyl ether*
Tricyclo[3,3,1,$0^{2,8}$]nona-3,6-dien-9-one†

C_9H_8OS
2,3-Dihydrobenzo[b]thiin-3-one
3-Hydroxybenzo[*b*]thiophene, *Me ether*

$C_9H_8O_2$
Acrylic Acid, *Phenyl ester*
Atropic Acid
Benzoylacetaldehyde
Chromanone
Cinnamic Acid
Cyclo-octatetraenecarboxylic Acid
2,2′-Difurylmethane†
3,4-Dihydrocoumarin
Drosophillin E†
4-Hydroxybenzofuran, *Me ether*
5-Hydroxybenzofuran, *Me ether*
6-Hydroxybenzofuran, *Me ether*
o-Hydroxycinnamaldehyde
p-Hydroxycinnamaldehyde
2-Hydroxy-1-indanone
4-Hydroxy-1-indanone
5-Hydroxy-1-indanone
6-Hydroxy-1-indanone
7-Hydroxy-2-indanone
2-Hydroxymethyl-benzofuran
5-Hydroxymethyl-benzofuran
3-Hydroxy-2-methylbenzofuran
3-Hydroxy-5-methylbenzofuran
3-Hydroxy-6-methylbenzofuran
3-Hydroxy-7-methylbenzofuran
6-Hydroxy-3-methylbenzofuran
1-Isochromanone
3-Isochromanone
3,4-Methylenedioxystyrene
3-Methylphthalide
5-Methylphthalide
1-Phenylpropane-1,2-dione
p-Toluylformaldehyde

$C_9H_8O_2S$
2-Acetyl-3-hydroxy-5-prop-1-ynylthiophene
3-*o*-Mercaptophenylacrylic Acid
3-*p*-Mercaptophenylacrylic Acid

$C_9H_8O_3$
Acetophenone-*o*-carboxylic Acid
Acetophenone-*m*-carboxylic Acid
Acetophenone-*p*-carboxylic Acid
Acetylpiperone
O-Acetylsalicylaldehyde
Benzoylacetic Acid
Benzoylformic Acid, *Me ester*
Bicyclo[2,2,1]hept-5-ene-2,3-dicarboxylic Acid, *Anhydride*
o-Coumaric Acid
m-Coumaric Acid
p-Coumaric Acid
Coumarinic Acid
2,4-Diformylphenol, *Me ether*
3,4-Dihydroxybenzaldehyde, *Ethylene ether*
2-Formylbenzoic Acid, *Me ester*
3-Formylbenzoic Acid, *Me ester*
4-Formylbenzoic Acid, *Me ester*
Formylphenylacetic Acid
3-Hydroxy-3-methylphthalide†
3-Hydroxy-4-methylphthalide
3-Hydroxy-6-methylphthalide
4-Hydroxy-3-methylphthalide
5-Hydroxy-6-methylphthalide
7-Hydroxy-3-methylphthalide
7-Hydroxy-6-methylphthalide
9-Hydroxynon-7-ene-3,5-diynoic Acid†
4-Hydroxyphthalide, *Me ether*†
5-Hydroxyphthalide, *Me ether*†
6-Hydroxyphthalide, *Me ether*†
7-Hydroxyphthalide, *Me ether*†
3,4-Methylenedioxyphenylacetaldehyde
Nona-4,6,8-triyne-1,2,3-triol
5-Norbornene-2,3-dicarboxylic anhydride†
2-Phenyloxiran-1-carboxylic Acid
Phenylpyruvic Acid
m-Toluylformic Acid
p-Toluylformic Acid

$C_9H_8O_3S$
Thio-oxalic Acid, S-p-*Tolyl ester*

$C_9H_8O_4$
2-Acetyl-6-hydroxybenzoic Acid
O-Acetylsalicyclic Acid
Caffeic Acid
2-Carboxyphenylacetic Acid
3-Carboxyphenylacetic Acid
4-Carboxyphenylacetic Acid
Cycloheptatriene-1,6-dicarboxylic Acid†
Dehydrocarolic Acid
2,4-Diformyl-6-methoxyphenol

$C_9H_8O_4$ (*continued*)
2,5-Dihydroxycinnamic Acid
3,5-Dihydroxycinnamic Acid
2,4-Dihydroxy-6-methylisophthalaldehyde
4,6-Dihydroxy-2-methylisophthalaldehyde
3-Formyl-2-hydroxybenzoic Acid, *Me ester*
4-Formyl-3-hydroxybenzoic Acid, *Me ester*
5-Formyl-2-hydroxybenzoic Acid, *Me ester*
2-Formylphenoxyacetic Acid
3-Formylphenoxyacetic Acid
4-Formylphenoxyacetic Acid
3-Hydroxybenzoic Acid, O-*Ac*
4-Hydroxybenzoic Acid, O-*Ac*
4-Hydroxybenzoylformic Acid, *Me ether*
5-Hydroxy-7-methoxyphthalide†
7-Hydroxy-5-methoxyphthalide†
2-Hydroxyphenylpyruvic Acid
3-Hydroxyphenylpyruvic Acid
4-Hydroxyphenylpyruvic Acid
Isophthalic Acid, *Mono-Me ester*
3,4-Methylenedioxyphenylacetic Acid
3,4-Methylenedioxybenzoic Acid, *Me ester*
2-Methylisophthalic Acid
4-Methylisophthalic Acid
5-Methylisophthalic Acid
3-Methylphthalic Acid
4-Methylphthalic Acid
Methylterephthalic Acid
Myristicinaldehyde
Phenylmalonic Acid
Phthalic Acid, *Me ester*
Terephthalic Acid, *Mono-Me ester*
Tetracylo[2,2,1,$0^{2,6}$,$0^{3,5}$]heptane-2,3-dicarboxylic Acid
Umbellic Acid

$C_9H_8O_4S$
S-o-Carboxyphenylthioglycollic Acid
S-p-Carboxyphenylthioglycollic Acid

$C_9H_8O_5$
5-Acetyl-2,4-dihydroxybenzoic Acid
5-Carboxyvanillin†
3,4-Dihydroxybenzoylformic Acid, 3-*Me ether*
Flavipin
2-Formyl-5,6-dihydroxybenzoic Acid, 5-*Me ether*
3-Formyl-4,6-dihydroxy-*o*-toluic Acid
5-Formyl-4,6-dihydroxy-*o*-toluic Acid
3-Formyl-4-hydroxy-5-methoxybenzoic Acid
5-Formyl-2-hydroxy-3-methoxybenzoic Acid
Furan-2,5-dicarboxylic Acid, *Mono-allyl ester*
2-Hydroxyisophthalic Acid, *Mono-Me ester*
2-Hydroxyisophthalic Acid, *Me ether*
4-Hydroxyisophthalic Acid, *Me ether*
5-Hydroxyisophthalic Acid, *Me ether*
2-Hydroxy-4-methylisophthalic Acid
2-Hydroxy-5-methylisophthalic Acid
4-Hydroxy-5-methylisophthalic Acid
4-Hydroxy-6-methylisophthalic Acid
5-Hydroxy-4-methylisophthalic Acid
5-Hydroxy-3-methylphthalic Acid
5-Hydroxy-4-methylphthalic Acid
3-Hydroxy-5-methylphthalic Acid
2-Hydroxy-6-methylterephthalic Acid†
5-Hydroxy-2-methylterephthalic Acid
3-Hydroxyphthalic Acid, 1-*Me ester*
3-Hydroxyphthalic Acid, 2-*Me ester*
3-Hydroxyphthalic Acid, *Me ether*
4-Hydroxyphthalic Acid, 1-*Me ester*
4-Hydroxyphthalic Acid, 2-*Me ester*
4-Hydroxyphthalic Acid, *Me ether*
Hydroxyterephthalic Acid, 1-*Me ester*
Hydroxyterephthalic Acid, 4-*Me ester*
Hydroxyterephthalic Acid, *Me ether*
2-Hydroxy-6-methylterephthalic Acid†
3,4-Methylenedioxymandelic Acid
Myristicic Acid
3-(2,3,4-Trihydroxyphenyl)acrylic Acid
3-(2,4,5-Trihydroxyphenyl)acrylic Acid
3-(2,4,6-Trihydroxyphenyl)acrylic Acid
3-(3,4,5-Trihydroxyphenyl)acrylic Acid

$C_9H_8O_6$
Barbatol-carboxylic Acid
Chelidonic Acid, *Di-Me ester*
Chelidonic Acid, *Et ester*
2,6-Dimethyl-γ-pyrone-3,5-dicarboxylic Acid

$C_9H_8O_7$
4,6-Dihydroxy-3-methoxyphthalic Acid
3-Hydroxy-4-oxo-4*H*-pyran-2,6-dicarboxylic Acid, *Di-Me ester*
3-Hydroxy-4-oxo-4*H*-pyran-2,6-dicarboxylic Acid, *Mono-Et ester*
3-Hydroxy-4-oxo-4*H*-pyran-2,6-dicarboxylic Acid, *Et ether*

C_9H_8S
Isothiochromene
5-Methylbenzo[*b*]thiophene
6-Methylbenzo[*b*]thiophene

C_9H_9 (ion)
Cyclononatetraenide†

C_9H_9Br
1-Bromo-2-phenylpropene
1-Bromo-3-phenylpropene
2-Bromo-3-phenylpropene
3-Bromo-1-phenylpropene
3-Bromo-3-phenylpropene
2-*m*-Bromophenylpropene
1-*p*-Bromophenylpropene
2-*p*-Bromophenylpropene
3-*p*-Bromophenylpropene

C_9H_9BrO
α-Bromo-2-methylacetophenone
α-Bromo-4-methylacetophenone
α-Bromopropiophenone
β-Bromopropiophenone
2-Bromopropiophenone
3-Bromopropiophenone
4-Bromopropiophenone

$C_9H_9BrO_2$
o-Bromobenzoic Acid, *Et ester*
m-Bromobenzoic Acid, *Et ester*
p-Bromobenzoic Acid, *Et ester*
2-Bromo-3,5-dimethylbenzoic Acid
2-Bromo-4,5-dimethylbenzoic Acid
2-Bromo-4,6-dimethylbenzoic Acid
4-Bromo-2,5-dimethylbenzoic Acid
4-Bromo-2,6-dimethylbenzoic Acid
4-Bromo-3,5-dimethylbenzoic Acid
5-Bromo-2,4-dimethylbenzoic Acid
2-Bromo-4-hydroxybenzaldehyde, *Et ether*

$C_9H_9BrO_2$ (*continued*)
5-Bromo-2-hydroxybenzaldehyde, *Et ether*
α-Bromophenylacetic Acid, *Me ester*
2-Bromo-2-phenylpropionic Acid
2-Bromo-3-phenylpropionic Acid
3-Bromo-2-phenylpropionic Acid
3-Bromo-3-phenylpropionic Acid
3-*o*-Bromophenylpropionic Acid
3-*m*-Bromophenylpropionic Acid
3-*p*-Bromophenylpropionic Acid
2-Bromopropionic Acid, *Phenyl ester*
α-Bromo-*p*-toluic Acid, *Me ester*
3-Bromo-*o*-toluic Acid, *Me ester*
5-Bromo-*o*-toluic Acid, *Me ester*
6-Bromo-*o*-toluic Acid, *Me ester*

$C_9H_9BrO_3$
5-Bromo-2,3-dihydroxybenzaldehyde, *Di-Me Ether*
2-Bromo-3,4-dimethoxybenzaldehyde
2-Bromo-4,5-dimethoxybenzaldehyde
3-Bromo-4,5-dimethoxybenzaldehyde
3-Bromo-4-hydroxybenzoic Acid, *Et ester*
4-Bromo-3-hydroxybenzoic Acid, *Et ester*
5-Bromo-2-hydroxybenzoic Acid, *Me ether, Me ester*
5-Bromo-2-hydroxybenzoic Acid, *Et ester*
5-Bromo-2-hydroxybenzoic Acid, *Et ether*
3-Bromo-4-hydroxyphenylacetic Acid, *Me ester*
3-Bromo-4-hydroxyphenylacetic Acid, *Me ether*
5-Bromo-2-hydroxy-*m*-toluic Acid, *Me ester*
3-Bromo-2-hydroxy-*p*-toluic Acid, *Me ether*
5-Bromo-2-hydroxy-*p*-toluic Acid, *Me ester*
3-Bromo-4-methoxybenzoic Acid, *Me ester*

$C_9H_9BrO_4$
5-Bromo-2,4-dihydroxybenzoic Acid, *Et ester*
5-Bromo-2,4-dihydroxybenzoic Acid, 4-*Me ether*
5-Bromo-2,4-dihydroxybenzoic Acid, *Di-Me ether*
2-Bromo-3,4-dimethoxybenzoic Acid
2-Bromo-4,5-dimethoxybenzoic Acid
3-Bromo-4,5-dimethoxybenzoic Acid
3-Bromo-4-hydroxy-5-methoxybenzoic Acid, *Me ester*

$C_9H_9Br_2NO$
2,3-Dibromo-3-phenylpropionic Acid, *Amide*

$C_9H_9Br_2NO_2$
2-Amino-3,5-dibromobenzoic Acid, *Et ester*
4-Amino-3,5-dibromobenzoic Acid, *Et ester*

$C_9H_9Br_3$
1,2,4-Tribromo-3,5,6-trimethylbenzene

$C_9H_9Br_3O$
2,4,6-Tribromo-*m*-cresol, *Et ether*
2,4,6-Tribromophenol, *Propyl ether*

$C_9H_9Br_3O_2$
2,4,6-Tribromoresorcinol, *Me-Et ether*

$C_9H_9Br_3O_3$
Tribromophloroglucinol, *Tri-Me ether*
Tribromopyrogallol, *Tri-Me ether*

C_9H_9Cl
α-Chloro-*p*-methylstyrene
β-Chloro-*p*-methylstyrene
1-Chloro-1-phenylpropene
1-Chloro-2-phenylpropene
1-Chloro-3-phenylpropene
2-Chloro-1-phenylpropene
2-Chloro-3-phenylpropene
3-Chloro-1-phenylpropene

$C_9H_9ClHgO_2$
Mercuri-*o*-carboxyphenyl chloride, *Et ester*
Mercuri-*m*-carboxyphenyl chloride, *Et ester*
Mercuri-*p*-carboxyphenyl chloride, *Et ester*

$C_9H_9ClN_2$
3-Chloroindazole, 1-*Et*

$C_9H_9ClN_2O_3$
4-Chloro-3-nitrobenzoic Acid, *Di-Me-amide*
5-Chloro-2-nitrobenzoic Acid, *Di-Me-amide*

C_9H_9ClO
α-Chloro-2-methylacetophenone
α-Chloro-4-methylacetophenone
2-Chloro-4-methylacetophenone
2-Chloro-5-methylacetophenone
3-Chloro-4-methylacetophenone
4-Chloro-2-methylacetophenone
4-Chloro-3-methylacetophenone
γ-Chloropropiophenone
2-Chloropropiophenone
3-Chloropropiophenone
4-Chloropropiophenone
2,3-Dimethylbenzoic Acid, *Chloride*
2,4-Dimethylbenzoic Acid, *Chloride*
2,6-Dimethylbenzoic Acid, *Chloride*
3,5-Dimethylbenzoic Acid, *Chloride*
o-Ethylbenzoic Acid, *Chloride*
2-Phenylpropionic Acid, *Chloride*
3-Phenylpropionic Acid, *Chloride*

$C_9H_9ClO_2$
Chloroacetic Acid, *Benzyl ester*
o-Chlorobenzoic Acid, *Et ester*
m-Chlorobenzoic Acid, *Et ester*
p-Chlorobenzoic Acid, *Et ester*
2-Chloro-4,6-dimethylbenzoic Acid
4-Chloro-3,5-dimethylbenzoic Acid
5-Chloro-2,4-dimethylbenzoic Acid
Chloroformic Acid, *Phenylethyl ester*
α-Chloro-2-hydroxyacetophenone, *Me ether*
α-Chloro-4-hydroxyacetophenone, *Me ether*
5-Chloromethyl-2-hydroxybenzaldehyde, *Me ether*
2-Chloro-2-phenylpropionic Acid
2-Chloro-3-phenylpropionic Acid
3-Chloro-2-phenylpropionic Acid
3-Chloro-3-phenylpropionic Acid
3-*o*-Chlorophenylpropionic Acid
3-*m*-Chlorophenylpropionic Acid
3-*p*-Chlorophenylpropionic Acid
α-Chlorophenylacetic Acid, *Me ester*
o-Chlorophenylacetic Acid, *Me ester*
m-Ethoxybenzoic Acid, *Chloride*
p-Ethoxybenzoic Acid, *Chloride*
p-Hydroxymethyl-benzoic Acid, *Me ether, Chloride*
Mandelic Acid, *Me ether, Chloride*
p-Methoxyphenylacetic Acid, *Chloride*
2-Phenoxypropionic Acid, *Chloride*

$C_9H_9ClO_2S$
Indane-4-sulphonic Acid, *Chloride*

$C_9H_9ClO_3$
2-Chloro-3-hydroxybenzoic Acid, *Et ester*
3-Chloro-2-hydroxybenzoic Acid, *Et ester*
3-Chloro-4-hydroxybenzoic Acid, *Et ester*
5-Chloro-2-hydroxybenzoic Acid, *Et ester*
5-Chloro-2-hydroxybenzoic Acid, *Et ether*
3-Chloro-2-hydroxy-3-phenylpropionic Acid
α-Chloro-4-hydroxy-*m*-toluic Acid, *Me ester*
3-Chloro-6-hydroxy-*p*-toluic Acid, *Me ester*
3-Chloro-6-hydroxy-*p*-toluic Acid, *Me ether*
p-Chloromandelic Acid, *Me ether*
3-Chloro-4-methoxybenzoic Acid, *Me ester*
5-Chloromethyl-2-hydroxybenzoic Acid, *Me ester*
2-Chloro-4-methylphenoxyacetic Acid
2-Chloro-6-methylphenoxyacetic Acid
4-Chloro-2-methylphenoxyacetic Acid
o-Chlorophenoxyacetic Acid, *Me ester*
p-Chlorophenoxyacetic Acid, *Me ester*
2,5-Dimethoxybenzoic Acid, *Chloride*
3,4-Dimethoxybenzoic Acid, *Chloride*
3,5-Dimethoxybenzoic Acid, *Chloride*

$C_9H_9ClO_4$
4-Chloro-5-hydroxy-2-methylphenoxyacetic Acid†

$C_9H_9Cl_2NO_4S$
N-Dichloro-*p*-sulphonamidobenzoic Acid, *Et ester*

$C_9H_9Cl_3O$
2,4,6-Trichloro-*m*-cresol, *Et ether*

$C_9H_9Cl_3O_3$
Trichlorophloroglucinol, *Tri-Me ether*
Trichloropyrogallol, *Tri-Me ether*

$C_9H_9Cl_6O_3$
2,4,6-Tri-(1,1-dichloroethyl)-1,3,5-triazine

$C_9H_9FO_2$
o-Fluorobenzoic Acid, *Et ester*
m-Fluorobenzoic Acid, *Et ester*
p-Fluorobenzoic Acid, *Et ester*
3-*m*-Fluorophenylpropionic Acid
3-*p*-Fluorophenylpropionic Acid

$C_9H_9HgNaO_2S$
Thiomersal

$C_9H_9IO_2$
o-Iodobenzoic Acid, *Et ester*
m-Iodobenzoic Acid, *Et ester*
p-Iodobenzoic Acid, *Et ester*
2-Iodo-4,6-dimethylbenzoic Acid
3-Iodo-2,4-dimethylbenzoic Acid
4-Iodo-2,6-dimethylbenzoic Acid
p-Iodomethylbenzoic Acid, *Me ester*
2-Iodo-3-methylbenzoic Acid, *Me ester*
3-Iodo-4-methylbenzoic Acid, *Me ester*
3-Iodo-3-phenylpropionic Acid
3-*o*-Iodophenylpropionic Acid
3-*m*-Iodophenylpropionic Acid
3-*p*-Iodophenylpropionic Acid

$C_9H_9IO_3$
1,3-Dihydro-1-hydroxy-3-oxo-1,2-benziodoxole, 1-*Et ether*†
2-Hydroxy-4-iodobenzoic Acid, *Et ester*
2-Hydroxy-5-iodobenzoic Acid, *Et ester*
3-Hydroxy-2-iodobenzoic Acid, *Me ether*, *Me ester*
4-Hydroxy-3-iodobenzoic Acid, *Et ether*
4-Hydroxy-2-iodo-3-methoxybenzaldehyde, *Me ether*
3-Iodo-4-methoxybenzoic Acid, *Me ester*

$C_9H_9I_2NO$
2,4-Di-iodo-*m*-toluidine, N-*Ac*
2,5-Di-iodo-*m*-toluidine, N-*Ac*
4,5-Di-iodo-*m*-toluidine, N-*Ac*
4,6-Di-iodo-*m*-toluidine, N-*Ac*
5,6-Di-iodo-*m*-toluidine, N-*Ac*
2,5-Di-iodo-*p*-toluidine, N-*Ac*
2,6-Di-iodo-*p*-toluidine, N-*Ac*

$C_9H_9I_2NO_2$
2-Amino-3,5-di-iodobenzoic Acid, *Et ester*
2-Amino-4,5-di-iodobenzoic Acid, *Et ester*
4-Amino-3,5-di-iodobenzoic Acid, *Et ester*

$C_9H_9I_2NO_3$
3,5-Di-iodotyrosine
Iodogorgoic Acid

$C_9H_9I_3O$
2,4,6-Tri-iodophenol, *Propyl ether*

C_9H_9N
1,2-Dihydroquinoline
2,4-Dimethylbenzoic Acid, *Nitrile*
2,5-Dimethylbenzoic Acid, *Nitrile*
2,6-Dimethylbenzoic Acid, *Nitrile*
3,4-Dimethylbenzoic Acid, *Nitrile*
o-Ethylbenzoic Acid, *Nitrile*
m-Ethylbenzoic Acid, *Nitrile*
1-Methylindole
2-Methylindole
3-Methylindole
4-Methylindole
5-Methylindole
6-Methylindole
7-Methylindole
2-Methylindolizine
3-Methylindolizine
1-Methyl-1*H*-1-pyrindine†
3-Phenyl-1-azabicyclo[1,1,0]butane†
2-Phenylpropionic Acid, *Nitrile*
3-Phenylpropionic Acid, *Nitrile*
9α*H*-Quinolizine
o-Tolylacetic Acid, *Nitrile*
m-Tolylacetic Acid, *Nitrile*
p-Tolylacetic Acid, *Nitrile*

C_9H_9NO
Atropic Acid, *Amide*
Cinnamic Acid, *Amide*
N-(2-Furfuryl)pyrrole†
2-Hydroxy-4,6-dimethylbenzoic Acid, *Nitrile*
4-Hydroxy-2,5-dimethylbenzoic Acid, *Nitrile*
4-Hydroxy-3,5-dimethylbenzoic Acid, *Nitrile*
4-Hydroxyindole, *Me ether*
5-Hydroxyindole, *Me ether*
6-Hydroxyindole, *Me ether*
7-Hydroxyindole, *Me ether*
3-Hydroxy-2-methylindole
2-Hydroxyphenylacetic Acid, *Me ether*, *Nitrile*
3-Hydroxyphenylacetic Acid, *Me ether*, *Nitrile*
2-Hydroxy-3-phenylpropionic Acid, *Nitrile*
3-*o*-Hydroxyphenylpropionic Acid, *Nitrile*
3-*p*-Hydroxyphenylpropionic Acid, *Nitrile*

C_9H_9NO (*continued*)
6-Hydroxy-*m*-toluic Acid, *Me ether*, *Nitrile*
2-Indolylmethanol
3-Indolylmethanol
4-Indolylmethanol
Mandelic Acid, *Me ether*, *Nitrile*
p-Methoxyphenylacetic Acid, *Nitrile*
1-Methyloxindole
3-Methyloxindole
4-Methyloxindole
5-Methyloxindole
6-Methyloxindole
7-Methyloxindole
2-Phenyl-2-oxazoline
Phthalimidine, N-*Me*
1,2,3,4-Tetrahydro-2-oxoquinoline

C_9H_9NOS
2-Hydroxybenzothiazole, *Et ether*
5-Phenyl-2-oxazolidinethione

$C_9H_9NO_2$
p-Acetamidobenzaldehyde
2-Acetoacetylpyridine
3-Acetoacetylpyridine
4-Acetoacetylpyridine
Acetophenone-*o*-carboxylic Acid, *Amide*
o-Aminocinnamic Acid
m-Aminocinnamic Acid
p-Aminocinnamic Acid
Benzamidoacetaldehyde
Benzomorpholone, N-*Me*
Benzoxazolin-2-one, N-*Et*
Bicyclo[2,2,1]hept-5-ene-2,3-dicarboxylic Acid, *Imide*
Cinnamohydroxamic Acid
1,2-Dihydroxyindole, 1-*Me ether*
α,2-Dihydroxyphenylacetic Acid, 2-*Me ether*, *Nitrile*
α,4-Dihydroxyphenylacetic Acid, 4-*Me ether*, *Nitrile*
2,3-Dihydroxy-2-phenylpropionic Acid, *Nitrile*
5,6-Dihydroxyskatole†
2,5-Dimethoxybenzoic Acid, *Nitrile*
2,6-Dimethoxybenzoic Acid, *Nitrile*
Epinochrome
Gentianidine†
Indoline-2-carboxylic Acid†
3,4-Methylenedioxybenzaldehyde, *Methylimide*
4-Nitroindane
5-Nitroindane
Phenylpyruvic Acid, *Amide*
2-Pyridineacrylic Acid, *Me ester*
p-Toluylformic Acid, *Amide*

$C_9H_9NO_2S$
o-Nitrobenzenethiol, S-*Allyl ether*
p-Nitrobenzenethiol, S-*Allyl ether*

$C_9H_9NO_3$
N-Acetylanthranilic Acid
Adrenochrome
Adrenolutin
2-Amino-3-methylbenzoylformic Acid
2-Amino-5-methylbenzoylformic Acid
4-Amino-3-methylbenzoylformic Acid
2-Carboxyphenylacetic Acid, α-*Amide*
2-Carboxyphenylacetic Acid, 2-*Amide*
4-Carboxyphenylacetic Acid, α-*Amide*
4-Carboxyphenylacetic Acid, 4-*Amide*
2,4-Dimethyl-5-nitrobenzaldehyde
3,5-Dimethyl-2-nitrobenzaldehyde
3,6-Dimethyl-2-nitrobenzaldehyde
N-Formyl-*N*-phenylglycine
2-Formylphenylglycine
Hippuric Acid
4-Hydroxybenzoylformic Acid, *Me ether*, *Amide*
3-Hydroxy-β-nitrostyrene, *Me ether*
4-Hydroxy-β-nitrostyrene, *Me ether*
4-Hydroxy-3-nitrostyrene, *Me ether*
Isophthalamic Acid, *Me ester*
Isophthalic Acid, *Mono-Me ester*, *Amide*
Malonanilic Acid
3,4-Methylenedioxyphenylacetic Acid, *Amide*
4-Methyl-2-nitroacetophenone
4-Methyl-3-nitroacetophenone
N-Methyloxanilic Acid
3′-Methyloxanilic Acid
o-Nitrophenol, *Allyl ether*
p-Nitrophenol, *Allyl ether*
o-Nitrophenylacetone
p-Nitrophenylacetone
o-Nitropropiophenone
m-Nitropropiophenone
p-Nitropropiophenone
o-Nitrosobenzoic Acid, *Et ester*
m-Nitrosobenzoic Acid, *Et ester*
p-Nitrosobenzoic Acid, *Et ester*
Oxanilic Acid, *Me ester*
Terephthalamic Acid, *Me ester*

$C_9H_9NO_3S$
Saccharin, N-*Et*
Sulphazone, *Me ether*

$C_9H_9NO_4$
4-Aminoisophthalic Acid, 1-*Me-ester*
4-Aminoisophthalic Acid, 3-*Me ester*
4-Aminoisophthalic Acid, N-*Me*
3-Aminophthalic Acid, 1-*Me ester*
4-Aminophthalic Acid, 1-*Me ester*
Aminoterephthalic Acid, 1-*Me ester*
Aminoterephthalic Acid, 4-*Me ester*
Aminoterephthalic Acid, *N-Me*
Anilinomalonic Acid
Apophyllenic Acid, *Me ester*
m-Carboxyphenylglycine†
2,4-Dimethyl-3-nitrobenzoic Acid
2,4-Dimethyl-5-nitrobenzoic Acid
2,4-Dimethyl-6-nitrobenzoic Acid
2,5-Dimethyl-4-nitrobenzoic Acid
3,5-Dimethyl-2-nitrobenzoic Acid
3,5-Dimethyl-4-nitrobenzoic Acid
2,6-Dimethylpyridine-3,4-dicarboxylic Acid
2,4-Dimethylpyridine-3,5-dicarboxylic Acid
2,6-Dimethylpyridine-3,5-dicarboxylic Acid
2-Ethyl-4-nitrobenzoic Acid
2-Ethyl-5-nitrobenzoic Acid
4-Ethyl-3-nitrobenzoic Acid
2-Hydroxybenzamidoacetic Acid
3-Hydroxybenzamidoacetic Acid
4-Hydroxybenzamidoacetic Acid
2-Hydroxy-4,5-dimethyl-3-nitrobenzaldehyde
2-Hydroxy-5,6-dimethyl-3-nitrobenzaldehyde
4-Hydroxy-2,5-dimethyl-3-nitrobenzaldehyde

$C_9H_9NO_4$ (*continued*)

2-Hydroxyisophthalic Acid, *Mono-Me ester, Amide*
2-Hydroxy-5-nitrobenzaldehyde, *Et ether*
4-Hydroxy-3-nitrobenzaldehyde, *Et ether*
5-Hydroxy-2-nitrobenzaldehyde, *Et ether*
2-Hydroxy-5-nitrosobenzoic Acid, *Et ester*
Isatoic Acid, 2-*Me ester*
3,4-Methylenedioxymandelic Acid, *Amide*
N-Methylisatoic Acid
3-Methyl-4-nitrophenylacetic Acid
4-Methyl-3-nitrophenylacetic Acid
5-Methyl-2-nitrophenylacetic Acid
Myristicic Acid, *Amide*
o-Nitrobenzoic Acid, *Et ester*
m-Nitrobenzoic Acid, *Et ester*
p-Nitrobenzoic Acid, *Et ester*
o-Nitrophenoxyacetone
p-Nitrophenoxyacetone
o-Nitrophenylacetic Acid, *Me ester*
p-Nitrophenylacetic Acid, *Me ester*
2-*o*-Nitrophenylpropionic Acid
2-*m*-Nitrophenylpropionic Acid
2-*p*-Nitrophenylpropionic Acid
3-*o*-Nitrophenylpropionic Acid
3-*m*-Nitrophenylpropionic Acid
3-*p*-Nitrophenylpropionic Acid
3-Nitro-*o*-toluic Acid, *Me ester*
5-Nitro-*o*-toluic Acid, *Me ester*
6-Nitro-*o*-toluic Acid, *Me ester*
2-Nitro-*m*-toluic Acid, *Me ester*
4-Nitro-*m*-toluic Acid, *Me ester*
5-Nitro-*m*-toluic Acid, *Me ester*
6-Nitro-*m*-toluic Acid, *Me ester*
3-Nitro-*p*-toluic Acid, *Me ester*
Pencolide
N-Phenylglycine-*o*-carboxylic Acid
N-Phenylglycine-*m*-carboxylic Acid
N-Phenylglycine-*p*-carboxylic Acid
Pyridine-2,6-diacetic Acid
Pyridine-2,3-dicarboxylic Acid, *Di-Me ester*
Pyridine-2,3-dicarboxylic Acid, 1-*Et ester*
Pyridine-2,4-dicarboxylic Acid, *Di-Me ester*
Pyridine-2,5-dicarboxylic Acid, *Di-Me ester*
Pyridine-2,6-dicarboxylic Acid, *Di-Me ester*
Pyridine-2,6-dicarboxylic Acid, *Mono-Et ester*
Pyridine-3,4-dicarboxylic Acid, *Di-Me ester*
Pyridine-3,4-dicarboxylic Acid, 4-*Et ester*
Pyridine-3,5-dicarboxylic Acid, *Di-Me ester*
Pyridine-3,5-dicarboxylic Acid, *Mono-Et ester*

$C_9H_9NO_5$

(3-Carboxy-4-hydroxyphenyl)glycine†
2,3-Dihydroxybenzoylglycine
2,4-Dihydroxy-7-methoxy-1,4-benzoxazolin-3-one
2,4-Dihydroxy-5-nitroacetophenone, 4-*Me ether*
2,3-Dihydroxy-5-nitrobenzaldehyde, 3-*Et ether*
2,3-Dimethoxy-5-nitrobenzaldehyde
2,3-Dimethoxy-6-nitrobenzaldehyde
2,4-Dimethoxy-3-nitrobenzaldehyde
2,4-Dimethoxy-5-nitrobenzaldehyde
2,5-Dimethoxy-3-nitrobenzaldehyde
3,4-Dimethoxy-2-nitrobenzaldehyde
3,4-Dimethoxy-5-nitrobenzaldehyde
3,6-Dimethoxy-2-nitrobenzaldehyde
4,5-Dimethoxy-2-nitrobenzaldehyde
Ekapterin
2-Hydroxy-3-nitrobenzoic Acid, *Et ether*
2-Hydroxy-3-nitrobenzoic Acid, *Et ester*
2-Hydroxy-4-nitrobenzoic Acid, *Me ether, Me ester*
2-Hydroxy-4-nitrobenzoic Acid, *Et ether*
2-Hydroxy-4-nitrobenzoic Acid, *Et ester*
2-Hydroxy-5-nitrobenzoic Acid, *Me ether, Me ester*
2-Hydroxy-5-nitrobenzoic Acid, *Et ether*
2-Hydroxy-5-nitrobenzoic Acid, *Et ester*
3-Hydroxy-2- nitrobenzoic Acid, *Et ester*
3-Hydroxy-4-nitrobenzoic Acid, *Et ether*
3-Hydroxy-4-nitrobenzoic Acid, *Et ester*
4-Hydroxy-2-nitrobenzoic Acid, *Et ester*
4-Hydroxy-3-nitrobenzoic Acid, *Et ether*
Hydroxy-*o*-nitrophenylacetic Acid, *Me ester*
Hydroxy-*o*-nitrophenylacetic Acid, *Me ether*
Hydroxy-*m*-nitrophenylacetic Acid, *Me ester*
Hydroxy-*p*-nitrophenylacetic Acid, *Me ester*
2-Hydroxy-3-*o*-nitrophenylpropionic Acid
3-Hydroxy-3-*o*-nitrophenylpropionic Acid
3-Hydroxy-3-*m*-nitrophenylpropionic Acid
3-Hydroxy-3-*p*-nitrophenylpropionic Acid
3-(4-Hydroxy-3-nitrophenyl)propionic Acid
2-Hydroxy-5-nitro-*p*-toluic Acid, *Me ether*
2-Hydroxy-5-nitro-*p*-toluic Acid, *Me ester*
4-Hydroxy-5-nitro-*m*-toluic Acid, *Me ester*
4-Hydroxypyridine-2,6-dicarboxylic Acid, *Di-Me ester*
4-Methoxy-3-nitrobenzoic Acid, *Me ester*
4-Methoxy-2-nitrophenylacetic Acid
4-Methoxy-3-nitrophenylacetic Acid
o-Nitrophenoxyacetic Acid, *Me ester*
m-Nitrophenoxyacetic Acid, *Me ester*
p-Nitrophenoxyacetic Acid, *Me ester*
2-*o*-Nitrophenoxypropionic Acid
2-*m*-Nitrophenoxypropionic Acid
2-*p*-Nitrophenoxypropionic Acid
3-*o*-Nitrophenoxypropionic Acid
3-*m*-Nitrophenoxypropionic Acid
3-*p*-Nitrophenoxypropionic Acid

$C_9H_9NO_6$

2,3-Dihydroxy-5-nitrobenzoic Acid, 3-*Et ether*
2,5-Dihydroxy-4-nitrobenzoic Acid, 5-*Me ether, Me ester*
2,5-Dihydroxy-6-nitrobenzoic Acid, 5-*Me ether, Me ester*
3,4-Dihydroxy-2-nitrophenylacetic Acid, 4-*Me ether*
2,6-Dihydroxypyridine-3,5-dicarboxylic Acid, *Et ether*
2,3-Dimethoxy-4-nitrobenzoic Acid
2,3-Dimethoxy-5-nitrobenzoic Acid
2,3-Dimethoxy-6-nitrobenzoic Acid
2,4-Dimethoxy-3-nitrobenzoic Acid
2,4-Dimethoxy-5-nitrobenzoic Acid
2,5-Dimethoxy-3-nitrobenzoic Acid
2,5-Dimethoxy-4-nitrobenzoic Acid
2,6-Dimethoxy-3-nitrobenzoic Acid
3,4-Dimethoxy-2-nitrobenzoic Acid
3,4-Dimethoxy-5-nitrobenzoic Acid
3,5-Dimethoxy-2-nitrobenzoic Acid
3,5-Dimethoxy-4-nitrobenzoic Acid

$C_9H_9NO_6$ (*continued*)
3,6-Dimethoxy-2-nitrobenzoic Acid
4,5-Dimethoxy-2-nitrobenzoic Acid
4-Hydroxy-3-methoxy-5-nitrobenzoic Acid, *Me ester*
5-Hydroxy-4-methoxy-2-nitrobenzoic Acid, *Me ester*
4-Hydroxy-3-methoxy-2-nitrophenylacetic Acid
4-Hydroxy-3-methoxy-5-nitrophenylacetic Acid
4-Hydroxy-5-methoxy-2-nitrophenylacetic Acid
2-Hydroxy-4-nitrobenzoic Acid, *Ac*
3-*m*-Nitrophenylglyceric Acid
3-*p*-Nitrophenylglyceric Acid
Stizolobic Acid †
Stizolobinic Acid †

$C_9H_9NO_6S$
3-Sulphophthalic Acid, *Me ester*

$C_9H_9NO_7S$
2-Nitro-4-sulphobenzoic Acid, *Di-Me ester*

$C_9H_9NS_2$
2-Mercaptobenzothiazole, *Et ether*

$C_9H_9N_3$
1,3-Diaminoisoquinoline †
3,4-Diaminoquinoline
3,5-Diaminoquinoline
4,6-Diaminoquinoline
4,8-Diaminoquinoline
5,6-Diaminoquinoline
5,7-Diaminoquinoline
5,8-Diaminoquinoline
6,8-Diaminoquinoline
7,8-Diaminoquinoline
1-Methyl-2-phenyl-1,3,4-triazole
2-Methyl-1-phenyl-1,3,4-triazole
2-Methyl-5-phenyl-1,3,4-triazole
4-Methyl-1-phenyl-1,2,3-triazole
4-Methyl-2-phenyl-1,2,3-triazole
5-Methyl-1-phenyl-1,2,3-triazole
3-Methyl-1-phenyl-1,2,4-triazole
5-Methyl-1-phenyl-1,2,4-triazole
3-Methyl-1-phenyl-1,2,5-triazole
2-Quinolylhydrazine
5-Quinolylhydrazine
6-Quinolylhydrazine
8-Quinolylhydrazine

$C_9H_9N_3OS$
Benzothiazole-2-acetic Acid, *Hydrazide*

$C_9H_9N_3O_2$
Benzylidenebiuret
4-Dimethylamino-3-nitrobenzoic Acid, *Nitrile*
1,2-Dimethyl-5-nitrobenzimidazole
1,2-Dimethyl-6-nitrobenzimidazole
2,5-Dimethyl-6-nitrobenzimidazole
2,5-Dimethyl-7-nitrobenzimidazole
2,7-Dimethyl-5-nitrobenzimidazole
4,7-Dimethyl-6-nitrobenzimidazole
5,6-Dimethyl-4-nitroindazole
5,6-Dimethyl-7-nitroindazole
6-Nitroindazole, N-*Et*
1-Phenylurazole, 2-*Me*
1-Phenylurazole, 4-*Me*
1-Phenylurazole, 3-*Me ether*

$C_9H_9N_3O_2S_2$
2′-(2-Aminoethyl)-2,4′-bithiazole-4-carboxylic Acid †
Sulphathiazole

$C_9H_9N_3O_3$
2-Amino-3-nitrocinnamic Acid, *Amide*

$C_9H_9N_3O_4$
1,2,3,4-Tetrahydro-6,8-dinitroquinoline

$C_9H_9N_3O_6$
2-Amino-3,5-dinitrobenzoic Acid, *Et ester*
2-Amino-3,5-dinitrobenzoic Acid, N-*Di-Me*
4-Amino-3,5-dinitrobenzoic Acid, *Et ester*
4-Amino-3,5-dinitrobenzoic Acid, N-*Me, Me ester*
4-Amino-3,5-dinitrobenzoic Acid, N-*Di-Me*
Triazine-tricarboxylic Acid, *Tri-Me ester*
1,2,4-Trimethyl-3,5,6-trinitrobenzene
1,3,5-Trimethyl-2,4,6-trinitrobenzene

$C_9H_9N_3O_7$
3,5-Dimethyl-2,4,6-trinitrophenol, *Me ether*
2,4,6-Trinitro-*m*-cresol, *Et ether*

$C_9H_9N_3O_8$
Styphnic Acid, *Me-Et ether*

$C_9H_9N_3S$
Amiphenazole

$C_9H_9N_5O_5$
Erythropterin

$C_9H_9N_5O_8$
N,2,4,6-Tetranitroaniline, N-*Propyl*
N,2,4,6-Tetranitroaniline, N-*Isopropyl*

C_9H_{10}
Allylbenzene
Bicyclo[6,1,0]nonatriene †
Cyclononatetraene Anion Radical †
Cyclononatetraene Dianion †
Indane
6-Isopropenylfulvene †
Methylcyclo-octatetraene
α-Methylstyrene
β-Methylstyrene
o-Methylstyrene
m-Methylstyrene
p-Methylstyrene
6-Methyl-6-vinylfulvene †
Pentacyclo[4,3,0,$0^{2,5}$,$0^{3,8}$,$0^{4,7}$]nonane †
Phenylcyclopropane
6-[1-Propenyl]fulvene †
Tricyclo[3,3,1,$0^{2,8}$]nona-3,6-diene †

$C_9H_{10}BrNO$
2-Bromo-4,6-dimethylbenzoic Acid, *Amide*
4-Bromo-2,5-dimethylbenzoic Acid, *Amide*
5-Bromo-2,4-dimethylbenzoic Acid, *Amide*

$C_9H_{10}BrNO_2$
2-Amino-5-bromobenzoic Acid, *Et ester*
4-Amino-2-bromobenzoic Acid, *Et ester*
3-(3-Amino-4-bromophenyl)propionic Acid
3-(4-Amino-3-bromophenyl)propionic Acid
3-Bromo-*p*-anisidine, N-*Ac*

$C_9H_{10}BrNO_3$
4-Bromo-6-nitro-*m*-cresol, *Et ether*

$C_9H_{10}Br_2$
2,4-Dibromo-1,3,5-trimethylbenzene
4,6-Dibromo-2,5-xylenol, *Me ether*

$C_9H_{10}ClNO$
4-Amino-2-chlorobenzaldehyde, N-*Di-Me*
4-Amino-2-chlorobenzaldehyde, N-*Et*
2-Chloro-4,6-dimethylbenzoic Acid, *Amide*
3-*o*-Chlorophenylpropionic Acid, *Amide*
p-Dimethylaminobenzoic Acid, *Chloride*

$C_9H_{10}ClNO_2$
2-Amino-4-chlorobenzoic Acid, *Et ester*
4-Amino-2-chlorobenzoic Acid, *Et ester*

$C_9H_{10}ClNO_4$
2-Amino-5-chloro-3,4-dimethoxybenzoic Acid †

$C_9H_{10}Cl_2N_2O$
Diuron

$C_9H_{10}Cl_2N_2O_2$
N-(3,4-Dichlorophenyl)-*N′*-methoxy-*N′* methylurea †

$C_9H_{10}Cl_2O$
2,6-Dichloro-*p*-cresol, *Et ether*

$C_9H_{10}Cl_2O_2$
2,5-Dichlorobenzaldehyde, *Di-Me acetal*
6-Methyl-4-cyclohexene-1,3-dicarboxylic Acid, *Dichloride*

$C_9H_{10}Cl_2O_4S_2$
2,4,6-Trimethylbenzene-1,3-disulphonic Acid, *Dichloride*

$C_9H_{10}FNO$
3-*m*-Fluorophenylpropionic Acid, *Amide*

$C_9H_{10}FNO_3$
2-Fluorotyrosine
3-Fluorotyrosine

$C_9H_{10}HgO_2$
Mercuri-benzyl acetate
Mercuri-*o*-tolyl acetate
Mercuri-*m*-tolyl acetate
Mercuri-*p*-tolyl acetate

$C_9H_{10}HgO_3$
Mercuri-*o*-hydroxyphenyl acetate, *Me ether*
Mercuri-*p*-hydroxyphenyl acetate, *Me ether*
Mercuri-2-hydroxy-*m*-tolyl acetate
Mercuri-4-hydroxy-*m*-tolyl acetate

$C_9H_{10}INO_2$
2-Amino-4-iodobenzoic Acid, N-*Et*
2-Amino-5-iodobenzoic Acid, N-*Et*
4-Amino-3-iodobenzoic Acid, N-*Di-Me*
α-Amino-β-4-iodophenylpropionic Acid

$C_9H_{10}I_2O$
2,4-Di-iodophenol, *Propyl ether*

$C_9H_{10}N_2$
2-Anilinopropionic Acid, *Nitrile*
Apoharmine, N-*Me*
p-Dimethylaminobenzoic Acid, *Nitrile*
1,2-Dimethylbenzimidazole
1,5-Dimethylbenzimidazole
1,6-Dimethylbenzimidazole
2,6-Dimethylbenzimidazole
5,6-Dimethylbenzimidazole
5,7-Dimethylbenzimidazole
1,3-Dimethylindazole
1,5-Dimethylindazole
2,3-Dimethylindazole
2,5-Dimethylindazole
5,7-Dimethylindazole
o-Ethylaminobenzoic Acid, *Nitrile*
1-Ethylindazole
2-Ethylindazole
3-Indolylmethylamine
Myosmine †
1-Phenylpyrazoline
3-Phenylpyrazoline
4-Phenylpyrazoline
5-Phenylpyrazoline
N-Phenylsarcosine, *Nitrile*
N-*p*-Tolylglycine, *Nitrile*

$C_9H_{10}N_2O$
3-Amino-3,4-dihydro-2-quinolone
7-Amino-3,4-dihydro-2-quinolone
2-Amino-6-hydroxybenzoic Acid, *Et ether, Nitrile*
Cinnamic Acid, *Hydrazide*
2-Hydroxyphenylglycine, *Et ester*, *Nitrile*
Indoline-2-carboxylic Acid, *Amide* †

$C_9H_{10}N_2O_2$
N-Acetylanthranilic Acid, *Amide*
1-Amino-6-nitroindane
5-Amino-4-nitroindane
Benzylidenehydrazinoacetic Acid
4-Carboxyphenylacetic Acid, *Diamide*
3,4-Diaminocinnamic Acid
2,3-Dihydro-2-methyl-5-nitroindole
2,3-Dihydro-2-methyl-6-nitroindole
2,3-Dihydro-2-methyl-7-nitroindole
2,3-Dihydro-3-methyl-6-nitroindole
Hippuric Acid, *Amide*
Malonamide, N-*Phenyl*
4-Methylphthalic Acid, *Diamide*
Phenacetylurea
Phenylglyoxime, N-*Me*
Phenylglyoxime, 1-*Me ether*
Phenylmalonic Acid, *Diamide*
1,2,3,4-Tetrahydro-6-nitroquinoline
1,2,3,4-Tetrahydro-7-nitroquinoline
1,2,3,4-Tetrahydro-8-nitroquinoline
p-Tolyglyoxime

$C_9H_{10}N_2O_2S_2$
Aureothricin

$C_9H_{10}N_2O_3$
Allophanic Acid, *Benzyl ester*
o-Aminobenzamidoacetic Acid
m-Aminobenzamidoacetic Acid
p-Aminobenzamidoacetic Acid
2-Amino-5-nitrobenzaldehyde, N-*Di-Me*
4-Amino-3-nitrobenzaldehyde, N-*Di-Me*
2-Amino-5-nitrosobenzoic Acid, *Et ester*
p-Amino-oxanilic Acid, *Me ester*
2,4-Dimethyl-3-nitrobenzoic Acid, *Amide*
2,4-Dimethyl-5-nitrobenzoic Acid, *Amide*
2,6-Dimethylpyridine-3,4-dicarboxylic Acid, 4-*Amide*
2-Ethylamino-5-nitrosobenzoic Acid
o-(*N*-Ethyl-*N*-nitrosoamino)benzoic Acid
Ethyl phenylaminoformate, N-*Nitroso*

$C_9H_{10}N_2O_3$ (*continued*)
2-*N*-Methylamino-5-nitrosobenzoic Acid, *Me ester*
2-(*N*-Methyl-*N*-nitrosoamino)benzoic Acid, *Me ester*
o-Nitrobenzamide, N-*Di-Me*
m-Nitrobenzamide, N-*Et*
p-Nitrobenzamide, N-*Et*
p-Nitrophenylacetic Acid, *Methylamide*
3-*o*-Nitrophenylpropionic Acid, *Amide*
3-*p*-Nitrophenylpropionic Acid, *Amide*
3-Nitro-*o*-toluic Acid, *Methylamide*
5-Nitro-*o*-toluic Acid, *Methylamide*
2-Nitro-*m*-toluic Acid, *Methylamide*
3-Nitro-*p*-toluic Acid, *Methylamide*
N-Phenylglycine-*o*-carboxylic Acid, α-*Amide*
2-Phenylhydantoic Acid
5-Phenylhydantoic Acid

$C_9H_{10}N_2O_4$
2-Amino-3,5-dimethyl-6-nitrobenzoic Acid
5-Amino-2,4-dimethyl-3-nitrobenzoic Acid
2-Amino-3-nitrobenzoic Acid, *Et ester*
2-Amino-4-nitrobenzoic Acid, *Et ester*
2-Amino-4-nitrobenzoic Acid, N-*Et*
2-Amino-5-nitrobenzoic Acid, *Et ester*
3-Amino-2-nitrobenzoic Acid, *Et ester*
3-Amino-4-nitrobenzoic Acid, *Et ester*
4-Amino-2-nitrobenzoic Acid, *Et ester*
4-Amino-3-nitrobenzoic Acid, *Et ester*
4-Amino-3-nitrobenzoic Acid, N-*Et*
5-Amino-2-nitrobenzoic Acid, *Et ester*
5-Amino-3-nitrobenzoic Acid, *Et ester*
5-Amino-3-nitrobenzoic Acid, N-*Et*
4-Amino-2-nitrophenylacetic Acid, *Me ester*
2-Amino-3-(4-nitrophenyl)propionic Acid
3-Amino-3-(2-nitrophenyl)propionic Acid
3-Amino-3-(3-nitrophenyl)propionic Acid
3-Amino-3-(4-nitrophenyl)propionic Acid
3-(4-Amino-2-nitrophenyl)propionic Acid
3-(4-Amino-3-nitrophenyl)propionic Acid
4-Amino-6-nitro-*m*-toluic Acid, *Me ester*
6-Amino-4-nitro-*m*-toluic Acid, *Me ester*
4-Dimethylamino-3-nitrobenzoic Acid
Ethyl *o*-nitrophenylaminoformate
Ethyl *m*-nitrophenylaminoformate
Ethyl *p*-nitrophenylaminoformate
2-Hydroxy-3-*o*-nitrophenylpropionic Acid, *Amide*
3-Hydroxy-3-*o*-nitrophenylpropionic Acid, *Amide*
3-Hydroxy-3-*p*-nitrophenylpropionic Acid, *Amide*
4-Methoxy-3-nitrophenylacetic Acid, *Amide*
4-Methylamino-3-nitrobenzoic Acid, *Me ester*
3,4-Methylenedioxy-6-nitroaniline, N-*Di-Me*
3,4-Methylenedioxy-6-nitroaniline, N-*Et*
N-(2-Nitro-4-tolyl)glycine
N-(3-Nitro-2-tolyl)glycine
N-(3-Nitro-4-tolyl)glycine
N-(4-Nitro-2-tolyl)glycine
N-(4-Nitro-3-tolyl)glycine
N-(5-Nitro-2-tolyl)glycine
1,3,5-Trimethyl-2,4-dinitrobenzene

$C_9H_{10}N_2O_4S$
2,4-Dinitrobenzenethiol, *Isopropyl ether*

$C_9H_{10}N_2O_5$
3,4-Dimethoxy-2-nitrobenzoic Acid, *Amide*
3,5-Dimethoxy-2-nitrobenzoic Acid, *Amide*
4,6-Dinitro-*o*-cresol, *Et ether*
4,6-Dinitro-*m*-cresol, *Et ether*
2,6-Dinitro-*p*-cresol, *Et ether*
4,6-Dinitro-2,5-xylenol, *Me ether*

$C_9H_{10}N_2O_6$
4,6-Dinitroguaiacol, *Et ether*
2,4-Dinitroresorcinol, 1-*Et*-3-*Me ether*

$C_9H_{10}N_2O_6S$
p-Methoxycarbonylaminobenzenesulphonylcarbamic Acid†

$C_9H_{10}N_2O_7$
1,2,3-Trihydroxy-4,5-dinitrobenzene, *Tri-Me ether*
1,2,3-Trihydroxy-4,6-dinitrobenzene, *Tri-Me ether*

$C_9H_{10}N_2S_2$
2-Benzothiazolesulphenamide, N-*Ethyl*

$C_9H_{10}N_4$
5-Amino-1-phenyl-1,2,3-triazole, 5-N-*Me*

$C_9H_{10}N_4O_2S_2$
Sulphamethizole

$C_9H_{10}N_4O_5$
2′-Deoxy-5-diazouridine†

$C_9H_{10}N_4O_6$
5-Diazouridine†
3,5-Dimethyl-2,4,6-trinitroaniline, N-*Me*

$C_9H_{10}O$
Allyl phenyl Ether
Anol
Chavicol
Chroman
Cinnamyl Alcohol
2,3-Dihydro-2-methylbenzofuran
2,3-Dihydro-5-methylbenzofuran
2,3-Dimethylbenzaldehyde
2,4-Dimethylbenzaldehyde
2,5-Dimethylbenzaldehyde
2,6-Dimethylbenzaldehyde
3,4-Dimethylbenzaldehyde
3,5-Dimethylbenzaldehyde
o-Ethylbenzaldehyde
m-Ethylbenzaldehyde
p-Ethylbenzaldehyde
1-Hydroxyindane
4-Hydroxyindane
5-Hydroxyindane
2-Hydroxystyrene, *Me ether*
3-Hydroxystyrene, *Me ether*
4-Hydroxystyrene, *Me ether*
Isochroman
2-Methylacetophenone
3-Methylacetophenone
4-Methylacetophenone
Methyl styryl Ether
1-Phenylallyl Alcohol
2-Phenylpropanal
3-Phenylpropanal
1-Phenyl-2-propanone
1-Phenylvinyl Alcohol, *Me ether*

$C_9H_{10}O$ (*continued*)

2-Phenylvinyl Alcohol, *Me ether*
o-Propenylphenol
m-Propenylphenol
Propiophenone
o-Tolylacetaldehyde
m-Tolylacetaldehyde
p-Tolylacetaldehyde

$C_9H_{10}OS$

Thioacetic Acid, O-*Benzyl ester*
Thiobenzoic Acid, S-*Et ester*

$C_9H_{10}OS_2$

2-Hydroxybenzenethionothiolic Acid, *Me ester, Me ether*

$C_9H_{10}O_2$

3-Allylcatechol
4-Allylcatechol
Benzyl acetate
2,3-Dimethylbenzoic Acid
2,4-Dimethylbenzoic Acid
2,5-Dimethylbenzoic Acid
2,6-Dimethylbenzoic Acid
3,4-Dimethylbenzoic Acid
3,5-Dimethylbenzoic Acid
Ethyl benzoate
o-Ethylbenzoic Acid
m-Ethylbenzoic Acid
p-Ethylbenzoic Acid
2-Ethyl-5-methyl-1,4-benzoquinone
2-Ethyl-6-methyl-1,4-benzoquinone
3-Ethyltropolone†
4-Ethyltropolone†
5-Ethyltropolone†
Guaiacol, *Vinyl ether*
2-Hydroxyacetophenone, *Me ether*
3-Hydroxyacetophenone, *Me ether*
4-Hydroxyacetophenone, *Me ether*
α-Hydroxyacetophenone, *Me ether*
m-Hydroxybenzaldehyde, *Et ether*
p-Hydroxybenzaldehyde, *Et ether*
p-Hydroxycinnamyl Alcohol
2-Hydroxy-3,5-dimethylbenzaldehyde
2-Hydroxy-3,6-dimethylbenzaldehyde
2-Hydroxy-4,5-dimethylbenzaldehyde
2-Hydroxy-4,6-dimethylbenzaldehyde
2-Hydroxy-5,6-dimethylbenzaldehyde
4-Hydroxy-2,3-dimethylbenzaldehyde
4-Hydroxy-2,5-dimethylbenzaldehyde
4-Hydroxy-2,6-dimethylbenzaldehyde
4-Hydroxy-3,5-dimethylbenzaldehyde
α-Hydroxy-4-methylacetophenone
2-Hydroxy-3-methylacetophenone
2-Hydroxy-4-methylacetophenone
2-Hydroxy-5-methylacetophenone
4-Hydroxy-2-methylacetophenone
4-Hydroxy-3-methylacetophenone
p-Hydroxymethyl-benzaldehyde, *Me ether*
2-Hydroxymethyloxiran, *Phenyl ether*
1-Hydroxymethyl-2-phenyloxiran
2-Hydroxyphenylacetaldehyde, *Me ether*
3-Hydroxyphenylacetaldehyde, *Me ether*
2-Hydroxy-2-phenylacetaldehyde, *Me ether*
1-Hydroxy-3-phenylacetone
p-Hydroxyphenylacetone
1-Hydroxy-1-phenyl-2-propanone
β-Hydroxypropiophenone
γ-Hydroxypropiophenone
2-Hydroxypropiophenone
3-Hydroxypropiophenone
4-Hydroxypropiophenone
4-Hydroxy-*o*-tolualdehyde, *Me ether*
6-Hydroxy-*o*-tolualdehyde, *Me ether*
2-Hydroxy-*m*-tolualdehyde, *Me ether*
4-Hydroxy-*m*-tolualdehyde, *Me ether*
6-Hydroxy-*m*-tolualdehyde, *Me ether*
2-Hydroxy-*p*-tolualdehyde, *Me ether*
6-Isochromanol
4-Methoxyphenylacetaldehyde
2-Methoxy-4-vinylphenol
4-Methyl-4-cyclohexene-1,2-dicarboxylic Acid, *Di-nitrile*
Non-2-ene-4,6-diyne-1,9-diol
Phenoxyacetone
Phenylacetic Acid, *Me ester*
2-Phenylpropionic Acid
3-Phenylpropionic Acid
Propionic Acid, *Phenyl ester*
Salicylaldehyde, *Et ether*
o-Toluic Acid, *Me ester*
m-Toluic Acid, *Me ester*
p-Toluic Acid, *Me ester*
o-Tolylacetic Acid
m-Tolylacetic Acid
p-Tolylacetic Acid
2,3,5-Trimethyl-1,4-benzoquinone

$C_9H_{10}O_2S$

2,3-Dihydro-3*H*-benzo[*b*]thiin, S-*dioxide*
o-Mercaptobenzoic Acid, *Et ether*
m-Mercaptobenzoic Acid, *Et ester*
p-Mercaptobenzoic Acid, S-*Et*
2-Mercapto-3-phenylpropionic Acid
3-Mercapto-3-phenylpropionic Acid
3-*o*-Mercaptophenylpropionic Acid
6-Mercapto-*m*-toluic Acid, S-*Me*
α-Mercaptophenylacetic Acid, *Me ester*†
o-Methylmercaptobenzoic Acid, *Me ester*
Phenylmercaptoacetic Acid, *Me ester*
Thioglycollic Acid, S-*Benzyl*

$C_9H_{10}O_2S_2$

Thiolbenzenesulphonic Acid, *Allyl ester*

$C_9H_{10}O_3$

Acetovanillone
Acetylquinol, 5-*Me ether*
Anisic Acid, *Me ester*
Bourbonal
Caffeyl Alcohol
α,4-Dihydroxyacetophenone, 4-*Me ether*
2,3-Dihydroxybenzaldehyde, *Di-Me ether*
2,4-Dihydroxybenzaldehyde, *Di-Me ether*
2,4-Dihydroxybenzaldehyde, 4-*Et ether*
2,4-Dihydroxy-3,6-dimethylbenzaldehyde
2,4-Dihydroxy-3-methylacetophenone
2,4-Dihydroxy-5-methylacetophenone
2,4-Dihydroxy-6-methylacetophenone
2,6-Dihydroxy-3-methylacetophenone
2,6-Dihydroxy-4-methylacetophenone
4,5-Dihydroxy-2-methylacetophenone
1,2-Dihydroxynona-3,5-diyn-7-one†
β,γ-Dihydroxypropiophenone

$C_9H_{10}O_3$ *(continued)*
2,3-Dihydroxypropiophenone
2,4-Dihydroxypropiophenone
2,5-Dihydroxypropiophenone
2,6-Dihydroxypropiophenone
3,4-Dihydroxypropiophenone
3,5-Dihydroxypropiophenone
4,5-Dihydroxy-*o*-tolualdehyde, 5-*Me ether*
4,6-Dihydroxy-*o*-tolualdehyde, 3-*Me ether*
4,6-Dihydroxy-*o*-tolualdehyde, 5-*Me ether*
3,4-Dimethoxybenzaldehyde
o-Ethoxybenzoic Acid
m-Ethoxybenzoic Acid
p-Ethoxybenzoic Acid
Ethyl salicylate
3-α-Furylacrylic Acid, *Et ester*
Gallacetonin
3-Hydroxybenzoic Acid, *Et ester*
4-Hydroxybenzoic Acid, *Et ester*
2-Hydroxy-3,5-dimethylbenzoic Acid
2-Hydroxy-3,6-dimethylbenzoic Acid
2-Hydroxy-4,5-dimethylbenzoic Acid
2-Hydroxy-4,6-dimethylbenzoic Acid
2-Hydroxy-5,6-dimethylbenzoic Acid
3-Hydroxy-4,5-dimethylbenzoic Acid
3-Hydroxy-4,6-dimethylbenzoic Acid
4-Hydroxy-2,5-dimethylbenzoic Acid
4-Hydroxy-2,6-dimethylbenzoic Acid
4-Hydroxy-3,5-dimethylbenzoic Acid
2-Hydroxy-4-methoxyacetophenone
4-Hydroxy-2-methoxyacetophenone
4-Hydroxy-3-methoxyphenylacetaldehyde
3-Hydroxy-4-methoxy-*o*-tolualdehyde
4-Hydroxy-3-methoxy-*o*-tolualdehyde
4-Hydroxy-5-methoxy-*o*-tolualdehyde
6-Hydroxy-4-methoxy-*o*-tolualdehyde
6-Hydroxy-5-methoxy-*o*-tolualdehyde
4-Hydroxy-5-methoxy-*m*-tolualdehyde
6-Hydroxy-5-methoxy-*m*-tolualdehyde
4-Hydroxy-6-methoxy-*o*-toluic Aldehyde
o-Hydroxymethyl-benzoic Acid, *Me ether*
p-Hydroxymethyl-benzoic Acid, *Me ether*
2-Hydroxymethyl-*p*-toluic Acid
3-Hydroxymethyl-*p*-toluic Acid
4-Hydroxymethyl-*o*-toluic Acid
2-Hydroxyphenylacetic Acid, *Me ester*
2-Hydroxyphenylacetic Acid, *Me ether*
3-Hydroxyphenylacetic Acid, *Me ether*
4-Hydroxyphenylacetic Acid, *Me ester*
2-*p*-Hydroxyphenylpropionic Acid ★†
2-Hydroxy-2-phenylpropionic Acid
2-Hydroxy-3-phenylpropionic Acid
3-*o*-Hydroxyphenylpropionic Acid
3-*m*-Hydroxyphenylpropionic Acid
3-*p*-Hydroxyphenylpropionic Acid
3-Hydroxy-2-phenylpropionic Acid
3-Hydroxy-3-phenylpropionic Acid
4-(2-Hydroxypropionyl)phenol
4-(3-Hydroxypropionyl)phenol
3-Hydroxy-*o*-toluic Acid, *Me ester*
3-Hydroxy-*o*-toluic Acid, *Me ether*
4-Hydroxy-*o*-toluic Acid, *Me ether*
5-Hydroxy-*o*-toluic Acid, *Me ester*
5-Hydroxy-*o*-toluic Acid, *Me ether*
6-Hydroxy-*o*-toluic Acid, *Me ester*
6-Hydroxy-*o*-toluic Acid, *Me ether*
2-Hydroxy-*m*-toluic Acid, *Me ester*
2-Hydroxy-*m*-toluic Acid, *Me ether*
4-Hydroxy-*m*-toluic Acid, *Me ether*
5-Hydroxy-*m*-toluic Acid, *Me ester*
5-Hydroxy-*m*-toluic Acid, *Me ether*
6-Hydroxy-*m*-toluic Acid, *Me ester*
6-Hydroxy-*m*-toluic Acid, *Me ether*
2-Hydroxy-*p*-toluic Acid, *Me ester*
2-Hydroxy-*p*-toluic Acid, *Me ether*
3-Hydroxy-*p*-toluic Acid, *Me ether*
5-Hydroxytoluquinone, *Et ether*
2-Hydroxy-2-*m*-tolylacetic Acid
2-Hydroxy-2-*p*-tolylacetic Acid
Ipomeanine
Mandelic Acid, *Me ether*
Mandelic Acid, *Me ester*
o-Methoxybenzoic Acid, *Me ester*
m-Methoxybenzoic Acid, *Me ester*
p-Methoxyphenylacetic Acid
4-Methyl-1-cyclohexene-1,2-dicarboxylic Acid, *Anhydride*
1-Methyl-2-cyclohexene-1,2-dicarboxylic Acid, *Anhydride*
1-Methyl-4-cyclohexene-1,2-dicarboxylic Acid, *Anhydride*
3-Methyl-4-cyclohexene-1,2-dicarboxylic Acid, *Anhydride*
4-Methyl-4-cyclohexene-1,2-dicarboxylic Acid, *Anhydride*
2-(3,4-Methylenedioxyphenyl)ethanol
Non-4-ene-6,8-diyne-1,2,3-triol
Phenoxyacetic Acid, *Me ester*
2-Phenoxypropionic Acid
3-Phenoxypropionic Acid
Quinol, *Me ether*, *Ac*

$C_9H_{10}O_3S$
p-Ethylsulphonylbenzaldehyde
Indane-4-sulphonic Acid
Indane-5-sulphonic Acid

$C_9H_{10}O_4$
3-Acetyl-2-hydroxy-6-methylpyran-4-one, *Me ester*
Bicyclo[2,2,1]hept-5-ene-1,4-dicarboxylic Acid
Bicyclo[2,2,1]hept-5-ene-2,2-dicarboxylic Acid
Bicyclo[2,2,1]hept-5-ene-2,3-dicarboxylic Acid
Carolic Acid
1,4-Cyclohexadiene-1,4-dicarboxylic Acid, *Mono-Me ester*
2,6-Cyclohexadiene-1,2-dicarboxylic Acid, *Mono-Me ester*
2,3-Dihydroxybenzoic Acid, *Et ester*
2,3-Dihydroxybenzoic Acid, 2-*Me ether*, *Me ester*
2,3-Dihydroxybenzoic Acid, 3-*Me ether*, *Me ester*
2,4-Dihydroxybenzoic Acid, 4-*Me ether*, *Me ester*
2,4-Dihydroxybenzoic Acid, 2-*Et ether*
2,5-Dihydroxybenzoic Acid, *Et ester*
2,5-Dihydroxybenzoic Acid, 5-*Me ether*, *Me ester*
2,5-Dihydroxybenzoic Acid, 5-*Et ether*
3,4-Dihydroxybenzoic Acid, *Et ester*
3,5-Dihydroxybenzoic Acid, *Et ester*
2,4-Dihydroxy-5,6-dimethylbenzoic Acid †

$C_9H_{10}O_4$ (*continued*)
α,2-Dihydroxyphenylacetic Acid, 2-*Me ether*
α,3-Dihydroxyphenylacetic Acid, 3-*Me ether*
α,4-Dihydroxyphenylacetic Acid, 4-*Me ether*
2,3-Dihydroxyphenylacetic Acid, 3-*Me ether*
2,5-Dihydroxyphenylacetic Acid, *Me ester*
2,5-Dihydroxyphenylacetic Acid, *Me ether*
2,3-Dihydroxy-2-phenylpropionic Acid
2,3-Dihydroxy-3-phenylpropionic Acid
2-(3,4-Dihydroxyphenyl)propionic Acid
3-(2,4-Dihydroxyphenyl)propionic Acid
3-(2,6-Dihydroxyphenyl)propionic Acid
3-(3,4-Dihydroxyphenyl)propionic Acid
3-(3,5-Dihydroxyphenyl) propionic Acid
3,5-Dihydroxy-*o*-toluic Acid, *Me ester*
4,6-Dihydroxy-*o*-toluic Acid, *Me ester*
2,4-Dihydroxy-*m*-toluic Acid, *Me ester*
5,6-Dihydroxy-*m*-toluic Acid, 5-*Me ether*
2,6-Dihydroxy-*p*-toluic Acid, *Me ester*
2,6-Dihydroxy-*p*-toluic Acid, 3-*Me ether*
3,5-Dihydroxy-*p*-toluic Acid, *Me ester*
2,3-Dimethoxybenzoic Acid
2,4-Dimethoxybenzoic Acid
2,5-Dimethoxybenzoic Acid
2,6-Dimethoxybenzoic Acid
3,4-Dimethoxybenzoic Acid
3,5-Dimethoxybenzoic Acid
1,4-Dimethoxy-2,3-methylenedioxybenzene
4,6-Dimethyl-2-oxo-2*H*-pyran-5-carboxylic Acid, *Me ester*
2-Ethyl-4,6-dihydroxybenzoic Acid †
Fumigatin, *Me ether*
α-Furoylacetic Acid, *Et ester*
3-α-Furylacrylic Acid, *Methoxymethyl ester*
4-Hydroxy-3,5-dimethoxybenzaldehyde
3-Hydroxy-4-methoxybenzoic Acid, *Me ester*
4-Hydroxy-3-methoxyphenylacetic Acid
4-Hydroxy-6-methoxy-*o*-toluic Acid
6-Hydroxy-4-methoxy-*o*-toluic Acid
2-Hydroxy-4-methoxy-*m*-toluic Acid
2-Hydroxy-6-methoxy-*p*-toluic Acid
3-Hydroxy-5-methoxy-*p*-toluquinone, *Me ether* †
4-Hydroxy-6-(2-oxopropyl)-2*H*-pyran-2-one, 4-*Me ether* †
Mesotan
Nortricyclene-2,3-dicarboxylic Acid
Palasonin †
α,2,4-Trihydroxyacetophenone, α-*Me ether*
α,2,4-Trihydroxyacetophenone, 4-*Me ether*
2,3,4-Trihydroxyacetophenone, 2-(or 4)-*Me ether*
2,3,4-Trihydroxyacetophenone, 3-*Me ether*
2,3,4-Trihydroxyacetophenone, 4-*Me ether*
2,4,5-Trihydroxyacetophenone, 4-*Me ether*
2,4,5-Trihydroxyacetophenone, 5-*Me ether*
2,4,6-Trihydroxyacetophenone, 2-*Me ether*
2,4,6-Trihydroxyacetophenone, 4-*Me ether*
2,3,4-Trihydroxybenzaldehyde, 3,4-*Di-Me ether*
2,3,5-Trihydroxybenzaldehyde, 2,3-*Di-Me ether*
2,4,5-Trihydroxybenzaldehyde, 4,5-*Di-Me ether*
2,4,6-Trihydroxybenzaldehyde, 4,6-*Di-Me ether*
3,4,5-Trihydroxybenzaldehyde, 3,5-*Di-Me ether*
2,3,4-Trihydroxypropiophenone
2,4,5-Trihydroxypropiophenone
2,4,6-Trihydroxypropiophenone
3,4,5-Trihydroxypropiophenone
Vanillic Acid, *Me ester*

$C_9H_{10}O_5$
2,5-Dimethylfuran-3,4-dicarboxylic Acid, *Me ester*
Gallic Acid, *Et ester*
Gallic Acid, 3-*Me ether*, *Me ester*
Gallic Acid, 4-*Me ether*, *Me ester*
Gallic Acid, 3,4-*Di-Me ether*
4-Hydroxy-3,5-dimethoxybenzoic Acid
Methronic Acid, *Mono-Me ester*
2-Methylfuran-3,4-dicarboxylic Acid, *Et ester*
5-Methylfuran-2,3-dicarboxylic Acid, *Et ester*
α,2,4,6-Tetrahydroxyacetophenone, α-*Me ether*
α,3,4,5-Tetrahydroxyacetophenone, α-*Me ether*
2,3,4-Trihydroxybenzoic Acid, *Me ester*, 4-*Me ether*
2,3,4-Trihydroxybenzoic Acid, *Et ester*
2,3,4-Trihydroxybenzoic Acid, 2,3-*Di-Me ether*
2,3,4-Trihydroxybenzoic Acid, 3,4-*Di-Me ether*
2,3,5-Trihydroxybenzoic Acid, 2,3-*Di-Me ether*
2,3,6-Trihydroxybenzoic Acid, 2,3-*Di-Me ether*
2,4,5-Trihydroxybenzoic Acid, 4,5-*Di-Me ether*
2,4,6-Trihydroxybenzoic Acid, *Et ester*
2,4,6-Trihydroxybenzoic Acid, 4-*Me ether*, *Me ester*
2,4,6-Trihydroxybenzoic Acid, 2,6-*Di-Me ether*
2,4,6-Trihydroxybenzoic Acid, 4,6-*Di-Me ether*
4,5,6-Trihydroxy-*o*-toluic Acid, *Me ester*
2,4,6-Trihydroxy-*m*-toluic Acid, *Me ester*
2,4,6-Trihydroxy-*m*-toluic Acid, 4-*Me ether*

$C_9H_{10}O_5S$
m-Sulphobenzoic Acid, *Di-Me ester*
p-Sulphobenzoic Acid, *Di-Me ester*

$C_9H_{10}O_6$
Carolinic Acid
5,6-Dihydroxy-4-oxopyran-2-carboxylic Acid, *Di-Me ether*, *Me ester*
Rubiginic Acid, *Me ester*, *Di-Me ether*
2,3,4,5-Tetrahydroxybenzoic Acid, 2,5-*Di-Me ether*
2,3,4,5-Tetrahydroxybenzoic Acid, 3,5-*Di-Me ether*

$C_9H_{10}S$
2,3-Dihydro-3*H*-benzo[*b*]thiin
Isothiochroman

$C_9H_{10}S_2$
Dithiobenzoic Acid, *Et ester*

$C_9H_{11}Br$
α-Bromo-3,5-dimethyltoluene
β-Bromoisopropylbenzene
o-Bromoisopropylbenzene
m-Bromoisopropylbenzene
p-Bromoisopropylbenzene
1-Bromo-2,4,6-trimethylbenzene
3-Bromo-1,2,4-trimethylbenzene
5-Bromo-1,2,4-trimethylbenzene
6-Bromo-1,2,4-trimethylbenzene

$C_9H_{11}BrHg$
Mercuri-2,4,6-trimethylphenyl bromide

$C_9H_{11}BrN_2O$
N-Methyl-4-nitrosoaniline, N-β-*Bromoethyl*

$C_9H_{11}BrN_2O_2$
2-Bromo-5-nitro-*p*-toluidine, N-*Di-Me*
Pyridestigmine Bromide

$C_9H_{11}BrN_2O_5$
5-Bromodeoxyuridine †

$C_9H_{11}BrN_2O_6$
5-Bromouridine †

$C_9H_{11}BrO$
m-Bromobenzyl Alcohol, *Et ether*
p-Bromobenzyl Alcohol, *Et ether*
4-Bromo-*o*-cresol, *Et ether*
2-Bromo-*p*-cresol, *Et ether*
6-Bromo-2,4-xylenol, *Me ether*
6-Bromo-3,4-xylenol, *Me ether*

$C_9H_{11}BrO_2$
4-Bromocatechol, 1-*Me*-2-*Et ether*
4-Bromo-5-methylresorcinol, *Di-Me ether*
Guaiacol, β-*Bromoethyl ether*

$C_9H_{11}Cl$
α-Chloro-2,4-dimethyltoluene
α-Chloro-2,5-dimethyltoluene
α-Chloro-3,4-dimethyltoluene
α-Chloro-3,5-dimethyltoluene
2-Chloroisopropylbenzene
p-Chloroisopropylbenzene
1-Chloropropylbenzene
2-Chloropropylbenzene
3-Chloropropylbenzene
o-Chloropropylbenzene
p-Chloropropylbenzene
1-Chloro-2,3,5-trimethylbenzene
1-Chloro-2,4,5-trimethylbenzene
1-Chloro-2,3,6-trimethylbenzene
1-Chloro-2,4,6-trimethylbenzene

$C_9H_{11}ClHg$
Mercuri-2,4,6-trimethylphenyl chloride

$C_9H_{11}ClN_2O$
N-Methyl-4-nitrosoaniline, N-β-*Chloroethyl*
Monuron

$C_9H_{11}ClN_2O_2$
N-(4-Chlorophenyl)-*N'*-methoxy-*N'*-methylurea †

$C_9H_{11}ClO$
o-Chlorobenzyl Alcohol, *Et ether*
m-Chlorobenzyl Alcohol, *Et ether*
p-Chlorobenzyl Alcohol, *Et ether*
5-Chloro-*o*-cresol, *Et ether*
4-Chloro-*m*-cresol, *Et ether*
2-Chloro-*p*-cresol, *Et ether*

$C_9H_{11}ClO_2$
5-Chloro-2-methoxyphenol, *Et ether*
2-(4-Chloro-*o*-tolyloxy)ethanol

$C_9H_{11}ClO_2S$
2,4,6-Trimethylbenzenesulphonic Acid, *Chloride*

$C_9H_{11}ClO_3$
5-Chloropyrogallol, *Tri-Me ether*
Chlorphenesin

$C_9H_{11}ClO_4$
2,3-Dimethylbutane-1,2,3-tricarboxylic Acid, *Chloride*

$C_9H_{11}F$
p-Fluoropropylbenzene
1-Fluoro-2,4,6-trimethylbenzene

$C_9H_{11}FO_2S$
2,4,6-Trimethylbenzenesulphonic Acid, *Fluoride*

$C_9H_{11}F_7O_2$
Heptafluorobutyric Acid, *Pentyl ester*

$C_9H_{11}HgI$
Mercuri-2,4,6-trimethylphenyl iodide

$C_9H_{11}HgNO_2$
Mercuri-4-amino-*o*-tolyl acetate
Mercuri-4-amino-*m*-tolyl acetate
Mercuri-*p*-methylaminophenyl acetate

$C_9H_{11}I$
p-Iodoisopropylbenzene
5-Iodo-1,2,4-trimethylbenzene

$C_9H_{11}N$
1-Aminoindane †
2-Aminoindane †
4-Aminoindane
5-Aminoindane
N-Benzylidene-ethylamine
Cinnamylamine
2,3-Dihydro-1-methylindole
2,3-Dihydro-2-methylindole
2,3-Dihydro-3-methylindole
2,3-Dihydro-4-methylindole
2,3-Dihydro-7-methylindole
N-Phenylazetidine †
2-Phenylcyclopropylamine †
1,2,3,4-Tetrahydroisoquinoline
5,6,7,8-Tetrahydroisoquinoline
1,2,3,4-Tetrahydroquinoline
5,6,7,8-Tetrahydroquinoline

$C_9H_{11}NO$
Acet-*o*-toluidide
Acet-*m*-toluidide
Acet-*p*-toluidide
2-Amino-3,5-dimethylbenzaldehyde
2-Amino-2-hydroxyindane
α-Aminopropiophenone
β-Aminopropiophenone
o-Aminopropiophenone
m-Aminopropiophenone
p-Aminopropiophenone
o-Dimethylaminobenzaldehyde
m-Dimethylaminobenzaldehyde
p-Dimethylaminobenzaldehyde
2,3-Dimethylbenzoic Acid, *Amide*
2,4-Dimethylbenzoic Acid, *Amide*
2,5-Dimethylbenzoic Acid, *Amide*
2,6-Dimethylbenzoic Acid, *Amide*
3,4-Dimethylbenzoic Acid, *Amide*
3,5-Dimethylbenzoic Acid, *Amide*
p-Ethylaminobenzaldehyde
N-Ethylbenzamide
Ethyl benzimidate
o-Ethylbenzoic Acid, *Amide*
p-Ethylbenzoic Acid, *Amide*
3-Hydroxy-3-phenylazetidine †
7-Hydroxy-1,2,3,4-tetrahydroisoquinoline
N-Methylacetanilide

$C_9H_{11}NO$ (*continued*)

α-Methylaminoacetophenone
p-Methylaminoacetophenone
4-Methylamino-*m*-toluic Aldehyde
2-Phenylpropionic Acid, *Amide*
3-Phenylpropionic Acid, *Amide*
1-(2-Pyridyl)-1-butanone
1-(3-Pyridyl)-1-butanone
1-(4-Pyridyl)-1-butanone
1,2,3,4-Tetrahydro-4-hydroxyquinoline
1,2,3,4-Tetrahydro-5-hydroxyquinoline
1,2,3,4-Tetrahydro-6-hydroxyquinoline
1,2,3,4-Tetrahydro-7-hydroxyquinoline
1,2,3,4-Tetrahydro-8-hydroxyquinoline
5,6,7,8-Tetrahydro-3-hydroxyquinoline
5,6,7,8-Tetrahydro-4-hydroxyquinoline
o-Toluamide, N-*Me*
m-Toluamide, N-*Me*
p-Toluamide, N-*Me*
o-Tolylacetic Acid, *Amide*
m-Tolylacetic Acid, *Amide*
p-Tolylacetic Acid, *Amide*
Venoterpine†

$C_9H_{11}NOS$

Ethyl phenylaminothioformate

$C_9H_{11}NO_2$

m-Aminobenzoic Acid, *Et ester*
2-Amino-3,5-dimethylbenzoic Acid
4-Amino-2,6-dimethylbenzoic Acid
4-Amino-3,5-dimethylbenzoic Acid
o-β-Aminoethylbenzoic Acid
p-β-Aminoethylbenzoic Acid
4-Amino-2-ethylbenzoic Acid
5-Amino-2-ethylbenzoic Acid
1-(2-Aminoethyl)-3,4-methylenedioxybenzene
α-Amino-4-hydroxyacetophenone, *Me ether*
3-Amino-4-hydroxyacetophenone, *Me ether*
α-Aminophenylacetic Acid, *Me ester*
2-*p*-Aminophenylpropionic Acid
3-*m*-Aminophenylpropionic Acid
3-*p*-Aminophenylpropionic Acid
2-Amino-2-phenylpropionic Acid
3-Amino-2-phenylpropionic Acid
3-Amino-3-phenylpropionic Acid
4-Amino-*m*-toluic Acid, *Me ester*
4-Amino-*m*-toluic Acid, N-*Me*
6-Amino-*m*-toluic Acid, *Me ester*
6-Amino-*m*-toluic Acid, N-*Me*
2-Anilinopropionic Acid
3-Anilinopropionic Acid
Anthranilic Acid, *Et ester*
Benzocaine
Benzylglycine
o-Dimethylaminobenzoic Acid
m-Dimethylaminobenzoic Acid
p-Dimethylaminobenzoic Acid
4-Dimethylamino-2-hydroxybenzaldehyde
4,6-Dimethylpyridine-2-carboxylic Acid, *Me ester*
m-Ethoxybenzoic Acid, *Amide*
N-Ethoxycarbonylazepine
o-Ethylaminobenzoic Acid
m-Ethylaminobenzoic Acid
p-Ethylaminobenzoic Acid
Ethyl phenylaminoformate
2-Hydroxy-2-phenylpropionic Acid, *Amide*
2-Hydroxy-3-phenylpropionic Acid, *Amide*
3-*o*-Hydroxyphenylpropionic Acid, *Amide*
3-*p*-Hydroxyphenylpropionic Acid, *Amide*
3-Hydroxy-2-phenylpropionic Acid, *Amide*
3-Hydroxy-3-phenylpropionic Acid, *Amide*
4-Hydroxy-*m*-toluic Acid, *Me ether*, *Amide*
6-Hydroxy-*m*-toluic Acid, *Me ether*, *Amide*
Lactic Acid, *Anilide*
Mandelamide, *Me ether*
Mandelamide, N-*Me*
Methoxyacetic Acid, *Anilide*
p-Methoxyphenylacetic Acid, *Amide*
o-Methylaminobenzoic Acid, *Me ester*
m-Methylaminobenzoic Acid, *Me ester*
p-Methylaminobenzoic Acid, *Me ester*
2-Methylamino-2-phenylacetic Acid
p-Methylaminophenylacetic Acid
2-Methylamino-*m*-toluic Acid
4-Methylamino-*m*-toluic Acid
6-Methylamino-*m*-toluic Acid
N-Methylphenylcarbamic Acid, *Me ester*
4-Methylpyridine-2-carboxylic Acid, *Et ester*
6-Methylpyridine-2-carboxylic Acid, *Et ester*
2-Methylpyridine-3-carboxylic Acid, *Et ester*
4-Methylpyridine-3-carboxylic Acid, *Et ester*
6-Methylpyridine-3-carboxylic Acid, *Et ester*
Nitromesitylene
α-Nitromesitylene
o-Nitropropylbenzene
m-Nitropropylbenzene
p-Nitropropylbenzene
2-Phenoxypropionic Acid, *Amide*
β-Phenylalanine
N-Phenylglycine, *Me ester*
N-Phenylsarcosine
2-Pyridineacetic Acid, *Et ester*
3-Pyridineacetic Acid, *Et ester*
4-Pyridineacetic Acid, *Et ester*
Pyridine-3-carboxylic Acid, *Propyl ester*
N-*o*-Tolylglycine
N-*m*-Tolylglycine
N-*p*-Tolylglycine
1,2,4-Trimethyl-3-nitrobenzene
1,2,4-Trimethyl-5-nitrobenzene
1,2,4-Trimethyl-6-nitrobenzene

$C_9H_{11}NO_2S$

2-Amino-3-mercapto-3-phenylpropionic Acid
Indane-4-sulphonic Acid, *Amide*
Indane-5-sulphonic Acid, *Amide*
o-Nitrobenzenethiol, S-*Propyl ether*
p-Nitrobenzenethiol, S-*Propyl ether*
p-Nitrobenzenethiol, S-*Isopropyl ether*
S-Phenylcysteine

$C_9H_{11}NO_3$

Adrenalone
2-Amino-4-hydroxybenzoic Acid, *Me ether*, *Me ester*
2-Amino-4-hydroxybenzoic Acid, *Et ether*
2-Amino-5-hydroxybenzoic Acid, *Et ester*
3-Amino-2-hydroxybenzoic Acid, *Et ester*
3-Amino-4-hydroxybenzoic Acid, *Et ether*
4-Amino-3-hydroxybenzoic Acid, *Et ester*
5-Amino-2-hydroxybenzoic Acid, *Et ester*
2-Amino-3-hydroxy-3-phenylpropionic Acid

$C_9H_{11}NO_3$ (*continued*)
3-Amino-2-hydroxy-3-phenylpropionic Acid
4-Aminophenoxyacetic Acid, *Me ester*
3-*p*-Aminophenyl-2-hydroxypropionic Acid
Bicyclo[2,2,1]hept-5-ene-2,3-dicarboxylic Acid, *Mono-amide*
α,4-Dihydroxyphenylacetic Acid, 4-*Me ether, Amide*
2,3-Dihydroxy-3-phenylpropionic Acid, *Amide*
2,5-Dimethoxybenzoic Acid, *Amide*
2,6-Dimethoxybenzoic Acid, *Amide*
3,4-Dimethoxybenzoic Acid, *Amide*
3,5-Dimethoxybenzoic Acid, *Amide*
2,4-Dimethyl-5-nitrophenol, *Me ether*
2,4-Dimethyl-6-nitrophenol, *Me ether*
2,5-Dimethyl-3-nitrophenol, *Me ether*
2,6-Dimethyl-4-nitrophenol, *Me ether*
3,5-Dimethyl-2-nitrophenol, *Me ether*
3,5-Dimethyl-4-nitrophenol, *Me ether*
4,5-Dimethyl-2-nitrophenol, *Me ether*
Ethyl 2-hydroxyphenylcarbamate
Ethyl 3-hydroxyphenylcarbamate
Ethyl 4-hydroxyphenylcarbamate
2-Hydroxyphenylglycine, *Me ether*
4-Hydroxyphenylglycine, *Me ether*
4-Hydroxyphenylglycine, *Me ester*
6-Hydroxypyridine-3-carboxylic Acid, *Me ether, Et ester*
Lactic Acid, p-*Hydroxyanilide*
3-Methoxy-2-methylaminobenzoic Acid
4-Nitro-*o*-cresol, *Et ether*
5-Nitro-*o*-cresol, *Et ether*
6-Nitro-*o*-cresol, *Et ether*
4-Nitro-*m*-cresol, *Et ether*
6-Nitro-*m*-cresol, *Et ether*
2-Nitro-*p*-cresol, *Et ether*
2-Nitro-1-phenylethanol, *Me ether*
2,4,6-Trimethyl-3-nitrophenol
3,4,5-Trimethyl-2-nitrophenol
Tyrosine
o-Tyrosine
m-Tyrosine

$C_9H_{11}NO_3S$
p-Formylbenzenesulphonic Acid, *Dimethylamide*

$C_9H_{11}NO_4$
2-Amino-3,5-dihydroxybenzoic Acid, 3,5-*Di-Me ether*
2-Amino-5,6-dihydroxybenzoic Acid, 5,6-*Di-Me ether*
3-Amino-2,5-dihydroxybenzoic Acid, 2,5-*Di-Me ether*
3-Amino-5,6-dihydroxybenzoic Acid, 5,6-*Di-Me ether*
4-Amino-3,5-dihydroxybenzoic Acid, 3,5-*Di-Me ether*
2-Amino-3,4-dimethoxybenzoic Acid
2-Amino-4,5-dimethoxybenzoic Acid
5-Amino-3,4-dimethoxybenzoic Acid
1,2-Dihydroxy-4-nitro-5-propylbenzene
α,4-Dihydroxy-3-nitrotoluene, 4-*Et ether*
3,5-Dihydroxy-2-nitrotoluene, 3-*Et ether*
3,5-Dihydroxy-2-nitrotoluene, 5-*Et ether*
β-[2,4-Dihydroxyphenyl]alanine
β-[2,5-Dihydroxyphenyl]alanine
β-[3,4-Dihydroxyphenyl]alanine
β-[3,5-Dihydroxyphenyl]alanine
1,4-Dihydroxy-2,3,5-trimethyl-6-nitrobenzene
3,4-Dimethoxy-2-nitrotoluene
3,4-Dimethoxy-5-nitrotoluene
4,5-Dimethoxy-2-nitrotoluene
3,5-Dimethylpyrrole-2,4-dicarboxylic Acid, 4-*Me ester*
o-Nitrobenzaldehyde, *Di-Me acetal*
m-Nitrobenzaldehyde, *Di-Me acetal*
p-Nitrobenzaldehyde, *Di-Me acetal*
4-Nitrocatechol, 1-*Me ether*, 2-*Et ether*
4-Nitrocatechol, 2-*Me ether*, 1-*Et ether*
1-*o*-Nitrophenyl-1,3-propanediol
Pyrrole-2,5-dicarboxylic Acid, N-*Me, Di-Me ester*

$C_9H_{11}NO_4S$
o-Sulphobenzoic Acid, 1-*Et ester, Amide*
o-Sulphobenzoic Acid, 1-*Ethylamide*
p-Sulphobenzoic Acid, 1-*Et ester, Amide*

$C_9H_{11}NO_5$
2,3-Dihydroxy-5-nitrobenzyl Alcohol, 2,3-*Di-Me ether*
2,5-Dimethyl-4-nitrofuran-3-carboxylic Acid, *Et ester*
Tartramidic Acid, *Et ester*
1,2,3-Trihydroxy-4-nitrobenzene, *Tri-Me ether*
1,2,3-Trihydroxy-5-nitrobenzene, *Tri-Me ether*
1,2,4-Trihydroxy-5-nitrobenzene, *Tri-Me ether*
1,3,5-Trihydroxy-2-nitrobenzene, *Tri-Me ether*

$C_9H_{11}NO_5S$
4-Hydroxy-3-sulphobenzoic Acid, *Sulphonamide, Et ether*

$C_9H_{11}NO_6$
Showdomycin†

$C_9H_{11}NS_2$
Ethyl phenylaminodithioformate

$C_9H_{11}N_3$
2-β-Phenylhydrazinopropionic Acid, *Nitrile*

$C_9H_{11}N_3O$
Phenylhydrazidine

$C_9H_{11}N_3O_2$
Anilinomalonic Acid, *Diamide*
2,6-Dimethylpyridine-3,4-dicarboxylic Acid, *Diamide*
o-(*N*-Ethyl-*N*-nitrosoamino)benzoic Acid, *Amide*
N-Phenylglycine-*o*-carboxylic Acid, *Diamide*
N-Phenylglycine-*m*-carboxylic Acid, *Diamide*
2-Phenylhydantoic Acid, *Amide*

$C_9H_{11}N_3O_3$
3-Amino-4-nitrobenzoic Acid, N-*Me, methylamide*
3, 4-Dimethyl-3-nitrobenzoic Acid, *Amide*

$C_9H_{11}N_3O_5$
4-Amino-2,6-dinitrophenol, N-*Me*, N-*Et*
4-Amino-2,6-dinitrophenol, N-*Di-Me, Me ether*

$C_9H_{11}N_3O_{10}S$
Trimethyloxonium 2,4,6-Trinitrobenzenesulphonate†

$C_9H_{11}N_5O_2$
Hippuric Acid, *Hydrazide*

$C_9H_{11}N_5O_3$
Biopterin
Ichthyopterin

$C_9H_{11}N_5O_4$
Neopterin†

$C_9H_{11}N_5O_5$
8-Azainosine

$C_9H_{11}N_5O_6$
8-Azaxanthine-7-xylopyranoside
8-Azaxanthine-9-xylopyranoside
8-Azaxanthosine

C_9H_{12}
Cumene
1,3,5-Cyclononatriene†
1,4,7-Cyclononatriene★†
o-Ethyltoluene
m-Ethyltoluene
p-Ethyltoluene
1,4-Nonadi-yne
1,5-Nonadi-yne
1,8-Nonadi-yne
2,7-Nonadi-yne
Propylbenzene
Tetracyclo[3,3,1,$0^{2,4}0^{3,7}$]nonane†
Triasterane†
1,2,3-Trimethylbenzene
1,2,4-Trimethylbenzene
1,3,5-Trimethylbenzene

$C_9H_{12}Al$
Triallylaluminium

$C_9H_{12}ClN$
α-Amino-2-chlorotoluene, N-*Di-Me*

$C_9H_{12}NO_6P$
Tyrosine-*O*-phosphate†

$C_9H_{12}N_2$
2-2′-Pyridyl-pyrrolidine
2-3′-Pyridyl-pyrrolidine
1,2,3,4-Tetrahydroquinoxaline, N-*Me*

$C_9H_{12}N_2O$
2-Anilinopropionic Acid, *Amide*
o-Dimethylaminobenzoic Acid, *Amide*
p-Dimethylaminobenzoic Acid, *Amide*
o-Ethylaminobenzoic Acid, *Amide*
1-Ethyl-1-phenylurea
1-Ethyl-3-phenylurea
Fenuron
Kynuramine
2-Methylamino-2-phenylacetic Acid, *Amide*
N-Methyl-4-nitrosoaniline, N-*Et*
N^1-Methyl-N^1-*p*-tolylurea
p-Nitrosoaniline, N-*Propyl*
4-Nitroso-*o*-toluidine, N-*Et*
4-Nitroso-*m*-toluidine, N-*Di-Me*
β-Phenylalanine, *Amide*
N-Phenylsarcosine, *Amide*
3-Phenylpropionic Acid, *Hydrazide*
N-*o*-Tolylglycine, *Amide*
N-*p*-Tolylglycine, *Amide*

$C_9H_{12}N_2OS$
2-Hydroxyphenylthiourea, *Et ether*
3-Hydroxyphenylthiourea, *Et ether*
4-Hydroxyphenylthiourea, *Et ether*

$C_9H_{12}N_2O_2$
2-Amino-6-hydroxybenzoic Acid, *Et ether, Amide*
o-Aminophenylurethane
p-Aminophenylurethane
Butane-2,2,3-tricarboxylic Acid, 2,3-*Dinitrile*, 2-*Et ester*
2,5-Diaminobenzoic Acid, *Et ester*
3,4-Diaminobenzoic Acid, *Et ester*
3,5-Diaminobenzoic Acid, *Et ester*
3-(3,4-Diaminophenyl)propionic Acid
3,5-Diamino-*p*-toluic Acid, *Me ester*
2,4-Dimethyl-6-nitroaniline, N-*Me*
N-Dimethyl-3-nitro-*o*-toluidine
N-Dimethyl-4-nitro-*o*-toluidine
N-Dimethyl-5-nitro-*o*-toluidine
N-Dimethyl-4-nitro-*m*-toluidine
N-Dimethyl-5-nitro-*m*-toluidine
N-Dimethyl-6-nitro-*m*-toluidine
N-Dimethyl-2-nitro-*p*-toluidine
N-Dimethyl-3-nitro-*p*-toluidine
Dulcin
N-Ethyl-4-nitro-*o*-toluidine
N-Ethyl-5-nitro-*o*-toluidine
N-Ethyl-6-nitro-*m*-toluidine
N-Ethyl-2-nitro-*p*-toluidine
N-Ethyl-3-nitro-*p*-toluidine
4-Ethyl-5-nitro-*o*-toluidine
4-Ethyl-6-nitro-*o*-toluidine
5-Ethyl-3-nitro-*o*-toluidine
4-Ethyl-6-nitro-*m*-toluidine
2-Hydroxyphenylglycine, *Et ester*, *Amide*
o-Nitrobenzylamine, N-*Et*
p-Nitrobenzylamine, N-*Et*
p-Nitrobenzylamine, N-*Di-Me*
2-Nitro-*N*-propylaniline
4-Nitro-*N*-propylaniline
Phenylhydrazine-α-carboxylic Acid, *Et ester*
Phenylhydrazine-β-carboxylic Acid, *Et ester*
2-β-Phenylhydrazinopropionic Acid
2,4,5-Trimethyl-3-nitroaniline
2,4,5-Trimethyl-6-nitroaniline
2,4,6-Trimethyl-3-nitroaniline

$C_9H_{12}N_2O_3$
2-Amino-4-nitrophenol, *Propyl ether*
Dormin
2-Hydroxy-5-nitropyridine, *Butyl ether*
2-Methoxy-4-nitroaniline, N-*Di-Me*
2-Methoxy-5-nitroaniline, N-*Di-Me*
4-Methoxy-3-nitroaniline, N-*Di-Me*
Tetrahydropyran-2,6-dicarboxylic Acid, *Diamide*

$(C_9H_{12}N_2O_3)_n$
Actinoleukin

$C_9H_{12}N_2O_4$
2-Amino-5-nitroquinol, 1-*Me*-4-*Et ether*

$C_9H_{12}N_2O_4S$
3-Nitrotoluene-α-sulphonic Acid, *Dimethylamide*
3-Nitrotoluene-α-sulphonic Acid, *Ethylamide*
4-Nitrotoluene-α-sulphonic Acid, *Dimethylamide*
4-Nitrotoluene-α-sulphonic Acid, *Ethylamide*
2-Nitrotoluene-4-sulphonic Acid, *Ethylamide*

$C_9H_{12}N_2O_6$
Pseudouridine
Pseudouridine A_F†
Pseudouridine A_S†
Pseudouridine B†
Pseudouridine C★†
Uridine

$C_9H_{12}N_2O_6S_2$
2,4-Disulphobenzoic Acid, 1-*Et ester*

$C_9H_{12}N_2O_8$
Orotidine

$C_9H_{12}N_2S$
N^1-Methyl-N^1-*o*-tolylthiourea
N^1-Methyl-N^2-*o*-tolylthiourea
N^1-Methyl-N^2-*p*-tolylthiourea
Phenylthiourea, 3-N-*Di-Me*
Phenylthiourea, 3-N-*Et*

$C_9H_{12}N_3O_7P$
Cytidine-2′,3′-phosphate

$C_9H_{12}N_4O_2$
Methylcaffeine
Theobromine, N-*Et*

$C_9H_{12}N_4O_3$
Tetramethyluric Acid

$C_9H_{12}N_6$
2,4,6-Tri(1-aziridinyl)-*s*-triazine

$C_9H_{12}N_6O_4$
8-Aza-adenosine

$C_9H_{12}N_6O_5$
8-Azaguanosine

$C_9H_{12}O$
Benzyl ethyl Ether
3-*tert*-Butylcyclopentadienone†
o-Cresol, *Et ether*
m-Cresol, *Et ether*
p-Cresol, *Et ether*
2,3-Dimethylbenzyl Alcohol
2,4-Dimethylbenzyl Alcohol
2,5-Dimethylbenzyl Alcohol
3,4-Dimethylbenzyl Alcohol
3,5-Dimethylbenzyl Alcohol
2,3-Dimethylphenol, *Me ether*
2,4-Dimethylphenol, *Me ether*
2,5-Dimethylphenol, *Me ether*
2,6-Dimethylphenol, *Me ether*
3,4-Dimethylphenol, *Me ether*
3,5-Dimethylphenol, *Me ether*
m-Ethylbenzyl Alcohol
p-Ethylbenzyl Alcohol
α-Hydroxy-*o*-xylene, *Me ether*
o-Isopropylphenol
m-Isopropylphenol
p-Isopropylphenol
Isopropyl phenyl Ether
Methyl phenethyl Ether
1-Phenyl-1-propanol
2-Phenyl-1-propanol
3-Phenyl-1-propanol
1-Phenyl-2-propanol
2-Phenyl-2-propanol
o-Propylphenol
m-Propylphenol
p-Propylphenol
1-*m*-Tolylethanol
1-*p*-Tolylethanol
2,6,6-Trimethyl-2,4-cyclohexadienone†
2,3,4-Trimethylphenol
2,3,5-Trimethylphenol
2,4,5-Trimethylphenol
2,4,6-Trimethylphenol
3,4,5-Trimethylphenol

$C_9H_{12}OS$
Ethyl *p*-tolyl Sulphoxide†
p-Hydroxybenzenethiol, S-*Me*, *Et ether*
p-Hydroxybenzenethiol, S-*Et*, *Me ether*
p-Hydroxybenzenethiol, S-*Propyl*

$C_9H_{12}O_2$
Benzyl hydroxyethyl Ether
Catechol, *Propyl ether*
Cyclohexylpropiolic Acid
α,β-Dihydroxyethylbenzene, *Me ether*
1,2-Dihydroxy-4-isopropylbenzene
1,3-Dihydroxy-2-isopropylbenzene
1,3-Dihydroxy-4-isopropylbenzene
1,3-Dihydroxy-5-isopropylbenzene
1,4-Dihydroxy-2-isopropylbenzene
1,2-Dihydroxy-3-propylbenzene
1,2-Dihydroxy-4-propylbenzene
1,3-Dihydroxy-2-propylbenzene
1,3-Dihydroxy-4-propylbenzene
1,3-Dihydroxy-5-propylbenzene
1,4-Dihydroxy-2-propylbenzene
α,2-Dihydroxytoluene, *Di-Me ether*
α,2-Dihydroxytoluene, α-*Et ether*
α,2-Dihydroxytoluene, 2-*Et ether*
2,3-Dihydroxytoluene, *Di-Me ether*
2,4-Dihydroxytoluene, *Di-Me ether*
2,6-Dihydroxytoluene, *Di-Me ether*
3,4-Dihydroxytoluene, 3-*Et ether*
3,5-Dihydroxytoluene, *Di-Me ether*
1,3-Dihydroxy-2,4,5-trimethylbenzene
1,3-Dihydroxy-2,4,6-trimethylbenzene
1,5-Dihydroxy-2,3,4-trimethylbenzene
2,5-Dihydroxy-*m*-xylene, 5-*Me ether*
4,5-Dihydroxy-*m*-xylene, 4-*Me ether*
4,6-Dihydroxy-*m*-xylene, 6-*Me ether*
2,5-Dihydroxy-*p*-xylene, *Me ether*
2,6-Dihydroxy-*p*-xylene, *Me ether*
2,3-Dimethoxytoluene
3,4-Dimethoxytoluene
2-Ethyl-5-methylquinol
2-Ethyl-6-methylquinol
Guaiacol, *Et ether*
2-Hydroxyethyl *o*-tolyl Ether
2-Hydroxyethyl *m*-tolyl Ether
2-Hydroxyethyl *p*-tolyl Ether
2-(4-Hydroxyphenyl)ethanol, 4-*Me ether*
2-*o*-Hydroxyphenyl-2-propanol
2-*m*-Hydroxyphenyl-2-propanol
2-α-Hydroxypropylphenol
2-γ-Hydroxypropylphenol
3-α-Hydroxypropylphenol
4-α-Hydroxypropylphenol
4-β-Hydroxypropylphenol
4-γ-Hydroxypropylphenol
p-Methoxybenzyl Alcohol, *Me ether*

$C_9H_{12}O_2$ (*continued*)
Methylquinol, *Di-Me ether*
Methylquinol, 5-*Et ether*
Norbornene-7-carboxylic Acid, *Me ester*
1-Phenyl-1,2-propanediol
2-Phenyl-1,2-propanediol
3-Phenyl-1,2-propanediol
1-Phenyl-1,3-propanediol
2-Phenyl-1,3-propanediol
1,3-Propanediol, *Phenyl ether*
Resorcinol, *Me-Et ether*
Resorcinol, *Mono-propyl ether*
Terephthalyl Alcohol, *Mono-Me ether*
2,3,5-Trimethylquinol

$C_9H_{12}O_2S$
Phenyl propyl sulphone
Thiophene-2-carboxylic Acid, *Butyl ester*
Toluene-*p*-sulphinic Acid, *Et ester*

$C_9H_{12}O_2S_2$
4-Acetyl-2-(1-acetylethylidene)-4-methyl-1,3-dithietan†
Thiolbenzenesulphonic Acid, *Propyl ester*

$C_9H_{12}O_3$
2-Carboxy-3-methylcyclopentylacetic Acid, *Anhydride*†
2,5-Diacetylcyclopentanone†
1-(2,5-Dihydroxyphenyl)propan-2-ol†
1,2-Dimethylcyclopentane-1,3-dicarboxylic Acid, *Anhydride*
2,2-Dimethylcyclopentane-1,3-dicarboxylic Acid, *Anhydride*
2,4-Dimethylfuran-3-carboxylic Acid, *Et ester*
2,5-Dimethylfuran-3-carboxylic Acid, *Et ester*
Filicinic Acid, *Me ether*
Furan-2-carboxylic Acid, *Butyl ester*
Furan-2-carboxylic Acid, *Isobutyl ester*
2-Furylacetic Acid, *Propyl ester*
2-Furylacetic Acid, *Isopropyl ester*
3-(2-Furyl)propionic Acid, *Et ester*
5-(2-Furyl)valeric Acid
Glycerol, 1-*Phenyl ether*
Guaiacol, β-*Hydroxyethyl ether*
α-Hydroxy-3,4-dimethoxytoluene
Iridol
Metacrolein
1-Methylcyclohexane-1,2-dicarboxylic Acid, *Anhydride*
2-Methyl-4-oxo-2-cyclohexene-1-carboxylic Acid, *Me ester*
Nona-3,5-diyne-1,2,7-triol†
1-Phenylglycerol
2-Phenylglycerol
Phloroglucinol, *Tri-Me ether*
Pyrogallol, *Tri-Me ether*
1,2,4-Trihydroxybenzene, 1,2,4-*Tri-Me ether*
2,3,5-Trihydroxytoluene, 2,3-*Di-Me ether*
2,4,5-Trihydroxytoluene, 4-*Et ether*
2,4,6-Trihydroxytoluene, 2,4-*Di-Me ether*
2,4,6-Trihydroxytoluene, 4-*Et ether*
2,4,6-Trihydroxytoluene, 6-*Et ether*
1,3,5-Trihydroxy-2,4,6-trimethylbenzene
1,3,5-Trishydroxymethylbenzene†

$C_9H_{12}O_3S$
Benzenesulphonic Acid, *Propyl este*
Toluene-*p*-sulphonic Acid, *Et ester*
2,4,6-Trimethylbenzenesulphonic Acid

$C_9H_{12}O_4$
Antiarol
1-Cyclohexene-1,3-dicarboxylic Acid, *Mono-Me ester*
1-Cyclopentene-1,2-dicarboxylic Acid, *Di-Me ester*
Diallylmalonic Acid
α,4-Dihydroxy-3,5-dimethoxytoluene
3-(1,2-Epoxypropyl)-5 6-dihydro-5-hydroxy-6-methylpyran-2-one†
Genipic Acid†
2-Isobutylidene-5,5-dimethyl-4,6-dioxo-*m*-dioxan†
Isosantenenic Acid
3-Methyl-3-cyclohexene-1,1-dicarboxylic Acid
4-Methyl-1-cyclohexene-1,2-dicarboxylic Acid
1-Methyl-2-cyclohexene-1,2-dicarboxylic Acid
4-Methyl-3-cyclohexene-1,2-dicarboxylic Acid
5-Methyl-3-cyclohexene-1,2-dicarboxylic Acid
6-Methyl-3-cyclohexene-1,2-dicarboxylic Acid
1-Methyl-4-cyclohexene-1,2-dicarboxylic Acid
3-Methyl-4-cyclohexene-1,2-dicarboxylic Acid
4-Methyl-4-cyclohexene-1,2-dicarboxylic Acid
4-Methyl-3-cyclohexene-1,3-dicarboxylic Acid
6-Methyl-4-cyclohexene-1,3-dicarboxylic Acid
1-Methyl-2-cyclopentene-1,2-dicarboxylic Acid, 2-*Me ester*
3-Methylmuconic Acid, *Di-Me ester*

$C_9H_{12}O_4S$
5-Hydroxy-2,4-dimethylbenzenesulphonic Acid, *Me ether*

$C_9H_{12}O_5$
2,3-Dimethylbutane-1,2,3-tricarboxylic Acid, 2,3-*Anhydride*

$C_9H_{12}O_6$
Aconitic Acid, *Tri-Me ester*
2,6-Dioxopimelic Acid, *Di-Me ester*
3-Methylcyclopropane-1,1,2-tricarboxylic Acid, *Mono-Et ester*
cis-2-Pentene-2,3,5-tricarboxylic Acid, *Mono-Me ester*

$C_9H_{12}O_6S_2$
2,4,6-Trimethylbenzene-1,3-disulphonic Acid

$C_9H_{12}O_8$
Methane-tetracarboxylic Acid, *Tetra-Me ester*
Methane-tetracetic Acid

$C_9H_{12}S$
α-Toluenethiol, *Et ether*
o-Toluenethiol, S-*Et*
p-Toluenethiol, S-*Et*

$C_9H_{12}S_3$
Benzene-1,3,5-trithiol, *Tri-Me ether*

$C_9H_{13}ClO$
2-Methyl-1-cyclohexenyl-1-acetic Acid, *Chloride*
3-Methyl-1-cyclohexenyl-1-acetic Acid, *Chloride*
2,3,3-Trimethyl-1-cyclopentene-1-carboxylic Acid, *Chloride*

$C_9H_{13}ClO_4$
3-Chloroglutaconic Acid, *Di-Et ester*

$C_9H_{13}Cl_3O_2$
2,2,3-Trichloropropionic Acid, *Cyclohexyl ester*

$C_9H_{13}N$
exo-cis-Bicyclo[3,3,0]octane-2-carboxylic Acid, *Nitrile*†
2-Butylpyridine
2-(2-Butyl)pyridine
2-*tert*-Butylpyridine
3-Butylpyridine
3-*tert*-Butylpyridine
4-Butylpyridine
4-(2-Butyl)pyridine
4-*tert*-Butylpyridine
2,1′-Cyclohexenylpropionic Acid, *Nitrile*
2,4-Diethylpyridine
2,6-Diethylpyridine
3,4-Diethylpyridine
2,4-Dimethylbenzylamine
3,5-Dimethylbenzylamine
N-Dimethylbenzylamine
N-Dimethyl-*o*-toluidine
N-Dimethyl-*m*-toluidine
N-Dimethyl-*p*-toluidine
N-Ethylbenzylamine
2-Ethyl-3,5-dimethylpyridine
2-Ethyl-4,6-dimethylpyridine
4-Ethyl-2,6-dimethylpyridine
4-Ethyl-3,5-dimethylpyridine
N-Ethyl-*N*-methylaniline
N-Ethyl-*o*-toluidine
N-Ethyl-*m*-toluidine
N-Ethyl-*p*-toluidine
4-Ethyl-*o*-toluidine
5-Ethyl-*o*-toluidine
4-Ethyl-*m*-toluidine
6-Ethyl-*m*-toluidine
N-Isopropylaniline
o-Isopropylaniline
p-Isopropylaniline
2-Isopropyl-5-methylpyridine
5-Isopropyl-2-methylpyridine
2-Methyl-1-cyclohexenyl-1-acetic Acid, *Nitrile*
3-Methyl-1-cyclohexenyl-1-acetic Acid, *Nitrile*
4-Methyl-1-cyclohexenyl-1-acetic Acid, *Nitrile*
Norcampholenic Acid, *Nitrile*
1-Phenyl-1-propylamine
2-Phenyl-1-propylamine
3-Phenyl-1-propylamine
1-Phenyl-2-propylamine★†
2-Phenyl-2-propylamine
N-Propylaniline
o-Propylaniline
m-Propylaniline
p-Propylaniline
4,5,6,7-Tetrahydro-2-methylindole
4,5,6,7-Tetrahydro-3-methylindole
2,3,4,5-Tetramethylpyridine
2,3,4,6-Tetramethylpyridine
2,3,5,6-Tetramethylpyridine
N,2,3-Trimethylaniline
N,2,4-Trimethylaniline
N,2,5-Trimethylaniline
N,2,6-Trimethylaniline
2,3,5-Trimethylaniline
2,3,6-Trimethylaniline
2,4,5-Trimethylaniline
2,4,6-Trimethylaniline
3,4,5-Trimethylaniline
2,3,3-Trimethyl-1-cyclopentene-1-carboxylic Acid, *Nitrile*
2,2,3-Trimethyl-3-cyclopentene-1-carboxylic Acid, *Nitrile*

$C_9H_{13}NO$
o-Aminobenzyl Alcohol, *Et ether*
4-Amino-*o*-cresol, *Et ether*
5-Amino-*o*-cresol, *Et ether*
4-Amino-*m*-cresol, *Et ether*
2-Amino-*p*-cresol, *Et ether*
p-Aminophenol, *Propyl ether*
p-Aminophenol, N-*Isopropyl*
6-Amino-2,4-xylenol, *Me ether*
3-Amino-2,5-xylenol, *Me ether*
4-Amino-2,6-xylenol, *Me ether*
2-Amino-3,5-xylenol, *Me ether*
4-Amino-3,5-xylenol, *Me ether*
9-Azatricyclo[4,3,1,$0^{4,9}$]decan-7-one†
o-Dimethylaminophenol, *Me ether*
m-Dimethylaminophenol, *Me ether*
p-Dimethylaminophenol, *Me ether*
o-Ethylaminophenol, *Me ether*
Halostachine
2-Hydroxybenzylamine, *Et ether*
4-Hydroxy-2,6-dimethylpyridine, *Et ether*
N-2-Hydroxyethyl-*N*-methylaniline
N-2-Hydroxyethyl-*o*-toluidine
N-2-Hydroxyethyl-*m*-toluidine
N-2-Hydroxyethyl-*p*-toluidine
4-Hydroxymethyl-*N*-dimethylaniline
N-(3-Hydroxypropyl)aniline
2-(4-Methoxyphenyl)ethylamine
o-β-Methylaminoethylphenol
m-β-Methylaminoethylphenol
p-β-Methylaminoethylphenol
N-Methyl-*o*-phenetidine
N-Methyl-*m*-phenetidine
N-Methyl-*p*-phenetidine
Norephedrine
Nor-ψ-ephedrine
Paredrine
Phenylalaninol†
Pyrrole-2-aldehyde, N-*Butyl*
Tyramine, *Me ether*
Tyramine, N-*Me*

$C_9H_{13}NO_2$
5-Amino-2-hydroxybenzyl Alcohol, α-*Et ether*
4-Aminoresorcinol, 1-*Me*-3-*Et ether*
4-Aminoresorcinol, 3-*Me*-1-*Et ether*
Anhydroecgonine
2-Azaspiro[4,5]decane-1,3-dione
1,4-Cyclohexadiene-1-alanine†
2,3-Dihydroxybenzylamine, *Di-Me ether*
3,4-Dihydroxybenzylamine, *Di-Me ether*
2-(3,4-Dihydroxyphenyl)ethylamine, N-*Me*†
2-(3,4-Dihydroxyphenyl)ethylamine, 3-*Me ether*†
2,5-Dihydroxypyridine, *Di-Et ether*
2,6-Dihydroxypyridine, *Di-Et ether*

$C_9H_{13}NO_2$ *(continued)*
3,4-Dimethoxybenzylamine
3,4-Dimethylglutaconic Acid, *Et ester nitrile*
3,4-Dimethylpyrrole-2-carboxylic Acid, *Et ester*
3,5-Dimethylpyrrole-2-carboxylic Acid, *Et ester*
4,5-Dimethylpyrrole-2-carboxylic Acid, *Et ester*
2,4-Dimethylpyrrole-3-carboxylic Acid, *Et ester*
2,5-Dimethylpyrrole-3-carboxylic Acid, *Et ester*
4,5-Dimethylpyrrole-3-carboxylic Acid, *Et ester*
β-(2,4-Dimethyl-3-pyrryl)propionic Acid
Epinine
Ethinamate
3-Ethyl-4-methylpyrrole-2-carboxylic Acid, *Me ester*
4-Ethyl-3-methylpyrrole-2-carboxylic Acid, *Me ester*
Metaraminol
2′-Methylaminoethylquinol
2-Methyl-1-butene-1,1-dicarboxylic Acid, *Et ester*, *Nitrile*
Phenylephrine
Pyrrole-2-carboxylic Acid, *Butyl ester*
Pyrrole-2-carboxylic Acid, *Isobutyl ester*
Sympathol

$C_9H_{13}NO_2S$
2,4,6-Trimethylbenzenesulphonic Acid, *Amide*
m-Xylene-4-sulphonic Acid, *Methylamide*

$C_9H_{13}NO_3$
Acetylethylmalonic Acid, *Et ester-nitrile*
Adermin, 3-*Me ether*
Adrenaline
Butyrylmalonic Acid, *Et ester nitrile*
4-Hydroxy-3,5-dimethoxybenzylamine†
4-Hydroxymethylene-2-pentyl-5-oxazolone
Tetrahydropyran-4,4-dicarboxylic Acid, *Mononitrile, Et ester*

$C_9H_{13}NO_3S$
5-Hydroxy-2,4-dimethylbenzenesulphonic Acid, *Amide*

$C_9H_{13}NO_4$
2-Amino-3-hydroxy-3,2′-furylpropionic Acid, *Et ester*
Cyanosuccinic Acid, *Di-Et ester*
2-Pentenylpenaldic Acid

$C_9H_{13}NO_5$
γ-L-Glutamyl-β-cyano-L-alanine

$C_9H_{13}NS$
o-Aminobenzenethiol, *Tri-Me*
m-Aminobenzenethiol, *Tri-Me*
p-Aminobenzenethiol, *Tri-Me*

$C_9H_{13}N_3O$
Iproniazid
2-β-Phenylhydrazinopropionic Acid, *Amide*
2-Phenylsemicarbazide, 4-*Et*
4-Phenylsemicarbazide, 1,1-*Di-Me*

$C_9H_{13}N_3O_2$
Aminometradine
Amisometradine
2,4-Diamino-5-nitrotoluene, 4-N-*Di-Me*

$C_9H_{13}N_3O_4$
Deoxycytidine (given in error as $C_9H_{13}N_3O_5$)

$C_9H_{13}N_3O_5$
Cytidine

C_9H_{14}
Bicyclo[3,3,1]non-1-ene†
Cyclononyne
Cyclosantene
2,3-Dimethyl-2-norbornene
1-Nonen-3-yne
2-Nonen-4-yne
Tricyclo[3,3,1,$0^{3,7}$]nonane†
Tricyclo[4,2,1,$0^{3,7}$]nonane
Tricyclo[4,3,0,$0^{3,7}$]nonane†
Tricyclo[4,3,0,$0^{3,8}$]nonane†

$C_9H_{14}Br_2O_2$
2,3-Dibromopropionic Acid, *Cyclohexyl ester*

$C_9H_{14}Cl_2O_2$
Azelaic Acid, *Dichloride*
2,3-Dichloropropionic Acid, *Cyclohexyl ester*
Dipropylmalonic Acid, *Dichloride*
2-Isopropylbutane-1,4-dicarboxylic Acid, *Dichloride*

$C_9H_{14}N_2$
o-Aminobenzylamine, αN-*Di-Me*
m-Aminobenzylamine, αN-*Di-Me*
p-Aminobenzylamine, αN-*Di-Me*
Azelaic Acid, *Dinitrile*
2,4-Diaminotoluene, 2-N-*Et*
2,4-Diaminotoluene, 4-N-*Et*
2,5-Diaminotoluene, 2-N-*Et*
3,4-Diaminotoluene, 3,4-N-*Di-Me*
3,4-Diaminotoluene, 3-N-*Et*
3,4-Diaminotoluene, 4-N-*Et*
1,3-Diaza-adamantane
2-*N*-Dimethyl-2,4-toluenediamine
4-*N*-Dimethyl-2,4-toluenediamine
2-*N*-Dimethyl-2,5-toluenediamine
5-*N*-Dimethyl-2,5-toluenediamine
4-*N*-Dimethyl-3,4-toluenediamine
3-*N*-Dimethyl-3,5-toluenediamine
Dipropylmalonic Acid, *Dinitrile*
2-Ethyl-5-methyladipic Acid, *Dinitrile*
α-Methylphenethylhydrazine†
m-Phenylenediamine, N,N′-*Tri-Me*
p-Phenylenediamine, N,N′-*Tri-Me*
p-Phenylenediamine, N-*Propyl*
3-Propyladipic Acid, *Dinitrile*

$C_9H_{14}N_2O_2$
2-Amino-1-*p*-aminophenyl-1,3-propanediol†
3-Butyl-6-methyluracil
Diallylmalonic Acid, *Diamide*
1-Ethyl-5-methylpyrazole-3-carboxylic Acid, *Et ester*
2-Ethyl-5-methylpyrazole-3-carboxylic Acid, *Et ester*

$C_9H_{14}N_2O_3$
Gemonal

$C_9H_{14}N_2O_4$
2,4-Dioxo-3-imidazolidylacetic Acid, *Isobutyl ester*

$C_9H_{14}N_2O_4S$
Thiostreptine†

$C_9H_{14}N_2O_4S_2$
2,4,6-Trimethylbenzene-1,3-disulphonic Acid, *Diamide*

$C_9H_{14}N_2O_8$
Aspergillomarasmine B†

$C_9H_{14}N_2O_{12}P_2$
Pseudouridine C 3′,5′-Diphosphate†
Pseudouridine C 5′-Diphosphate†

$C_9H_{14}N_3O_5$
Azathymidine

$C_9H_{14}N_3O_8P$
Cytidine-5′-phosphate
Cytidylic Acid *a*
Cytidylic Acid *b*

$C_9H_{14}N_4O_3$
Carnosine

$C_9H_{14}N_4O_5$
5-Amino-4-imidazolecarboxamide riboside

$C_9H_{14}O$
2-Allylcyclohexanone
Apocitral
Camphenilone
Cyclohexenylacetone
Dicyclobutyl Ketone
2,6-Dimethylhepta-1,5-dien-4-one†
2,6-Dimethyl-2,5-heptadien-4-one
1,7-Dimethyl-2-norbornanone
2,2-Dimethyl-5-oxobicyclo[2,2,1]heptane
2-Ethyl-3-methylcyclohex-2-en-1-one
3-Ethyl-2-methylcyclohex-2-en-1-one
3-Ethyl-4-methylcyclohex-2-en-1-one
5-Ethyl-3-methylcyclohex-2-en-1-one
6-Ethyl-3-methylcyclohex-2-en-1-one
5-Isopropylbicyclo[3,1,0]hexan-2-one
4-Isopropyl-2-cyclohexen-1-one
4-Isopropyl-3-cyclohexen-1-one
2-Isopropylidene-5-methylcyclopentanone
2-Isopropyl-5-methyl-2-cyclopentene-1-one
Isosantenone
2,6-Nonadienal
Nopinone
2-Pentylfuran†
2,4,4-Trimethyl-2-cyclohexen-1-one
2,5,5-Trimethyl-3-cyclohexen-1-one
3,4,4-Trimethyl-2-cyclohexen-1-one
3,4,6-Trimethyl-2-cyclohexen-1-one
3,5,5-Trimethyl-2-cyclohexen-1-one
3,6,6-Trimethyl-2-cyclohexen-1-one
4,4,6-Trimethyl-2-cyclohexen-1-one
5,6,6-Trimethyl-2-cyclohexen-1-one
2,2,3-Trimethyl-3-cyclohexen-1-one

$C_9H_{14}O_2$
Bicyclo[2,2,2]octane-1-carboxylic Acid
exo-cis-Bicyclo[3,3,0]octane-2-carboxylic Acid†
Castelamarin
Cyclobutane-carboxylic Acid, *Cyclobutyl ester*
1-Cyclohexene-1-carboxylic Acid, *Et ester*
3-Cyclohexene-1-carboxylic Acid, *Et ester*
2-1′-Cyclohexenylpropionic Acid
Cyclohexylideneacetic Acid, *Me ester*
2-Cyclohexylidenepropionic Acid
1,2-Cyclononanedione
4,4-Dimethyl-2-pentynoic Acid, *Et ester*
2-Heptynoic Acid, *Et ester*
2-Hydroxy-1-methylcyclohexaneacetic Acid, *Lactone*
2-Hydroxy-2-methylcyclohexaneacetic Acid, *Lactone*
2-Hydroxy-4-methylcyclohexaneacetic Acid, *Lactone*
2-Hydroxymethyl-3-methylcyclopentylacetic Acid, *Lactone*†
3-Methyl-2-cyclohexene-1-carboxylic Acid, *Me ester*
2-Methyl-1-cyclohexenyl-1-acetic Acid
3-Methyl-1-cyclohexenyl-1-acetic Acid
4-Methyl-1-cyclohexenyl-1-acetic Acid
1-Methyl-2-cyclopentene-1-carboxylic Acid, *Et ester*
2-Methyl-1-cyclopentene-1-acetic Acid, *Me ester*
2-Methyl-2,4-hexadienoic Acid, *Et ester*
3-Methyl-2,4-hexadienoic Acid, *Et ester*
4-Methyl-2,4-hexadienoic Acid, *Et ester*
5-Methyl-2,4-hexadienoic Acid, *Et ester*
3-Methyl-2-hexenoic Acid, *Me ester*
2-Nonynoic Acid
3-Nonynoic Acid
4-Nonynoic Acid
5-Nonynoic Acid
6-Nonynoic Acid
7-Nonynoic Acid
8-Nonynoic Acid
Norcampholenic Acid
2-Octynoic Acid, *Me ester*
2,4-Pentadienoic Acid, *Butyl ester*
1,2,3-Trimethyl-2-cyclopentene-1-carboxylic Acid
2,3,3-Trimethyl-1-cyclopentene-1-carboxylic Acid
2,2,3-Trimethyl-3-cyclopentene-1-carboxylic Acid
4,5,5-Trimethyl-1-cyclopentene-1-carboxylic Acid

$C_9H_{14}O_3$
3-Acetylacrylic Acid, *Butyl ester*
3-Acetyl-2,2-dimethylcyclobutane-1-carboxylic Acid
1,3-Dimethyl-2-oxocyclohexanecarboxylic Acid
1,3-Dimethyl-5-oxocyclohexanecarboxylic Acid
1,4-Dimethyl-2-oxocyclohexanecarboxylic Acid
2,2-Dimethyl-4-oxocyclohexanecarboxylic Acid
2,3-Dimethyl-4-oxocyclohexanecarboxylic Acid
2,4-Dimethyl-6-oxocyclohexanecarboxylic Acid
2,5-Dimethyl-4-oxocyclohexanecarboxylic Acid
2,6-Dimethyl-4-oxocyclohexanecarboxylic Acid
3,3-Dimethyl-2-oxocyclohexanecarboxylic Acid
4,4-Dimethyl-2-oxocyclohexanecarboxylic Acid
4,4-Dimethyl-3-oxocyclohexanecarboxylic Acid
4,5-Dimethyl-2-oxocyclohexanecarboxylic Acid
β-Formylcrotonic Acid, *Butyl ester*†
2-Isopropylbutane-1,3-dicarboxylic Acid, *Anhydride*
2-Isopropyl-2-methylpropane-1,3-dicarboxylic Acid, *Anhydride*
2-Isopropyl-2-methylglutaric Acid, *Anhydride*†
1-Methyl-2-oxo-1-cyclohexaneacetic Acid
1-Methyl-3-oxo-1-cyclohexaneacetic Acid

$C_9H_{14}O_3$ *(continued)*
3-Methyl-2-oxo-1-cyclohexaneacetic Acid
4-Methyl-2-oxo-1-cyclohexaneacetic Acid
4-Methyl-3-oxo-1-cyclohexaneacetic Acid
5-Methyl-2-oxo-1-cyclohexaneacetic Acid
5-Methyl-3-oxo-1-cyclohexaneacetic Acid
1-Methyl-2-oxo-1-cyclohexanecarboxylic Acid, *Me ester*
4-Methyl-3-oxo-1-cyclohexanecarboxylic Acid, *Me ester*
1-Methyl-4-oxo-1-cyclohexanecarboxylic Acid, *Me ester*
1-Methyl-2-oxocyclopentane-1-carboxylic Acid, *Et ester*
1-Methyl-3-oxocyclopentane-1-carboxylic Acid, *Et ester*
2-Methyl-3-oxocyclopentane-1-carboxylic Acid, *Et ester*
3-Methyl-2-oxocyclopentane-1-carboxylic Acid, *Et ester*
4-Methyl-2-oxocyclopentane-1-carboxylic Acid, *Et ester*
4-Methyl-3-oxocyclopentane-1-carboxylic Acid, *Et ester*
5-Methyl-3-oxocyclopentane-1-carboxylic Acid, *Et ester*
2-Oxocyclohexanecarboxylic Acid, *Et ester*
3-Oxocyclohexanecarboxylic Acid, *Et ester*
4-Oxocyclohexanecarboxylic Acid, *Et ester*
2-2′-Oxocyclohexanepropionic Acid
2-3′-Oxocyclohexanepropionic Acid
Pyruvic Acid, *Cyclohexyl ester*
Umbellulonic Acid

$C_9H_{14}O_4$
4-Acetyl-5-oxohexanoic Acid, *Me ester*
Allylisopropylmalonic Acid
1-Carboxycyclohexylacetic Acid
3-Carboxycyclohexylacetic Acid
3-Carboxy-2,2-dimethylcyclobutylacetic Acid
4-Carboxy-2,2-dimethylcyclobutylacetic Acid
3-(2-Carboxyethyl)-4,4-dimethylbutyrolactone
2-Carboxy-3-methylcyclopentylacetic Acid†
2-Carboxy-6-methyl-4-heptalactone†
α-Carboxy-γ-octalactone†
Citraconic Acid, *Di-Et ester*
Cycloheptane-1,1-dicarboxylic Acid
Cyclohexane-1,2-dicarboxylic Acid, *Mono-Me ester*
Cyclohexane-1,3-dicarboxylic Acid, *Mono-Me ester*
Cyclohexane-1,4-dicarboxylic Acid, *Mono-Me ester*
Cyclohexylmalonic Acid
Cyclopentane-1,2-dicarboxylic Acid, *Di-Me ester*
Cyclopentane-1,3-dicarboxylic Acid, *Di-Me ester*
Cyclopropane-1,1-dicarboxylic Acid, *Di-Et ester*
2,2-Diacetylpropionic Acid, *Et ester*
1,2-Dimethylcyclopentane-1,3-dicarboxylic Acid
2,2-Dimethylcyclopentane-1,3-dicarboxylic Acid
3,3-Dimethylcyclopropane-1,2-dicarboxylic Acid, *Di-Me ester*
3,3-Dimethylcyclopropane-1,2-dicarboxylic Acid, *Et ester*
2,2-Dimethyl-5-oxo-oxolan-3-acetic Acid, *Me ester*
2,2-Dimethyl-5-oxo-oxolan-3-carboxylic Acid, *Et ester*
2,5-Dioxohexane-3-carboxylic Acid, *Et ester*
4-Ethylglutaconic Acid, *Di-Me ester*
2-Ethyl-3-methylmaleic Acid, *Di-Me ester*
4-Ethyltetrahydro-5-oxo-3-furoic Acid, *Et ester*
4-Ethyltetrahydro-5-oxofuran-3-carboxylic Acid, *Et ester*
Glutaconic Acid, *Di-Et ester*
3-Hydroxy-2,2,3-trimethyladipic Acid, *Lactone*
Mesaconic Acid, *Di-Et ester*
2-Methyl-1-butene-1,1-dicarboxylic Acid, *Et ester*
3-Methyl-1-butene-1,1-dicarboxylic Acid, *Et ester*
3-Methyl-2-butene-1,2-dicarboxylic Acid, *Di-Me ester*
2-Methyl-1-butene-1,4-dicarboxylic Acid, 1-*Et ester*
2-Methyl-2-butene-1,4-dicarboxylic Acid, *Di-Me ester*
2-Methylcyclohexane-1,1-dicarboxylic Acid
3-Methylcyclohexane-1,1-dicarboxylic Acid
4-Methylcyclohexane-1,1-dicarboxylic Acid
1-Methylcyclohexane-1,2-dicarboxylic Acid
4-Methylcyclohexane-1,2-dicarboxylic Acid
5-Methyl-1-hexene-1,2-dicarboxylic Acid
2-Propene-1,2-dicarboxylic Acid, *Di-Et ester*

$C_9H_{14}O_5$
Acetonedicarboxylic Acid, *Di-Et ester*
3-Acetylglutaric Acid, *Di-Me ester*
Acetylmalonic Acid, *Di-Et ester*
Acetylmalonic Acid, *Mono-isobutyl ester*
2-Acetyl-2-methylsuccinic Acid, *Di-Me ester*
2-Acetyl-3-methylsuccinic Acid, *Di-Me ester*
2-Acetyl-3-methylsuccinic Acid, *Mono-Et ester*
Formylsuccinic Acid, *Di-Et ester*
2-Oxalopropionic Acid, *Di-Et ester*
4-Oxo-1,7-heptanedioic Acid, *Di-Me ester*
4-Oxo-1,7-heptanedioic Acid, *Mono-Et ester*
5-Oxononane-1,9-dioic Acid
Tetrahydropyran-2,6-dicarboxylic Acid, *Di-Me ester*

$C_9H_{14}O_6$
Aldgarose†
3-Carboxymethyl-2-methylbutane-2,4-dicarboxylic Acid
2,3-Dimethylbutane-1,2,3-tricarboxylic Acid
1,3-Dioxolan-4,5-dicarboxylic Acid, *Di-Et ester*
Glyceric Acid, *Et ester*, *di-Ac*
Hexane-1,1,6-tricarboxylic Acid
Hexane-1,2,2-tricarboxylic Acid
Hexane-1,2,3-tricarboxylic Acid
Hexane-1,2,6-tricarboxylic Acid
Hexane-1,3,3-tricarboxylic Acid
Hexane-1,3,6-tricarboxylic Acid
Hexane-1,4,4-tricarboxylic Acid
Hexane-1,4,5-tricarboxylic Acid
Hexane-2,3,5-tricarboxylic Acid
Hexane-2,4,4-tricarboxylic Acid
Methane-tricarboxylic Acid, *Mono-Me*, *Di-Et ester*
Triacetin
Tricarballylic Acid, *Tri-Me ester*

$C_9H_{14}O_7$
Trimethyl citrate

$C_9H_{14}S$
Thia-adamantane

$C_9H_{15}BO_3$
Triallyl borate

$C_9H_{15}BrO_2$
1-Bromocyclohexane-carboxylic Acid, *Et ester*
3-Bromocyclohexane-carboxylic Acid, *Et ester*

$C_9H_{15}BrO_4$
3-Bromoglutaric Acid, *Di-Et ester*

$C_9H_{15}ClO$
2-Cyclohexylpropionic Acid, *Chloride*
1,3-Dimethylcyclohexanecarboxylic Acid, *Chloride*
2,6-Dimethylcyclohexanecarboxylic Acid, *Chloride*
3,4-Dimethylcyclohexanecarboxylic Acid, *Chloride*
2-Methylcyclohexaneacetic Acid, *Chloride*
3-Methylcyclohexaneacetic Acid, *Chloride*
4-Methylcyclohexaneacetic Acid, *Chloride*
8-Nonenoic Acid, *Chloride*
2,2,3-Trimethylcyclopentane-1-carboxylic Acid, *Chloride*

$C_9H_{15}ClO_2$
3-Chloro-2-ethylcrotonic Acid, *Propyl ester*

$C_9H_{15}ClO_3$
Pimelic Acid, *Et ester*, *Chloride*

$C_9H_{15}ClO_4$
2-Chloro-2-ethylmalonic Acid, *Di-Et ester*
2-Chloroglutaric Acid, *Di-Et ester*

$C_9H_{15}N$
1-Aza-adamantane
2-Cyclohexylpropionic Acid, *Nitrile*
2,4-Diethyl-3-methylpyrrole
2,5-Diethyl-1-methylpyrrole
2,5-Diethyl-3-methylpyrrole
3,4-Diethyl-2-methylpyrrole
Isonorlupinene
N-Methylgranatenine
2-Methyloctanoic Acid, *Nitrile*
3-Methyloctanoic Acid, *Nitrile*
Phyllopyrrole
Pyrrole, N-3-*Methylbutyl*
Triallylamine

$C_9H_{15}NO$
9-Aza-9-methylbicyclo[3,3,1]nonan-3-one
exo-cis-Bicyclo[3,3,0]octane-2-carboxylic Acid, *Amide*†
1-Methoxymethyl-1,2-dehydro-8α-pyrrolizidine†
2-Methyl-1-cyclohexenyl-1-acetic Acid, *Amide*
3-Methyl-1-cyclohexenyl-1-acetic Acid, *Amide*
4-Methyl-1-cyclohexenyl-1-acetic Acid, *Amide*
1-Methyl-9-azabicyclo[3,3,1]nonan-3-one†
9-Methyl-9-azabicyclo[3,3,1]nonan-3-one★†
2-Nonynoic Acid, *Amide*
3-Nonynoic Acid, *Amide*
4-Nonynoic Acid, *Amide*
5-Nonynoic Acid, *Amide*
6-Nonynoic Acid, *Amide*
7-Nonynoic Acid, *Amide*
8-Nonynoic Acid, *Amide*
3-Oxononanoic Acid, *Nitrile*
Supinidine
Tetrahydro-2,6,6-trimethylpyran-2-carboxylic Acid *Nitrile*
1,2,3-Trimethyl-2-cyclopentene-1-carboxylic Acid, *Amide*
2,3,3-Trimethyl-1-cyclopentene-1-carboxylic Acid, *Amide*
2,2,3-Trimethyl-3-cyclopentene-1-carboxylic Acid, *Amide*
4,5,5-Trimethyl-1-cyclopentene-1-carboxylic Acid, *Amide*

$C_9H_{15}NO_2$
Diethylmalonic Acid, *Et ester*
3,3-Dimethylglutaric Acid, *Et ester*, *Nitrile*
1β,2β-Epoxy-1α-methoxymethyl-8α-pyrrolizidine†
7α-Hydroxy-1-methoxymethyl-1,2-dehydro-8α-pyrrolizidine†
7β-Hydroxy-1-methoxymethyl-1,2-dehydro-8α-pyrrolizidine†
Isobutylmalonic Acid, *Et ester*, *Nitrile*
2-Isopropylbutane-1,3-dicarboxylic Acid, *Imide*
Meroquinene
2-Methylhexane-3,3-dicarboxylic Acid, *Mononitrile*
Methylpropylmalonic Acid, *Et ester*, *Nitrile*
Pyrrolizidine-1-carboxylic Acid, *Me ester*†
8*H*β-Pyrrolizidine-1α-carboxylic Acid, *Me ester*†
Retronecine, *Me ether*
Succinimide, N-3-*Methylbutyl*

$C_9H_{15}NO_3$
Actinobolamine†
Alloecgonine
Cyclohexanecarbonylglycine
Ecgonine
α-Ecgonine
ψ-Ecgonine
trans-3-Methyl-1-butene-1,2-dicarboxylic Acid, *Et ester-amide*
Nor-ψ-ecgonine, *Me ester*
Propylfumaric Acid, *Et ester-amide*

$C_9H_{15}NO_3S$
2-(5-Carboxypentyl)-4-thiazolidone

$C_9H_{15}NO_4$
4-Carboxy-1-methylpiperidine-2-acetic Acid
Pentylpenaldic Acid
Piperidine-2,3-dicarboxylic Acid, *Di-Me ester*
Piperidine-2,6-dicarboxylic Acid, *Di-Me ester*

$C_9H_{15}NO_5$
Acetamidomalonic Acid, *Di-Et ester*†

$C_9H_{15}NO_6$
Tri-(carboxymethyl)amine, *Tri-Me ester*

$C_9H_{15}N_3$
4-Amino-2,6-diethyl-5-methylpyrimidine

$C_9H_{15}N_3O_2$
Hercynin

$C_9H_{15}N_3O_2S$
Ergothioneine

$C_9H_{15}N_3O_3$
Cyanuric Acid, O-*Tri-Et ester*
$C_9H_{15}N_3O_6S$
Asparthione
$C_9H_{15}N_3O_7$
Lycomarasmine
$C_9H_{15}N_4O_8P$
5-Amino-1-β-D-ribofuranosylimidazole-4-carboxyamide-5′-phosphate †
$C_9H_{15}NaO_9$
Digeneaside
$C_9H_{15}O_4P$
Phosphoric Acid, *Triallyl ester*
$C_9H_{15}O_5P$
Phosphonoacetic Acid, 3,3-*Diallyl-1-Me ester*
C_9H_{16}
Allylcyclohexane
Bicyclo[3,3,1]nonane
Cyclononene
Bicyclo[6,1,0]nonane †
7,7-Dimethylbicyclo[2,2,1]heptane
2,6-Dimethyl-1,3-heptadiene
2,6-Dimethyl-1,5-heptadiene
2,4-Dimethyl-2,4-heptadiene
2,6-Dimethyl-2,4-heptadiene
2,6-Dimethyl-2,5-heptadiene
3,5-Dimethyl-2,4-heptadiene
3,6-Dimethyl-2,4-heptadiene
1-Ethylidene-2-methylcyclohexane
1-Ethylidene-3-methylcyclohexane
1-Ethylidene-4-methylcyclohexane
1-Ethyl-3-methylcyclohexene
1-Ethyl-4-methylcyclohexene
1-Ethyl-5-methylcyclohexene
1-Isopropyl-3-methylcyclopentene
4-Isopropyl-1-methylcyclopentene
7-Methyl-2,4-octadiene
4-Methyl-3,5-octadiene
1,8-Nonadiene
2,7-Nonadiene
1-Nonyne
2-Nonyne
3-Nonyne
Nopinane
1,2,3,3-Tetramethylcyclopentene
1,2,3-Trimethylcyclohexene
1,3,5-Trimethylcyclohexene
1,4,4-Trimethylcyclohexene
1,4,5-Trimethylcyclohexene
1,5,5-Trimethylcyclohexene
1,5,6-Trimethylcyclohexene
1,6,6-Trimethylcyclohexene
$C_9H_{16}BrClO$
2-Bromononanoic Acid, *Chloride*
9-Bromononanoic Acid, *Chloride*
$C_9H_{16}BrNO$
2,2,6,6-Tetramethyl-4-piperidone, N-*Bromo*
$C_9H_{16}ClNO$
1-Chlorocyclohexane-1-carboxylic Acid, *Et-amide*
$C_9H_{16}N_2$
1,8-Diazabicyclo[5,4,0]undec-7-ene †
2-Methyl-1,3′-methylbutylimidazole
5-Methyl-4,2′-methylbutylimidazole
$C_9H_{16}N_2O$
Dipropylmalonic Acid, *Mononitrile*, *Amide*
Loline, N-*Me* †
9-Methyl-3,9-diazabicyclo[3,3,1]nonane-7-one
$C_9H_{16}N_2O_2$
Cyclohexanecarbonylglycine, *Amide*
Ecgonine, *Amide*
$C_9H_{16}N_2O_4S$
γ-L-Glutamyl-*S*-methylcysteine
$C_9H_{16}N_2O_5$
γ-L-Glutamyl-β-aminoisobutyric Acid
γ-L-Glutamyl-γ-aminobutyric Acid †
$C_9H_{16}N_2O_5S$
N-Acetylcystathionine †
Aletheine, N-*Di-Ac*
$C_9H_{16}N_4O_6$
Uroxanic Acid, *Di-Et ester*
$C_9H_{16}O$
1-Acetyl-1-methylcyclohexane
1-Acetyl-2-methylcyclohexane
1-Acetyl-3-methylcyclohexane
1-Acetyl-4-methylcyclohexane
1-Allylcyclohexanol
Camphenilol
Cryptol
Cyclohexyl ethyl Ketone
Cyclononanone
5,5-Dimethylbicyclo[2,2,1]heptan-2-ol
1,7-Dimethyl-2-norbornanol
2,3-Dimethyl-2-norbornanol
2-Ethyl-2-methylcyclohexanone
2-Ethyl-5-methylcyclohexanone
3-Ethyl-5-methylcyclohexanone
1,6-Heptadien-4-ol, *Et ether*
2-Isopropylcyclohexanone
3-Isopropylcyclohexanone
4-Isopropylcyclohexanone
2-Isopropyl-2-methylcyclopentanone
2-Isopropyl-4-methylcyclopentanone
2-Isopropyl-5-methylcyclopentanone
3-Isopropyl-5-methylcyclopentanone
6-Methyl-5-methylene-2-heptanone
5-Methyl-1-octen-3-one
2-Methyl-4-octen-3-one
4-Methyl-4-octen-3-one
7-Methyl-5-octen-3-one
7-Methyl-5-octen-4-one
2-Methyl-2-octen-4-one
2-Methyl-5-octen-4-one
2-Methyl-7-octen-4-one
2,6-Nonadien-1-ol
1,6-Nonadien-3-ol
2-Nonen-1-al
2-Nonyn-1-ol
4-Nonyn-1-ol
4-Nonyn-2-ol
1-Nonyn-3-ol
1-Nonyn-4-ol
Nopinol
Norisoborneol
2-Propylcyclohexanone

$C_9H_{16}O$ (*continued*)
4-Propylcyclohexanone
2,2,3-Trimethylcyclohexanone
2,2,4-Trimethylcyclohexanone
2,2,5-Trimethylcyclohexanone
2,2,6-Trimethylcyclohexanone
2,3,3-Trimethylcyclohexanone
2,3,6-Trimethylcyclohexanone
2,4,4-Trimethylcyclohexanone
2,4,5-Trimethylcyclohexanone
2,4,6-Trimethylcyclohexanone
2,5,5-Trimethylcyclohexanone
3,3,4-Trimethylcyclohexanone
3,3,5-Trimethylcyclohexanone
3,4,4-Trimethylcyclohexanone
1,3,5-Trimethyl-2-cyclohexen-1-ol
2,2,5-Trimethyl-3-cyclohexen-1-ol
2,4,4-Trimethyl-2-cyclohexen-1-ol
3,5,5-Trimethyl-2-cyclohexen-1-ol

$C_9H_{16}O_2$
Angelic Acid, *Isobutyl ester*
Brevicomin†
Crotonic Acid, *Pentyl ester*
Cycloheptane-carboxylic Acid, *Me ester*
Cyclohexanecarboxylic Acid, *Et ester*
Cyclohexylacetic Acid, *Me ester*
2-Cyclohexylpropionic Acid
3-Cyclohexylpropionic Acid
Cyclo-octanecarboxylic Acid
2-Cyclopentylpropionic Acid, *Me ester*
1,3-Dimethylcyclohexanecarboxylic Acid
1,4-Dimethylcyclohexanecarboxylic Acid
2,2-Dimethylcyclohexanecarboxylic Acid
2,4-Dimethylcyclohexanecarboxylic Acid
2,5-Dimethylcyclohexanecarboxylic Acid
2,6-Dimethylcyclohexanecarboxylic Acid
3,4-Dimethylcyclohexanecarboxylic Acid
3,5-Dimethylcyclohexanecarboxylic Acid
2,2-Dimethyl-6-oxoheptanal
3,4-Dimethyl-3-pentenoic Acid, *Et ester*
2-Heptenoic Acid, *Et ester*
5-Hydroxynonanoic Acid, *Lactone*†
3-Hydroxy-2,2,5-trimethylhexanoic Acid, *Lactone*
2-Methylacrylic Acid, neo-*Pentyl ester*
2-Methylcyclohexaneacetic Acid
3-Methylcyclohexaneacetic Acid
4-Methylcyclohexaneacetic Acid
1-Methylcyclohexane-1-carboxylic Acid, *Me ester*
2-Methylcyclohexane-1-carboxylic Acid, *Me ester*
3-Methylcyclohexane-1-carboxylic Acid, *Me ester*
4-Methylcyclohexane-1-carboxylic Acid, *Me ester*
2-Methylcyclopentane-1-carboxylic Acid, *Et ester*
2-Methyl-2-hexenoic Acid, *Et ester*
3-Methyl-2-hexenoic Acid, *Et ester*
5-Methyl-2-hexenoic Acid, *Et ester*
2-Methyl-3-hexenoic Acid, *Et ester*
3-Methyl-3-hexenoic Acid, *Et ester*
4-Methyl-3-hexenoic Acid, *Et ester*
5-Methyl-4-hexenoic Acid, *Et ester*
2,4-Nonadione
2,8-Nonadione
3,4-Nonadione
2-Nonenoic Acid
3-Nonenoic Acid
8-Nonenoic Acid
7-Octenoic Acid, *Me ester*
1,2,3-Trimethylcyclopentane-1-carboxylic Acid
2,2,3-Trimethylcyclopentane-1-carboxylic Acid

$C_9H_{16}O_3$
Acetoacetic Acid, *Isopentyl ester*
Azelaialdehydic Acid
2,2-Dimethyl-6-oxoheptanoic Acid
4,4-Dimethyl-6-oxoheptanoic Acid
3,3-Dimethyl-5-oxohexanoic Acid, *Me ester*
2-Ethyl-2-methylacetoacetic Acid, *Et ester*
3-Formyl-2-methylpropionic Acid, *Butyl ester*
2-Hexanone-3-carboxylic Acid, *Et ester*
2-Hydroxycycloheptanecarboxylic Acid, *Me ester*†
1-Hydroxycyclohexanecarboxylic Acid, *Et ester*
2-Hydroxycyclohexanecarboxylic Acid, *Et ester*
3-Hydroxycyclohexanecarboxylic Acid, *Et ester*
4-Hydroxycyclohexanecarboxylic Acid, *Et ester*
2,2′-Hydroxycyclohexanepropionic Acid
1-Hydroxy-3,5-dimethylcyclohexanecarboxylic Acid
3-Hydroxy-4,4-dimethylcyclohexanecarboxylic Acid
4-Hydroxy-2,4-dimethylcyclohexanecarboxylic Acid
4-Hydroxy-2,6-dimethylcyclohexanecarboxylic Acid
1-Hydroxy-2-methylcyclohexaneacetic Acid
2-Hydroxy-1-methylcyclohexaneacetic Acid
2-Hydroxy-2-methylcyclohexaneacetic Acid
1-Hydroxy-3-methylcyclohexaneacetic Acid
1-Hydroxy-4-methylcyclohexaneacetic Acid
2-Hydroxy-4-methylcyclohexaneacetic Acid
2-Hydroxy-1-methylcyclohexane-1-carboxylic Acid, *Me ether*
1-Hydroxy-3-methyl-1-cyclopentaneacetic Acid, *Me ester*
2-Hydroxy-1-methylcyclopentane-1-carboxylic Acid, *Et ester*
2-Hydroxy-5-methylcyclopentane-1-carboxylic Acid, *Et ester*
2-Hydroxy-3-methylcyclopentane-1-carboxylic Acid, *Et ester*
3-Hydroxy-5-methylcyclopentane-1-carboxylic Acid, *Et ester*
Levulinic Acid, *Butyl ester*
Levulinic Acid, 2-*Butyl ester*
Levulinic Acid, *Isobutyl ester*
Mesitonic Acid, *Et ester*
2-Methyl-5-oxohexanoic Acid, *Et ester*
3-Methyl-5-oxohexanoic Acid, *Et ester*
4-Methyl-5-oxohexanoic Acid, *Et ester*
2-Oxocyclo-octane-1-carboxylic Acid,†
2-Oxoheptanoic Acid, *Et ester*
3-Oxoheptanoic Acid, *Et ester*
5-Oxoheptanoic Acid, *Et ester*
6-Oxoheptanoic Acid, *Et ester*
2-Oxononanoic Acid
3-Oxononanoic Acid

$C_9H_{16}O_3$ (*continued*)
4-Oxononanoic Acid
5-Oxononanoic Acid
6-Oxononanoic Acid
7-Oxononanoic Acid
8-Oxononanoic Acid
3-Oxo-octanoic Acid, *Me ester*
4-Oxo-octanoic Acid, *Me ester*
6-Oxo-octanoic Acid, *Me ester*
2-Oxovaleric Acid, *Isobutyl ester*
Tetrahydro-2,6,6-trimethylpyran-2-carboxylic Acid

$C_9H_{16}O_4$
Azelaic Acid
tert-Butyl ethyl malonate†
(2-Butyl)malonic Acid, *Di-Me ester*
tert-Butylmalonic Acid, *Mono-Et ester*
3-Carboxymethyl-5-methylhexanoic Acid
Diethylmalonic Acid, *Di-Me ester*
2,2-Dimethylglutaric Acid, *Di-Me ester*
3,3-Dimethylglutaric Acid, *Di-Me ester*
3,3-Dimethylglutaric Acid, *Mono-Et ester*
Dimethylmalonic Acid, *Di-Me ester*
Dipropylmalonic Acid
2-Ethylglutaric Acid, *Di-Me ester*
Ethylmalonic Acid, *Di-Et ester*
2-Ethyl-4-methyladipic Acid
2-Ethyl-5-methyladipic Acid
3-Ethyl-4-methyladipic Acid
3-Ethyl-3-methylglutaric Acid, *Me ester*
2-Ethyl-2-methylsuccinic Acid, *Di-Me ester*
Glucal, 3,4,6-*Tri-Me ether*
Glutaric Acid, *Di-Et ester*
Heptane-1,1-dicarboxylic Acid
Heptane-1,2-dicarboxylic Acid
Heptane-1,3-dicarboxylic Acid
Heptane-1,4-dicarboxylic Acid
Heptane-1,5-dicarboxylic Acid
Heptane-2,4-dicarboxylic Acid
Heptane-2,6-dicarboxylic Acid
Heptane-3,3-dicarboxylic Acid
Heptane-3,4-dicarboxylic Acid
Heptane-3,5-dicarboxylic Acid
Isobutylmalonic Acid, *Di-Me ester*
2-Isopropylbutane-1,3-dicarboxylic Acid
2-Isopropylbutane-1,4-dicarboxylic Acid
2-Isopropyl-2-methylglutaric Acid†
2-Isopropyl-2-methylpropane-1,3-dicarboxylic Acid
Malonic Acid, *Dipropyl ester*
2-Methyladipic Acid, *Di-Me ester*
3-Methyladipic Acid, *Mono-Et ester*
5-Methylhexane-1,1-dicarboxylic Acid
5-Methylhexane-1,2-dicarboxylic Acid
5-Methylhexane-1,3-dicarboxylic Acid
5-Methylhexane-1,4-dicarboxylic Acid
5-Methylhexane-2,2-dicarboxylic Acid
5-Methylhexane-2,3-dicarboxylic Acid
2-Methylhexane-3,3-dicarboxylic Acid
3-Methyloctanedioic Acid
4-Methyloctanedioic Acid
Methylpropylmalonic Acid, *Di-Me ester*
Methylsuccinic Acid, *Di-Et ester*
Octanedioic Acid, *Me ester*
Pimelic Acid, *Di-Me ester*
Pimelic Acid, *Et ester*
3-Propyladipic Acid
Propylsuccinic Acid, *Di-Me ester*
Succinic Acid, *Pentyl ester*
Tetramethylsuccinic Acid, *Mono-Me ester*
2,2,4-Trimethyladipic Acid
2,2,5-Trimethyladipic Acid
3,3,4-Trimethyladipic Acid
Trimethylsuccinic Acid, β-*Et ester*

$C_9H_{16}O_5$
3,6-Anhydro-D-glucose, *Tri-Me ether*
Chromose B†
5-Hydroxyazelaic Acid
3-Hydroxy-2,4-dimethylglutaric Acid, *Et ester*
3-Hydroxyglutaric Acid, *Di-Et ester*
Hydroxymethyl-malonic Acid, *Me ether, Di-Et ester*
Hydroxymethyl-succinic Acid, *Di-Et ester*
2-Hydroxy-3-methylsuccinic Acid, *Di-Et ester*
3-Hydroxy-2,2,3-trimethyladipic Acid
5-Hydroxy-2,2,5-trimethyladipic Acid
2-Hydroxy-2,4,4-trimethyladipic Acid
3-Hydroxy-3,4,4-trimethyladipic Acid
Laevoglucosan, *Tri-Me*
Laevoglucosan, 2,3,4-*Tri-Me ether*
Methoxysuccinic Acid, *Di-Et ester*
Mycarose, 4-O-*Ac*
Propionic Acid, *Glycerol di-ester*

$C_9H_{16}O_6$
Gluconic Acid, γ-*Lactone*, 2,3,6-*Tri-Me ether*
Quinic Acid, *Et ester*

$C_9H_{16}O_7$
Galacturonic Acid, 3,4-*Di-Me ether, Methyl pyranoside*
Galacturonic Acid, 2,3,4-*Tri-Me ether*
Glucuronic Acid, 2,3,4-*Tri-Me ether*
Trihydroxyglutaric Acid, *Di-Me ester*, 1,2-*Di-Me ether*

$C_9H_{16}O_8$
Rhamno-octonic Acid, γ-*Lactone*

$C_9H_{17}BrO_2$
2-Bromoisobutyric Acid, 3-*Methylbutyl ester*
2-Bromononanoic Acid
9-Bromononanoic Acid

$C_9H_{17}ClO$
3-Methyl-2-propylvaleric Acid, *Chloride*
Nonanoic Acid, *Chloride*

$C_9H_{17}ClO_2$
2-Hydroxynonanoic Acid, *Chloride*

$C_9H_{17}Cl_3$
1,1,3-Trichlorononane†

$C_9H_{17}N$
1-Azaspiro[4,5]decane
Decahydroquinoline
2-Ethylquinuclidine
3-Ethylquinuclidine
N-Methyl-γ-coniceine
N-Methylgranatanine
1-Methylindolizidine
2-Methylindolizidine
3-Methylindolizidine
5-Methylindolizidine
Nonanoic Acid, *Nitrile*

$C_9H_{17}N$ (*continued*)
Norbornylamine
Norlupinane
Norlupinane-B
Octahydro-1-methylindole
Octahydro-2-methylindole
Pinidine†
3-Vinylpiperidine, N-*Et*

$C_9H_{17}NO$
2-Cyclohexylpropionic Acid, *Amide*
3-Cyclohexylpropionic Acid, *Amide*
1,3-Dimethylcyclohexanecarboxylic Acid, *Amide*
1,4-Dimethylcyclohexanecarboxylic Acid, *Amide*
2,4-Dimethylcyclohexanecarboxylic Acid, *Amide*
2,6-Dimethylcyclohexanecarboxylic Acid, *Amide*
3,5-Dimethylcyclohexanecarboxylic Acid, *Amide*
3α-Granatanol†
3β-Granatanol†
2-Hydroxymethyl-8-methyl-8-azabicyclo[3,2,1]-octane
Isopelletierine, N-*Me*
2-Methylcyclohexaneacetic Acid, *Amide*
4-Methylcyclohexaneacetic Acid, *Amide*
N-Methylgranatoline
2-Methyloctanoic Acid, *Amide*
3-Methyloctanoic Acid, *Amide*
6-Methyloctanoic Acid, *Amide*
7-Methyloctanoic Acid, *Amide*
2-Propionylpiperidine, N-*Me*
1,2,3,6-Tetramethyl-4-piperidone
2,2,6,6-Tetramethyl-4-piperidone
1,2,3-Trimethylcyclopentane-1-carboxylic Acid, *Amide*
2,2,3-Trimethylcyclopentane-1-carboxylic Acid, *Amide*
2,2,6-Trimethyl-4-piperidone, N-*Me*
Tropine, *Me ether*

$C_9H_{17}NO_2$
1-Aminocyclohexanecarboxylic Acid, *Et ester*
2-Aminocyclohexanecarboxylic Acid, *Et ester*
3-Aminocyclohexanecarboxylic Acid, *Et ester*
4-Aminocyclohexanecarboxylic Acid, *Et ester*
3-Cyclohexylalanine
3-Ethyl-4-piperidylacetic Acid
N-Methylgranatoline, N-*Oxide*
3-Oxononanoic Acid, *Amide*
3-Piperidineacetic Acid, *Et ester*
4-Piperidineacetic Acid, *Et ester*
Piperidinoacetic Acid, *Et ester*
Stachydrine, *Et ester*
Tetrahydro-2,6,6-trimethylpyran-2-carboxylic Acid *Amide*
2,2,6,6-Tetramethyl-4-piperidone, N-*OH*

$C_9H_{17}NO_3$
Azelaic Acid, *Monoamide*
Diethylmalonic Acid, *Mono-amide*, *Et ester*
N-Formyl-isoleucine, *Et ester*
3-(4-Hydroxycyclohexyl)-alanine
2-Methylhexane-3,3-dicarboxylic Acid, *Mono-amide*

$C_9H_{17}NO_4$
L-Aspartic Acid, β-*Isoamyl ester*
Glutamic Acid, *Di-Et ester*
N-Methylaspartic Acid, *Di-Et ester*

$C_9H_{17}NO_5$
Pantothenic Acid

$C_9H_{17}NO_6$
Galacturonic Acid, 2,3-*Di-Me ether*, *Methyl pyranoside*, *Amide*
Galacturonic Acid, 3,4-*Di-Me ether*, *Methyl pyranoside*, *Amide*

$C_9H_{17}NO_7$
Muramic Acid

$C_9H_{17}NS$
n-Octyl thiocyanate

$C_9H_{17}N_3O_4$
Diglycyl-valine

$C_9H_{17}O_5P$
Phosphonoacetic Acid, 3,3-*Di-Et*-1-*allyl ester*

C_9H_{18}
Citronellene
Cyclononane
1-Ethyl-2-methylcyclohexane
1-Ethyl-3-methylcyclohexane
1-Ethyl-4-methylcyclohexane
Isopropylcyclohexane
1-Isopropyl-2-methylcyclopentane
1-Isopropyl-3-methylcyclopentane
2-Methyl-1-octene
3-Methyl-2-octene
4-Methyl-2-octene
7-Methyl-3-octene
2-Methyl-4-octene
1-Nonene
2-Nonene
3-Nonene
4-Nonene
Propylcyclohexane
1,1,3-Trimethylcyclohexane
1,1,4-Trimethylcyclohexane
1,2,3-Trimethylcyclohexane
1,2,4-Trimethylcyclohexane
1,3,5-Trimethylcyclohexane

$C_9H_{18}BrN$
5-Azoniaspiro[4,5]decane, *Bromide*

$C_9H_{18}Br_2$
1,2-Dibromononane
1,4-Dibromononane
1,9-Dibromononane

$C_9H_{18}ClN$
5-Azoniaspiro[4,5]decane, *Chloride*

$C_9H_{18}Cl_2$
1,9-Dichlorononane

$C_9H_{18}HgO_2$
Mercuri-heptyl acetate

$C_9H_{18}IN$
5-Azoiaspiro[4,5]decane, *Iodide*

$C_9H_{18}N_2$
3-Amino-9-azabicyclo[3,3,1]nonane, Imino-N-*Me*
2,8-Diazaspiro[5,5]undecane
N,*N*-Dibutycyanamide

$C_9H_{18}N_2O$
2,3-Di-*tert*-butyl-2,3-diazacyclopropanone†

$C_9H_{18}N_2O_2$
Azelaic Acid, *Diamide*
Dipropylmalonic Acid, *Diamide*
2-Isopropylbutane-1,4-dicarboxylic Acid, *Diamide*
Malonamide, N,N'-*Dipropyl*
Piperazine-*N*-carboxylic Acid, *Butyl ester*

$C_9H_{18}N_2O_3$
L-Alanyl-D-leucine
L-Alanyl-L-leucine
DL-Alanyl-leucine
Leucylalanine
Leucylglycine, *Me ester*

$C_9H_{18}N_2O_4$
Meprobamate

$C_9H_{18}N_4O_4$
N^{α}-1-Carboxyethylarginine
Octopin

$C_9H_{18}O$
Cyclononanol
Cyclo-octylmethanol
2,6-Dimethyl-3-heptanone
2,6-Dimethyl-4-heptanone
3,5-Dimethyl-4-heptanone
4,6-Dimethyl-1-hepten-4-ol
1-Ethyl-2-methylcyclohexanol
1-Ethyl-3-methylcyclohexanol
1-Ethyl-4-methylcyclohexanol
2-Ethyl-4-methylcyclohexanol
2-Ethyl-5-methylcyclohexanol
2-Ethyl-6-methylcyclohexanol
4-Ethyl-2-methylcyclohexanol
4-Ethyl-3-methylcyclohexanol
1-Hepten-2-ol, *Et ether*
1-Isopropylcyclohexanol
2-Isopropylcyclohexanol
4-Isopropylcyclohexanol
2-Isopropyl-5-methylcyclopentanol
3-Isopropyl-1-methylcyclopentanol
6-Methyl-5-hepten-2-ol, *Me ether*
3-Methyl-2-octanone
4-Methyl-2-octanone
5-Methyl-2-octanone
7-Methyl-2-octanone
2-Methyl-3-octanone
4-Methyl-3-octanone
5-Methyl-3-octanone
7-Methyl-3-octanone
3-Methyl-4-octanone
6-Methyl-4-octanone
7-Methyl-4-octanone
7-Methyl-2-octen-1-ol
3-Methyl-6-octen-1-ol
2-Methyl-3-octen-2-ol
3-Methyl-4-octen-3-ol
4-Methyl-1-octen-4-ol
2-Methyl-2-octen-4-ol
7-Methyl-2-octen-4-ol
5-Methyl-5-octen-4-ol
Nonanal
2-Nonanone
3-Nonanone
4-Nonanone
5-Nonanone
1-Nonen-3-ol
6-Nonen-3-ol
8-Nonen-1-ol
Non-3-en-1-ol †
1-Octen-2-ol, *Me ether*
1-Propylcyclohexanol
2-Propylcyclohexanol
4-Propylcyclohexanol
2,2,4,4-Tetramethyl-3-pentanone
1,2,2-Trimethylcyclohexanol
1,2,6-Trimethylcyclohexanol
1,3,3-Trimethylcyclohexanol
1,3,5-Trimethylcyclohexanol
1,4,4-Trimethylcyclohexanol
2,2,3-Trimethylcyclohexanol
2,2,5-Trimethylcyclohexanol
2,2,6-Trimethylcyclohexanol
2,3,3-Trimethylcyclohexanol
2,3,5-Trimethylcyclohexanol
2,3,6-Trimethylcyclohexanol
2,4,4-Trimethylcyclohexanol
2,4,5-Trimethylcyclohexanol
2,4,6-Trimethylcyclohexanol
2,5,5-Trimethylcyclohexanol
3,3,5-Trimethylcyclohexanol

$C_9H_{18}O_2$
2,5-Dimethylhexanoic Acid, *Me ester*
2-Ethylhexanoic Acid, *Me ester*
2-Ethyl-2-methylhexanoic Acid
2-Ethyl-3-methylhexanoic Acid
2-Ethyl-5-methylhexanoic Acid
3-Ethyl-3-methylhexanoic Acid
2-Ethylvaleric Acid, *Et ester*
Heptanoic Acid, *Et ester*
Hexanoic Acid, *Propyl ester*
2-Hexanol, *Propionyl*
5-Hydroxypentanal, *Isobutyl ether*
2-Hydroxy-3-pentanone, *Butyl ether*
2-Isobutylvaleric Acid
Isopentyl isobutyrate
Isovaleric Acid, *Isobutyl ester*
2-Methyl-1-butyl butyrate
3-Methyl-1-butyl butyrate
2-Methyl-2-butyl butyrate
2-Methyl-1-butyl isobutyrate
2-Methyl-2-butyl isobutyrate
2-Methylbutyric Acid, *Butyl ester*
2-Methylbutyric Acid, 2-*Butyl ester*
2-Methylbutyric Acid, *Isobutyl ester*
5-Methylheptanoic Acid, *Me ester*
4-Methylhexane-3-carboxylic Acid, *Me ester*
2-Methylhexanoic Acid, *Et ester*
3-Methylhexanoic Acid, *Et ester*
4-Methylhexanoic Acid, *Et ester*
5-Methylhexanoic Acid, *Et ester*
2-Methyloctanoic Acid
3-Methyloctanoic Acid
4-Methyloctanoic Acid
5-Methyloctanoic Acid
6-Methyloctanoic Acid
7-Methyloctanoic Acid
2-Methyl-2-propylvaleric Acid
3-Methyl-2-propylvaleric Acid
Nonanoic Acid
Octanoic Acid, *Me ester*

$C_9H_{18}O_2$ *(continued)*
2-Octanol, *Formyl*
1-Penten-3-one, *Diethylacetal*
Pentyl butyrate
Valeric Acid, *Butyl ester*
Valeric Acid, sec-*Butyl ester*

$C_9H_{18}O_3$
Butyric Acid, *Butoxymethyl ester*
Dibutyl carbonate
Di-2-butyl carbonate
Di-isobutyl carbonate
3-Ethyl-3-hydroxyvaleric Acid, *Et ester*
3-Hydroxy-2,2-dimethylpropionic Acid, *Et ether*, *Et ester*
2-Hydroxynonanoic Acid
3-Hydroxynonanoic Acid
4-Hydroxynonanoic Acid
5-Hydroxynonanoic Acid†
7-Hydroxynonanoic Acid
9-Hydroxynonanoic Acid
8-Hydroxyoctanoic Acid, *Me ester*
3-Hydroxy-2,2,3-trimethylbutyric Acid, *Et ester*
3-Hydroxy-2,2,5-trimethylhexanoic Acid
2-Hydroxyvaleric Acid, *Et ether*, *Et ester*
5-Hydroxyvaleric Acid, *Et ether*, *Et ester*
Levulinic Aldehyde, *Di-Et acetal*
Propanal, *Parapropanal*

$C_9H_{18}O_4$
Cladinose, *Me-glycoside*
2,3-Dihydroxybutyric Acid, *Pentyl ester*
2,3-Dihydroxy-2-methylpropionic Acid, *Pentyl ester*
Formylacetic Acid, *Et ester*
Glycerol, 1-*Hexanoyl*

$C_9H_{18}O_5$
Fucose, 2,3,4-*Tri-Me ether*
Rhamnose, 2,3,4-*Tri-Me ether*
Thevetose, *Ethyl-β-thevetoside*

$C_9H_{18}O_6$
2,3-Di-*O*-methylgalactose, *Methyl galactoside*
2,4-Di-*O*-methylgalactose, *Methyl galactoside*
2,6-Di-*O*-methylgalactose, *Methyl galactoside*
3,4-Di-*O*-methylgalactose, *Methyl galactoside*
4,6-Di-*O*-methylgalactose, *Methyl galactoside*
2,3-Di-*O*-methylglucose, *Methyl glucoside*
2,4-Di-*O*-methylglucose, *Methyl glucoside*
2,6-Di-*O*-methylglucose, *Methyl glucoside*
3,4-Di-*O*-methylglucose, *Methyl glucoside*
3,6-Di-*O*-methylglucose, *Methyl glucoside*
4,6-Di-*O*-methylglucose, *Methyl glucoside*
Fructose, 1,3,4-*Tri-Me ether*
Fructose, 1,3,6-*Tri-Me ether*
Fructose, 3,4,6-*Tri-Me ether*
Rhamnonic Acid, 2,3,5-*Tri-Me ether*
2,3,4-Tri-*O*-methylgalactose
2,3,5-Tri-*O*-methylgalactose
2,3,6-Tri-*O*-methylgalactose
2,4,6-Tri-*O*-methylgalactose
3,4,6-Tri-*O*-methylgalactose
2,3,4-Tri-*O*-methylglucose
2,3,5-Tri-*O*-methylglucose
2,3,6-Tri-*O*-methylglucose
2,4,6-Tri-*O*-methylglucose
3,4,6-Tri-*O*-methylglucose
3,5,6-Tri-*O*-methylglucose

$C_9H_{18}O_7$
Galactonic Acid, 2,3,4-*Tri-Me ether*
Galactonic Acid, 2,3,5-*Tri-Me ether*
Galactonic Acid, 2,3,6-*Tri-Me ether*
α-*Rhamno*hexonic Acid, *Et ester*

$C_9H_{18}O_8$
Floridoside
O-β-D-Galactofuranosyl(1→1)-D-glycerol†
O-α-D-Galactopyranosyl-(1→2)-glycerol†
Gluco-octose, *Methylglycoside*
Rhamno-octose

$C_9H_{18}O_9$
Gluco-nonose
Kurrin
D-*erythro*-L-*gluco*-Nonulose†
Rhamno-octonic Acid

$C_9H_{19}BrHg$
Mercuri-nonyl bromide

$C_9H_{19}ClO_2$
2-Chloropropanal, *Dipropyl acetal*

$C_9H_{19}N$
1-Butylpiperidine
1-(2-Butyl)piperidine
1-*tert*-Butylpiperidine
2-Butylpiperidine
2-(2-Butyl)piperidine
3-Butylpiperidine
4-Butylpiperidine
4-(2-Butyl)piperidine
Cyclopentamine
2,2-Diethyl-1-methylpyrrolidine
N-Dipropylallylamine
1-Ethyl-2,6-dimethylpiperidine
2-Ethyl-3,5-dimethylpiperidine
3-Ethyl-1,2-dimethylpiperidine
4-Ethyl-1,4-dimethylpiperidine
4-Ethyl-2,6-dimethylpiperidine
2-Isopropylpiperidine, N-*Me*
N-Methylconiine
Octinum
Perhydroazecine
Phyllopyrrolidine
Piperidine, N-*Butyl*
N-Propylcyclohexylamine
1-Propylcyclohexylamine
2-Propylcyclohexylamine
Pyrrolidine, N-*Pentyl*
1,2,3,6-Tetramethylpiperidine
1,2,5,5-Tetramethylpiperidine
2,2,4,6-Tetramethylpiperidine
2,2,6,6-Tetramethylpiperidine

$C_9H_{19}NO$
5-Azoniaspiro[4,5]decane(hydroxide)
4-Hydroxy-1,2,3,6-tetramethylpiperidine
4-Hydroxy-2,2,6,6-tetramethylpiperidine
4-Hydroxy-2,2,6-trimethylpiperidine, N-*Me*
N-Methylconhydrine
3-Methyl-2-propylvaleric Acid, *Amide*
Nonanoic Acid, *Amide*

$C_9H_{19}NO_2$
2-Aminohexanoic Acid, *Isopropyl ester*
2-Aminononanoic Acid
3-Aminononanoic Acid

$C_9H_{19}NO_2$ (*continued*)
6-Aminononanoic Acid
9-Aminononanoic Acid
2-Amino-octanoic Acid, *Me ester*
Butylcarbamic Acid, *Bu ester*
6-Dimethylaminohexanoic Acid, *Me ester*
Piperidinoacetaldehyde, *Di-Me acetal*
Valine, *Butyl ester*

$C_9H_{19}NO_3$
Desosamine, *Me glycoside*

$C_9H_{19}NO_4$
Acetylcholine, *Acetate*
Amosamine, α-*Me amosaminide* †

$C_9H_{19}NO_5$
Glucosamine, 3,4,6-*Tri-Me ether*

$C_9H_{19}NO_6$
Galactonic Acid, 2,3,4-*Tri-Me ether*, *Amide*
Gluconic Acid, 2,4,6-*Tri-Me ether*, *Amide*
Glucosaminic Acid, 3,4,6-*Tri-Me ether*

$C_9H_{19}NS_2$
N-Dibutyldithiocarbamic Acid

$C_9H_{19}N_3O_4$
N^{ϵ}-(2-Amino-2-carboxyethyl)-L-lysine †

$C_9H_{19}N_5O$
Amidinomycin †

$C_9H_{19}N_5O_3$
L-Alanyl-L-arginine
L-Arginyl-L-alanine
L-Arginyl-D-alanine

C_9H_{20}
2,2-Dimethylheptane
2,3-Dimethylheptane
2,4-Dimethylheptane
2,5-Dimethylheptane
2,6-Dimethylheptane
3,3-Dimethylheptane
3,4-Dimethylheptane
3,5-Dimethylheptane
4,4-Dimethylheptane
4-Ethylheptane
2-Methyloctane
3-Methyloctane
4-Methyloctane
Nonane
2,2,4,4-Tetramethylpentane

$C_9H_{20}N_2$
1,2,2,5,5-Pentamethylpiperazine
2,3,4,5,6-Pentamethylpiperazine
2,3,5,6-Tetramethylpiperazine, N-*Me*

$C_9H_{20}N_2O$
1,1-Di-isobutylurea
1,3-Di-isobutylurea
1,1-Dibutylurea
1,3-Dibutylurea

$C_9H_{20}N_2O_2$
6-*N*-Trimethyl-L-lysine betaine †

$C_9H_{20}N_2S$
1,3-Dibutylthiourea

$C_9H_{20}O$
Butyl 2-methylbutyl Ether
2,2-Dimethyl-1-heptanol
2,5-Dimethyl-1-heptanol
2,4-Dimethyl-2-heptanol
2,5-Dimethyl-2-heptanol
2,6-Dimethyl-2-heptanol
4,6-Dimethyl-2-heptanol
5,6-Dimethyl-2-heptanol
2,2-Dimethyl-3-heptanol
2,3-Dimethyl-3-heptanol
2,6-Dimethyl-3-heptanol
3,5-Dimethyl-3-heptanol
3,6-Dimethyl-3-heptanol
2,2-Dimethyl-4-heptanol
2,4-Dimethyl-4-heptanol
2,6-Dimethyl-4-heptanol
3,3-Dimethyl-4-heptanol
3,5-Dimethyl-4-heptanol
3,5-Dimethyl-3-hexanol, *Me ether*
2,4-Dimethyl-2-pentanol, *Et ether*
4-Ethyl-4-heptanol
Ethyl heptyl Ether
2-Ethyl-2-methyl-1-hexanol
2-Ethyl-3-methyl-1-hexanol
3-Ethyl-2-methyl-2-hexanol
3-Ethyl-2-methyl-3-hexanol
4-Ethyl-2-methyl-4-hexanol
Isobutyl 2-methylbutyl Ether
6-Methyl-2-heptanol, *Me ether*
5-Methyl-2-hexanol, *Et ether*
2-Methyl-1-octanol
3-Methyl-1-octanol
4-Methyl-1-octanol
5-Methyl-1-octanol
6-Methyl-1-octanol
7-Methyl-1-octanol
2-Methyl-2-octanol
3-Methyl-2-octanol
7-Methyl-2-octanol
2-Methyl-3-octanol
3-Methyl-3-octanol
4-Methyl-3-octanol
6-Methyl-3-octanol
2-Methyl-4-octanol
4-Methyl-4-octanol
5-Methyl-4-octanol
7-Methyl-4-octanol
Methyl octyl Ether
1-Nonanol
2-Nonanol
3-Nonanol
4-Nonanol
5-Nonanol
2-Octanol, *Me ether*
2,2,4,4-Tetramethyl-3-pentanol
2,4,4-Trimethyl-3-hexanol
3,5,5-Trimethylhexanol

$C_9H_{20}OS_2$
5-Hydroxypentanal, *Di-Et mercaptal*

$C_9H_{20}O_2$
Biformyne
2,2-Dimethylpropanal, *Di-Et acetal*
1,7-Heptanediol, *Di-Me ether*
3-Methyl-2-butanone, *Di-Et acetal*
1,9-Nonanediol
1,5-Pentanediol, *Monobutyl ether*
3-Pentanone, *Di-Et acetal*
n-Valeraldehyde, *Di-Et acetal*

$C_9H_{20}O_3$
Glycerol, *Tri-Et ether*
Glycerol, 1,3-*Dipropyl ether*
Glycerol, 1,3-*Di-isopropyl ether*
1,2,3-Nonanetriol
Orthoacetic Acid, *Me-di-3-methylbutyl ester*
Orthoformic Acid, *Et-dipropyl ester*
Orthopropionic Acid, *Tri-Et ester*

$C_9H_{20}O_3S$
Nonane-1-sulphonic Acid

$C_9H_{20}O_4$
Ethyl orthocarbonate †
Orthocarbonic Acid, *Tetra-Et ester*

$C_9H_{20}O_4S_2$
Ethylsulphonal

$C_9H_{20}O_5$
Rhamnitol, 3,4,5-*Tri-Me ether*

$C_9H_{21}AlO_3$
Aluminium isopropoxide

$C_9H_{21}BO_3$
Tripropyl borate

$C_9H_{21}N$
1-Amino-2-methyloctane
1-Aminononane
2-Aminononane
N-Isobutyl-3-methylbutylamine
Methyldi-(2-butyl)amine †
Tripropylamine

$C_9H_{21}NO$
1-Diethylamino-2-propanol, *Et ether*
2-Heptylaminoethanol

$C_9H_{21}NO_3$
Muscarine

$C_9H_{21}O_4P$
Phosphoric Acid, *Tripropyl ester*

$C_9H_{22}AuCl_4NO_2$
Muscaridine, *Chloroaurate*

$C_9H_{22}NO_2$ (ion)
Muscaridine

$C_9H_{22}N_5O_4$
Actinorubin (or $C_6H_{14}N_2O_3$)

$C_9H_{22}O_4P_2S_4$
Ethion

$C_9H_{23}NO_2$
Leucine-choline

$C_9H_{23}N_2OP$
Methylphosphonic Acid, *Di-diethylamide*

$C_9H_{24}B_3N_3$
N,N,',N''-Tri-isopropylborazole
N,N',N''-Tripropylborazole

$C_9H_{24}I_2N_2O$
Endoiodin

$C_9H_{24}N_4$
Tris-(3-aminopropyl)ammonia †

$C_9H_{24}O_3Si_3$
2,4,6-Triethyl-2,4,6-trimethylcyclotrisiloxane

C_9N_6
Diazotetracyanocyclopentadiene †

C_{10}

$C_{10}Br_8$
Octabromofulvalene †

$C_{10}Cl_8$
Octachlorofulvalene †
Octachloronaphthalene

$C_{10}D_8$
Octadeuteronaphthalene

$C_{10}HF_{19}O_2$
Perfluorodecanoic Acid

$C_{10}H_2O_6$
Benzene-1,2,4,5-tetracarboxylic Acid, *Di-anhydride*

$C_{10}H_3Br_3O_2$
3,4,6-Tribromo-1,2-naphthoquinone
3,5,6-Tribromo-1,2-naphthoquinone

$C_{10}H_4Br_2O_2$
3,4-Dibromo-1,2-naphthoquinone
3,6-Dibromo-1,2-naphthoquinone
4,6-Dibromo-1,2-naphthoquinone
2,3-Dibromo-1,4-naphthoquinone
5,8-Dibromo-1,4-naphthoquinone

$C_{10}H_4Br_4$
1,2,3,4-Tetrabromonaphthalene
1,2,6,8-Tetrabromonaphthalene
1,3,5,8-Tetrabromonaphthalene
1,3,6,8-Tetrabromonaphthalene
1,4,6,7-Tetrabromonaphthalene

$C_{10}H_4Br_4O$
1,3,4,6-Tetrabromo-2-naphthol
1,3,5,6-Tetrabromo-2-naphthol
1,3,6,7-Tetrabromo-2-naphthol

$C_{10}H_4Br_5NO$
2-(3,5-Dibromo-2-hydroxyphenyl)-3,4,5-tribromopyrrole †

$C_{10}H_4ClN_3O_6$
1-Chloro-2,4,5-trinitronaphthalene
1-Chloro-2,4,8-trinitronaphthalene
2-Chloro-1,6,8-trinitronaphthalene

$C_{10}H_4Cl_2O_2$
3,4-Dichloro-1,2-naphthoquinone
5,8-Dichloro-1,2-naphthoquinone
2,3-Dichloro-1,4-naphthoquinone
2,6-Dichloro-1,4-naphthoquinone
2,7-Dichloro-1,4-naphthoquinone
5,6-Dichloro-1,4-naphthoquinone
5,8-Dichloro-1,4-naphthoquinone
1,5-Dichloro-2,6-naphthoquinone

$C_{10}H_4Cl_4$
1,2,3,4-Tetrachloronaphthalene
1,2,3,5-Tetrachloronaphthalene
1,2,3,7-Tetrachloronaphthalene
1,2,4,6-Tetrachloronaphthalene
1,2,5,6-Tetrachloronaphthalene
1,3,5,7-Tetrachloronaphthalene
1,3,5,8-Tetrachloronaphthalene
1,3,6,7-Tetrachloronaphthalene
1,4,6,7-Tetrachloronaphthalene

$C_{10}H_4Cl_4O_8S_4$
Naphthalene-1,3,5,7-tetrasulphonic Acid, *Tetrachloride*

$C_{10}H_4Cl_6O_2$
Resorcinol, *Di-trichlorovinyl ether*

$C_{10}H_4F_4$
1,2,3,4-Tetrafluoronaphthalene†

$C_{10}H_4N_4O_8$
1,2,4,7-Tetranitronaphthalene
1,2,5,8-Tetranitronaphthalene
1,2,6,8-Tetranitronaphthalene
1,3,5,7-Tetranitronaphthalene
1,3,5,8-Tetranitronaphthalene
1,3,6,8-Tetranitronaphthalene
1,4,5,8-Tetranitronaphthalene

$C_{10}H_4N_4O_9$
2,4,5,7-Tetranitro-1-naphthol

$C_{10}H_4O_4$
2-Decene-4,6,8-triynedioic Acid

$C_{10}H_5Br_2NO_2$
1,2-Dibromo-6-nitronaphthalene
1,2-Dibromo-7-nitronaphthalene
1,3-Dibromo-7-nitronaphthalene
1,4-Dibromo-2-nitronaphthalene
1,4-Dibromo-6-nitronaphthalene
1,4-Dibromo-8-nitronaphthalene
2,4-Dibromo-1-nitronaphthalene

$C_{10}H_5Br_2NO_3$
2,4-Dibromo-6-nitro-1-naphthol
3,6-Dibromo-1-nitro-2-naphthol

$C_{10}H_5BrN_2O_4$
1-Bromo-2,4-dinitronaphthalene
1-Bromo-4,8-dinitronaphthalene
1-Bromo-5,8-dinitronaphthalene
2-Bromo-1,3-dinitronaphthalene
2-Bromo-1,7-dinitronaphthalene

$C_{10}H_5BrO_2$
3-Bromo-1,2-naphthoquinone
4-Bromo-1,2-naphthoquinone
6-Bromo-1,2-naphthoquinone
2-Bromo-1,4-naphthoquinone
5-Bromo-1,4-naphthoquinone

$C_{10}H_5BrO_3$
2-Bromo-3-hydroxy-1,4-naphthoquinone
2-Bromo-5-hydroxy-1,4-naphthoquinone
6-Bromo-2-hydroxy-1,4-naphthoquinone

$C_{10}H_5Br_3$
1,2,4-Tribromonaphthalene
1,2,6-Tribromonaphthalene
1,3,6-Tribromonaphthalene
1,3,8-Tribromonaphthalene
1,4,5-Tribromonaphthalene
1,4,6-Tribromonaphthalene

$C_{10}H_5ClN_2O_4$
1-Chloro-2,4-dinitronaphthalene
1-Chloro-4,5-dinitronaphthalene
2-Chloro-1,8-dinitronaphthalene
3-Chloro-1,8-dinitronaphthalene
4-Chloro-1,5-dinitronaphthalene
6-Chloro-1,3-dinitronaphthalene

$C_{10}H_5ClO_2$
2-Chloro-1,4-naphthoquinone
3-Chloro-1,2-naphthoquinone
4-Chloro-1,2-naphthoquinone
5-Chloro-1,4-naphthoquinone
6-Chloro-1,4-naphthoquinone

$C_{10}H_5ClO_3$
2-Chloro-3-hydroxy-1,4-naphthoquinone
6-Chloro-3-hydroxy-1,4-naphthoquinone
Coumarin-3-carboxylic Acid, *Chloride*

$C_{10}H_5Cl_2NO$
2-Chloroquinoline-4-carboxylic Acid, *Chloride*

$C_{10}H_5Cl_2NO_2$
1,3-Dichloro-8-nitronaphthalene
1,4-Dichloro-2-nitronaphthalene
1,4-Dichloro-5-nitronaphthalene
1,5-Dichloro-3-nitronaphthalene
1,5-Dichloro-4-nitronaphthalene
1,6-Dichloro-4-nitronaphthalene

$C_{10}H_5Cl_2NO_6S_2$
4-Nitronaphthalene-2,6-disulphonic Acid, *Dichloride*
4-Nitronaphthalene-2,7-disulphonic Acid, *Dichloride*

$C_{10}H_5Cl_3$
1,2,3-Trichloronaphthalene
1,2,4-Trichloronaphthalene
1,2,5-Trichloronaphthalene
1,2,6-Trichloronaphthalene
1,2,7-Trichloronaphthalene
1,2,8-Trichloronaphthalene
1,3,5-Trichloronaphthalene
1,3,6-Trichloronaphthalene
1,3,7-Trichloronaphthalene
1,3,8-Trichloronaphthalene
1,4,5-Trichloronaphthalene
1,4,6-Trichloronaphthalene
1,6,7-Trichloronaphthalene
2,3,6-Trichloronaphthalene

$C_{10}H_5Cl_3O$
2,3,4-Trichloro-1-naphthol
1,3,4-Trichloro-2-naphthol
1,4,5-Trichloro-2-naphthol

$C_{10}H_5Cl_3O_6S_3$
Naphthalene-1,3,5-trisulphonic Acid, *Trichloride*
Naphthalene-1,3,6-trisulphonic Acid, *Trichloride*
Naphthalene-1,3,7-trisulphonic Acid, *Trichloride*
Naphthalene-1,4,5-trisulphonic Acid, *Trichloride*
Naphthalene-2,3,6-trisulphonic Acid, *Trichloride*

$C_{10}H_5Cl_3O_7S_3$
1-Naphthol-2,4,7-trisulphonic Acid, *Chloride*

$C_{10}H_5Cl_4N$
2,3,4,5-Tetrachloropyrrole, N-*Phenyl*

$C_{10}H_5Cl_7$
Heptachlor

$C_{10}H_5F_3O_7S_3$
2-Naphthol-3,6,8-trisulphonic Acid, *Trifluoride*

$C_{10}H_5F_7O_2$
Heptafluorobutyric Acid, *Phenyl ester*

$C_{10}H_5IN_2O_4$
1-Iodo-3,4-dinitronaphthalene
1-Iodo-4,8-dinitronaphthalene

$C_{10}H_5NO_2$
Coumarin-3-carboxylic Acid, *Nitrile*

$C_{10}H_5NO_4$
3-Nitro-1,2-naphthoquinone

$C_{10}H_5NO_5$
6-Hydroxy-5-nitro-1,3-naphthoquinone
7-Hydroxy-8-nitro-1,2-naphthoquinone
2-Hydroxy-3-nitro-1,4-naphthoquinone

$C_{10}H_5N_3O_2$
o-Nitrobenzylidenemalonic Acid, *Dinitrile*

$C_{10}H_5N_3O_6$
1,2,3-Trinitronaphthalene
1,2,4-Trinitronaphthalene
1,2,5-Trinitronaphthalene
1,3,5-Trinitronaphthalene
1,3,6-Trinitronaphthalene
1,3,8-Trinitronaphthalene
1,4,5-Trinitronaphthalene

$C_{10}H_5N_3O_7$
2,4,5-Trinitro-1-naphthol
2,4,7-Trinitro-1-naphthol
2,4,8-Trinitro-1-naphthol
1,5,8-Trinitro-2-naphthol
1,6,8-Trinitro-2-naphthol

$C_{10}H_5N_5O_8$
2,4,5,7-Tetranitro-1-naphthylamine
2,4,5,8-Tetranitro-1-naphthylamine

$C_{10}H_6$
2,4,6,8-Decatetrayne
1,2-Diethynylbenzene
1,3-Diethynylbenzene
1,4-Diethynylbenzene

$C_{10}H_6BrCl$
1-Bromo-2-chloronaphthalene
1-Bromo-4-chloronaphthalene
1-Bromo-5-chloronaphthalene
1-Bromo-6(or 7)-chloronaphthalene
1-Bromo-8-chloronaphthalene
2-Bromo-1-chloronaphthalene
6-Bromo-1-chloronaphthalene

$C_{10}H_6BrClO$
2-Bromo-4-chloro-1-naphthol
4-Bromo-2-chloro-1-naphthol
6-Bromo-1-chloro-2-naphthol

$C_{10}H_6BrClO_2S$
4-Chloronaphthalene-1-sulphonic Acid, *Bromide*
5-Chloronaphthalene-1-sulphonic Acid, *Bromide*
7-Chloronaphthalene-1-sulphonic Acid, *Bromide*
6-Chloronaphthalene-2-sulphonic Acid, *Bromide*

$C_{10}H_6BrF$
1-Bromo-2-fluoronaphthalene
1-Bromo-4-fluoronaphthalene

$C_{10}H_6BrNO_2$
1-Bromo-2-nitronaphthalene
1-Bromo-3-nitronaphthalene
1-Bromo-4-nitronaphthalene
1-Bromo-5-nitronaphthalene
1-Bromo-6-nitronaphthalene
1-Bromo-7-nitronaphthalene
1-Bromo-8-nitronaphthalene
2-Bromo-1-nitronaphthalene
2-Bromo-3-nitronaphthalene
2-Bromo-6-nitronaphthalene
2-Bromo-7-nitronaphthalene
3-Bromo-1-nitronaphthalene
6-Bromo-1-nitronaphthalene
7-Bromo-1-nitronaphthalene
8-Bromoquinoline-5-carboxylic Acid
3-Bromoquinoline-6-carboxylic Acid
3-Bromoquinoline-8-carboxylic Acid

$C_{10}H_6BrNO_3$
2-Bromo-4-nitro-1-naphthol
3-Bromo-1-nitro-2-naphthol
4-Bromo-2-nitro-1-naphthol
5-Bromo-2-nitro-1-naphthol
6-Bromo-1-nitro-2-naphthol

$C_{10}H_6BrNO_4$
Propargyl 2-Bromo-3-nitrobenzoate†

$C_{10}H_6Br_2$
1,2-Dibromonaphthalene
1,3-Dibromonaphthalene
1,4-Dibromonaphthalene
1,5-Dibromonaphthalene
1,6-Dibromonaphthalene
1,7-Dibromonaphthalene
1,8-Dibromonaphthalene
2,3-Dibromonaphthalene
2,6-Dibromonaphthalene
2,7-Dibromonaphthalene

$C_{10}H_6Br_2O$
2,4-Dibromo-1-naphthol
1,3-Dibromo-2-naphthol
1,6-Dibromo-2-naphthol
3,6-Dibromo-2-naphthol
3,7-Dibromo-2-naphthol
4,6-Dibromo-2-naphthol
5,8-Dibromo-2-naphthol

$C_{10}H_6Br_2O_2$
1,3-Dibromo-2,4-dihydroxynaphthalene
1,4-Dibromo-2,3-dihydroxynaphthalene
1,5-Dibromo-2,6-dihydroxynaphthalene
1,5-Dibromo-4,8-dihydroxynaphthalene
2,3-Dibromo-1,4-dihydroxynaphthalene
2,3-Dibromo-6,7-dihydroxynaphthalene
2,6-Dibromo-1,5-dihydroxynaphthalene
2,7-Dibromo-3,6-dihydroxynaphthalene
3,6-Dibromo-1,2-dihydroxynaphthalene

$C_{10}H_6Br_2O_4S_2$
Naphthalene-2,7-disulphonic Acid, *Dibromide*

$C_{10}H_6Br_3N$
1,3,4-Tribromo-2-naphthylamine
1,3,6-Tribromo-2-naphthylamine
3,4,7-Tribromo-2-naphthylamine

$C_{10}H_6ClF$
1-Chloro-4-fluoronaphthalene
1-Chloro-5-fluoronaphthalene
1-Chloro-8-fluoronaphthalene

$C_{10}H_6ClFO_2S$
4-Fluoronaphthalene-1-sulphonic Acid, *Chloride*
5-Fluoronaphthalene-1-sulphonic Acid, *Chloride*
6-Fluoronaphthalene-2-sulphonic Acid, *Chloride*

$C_{10}H_6ClI$
1-Chloro-2-iodonaphthalene
1-Chloro-3-iodonaphthalene
1-Chloro-4-iodonaphthalene
1-Chloro-5-iodonaphthalene
1-Chloro-7-iodonaphthalene
1-Chloro-8-iodonaphthalene
2-Chloro-1-iodonaphthalene
2-Chloro-5-iodonaphthalene
2-Chloro-6-iodonaphthalene
2-Chloro-7-iodonaphthalene
7-Chloro-1-iodonaphthalene

$C_{10}H_6ClNO$
o-Carboxycinnamic Acid, 2-*Nitrile*, *Chloride*
Quinoline-2-carboxylic Acid, *Chloride*
Quinoline-4-carboxylic Acid, *Chloride*

$C_{10}H_6ClNO_2$
2-Chloro-α-cyanocinnamic Acid
4-Chloro-α-cyanocinnamic Acid
2-Chloro-1,4-naphthoquinone, 4-*Oxime*
3-Chloro-1,2-naphthoquinone, 1-*Oxime*
1-Chloro-2-nitronaphthalene
1-Chloro-3-nitronaphthalene
1-Chloro-4-nitronaphthalene
1-Chloro-5-nitronaphthalene
1-Chloro-6-nitronaphthalene
1-Chloro-8-nitronaphthalene
2-Chloro-1-nitronaphthalene
2-Chloro-3-nitronaphthalene
2-Chloro-6-nitronaphthalene
2-Chloro-7-nitronaphthalene
3-Chloro-1-nitronaphthalene
6-Chloro-1-nitronaphthalene
7-Chloro-1-nitronaphthalene
4-Chloroquinoline-2-carboxylic Acid
7-Chloroquinoline-2-carboxylic Acid
2-Chloroquinoline-3-carboxylic Acid
4-Chloroquinoline-3-carboxylic Acid
2-Chloroquinoline-4-carboxylic Acid
3-Chloroquinoline-4-carboxylic Acid
5-Chloroquinoline-4-carboxylic Acid
6-Chloroquinoline-4-carboxylic Acid
7-Chloroquinoline-4-carboxylic Acid
8-Chloroquinoline-4-carboxylic Acid
6-Hydroxyquinoline-4-carboxylic Acid, *Chloride*
2-3′-Indolylglyoxylic Acid, *Chloride*

$C_{10}H_6ClNO_3$
2-Chloro-3-nitro-1-naphthol
2-Chloro-4-nitro-1-naphthol
2-Chloro-6-nitro-1-naphthol
4-Chloro-2-nitro-1-naphthol
Phthalimidoacetic Acid, *Chloride*

$C_{10}H_6ClNO_4S$
4-Nitronaphthalene-1-sulphonic Acid, *Chloride*
5-Nitronaphthalene-1-sulphonic Acid, *Chloride*
6-Nitronaphthalene-1-sulphonic Acid, *Chloride*
7-Nitronaphthalene-1-sulphonic Acid, *Chloride*
8-Nitronaphthalene-1-sulphonic Acid, *Chloride*
1-Nitronaphthalene-2-sulphonic Acid, *Chloride*
4-Nitronaphthalene-2-sulphonic Acid, *Chloride*
5-Nitronaphthalene-2-sulphonic Acid, *Chloride*
8-Nitronaphthalene-2-sulphonic Acid, *Chloride*

$C_{10}H_6ClNO_4S_2$
1,8-Naphthosultam-4-sulphonic Acid, *Chloride*

$C_{10}H_6Cl_2$
1,2-Dichloronaphthalene
1,3-Dichloronaphthalene
1,4-Dichloronaphthalene
1,5-Dichloronaphthalene
1,6-Dichloronaphthalene
1,7-Dichloronaphthalene
1,8-Dichloronaphthalene
2,3-Dichloronaphthalene
2,6-Dichloronaphthalene
2,7-Dichloronaphthalene

$C_{10}H_6Cl_2N_2O_2$
Pyrrolnitrin †

$C_{10}H_6Cl_2O$
2,3-Dichloro-1-naphthol
2,4-Dichloro-1-naphthol
5,7-Dichloro-1-naphthol
5,8-Dichloro-1-naphthol
6,7-Dichloro-1-naphthol
7,8-Dichloro-1-naphthol
1,3-Dichloro-2-naphthol
1,4-Dichloro-2-naphthol
1,5-Dichloro-2-naphthol
1,6-Dichloro-2-naphthol
1,7-Dichloro-2-naphthol
1,8-Dichloro-2-naphthol
3,4-Dichloro-2-naphthol
3,6-Dichloro-2-naphthol
4,8-Dichloro-2-naphthol
5,6-Dichloro-2-naphthol
5,8-Dichloro-2-naphthol

$C_{10}H_6Cl_2O_2$
Benzylidenemalonic Acid, *Dichloride*
1,2-Dichloro-3,4-dihydroxynaphthalene
1,3-Dichloro-2,4-dihydroxynaphthalene
1,4-Dichloro-2,3-dihydroxynaphthalene
1,5-Dichloro-2,6-dihydroxynaphthalene
1,5-Dichloro-4,8-dihydroxynaphthalene
1,8-Dichloro-2,7-dihydroxynaphthalene
2,3-Dichloro-1,4-dihydroxynaphthalene

$C_{10}H_6Cl_2O_2S$
2-Chloronaphthalene-1-sulphonic Acid, *Chloride*
4-Chloronaphthalene-1-sulphonic Acid, *Chloride*
5-Chloronaphthalene-1-sulphonic Acid, *Chloride*
6-Chloronaphthalene-1-sulphonic Acid, *Chloride*
7-Chloronaphthalene-1-sulphonic Acid, *Chloride*
8-Chloronaphthalene-1-sulphonic Acid, *Chloride*
1-Chloronaphthalene-2-sulphonic Acid, *Chloride*
4-Chloronaphthalene-2-sulphonic Acid, *Chloride*
5-Chloronaphthalene-2-sulphonic Acid, *Chloride*
6-Chloronaphthalene-2-sulphonic Acid, *Chloride*
7-Chloronaphthalene-2-sulphonic Acid, *Chloride*
8-Chloronaphthalene-2-sulphonic Acid, *Chloride*

$C_{10}H_6Cl_2O_4S_2$
Naphthalene-1,3-disulphonic Acid, *Dichloride*
Naphthalene-1,4-disulphonic Acid, *Dichloride*
Naphthalene-1,5-disulphonic Acid, *Dichloride*
Naphthalene-1,6-disulphonic Acid, *Dichloride*
Naphthalene-1,7-disulphonic Acid, *Dichloride*
Naphthalene-2,6-disulphonic Acid, *Dichloride*
Naphthalene-2,7-disulphonic Acid, *Dichloride*

$C_{10}H_6Cl_2O_5S_2$
1-Naphthol-2,4-disulphonic Acid, *Dichloride*
1-Naphthol-3,6-disulphonic Acid, *Dichloride*

$C_{10}H_6Cl_2O_5S_2$ (*continued*)
1-Naphthol-3,8-disulphonic Acid, *Dichloride*
1-Naphthol-4,7-disulphonic Acid, *Dichloride*
1-Naphthol-4,8-disulphonic Acid, *Dichloride*
2-Naphthol-1,6-disulphonic Acid, *Dichloride*
2-Naphthol-1,7-disulphonic Acid, *Dichloride*
2-Naphthol-3,6-disulphonic Acid, *Dichloride*
2-Naphthol-6,8-disulphonic Acid, *Dichloride*

$C_{10}H_6Cl_4O_4$
Tetrachlorophthalic Acid, *Di-Me ester*
Tetrachlorophthalic Acid, *Mono-Et ester*
Tetrachloroterephthalic Acid, *Di-Me ester*

$C_{10}H_6Cl_6O$
1-Hydroxychlordene

$C_{10}H_6FNO_2$
1-Fluoro-4-nitronaphthalene

$C_{10}H_6F_2O_4S_2$
Naphthalene-1,5-disulphonic Acid, *Difluoride*

$C_{10}H_6F_7NO$
Heptafluorobutyric Acid, *Anilide*

$C_{10}H_6INO_2$
1-Iodo-2-nitronaphthalene
2-Iodo-1-nitronaphthalene
3-Iodo-1-nitronaphthalene
3-Iodo-2-nitronaphthalene
4-Iodo-1-nitronaphthalene
4-Iodo-2-nitronaphthalene
5-Iodo-1-nitronaphthalene
5-Iodo-2-nitronaphthalene
6-Iodo-2-nitronaphthalene
7-Iodo-2-nitronaphthalene
8-Iodo-1-nitronaphthalene

$C_{10}H_6INO_3$
2-Iodo-4-nitro-1-naphthol
2-Iodo-6-nitro-1-naphthol
4-Iodo-2-nitro-1-naphthol
4-Iodo-6-nitro-1-naphthol

$C_{10}H_6I_2$
1,2-Di-iodonaphthalene
1,4-Di-iodonaphthalene
1,8-Di-iodonaphthalene

$C_{10}H_6I_4O_4$
Tetraiodoterephthalic Acid, *Di-Me ester*

$C_{10}H_6N_2$
Benzylidenemalonic Acid, *Dinitrile*
1-Cyanoisoquinoline
5(or 8)-Cyanoisoquinoline
Quinoline-2-carboxylic Acid, *Nitrile*
Quinoline-3-carboxylic Acid, *Nitrile*
Quinoline-4-carboxylic Acid, *Nitrile*
Quinoline-5-carboxylic Acid, *Nitrile*
Quinoline-6-carboxylic Acid, *Nitrile*
Quinoline-7-carboxylic Acid, *Nitrile*
Quinoline-8-carboxylic Acid, *Nitrile*

$C_{10}H_6N_2O$
5-Aminoquinoline-4-carboxylic Acid, *Lactam*
2-Hydroxyquinoline-3-carboxylic Acid, *Nitrile*
7-Hydroxyquinoline-4-carboxylic Acid, *Nitrile*
6-Hydroxyquinoline-5-carboxylic Acid, *Nitrile*
Naphth[1,2-*d*]oxadiazole
Naphth[2,1-*d*]oxadiazole

$C_{10}H_6N_2O_2$
2,4-Dihydroxyquinoline-3-carboxylic Acid, *Nitrile*
Indole-2,6-dicarboxylic Acid, 6-*Nitrile*
Pyrocoll

$C_{10}H_6N_2O_4$
1,2-Dinitronaphthalene
1,3-Dinitronaphthalene
1,4-Dinitronaphthalene
1,5-Dinitronaphthalene
1,6-Dinitronaphthalene
1,7-Dinitronaphthalene
1,8-Dinitronaphthalene
2,3-Dinitronaphthalene
2,6-Dinitronaphthalene
2,7-Dinitronaphthalene
o-Nitrobenzylidenemalonic Acid, *Mononitrile*
m-Nitrobenzylidenemalonic Acid, *Mononitrile*
p-Nitrobenzylidenemalonic Acid, *Mononitrile*
5-Nitro-4-nitroso-1-naphthol
7-Nitro-4-nitroso-1-naphthol
8-Nitro-4-nitroso-1-naphthol
5-Nitroquinoline-2-carboxylic Acid
8-Nitroquinoline-2-carboxylic Acid
3-Nitroquinoline-4-carboxylic Acid
5-Nitroquinoline-4-carboxylic Acid
6-Nitroquinoline-4-carboxylic Acid
8-Nitroquinoline-4-carboxylic Acid

$C_{10}H_6N_2O_5$
2,4-Dinitro-1-naphthol
4,5-Dinitro-1-naphthol
4,6-Dinitro-1-naphthol
4,8-Dinitro-1-naphthol
1,5-Dinitro-2-naphthol
1,6-Dinitro-2-naphthol
1,8-Dinitro-2-naphthol
4,5-Dinitro-2-naphthol
4,8-Dinitro-2-naphthol
5,8-Dinitro-2-naphthol
4-Hydroxy-6-nitroquinoline-3-carboxylic Acid
4-Hydroxy-7-nitroquinoline-3-carboxylic Acid
4-Hydroxy-8-nitroquinoline-3-carboxylic Acid
8-Hydroxy-5-nitroquinoline-7-carboxylic Acid

$C_{10}H_6N_2O_6$
1,4-Dihydroxy-2,3-dinitronaphthalene
1,5-Dihydroxy-2,4-dinitronaphthalene
1,5-Dihydroxy-2,6-dinitronaphthalene
1,8-Dihydroxy-2,4-dinitronaphthalene
2,6-Dihydroxy-1,5-dinitronaphthalene
2,7-Dihydroxy-1,8-dinitronaphthalene

$C_{10}H_6N_2O_8S$
Flavianic Acid

$C_{10}H_6N_2S$
Naphtho[1,8-*cd*][1,2,6]thiadiazine†

$C_{10}H_6N_4O_2$
Alloxazine

$C_{10}H_6N_4O_6$
2,3,4-Trinitro-1-naphthylamine
2,4,5-Trinitro-1-naphthylamine
2,4,6-Trinitro-1-naphthylamine
1,6,8-Trinitro-2-naphthylamine

$C_{10}H_6N_6O_4S_2$
Naphthalene-1,5-disulphonic Acid, *Diazide*

$C_{10}H_6O$
2-Decene-4,6,8-triyn-1-al

$C_{10}H_6O_2$
2-Decene-4,6,8-triynoic Acid
8-Decene-2,4,6-triynoic Acid
1,2-Naphthoquinone
1,4-Naphthoquinone
Non-2-ene-4,6,8-triynoic Acid, *Me ester*
Phenylcyclobutadienoquinone

$C_{10}H_6O_3$
Diatetryne III
10-Hydroxydec-2-ene-4,6,8-triynoic Acid†
5-Hydroxy-1,4-naphthaquinone

$C_{10}H_6O_3S$
Naphthosultone

$C_{10}H_6O_4$
Ayapin
5-Benzylidene-1,3-dioxolan-2,4-dione†
Coumarin-3-carboxylic Acid
2,8-Decadiene-4,6-diynedioic Acid
2,4,6-Decatriynedioic Acid
Di-2-furylglyoxal
2,3-Dihydroxy-1,4-naphthoquinone
2,5-Dihydroxy-1,4-naphthoquinone
2,6-Dihydroxy-1,4-naphthoquinone
2,7-Dihydroxy-1,4-naphthoquinone
5,8-Dihydroxy-1,4-naphthoquinone
1,3-Dioxoindane-2-carboxylic Acid
Phthaloylacetic Acid

$C_{10}H_6O_5$
Furan-2-carboxylic Acid, *Anhydride*
3-Hydroxy-2*H*-pyran-2-one, *Anhydride*
3,4-Methylenedioxyhomophthalic Acid, *Anhydride*
Naphthopurpurin
2,5,7-Trihydroxy-1,4-naphthoquinone

$C_{10}H_6O_5S_2$
Naphthalene-1,2-disulphonic Acid, *Anhydride*
Naphthalene-1,8-disulphonic Acid, *Anhydride*

$C_{10}H_6O_6$
Cotarnic Acid, *Anhydride*
5-Hydroxytoluene-2,3,6-tricarboxylic Acid, *Anhydride*†
5-Methoxy-3,4-methylenedioxyphthalic Acid, *Anhydride*
Spinochrome B†
2,3,5,8-Tetrahydroxy-1,4-naphthoquinone
2,5,7,8-Tetrahydroxy-1,4-naphthaquinone†

$C_{10}H_6O_6S_2$
Naphthosultone-3-sulphonic Acid
Naphthosultone-4-sulphonic Acid

$C_{10}H_6O_7$
Spinochrome D†

$C_{10}H_6O_8$
Benzene-1,2,3,4-tetracarboxylic Acid
Benzene-1,2,3,5-tetracarboxylic Acid
Benzene-1,2,4,5-tetracarboxylic Acid
Spinochrome E

$C_{10}H_6O_{10}$
3,6-Dihydroxybenzene-1,2,4,5-tetracarboxylic Acid

$C_{10}H_6S_2$
Thieno[3,2-*b*]thianaphthene

$C_{10}H_7Br$
1-Bromonaphthalene
2-Bromonaphthalene

$C_{10}H_7BrHg$
Mercuri-1-naphthyl bromide

$C_{10}H_7BrN_2O_2$
2-Bromo-4-nitro-1-naphthylamine
3-Bromo-2-nitro-1-naphthylamine
3-Bromo-4-nitro-1-naphthylamine
4-Bromo-2-nitro-1-naphthylamine
4-Bromo-3-nitro-1-naphthylamine
4-Bromo-5-nitro-1-naphthylamine
4-Bromo-8-nitro-1-naphthylamine
5-Bromo-2-nitro-1-naphthylamine
5-Bromo-4-nitro-1-naphthylamine
6-Bromo-1-nitro-2-naphthylamine

$C_{10}H_7BrO$
2-Bromo-1-naphthol
3-Bromo-1-naphthol
4-Bromo-1-naphthol
5-Bromo-1-naphthol
6-Bromo-1-naphthol
7-Bromo-1-naphthol
8-Bromo-1-naphthol
1-Bromo-2-naphthol
3-Bromo-2-naphthol
4-Bromo-2-naphthol
5(or 8-)Bromo-2-naphthol
6-Bromo-2-naphthol
7-Bromo-2-naphthol

$C_{10}H_7BrO_2S$
Naphthalene-1-sulphonic Acid, *Bromide*
Naphthalene-2-sulphonic Acid, *Bromide*

$C_{10}H_7BrS$
4-Bromonaphthalene-1-thiol

$C_{10}H_7Br_2N$
2-Dibromomethyl-quinoline
6-Dibromomethyl-quinoline
6,8-Dibromo-2-methylquinoline
2,4-Dibromo-1-naphthylamine
1,3-Dibromo-2-naphthylamine
1,4-Dibromo-2-naphthylamine
1,6-Dibromo-2-naphthylamine
3,6-Dibromo-2-naphthylamine
5,8-Dibromo-2-naphthylamine
7,8-Dibromo-2-naphthylamine

$C_{10}H_7Br_2NO$
5,7-Dibromo-8-hydroxyquinoline, *Me ether*

$C_{10}H_7Cl$
1-Chloronaphthalene
2-Chloronaphthalene

$C_{10}H_7ClHg$
Mercuri-1-naphthyl chloride
Mercuri-2-naphthyl chloride

$C_{10}H_7ClN_2$
Naphthalene-1-diazonium chloride
Naphthalene-2-diazonium chloride

$C_{10}H_7ClN_2O$
2-Chloroquinoline-3-carboxylic Acid, *Amide*
2-Chloroquinoline-4-carboxylic Acid, *Amide*
6-Chloroquinoline-4-carboxylic Acid, *Amide*

$C_{10}H_7ClN_2O_2$
2-Chloro-4-methyl-6-nitroquinoline
2-Chloro-6-methyl-8-nitroquinoline
4-Chloro-2-methyl-3-nitroquinoline
8-Chloromethyl-5-nitroquinoline
2-Chloro-4-nitro-1-naphthylamine
3-Chloro-2-nitro-1-naphthylamine
4-Chloro-2-nitro-1-naphthylamine
4-Chloro-3-nitro-1-naphthylamine
8-Chloro-4-nitro-1-naphthylamine

$C_{10}H_7ClO$
2-Chloro-1-naphthol
3-Chloro-1-naphthol
4-Chloro-1-naphthol
5-Chloro-1-naphthol
6-Chloro-1-naphthol
7-Chloro-1-naphthol
8-Chloro-1-naphthol
1-Chloro-2-naphthol
3-Chloro-2-naphthol
4-Chloro-2-naphthol
5-Chloro-2-naphthol
6-Chloro-2-naphthol
7-Chloro-2-naphthol
8-Chloro-2-naphthol

$C_{10}H_7ClO_2S$
Naphthalene-1-sulphonic Acid, *Chloride*
Naphthalene-2-sulphonic Acid, *Chloride*

$C_{10}H_7ClO_3S$
2-Chloronaphthalene-1-sulphonic Acid
4-Chloronaphthalene-1-sulphonic Acid
5-Chloronaphthalene-1-sulphonic Acid
6-Chloronaphthalene-1-sulphonic Acid
7-Chloronaphthalene-1-sulphonic Acid
8-Chloronaphthalene-1-sulphonic Acid
1-Chloronaphthalene-2-sulphonic Acid
4-Chloronaphthalene-2-sulphonic Acid
5-Chloronaphthalene-2-sulphonic Acid
6-Chloronaphthalene-2-sulphonic Acid
7-Chloronaphthalene-2-sulphonic Acid
8-Chloronaphthalene-2-sulphonic Acid
2-Naphthol-1-sulphonic Acid, *Chloride*

$C_{10}H_7ClS$
4-Chloronaphthalene-1-thiol
8-Chloronaphthalene-1-thiol

$C_{10}H_7Cl_2N$
2,4-Dichloro-3-methylquinoline
2,4-Dichloro-5-methylquinoline
3,4-Dichloro-2-methylquinoline
4,5-Dichloro-2-methylquinoline
4,5-Dichloro-3-methylquinoline
4,6-Dichloro-3-methylquinoline
4,7-Dichloro-2-methylquinoline
4,7-Dichloro-3-methylquinoline
4,7-Dichloro-8-methylquinoline
4,8-Dichloro-2-methylquinoline
4,8-Dichloro-3-methylquinoline
5,8-Dichloro-2-methylquinoline
2,4-Dichloro-1-naphthylamine
4,7-Dichloro-1-naphthylamine
4,8-Dichloro-1-naphthylamine
5,7-Dichloro-1-naphthylamine
5,8-Dichloro-1-naphthylamine
1,3-Dichloro-2-naphthylamine
1,4-Dichloro-2-naphthylamine
1,5-Dichloro-2-naphthylamine
1,6-Dichloro-2-naphthylamine
1,8-Dichloro-2-naphthylamine
4,8-Dichloro-2-naphthylamine
5,8-Dichloro-2-naphthylamine

$C_{10}H_7Cl_2NO$
4,7-Dichloro-8-hydroxyquinoline, *Me ether*

$C_{10}H_7Cl_2NO_6$
2,5-Dichloro-6-nitroterephthalic Acid, *Di-Me ester*

$C_{10}H_7Cl_7$
β-Dihydroheptachlor†

$C_{10}H_7F$
1-Fluoronaphthalene
2-Fluoronaphthalene

$C_{10}H_7FO_2S$
Naphthalene-1-sulphonic Acid, *Fluoride*
Naphthalene-2-sulphonic Acid, *Fluoride*

$C_{10}H_7FO_3S$
4-Fluoronaphthalene-1-sulphonic Acid
5-Fluoronaphthalene-1-sulphonic Acid
6-Fluoronaphthalene-2-sulphonic Acid

$C_{10}H_7HgI$
Mercuri-1-naphthyl iodide

$C_{10}H_7I$
1-Iodonaphthalene
2-Iodonaphthalene

$C_{10}H_7IO$
3-Iodo-1-naphthol
5-Iodo-1-naphthol
1-Iodo-2-naphthol
3-Iodo-2-naphthol
4-Iodo-2-naphthol

$C_{10}H_7IN_2O_2$
2-Iodo-3-nitro-1-naphthylamine
2-Iodo-4-nitro-1-naphthylamine
2-Iodo-5-nitro-1-naphthylamine
3-Iodo-2-nitro-1-naphthylamine
4-Iodo-2-nitro-1-naphthylamine
4-Iodo-6-nitro-1-naphthylamine
3-Iodo-1-nitro-2-naphthylamine

$C_{10}H_7N$
Cyclazine
Cycl[3,2,2]azine†
Indene-3-carboxylic Acid, *Nitrile*

$C_{10}H_7NO$
3-Hydroxyindene-2-carboxylic Acid, *Nitrile*
2-Hydroxyindene-3-carboxylic Acid, *Nitrile*
Isoquinoline-1-aldehyde
1-Nitrosonaphthalene
Quinoline-2-aldehyde
Quinoline-3-aldehyde
Quinoline-4-aldehyde
Quinoline-5-aldehyde
Quinoline-6-aldehyde
Quinoline-7-aldehyde
Quinoline-8-aldehyde

$C_{10}H_7NO_2$
4-Amino-1,2-naphthoquinone
2-Amino-1,4-naphthoquinone
5-Amino-1,4-naphthoquinone

$C_{10}H_7NO_2$ *(continued)*
o-Carboxycinnamic Acid, 2-*Nitrile*
α-Cyanocinnamic Acid
6-Hydroxyquinoline-5-aldehyde
8-Hydroxyquinoline-5-aldehyde
8-Hydroxyquinoline-7-aldehyde
7-Hydroxyquinoline-8-aldehyde
Isoquinoline-1-carboxylic Acid
Maleanil
1-Nitronaphthalene
2-Nitronaphthalene
2-Nitroso-1-naphthol
4-Nitroso-1-naphthol
1-Nitroso-2-naphthol
Phthalimide, N-*Vinyl*
Quinoline-2-carboxylic Acid
Quinoline-3-carboxylic Acid
Quinoline-4-carboxylic Acid
Quinoline-5-carboxylic Acid
Quinoline-6-carboxylic Acid
Quinoline-7-carboxylic Acid
Quinoline-8-carboxylic Acid

$C_{10}H_7NO_2S$
1,8-Naphthosultam

$C_{10}H_7NO_3$
Coumarin-3-carboxylic Acid, *Amide*
1-Hydroxyisoquinoline-3-carboxylic Acid
6-Hydroxyisoquinoline-3-carboxylic Acid
1-Hydroxyisoquinoline-4-carboxylic Acid
3-Hydroxy-6-nitrophthalic Acid, *Dinitrile*, *Et ether*
4-Hydroxyquinoline-2-carboxylic Acid
2-Hydroxyquinoline-3-carboxylic Acid
4-Hydroxyquinoline-3-carboxylic Acid
2-Hydroxyquinoline-4-carboxylic Acid
3-Hydroxyquinoline-4-carboxylic Acid
6-Hydroxyquinoline-4-carboxylic Acid
7-Hydroxyquinoline-4-carboxylic Acid
8-Hydroxyquinoline-4-carboxylic Acid
6-Hydroxyquinoline-5-carboxylic Acid
8-Hydroxyquinoline-5-carboxylic Acid
5-Hydroxyquinoline-6-carboxylic Acid
8-Hydroxyquinoline-6-carboxylic Acid
8-Hydroxyquinoline-7-carboxylic Acid
3-Hydroxyquinoline-8-carboxylic Acid
4-Hydroxyquinoline-8-carboxylic Acid
2-3′-Indolylglyoxylic Acid
2-Nitro-1-naphthol
3-Nitro-1-naphthol
4-Nitro-1-naphthol
5-Nitro-1-naphthol
6-Nitro-1-naphthol
8-Nitro-1-naphthol
1-Nitro-2-naphthol
4-Nitro-2-naphthol
5-Nitro-2-naphthol
6-Nitro-2-naphthol
7-Nitro-2-naphthol
8-Nitro-2-naphthol

$C_{10}H_7NO_4$
1,2-Dihydroxy-3-nitronaphthalene
4,8-Dihydroxyquinoline-2-carboxylic Acid
6,7-Dihydroxyquinoline-2-carboxylic Acid
2,4-Dihydroxyquinoline-3-carboxylic Acid
2,6-Dihydroxyquinoline-4-carboxylic Acid
2,4-Dihydroxyquinoline-6-carboxylic Acid
Indole-1,3-dicarboxylic Acid
Indole-2,3-dicarboxylic Acid
Indole-2,5-dicarboxylic Acid
Indole-2,6-dicarboxylic Acid
Indole-3,6-dicarboxylic Acid
7-Methyl-6-nitrocoumarin
7-Methyl-8-nitrocoumarin
o-Nitrophenylpropiolic Acid, *Me ester*
Phthalimidoacetic Acid

$C_{10}H_7NO_4S$
4-Nitronaphthalene-1-sulphinic Acid
5-Nitronaphthalene-1-sulphinic Acid
5-Nitronaphthalene-2-sulphinic Acid
8-Nitronaphthalene-1-sulphinic Acid
1-Nitronaphthalene-2-sulphinic Acid
8-Nitronaphthalene-2-sulphinic Acid

$C_{10}H_7NO_5S$
2-Nitronaphthalene-1-sulphonic Acid
4-Nitronaphthalene-1-sulphonic Acid
5-Nitronaphthalene-1-sulphonic Acid
6-Nitronaphthalene-1-sulphonic Acid
7-Nitronaphthalene-1-sulphonic Acid
8-Nitronaphthalene-1-sulphonic Acid
1-Nitronaphthalene-2-sulphonic Acid
4-Nitronaphthalene-2-sulphonic Acid
5-Nitronaphthalene-2-sulphonic Acid
8-Nitronaphthalene-2-sulphonic Acid

$C_{10}H_7NO_5S_2$
1,8-Naphthosultam-4-sulphonic Acid

$C_{10}H_7NO_6$
3-(4,5-Methylenedioxy-2-nitrophenyl)acrylic Acid
o-Nitrobenzylidenemalonic Acid
m-Nitrobenzylidenemalonic Acid
p-Nitrobenzylidenemalonic Acid

$C_{10}H_7NO_6S$
2-Nitro-1-naphthol-4-sulphonic Acid
2-Nitro-1-naphthol-7-sulphonic Acid
1-Nitro-2-naphthol-6-sulphonic Acid
6-Nitro-2-naphthol-4-sulphonic Acid
6-Nitro-2-naphthol-8-sulphonic Acid
8-Nitro-2-naphthol-6-sulphonic Acid

$C_{10}H_7NO_7$
3,4-Dimethoxy-6-nitrobenzene-1,2-dicarboxylic Acid, *Anhydride*

$C_{10}H_7NO_8S_2$
3-Nitronaphthalene-1,5-disulphonic Acid
4-Nitronaphthalene-1,5-disulphonic Acid
8-Nitronaphthalene-1,6-disulphonic Acid
4-Nitronaphthalene-2,6-disulphonic Acid
4-Nitronaphthalene-2,7-disulphonic Acid

$C_{10}H_7NO_{11}S_3$
8-Nitronaphthalene-1,3,6-trisulphonic Acid

$C_{10}H_7NS$
4*H*-Thieno[3,2-*b*]indole

$C_{10}H_7N_3$
1-Naphthyl azide
2-Naphthyl azide

$C_{10}H_7N_3O_3$
o-Nitrobenzylidenemalonic Acid, *Mononitrile*, *Amide*

$C_{10}H_7N_3O_4$
2,3-Dinitro-1-naphthylamine
2,4-Dinitro-1-naphthylamine
2,6-Dinitro-1-naphthylamine
3,5-Dinitro-1-naphthylamine
3,6-Dinitro-1-naphthylamine
4,5-Dinitro-1-naphthylamine
3,8-Dinitro-1-naphthylamine
4,6-Dinitro-1-naphthylamine
4,8-Dinitro-1-naphthylamine
1,5-Dinitro-2-naphthylamine
1,6-Dinitro-2-naphthylamine
1,7-Dinitro-2-naphthylamine
1,8-Dinitro-2-naphthylamine
4,5-Dinitro-2-naphthylamine
5,8-Dinitro-2-naphthylamine

$C_{10}H_7N_3O_5$
2-Hydroxy-6,8-dinitroquinoline, *Me ether*

$C_{10}H_7N_3S$
Thiabendazole†

$C_{10}H_8$
Azulene
Benzobicyclo[2,2,0]hexa-2,5-diene†
Benzofulvene
Cyclodeca-1,3,7-triyne†
Fulvalene
Naphthalene

$C_{10}H_8BrN$
2-Bromomethylquinoline
2-Bromo-4-methylquinoline
2-Bromo-6-methylquinoline
3-Bromomethylquinoline
3-Bromo-2-methylquinoline
3-Bromo-4-methylquinoline
4-Bromomethylquinoline
5-Bromo-8-methylquinoline
6-Bromo-2-methylquinoline
6-Bromo-8-methylquinoline
7-Bromo-8-methylquinoline
8-Bromomethylquinoline
8-Bromo-6-methylquinoline
2-Bromo-1-naphthylamine
3-Bromo-1-naphthylamine
4-Bromo-1-naphthylamine
5-Bromo-1-naphthylamine
7-Bromo-1-naphthylamine
8-Bromo-1-naphthylamine
1-Bromo-2-naphthylamine
3-Bromo-2-naphthylamine
4-Bromo-2-naphthylamine
5-Bromo-2-naphthylamine
6-Bromo-2-naphthylamine
7-Bromo-2-naphthylamine
8-Bromo-2-naphthylamine

$C_{10}H_8BrNO$
4-Bromo-2-hydroxyquinoline, *Me ether*
4-Bromo-6-hydroxyquinoline, *Me ether*
5-Bromo-6-hydroxyquinoline, *Me ether*
5-Bromo-8-hydroxyquinoline, *Me ether*
7-Bromo-8-hydroxyquinoline, *Me ether*

$C_{10}H_8BrNO_2$
Phthalimide, N-2-*Bromoethyl*

$C_{10}H_8Br_2O_2$
α,β-Dibromocinnamic Acid, *Me ester*

$C_{10}H_8Br_2O_4$
4,6-Dibromoisophthalic Acid, *Di-Me ester*
3,4-Dibromophthalic Acid, *Di-Me ester*
4,5-Dibromophthalic Acid, *Di-Me ester*

$C_{10}H_8ClN$
2-Chloromethylquinoline
6-Chloromethylquinoline
8-Chloromethylquinoline
2-Chloro-3-methylquinoline
2-Chloro-4-methylquinoline
2-Chloro-6-methylquinoline
2-Chloro-8-methylquinoline
3-Chloro-2-methylquinoline
3-Chloro-4-methylquinoline
3-Chloro-6-methylquinoline
3-Chloro-8-methylquinoline
4-Chloro-2-methylquinoline
4-Chloro-3-methylquinoline
4-Chloro-7-methylquinoline
5-Chloro-2-methylquinoline
5-Chloro-4-methylquinoline
6-Chloro-2-methylquinoline
6-Chloro-3-methylquinoline
6-Chloro-4-methylquinoline
6-Chloro-8-methylquinoline
7-Chloro-2-methylquinoline
7-Chloro-3-methylquinoline
7-Chloro-4-methylquinoline
7-Chloro-8-methylquinoline
8-Chloro-2-methylquinoline
8-Chloro-4-methylquinoline
8-Chloro-5-methylquinoline
8-Chloro-6-methylquinoline
2-Chloro-1-naphthylamine
3-Chloro-1-naphthylamine
4-Chloro-1-naphthylamine
5-Chloro-1-naphthylamine
6-Chloro-1-naphthylamine
7-Chloro-1-naphthylamine
8-Chloro-1-naphthylamine
1-Chloro-2-naphthylamine
4-Chloro-2-naphthylamine

$C_{10}H_8ClNO$
1-Chloro-3-hydroxyisoquinoline, *Me ether*
1-Chloro-4-hydroxyisoquinoline, *Me ether*
3-Chloro-1-hydroxyisoquinoline, *Me ether*
2-Chloro-6-hydroxy-4-methylquinoline
2-Chloro-7(?)-hydroxy-4-methylquinoline
4-Chloro-6-hydroxy-2-methylquinoline
4-Chloro-7-hydroxy-2-methylquinoline
5-Chloro-8-hydroxy-2-methylquinoline
7-Chloro-4-hydroxy-2-methylquinoline
8-Chloro-2-hydroxy-4-methylquinoline
8-Chloro-4-hydroxy-2-methylquinoline
2-Chloro-6-hydroxyquinoline, *Me ether*
2-Chloro-8-hydroxyquinoline, *Me ether*
3-Chloro-6-hydroxyquinoline, *Me ether*
3-Chloro-7-hydroxyquinoline, *Me ether*
3-Chloro-8-hydroxyquinoline, *Me ether*
4-Chloro-6-hydroxyquinoline, *Me ether*

$C_{10}H_8ClNO_2$
4-Chloroindole-3-acetic Acid†
1-Hydroxyindole-2-carboxylic Acid, *Me ether, Chloride*
Phthalimide, N-2-*Chloroethyl*

$C_{10}H_8ClNO_2S$
2-Chloronaphthalene-1-sulphonic Acid, *Amide*
4-Chloronaphthalene-1-sulphonic Acid, *Amide*
5-Chloronaphthalene-1-sulphonic Acid, *Amide*
6-Chloronaphthalene-1-sulphonic Acid, *Amide*
7-Chloronaphthalene-1-sulphonic Acid, *Amide*
8-Chloronaphthalene-1-sulphonic Acid, *Amide*
1-Chloronaphthalene-2-sulphonic Acid, *Amide*
4-Chloronaphthalene-2-sulphonic Acid, *Amide*
5-Chloronaphthalene-2-sulphonic Acid, *Amide*
6-Chloronaphthalene-2-sulphonic Acid, *Amide*
7-Chloronaphthalene-2-sulphonic Acid, *Amide*
8-Chloronaphthalene-2-sulphonic Acid, *Amide*
2-Methylquinoline-3-sulphonic Acid, *Chloride*

$C_{10}H_8ClNO_3$
5-Chloro-6-methoxy-1-methylisatin †

$C_{10}H_8ClNO_4$
α-Chloro-4-nitrocinnamic Acid, *Me ester*
β-Chloro-3-nitrocinnamic Acid, *Me ester*

$C_{10}H_8ClNO_5$
3-Nitrophthalic Acid, 1-*Et ester*, *Chloride*
3-Nitrophthalic Acid, 2-*Et ester*, *Chloride*

$C_{10}H_8ClNO_6$
3-Nitro-opianic Acid, *Chloride*

$C_{10}H_8Cl_2O_2$
Benzylmalonic Acid, *Dichloride*
α,β-Dichlorocinnamic Acid, *Me ester*
Methylphenylmalonic Acid, *Dichloride*
Phenylsuccinic Acid, *Dichloride*

$C_{10}H_8Cl_2O_4$
2,5-Dichloroterephthalic Acid, *Di-Me ester*

$C_{10}H_8Cl_3NO_5S$
α-Chloro-α-chlorosulphenyl-4-nitro-2,5-dimethoxyphenylacetyl chloride †

$C_{10}H_8Cl_4$
Naphthalene 1,2,3,4-Tetrachloride †

$C_{10}H_8FN$
4-Fluoro-1-naphthylamine

$C_{10}H_8FNO_2S$
4-Fluoronaphthalene-1-sulphonic Acid, *Amide*
5-Fluoronaphthalene-1-sulphonic Acid, *Amide*
6-Fluoronaphthalene-2-sulphonic Acid, *Amide*

$C_{10}H_8HgO$
Mercuri-1-naphthyl hydroxide

$C_{10}H_8IN$
6-Iodo-2-methylquinoline
2-Iodo-1-naphthylamine
3-Iodo-1-naphthylamine
4-Iodo-1-naphthylamine
5-Iodo-1-naphthylamine
8-Iodo-1-naphthylamine
1-Iodo-2-naphthylamine
3-Iodo-2-naphthylamine
4-Iodo-2-naphthylamine
6-Iodo-2-naphthylamine
7-Iodo-2-naphthylamine

$C_{10}H_8INO$
6-Hydroxy-4-iodoquinoline, *Me ether*

$C_{10}H_8INO_2$
Phthalimide, N-2-*Iodoethyl*

$C_{10}H_8I_2O_2$
α,β-Di-iodocinnamic Acid, *Me ester*

$C_{10}H_8I_2O_4$
Iodoacetic Acid, *Catechol ester*

$C_{10}H_8I_3NO_3$
3-Amino-2,4,6-tri-iodobenzoic Acid, N-*Propionyl*

$C_{10}H_8NO_3$
2-Methyl-4-nitroquinoline, N-*Oxide*
2-Methyl-5-nitroquinoline, N-*Oxide*
2-Methyl-8-nitroquinoline, N-*Oxide*

$C_{10}H_8N_2$
Benzene-1,2-diacetic Acid, *Dinitrile*
Benzene-1,3-diacetic Acid, *Dinitrile*
Benzene-1,4-diacetic Acid, *Dinitrile*
2,2′-Bipyridyl
2,3′-Bipyridyl
2,4′-Bipyridyl
3,3′-Bipyridyl
4,4′-Bipyridyl
3-Indolylacetonitrile
4-Indolylacetonitrile
2-Methylindole-3-carboxylic Acid, *Nitrile*
Methylphenylmalonic Acid, *Dinitrile*
3-Phenylpyridazine
4-Phenylpyridazine
Phenylsuccinic Acid, *Dinitrile*

$C_{10}H_8N_2O$
α-Cyanocinnamic Acid, *Amide*
2-Nitroso-1-naphthylamine
4-Nitroso-1-naphthylamine
1-Nitroso-2-naphthylamine
3-Phenoxypyridazine †
Quinoline-2-carboxylic Acid, *Amide*
Quinoline-3-carboxylic Acid, *Amide*
Quinoline-4-carboxylic Acid, *Amide*
Quinoline-6-carboxylic Acid, *Amide*

$C_{10}H_8N_2O_2$
8-Aminoquinoline-2-carboxylic Acid
2-Aminoquinoline-3-carboxylic Acid
3-Aminoquinoline-4-carboxylic Acid
5-Aminoquinoline-4-carboxylic Acid
5-Aminoquinoline-6-carboxylic Acid
2,5-Diamino-1,4-naphthoquinone
2,7-Diamino-1,4-naphthoquinone
2,8-Diamino-1,4-naphthoquinone
Di-2-pyrrylglyoxal
Hydantoin, 5-*Benzylidene*
2-Hydroxyquinoline-3-carboxylic Acid, *Amide*
6-Hydroxyquinoline-4-carboxylic Acid, *Amide*
6-Hydroxyquinoline-5-carboxylic Acid, *Amide*
2-3′-Indolylglyoxylic Acid, *Amide*
2-Methyl-4-nitroquinoline
2-Methyl-5-nitroquinoline
2-Methyl-6-nitroquinoline
2-Methyl-8-nitroquinoline
3-Methyl-6-nitroquinoline
4-Methyl-3-nitroquinoline
4-Methyl-8-nitroquinoline
5-Methyl-6-nitroquinoline
5-Methyl-8-nitroquinoline
6-Methyl-5-nitroquinoline
6-Methyl-8-nitroquinoline

$C_{10}H_8N_2O_2$ (*continued*)
7-Methyl-6-nitroquinoline
7-Methyl-8-nitroquinoline
8-Methyl-5-nitroquinoline
8-Methyl-6-nitroquinoline
2-Nitro-1-naphthylamine
3-Nitro-1-naphthylamine
4-Nitro-1-naphthylamine
5-Nitro-1-naphthylamine
6-Nitro-1-naphthylamine
7-Nitro-1-naphthylamine
8-Nitro-1-naphthylamine
1-Nitro-2-naphthylamine
3-Nitro-2-naphthylamine
4-Nitro-2-naphthylamine
5-Nitro-2-naphthylamine
6-Nitro-2-naphthylamine
7-Nitro-2-naphthylamine
8-Nitro-2-naphthylamine
3-Phenylpyrazole-1-carboxylic Acid
4-Phenylpyrazole-1-carboxylic Acid
5-Phenylpyrazole-1-carboxylic Acid
1-Phenylpyrazole-3-carboxylic Acid
4-Phenylpyrazole-3-carboxylic Acid
5(3)-Phenylpyrazole-3(5)-carboxylic Acid
5-Phenylpyrazole-4-carboxylic Acid
1-Phenylpyrazole-4-carboxylic Acid
1-Phenylpyrazole-5-carboxylic Acid
1-Phenyluracil
4-Phenyluracil
5-Phenyluracil
Quinoxaline-2-carboxylic Acid, *Me ester* †

$C_{10}H_8N_2O_3$
7-Amino-6-methoxyquinoline-5,8-dione †
4-Amino-2-nitro-1-naphthol
8-Amino-5-nitro-1-naphthol
7-Amino-8-nitro-2-naphthol
2-Hydroxy-4-methyl-5-nitroquinoline
2-Hydroxy-4-methyl-6-nitroquinoline
2-Hydroxy-4-methyl-8-nitroquinoline
2-Hydroxy-6-methyl-5-nitroquinoline
2-Hydroxy-6-methyl-8-nitroquinoline
2-Hydroxy-7-methyl-8-nitroquinoline
2-Hydroxy-8-methyl-5-nitroquinoline
2-Hydroxy-8-methyl-6-nitroquinoline
3-Hydroxy-2-methyl-5-nitroquinoline
4-Hydroxy-2-methyl-3-nitroquinoline
4-Hydroxy-2-methyl-6-nitroquinoline
4-Hydroxy-2-methyl-8-nitroquinoline
4-Hydroxy-7-methyl-3-nitroquinoline
5-Hydroxy-6-methyl-8-nitroquinoline
5-Hydroxy-8-methyl-6-nitroquinoline
6-Hydroxy-2-methyl-8-nitroquinoline
6-Hydroxy-3-methyl-8-nitroquinoline
6-Hydroxy-5-methyl-8-nitroquinoline
8-Hydroxy-2-methyl-5-nitroquinoline
8-Hydroxy-5-methyl-7-nitroquinoline
8-Hydroxy-7-methyl-5-nitroquinoline
2-Hydroxy-5-nitroquinoline, *Me ether*
2-Hydroxy-5-nitroquinoline, N-*Me*
2-Hydroxy-6-nitroquinoline, *Me ether*
2-Hydroxy-6-nitroquinoline, N-*Me*
2-Hydroxy-7-nitroquinoline, N-*Me*
2-Hydroxy-8-nitroquinoline, N-*Me*
4-Hydroxy-3-nitroquinoline, *Me ether*
4-Hydroxy-8-nitroquinoline, *Me ether*
5-Hydroxy-8-nitroquinoline, *Me ether*
6-Hydroxy-5-nitroquinoline, *Me ether*
6-Hydroxy-7-nitroquinoline, *Me ether*
6-Hydroxy-8-nitroquinoline, *Me ether*
7-Hydroxy-8-nitroquinoline, *Me ether*
8-Hydroxy-5-nitroquinoline, *Me ether*
8-Hydroxy-6-nitroquinoline, *Me ether*
1-Phenylbarbituric Acid
5-Phenylbarbituric Acid
1-Phenyl-4-pyrazolone-3-carboxylic Acid
1-Phenyl-5-pyrazolone-3-carboxylic Acid
1-Phenyl-5-pyrazolone-4-carboxylic Acid
2-Phenyl-5-pyrazolone-4-carboxylic Acid

$C_{10}H_8N_2O_4S$
4-Nitronaphthalene-1-sulphonic Acid, *Amide*
5-Nitronaphthalene-1-sulphonic Acid, *Amide*
6-Nitronaphthalene-1-sulphonic Acid, *Amide*
7-Nitronaphthalene-1-sulphonic Acid, *Amide*
8-Nitronaphthalene-1-sulphonic Acid, *Amide*
1-Nitronaphthalene-2-sulphonic Acid, *Amide*
4-Nitronaphthalene-2-sulphonic Acid, *Amide*
5-Nitronaphthalene-2-sulphonic Acid, *Amide*
8-Nitronaphthalene-2-sulphonic Acid, *Amide*

$C_{10}H_8N_2O_5S$
4-Amino-8-nitronaphthalene-1-sulphonic Acid
5-Amino-8-nitronaphthalene-1-sulphonic Acid
7-Amino-3-nitronaphthalene-1-sulphonic Acid
1-Amino-5-nitronaphthalene-2-sulphonic Acid
5-Amino-8-nitronaphthalene-2-sulphonic Acid
6-Amino-4-nitronaphthalene-2-sulphonic Acid
8-Amino-5-nitronaphthalene-2-sulphonic Acid

$C_{10}H_8N_2O_6$
2,4-Dinitrocinnamic Acid, *Me ester*

$C_{10}H_8N_2O_6S$
2-Amino-4-nitro-1-naphthol-7-sulphonic Acid
3-Amino-5-nitro-2-naphthol-7-sulphonic Acid
4-Amino-2-nitro-1-naphthol-7-sulphonic Acid

$C_{10}H_8N_2O_7$
2-(2,4-Dinitrophenyl)acetoacetic Acid †

$C_{10}H_8N_2O_8$
2,4-Dinitroisophthalic Acid, *Di-Me ester*
4,6-Dinitroisophthalic Acid, *Mono-Et ester*
3,5-Dinitrophthalic Acid, *Et ester*
4,5-Dinitrophthalic Acid, *Di-Me ester*
2,6-Dinitroterephthalic Acid, 4-*Et ester*

$C_{10}H_8N_4O_4$
Indigoidin †

$C_{10}H_8N_4O_5$
Picrolonic Acid

$C_{10}H_8O$
1-Benzoxepin †
3-Benzoxepin
Cyclo-octa[*c*]furan †
2-Decene-4,6,8-triyn-1-ol
1,2-Dihydro-1,2-epoxynaphthalene †
1,6-Epoxy[10]annulene †
2-Hydroxyazulene †
4-Hydroxyazulene †
6-Hydroxyazulene †
Indene-2-aldehyde
2-Methylindone

$C_{10}H_8O$ (*continued*)
3-Methylindone
1-Naphthol
2-Naphthol

$C_{10}H_8OS$
2-Hydroxy-1-naphthalenethiol
5-Hydroxy-1-naphthalenethiol
8-Hydroxy-1-naphthalenethiol
1-Hydroxy-2-naphthalenethiol
6-Hydroxy-2-naphthalenethiol
7-Hydroxy-2-naphthalenethiol

$C_{10}H_8O_2$
2-Acetylbenzofuran
3-Benzoylpropionic Acid, *Lactone*
2,8-Decadiene-4,6-diynoic Acid
2-Decene-4,6,8-triyne-1,10-diol
Dicyclopentadienone†
1,2-Dihydroxynaphthalene
1,3-Dihydroxynaphthalene
1,4-Dihydroxynaphthalene
1,5-Dihydroxynaphthalene
1,6-Dihydroxynaphthalene
1,7-Dihydroxynaphthalene
1,8-Dihydroxynaphthalene
2,3-Dihydroxynaphthalene
2,6-Dihydroxynaphthalene
2,7-Dihydroxynaphthalene
Indene-1-carboxylic Acid†
Indene-2-carboxylic Acid
Indene-3-carboxylic Acid
Indene-5(or 6)-carboxylic Acid
2-Methylchromone
3-Methylchromone
6-Methylchromone
7(or 5)-Methylchromone
8-Methylchromone
3-Methylcoumarin
4-Methylcoumarin
5-Methylcoumarin
6-Methylcoumarin
7-Methylcoumarin
8-Methylcoumarin
Nona-3,4-diene-6,8-diynoic Acid, *Me ester*
4-Phenylbuta-2,3-dienoic Acid†
3-Phenyl-2-propynoic Acid, *Me ester*

$C_{10}H_8O_2S$
Benzo[*b*]thiophene-2-carboxylic Acid, *Me ester*
Benzo[*b*]thiophene-3-carboxylic Acid, *Me ester*
Naphthalene-1-sulphinic Acid
Naphthalene-2-sulphinic Acid

$C_{10}H_8O_3$
Benzofuran-2-carboxylic Acid, *Me ester*
3-Benzoylacrylic Acid
3,4-Dimethylphthalic Acid, *Anhydride*
3,5-Dimethylphthalic Acid, *Anhydride*
3,6-Dimethylphthalic Acid, *Anhydride*
4,5-Dimethylphthalic Acid, *Anhydride*
Erythrocentaurin
4-Formylcinnamic Acid
Herniarin★†
3-Hydroxycoumarin, *Me ether*
4-Hydroxycoumarin, *Me ether*
5-Hydroxycoumarin, *Me ether*
3-Hydroxyindene-2-carboxylic Acid
2-Hydroxyindene-3-carboxylic Acid
5-Hydroxy-2-methylchromone
7-Hydroxy-4-methylcoumarin
7-Hydroxy-5-methylcoumarin
7-Hydroxy-6-methylcoumarin
7-Hydroxy-8-methylcoumarin
8-Hydroxy-3-methylisochroman-1-one
2-Hydroxyphenylpropiolic Acid, *Me ether*
3-Hydroxyphenylpropiolic Acid, *Me ether*
4-Hydroxyphenylpropiolic Acid, *Me ether*
3-Methylbenzofuran-2-carboxylic Acid
6-Methylbenzofuran-2-carboxylic Acid
2-Methylbenzofuran-7-carboxylic Acid
5,6-Methylenedioxy-1-indanone
3-(3,4-Methylenedioxyphenyl)acrolein
Phenylsuccinic Acid, *Anhydride*
1,2,3-Trihydroxynaphthalene
1,2,4-Trihydroxynaphthalene
1,2,7-Trihydroxynaphthalene
1,3,6-Trihydroxynaphthalene
1,4,5-Trihydroxynaphthalene
1,4,6-Trihydroxynaphthalene
1,6,7-Trihydroxynaphthalene

$C_{10}H_8O_3S$
3-Hydroxybenzo[*b*]thiophene-2-carboxylic Acid, *Me ether*
3-Hydroxybenzo[*b*]thiophene-2-carboxylic Acid, *Me ester*
Naphthalene-1-sulphonic Acid
Naphthalene-2-sulphonic Acid

$C_{10}H_8O_4$
Aesculetin, 7-*Me ether*
Alginetin
Anemonin
Benzoylpyruvic Acid
Benzylidenemalonic Acid
o-Carboxycinnamic Acid
m-Carboxycinnamic Acid
p-Carboxycinnamic Acid
Cubanedicarboxylic Acid†
Daphnetin, 7-*Me ether*
Daphnetin, 8-*Me ether*
2-Decene-4,6-diynedioic Acid
1,2-Difuryl-2-oxoethanol
3,7-Dihydroxychromone, 7-*Me ether*
5,7-Dihydroxychromone, 7-*Me ether*
4,7-Dihydroxycoumarin, 4-*Me ether*
2,10-Dihydroxydeca-4,6,8-triynoic Acid
2,10-Dihydroxydec-4-ene-6,8-diynoic Acid
6,7-Dihydroxy-4-methylcoumarin
Fumaric Acid, *Mono-phenyl ester*
2-Hydroxy-3,4-methylenedioxy-6-vinylbenzaldehyde
Maleic Acid, *Monophenyl ester*
3-(2,3-Methylenedioxyphenyl)acrylic Acid
3-(3,4-Methylenedioxyphenyl)acrylic Acid
Phenylfumaric Acid
Phenylmaleic Acid
Phthalide-3-acetic Acid
Phthalide-3-carboxylic Acid, *Me ester*
Scopoletin
1,2,3,4-Tetrahydroxynaphthalene
1,2,4,5-(or 8)-Tetrahydroxynaphthalene
1,2,4,6-Tetrahydroxynaphthalene
1,2,4,7-Tetrahydroxynaphthalene

$C_{10}H_8O_4$ (*continued*)
- 1,3,6,8-Tetrahydroxynaphthalene†
- 1,4,5,8-Tetrahydroxynaphthalene

$C_{10}H_8O_4S$
- 1-Naphthol-2-sulphonic Acid
- 1-Naphthol-3-sulphonic Acid
- 1-Naphthol-4-sulphonic Acid
- 1-Naphthol-5-sulphonic Acid
- 1-Naphthol-6-sulphonic Acid
- 1-Naphthol-7-sulphonic Acid
- 1-Naphthol-8-sulphonic Acid
- 2-Naphthol-1-sulphonic Acid
- 2-Naphthol-3-sulphonic Acid
- 2-Naphthol-4-sulphonic Acid
- 2-Naphthol-5-sulphonic Acid
- 2-Naphthol-6-sulphonic Acid
- 2-Naphthol-7-sulphonic Acid
- 2-Naphthol-8-sulphonic Acid

$C_{10}H_8O_4S_2$
- Naphthalene-1,4-disulphinic Acid
- Naphthalene-1,5-disulphinic Acid
- Naphthalene-1,6-disulphinic Acid
- Naphthalene-1,7-disulphinic Acid
- Naphthalene-2,6-disulphinic Acid
- Naphthalene-2,7-disulphinic Acid

$C_{10}H_8O_5$
- Benzoylmalonic Acid
- Di-2-furylglycollic Acid
- 3,5-Dihydroxyphthalic Acid, *Di-Me ether, Anhydride*
- 3,6-Dihydroxyphthalic Acid, *Di-Me ether, Anhydride*
- 3,4-Dimethoxyphthalic Acid, *Anhydride*
- Fraxetin
- Metahemipinic Acid, *Anhydride*
- Phthalonic Acid, α-*Me ester*

$C_{10}H_8O_6$
- Benzene-1,2,3-tricarboxylic Acid, 2-*Me ester*
- Benzene-1,2,4-tricarboxylic Acid, 1-*Me ester*
- Benzene-1,2,4-tricarboxylic Acid, 2-*Me ester*
- Benzene-1,2,4-tricarboxylic Acid, 4-*Me ester*
- Benzene-1,3,5-tricarboxylic Acid, *Mono-Me ester*
- Chorismic Acid†
- 2,4-Dihydroxy-6-pyruvylbenzoic Acid
- 3,4-Methylenedioxyhomophthalic Acid
- 4,5-Methylenedioxyhomophthalic Acid
- 4,5-Methylenedioxyphthalic Acid, *Mono-Me ester*
- Toluene-2,3,6-tricarboxylic Acid
- Toluene-2,4,5-tricarboxylic Acid
- Toluene-2,4,6-tricarboxylic Acid

$C_{10}H_8O_6S_2$
- Naphthalene-1,2-disulphonic Acid
- Naphthalene-1,3-disulphonic Acid
- Naphthalene-1,4-disulphonic Acid
- Naphthalene-1,5-disulphonic Acid
- Naphthalene-1,6-disulphonic Acid
- Naphthalene-1,7-disulphonic Acid
- Naphthalene-1,8-disulphonic Acid
- Naphthalene-2,6-disulphonic Acid
- Naphthalene-2,7-disulphonic Acid

$C_{10}H_8O_7$
- Cotarnic Acid
- 4-Hydroxybenzene-1,2,3-tricarboxylic Acid, *Me ether*
- 5-Hydroxybenzene-1,2,3-tricarboxylic Acid, *Me ether*
- 3-Hydroxybenzene-1,2,4-tricarboxylic Acid, *Me ether*
- 5-Hydroxybenzene-1,2,4-tricarboxylic Acid, *Me ether*
- 6-Hydroxybenzene-1,2,4-tricarboxylic Acid, *Me ether*
- 2-Hydroxybenzene-1,3,5-tricarboxylic Acid, *Me ether*
- 5-Hydroxytoluene-2,3,4-tricarboxylic Acid★†
- 5-Hydroxytoluene-2,3,6-tricarboxylic Acid†
- 3-Hydroxytoluene-2,4,5-tricarboxylic Acid
- 3-Hydroxytoluene-2,4,6-tricarboxylic Acid
- 5-Methoxy-3,4-methylenedioxyphthalic Acid

$C_{10}H_8O_7S_2$
- 1-Naphthol-2,4-disulphonic Acid
- 1-Naphthol-2,5-disulphonic Acid
- 1-Naphthol-2,7-disulphonic Acid
- 1-Naphthol-3,6-disulphonic Acid
- 1-Naphthol-3,7-disulphonic Acid
- 1-Naphthol-3,8-disulphonic Acid
- 1-Naphthol-4,7-disulphonic Acid
- 1-Naphthol-4,8-disulphonic Acid
- 1-Naphthol-5,8-disulphonic Acid
- 1-Naphthol-6,8-disulphonic Acid
- 2-Naphthol-1,4-disulphonic Acid
- 2-Naphthol-1,5-disulphonic Acid
- 2-Naphthol-1,6-disulphonic Acid
- 2-Naphthol-1,7-disulphonic Acid
- 2-Naphthol-3,6-disulphonic Acid
- 2-Naphthol-3,7-disulphonic Acid
- 2-Naphthol-4,7-disulphonic Acid
- 2-Naphthol-4,8-disulphonic Acid
- 2-Naphthol-6,8-disulphonic Acid

$C_{10}H_8O_8S_2$
- 4,5-Dihydroxynaphthalene-2,7-disulphonic Acid

$C_{10}H_8O_9S_3$
- Naphthalene-1,3,5-trisulphonic Acid
- Naphthalene-1,3,6-trisulphonic Acid
- Naphthalene-1,3,7-trisulphonic Acid
- Naphthalene-1,4,5-trisulphonic Acid
- Naphthalene-2,3,6-trisulphonic Acid

$C_{10}H_8O_{10}S_3$
- 1-Naphthol-2,4,7-trisulphonic Acid
- 1-Naphthol-2,4,8-trisulphonic Acid
- 1-Naphthol-3,6,8-trisulphonic Acid
- 1-Naphthol-4,6,8-trisulphonic Acid
- 2-Naphthol-1,3,6-trisulphonic Acid
- 2-Naphthol-1,3,7-trisulphonic Acid
- 2-Naphthol-3,6,7-trisulphonic Acid
- 2-Naphthol-3,6,8-trisulphonic Acid

$C_{10}H_8O_{12}S_4$
- Naphthalene-1,3,5,7-tetrasulphonic Acid

$C_{10}H_8O_{13}S_4$
- 1-Naphthol-2,3,4,6-tetrasulphonic Acid
- 2-Naphthol-1,3,6,7-tetrasulphonic Acid

$C_{10}H_8S$
- 1-Naphthalenethiol
- 2-Naphthalenethiol
- 2-Phenylthiophene
- 3-Phenylthiophene

$C_{10}H_8S_2$
1,5-Naphthalenedithiol
1,8-Naphthalenedithiol
2,6-Naphthalenedithiol
2,7-Naphthalenedithiol

$C_{10}H_8Se$
1-Selenylnaphthalene

$C_{10}H_9BrO_2$
α-Bromocinnamic Acid, *Me ester*
β-Bromocinnamic Acid, *Me ester*
p-Bromocinnamic Acid, *Me ester*

$C_{10}H_9BrO_4$
2-Bromoisophthalic Acid, *Di-Me ester*
3-Bromophthalic Acid, *Di-Me ester*
4-Bromophthalic Acid, *Di-Me ester*
3-Bromophthalic Acid, *Mono-Et ester*
Bromoterephthalic Acid, *Di-Me ester*

$C_{10}H_9Br_5O$
Pentabromophenol, n-*Butyl ether*
Pentabromophenol, 2-*Butyl ether*
Pentabromophenol, *Isobutyl ether*

$C_{10}H_9Cl$
1-Chlorocyclodeca-1,3,8-trien-6-yne†

$C_{10}H_9ClO$
Indane-2-carboxylic Acid, *Chloride*
Indane-5-carboxylic Acid, *Chloride*
2-Methyl-3-phenylacrylic Acid, *Chloride*
4-Phenyl-3-butenoic Acid, *Chloride*

$C_{10}H_9ClO_2$
α-Chlorocinnamic Acid, *Me ester*
β-Chlorocinnamic Acid, *Me ester*
o-Chlorocinnamic Acid, *Me ester*
3-*p*-Methoxyphenylacrylic Acid, *Chloride*

$C_{10}H_9ClO_3$
3-(3,4-Methylenedioxyphenyl)propionic Acid, *Chloride*

$C_{10}H_9ClO_4$
4-Chlorophthalic Acid, *Di-Me ester*
Chloroterephthalic Acid, *Di-Me ester*
2-Formyl-5,6-dimethoxybenzoic Acid, ψ-*Chloride*

$C_{10}H_9Cl_3O$
3-(2,3-Dichlorophenyl)butyric Acid, *Chloride*
3-(2,5-Dichlorophenyl)butyric Acid, *Chloride*

$C_{10}H_9Cl_3O_2$
Chloralacetophenone
α,α,α-Trichloro-*o*-toluic Acid, *Et ester*
α,α,α-Trichloro-*p*-toluic Acid, *Et ester*

$C_{10}H_9Cl_5O$
Pentachlorophenol, *Butyl ether*

$C_{10}H_9IO_2$
2-Iodo-3-phenylacrylic Acid, *Me ester*
3-*o*-Iodophenylacrylic Acid, *Me ester*

$C_{10}H_9IO_4$
2-Iodobenzene-1,3-diacetic Acid†
2-Iodoisophthalic Acid, *Di-Me ester*
5-Iodoisophthalic Acid, *Di-Me ester*
3-Iodophthalic Acid, *Di-Me ester*
4-Iodophthalic Acid, *Di-Me ester*
Iodoterephthalic Acid, *Di-Me ester*

$C_{10}H_9I_2NO_5$
3,5-Di-iodo-1-methyl-4-pyridone-2,6-dicarboxylic Acid, *Di-Me ester*

$C_{10}H_9N$
1,6-Iminocyclodecapentaene†
1-Methylisoquinoline
3-Methylisoquinoline
4-Methylisoquinoline
5-Methylisoquinoline
6-Methylisoquinoline
7-Methylisoquinoline
8-Methylisoquinoline
2-Methylquinoline
3-Methylquinoline
4-Methylquinoline
5-Methylquinoline
6-Methylquinoline
7-Methylquinoline
8-Methylquinoline
1-Naphthylamine
2-Naphthylamine
4-Phenyl-3-butenoic Acid, *Nitrile*
2-Phenylcrotonic Acid, *Nitrile*
1-Phenylcyclopropane-1-carboxylic Acid, *Nitrile*
1-Phenylpyrrole
2-Phenylpyrrole
N-Pyridinium cyclopentadienide†
3-*m*-Tolylacrylic Acid, *Nitrile*
3-*p*-Tolylacrylic Acid, *Nitrile*

$C_{10}H_9NO$
3-Acetylindole
2-Amino-1-naphthol
4-Amino-1-naphthol
5-Amino-1-naphthol
6-Amino-1-naphthol
7-Amino-1-naphthol
8-Amino-1-naphthol
1-Amino-2-naphthol
3-Amino-2-naphthol
4-Amino-2-naphthol
5-Amino-2-naphthol
6-Amino-2-naphthol
7-Amino-2-naphthol
8-Amino-2-naphthol
3-Benzoylpropionic Acid, *Nitrile*
Carbostyril, 2-*Me ether*
2,4-Dimethylbenzoylformic Acid, *Nitrile*
Echinopsine
1-Hydroxyisoquinoline, *Me ether*
1-Hydroxyisoquinoline, N-*Me*
7-Hydroxyisoquinoline, *Me ether*
2-Hydroxy-3-methylquinoline
2-Hydroxy-4-methylquinoline
2-Hydroxy-6-methylquinoline
2-Hydroxy-8-methylquinoline
3-Hydroxy-2-methylquinoline
3-Hydroxy-6-methylquinoline
4-Hydroxy-2-methylquinoline
4-Hydroxy-3-methylquinoline
4-Hydroxy-7-methylquinoline
5-Hydroxy-2-methylquinoline
5-Hydroxy-8-methylquinoline
6-Hydroxy-2-methylquinoline
6-Hydroxy-4-methylquinoline
6-Hydroxy-8-methylquinoline

$C_{10}H_9NO$ (*continued*)
7-Hydroxy-6-methylquinoline
8-Hydroxy-1-methylquinoline
8-Hydroxy-2-methylquinoline
8-Hydroxy-3-methylquinoline
8-Hydroxy-5-methylquinoline
8-Hydroxy-6-methylquinoline
8-Hydroxy-7-methylquinoline
2-Hydroxymethyl-quinoline
2-Hydroxy-4-phenyl-3-butenoic Acid, *Nitrile*
4-Hydroxyquinoline, *Me ether*
5-Hydroxyquinoline, *Me ether*
6-Hydroxyquinoline, *Me ether*
7-Hydroxyquinoline, *Me ether*
8-Hydroxyquinoline, *Me ether*
Indene-3-carboxylic Acid, *Amide*
3-*p*-Methoxyphenylacrylic Acid, *Nitrile*
2-Methylindole, N-*Formyl*
2-Methylindole-3-aldehyde
5-Methylindole-3-aldehyde
7-Methylindole-3-aldehyde
N-Methyl-α-quinolone
2-Methyl-4(1*H*)-quinolone
N-1-Naphthylhydroxylamine
N-2-Naphthylhydroxylamine
3-Oxo-2-phenylbutryic Acid, *Nitrile*

$C_{10}H_9NO_2$
O-Benzoyl-lactic Acid, *Nitrile*
3-*o*-Carboxyphenylpropionic Acid, 2-*Nitrile*
3-*p*-Carboxyphenylpropionic Acid, 4-*Nitrile*
o-Cyanobenzoic Acid, *Et ester*
m-Cyanobenzoic Acid, *Et ester*
p-Cyanobenzoic Acid, *Et ester*
2-Cyano-3-phenylpropionic Acid
3-Cyano-3-phenylpropionic Acid
1,4-Dihydroxyisoquinoline, 4-*Me ether*
2,3-Dihydroxy-4-methylquinoline
2,3-Dihydroxy-6-methylquinoline
2,4-Dihydroxy-5-methylquinoline
2,7-Dihydroxy-4-methylquinoline
4,6-Dihydroxy-2-methylquinoline
4,7-Dihydroxy-2-methylquinoline
4,8-Dihydroxy-2-methylquinoline
5,6-Dihydroxy-2-methylquinoline
6,7-Dihydroxy-2-methylquinoline
7,8-Dihydroxy-2-methylquinoline
2,3-Dihydroxy-1-naphthylamine
2,4-Dihydroxy-1-naphthylamine
2,7-Dihydroxy-1-naphthylamine
3,4-Dihydroxy-1-naphthylamine
5,8-Dihydroxy-1-naphthylamine
1,4-Dihydroxy-2-naphthylamine
1,5-Dihydroxy-2-naphthylamine
1,6-Dihydroxy-2-naphthylamine
3,4-Dihydroxy-2-naphthylamine
2,6-Dihydroxyquinoline, 6-*Me ether*
1,5-Dimethylisatin
1,7-Dimethylisatin
4,5-Dimethylisatin
4,6-Dimethylisatin
4,7-Dimethylisatin
5,6-Dimethylisatin
5,7-Dimethylisatin
6,7-Dimethylisatin
3,4-Dimethylphthalic Acid, *Imide*
1-Ethylisatin
4-Ethylisatin
5-Ethylisatin
6-Ethylisatin
Gentianine
4-Hydroxybenzoylformic Acid, *Et ether*, *Nitrile*
Indole-2-carboxylic Acid, *Me ester*
Indole-2-carboxylic Acid, N-*Me*
Indole-4-carboxylic Acid, *Me ester*
Indole-6-carboxylic Acid, *Me ester*
1-Indolylacetic Acid
2-Indolylacetic Acid
3-Indolylacetic Acid
4-Indolylacetic Acid
6-Methylaminocoumarin
3-Methylbenzofuran-2-carboxylic Acid, *Amide*
3-Methylindole-1-carboxylic Acid
1-Methylindole-2-carboxylic Acid
3-Methylindole-2-carboxylic Acid
5-Methylindole-2-carboxylic Acid
6-Methylindole-2-carboxylic Acid
7-Methylindole-2-carboxylic Acid
2-Methylindole-3-carboxylic Acid
4-Methylindole-3-carboxylic Acid
5-Methylindole-3-carboxylic Acid
7-Methylindole-3-carboxylic Acid
3-Methylphthalic Acid, 1-*Me ester* 2-*nitrile*
4-Methylphthalic Acid, 1-*Me ester* 2-*nitrile*
1-(*p*-Nitrophenyl)-1,3-butadiene†
Phenylsuccinic Acid, *Imide*
Phthalimide, N-*Et*
o-Propionylbenzoic Acid, *Anhydro-oxime*
Succinanil

$C_{10}H_9NO_2S$
Benzothiazole-2-carboxylic Acid, *Et ester*
Benzothiazole-6-carboxylic Acid, *Et ester*
S-*o*-Carboxyphenylthioglycollic Acid, 2-*Nitrile*, *Me ester*
6-Methylbenzothiazole-2-carboxylic Acid, *Me ester*
Naphthalene-1-sulphonic Acid, *Amide*
Naphthalene-2-sulphonic Acid, *Amide*

$C_{10}H_9NO_3$
Anthroxanic Acid, *Et ester*
Benzoylpyruvic Acid, *Amide*
5-Formyl-3,4-dimethoxybenzoic Acid, *Nitrile*
5-Hydroxyindole-3-acetic Acid†
1-Hydroxyindole-2-carboxylic Acid, *Me ester*
1-Hydroxyindole-2-carboxylic Acid, *Me ether*
3-Hydroxyindole-2-carboxylic Acid, *Me ester*
4-Hydroxyindole-2-carboxylic Acid, *Me ether*
4-Hydroxyindole-2-carboxylic Acid, *Me ester*
5-Hydroxyindole-2-carboxylic Acid, *Me ether*
5-Hydroxyindole-2-carboxylic Acid, *Me ester*
6-Hydroxyindole-2-carboxylic Acid, *Me ether*
7-Hydroxyindole-2-carboxylic Acid, *Me ether*
3-Hydroxy-5-methylindole-2-carboxylic Acid
3-Hydroxy-7-methylindole-2-carboxylic Acid
Malanil
Maleanilic Acid
3-(3,4-Methylenedioxyphenyl)acrylic Acid, *Amide*
3-Methyl-6-nitroindan-1-one
4-*o*-Nitrophenyl-3-buten-2-one
4-*m*-Nitrophenyl-3-buten-2-one

$C_{10}H_9NO_3$ (*continued*)
4-*p*-Nitrophenyl-3-buten-2-one
1-Nitro-4-phenyl-3-buten-2-one
3-*p*-Nitrosophenylacrylic Acid, *Me ester*
Phthaloxime, O-*Et*
Succinimide, N-p-*Hydroxyphenyl*
1,2,3,4-Tetrahydro-4-oxoquinoline-2-carboxylic Acid †

$C_{10}H_9NO_3S$
4-Methylquinoline-2-sulphonic Acid
2-Methylquinoline-3-sulphonic Acid
2-Methylquinoline-4-sulphonic Acid
2-Methylquinoline-5-sulphonic Acid
6-Methylquinoline-5-sulphonic Acid
8-Methylquinoline-5-sulphonic Acid
2-Methylquinoline-6-sulphonic Acid
4-Methylquinoline-6-sulphonic Acid
8-Methylquinoline-6-sulphonic Acid
6-Methylquinoline-7-sulphonic Acid
2-Methylquinoline-8-sulphonic Acid
6-Methylquinoline-8-sulphonic Acid
1-Naphthol-8-sulphonic Acid, *Amide*
2-Naphthol-3-sulphonic Acid, *Amide*
2-Naphthol-6-sulphonic Acid, *Amide*
1-Naphthylamine-2-sulphonic Acid
1-Naphthylamine-3-sulphonic Acid
1-Naphthylamine-4-sulphonic Acid
1-Naphthylamine-5-sulphonic Acid
1-Naphthylamine-6-sulphonic Acid
1-Naphthylamine-7-sulphonic Acid
1-Naphthylamine-8-sulphonic Acid
2-Naphthylamine-1-sulphonic Acid
2-Naphthylamine-4-sulphonic Acid
2-Naphthylamine-5-sulphonic Acid
2-Naphthylamine-6-sulphonic Acid
2-Naphthylamine-7-sulphonic Acid
2-Naphthylamine-8-sulphonic Acid
Quinoline-8-sulphonic Acid, *Me ester*

$C_{10}H_9NO_4$
5,6-Dihydroxyisatin, *Di-Me ether*
6,7-Dihydroxyisatin, *Di-Me ether*
2,5-Dimethoxy-3,4-methylenedioxybenzoic Acid, *Nitrile*
3,4-Dimethoxyphthalic Acid, *Imide*
3,4-Dimethoxyphthalic Acid, 1-*Nitrile*
3-(2-Hydroxy-3-nitrophenyl)acrolein, *Me ether*
2-Methyl-3-*o*-nitrophenylacrylic Acid
2-Methyl-3-*m*-nitrophenylacrylic Acid
2-Methyl-3-*p*-nitrophenylacrylic Acid
3-(3-Methyl-2-nitrophenyl)acrylic Acid
3-(3-Methyl-4-nitrophenyl)acrylic Acid
3-(3-Methyl-6-nitrophenyl)acrylic Acid
3-(4-Methyl-3-nitrophenyl)acrylic Acid
o-Nitrobenzoylacetone
m-Nitrobenzoylacetone
p-Nitrobenzoylacetone
5-Nitroindane-2-carboxylic Acid
3-*o*-Nitrophenylacrylic Acid, *Me ester*
3-*m*-Nitrophenylacrylic Acid, *Me ester*
3-*p*-Nitrophenylacrylic Acid, *Me ester*
3-*p*-Nitrophenylcrotonic Acid

$C_{10}H_9NO_4S$
1-Amino-2-naphthol-4-sulphonic Acid
1-Amino-2-naphthol-6-sulphonic Acid
1-Amino-5-naphthol-7-sulphonic Acid
1-Amino-8-naphthol-4-sulphonic Acid
2-Amino-5-naphthol-7-sulphonic Acid
2-Amino-8-naphthol-6-sulphonic Acid

$C_{10}H_9NO_5$
2-Acetyl-3-nitrobenzoic Acid, *Me ester*
2-Acetyl-5-nitrobenzoic Acid, *Me ester*
2-Acetyl-6-nitrobenzoic Acid, *Me ester*
4-Acetyl-2-nitrobenzoic Acid, *Me ester*
3-(2-Hydroxy-4-methyl-5-nitrophenyl)acrylic Acid
α-Hydroxy-2-nitrocinnamic Acid, *Me ether*
α-Hydroxy-2-nitrocinnamic Acid, *Me ester*
3-(2-Hydroxy-3-nitrophenyl)acrylic Acid, *Me ether*
3-(2-Hydroxy-3-nitrophenyl)acrylic Acid, *Me ester*
3-(2-Hydroxy-5-nitrophenyl)acrylic Acid, *Me ether*
3-(2-Hydroxy-5-nitrophenyl)acrylic Acid, *Me ester*
3-(3-Hydroxy-4-nitrophenyl)acrylic Acid, *Me ether*
3-(3-Hydroxy-4-nitrophenyl)acrylic Acid, *Me ester*
3-(3-Hydroxy-6-nitrophenyl)acrylic Acid, *Me ether*
3-(4-Hydroxy-3-nitrophenyl)acrylic Acid, *Me ester*
3-(4-Hydroxy-3-nitrophenyl)acrylic Acid, *Me ether*
Kynuric Acid, α-*Me ester*
p-Nitrobenzoylacetic Acid, *Me ester*
o-Nitrobenzoylformic Acid, *Et ester*
3-*o*-Nitrophenyloxiran-2-carboxylic Acid, *Me ester*
3-*m*-Nitrophenyloxiran-2-carboxylic Acid, *Me ester*
3-*o*-Nitrophenylpyruvic Acid, *Me ester*
3-*p*-Nitrophenylpyruvic Acid, *Me ester*

$C_{10}H_9NO_6$
o-Nitrobenzylmalonic Acid
m-Nitrobenzylmalonic Acid
p-Nitrobenzylmalonic Acid
2-Nitroisophthalic Acid, *Di-Me ester*
4-Nitroisophthalic Acid, *Di-Me ester*
5-Nitroisophthalic Acid, *Di-Me ester*
3-Nitrophthalic Acid, *Di-Me ester*
3-Nitrophthalic Acid, 1-*Et ester*
3-Nitrophthalic Acid, 2-*Et ester*
4-Nitrophthalic Acid, *Di-Me ester*
4-Nitrophthalic Acid, 1-*Et ester*
4-Nitrophthalic Acid, 2-*Et ester*
2-Nitroterephthalic Acid, *Di-Me ester*
2-Nitroterephthalic Acid, 1-*Et ester*
2-Nitroterephthalic Acid, 4-*Et ester*

$C_{10}H_9NO_6S_2$
1-Naphthol-3,8-disulphonic Acid, 8-*Amide*
1-Naphthylamine-2,4-disulphonic Acid
1-Naphthylamine-2,5-disulphonic Acid
1-Naphthylamine-2,7-disulphonic Acid
1-Naphthylamine-2,8-disulphonic Acid
1-Naphthylamine-3,5-disulphonic Acid
1-Naphthylamine-3,6-disulphonic Acid
1-Naphthylamine-3,7-disulphonic Acid

$C_{10}H_9NO_6S_2$ (*continued*)
1-Naphthylamine-3,8-disulphonic Acid
1-Naphthylamine-4,6-disulphonic Acid
1-Naphthylamine-4,7-disulphonic Acid
1-Naphthylamine-4,8-disulphonic Acid
1-Naphthylamine-5,8-disulphonic Acid
2-Naphthylamine-1,5-disulphonic Acid
2-Naphthylamine-1,6-disulphonic Acid
2-Naphthylamine-1,7-disulphonic Acid
2-Naphthylamine-3,6-disulphonic Acid
2-Naphthylamine-3,7-disulphonic Acid
2-Naphthylamine-4,7-disulphonic Acid
2-Naphthylamine-4,8-disulphonic Acid
2-Naphthylamine-5,7-disulphonic Acid
2-Naphthylamine-6,8-disulphonic Acid

$C_{10}H_9NO_7$
4-Hydroxy-5-nitroisophthalic Acid, *Di-Me ester*
2-Nitromyristicic Acid, *Me ester*
3-Nitro-opianic Acid

$C_{10}H_9NO_7S_2$
1-Amino-2-naphthol-3,6-disulphonic Acid
1-Amino-8-naphthol-2,4-disulphonic Acid
1-Amino-8-naphthol-3,6-disulphonic Acid
1-Amino-8-naphthol-4,6-disulphonic Acid
2-Amino-8-naphthol-3,6-disulphonic Acid

$C_{10}H_9NO_8$
3,4-Dimethoxy-6-nitrobenzene-1,2-dicarboxylic Acid

$C_{10}H_9NO_9S_3$
1-Naphthylamine-2,4,6-trisulphonic Acid
1-Naphthylamine-2,4,7-trisulphonic Acid
1-Naphthylamine-2,5,7-trisulphonic Acid
1-Naphthylamine-3,5,7-trisulphonic Acid
1-Naphthylamine-3,6,8-trisulphonic Acid
1-Naphthylamine-4,6,8-trisulphonic Acid
2-Naphthylamine-1,3,7-trisulphonic Acid
2-Naphthylamine-1,5,7-trisulphonic Acid
2-Naphthylamine-3,5,7-trisulphonic Acid
2-Naphthylamine-3,6,7-trisulphonic Acid
2-Naphthylamine-3,6,8-trisulphonic Acid
2-Naphthylamine-4,6,8-trisulphonic Acid

$C_{10}H_9NO_{12}S_4$
2-Naphthylamine-1,3,6,7-tetrasulphonic Acid

$C_{10}H_9NS$
4-Amino-1-naphthalenethiol
8-Amino-2-naphthalenethiol
2-Quinolinethiol, S-*Me*

$C_{10}H_9N_3$
2-Aminopyrimidine, N-*Phenyl*
4-Aminopyrimidine, N-*Phenyl*
2,2′-Dipyridylamine
2,3′-Dipyridylamine
2,4′-Dipyridylamine
3,3′-Dipyridylamine
3,4′-Dipyridylamine
4,4′-Dipyridylamine

$C_{10}H_9N_3O$
3-Phenylpyrazole-1-carboxylic Acid, *Amide*
5-Phenylpyrazole-1-carboxylic Acid, *Amide*

$C_{10}H_9N_3O_2$
4-Amino-2-methyl-3-nitroquinoline

$C_{10}H_9N_3O_3$
m-Nitrobenzylmalonic Acid, *Amide-nitrile*

$C_{10}H_9N_3O_4$
Alloxanic Acid, *Anilide*
Anil-alloxan
6-Nitroindazole-1-carboxylic Acid, *Et ester*
6-Nitroindazole-3-carboxylic Acid, *Et ester*

$C_{10}H_9N_3O_6$
1,2,3,4-Tetrahydro-6,8-dinitroquinoline-1-carboxylic Acid

$C_{10}H_9N_3O_6S_2$
4-Nitronaphthalene-2,6-disulphonic Acid, *Diamide*
4-Nitronaphthalene-2,7-disulphonic Acid, *Diamide*

$C_{10}H_9N_3O_8$
2,4,6-Trinitrobenzoic Acid, *Propyl ester*
2,4,6-Trinitrobenzoic Acid, *Isopropyl ester*
2,3,6-Trinitro-*p*-toluic Acid, *Et ester*

$C_{10}H_9P$
1-Phenylphosphole†

$C_{10}H_{10}$
Benzo[1,2:4,5]dicyclobutene†
Bicyclo[4,2,2]deca-2,4,7,9-tetraene†
Bicyclo[6,2,0]deca-1,3,5,7-tetraene†
Bicyclo[6,2,0]deca-2,4,6,9-tetraene†
Cyclodeca-1,3,5,7,9-pentaene†
Deca-1,5,9-triyne†
1,2-Dihydrofulvalene
1,2-Dihydronaphthalene
1,4-Dihydronaphthalene
9,10-Dihydronaphthalene†
7,8-Dimethylenecyclo-octa-1,3,5-triene†
1,2-Divinylbenzene
1,3-Divinylbenzene
1,4-Divinylbenzene
2-Methylindene
3-Methylindene
Pentacyclo[4,4,0,$0^{2,5}$,$0^{3,8}$,$0^{4,7}$]deca-9-ene†
4-Phenyl-1,2-butadiene
1-Phenyl-1,3-butadiene
2-Phenyl-1,3-butadiene
1-Phenyl-1-butyne
4-Phenyl-1-butyne
Tetracyclo[4,4,0,$0^{2,8}$,$0^{5,7}$]deca-3,9-diene†
Tricyclo[3,3,2,$0^{2,8}$]deca-3,6,9-triene†
Tricyclo[4,2,2,$0^{2,5}$]deca-3,7,9-triene†
Tricyclo[5,2,1,$0^{4,10}$]deca-2,5,8-triene†
Tricyclo[5,3,0,$0^{4,8}$]deca-2,5,9-triene†
8-Vinylheptafulvene†

$C_{10}H_{10}BrNO_4$
2-Bromobutyric Acid, o-*Nitrophenyl ester*
2-Bromobutyric Acid, m-*Nitrophenyl ester*
2-Bromobutyric Acid, p-*Nitrophenyl ester*
2-Bromo-5-nitro-*p*-toluic Acid, *Et ester*

$C_{10}H_{10}Br_2O_2$
4-Allyl-3,6-dibromoguaiacol
2,2-Dibromo-4-phenylbutyric Acid
2,3-Dibromo-3-phenylbutyric Acid
3,4-Dibromo-4-phenylbutyric Acid
2,3-Dibromo-3-phenylpropionic Acid, *Me ester*
3,6-Dibromo-4-propenylguaiacol

$C_{10}H_{10}Br_2O_2$ (*continued*)
3,5-Dibromo-*p*-toluic Acid, *Et ester*
3,6-Dibromo-*p*-toluic Acid, *Et ester*

$C_{10}H_{10}Br_2O_3$
3,5-Dibromo-2-hydroxybenzoic Acid, *Et ether, Me ester*
3,5-Dibromo-4-methoxybenzoic Acid, *Et ester*

$C_{10}H_{10}Br_2O_5$
2,6-Dibromo-3,4,5-trihydroxybenzoic Acid, *Tri-Me ether*

$C_{10}H_{10}Br_4O_5$
3,5-Dibromolaevulinic Acid, *Anhydride*

$C_{10}H_{10}ClNO_3$
2-*p*-Nitrophenylbutyric Acid, *Chloride*
3-*p*-Nitrophenylbutyric Acid, *Chloride*

$C_{10}H_{10}ClNO_4$
2-Chloro-5-nitro-*p*-toluic Acid, *Et ester*

$C_{10}H_{10}ClNO_6S$
4-Nitro-6-sulpho-*m*-toluic Acid, 6-*Chloride, Et ester*

$C_{10}H_{10}Cl_2$
1,6-Dichlorocyclodeca-1,3,6,8-tetraene†

$C_{10}H_{10}Cl_2N_2O_4$
2,5-Dichloro-4-isopropyl-3,6-dinitrotoluene

$C_{10}H_{10}Cl_2O_2$
α,α-Dichlorophenylacetic Acid, *Et ester*
2,2-Dichloro-4-phenylbutyric Acid
3-(2,3-Dichlorophenyl)butyric Acid
3-(2,5-Dichlorophenyl)butyric Acid
4-(2,5-Dichlorophenyl)butyric Acid
2,3-Dichloro-3-phenylpropionic Acid, *Me ester*
3,3-Dichloro-2-phenylpropionic Acid, *Me ester*
α,α-Dichloro-*o*-toluic Acid, *Et ester*
α,α-Dichloro-*p*-toluic Acid, *Et ester*

$C_{10}H_{10}Cl_2O_3$
4-(2,4-Dichlorophenoxy)butyric Acid

$C_{10}H_{10}Cl_2O_4$
2,4-Dichloro-5-hydroxyphenoxyacetic Acid, *Me ether, Me ester*†
2,5-Dichloro-4-hydroxyphenoxyacetic Acid, *Me ether, Me ester*†

$C_{10}H_{10}Cl_4O_2$
Tetrachloroquinol, *Di-Et ether*
Tetrachlororesorcinol, *Di-Et ether*

$C_{10}H_{10}Co$
Dicyclopentadienyl cobalt

$C_{10}H_{10}Fe$
Ferrocene

$C_{10}H_{10}FeO$
Hydroxyferrocene†

$C_{10}H_{10}N_2$
2-Aminomethyl-quinoline
4-Aminomethyl-quinoline
2-Amino-4-methylquinoline
2-Amino-8-methylquinoline
3-Amino-2-methylquinoline
3-Amino-4-methylquinoline
4-Amino-2-methylquinoline
4-Amino-7-methylquinoline
5-Amino-2-methylquinoline
5-Amino-4-methylquinoline
5-Amino-6-methylquinoline
5-Amino-8-methylquinoline
6-Amino-2-methylquinoline
6-Amino-4-methylquinoline
6-Amino-8-methylquinoline
7-Amino-2-methylquinoline
7-Amino-8-methylquinoline
8-Amino-2-methylquinoline
8-Amino-4-methylquinoline
8-Amino-5-methylquinoline
8-Amino-6-methylquinoline
8-Amino-7-methylquinoline
3-Anilinocrotonic Acid, *Nitrile*
2,4-Dimethylquinazoline
2,6-Dimethylquinazoline
2,3-Dimethylquinoxaline
2,6(or 2,7)-Dimethylquinoxaline
6,7-Dimethylquinoxaline
1-Methyl-2-phenylimidazole
2-Methyl-1-phenylimidazole
2-Methyl-4-phenylimidazole
4-Methyl-2-phenylimidazole
4-Methyl-5-phenylimidazole
1-Methyl-3-phenylpyrazole
1-Methyl-5-phenylpyrazole
3-Methyl-1-phenylpyrazole
4-Methyl-1-phenylpyrazole
5-Methyl-1-phenylpyrazole
5(3)-Methyl-3(5)-phenylpyrazole
1,2-Naphthylenediamine
1,3-Naphthylenediamine
1,4-Naphthylenediamine
1,5-Naphthylenediamine
1,6-Naphthylenediamine
1,7-Naphthylenediamine
1,8-Naphthylenediamine
2,3-Naphthylenediamine
2,6-Naphthylenediamine
2,7-Naphthylenediamine
1-Naphthylhydrazine
2-Naphthylhydrazine
2,2′-Nicotyrine
3,2′-Nicotyrine

$C_{10}H_{10}N_2O$
2-Acetylbenzimidazole, 1-N-*Me*†
4-Amino-6-hydroxyquinoline, *Me ether*
5-Amino-6-hydroxyquinoline, *Me ether*
5-Amino-8-hydroxyquinoline, *Me ether*
6-Amino-2-hydroxyquinoline, *Me ether*
6-Amino-8-hydroxyquinoline, *Me ether*
8-Amino-4-hydroxyquinoline, *Me ether*
8-Amino-6-hydroxyquinoline, *Me ether*
8-Amino-7-hydroxyquinoline, *Me ether*
2-Cyano-3-phenylpropionic Acid, *Amide*
2,4-Diamino-1-naphthol
2,6-Diamino-1-naphthol
2,8-Diamino-1-naphthol
3,4-Diamino-1-naphthol
4,5-Diamino-1-naphthol
4,8-Diamino-1-naphthol
1,3-Diamino-2-naphthol
1,4-Diamino-2-naphthol
1,6-Diamino-2-naphthol
1,7-Diamino-2-naphthol

$C_{10}H_{10}N_2O$ (*continued*)
3,4-Diamino-2-naphthol
7,8-Diamino-2-naphthol
Glomerine†
1-Methyl-5-phenyl-4-pyrazolin-3-one
4-Methyl-1-phenyl-4-pyrazolin-3-one
5-Methyl-1-phenyl-4-pyrazolin-3-one
1-Methyl-3-phenyl-2-pyrazolin-5-one
2-Methyl-1-phenyl-3-pyrazolin-5-one
3-Methyl-1-phenyl-2-pyrazolin-5-one
4-Methyl-1-phenyl-2-pyrazolin-5-one
4-Methyl-3-phenyl-2-pyrazolin-5-one
Phenaceturic Acid, *Nitrile*
Phenylcyanoacetic Acid, *Methylamide*
Phthalazone, N-*Et*

$C_{10}H_{10}N_2OS$
2-Thiohydantoin, 3-N-*Benzyl*

$C_{10}H_{10}N_2O_2$
Benzylidenemalonic Acid, *Diamide*
2,3-Dimethyl-4-nitroindole
2,3-Dimethyl-5-nitroindole
2,3-Dimethyl-6-nitroindole
2,3-Dimethyl-7-nitroindole
1-Hydroxyindole-2-carboxylic Acid, *Me ether, Amide*
2,2′-Indazolylpropionic Acid
3,1′-Indazolylpropionic Acid
3,2′-Indazolylpropionic Acid
Indoxyl, N-*Nitroso, Et ether*
4-Isopropyl-3-nitrobenzoic Acid, *Nitrile*
7-Methylindazole-1-carboxylic Acid, *Me ester*
5-Methylindazole-2-carboxylic Acid, *Me ester*
7-Methylindazole-2-carboxylic Acid, *Me ester*
1-Methylindazole-3-carboxylic Acid, *Me ester*
2-Methylindazole-3-carboxylic Acid, *Me ester*
6-Methylindazole-3-carboxylic Acid, *Me ester*
1-Methyl-3-phenylhydantoin
3-Methyl-1-phenylhydantoin
3-Methyl-5-phenylhydantoin
5-Methyl-1-phenylhydantoin
5-Methyl-3-phenylhydantoin
5-Methyl-5-phenylhydantoin
3-Nitroindole, N-*Et*
N-Phenylglycine-*o*-carboxylic Acid, *α-Nitrile, β-Me ester*
5-Phenylhydantoin, N-*Me*

$C_{10}H_{10}N_2O_2S$
3-(2-Benzothiazolyl)alanine
1-Naphthylamine-3-sulphonic Acid, *Amide*
1-Naphthylamine-4-sulphonic Acid, *Amide*
1-Naphthylamine-5-sulphonic Acid, *Amide*
1-Naphthylamine-6-sulphonic Acid, *Amide*
1-Naphthylamine-7-sulphonic Acid, *Amide*
2-Naphthylamine-5-sulphonic Acid, *Amide*

$C_{10}H_{10}N_2O_3$
3,3-Dimethyl-4(or 6-)nitro-oxindole
3,3-Dimethyl-5-nitro-oxindole
3,3-Dimethyl-7-nitro-oxindole
Opiazone
2-Pyruvoylaminobenzamide†

$C_{10}H_{10}N_2O_3S$
1,2-Naphthylenediamine-3-sulphonic Acid
1,2-Naphthylenediamine-4-sulphonic Acid
1,2-Naphthylenediamine-5-sulphonic Acid
1,2-Naphthylenediamine-6-sulphonic Acid
1,2-Naphthylenediamine-7-sulphonic Acid
1,3-Naphthylenediamine-5-sulphonic Acid
1,3-Naphthylenediamine-6-sulphonic Acid
1,4-Naphthylenediamine-2-sulphonic Acid
1,4-Naphthylenediamine-5-sulphonic Acid
1,4-Naphthylenediamine-6-sulphonic Acid
1,5-Naphthylenediamine-2-sulphonic Acid
1,5-Naphthylenediamine-4-sulphonic Acid
1,6-Naphthylenediamine-4-sulphonic Acid
1,8-Naphthylenediamine-4-sulphonic Acid

$C_{10}H_{10}N_2O_4$
2,6-Dihydroxy-3-nitrobenzoic Acid, *Nitrile, 6-Me-2-Et ether*
3,4-Dimethoxy-2-nitrophenylacetic Acid, *Nitrile*

$C_{10}H_{10}N_2O_4S_2$
Naphthalene-1,3-disulphonic Acid, *Diamide*
Naphthalene-1,4-disulphonic Acid, *Diamide*
Naphthalene-1,5-disulphonic Acid, *Diamide*
Naphthalene-1,6-disulphonic Acid, *Diamide*
Naphthalene-1,7-disulphonic Acid, *Diamide*
Naphthalene-2,6-disulphonic Acid, *Diamide*
Naphthalene-2,7-disulphonic Acid, *Diamide*

$C_{10}H_{10}N_2O_5$
4-Amino-5-nitro-*o*-toluic Acid, N-*Ac*
4-Amino-6-nitro-*m*-toluic Acid, N-*Ac*
6-Amino-4-nitro-*m*-toluic Acid, N-*Ac*
2-Nitro-oxanilic Acid, *Et ester*
3-Nitro-oxanilic Acid, *Et ester*
4-Nitro-oxanilic Acid, *Et ester*

$C_{10}H_{10}N_2O_6$
2-Amino-6-nitrobenzoic Acid, N-*Carbethoxyl*
4-Amino-6-nitroisophthalic Acid, *Di-Me ester*
2-Amino-5-nitroterephthalic Acid, *Di-Me ester*
3,5-Dinitrobenzoic Acid, n-*Propyl ester*
2,4-Dinitrophenylacetic Acid, *Et ester*
4,5-Dinitro-*o*-toluic Acid, *Et ester*
5,6-Dinitro-*o*-toluic Acid, *Et ester*
4,6-Dinitro-*m*-toluic Acid, *Et ester*
3,5-Dinitro-*p*-toluic Acid, *Et ester*
3-Nitro-opianic Acid, *Amide*

$C_{10}H_{10}N_2O_6S_2$
1,8-Naphthylenediamine-3,6-disulphonic Acid
1,8-Naphthylenediamine-4,5-disulphonic Acid

$C_{10}H_{10}N_2O_7$
2-Hydroxy-3,5-dinitrobenzoic Acid, *Me ether, Et ester*
2-Hydroxy-3,5-dinitrobenzoic Acid, *Et ether, Me ester*
2-Hydroxy-3,5-dinitrobenzoic Acid, *Propyl ester*
2-Hydroxy-3,5-dinitrobenzoic Acid, *Isopropyl ester*
4-Methoxy-3,5-dinitrobenzoic Acid, *Et ester*

$C_{10}H_{10}N_2O_8$
3,4-Dimethoxy-2,6-dinitrobenzoic Acid, *Me ester*
4,5-Dimethoxy-2,3-dinitrobenzoic Acid, *Me ester*

$C_{10}H_{10}N_2S_2$
5-Ethyl-4-phenyl-1,3,4-thiadiazole-2-thiol mesoionic didehydro derivative†

$C_{10}H_{10}N_4O_2S$
Sulphadiazine
Sulphapyrazine
$C_{10}H_{10}N_4O_5$
1,2,3,4-Tetrahydro-6,8-dinitroquinoline-1-carboxylic Acid, *Amide*
$C_{10}H_{10}N_4O_6$
5,5′-Bibarbituric Acid, *Di-Me deriv.*
$C_{10}H_{10}Ni$
Dicyclopentadienyl nickel
$C_{10}H_{10}O$
Crotonophenone
Cyclodecapentaene-1,6-epoxide †
2,8-Decadiene-4,6-diyn-1-ol
Deca-3,4-diene-6,8-diyn-1-ol †
Deca-4,5-diene-7,9-diyn-1-ol †
Deca-4,6,8-triyn-1-ol †
3,4-Dihydro-1(2*H*)-naphthalenone
3,4-Dihydro-2(1*H*)-naphthalenone
2,5-Dimethylbenzofuran
2,6-Dimethylbenzofuran
2,7-Dimethylbenzofuran
3,5-Dimethylbenzofuran
3,6-Dimethylbenzofuran
3,7-Dimethylbenzofuran
4,6-Dimethylbenzofuran
4,7-Dimethylbenzofuran
5,6-Dimethylbenzofuran
5,7-Dimethylbenzofuran
6,7-Dimethylbenzofuran
Indane-1-aldehyde
Indane-2-aldehyde
Indane-4-aldehyde
Indane-5-aldehyde
1-Indenylmethanol
2-Methyl-3-chromene †
2-Methyl-1-indanone
3-Methyl-1-indanone
4-Methyl-1-indanone
6-Methyl-1-indanone
1-Methyl-2-indanone
2-Methyl-3-phenylacrolein
3-Methyl-3-phenylacrolein
4-Phenyl-3-buten-2-one
2-Phenyl-3-butyn-2-ol
2-Phenylcrotonaldehyde
4-Phenylcrotonaldehyde
3-Phenyl-2-propyn-1-ol, *Me ether*
3-*p*-Tolylacrolein
Tricyclo[3,3,2,$0^{2,8}$]deca-3,6-dien-9-one †
$C_{10}H_{10}OS$
3-Hydroxybenzo[*b*]thiophene, *Et ether*
$C_{10}H_{10}O_2$
Acrylic Acid, *Benzyl ester*
Benzoic Acid, *Allyl ester*
Benzoylacetone
2-Benzoylpropanal
2-Benzylacrylic Acid
Deca-4,5-diene-7,9-diyne-1,3-diol †
2-Decene-4,6-diynoic Acid
1,3-Diacetylbenzene
1,4-Diacetylbenzene
1,2-Dihydronaphthalene-1,2-diol †
β-Dolabrin
6-Hydroxybenzofuran, *Et ether*
o-Hydroxycinnamaldehyde, *Me ether*
p-Hydroxycinnamaldehyde, *Me ether*
4-Hydroxy-1-indanone, *Me ether*
5-Hydroxy-1-indanone, *Me ether*
6-Hydroxy-1-indanone, *Me ether*
7-Hydroxy-2-indanone, *Me ether*
3-Hydroxy-5-methylbenzofuran, *Me ether*
6-Hydroxy-3-methylbenzofuran, *Me ether*
3-Hydroxy-4-methylphthalide, *Me ether*
3-Hydroxy-6-methylphthalide, *Me ether*
4-Hydroxy-3-methylphthalide, *Me ether*
5-Hydroxy-6-methylphthalide, *Me ether*
7-Hydroxy-3-methylphthalide, *Me ether*
7-Hydroxy-6-methylphthalide, *Me ether*
4-(2-Hydroxyphenyl)-3-buten-2-one
4-(3-Hydroxyphenyl)-3-buten-2-one
4-(4-Hydroxyphenyl)-3-buten-2-one
4-Hydroxy-2-phenylbutyric Acid, *Lactone*
4-Hydroxy-4-phenylbutyric Acid, *Lactone*
Indane-1-carboxylic Acid
Indane-2-carboxylic Acid
Indane-4-carboxylic Acid
Indane-5-carboxylic Acid
Isosafrole
2-Methyl-4-chromanone
3-Methyl-4-chromanone
6-Methyl-4-chromanone
7-Methyl-4-chromanone
8-Methyl-4-chromanone
Methyl cinnamate
3,4-Methylenedioxyallylbenzene
2-Methyl-3-phenylacrylic Acid
4-Phenyl-3-butenoic Acid
2-Phenylcrotonic Acid
3-Phenylcrotonic Acid
4-Phenylcrotonic Acid
1-Phenylcyclopropane-1-carboxylic Acid
2-Phenylcyclopropane-1-carboxylic Acid
Salicylaldehyde, *Allyl ether*
2,3,4,5-Tetrahydro-5-oxo-1-benzoxepin
3-*o*-Tolylacrylic Acid
3-*m*-Tolylacrylic Acid
3-*p*-Tolylacrylic Acid
$C_{10}H_{10}O_2S$
3-*o*-Mercaptophenylacrylic Acid, S-*Me*
3-*o*-Mercaptophenylacrylic Acid, *Me ester*
3-*p*-Mercaptophenylacrylic Acid, S-*Me ether*
6-(2-Thienyl)-2,4-hexadienoic Acid †
$C_{10}H_{10}O_2S_2$
2,2′-Dithenyl Sulphone †
2,3′-Dithenyl Sulphone †
3,3′-Dithenyl Sulphone †
$C_{10}H_{10}O_3$
Acetophenone-*p*-carboxylic Acid, *Me ester*
2-Allyloxybenzoic Acid
4-Allyloxybenzoic Acid
Benzoylacetic Acid, *Me ester*
Benzoylformic Acid, *Et ester*
2-Benzoylpropionic Acid
3-Benzoylpropionic Acid
Benzylpyruvic Acid
1-*o*-Carboxyphenyl-2-propanone
m-Coumaric Acid, *Me ester*

$C_{10}H_{10}O_3$ *(continued)*
p-Coumaric Acid, *Me ester*
Dihydro-7-hydroxy-4,6-dimethylisobenzofuran
2,4-Dimethylbenzoylformic Acid
2,5-Dimethylbenzoylformic Acid
3,4-Dimethylbenzoylformic Acid
α-Dolabrinol
3-Ethyl-7-hydroxyisobenzofuran-1-one
2-Formylbenzoic Acid, *Et ester*
3-Formylbenzoic Acid, *Et ester*
Formylphenylacetic Acid, *Me ester*
10-Hydroxydec-2-ene-4,6-diynoic Acid
4-Hydroxy-3-methoxycinnamaldehyde
4-Hydroxymethyl-cinnamic Acid
3-(2-Hydroxy-4-methylphenyl)acrylic Acid
3-(2-Hydroxy-5-methylphenyl)acrylic Acid
3-(4-Hydroxy-2-methylphenyl)acrylic Acid
9-Hydroxynon-7-ene-3,5-diynoic Acid, *Me ester*†
2-Hydroxy-4-phenyl-3-butenoic Acid
4-*p*-Hydroxyphenyl-3-butenoic Acid
3-*o*-Hydroxyphenylcrotonic Acid
3-*m*-Hydroxyphenylcrotonic Acid
3-*p*-Hydroxyphenylcrotonic Acid
3-*o*-Hydroxyphenyl-2-methylacrylic Acid
3-*m*-Hydroxyphenyl-2-methylacrylic Acid
3-*p*-Hydroxyphenyl-2-methylacrylic Acid
4-Hydroxyphthalide, *Et ether*†
5-Hydroxyphthalide, *Et ether*†
6-Hydroxyphthalide, *Et ether*†
3-*o*-Methoxyphenylacrylic Acid
3-*m*-Methoxyphenylacrylic Acid
3-*p*-Methoxyphenylacrylic Acid
1-*p*-Methoxyphenyl-1,2-propanedione
3,4-Methylenedioxypropiophenone
Ochracin
3-Oxo-2-phenylbutyric Acid
3-Oxo-4-phenylbutyric Acid
2-Phenyloxiran-1-carboxylic Acid, *Me ester*
Phenylpyruvic Acid, *Me ester*
o-Propionylbenzoic Acid
Pyruvic Acid, *Benzyl ester*
Salicylic Acid, *Allyl ester*
m-Toluylformic Acid, *Me ester*

$C_{10}H_{10}O_4$
Acetylenedicarboxylic Acid, *Di-allyl ester*
2-Acetyl-6-hydroxybenzoic Acid, *Me ether*
O-Acetylsalicylic Acid, *Me ester*
Benzene-1,2-diacetic Acid
Benzene-1,3-diacetic Acid
Benzene-1,4-diacetic Acid
O-Benzoyl-lactic Acid
Benzylmalonic Acid
2-Carboxyphenylacetic Acid, 2-*Me ester*
3-*o*-Carboxyphenylpropionic Acid
3-*m*-Carboxyphenylpropionic Acid
3-*p*-Carboxyphenylpropionic Acid
4,6-Diacetylresorcinol
2,4-Diformyl-6-methoxyphenol, *Me ether*
2,5-Dihydroxycinnamic Acid, 2-*Me ether*
2,10-Dihydroxydec-4-ene-6,8-diynoic Acid
5,7-Dihydroxyphthalide, *Di-Me ether*†
4,5-Dimethylisophthalic Acid
4,6-Dimethylisophthalic Acid
3,4-Dimethylphthalic Acid
3,5-Dimethylphthalic Acid
3,6-Dimethylphthalic Acid
4,5-Dimethylphthalic Acid
2,5-Dimethylterephthalic Acid
2,6-Dimethylterephthalic Acid
Eugentiogenin
3-Formyl-2-hydroxybenzoic Acid, *Et ester*
5-Formyl-2-hydroxybenzoic Acid, *Et ester*
2-Formylphenoxyacetic Acid, *Me ester*
4-(2-Furyl)-2-oxo-3-butenoic Acid, *Et ester*
Gentiogenin
Glyceraldehyde, 2-*Benzoyl*
2-Hydroxybenzoylformic Acid, *Et ester*
4-Hydroxybenzoylformic Acid, *Et ether*
2-Hydroxy-3-methoxycinnamic Acid†
3-Hydroxy-4-methoxycinnamic Acid
4-Hydroxy-3-methoxycinnamic Acid
2-Hydroxy-4-oxo-phenylbutyric Acid
2-Hydroxyphenylpyruvic Acid, *Me ether*
3-Hydroxyphenylpyruvic Acid, *Me ether*
Isophthalic Acid, *Di-Me ester*
Isophthalic Acid, *Mono-Et ester*
Meconine
m-Meconine
ψ-Meconine
p-Methoxybenzoylacetic Acid
3,4-Methylenedioxybenzoic Acid, *Et ester*
3,4-Methylenedioxyphenylacetic Acid, *Me ester*
2-(3,4-Methylenedioxyphenyl)propionic Acid
3-(3,4-Methylenedioxyphenyl)propionic Acid
Methylphenylmalonic Acid
Methylterephthalic Acid, 1-*Me ester*
Methylterephthalic Acid, 4-*Me ester*
2,6-Octadien-4-ynedioic Acid, *Di-Me ester*
Phenylmalonic Acid, *Me ester*
Phenylsuccinic Acid
Phthalic Acid, *Di-Me ester*
Phthalic Acid, *Et ester*
Quadrilineatin
Salacetol
Succinic Acid, *Phenyl ester*
Terephthalic Acid, *Di-Me ester*
Terephthalic Acid, *Mono-Et ester*
Umbellic Acid, *Me ester*
Umbellic Acid, 4-*Me ether*

$C_{10}H_{10}O_4S$
S-*o*-Carboxyphenylglycollic Acid, 2-*Me ester*

$C_{10}H_{10}O_5$
5-Acetyl-2,4-dihydroxybenzoic Acid, *Me ester*
Curvulinic Acid†
1,3-Diacetyl-2,4,5-trihydroxybenzene
1,5-Diacetyl-2,3,4-trihydroxybenzene
2,4-Dihydroxybenzoylformic Acid, *Di-Me ether*
2,4-Dihydroxybenzoylformic Acid, 4-*Et ether*
2,5-Dihydroxybenzoylformic Acid, *Di-Me ether*
2,6-Dihydroxybenzoylformic Acid, *Di-Me ether*
3,4-Dihydroxybenzoylformic Acid, *Di-Me ether*
2,3-Dimethoxy-4,5-methylenedioxybenzaldehyde

$C_{10}H_{10}O_5$ (*continued*)
- 2,5-Dimethoxy-3,4-methylenedioxybenzaldehyde
- Flavipin, *Me ether*
- 2-Formyl-5,6-dihydroxybenzoic Acid, 5-*Me ether*, ψ-*Me ester*
- 5-Formyl-4,6-dihydroxy-*o*-toluic Acid, *Me ester*
- 2-Formyl-3,4-dimethoxybenzoic Acid
- 2-Formyl-4,5-dimethoxybenzoic Acid
- 2-Formyl-5,6-dimethoxybenzoic Acid
- 5-Formyl-3,4-dimethoxybenzoic Acid
- 3-Formyl-4-hydroxy-5-methoxybenzoic Acid, *Me ester*
- 5-Formyl-2-hydroxy-3-methoxybenzoic Acid, *Me ether*
- Glyceric Acid, 2-*Benzoyl*
- 2-Hydroxyisophthalic Acid, *Di-Me ester*
- 4-Hydroxyisophthalic Acid, *Di-Me ester*
- 4-Hydroxyisophthalic Acid, *Mono-Et ester*
- 5-Hydroxyisophthalic Acid, *Di-Me ester*
- 2-Hydroxy-5-methylisophthalic Acid, *Me ether*
- 4-Hydroxy-6-methylisophthalic Acid, *Me ether*
- 5-Hydroxy-3-methylphthalic Acid, *Me ether*
- 3-Hydroxy-5-methylphthalic Acid, *Me ether*
- 2-Hydroxy-6-methylterephthalic Acid, *Me ether*★†
- 3-Hydroxyphthalic Acid, *Di-Me ester*
- 4-Hydroxyphthalic Acid, *Di-Me ester*
- 4-Hydroxyphthalic Acid, 1-*Et ester*
- 4-Hydroxyphthalic Acid, 2-*Et ester*
- 4-Hydroxyphthalic Acid, *Et ether*
- Hydroxyterephthalic Acid, *Di-Me ester*
- Hydroxyterephthalic Acid, *Et ether*
- 2-Phenylmalic Acid
- 3-Phenylmalic Acid
- Stipitatic Acid, *Et ester*
- 3-(2,4,5-Trihydroxyphenyl)acrylic Acid, *Me ester*
- 3-(2,4,5-Trihydroxyphenyl)acrylic Acid, 5-*Me ether*
- 3-(2,4,6-Trihydroxyphenyl)acrylic Acid, *Me ester*

$C_{10}H_{10}O_6$
- Carlic Acid
- 2,4-Dihydroxyisophthalic Acid, *Et ester*
- 4,6-Dihydroxyisophthalic Acid, *Di-Me ester*
- 4,6-Dihydroxyisophthalic Acid, *Di-Me ether*
- 3,5-Dihydroxyphthalic Acid, *Di-Me ether*
- 3,6-Dihydroxyphthalic Acid, *Di-Me ester*
- 3,6-Dihydroxyphthalic Acid, *Di-Me ether*
- 2,3-Dihydroxyterephthalic Acid, *Di-Me ester*
- 2,3-Dihydroxyterephthalic Acid, *Di-Me ether*
- 2,5-Dihydroxyterephthalic Acid, *Mono-Et ester*
- 2,5-Dihydroxyterephthalic Acid, *Di-Me ether*
- 2,6-Dihydroxyterephthalic Acid, *Di-Me ester*
- 2,6-Dihydroxyterephthalic Acid, *Di-Me ether*
- 4,5-Dimethoxyisophthalic Acid
- 2,3-Dimethoxy-4,5-methylenedioxybenzoic Acid
- 2,5-Dimethoxy-3,4-methylenedioxybenzoic Acid
- 3,4-Dimethoxyphthalic Acid
- Metahemipinic Acid
- Prephenic Acid

$C_{10}H_{10}O_7$
- Furan-2,3,4-tricarboxylic Acid, *Tri-Me ester*
- Furan-2,3,5-tricarboxylic Acid, *Tri-Me ester*
- 3-Hydroxy-4-oxo-4*H*-pyran-2,6-dicarboxylic Acid, *Mono-propyl ester*
- 2,4,6-Trihydroxyisophthalic Acid, *Di-Me ester*
- 3,4,5-Trihydroxyphthalic Acid, 3,5-*Di-Me ether*

$C_{10}H_{10}Ru$
- Ruthenocene

$C_{10}H_{10}S_3$
- 2,2′-Dithenyl Sulphide†
- 2,3′-Dithenyl Sulphide†
- 3,3′-Dithenyl Sulphide†

$C_{10}H_{10}Ti$
- Dicyclopentadienyl titanium

$C_{10}H_{11}AsCl_2O_2$
- Benzenearsonous Acid, *Di-2-chloroethyl ester*

$C_{10}H_{11}AsO_2$
- Benzenearsonous Acid, *Ethylene ester*

$C_{10}H_{11}Br$
- 5-Bromo-1,2,3,4-tetrahydronaphthalene
- 6-Bromotetrahydronaphthalene

$C_{10}H_{11}BrN_2$
- 3-Bromo-4-methylazobenzene
- 4-Bromo-3-methylazobenzene
- 4′-Bromo-4-methylazobenzene

$C_{10}H_{11}BrO$
- 2-Bromo-1,2,3,4-tetrahydro-1-naphthol
- 1-Bromo-5,6,7,8-tetrahydro-2-naphthol
- 3-Bromo-1,2,3,4-tetrahydro-2-naphthol

$C_{10}H_{11}BrO_2$
- 2-Bromo-4,6-dimethylbenzoic Acid, *Me ester*
- 4-Bromo-2,6-dimethylbenzoic Acid, *Me ester*
- 3-Bromo-4-isopropylbenzoic Acid
- 3-Bromo-2-isopropyl-5-methyl-1,4-benzoquinone
- 3-Bromo-5-isopropyl-2-methyl-1,4-benzoquinone
- α-Bromophenylacetic Acid, *Et ester*
- *o*-Bromophenylacetic Acid, *Et ester*
- *m*-Bromophenylacetic Acid, *Et ester*
- *p*-Bromophenylacetic Acid, *Et ester*
- 2-Bromo-3-phenylbutyric Acid
- 2-Bromo-4-phenylbutyric Acid
- 3-Bromo-2-phenylbutyric Acid
- 3-Bromo-4-phenylbutyric Acid
- 4-Bromo-4-phenylbutyric Acid
- 3-Bromo-3-phenylpropionic Acid, *Me ester*
- 3-*p*-Bromophenylpropionic Acid, *Me ester*
- α-Bromo-*m*-toluic Acid, *Et ester*
- α-Bromo-*p*-toluic Acid, *Et ester*
- 4-Bromo-*m*-toluic Acid, *Et ester*

$C_{10}H_{11}BrO_3$
- α-Bromo-2,4-dihydroxyacetophenone, *Di-Me ether*
- α-Bromo-3,4-dihydroxyacetophenone, *Di-Me ether*
- 5-Bromo-2-hydroxybenzoic Acid, *Et ether*, *Me ester*
- 5-Bromo-2-hydroxy-*m*-toluic Acid, *Et ester*
- 5-Bromo-2-hydroxy-*p*-toluic Acid, *Me ester*, *Me ether*

$C_{10}H_{11}BrO_3$ *(continued)*
3-Bromo-4-methoxybenzoic Acid, *Et ester*
o-Bromophenoxyacetic Acid, *Et ester*
p-Bromophenoxyacetic Acid, *Et ester*
Methyl salicylate, β-*Bromoethyl ether*

$C_{10}H_{11}BrO_4$
2-Bromo-3,4-dimethoxybenzoic Acid, *Me ester*
3-Bromo-4,5-dimethoxybenzoic Acid, *Me ester*

$C_{10}H_{11}Br_2NO$
2,2-Dibromo-4-phenylbutyric Acid, *Amide*

$C_{10}H_{11}Br_3O_2$
2,4,6-Tribromoresorcinol, *Di-Et ether*

$C_{10}H_{11}Br_3O_3$
Tribromophloroglucinol, *Di-Et ether*

$C_{10}H_{11}Cl$
α-Chloro-2,4-dimethylstyrene
β-Chloro-α,4-dimethylstyrene

$C_{10}H_{11}ClO$
2-Benzylpropionic Acid, *Chloride*
2-Chloro-1,2,3,4-tetrahydro-1-naphthol
3-Chloro-1,2,3,4-tetrahydro-2-naphthol
2,4-Dimethylphenylacetic Acid, *Chloride*
p-Isopropylbenzoic Acid, *Chloride*
2-Methyl-2-phenylpropionic Acid, *Chloride*
2-Phenylbutyric Acid, *Chloride*
3-Phenylbutyric Acid, *Chloride*
o-Propylbenzoic Acid, *Chloride*
2,4,6-Trimethylbenzoic Acid, *Chloride*

$C_{10}H_{11}ClO_2$
α-Chloro-4-hydroxyacetophenone, *Et ether*
2-Chloro-3-isopropyl-6-methyl-1,4-benzoquinone
3-Chloro-5-isopropyl-2-methyl-1,4-benzoquinone
α-Chlorophenylacetic Acid, *Et ester*
o-Chlorophenylacetic Acid, *Et ester*
m-Chlorophenylacetic Acid, *Et ester*
p-Chlorophenylacetic Acid, *Et ester*
4-Chloro-*o*-toluic Acid, *Et ester*
α-Chloro-*m*-toluic Acid, *Et ester*
4-Chloro-*m*-toluic Acid, *Et ester*
p-Hydroxymethyl-benzoic Acid, *Et ether, Chloride*
2-Methyl-2-phenoxypropionic Acid, *Chloride*
3-*m*-Methoxyphenylpropionic Acid, *Chloride*
3-*p*-Methoxyphenylpropionic Acid, *Chloride*
2-Phenoxybutyric Acid, *Chloride*
p-Propoxybenzoic Acid, *Chloride*

$C_{10}H_{11}ClO_3$
α-Chloro-2,4-dihydroxyacetophenone, *Di-Me ether*
α-Chloro-4-hydroxy-*m*-toluic Acid, *Et ester*
3-Chloro-6-hydroxy-*p*-toluic Acid, *Et ester*
3-Chloro-6-hydroxy-*p*-toluic Acid, *Me ether, Me ester*
3-Chloro-6-hydroxy-*p*-toluic Acid, *Et ether*
5-Chloromethyl-2-hydroxybenzoic Acid, *Et ester*
o-Chlorophenoxyacetic Acid, *Et ester*
p-Chlorophenoxyacetic Acid, *Et ester*
3,4-Dimethoxyphenylacetic Acid, *Chloride*

$C_{10}H_{11}ClO_4$
4-Chloro-5-hydroxy-2-methylphenoxyacetic Acid, *Me ether*†
Gallic Acid, *Tri-Me ether, Chloride*
2,3,4-Trihydroxybenzoic Acid, 2,3,4-*Tri-Me ether, Chloride*
2,4,5-Trimethoxybenzoic Acid, *Chloride*

$C_{10}H_{11}ClO_7$
D-Glucurone, *Di-Ac, Chloride*

$C_{10}H_{11}Cl_3O_2$
Trichloroquinol, *Di-Et ether*

$C_{10}H_{11}FeN$
Ferrocenylamine

$C_{10}H_{11}IO$
2-Iodo-4,6-dimethylacetophenone

$C_{10}H_{11}IO_2$
2-Iodo-3-isopropyl-6-methylbenzoquinone
2-Iodo-6-isopropyl-3-methylbenzoquinone
o-Iodomethylbenzoic Acid, *Et ester*
3-Iodo-4-methylbenzoic Acid, *Et ester*

$C_{10}H_{11}IO_3$
3-Iodo-4-methoxybenzoic Acid, *Et ester*

$C_{10}H_{11}I_2NO_3$
3,5-Di-iodotyrosine, *Me ester*
Iodogorgoic Acid, *Me ester*

$C_{10}H_{11}N$
3,4-Dihydro-1-methylisoquinoline
3,4-Dihydro-3-methylisoquinoline
3,4-Dihydro-4-methylisoquinoline
3,4-Dihydro-5-methylisoquinoline
1,2-Dimethylindole
1,3-Dimethylindole
1,4-Dimethylindole
1,5-Dimethylindole
1,6-Dimethylindole
1,7-Dimethylindole
2,3-Dimethylindole
2,4-Dimethylindole
2,5-Dimethylindole
2,6-Dimethylindole
2,7-Dimethylindole
3,5-Dimethylindole
3,6-Dimethylindole
3,7-Dimethylindole
4,6-Dimethylindole
4,7-Dimethylindole
5,6-Dimethylindole
5,7-Dimethylindole
6,7-Dimethylindole
2,4-Dimethylphenylacetic Acid, *Nitrile*
2,5-Dimethylphenylacetic Acid, *Nitrile*
3,5-Dimethylphenylacetic Acid, *Nitrile*
1-Ethylindole
2-Ethylindole
3-Ethylindole
p-Isopropylbenzoic Acid, *Nitrile*
2-Methyl-2-phenylpropionic Acid, *Nitrile*
N-Methyl-*N*-2-propynylaniline
2-Phenylbutyric Acid, *Nitrile*
4-Phenylbutyric Acid, *Nitrile*
1-Phenyl-2-pyrroline
2-Phenyl-2-pyrroline

$C_{10}H_{11}N$ (*continued*)
1-Phenyl-3-pyrroline
2-Phenyl-3-pyrroline
o-Propylbenzoic Acid, *Nitrile*
p-Propylbenzoic Acid, *Nitrile*
2-*p*-Tolylpropionic Acid, *Nitrile*
3-*p*-Tolylpropionic Acid, *Nitrile*
2,4,5-Trimethylbenzoic Acid, *Nitrile*
2,4,6-Trimethylbenzoic Acid, *Nitrile*

$C_{10}H_{11}NO$
Abikoviromycin†
Boschniakine†
1,3-Dimethyloxindole
1,7-Dimethyloxindole
3,3-Dimethyloxindole
3,6-Dimethyloxindole
4,7-Dimethyloxindole
5,7-Dimethyloxindole
4-Hydroxy-2,6-dimethylbenzoic Acid, *Me ether*, *Nitrile*
3β-Hydroxyethylindole
α-Hydroxyisopropylbenzoic Acid, *Nitrile*
2-Hydroxyphenylacetic Acid, *Et ether*, *Nitrile*
4-Hydroxyphenylacetic Acid, *Et ether*, *Nitrile*
Indane-1-carboxylic Acid, *Amide*
Indane-2-carboxylic Acid, *Amide*
Indane-4-carboxylic Acid, *Amide*
Indane-5-carboxylic Acid, *Amide*
Mandelic Acid, *Et ether*, *Nitrile*
3-*p*-Methoxyphenylpropionic Acid, *Nitrile*
3-Methyl-2-indolemethanol
2-Methyl-3-indolemethanol
2-Methyl-3-phenylacrylic Acid, *Amide*
2-Phenoxybutyric Acid, *Nitrile*
4-Phenoxybutyric Acid, *Nitrile*
4-Phenyl-3-butenoic Acid, *Amide*
2-Phenylcrotonic Acid, *Amide*
3-Phenylcrotonic Acid, *Amide*
1-Phenylcyclopropane-1-carboxylic Acid, *Amide*
2-Phenylcyclopropane-1-carboxylic Acid, *Amide*
1-Phenyl-2-pyrrolidone
4-Phenyl-2-pyrrolidone
5-Phenyl-2-pyrrolidone
Phthalimidine, N-*Et*
Physostigmol
1,2,3,4-Tetrahydro-2-oxoquinoline, N-*Me*
3-*p*-Tolylacrylic Acid, *Amide*

$C_{10}H_{11}NO_2$
Acetoacetanilide
p-Aminobenzoic Acid, *Allyl ester*
o-Aminocinnamic Acid, *Me ester*
m-Aminocinnamic Acid, *Me ester*
p-Aminocinnamic Acid, *Me ester*
3-Anilinocrotonic Acid
2-Benzoylpropionic Acid, *Amide*
3-Benzoylpropionic Acid, *Amide*
Benzylpenilloaldehyde
Bicyclo[2,2,1]hept-5-ene-2,3-dicarboxylic Acid, *Methylimide*
Boschniakinic Acid†
Dehydrofusaric Acid
Diacetanilide
2,3-Dihydro-4-methylindole-3-carboxylic Acid
2,3-Dihydro-7-methylindole-3-carboxylic Acid
5,6-Dihydroxyindole, *Di-Me ether*
3,4-Dimethoxyphenylacetic Acid, *Nitrile*
5-Hydroxy-3β-hydroxyethylindole★†
2-Hydroxy-4-phenyl-3-butenoic Acid, *Amide*
Indoline-2-carboxylic Acid, *Me ester*†
4-Methoxy-2,6-dimethylbenzonitrile, *N*-Oxide†
3-*o*-Methoxyphenylacrylic Acid, *Amide*
3-*p*-Methoxyphenylacrylic Acid, *Amide*
3-Methylamino-3-phenylacrylic Acid
o-Propionylbenzoic Acid, *Amide*
2-Pyridineacrylic Acid, *Et ester*
3-Pyridineacrylic Acid, *Et ester*
4-Pyridineacrylic Acid, *Et ester*
1,2,3,4-Tetrahydroisoquinoline-2-carboxylic Acid
1,2,3,4-Tetrahydroisoquinoline-3-carboxylic Acid
5,6,7,8-Tetrahydroisoquinoline-3-carboxylic Acid
1,2,3,4-Tetrahydro-5-nitronaphthalene
1,2,3,4-Tetrahydro-6-nitronaphthalene
1,2,3,4-Tetrahydroquinoline-2-carboxylic Acid
1,2,3,4-Tetrahydroquinoline-4-carboxylic Acid
5,6,7,8-Tetrahydroquinoline-4-carboxylic Acid
1,2,3,4-Tetrahydroquinoline-5-carboxylic Acid
1,2,3,4-Tetrahydroquinoline-6-carboxylic Acid
1,2,3,4-Tetrahydroquinoline-7-carboxylic Acid
1,2,3,4-Tetrahydroquinoline-8-carboxylic Acid

$C_{10}H_{11}NO_2S$
Veratryl Isothiocyanate†

$C_{10}H_{11}NO_3$
N-Acetylanthranilic Acid, *Me ester*
N-Acetylanthranilic Acid, N-*Me*
O-Benzoyl-lactic Acid, *Amide*
2,4-Dimethyl-3-nitroacetophenone
2,4-Dimethyl-5-nitroacetophenone
2,6-Dimethylterephthalic Acid, 4-*Amide*
Gallic Acid, *Tri-Me ether*, *Nitrile*
Gentioflavine†
Hippuric Acid, *Me ester*
p-Hydroxybenzylpenilloaldehyde
4-Isopropyl-3-nitrobenzaldehyde
Malonanilic Acid, *Me ester*
2-(3,4-Methylenedioxyphenyl)propionic Acid, *Amide*
3-(3,4-Methylenedioxyphenyl)propionic Acid, *Amide*
Methylmalonic Acid, *Monoanilide*
N-Methyloxanilic Acid, *Me ester*
4-*o*-Nitrophenyl-2-butanone
4-*p*-Nitrophenyl-2-butanone
o-Nitrosobenzoic Acid, *Propyl ester*
o-Nitrosobenzoic Acid, *Isopropyl ester*
Oxanilic Acid, *Et ester*
Phenaceturic Acid
N-Phenylglycine, N-*Ac*

$C_{10}H_{11}NO_3$ *(continued)*
Phenylsuccinic Acid, α-*Amide*
Phenylsuccinic Acid, β-*Amide*
Succinanilic Acid
5,6,7,8-Tetrahydro-2-nitro-1-naphthol
5,6,7,8-Tetrahydro-4-nitro-1-naphthol
1,2,3,4-Tetrahydro-7-nitro-1-naphthol
5,6,7,8-Tetrahydro-3-nitro-2-naphthol
5,6,7,8-Tetrahydro-4-nitro-2-naphthol
2,4,5-Trimethoxybenzoic Acid, *Nitrile*

$C_{10}H_{11}NO_4$
2-Aminoisophthalic Acid, N-*Di-Me*
4-Aminoisophthalic Acid, *Di-Me ester*
4-Aminoisophthalic Acid, 1-*Et ester*
4-Aminoisophthalic Acid, N-*Me*, 1-*Me ester*
4-Aminoisophthalic Acid, N-*Di-Me*
5-Aminoisophthalic Acid, *Di-Me ester*
3-Aminophthalic Acid, *Di-Me ester*
4-Aminophthalic Acid, *Di-Me ester*
Aminoterephthalic Acid, *Di-Me ester*
Aminoterephthalic Acid, N-*Di-Me*
Apophyllenic Acid, *Et ester*
m-Carboxyphenylalanine†
3,5-Dimethyl-2-nitrophenylacetic Acid
2,6-Dimethylpyridine-3,4-dicarboxylic Acid, 3-*Me ester*
2,6-Dimethylpyridine-3,4-dicarboxylic Acid, 4-*Me ester*
Isatoic Acid, *Di-Me ester*
Isatoic Acid, 2-*Et ester*
4-Isopropyl-2-nitrobenzoic Acid
4-Isopropyl-3-nitrobenzoic Acid
Malic Acid, *Monoanilide*
p-Methoxybenzamidoacetic Acid
N-Methylisatoic Acid, *Me ester*
6-Methylpyridine-2,4-dicarboxylic Acid, *Di-Me ester*
6-Methylpyridine-3,4-dicarboxylic Acid, *Di-Me ester*
2-Methylpyridine-3,5-dicarboxylic Acid, *Di-Me ester*
o-Nitrobenzoic Acid, *Propyl ester*
p-Nitrobenzoic Acid, *Propyl ester*
p-Nitrobenzoic Acid, *Isopropyl ester*
6-Nitroeugenol
6-Nitroisoeugenol
o-Nitrophenylacetic Acid, *Et ester*
p-Nitrophenylacetic Acid, *Et ester*
2-*p*-Nitrophenylbutyric Acid
3-*p*-Nitrophenylbutyric Acid
4-*p*-Nitrophenylbutyric Acid
3-*o*-Nitrophenylpropionic Acid, *Me ester*
4-Nitro-*m*-toluic Acid, *Et ester*
3-Nitro-*p*-toluic Acid, *Et ester*
N-Phenylglycine-*o*-carboxylic Acid, α-*Me ester*
N-Phenylglycine-*o*-carboxylic Acid, β-*Me ester*
N-Phenylglycine-*o*-carboxylic Acid, N-*Me*
2-Phenyliminodiacetic Acid
N-Phenyliminodiacetic Acid
Pyridine-2,3-dicarboxylic Acid, 1-*Me*-2-*Et ester*
Pyridine-2,3-dicarboxylic Acid, 2-*Me*-1-*Et ester*

$C_{10}H_{11}NO_5$
3-(3-Carboxy-4-hydroxyphenyl)-L-alanine
2,4-Dihydroxy-5-nitroacetophenone, *Di-Me ether*
3,4-Dihydroxy-α-nitroacetophenone, *Di-Me ether*
2,3-Dihydroxy-5-nitrobenzaldehyde, 2-*Me*-3-*Et ether*
2,3-Dihydroxy-5-nitrobenzaldehyde, 3-*Me*-2-*Et ether*
2,3-Dihydroxy-6-nitrobenzaldehyde, 3-*Me*-2-*Et ether*
2,4-Dihydroxy-3-nitrobenzaldehyde, 4-*Me*-2-*Et ether*
2,4-Dihydroxy-5-nitrobenzaldehyde, 4-*Me*-2-*Et ether*
3,4-Dimethoxyphthalic Acid, 1-*Amide*
3,4-Dimethoxyphthalic Acid, 2-*Amide*
Flavensomycinic Acid
3-Hydroxy-4-methoxy-2-nitrobenzaldehyde, *Et ether*
4-Hydroxy-3-methoxy-2-nitrobenzaldehyde, *Et ether*
4-Hydroxy-5-methoxy-2-nitrobenzaldehyde, *Et ether*
2-Hydroxy-4-nitrobenzoic Acid, *Me ether*, *Et ester*
2-Hydroxy-4-nitrobenzoic Acid, *Propyl ether*
2-Hydroxy-4-nitrobenzoic Acid, *Propyl ester*
Hydroxy-*o*-nitrophenylacetic Acid, *Et ester*
Hydroxy-*m*-nitrophenylacetic Acid, *Et ester*
Hydroxy-*p*-nitrophenylacetic Acid, *Et ester*
3-Hydroxy-3-*o*-nitrophenylpropionic Acid, *Me ester*
3-Hydroxy-3-*p*-nitrophenylpropionic Acid, *Me ester*
3-(4-Hydroxy-3-nitrophenyl)propionic Acid, *Me ester*
2-Hydroxy-5-nitro-*m*-toluic Acid, *Et ester*
2-Hydroxy-5-nitro-*p*-toluic Acid, *Et ether*
6-Hydroxy-5-nitro-*m*-toluic Acid, *Et ester*
2-Hydroxy-3-nitro-*p*-toluic Acid, *Et ester*
4-Methoxy-3-nitrobenzoic Acid, *Et ester*
4-Methoxy-3-nitrophenylacetic Acid, *Me ester*
1,2-*p*-Nitrobenzylideneglycerol
1,3-*p*-Nitrobenzylideneglycerol
o-Nitrophenoxyacetic Acid, *Et ester*
m-Nitrophenoxyacetic Acid, *Et ester*
p-Nitrophenoxyacetic Acid, *Et ester*
2-*m*-Nitrophenoxypropionic Acid, *Me ester*
Tartranilic Acid

$C_{10}H_{11}NO_6$
2,3-Dihydroxy-5-nitrobenzoic Acid, 2-*Me*-3-*Et ether*
2,3-Dihydroxy-5-nitrobenzoic Acid, 3-*Me*-2-*Et ether*
2,3-Dihydroxy-6-nitrobenzoic Acid, 3-*Me*-2-*Et ether*
2,3-Dimethoxy-5-nitrobenzoic Acid, *Me ester*
2,3-Dimethoxy-6-nitrobenzoic Acid, *Me ester*
2,4-Dimethoxy-3-nitrobenzoic Acid, *Me ester*
2,4-Dimethoxy-5-nitrobenzoic Acid, *Me ester*
2,5-Dimethoxy-3-nitrobenzoic Acid, *Me ester*
3,4-Dimethoxy-2-nitrobenzoic Acid, *Me ester*
3,4-Dimethoxy-5-nitrobenzoic Acid, *Me ester*
3,6-Dimethoxy-2-nitrobenzoic Acid, *Me ester*
4,5-Dimethoxy-2-nitrobenzoic Acid, *Me ester*
3,4-Dimethoxy-2-nitrophenylacetic Acid
3,4-Dimethoxy-5-nitrophenylacetic Acid

$C_{10}H_{11}NO_6$ *(continued)*
4,5-Dimethoxy-2-nitrophenylacetic Acid
4-Hydroxy-3-methoxy-5-nitrophenylacetic Acid, *Me ester*
3-*m*-Nitrophenylglyceric Acid, *Me ester*

$C_{10}H_{11}NO_6S$
2-Sulphophthalic Acid, *Me ester, Di-Et ester*

$C_{10}H_{11}NO_7$
2-Nitrogallic Acid, *Tri-Me ether*

$C_{10}H_{11}NO_7S$
4-Nitro-2-sulphobenzoic Acid, 1-*Me*-2-*Et ester*
4-Nitro-2-sulphobenzoic Acid, 1-*Et*-2-*Me ester*
4-Nitro-6-sulpho-*m*-toluic Acid, *Di-Me ester*
6-Nitro-4-sulpho-*m*-toluic Acid, *Di-Me ester*

$C_{10}H_{11}NS_2$
Phenyldithiocarbamic Acid, *Allyl ester*

$C_{10}H_{11}N_3$
4-Amino-3-methyl-1-phenylpyrazole
5-Amino-3-methyl-1-phenylpyrazole
3,4-Diamino-2-methylquinoline
4,6-Diamino-2-methylquinoline
4,8-Diamino-2-methylquinoline
1,2,4-Triaminonaphthalene
1,2,5-Triaminonaphthalene
1,2,6-Triaminonaphthalene
1,2,7-Triaminonaphthalene
1,3,6-Triaminonaphthalene
1,3,7-Triaminonaphthalene
1,3,8-Triaminonaphthalene
1,4,5-Triaminonaphthalene

$C_{10}H_{11}N_3O_2$
β-1-Benzimidazolylalanine
β-2-Benzimidazolylalanine
1-Phenylurazole, 2,4-*Di-Me*
1-Phenylurazole, 2-*Et*
1-Phenylurazole, 3(5)-*Et ether*

$C_{10}H_{11}N_3O_2S_2$
2′-(2-Aminoethyl)-2,4′-bithiazole-4-carboxylic Acid, *Me ester*†
Sulphamethylthiazole

$C_{10}H_{11}N_3O_4$
o-Nitrobenzylmalonic Acid, *Diamide*
m-Nitrobenzylmalonic Acid, *Diamide*

$C_{10}H_{11}N_3O_8$
Styphnic Acid, *Di-Et ether*
3,4,5-Trinitrocatechol, *Di-Et ether*

$C_{10}H_{11}N_5O_8$
N,2,4,6-Tetranitroaniline, N-*Butyl*
N,2,4,6-Tetranitroaniline, N-*Isobutyl*

$C_{10}H_{12}$
1,2-Dimethylcyclo-octatetraene
2,4-Dimethylstyrene
2,5-Dimethylstyrene
3,4-Dimethylstyrene
3,5-Dimethylstyrene
Ethylcyclo-octatetraene
4,7-Methylene-4,7,8,9-tetrahydroindene
1-Methylindane
2-Methylindane
4-Methylindane
5-Methylindane
2-Methyl-1-phenyl-1-propene
Pentacyclo[4,4,$0^{2,5}$,$0^{3,8}$,$0^{4,7}$]decane†
1-Phenyl-1-butene
2-Phenyl-1-butene
4-Phenyl-1-butene
1-Phenyl-2-butene
2-Phenyl-2-butene
Phenylcyclobutane
1,2,9,10-Tetrahydronaphthalene†
Tetrahydronaphthalene
Tetravinylethylene†
1-*p*-Tolylpropene
2-*o*-Tolylpropene
2-*m*-Tolylpropene
2-*p*-Tolylpropene

$C_{10}H_{12}BrN$
4-Bromo-5,6,7,8-tetrahydro-1-naphthylamine
1-Bromo-5,6,7,8-tetrahydro-2-naphthylamine
4-Bromo-5,6,7,8-tetrahydro-2-naphthylamine

$C_{10}H_{12}BrNO$
2-Bromobutyric Acid, *Anilide*
3-Bromo-4-isopropylbenzoic Acid, *Amide*

$C_{10}H_{12}BrNO_2$
5-Bromo-*o*-phenetidine, N-*Ac*
2-Bromo-*p*-phenetidine, N-*Ac*
3-Bromo-*p*-phenetidine, N-*Ac*
N-*o*-Bromophenylglycine, *Et ester*
N-*p*-Bromophenylglycine, *Et ester*

$C_{10}H_{12}Br_2O$
2,4-Dibromo-6-isopropyl-*m*-cresol

$C_{10}H_{12}Br_2O_2$
2,6-Dibromoresorcinol, *Di-Et ether*

$C_{10}H_{12}Br_2O_3$
2,3-Dibromo-4,5-dihydroxybenzyl methyl ether, *Di-Me ether*

$C_{10}H_{12}ClNO_2$
5-Chloro-*o*-phenetidine, N-*Ac*
6-Chloro-*m*-phenetidine, N-*Ac*
2-Chloro-*p*-phenetidine, N-*Ac*
3-Chloro-*p*-phenetidine, N-*Ac*
o-Chlorophenylglycine, *Et ester*
m-Chlorophenylcarbamic Acid, *Isopropyl ester*
Isopropyl *N*-(3-chlorophenyl)carbamate
o-Tolylcarbamic Acid, β-*Chloroethyl ester*
p-Tolylcarbamic Acid, β-*Chloroethyl ester*

$C_{10}H_{12}ClNO_4$
2-Amino-5-chloro-3,4-dimethoxybenzoic Acid, *Me ester*†
6-Chloro-4-nitroresorcinol, *Di-Et ether*

$C_{10}H_{12}Cl_2$
2,5-Dichloro-4-isopropyltoluene

$C_{10}H_{12}FNO_3$
2-Fluorotyrosine, *Me ether*

$C_{10}H_{12}FN_5O_4$
8-Fluoroadenosine†

$C_{10}H_{12}F_3N_3O_4S_2$
Bendrofluazide

$C_{10}H_{12}HgO_3$
Mercuri-*p*-hydroxyphenyl acetate, *Et ether*

$C_{10}H_{12}INO_2$
4-Ethoxy-3-iodoacetanilide

$C_{10}H_{12}N_2$
2-β-Aminoethylindole
3-Amino-4-isopropylbenzoic Acid, *Nitrile*
Anatabine
2-Anilinobutyric Acid, *Nitrile*
2-Anilino-2-methyl-propionic Acid, *Nitrile*
2-Benzylimidazoline
1-Ethyl-3-methylindazole
1-Ethyl-5-methylindazole
2-Ethyl-3-methylindazole
2-Ethyl-5-methylindazole
Isonicoteine
1-Methyl-3-phenylpyrazoline
1-Methyl-5-phenylpyrazoline
3-Methyl-1-phenylpyrazoline
3-Methyl-5-phenylpyrazoline
4-Methyl-1-phenylpyrazoline
5-Methyl-1-phenylpyrazoline
5-Methyl-3-phenylpyrazoline
N-Phenylglycine, N-*Et*, *Nitrile*
Tryptamine

$C_{10}H_{12}N_2O$
Acetoacetanilide, *Oxime*
3-(2-Aminoethyl)-4-hydroxyindole†
3-(2-Aminoethyl)-5-hydroxyindole★†
3-(2-Aminoethyl)-7-hydroxyindole
Cotinine†
1,2,3,4-Tetrahydroisoquinoline-2-carboxylic Acid, *Amide*

$C_{10}H_{12}N_2O_2$
N-Acetylanthranilic Acid, N-*Me*, *Amide*
Benzene-1,2-diacetic Acid, *Diamide*
Benzene-1,4-diacetic Acid, *Diamide*
Benzylmalonic Acid, *Diamide*
2,3-Dihydro-2-methyl-5-nitroindole, N-*Me*
Malonamide, N-*Benzyl*
o-Nitrobenzylamine, N-*Allyl*
p-Nitrobenzylamine, N-*Allyl*
Phenaceturic Acid, *Amide*
Phenylglyoxime, O,N-*Di-Me*
Phenylglyoxime, *Di-Me ether*
Phenylsuccinic Acid, *Diamide*
Succinanilic Acid, *Amide*
1,2,3-4,Tetrahydro-2-methyl-6-nitroquinoline
5,6,7,8-Tetrahydro-2-nitro-1-naphthylamine
5,6,7,8-Tetrahydro-3-nitro-1-naphthylamine
5,6,7,8-Tetrahydro-4-nitro-1-naphthylamine
5,6,7,8-Tetrahydro-1-nitro-2-naphthylamine
5,6,7,8-Tetrahydro-3-nitro-2-naphthylamine
5,6,7,8-Tetrahydro-4-nitro-2-naphthylamine

$C_{10}H_{12}N_2O_3$
o-Aminobenzamidoacetic Acid, *Me ester*
p-Aminobenzamidoacetic Acid, *Me ester*
5-Amino-2-nitroacetophenone, N-*Di-Me*
p-Amino-oxanilic Acid, *Et ester*
5,5-Diallylbarbituric Acid
2-Ethylamino-5-nitrosobenzoic Acid, *Me ester*
Isatoic Acid, *Amide*
Kynurenine
2-*N*-Methylamino-5-nitrosobenzoic Acid, *N-Carbethoxyl*
p-Nitrophenylacetic Acid, *Dimethylamide*
p-Nitrophenylacetic Acid, *Ethylamide*
3-*p*-Nitrophenylpropionic Acid, *Methylamide*
3-Nitro-*o*-toluic Acid, *Dimethylamide*
5-Nitro-*o*-toluic Acid, *Dimethylamide*
2-Nitro-*m*-toluic Acid, *Dimethylamide*
3-Nitro-*p*-toluic Acid, *Dimethylamide*
N-Phenylglycine-*o*-carboxylic Acid, β-*Me ester*, α-*Amide*

$C_{10}H_{12}N_2O_4$
2-Amino-5-nitrobenzoic Acid, N-*Me*, *Et ester*
4-Amino-2-nitrophenylacetic Acid, *Et ester*
4,6-Diaminoisophthalic Acid, *Di-Me ester*
4,6-Diaminoisophthalic Acid, *Mono-Et ester*
2,5-Diaminoterephthalic Acid, *Di-Me ester*
2,6-Diaminoterephthalic Acid, *Di-Me ester*
4-Dimethylamino-3-nitrobenzoic Acid, *Me ester*
Ethyl 2-methyl-4-nitrophenylaminoformate
Ethyl 2-methyl-5-nitrophenylaminoformate
Ethyl 2-methyl-6-nitrophenylaminoformate
Ethyl 4-methyl-2-nitrophenylaminoformate
Ethyl 4-methyl-3-nitrophenylaminoformate
3-Hydroxykynurenine†
3-(4-Hydroxy-3-nitrophenyl)propionic Acid, *Me ester*, *Amide*
4-Methylamino-3-nitrobenzoic Acid, *Et ester*
3,4-Methylenedioxy-6-nitroaniline, N-*Propyl*
o-Nitrophenylcarbamic Acid, *Isopropyl ester*
m-Nitrophenylcarbamic Acid, *Isopropyl ester*
p-Nitrophenylcarbamic Acid, *Isopropyl ester*
N-o-Nitrophenylglycine, *Et ester*
N-m-Nitrophenylglycine, *Et ester*
N-p-Nitrophenylglycine, *Et ester*
N-(4-Nitro-2-tolyl)glycine, *Me ester*
N-(5-Nitro-2-tolyl)glycine, *Me ester*
Tartranilic Acid, *Amide*
1,2,3,4-Tetramethyl-5,6-dinitrobenzene
1,2,3,5-Tetramethyl-4,6-dinitrobenzene
1,2,4,5-Tetramethyl-3,6-dinitrobenzene

$C_{10}H_{12}N_2O_4S$
2,4-Dinitrobenzenethiol, *Isobutyl ether*

$C_{10}H_{12}N_2O_5$
Dinoseb†

$C_{10}H_{12}N_2O_5S$
7-Aminocephalosporanic Acid

$C_{10}H_{12}N_2O_6$
3,5-Dinitrocatechol, *Di-Et ether*
2,3-Dinitroquinol, *Di-Et ether*
2,5-Dinitroquinol, *Di-Et ether*
2,4-Dinitroresorcinol, *Di-Et ether*
4,6-Dinitroresorcinol, *Di-Et ether*
2-Nitrogallic Acid, *Tri-Me ether*, *Amide*

$C_{10}H_{12}N_2O_6S$
p-Methoxycarbonylaminobenzenesulphonylcarbamic Acid, *Me ester*†

$C_{10}H_{12}N_2O_7S$
Rhodnitin†

$C_{10}H_{12}N_2S_2$
2-Benzothiazolesulphenamide, N-*Isopropyl*

$C_{10}H_{12}N_4O_2S_2$
N'-(5-Ethyl-1,3,4-thiadiazol-2-yl)sulphanilamide

$C_{10}H_{12}N_4O_4$
Cytosinine†
Nebularine

$C_{10}H_{12}N_4O_4S_2$
Naphthalene-1,5-disulphonic Acid, *Dihydrazide*
$C_{10}H_{12}N_4O_5$
Formycin B†
Inosine
3-Isoinosine†
$C_{10}H_{12}N_4O_6$
3,5-Dimethyl-2,4,6-trinitroaniline, N-*Et*
Xanthosine
$C_{10}H_{12}N_4O_7$
9-β-D-Ribofuranosyluric Acid†
3-Ribosyluric Acid
$C_{10}H_{12}O$
Anethole
Benzyl Alcohol, *Allyl ether*
2-Benzylpropanal
Butyrophenone
p-Cresol, *Allyl ether*
8,9-Dehydroadamantan-2-one†
2,4-Dihydroxyacetophenone, 4-*Et ether*
2,4-Dimethylacetophenone
2,5-Dimethylacetophenone
3,4-Dimethylacetophenone
3,5-Dimethylacetophenone
Esdragol
β-Ethoxystyrene
1-Hydroxyindane, *Me ether*
4-Hydroxyindane, *Me ether*
5-Hydroxyindane, *Me ether*
4-Hydroxystyrene, *Et ether*
Isobutyrophenone
p-Isopropylbenzaldehyde
2-Methylchroman
3-Methylchroman
6-Methylchroman
7-Methylchroman
8-Methylchroman
2-Methyl-2-phenylpropanal
Nezukone†
1-Phenylallyl Alcohol, *Me ether*
2-Phenylbutanal
3-Phenylbutanal
4-Phenylbutanal
1-Phenyl-2-butanone
4-Phenyl-2-butanone
1-Phenyl-2-buten-1-ol
1-Phenyl-3-buten-1-ol
3-Phenyl-2-buten-1-ol
4-Phenyl-2-buten-1-ol
3-Phenyl-3-buten-1-ol
4-Phenyl-3-buten-1-ol
2-Phenyl-3-buten-2-ol
4-Phenyl-3-buten-2-ol
1-Phenylvinyl Alcohol, *Et ether*
2-Phenylvinyl Alcohol, *Et ether*
4-Propenyl-*o*-cresol
6-Propenyl-*o*-cresol
4-Propenyl-*m*-cresol
2-Propenyl-*p*-cresol
o-Propenylphenol, *Me ether*
m-Propenylphenol, *Me ether*
o-Propionyltoluene
m-Propionyltoluene
p-Propionyltoluene
1,2,3,4-Tetrahydro-1-naphthol
5,6,7,8-Tetrahydro-1-naphthol
1,2,3,4-Tetrahydro-2-naphthol
5,6,7,8-Tetrahydro-2-naphthol
o-Tolylacetone
m-Tolylacetone
p-Tolylacetone
3-*o*-Tolylpropanal
3-*p*-Tolylpropanal
1-*p*-Tolylcyclopropanol†
2,3,4-Trimethylbenzaldehyde
2,3,6-Trimethylbenzaldehyde
2,4,5-Trimethylbenzaldehyde
2,4,6-Trimethylbenzaldehyde
3,4,5-Trimethylbenzaldehyde
$C_{10}H_{12}OS$
Dimethylsulphonium Phenacylide†
Phenacylidenedimethylsulphurane†
Phenylthiolacetic Acid, *Et ester*
Thiobenzoic Acid, S-*Propyl ester*
Thiobutyric Acid, S-*Phenyl ester*
Thio-*o*-toluic Acid, S-*Et ester*
Thio-*p*-toluic Acid, S-*Et ester*
Thio-*p*-toluic Acid, O-*Et ester*
$C_{10}H_{12}O_2$
2-Allyl-6-methoxyphenol
Benzoic Acid, *Propyl ester*
Benzoic Acid, *Isopropyl ester*
1-Benzoyl-1-propanol
3-Benzoyl-1-propanol
1-Benzoyl-2-propanol
2-Benzylpropionic Acid
Butyric Acid, *Phenyl ester*
Chavibetol
α,2-Dihydroxytoluene, 2-*Allyl ether*
2,4-Dimethylbenzoic Acid, *Me ester*
2,3-Dimethylphenylacetic Acid
2,4-Dimethylphenylacetic Acid
2,5-Dimethylphenylacetic Acid
3,5-Dimethylphenylacetic Acid
Egomaketone
4-Ethyl-α-hydroxyacetophenone
Eugenol
Evodone
Guaiacol, *Allyl Ether*
2-Hydroxyacetophenone, *Et ether*
3-Hydroxyacetophenone, *Et ether*
4-Hydroxyacetophenone, *Et ether*
α-Hydroxyacetophenone, *Et ether*
3-Hydroxy-2-butanone, *Phenyl ether*
o-Hydroxybutyrophenone
m-Hydroxybutyrophenone
p-Hydroxybutyrophenone
2-Hydroxy-4,5-dimethylbenzaldehyde, *Me ether*
2-Hydroxy-4,6-dimethylbenzaldehyde, *Me ether*
4-Hydroxy-2,6-dimethylbenzaldehyde, *Me ether*
4-Hydroxy-3,5-dimethylbenzaldehyde, *Me ether*
2-Hydroxy-4-methylacetophenone, *Me ether*
2-Hydroxy-5-methylacetophenone, *Me ether*
4-Hydroxy-2-methylacetophenone, *Me ether*
p-Hydroxyethyl-benzaldehyde, *Et ether*
1-Hydroxymethyl-2-phenyloxiran, *Me ether*

$C_{10}H_{12}O_2$ (*continued*)

2-Hydroxy-3-methylpropiophenone
2-Hydroxy-4-methylpropiophenone
2-Hydroxy-5-methylpropiophenone
2-Hydroxy-6-methylpropiophenone
4-Hydroxy-2-methylpropiophenone
4-Hydroxy-3-methylpropiophenone
1-Hydroxy-3-phenylacetone, *Me ether*
p-Hydroxyphenylacetone, *Me ether*
1-Hydroxy-1-phenyl-2-propanone, *Me ether*
β-Hydroxypropiophenone, *Me ether*
γ-Hydroxypropiophenone, *Me ether*
2-Hydroxypropiophenone, *Me ether*
3-Hydroxypropiophenone, *Me ether*
4-Hydroxypropiophenone, *Me ether*
3-(2-Hydroxy-2-propyl)benzoic Acid
4-Hydroxy-*o*-tolualdehyde, *Et ether*
4-Hydroxy-*m*-tolualdehyde, *Et ether*
6-Hydroxy-*m*-tolualdehyde, *Et ether*
2-Hydroxy-3,4,6-trimethylbenzaldehyde
2-Hydroxy-3,5,6-trimethylbenzaldehyde
2-Hydroxy-4,5,6-trimethylbenzaldehyde
Isoegomaketone†
o-Isopropylbenzoic Acid
m-Isopropylbenzoic Acid
p-Isopropylbenzoic Acid
2-Isopropyl-5-methyl-1,4-benzoquinone
p-Methoxycinnamyl Alcohol†
p-Methoxyphenylacetone
2-Methoxy-4-propenylphenol
2-Methoxy-5-propylphenol
2-Methyl-2-phenylpropionic Acid
Phenylacetic Acid, *Et ester*
2-Phenylbutyric Acid
3-Phenylbutyric Acid
4-Phenylbutyric Acid
4-Phenyl-*m*-dioxane†
2-Phenylpropionic Acid, *Me ester*
3-Phenylpropionic Acid, *Me ester*
Propionic Acid, *Benzyl ester*
o-Propylbenzoic Acid
m-Propylbenzoic Acid
p-Propylbenzoic Acid
Tetrahydronaphthalene, α-*Hydroperoxide*
2,3,5,6-Tetramethyl-1,4-benzoquinone
α-Thujaplicin
β-Thujaplicin
γ-Thujaplicin
Thujic Acid
o-Toluic Acid, *Et ester*
m-Toluic Acid, *Et ester*
p-Toluic Acid, *Et ester*
m-Tolylacetic Acid, *Me ester*
2-*p*-Tolylpropionic Acid
3-*o*-Tolylpropionic Acid
3-*m*-Tolylpropionic Acid
3-*p*-Tolylpropionic Acid
2,3,4-Trimethylbenzoic Acid
2,3,5-Trimethylbenzoic Acid
2,3,6-Trimethylbenzoic Acid
2,4,5-Trimethylbenzoic Acid
2,4,6-Trimethylbenzoic Acid
3,4,5-Trimethylbenzoic Acid

$C_{10}H_{12}O_2S$

3-Mercapto-3-phenylpropionic Acid, *Me ester*
Phenylmercaptoacetic Acid, *Et ester*

$C_{10}H_{12}O_3$

Acetoveratrone
Acetylquinol, 2,5-*Di-Me ether*
Acetylquinol, 5-*Et ether*
Anisic Acid, *Et ester*
1,2-*O*-Benzylidenglycerol
1,3-*O*-Benzylideneglycerol
Coniferyl Alcohol
α,4-Dihydroxyacetophenone, 4-*Et ether*
2,3-Dihydroxyacetophenone, *Di-Me ether*
2,4-Dihydroxyacetophenone, *Di-Me ether*
2,6-Dihydroxyacetophenone, *Di-Me ether*
3,5-Dihydroxyacetophenone, *Di-Me ether*
3,4-Dihydroxybenzaldehyde, 3-*Propyl ether*
2,4-Dihydroxybutyrophenone
2,6-Dihydroxybutyrophenone
2,4-Dihydroxy-3,5-dimethylacetophenone
2,4-Dihydroxy-3-methylacetophenone, 4-*Me ether*
2,4-Dihydroxy-6-methylacetophenone, 4-*Me ether*
2,4-Dihydroxy-6-methylacetophenone, 2-*Me ether*
4,5-Dihydroxy-2-methylacetophenone, 4-*Me ether*
2,4-Dihydroxypropiophenone, 4-*Me ether*
2,5-Dihydroxypropiophenone, 5-*Me ether*
4,5-Dihydroxy-*o*-tolualdehyde, 4-*Et ether*
4,6-Dihydroxy-*o*-tolualdehyde, *Di-Me ether*
4,6-Dihydroxy-*m*-tolualdehyde, *Di-Me ether*
2,6-Dihydroxy-*p*-tolualdehyde, *Di-Me ether*
3,4-Dimethoxyphenylacetaldehyde
3,4-Dimethoxy-*o*-toluic Aldehyde
4,5-Dimethoxy-*o*-toluic Aldehyde
Ethoxyacetic Acid, *Phenyl ester*
o-Ethoxybenzoic Acid, *Me ester*
3-α-Furylacrylic Acid, *Propyl ester*
2-Hydroxy-4,5-dimethylbenzoic Acid, *Me ether*
2-Hydroxy-4,5-dimethylbenzoic Acid, *Me ester*
2-Hydroxy-4,6-dimethylbenzoic Acid, *Me ether*
3-Hydroxy-4,5-dimethylbenzoic Acid, *Me ester*
3-Hydroxy-4,5-dimethylbenzoic Acid, *Me ether*
4-Hydroxy-2,5-dimethylbenzoic Acid, *Me ether*
4-Hydroxy-3,5-dimethylbenzoic Acid, *Me ester*
2-Hydroxy-4-isopropylbenzoic Acid
3-Hydroxy-4-isopropylbenzoic Acid
α-Hydroxyisopropylbenzoic Acid
3-Hydroxy-4-methoxybenzaldehyde, *Et ether*
2-Hydroxy-4-methoxy-3,6-dimethylbenzaldehyde
4-Hydroxy-2-methoxy-3,6-dimethylbenzaldehyde
3-Hydroxy-4-methoxy-*o*-tolualdehyde, *Me ether*
4-Hydroxy-6-methoxy-*o*-toluic Aldehyde, *Me ether*
o-Hydroxymethyl-benzoic Acid, *Et ether*
p-Hydroxymethyl-benzoic Acid, *Et ester*
2-Hydroxy-2-methyl-3-phenylpropionic Acid
3-Hydroxy-2-methyl-3-phenylpropionic Acid
2-Hydroxyphenylacetic Acid, *Et ether*
3-Hydroxyphenylacetic Acid, *Et ether*
4-Hydroxyphenylacetic Acid, *Et ether*
4-Hydroxyphenylacetic Acid, *Et ester*
2-Hydroxy-2-phenylbutyric Acid
2-Hydroxy-4-phenylbutyric Acid

$C_{10}H_{12}O_3$ (*continued*)
3-Hydroxy-2-phenylbutyric Acid
3-Hydroxy-3-phenylbutyric Acid
3-Hydroxy-4-phenylbutyric Acid
4-Hydroxy-2-phenylbutyric Acid
4-Hydroxy-4-phenylbutyric Acid
2-*p*-Hydroxyphenylbutyric Acid
4-*o*-Hydroxyphenylbutyric Acid
4-*p*-Hydroxyphenylbutyric Acid
2-*p*-Hydroxyphenylpropionic Acid, *Me ether*
2-Hydroxy-3-phenylpropionic Acid, *Me ester*
3-*p*-Hydroxyphenylpropionic Acid, *Me ester*
3-Hydroxy-2-phenylpropionic Acid, *Me ester*
3-Hydroxy-2-phenylpropionic Acid, *Me ether*
3-Hydroxy-3-phenylpropionic Acid, *Me ester*
3-Hydroxy-3-phenylpropionic Acid, *Me ether*
3-Hydroxy-*o*-toluic Acid, *Me ester*, *Me ether*
3-Hydroxy-*o*-toluic Acid, *Et ester*
3-Hydroxy-*o*-toluic Acid, *Et ether*
4-Hydroxy-*o*-toluic Acid, *Et ester*
4-Hydroxy-*o*-toluic Acid, *Me ether*, *Me ester*
4-Hydroxy-*o*-toluic Acid, *Et ether*
5-Hydroxy-*o*-toluic Acid, *Me ester*, *Me ether*
5-Hydroxy-*o*-toluic Acid, *Et ester*
6-Hydroxy-*o*-toluic Acid, *Me ester*, *Me ether*
6-Hydroxy-*o*-toluic Acid, *Et ether*
2-Hydroxy-*m*-toluic Acid, *Et ester*
2-Hydroxy-*m*-toluic Acid, *Me ester*, *Me ester*
4-Hydroxy-*m*-toluic Acid, *Et ether*
4-Hydroxy-*m*-toluic Acid, *Me ether*, *Me ester*
4-Hydroxy-*m*-toluic Acid, *Et ester*
5-Hydroxy-*m*-toluic Acid, *Me ether*, *Me ester*
6-Hydroxy-*m*-toluic Acid, *Et ester*
6-Hydroxy-*m*-toluic Acid, *Me ether*, *Me ester*
2-Hydroxy-*p*-toluic Acid, *Et ester*
2-Hydroxy-*p*-toluic Acid, *Me ether*, *Me ester*
3-Hydroxy-*p*-toluic Acid, *Et ether*
2-Hydroxy-2-*p*-tolylacetic Acid, *Me ester*
2-Hydroxy-3,4,6-trimethylbenzoic Acid
2-Hydroxy-3,5,6-trimethylbenzoic Acid
4-Isobutyrylresorcinol
o-Isopropoxybenzoic Acid
m-Isopropoxybenzoic Acid
p-Isopropoxybenzoic Acid
Lubanol
Mandelic Acid, *Me ether*, *Me ester*
Mandelic Acid, *Et ether*
Mandelic Acid, *Et ester*
o-Methoxybenzaldehyde, *Ethylene acetal*
o-Methoxybenzoic Acid, *Et ester*
m-Methoxybenzoic Acid, *Et ester*
p-Methoxybenzyl acetate
3-*p*-Methoxyphenylacetic Acid, *Me ester*
p-Methoxyphenylpropionic Acid
3-*o*-Methoxyphenylpropionic Acid
3-*m*-Methoxyphenylpropionic Acid
2-Methyl-2-phenoxypropionic Acid
2-Methyl-3-phenoxypropionic Acid
Phenoxyacetic Acid, *Et ester*
2-Phenoxybutyric Acid
4-Phenoxybutyric Acid
3-Phenoxypropionic Acid, *Me ester*
o-Propoxybenzoic Acid
m-Propoxybenzoic Acid
p-Propoxybenzoic Acid
Salicylic Acid, *Propyl ether*
α-Thujaplicinol
Vanillin, *Et ether*

$C_{10}H_{12}O_3S$
α-Vinylbenzenesulphonic Acid, *Et ester*

$C_{10}H_{12}O_4$
3-Acetyl-2-hydroxy-6-methylpyran-4-one, *Et ester*
Aurantiogliocladin
Batatic Acid
Cantharidin
Curvulol†
1,3-Cyclohexadiene-1,4-dicarboxylic Acid, *Di-Me ester*
1,4-Cyclohexadiene-1,4-dicarboxylic Acid, *Di-Me ester*
1,4-Cyclohexadiene-1,4-dicarboxylic Acid, *Mono-Et ester*
2,5-Cyclohexadiene-1,4-dicarboxylic Acid, *Di-Me ester*
2,6-Cyclohexadiene-1,2-dicarboxylic Acid, *Mono-Et ester*
2,3-Dihydroxybenzoic Acid, 3-*Me ether*, *Et ester*
2,3-Dihydroxybenzoic Acid, 3-*Me ether*, 2-*Et ether*
2,4-Dihydroxybenzoic Acid, 4-*Me ether*, *Et ester*
2,4-Dihydroxybenzoic Acid, 4-*Et ether*, *Me ester*
2,5-Dihydroxy-1,4-benzoquinone, *Di-Et ether*
α,2-Dihydroxyphenylacetic Acid, 2-*Et ether*
α,4-Dihydroxyphenylacetic Acid, *Et ester*
2,4-Dihydroxyphenylacetic Acid, *Di-Me ether*
2,5-Dihydroxyphenylacetic Acid, *Et ester*
2,3-Dihydroxy-3-phenylpropionic Acid, *Me ester*
3-(3,4-Dihydroxyphenyl)propionic Acid, 3-*Me ether*
3-(3,4-Dihydroxyphenyl)propionic Acid, 4-*Me ether*
3-(3,5-Dihydroxyphenyl)propionic Acid, 3-*Me ether*
2,4-Dihydroxy-6-propylbenzoic Acid
3,5-Dihydroxy-*o*-toluic Acid, *Et ester*
4,6-Dihydroxy-*o*-toluic Acid, *Et ester*
5,6-Dihydroxy-*m*-toluic Acid, 5-*Me ether*, *Me ester*
2,5-Dihydroxy-*p*-toluic Acid, *Et ester*
2,6-Dihydroxy-*p*-toluic Acid, *Et ester*
2,6-Dihydroxy-*p*-toluic Acid, *Di-Me ether*
3,5-Dihydroxy-*p*-toluic Acid, *Di-Me ether*
2,3-Dimethoxybenzoic Acid, *Me ester*
2,4-Dimethoxybenzoic Acid, *Me ester*
2,6-Dimethoxybenzoic Acid, *Me ester*
3,4-Dimethoxybenzoic Acid, *Me ester*
3,5-Dimethoxybenzoic Acid, *Me ester*
2,3-Dimethoxyphenylacetic Acid
3,4-Dimethoxyphenylacetic Acid
4,5-Dimethoxy-*o*-toluic Acid
4,6-Dimethyl-2-oxo-2*H*-pyran-5-carboxylic Acid, *Et ester*
α-Furoylacetic Acid, *Propyl ester*
Glycerol, 1-*Benzoyl*
Guaiacol, *Mono-guaiacol carbonate*, *Et ester*
3-Hydroxy-4-methoxybenzoic Acid, *Et ester*

$C_{10}H_{12}O_4$ (*continued*)
3-Hydroxy-4-methoxybenzoic Acid, *Et ether*
2-Hydroxy-4-methoxy-3,6-dimethylbenzoic Acid
4-Hydroxy-2-methoxy-3,6-dimethylbenzoic Acid
2-Hydroxy-2-*p*-methoxyphenylpropionic Acid
4-Hydroxy-6-methoxy-*o*-toluic Acid, *Me ester*
6-Hydroxy-4-methoxy-*o*-toluic Acid, *Me ether*
2-Hydroxy-4-methoxy-*m*-toluic Acid, *Me ester*
2-Hydroxy-6-methoxy-*p*-toluic Acid, *Me ester*
Hygrophyllinecic Acid, *Di-lactone*†
Isocantharidin
Maleic Acid, *Diallyl ester*
3,4-Methylenedioxybenzaldehyde, *Di-Me acetal*
2,4,6-Octatrienedioic Acid, *Di-Me ester*
Sparassol
α,2,4-Trihydroxyacetophenone, α-4-*Di-Me ether*
α,2,4-Trihydroxyacetophenone, 2,4-*Di-Me ether*
α,2,4-Trihydroxacetophenone, α-*Et ether*
α,3,4-Trihydroxyacetophenone, 3,4-*Di-Me ether*
2,3,4-Trihydroxyacetophenone, 2,3-*Di-Me ether*
2,3,4-Trihydroxyacetophenone, 2,4-*Di-Me ether*
2,3,4-Trihydroxyacetophenone, 3-(or 4)-*Et ether*
2,3,4-Trihydroxyacetophenone, 3,4-*Di-Me ether*
2,4,5-Trihydroxyacetophenone, 4,5-*Di-Me ether*
2,4,6-Trihydroxyacetophenone, 2,4-*Di-Me ether*
2,4,6-Trihydroxyacetophenone, 2,6-*Di-Me ether*
2,3,4-Trihydroxybenzaldehyde, 2,3,4-*Tri-Me ether*
2,3,5-Trihydroxybenzaldehyde, 2,3,5-*Tri-Me ether*
2,4,5-Trihydroxybenzaldehyde, 4-*Me*-5-*Et ether*
2,4,5-Trihydroxybenzaldehyde, 5-*Me*-4-*Et ether*
2,4,6-Trihydroxybenzaldehyde, 2,4,6-*Tri-Me ether*
3,4,5-Trihydroxybenzaldehyde, *Tri-Me ether*
2,3,4-Trihydroxybutyrophenone
2,4,6-Trihydroxybutyrophenone
3,4,5-Trihydroxybutyrophenone
2,4,6-Trihydroxyisobutyrophenone
2,3,6-Trihydroxy-*p*-toluic aldehyde, 2,6-*Di-Me ether*
2,4,5-Trimethoxybenzaldehyde
Vanillic Acid, *Et ester*
Vanillic Acid, *Et ether*

$C_{10}H_{12}O_4S$
Phenylsulphonylacetic Acid, *Et ester*
Thiophene-2,4-dicarboxylic Acid, *Di-Et ester*
Thiophene-2,5-dicarboxylic Acid, *Di-Et ester*
Thiophene-3,4-dicarboxylic Acid, *Di-Et ester*

$C_{10}H_{12}O_5$
Antibiotic U-13,933†
2,3-Dihydroxymandelic Acid, 2,3-*Di-Me ether*
2,5-Dihydroxymandelic Acid, 2,5-*Di-Me ether*
3,4-Dihydroxymandelic Acid, *Et ester*
3,4-Dihydroxymandelic Acid, 3,4-*Di-Me ether*
2,5-Dimethylfuran-3,4-dicarboxylic Acid, *Di-Me ester*
2,5-Dimethylfuran-3,4-dicarboxylic Acid, *Et ester*
Furan-2,3-dicarboxylic Acid, *Di-Et ester*
Furan-2,4-dicarboxylic Acid, *Di-Et ester*
Furan-3,4-dicarboxylic Acid, *Di-Et ester*
Gallic Acid, *Propyl ester*
Gallic Acid, *Isopropyl ester*
Gallic Acid, 3,4-*Di-Me ether*, *Me ester*
Gallic Acid, *Tri-Me ether*
4-Hydroxy-3,5-dimethoxybenzoic Acid, *Me ester*
4-Hydroxy-3,5-dimethoxyphenylacetic Acid
Iridic Acid
Methronic Acid, *Di-Me ester*
Methronic Acid, *Mono-Et ester*
Spinulosin, *Di-Me ether*
α,3,4,5-Tetrahydroxyacetophenone, 3,5-*Di-Me ether*
2,3,4,5-Tetrahydroxyacetophenone, 3,4-*Di-Me ether*
2,3,4,6-Tetrahydroxyacetophenone, 3,4-*Di-Me ether*
2,3,4-Trihydroxybenzoic Acid, *Me ester*, *Di-Me ether*
2,3,4-Trihydroxybenzoic Acid, 2,3,4-*Tri-Me ether*
2,3,5-Trihydroxybenzoic Acid, 2,3,5-*Tri-Me ether*
2,3,6-Trihydroxybenzoic Acid, 2,3,6-*Tri-Me ether*
2,4,6-Trihydroxybenzoic Acid, 2,6-*Di-Me ether*, *Me ester*
2,4,6-Trihydroxybenzoic Acid, 4,6-*Di-Me ether*, *Me ester*
2,4,6-Trihydroxybenzoic Acid, 2,4,6-*Tri-Me ether*
2,4,6-Trihydroxy-*m*-toluic Acid, 4-*Me ether*, *Me ester*
2,4,5-Trimethoxybenzoic Acid

$C_{10}H_{12}O_6$
Carlosic Acid
2,5-Dioxocyclohexane-1,4-dicarboxylic Acid, *Di-Me ester*
2,5-Dioxocyclohexane-1,4-dicarboxylic Acid, *Et ester*
Phorbic Acid, *Di-lactone*, *Et ester*
2,3,4,5-Tetrahydroxybenzoic Acid, 3,4,5-*Tri-Me ether*

$C_{10}H_{12}O_7$
Azaleintin

$C_{10}H_{13}AsO_2$
Benzenearsonous Acid, *Di-Et ester*

$C_{10}H_{13}Br$
2-Bromo-4-isopropyltoluene
3-Bromo-4-isopropyltoluene
p-(2-Bromo-2-propyl)toluene

$C_{10}H_{13}BrN_2O_3$
Noctal

$C_{10}H_{13}BrO$
4-Bromo-5-isopropyl-*o*-cresol
6-Bromo-2,4-xylenol, *Et ether*

$C_{10}H_{13}Cl$
p-(1-Butyl)chlorobenzene
p-(*tert*-Butyl)chlorobenzene
(1-Chloro-1-butyl)benzene
(3-Chloro-1-butyl)benzene
(4-Chloro-1-butyl)benzene
(4-Chloro-2-butyl)benzene
2-Chloro-4-isopropyltoluene
3-Chloro-4-isopropyltoluene
3-Chloro-5-isopropyltoluene

$C_{10}H_{13}ClN_2O_3S$
Chlorpropamide

$C_{10}H_{13}ClO$
4-Chloro-6-isopropyl-*m*-cresol
Shonanic Acid, *Chloride*
Teresantalic Acid, *Chloride*
Tricyclenic Acid, *Chloride*

$C_{10}H_{13}ClO_3$
1-Methyl-4-cyclohexene-1,2-dicarboxylic Acid, *Chloride*

$C_{10}H_{13}ClO_3S$
5-Hydroxy-2,4-dimethylbenzenesulphonic Acid, *Et ether*, *Chloride*

$C_{10}H_{13}Cl_2I$
2-Iodo-4-isopropyltoluene, *Dichloride*
3-Iodo-4-isopropyltoluene, *Dichloride*

$C_{10}H_{13}Cl_2N$
NN-Di-(2-chloroethyl)aniline †

$C_{10}H_{13}HgNO_2$
Mercuri-*p*-dimethylaminophenylacetate
Mercuri-*p*-ethylaminophenyl acetate

$C_{10}H_{13}I$
2-Iodo-4-isopropyltoluene
3-Iodo-4-isopropyltoluene

$C_{10}H_{13}IO$
5-Hydroxy-2-iodo-4-isopropyltoluene

$C_{10}H_{13}N$
Actinidine †
N-Allyl-*N*-methylaniline
1-Aminoindane, N-*Me* †
1-Amino-2-methylindane
2-Amino-1-methylindane
2-Amino-2-methylindane
Benzylamine, N-*Allyl*
4-Isopropenyl-1-cyclohexene-1-carboxylic Acid, *Nitrile*
Myrtenic Acid, *Nitrile*
1-Phenylpyrrolidine
2-Phenylpyrrolidine
3-Phenylpyrrolidine
1,2,3,4-Tetrahydro-1-methylisoquinoline
1,2,3,4-Tetrahydro-2-methylisoquinoline
1,2,3,4-Tetrahydro-3-methylisoquinoline
1,2,3,4-Tetrahydro-6-methylisoquinoline
1,2,3,4-Tetrahydro-7-methylisoquinoline
1,2,3,4-Tetrahydro-1-methylquinoline
1,2,3,4-Tetrahydro-2-methylquinoline
1,2,3,4-Tetrahydro-3-methylquinoline
1,2,3,4-Tetrahydro-4-methylquinoline
1,2,3,4-Tetrahydro-5-methylquinoline
1,2,3,4-Tetrahydro-6-methylquinoline
1,2,3,4-Tetrahydro-7-methylquinoline
1,2,3,4-Tetrahydro-8-methylquinoline
5,6,7,8-Tetrahydro-2-methylquinoline
5,6,7,8-Tetrahydro-3-methylquinoline
5,6,7,8-Tetrahydro-4-methylquinoline
5,6,7,8-Tetrahydro-8-methylquinoline
1,2,3,4-Tetrahydro-1-naphthylamine
5,6,7,8-Tetrahydro-1-naphthylamine
1,2,3,4-Tetrahydro-2-naphthylamine
5,6,7,8-Tetrahydro-2-naphthylamine
Tricyclenic Acid, *Nitrile*

$C_{10}H_{13}NO$
N-Allyl-*N*-methylaniline, N-*Oxide*
4-Aminobutyrophenone
2-Benzylpropionic Acid, *Amide*
Butyric Acid, *Anilide*
2,4-Dimethylphenylacetic Acid, *Amide*
2,5-Dimethylphenylacetic Acid, *Amide*
3,5-Dimethylphenylacetic Acid, *Amide*
N-Ethylaniline, N-*Ac*
o-Ethylaniline, N-*Ac*
m-Ethylaniline, N-*Ac*
p-Ethylaniline, N-*Ac*
Isobutyric Acid, *Anilide*
p-Isopropylbenzoic Acid, *Amide*
β-Methylaminopropiophenone
γ-Methylaminopropiophenone
2-Methyl-2-phenylpropionic Acid, *Amide*
N-Methyl-*o*-toluidine, N-*Ac*
N-Methyl-*m*-toluidine, N-*Ac*
N-Methyl-*p*-toluidine, N-*Ac*
2-Phenylbutyric Acid, *Amide*
3-Phenylbutyric Acid, *Amide*
4-Phenylbutyric Acid, *Amide*
o-Propylbenzoic Acid, *Amide*
Tecostidine †
1,2,3,4-Tetrahydro-8-hydroxy-1-methylquinoline
1,2,3,4-Tetrahydro-8-hydroxy-2-methylquinoline
1,2,3,4-Tetrahydro-8-hydroxyquinoline, *Me ether*
5,6,7,8-Tetrahydro-4-hydroxyquinoline, *Me ether*
1,2,3,4-Tetrahydro-6-methoxyquinoline
o-Toluamide, N-*Di-Me*
m-Toluamide, N-*Di-Me*
p-Toluamide, N-*Di-Me*
p-Toluamide, N-*Et*
2-*p*-Tolylpropionic Acid, *Amide*
3-*p*-Tolylpropionic Acid, *Amide*
2,4,5-Trimethylbenzoic Acid, *Amide*
2,4,6-Trimethylbenzoic Acid, *Amide*

$C_{10}H_{13}NOS$
Benzylthiourethane
Phenylthiocarbamic Acid, O-*Propyl ester*
Phenylthiocarbamic Acid, *Isobutyl ester*

$C_{10}H_{13}NO_2$
N-Acetyltyramine
2-Amino-3,5-dimethylbenzoic Acid, *Me ester*
4-Amino-3,5-dimethylbenzoic Acid, *Me ester*
α-Amino-4-hydroxyacetophenone, N-*Di-Me*
4-Amino-2-hydroxyacetophenone, N-*Di-Me*

$C_{10}H_{13}NO_2$ (*continued*)

5-Amino-2-hydroxyacetophenone, *Et ether*
3-Amino-4-isopropylbenzoic Acid
2-Amino-2-methyl-3-phenylpropionic Acid
3-Amino-2-methyl-3-phenylpropionic Acid
α-Aminophenylacetic Acid, *Et ester*
2-*p*-Aminophenylbutyric Acid
3-*p*-Aminophenylbutyric Acid
4-*o*-Aminophenylbutyric Acid
4-*p*-Aminophenylbutyric Acid
2-Amino-1-phenylbutyric Acid
2-Amino-3-phenylbutyric Acid
2-Amino-4-phenylbutyric Acid
3-Amino-3-phenylbutyric Acid
4-Amino-4-phenylbutyric Acid
2-Amino-*m*-toluic Acid, N-*Et*
2-Anilinobutyric Acid
3-Anilinobutyric Acid
2-Anilino-2-methylpropionic Acid
3-Anilino-2-methylpropionic Acid
3-Anilinopropionic Acid, *Me ester*
Benzylurethane
Bicyclo[2,2,2]octane-1,4-dicarboxylic Acid, *Mono-nitrile*
1-Butyl-2-nitrobenzene
1-Butyl-3-nitrobenzene
1-Butyl-4-nitrobenzene
1-(2-Butyl)-2-nitrobenzene
1-(2-Butyl)-4-nitrobenzene
1-*tert*-Butyl-2-nitrobenzene
1-*tert*-Butyl-3-nitrobenzene
1-*tert*-Butyl-4-nitrobenzene
ω-Camphanic Acid, *Nitrile*
5,6-Dihydroxyindoline, 5,6-*Di-Me ether*†
o-Dimethylaminobenzoic Acid, *Me ester*
m-Dimethylaminobenzoic Acid, *Me ester*
p-Dimethylaminobenzoic Acid, *Me ester*
p-Dimethylaminophenylacetic Acid†
α,α-Dimethyl-β-nitroethylbenzene
3,5-Dimethylpyridine-2-carboxylic Acid, *Et ester*
4,6-Dimethylpyridine-2-carboxylic Acid, *Et ester*
2,4-Dimethylpyridine-3-carboxylic Acid, *Et ester*
2,5-Dimethylpyridine-3-carboxylic Acid, *Et ester*
2,6-Dimethylpyridine-3-carboxylic Acid, *Et ester*
o-Ethylaminobenzoic Acid, *Me ester*
p-Ethylaminobenzoic Acid, *Me ester*
p-Ethylaminophenol, *Acetate*
Fusaric Acid
α-Hydroxyisopropylbenzoic Acid, *Amide*
p-Hydroxymethyl-benzoic Acid, *Et ether*, *Amide*
4-Hydroxyphenylacetic Acid, *Et ether*, *Amide*
2-Hydroxy-2-phenylbutyric Acid, *Amide*
2-Hydroxy-4-phenylbutyric Acid, *Amide*
4-Hydroxy-4-phenylbutyric Acid, *Amide*
4-Hydroxy-*m*-toluic Acid, *Et ether*, *Amide*
6-Hydroxy-*m*-toluic Acid, *Amide*, *Et ether*
1-Isopropyl-3-methyl-4-nitrobenzene
1-Isopropyl-4-methyl-2-nitrobenzene
1-Isopropyl-4-methyl-3-nitrobenzene
5-Isopropyl-4-nitroso-*o*-cresol
6-Isopropyl-4-nitroso-*m*-cresol
Lactic Acid, o-*Toluidide*
Lactic Acid, p-*Toluidide*
Lactic Acid, N-*Methylanilide*
Mandelamide, N-*Di-Me*
Mandelamide, N-*Et*
Mandelamide, *Et ether*
3-*o*-Methoxyphenylpropionic Acid, *Amide*
3-*p*-Methoxyphenylpropionic Acid, *Amide*
o-Methylaminobenzoic Acid, *Et ester*
m-Methylaminobenzoic Acid, *Et ester*
2-Methylamino-3-phenylpropionic Acid
3-Methylamino-3-phenylpropionic Acid
1-Methyl-2-cyclopentene-1,2-dicarboxylic Acid, 2-*Et ester*-3-*nitrile*
2-Methyl-1-nitro-1-phenylpropane
2-Methyl-2-phenoxypropionic Acid, *Amide*
N-Methylphenylcarbamic Acid, *Et ester*
1-Nitro-1-phenylbutane
1-Nitro-3-phenylbutane
1-Nitro-4-phenylbutane
o-Phenetidine, N-*Ac*
m-Phenetidine, N-*Ac*
p-Phenetidine, N-*Ac*
2-Phenoxybutyric Acid, *Amide*
4-Phenoxybutyric Acid, *Amide*
β-Phenylalanine, *Me ester*
Phenylcarbamic Acid, *Propyl ester*
Phenylcarbamic Acid, *Isopropyl ester*
N-Phenylglycine, *Et ester*
N-Phenylglycine, N-*Et*
N-Phenylsarcosine, *Me ester*
Propaesin
p-Propoxybenzoic Acid, *Amide*
1,2,4,5-Tetramethyl-3-nitrobenzene
o-Tolylcarbamic Acid, *Et ester*
m-Tolylcarbamic Acid, *Et ester*
p-Tolylcarbamic Acid, *Et ester*
2,4,5-Trimethyl-α-nitrotoluene

$C_{10}H_{13}NO_2S$

Cysteine, *Benzyl ether*
p-Nitrobenzenethiol, S-*Butyl ether*
p-Vinylbenzenesulphonic Acid, *Dimethylamide*

$C_{10}H_{13}NO_3$

4-Aminophenoxyacetic Acid, *Et ester*
4-Butyl-2-nitrophenol
Damascenine
1,3-Dihydroxy-4-nitrosobenzene, *Di-Et ether*
α,2-Dihydroxyphenylacetic Acid, 2-*Et ether*, *Amide*
2,3-Dimethoxyphenylacetic Acid, *Amide*
3,4-Dimethoxyphenylacetic Acid, *Amide*
2,5-Dimethyl-4-nitrophenol, *Et ether*
Ethyl 2-hydroxyphenylcarbamate, *Me ether*
Ethyl 4-hydroxyphenylcarbamate, *Me ether*
3-Hydroxy-4-methoxybenzoic Acid, *Amide*
2-Hydroxyphenylglycine, *Et ether*
4-Hydroxyphenylglycine, *Et ester*
3-Isopropyl-6-methyl-2-nitrophenol
5-Isopropyl-2-methyl-3-nitrophenol
5-Isopropyl-2-methyl-4-nitrophenol
6-Isopropyl-2-nitro-*m*-cresol
6-Isopropyl-4-nitro-*m*-cresol
Lactic Acid, p-*Anisidide*
6-Nitro-*o*-cresol, *Propyl ether*
o-Nitrophenol, *Isobutyl ether*

$C_{10}H_{13}NO_3$ (*continued*)
p-Nitrophenol, *Isobutyl ether*
2-Nitro-1-phenylethanol, *Et ether*
Surinamine
2,3,5,6-Tetramethyl-4-nitrophenol
Tyrosine, *Me ester*
Tyrosine, *Me ether*
o-Tyrosine, *Me ester*
o-Tyrosine, *Me ether*

$C_{10}H_{13}NO_4$
2-Amino-3,4-dihydroxybenzoic Acid, 3-*Me*-4-*ether*
2-Amino-4,5-dimethoxybenzoic Acid, *Me ester*
5-Amino-3,4-dimethoxybenzoic Acid, *Me ester*
β-4-Carboxymethyl-2-methyl-3-pyrrylpropionic Acid
1,3-Dihydroxy-2-nitrobenzene, *Di-Et ether*
1,3-Dihydroxy-4-nitrobenzene, *Di-Et ether*
1,2-Dihydroxy-4-nitro-5-propylbenzene, 3-*Me ether*
β-3,4-Dihydroxyphenyl alanine, 3-*Me ether*
4,6-Dihydroxypyridine-2-carboxylic Acid, *Di-Et ether*
4,5-Dimethylpyrrole-2,3-dicarboxylic Acid, 3-*Et ester*
3,5-Dimethylpyrrole-2,4-dicarboxylic Acid, *Di-Me ester*
3,5-Dimethylpyrrole-2,4-dicarboxylic Acid, 2-*Et ester*
3,5-Dimethylpyrrole-2,4-dicarboxylic Acid, 4-*Et ester*
Gallic Acid, *Tri-Me ether*, *Amide*
Iridic Acid, *Amide*
4-Nitrocatechol, *Di-Et ether*
2-Nitroquinol, *Di-Et ether*
Pyrrole-2,5-dicarboxylic Acid, *Di-Et ester*
Pyrrole-3,4-dicarboxylic Acid, *Di-Et ester*
Presinol †
2,3,4-Trihydroxybenzoic Acid, 2,3,4-*Tri-Me ether*, *Amide*
1,3,5-Trihydroxy-2-nitrosobenzene, 1,3-*Di-Et ether*
1,3,5-Trihydroxy-2-nitrosobenzene, 1,5-*Di-Et ether*
2,4,5-Trimethoxybenzoic Acid, *Amide*

$C_{10}H_{13}NO_4S$
p-Sulphobenzoic Acid, 4-*Isopropylamide*

$C_{10}H_{13}NO_5$
2,3-Dihydroxy-5-nitrobenzyl Alcohol, 3-*Me*-2-*Et ether*

$C_{10}H_{13}NO_6$
3-Deazauridine †

$C_{10}H_{13}NS$
1,4-Thiazan, N-*Phenyl*

$C_{10}H_{13}NS_2$
Phenyldithiocarbamic Acid, *Propyl ester*

$C_{10}H_{13}N_2O_{11}P$
Orotidine-5′-phosphate

$C_{10}H_{13}N_3$
2-Methyl-2-β-phenylhydrazinopropionic Acid, *Nitrile*
2-β-Phenylhydrazinobutyric Acid, *Nitrile*
1,2,3,4-Tetrahydro-1-phenylisoquinoline, N-*Me*

$C_{10}H_{13}N_3O_2$
Guanoxan †
2-Phenyliminodiacetic Acid, *Diamide*
N-Phenyliminodiacetic Acid, *Diamide*

$C_{10}H_{13}N_5$
6-(3-Methylbut-2-enylamino)purine †
6-(3-Methylbut-3-enylamino)purine †
Triacanthine

$C_{10}H_{13}N_5O$
2′,3′,5′-Trideoxyadenosine †
Zeatin †

$C_{10}H_{13}N_5O_2$
2′,3′-Dideoxyadenosine †
2′,5′-Dideoxyadenosine †

$C_{10}H_{13}N_5O_3$
Cordycepin (2′-Deoxyadenosine)
Cyclodeoxyadenosine
2′-Deoxyadenosine †
3′-Deoxyadenosine †
9-(3-Deoxy-3-*C*-hydroxymethyl-β-D-*erythro*-furanosyl)adenine †
3α-(2′-Deoxy-D-ribofuranosyl)adenine †
3β-(2′-Deoxy-D-ribofuranosyl)adenine †

$C_{10}H_{13}N_5O_3S$
2′-Deoxythioguanosine
3′-Thio-adenosine †

$C_{10}H_{13}N_5O_4$
Adenosine
9β-D-Arabinofuranosyladenine †
Cycloadenosine
Formycin †
Guanine deoxyriboside
3-Isoadenosine †
9(β-D-Lyxofuranosyl)adenine †
3-β-D-Ribofuranosyladenine
7-D-Ribofuranosyladenine †
9-(β-D-Xylofuranosyl)-adenine †

$C_{10}H_{13}N_5O_5$
9β-D-Arabinofuranosylguanine †
Crotonoside
Guanosine

$C_{10}H_{13}N_5O_7$
8-Azaxanthine-7-glucopyranoside
8-Azaxanthine-9-glucopyranoside

$C_{10}H_{13}N_5O_{10}P_2$
Adenosine-2′,5′-diphosphate
Adenosine-3′,5′-diphosphate

$C_{10}H_{14}$
Butylbenzene
2-Butylbenzene
tert-Butylbenzene
Cosmene
Cyclodeca-1,2,3-triene †
2,4-Dehydroadamantane †
o-Diethylbenzene
m-Diethylbenzene
p-Diethylbenzene
Hexahydronaphthalene
1,2,5,6,9,10-Hexahydronaphthalene †
Isobutylbenzene

$C_{10}H_{14}$ (*continued*)
5-Isopropylidene-2-methyl-1,3-cyclohexadiene
o-Isopropyltoluene
m-Isopropyltoluene
p-Isopropyltoluene
o-Propyltoluene
m-Propyltoluene
p-Propyltoluene
1,2,3,4-Tetramethylbenzene
1,2,3,5-Tetramethylbenzene
1,2,4,5-Tetramethylbenzene
4,7,7-Trimethyltricyclo[2,2,1,$0^{2,6}$]heptane-3-one

$C_{10}H_{14}BrClO_3S$
3-Bromocamphor-8-sulphonic Acid, *Chloride*
3-Bromo-(+)-camphor-10-sulphonic Acid, *Chloride*
10-Bromocamphor-3-sulphonic Acid, *Chloride*

$C_{10}H_{14}BrN$
3-Bromodiethylaniline
4-Bromodiethylaniline

$C_{10}H_{14}Br_2O$
3,3-Dibromo-(+)-camphor
3,6-Dibromo(+)-camphor
3,8-Dibromo-(+)-camphor
3,10-Dibromo-(+)-camphor

$C_{10}H_{14}Br_2O_3S$
3-Bromocamphor-8-sulphonic Acid, *Bromide*
3-Bromo-(+)-camphor-10-sulphonic Acid, *Bromide*

$C_{10}H_{14}Br_2O_4$
1,2-Dibromocyclohexane-1,4-dicarboxylic Acid, *Di-Me ester*
1,4-Dibromocyclohexane-1,4-dicarboxylic Acid, *Di-Me ester*
2,3-Dibromocyclohexane-1,4-dicarboxylic Acid, *Di-Me ester*
2,5-Dibromocyclohexane-1,4-dicarboxylic Acid, *Di-Me ester*

$C_{10}H_{14}CaO_{12}$
D-Lyxuronic Acid, *Ca salt*

$C_{10}H_{14}ClHgN$
Mercuri-*p*-diethylaminophenyl chloride

$C_{10}H_{14}ClN$
m-Chloroaniline, N-*Di-Et*

$C_{10}H_{14}ClNO$
1,2,2-Trimethylcyclopentane-1,3-dicarboxylic Acid, α-*Nitrile*, *Chloride*
1,2,2-Trimethylcyclopentane-1,3-dicarboxylic Acid, β-*Nitrile*, *Chloride*

$C_{10}H_{14}Cl_2O$
3,3-Dichloro-(+)-camphor
3,8-Dichloro-(+)-camphor

$C_{10}H_{14}Cl_2O_2$
Camphenic Acid, *Dichloride*
1,2,2-Trimethylcyclopentane-1,3-dicarboxylic Acid, *Dichloride*

$C_{10}H_{14}IN$
N-Diethyl-*p*-iodoaniline

$C_{10}H_{14}NO_5PS$
Parathion

$C_{10}H_{14}NO_6P$
Diethyl *p*-nitrophenylphosphate†

$C_{10}H_{14}N_2$
Anabasine
Isoneonicotine
Isonicotine
Nicotidine
Nicotimine
Nicotine
γ-Nicotine†
N-Phenylpiperazine
2-Phenylpiperazine
1-4′-Pyridyl-piperidine
2-2′-Pyridyl-piperidine
4-3′-Pyridyl-piperidine
p-Tolamidine, α,β-N-*Di-Me*
p-Tolamidine, β-N-*Et*

$C_{10}H_{14}N_2O$
2-Anilinobutyric Acid, *Amide*
2-Anilino-2-methylpropionic Acid, *Amide*
Diazocamphor
N-Diethyl-4-nitrosoaniline
Fusaric Acid, *Amide*
N-Methyl-4-nitrosoaniline, N-*Propyl*
p-Nitrosoaniline, N-*Butyl*
N-Phenylglycine, N-*Et*, *Amide*

$C_{10}H_{14}N_2OS$
4-Hydroxyphenylthiourea, *Propyl ether*

$C_{10}H_{14}N_2O_2$
2,4-Diaminophenylacetic Acid, *Et ester*
N-Diethyl-2-nitroaniline
N-Diethyl-3-nitroaniline
N-Diethyl-4-nitroaniline
2,4-Dimethyl-5-nitroaniline, N-*Di-Me*
4-Hydroxyphenylurea, *Propyl ether*
Isopilocarpidine
2-Isopropyl-5-methyl-1,4-benzoquinone, 1-, 4-*Dioxime*
2-Methylbutane-1,1,2-tricarboxylic Acid, 1-*Mono-Et ester*, 1,2-*Di-nitrile*
3-Methylbutane-2,2,3-tricarboxylic Acid, 3-*Et ester*, *Di-nitrile*
2-Methyl-2-α-phenylhydrazinopropionic Acid
2-Methyl-2-β-phenylhydrazinopropionic Acid
p-Nitrobenzylamine, N-*Propyl*
2-α-Phenylhydrazinoacetic Acid, *Et ester*
2-β-Phenylhydrazinoacetic Acid, *Et ester*
2-β-Phenylhydrazinobutyric Acid
3-α-Phenylhydrazinobutyric Acid
3-β-Phenylhydrazinobutyric Acid
Pilocarpidine
Proline, *Anhydride*
p-Propoxybenzoic Acid, *Hydrazide*
4,5,6,7-Tetrahydroindazole-1-carboxylic Acid, *Et ester*
4,5,6,7-Tetrahydroindazole-2-carboxylic Acid, *Et ester*
4,5,6,7-Tetrahydroindazole-3-carboxylic Acid, *Et ester*

$C_{10}H_{14}N_2O_2S$
Pyrrothine, *Isobutyryl*
Veratrylthiourea†

$C_{10}H_{14}N_2O_3$
4-Amino-5-isopropyl-6-nitro-*o*-cresol

$C_{10}H_{14}N_2O_3$ (*continued*)
4-Amino-6-isopropyl-2-nitro-*m*-cresol
2-Amino-4-nitrophenol, *Butyl ether*
Aprobarbital

$C_{10}H_{14}N_2O_3S$
α-Dehydrobiotin †

$C_{10}H_{14}N_2O_4$
2-Methylimidazole-4,5-dicarboxylic Acid, *Di-Et ester*
Porphobilinogen †

$C_{10}H_{14}N_2O_5$
Thymidine

$C_{10}H_{14}N_2O_6$
3-Methyluridine
Uridine, N-*Me*

$C_{10}H_{14}N_4O_2$
Theobromine, N-*Propyl*
Theophylline, 7-N-*Propyl*

$C_{10}H_{14}N_4O_4$
Diprophylline

$C_{10}H_{14}N_4O_5$
α-L-Aspartyl-L-histidine
β-L-Aspartyl-L-histidine

$C_{10}H_{14}N_4O_6P$
Adenylic Acid a
Adenylic Acid b

$C_{10}H_{14}N_5O_7P$
Adenylic Acid
Guanosine-5′-phosphate
Inosinic Acid

$C_{10}H_{14}N_5O_8P$
Guanylic Acid

$C_{10}H_{14}N_6O_3$
3′-Amino-3′-deoxyadenosine
9-(2-Amino-2-deoxy-α-D-ribofuranosyl)adenine †
9-(2-Amino-2-deoxy-β-D-ribofuranosyl)adenine †

$C_{10}H_{14}O$
Adamantan-2-one †
Benzyl isopropyl Ether
o-Butylphenol
m-Butylphenol
p-Butylphenol
o-2-Butylphenol
p-2-Butylphenol
m-*tert*-Butylphenol
p-*tert*-Butylphenol
tert-Butyl phenyl Ether
6-Camphenone
Carvacrol
Carvone
Chrysanthenone
Cinerone
o-Cresol, *Propyl ether*
m-Cresol, *Propyl ether*
p-Cresol, *Propyl ether*
Deca-2,4,6,8-tetraenol †
2,3-Dimethylphenol, *Et ether*
2,4-Dimethylphenol, *Et ether*
2,5-Dimethylphenol, *Et ether*
2,6-Dimethylphenol, *Et ether*
3,4-Dimethylphenol, *Et ether*
3,5-Dimethylphenol, *Et ether*
Eucarvone
Filifolone †
3,4,5,6,7,8-Hexahydro-1(2*H*)-naphthalenone
3,4,5,6,7,8-Hexahydro-2(1*H*)-naphthalenone
4,4α,5,6,7,8-Hexahydro-2(3*H*)-naphthalenone
α-Hydroxy-*o*-xylene, *Et ether*
α-Hydroxy-*m*-xylene, *Et ether*
α-Hydroxy-*p*-xylene, *Et ether*
Isobutyl phenyl Ether
4-Isopropenyl-1-cyclohexene-1-aldehyde
6-Isopropenyl-2-methyl-2-cyclohexen-1-one
p-Isopropylbenzyl Alcohol
3-Isopropyl-*o*-cresol
4-Isopropyl-*o*-cresol
6-Isopropyl-*o*-cresol
2-Isopropyl-*m*-cresol
4-Isopropyl-*m*-cresol
5-Isopropyl-*m*-cresol
6-Isopropyl-*m*-cresol
2-Isopropyl-*p*-cresol
3-Isopropyl-*p*-cresol
4-Isopropyl-1,5-cyclohexadiene-1-aldehyde
o-Isopropylphenol, *Me ether*
m-Isopropylphenol, *Me ether*
p-Isopropylphenol, *Me ether*
Isopropyl *m*-tolyl Ether
Menthofuran †
2-Methyl-1-phenyl-1-propanol
2-Methyl-2-phenylpropanol
2-Methyl-1-phenyl-2-propanol
Myrtenal
Perillene
1-Phenyl-1-butanol
2-Phenyl-1-butanol
3-Phenyl-1-butanol
4-Phenyl-1-butanol
2-Phenyl-2-butanol
4-Phenyl-2-butanol
1-Phenylethanol, *Et ether*
3-Phenol-1-propanol, *Me ether*
1-Phenyl-2-propanol, *Me ether*
2(10)-Pinen-3-one
2-Pinen-4-one
Piperitenone †
o-Propylphenol, *Me ether*
m-Propylphenol, *Me ether*
p-Propylphenol, *Me ether*
2,3,4,5-Tetramethylphenol
2,3,4,6-Tetramethylphenol
2,3,5,6-Tetramethylphenol
1-*m*-Tolyl-1-propanol
1-*p*-Tolyl-1-propanol
3-*o*-Tolyl-1-propanol
3-*m*-Tolyl-1-propanol
3-*p*-Tolyl-1-propanol
1-*m*-Tolyl-2-propanol
Tricyclal
α,α,2-Trimethylbenzyl Alcohol
α,α,3-Trimethylbenzyl Alcohol
α,α,4-Trimethylbenzyl Alcohol
2,3,4-Trimethylbenzyl Alcohol
2,3,6-Trimethylbenzyl Alcohol
2,4,5-Trimethylbenzyl Alcohol
2,4,6-Trimethylbenzyl Alcohol
2,6,6-Trimethyl-1,3-cyclohexadien-1-ol

$C_{10}H_{14}O$ (*continued*)
2,3,5-Trimethylphenol, *Me ether*
2,4,5-Trimethylphenol, *Me ether*
2,4,6-Trimethylphenol, *Me ether*
Umbellulone

$C_{10}H_{14}OS$
o-Hydroxybenzenethiol, *Di-Et ether*
p-Hydroxybenzenethiol, S-*Et*, *Et ether*
p-Hydroxybenzenethiol, S-*Propyl*, *Me ether*
p-Hydroxybenzenethiol, S-*Butyl*

$C_{10}H_{14}OS_2$
3-Hydroxy-1,2-propanedithiol★, 3-*Benzyl ether*†

$C_{10}H_{14}O_2$
Betuligenol
Camphorquinone
Catechol, *Di-Et ether*
Catechol, n-*Butyl ether*
Chamic Acid
Chaminic Acid
Cinerolone
Coerulignol
Cyclohexylpropiolic Acid, *Me ester*
2,3-Diacetylnaphthalene†
α,β-Dihydroxyethylbenzene, *Et ether*
1,2-Dihydroxy-3-propylbenzene, 1-*Me ether*
3,4-Dihydroxytoluene, 3-*Me*, 4-*Et ether*
α,2-Dihydroxy-3,5,6-trimethyltoluene
2,5-Dihydroxy-*p*-xylene, *Et ether*
Dolichodial
Elsholtzione
Guaiacol, *Propyl ether*
2-Hydroxyethyl phenyl Ether, *Et ether*
2-(2-Hydroxyphenyl)ethanol, 2-*Et ether*
2-(4-Hydroxyphenyl)ethanol, 4-*Et ether*
2-*o*-Hydroxyphenyl-2-propanol, *Me ether*
2-*m*-Hydroxyphenyl-2-propanol, *Me ether*
2-α-Hydroxypropylphenol, 1-*Me ether*
4-α-Hydroxypropylphenol, 1-*Me ether*
4-β-Hydroxypropylphenol, *Me ether*
4-γ-Hydroxypropylphenol, 1′-*Me ether*
4-γ-Hydroxypropylphenol, 4-*Me ether*
4-Isobutylresorcinol
4-Isopropenyl-1-cyclohexene-1-carboxylic Acid
2-Isopropyl-5-methylquinol
Isoteresentalic Acid
Myrtenic Acid
Neonepetalactone†
Nepetalactone†
1-Phenyl-1,2-butanediol
2-Phenyl-1,2-butanediol
4-Phenyl-1,2-butanediol
1-Phenyl-1,3-butanediol
3-Phenyl-1,3-butanediol
1-Phenyl-1,4-butanediol
Quinol, *Di-Et ether*
Resorcinol, *Di-Et ether*
Resorcinol, *Me-Propyl ether*
Resorcinol, *Monobutyl ether*
Rhododendrol
Rotundifolone†
Shonanic Acid
Terephthalyl Alcohol, *Di-Me ether*
Terephthalyl Alcohol, *Mono-Et ether*
Teresantalic Acid
2,3,5,6-Tetramethylquinol
Tricyclenic Acid
2,6,6-Trimethyl-1,3-cyclohexadiene-1-carboxylic Acid
3,5,6-Trimethylguaiacol
4,5,6-Trimethylguaiacol

$C_{10}H_{14}O_2S$
2-Acetyl-3-hydroxythiophene, 3-tert-*Bu ether*†
2-Thiopheneacetic Acid, *Butyl ester*
Thiophene-2-carboxylic Acid, *Amyl ester*

$C_{10}H_{14}O_2S_2$
Thiolbenzenesulphonic Acid, *Butyl ester*

$C_{10}H_{14}O_3$
Barnol†
1-*O*-Benzyl-L-glycerol†
α,4-Dihydroxy-3-methoxytoluene, 4-*Et ether*
5-Ethyl-2-methylfuran-3-carboxylic Acid, *Et ester*
Exogonic Acid
Filicinic Acid, *Et ether*
Furan-2-carboxylic Acid, *Pentyl ester*
2-Furylacetic Acid, *Butyl ester*
2,4,5,6,7,7*a*-Hexahydro-7α-hydroxy-3,6-dimethylbenzofuran-2-one†
1-*p*-Methoxyphenyl-1,2-propanediol
4-Methyl-6-oxo-1-cyclohexane-1-carboxylic Acid, *Et ester*
2-Methyl-4-oxo-2-cyclohexene-1-carboxylic Acid, *Et ester*
1-Methyl-2-oxo-2-cyclohexene-1-carboxylic Acid, *Et ester*
Myanesin
Norcamphor-3-carboxylic Acid, *Et ester*
Orthobenzoic Acid, *Tri-Me ester*
1-Phenylglycerol, 1-*Me ether*
1-Phenylglycerol, 2-*Me ether*
1-Phenylglycerol, 3-*Me ether*
Phloroglucinol, *Di-Et ether*
Pyrogallol, 1,3-*Di-Et ether*
Ramulosin†
1,1,3,3-Tetramethylcyclohexane-2,4,6-trione
1,2,4-Trihydroxybenzene, 1,2-*Di-Et ether*
1,3,5-Trihydroxy-2,4,6-trimethylbenzene, *Mono-Me ether*
1,2,2-Trimethylcyclopentane-1,3-dicarboxylic Anhydride
2,4,5-Trimethylfuran-3-carboxylic Acid, *Et ester*

$C_{10}H_{14}O_3S$
Toluene-*p*-sulphonic Acid, *Propyl ester*

$C_{10}H_{14}O_4$
Acetylenedicarboxylic Acid, *Di-isopropyl ester*
Adipic Acid, *Divinyl ester*
Antiarol, *Me ether*
Bicyclo[2,2,2]octane-1,4-dicarboxylic Acid
π-Camphanic Acid
ω-Camphanic Acid
2-Carboxy-7-noneno-4-lactone†
Chrysanthemum-dicarboxylic Acid
1-Cyclohexene-1,3-dicarboxylic Acid, *Di-Me ester*
1-Cyclohexene-1,3-dicarboxylic Acid, *Mono-Et ester*
1-Cyclohexene-1,4-dicarboxylic Acid, *Di-Me ester*

$C_{10}H_{14}O_4$ (*continued*)
2-Cyclohexene-1,4-dicarboxylic Acid, *Di-Me ester*
3-Cyclohexene-1,3-dicarboxylic Acid, *Di-Me ester*
4-Cyclohexene-1,2-dicarboxylic Acid, *Di-Me ester*
Fumigatol, *Et ether*†
Genipic Acid, *Me ester*†
Gliorosein
Guaiacol, 1-*Monoglyceryl ether*
Iso-oxocamphoric Acid, *Lactone*
1-Methyl-2-cyclohexene-1,2-dicarboxylic Acid, 1-*Mono-Me ester*
1-Methyl-2-cyclohexene-1,2-dicarboxylic Acid, 2-*Mono-Me ester*
1-Methyl-4-cyclohexene-1,2-dicarboxylic Acid, 1-*Mono-Me ester*
1-Methyl-4-cyclohexene-1,2-dicarboxylic Acid, 2-*Mono-Me ester*
3-Methylenecylopropane-1,2-dicarboxylic Acid, *Di-Et ester*
Muconic Acid, *Di-Et ester*
Succinic Acid, *Diallyl ester*
Tetra-acetylethane†
Tetrahydro-2,6,6-trimethylpyran-2,5-dicarboxylic Acid, *Anhydride*
1,2,3,4-Tetrahydroxybenzene, *Tetra-Me ether*
1,2,3,5-Tetrahydroxybenzene, 1,3-*Di-Me ether*, 2-*Et ether*
1,2,4,5-Tetrahydroxybenzene, 1,4-*Di-Et ether*
α,2,3,5-Tetrahydroxytoluene, 2,3,5-*Tri-Me ether*
α,2,4,6-Tetrahydroxytoluene, 2,4,6-*Tri-Me ether*
α,3,4,5-Tetrahydroxytoluene, 3,4,5-*Tri-Me ether*

$C_{10}H_{14}O_4S$
5-Hydroxy-2,4-dimethylbenzenesulphonic Acid, *Et ether*

$C_{10}H_{14}O_5$
Acetomycin
4-Formylglutaconic Acid, *Di-Et ester*
Guaiacyl Glycerol†
Heptane-1,3,7-tricarboxylic Acid, *Anhydride*
Hygrophyllinecic Acid, *Mono-lactone*†
α-Longinecic Acid
Sceleranecic Acid
Seneciphyllic Acid†

$C_{10}H_{14}O_6$
2,3-Diacetylsuccinic Acid, *Di-Me ester*
2,3-Diacetylsuccinic Acid, *Et ester*
2,2-Dimethylcyclopentane-1,1,3-tricarboxylic Acid
3,4-Dioxoadipic Acid, *Di-Et ester*
1-Methylcyclopropane-1,2,3-tricarboxylic Acid, *Tri-Me ester*
cis-2-Pentene-2,3,5-tricarboxylic Acid, *Mono-Et ester*
Riddellic Acid
Tartaric Acid, *Diallyl ester*

$C_{10}H_{14}O_8$
Ethane-tetracarboxylic Acid, *Tetra-Me ester*
Ethane-tetracarboxylic Acid, sym.-*Di-Et ester*
Hexane-1,1,3,3-tetracarboxylic Acid
Hexane-1,1,4,4-tetracarboxylic Acid
Hexane-1,1,6,6-tetracarboxylic Acid
Hexane-1,2,3,6-tetracarboxylic Acid
Hexane-1,2,4,6-tetracarboxylic Acid
Hexane-1,2,5,6-tetracarboxylic Acid
Hexane-1,3,4,6-tetracarboxylic Acid
Hexane-2,2,3,4-tetracarboxylic Acid
Hexane-2,2,5,5-tetracarboxylic Acid,
Succinperoxide, *Di-Me ester*

$C_{10}H_{14}S$
2-Isopropyl-5-methylbenzenethiol
5-Isopropyl-2-methylbenzenethiol
p-Toluenethiol, S-*Isopropyl*

$C_{10}H_{14}S_2$
Benzene-1,4-dithiol, *Di-Et ether*

$C_{10}H_{15}As$
Phenylarsine, As-*Di-Et*

$C_{10}H_{15}AsO_3$
Benzenearsonic Acid, *Di-Et ester*

$C_{10}H_{15}BrO$
3-Bromocamphor
5-Bromocamphor
8-Bromocamphor
10-Bromocamphor

$C_{10}H_{15}BrO_3S$
Camphor-8-sulphonic Acid, *Bromide*
Camphor-10(or 6-)-sulphonic Acid, *Bromide*

$C_{10}H_{15}BrO_4S$
3-Bromocamphor-8-sulphonic Acid
3-Bromo-(+)-camphor-10-sulphonic Acid
10-Bromocamphor-3-sulphonic Acid

$C_{10}H_{15}ClN_2$
4-Amino-3-chloro-*N*-diethylaniline

$C_{10}H_{15}ClO$
3-Chlorocamphor
4-Chlorocamphor
5-Chlorocamphor
6-Chlorocamphor
8-Chlorocamphor
10-Chlorocamphor
2,6-Decadienoic Acid, *Chloride*

$C_{10}H_{15}ClO_3S$
Camphor-3-sulphonic Acid, *Chloride*
Camphor-8-sulphonic Acid, *Chloride*
Camphor-10(or 6-)sulphonic Acid, *Chloride*

$C_{10}H_{15}N$
α-Aminobutylbenzene
γ-Aminobutylbenzene
δ-Aminobutylbenzene
α-Aminoisobutylbenzene
o-*n*-Butylaniline
m-*n*-Butylaniline
p-*n*-Butylaniline
o-2-Butylaniline
m-2-Butylaniline
p-2-Butylaniline
o-*tert*-Butylaniline
m-*tert*-Butylaniline
p-*tert*-Butylaniline
N-Butylaniline
N-2-Butylaniline
N-*tert*-Butylaniline
N-Diethylaniline
2,3-Dimethylaniline, N-*Et*

$C_{10}H_{15}N$ (*continued*)
2,5-Dimethylaniline, N-*Et*
2,6-Dimethylaniline, N-*Et*
N-Dimethyl-2,3-xylidine
N-Dimethyl-2,4-xylidine
N-Dimethyl-2,5-xylidine
N-Dimethyl-2,6-xylidine
N-Dimethyl-3,4-xylidine
N-Dimethyl-3,5-xylidine
N-Ethyl-*N*-methylbenzylamine
N-Ethyl-*N*-methyl-*p*-toluidine
Geranic Acid, *Nitrile*
N-Isobutylaniline
o-Isobutylaniline
p-Isobutylaniline
p-Isopropylbenzylamine
2-Isopropylidene-5-methylcyclopentane-1-carboxylic Acid, *Nitrile*
3-Isopropylidene-1-methylcyclopentane-1-carboxylic Acid, *Nitrile*
N-Isopropyl-*N*-methylaniline
4-Isopropyl-*N*-methylaniline
5-Isopropyl-*o*-toluidine
6-Isopropyl-*m*-toluidine
N-Isopropyl-*p*-toluidine
2-Methyl-2-(3-methyl-2-cyclopentenyl)propionic Acid, *Nitrile*
Methylpropylaniline
Pervitin
2-Phenyl-1-butylamine
3-Phenyl-1-butylamine
1-Phenyl-1-propylamine, N-*Me*
2-Phenyl-1-propylamine, N-*Me*
3-Phenyl-1-propylamine, N-*Me*
N-Propyl-*o*-toluidine
N-propyl-*p*-toluidine
2-Propyl-*p*-toluidine
2,3,4,5-Tetramethylaniline
2,3,4,6-Tetramethylaniline
2,3,5,6-Tetramethylaniline
2,4,5-Trimethylbenzylamine
2,4,6-Trimethylbenzylamine
3,4,5-Trimethylbenzylamine
2,2,3-Trimethyl-3-cyclopentenylacetic Acid, *Nitrile*
2,3,3-Trimethyl-1-cyclopentenylacetic Acid, *Nitrile*

$C_{10}H_{15}NO$
2-Aminoadamantan-1-ol†
2-Amino-6-isopropyl-*m*-cresol
4-Amino-6-isopropyl-*m*-cresol
p-Aminophenol, N-sec-*Butyl*
6-Amino-2,4-xylenol, *Et ether*
4-Amino-2,5-xylenol, *Et ether*
4-Amino-3,5-xylenol, N-*Et*
o-Diethylaminophenol
m-Diethylaminophenol
p-Diethylaminophenol
m-Dimethylaminophenol, *Et ether*
p-Dimethylaminophenol, *Et ether*
Ephedrine
ψ-Ephedrine
o-Ethylaminophenol, *Et ether*
Hordenine
4-Isopropenyl-1-cyclohexene-1-carboxylic Acid, *Amide*
o-β-Methylaminoethylphenol, *Me ether*
m-β-Methylaminoethylphenol, *Me ether*
p-β-Methylaminoethylphenol, *Me ether*
p-(2-Methylaminopropyl)phenol†
Pyrrole-2-aldehyde, N-3-*Methylbutyl*
Shonanic Acid, *Amide*
Tricyclenic Acid, *Amide*
Tyramine, N-*Et*

$C_{10}H_{15}NO_2$
α-Aminomethyl-*p*-hydroxybenzyl Alcohol, N-*Et*
α-Aminomethyl-*p*-hydroxybenzyl Alcohol, N-*Di-Me*
Aminoquinol, *Di-Et ether*
2-Aminoresorcinol, *Di-Et ether*
4-Aminoresorcinol, *Di-Et ether*
Anhydroecgonine, *Me ester*
2,4-Dihydroxy-6-methylpyridine, *Di-Et ether*
2-(3,4-Dihydroxyphenyl)ethylamine, *Di-Me ether*†
3,4-Dimethoxyphenethylamine
Effortil
4-Ethyl-3-methylglutaconic Acid, *Et ester-nitrile*
3-Ethyl-4-methylpyrrole-2-carboxylic Acid, *Et ester*
3-Ethyl-5-methylpyrrole-2-carboxylic Acid, *Et ester*
4-Ethyl-3-methylpyrrole-2-carboxylic Acid, *Et ester*
5-Ethyl-3-methylpyrrole-2-carboxylic Acid, *Et ester*
5-Ethyl-4-methylpyrrole-2-carboxylic Acid, *Et ester*
Isonitrosocamphor
Norecgonidine, *Et ester*
Pyridine-4-aldehyde, *Di-Et acetal*
Pyrrole-2-carboxylic Acid, 3-*Methylbutyl ester*
1,2,2-Trimethylcyclopentane-1,3-dicarboximide
1,2,2-Trimethylcyclopentane-1,3-dicarboxylic Acid, α-*Nitrile*
1,2,2-Trimethylcyclopentane-1,3-dicarboxylic Acid, β-*Nitrile*

$C_{10}H_{15}NO_2S$
2,4,6-Trimethylbenzenesulphonic Acid, *Methylamide*
m-Xylene-4-sulphonic Acid, *Dimethylamide*

$C_{10}H_{15}NO_3$
Adermin, 4-*Et ether*
α-Amino-3,4,5-trihydroxytoluene, *Tri-Me ether*
π-Camphanic Acid, *Amide*
ω-Camphanic Acid, *Amide*
4-Hydroxy-3,5-dimethoxybenzylamide, O-*Me ether*†
4-Hydroxymethylene-2-pentyl-5-oxazolone, *Me ether*
3-Nitro-(+)-camphor
Tenuazonic Acid

$C_{10}H_{15}NO_3S$
N-Diethylaniline-*m*-sulphonic Acid
5-Hydroxy-2,4-dimethylbenzenesulphonic Acid, *Et ether*, *Amide*

$C_{10}H_{15}NO_4$
3-Carboxymethyl-4-isopropenylpyrrolidine-2-carboxylic Acid
2-Pentenylpenaldic Acid, *Me ester*

$C_{10}H_{15}NS$
5-Amino-*o*-toluenethiol, N,S-*Tri-Me*

$C_{10}H_{15}N_2O$
1-Nitrocamphene †

$C_{10}H_{15}N_3O$
2-Methyl-2-α-phenylhydrazinopropionic Acid, *Amide*
2-Methyl-2-β-phenylhydrazinopropionic Acid, *Amide*
2-β-Phenylhydrazinobutyric Acid, *Amide*

$C_{10}H_{15}N_3O_4$
Deoxy-5-methylcytidine

$C_{10}H_{15}N_3O_5$
1-Methylcytidine
5-Methylcytidine

$C_{10}H_{15}N_3O_8$
Convicine †

$C_{10}H_{15}N_4O_9P$
Xanthylic Acid

$C_{10}H_{15}N_5$
1-Phenethyldiguanide

$C_{10}H_{15}N_5O$
Dihydrozeatin †

$C_{10}H_{15}N_5O_{10}P_2$
Adenosine diphosphate

$C_{10}H_{15}N_5O_{11}P_2$
Guanosine diphosphate

$C_{10}H_{15}O_2PS_2$
O,O-Dimethyl-*S*-α-methylbenzylphosphorothidothionate

$C_{10}H_{15}O_3P$
Phenylphosphonic Acid, *Di-Et ester*

$C_{10}H_{15}O_3PS$
O,O-Diethyl-*S*-phenylphosphorothiolate

$C_{10}H_{16}$
Adamantane
Allo-ocimene
2-Bornene
Camphene
3-Carene
4-Carene
Cyclodecyne
4,4-Dimethylbicyclo[3,2,1]oct-2-ene
2,2-Dimethyl-5-methylenebicyclo[2,2,1]heptane
7,7-Dimethyl-2-methylenebicyclo[2,2,1]heptane
2,5-Dimethyl-3-vinylhexa-1,4-diene †
Fenchelene
Geraniene
1-Isopropenyl-2-methylcyclohexene
1-Isopropenyl-3-methylcyclohexene
1-Isopropenyl-4-methylcyclohexene
1-Isopropenyl-5-methylcyclohexene
3-Isopropenyl-1-methylcyclohexene
3-Isopropenyl-2-methylcyclohexene
3-Isopropenyl-4-methylcyclohexene
4-Isopropenyl-1-methylcyclohexene
4-Isopropenyl-3-methylcyclohexene
4-Isopropenyl-5-methylcyclohexene
5-Isopropenyl-1-methylcyclohexene
1-Isopropenyl-4-methylenecyclohexane
3-Isopropylidene-6-methylcyclohexene
4-Isopropylidene-1-methylcyclohexene
1-Isopropylidene-4-methylenecyclohexane
1-Isopropyl-4-methyl-1,3-cyclohexadiene
5-Isopropyl-2-methyl-1,3-cyclohexadiene
2-Isopropyl-5-methyl-1,3-cyclohexadiene
1-Isopropyl-4-methyl-1,4-cyclohexadiene
1-Isopropyl-4-methylenecyclohexene
3-Isopropyl-6-methylenecyclohexene
Mycelene
Myrcene
Ocimene
1,2,3,4,4*a*,5,6,7-Octahydronaphthalene
1,2,3,4,4*a*,5,6,8*a*-Octahydronaphthalene
1,2,3,4,4*a*,5,8,8*a*-Octahydronaphthalene
1,2,3,4,5,6,7,8-Octahydronaphthalene
α-Pinene
β-Pinene
4(10)-Thujene
Tricyclene
Tricyclo[4,4,0,$0^{3,8}$]decane †
Tricyclo[4,3,1,$0^{3,8}$]decane †
Tricyclo[5,2,1,$0^{4,10}$]decane †
1,5,5-Trimethylbicyclo[2,2,1]hept-2-ene
2,5,5-Trimethylbicyclo[2,2,1]hept-2-ene
2,7,7-Trimethylbicyclo[2,2,1]hept-2-ene
3,6,6-Trimethylbicyclo[3,1,1]hept-2-ene
3,3,4-Trimethyltricyclo[2,2,1,$0^{2,6}$]heptane

$C_{10}H_{16}BrNO$
Edrophonium bromide

$C_{10}H_{16}BrNO_3S$
3-Bromocamphor-8-sulphonic Acid, *Amide*
3-Bromo-(+)-camphor-10-sulphonic Acid, *Amide*
10-Bromocamphor-3-sulphonic Acid, *Amide*

$C_{10}H_{16}Br_2O_3$
4-Bromovaleric Acid, *Anhydride*

$C_{10}H_{16}Br_2O_4$
2,5-Dibromoadipic Acid, *Di-Et ester*

$C_{10}H_{16}ClNO$
Edrophonium chloride

$C_{10}H_{16}ClO_7$
Erythronic Acid, *Tri-Ac Chloride*

$C_{10}H_{16}Cl_2O_2$
Decanedioic Acid, *Dichloride*
6-Methylheptane-2,5-dicarboxylic Acid, *Dichloride*

$C_{10}H_{16}CuN_4O_4$
Anserine, *Cu deriv.*

$C_{10}H_{16}I_2O_3$
Exogonic Acid, "*Di-hydriodide*"

$C_{10}H_{16}KNO_9S_2$
Sinigrin

$C_{10}H_{16}N_2$
o-Amino-*N*-diethylaniline
m-Amino-*N*-diethylaniline
p-Amino-*N*-diethylaniline
Convolvicine
Decanedioic Acid, *Dinitrile*
α,β-Diethylphenylhydrazine
β,β-Diethylphenylhydrazine
Lupinic Acid, *Nitrile*

$C_{10}H_{16}N_2$ (*continued*)
o-Phenylenediamine, N,N'-*Tetra-Me*
m-Phenylenediamine, N,N'-*Tetra-Me*
p-Phenylenediamine, N,N'-*Tetra-Me*

$C_{10}H_{16}N_2O$
3-Carbamoyl-1,2,2-trimethylcyclopentane-carboxylic Acid, *Nitrile*
3-Carbamoyl-2,2,3-trimethylcyclopentane-carboxylic Acid, *Nitrile*

$C_{10}H_{16}N_2O_2$
Decorticasine

$C_{10}H_{16}N_2O_3$
5-(2-Butyl)-5-ethylbarbituric Acid
Neonal
Proponal

$C_{10}H_{16}N_2O_3S$
Biotin

$C_{10}H_{16}N_2O_4$
2,4-Dioxo-3-imidazolidylacetic Acid, 3-*Methylbutyl ester*
Hexahydro-2-oxo-1*H*-furo[3,4]iminazole-4-valeric Acid

$C_{10}H_{16}N_2O_7$
α-L-Glutamyl-L-glutamic Acid†
γ-L-Glutamyl-L-glutamic Acid†

$C_{10}H_{16}N_2O_8$
Ethylenediaminetetra-acetic Acid

$C_{10}H_{16}N_2O_9S$
S-Sulphoglutathione

$C_{10}H_{16}N_2O_{11}P_2$
Thymidine-5'-pyrophosphate

$C_{10}H_{16}N_2O_{13}$
3-Nitro-opianic Acid, *Anhydride*

$C_{10}H_{16}N_4O_3$
Anserine
Balenine†
Ophidine†

$C_{10}H_{16}N_4O_7$
Vicine

$C_{10}H_{16}N_5O_{13}P_3$
Adenosine triphosphate
Adenosine-2',3',5'-triphosphate

$C_{10}H_{16}N_5O_{14}P_3$
Guanosine triphosphate

$C_{10}H_{16}O$
Allocyclocitral
Anthemol
1-Camphenol
6-Camphenol
(+)-Camphor
(−)-Camphor
(±)-Camphor
epi-Camphor
Carone
Carvenone
Carveol
Citral
α-Cyclocitral
β-Cyclocitral
4-Cyclohexyl-3-buten-2-one
β-Cyclolavandulal†
2,4-Decadienal
2,6-Decadienal
1-Decalone
2,2-Dimethyl-6-methylene-3-oxabicyclo[3,3,0]-octane†
8,8-Dimethyl-2-methylene-6-oxabicyclo[4,2,1]-octane†
2,5-Dimethyl-4-vinylhexa-2,5-dien-1-ol†
Fenchone
1-Hydroxymethyl-4-isopropenylcyclohexene
2-Isopropenyl-5-methylcyclohexanone
5-Isopropenyl-2-methylcyclohexanone
4-Isopropyl-1-cyclohexene-1-aldehyde
4-Isopropyl-2,3-dimethyl-2-cyclopenten-1-one
2-Isopropylidene-5-methyl-1-cyclohexanone
6-Isopropylidene-3-methyl-1-cyclohexen-1-ol
6-Isopropylidene-3-methyl-2-cyclohexen-1-ol
2-Isopropyl-3-methyl-2-cyclohexen-1-one
2-Isopropyl-5-methyl-2-cyclohexen-1-one
4-Isopropyl-5-methyl-2-cyclohexen-1-one
5-Isopropyl-2-methyl-2-cyclohexen-1-one
5-Isopropyl-3-methyl-2-cyclohexen-1-one
6-Isopropyl-2-methyl-2-cyclohexen-1-one
6-Isopropyl-3-methyl-2-cyclohexen-1-one
5-Isopropyl-2-methylenecyclohexanone
Matabiether†
2-Methyl-6-methyleneocta-3,7-dien-2-ol†
2-Methyl-6-methyleneocta-2,7-dien-4-ol†
Methylnopinone
Myrtenol
α-Naphthanone
β-Naphthanone
Octahydro-1-naphthol
3-Pinanone
4-Pinanone
2(10)-Pinen-3-ol
2-Pinen-4-ol
Pinol
Tagetone
Teresantalol
4(10)-Thujen-3-ol
Thujone
1,5,5-Trimethylbicyclo[2,2,1]heptan-2-one
2,2,5-Trimethyl-3-oxonorbornane
2,2,5-Trimethylcyclohept-4-enone†

$C_{10}H_{16}O_2$
Allocyclogeranic Acid
2-Allyl-4-pentenoic Acid, *Et ester*
Ascaridol
Buchu-camphor
α-Campholide
Chrysanthemum-monocarboxylic Acid
Cyclodecane-1,2-dione
Cyclodecane-1,6-dione
2-1'-Cyclohexenylpropionic Acid, *Me ester*
Cyclohexylideneacetic Acid, *Et ester*
β-Cyclolavandulic Acid†
2-1'-Cyclopentenylpropionic Acid, *Et ester*
2-2'-Cyclopentenylpropionic Acid, *Et ester*
2,4-Decadienoic Acid
2,6-Decadienoic Acid
4,6-Decadienoic Acid

$C_{10}H_{16}O_2$ (*continued*)

Dec-2-en-5-olide †
Dihydronepetalactone †
3,7-Dimethyl-2,4-octadienoic Acid
2-(2-Ethylidenecyclobutyl)-2-methylpropionic Acid
Geranic Acid★ †
2-Hydroxyepicamphor †
1-Hydroxymethyl-7,7-dimethylbicyclo[2,2,1]-heptan-2-one
7-Hydroxymethyl-1,7-dimethylbicyclo[2,2,1]-heptan-2-one
5-(1-Hydroxy-1-methylethyl)-2-methyl-2-cyclohexen-1-one
3-Hydroxy-1,7,7-trimethylbicyclo[2,2,1]heptan-2-one★ †
4-Hydroxy-1,7,7-trimethylbicyclo[2,2,1]heptan-2-one
5-Hydroxy-1,7,7-trimethylbicyclo[2,2,1]heptan-2-one
6-Hydroxy-1,7,7-trimethylbicyclo[2,2,1]heptan-2-one
Iridodial
Iridomyrmecin
Isodihydronepetalactone †
Isoiridomyrmecin
4-Isopropyl-1-cyclohexene-1-carboxylic Acid
2-Isopropylidene-5-methylcyclopentane-1-carboxylic Acid
3-Isopropylidene-1-methylcyclopentane-1-carboxylic Acid
5-Isopropyl-2-methyl-1,3-cyclohexanedione
2-Methyl-1-cyclohexene-1-carboxylic Acid, *Et ester*
4-Methyl-1-cyclohexene-1-carboxylic Acid, *Et ester*
5-Methyl-1-cyclohexene-1-carboxylic Acid, *Et ester*
3-Methyl-2-cyclohexene-1-carboxylic Acid, *Et ester*
5-Methyl-2-cyclohexene-1-carboxylic Acid, *Et ester*
6-Methyl-2-cyclohexene-1-carboxylic Acid, *Et ester*
2-Methyl-3-cyclohexene-1-carboxylic Acid, *Et ester*
3-Methyl-3-cyclohexene-1-carboxylic Acid, *Et ester*
4-Methyl-3-cyclohexene-1-carboxylic Acid, *Et ester*
5-Methyl-3-cyclohexene-1-carboxylic Acid, *Et ester*
6-Methyl-3-cyclohexene-1-carboxylic Acid, *Et ester*
3-Methyl-1-cyclohexenyl-1-acetic Acid, *Me ester*
4-Methyl-1-cyclohexenyl-1-acetic Acid, *Me ester*
3-Methyl-1-cyclopentene-1-acetic Acid, *Et ester*
2-Methyl-2-(3-methyl-2-cyclopentenyl)propionic Acid
4-Methyl-2-pentynoic Acid, *Isobutyl ester*
2-Nonynoic Acid, *Me ester*
4-Nonynoic Acid, *Me ester*
5-Nonynoic Acid, *Me ester*
6-Nonynoic Acid, *Me ester*
7-Nonynoic Acid, *Me ester*
8-Nonynoic Acid, *Me ester*
2-Octynoic Acid, *Et ester*
cis-5-Pent-2′-enylpentanolide-(5,1)
β-Pulegenic Acid
Pyrocin
2,6,6-Trimethyl-1-cyclohexene-1-carboxylic Acid
5,6,6-Trimethyl-1-cyclohexene-1-carboxylic Acid
2,3,6-Trimethyl-2-cyclohexene-1-carboxylic Acid
2,6,6-Trimethyl-2-cyclohexene-1-carboxylic Acid
1,2,2-Trimethyl-3-cyclohexene-1-carboxylic Acid
1,3,4-Trimethyl-3-cyclohexene-1-carboxylic Acid
2,2,6-Trimethyl-3-cyclohexene-1-carboxylic Acid
2,2,4-Trimethyl-3-cyclohexene-1-carboxylic Acid
2,6,6-Trimethyl-3-cyclohexene-1-carboxylic Acid
3,4,6-Trimethyl-3-cyclohexene-1-carboxylic Acid
2,3,3-Trimethyl-1-cyclopentene-1-carboxylic Acid, *Me ester*
2,2,3-Trimethyl-3-cyclopentene-1-carboxylic Acid, *Me ester*
2,2,3-Trimethyl-3-cyclopentenylacetic Acid
2,3,3-Trimethyl-1-cyclopentenylacetic Acid

$C_{10}H_{16}O_3$

3-Acetyl-2,2-dimethylcyclobutaneacetic Acid
3-Acetyl-2,2-dimethylcyclobutane-1-carboxylic Acid, *Me ester*
Decanedioic Acid, *Anhydride*
β-Formylcrotonic Acid, 3-*Methylbutyl ester* †
10-Hydroxy-2-decynoic Acid †
2-Methylheptane-3,4-dicarboxylic Acid, *Anhydride*
4-Methyl-3-oxo-1-cyclohexaneacetic Acid, *Me ester*
1-Methyl-2-oxo-1-cyclohexanecarboxylic Acid, *Et ester*
2-Methyl-4-oxo-1-cyclohexanecarboxylic Acid, *Et ester*
3-Methyl-2-oxo-1-cyclohexanecarboxylic Acid, *Et ester*
4-Methyl-2-oxo-1-cyclohexanecarboxylic Acid, *Et ester*
4-Methyl-3-oxo-1-cyclohexanecarboxylic Acid, *Et ester*
5-Methyl-2-oxo-1-cyclohexanecarboxylic Acid, *Et ester*
5-Methyl-3-oxo-1-cyclohexanecarboxylic Acid, *Et ester*
4-Methyl-2-oxocyclopentane-1-carboxylic Acid, *Propyl ester*
Nepetalic Acid †
Nopinic Acid
Oleuropeic Acid †
9-Oxo-2-decenoic Acid
2,4,5-Trihydroxytoluene, *Tri-Me ether*
2,4,6-Trihydroxytoluene, 2,4,6-*Tri-Me ether*

$C_{10}H_{16}O_4$
Allylmalonic Acid, *Di-Et ester*
1-Butene-1,3-dicarboxylic Acid, *Di-Et ester*
2-Butene-1,4-dicarboxylic Acid, *Di-Et ester*
Camphenic Acid
α-Carboxy-γ-nonalactone†
α-Carboxy-δ-nonalactone†
Cyclobutane-1,1-dicarboxylic Acid, *Di-Et ester*
Cyclobutane-1,2-dicarboxylic Acid, *Di-Et ester*
Cyclohexane-1,2-dicarboxylic Acid, *Di-Me ester*
Cyclohexane-1,3-dicarboxylic Acid, *Di-Me ester*
Cyclohexane-1,4-dicarboxylic Acid, *Di-Me ester*
2-Decenedioic Acid
2,2-Diacetylbutyric Acid, *Et ester*
2,3-Diacetylbutyric Acid, *Et ester*
2,2-Dimethylcyclobutane-1,3-dicarboxylic Acid
Dimethylfumaric Acid, *Di-Et ester*
2,4-Dimethylglutaconic Acid, *Me-Et ester*
Dimethylmaleic Acid, *Di-Et ester*
2,2-Dimethyl-5-oxo-oxolan-3-acetic Acid, *Et ester*
Ethylfumaric Acid, *Di-Et ester*
Maleic Acid, *Dipropyl ester*
Maleic Acid, *Di-isopropyl ester*
1-Methylcyclohexane-1,2-dicarboxylic Acid, 1-*Me ester*
1-Methylcyclohexane-1,2-dicarboxylic Acid, 2-*Me ester*
2-Methylcyclopropane-1,1-dicarboxylic Acid, *Di-Et ester*
3-Methylcyclopropane-1,2-dicarboxylic Acid, *Di-Et ester*
3-Methyl-2,5-dioxohexane-3-carboxylic Acid, *Et ester*
6-Methyl-3-heptene-1,3-dicarboxylic Acid
6-Methyl-2-heptene-2,3-dicarboxylic Acid
6-Methyl-1-heptene-4,7-dicarboxylic Acid
2-Methylpropene-1,3-dicarboxylic Acid, *Di-Et ester*
Succinic Acid, *Cyclohexyl ester*
1,2,2-Trimethylcyclopentane-1,3-dicarboxylic Acid
4,4,5-Trimethylcyclopentane-1,3-dicarboxylic Acid

$C_{10}H_{16}O_4S$
Camphor-3-sulphonic Acid
Camphor-8-sulphonic Acid
Camphor-10(or 6-)-sulphonic Acid
Tetrahydrothiophene-2,5-dicarboxylic Acid, *Di-Et ester*

$C_{10}H_{16}O_5$
3-Acetonylglutaric Acid, *Di-Me ester*†
Acetylmalonic Acid, *Mono-isoamyl ester*
Acetylsuccinic Acid, *Di-Et ester*
6-Hydroxy-5-methyl-2-heptene-3,6-dicarboxylic Acid
Iso-oxocamphoric Acid
3-Oxalobutyric Acid, *Di-Et ester*
2-Oxohexanedioic Acid, *Di-Et ester*
3-Oxohexanedioic Acid, *Di-Et ester*
Tetrahydro-2,6,6-trimethylpyran-2,5-dicarboxylic Acid

$C_{10}H_{16}O_6$
Acetoxysuccinic Acid, *Di-Et ester*
Butane-1,2,4-tricarboxylic Acid, *Tri-Me ester*
2,3-Dimethylbutane-1,2,3-tricarboxylic Acid, *Mono-Me ester*
Heptane-1,1,3-tricarboxylic Acid
Heptane-1,1,7-tricarboxylic Acid
Heptane-1,2,2-tricarboxylic Acid
Heptane-1,3,3-tricarboxylic Acid
Heptane-1,3,7-tricarboxylic Acid
Heptane-1,4,7-tricarboxylic Acid
Heptane-1,5,5-tricarboxylic Acid
Heptane-2,4,4-tricarboxylic Acid
Heptane-2,4,6-tricarboxylic Acid
Hygrophyllinecic Acid†
Isatinecic Acid
Jaconecic Acid
Methane-tricarboxylic Acid, *Tri-Et ester*
2-Methylbutane-1,1,2-tricarboxylic Acid, 1-*Mono-Et ester*
3-Methylbutane-2,2,3-tricarboxylic Acid, 3-*Et ester*
Retronecic Acid

$C_{10}H_{16}O_7$
Trimethyl citrate, *Me ether*

$C_{10}H_{17}Br$
2-Bromo-1,7,7-trimethylbicyclo[2,2,1]heptane

$C_{10}H_{17}Cl$
2-Chlorobornane
1-Chlorodecahydronaphthalene
2-Chlorodecahydronaphthalene
9-Chlorodecahydronaphthalene

$C_{10}H_{17}ClO$
2-Decenoic Acid, *Chloride*
2-Isopropyl-5-methylcyclopentane-1-carboxylic Acid, *Chloride*
3-Isopropyl-1-methylcyclopentane-1-carboxylic Acid, *Chloride*
1,2,2,3-Tetramethylcyclopentane-1-carboxylic Acid *Chloride*
2,2,3-Trimethylcyclopentylacetic Acid, *Chloride*

$C_{10}H_{17}ClO_3$
Octanedioic Acid, *Et ester*, *Chloride*

$C_{10}H_{17}N$
1-Aminoadamantane†
2-Aminoadamantane†
Camphenamine
Citronellic Acid, *Nitrile*
Dicadylamine
2-Ethyl-4-methyl-3-propylpyrrole
2-Isopropyl-5-methylcyclopentane-1-carboxylic Acid, *Nitrile*
3-Isopropyl-1-methylcyclopentane-1-carboxylic Acid, *Nitrile*
1,2,2,3-Tetramethylcyclopentane-1-carboxylic Acid, *Nitrile*
2,2,3,3-Tetramethylcyclopentane-1-carboxylic Acid, *Nitrile*

$C_{10}H_{17}NO$
Allocyclogeranic Acid, *Amide*
3-Aminocamphor
4-Aminocamphor

$C_{10}H_{17}NO$ (*continued*)
2-Isopropylidene-5-methylcyclopentane-1-carboxylic Acid, *Amide*
3-Isopropylidene-1-methylcyclopentane-1-carboxylic Acid, *Amide*
Lupinal
2-Methyl-2-(3-methyl-2-cyclopentenyl)propionic Acid, *Amide*
β-Pulegenic Acid, *Amide*
2,6,6-Trimethyl-2-cyclohexene-1-carboxylic Acid, *Amide*
1,2,2-Trimethyl-3-cyclohexene-1-carboxylic Acid, *Amide*
2,2,3-Trimethyl-3-cyclopentenylacetic Acid, *Amide*
2,3,3-Trimethyl-1-cyclopentenylacetic Acid, *Amide*

$C_{10}H_{17}NO_2$
2-Isopropylpropane-1,3-dicarboxylic Acid, *Et ester-nitrile*
Lupinic Acid
Methyprylone
Pyrrolizidine-1-carboxylic Acid, *Et ester*†
2-*p*-Tolylquinoline-4-carboxylic Acid, *Et ester*

$C_{10}H_{17}NO_3$
3-Carbamoyl-1,2,2-trimethylcyclopentanecarboxylic Acid
3-Carbamoyl-2,2,3-trimethylcyclopentanecarboxylic Acid
Cyclohexanecarbonylglycine, *Me ester*
Ecgonine, *Me ester*
α-Ecgonine, *Me ester*
ψ-Ecgonine, *Me ester*
Isobutylfumaric Acid, *Et ester-amide*
4-Nitromenthone
8-Nitromenthone
Nor-ψ-ecgonine, *Et ester*

$C_{10}H_{17}NO_3S$
Camphor-3-sulphonic Acid, *Amide*
Camphor-8-sulphonic Acid, *Amide*
Camphor-10(or 6-)sulphonic Acid, *Amide*
2-(5-Carboxypentyl)-4-thiazolidone, *Me ester*

$C_{10}H_{17}NO_4$
Lobelinic Acid
N-Methylpiperidine-2,6-dicarboxylic Acid, *Di-Me ester*
Pentylpenaldic Acid, *Me ester*
Tropinic Acid, *Di-Me ester*

$C_{10}H_{17}NO_6$
Linamarin

$C_{10}H_{17}N_2O$
4-Methylpentane-1,1-dicarboxylic Acid, *Et ester-nitrile*

$C_{10}H_{17}N_3O_6$
γ-Glutamylalanylglycine
γ-L-Glutamyl-L-glutamic Acid, *γ'-Amide*†

$C_{10}H_{17}N_3O_6S$
Glutathione

$C_{10}H_{17}N_3O_8$
Aspergillomarasmine A†

$C_{10}H_{17}N_7O_4$
Saxitoxin

$C_{10}H_{17}O_3P$
1,3,5,7-Tetramethyl-2,4,8-trioxa-6-phospha-adamantane

$C_{10}H_{17}O_5P$
Phosphonoacetic Acid, 3,3-*Diallyl*-1-*Et ester*

$C_{10}H_{18}$
Bornane
Carane
Cyclodecene
Decahydronaphthalene
1-Decyne
3-Decyne
4-Decyne
5-Decyne
3,7-Dimethyl-1,6-octadiene
2,7-Dimethyl-1,7-(or 2,7-)octadiene
4,4-Dimethyl-1,7-octadiene
2,4-Dimethyl-2,4-octadiene
3,7-Dimethyl-2,4-octadiene
2,6-Dimethyl-2,5-octadiene
2,6-Dimethyl-2,6-octadiene
2,7-Dimethyl-2,6-octadiene
3,6-Dimethyl-2,6-octadiene
4,5-Dimethyl-2,6-octadiene
2,6-Dimethyl-2,7-octadiene
3,6-Dimethyl-3,5-octadiene
Hexamethylbicyclo[1,1,0]butane†
1-Isopropenyl-3-methylcyclohexane
1-Isopropenyl-4-methylcyclohexane
1-Isopropylidene-2-methylcyclohexane
1-Isopropylidene-3-methylcyclohexane
1-Isopropylidene-4-methylcyclohexane
1-Isopropyl-2-methylcyclohexene
1-Isopropyl-4-methylcyclohexene
1-Isopropyl-5-methylcyclohexene
3-Isopropyl-1-methylcyclohexene†
3-Isopropyl-5-methylcyclohexene
3-Isopropyl-6-methylcyclohexene
4-Isopropyl-1-methylcyclohexene
5-Isopropyl-1-methylcyclohexene★†
Pinane
2,3,4,5-Tetramethylhexa-1,4-diene†
2,2,5,5-Tetramethylhex-3-yne†
Thujane
1,3,3-Trimethylbicyclo[2,2,1]heptane
2,2,3-Trimethylbicyclo[2,2,1]heptane
2,7,7-Trimethylbicyclo[2,2,1]heptane
1,4,4-Trimethylcycloheptene

$C_{10}H_{18}Br_2O_2$
3-Decenoic Acid, *Dibromo*

$C_{10}H_{18}NO_2S$
5-(2-Butyl)-5-ethyl-1-methylthiobarbituric Acid

$C_{10}H_{18}N_2O$
Lupinic Acid, *Amide*

$C_{10}H_{18}N_2O_2$
Camphenic Acid, *Diamide*
Slaframine†
1,2,2-Trimethylcyclopentane-1,3-dicarboxylic Acid, *Diamide*

$C_{10}H_{18}N_2O_2S_2$
Methionine, *Anhydride*

$C_{10}H_{18}N_2O_3$
4-Methyl-2-oxo-5-imidazolidinehexanoic Acid

$C_{10}H_{18}N_2O_4$
2,2′-Azodi-2-methylpropionic Acid, *Di-Me ester*
Piperazine-*N,N′*-dicarboxylic Acid, *Di-Et ester*

$C_{10}H_{18}N_2O_4S$
γ-L-Glutamyl-L-methionine

$C_{10}H_{18}N_2O_5$
γ-L-Glutamyl-L-valine

$C_{10}H_{18}N_2O_7$
Succinamide, N-D-*Glucoside*

$C_{10}H_{18}N_4O_5$
Biuret Base

$C_{10}H_{18}N_8O_2$
Leucoporphyrindene

$C_{10}H_{18}O$
Allocyclogeraniol
Borneol
(−)-*epi*-Borneol
2-*tert*-Butylcyclohexanone†
3-*tert*-Butylcyclohexanone†
4-*tert*-Butylcyclohexanone†
Cineole
m-Cineole
1,4-Cineole
Citronellal
Cyclodecanone
Cyclohexyl isopropyl Ketone
Cyclohexyl propyl Ketone
β-Cyclolavandulol†
Decahydro-1-naphthol
Decahydro-2-naphthol
2-Decenal
3-Decenal
6-Decenal
2-Decen-2-one
2,2-Dimethyl-5-(2-methylpropenyl)-tetrahydrofuran†
3-Ethyl-4-methyl-3-hepten-2-one
3-Ethyl-4-methyl-4-hepten-2-one
3-Ethyl-4-methyl-4-hepten-3-one
4-Ethyl-2-methyl-2-hepten-6-one
5-Ethyl-4-methyl-5-hepten-3-one
Geraniol
1-(1-Hydroxy-1-methylethyl)-2-methylcyclohexene
1-(1-Hydroxy-1-methylethyl)-3-methylcyclohexene
1-(1-Hydroxy-1-methylethyl)-4-methylcyclohexene
1-(1-Hydroxy-1-methylethyl)-5-methylcyclohexene
3-(1-Hydroxy-1-methylethyl)-1-methylcyclohexene
3-(1-Hydroxy-1-methylethyl)-4-methylcyclohexene
3-(1-Hydroxy-1-methylethyl)-5-methylcyclohexene
4-(1-Hydroxy-1-methylethyl)-1-methylcyclohexene
4-(1-Hydroxy-1-methylethyl)-3-methylcyclohexene
6-(1-Hydroxy-1-methylethyl)-1-methylcyclohexene
4-(1-Hydroxy-1-methylethyl)-5-methylcyclohexene
5-(1-Hydroxy-1-methylethyl)-1-methylcyclohexene
5-(1-Hydroxy-1-methylethyl)-3-methylcyclohexene
2-Hydroxymethyl-1,3,3-trimethylcyclohexene
3-Hydroxymethyl-2,4,4-trimethylcyclohexene
Isoborneol
epi-Isoborneol
2-2′-Isobutenyl-4-methyltetrahydropyran
2-Isobutyl-4-methylcyclopentanone
2-Isobutyl-5-methylcyclopentanone
Isogeraniol
Isomyrtanol
2-Isopropenyl-5-methylcyclohexanol
4-Isopropenyl-1-methylcyclohexanol
5-Isopropenyl-2-methylcyclohexanol
2-Isopropylidene-5-methylcyclohexanol
4-Isopropylidene-1-methylcyclohexanol
5-Isopropylidene-2-methylcyclohexanol
2-Isopropyl-3-methylcyclohexanone
2-Isopropyl-4-methylcyclohexanone
2-Isopropyl-5-methylcyclohexanone
2-Isopropyl-6-metbylcyclohexanone
3-Isopropyl-5-methylcyclohexanone
4-Isopropyl-3-methylcyclohexanone
5-Isopropyl-2-methylcyclohexanone
3-Isopropyl-6-methyl-2-cyclohexen-1-ol
4-Isopropyl-1-methyl-2-cyclohexen-1-ol
5-Isopropyl-2-methyl-2-cyclohexen-1-ol
6-Isopropyl-3-methyl-2-cyclohexen-1-ol
1-Isopropyl-4-methyl-3-cyclohexen-1-ol
4-Isopropyl-1-methyl-3-cyclohexen-1-ol
Lavandulol
Linalool
2-Methyl-6-methyleneoct-7-en-4-ol†
6-Methyl-5-nonen-4-one
6-Methyl-6-nonen-4-one
Methylnopinol
2-Methyl-6-propylcyclohexanone
4-Methyl-2-propylcyclohexanone
5-Methyl-2-propylcyclohexanone
Myrtanol
Nerol
3-Pinanol
4-Pinanol
1(7)-Terpinen-4-ol†
δ-Terpineol†
3-Thujanol
Thujan-4-ol†
1,3,3-Trimethylbicyclo[2,2,1]heptan-2-ol
1,5,5-Trimethylbicyclo[2,2,1]heptan-2-ol
2,2,3-Trimethylcycloheptanone
2,2,6-Trimethylcycloheptanone
2,3,7-Trimethylcycloheptanone
2,6,6-Trimethylcycloheptanone
2,2,6-Trimethyl-6-vinyltetrahydropyran†
Yomogi Alcohol A†

$C_{10}H_{18}O_2$
Angelic Acid, *Isoamyl ester*
Camphene Glycol
Citronellic Acid
Cycloheptane-carboxylic Acid, *Et ester*
Cyclohexylacetic Acid, *Et ester*

$C_{10}H_{18}O_2$ (*continued*)
2-Decenoic Acid
3-Decenoic Acid
4-Decenoic Acid
8-Decenoic Acid
9-Decenoic Acid
Dihydrochrysanthemum monocarboxylic Acid
2,6-Dimethylcyclohexanecarboxylic Acid, *Me ester*
1,4-Diphenylbutane-2,3-diol
1,5-Hexadiene-3,4-diol, *Di-Et ether*
2-Hydroxycyclodecanone†
4-Hydroxydecanoic Acid, *Lactone*
5-Hydroxydecanoic Acid, *Lactone*
10-Hydroxydecanoic Acid, *Lactone*
3-(1-Hydroxy-1-methylethyl)-6-methylcyclohexanone
5-(1-Hydroxy-1-methylethyl)-2-methyl-2-cyclohexen-1-ol
5-(10-Hydroxy-2-methylethyl)-2-methyl-2-cyclohexen-1-ol
Isoneomatatabiol†
2-Isopropyl-1-methylcyclopentane-1-carboxylic Acid
3-Isopropyl-1-methylcyclopentane-1-carboxylic Acid
1-Isopropyl-4-methyl-2-oxocyclohexanol
3-Isopropyl-6-methyl-2-oxocyclohexanol
4-Isopropyl-1-methyl-2-oxocyclohexanol
2-Methylcrotonic Acid, 3-*Methylbutyl ester*
1-Methylcyclohexane-1-carboxylic Acid, *Et ester*
2-Methylcyclohexane-1-carboxylic Acid, *Et ester*
3-Methylcyclohexane-1-carboxylic Acid, *Et ester*
4-Methylcyclohexane-1-carboxylic Acid, *Et ester*
2-Methyloctanoic Acid, *Me ester*
6-Methyloctanoic Acid, *Me ester*
Neomatatabiol†
7-Octenoic Acid, *Et ester*
1,2,2,3-Tetramethylcyclopentane-1-carboxylic Acid
2,2,3,3-Tetramethylcyclopentane-1-carboxylic Acid
2,2,3-Trimethylcyclopentenylacetic Acid
Uroterpenol†

$C_{10}H_{18}O_3$
Ascaridol, *Glycol*
Azelaialdehydic Acid, *Me ester*
2,2-Dimethyl-3-oxohexanoic Acid, *Et ester*
3,3-Dimethyl-5-oxohexanoic Acid, *Et ester*
2,2-Dimethylpropionic Acid, *Anhydride*
2-Ethyl-3-oxo-hexanoic Acid, *Et ester*
9-Formylnonanoic Acid
9-Hydroxydec-2-enoic Acid
10-Hydroxydec-2-enoic Acid
1-Hydroxy-3-methylcyclohexaneacetic Acid, *Me ester*
1-Hydroxy-2-methylcyclohexane-1-carboxylic Acid, *Et ether*
1-Hydroxy-2-methylcyclohexane-1-carboxylic Acid, *Et ester*
2-Hydroxy-4-methylcyclohexane-1-carboxylic Acid, *Et ester*
2-Hydroxy-5-methylcyclohexane-1-carboxylic Acid, *Et ester*
1-Hydroxy-3-methyl-1-cyclopentaneacetic Acid, *Et ester*
5-Hydroxy-4-octanone, *Ac*
Isovaleric Acid, *Anhydride*
Levulinic Acid, *Pentyl ester*
2-Methylbutyric Acid, *Anhydride*
2-Methyl-3-oxovaleric Acid, *Isobutyl ester*
Mullilam Diol†
2-Oxodecanoic Acid
3-Oxodecanoic Acid
4-Oxodecanoic Acid
5-Oxodecanoic Acid
6-Oxodecanoic Acid
8-Oxodecanoic Acid
9-Oxodecanoic Acid
3-Oxononanoic Acid, *Me ester*
7-Oxononanoic Acid, *Me ester*
3-Oxo-octanoic Acid, *Et ester*
4-Oxo-octanoic Acid, *Et ester*
6-Oxo-octanoic Acid, *Et ester*
7-Oxo-octanoic Acid, *Et ester*
2-Oxopentane-3-carboxylic Acid, *Isobutyl ester*
2-Oxocyclo-octane-1-carboxylic Acid, *Me ester*†
Pinol Glycol
Tetrahydro-2,6,6-trimethylpyran-2-carboxylic Acid, *Me ester*
Valeric Acid, *Anhydride*

$C_{10}H_{18}O_4$
Adipic Acid, *Di-Et ester*
Butylmethylmalonic Acid, *Di-Me ester*
Butylsuccinic Acid, *Mono-Et ester*
Decanedioic Acid
2,3-Diethylsuccinic Acid, *Di-Me ester*
2,2-Dimethylsuccinic Acid, *Di-Et ester*
2,3-Dimethylsuccinic Acid, *Di-Et ester*
2-Ethylbutane-1,1-dicarboxylic Acid, *Mono-Et ester*
Ethylmethylmalonic Acid, *Di-Et ester*
Ethylsuccinic Acid, *Di-Et ester*
Hexane-3,3-dicarboxylic Acid, *Di-Me ester*
8-Hydroxyoctanoic Acid, *Ac*
2-Isobutyl-4-methylglutaric Acid
3-Isobutyl-3-methylglutaric Acid
Isohexylsuccinic Acid
Isopropylmalonic Acid, *Di-Et ester*
2-Isopropylpropane-1,3-dicarboxylic Acid, *Di-Me ester*
3-Methylazelaic Acid
4-Methylazelaic Acid
5-Methylazelaic Acid
2-Methylheptane-1,1-dicarboxylic Acid
6-Methylheptane-2,3-dicarboxylic Acid
6-Methylheptane-2,5-dicarboxylic Acid
2-Methylheptane-3,4-dicarboxylic Acid
2-Methylheptane-4,4-dicarboxylic Acid
4-Methyl-2-propyladipic Acid
Nonactinic Acid
Octanedioic Acid, *Di-Me ester*
Octanedioic Acid, *Et ester*

$C_{10}H_{18}O_4$ *(continued)*
Oxalic Acid, *Dibutyl ester*
Propylmalonic Acid, *Di-Et ester*
Succinic Acid, *Dipropyl ester*
Tetramethylsuccinic Acid, *Di-Me ester*
Tetramethylsuccinic Acid, *Mono-Et ester*
2,2,4-Trimethylglutaric Acid, *Di-Me ester*
2,2,3-Trimethylpimelic Acid
2,2,6-Trimethylpimelic Acid
2,3,6-Trimethylpimelic Acid

$C_{10}H_{18}O_4Si$
Triallyl methyl orthosilicate

$C_{10}H_{18}O_5$
2-Hydroxyadipic Acid, *Di-Et ester*
α-Longinecic Acid, *Tetrahydro*
Malic Acid, *Dipropyl ester*
Malic Acid, *Di-isopropyl ester*
Malic Acid, *Et ether*, *Di-Et ester*

$C_{10}H_{18}O_6$
Dimethoxysuccinic Acid, *Di-Et ester*
Gluconic Acid, γ-*Lactone*, *Tetra-Me ether*
Tartaric Acid, *Dipropyl ester*
Tartaric Acid, *Di-isopropyl ester*

$C_{10}H_{18}O_7$
Galacturonic Acid, 2,3-*Di-Me ether*, *Methyl pyranoside*, *Me ester*
Galacturonic Acid, 2,3,4-*Tri-Me ether*, *Methyl-pyranoside*
Glucuronic Acid, 2,3,4-*Tri-Me ether*, β-*Methyl glycoside*

$C_{10}H_{18}O_8$
Allomucic Acid, *Di-Et ester*
Mucic Acid, *Di-Et ester*
Saccharic Acid, *Di-Et ester*

$C_{10}H_{18}O_9$
β-D-Ribofuranosyl-β-D-ribofuranoside†
Xylobiose

$C_{10}H_{19}Br$
1-Bromo-1-isopropyl-4-methylcyclohexane
1-Bromo-3-isopropyl-5-methylcyclohexane
1-Bromo-2-isopropyl-5-methylcyclohexane

$C_{10}H_{19}BrO_2$
2-Bromodecanoic Acid
4-Bromodecanoic Acid
10-Bromodecanoic Acid
2-Bromo-octanoic Acid, *Et ester*
7-Bromo-octanoic Acid, *Et ester*
8-Bromo-octanoic Acid, *Et ester*
2-Bromo-2-propylvaleric Acid, *Et ester*
2-Octanol, *Bromoacetyl*

$C_{10}H_{19}Cl$
1-Chloro-1-isopropyl-4-methylcyclohexane
1-Chloro-2-isopropyl-5-methylcyclohexane
1-Chloro-3-isopropyl-5-methylcyclohexane

$C_{10}H_{19}ClO$
Decanoic Acid, *Chloride*
2-Methylnonanoic Acid, *Chloride*

$C_{10}H_{19}Cl_3O_2$
Chloral, *Di-isobutyl acetal*

$C_{10}H_{19}I$
1-Iodo-2-isopropyl-5-methylcyclohexane
1-Iodo-3-isopropyl-5-methylcyclohexane

$C_{10}H_{19}N$
2-Aminobornane
9-Aminodecahydronaphthalene
2-(2-Aminoethyl)-1,5,5-trimethylcyclopentene
4-(2-Aminoethyl)-1,5,5-trimethylcyclopentene
2-Amino-1,3,3-trimethylbicyclo[2,2,1]heptane
1-Azabicyclo[5,4,0]hendecane
Camphidine
Decahydro-*N*-methylquinoline
Decahydro-2-methylquinoline
Decahydro-1-naphthylamine
Decahydro-2-naphthylamine
Decanoic Acid, *Nitrile*
5,7-Dimethyl-octahydroindolizine†
1-Ethylindolizidine
2-Ethylindolizidine
3-Ethylindolizidine
5-Ethylindolizidine
6-Ethylindolizidine
Geranylamine
1-Lupinane
2-Lupinane
2-Methylnonanoic Acid, *Nitrile*
Pyrrolidine, N-*Cyclohexyl*
3-Thujylamine

$C_{10}H_{19}NO$
Citronellic Acid, *Amide*
2-Decenoic Acid, *Amide*
1,3-Di-*tert*-butylaziridin-2-one†
Isolupinine
2-Isopropyl-5-methylcyclopentane-1-carboxylic Acid, *Amide*
3-Isopropyl-1-methylcyclopentane-1-carboxylic Acid, *Amide*
1-Lupinine
2-Lupinine
3-Lupinine
4-Lupinine
1,2,2,3-Tetramethylcyclopentane-1-carboxylic Acid, *Amide*
2,2,3,3-Tetramethylcyclopentane-1-carboxylic Acid, *Amide*
2,2,6,6-Tetramethyl-4-piperidone, N-*Me*
2,2,3-Trimethylcyclopentylacetic Acid, *Amide*
Virgilidine

$C_{10}H_{19}NOS_2$
8-Methylsulphinyloctyl isothiocyanate†

$C_{10}H_{19}NO_2$
3-Cyclohexylalanine, *Me ester*
3-Ethyl-4-piperidylacetic Acid, *Me ester*
Heptylpenilloaldehyde
1-Methyl-4-(2-nitro-2-propyl)cyclohexane
Valeric Acid, *Imide*

$C_{10}H_{19}NO_3$
Decanedioic Acid, *Monoamide*

$C_{10}H_{19}NO_4$
3,3′-Iminodibutyric Acid, *Di-Me ester*
2,2′-Iminodipropionic Acid, *Di-Et ester*
2,3′-Iminodipropionic Acid, *Di-Et ester*
3,3′-Iminodipropionic Acid, *Di-Et ester*
N-Methylglutamic Acid, *Di-Et ester*

$C_{10}H_{19}NO_5$
Pantothenic Acid, *Me-ester*
γ-Pantothenic Acid†

$C_{10}H_{19}NO_6$
Galacturonic Acid, 2,3,4-*Tri-Me ether, Methylpyranoside, Amide*
Galacturonic Acid, 2,3,5-*Tri-Me ether, Methylpyranoside, Amide*
Glucuronic Acid, 2,3,4-*Tri-Me ether, Methyl glycoside,* (α)-*Amide*

$C_{10}H_{19}NO_8S_2$
Glucosisymbrin

$C_{10}H_{19}NO_9S_2$
Glucoputranjivin

$C_{10}H_{19}N_3O_4$
Diglycyl-leucine
Glycyl-leucyl-glycine
Leucyl-β-asparagine

$C_{10}H_{19}O_6PS_2$
Malathion †

$C_{10}H_{20}$
1-Decene
4-Decene
5-Decene
1-Isopropyl-2-methylcyclohexane
1-Isopropyl-3-methylcyclohexane
1-Isopropyl-4-methylcyclohexane
1-Methyl-1-propylcyclohexane
1-Methyl-2-propylcyclohexane
1-Methyl-3-propylcyclohexane
1-Methyl-4-propylcyclohexane
1,1,3,4-Tetramethylcyclohexane
1,1,3,5-Tetramethylcyclohexane
1,2,3,4-Tetramethylcyclohexane
1,2,3,5-Tetramethylcyclohexane
1,2,4,5-Tetramethylcyclohexane
3,3,5,5-Tetramethylcyclohexane
1,1,2-Trimethylcycloheptane
1,1,4-Trimethylcycloheptane

$C_{10}H_{20}BrN$
5-Azoniaspiro[4,6]undecane, *Bromide*
6-Azoniaspiro[5,5]undecane, *Bromide*

$C_{10}H_{20}Br_2$
1,5-Dibromodecane
1,10-Dibromodecane
5,6-Dibromodecane

$C_{10}H_{20}ClN$
5-Azoniaspiro[4,6]undecane, *Chloride*
6-Azoniaspiro[5,5]undecane, *Chloride*

$C_{10}H_{20}Cl_2$
1,10-Dichlorodecane

$C_{10}H_{20}N_2$
2,2′-Bipiperidyl
2,3′-Bipiperidyl
2,4′-Bipiperidyl
3,3′-Bipiperidyl
3,4′-Bipiperidyl
4,4′-Bipiperidyl

$C_{10}H_{20}N_2O_2$
Adipic Acid, *Di-*N-*dimethylamide*
Decanedioic Acid, *Diamide*
6-Methylheptane-2,5-dicarboxylic Acid, *Diamide*

$C_{10}H_{20}N_2O_3$
Allophanic Acid, *Octyl ester*
D-Methionyl-D-methionine
Valine, *Anhydride*

$C_{10}H_{20}N_2O_4$
2,5-Diaminoadipic Acid, *Di-Et ester*
Ethylenedicarbamic Acid, *Di-Et ester*
2,2′-Hydrazodi-(2-methylpropionic) Acid, *Di-Me ester*
2,2′-Hydrazodipropionic Acid, *Di-Et ester*

$C_{10}H_{20}N_2O_4S_2$
Cystine, *Di-Et ester*
4,4′-Diamino-4,4′-dicarboxy-1,1′-dibutyldi-sulphide
Penicillamine, *Disulphide*
Tetramethylene-*S,S′*-dicystine

$C_{10}H_{20}N_2S_4$
Antietil

$C_{10}H_{20}N_4O_8$
2,3,4,5-Tetramethylhexane dinitrosate

$C_{10}H_{20}O$
Androl
2-*tert*-Butylcyclohexanol †
3-*tert*-Butylcyclohexanol †
4-*tert*-Butylcyclohexanol †
Carvomenthol
Citronellol
Cyclodecanol
Decanal
2-Decanone
3-Decanone
4-Decanone
Dec-3-en-1-ol †
2,6-Dimethyl-4-octanone
3-Ethyl-6-methyl-2-heptanone
4-Ethyl-3-methyl-2-heptanone
4-Ethyl-6-methyl-2-heptanone
5-Ethyl-2-methyl-4-heptanone
5-Ethyl-6-methyl-2-heptanone
1-(1-Hydroxy-1-methylethyl)-2-methylcyclohexane
1-(1-Hydroxy-1-methylethyl)-3-methylcyclohexane
1-(1-Hydroxy-1-methylethyl)-4-methylcyclohexane
1-(2-Hydroxy-1-methylethyl)-4-methylcyclohexane
1-Isopropyl-2-methylcyclohexanol
1-Isopropyl-3-methylcyclohexanol
1-Isopropyl-4-methylcyclohexanol
2-Isopropyl-5-methylcyclohexanol
2-Isopropyl-5-methylcyclohexanol
3-Isopropyl-5-methylcyclohexanol
4-Isopropyl-1-methylcyclohexanol
4-Isopropyl-2-methylcyclohexanol
4-Methyl-2-nonanone
7-Methyl-2-nonanone
2-Methyl-3-nonanone
4-Methyl-3-nonanone
2-Methyl-4-nonanone
6-Methyl-4-nonanone
2-Methyl-5-nonanone
3-Methyl-5-nonanone
2-Methyl-1-propylcyclohexanol
2-Methyl-2-propylcyclohexanol

$C_{10}H_{20}O$ (*continued*)
2-Methyl-6-propylcyclohexanol
3-Methyl-1-propylcyclohexanol
4-Methyl-1-propylcyclohexanol
4-Methyl-2-propylcyclohexanol
5-Methyl-2-propylcyclohexanol
4-Propyl-1-hepten-4-ol
2,6,6-Trimethylcycloheptanol

$C_{10}H_{20}O_2$
Decanoic Acid
1-(1,2-Dihydroxy-1-methylethyl)-4-methylcyclohexane
3,3-Dimethylbutyric Acid, *Butyl ester*
2,5-Dimethylhexanoic Acid, *Et ester*
2-Ethylhexanoic Acid, *Et ester*
3-Ethylhexanoic Acid, *Et ester*
Hexanoic Acid, *Butyl ester*
2-Hexanol, *Butyryl*
5-Hydroxy-2,7-dimethyl-4-octanone
1-(1-Hydroxy-1-methylethyl)-3-methylcyclohexanol
1-(1-Hydroxy-1-methylethyl)-4-methylcyclohexanol
1-(2-Hydroxy-1-methylethyl)-4-methylcyclohexanol
2-(1-Hydroxy-1-methylethyl)-5-methylcyclohexanol
3-(1-Hydroxy-1-methylethyl)-1-methylcyclohexanol
3-(1-Hydroxy-1-methylethyl)-6-methylcyclohexanol
4-(1-Hydroxy-1-methylethyl)-1-methylcyclohexanol
5-(2-Hydroxy-1-methylethyl)-2-methylcyclohexanol
Isopentyl-3-methylbutyrate
1-Isopropyl-4-methyl-1,2-cyclohexanediol
1-Isopropyl-4-methyl-1,3-cyclohexanediol
1-Isopropyl-4-methyl-1,4-cyclohexanediol
2-Isopropyl-5-methyl-1,4-cyclohexanediol★ †
3-Isopropyl-6-methyl-1,2-cyclohexanediol
Isovaleric Acid, *Isopentyl ester*
Ketene, *Di-isobutyl acetal*
2-Methylbutyric Acid, 3-*Methylbutyl ester*
4-Methylheptanoic Acid, *Et ester*
5-Methylheptanoic Acid, *Et ester*
6-Methylheptanoic Acid, *Et ester*
3-Methylhexane-3-carboxylic Acid, *Et ester*
5-Methylhexane-3-carboxylic Acid, *Et ester*
2-Methylnonanoic Acid
3-Methylnonanoic Acid
4-Methylnonanoic Acid
5-Methylnonanoic Acid
7-Methylnonanoic Acid
3-Methyl-2-propylvaleric Acid, *Me ester*
4-Methylvaleric Acid, *Isobutyl ester*
Nonanoic Acid, *Me ester*
Octanoic Acid, *Et ester*
2-Octanol, *Ac*
2-Propylvaleric Acid, *Et ester*
Valeric Acid, *Pentyl ester*

$C_{10}H_{20}O_3$
2-Hydroxydecanoic Acid
4-Hydroxydecanoic Acid
5-Hydroxydecanoic Acid

10-Hydroxydecanoic Acid
2-Hydroxyhexanoic Acid, *Et ether*, *Et ester*
2-Hydroxymethylnonanoic Acid
3-Hydroxy-4-methylvaleric Acid, *Et ether*, *Et ester*
9-Hydroxynonanoic Acid, *Me ester*
2-Hydroxyoctanoic Acid, *Et ester*
3-Hydroxyoctanoic Acid, *Et ester*
2-Hydroxy-2-propylvaleric Acid, *Et ester*
3-Hydroxy-2,2,3-trimethylvaleric Acid, *Et ester*
3-Hydroxy-2,2,4-trimethylvaleric Acid, *Et ester*
3-Hydroxy-2,3,4-trimethylvaleric Acid, *Et* ester
3-Hydroxy-3,4,4-trimethylvaleric Acid, *Et ester*

$C_{10}H_{20}O_4Si$
Diallyl diethyl orthosilicate

$C_{10}H_{20}O_5$
Mannitan, *Di-Et ether*
Sorbitan, *Tetra-Me ether*
Styracititol, *Tetra-Me ether*

$C_{10}H_{20}O_6$
Fructose, 1,3,4,5-*Tetra-Me ether*
Glucose, *Butylglucoside*
Mannose, *Tetra-Me ether*
Sorbose, 1,3,4,5-*Tetra-Me ether*
Sorbose, 1,3,4,6-*Tetra-Me ether*
Tagatose, *Tetra-Me ether*
2,3,4,6-Tetra-*O*-methylgalactose
2,3,5,6-Tetra-*O*-methylgalactose
2,3,4,6-Tetra-*O*-methylglucose
2,3,5,6-Tetra-*O*-methylglucose

$C_{10}H_{20}O_7$
Galactonic Acid, 2,3,4,6-*Tetra-Me ether*
Gluconic Acid, 2,3,4-*Tri-Me ether*, *Me ester*
Mannonic Acid, 2,3,4,5-*Tetra-Me ether*

$C_{10}H_{20}O_9$
D-Mannopyranosyl-1-*meso*-erythritol

$C_{10}H_{21}Br$
1-Bromodecane

$C_{10}H_{21}BrHg$
Mercuri-decyl bromide

$C_{10}H_{21}BrO_2$
Bromoacetaldehyde, *Dibutyl acetal*
Bromoacetaldehyde, *Di-isobutyl acetal*

$C_{10}H_{21}Cl$
1-Chlorodecane

$C_{10}H_{21}ClO_2$
Chloroacetaldehyde, *Dibutyl acetal*

$C_{10}H_{21}I$
1-Iododecane

$C_{10}H_{21}N$
Azaundecane
Cyclohexylamine
1-Cyclohexyl-2-methylaminopropane †
1,3-Diethyl-2-methylpiperidine
2-Isopropyl-5-methylcyclohexylamine
1,2,2,6,6-Pentamethylpiperidine
Piperidine, N-*Pentyl*

$C_{10}H_{21}NO$
5-Azoniaspiro[4,6]undecane (hydroxide)
6-Azoniaspiro[5,5]undecane (hydroxide)
Decanoic Acid, *Amide*
4-Hydroxy-2,2,6,6-tetramethylpiperidine, N-*Me*
2-Methylnonanoic Acid, *Amide*
7-Methylnonanoic Acid, *Amide*

$C_{10}H_{21}NO_2$
2-Aminodecanoic Acid
3-Aminodecanoic Acid
10-Aminodecanoic Acid
2-Aminohexanoic Acid, N-*Di-Et*
2-Amino-octanoic Acid, *Et ester*
3-Amino-octanoic Acid, *Et ester*
6-Dimethylaminohexanoic Acid, *Et ester*
Piperidine-2-aldehyde, *Di-Et acetal*
Piperidine-3-aldehyde, *Di-Et acetal*

$C_{10}H_{21}NO_3$
Desosamine, *Et glycoside*

$C_{10}H_{21}NO_5$
Glucosamine, 3,4,6-*Tri-Me ether*, *Methyl glycoside*

$C_{10}H_{21}NO_6$
Gluconic Acid, 2,3,4-*Tri-Me ether*, *Me ester* α-*form*
Gluconic Acid, 2,3,4,6-*Tetra-Me ether*

$C_{10}H_{21}NO_{10}S_3$
Glucoraphenin

$C_{10}H_{21}N_3O$
Diethylcarbamazine

$C_{10}H_{21}O_5P$
Phosphonoacetic Acid, 3,3-*Dipropyl*-1-*ethyl ester*

$C_{10}H_{22}$
Decane
2,3-Dimethyloctane
2,4-Dimethyloctane
2,5-Dimethyloctane
2,6-Dimethyloctane
2,7-Dimethyloctane
3,6-Dimethyloctane
4,4-Dimethyloctane
4,5-Dimethyloctane
2-Methylnonane
3-Methylnonane
4-Methylnonane
5-Methylnonane

$C_{10}H_{22}Cl_2N_2O_4$
Mannomustine

$C_{10}H_{22}N_2$
Piperazine, N,N-*Dipropyl*
2,3,5,6-Tetramethylpiperazine, N,N-*Di-Me*

$C_{10}H_{22}N_2O_4Pb$
Lead dipentyldinitrate

$C_{10}H_{22}O$
1-Decanol
2-Decanol
3-Decanol
Di-3-methylbutyl Ether
3,7-Dimethyl-1-octanol
4,7-Dimethyl-1-octanol
7,7-Dimethyl-1-octanol
2,6-Dimethyl-2-octanol
2,7-Dimethyl-2-octanol
2,2-Dimethyl-3-octanol
2,3-Dimethyl-3-octanol
2,7-Dimethyl-3-octanol
3,6-Dimethyl-3-octanol
3,7-Dimethyl-3-octanol
2,4-Dimethyl-4-octanol
2,5-Dimethyl-4-octanol
2,6-Dimethyl-4-octanol
2,7-Dimethyl-4-octanol
4,7-Dimethyl-4-octanol
Dipentyl Ether
3-Ethyl-2-methyl-3-heptanol
3-Ethyl-6-methyl-3-heptanol
5-Ethyl-2-methyl-3-heptanol
5-Ethyl-4-methyl-3-heptanol
2-Methyl-1-nonanol
3-Methyl-1-nonanol
4-Methyl-1-nonanol
5-Methyl-1-nonanol
2-Methyl-2-nonanol
2-Methyl-3-nonanol
2-Methyl-4-nonanol
4-Methyl-4-nonanol
5-Methyl-5-nonanol
2-Nonanol, *Me ether*
1-Octanol, *Et ether*
2-Octanol, *Et ether*
2,4,6-Trimethyl-4-heptanol

$C_{10}H_{22}OS$
Di-2-methylbutyl sulphoxide
Di-3-methylbutyl sulphoxide
Dipentyl Sulphoxide †

$C_{10}H_{22}OS_2$
5-Hydroxypentanal, *Di-Et mercaptal*, *Me ether*

$C_{10}H_{22}O_2$
Acetaldehyde, *Dibutyl acetal*
Acetaldehyde, *Di-isobutyl acetal*
1,4-Butanediol, *Dipropyl ether*
1,4-Butanediol, *Di-isopropyl ether*
1,2-Decanediol
1,3-Decanediol
1,4-Decanediol
1,5-Decanediol
1,10-Decanediol
2,9-Decanediol
3,8-Decanediol
4,7-Decanediol
Di-2-methyl-2-butyl peroxide
1,6-Hexanediol, *Di-Et ether*
2-Hexanone, *Di-Et ketal*
1,8-Octanediol, *Di-Me ether*

$C_{10}H_{22}O_2S$
Di-2-methylbutyl sulphone
Di-3-methylbutyl sulphone
Dipentyl Sulphone †

$C_{10}H_{22}O_3$
1,3,3-Butanetriol, *Tri-Et ether*
Di-(2-hydroxyethyl) Ether, *Dipropyl ether*
Orthoacetic Acid, *Et-dipropyl ester*
Orthoacetic Acid, *Di-isobutyl ester*
Orthoformic Acid, *Tripropyl ester*

$C_{10}H_{22}O_4$
Glyoxal, *Tetra-Et acetal*
$C_{10}H_{22}O_5$
Rhamnitol, 1,3,4,5-*Tetra-Me ether*
$C_{10}H_{22}O_6$
Mannitol, 1,2,3,4-*Tetra-Me ether*
Sorbitol, *Tetra-Me ether*
$C_{10}H_{22}O_7$
Dipentaerythritol
$C_{10}H_{22}S$
Di-2-methylbutyl sulphide
Di-3-methylbutyl sulphide
Dipentyl sulphide†
$C_{10}H_{22}S_2$
Di-2-methylbutyl disulphide
Di-3-methylbutyl disulphide
Dipentyl disulphide†
$C_{10}H_{23}AsO_3$
Butyl ethyl arsenite
$C_{10}H_{23}N$
Di-(3-methylbutyl)amine
Dipentylamine
$C_{10}H_{23}NO$
2-Di-isobutylaminoethanol
$C_{10}H_{23}O_3PS$
Di-*O*,*O*-butyl-*S*-ethylphosphorothiolate
$C_{10}H_{24}N_2$
1,10-Diaminodecane
$C_{10}H_{24}Pb$
Lead dibutyldimethyl
$C_{10}H_{25}BrMgO_2$
Ethylmagnesium bromide dietherate†
$C_{10}H_{25}Sb$
Pentaethylstibine
$C_{10}H_{26}N_4$
Spermine
$C_{10}H_{30}O_5Si_5$
2,4,6,8,10-Pentaethylcyclopentasiloxane
$C_{10}N_6$
Hexacyano-1,3-butadiene†
$C_{10}O_8$
1,4-Benzoquinone-tetracarboxylic Acid, *Dianhydride*†
$C_{10}O_{10}$
Tetrahydroxy-*p*-benzoquinone bisoxalate†

C_{11}

$C_{11}H_4N_2O_7$
4,5-Dinitronaphthalene-1,8-dicarboxylic Acid, *Anhydride*
$C_{11}H_5Br_2N$
1,6-Dibromo-2-naphthoic Acid, *Nitrile*
$C_{11}H_5Cl_2N$
5,8-Dichloro-2-naphthoic Acid, *Nitrile*
$C_{11}H_5NO_3$
Quinoline-2,3-dicarboxylic Acid, *Anhydride*
$C_{11}H_6BrClO$
7-Bromo-1-naphthoic Acid, *Chloride*
$C_{11}H_6BrN$
4-Bromo-1-naphthoic Acid, *Nitrile*
5-Bromo-1-naphthoic Acid, *Nitrile*
$C_{11}H_6BrNO_4$
5-Bromo-8-nitro-1-naphthoic Acid
8-Bromo-3-nitro-1-naphthoic Acid
8-Bromo-5-nitro-1-naphthoic Acid
$C_{11}H_6Br_2O_2$
1,6-Dibromo-2-naphthoic Acid
4,7-Dibromo-2-naphthoic Acid
5,8-Dibromo-2-naphthoic Acid
$C_{11}H_6Br_4O$
1,3,4,6-Tetrabromo-2-naphthol, *Me ether*
$C_{11}H_6Br_5NO$
2-(3,5-Dibromo-2-hydroxyphenyl)-3,4,5-tribromopyrrole, *Me ether*†
$C_{11}H_6ClIO$
7-Iodo-1-naphthoic Acid, *Chloride*
$C_{11}H_6ClN$
4-Chloro-1-naphthoic Acid, *Nitrile*
5-Chloro-1-naphthoic Acid, *Nitrile*
5-Chloro-2-naphthoic Acid, *Nitrile*
$C_{11}H_6ClNO_3$
3-Nitro-1-naphthoic Acid, *Chloride*
4-Nitro-1-naphthoic Acid, *Chloride*
5-Nitro-1-naphthoic Acid, *Chloride*
6-Nitro-1-naphthoic Acid, *Chloride*
5-Nitro-2-naphthoic Acid, *Chloride*
$C_{11}H_6Cl_2O$
5-Chloro-1-naphthoic Acid, *Chloride*
6-Chloro-1-naphthoic Acid, *Chloride*
7-Chloro-1-naphthoic Acid, *Chloride*
3-Chloro-2-naphthoic Acid, *Chloride*
5-Chloro-2-naphthoic Acid, *Chloride*
$C_{11}H_6Cl_2O_2$
4-Chloro-3-hydroxy-2-naphthoic Acid, *Chloride*
3,6-Dichloro-1-naphthoic Acid
5,8-Dichloro-1-naphthoic Acid
4,5-Dichloro-2-naphthoic Acid
5,8-Dichloro-2-naphthoic Acid
$C_{11}H_6F_{14}O_4$
Heptafluorobutyric Acid, *Trimethylene ester*
$C_{11}H_6N_2O_2$
3-Nitro-1-naphthoic Acid, *Nitrile*
4-Nitro-1-naphthoic Acid, *Nitrile*
5-Nitro-1-naphthoic Acid, *Nitrile*
8-Nitro-1-naphthoic Acid, *Nitrile*
1-Nitro-2-naphthoic Acid, *Nitrile*
5-Nitro-2-naphthoic Acid, *Nitrile*
8-Nitro-2-naphthoic Acid, *Nitrile*
Quinoline-2,3-dicarboxylic Acid, *Imide*
Quinoline-2,4-dicarboxylic Acid, 2-*Nitrile*
$C_{11}H_6N_2O_3$
3-Nitronaphthastyril
4-Nitronaphthastyril
5-Nitronaphthastyril
$C_{11}H_6N_2O_6$
3,6-Dinitro-1-naphthoic Acid
4,5-Dinitro-1-naphthoic Acid

$C_{11}H_6N_2O_6$ (*continued*)
5,8-Dinitro-1-naphthoic Acid
6,8-Dinitro-1-naphthoic Acid
4,5-Dinitro-2-naphthoic Acid
4,7-Dinitro-2-naphthoic Acid

$C_{11}H_6O_2$
8-Hydroxy-1-naphthoic Acid, *Lactone*

$C_{11}H_6O_3$
Angelicin
4*H*-Furo[3,2-*c*]benzopyran-4-one
Psoralene

$C_{11}H_6O_4$
Bergaptol
Xanthotoxol

$C_{11}H_6O_6$
Phthaloylmalonic Acid

$C_{11}H_6O_{10}$
Benzenepentacarboxylic Acid

$C_{11}H_7BrO_2$
3-Bromo-1-naphthoic Acid
4-Bromo-1-naphthoic Acid
5-Bromo-1-naphthoic Acid
7-Bromo-1-naphthoic Acid
8-Bromo-1-naphthoic Acid
1-Bromo-2-naphthoic Acid
3-Bromo-2-naphthoic Acid

$C_{11}H_7BrO_3$
3-Bromo-4-hydroxy-1-naphthoic Acid
4-Bromo-3-hydroxy-1-naphthoic Acid
6-Bromo-2-hydroxy-1-naphthoic Acid
4-Bromo-1-hydroxy-2-naphthoic Acid
4-Bromo-3-hydroxy-2-naphthoic Acid
6-Bromo-3-hydroxy-2-naphthoic Acid
7-Bromo-3-hydroxy-2-naphthoic Acid

$C_{11}H_7ClO$
2-Chloro-1-naphthaldehyde
4-Chloro-1-naphthaldehyde
6-Chloro-1-naphthaldehyde
7-Chloro-1-naphthaldehyde
3-Chloro-2-naphthaldehyde
1-Naphthoic Acid, *Chloride*
2-Naphthoic Acid, *Chloride*

$C_{11}H_7ClO_2$
Chloroformic Acid, 1-*Naphthyl ester*
2-Chloro-1-naphthoic Acid
4-Chloro-1-naphthoic Acid
5-Chloro-1-naphthoic Acid
6-Chloro-1-naphthoic Acid
7-Chloro-1-naphthoic Acid
8-Chloro-1-naphthoic Acid
1-Chloro-2-naphthoic Acid
3-Chloro-2-naphthoic Acid
4-Chloro-2-naphthoic Acid
5-Chloro-2-naphthoic Acid
6-Chloro-2-naphthoic Acid
8-Chloro-2-naphthoic Acid
1-Hydroxy-2-naphthoic Acid, *Chloride*
3-Hydroxy-2-naphthoic Acid, *Chloride*

$C_{11}H_7ClO_3$
4-Chloro-1-hydroxy-2-naphthoic Acid
4-Chloro-3-hydroxy-2-naphthoic Acid
5-Chloro-3-hydroxy-2-naphthoic Acid
5-Chloro-6-hydroxy-2-naphthoic Acid
6-Chloro-3-hydroxy-2-naphthoic Acid
7-Chloro-3-hydroxy-2-naphthoic Acid
7-Chloro-4-hydroxy-2-naphthoic Acid
4-Methylcoumarin-3-carboxylic Acid, *Chloride*

$C_{11}H_7Cl_2NO_2$
1-Methylindole-2,3-dicarboxylic Acid, *Dichloride*

$C_{11}H_7Cl_2NO_3$
Pyoluteorin

$C_{11}H_7HgN$
Mercuri-1-naphthyl cyanide

$C_{11}H_7IO_2$
4-Iodo-1-naphthoic Acid
5-Iodo-1-naphthoic Acid
7-Iodo-1-naphthoic Acid
8-Iodo-1-naphthoic Acid
3-Iodo-2-naphthoic Acid
5-Iodo-2-naphthoic Acid

$C_{11}H_7N$
1-Naphthonitrile
2-Naphthonitrile

$C_{11}H_7NO$
Furo[3,2-*c*]quinoline
3-Hydroxy-2-naphthoic Acid, *Nitrile*
Naphthastyril
Naphth[1,2-*d*]oxazole
Naphth[2,1-*d*]oxazole
1-Naphthyl isocyanate
2-Naphthyl isocyanate

$C_{11}H_7NO_2$
4-Methylcoumarin-3-carboxylic Acid, *Nitrile*

$C_{11}H_7NO_3$
1-Methylindole-2,3-dicarboxylic Acid, *Anhydride*
4-Nitro-1-naphthaldehyde
5-Nitro-1-naphthaldehyde
8-Nitro-1-naphthaldehyde
1-Nitro-2-naphthaldehyde

$C_{11}H_7NO_4$
2-Hydroxy-6-nitro-1-naphthaldehyde
2-Methyl-3-nitro-1,4-naphthoquinone
3-Nitro-1-naphthoic Acid
4-Nitro-1-naphthoic Acid
5-Nitro-1-naphthoic Acid
6-Nitro-1-naphthoic Acid
8-Nitro-1-naphthoic Acid
1-Nitro-2-naphthoic Acid
5-Nitro-2-naphthoic Acid
6-Nitro-2-naphthoic Acid
7-Nitro-2-naphthoic Acid
8-Nitro-2-naphthoic Acid
Quinoline-2,3-dicarboxylic Acid
Quinoline-2,4-dicarboxylic Acid
Quinoline-2,6-dicarboxylic Acid
Quinoline-3,7-dicarboxylic Acid
Quinoline-5,6-dicarboxylic Acid
Quinoline-5,8-dicarboxylic Acid
Quinoline-6,7-dicarboxylic Acid
Quinoline-7,8-dicarboxylic Acid

$C_{11}H_7NO_5$
2-Hydroxy-6-nitro-1-naphthoic Acid
4-Hydroxy-3-nitro-1-naphthoic Acid

$C_{11}H_7NO_5$ (*continued*)
8-Hydroxy-5-nitro-1-naphthoic Acid
1-Hydroxy-4-nitro-2-naphthoic Acid
3-Hydroxy-4-nitro-2-naphthoic Acid

$C_{11}H_7NS$
Thianaphtheno[2,3-*c*]pyridine
Thianaphtheno[3,2-*c*]pyridine

$C_{11}H_7NS_2$
2-Mercaptonaphtho[1,2-*d*]thiazole
2-Mercaptonaphtho[2,1-*d*]thiazole

$C_{11}H_7N_3O_7$
2,4,5-Trinitro-1-naphthol, *Me ether*
1,6,8-Trinitro-2-naphthol, *Me ether*

$C_{11}H_8$
5*H*-Benzocyclohepten-5-one†
7*H*-Benzocyclohepten-7-one†

$C_{11}H_8BrClO$
6-Bromo-1-chloro-2-naphthol, *Me ether*

$C_{11}H_8BrNO$
5-Bromo-1-naphthoic Acid, *Amide*
7-Bromo-1-naphthoic Acid, *Amide*
8-Bromo-1-naphthoic Acid, *Amide*

$C_{11}H_8BrNO_3$
4-Bromo-2-nitro-1-naphthol, *Me ether*
6-Bromo-1-nitro-2-naphthol, *Me ether*

$C_{11}H_8Br_2O$
1,6-Dibromo-2-naphthol, *Me ether*
4,6-Dibromo-2-naphthol, *Me ether*
5,8-Dibromo-2-naphthol, *Me ether*

$C_{11}H_8ClN$
2-Chloro-6-phenylpyridine
4-Chloro-2-phenylpyridine

$C_{11}H_8ClNO$
4-Chloro-1-naphthoic Acid, *Amide*
5-Chloro-1-naphthoic Acid, *Amide*
6-Chloro-1-naphthoic Acid, *Amide*
7-Chloro-1-naphthoic Acid, *Amide*
8-Chloro-1-naphthoic Acid, *Amide*
3-Chloro-2-naphthoic Acid, *Amide*
5-Chloro-2-naphthoic Acid, *Amide*
6-Chloro-2-naphthoic Acid, *Amide*
3-Methylquinoline-4-carboxylic Acid, *Chloride*

$C_{11}H_8ClNO_2$
4-Chloro-α-cyanocinnamic Acid, *Me ester*
4-Chloro-3-hydroxy-2-naphthoic Acid, *Amide*
2-Chloroquinoline-4-carboxylic Acid, *Me ester*
6-Chloroquinoline-4-carboxylic Acid, *Me ester*
6-Methoxyquinoline-4-carboxylic Acid, *Chloride*

$C_{11}H_8ClNO_3$
2-Phthalimidopropionic Acid, *Chloride*
3-Phthalimidopropionic Acid, *Chloride*

$C_{11}H_8Cl_2O$
2,4-Dichloro-1-naphthol, *Me ether*
4,8-Dichloro-2-naphthol, *Me ether*

$C_{11}H_8Cl_2O_2$
Indane-2,2-dicarboxylic Acid, *Dichloride*

$C_{11}H_8F_7NO$
Heptafluorobutyric Acid, *Benzylamide*

$C_{11}H_8INO$
7-Iodo-1-naphthoic Acid, *Amide*
3-Iodo-2-naphthoic Acid, *Amide*
5-Iodo-2-naphthoic Acid, *Amide*

$C_{11}H_8I_3NO_3$
3-Amino-2,4,6-tri-iodobenzoic Acid, N-*Ethoxalyl*

$C_{11}H_8N_2$
4-Amino-1-naphthoic Acid, *Nitrile*
5-Amino-1-naphthoic Acid, *Nitrile*
1-Amino-2-naphthoic Acid, *Nitrile*
4-Amino-2-naphthoic Acid, *Nitrile*
5-Amino-2-naphthoic Acid, *Nitrile*
7-Amino-2-naphthoic Acid, *Nitrile*
8-Amino-2-naphthoic Acid, *Nitrile*
Benz[*e*]indazole
Benz[*g*]indazole
α-Carboline
β-Carboline
γ-Carboline
1,5-Diazafluorene
1,8-Diazafluorene
Imidazo[5,1-*a*]isoquinoline†
2-Methylquinoline-3-carboxylic Acid, *Nitrile*
8-Methylquinoline-4-carboxylic Acid, *Nitrile*
2-Methylquinoline-5-carboxylic Acid, *Nitrile*
Naphth[1,2-*d*]imidazole
Perimidine (incorrectly given as Naphth[1,8-*de*]-imidazole)
Pyrrolo[1,2-*a*]quinoxaline†

$C_{11}H_8N_2O$
2-Hydroxy-4-methylquinoline-3-carboxylic Acid, *Nitrile*
2-Hydroxyquinoline-3-carboxylic Acid, *Me ether*, *Nitrile*
6-Methoxyquinoline-4-carboxylic Acid, *Nitrile*
Perimidone (incorrectly given as Naphth[1,8-*de*]-imidazol-2-one)

$C_{11}H_8N_2O_2$
2-*o*-Nitrophenylpyridine
2-*m*-Nitrophenylpyridine
2-*p*-Nitrophenylpyridine
3-*o*-Nitrophenylpyridine
3-*m*-Nitrophenylpyridine
3-*p*-Nitrophenylpyridine
4-*o*-Nitrophenylpyridine
4-*m*-Nitrophenylpyridine
4-*p*-Nitrophenylpyridine
6-Phenylpyridazine-3-carboxylic Acid
5-Phenylpyridazine-4-carboxylic Acid

$C_{11}H_8N_2O_3$
3-Nitro-1-naphthoic Acid, *Amide*
4-Nitro-1-naphthoic Acid, *Amide*
5-Nitro-1-naphthoic Acid, *Amide*
6-Nitro-1-naphthoic Acid, *Amide*
8-Nitro-1-naphthoic Acid, *Amide*
5-Nitro-2-naphthoic Acid, *Amide*
8-Nitro-2-naphthoic Acid, *Amide*
Quinoline-2,3-dicarboxylic Acid, *Amide*

$C_{11}H_8N_2O_3S_2$
Luciferin

$C_{11}H_8N_2O_4$
1-Methyl-2,4-dinitronaphthalene
1-Methyl-4,5-dinitronaphthalene

$C_{11}H_8N_2O_4$ (*continued*)
1-Methyl-4,8-dinitronaphthalene
2-Methyl-1,5-dinitronaphthalene
2-Methyl-5-nitroquinoline-3-carboxylic Acid
2-Methyl-8-nitroquinoline-3-carboxylic Acid
o-Nitrobenzylidenemalonic Acid, *Mononitrile, Me ester*
m-Nitrobenzylidenemalonic Acid, *Mononitrile, Me ester*
5-Nitrofuran-2-carboxylic Acid, *Anilide*
4-Phenylpyrazole-1,3(5)-dicarboxylic Acid
3-Phenylpyrazole-1,4-dicarboxylic Acid
3(5)-Phenylpyrazole-1,5(3)-dicarboxylic Acid
1-Phenylpyrazole-3,4-dicarboxylic Acid
5-Phenylpyrazole-3,4-dicarboxylic Acid
1-Phenylpyrazole-3,5-dicarboxylic Acid
4-Phenylpyrazole-3,5-dicarboxylic Acid
1-Phenylpyrazole-4,5-dicarboxylic Acid

$C_{11}H_8N_2O_5$
2,4-Dinitro-1-naphthol, *Me ether*
4,5-Dinitro-1-naphthol, *Me ether*
1,6-Dinitro-2-naphthol, *Me ether*
1,8-Dinitro-2-naphthol, *Me ether*
8-Hydroxy-5-nitroquinoline-7-carboxylic Acid, *Me ester*

$C_{11}H_8N_2O_6$
1,8-Dihydroxy-2,4-dinitronaphthalene, 1-*Me ether*

$C_{11}H_8N_4O_2$
10-Methylisoalloxazine†

$C_{11}H_8N_4O_3$
3-Phenyluric Acid
9-Phenyluric Acid

$C_{11}H_8O$
1-Naphthaldehyde
2-Naphthaldehyde

$C_{11}H_8OS$
2-Benzoylthiophene
3-Benzoylthiophene

$C_{11}H_8O_2$
Azulene-1-carboxylic Acid
Azulene-5-carboxylic Acid
Azulene-6-carboxylic Acid
3,4-Benzotropolone
4,5-Benzotropolone
2-Benzoylfuran
2-Decene-4,6,8-triynoic Acid, *Me ester*
8-Decene-2,4,6-triynoic Acid, *Me ester*
Drosophilin C†
Drosophilin D†
2-Hydroxy-1-naphthaldehyde
3-Hydroxy-1-naphthaldehyde
4-Hydroxy-1-naphthaldehyde
5-Hydroxy-1-naphthaldehyde
1-Hydroxy-2-naphthaldehyde
3-Hydroxy-2-naphthaldehyde
4-Hydroxy-2-naphthaldehyde
6-Hydroxy-2-naphthaldehyde
3-Methyl-1,2-naphthoquinone
4-Methyl-1,2-naphthoquinone
5-Methyl-1,2-naphthoquinone
6-Methyl-1,2-naphthoquinone
2-Methyl-1,4-naphthoquinone★†
5-Methyl-1,4-naphthoquinone
6-Methyl-1,4-naphthoquinone
1-Naphthoic Acid
2-Naphthoic Acid
Nemotin
6-Phenyl-2*H*-pyran-2-one
2-Phenyl-4*H*-pyran-4-one

$C_{11}H_8O_3$
3-Acetylcoumarin
2,3-Dihydroxy-1-naphthaldehyde
2,4-Dihydroxy-1-naphthaldehyde
2,5-Dihydroxy-1-naphthaldehyde
2,6-Dihydroxy-1-naphthaldehyde
2,7-Dihydroxy-1-naphthaldehyde
2,8-Dihydroxy-1-naphthaldehyde
3,4-Dihydroxy-1-naphthaldehyde
4,5-Dihydroxy-1-naphthaldehyde
4,6-Dihydroxy-1-naphthaldehyde
4,7-Dihydroxy-1-naphthaldehyde
4,8-Dihydroxy-1-naphthaldehyde
1,4-Dihydroxy-2-naphthaldehyde
1,5-Dihydroxy-2-naphthaldehyde
1,8-Dihydroxy-2-naphthaldehyde
Furan-2-carboxylic Acid, *Phenyl ester*
10-Hydroxydec-2-ene-4,6,8-triynoic Acid, *Me ester*†
2-Hydroxy-1-naphthoic Acid
3-Hydroxy-1-naphthoic Acid
4-Hydroxy-1-naphthoic Acid
5-Hydroxy-1-naphthoic Acid
6-Hydroxy-1-naphthoic Acid
7-Hydroxy-1-naphthoic Acid
8-Hydroxy-1-naphthoic Acid
1-Hydroxy-2-naphthoic Acid
3-Hydroxy-2-naphthoic Acid
4-Hydroxy-2-naphthoic Acid
5-Hydroxy-2-naphthoic Acid
6-Hydroxy-2-naphthoic Acid
7-Hydroxy-2-naphthoic Acid
8-Hydroxy-2-naphthoic Acid
Phenylcitraconic Acid, *Anhydride*
Phenylitaconic Acid, *Anhydride*
3-Phenyl-2-pentenedioic Acid, *Anhydride*
Phthiocol
Plumbagin

$C_{11}H_8O_4$
Coumarin-3-carboxylic Acid, *Me ester*
2,7-Dihydroxy-1-naphthoic Acid
3,4-Dihydroxy-1-naphthoic Acid
4,5-Dihydroxy-1-naphthoic Acid
4,8-Dihydroxy-1-naphthoic Acid
6,7-Dihydroxy-1-naphthoic Acid
1,3-Dihydroxy-2-naphthoic Acid
1,4-Dihydroxy-2-naphthoic Acid
1,5-Dihydroxy-2-naphthoic Acid
1,7-Dihydroxy-2-naphthoic Acid
1,8-Dihydroxy-2-naphthoic Acid
3,4-Dihydroxy-2-naphthoic Acid
3,5-Dihydroxy-2-naphthoic Acid
3,6-Dihydroxy-2-naphthoic Acid
3,7-Dihydroxy-2-naphthoic Acid
2,3-Dihydroxy-1,4-naphthoquinone, *Me ether*
Droserone
Indene-2,3-dicarboxylic Acid
Indene-3,6-dicarboxylic Acid

$C_{11}H_8O_4$ (*continued*)
3-Methylchromone-2-carboxylic Acid
4-Methylcoumarin-3-carboxylic Acid
5-Methylcoumarin-3-carboxylic Acid
6-Methylcoumarin-3-carboxylic Acid
7-Methylcoumarin-3-carboxylic Acid
8-Methylcoumarin-3-carboxylic Acid
6-Methylcoumarin-4-carboxylic Acid
7-Methylcoumarin-4-carboxylic Acid

$C_{11}H_8O_5$
6-Formyl-7-hydroxy-5-methoxycoumarin
Oosponol †
Purpurogallin
2,5,8-Trihydroxy-3-methyl-1,4-naphthoquinone
6,7,8-Trihydroxy-1-naphthoic Acid

$C_{11}H_8O_6$
Glauconin
5-Hydroxytoluene-2,3,4-tricarboxylic Acid, *Me ether*, *Anhydride* †
5-Hydroxytoluene-2,3,6-tricarboxylic Acid, *Me ether*, *Anhydride* †
3-Hydroxytoluene-2,4,5-tricarboxylic Acid, *Me ester*, *Anhydride*

$C_{11}H_9Br$
1-Bromomethylnaphthalene
1-Bromo-2-methylnaphthalene
1-Bromo-3-methylnaphthalene
1-Bromo-4-methylnaphthalene
1-Bromo-5-methylnaphthalene
1-Bromo-6-methylnaphthalene
1-Bromo-7-methylnaphthalene
1-Bromo-8-methylnaphthalene
2-Bromomethylnaphthalene
2-Bromo-1-methylnaphthalene
3-Bromo-1-methylnaphthalene
7-Bromo-1-methylnaphthalene

$C_{11}H_9BrNO$
5-Bromo-6-hydroxyquinoline, *Et ether*
5-Bromo-8-hydroxyquinoline, *Et ether*

$C_{11}H_9BrO$
3-Bromo-2-hydroxy-1-methylnaphthalene
6-Bromo-2-hydroxy-1-methylnaphthalene
5-Bromo-1-naphthol, *Me ether*
1-Bromo-2-naphthol, *Me ether*
3-Bromo-2-naphthol, *Me ether*
4-Bromo-2-naphthol, *Me ether*
6-Bromo-2-naphthol, *Me ether*

$C_{11}H_9BrO_6$
6-Bromobenzene-1,2,4-tricarboxylic Acid, 2,4-*Di-Me ester*

$C_{11}H_9Br_2NO_2$
Phthalimide, N-2,3-*Dibromopropyl*

$C_{11}H_9Cl$
1-Chloromethylnaphthalene
2-Chloromethylnaphthalene
2-Chloro-1-methylnaphthalene

$C_{11}H_9ClN_2O$
5-Methyl-3-phenylpyrazole-1-carboxylic Acid, *Chloride*
5-Methyl-1-phenylpyrazole-3-carboxylic Acid, *Chloride*
5-Methyl-1-phenylpyrazole-4-carboxylic Acid, *Chloride*
3-Methyl-1-phenylpyrazole-5-carboxylic Acid, *Chloride*

$C_{11}H_9ClO_2S$
2-Methylnaphthalene-1-sulphonic Acid, *Chloride*
4-Methylnaphthalene-1-sulphonic Acid, *Chloride*
7-Methylnaphthalene-1-sulphonic Acid, *Chloride*
4-Methylnaphthalene-2-sulphonic Acid, *Chloride*
5-Methylnaphthalene-2-sulphonic Acid, *Chloride*
6-Methylnaphthalene-2-sulphonic Acid, *Chloride*

$C_{11}H_9ClO_3$
3-Chloro-1-naphthol, *Me ether*
1-Chloro-2-naphthol, *Me ether*
3-Chloro-2-naphthol, *Me ether*
4-Chloro-2-naphthol, *Me ether*

$C_{11}H_9ClO_3S$
4-Chloronaphthalene-1-sulphonic Acid, *Me ester*
5-Chloronaphthalene-1-sulphonic Acid, *Me ester*
7-Chloronaphthalene-1-sulphonic Acid, *Me ester*
8-Chloronaphthalene-1-sulphonic Acid, *Me ester*
6-Chloronaphthalene-2-sulphonic Acid, *Me ester*
7-Chloronaphthalene-2-sulphonic Acid, *Me ester*
2-Naphthol-3-sulphonic Acid, *Me ether*, *Chloride*
2-Naphthol-6-sulphonic Acid, *Me ether*, *Chloride*
2-Naphthol-8-sulphonic Acid, *Me ether*, *Chloride*

$C_{11}H_9ClO_4$
Sordidone †

$C_{11}H_9Cl_2NO_2$
Phthalimide, N-2,3-*Dichloropropyl*

$C_{11}H_9F$
1-Fluoro-2-methylnaphthalene

$C_{11}H_9IO$
5-Iodo-1-naphthol, *Me ether*
3-Iodo-2-naphthol, *Me ether*
4-Iodo-2-naphthol, *Me ether*

$C_{11}H_9N$
1,2-Dihydrocyclobuta[*b*]quinoline †
3-Indenylacetic Acid, *Nitrile*
5-Phenyl-2,4-pentadienoic Acid, *Nitrile*
2-Phenylpyridine
3-Phenylpyridine
4-Phenylpyridine
9*H*-Pyrrolo[1,2-*a*]indole †
2-Vinylquinoline

$C_{11}H_9NO$
2-Benzoylpyrrole
Furfurylideneaniline
3-Hydroxyindene-2-carboxylic Acid, *Nitrile*, *Me ether*
2-Hydroxyindene-3-carboxylic Acid, *Nitrile*, *Me ether*
3-Hydroxypyridine, *Phenyl ether*
2-Methylquinoline-5-aldehyde

$C_{11}H_9NO$ (*continued*)
2-Methylquinoline-6-aldehyde
1-Phenyl-α-pyridone
6-Phenyl-α-pyridone
1-Phenyl-γ-pyridone
6-Phenyl-γ-pyridone
5-Phenylpyrrole-2-aldehyde
2-Quinolineacetaldehyde

$C_{11}H_9NO_2$
2-Amino-1-naphthoic Acid
3-Amino-1-naphthoic Acid
4-Amino-1-naphthoic Acid
5-Amino-1-naphthoic Acid
6-Amino-1-naphthoic Acid
7-Amino-1-naphthoic Acid
8-Amino-1-naphthoic Acid
1-Amino-2-naphthoic Acid
3-Amino-2-naphthoic Acid
4-Amino-2-naphthoic Acid
5-Amino-2-naphthoic Acid
6-Amino-2-naphthoic Acid
7-Amino-2-naphthoic Acid
8-Amino-2-naphthoic Acid
α-Cyanocinnamic Acid, *Me ester*
Furan-3-carboxylic Acid, *Anilide*
4-Hydroxy-2-methylquinoline-3-aldehyde
3-Hydroxy-1-naphthoic Acid, *Amide*
1-Hydroxy-2-naphthoic Acid, *Amide*
3-Hydroxy-2-naphthoic Acid, *Amide*
8-Hydroxyquinoline-5-aldehyde, *Me ether*
3-3′-Indolylacrylic Acid
2-Methyl-1,4-naphthoquinone, 4-*Mono-oxime*†
1-Methyl-2-nitronaphthalene
1-Methyl-4-nitronaphthalene
1-Methyl-5-nitronaphthalene
1-Methyl-6-nitronaphthalene
1-Methyl-7-nitronaphthalene
1-Methyl-8-nitronaphthalene
2-Methyl-1-nitronaphthalene
2-Methyl-3-nitronaphthalene
2-Methyl-6-nitronaphthalene
2-Methyl-7-nitronaphthalene
2-Methyl-8-nitronaphthalene
6-Methyl-1-nitronaphthalene
3-Methylquinoline-2-carboxylic Acid
4-Methylquinoline-2-carboxylic Acid
2-Methylquinoline-3-carboxylic Acid
2-Methylquinoline-4-carboxylic Acid
3-Methylquinoline-4-carboxylic Acid
7-Methylquinoline-4-carboxylic Acid
8-Methylquinoline-4-carboxylic Acid
2-Methylquinoline-5-carboxylic Acid
8-Methylquinoline-5-carboxylic Acid
2-Methylquinoline-6-carboxylic Acid
4-Methylquinoline-6-carboxylic Acid
2-Methylquinoline-8-carboxylic Acid
5-Methylquinoline-8-carboxylic Acid
6-Methylquinoline-8-carboxylic Acid
N-1-Naphthoylhydroxylamine
N-2-Naphthoylhydroxylamine
1-Nitromethylnaphthalene
2-Nitromethylnaphthalene
2-Nitroso-1-naphthol, *Me ether*
4-Nitroso-1-naphthol, *Me ether*
1-Nitroso-2-naphthol, *Me ether*
3-Phenyl-2-pentenedioic Acid, *Mononitrile*
3-Phenyl-2-pentenedioic Acid, *Imide*
1-Phenylpyrrole-2-carboxylic Acid
2-Phenylpyrrole-3-carboxylic Acid
5-Phenylpyrrole-3-carboxylic Acid
Phthalimide, N-*Allyl*
Phthalimide, N-*Cyclopropyl*
Phthalimide, N-*Propenyl*
2-Quinolineacetic Acid
3-Quinolineacetic Acid
4-Quinolineacetic Acid
6-Quinolineacetic Acid
8-Quinolineacetic Acid
Quinoline-2-carboxylic Acid, *Me ester*
Quinoline-3-carboxylic Acid, *Me ester*
Quinoline-4-carboxylic Acid, *Me ester*
Quinoline-6-carboxylic Acid, *Me ester*
Quinoline-8-carboxylic Acid, *Me ester*

$C_{11}H_9NO_2S$
1,8-Naphthosultam, N-*Me*

$C_{11}H_9NO_3$
3-Amino-4-hydroxy-1-napthoic Acid
4-Amino-3-hydroxy-1-napthoic Acid
4-Amino-1-hydroxy-2-napthoic Acid
4-Amino-3-hydroxy-2-napthoic Acid
7-Amino-3-hydroxy-2-napthoic Acid
Anibine†
Benzoylmalonic Acid, *Me-ester-nitrile*
2-Benzyl-4-hydroxymethylene-5-oxazolone
3-Formyl-2-hydroxy-4-methoxyquinoline
1-Hydroxyisoquinoline-3-carboxylic Acid, N-*Me*
1-Hydroxyisoquinoline-4-carboxylic Acid, N-*Me*
2-Hydroxy-4-methylquinoline-3-carboxylic Acid
4-Hydroxy-2-methylquinoline-3-carboxylic Acid
2-Hydroxy-3-methylquinoline-4-carboxylic Acid
3-Hydroxy-2-methylquinoline-4-carboxylic Acid
2-Hydroxy-4-methylquinoline-8-carboxylic Acid
4-Hydroxyquinoline-2-carboxylic Acid, *Me ester*
4-Hydroxyquinoline-2-carboxylic Acid, *Me ether*
2-Hydroxyquinoline-3-carboxylic Acid, *Me ester*
2-Hydroxyquinoline-3-carboxylic Acid, *Me ether*
4-Hydroxyquinoline-3-carboxylic Acid, *Me ether*
2-Hydroxyquinoline-4-carboxylic Acid, *Me ester*
7-Hydroxyquinoline-4-carboxylic Acid, *Me ether*
8-Hydroxyquinoline 5-carboxylic Acid, *Me ether*
4-Hydroxyquinoline-8-carboxylic Acid, *Me ester*
3-3′-Indolylpyruvic Acid
6-Methoxyquinoline-4-carboxylic Acid
4-Methylcoumarin-3-carboxylic Acid, *Amide*
6-Methyl-1-nitro-2-naphthol
2-Nitro-1-naphthol, *Me ether*
4-Nitro-1-naphthol, *Me ether*
5-Nitro-1-naphthol, *Me ether*
8-Nitro-1-naphthol, *Me ether*
1-Nitro-2-naphthol, *Me ether*
4-Nitro-2-naphthol, *Me ether*
6-Nitro-2-naphthol, *Me ether*
8-Nitro-2-naphthol, *Me ether*★†

$C_{11}H_9NO_4$
4,8-Dihydroxyquinoline-2-carboxylic Acid, *Me ester*
2,4-Dihydroxyquinoline-3-carboxylic Acid, *Me ester*

$C_{11}H_9NO_4$ (*continued*)
Indole-2,3-dicarboxylic Acid, *Mono-Me ester*
Isatogenic Acid, *Et ester*
1-Methylindole-2,3-dicarboxylic Acid
3-Methylindole-2,5-dicarboxylic Acid
o-Nitrophenylpropiolic Acid, *Et ester*
p-Nitrophenylpropiolic Acid, *Et ester*
2-Phthalimidopropionic Acid
3-Phthalimidopropionic Acid

$C_{11}H_9NO_5$
7-Hydroxy-4,8-dimethyl-6-nitrocoumarin †

$C_{11}H_9NO_5S$
4-Nitronaphthalene-1-sulphonic Acid, *Me ester*
5-Nitronaphthalene-1-sulphonic Acid, *Me ester*
8-Nitronaphthalene-1-sulphonic Acid, *Me ester*

$C_{11}H_9NO_6$
2-Methylcarbamoyl-4,5-methylenedioxybenzoylformic Acid
3-(4,5-Methylenedioxy-2-nitrophenyl)acrylic Acid, *Me ester*
5-Nitroindane-2,2-dicarboxylic Acid

$C_{11}H_9N_2$ (ion)
Dipyrido[1,2-*c*:2′,1′-*e*]imidazolium cation †

$C_{11}H_9N_3O_2$
Anthranilinodiacetic Acid, α-*Dinitrile*
Quinoline-2,3-dicarboxylic Acid, *Diamide*

$C_{11}H_9N_3O_4$
4,8-Dinitro-1-naphthylamine, N-*Me*
5-Isonitrosobarbituric Acid, 5-O-*Benzyl ether*

$C_{11}H_{10}$
5*H*-Benzocycloheptene †
7*H*-Benzocycloheptene †
7*H*-Cycloheptabenzene
1,6-Methanocyclodecapentaene †
1-Methylazulene
2-Methylazulene
4-Methylazulene
5-Methylazulene
6-Methylazulene
1-Methylnaphthalene
2-Methylnaphthalene

$C_{11}H_{10}BrNO_2$
Phthalimide, N-2-*Bromopropyl*

$C_{11}H_{10}ClN$
2-Chloro-3,4-dimethylquinoline
2-Chloro-4,6-dimethylquinoline
2-Chloro-4,8-dimethylquinoline
2-Chloro-6,8-dimethylquinoline
3-Chloro-2,4-dimethylquinoline
3-Chloro-2,7-dimethylquinoline
4-Chloro-2,3-dimethylquinoline
4-Chloro-2,5-dimethylquinoline
4-Chloro-3,5-dimethylquinoline
4-Chloro-3,6-dimethylquinoline
4-Chloro-3,8-dimethylquinoline
4-Chloro-6,7-dimethylquinoline
5-Chloro-4,6-dimethylquinoline

$C_{11}H_{10}ClNO$
3-Chloro-1-hydroxyisoquinoline, *Et ether*
4-Chloro-6-hydroxy-2-methylquinoline, *Me ether*
4-Chloro-7-hydroxy-2-methylquinoline, *Me ether*
5-Chloro-8-hydroxy-2-methylquinoline, *Me ether*
8-Chloro-2-hydroxy-4-methylquinoline, *Me ether*
4-Chloro-2-hydroxyquinoline, *Et ether*
5-Chloro-6-hydroxyquinoline, *Et ether*

$C_{11}H_{10}ClNO_2$
4-Chloroindole-3-acetic Acid, *Me ester* †
Mesacon-α-anilic Acid, *Chloride*
1-Phenyl-5-pyrrolidone-3-carboxylic Acid, *Chloride*
Phthalimide, N-2-*Chloropropyl*

$C_{11}H_{10}ClNO_4$
α-Chloro-4-nitrocinnamic Acid, *Et ester*
5-Chloro-2-nitrocinnamic Acid, *Et ester*

$C_{11}H_{10}ClNO_5$
3-Nitrophthalic Acid, 1-*Propyl ester*, *Chloride*
3-Nitrophthalic Acid, 2-*Propyl ester*, *Chloride*

$C_{11}H_{10}Cl_2O_2$
3-Phenylglutaric Acid, *Dichloride*

$C_{11}H_{10}Cl_2O_4$
5,7-Dichloro-3,4-dihydro-8-hydroxy-6-methoxy-3-methylisocoumarin †

$C_{11}H_{10}FeO_2$
Ferrocenecarboxylic Acid

$C_{11}H_{10}INO_2$
Phthalimide, N-3-*Iodopropyl*

$C_{11}H_{10}I_2O_2$
α,β-Di-iodocinnamic Acid, *Et ester*

$C_{11}H_{10}I_3NO_3$
3-Amino-2,4,6-tri-iodobenzoic Acid, N-*Ac*, *Et ester*
3-Amino-2,4,6-tri-iodobenzoic Acid, N-*Butyryl*

$C_{11}H_{10}N_2$
2-*m*-Aminophenylpyridine
3-*m*-Aminophenylpyridine
4-*m*-Aminophenylpyridine
2-*p*-Aminophenylpyridine
5-Amino-2-phenylpyridine
2-Anilinopyridine
3-3′-Indolylpropionic Acid, *Nitrile*
2,4,6-Trimethylisophthalic Acid, *Dinitrile*

$C_{11}H_{10}N_2O$
1-Acetyl-4-phenylpyrazole
3-Acetyl-1-phenylpyrazole
4-Acetyl-1-phenylpyrazole
4-Amino-1-naphthoic Acid, *Amide*
3-Amino-2-naphthoic Acid, *Amide*
Cinnamoylglycine, *Nitrile*
2-Methylquinoline-3-carboxylic Acid, *Amide*
2-Methylquinoline-4-carboxylic Acid, *Amide*
3-Methylquinoline-4-carboxylic Acid, *Amide*
7-Methylquinoline-4-carboxylic Acid, *Amide*
1-Naphthoylhydrazine
2-Naphthoylhydrazine
α-Naphthylurea
β-Naphthylurea
4-Nitroso-1-naphthylamine, N-*Me*
1-Nitroso-2-naphthylamine, N-*Me*

$C_{11}H_{10}N_2O$ (*continued*)
- 2-Quinolineacetic Acid, *Nitrile*
- 4-Quinolineacetic Acid, *Nitrile*
- 8-Quinolineacetic Acid, *Nitrile*

$C_{11}H_{10}N_2O_2$
- 8-Aminoquinoline-2-carboxylic Acid, *Me ester*
- 2-Aminoquinoline-3-carboxylic Acid, *Me ester*
- 5-Aminoquinoline-6-carboxylic Acid, *Me ester*
- 1,3-Diamino-2-naphthoic Acid
- 1,4-Diamino-2-naphthoic Acid
- 4,5-Diamino-2-naphthoic Acid
- 2-Hydroxy-3-methylquinoline-4-carboxylic Acid, *Amide*
- 6-Methoxyquinoline-4-carboxylic Acid, *Amide*
- *N*-Methyl-2-nitro-1-naphthylamine
- *N*-Methyl-4-nitro-1-naphthylamine
- *N*-Methyl-8-nitro-1-naphthylamine
- 2-Methyl-4-nitro-1-naphthylamine
- 4-Methyl-2-nitro-1-naphthylamine
- 4-Methyl-3-nitro-1-naphthylamine
- 5-Methyl-2-nitro-1-naphthylamine
- 5-Methyl-4-nitro-1-naphthylamine
- 6-Methyl-2-nitro-1-naphthylamine
- 6-Methyl-4-nitro-1-naphthylamine
- 6-Methyl-5-nitro-1-naphthylamine
- 7-Methyl-2-nitro-1-naphthlyamine
- 7-Methyl-4-nitro-1-naphthylamine
- 7-Methyl-8-nitro-1-naphthylamine
- 8-Methyl-2-nitro-1-naphthylamine
- 8-Methyl-4-nitro-1-naphthylamine
- *N*-Methyl-1-nitro-2-naphthylamine
- *N*-Methyl-6-nitro-2-naphthylamine
- 1-Methyl-4-nitro-2-naphthylamine
- 3-Methyl-5-phenylpyrazole-1-carboxylic Acid
- 5-Methyl-3-phenylpyrazole-1-carboxylic Acid
- 1-Methyl-4-phenylpyrazole-3-carboxylic Acid
- 1-Methyl-5-phenylpyrazole-3-carboxylic Acid
- 5-Methyl-1-phenylpyrazole-3-carboxylic Acid
- 3-Methyl-1-phenylpyrazole-4-carboxylic Acid
- 3-Methyl-5-phenylpyrazole-4-carboxylic Acid
- 5-Methyl-1-phenylpyrazole-4-carboxylic Acid
- 1-Methyl-3-phenylpyrazole-5-carboxylic Acid
- 1-Methyl-4-phenylpyrazole-5-carboxylic Acid
- 3-Methyl-1-phenylpyrazole-5-carboxylic Acid
- 4-Methyl-3(5)-phenylpyrazole-5(3)-carboxylic Acid
- 5-Phenylpyrazole-1-carboxylic Acid, *Me ester*
- 1-Phenylpyrazole-3-carboxylic Acid, *Me ester*
- 4-Phenylpyrazole-3-carboxylic Acid, *Me ester*
- 5(3)-Phenylpyrazole-3(5)-carboxylic Acid, *Me ester*
- 1-Phenylpyrazole-4-carboxylic Acid, *Me ester*
- 5-Phenylpyrazole-4-carboxylic Acid, *Me ester*
- 1-Phenylpyrazole-5-carboxylic Acid, *Me ester*
- Vasicinone

$C_{11}H_{10}N_2O_2S_2$
- *N*-(*p*-Acetamidophenyl)rhodanine†

$C_{11}H_{10}N_2O_3$
- 4-Amino-2-nitro-1-naphthol, *Me ether*
- 8-Amino-5-nitro-1-naphthol, *Me ether*
- 7-Amino-8-nitro-2-naphthol, *Me ether*
- 2-Hydroxy-3,4-dimethyl-6-nitroquinoline
- 2-Hydroxy-4,6-dimethyl-3-nitroquinoline
- 2-Hydroxy-4,7-dimethyl-6-nitroquinoline
- 2-Hydroxy-4,7-dimethyl-8-nitroquinoline
- 2-Hydroxy-4,8-dimethyl-6-nitroquinoline
- 4-Hydroxy-2,3-dimethyl-6-nitroquinoline
- 2-Hydroxy-4-methyl-6-nitroquinoline, N-*Me*
- 2-Hydroxy-6-methyl-5-nitroquinoline, N-*Me*
- 2-Hydroxy-6-methyl-8-nitroquinoline, N-*Me*
- 2-Hydroxy-7-methyl-8-nitroquinoline, N-*Me*
- 2-Hydroxy-8-methyl-5-nitroquinoline, N-*Me*
- 4-Hydroxy-2-methyl-3-nitroquinoline, *Me ether*
- 4-Hydroxy-2-methyl-8-nitroquinoline, *Me ether*
- 6-Hydroxy-2-methyl-8-nitroquinoline, *Me ether*
- 6-Hydroxy-3-methyl-8-nitroquinoline, *Me ether*
- 6-Hydroxy-5-methyl-8-nitroquinoline, *Me ether*
- 2-Hydroxy-5-nitroquinoline, N-*Et*
- 2-Hydroxy-6-nitroquinoline, *Et ether*
- 2-Hydroxy-6-nitroquinoline, N-*Et*
- 2-Hydroxy-7-nitroquinoline, N-*Et*
- 2-Hydroxy-8-nitroquinoline, N-*Et*
- 3-Hydroxy-4(?)-nitroquinoline, *Et ether*
- 4-Hydroxy-3-nitroquinoline, *Et ether*
- 6-Hydroxy-5-nitroquinoline, *Et ether*
- 6-Hydroxy-8-nitroquinoline, *Et ether*
- 8-Hydroxy-5-nitroquinoline, *Et ether*
- 1-Methylindole-2,3-dicarboxylic Acid, *Monoamide*
- 1-Phenyl-4-pyrazolone-3-carboxylic Acid, *Me ester*
- 1-Phenyl-5-pyrazolone-3-carboxylic Acid, *Me ester*

$C_{11}H_{10}N_2O_4$
- Anthranilinodiacetic Acid, α-*Mononitrile*
- 4,6-Dihydroxy-3-nitroquinoline, *Di-Me ether*
- 5,6-Dihydroxy-8-nitroquinoline, *Di-Me ether*
- 6-Nitroindole-3-carboxylic Acid, *Et ester*
- 2-Nitrotoluene-α,4-dicarboxylic Acid, 4-*Nitrile*, *Et ester*
- Succinimide, N-o-*Nitrobenzyl*
- Succinimide, N-p-*Nitrobenzyl*

$C_{11}H_{10}N_2O_4S$
- 8-Nitronaphthalene-1-sulphonic Acid, *Me-amide*

$C_{11}H_{10}N_2O_6$
- 2,4-Dinitrocinnamic Acid, *Et ester*

$C_{11}H_{10}N_2S$
- α-Naphthylthiourea
- β-Naphthylthiourea

$C_{11}H_{10}N_2S$ (di-ion)
- 6*H*-Dipyrido[2,1-*b*:1′,2′-*e*]-1,3,5-thiadiadiazinium

$C_{11}H_{10}N_4O_2$
- 1-Phenylpyrazole-3,5-dicarboxylic Acid, *Diamide*
- 1-Phenylpyrazole-4,5-dicarboxylic Acid, *Diamide*

$C_{11}H_{10}N_4O_4$
- Phenyl-ψ-uric Acid

$C_{11}H_{10}O$
- 2-Hydroxyazulene, *Me ether*†
- 6-Hydroxyazulene, *Me ether*†
- 2-Methyl-1-naphthol
- 3-Methyl-1-naphthol
- 4-Methyl-1-naphthol

$C_{11}H_{10}O$ (*continued*)
5-Methyl-1-naphthol
6-Methyl-1-naphthol
7-Methyl-1-naphthol
1-Methyl-2-naphthol
3-Methyl-2-naphthol
4-Methyl-2-naphthol
5-Methyl-2-naphthol
6-Methyl-2-naphthol
8-Methyl-2-naphthol
Methyl 1-naphthyl Ether
Methyl 2-naphthyl Ether
1-Naphthylmethanol
2-Naphthylmethanol
1-Phenyl-1-penten-4-yn-3-ol †

$C_{11}H_{10}OS$
2-Hydroxy-1-naphthalenethiol, O-*Me ether*

$C_{11}H_{10}O_2$
2-Allyl-4-pentenoic Acid, *Allyl ester*
2,8-Decadiene-4,6-diynoic Acid, *Me ester*
1,4-Dihydro-1-naphthoic Acid
3,4-Dihydro-1-naphthoic Acid
1,2-Dihydro-2-naphthoic Acid
1,4-Dihydro-2-naphthoic Acid
3,4-Dihydro-2-naphthoic Acid
1,2-Dihydroxy-4-methylnaphthalene
1,4-Dihydroxy-2-methylnaphthalene
1,4-Dihydroxy-5-methylnaphthalene
1,6-Dihydroxy-4-methylnaphthalene
2,3-Dihydroxy-1-methylnaphthalene
2,6-Dihydroxy-1-methylnaphthalene
2,7-Dihydroxy-1-methylnaphthalene
1,2-Dihydroxynaphthalene, 1-*Me ether*
2-Ethylchromone
3-Ethylcoumarin
4-Ethylcoumarin
Furfuryl phenyl Ether
1-Hydroxymethyl-2-naphthol
Indene-3-carboxylic Acid, *Me ester*
3-Indenylacetic Acid
7-Indenylacetic Acid
4-Methoxy-1-naphthol †
3-Methylindene-2-carboxylic Acid
1-Methylindene-3-carboxylic Acid
3-Methylnaphthalene-1,8-diol †
5-Phenyl-2,4-pentadienoic Acid
3-Phenyl-2-propynoic Acid, *Et ester*
Phyllomerol

$C_{11}H_{10}O_2S$
Benzo[*b*]thiophene-2-carboxylic Acid, *Et ester*
Benzo[*b*]thiophene-3-carboxylic Acid, *Et ester*
Naphthalene-1-sulphinic Acid, *Me ester*
Naphthalene-2-sulphinic Acid, *Me ester*

$C_{11}H_{10}O_3$
Benzofuran-2-carboxylic Acid, *Et ester*
3-Benzoylacrylic Acid, *Me ester*
Benzylsuccinic Acid, *Anhydride*
4-Formylcinnamic Acid, *Me ester*
4-Hydroxycoumarin, *Et ether*
7-Hydroxy-4,8-dimethylcoumarin †
7-Hydroxy-4-methylcoumarin, *Me ether*
7-Hydroxy-5-methylcoumarin, *Me ether*
7-Hydroxy-8-methylcoumarin, *Me ether*
2-Hydroxyphenylpropiolic Acid, *Et ether*
3-Hydroxyphenylpropiolic Acid, *Et ether*
8-Methoxy-4-methylcoumarin †
3-Methylbenzofuran-2-carboxylic Acid, *Me ester*
4-(3,4-Methylenedioxyphenyl)-3-buten-2-one
Nemotinic Acid
Oospolactone
2-Phenylglutaric Acid, *Anhydride*
3-Phenylglutaric Acid, *Anhydride*
Psilotinin †
1,2,7-Trihydroxynaphthalene, 7-*Me ether*
Umbelliferone, *Et ether*

$C_{11}H_{10}O_3S$
3-Hydroxybenzo[*b*]thiophene-2-carboxylic Acid, *Et ether*
3-Hydroxybenzo[*b*]thiophene-2-carboxylic Acid, *Et ester*
2-Methylnaphthalene-1-sulphonic Acid
4-Methylnaphthalene-1-sulphonic Acid
5-Methylnaphthalene-1-sulphonic Acid
7-Methylnaphthalene-1-sulphonic Acid
4-Methylnaphthalene-2-sulphonic Acid
5-Methylnaphthalene-2-sulphonic Acid
6-Methylnaphthalene-2-sulphonic Acid
Naphthalene-1-sulphonic Acid, *Me ester*
Naphthalene-2-sulphonic Acid, *Me ester*
2-Naphthylmethane-sulphonic Acid

$C_{11}H_{10}O_4$
Aesculetin, 7-*Et ether*
Aesculetin, *Di-Me ether*
Anhydrosepedonin †
Benzoylpyruvic Acid, *Me ester*
Benzylfumaric Acid
Citropten
Daphnetin, 7,8-*Di-Me ether*
Daphnetin, 7-*Et ether*
Dihydrogladiolic Acid, *Lactone*
3,7-Dihydroxychromone, *Di-Me ether*
5,7-Dihydroxychromone, *Di-Me ether*
7,8-Dihydroxychromone, *Di-Me ether*
4,7-Dihydroxycoumarin, *Di-Me ether*
Eugenin
Eugenitol
Fumaric Acid, *Mono-benzyl ester*
2-Hydroxy-3,4-methylenedioxy-6-vinylbenzaldehyde, *Me ether*
Indane-1,2-dicarboxylic Acid
Indane-2,2-dicarboxylic Acid
Isoeugenitol
Mesaconic Acid, α-*Phenyl ester*
3-(3,4-Methylenedioxyphenyl)acrylic Acid, *Me ester*
Phenylaticonic Acid
Phenylcitraconic Acid
1-Phenylcyclopropane-1,2-dicarboxylic Acid
3-Phenylcyclopropane-1,2-dicarboxylic Acid
Phenylitaconic Acid
3-Phenyl-2-pentenedioic Acid
4-Phenyl-2-pentenedioic Acid
5-Phenyl-4-pentenedioic Acid
Phthalide-3-acetic Acid, *Me ester*
Scopoletin, *Me ether*
Tetrahydro-5-oxo-2-phenylfuran-3-carboxylic Acid
Undec-2-ene-4,6-diyne-1,11-dioic Acid †

$C_{11}H_{10}O_5$
Cyclopolide
Fraxidin
Fraxinol
Gladiolic Acid
Isofraxidin
Oospoglycol†
Phthalonic Acid, *Di-Me ester*
Reticulol†
Rosellinic Acid†

$C_{11}H_{10}O_6$
Benzene-1,2,3-tricarboxylic Acid, 1,3-*Di-Me ester*
Benzene-1,2,4-tricarboxylic Acid, 1,2-*Di-Me ester*
Cyclopaldic Acid
4,5-Methylenedioxyphthalic Acid, *Di-Me ester*
3,4,5-Trihydroxyphthalic Acid, *Tri-Me ether, Anhydride*
Ustic Acid, *Anhydroustic Acid*

$C_{11}H_{10}O_7$
5-Hydroxytoluene-2,3,4-tricarboxylic Acid, *Me ether*†
5-Hydroxytoluene-2,3,6-tricarboxylic Acid, *Me ether*†
3-Hydroxytoluene-2,4,5-tricarboxylic Acid, *Me ether*

$C_{11}H_{10}S$
Methyl 1-naphthyl sulphide
Methyl 2-naphthyl sulphide

$C_{11}H_{11}BrN_2O$
4-Bromoantipyrine
p-Bromoantipyrine

$C_{11}H_{11}BrO_2$
α-Bromocinnamic Acid, *Et ester*
β-Bromocinnamic Acid, *Et ester*

$C_{11}H_{11}Br_5O$
Pentabromophenol, 3-*Methylbutyl ether*

$C_{11}H_{11}ClN_2O$
Tryptophane, *Chloride*

$C_{11}H_{11}ClO$
α-Ethylcinnamic Acid, *Chloride*
1-Methylindane-2-carboxylic Acid, *Chloride*
5,6,7,8-Tetrahydro-2-naphthoic Acid, *Chloride*

$C_{11}H_{11}ClO_2$
α-Chlorocinnamic Acid, *Et ester*
β-Chlorocinnamic Acid, *Et ester*
o-Chlorocinnamic Acid, *Et ester*

$C_{11}H_{11}ClO_4$
7-Chloro-3,4-dihydro-8-hydroxy-6-methoxy-3-methylisocoumarin†

$C_{11}H_{11}Cl_2NO_2$
m-Chlorophenylcarbamic Acid, 4-*Chlorobut-2-enyl ester*†

$C_{11}H_{11}Cl_4NO_2$
Chlorbetamide

$C_{11}H_{11}FO_2$
2-Fluorocinnamic Acid, *Et ester*

$C_{11}H_{11}IN_2O$
4-Iodo-2,3-dimethyl-5-oxo-1-phenyl-Δ^3-pyrazoline

$C_{11}H_{11}IO_2$
3-*o*-Iodophenylacrylic Acid, *Et ester*
3-*m*-Iodophenylacrylic Acid, *Et ester*
3-*p*-Iodophenylacrylic Acid, *Et ester*

$C_{11}H_{11}N$
2,3-Dimethylquinoline
2,4-Dimethylquinoline
2,5-Dimethylquinoline
2,6-Dimethylquinoline
2,7-Dimethylquinoline
2,8-Dimethylquinoline
3,4-Dimethylquinoline
3,5-Dimethylquinoline
3,6-Dimethylquinoline
3,7-Dimethylquinoline
3,8-Dimethylquinoline
4,5-Dimethylquinoline
4,6-Dimethylquinoline
4,7-Dimethylquinoline
4,8-Dimethylquinoline
5,6-Dimethylquinoline
5,7-Dimethylquinoline
5,8-Dimethylquinoline
6,7-Dimethylquinoline
6,8-Dimethylquinoline
7,8-Dimethylquinoline
1-Ethylisoquinoline
3-Ethylisoquinoline
4-Ethylisoquinoline
2-Ethylquinoline
3-Ethylquinoline
4-Ethylquinoline
6-Ethylquinoline
7-Ethylquinoline
8-Ethylquinoline
1,6-Iminocyclodecapentaene, N-*Me*†
N-Methyl-1-naphthylamine
2-Methyl-1-naphthylamine
3-Methyl-1-naphthylamine
4-Methyl-1-naphthylamine
5-Methyl-1-naphthylamine
6-Methyl-1-naphthylamine
7-Methyl-1-naphthylamine
8-Methyl-1-naphthylamine
N-Methyl-2-naphthylamine
1-Methyl-2-naphthylamine
3-Methyl-2-naphthylamine
4-Methyl-2-naphthylamine
5-Methyl-2-naphthylamine
6-Methyl-2-naphthylamine
7-Methyl-2-naphthylamine
8-Methyl-2-naphthylamine
2-Methyl-1-phenylpyrrole
2-Methyl-4-phenylpyrrole
2-Methyl-5-phenylpyrrole
2-Methyl-5-phenylpyrrole
1-Naphthylmethylamine
2-Naphthylmethylamine
5,6,7,8-Tetrahydro-1-naphthoic Acid, *Nitrile*

$C_{11}H_{11}NO$
2-Amino-1-naphthol, *Me ether*
4-Amino-1-naphthol, *Me ether*
8-Amino-1-naphthol, *Me ether*
1-Amino-2-naphthol, *Me ether*
6-Amino-2-naphthol, *Me ether*

$C_{11}H_{11}NO$ (*continued*)
8-Amino-2-naphthol, *Me ether*
2-Benzoylbutyric Acid, *Nitrile*
2-Benzylacetoacetic Acid, *Nitrile*
Carbostyril, 2-*Et ether*
Carbostyril, N-*Ethyl-α-quinoline*
3,4-Dihydro-1-methyl-3,4-methylenequinolin-2-one†
1,2-Dihydro-2-naphthoic Acid, *Amide*
1,4-Dihydro-2-naphthoic Acid, *Amide*
3,4-Dihydro-2-naphthoic Acid, *Amide*
1,2-Dimethyl-4-quinolone†
2-Hydroxy-3,4-dimethylquinoline
2-Hydroxy-4,6-dimethylquinoline
2-Hydroxy-4,7-dimethylquinoline
2-Hydroxy-4,8-dimethylquinoline
2-Hydroxy-6,8-dimethylquinoline
4-Hydroxy-2,3-dimethylquinoline
4-Hydroxy-2,5-dimethylquinoline
4-Hydroxy-2,6-dimethylquinoline
4-Hydroxy-2,8-dimethylquinoline
4-Hydroxy-6,8-dimethylquinoline
5-Hydroxy-2,4-dimethylquinoline
5-Hydroxy-6,8-dimethylquinoline
6-Hydroxy-2,4-dimethylquinoline
7-Hydroxy-2,4-dimethylquinoline
8-Hydroxy-2,3-dimethylquinoline
8-Hydroxy-2,4-dimethylquinoline
1-Hydroxyisoquinoline, *Et ether*
7-Hydroxyisoquinoline, *Et ether*
2-Hydroxy-4-methylquinoline, *Me ether*
3-Hydroxy-2-methylquinoline, *Me ether*
4-Hydroxy-2-methylquinoline, *Me ether*
5-Hydroxy-2-methylquinoline, *Me ether*
5-Hydroxy-8-methylquinoline, *Me ether*
6-Hydroxy-2-methylquinoline, *Me ether*
6-Hydroxy-4-methylquinoline, *Me ether*
8-Hydroxy-2-methylquinoline, *Me ether*
4-Hydroxyquinoline, *Et ether*
6-Hydroxyquinoline, *Et ether*
8-Hydroxyquinoline, *Et ether*
2-Methylindole-3-aldehyde, N-*Me*
2-Methyl-5-*p*-tolyloxazole
5-Phenyl-2,4-pentadienoic Acid, *Amide*

$C_{11}H_{11}NO_2$
o-Aminophenylpropiolic Acid, *Et ester*
m-Aminophenylpropiolic Acid, *Et ester*
2-Cyano-3-phenylpropionic Acid, *Me ester*
3-Cyano-3-phenylpropionic Acid, *Me ester*
6,7-Dihydroxyisoquinoline, *Di-Me ether*
4,6-Dihydroxy-2-methylquinoline, 6-*Me ether*
4,8-Dihydroxy-2-methylquinoline, 8-*Me ether*
6,7-Dihydroxy-2-methylquinoline, 6-*Me ether*
2,7-Dihydroxy-1-naphthylamine, 7-*Me ether*
1,5-Dihydroxy-2-naphthylamine, 5-*Me ether*
2,4-Dihydroxyquinoline, 2-*Et ether*
5,7-Dimethylisatin, *Me ether*
Indole-2-carboxylic Acid, *Et ester*
Indole-4-carboxylic Acid, *Et ester*
3-Indolylacetic Acid, *Me ester*
3-3′-Indolylpropionic Acid
1-Methylindole-2-carboxylic Acid, *Me ester*
3-Methylindole-2-carboxylic Acid, *Me ester*
2-Methylindole-3-carboxylic Acid, *Me ester*
Phensoximide
Phenylcyanoacetic Acid, *Et ester*
3-Phenylglutaric Acid, *Imide*
Phenylsuccinic Acid, β-*Me ester*, α-*Nitrile*
Phthalimide, N-*Propyl*
Succinimide, N-*Benzyl*
Succinimide, N-o-*Tolyl*
Succinimide, N-m-*Tolyl*
Succinimide, N-p-*Tolyl*

$C_{11}H_{11}NO_2S$
2-Methylbenzothiazole-6-carboxylic Acid, *Et ester*
2-Methylnaphthalene-1-sulphonic Acid, *Amide*
4-Methylnaphthalene-1-sulphonic Acid, *Amide*
5-Methylnaphthalene-1-sulphonic Acid, *Amide*
7-Methylnaphthalene-1-sulphonic Acid, *Amide*
4-Methylnaphthalene-2-sulphonic Acid, *Amide*
5-Methylnaphthalene-2-sulphonic Acid, *Amide*
6-Methylnaphthalene-2-sulphonic Acid, *Amide*

$C_{11}H_{11}NO_3$
Cinnamoylglycine
Edulitine†
5-Hydroxyindole-3-acetic Acid, *Me ether*†
1-Hydroxyindole-1-carboxylic Acid, *Et ester*
1-Hydroxyindole-1-carboxylic Acid, *Me ether*, *Me ester*
3-Hydroxyindole-2-carboxylic Acid, *Et ester*
3-Hydroxyindole-2-carboxylic Acid, 3-*Et ether*
3-3′-Indolyl-lactic Acid
Maleanilic Acid, *Me ester*
Maleic Acid, *Mono-o-toluidide*
Mesacon-α-anilic Acid
Mesacon-β-anilic Acid
Mesaconic Acid, α-*Phenyl ester*, β-*Amide*
3-*m*-Nitrosophenylacrylic Acid, *Et ester*
3-*p*-Nitrosophenylacrylic Acid, *Et ester*
Norcotarnine
Oxyhydrastinine
3-Phenyl-2-pentenedioic Acid, *Monoamide*
1-Phenyl-3-pyrrolidone-2-carboxylic Acid
1-Phenyl-3-pyrrolidone-4-carboxylic Acid
3-Phenyl-5-pyrrolidone-2-carboxylic Acid
1-Phenyl-2-pyrrolidone-3-carboxylic Acid
1-Phenyl-5-pyrrolidone-3-carboxylic Acid
Succinimide, N-p-*Hydroxyphenyl*, *Me ether*
1,2,3,4-Tetrahydro-4-oxoquinoline-2-carboxylic Acid, *Me ester*†

$C_{11}H_{11}NO_3S$
2-Naphthol-3-sulphonic Acid, *Me ether*, *Amide*
2-Naphthol-6-sulphonic Acid, *Me ether*, *Amide*
1-Naphthylamine-4-sulphonic Acid, N-*Me*
Quinoline-8-sulphonic Acid, *Et ester*

$C_{11}H_{11}NO_4$
Benzylpenaldic Acid
5,6-Dihydroxyindole-2-carboxylic Acid, *Di-Me ether*
5,6-Dihydroxyindole-2-carboxylic Acid, *Et ester*
2-Methyl-3-*p*-nitrophenylacrylic Acid, *Me ester*
3-(3-Methyl-2-nitrophenyl)acrylic Acid, *Me ester*
3-(3-Methyl-4-nitrophenyl)acrylic Acid, *Me ester*
3-(3-Methyl-6-nitrophenyl)acrylic Acid, *Me ester*
3-(4-Methyl-3-nitrophenyl)acrylic Acid, *Me ester*
3-*o*-Nitrophenylacrylic Acid, *Et ester*

$C_{11}H_{11}NO_4$ (*continued*)
3-*m*-Nitrophenylacrylic Acid, *Et ester*
3-*p*-Nitrophenylacrylic Acid, *Et ester*
3-*p*-Nitrophenylcrotonic Acid, *Me ester*

$C_{11}H_{11}NO_4S$
N-Ethyl-5-phenylisoxazolium-3′-sulphonate†

$C_{11}H_{11}NO_5$
2-Acetyl-3-nitrobenzoic Acid, *Et ester*
2-Acetyl-4-nitrobenzoic Acid, *Et ester*
2-Acetyl-5-nitrobenzoic Acid, *Et ester*
p-Hydroxybenzylpenaldic Acid
3-(2-Hydroxy-4-methyl-5-nitrophenyl)acrylic Acid, *Me ester*
α-Hydroxy-2-nitrocinnamic Acid, *Me ether*, *Me ester*
α-Hydroxy-2-nitrocinnamic Acid, *Et ether*
α-Hydroxy-2-nitrocinnamic Acid, *Et ester*
α-Hydroxy-4-nitrocinnamic Acid, *Et ester*
3-(2-Hydroxy-3-nitrophenyl)acrylic Acid, *Me ether*, *Me ester*
3-(2-Hydroxy-5-nitrophenyl)acrylic Acid, *Me ether*, *Me ester*
3-(2-Hydroxy-5-nitrophenyl)acrylic Acid, *Et ether*
3-(4-Hydroxy-3-nitrophenyl)acrylic Acid, *Et ester*
Kynuric Acid, α-*Et ester*
4-Methoxy-3-nitrobenzoic Acid, *Allyl ester*
o-Nitrobenzoylacetic Acid, *Et ester*
m-Nitrobenzoylacetic Acid, *Et ester*
p-Nitrobenzoylacetic Acid, *Et ester*
3-*m*-Nitrophenyloxiran-2-carboxylic Acid, *Et ester*
3-*o*-Nitrophenylpyruvic Acid, *Et ester*

$C_{11}H_{11}NO_6$
Anthranilinodiacetic Acid
α,α-Diacetoxy-2-nitrotoluene
α,α-Diacetoxy-3-nitrotoluene
α,α-Diacetoxy-4-nitrotoluene
α,3-Dihydroxy-2-nitrocinnamic Acid, *Di-Me ether*
2,3-Dihydroxy-5-nitrocinnamic Acid, *Di-Me ether*
2,3-Dihydroxy-6-nitrocinnamic Acid, *Di-Me ether*
3,4-Dihydroxy-2-nitrocinnamic Acid, *Di-Me ether*
4,5-Dihydroxy-2-nitrocinnamic Acid, *Di-Me ether*
2-*o*-Nitrophenylglutaric Acid
2-*p*-Nitrophenylglutaric Acid
3-*o*-Nitrophenylglutaric Acid
3-*m*-Nitrophenylglutaric Acid
3-*p*-Nitrophenylglutaric Acid
2-Nitro-3-phenylglutaric Acid
3-Nitrophthalic Acid, 1-*Propyl ester*
3-Nitrophthalic Acid, 2-*Propyl ester*
3-Nitrophthalic Acid, 2-*Isopropyl ester*
4-Amino-2-hydroxyquinoline, *Et ether*
2-Nitroterephthalic Acid, 1-*Propyl ester*
4-Nitrotoluene-α,2-dicarboxylic Acid, *Di-Me ester*
2-Nitrotoluene-α,4-dicarboxylic Acid, *Di-Me ester*
Pyridine-2,3,4-tricarboxylic Acid, *Tri-Me ester*

$C_{11}H_{11}NO_7$
2-Nitromyristicic Acid, *Et ester*
3-Nitro-opianic Acid, *Me ester*
o-Nitrophenoxymalonic Acid, *Di-Me ester*
m-Nitrophenoxymalonic Acid, *Di-Me ester*
p-Nitrophenoxymalonic Acid, *Di-Me ester*

$C_{11}H_{11}NO_8$
3,4-Dihydroxy-6-nitrophthalic Acid, 4-*Me ether*, *Di-Me ester*

$C_{11}H_{11}NS$
4-Amino-1-naphthalenethiol, S-*Me*
2-Methyl-5-*p*-tolylthiazole
2-Quinolinethiol, S-*Et*
8-Quinolinethiol, S-*Et*

$C_{11}H_{11}N_3$
Benzyliminodiacetic Acid, *Dinitrile*

$C_{11}H_{11}N_3O$
5-Methyl-3-phenylpyrazole-1-carboxylic Acid, *Amide*
5-Methyl-1-phenylpyrazole-3-carboxylic Acid, *Amide*
3-Methyl-1-phenylpyrazole-5-carboxylic Acid, *Amide*
1-α-Naphthylsemicarbazide
1-β-Naphthylsemicarbazide
4-α-Naphthylsemicarbazide
4-β-Naphthylsemicarbazide

$C_{11}H_{11}N_3O_2$
1-Methylindole-2,3-dicarboxylic Acid, *Diamide*

$C_{11}H_{11}N_3O_2S$
Sulphapyridine

$C_{11}H_{11}N_3O_4$
6-Nitroindazole-3-carboxylic Acid, *Et ester*, N-*Me*

$C_{11}H_{11}N_3O_6$
1,2,3,4-Tetrahydro-6,8-dinitroquinoline-1-carboxylic Acid, *Me ester*

$C_{11}H_{11}N_3O_8$
2,4,6-Trinitrobenzoic Acid, *Butyl ester*
2,4,6-Trinitrobenzoic Acid, *Isobutyl ester*

$C_{11}H_{11}N_3S$
1-α-Naphthylthiosemicarbazide
1-β-Naphthylthiosemicarbazide
4-α-Naphthylthiosemicarbazide

$C_{11}H_{11}N_5O_3$
N-Benzoylglycylglycine, *Azide*

$C_{11}H_{11}N_5O_3S_2$
Urothione†

$C_{11}H_{12}$
1,2-Dihydro-3-methylnaphthalene
1,2-Dihydro-4-methylnaphthalene
5-Phenyl-1,2-pentadiene
1-Phenyl-1,3-pentadiene
2-Phenyl-1,3-pentadiene
3-Phenyl-1,3-pentadiene

$C_{11}H_{12}AsN$
8-Dimethylarsinoquinoline†

$C_{11}H_{12}Br_2N_2O_5$
Bromamphenicol

$C_{11}H_{12}Br_2O_2$
4-Allyl-3,6-dibromoguaiacol, *Me ether*
2,3-Dibromo-3-phenylbutyric Acid, *Me ester*
2,3-Dibromo-3-phenylpropionic Acid, *Et ester*
Zebromal

$C_{11}H_{12}Br_2O_3$
3,5-Dibromo-4-hydroxybenzoic Acid, *Butyl ester*

$C_{11}H_{12}ClNO_3S$
Chlormezanone

$C_{11}H_{12}Cl_2N_2O_5$
Chloramphenicol

$C_{11}H_{12}Cl_2O_2$
2,3-Dichloro-3-phenylpropionic Acid, *Et ester*

$C_{11}H_{12}Cl_2O_3$
2,4-Dichlorophenoxyacetic Acid, *Propyl ester*

$C_{11}H_{12}INO_3$
o-Iodobenzamidoacetic Acid, *Et ester*
p-Iodobenzamidoacetic Acid, *Et ester*

$C_{11}H_{12}I_3NO_2$
Iopanoic Acid

$C_{11}H_{12}N_2$
4-Amino-2,3-dimethylquinoline
5-Amino-2,6-dimethylquinoline
5-Amino-4,6-dimethylquinoline
5-Amino-6,8-dimethylquinoline
6-Amino-2,4-dimethylquinoline
6-Amino-5,7-dimethylquinoline
7(or 5)-Amino-2,4-dimethylquinoline
7-Amino-2,8-dimethylquinoline
8-Amino-2,3-dimethylquinoline
8-Amino-5,6-dimethylquinoline
6-Aminoquinoline, N-*Di-Me*
1,2-Diamino-4-methylnaphthalene
1,2-Diamino-5-methylnaphthalene
1,2-Diamino-7-methylnaphthalene
1,3-Diamino-4-methylnaphthalene
1,3-Diamino-5-methylnaphthalene
1,3-Diamino-6-methylnaphthalene
1,3-Diamino-7-methylnaphthalene
1,4-Diamino-2-methylnaphthalene
1,5-Diamino-2-methylnaphthalene
1,5-Diamino-4-methylnaphthalene
1,8-Diamino-2-methylnaphthalene
1,8-Diamino-4-methylnaphthalene
3,4-Dimethyl-1-phenylpyrazole
3,5-Dimethyl-1-phenylpyrazole
4,5-Dimethyl-1-phenylpyrazole
3-Methyl-1-*p*-tolylpyrazole
3-Methyl-5-*p*-tolylpyrazole
5-Methyl-1-*p*-tolylpyrazole

$C_{11}H_{12}N_2O$
4-Amino-2-hydroxyquinoline, *Et ether*
5-Amino-6-hydroxyquinoline, *Et ether*
5-Amino-8-hydroxyquinoline, *Et ether*
8-Amino-6-hydroxyquinoline, *Et ether*
Antipyrine
7,8-Diamino-2-naphthol, *Me ether*
1,2-Dimethyl-5-phenyl-3-pyrazolone
1,4-Dimethyl-2-phenyl-3-pyrazolone
2,5-Dimethyl-1-phenyl-3-pyrazolone
2,5-Dimethyl-4-phenyl-3-pyrazolone
4,5-Dimethyl-1-phenyl-3-pyrazolone
4,5-Dimethyl-2-phenyl-3-pyrazolone
Homoglomerine†
Isovasicine
3-Methyl-1-*o*-tolyl-2-pyrazolin-5-one
3-Methyl-1-*p*-tolyl-2-pyrazolin-5-one
5-Methyl-1-*o*-tolyl-4-pyrazolin-3-one
5-Methyl-1-*p*-tolyl-4-pyrazolin-3-one
Peganine

$C_{11}H_{12}N_2O_2$
Ethotoin
6-Hydroxypeganin
2-Indazolylacetic Acid, *Et ester*
2,2′-Indazolylbutyric Acid
β-5-Indole-α-alanine†
Mesacon-α-anilic Acid, *Amide*
5-Methylindazole-2-carboxylic Acid, *Et ester*
2-Methyl-3-nitroindole, N-*Et*
Nirvanol
N-Phenylglycine-*o*-carboxylic Acid, β-*Et ester*, α-*Nitrile*
Tryptophane

$C_{11}H_{12}N_2O_2S$
2-Methylaminobenzothiazole-6-carboxylic Acid, *Et ester*

$C_{11}H_{12}N_2O_3$
3,3-Dimethyl-5-nitro-oxindole, N-*Me*
2-Hydroxytryptophan†
5-Hydroxytryptophan†
7-Hydroxytryptophan†

$C_{11}H_{12}N_2O_4$
2-Amino-4-nitrocinnamic Acid, *Et ester*
N-Benzoylglycylglycine
2,6-Dihydroxy-3-nitrobenzoic Acid, *Nitrile*, *Di-Et ether*

$C_{11}H_{12}N_2O_5$
o-Nitrobenzamidoacetic Acid, *Et ester*
m-Nitrobenzamidoacetic Acid, *Et ester*
p-Nitrobenzamidoacetic Acid, *Et ester*

$C_{11}H_{12}N_2O_6$
3,5-Dinitrobenzoic Acid, n-*Butyl ester*

$C_{11}H_{12}N_2O_7$
2-Hydroxy-3,5-dinitrobenzoic Acid, *Et ether*, *Et ester*
2-Hydroxy-3,5-dinitrobenzoic Acid, *Butyl ester*
4-Hydroxy-3,5-dinitrobenzoic Acid, n-*Butyl ester*
4-Hydroxy-3,5-dinitrobenzoic Acid, *Et ether*, *Et ester*

$C_{11}H_{12}N_2S$
Tetramisole†

$C_{11}H_{12}N_4O_2S$
Sulphamerazine

$C_{11}H_{12}N_4O_3S$
Sulphamethoxypyridazine

$C_{11}H_{12}O$
2-Acetylindane
Cyclobutyl phenyl Ketone
3,4-Dihydro-1-methylnaphthalen-2-(1*H*)-one†
2,2-Dimethyl-3-chromene†
3-Methyl-1-phenyl-2-buten-1-one
3-Methyl-4-phenyl-3-buten-2-one
2-Phenylcyclopentanone
3-Phenylcyclopentanone

$C_{11}H_{12}O$ (*continued*)
1-Phenyl-1-penten-3-one
5,6,7,8-Tetrahydro-2-naphthaldehyde
4-*o*-Tolyl-3-buten-2-one
4-*p*-Tolyl-3-buten-2-one
Undeca-5,6-diene-8,10-diyn-1-ol †
Undeca-5,7,9-triyn-1-ol †

$C_{11}H_{12}O_2$
Atropic Acid, *Et ester*
2-Benzoylbutanal
2-Decene-4,6-diynoic Acid, *Me ester*
Ethyl cinnamate
α-Ethylcinnamic Acid
β-Ethylcinnamic Acid
3-Hydroxy-5-methylbenzofuran, *Et ether*
6-Hydroxy-3-methylbenzofuran, *Et ether*
4-(2-Hydroxyphenyl)-3-buten-2-one, *Me ether*
4-(3-Hydroxyphenyl)-3-buten-2-one, *Me ether*
4-Hydroxy-4-phenylvaleric Acid, *Lactone*
4-Hydroxy-5-phenylvaleric Acid, *Lactone*
Indane-1-carboxylic Acid, *Me ester*
Indane-2-carboxylic Acid, *Me ester*
Indane-4-carboxylic Acid, *Me ester*
Isotubanol
4-*p*-Methoxyphenyl-3-buten-2-one
7-Methylindane-1-carboxylic Acid
1-Methylindane-2-carboxylic Acid
1-Methylindane-4-carboxylic Acid
6-Methylindane-4-carboxylic Acid
4-Methylindane-7-carboxylic Acid
2-Methyl-3-phenylacrylic Acid, *Me ester*
4-Phenyl-3-butenoic Acid, *Me ester*
3-Phenylcrotonic Acid, *Me ester*
1-Phenyl-1,3-pentanedione
1-Phenyl-2,3-pentanedione
1-Phenyl-1,4-pentanedione
1-Phenyl-2,4-pentanedione
3-Phenyl-2,4-pentanedione
3-Propylphthalide
1,2,3,4-Tetrahydro-1-naphthoic Acid
5,6,7,8-Tetrahydro-1-naphthoic Acid
1,2,3,4-Tetrahydro-2-naphthoic Acid
5,6,7,8-Tetrahydro-2-naphthoic Acid
3-*p*-Tolylacrylic Acid, *Me ester*
Tubanol
Undeca-5,6-diene-8,10-diyne-1,3-diol †
Undeca-5,6-diene-8,10-diyne-1,4-diol †

$C_{11}H_{12}O_3$
Acetoacetic Acid, *Benzyl ester*
Acetophenone-*o*-carboxylic Acid, *Et ester*
Acetophenone-*p*-carboxylic Acid, *Et ester*
2-Allyloxybenzoic Acid, *Me ester*
4-Allyloxybenzoic Acid, *Me ester*
Benzoylacetic Acid, *Et ester*
2-Benzoylbutyric Acid
3-Benzoylbutyric Acid
4-Benzoylbutyric Acid
Benzoylformic Acid, *Propyl ester*
2-Benzoyl-2-methylpropionic Acid
3-Benzoyl-2-methylpropionic Acid
2-Benzoylpropionic Acid, *Me ester*
3-Benzoylpropionic Acid, *Me ester*
2-Benzylacetoacetic Acid
2,6-Diacetyl-*p*-cresol
4,6-Diacetyl-*m*-cresol
α-Ethyl-*o*-hydroxy-*trans*-cinnamic Acid
Formylphenylacetic Acid, *Et ester*
Furethrolone
1-Hydroxy-2-indanylacetic Acid
3-(4-Hydroxy-2-methylphenyl)acrylic Acid, *Me ether*
3-*o*-Hydroxyphenylcrotonic Acid, *Me ether*
3-*m*-Hydroxyphenylcrotonic Acid, *Me ether*
3-*p*-Hydroxyphenylcrotonic Acid, *Me ether*
3-*o*-Hydroxyphenyl-2-methylacrylic Acid, *Me ether*
3-*m*-Hydroxyphenyl-2-methylacrylic Acid, *Me ether*
3-*p*-Hydroxyphenyl-2-methylacrylic Acid, *Me ether*
3-Methoxy-4,5-methylenedioxy-1-propenylbenzene
3-*o*-Methoxyphenylacrylic Acid, *Me ester*
3-*m*-Methoxyphenylacrylic Acid, *Me ester*
3-*p*-Methoxyphenylacrylic Acid, *Me ester*
2-Methoxy-4-propenylphenol, *Formyl*
3,4-Methylenedioxybutyrophenone
4-(3,4-Methylenedioxyphenyl)-2-butanone
5-Methylmellein †
Methyl salicylate, *Allyl ether*
Myristicin
Ochracin, *Me ether*
3-Oxo-4-phenylbutyric Acid, *Me ester*
4-Oxo-2-phenylvaleric Acid
4-Oxo-3-phenylvaleric Acid
4-Oxo-5-phenylvaleric Acid
2-Phenyloxiran-1-carboxylic Acid, *Et ester*
Phenylpyruvic Acid, *Et ester*
o-Propionylbenzoic Acid, *Me ester*
m-Toluylformic Acid, *Et ester*
p-Toluylformic Acid, *Et ester*
2,4,5-Trimethylbenzoylformic Acid
2,4,5-Trimethylbenzoylformic Acid
2,4,6-Trimethylbenzoylformic Acid

$C_{11}H_{12}O_4$
O-Acetylsalicyclic Acid, *Et ester*
Benzylsuccinic Acid
Caffeic Acid, *Et ester*
Caffeic Acid, *Di-Me ether*
2-Carboxyphenylacetic Acid, 2-*Et ester*
2-Carboxyphenylacetic Acid, α-*Et ester*
Crocatone †
α,α-Diacetoxytoluene
4,6-Diacetylresorcinol, *Mono-Me ether*
3,4-Dihydro-8-hydroxy-6-methoxy-3-methylisocoumarin †
2,5-Dihydroxycinnamic Acid, *Di-Me ether*
3,5-Dihydroxycinnamic Acid, *Di-Me ether*
2,6-Dimethylterephthalic Acid, 1-*Me ester*
2,6-Dimethylterephthalic Acid, 4-*Me ester*
2-Formylphenoxyacetic Acid, *Et ester*
3-Formylphenoxyacetic Acid, *Et ester*
4-Formylphenoxyacetic Acid, *Et ester*
Glycerol 1,2-Carbonate, 3-*Benzyl ether* †
4-Hydroxy-3-methoxycinnamic Acid, *Me ester*
4-Hydroxy-6-methoxy-*o*-toluic Aldehyde, O-*Ac*
p-Methoxybenzoylacetic Acid, *Me ester*
3,4-Methylenedioxyphenylacetic Acid, *Et ester*
4-Methylisophthalic Acid, *Di-Me ester*
5-Methylisophthalic Acid, *Di-Me ester*

$C_{11}H_{12}O_4$ (*continued*)
Methylterephthalic Acid, *Di-Me ester*
1-(4-Pentenoyl)-2,3,6-trihydroxybenzene
2-Phenylglutaric Acid
3-Phenylglutaric Acid
Phenylmalonic Acid, *Di-Me ester*
Phenylmalonic Acid, *Et ester*
Phenylsuccinic Acid, α-*Me ester*
Phenylsuccinic Acid, β-*Me ester*
Phthalic Acid, *Me-Et ester*
Succinic Acid, *Benzyl ester*
Terephthalic Acid, *Mono-propyl ester*
Tetracyclo[2,2,1,$0^{2,6}$,$0^{3,5}$]heptane-2,3-dicarboxylic Acid, *Di-Me ester*
2,4,6-Trimethylisophthalic Acid
Umbellic Acid, *Et ester*
Umbellic Acid, *Di-Me ether*

$C_{11}H_{12}O_4S$
S-o-Carboxyphenylthioglycollic Acid, *Di-Me ester*
S-o-Carboxyphenylthioglycollic Acid, 2-*Et ester*
S-*p*-Carboxyphenylthioglycollic Acid, *Di-Me ester*

$C_{11}H_{12}O_5$
5-Acetyl-2,4-dihydroxybenzoic Acid, *Et ester*
5-Acetyl-2,4-dihydroxybenzoic Acid, *Di-Me ether*
Curvulinic Acid, *Me ester*†
Dihydrogladiolic Acid
Elenolide
2-Formyl-5,6-dihydroxybenzoic Acid, 5-*Me ether*, ψ-*Et ester*
3-Formyl-4,6-dihydroxy-*o*-toluic Acid, *Et ester*
5-Formyl-4,6-dihydroxy-*o*-toluic Acid, *Et ester*
2-Formyl-4,5-dimethoxybenzoic Acid, α-*Me ester*
2-Formyl-5,6-dimethoxybenzoic Acid, α-*Me ester*
Guaiacol, *Mono-guaiacol succinate*
3-(4-Hydroxy-3,5-dimethoxyphenyl)acrylic Acid
2-Hydroxy-5-methylisophthalic Acid, *Di-Me ester*
4-Hydroxy-5-methylisophthalic Acid, *Di-Me ester*
4-Hydroxy-6-methylisophthalic Acid, *Di-Me ester*
4-Hydroxy-6-methylisophthalic Acid, 3-*Et ester*
3-Hydroxyphthalic Acid, *Me ether*, *Di-Me ester*
4-Hydroxyphthalic Acid, *Me ether*, *Di-Me ester*
Hydroxyterephthalic Acid, *Me-ether*, *Di-Me ester*
3,4-Methylenedioxymandelic Acid, *Et ester*
Myristicic Acid, *Et ester*
Sepedonin†

$C_{11}H_{12}O_6$
Chelidonic Acid, *Di-Et ester*
Curvulic Acid†
Cyclopolic Acid
4,5-Dimethoxyisophthalic Acid, 1-*Me ester*
2,5-Dimethoxy-3,4-methylenedioxybenzoic Acid, *Me ester*
3,4-Dimethoxyphthalic Acid, 1-*Me ester*
4-Hydroxy-6-methoxy-*o*-toluic Acid, O-*Carbomethoxyl*

$C_{11}H_{12}O_7$
2-Carboxy-4,5-dimethoxyphenoxyacetic Acid
3-Hydroxy-4-oxo-4*H*-pyran-2,6-dicarboxylic Acid, *Di-Et ester*
Piscidic Acid†
2,4,6-Trihydroxyisophthalic Acid, *Tri-Me ether*
4,5,6-Trihydroxyisophthalic Acid, *Tri-Me ether*
3,4,5-Trihydroxyphthalic Acid, *Tri-Me ether*
3,4,6-Trihydroxyphthalic Acid, *Tri-Me ether*
Ustic Acid

$C_{11}H_{13}AsO_2$
Benzenearsonous Acid, *Trimethylene ester*

$C_{11}H_{13}BrO$
2-Bromo-1,2,3,4-tetrahydro-1-naphthol, *Me ether*
3-Bromo-4-hydroxybenzoic Acid, *Butyl ester*

$C_{11}H_{13}BrO_2$
2-Bromobutyric Acid, o-*Tolyl ester*
2-Bromo-4,6-dimethylbenzoic Acid, *Et ester*
α-Bromophenylacetic Acid, *Propyl ester*
2-Bromo-3-phenylpropionic Acid, *Et ester*
3-Bromo-2-phenylpropionic Acid, *Et ester*
3-Bromo-3-phenylpropionic Acid, *Et ester*

$C_{11}H_{13}BrO_3$
2-Bromobutyric Acid, *Guaiacol ester*

$C_{11}H_{13}ClO$
2,2-Dimethyl-3-phenylpropionic Acid, *Chloride*
3-(2,4-Dimethylphenyl)propionic Acid, *Chloride*
3-(2,5-Dimethylphenyl)propionic Acid, *Chloride*
5-Isopropyl-*o*-toluic Acid, *Chloride*
6-Isopropyl-*m*-toluic Acid, *Chloride*
2-Methyl-3-phenylbutyric Acid, *Chloride*
2-Methyl-4-phenylbutyric Acid, *Chloride*
3-Methyl-4-phenylbutyric Acid, *Chloride*
2-Methyl-3-*p*-tolylpropionic Acid, *Chloride*
4-Phenylvaleric Acid, *Chloride*
2,3,4,6-Tetramethylbenzoic Acid, *Chloride*
2,3,5,6-Tetramethylbenzoic Acid, *Chloride*
3-*p*-Tolylbutyric Acid, *Chloride*

$C_{11}H_{13}ClO_2$
α-Chlorophenylacetic Acid, *Propyl ester*
2-Chloro-2-phenylpropionic Acid, *Et ester*
2-*p*-Hydroxyphenylbutyric Acid, *Me ether*, *Chloride*
4-*p*-Hydroxyphenylbutyric Acid, *Me ether*, *Chloride*
4-*p*-Hydroxyphenylvaleric Acid, *Chloride*

$C_{11}H_{13}ClO_3$
3-Chloro-2-hydroxy-2-methylpropionic Acid, *Benzyl ester*
3-Chloro-6-hydroxy-*p*-toluic Acid, *Propyl ester*
3-Chloro-6-hydroxy-*p*-toluic Acid, *Propyl ether*
3-Chloro-6-hydroxy-*p*-toluic Acid, *Isopropyl ether*
3-Chloro-6-hydroxy-*p*-toluic Acid, *Et ether*, *Me ester*

$C_{11}H_{13}ClO_4$
4-Chloro-5-hydroxy-2-methyl phenoxyacetic Acid, *Me ether*, *Me ester*†

$C_{11}H_{13}ClO_5$
2,3,4,6-Tetrahydroxybenzoic Acid, *Tetra-Me ether*, *Chloride*

$C_{11}H_{13}N$

3-(2,4-Dimethylphenyl)propionic Acid, *Nitrile*
1-Ethyl-5-methylindole
2-Ethyl-3-methylindole
3-Ethyl-2-methylindole
2-Isopropylindole
3-Isopropylindole
Lilolidine
2-Methyl-2-phenylbutyric Acid, *Nitrile*
3-Methyl-2-phenylbutyric Acid, *Nitrile*
3-Methyl-4-phenylbutyric Acid, *Nitrile*
2-Phenylvaleric Acid, *Nitrile*
4-Phenylvaleric Acid, *Nitrile*
5-Phenylvaleric Acid, *Nitrile*
2-Propylindole
3-Propylindole
1,2,5,6-Tetrahydro-1-phenylpyridine
1,4,5,6-Tetrahydro-2-phenylpyridine
1,2,3,6-Tetrahydro-4-phenylpyridine
2,3,4,6-Tetramethylbenzoic Acid, *Nitrile*
2,3,5,6-Tetramethylbenzoic Acid, *Nitrile*
1,2,3-Trimethylindole
1,2,5-Trimethylindole
2,3,4(or 2,3,6)-Trimethylindole
2,3,5-Trimethylindole
2,3,7-Trimethylindole
2,4,7-Trimethylindole
2,5,7-Trimethylindole
2,4,5-Trimethylphenylacetic Acid, *Nitrile*

$C_{11}H_{13}NO$

α-Ethylcinnamic Acid, *Amide*
β-Ethylcinnamic Acid, *Amide*
2-Hydroxy-2-(4-isopropylphenyl)acetic Acid, *Nitrile*
4-Hydroxy-5-isopropyl-*o*-toluic Acid, *Nitrile*
2-Hydroxy-2-(2,4,6-trimethylphenyl)acetic Acid, *Nitrile*
3-3′-Indolylpropanol
1-Methylindane-2-carboxylic Acid, *Amide*
Physostigmol, *Me ether*
1,2,3,4-Tetrahydro-1-naphthoic Acid, *Amide*
5,6,7,8-Tetrahydro-1-naphthoic Acid, *Amide*
5,6,7,8-Tetrahydro-2-naphthoic Acid, *Amide*

$C_{11}H_{13}NOS$

Dimethyldithiocarbamic Acid, *Phenacyl ester*

$C_{11}H_{13}NO_2$

o-Aminocinnamic Acid, *Et ester*
m-Aminocinnamic Acid, *Et ester*
p-Aminocinnamic Acid, *Et ester*
3-Anilinocrotonic Acid, *Me ester*
2-Benzoylbutyric Acid, *Amide*
2-Benzylacetoacetic Acid, *Amide*
Bicyclo[2,2,1]hept-5-ene-2,3-dicarboxylic Acid, *Ethylimide*
5-Hydroxy-3β-hydroxyethylindole, 5-*Me ether*†
Latumcidin
Mandelamide, *Allyl ether*
2-Methylacetoacetic Acid, *Anilide*
4-Oxo-2-phenylvaleric Acid, *Amide*
1,2,3,4-Tetrahydro-6,7-methylenedioxy-*N*-methylisoquinoline
1,2,3,4-Tetrahydroquinoline-2-carboxylic Acid, *Me ester*
1,2,3,4-Tetrahydroquinoline-4-carboxylic Acid, *Me ester*
1,2,3,4-Tetrahydroquinoline-4-carboxylic Acid, N-*Me*
1,2,3,4-Tetrahydroquinoline-5-carboxylic Acid, N-*Me*
1,2,3,4-Tetrahydroquinoline-6-carboxylic Acid, N-*Me*
1,2,3,4-Tetrahydroquinoline-7-carboxylic Acid, N-*Me*
1,2,3,4-Tetrahydroquinoline-8-carboxylic Acid, N-*Me*

$C_{11}H_{13}NO_3$

N-Acetylanthranilic Acid, *Et ester*
Acetylglycine, *Benzyl ester*
4-Amino-3-methylbenzoylformic Acid, N-*Et*
Corydaldine
N-Formyl-*N*-phenylglycine, *Et ester*
Hippuric Acid, *Et ester*
Malonanilic Acid, *Et ester*
2-β-Methylaminoethyl-4,5-methylenedioxy-benzaldehyde
3-(3,4-Methylenedioxyphenyl)propionic Acid, *Methylamide*
Methylmalonic Acid, *Mono-Me ester*, *Anilide*
Methylmalonic Acid, *Mono*-p-*toluidide*
o-Nitrosobenzoic Acid, *Isobutyl ester*
Oxanilic Acid, *Propyl ester*
Oxanilic Acid, *Isopropyl ester*
Phenaceturic Acid, *Me ester*
Phenylpyruvic Acid, *Me ester*, *Oxime*
Phenylsuccinic Acid, α-*Me ester*, β-*Amide*
Succinanilic Acid, *Me ester*

$C_{11}H_{13}NO_4$

4-Aminoisophthalic Acid, N-*Me*, *Di-Me ester*
4-Aminoisophthalic Acid, N-*Di-Me*, 1-*Me ester*
4-Aminoisophthalic Acid, N-*Di-Me*, 3-*Me ester*
Anilinomalonic Acid, *Di-Me ester*
2-Anilino-2-methylsuccinic Acid
Benzyliminodiacetic Acid
2,4-Dimethyl-5-nitrobenzoic Acid, *Et ester*
3,5-Dimethyl-2-nitrobenzoic Acid, *Et ester*
3,5-Dimethyl-4-nitrobenzoic Acid, *Et ester*
2,6-Dimethylpyridine-3,4-dicarboxylic Acid, *Di-Me ester*
2,6-Dimethylpyridine-3,4-dicarboxylic Acid, 3-*Et ester*
2-Hydroxybenzamidoacetic Acid, *Et ester*
4-Isopropyl-3-nitrobenzoic Acid, *Me ester*
Malic Acid, *Mono*-o-*toluidide*
Malic Acid, *Mono*-p-*toluidide*
N-Methylisatoic Acid, *Et ester*
o-Nitrobenzoic Acid, n-*Butyl ester*
m-Nitrobenzoic Acid, n-*Butyl ester*
p-Nitrobenzoic Acid, n-*Butyl ester*
p-Nitrophenylacetic Acid, *Propyl ester*
3-*p*-Nitrophenylbutyric Acid, *Me ester*
4-*p*-Nitrophenylbutyric Acid, *Me ester*
3-*p*-Nitrophenylpropionic Acid, *Et ester*
3-Phenylglutamic Acid
4-Phenylglutamic Acid
N-Phenylglycine-*o*-carboxylic Acid, *Di-Me ester*
N-Phenylglycine-*o*-carboxylic Acid, α-*Et ester*
N-Phenylglycine-*o*-carboxylic Acid, β-*Et ester*
N-Phenylglycine-*o*-carboxylic Acid, N-*Et*
Pyridine-2,6-diacetic Acid, *Di-Me ester*

$C_{11}H_{13}NO_4$ *(continued)*
Pyridine-2,3-dicarboxylic Acid, *Di-Et ester*
2,3,4,6-Tetrahydroxybenzoic Acid, *Nitrile*

$C_{11}H_{13}NO_5$
2,3-Dihydroxy-5-nitrobenzaldehyde, *Di-Et ether*
2,3-Dihydroxy-6-nitrobenzaldehyde, *Di-Et ether*
2-Hydroxy-3-nitrobenzoic Acid, *Et ether, Et ester*
2-Hydroxy-4-nitrobenzoic Acid, *Propyl ether, Me ester*
2-Hydroxy-4-nitrobenzoic Acid, *Butyl ester*
2-Hydroxy-5-nitrobenzoic Acid, *Et ether, Et ester*
2-Hydroxy-5-nitrobenzoic Acid, *Butyl ester*
3-Hydroxy-2-nitrobenzoic Acid, *Et ester, Et ether*
3-Hydroxy-4-nitrobenzoic Acid, *Et ether, Et ester*
4-Hydroxy-3-nitrobenzoic Acid, *Et ether, Et ester*
4-Hydroxy-3-nitrobenzoic Acid, *Me ester, Propyl ether*
4-Hydroxy-3-nitrobenzoic Acid, *Butyl ester*
3-Hydroxy-3-*m*-nitrophenylpropionic Acid, *Et ester*
3-Hydroxy-3-*p*-nitrophenylpropionic Acid, *Et ester*
3-(4-Hydroxy-3-nitrophenyl)propionic Acid, *Et ester*
4-Hydroxypyridine-2,6-dicarboxylic Acid, *Di-Et ester*
4-Methoxy-3-nitrobenzoic Acid, *Propyl ester*
4-Methoxy-3-nitrophenylacetic Acid, *Et ester*
1,2-*p*-Nitrobenzylideneglycerol, *Me ether*
1,3-*p*-Nitrobenzylideneglycerol, *Me ether*
2-*o*-Nitrophenoxypropionic Acid, *Et ester*
2-*m*-Nitrophenoxypropionic Acid, *Et ester*
2-*p*-Nitrophenoxypropionic Acid, *Et ester*
Serine, N-*Benzyloxycarbonyl*
Tartranilic Acid, *Me ester*

$C_{11}H_{13}NO_5S$
p-Sulphobenzoic Acid, 4-*Morpholide*

$C_{11}H_{13}NO_6$
2,3-Dihydroxy-5-nitrobenzoic Acid, *Di-Et ether*
2,6-Dihydroxypyridine-3,4-dicarboxylic Acid, *Di-Et ester*
2,6-Dihydroxypyridine-3,5-dicarboxylic Acid, *Di-Et ester*
2,3-Dimethoxy-4-nitrobenzoic Acid, *Et ester*
2,3-Dimethoxy-5-nitrobenzoic Acid, *Et ester*
3,5-Dimethoxy-2-nitrobenzoic Acid, *Et ester*
3,5-Dimethoxy-4-nitrobenzoic Acid, *Et ester*
4,5-Dimethoxy-2-nitrobenzoic Acid, *Et ester*
3-*m*-Nitrophenylglyceric Acid, *Et ester*
2-Nitrogallic Acid, *Tri-Me ether, Me ester*
Xylose★, o-*Nitrophenyl-β-D-xylopyranoside*†

$C_{11}H_{13}NO_7S$
4-Nitro-2-sulphobenzoic Acid, *Di-Et ester*

$C_{11}H_{13}N_3$
Iminopyrine

$C_{11}H_{13}N_3O$
4-Aminoantipyrine
5-Benzylcreatinine
Tryptophane, *Amide*

$C_{11}H_{13}N_3O_2$
1-Phenylurazole, 2-*Me*-4-*Et*

$C_{11}H_{13}N_3O_3$
N-Benzoylglycylglycine, *Amide*

$C_{11}H_{13}N_3O_3S$
Sulphisoxazole

$C_{11}H_{13}N_3O_6$
3-Butyl-2,4,6-trinitrotoluene
3-*tert*-Butyl-2,4,6-trinitrotoluene

$C_{11}H_{13}N_5O_4$
Angustmycin A

$C_{11}H_{14}$
2-Methyl-1-phenyl-1-butene
3-Methyl-1-phenyl-1-butene
3-Methyl-2-phenyl-1-butene
2-Methyl-3-phenyl-2-butene
2-Methyl-4-phenyl-2-butene
3-Methyl-1-phenyl-2-butene
Phenylcyclopentane
1-Phenyl-1-pentene
2-Phenyl-1-pentene
3-Phenyl-1-pentene
5-Phenyl-1-pentene
1-Phenyl-2-pentene
2-Phenyl-2-pentene
3-Phenyl-2-pentene
5-Phenyl-2-pentene
Propylcyclo-octatetrene★†
1,2,3,4-Tetrahydro-1-methylnaphthalene
1,2,3,4-Tetrahydro-5-methylnaphthalene
1,2,3,4-Tetrahydro-6-methylnaphthalene

$C_{11}H_{14}ClNO$
4-Amino-2-chlorobenzaldehyde, N-*Di-Et*

$C_{11}H_{14}ClNO_2$
o-Tolylcarbamic Acid, *γ-Chloropropyl ester*
p-Tolylcarbamic Acid, *γ-Chloropropyl ester*

$C_{11}H_{14}Cl_2N_2O_3$
Threo-1-*p*-Aminophenyl-2-dichloroacetamido-1,3-propanediol†

$C_{11}H_{14}INO_2$
α-Amino-β-4-iodophenylpropionic Acid, *Et ester*

$C_{11}H_{14}N_2$
2-Anilino-3-methyl-butyric Acid, *Nitrile*
2-Anilinovaleric Acid, *Nitrile*
o-Diethylaminobenzoic Acid, *Nitrile*
p-Diethylaminobenzoic Acid, *Nitrile*
Dipterine
Donaxine
3-3′-Indolylpropylamine
2-Methyltryptamine†
2-3′-Pyridyl-pyrrolidine, N-*Et*

$C_{11}H_{14}N_2O$
3-(2-Aminoethyl)-5-hydroxyindole, *Me ether*
Cytisine
Donaxine★, N-*Oxide*†
Surinamine, *Me ether-nitrile*

$C_{11}H_{14}N_2O_2$
Benzylidenediacetamide
5,6,7,8-Tetrahydro-3-nitro-2-naphthylamine, N-*Me*

$C_{11}H_{14}N_2O_3$
o-Aminobenzamidoacetic Acid, *Et ester*
m-Aminobenzamidoacetic Acid, *Et ester*
p-Aminobenzamidoacetiic Acid *Et ester*
2-Ethylamino-5-nitrosobenzoic Acid, *Et ester*
3-*p*-Nitrophenylpropionic Acid, *Dimethylamide*
N-Phenylglycine-*o*-carboxylic Acid, β-*Et ester*, α-*Amide*
2-Phenylhydantoic Acid, *Et ester*
5-Phenylhydantoic Acid, *Et ester*

$C_{11}H_{14}N_2O_4$
2-Amino-4-nitrobenzoic Acid, N-*Et*, *Et ester*
4-Amino-3-nitrobenzoic Acid, N-*Et*, *Et ester*
4-Amino-3-nitrobenzoic Acid, N-*Di-Et*
4-Dimethylamino-3-nitrobenzoic Acid, *Et ester*
Glycyltyrosine
3,4-Methylenedioxy-6-nitroaniline, N-*Butyl*
o-Nitrophenylcarbamic Acid, *Isobutyl ester*
p-Nitrophenylcarbamic Acid, *Isobutyl ester*
N-(2-Nitro-4-tolyl)glycine, *Et ester*
N-(4-Nitro-2-tolyl)glycine, *Et ester*
N-(5-Nitro-2-tolyl)glycine, *Et ester*

$C_{11}H_{14}N_2O_4S$
S-(1-Acetamido-4-hydroxyphenyl)cysteine†

$C_{11}H_{14}N_2O_5$
Antimycic Acid

$C_{11}H_{14}N_2S$
2-Isobutylaminobenzothiazole
1,4,5,6-Tetrahydro-1-methyl-2-[2-(2-thienyl)-vinyl]pyrimidine†

$C_{11}H_{14}N_4O_4$
4-Amino-1-β-D-ribofuranosylimidazo[4,5-*c*]-pyridine†
Cytosinine, *Me ester*
Tubercidin†

$C_{11}H_{14}O$
Chavicol, *Et ether*
2,2-Dimethylchroman†
2,2-Dimethyl-1-phenyl-1-propane
1-Hydroxyindane, *Et ether*
5-Hydroxyindane, *Et ether*
p-Isopropylacetophenone
4-Isopropyl-*o*-toluic Aldehyde
5-Isopropyl-*o*-toluic Aldehyde
4-Isopropyl-*m*-toluic Aldehyde
Isopropyl *o*-tolyl Ketone
Isopropyl *m*-tolyl Ketone
Isopropyl *p*-tolyl Ketone
Isovalerophenone
2-Methylbutyrophenone
3-Methylbutyrophenone
4-Methylbutyrophenone
4-Methyl-4-phenylbutanal
2-Methyl-1-phenyl-1-butanone
3-Methyl-1-phenyl-2-butanone
1-Phenylallyl Alcohol, *Et ether*
1-Phenylcyclopentanol
3-Phenylcyclopentanol
1-Phenyl-2-pentanone
3-Phenyl-2-pentanone
4-Phenyl-2-pentanone
5-Phenyl-2-pentanone
1-Phenyl-3-pentanone
2-Phenyl-3-pentanone
2-Phenylvaleraldehyde
5-Phenylvaleraldehyde
2-Phenylvinyl Alcohol, *Propyl ether*
4-Propenyl-*o*-cresol, *Me ether*
4-Propenyl-*m*-cresol, *Me ether*
2-Propenyl-*p*-cresol, *Me ether*
o-Propenylphenol, *Et ether*
m-Propenylphenol, *Et ether*
p-Propylacetophenone
1,2,3,4-Tetrahydro-1-methyl-1-naphthol
1,2,3,4-Tetrahydro-2-methyl-1-naphthol
1,2,3,4-Tetrahydro-6-methyl-1-naphthol
5,6,7,8-Tetrahydro-3-methyl-1-naphthol
5,6,7,8-Tetrahydro-4-methyl-1-naphthol
1,2,3,4-Tetrahydro-6-methyl-2-naphthol
5,6,7,8-Tetrahydro-1-methyl-2-naphthol
5,6,7,8-Tetrahydro-3-methyl-2-naphthol
5,6,7,8-Tetrahydro-4-methyl-2-naphthol
5,6,7,8-Tetrahydro-2-naphthol, *Me ether*
2,3,4,5-Tetramethylbenzaldehyde
2,3,4,6-Tetramethylbenzaldehyde
2,3,5,6-Tetramethylbenzaldehyde
2,4,5-Trimethylacetophenone
2,4,6-Trimethylacetophenone
3,4,5-Trimethylacetophenone
Valerophenone

$C_{11}H_{14}OS$
Ethylmethylsulphonium phenacylide†
Thiobenzoic Acid, S-*Butyl ester*
Thiobenzoic Acid, S-2-*Butyl ester*
Thiobenzoic Acid, S-*Isobutyl ester*
Thiobenzoic Acid, S-tert-*Butyl ester*

$C_{11}H_{14}O_2$
Actinidiolide†
Benzoic Acid, *Butyl ester*
Benzoic Acid, 2-*Butyl ester*
Benzoic Acid, tert-*Butyl ester*
Benzoic Acid, *Isobutyl ester*
1-Benzoyl-2-propanol, *Me ether*
2-Benzylpropionic Acid, *Me ester*
5-*tert*-Butyl-2-hydroxybenzaldehyde
1-Butyl-3,4-methylenedioxybenzene†
Butyric Acid, *Benzyl ester*
3,5-Dimethylbenzoic Acid, *Et ester*
2,4-Dimethylphenylacetic Acid, *Me ester*
2,5-Dimethylphenylacetic Acid, *Me ester*
2,2-Dimethyl-3-phenylpropionic Acid
3-(2,3-Dimethylphenyl)propionic Acid
3-(2,4-Dimethylphenyl)propionic Acid
3-(2,5-Dimethylphenyl)propionic Acid
3-(3,5-Dimethylphenyl)propionic Acid
o-Ethylbenzoic Acid, *Et ester*
p-Ethylbenzoic Acid, *Et ester*
Eugenol, *Me ether*
p-Hydroxybutyrophenone, *Me ether*
4-Hydroxy-2,6-dimethylbenzaldehyde, *Et ether*
4-Hydroxy-3,5-dimethylbenzaldehyde, *Et ether*
2-Hydroxy-6-isopropyl-*m*-tolualdehyde
4-Hydroxy-6-isopropyl-*m*-tolualdehyde
4-Hydroxy-5-isopropyl-*o*-toluic Aldehyde

$C_{11}H_{14}O_2$ (*continued*)
6-Hydroxy-5-isopropyl-*o*-toluic Aldehyde
4-Hydroxy-3-methoxy-5-propenyltoluene
α-Hydroxy-4-methylacetophenone, *Et ether*
2-Hydroxy-4-methylacetophenone, *Et ether*
4-Hydroxy-2-methylacetophenone, *Et ether*
2-Hydroxy-3-methylbutyrophenone
2-Hydroxy-4-methylbutyrophenone
2-Hydroxy-5-methylbutyrophenone
4-Hydroxy-2-methylbutyrophenone
4-Hydroxy-3-methylbutyrophenone
2-Hydroxy-5-methylpropiophenone, *Me ether*
2-Hydroxy-6-methylpropiophenone, *Me ether*
4-Hydroxy-2-methylpropiophenone, *Me ether*
4-Hydroxy-3-methylpropiophenone, *Me ether*
5-Hydroxy-2-pentanone, *Phenyl ether*
1-Hydroxy-3-phenylacetone, *Et ether*
β-Hydroxypropiophenone, *Et ether*
γ-Hydroxypropiophenone, *Et ether*
4-Hydroxypropiophenone, *Et ether*
Isobutyric Acid, *Benzyl ester*
4-Isobutyryl-*o*-cresol
6-Isobutyryl-*o*-cresol
6-Isobutyryl-*m*-cresol
2-Isobutyryl-*p*-cresol
p-Isopropylbenzoic Acid, *Me ester*
5-Isopropyl-*o*-toluic Acid
6-Isopropyl-*m*-toluic Acid
Isosafroeugenol
Isovaleric Acid, *Phenyl ester*
β-Hydroxyvalerophenone
ω-Hydroxyvalerophenone
m-Hydroxyvalerophenone
p-Hydroxyvalerophenone
2-Methoxy-4-propenylphenol, 1-*Me ether*
2-(3-Methylbutyryl)phenol
3-(3-Methylbutyryl)phenol
4-(3-Methylbutyryl)phenol
2-Methyl-2-phenylbutyric Acid,
2-Methyl-3-phenylbutyric Acid
2-Methyl-4-phenylbutyric Acid
3-Methyl-2-phenylbutyric Acid
3-Methyl-3-phenylbutyric Acid
3-Methyl-4-phenylbutyric Acid
2-Methyl-2-phenylpropionic Acid, *Me ester*
2-Methyl-3-*o*-tolylpropionic Acid
2-Methyl-3-*p*-tolylpropionic Acid
Phenylacetic Acid, *Propyl ester*
2-Phenylbutyric Acid, *Me ester*
3-Phenylbutyric Acid, *Me ester*
2-Phenylpropionic Acid, *Et ester*
3-Phenylpropionic Acid, *Et ester*
2-Phenylvaleric Acid
3-Phenylvaleric Acid
4-Phenylvaleric Acid
5-Phenylvaleric Acid
2-Propyl-*p*-toluic Acid
Pyrethrolone
2,3,4,5-Tetramethylbenzoic Acid
2,3,4,6-Tetramethylbenzoic Acid
2,3,5,6-Tetramethylbenzoic Acid
α-Thujaplicin, *Me ether*
m-Toluic Acid, *Propyl ester*
m-Tolylacetic Acid, *Et ester*
p-Tolylacetic Acid, *Et ester*
3-*p*-Tolylbutyric Acid
4-*o*-Tolylbutyric Acid
4-*p*-Tolylbutyric Acid
3-*p*-Tolylpropionic Acid, *Me ester*
2,4,6-Trimethylbenzoic Acid, *Me ester*
2,4,5-Trimethylphenylacetic Acid
2,4,6-Trimethylphenylacetic Acid

$C_{11}H_{14}O_2S$
o-Mercaptobenzoic Acid, *Et ether*, *Et ester*
Phenylmercaptoacetic Acid, *Propyl ester*

$C_{11}H_{14}O_3$
Acetovanillone, *Et ether*
4-Allyl-2,6-dimethoxyphenol†
Anisic Acid, N-*Propyl ester*
1,2-*O*-Benzylideneglycerol, 3-*Me ether*
1,3-*O*-Benzylideneglycerol, 2-*Me ether*
m-Butoxybenzoic Acid,
p-Butoxybenzoic Acid
p-2-Butoxybenzoic Acid
Coniferyl Alcohol, γ-*Me ether*
2,4-Dihydroxyacetophenone, 2-*Me*, 4-*Et ether*
2,4-Dihydroxyacetophenone, 4-*Propyl ether*
2,4-Dihydroxybenzaldehyde, *Di-Et ether*
2,5-Dihydroxybenzaldehyde, *Di-Et ether*
3,4-Dihydroxybenzaldehyde, 3-*Isobutyl ether*
2,4-Dihydroxy-6-methylacetophenone, *Di-Me ether*
2,6-Dihydroxy-4-methylacetophenone, *Di-Me ether*
4,5-Dihydroxy-2-methylacetophenone, *Di-Me ether*
2,4-Dihydroxypropiophenone, *Di-Me ether*
2,4-Dihydroxypropiophenone, 4-*Et ether*
2,5-Dihydroxypropiophenone, *Di-Me ether*
2,6-Dihydroxypropiophenone, *Di-Me ether*
3,4-Dihydroxypropiophenone, *Di-Me ether*
3,5-Dihydroxypropiophenone, *Di-Me ether*
4,6-Dihydroxy-*o*-tolualdehyde, 5-*Me*, 3-*Et ether*
Ethoxyacetic Acid, *Benzyl ester*
o-Ethoxybenzoic Acid, *Et ester*
m-Ethoxybenzoic Acid, *Et ester*
p-Ethoxybenzoic Acid, *Et ester*
3-α-Furylacrylic Acid, *Butyl ester*
3-α-Furylacrylic Acid, *Isobutyl ester*
2-Hydroxy-4,6-dimethylbenzoic Acid, *Me ether*, *Me ester*
3-Hydroxy-4,5-dimethylbenzoic Acid, *Et ester*
3-Hydroxy-4,5-dimethylbenzoic Acid, *Et ether*
4-Hydroxy-2,6-dimethylbenzoic Acid, *Et ester*
4-Hydroxy-3,5-dimethylbenzoic Acid, *Et ester*
2-Hydroxy-2-(4-isopropylphenyl)acetic Acid
4-Hydroxy-5-isopropyl-*o*-toluic Acid
6-Hydroxy-5-isopropyl-*o*-toluic Acid
2-Hydroxy-5-isopropyl-*m*-toluic Acid
2-Hydroxy-6-isopropyl-*m*-toluic Acid
4-Hydroxy-6-isopropyl-*m*-toluic Acid
4-(3-Hydroxy-4-methoxyphenyl)-2-butanone
p-Hydroxymethyl-benzoic Acid, *Propyl ester*
2-Hydroxy-4-phenylbutyric Acid, *Me ester*
3-Hydroxy-3-phenylbutyric Acid, *Me ester*
2-*p*-Hydroxyphenylbutyric Acid, *Me ether*
4-*o*-Hydroxyphenylbutyric Acid, *Me ether*
4-*p*-Hydroxyphenylbutyric Acid, *Me ether*
2-*p*-Hydroxyphenylpropionic Acid, *Et ether*
2-Hydroxy-2-phenylpropionic Acid, *Et ester*
2-Hydroxy-2-phenylpropionic Acid, *Et ether*

$C_{11}H_{14}O_3$ (*continued*)
2-Hydroxy-3-phenylpropionic Acid, *Et ester*
3-*o*-Hydroxyphenylpropionic Acid, *Et ester*
3-*o*-Hydroxyphenylpropionic Acid, *Et ether*
3-*p*-Hydroxyphenylpropionic Acid, *Et ether*
3-Hydroxy-3-phenylpropionic Acid, *Et ester*
3-Hydroxy-3-phenylpropionic Acid, *Et ether*
3-Hydroxy-2-phenylvaleric Acid
3-Hydroxy-3-phenylvaleric Acid
3-Hydroxy-5-phenylvaleric Acid
4-Hydroxy-4-phenylvaleric Acid
4-Hydroxy-5-phenylvaleric Acid
4-*p*-Hydroxyphenylvaleric Acid
5-*p*-Hydroxyphenylvaleric Acid
3-Hydroxy-*o*-toluic Acid, *Me ester*, *Et ether*
6-Hydroxy-*o*-toluic Acid, *Me ester*, *Et ether*
2-Hydroxy-2-*p*-tolylacetic Acid, *Et ester*
2-Hydroxy-2-(2,4,5-trimethylphenyl)acetic Acid
2-Hydroxy-2-(2,4,6-trimethylphenyl)acetic Acid
Isobutoxybenzoic Acid
o-Isopropoxybenzoic Acid, *Me ester*
Isopygmaein
Mandelic Acid, *Propyl ester*
Mandelic Acid, *Me ether*, *Et ester*
Mandelic Acid, *Et ether*, *Me ester*
Mandelic Acid, *Propyl ether*
Mandelic Acid, *Isopropyl ether*
p-Methoxyphenylacetic Acid, *Et ester*
3-*o*-Methoxyphenylpropionic Acid, *Me ester*
3-*p*-Methoxyphenylpropionic Acid, *Me ester*
4-(3-Methylbutyryl)resorcincol
2-Methyl-3-phenoxypropionic Acid, *Me ester*
Methyl salicylate, *Propyl ether*
Methyl salicylate, *Isopropyl ether*
2-Phenoxypropionic Acid, *Et ester*
3-Phenoxypropionic Acid, *Et ester*
o-Propoxybenzoic Acid, *Me ester*
m-Propoxybenzoic Acid, *Me ester*
p-Propoxybenzoic Acid, *Me ester*
Pygmaein
Vanillin, *Propyl ether*
Vanillin, *Isopropyl ether*
Zingerone

$C_{11}H_{14}O_3S$
p-Butylsulphonylbenzaldehyde

$C_{11}H_{14}O_4$
Bicyclo[2,2,1]hept-5-ene-1,4-dicarboxylic Acid, *Di-Me ester*
Bicyclo[2,2,1]hept-5-ene-2,3-dicarboxylic Acid, *Mono-Et ester*
Canadensolide†
2,4-Dihydroxybenzoic Acid, *Di-Et ether*
2,6-Dihydroxybenzoic Acid, *Di-Et ether*
3,4-Dihydroxybenzoic Acid, *Di-Et ether*
3,5-Dihydroxybenzoic Acid, *Di-Et ether*
α,2-Dihydroxyphenylacetic Acid, 2-*Et ether*, *Me ester*
α,2-Dihydroxyphenylacetic Acid, 2-*Me ether*, *Et ester*
α,3-Dihydroxyphenylacetic Acid, 3-*Me ether*, *Et ester*
α,4-Dihydroxyphenylacetic Acid, 4-*Me ether*, *Et ester*
2-(3,4-Dihydroxyphenyl)propionic Acid, *Di-Me ether*
3-(2,4-Dihydroxyphenyl)propionic Acid, *Di-Me ether*
3-(3,4-Dihydroxyphenyl)propionic Acid, *Di-Me ether*
3-(3,5-Dihydroxyphenyl)propionic Acid, *Di-Me ether*
2,4-Dihydroxy-6-propylbenzoic Acid, *Me ester*
3,5-Dihydroxy-*o*-toluic Acid, n-*Propyl ester*
5,6-Dihydroxy-*m*-toluic Acid, 5-*Me ether*, *Et ester*
2,6-Dihydroxy-*p*-toluic Acid, *Di-Me ether*, *Me ester*
2,4-Dimethoxybenzoic Acid, *Et ester*
2,5-Dimethoxybenzoic Acid, *Et ester*
2,6-Dimethoxybenzoic Acid, *Et ester*
3,4-Dimethoxybenzoic Acid, *Et ester*
3,5-Dimethoxybenzoic Acid, *Et ester*
3,4-Dimethoxyphenylacetic Acid, *Me ester*
4,5-Dimethoxy-*o*-toluic Acid, *Me ester*
4,6-Dimethyl-2-oxo-2*H*-pyran-5-carboxylic Acid, n-*Propyl ester*
2-Ethyl-4,6-dihydroxybenzoic Acid, *Et ester*†
α-Furoylacetic Acid, *Butyl ester*
3-α-Furylacrylic Acid, 2-*Ethoxyethyl ester*
Guaiacol, *Mono-guaiacol carbonate*, *Propyl ester*
2-Hydroxy-4-methoxy-3,6-dimethylbenzoic Acid, *Me ester*
2-Hydroxy-4-methoxy-3,6-dimethylbenzoic Acid, 2-*Me ether*
4-Hydroxy-2-methoxy-3,6-dimethylbenzoic Acid, *Me ester*
4-Hydroxy-3-methoxyphenylacetic Acid, *Et ester*
6-Hydroxy-4-methoxy-*o*-toluic Acid, *Et ester*
6-Hydroxy-4-methoxy-*o*-toluic Adid, *Et ether*
Nortricyclene-2,2-dicarboxylic Acid, *Di-Me ester*
Sparassol, 6-*Me ether*
Terrestric Acid
α,2,4-Trihydroxyacetophenone, *Tri-Me ether*
α,3,4-Trihydroxyacetophenone, *Tri-Me ether*
2,3,4-Trihydroxyacetophenone, *Tri-Me ether*
2,3,5-Trihydroxyacetophenone, 2,3,5-*Tri-Me ether*
2,4,5-Trihydroxyacetophenone, 2,4,5-*Tri-Me ether*
2,4,6-Trihydroxyacetophenone, *Tri-Me ether*
2,4,6-Trihydroxyacetophenone, 2-*Me*-4-*Et ether*
2,4,6-Trihydroxyacetophenone, 4-*Me*-2-*Et ether*
3,4,5-Trihydroxyacetophenone, 3,4,5-*Tri-Me ether*
2,4,5-Trihydroxybenzaldehyde, 2,4-*Di-Me*-5-*Et ether*
2,4,6-Trihydroxybutyrophenone, 2-*Me ether*
2,4,6-Trihydroxybutyrophenone, 4-*Me ether*
2,4,6-Trihydroxyisovalerophenone†
2,4,5-Trihydroxypropiophenone, 4,5-*Di-Me ether*
2,4,6-Trihydroxypropiophenone, 2,4-*Di-Me ether*
2,4,6-Trihydroxypropiophenone, 2,6-*Di-Me ether*
Vanillic Acid, *Propyl ester*
Vanillic Acid, *Isopropyl ester*

$C_{11}H_{14}O_5$
2,4-Dihydroxyacetophenone, 4-α-*Glycerol ether*
Gallic Acid, *Butyl ester*
Gallic Acid, *Isobutyl ester*
Gallic Acid, 3,4-*Di-Me ether*, 5-*Et ether*
Gallic Acid, *Tri-Me ether*, *Me ester*
Genipin
4-Hydroxy-3,5-dimethoxybenzoic Acid, *Et ester*
4-Hydroxy-3,5-dimethoxybenzoic Acid, 4-*Et ether*
Iridic Acid, *Me ester*
Iridic Acid, *Me ether*
α,2,4,6-Tetrahydroxyacetophenone, α,4,6-*Tri-Me ether*
α,3,4,5-Tetrahydroxyacetophenone, 3,4,5-*Tri-Me ether*
2,3,4,6-Tetrahydroxyacetophenone, 2,4,6-*Tri-Me ether*
2,3,4-Trihydroxybenzoic Acid, 2,3,4-*Tri-Me ether*, *Me ester*
2,4,5-Trihydroxybenzoic Acid, 2,4-*Di-Me*-5-*Et ether*
2,4,5-Trihydroxybenzoic Acid, 2,5-*Di-Me*-4-*Et ether*
2,4,6-Trihydroxybenzoic Acid, *Me*-2-*Et ether*
2,4,6-Trihydroxybenzoic Acid, 2,4,6-*Tri-Me ether*, *Me ester*
2,4,5-Trimethoxybenzoic Acid, *Me ester*
Verbenalol†

$C_{11}H_{14}O_5S$
o-Sulphobenzoic Acid, *Di-Et ester*
p-Sulphobenzoic Acid, *Di-Et ester*

$C_{11}H_{14}O_6$
2,5-Dioxocyclohexane-1,4-dicarboxylic Acid, *Propyl ester*
Genipinic Acid†
2,3,4,5-Tetrahydroxybenzoic Acid, 3,4,5-*Tri-Me ether*, *Me ester*
2,3,4,5-Tetrahydroxybenzoic Acid, *Tetra-Me ether*
2,3,4,6-Tetrahydroxybenzoic Acid, *Tetra-Me ether*

$C_{11}H_{14}O_6S$
2-Hydroxy-5-sulphobenzoic Acid, *Di-Et ester*

$C_{11}H_{14}O_7$
2,4,6-Trioxopimelic Acid, *Di-Et ester*

$C_{11}H_{14}O_8$
Methane-tetracarboxylic Acid, *Tetra-cyclohexyl ester*

$C_{11}H_{15}BrN_2O_3$
5-2′-Bromoallyl-5-2′-butylbarbituric Acid
Pronarcon

$C_{11}H_{15}BrO$
4-Bromo-5-isopropyl-*o*-cresol, *Me ether*

$C_{11}H_{15}ClO$
4-Chloro-6-isopropyl-*m*-cresol, *Me ether*
Dicadic Acid, *Chloride*

$C_{11}H_{15}ClO_2$
1-Adamantyl Chloroformate†
Phenaglycodol

$C_{11}H_{15}Cl_2N_5$
Chlorproguanil

$C_{11}H_{15}N$
N-Allyl-*N*-methyl-*o*-toluidine
N-Allyl-*N*-methyl-*p*-toluidine
1-Aminoindane, N-N-*Di-Me*†
2-Amino-2-methylindane, N-*Me*
1-Ethyl-1,2,3,4-tetrahydroquinoline
1-Methyl-2-phenylpyrrolidine
1-Methyl-3-phenylpyrrolidine
2-Methyl-1-phenylpyrrolidine
2-Methyl-4-phenylpyrrolidine
3-Methyl-3-phenylpyrrolidine
N-Phenylpiperidine
2-Phenylpiperidine
3-Phenylpiperidine
4-Phenylpiperidine
Pyrrolidine, N-*Benzyl*
Pyrrolidine, N-o-*Tolyl*
1,2,3,4-Tetrahydro-1,3-dimethylquinoline
1,2,3,4-Tetrahydro-1,6-dimethylquinoline
1,2,3,4-Tetrahydro-1,8-dimethylquinoline
1,2,3,4-Tetrahydro-2,2-dimethylquinoline
1,2,3,4-Tetrahydro-2,3-dimethylquinoline
1,2,3,4-Tetrahydro-2,4-dimethylquinoline
1,2,3,4-Tetrahydro-2,6-dimethylquinoline
1,2,3,4-Tetrahydro-2,8-dimethylquinoline
1,2,3,4-Tetrahydro-3,4-dimethylquinoline
1,2,3,4-Tetrahydro-4,6-dimethylquinoline
1,2,3,4-Tetrahydro-4,7-dimethylquinoline
1,2,3,4-Tetrahydro-4,8-dimethylquinoline
1,2,3,4-Tetrahydro-5,8-dimethylquinoline
1,2,3,4-Tetrahydro-6,8-dimethylquinoline
5,6,7,8-Tetrahydro-2,3-dimethylquinoline
5,6,7,8-Tetrahydro-2,4-dimethylquinoline
5,6,7,8-Tetrahydro-2,6-dimethylquinoline
5,6,7,8-Tetrahydro-2,7-dimethylquinoline
5,6,7,8-Tetrahydro-2,8-dimethylquinoline
1,2,3,4-Tetrahydroisoquinoline, N-*Et*
5,6,7,8-Tetrahydro-*N*-methyl-1-naphthylamine
5,6,7,8-Tetrahydro-4-methyl-1-naphthylamine
1,2,3,4-Tetrahydro-*N*-methyl-2-naphthylamine
5,6,7,8-Tetrahydro-*N*-methyl-2-naphthylamine
1,2,3,4-Tetrahydro-2-methylquinoline, N-*Me*
5,6,7,8-Tetrahydro-1-naphthylamine, N-*Me*
1,2,3,4-Tetrahydro-2-naphthylamine, N-*Me*
5,6,7,8-Tetrahydro-2-naphthylamine, N-*Me*

$C_{11}H_{15}NO$
3-Cyanocamphor
m-Diethylaminobenzaldehyde
p-Diethylaminobenzaldehyde
N-Diethylbenzamide
2,2-Dimethyl-3-phenylpropionic Acid, *Amide*
3-(2,4-Dimethylphenyl)propionic Acid, *Amide*
3-(3,5-Dimethylphenyl)propionic Acid, *Amide*
1-Ethyl-1,2,3,4-tetrahydro-8-hydroxyquinoline
5-Isopropyl-*o*-toluic Acid, *Amide*
6-Isopropyl-*m*-toluic Acid, *Amide*
β-Methylaminobutyrophenone
2-Methyl-2-phenylbutyric Acid, *Amide*
3-Methyl-2-phenylbutyric Acid, *Amide*
3-Methyl-4-phenylbutyric Acid, *Amide*
2-Methyl-3-*o*-tolylpropionic Acid, *Amide*
2-Methyl-3-*p*-tolylpropionic Acid, *Amide*
Phenmetrazine
2-Phenylvaleric Acid, *Amide*
5-Phenylvaleric Acid, *Amide*

$C_{11}H_{15}NO$ (*continued*)
1,2,3,4-Tetrahydro-8-hydroxy-1-methylquinoline, *Me ether*
1,2,3,4-Tetrahydro-8-hydroxy-2-methylquinoline, *Me ether*
1,2,3,4-Tetrahydro-5-hydroxyquinoline, *Et ether*
1,2,3,4-Tetrahydro-6-hydroxyquinoline, *Et ether*
1,2,3,4-Tetrahydro-8-hydroxyquinoline, *Et ether*
1,2,3,4-Tetrahydro-6-methoxyquinoline, N-*Me*
2,3,4,5-Tetramethylbenzoic Acid, *Amide*
2,3,4,6-Tetramethylbenzoic Acid, *Amide*
2,3,5,6-Tetramethylbenzoic Acid, *Amide*
2,4,5-Trimethylphenylacetic Acid, *Amide*
2,4,6-Trimethylphenylacetic Acid, *Amide*

$C_{11}H_{15}NO_2$
m-Aminobenzoic Acid, *Butyl ester*
4-Amino-2,6-dimethylbenzoic Acid, *Et ester*
4-Amino-3,5-dimethylbenzoic Acid, *Et ester*
α-Amino-4-hydroxyacetophenone, *Me ether*, N-*Di-Me*
3-Amino-4-isopropylbenzoic Acid, *Me ester*
3-*p*-Aminophenylpropionic Acid, *Et ester*
2-Amino-2-phenylpropionic Acid, *Et ester*
3-Amino-3-phenylpropionic Acid, *Et ester*
5-*o*-Aminophenylvaleric Acid
2-Amino-5-phenylvaleric Acid
3-Amino-3-phenylvaleric Acid
2-Anilino-3-methylbutyric Acid
2-Anilinopropionic Acid, *Et ester*
3-Anilinopropionic Acid, *Et ester*
2-Anilinovaleric Acid
Anthranilic Acid, *Butyl ester*
Anthranilic Acid, *Isobutyl ester*
Benzylglycine, *Et ester*
Bractamine
Butesin
p-Butylaminobenzoic Acid
2-*tert*-Butyl-6-methylpyridine-3-carboxylic Acid
2-*tert*-Butyl-6-methylpyridine-4-carboxylic Acid
3-*tert*-Butyl-5-nitrotoluene
4-*tert*-Butyl-2-nitrotoluene
5-*tert*-Butyl-2-nitrotoluene
Corypalline
Cycloform
o-Diethylaminobenzoic Acid
m-Diethylaminobenzoic Acid
p-Diethylaminobenzoic Acid
p-Dimethylaminobenzoic Acid, *Et ester*
3-Dimethylamino-3-phenylpropionic Acid
N-Dimethylglycine, *Benzyl ester*
o-Ethylaminobenzoic Acid, *Et ester*
p-Ethylaminobenzoic Acid, *Et ester*
Fusaric Acid, *Me ester*
2-Hydroxy-2-(4-isopropylphenyl)acetic Acid, *Amide*
2-Hydroxy-3-phenylpropionic Acid, *Et-amide*
3-Hydroxy-3-phenylpropionic Acid, *Et-amide*
Lactic Acid, *Anilide*, *Et ether*
Lactic Acid, N-*Ethylanilide*
Mandelamide, N-*Di-Me*, *Me ether*
2-Methylamino-2-phenylacetic Acid, *Et ester*
p-Methylaminophenylacetic Acid, *Et ester*
3-Methyl-1-nitro-1-phenylbutane
N-Methyl-*p*-phenetidine, N-*Ac*
β-Phenylalanine, *Et ester*
Phenylcarbamic Acid, *Butyl ester*
Phenylcarbamic Acid, *Isobutyl ester*
Phenylcarbamic Acid, tert-*Butyl ester*
N-Phenylsarcosine, *Et ester*
Pyridine-3-carboxylic Acid, *Isopentyl ester*
Salsoline
N-*o*-Tolylglycine, N-*Et*
N-*o*-Tolylglycine, *Et ester*
N-*m*-Tolylglycine, *Et ester*
N-*p*-Tolylglycine, *Et ester*

$C_{11}H_{15}NO_2S$
p-Nitrobenzenethiol, S-*Isopentyl ether*

$C_{11}H_{15}NO_3$
Anhalamine
Candidulin
3,4-Dimethoxybenzoic Acid, *Amide*, N-*Di-Me*
Ethyl-4-hydroxyphenylcarbamate, *Et ether*
Lactic Acid, p-*Phenetidide*
1-(3-Methoxy-4,5-methylenedioxyphenyl)2-propylamine†
p-Nitrophenol, 3-*Methylbutyl ether*
Surinamine, *Me ester*
Tyrosine, O,N-*Di-Me*
Tyrosine, *Me ester*, N-*Me*
Tyrosine, *Et ester*

$C_{11}H_{15}NO_3S$
p-Formylbenzenesulphonic Acid, *Diethylamide*

$C_{11}H_{15}NO_4$
2-Amino-4,5-dimethoxybenzoic Acid, *Et ester*
1,2-Dihydroxy-4-nitro-5-propylbenzene, *Di-Me ether*
β-[2,4-Dihydroxyphenyl]alanine, 2,4-*Di-Me ether*
β-[3,4-Dihydroxyphenyl]alanine, *Et ester*
β-[3,4-Dihydroxyphenyl]alanine, 3,4-*Di-Me ether*
o-Nitrobenzaldehyde, *Di-Et acetal*
m-Nitrobenzaldehyde, *Di-Et acetal*

$C_{11}H_{15}NO_5$
2,3-Dihydroxy-5-nitrobenzyl Alcohol, *Di-Et ether*
Methocarbamol

$C_{11}H_{15}NO_6S$
p-Sulphobenzoic Acid, 4-*Di-2-hydroxyethylamide*

$C_{11}H_{15}NS_2$
Diethyldithiocarbamic Acid, *Phenyl ester*

$C_{11}H_{15}N_3O_2$
Benzyliminodiacetic Acid, *Diamide*

$C_{11}H_{15}N_3O_5$
4-Amino-2,6-dinitrophenol, N-*Me*, N-*Et*, *Et ether*

$C_{11}H_{15}N_3O_8$
Polyoxin C†

$C_{11}H_{15}N_5O_3$
9-[β-DL-2α,3α-Dihydroxy-4β-(hydroxymethyl)-cyclopentyl]adenine†

$C_{11}H_{15}N_5O_3S$
5′-Methylthio-5′-deoxyadenosine
$C_{11}H_{15}N_5O_4$
Homoadenosine†
2′-*C*-Methyladenosine†
3′-*C*-Methyladenosine†
2′-*O*-Methyladenosine†
$C_{11}H_{15}N_5O_5$
2′-*O*-Methylguanosine†
Psicofuranine
$C_{11}H_{16}$
o-Butyltoluene
o-*tert*-Butyltoluene
m-Butyltoluene
m-2-Butyltoluene
m-*tert*-Butyltoluene
p-Butyltoluene
p-2-Butyltoluene
p-*tert*-Butyltoluene
p-Isobutyltoluene
4-Isopropyl-*o*-xylene
4-Isopropyl-*m*-xylene
2-Isopropyl-*p*-xylene
3′-Methylbutylbenzene
2-Methyl-2-phenylbutane
2-Methyl-3-phenylbutane
Pentamethylbenzene
Pentylbenzene
2-Phenylpentane
3-Phenylpentane
Tetracyclo[5,3,1,$0^{1,6}$,$0^{6,11}$]undecane†
$C_{11}H_{16}ClN_5$
Paludrine
$C_{11}H_{16}Cl_2O_2$
1-Nonene-1,9-dicarboxylic Acid, *Dichloride*
1,2,2,3-Tetramethyl-1,3-cyclopentanedicarboxylic Acid, *Dichloride*
$C_{11}H_{16}N_2$
1-Methyl-2,3′-pyridyl-piperidine
1-Methyl-3,2′-pyridyl-piperidine
$C_{11}H_{16}N_2O$
2-Anilino-3-methylbutyric Acid, *Amide*
2-Anilinovaleric Acid, *Amide*
p-Butylaminobenzoic Acid, *Amide*
2,4-Diaminobenzaldehyde, 2,4-N-*Tetra-Me*
p-Diethylaminobenzoic Acid, *Amide*
3-Methylamino-3-phenylpropionic Acid, *Methylamide*
4-Nitroso-*o*-toluidine, N-*Butyl*
$C_{11}H_{16}N_2O_2$
2,4-Diaminobenzoic Acid, n-*Butyl ester*
3,5-Diaminobenzoic Acid, n-*Butyl ester*
N-Diethyl-5-nitro-*o*-toluidine
N-Diethyl-2-nitro-*p*-toluidine
N-Diethyl-3-nitro-*p*-toluidine
Isopilocarpine
Neopilocarpine
o-Nitrobenzylamine, N-*Di-Et*
m-Nitrabenzylamine, N-*Di-Et*
p-Nitrobenzylamine, N-*Di-Et*
3-Nitro-*p*-toluidine, N-*Butyl*
2-β-Phenylhydrazinopropionic Acid, *Et ester*
Pilocarpine
$C_{11}H_{16}N_2O_3$
5-Allyl-5-isopropyl-1-methylbarbituric Acid
Delvinal
2-Hydroxy-5-nitrobenzylamine, N-*Di-Et*
$C_{11}H_{16}N_2O_5S$
Carbon Dioxide Biotin†
$C_{11}H_{16}N_2O_6$
Uridine, 2′,3′-*Di-Me ether*
$C_{11}H_{16}N_2S$
Phenylthiourea, 3-N-*Di-Et*
$C_{11}H_{16}N_4O_2$
Theobromine, N-*Butyl*
Theobromine, N-*Isobutyl*
$C_{11}H_{16}N_4O_5$
γ-L-Glutamyl-β-pyrazol-1-yl-L-alanine
$C_{11}H_{16}N_6O_4$
9-(2-Amino-2-deoxy-α-D-glucofuranosyl)-adenine†
9-(2-Amino-2-deoxy-β-D-glucofuranosyl)-adenine†
N^6-Dimethyl-8-aza-adenosine
$C_{11}H_{16}O$
Benzyl butyl Ether
Benzyl isobutyl Ether
4-Butyl-*o*-cresol
6-Butyl-*o*-cresol
6-2′-Butyl-*o*-cresol
3-*tert*-Butyl-*o*-cresol
4-*tert*-Butyl-*o*-cresol
5-*tert*-Butyl-*o*-cresol
6-*tert*-Butyl-*o*-cresol
6-Butyl-*m*-cresol
6-2′-Butyl-*m*-cresol
4-*tert*-Butyl-*m*-cresol
6-*tert*-Butyl-*m*-cresol
2-Butyl-*p*-cresol
2-*tert*-Butyl-*p*-cresol
m-Butylphenol, *Me ether*
p-Butylphenol, *Me ether*
p-2′-Butylphenol, *Me ether*
m-*tert*-Butylphenol, *Me ether*
p-*tert*-Butylphenol, *Me ether*
Carvacrol, *Me ether*
o-Cresol, n-*Butyl ether*
m-Cresol, n-*Butyl ether*
p-Cresol, n-*Butyl ether*
2,4-Dimethylbenzyl Alcohol, *Et ether*
Isopentyl phenyl Ether
3-Isopropyl-*o*-cresol, *Me ether*
4-Isopropyl-*o*-cresol, *Me ether*
6-Isopropyl-*o*-cresol, *Me ether*
4-Isopropyl-*m*-cresol, *Me ether*
6-Isopropyl-*m*-cresol, *Me ether*
2-Isopropyl-*p*-cresol, *Me ether*
o-Isopropylphenol, *Et ether*
p-Isopropylphenol, *Et ether*
o-3-Methyl-1-butylphenol
p-(2-Methyl-2-butyl)phenol
p-(3-Methyl-1-butyl)phenol
Methylenecamphor
3-Methyl-2,2′-pentenylcyclopenten-1-one
2-Methyl-1-phenyl-1-butanol
2-Methyl-2-phenyl-1-butanol

$C_{11}H_{16}O$ (*continued*)
3-Methyl-2-phenyl-1-butanol
2-Methyl-3-phenyl-1-butanol
2-Methyl-4-phenyl-1-butanol
3-Methyl-1-phenyl-1-butanol
2-Methyl-1-phenyl-2-butanol
2-Methyl-3-phenyl-3-butanol
4-Methylsafranal
Norecsantalal
Pentamethylphenol
Phenyl-1-butanol, *Me ether*
1-Phenyl-1-pentanol
3-Phenyl-1-pentanol
4-Phenyl-1-pentanol
5-Phenyl-1-pentanol
1-Phenyl-2-pentanol
2-Phenyl-2-pentanol
4-Phenyl-2-pentanol
5-Phenyl-2-pentanol
1-Phenyl-3-pentanol
2-Phenyl-3-pentanol
3-Phenyl-3-pentanol
3-Phenyl-1-propanol, *Et ether*
1-Phenyl-2-propanol, *Et ether*
o-Propylphenol, *Et ether*
m-Propylphenol, *Et ether*
p-Propylphenol, *Et ether*
2,4,6-Trimethylbenzyl Alcohol, *Me ether*
2,3,4-Trimethylphenethyl Alcohol
2,4,6-Trimethylphenethyl Alcohol
2,4,5-Trimethylphenol, *Et ether*

$C_{11}H_{16}OS$
p-Hydroxybenzenethiol, S-*Butyl*, *Me ether*
p-Hydroxybenzenethiol, S-n-*Pentyl*

$C_{11}H_{16}O_2$
Betuligenol, *Me ether*
2-*tert*-Butyl-4-methoxyphenol †
3-*tert*-Butyl-4-methoxyphenol †
Catechol, *Isopentyl ether*
Coerulignol, *Me ether*
Cyclohexylpropiolic Acid, *Et ester*
Dicadic Acid
1,2-Dihydroxy-4-isopropylbenzene, *Di-Me ether*
1,4-Dihydroxy-2-isopropylbenzene, *Di-Me ether*
1,2-Dihydroxy-3-propylbenzene, *Di-Me ether*
1,3-Dihydroxy-5-propylbenzene, *Di-Me ether*
1,4-Dihydroxy-2-propylbenzene, *Di-Me ether*
3,4-Dihydroxytoluene, *Di-Et ether*
3,5-Dihydroxytoluene, *Di-Et ether*
1,3-Dihydroxy-2,4,5-trimethylbenzene, *Di-Me ether*
α,2-Dihydroxy-3,5,6-trimethyltoluene, α-*Me ether*
Guaiacol, n-*Butyl ether*
3-Hydroxymethylenecamphor
2-α-Hydroxypropylphenol, 1-*Et ether*
4-α-Hydroxypropylphenol, 1-*Et ether*
2-Hydroxy-2,6,6-trimethylcyclohexylidene-1-acetic Acid Lactone †
Isoteresantalic Acid, *Me ester*
Jasmolone †
2-3′-Methylbutylresorcinol
4-3′-Methylbutylresorcinol
4-Methylcamphorquinone
Methylquinol, *Di-Et ether*
Norecsantalic Acid
2-Octynoic Acid, *Allyl ester*
1,5-Pentanediol, *Phenyl ether*
5-Pentylresorcinol
3-Phenylpropanal, *Di-Me acetal*
1-Phenyl-1,3-propanediol, *Di-Me ether*
Propiophenone, *Di-Me acetal*
Resorcinol, *Me-Isobutyl ether*
Resorcinol, *Monopentyl ether*
Shonanic Acid, *Me ester*
Teresantalic Acid, *Me ester*
2,2,7,7-Tetramethyl-1,6-dioxaspiro[4,4]nona-3,8-diene †
Tricyclenic Acid, *Me ester*
2,3,5-Trimethylquinol, *Di-Me ether*

$C_{11}H_{16}O_2S$
Toluene-*p*-sulphinic Acid, *Butyl ester*

$C_{11}H_{16}O_3$
Angustione
1-(2,5-Dihydroxyphenyl)propan-2-ol, *Di-Me ether* †
2,6-Dimethoxy-4-propylphenol
Exogonic Acid, *Me ester*
Furan-2-carboxylic Acid, *Hexyl ester*
5-(2-Furyl)valeric Acid, *Et ester*
α-Hydroxy-3,4-dimethoxytoluene, *Et ether*
Ligusticumic Acid
Loliolide †
1-*p*-Methoxyphenyl-1,2-propanediol, β-*Me ether*
3-Methylcyclohexane-1,1-diacetic Acid, *Anhydride*
4-Methylcyclohexane-1,1-diacetic Acid, *Anhydride*
Phloroglucinol, *Me ether*
1,1,3,3-Tetramethylcyclohexane-2,4,6-trione, *Me ether*
1,3,5-Trihydroxy-2,4,6-trimethylbenzene, *Mono-Et ether*

$C_{11}H_{16}O_3S$
Toluene-*p*-sulphonic Acid, *Butyl ester*
Toluene-*p*-sulphonic Acid, 2-*Butyl ester*
Toluene-*p*-sulphonic Acid, *Isobutyl ester*

$C_{11}H_{16}O_4$
Arjunetin
π-Camphanic Acid, *Me ester*
ω-Camphanic Acid, *Me ester*
1-Cyclopentene-1,2-dicarboxylic Acid, *Di-Et ester*
1-Cyclopentene-1,3-dicarboxylic Acid, *Di-Et ester*
4-Cyclopentene-1,3-dicarboxylic Acid, *Di-Et ester*
Diallylmalonic Acid, *Di-Me ester*
Dihydrocanadensolide †
1-Methyl-2-cyclohexene-1,2-dicarboxylic Acid, *Di-Me ester*
1-Methyl-4-cyclohexene-1,2-dicarboxylic Acid, *Di-Me ester*
3-Methyl-4-cyclohexene-1,2-dicarboxylic Acid, *Di-Me ester*
4-Methyl-4-cyclohexene-1,2-dicarboxylic Acid, *Di-Me ester*

$C_{11}H_{16}O_4$ (*continued*)
3-Methylmuconic Acid, *Di-Et ester*
Phoronic Acid, *Di-lactone*
Pyrethic Acid†
2,3,4,5-Tetrahydroxytoluene, *Tetra-Me ether*

$C_{11}H_{16}O_4S$
5-Hydroxy-2,4-dimethylbenzenesulphonic Acid, *Propyl ether*

$C_{11}H_{16}O_5$
Loganetin
Pentahydroxybenzene, *Penta-Me ether*

$C_{11}H_{16}O_6$
2,4-Dioxopimelic Acid, *Di-Et ester*
2,6-Dioxopimelic Acid, *Di-Et ester*
cis-2-Pentene-2,3,5-tricarboxylic Acid, *Tri-Me ester*

$C_{11}H_{16}O_7$
Oxalomalonic Acid, *Tri-Et ester*

$C_{11}H_{16}O_8$
Heptane-1,1,3,3-tetracarboxylic Acid
Heptane-1,1,5,5-tetracarboxylic Acid
Heptane-1,3,3,7-tetracarboxylic Acid
Heptane-1,4,4,7-tetracarboxylic Acid
Heptane-2,2,6,6-tetracarboxylic Acid
Heptane-3,3,5,5-tetracarboxylic Acid
Ranunculin

$C_{11}H_{16}S$
5-Isopropyl-2-methylbenzenethiol, S-*Me*

$C_{11}H_{17}As$
Phenylarsine, As-*Et-Propyl*

$C_{11}H_{17}BrO_4S$
3-Bromo-(+)-camphor-10-sulphonic Acid, *Me ester*

$C_{11}H_{17}Cl_3O_6$
α-Glucochloralose, *Tri-Me ether*
β-Glucochloralose, *Tri-Me ether*

$C_{11}H_{17}N$
γ-Aminobutylbenzene, N-*Me*
δ-Aminobutylbenzene, N-*Me*
N-Benzyldiethylamine
N-Diethyl-*o*-toluidine
N-Diethyl-*m*-toluidine
N-Diethyl-*p*-toluidine
N-Ethyl-*N*-isopropylaniline
N-Ethyl-*N*-propylaniline
N-Isobutylbenzylamine
N-Isobutyl-*N*-methylaniline
N-Isobutyl-*o*-toluidine
N-Isobutyl-*p*-toluidine
p-Isopropyl-*N*-methylbenzylamine
Mephentermine
N-Methyl-*N*-butylaniline
N-3-Methylbutylaniline
2,3,4,5,6-Pentamethylaniline
2-Phenyl-1-pentylamine
5-Phenyl-1-pentylamine
1-Phenyl-1-propylamine, N-*Et*
3-Phenyl-1-propylamine, N-*Di-Me*
3-Phenyl-1-propylamine, N-*Et*
o-Propylaniline, N-*Di-Me*
p-Propylaniline, N-*Di-Me*
1,2,2,3-Tetramethyl-3-cyclopentene-1-acetic Acid, *Nitrile*

$C_{11}H_{17}NO$
3-Aminomethylenecamphor
Dicadic Acid, *Amide*
Hordenine, *Me ether*
3-Hydroxymethylenecamphor, *Imino comp.*
5-Isopropyl-2-methyl-3-cyclohexanone-1-carboxylic Acid, *Nitrile*
Methoxyphenamine
N-Methylephedrine
N-Methyl-ψ-ephedrine
Tecomanine

$C_{11}H_{17}NO_2$
α-Aminomethyl-*p*-hydroxybenzyl Alcohol, N-*Propyl*
α-Aminomethyl-*p*-hydroxybenzyl Alcohol, N-*Isopropyl*
Anhydroecgonine, *Et ester*
3-Carboxy-2,2,3-trimethylcyclopentylacetonitrile
Cyclohexylmalonic Acid, *Et ester-nitrile*
3,4-Dimethoxy-*N*-methyl-α-phenethylamine
2,5-Dimethylpyrrole-3-carboxylic Acid, tert-*Butyl ester*
Ligusticumic Acid, *Amide*
2′-Methylaminoethylquinol, *Di-Me ether*
1,2,2-Trimethylcyclopentane-1,3-dicarboxylic Acid, α-*Nitrile*, β-*Me ester*

$C_{11}H_{17}NO_2S$
2,4,6-Trimethylbenzenesulphonic Acid, *Dimethylamide*

$C_{11}H_{17}NO_3$
4-Hydroxymethylene-2-pentyl-5-oxazolone, *Et ether*
Mescaline
Methoxamine

$C_{11}H_{17}NO_3S$
5-Hydroxy-2,4-dimethylbenzenesulphonic Acid, *Propyl ether*, *Amide*

$C_{11}H_{17}NO_4$
Butane-1,1,2-tricarboxylic Acid, 1-*Nitrile*-1,2-*di-Et ester*
Butane-1,1,4-tricarboxylic Acid, 4-*Nitrile*-1,1-*di-Et ester*
Butane-1,2,2-tricarboxylic Acid, 2-*Nitrile*-1-2-*di-Et ester*
Butane-2,2,3-tricarboxylic Acid, 2-*Nitrile*-2,3-*di-Et ester*
2-Pentenylpenaldic Acid

$C_{11}H_{17}N_2$ (ion)
Octahydrodipyrido[1,2-*a*:1′,2′-*c*]imidazol-10-ium†

$C_{11}H_{17}N_3$
1,1,3,3-Tetramethyl-2-phenylguanidine†

$C_{11}H_{17}N_3O_2$
4-Nitro-*m*-phenylenediamine, 1-N-*Di-Et*, 3-N-*Me*

$C_{11}H_{17}N_3O_3$
5-Ethyl-5-piperidinobarbituric Acid

$C_{11}H_{17}N_3O_3S$
Carbutamide

$C_{11}H_{17}N_3O_8$
Tetrodotoxin

$C_{11}H_{17}NaN_2O_2S$
Pentothal
$C_{11}H_{17}O_3PS$
S-Benzyl-*OO*-diethyl phosphorothiolate†
O-Benzyl-*OO*-diethyl phosphorothionate†
$C_{11}H_{17}O_3PSe$
Se-Benzyl-*OO*-diethyl phosphoroselenolate†
$C_{11}H_{17}O_5P$
Phosphonoacetic Acid, *Triallyl ester*
$C_{11}H_{18}$
Dictyopterone†
Tricyclo[4,3,1,1^{3,8}]undecane†
2,2,4-Trimethyl-3-methylenenorbonane
2,2,5-Trimethyl-3-methylenenorbonane
$C_{11}H_{18}Cl_2O_2$
Undecanedioic Acid, *Dichloride*
$C_{11}H_{18}NO$ (ion)
Candicine†
$C_{11}H_{18}NO_2$ (ion)
Coryneine†
$C_{11}H_{18}N_2$
o-Aminobenzylamine, αN-*Di-Et*
m-Aminobenzylamine, αN-*Di-Et*
p-Aminobenzylamine, αN-*Di-Et*
2,4-Diaminotoluene, N-*Tetra-Me*
2,4-Diaminotoluene, 2-N-*Di-Et*
2,4-Diaminotoluene, 4-N-*Butyl*
2,5-Diaminotoluene, 2,5-N-*Tetra-Me*
2,5-Diaminotoluene, 2-N-*Di-Et*
3,4-Diaminotoluene, 3,4-N-*Tetra-Me*
3,4-Diaminotoluene, 4-N-*Butyl*
N,N'-Diethyl-3,4-toluenediamine
Undecanedioic Acid, *Dinitrile*
$C_{11}H_{18}N_2O_3$
Amytal
Muta-aspergillic Acid
$C_{11}H_{18}N_2O_3S$
Biotin, *Me ester*
$C_{11}H_{18}N_2O_4$
L-Leucyl-D-glutamic Acid, *Anhydride*
$C_{11}H_{18}N_2O_5S$
γ-L-Glutamyl-*S*-allyl-L-cysteine
γ-L-Glutamyl-*S*-(prop-1-enyl)-L-cysteine
$C_{11}H_{18}N_2O_6S$
γ-L-Glutamyl-*S*-(prop-1-enyl)-L-cysteine sulphoxide
$C_{11}H_{18}N_4O_8$
Glycyl-L-aspartyl-L-serylglycine†
$C_{11}H_{18}O$
5-Isobutyl-1-methylcyclohexene-3-one
5-Isopropenyl-1,2-dimethyl-2-cyclohexen-1-ol
3-Methylcamphor
4-Methylcamphor
6-Methylcamphor
3-Methyl-2-pentyl-2-cyclopenten-1-one
Norecsantalol
1,8,8-Trimethylbicyclo[3,2,1]octan-2-one
5,8,8-Trimethylbicyclo[3,2,1]octan-2-one
$C_{11}H_{18}O_2$
Allocyclogeranic Acid, *Me ester*
Bicyclo[2,2,2]octane-1-carboxylic Acid, *Et ester*
Buchu-camphor, *Me ether*
2-1'-Cyclohexenylpropionic Acid, *Et ester*
2-Cyclohexylidenepropionic Acid, *Et ester*
1,2-Cycloundecanedione†
2,4-Decadienoic Acid, *Me ester*
4,6-Decadienoic Acid, *Me ester*
2,2-Dimethyl-3-isopropylidenecyclopropyl propionate
3,7-Dimethyl-2,6-nonadienoic Acid
Geranic Acid, *Me ester*★†
2-Hexynoic Acid, 3-*Methylbutyl ester*
3-Hydroxy-1,7,7-trimethylbicyclo[2,2,1]heptan-2-one, *Me ether*
Isoborneol, *Formyl*
3-Isopropylidene-2,2-dimethylcyclopropyl Propionate
2-Isopropylidene-5-methylcyclopentane-1-carboxylic Acid, *Me ester*
3-Isopropylidene-1-methylcyclopentane-1-carboxylic Acid, *Me ester*
2-Methyl-1-cyclohexenyl-1-acetic Acid, *Et ester*
3-Methyl-1-cyclohexenyl-1-acetic Acid, *Et ester*
4-Methyl-1-cyclohexenyl-1-acetic Acid, *Et ester*
2-Octynoic Acid, *Propyl ester*
2-Octynoic Acid, *Isopropyl ester*
1,2,2,3-Tetramethyl-3-cyclopentene-1-acetic Acid
2,2,7,7-Tetramethyl-1,6-dioxaspiro[4,4]non-3-ene†
2,3,6-Trimethyl-2-cyclohexene-1-carboxylic Acid, *Me ester*
1,2,2-Trimethyl-3-cyclohexene-1-carboxylic Acid, *Me ester*
1,3,4-Trimethyl-3-cyclohexene-1-carboxylic Acid, *Me ester*
2,3,3-Trimethyl-1-cyclopentene-1-carboxylic Acid, *Et ester*
2-Undecynoic Acid
6-Undecynoic Acid
9-Undecynoic Acid
10-Undecynoic Acid
$C_{11}H_{18}O_3$
3-Acetyl-2,2-dimethylcyclobutaneacetic Acid, *Me ester*
1,3-Dimethyl-2-oxocyclohexanecarboxylic Acid, *Et ester*
1,4-Dimethyl-2-oxocyclohexanecarboxylic Acid, *Et ester*
2,4-Dimethyl-6-oxocyclohexanecarboxylic Acid, *Et ester*
2,6-Dimethyl-4-oxocyclohexanecarboxylic Acid, *Et ester*
3,3-Dimethyl-2-oxocyclohexanecarboxylic Acid, *Et ester*
4,4-Dimethyl-2-oxocyclohexanecarboxylic Acid, *Et ester*
4,5-Dimethyl-2-oxocyclohexanecarboxylic Acid, *Et ester*
Gitaligenin
1-Isopropyl-3-methyl-2-cyclohexanone-1-carboxylic Acid

$C_{11}H_{18}O_3$ *(continued)*
1-Isopropyl-4-methyl-2-cyclohexanone-1-carboxylic Acid
1-Isopropyl-5-methyl-2-cyclohexanone-1-carboxylic Acid
5-Isopropyl-2-methyl-3-cyclohexanone-1-carboxylic Acid
1-Methyl-2-oxo-1-cyclohexaneacetic Acid, *Et ester*
1-Methyl-3-oxo-1-cyclohexaneacetic Acid, *Et ester*
3-Methyl-2-oxo-1-cyclohexaneacetic Acid, *Et ester*
4-Methyl-2-oxo-1-cyclohexaneacetic Acid, *Et ester*
5-Methyl-2-oxo-1-cyclohexaneacetic Acid, *Et ester*
5-Methyl-3-oxo-1-cyclohexaneacetic Acid, *Et ester*
2-(4-Methyl-2-oxo-1-cyclohexyl)-2-methylpropionic Acid
4-Methyl-2-oxocyclopentane-1-carboxylic Acid, *Isobutyl ester*
Nepetalic Acid, *Me ester* †
Umbellulonic Acid, *Et ester*

$C_{11}H_{18}O_4$
Allylmethylmalonic Acid, *Di-Et ester*
3-Carboxy-1-cyclopentylacetic Acid, 1-*Et ester*, α-*Me ester*
α-Carboxy-γ-decalactone †
α-Carboxy-δ-decalactone †
3-Carboxy-2,2-dimethylcyclobutylacetic Acid, *Di-Me ester*
4-Carboxy-2,2-dimethylcyclobutylacetic Acid, *Di-Me ester*
2-Carboxy-5,7-dimethyl-4-octalactone †
3-(2-Carboxyethyl)-4,4-dimethylbutyrolactone, *Et ester*
3-Carboxy-1,2,2-trimethylcyclopentylacetic Acid
3-Carboxy-2,2,3-trimethyl-1-cyclopentylacetic Acid
Cyclohexylmalonic Acid, *Di-Me ester*
Cyclopentane-1,2-dicarboxylic Acid, *Di-Et ester*
Cyclopentane-1,3-dicarboxylic Acid, *Di-Et ester*
1,2-Dimethylcyclopentane-1,3-dicarboxylic Acid, *Di-Me ester*
3,3-Dimethylcyclopropane-1,2-dicarboxylic Acid, *Di-Et ester*
2,4-Dimethylglutaconic Acid, *Di-Et ester*
3,4-Dimethylglutaconic Acid, *Di-Et ester*
4,4-Dimethylglutaconic Acid, *Di-Et ester*
4-Ethylglutaconic Acid, *Di-Et ester*
3-Hydroxy-2,2,3-trimethyladipic Acid, *Lactone, Et ester*
trans-3-Methyl-1-butene-1,2-dicarboxylic Acid, *Di-Et ester*
3-Methyl-2-butene-1,2-dicarboxylic Acid, *Di-Et ester*
3-Methyl-1-butene-2,3-dicarboxylic Acid, *Di-Et ester*
α-Methylcyclohexane-1,1-diacetic Acid
1-Methylcyclohexane-1,2-diacetic Acid
3-Methylcyclohexane-1,2-diacetic Acid
2-Methylcyclohexane-1,1-diacetic Acid
3-Methylcyclohexane-1,1-diacetic Acid
4-Methylcyclohexane-1,1-diacetic Acid
1-Nonene-1,9-dicarboxylic Acid
2-Nonene-1,9-dicarboxylic Acid
1,2,2,3-Tetramethyl-1,3-cyclopentanedicarboxylic Acid
1,2,2,5-Tetramethyl-1,3-cyclopentanedicarboxylic Acid
1,2,2-Trimethylcyclopentane-1,3-dicarboxylic Acid, α-*Me ester*
1,2,2-Trimethylcyclopentane-1,3-dicarboxylic Acid, β-*Me ester*

$C_{11}H_{18}O_4S$
Camphor-3-sulphonic Acid, *Me ester*
Camphor-10(or 6-)sulphonic Acid *Me ester*

$C_{11}H_{18}O_5$
Acetylethylmalonic Acid, *Di-Et ester*
2-Acetylglutaric Acid, *Di-Et ester*
3-Acetylglutaric Acid, *Di-Et ester*
2-Acetyl-2-methylsuccinic Acid, *Di-Et ester*
2-Acetyl-3-methylsuccinic Acid, *Di-Et ester*
Butyrylmalonic Acid, *Di-Et ester*
Heptanoylsuccinic Acid †
4-Oxo-1,7-heptanedioic Acid, *Di-Et ester*
5-Oxononane-1,9-dioic Acid, *Di-Me ester*
6-Oxoundecanedioic Acid †
Phoronic Acid
Tetrahydropyran-4,4-dicarboxylic Acid, *Di-Et ester*

$C_{11}H_{18}O_6$
Ascarylose, *Me ascaryloside*
Butane-1,1,4-tricarboxylic Acid, 1,1-*Di-Et ester*
2,3-Dimethylbutane-1,2,3-tricarboxylic Acid, *Mono-Et ester*
Methane-triacetic Acid, *Di-Et ester*
3-Methylbutane-1,2,3-tricarboxylic Acid, *Tri-Me ester*
6-Methylheptane-1,1,7-tricarboxylic Acid
6-Methylheptane-1,2,2-tricarboxylic Acid
Pentane-1,3,5-tricarboxylic Acid, *Tri-Me ester* †

$C_{11}H_{18}O_8$
1,2-Dihydroxyethane-1,1,2-tricarboxylic Acid, *Tri-Et ester*
Tuliposide A †

$C_{11}H_{18}O_9$
Tuliposide B †

$C_{11}H_{19}ClO$
10-Undecenoic Acid, *Chloride*

$C_{11}H_{19}ClO_3$
Decanedioic Acid, *Chloride*

$C_{11}H_{19}N$
2-Undecenoic Acid, *Nitrile*
10-Undecenoic Acid, *Nitrile*

$C_{11}H_{19}NO$
1,2,2,3-Tetramethyl-3-cyclopentene-1-acetic Acid, *Amide*

$C_{11}H_{19}NOS$
6-(2-Thienyl)-2,4-hexadienoic Acid, *Isobutylamide* †

$C_{11}H_{19}NO_2$
Dipropylmalonic Acid, *Nitrile*
Lupinic Acid, *Me ester*
Meroquinene, *Et ester*
2-Methylhexane-3,3-dicarboxylic Acid, *Mono-nitrile, Et ester*

$C_{11}H_{19}NO_2S_4$
Gerrardine†

$C_{11}H_{19}NO_3$
3-Carbamoyl-1,2,2-trimethylcyclopentane-carboxylic Acid, *Me ester*
3-Carbamoyl-1,2,2-trimethylcyclopentane-carboxylic Acid, N-*Me*
3-Carbamoyl-2,2,3-trimethylcyclopentane-carboxylic Acid, *Me ester*
3-Carbamoyl-2,2,3-trimethylcyclopentane-carboxylic Acid, N-*Me*
Cyclohexanecarbonylglycine, *Et ester*
1,2,2,3-Tetramethyl-1,3-cyclopentanedicarb-oxylic Acid, *Monoamide*

$C_{11}H_{19}NO_4$
Heptylpenaldic Acid
Pentylpenaldic Acid, *Et ester*
Piperidine-2,3-dicarboxylic Acid, *Di-Et ester*
Piperidine-2,6-dicarboxylic Acid, *Di-Et ester*

$C_{11}H_{19}NO_9$
N-Acetylneuraminic Acid

$C_{11}N_{19}NO_9S_2$
Gluconapin

$C_{11}H_{19}NO_{10}S_2$
Glucorapiferin
epi-Glucorapiferin†

$C_{11}H_{19}N_3O$
5-Butyl-2-dimethylamino-4-hydroxy-6-methylpyrimidine

$C_{11}H_{19}N_3O_6$
Ophthalmic Acid

$C_{11}H_{19}N_3O_6S$
γ-L-Glutamyl-L-cysteinyl-β-alanine†

$C_{11}H_{20}$
3-Methylene-1-decene
Spiro[5,5]undecane†
1,4,7,7-Tetramethylnorbornane
1-Undecyne★†
2-Undecyne★†
3-Undecyne†
4-Undecyne†
5-Undecyne★†

$C_{11}H_{20}ClN_5$
Chlorazin

$C_{11}H_{20}KNO_9S_2$
Glucochlearine

$C_{11}H_{20}KNO_{11}S_3$
Glucocheiroline

$C_{11}H_{20}N_2$
4,5-Di-*tert*-butylimidazole†

$C_{11}H_{20}N_2O_2$
1-Nonene-1,9-dicarboxylic Acid, *Diamide*

$C_{11}H_{20}N_2O_3$
4-Methyl-2-oxo-5-imidazolidinehexanoic Acid, *Me ester*

$C_{11}H_{20}N_2O_5$
γ-L-Glutamyl-L-isoleucine
γ-L-Glutamyl-L-leucine
L-Leucyl-D-glutamic Acid

$C_{11}H_{20}N_2O_6$
Saccharopine

$C_{11}H_{20}N_3O_3PS$
Pyrimithate†

$C_{11}H_{20}N_4O_2$
4,4,10,10-Tetramethyl-2,8-dioxo-1,3,7,9-tetra-aza-spiroundecane†

$C_{11}H_{20}N_4O_{11}P_2$
Cytidine diphosphate ethanolamine

$C_{11}H_{20}O$
Borneol, *Me ether*
Cycloundecanone
2,3-Di-*tert*-butylcyclopropanone†
Homomenthone
4-(1-Hydroxy-1-methylethyl)-1-methylcyclo-hexene, *Me ether*
Isoborneol, *Me ether*
1-Isopropyl-3,4-dimethylbicyclo[3,1,0]hexan-3-ol
Linalool, *Me ether*
3-Methyl-2,3′-methylbutylcyclopentanone
3-Methyl-2-pentylcyclopentanone
3-Methyl-4-pentylcyclopentanone
4-Methyl-2-pentylcyclopentanone
1,2,7,7-Tetramethylbicyclo[2,2,1]heptan-2-ol
3-Thujanol, *Me ether*

$C_{11}H_{20}O_2$
Citronellic Acid, *Me ester*
8-Decenoic Acid, *Me ester*
2-Ethylhexanoic Acid, *Allyl ester*
4-Ethyl-2-methyl-2-octenoic Acid†
4-Hydroxyundecanoic Acid, *Lactone*
2-Isopropyl-5-methylcyclopentane-1-carboxylic Acid, *Me ester*
3-Isopropyl-1-methylcyclopentane-1-carboxylic Acid, *Me ester*
3-Methyloctanoic Acid, *Et ester*
4-Methyloctanoic Acid, *Et ester*
5-Methyloctanoic Acid, *Et ester*
6-Methyloctanoic Acid, *Et ester*
7-Methyloctanoic Acid, *Et ester*
1,2,2,3-Tetramethylcyclopentane-1-carboxylic Acid, *Me ester*
2,2,3,3-Tetramethylcyclopentane-1-carboxylic Acid, *Me ester*
2,2,7,7-Tetramethyl-1,6-dioxaspiro[4,4]nonane†
2-Undecenoic Acid
6-Undecenoic Acid
9-Undecenoic Acid
10-Undecenoic Acid

$C_{11}H_{20}O_3$
Azelaialdehydic Acid, *Et ester*
2,2-Dimethyl-6-oxoheptanoic Acid, *Et ester*
9-Formylnonanoic Acid, *Me ester*
2-Hexanone-3-carboxylic Acid, *Isobutyl ester*

$C_{11}H_{20}O_3$ (*continued*)
- 4-Hydroxy-2,6-dimethylcyclohexanecarboxylic Acid, *Et ester*
- 1-Hydroxy-2-methylcyclohexaneacetic Acid, *Et ester*
- 2-Hydroxy-1-methylcyclohexaneacetic Acid, *Et ester*
- 2-Hydroxy-2-methylcyclohexaneacetic Acid, *Et ester*
- 1-Hydroxy-3-methylcyclohexaneacetic Acid, *Et ester*
- 2-Hydroxy-1-methylcyclohexane-1-carboxylic Acid, *Isopropyl ester*
- Levulinic Acid, *Hexyl ester*
- 2-Oxocyclo-octane-1-carboxylic Acid, *Et ester* †
- 3-Oxodecanoic Acid, *Me ester*
- 3-Oxononanoic Acid, *Et ester*
- 4-Oxononanoic Acid, *Et ester*
- 7-Oxononanoic Acid, *Et ester*
- 8-Oxononanoic Acid, *Et ester*
- 2-Oxopentane-3-carboxylic Acid, 3-*Methylbutyl ester*
- 2-Oxoundecanoic Acid
- 3-Oxoundecanoic Acid
- 4-Oxoundecanoic Acid
- 5-Oxoundecanoic Acid
- 9-Oxoundecanoic Acid
- 10-Oxoundecanoic Acid
- Tetrahydro-2,6,6-trimethylpyran-2-carboxylic Acid, *Et ester*

$C_{11}H_{20}O_4$
- Azelaic Acid, *Di-Me ester*
- Azelaic Acid, *Et ester*
- Butylmalonic Acid, *Di-Et ester*
- (2-Butyl)malonic Acid, *Di-Et ester*
- *tert*-Butylmalonic Acid, *Di-Et ester*
- Decanedioic Acid, *Me ester*
- Di-*tert*-butyl malonate †
- Diethylmalonic Acid, *Di-Et ester*
- 2,2-Dimethylglutaric Acid, *Di-Et ester*
- 2,4-Dimethylglutaric Acid, *Di-Et ester*
- 3,3-Dimethylglutaric Acid, *Di-Et ester*
- Dipropylmalonic Acid, *Et ester*
- 2-Ethylglutaric Acid, *Di-Et ester*
- 3-Ethylglutaric Acid, *Di-Et ester*
- 2-Ethyl-2-methylsuccinic Acid, *Di-Et ester*
- Homononactinic Acid
- Isopropylsuccinic Acid, *Di-Et ester*
- Malonic Acid, *Dibutyl ester*
- 2-Methyladipic Acid, *Di-Et ester*
- 3-Methyladipic Acid, *Di-Et ester*
- 3-Methylbutane-2,2-dicarboxylic Acid, *Di-Et ester*
- 5-Methylhexane-1,4-dicarboxylic Acid, *Di-Me ester*
- 5-Methylhexane-1,4-dicarboxylic Acid, *Mono-Et ester*
- Methylpropylmalonic Acid, *Di-Et ester*
- Nonane-1,1-dicarboxylic Acid
- Pimelic Acid, *Di-Et ester*
- Propylsuccinic Acid, *Di-Et ester*
- Trimethylsuccinic Acid, *Di-Et ester*
- Undecanedioic Acid

$C_{11}H_{20}O_4Si$
- Triallyl ethyl orthosilicate

$C_{11}H_{20}O_5$
- Butyric Acid, 1,3-*Glycerol dibutyrate*
- Di-isobutyrin
- 2-Hydroxyadipic Acid, *Di-Et ester*, *Me ether*
- Hydroxytrimethylsuccinic Acid, *Di-Et ester*
- Methoxysuccinic Acid, *Dipropyl ester*

$C_{11}H_{20}O_6$
- Mesoxalic Acid, *Di-Et acetal*, *Di-Et ester*

$C_{11}H_{20}O_7$
- Galacturonic Acid, 2,3,4-*Tri-Me ether*, *Me ester*

$C_{11}H_{20}O_{10}$
- Glucoxylose
- Isoprimeverose
- Primeverose
- Vicianose

$C_{11}H_{21}BrO_2$
- 2-Bromodecanoic Acid, *Me ester*
- 10-Bromodecanoic Acid, *Me ester*
- 2-Bromononanoic Acid, *Et ester*
- 9-Bromononanoic Acid, *Et ester*

$C_{10}H_{21}N$
- 1-Azabicyclo[5,5,0]dodecane
- 2-Butylpiperidine, N-*Me*
- Mecamylamine
- Undecanoic Acid, *Nitrile*
- Skytanthine

$C_{11}H_{21}NO$
- Azacyclododecan-2-one
- Hydroxyskytanthine I †
- Hydroxyskytanthin II †
- 11-Hydroxyundecanoic Acid, *Nitrile*
- 1-Lupinine, *Me ether*
- Tecostanine †
- 2,2,6,6-Tetramethyl-4-piperidone, N-*Et*
- 2,2,3-Trimethylcyclopentane-1-carboxylic Acid, *Ethylamide*
- 2-Undecenoic Acid, *Amide*
- 6-Undecenoic Acid, *Amide*
- 10-Undecenoic Acid, *Amide*

$C_{11}H_{21}NOS_2$
- (−)-9-Methylsulphinylnonyl isothiocyanate

$C_{11}H_{21}NO_2$
- 3-Cyclohexylalanine, *Et ester*
- 3-Ethyl-4-piperidylacetic Acid, *Et ester*

$C_{11}H_{21}NO_3$
- Dipropylmalonic Acid, *Et ester*, *Amide*
- 3-(4-Hydroxycyclohexyl)-alanine, *Et ester*
- Tyrosine, 3-*Methylbutyl ester*

$C_{11}H_{21}NO_5$
- Pantothenic Acid, *Et ester*

$C_{11}H_{21}NO_6$
- Galacturonic Acid, 2,3,4-*Tri-Me ether*, *Amide*

$C_{11}H_{21}NO_8S$
- 3-Methyl-3-butenylglucosinolate †

$C_{11}H_{21}NO_9S_2$
- Glucoibervirin

$C_{11}H_{21}NO_{10}S_2$
- Glucoconringin

$C_{11}H_{21}NO_{10}S_3$
- Glucoiberin

$C_{11}H_{21}N_3O_3$
Chaksine
Isochaksine

$C_{11}H_{21}N_3O_4$
Glycyl-D-alanyl-L-leucine
Glycyl-leucyl-alanine
L-Leucyl-D-glutamic Acid, *Amide*

$C_{11}H_{21}O_5P$
Phosphonoacetic Acid, 3,3-*Dipropyl*-1-*allyl ester*

$C_{11}H_{22}$
1-Undecene†
2-Undecene†
3-Undecene†
4-Undecene†
5-Undecene†

$C_{11}H_{22}BrN$
5-Azoniaspiro[4,7]dodecane, *Bromide*
6-Azoniaspiro[5,6]dodecane, *Bromide*

$C_{11}H_{22}Br_2$
1,2-Dibromoundecane
1,10-Dibromoundecane
1,11-Dibromoundecane
2,3-Dibromoundecane

$C_{11}H_{22}ClN$
6-Azoniaspiro[5,6]dodecane, *Chloride*

$C_{11}H_{22}NO_9S_2$
Glucosisaustricin†

$C_{11}H_{22}N_2O_2$
Malonamide, N,N′-*Dibutyl*
Undecanedioic Acid, *Diamide*

$C_{11}H_{22}N_2O_4S$
Pantetheine

$C_{11}H_{22}O$
Cycloundecanol
2,8-Dimethyl-5-nonanone
1-Isobutyl-3-methylcyclohexanol
2-Isopropyl-5-methylcyclohexanol, *Me ether*
3,3,5,5-Tetramethyl-4-heptanone
Undecanal
2-Undecanone
3-Undecanone
4-Undecanone
5-Undecanone
6-Undecanone
1-Undecen-3-ol
2-Undecen-4-ol
9-Undecen-1-ol
10-Undecen-1-ol
10-Undecen-2-ol

$C_{11}H_{22}O_2$
Decanoic Acid, *Me ester*
2-Ethyl-2-methylhexanoic Acid, *Et ester*
2-Ethyl-3-methylhexanoic Acid, *Et ester*
2-Ethyl-5-methylhexanoic Acid, *Et ester*
2-Hexanol, *Valeryl*
2-Methyldecanoic Acid
3-Methyldecanoic Acid
4-Methyldecanoic Acid
5-Methyldecanoic Acid
6-Methyldecanoic Acid
7-Methyldecanoic Acid
8-Methyldecanoic Acid
9-Methyldecanoic Acid
4-Methylvaleric Acid, 3-*Methylbutyl ester*
Nonanoic Acid, *Et ester*
Octanoic Acid, *Propyl ester*
2-Octanol, *Propionyl*
Undecanoic Acid

$C_{11}H_{22}O_3$
Di-3-methylbutyl carbonate
3-Ethyl-2-hydroxynonanoic Acid
2-Hydroxydecanoic Acid, *Me ester*
10-Hydroxydecanoic Acid, *Me ester*
2-Hydroxynonanoic Acid, *Et ester*
3-Hydroxynonanoic Acid, *Et ester*
7-Hydroxynonanoic Acid, *Et ester*
3-Hydroxy-2,2,5-trimethylhexanoic Acid, *Et ester*
2-Hydroxyundecanoic Acid
3-Hydroxyundecanoic Acid
4-Hydroxyundecanoic Acid
10-Hydroxyundecanoic Acid
11-Hydroxyundecanoic Acid

$C_{11}H_{22}O_4$
Glycerol, 1-*Octanoyl*

$C_{11}H_{22}O_6$
Fructose, *Me glycoside*, *Tetra-Me ether*
2,3,4,5,6-Penta-*O*-methylglucose

$C_{11}H_{22}O_7$
Gluconic Acid, *Penta-Me ether*
Mannonic Acid, 1,2,3,4,5-*Penta-Me ether*

$C_{11}H_{22}O_{10}$
Umbilicin

$C_{11}H_{23}N$
11-Amino-1-undecene
Azacyclododecane

$C_{11}H_{23}NO$
5-Azoniaspiro[4,7]dodecane (hydroxide)
6-Azoniaspiro[5,6]dodecane (hydroxide)
4-Hydroxy-2,2,6,6-tetramethylpiperidine,O,N-*Di-Me*
2-Methyldecanoic Acid, *Amide*
8-Methyldecanoic Acid, *Amide*
Undecanoic Acid, *Amide*

$C_{11}H_{23}NO_2$
10-Aminodecanoic Acid, *Me ester*
2-Aminovaleric Acid, *Et ester*, N-*Di-Et*
Piperidinoacetaldehyde, *Di-Et acetal*

$C_{11}H_{24}$
Undecane

$C_{11}H_{24}N_2O$
1,3-Di-(3-methylbutyl)urea

$C_{11}H_{24}O$
2,8-Dimethyl-5-nonanol
2-Nonanol, *Et ether*
1-Octanol, n-*Propyl ether*
2-Octanol, n-*Propyl ether*
1-Undecanol
2-Undecanol
3-Undecanol
5-Undecanol
6-Undecanol

$C_{11}H_{24}O_2$
1,7-Heptanediol, *Di-Et ether*
2-Heptanone, *Di-Et ketal*
2-Methyldecane-2,5-diol†
Undecane-1,3-diol†

$C_{11}H_{24}O_6$
Mannitol, 1,2,3,4,5-*Penta-Me ether*

$C_{11}H_{24}S$
Decyl methyl sulphide

$C_{11}H_{25}N$
1-Aminoundecane
2-Aminoundecane

$C_{11}H_{26}N_2$
1,3-Diaminopropane, 1,3-N-*Tetra-Et*

$C_{11}H_{28}Br_2N_2$
Pentamethonium Bromide

$C_{11}H_{28}I_2N_2$
Pentamethonium Iodide

C_{12}

$C_{12}Cl_{12}$
Perchloro-4,8-dimethylenetricyclo[3,3,2,$0^{1,5}$]-deca-2,6-diene†
Perchloro-3,4,7,8-tetramethylene-tricyclo-[4,2,0,$0^{2,5}$]octane†

$C_{12}F_8N_2$
Octafluorophenazine†

$C_{12}F_8S$
Octafluorodibenzothiophen†

$C_{12}H_2O_{10}$
Mellitic Acid, *Dianhydride*

$C_{12}H_4Br_6$
3,3′,4,4′,5,5′-Hexabromobiphenyl

$C_{12}H_4Br_6N_2O$
2,2′,4,4′,6,6′-Hexabromoazoxybenzene

$C_{12}H_4Cl_6$
2,2′,4,4′,6,6′-Hexachlorobiphenyl
3,3′,4,4′,5,5′-Hexachlorobiphenyl

$C_{12}H_4F_4O$
1,2,3,4-Tetrafluorodibenzofuran†

$C_{12}H_4N_2O_6$
5,6-Dinitroacenaphthenequinone

$C_{12}H_4N_6O_{12}$
2,2′,4,4′,6,6′-Hexanitrobiphenyl

$C_{12}H_4N_6O_{12}S$
Di-2,4,6-trinitrophenyl sulphide

$C_{12}H_4N_6O_{12}Se$
Di-2,4,6-trinitrophenyl selenide

$C_{12}H_4N_6O_{13}$
2,4,5-Trinitrophenyl 2,4,6-trinitrophenyl Ether

$C_{12}H_4N_6O_{16}$
3,3′,5,5′-Tetrahydroxy-2,2′,4,4′,6,6′-hexanitro-biphenyl

$C_{12}H_4N_8O_{12}$
2,2′,4,4′,6,6′-Hexanitroazobenzene

L

$C_{12}H_5BrO_2$
3-Bromoacenaphthenequinone
4-Bromoacenaphthenequinone
5-Bromoacenaphthenequinone

$C_{12}H_5BrO_3$
3-Bromonaphthalic Acid
4-Bromonaphthalic Acid

$C_{11}H_5Br_2ClO$
5,8-Dibromo-2-naphthoic Acid, *Chloride*

$C_{12}H_5NO_4$
5-Nitroacenaphthenequinone

$C_{12}H_5NO_5$
2-Nitronaphthalic Acid, *Anhydride*
3-Nitronaphthalic Acid, *Anhydride*
4-Nitronaphthalic Acid, *Anhydride*

$C_{12}H_5N_5O_8$
1,3,6,8-Tetranitrocarbazole

$C_{12}H_5N_7O_{12}$
Di-2,4,6-trinitrophenylamine

$C_{12}H_6$
1,5,9-Tridehydro[12]annulene†
1,3,5-Triethynylbenzene†

$C_{12}H_6Br_4$
2,2′,4,4′-Tetrabromobiphenyl
2,2′,6,6′-Tetrabromobiphenyl
2,3,4,4′-Tetrabromobiphenyl
2,3′,4,4′-Tetrabromobiphenyl
2,4,4′,5-Tetrabromobiphenyl
2,4,4′,6-Tetrabromobiphenyl
3,3′,4,4′-Tetrabromobiphenyl
3,3′,5,5′-Tetrabromobiphenyl

$C_{12}H_6Br_6N_2$
2,2′,4,4′,6,6-Hexabromohydrazobenzene

$C_{12}H_6Cl_2N_2O_4$
2,2′-Dichloro-4,4′-dinitrobiphenyl
2,2′-Dichloro-5,5′-dinitrobiphenyl
2,4′-Dichloro-3,3′-dinitrobiphenyl
3,3′-Dichloro-4,4′-dinitrobiphenyl
4,2′-Dichloro-3,3′-dinitrobiphenyl
4,4′-Dichloro-2,2′-dinitrobiphenyl
4,4′-Dichloro-2,3′-dinitrobiphenyl
4,4′-Dichloro-3,3′-dinitrobiphenyl
5,5′-Dichloro-2,2′-dinitrobiphenyl

$C_{12}H_6Cl_2O_2$
Naphthalene-1,4-dicarboxylic Acid, *Dichloride*
Naphthalene-1,5-dicarboxylic Acid, *Dichloride*
Naphthalene-1,8-dicarboxylic Acid, *Dichloride*

$C_{12}H_6Cl_4$
2,2′,4,4′-Tetrachlorobiphenyl
2,2′,5,5′-Tetrachlorobiphenyl
2,2′,6,6′-Tetrachlorobiphenyl
3,3′,4,4′-Tetrachlorobiphenyl
3,3′,5,5′-Tetrachlorobiphenyl

$C_{12}H_6Cl_6$
Mellitic Acid, *Hexa-chloride*

$C_{12}H_6N_2$
Acenaphthenequinone, *Azine*
Naphthalene-1,2-dicarboxylic Acid, *Dinitrile*
Naphthalene-1,3-dicarboxylic Acid, *Dinitrile*
Naphthalene-1,4-dicarboxylic Acid, *Dinitrile*
Naphthalene-1,5-dicarboxylic Acid, *Dinitrile*

$C_{12}H_6N_2$ (*continued*)
Naphthalene-1,6-dicarboxylic Acid, *Dinitrile*
Naphthalene-1,7-dicarboxylic Acid, *Dinitrile*
Naphthalene-1,8-dicarboxylic Acid, *Dinitrile*
Naphthalene-2,3-dicarboxylic Acid, *Dinitrile*
Naphthalene-2,6-dicarboxylic Acid, *Dinitrile*
Naphthalene-2,7-dicarboxylic Acid, *Dinitrile*

$C_{12}H_6N_2O_2$
4,7-Phenanthroline-5,6-dione

$C_{12}H_6N_2O_4$
4-Nitronaphthalic Acid, *Imide*

$C_{12}H_6N_2O_8$
3,6-Dinitronaphthalene-1,8-dicarboxylic Acid
4,5-Dinitronaphthalene-1,8-dicarboxylic Acid

$C_{12}H_6N_2S$
Acenaphtho[5,6-*cd*]-1,2,6-thiadiazine†

$C_{12}H_6N_2SSe$
Thianaphtheno[2,3-*e*]-2,1,3-benzoselenadiazole
Thianaphtheno[3,2-*e*]-2,1,3-benzoselenadiazole

$C_{12}H_6N_4$
4,5,9,10-Tetra-azapyrene†

$C_{12}H_6N_4O_2$
4,5,9,10-Tetra-azapyrene, *Di-oxide*†

$C_{12}H_6N_4O_8$
2,2′,4,4′-Tetranitrobiphenyl
2,2′,6,6′-Tetranitrobiphenyl
2′,3,4,4′-Tetranitrobiphenyl
3,3′,4,4′-Tetranitrobiphenyl
3,3′,5,5′-Tetranitrobiphenyl

$C_{12}H_6N_4S_4$
5,2′-4,4′-2′,5′-Quaterthiazole

$C_{12}H_6N_8O_{12}$
2,2′,4,4′6,6′-Hexanitrohydrazobenzene

$C_{12}H_6O_2$
Acenaphthenequinone

$C_{12}H_6O_3$
3-Hydroxyacenaphthenequinone
Naphthalene-1,2-dicarboxylic Acid, *Anhydride*
Naphthalene-1,8-dicarboxylic Acid, *Anhydride*
Naphthalene-2,3-dicarboxylic Acid, *Anhydride*
Naphtho[2,1-*b*]furan-1,2-dione

$C_{12}H_6O_4$
Benzene-1,3-dipropiolic Acid
Bicyclo[2,2,1]hept-5-ene-2,3-dicarboxylic Acid, *Mono-propyl ester*
3,6-Dihydroxy-1,2-acenaphthenequinone
3,8-Dihydroxy-1,2-acenaphthenequinone
2-Hydroxynaphthalene-1,8-dicarboxylic Acid, *Anhydride*
3-Hydroxynaphthalene-1,8-dicarboxylic Acid, *Anhydride*
4-Hydroxynaphthalene-1,8-dicarboxylic Acid, *Anhydride*

$C_{12}H_6O_5$
3,4-Dihydroxynaphthalic Anhydride
3 6-Dihydroxynaphthalic Anhydride

$C_{12}H_6O_8$
Galloflavin
Isogalloflavin

$C_{12}H_6O_{12}$
Mellitic Acid

$C_{12}H_7BrN_2O_2$
3-Bromo-6-nitrocarbazole

$C_{12}H_7BrO_4$
3-Bromonaphthalic Acid

$C_{12}H_7Br_2N$
1,3-Dibromocarbazole
3,6-Dibromocarbazole

$C_{11}H_7Br_2NO$
5,8-Dibromo-2-naphthoic Acid, *Amide*

$C_{12}H_7Br_2NO_2$
2,4′-Dibromo-4-nitrobiphenyl
2′,4-Dibromo-3-nitrobiphenyl
2,5-Dibromo-3′-nitrobiphenyl
2,5-Dibromo-4′-nitrobiphenyl
3,4-Dibromo-5-nitrobiphenyl
3,5-Dibromo-3′-nitrobiphenyl
3,5-Dibromo-4′-nitrobiphenyl
4,4′-Dibromo-2-nitrobiphenyl
4,4′-Dibromo-3-nitrobiphenyl
4,5-Dibromo-2-nitrobiphenyl

$C_{12}H_7Br_3$
2,2′,5-Tribromobiphenyl
2,3,4-Tribromobiphenyl
2,3′,4′-Tribromobiphenyl
2,3′,5-Tribromobiphenyl
2,3′,5′-Tribromobiphenyl
2,4,4′-Tribromobiphenyl
2,4,5-Tribromobiphenyl
2,4,6-Tribromobiphenyl
2,4′,5-Tribromobiphenyl
3,3′,5-Tribromobiphenyl
3,4,4′-Tribromobiphenyl
3,4′,5-Tribromobiphenyl

$C_{12}H_7Br_4N$
2,2′,4,4′-Tetrabromodiphenylamine

$C_{12}H_7ClN_2O_2$
3-Chloro-6-nitrocarbazole

$C_{12}H_7ClN_2O_5$
5-Chloro-2,4-dinitrophenol, *Phenyl ether*

$C_{12}H_7Cl_2N$
1,4-Dichlorocarbazole
2,7-Dichlorocarbazole
3,6-Dichlorocarbazole

$C_{12}H_7Cl_2NO_2$
2,4-Dichloro-3′-nitrobiphenyl
2,4′-Dichloro-4-nitrobiphenyl
2′,4-Dichloro-3-nitrobiphenyl
3,5-Dichloro-2′-nitrobiphenyl
3,5-Dichloro-3′-nitrobiphenyl
3,5-Dichloro-4′-nitrobiphenyl
4,4′-Dichloro-2-nitrobiphenyl

$C_{12}H_7Cl_2NO_3$
4,5-Dichloro-2-nitrophenol, *Phenyl ether*

$C_{12}H_7Cl_3$
2,3,5-Trichlorobiphenyl
2,4,6-Trichlorobiphenyl
2,4′,5-Trichlorobiphenyl
2′,3,4-Trichlorobiphenyl
2′,3,5-Trichlorobiphenyl
3,4′,5-Trichlorobiphenyl

$C_{12}H_7Cl_4N$
2,2′,4,4′-Tetrachlorodiphenylamine
2,3,4′,5-Tetrachlorodiphenylamine

$C_{12}H_7Cl_5O_2$
1,9,10,10,11-*exo*-Pentachloro-4,5-epoxypentacyclo[7,3,0,0^{2,6},0^{3,8},0^{7,11}]dodecan-12-one †

$C_{12}H_7I_2N$
2,7-Di-iodocarbazole
3,6-Di-iodocarbazole

$C_{12}H_7NO$
2-Azaphenalone †
1-Naphthylglyoxylic Acid, *Nitrile*

$C_{12}H_7NO_2$
Benz[*e*]isatin
Benz[*f*]isatin
Benz[*g*]isatin
2-Hydroxybenzo[*de*]quinolin-3-one
Naphthalene-1,2-dicarboxylic Acid, *Imide*
Naphthalene-1,8-dicarboxylic Acid, *Mononitrile*
Naphthalene-2,3-dicarboxylic Acid, *Mononitrile*
Naphthalene-2,3-dicarboxylic Acid, *Imide*
Naphthalene-2,6-dicarboxylic Acid, *Mononitrile*
Naphthalene-2,7-dicarboxylic Acid, *Mononitrile*

$C_{12}H_7NO_3$
3-Aminonaphthalic Acid, *Anhydride*
4-Aminonaphthalic Acid, *Anhydride*
2-Methylquinoline-3,4-dicarboxylic Acid, *Anhydride*
1-Nitrodibenzofuran
2-Nitrodibenzofuran
3-Nitrodibenzofuran
4-Nitrodibenzofuran
Resorufin

$C_{12}H_7NO_4$
Resazurin

$C_{12}H_7NO_6$
5-Nitronaphthalene-1,4-dicarboxylic Acid
2-Nitronaphthalic Acid
3-Nitronaphthalic Acid
4-Nitronaphthalic Acid

$C_{12}H_7N_3O_2$
1-Nitrophenazine
2-Nitrophenazine

$C_{12}H_7N_3O_3$
1-Nitrophenazine, 5-*Oxide*

$C_{12}H_7N_3O_6$
2,2′,4-Trinitrobiphenyl
2,3′,4-Trinitrobiphenyl
2,4,4′-Trinitrobiphenyl
2,4,6-Trinitrobiphenyl
3,3′,4-Trinitrobiphenyl
3,3′,5-Trinitrobiphenyl
3,4,4′-Trinitrobiphenyl

$C_{12}H_7N_3O_7$
2,4-Dinitrophenyl 2-nitrophenyl Ether
2,4-Dinitrophenyl 3-nitrophenyl Ether
2,4-Dinitrophenyl 4-nitrophenyl Ether
Phenyl 2,4,5-trinitrophenyl Ether
Phenyl 2,4,6-trinitrophenyl Ether

$C_{12}H_7N_5O_8$
2,2′,4,4′-Tetranitrodiphenylamine
2,2′,4,6-Tetranitrodiphenylamine
2,3′,4,6-Tetranitrodiphenylamine
2,4,4′,6-Tetranitrodiphenylamine

$C_{12}H_8$
Acenaphthylene
Biphenylene
Bisdehydro[12]annulene †
1,3,7,9-Cyclodecatetrayne †
s-Indacene †
1-Naphthylacetylene
2-Naphthylacetylene

$C_{12}H_8AsN$
Phenarsazine

$C_{12}H_8As_2$
Phenarsine

$C_{12}H_8BrCl$
2-Bromo-2′-chlorobiphenyl
3-Bromo-5-chlorobiphenyl
4′-Bromo-4-chlorobiphenyl

$C_{12}H_8BrF$
4-Bromo-4′-fluorobiphenyl

$C_{12}H_8BrI$
2-Bromo-2′-iodobiphenyl
Dibenzobromolium Iodide

$C_{12}H_8BrN$
1-Bromocarbazole
2-Bromocarbazole
3-Bromocarbazole
4-Bromocarbazole

$C_{12}H_8BrNO_2$
2-Bromo-2′-nitrobiphenyl
2-Bromo-3′-nitrobiphenyl
2-Bromo-4′-nitrobiphenyl
2-Bromo-5-nitrobiphenyl
3′-Bromo-2-nitrobiphenyl
3-Bromo-3′-nitrobiphenyl
3-Bromo-4′-nitrobiphenyl
4-Bromo-3-nitrobiphenyl
4-Bromo-4′-nitrobiphenyl
4′-Bromo-2-nitrobiphenyl
4′-Bromo-3-nitrobiphenyl

$C_{12}H_8BrNO_3$
3-Bromo-4-hydroxy-4′-nitrobiphenyl
3-Bromo-4-hydroxy-5-nitrobiphenyl

$C_{12}H_8BrNO_4$
8-Bromo-3-nitro-1-naphthoic Acid, *Me ester*
8-Bromo-5-nitro-1-naphthoic Acid, *Me ester*

$C_{12}H_8Br_2$
2,2′-Dibromobiphenyl
2,4-Dibromobiphenyl
2,4′-Dibromobiphenyl
2,5-Dibromobiphenyl
3,3′-Dibromobiphenyl
3,4-Dibromobiphenyl
4,4′-Dibromobiphenyl

$C_{12}H_8Br_2Hg$
Mercury di-*p*-bromophenyl

$C_{12}H_8Br_2N_2$
2,2′-Dibromoazobenzene

$C_{12}H_8Br_2N_2$ (*continued*)
2,4-Dibromoazobenzene
3,3′-Dibromoazobenzene
3,5-Dibromoazobenzene
4,4′-Dibromoazobenzene

$C_{12}H_8Br_2O$
3,4-Dibromo-5-hydroxybiphenyl
3,4′-Dibromo-4-hydroxybiphenyl
3,5-Dibromo-2-hydroxybiphenyl
3,5-Dibromo-4-hydroxybiphenyl
Di-4-bromophenyl Ether

$C_{12}H_8Br_2OS$
Di-4-bromophenyl sulphoxide

$C_{12}H_8Br_2O_2$
1,6-Dibromo-2-naphthoic Acid, *Me ester*
5,8-Dibromo-2-naphthoic Acid, *Me ester*

$C_{12}H_8Br_2O_2S$
Di-4-bromophenyl sulphone

$C_{12}H_8Br_2O_5S_2$
Bromobenzene-*p*-sulphonic Acid, *Anhydride*

$C_{12}H_8Br_2S$
Di-4-bromophenyl sulphide

$C_{12}H_8Br_4N_2$
2,2′,6,6′-Tetrabromobenzidine
3,3′,5,5′-Tetrabromobenzidine

$C_{12}H_8ClF$
4-Chloro-4′-fluorobiphenyl

$C_{12}H_8ClI$
2-Chloro-2′-iodobiphenyl
3-Chloro-3′-iodobiphenyl
4-Chloro-2′-iodobiphenyl
4-Chloro-4′-iodobiphenyl
Dibenzochloronium Iodide

$C_{12}H_8ClN$
1-Chlorocarbazole
2-Chlorocarbazole
3-Chlorocarbazole
4-Chlorocarbazole

$C_{12}H_8ClNO_2$
2-Chloro-2′-nitrobiphenyl
3-Chloro-3′-nitrobiphenyl
4-Chloro-4′-nitrobiphenyl
4′-Chloro-3-nitrobiphenyl
4-Chloropyridine-2-carboxylic Acid, *Phenyl ester*
5-Chloropyridine-3-carboxylic Acid, *Phenyl ester*

$C_{12}H_8ClNO_4$
(1-Nitro-2-naphthoxy)acetic Acid, *Chloride*

$C_{12}H_8ClNO_4S$
6-Nitroacenaphthene-3-sulphonic Acid, *Chloride*
6-Nitroacenaphthene-4-sulphonic Acid, *Chloride*
4′-Nitrobiphenyl-4-sulphonic Acid, *Chloride*

$C_{12}H_8ClNO_5S$
4-Chloro-2-nitrobenzenesulphonic Acid, *Phenyl ester*
4-Chloro-3-nitrobenzenesulphonic Acid, *Phenyl ester*

$C_{12}H_8ClN_3O_2$
3-Chloro-4′-nitroazobenzene
4-Chloro-4′-nitroazobenzene

$C_{12}H_8ClN_3O_4$
2-Chloro-2′,4′-dinitrodiphenylamine
3-Chloro-2′,4′-dinitrodiphenylamine
4-Chloro-2,6-dinitrodiphenylamine
4-Chloro-2′,4′-dinitrodiphenylamine
5-Chloro-2,4-dinitrodiphenylamine

$C_{12}H_8Cl_2$
1,2-Dichloroacenaphthene
5,6-Dichloroacenaphthene
2,2′-Dichlorobiphenyl
2,3-Dichlorobiphenyl
2,4′-Dichlorobiphenyl
2,5-Dichlorobiphenyl
3,3′-Dichlorobiphenyl
3,4-Dichlorobiphenyl
3,5-Dichlorobiphenyl
4,4′-Dichlorobiphenyl

$C_{12}H_8Cl_2Hg$
Mercury di-*o*-chlorophenyl
Mercury di-*p*-chlorophenyl

$C_{12}H_8Cl_2N_2$
2,2′-Dichloroazobenzene
2,4-Dichloroazobenzene
2,5-Dichloroazobenzene
3,3′-Dichloroazobenzene
4,4′-Dichloroazobenzene

$C_{12}H_8Cl_2N_2O$
2,2′-Dichloroazoxybenzene
3,3′-Dichloroazoxybenzene
4,4′-Dichloroazoxybenzene

$C_{12}H_8Cl_2N_2O_4S_2$
Azobenzene-3,3′-disulphonic Acid, *Dichloride*

$C_{12}H_8Cl_2N_2O_5S_2$
Azoxybenzene-3,3′-disulphonic Acid, *Dichloride*

$C_{12}H_8Cl_2OS$
Di-2-chlorophenyl Sulphoxide†
Di-3-chlorophenyl Sulphoxide†
Di-4-chlorophenyl Sulphoxide★†

$C_{12}H_8Cl_2O_2$
3,3′-Dichloro-2,2′-dihydroxybiphenyl
3,3′-Dichloro-4,4′-dihydroxybiphenyl
5,5′-Dichloro-2,2′-dihydroxybiphenyl
3,6-Dichloro-1-naphthoic Acid, *Me ester*
5,8-Dichloro-2-naphthoic Acid, *Me ester*
Di-2-chlorophenyl Sulphone†
Di-3-chlorophenyl Sulphone†
Di-4-chlorophenyl Sulphone★†

$C_{12}H_8Cl_2S$
2-Chlorophenyl 3-chlorophenyl sulphide
2-Chlorophenyl 4-chlorophenyl sulphide
3-Chlorophenyl 4-chlorophenyl sulphide
Di-2-chlorophenyl sulphide
Di-3-chlorophenyl sulphide
Di-4-chlorophenyl sulphide

$C_{12}H_8Cl_4N_2$
2,2′,5,5′-Tetrachlorobenzidine
2,2′,6,6′-Tetrachlorobenzidine
3,3′,5,5′-Tetrachlorobenzidine

$C_{12}H_8Cl_6$
Aldrin

$C_{12}H_8Cl_6O$
Endrin
HEOD★†

$C_{12}H_8Cl_6O_2$
1,2,3,4,10,10-Hexachloro-6,7-epoxy-1,4,4*a*,5,6,7,8,8*a*-octahydro-4*a*(or 5)-hydroxy-*exo*-1,4-*endo*-5,8-methanonaphthalene†

$C_{12}H_8FNO_2$
2-Fluoro-2′-nitrobiphenyl
2-Fluoro-4-nitrobiphenyl
2-Fluoro-4′-nitrobiphenyl
4-Fluoro-2-nitrobiphenyl
4-Fluoro-2′-nitrobiphenyl
4-Fluoro-4′-nitrobiphenyl

$C_{12}H_8F_2$
2,2′-Difluorobiphenyl
3,3′-Difluorobiphenyl
4,4′-Difluorobiphenyl

$C_{12}H_8Hg$
Dibenzomercurole

$C_{12}H_8HgI_2$
Mercury di-*p*-iodophenyl

$C_{12}H_8HgN_2O_4$
Mercury di-*o*-nitrophenyl
Mercury di-*p*-nitrophenyl

$C_{12}H_8I_2$
Dibenziodolium iodide

$C_{12}H_8N_2$
Benzo[*c*]cinnoline
Benzo[*f*]quinazoline
1,7-Phenanthroline
4,7-Phenanthroline
1,10-Phenanthroline
Phenazine
4-Phenyl-1,3-butadiene-1,1-dicarboxylic Acid, *Dinitrile*

$C_{12}H_8N_2O$
1-Hydroxyphenazine
2-Hydroxyphenazine
Perlolidine†

$C_{12}H_8N_2O_2$
2-Aminophenoxazin-3-one†
β-Carboline-1-carboxylic Acid†
1,6-Dihydroxyphenazine
2-Methylquinoline-3,4-dicarboxylic Acid, *Imide*
1-Nitrocarbazole
2-Nitrocarbazole
3-Nitrocarbazole
4-Nitrocarbazole

$C_{12}H_8N_2O_3$
3-Amino-2-nitrodibenzofuran
4-Amino-1-nitrodibenzofuran
4-Amino-3-nitrodibenzofuran
7-Amino-2-nitrodibenzofuran
8-Amino-1-nitrodibenzofuran
2-*o*-Nitrobenzoylpyridine
2-*m*-Nitrobenzoylpyridine
2-*p*-Nitrobenzoylpyridine
3-*p*-Nitrobenzoylpyridine
4-*o*-Nitrobenzoylpyridine
4-*m*-Nitrobenzoylpyridine
4-*p*-Nitrobenzoylpyridine
Pyridine-4-carboxylic Acid, *Anhydride*

$C_{12}H_8N_2O_4$
3,6-Dinitroacenaphthene
3,8-Dinitroacenaphthene
5,6-Dinitroacenaphthene
2,2′-Dinitrobiphenyl
2,3′-Dinitrobiphenyl
2,4-Dinitrobiphenyl
2,4′-Dinitrobiphenyl
3,3′-Dinitrobiphenyl
3,4-Dinitrobiphenyl
3,4′-Dinitrobiphenyl
4,4′-Dinitrobiphenyl
Iodinin

$C_{12}H_8N_2O_4S$
2,4-Dinitrophenyl phenyl sulphide
Di-2-nitrophenyl sulphide
Di-4-nitrophenyl sulphide
2-Nitrophenyl 4-nitrophenyl sulphide

$C_{12}H_8N_2O_4S_2$
Di-2-nitrophenyl disulphide
Di-3-nitrophenyl disulphide
Di-4-nitrophenyl disulphide

$C_{12}H_8N_2O_5$
2,4-Dinitrophenyl phenyl Ether
2,6-Dinitrophenyl phenyl Ether
3,4-Dinitrophenyl phenyl Ether
Di-2-nitrophenyl Ether
Di-3-nitrophenyl Ether
Di-4-nitrophenyl Ether
2-Hydroxy-3,5-dinitrobiphenyl
2-Hydroxy-4′,5-dinitrobiphenyl
2-Nitrophenyl-4-nitrophenyl Ether
3-Nitrophenyl-4-nitrophenyl Ether

$C_{12}H_8N_2O_6$
2,2′-Dihydroxy-3,3′-dinitrobiphenyl
2,2′-Dihydroxy-3,5′-dinitrobiphenyl
2,2′-Dihydroxy-5,5′-dinitrobiphenyl
4,4′-Dihydroxy-3,3′-dinitrobiphenyl
3,6-Dinitro-1-naphthoic Acid, *Me ester*
6,8-Dinitro-1-naphthoic Acid, *Me ester*
4,6-Dinitroresorcinol, *Phenyl ether*

$C_{12}H_8N_2O_6S$
2,4-Dinitrophenyl phenyl sulphone
Di-2-nitrophenyl sulphone
Di-3-nitrophenyl sulphone
Di-4-nitrophenyl sulphone
2-Nitrophenyl-3-nitrophenyl sulphone
3-Nitrophenyl 4-nitrophenyl sulphone

$C_{12}H_8N_4$
Dibenzo-1,3*a*,4,6*a*-tetra-azapentalene†
Dibenzo-1,3*a*,6,6*a*-tetra-azapentalene†

$C_{12}H_8N_4O_4$
2,2′-Dinitroazobenzene
2,3′-Dinitroazobenzene
2,4-Dinitroazobenzene
2,4′-Dinitroazobenzene
3,3′-Dinitroazobenzene
4,4′-Dinitroazobenzene

$C_{12}H_8N_4O_5$
2,2′-Dinitroazoxybenzene
2,4-Dinitroazoxybenzene
2,6-Dinitroazoxybenzene
3,3′-Dinitroazoxybenzene

$C_{12}H_8N_4O_5$ (*continued*)
3,5-Dinitroazoxybenzene
4,4′-Dinitroazoxybenzene

$C_{12}H_8N_4O_6$
4,5-Dinitronaphthalene-1,8-dicarboxylic Acid, *Diamide*
2,2′,4-Trinitrodiphenylamine
2,3′,4-Trinitrodiphenylamine
2,4,4′-Trinitrodiphenylamine
2,4,6-Trinitrodiphenylamine

$C_{12}H_8O$
Acenaphthenone
Dibenzofuran
Naphtho[1,2-*b*]furan
Naphtho[2,1-*b*]furan

$C_{12}H_8OS$
1-(2-Furyl)-4-(2-thienyl)but-1-en-3-yne †
Naphtho[2,1-*b*]thiophen-3-ol
Naphtho[1,2-*b*]thiophen-3-ol
Naphtho[2,3-*b*]thiophen-3-ol
Phenoxathiin

$C_{12}H_8O_2$
Biphenoquinone
Cyclo-octa[1,2-*c*,5,6-*c*′]difuran †
Dibenzo[1,4]dioxan
1-Hydroxydibenzofuran
2-Hydroxydibenzofuran
3-Hydroxydibenzofuran
4-Hydroxydibenzofuran
Naphthalaldehyde
Naphtho[1,2-*b*]furan-3(2*H*)-one
Naphtho[2,1-*b*]furan-2-one
Naphtho[2,1-*b*]furan-3(2*H*)-one
1-Naphthylglyoxal
2-Naphthylglyoxal
1-Oxaphenalen-3(2*H*)-one †

$C_{12}H_8O_2S_2$
Dibenzo-1,2-dithiin-5,5-dioxide
Thianthrene, 5,10-*Dioxide*

$C_{12}H_8O_3$
Benzofuril
1,2-Dihydroxydibenzofuran
1,4-Dihydroxydibenzofuran
1,9-Dihydroxydibenzofuran
2,8-Dihydroxydibenzofuran
2,9-Dihydroxydibenzofuran
3,7-Dihydroxydibenzofuran
3,8-Dihydroxydibenzofuran
4,6-Dihydroxydibenzofuran
4,8-Dihydroxydibenzofuran
1,2-Naphthaldehydic Acid
1,8-Naphthaldehydic Acid
1-Naphthylglyoxylic Acid
2-Naphthylglyoxylic Acid

$C_{12}H_8O_4$
Allobergaptene
Bergapten
Isobergapten
Naphthalene-1,2-dicarboxylic Acid
Naphthalene-1,3-dicarboxylic Acid
Naphthalene-1,4-dicarboxylic Acid
Naphthalene-1,5-dicarboxylic Acid
Naphthalene-1,6-dicarboxylic Acid
Naphthalene-1,7-dicarboxylic Acid
Naphthalene-1,8-dicarboxylic Acid
Naphthalene-2,3-dicarboxylic Acid
Naphthalene-2,6-dicarboxylic Acid
Naphthalene-2,7-dicarboxylic Acid
β-Sorigenin
Xanthotoxin

$C_{12}H_8O_4S_2$
Thianthrene, 5,5,10,10-*Tetraoxide*

$C_{12}H_8O_5$
1-Hydroxynaphthalene-2,4-dicarboxylic Acid
2-Hydroxynaphthalene-1,8-dicarboxylic Acid
3-Hydroxynaphthalene-1,8-dicarboxylic Acid
4-Hydroxynaphthalene-1,8-dicarboxylic Acid

$C_{12}H_8O_6$
Brevifolin

$C_{12}H_8O_7$
Purpurogallin-carboxylic Acid
Spinochrome A★ †
Spinochrome C †
Spinochrome S †

$C_{12}H_8S$
Dibenzothiophene
Naphtho[1,2-*b*]thiophen †
Naphtho[2,1-*b*]thiophen †
Naphtho[2,3-*b*]thiophen †

$C_{12}H_8S_2$
5-(3-Buten-1-ynyl)-2-2′-bithienyl †
Dibenzo-1,2-dithiin
Thianthrene

$C_{12}H_8S_3$
α-Terthienyl

$C_{12}H_8Se$
Dibenzoselenophene

$C_{12}H_9AsClN$
10-Chloro-5,10-dihydrophenarsazine

$C_{12}H_9Br$
1-Bromoacenaphthene
3-Bromoacenaphthene
5-Bromoacenaphthene
2-Bromobiphenyl
3-Bromobiphenyl
4-Bromobiphenyl

$C_{12}H_9BrN_2$
2-Bromoazobenzene
3-Bromoazobenzene
4-Bromoazobenzene

$C_{12}H_9BrO$
3-Bromo-2-hydroxybiphenyl
3-Bromo-4-hydroxybiphenyl
3-Bromo-5-hydroxybiphenyl
4-Bromo-4′-hydroxybiphenyl
5-Bromo-2-hydroxybiphenyl
p-Bromophenyl phenyl Ether

$C_{12}H_9BrO_2$
4-Bromo-1-naphthoic Acid, *Me ester*
7-Bromo-1-naphthoic Acid, *Me ester*
8-Bromo-1-naphthoic Acid, *Me ester*
3-Bromo-2-naphthoic Acid, *Me ester*

$C_{12}H_9BrO_3$
2-Bromo-3-hydroxy-1,4-naphthoquinone, *Et ether*
$C_{12}H_9BrS$
4-Bromophenyl phenyl sulphide
$C_{12}H_9Br_2N$
4,4′-Dibromodiphenylamine
$C_{12}H_9Br_2N$
2-Amino-3,5-dibromobiphenyl
2-Amino-4,4′-dibromobiphenyl
2-Amino-4,5-dibromobiphenyl
3-Amino-2′,4-dibromobiphenyl
3-Amino-4,5-dibromobiphenyl
3′-Amino-3,5-dibromobiphenyl
4-Amino-2,4′-dibromobiphenyl
4-Amino-2′,3-dibromobiphenyl
4-Amino-3,4′-dibromobiphenyl
4-Amino-3,5-dibromobiphenyl
4′-Amino-3,5-dibromobiphenyl
$C_{12}H_9BrN_2O_2$
4-Amino-3-bromo-2′-nitrobiphenyl
4-Amino-3-bromo-3′-nitrobipheny
4-Amino-3-bromo-4′-nitrobiphenyl
4-Amino-4′-bromo-3-nitrobiphenyl
4-Amino-5-bromo-3-nitrobiphenyl
5-Amino-2-bromo-3′-nitrobiphenyl
$C_{12}H_9Cl$
1-Chloroacenaphthene
3-Chloroacenaphthene
4-Chloroacenaphthene
5-Chloroacenaphthene
2-Chlorobiphenyl
3-Chlorobiphenyl
4-Chlorobiphenyl
$C_{12}H_9ClHgN_2O$
2-Chloromercuri-4-phenylazophenol†
$C_{12}H_9ClHgN_3O_3$
2-Chloromercuri-4-(*p*-nitrophenylazo)phenol†
$C_{12}H_9ClN_2$
3-Chloroazobenzene
4-Chloroazobenzene
$C_{12}H_9ClN_2O_2$
2-Chloro-2′-nitrodiphenylamine
3-Chloro-4′-nitrodiphenylamine
3′-Chloro-2-nitrodiphenylamine
4-Chloro-2-nitrodiphenylamine
4′-Chloro-2-nitrodiphenylamine
5-Chloro-2-nitrodiphenylamine
$C_{12}H_9ClN_2O_2S$
Azobenzene-3-sulphonic Acid, *Chloride*
$C_{12}H_9ClN_2O_4S$
4-Nitrodiphenylamine-2-sulphonic Acid, *Chloride*
$C_{12}H_9ClO$
1-Chloroacetylnaphthalene
2-Chloro-5-hydroxybiphenyl
3-Chloro-2-hydroxybiphenyl
3-Chloro-4-hydroxybiphenyl
4-Chloro-4′-hydroxybiphenyl
5-Chloro-2-hydroxybiphenyl
2-Methyl-1-naphthoic Acid, *Chloride*
4-Methyl-1-naphthoic Acid, *Chloride*
1-Naphthylacetic Acid, *Chloride*
2-Naphthylacetic Acid, *Chloride*
$C_{12}H_9ClOS_2$
5-(4-Chloro-3-hydroxybut-1-ynyl)-2,2′-bithienyl
$C_{12}H_9ClO_2$
α-Naphthoxyacetic Acid, *Chloride*
β-Naphthoxyacetic Acid, *Chloride*
$C_{12}H_9ClO_2S$
Acenaphthene-3-sulphonic Acid, *Chloride*
$C_{12}H_9ClO_2$
2-Chloro-1-naphthoic Acid, *Me ester*
5-Chloro-1-naphthoic Acid, *Me ester*
6-Chloro-1-naphthoic Acid, *Me ester*
7-Chloro-1-naphthoic Acid, *Me ester*
3-Chloro-2-naphthoic Acid, *Me ester*
5-Chloro-2-naphthoic Acid, *Me ester*
$C_{12}H_9ClO_3$
4-Chloro-1-hydroxy-2-naphthoic Acid, *Me ester*
4-Chloro-3-hydroxy-2-naphthoic Acid, *Me ester*
2-Chloro-3-hydroxy-1,4-naphthoquinone, *Et ether*
$C_{12}H_9Cl_2N$
2-Amino-3,5-dichlorobiphenyl
2-Amino-3′,5-dichlorobiphenyl
3-Amino-2′,4-dichlorobiphenyl
3-Amino-3′,5′-dichlorobiphenyl
4-Amino-3′,5′-dichlorobiphenyl
4-Amino-2,4′-dichlorobiphenyl
2,2′-Dichlorodiphenylamine
2,4-Dichlorodiphenylamine
2,4′-Dichlorodiphenylamine
2,5-Dichlorodiphenylamine
3,3′-Dichlorodiphenylamine
3,4-Dichlorodiphenylamine
3,4′-Dichlorodiphenylamine
3,5-Dichlorodiphenylamine
4,4′-Dichlorodiphenylamine
$C_{12}H_9Cl_2NO$
2-Amino-4-chlorophenol, o-*Chlorophenyl ether*
2-Amino-4-chlorophenol, p-*Chlorophenyl ether*
$C_{12}H_9Cl_2NO_3$
Pyoluteorin★, N-*Me*†
$C_{12}H_9Cl_2N_3$
4-Amino-2,3′-dichloroazobenzene
4-Amino-2′,3-dichloroazobenzene
2,2′-Dichlorodiazoaminobenzene
3,3′-Dichlorodiazoaminobenzene
4,4′-Dichlorodiazoaminobenzene
$C_{12}H_9F$
2-Fluorobiphenyl
3-Fluorobiphenyl
4-Fluorobiphenyl
$C_{12}H_9I$
3-Iodoacenaphthene
5-Iodoacenaphthene
2-Iodobiphenyl
3-Iodobiphenyl
4-Iodobiphenyl
$C_{12}H_9IO$
2-Iodophenyl phenyl Ether
3-Iodophenyl phenyl Ether
4-Iodophenyl phenyl Ether

$C_{12}H_9IO_2$
5-Iodo-1-naphthoic Acid, *Me ester*
7-Iodo-1-naphthoic Acid, *Me ester*
8-Iodo-1-naphthoic Acid, *Me ester*
3-Iodo-2-naphthoic Acid, *Me ester*
5-Iodo-2-naphthoic Acid, *Me ester*

$C_{12}H_9N$
1*H*-Benz[*f*]indole
1*H*-Benz[*g*]indole
3*H*-Benz[*e*]indole
Benz[*b*]indolizine
Benz[*f*]isoindole
Carbazole
2-Methyl-1-naphthoic Acid, *Nitrile*
1-Naphthylacetic Acid, *Nitrile*
2-Naphthylacetic Acid, *Nitrile*

$C_{12}H_9NO$
1-Aminodibenzofuran
2-Aminodibenzofuran
3-Aminodibenzofuran
4-Aminodibenzofuran
2-Benzoylpyridine
3-Benzoylpyridine
4-Benzoylpyridine
Benz[*e*]indol-2-one
Benz[*g*]indol-2-one
2-Hydroxycarbazole
3-Hydroxycarbazole
4-Hydroxycarbazole
3-Hydroxy-2-naphthoic Acid, *Me ether*, *Nitrile*
2-Methylnaphtho[1,2-*d*]oxazole
2-Methylnaphtho[2,1-*d*]oxazole
2-Methylnaphtho[2,3-*d*]oxazole
β-Naphthoxyacetic Acid, *Nitrile*
Phenoxazine

$C_{12}H_9NO_2$
2-Cyano-5-phenyl-2,4-pentadienoic Acid
ψ-Dictamnine
Dictamnine
1,7-Dihydroxycarbazole
1,8-Dihydroxycarbazole
2,3-Dihydroxycarbazole
Indophenol
1-Nitroacenaphthene
3-Nitroacenaphthene
4-Nitroacenaphthene
5-Nitroacenaphthene
2-Nitrobiphenyl
3-Nitrobiphenyl
4-Nitrobiphenyl
6-Phenylpyridine-2-carboxylic Acid
6-Phenylpyridine-3-carboxylic Acid
Pyridine-2-carboxylic Acid, *Phenyl ester*
Pyridine-3-carboxylic Acid, *Phenyl ester*
Pyridine-4-carboxylic Acid, *Phenyl ester*
2-Quinolineacrylic Acid
4-Quinolineacrylic Acid
8-Quinolineacrylic Acid

$C_{12}H_9NO_2S$
o-Nitrophenyl phenyl sulphide
p-Nitrophenyl phenyl sulphide
Phenothiazine 5,5-dioxide

$C_{12}H_9NO_3$
7-Hydroxy-4-methoxyfuro[2,3-*b*]quinoline†
2-Hydroxy-2′-nitrobiphenyl
2-Hydroxy-3-nitrobiphenyl
2-Hydroxy-3′-nitrobiphenyl
2-Hydroxy-4-nitrobiphenyl
2-Hydroxy-4′-nitrobiphenyl
2-Hydroxy-5-nitrobiphenyl
3-Hydroxy-2-nitrobiphenyl
3-Hydroxy-3′-nitrobiphenyl
3-Hydroxy-4-nitrobiphenyl
4-Hydroxy-2-nitrobiphenyl
4-Hydroxy-2′-nitrobiphenyl
4-Hydroxy-3-nitrobiphenyl
4-Hydroxy-4′-nitrobiphenyl
o-Nitrophenyl phenyl Ether
m-Nitrophenyl phenyl Ether
p-Nitrophenyl phenyl Ether

$C_{12}H_9NO_4$
5-Aminonaphthalene-2,3-dicarboxylic Acid
3-Aminonaphthalic Acid
4-Aminonaphthalic Acid
2,4′-Dihydroxy-4-nitrobiphenyl
2,5-Dihydroxy-3′-nitrobiphenyl
2,5-Dihydroxy-4-nitrobiphenyl
2,5-Dihydroxy-4′-nitrobiphenyl
2-Hydroxy-4-nitrophenyl phenyl Ether
2-Hydroxy-5-nitrophenyl phenyl Ether
4-Hydroxy-3-nitrophenyl phenyl Ether
2-Hydroxyphenyl 2-nitrophenyl Ether
2-Hydroxyphenyl-3-nitrophenyl Ether
2-Hydroxyphenyl 4-nitrophenyl Ether
4-Hydroxyphenyl 2-nitrophenyl Ether
6-Methyl-1-nitro-2-naphthoic Acid
6-Methyl-5-nitro-2-naphthoic Acid
2-Methylquinoline-3,4-dicarboxylic Acid
2-Methylquinoline-4,6-dicarboxylic Acid
3-Nitro-1-naphthoic Acid, *Me ester*
5-Nitro-1-naphthoic Acid, *Me ester*
8-Nitro-1-naphthoic Acid, *Me ester*
5-Nitro-2-naphthoic Acid, *Me ester*
Pyrrole-2,5-dicarboxylic Acid, N-*Phenyl*
Pyrrole-3,4-dicarboxylic Acid, N-*Phenyl*
Quinoline-2,3-dicarboxylic Acid, *Me ester*
Quinoline-6,7-dicarboxylic Acid, *Me ester*

$C_{12}H_9NO_4S$
o-Nitrophenyl phenyl sulphone
m-Nitrophenyl phenyl sulphone
p-Nitrophenyl phenyl sulphone

$C_{12}H_9NO_4S_2$
Benzene-*o*-disulphonic Acid, *Phenylimide*
Thiolbenzenesulphonic Acid, p-*Nitrophenyl ester*

$C_{12}H_9NO_5$
2-Hydroxy-6-nitro-1-naphthoic Acid, *Me ether*
1-Hydroxy-4-nitro-2-naphthoic Acid, *Me ester*
1-Hydroxy-4-nitro-2-naphthoic Acid, *Me ether*
3-Hydroxy-4-nitro-2-naphthoic Acid, *Me ester*
(1-Nitro-2-naphthoxy)acetic Acid

$C_{12}H_9NO_5S$
6-Nitroacenaphthene-3-sulphonic Acid
6-Nitroacenaphthene-4-sulphonic Acid
4′-Nitrobiphenyl-4-sulphonic Acid

$C_{12}H_9NS$
Phenothiazine

$C_{12}H_9NS_2$
2-Mercaptonaphtho[1,2-*d*]thiazole, S-*Me*

$C_{12}H_9NSe$
Selenazine

$C_{12}H_9N_3$
1-Aminophenazine
2-Aminophenazine
1-Methylacenaphtho[5,6-*de*]triazine†
2-Methylacenaphtho[5,6-*de*]triazine†
Normacrorine†
N-Pyridinium 2-benzimidazolide†

$C_{12}H_9N_3O$
1-Amino-2-hydroxyphenazine
2-Amino-3-hydroxyphenazine
8-Amino-2-hydroxyphenazine

$C_{12}H_9N_3O_2$
3-Amino-2-nitrocarbazole
3-Amino-4-nitrocarbazole
2-Nitroazobenzene
3-Nitroazobenzene
4-Nitroazobenzene

$C_{12}H_9N_3O_3$
2-Hydroxy-5-nitroazobenzene
4-Hydroxy-2′-nitroazobenzene
4-Hydroxy-3-nitroazobenzene
4-Hydroxy-3′-nitroazobenzene
4-Hydroxy-4′-nitroazobenzene
2-Nitroazoxybenzene
4-Nitroazoxybenzene

$C_{12}H_9N_3O_4$
2,4-Dihydroxy-2′-nitroazobenzene
2,4-Dihydroxy-3-nitroazobenzene
2,4-Dihydroxy-4′-nitroazobenzene
2,5-Dihydroxy-4′-nitroazobenzene
3,4-Dihydroxy-2′-nitroazobenzene
3,4-Dihydroxy-3′-nitroazobenzene
3,4-Dihydroxy-4′-nitroazobenzene
2,2′-Dinitrodiphenylamine
2,3′-Dinitrodiphenylamine
2,4-Dinitrodiphenylamine
2,4′-Dinitrodiphenylamine
2,6-Dinitrodiphenylamine
3,3′-Dinitrodiphenylamine
3,4′-Dinitrodiphenylamine
4,4′-Dinitrodiphenylamine

$C_{12}H_9N_3O_5$
2-Hydroxy-4,6-dinitrodiphenylamine
2′-Hydroxy-2,4-dinitrodiphenylamine
2′-Hydroxy-2,6-dinitrodiphenylamine
4-Hydroxy-2,2′-dinitrophenylamine
4′-Hydroxy-2,4-dinitrophenylamine
5-Hydroxy-2,4-dinitrophenylamine

$C_{12}H_9N_3O_5S$
3′-Nitroazobenzene-4-sulphonic Acid
4′-Nitroazobenzene-4-sulphonic Acid

$C_{12}H_9N_3O_7$
2,4,5-Trinitro-1-naphthol, *Et ether*
1,6,8-Trinitro-2-naphthol, *Et ether*

$C_{12}H_{10}$
Acenaphthene
1,2-Benzoheptafulvene
3,4-Benzoheptafulvene
Biphenyl
Heptalene
Hemi-Dewar Biphenyl†
1-Naphthylethylene
2-Naphthylethylene
6-Phenylfulvene†
Sesquifulvalene†

$C_{12}H_{10}AsCl$
Diphenylchloroarsine

$C_{12}H_{10}AsNO_2$
Phenarsazinic Acid

$C_{12}H_{10}As_2$
Arsenobenzene

$C_{12}H_{10}As_2O_2$
p-Arsenophenol

$C_{12}H_{10}BrClO$
6-Bromo-1-chloro-2-naphthol, *Et ether*

$C_{12}H_{10}BrN$
2-Amino-2′-bromobiphenyl
2-Amino-3′-bromobiphenyl
2-Amino-5-bromobiphenyl
3-Amino-2′-bromobiphenyl
3-Amino-4-bromobiphenyl
3-Amino-4′-bromobiphenyl
3-Amino-5-bromobiphenyl
4-Amino-2′-bromobiphenyl
4-Amino-3-bromobiphenyl
4-Amino-3′-bromobiphenyl
4-Amino-4′-bromobiphenyl

$C_{12}H_{10}BrNO$
5-Bromo-2-naphthylamine, N-*Ac*
6-Bromo-2-naphthylamine, N-*Ac*

$C_{12}H_{10}BrNO_2S$
1-(*p*-Bromobenzenesulphonyl) azepine†

$C_{12}H_{10}BrNO_3$
4-Bromo-2-nitro-1-naphthol, *Et ether*
6-Bromo-1-nitro-2-naphthol, *Et ether*

$C_{12}H_{10}Br_2N_2$
4,4′-Diamino-2,2′-dibromobiphenyl
4,4′-Diamino-2,6-dibromobiphenyl
4,4′-Diamino-3,3′-dibromobiphenyl
2,2′-Dibromohydrazobenzene
3,3′-Dibromohydrazobenzene
3,5-Dibromohydrazobenzene
4,4′-Dibromohydrazobenzene

$C_{12}H_{10}Br_2O$
1,6-Dibromo-2-naphthol, *Et ether*

$C_{12}H_{10}Br_2O_2$
1,5-Dibromo-2,6-dihydroxynaphthalene, *Di-Me ether*
2,6-Dibromo-1,5-dihydroxynaphthalene, *Di-Me ether*

$C_{12}H_{10}Cd$
Diphenylcadmium

$C_{12}H_{10}ClN$
2-Amino-4′-chlorobiphenyl
3-Amino-4′-chlorobiphenyl
4-Amino-3-chlorobiphenyl
3-Chlorodiphenylamine
4-Chlorodiphenylamine

$C_{12}H_{10}ClNO$
2-Amino-4-chlorophenol, *Phenyl ether*
1-Chloro-2-naphthylamine, N-*Ac*

$C_{12}H_{10}ClNO_2$
2-Chloro-α-cyanocinnamic Acid, *Et ester*
4-Chloro-α-cyanocinnamic Acid, *Et ester*
4-Chloroquinoline-3-carboxylic Acid, *Et ester*
2-Chloroquinoline-4-carboxylic Acid, *Et ester*
5-Chloroquinoline-4-carboxylic Acid, *Et ester*
6-Chloroquinoline-4-carboxylic Acid, *Et ester*
7-Chloroquinoline-4-carboxylic Acid, *Et ester*
8-Chloroquinoline-4-carboxylic Acid, *Et ester*

$C_{12}H_{10}ClNO_5S$
1-Nitro-2-naphthol-6-sulphonic Acid, *Et ether*, *Chloride*

$C_{12}H_{10}Cl_2N_2$
4,4′-Diamino-2,2′-dichlorobiphenyl
4,4′-Diamino-2,5-dichlorobiphenyl
4,4′-Diamino-3,3′-dichlorobiphenyl
2,2′-Dichlorohydrazobenzene
2,4-Dichlorohydrazobenzene
2,5-Dichlorohydrazobenzene
3,3′-Dichlorohydrazobenzene
4,4′-Dichlorohydrazobenzene

$C_{12}H_{10}Cl_2O_5S_2$
2-Naphthol-3,6-disulphonic Acid, *Et ether*, *Dichloride*
2-Naphthol-6,8-disulphonic Acid, *Et ether*, *Dichloride*

$C_{12}H_{10}Cl_4O_4$
Tetrachlorophthalic Acid, *Di-Et ester*

$C_{12}H_{10}FN$
2-Amino-2′-fluorobiphenyl
2-Amino-4-fluorobiphenyl
2-Amino-4′-fluorobiphenyl
4-Amino-2-fluorobiphenyl
4-Amino-2′-fluorobiphenyl
4-Amino-4′-fluorobiphenyl

$C_{12}H_{10}FeO_4$
1,1′-Ferrocenedicarboxylic Acid

$C_{12}H_{10}Hg$
Mercury diphenyl

$C_{12}H_{10}HgO_2$
Mercuri-1-naphthyl acetate

$C_{12}H_{10}HgO_3$
1-Mercuri-2-hydroxynaphthyl acetate

$C_{12}H_{10}IN$
2-Amino-2′-iodobiphenyl
4-Amino-4′-iodobiphenyl

$C_{12}H_{10}INO_3$
4-Iodo-2-nitro-1-naphthol, *Et ether*

$C_{12}H_{10}I_2N_2$
4,4′-Diamino-2,2′-di-iodobiphenyl †

$C_{12}H_{10}I_4O_4$
Tetraiodoterephthalic Acid, *Di-Et ester*

$C_{12}H_{10}N_2$
1-Aminocarbazole
2-Aminocarbazole
3-Aminocarbazole
4-Aminocarbazole
Azobenzene
2,4-Dimethylquinoline-3-carboxylic Acid, *Nitrile*
Harman
Isoharman
3-Methyl-β-carboline †
Naphth[1,2-*d*]imidazole, 1-N-*Me*

$C_{12}H_{10}N_2$ (di-ion)
Dipyrido[1,2-*a*:2′,1′-*c*]pyrazidinium †

$C_{12}H_{10}N_2O$
Azoxybenzene
2,3-Diaminodibenzofuran
2,8-Diaminodibenzofuran
3,7-Diaminodibenzofuran
3,8-Diaminodibenzofuran
Glucazidone
2-Hydroxyazobenzene
3-Hydroxyazobenzene
4-Hydroxyazobenzene
7-Methoxy-β-carboline
2-Quinolineacrylic Acid

$C_{12}H_{10}N_2O_2$
5-Amino-4-nitroacenaphthene
2-Amino-2′-nitrobiphenyl
2-Amino-3-nitrobiphenyl
2-Amino-4′-nitrobiphenyl
2-Amino-5-nitrobiphenyl
3-Amino-2′-nitrobiphenyl
3-Amino-3′-nitrobiphenyl
3-Amino-4-nitrobiphenyl
3-Amino-4′-nitrobiphenyl
4-Amino-2-nitrobiphenyl
4-Amino-2′-nitrobiphenyl
4-Amino-3-nitrobiphenyl
4-Amino-3′-nitrobiphenyl
4-Amino-4′-nitrobiphenyl
2,4-Dicyano-2-phenylbutyric Acid †
2,2′-Dihydroxyazobenzene
2,4-Dihydroxyazobenzene
2,4′-Dihydroxyazobenzene
2,5-Dihydroxyazobenzene
3,3′-Dihydroxyazobenzene
3,4-Dihydroxyazobenzene
3,4′-Dihydroxyazobenzene
4,4′-Dihydroxyazobenzene
Indole-2,6-dicarboxylic Acid, 6-*Nitrile*, 2-*Et ester*
Naphthalene-1,2-dicarboxylic Acid, *Diamide*
N-2-Naphthoylurea
2-Nitrodiphenylamine
3-Nitrodiphenylamine
4-Nitrodiphenylamine
Pyridoin

$C_{12}H_{10}N_2O_2S$
2-Amino-4′-nitrodiphenyl sulphide
3-Amino-4′-nitrodiphenyl sulphide
4-Amino-4′-nitrodiphenyl sulphide

$C_{12}H_{10}N_2O_3$
4′-Amino-2-hydroxy-4-nitrobiphenyl
2-Amino-2′-nitrodiphenyl Ether
2-Amino-4-nitrodiphenyl Ether
2-Amino-5-nitrodiphenyl Ether
4-Amino-3-nitrodiphenyl Ether
4-Amino-4′-nitrodiphenyl Ether
2,2′-Dihydroxyazoxybenzene

$C_{12}H_{10}N_2O_3$ (*continued*)
4-Hydroxy-2-nitro-*N*-phenylaniline
5-Hydroxy-2-nitro-*N*-phenylaniline
2-Hydroxy-5-nitropyridine, *Benzyl ether*
6-Hydroxy-8-nitroquinoline, *Allyl ether*
N-2-Hydroxyphenyl-2-nitroaniline
N-4-Hydroxyphenyl-2-nitroaniline
N-4-Hydroxyphenyl-4-nitroaniline

$C_{12}H_{10}N_2O_3S$
Azobenzene-4-sulphonic Acid

$C_{12}H_{10}N_2O_4$
2,6-Dimethyl-1,5-dinitronaphthalene†
o-Nitrobenzylidenemalonic Acid, *Mononitrile, Et ester*
m-Nitrobenzylidenemalonic Acid, *Mononitrile, Et ester*
p-Nitrobenzylidenemalonic Acid, *Mononitrile, Et ester*
5-Nitrofuran-2-carboxylic Acid, p-*Toluidide*
(1-Nitro-2-naphthoxy)acetic Acid, *Amide*

$C_{12}H_{10}N_2O_4S$
2-Nitrobenzenesulphonic Acid, *Anilide*
3-Nitrobenzenesulphonic Acid, *Anilide*
4-Nitrobenzenesulphonic Acid, *Anilide*
4′-Nitrobiphenyl-4-sulphonic Acid, *Amide*

$C_{12}H_{10}N_2O_5$
2,4-Dinitro-1-naphthol, *Et ether*
4,5-Dinitro-1-naphthol, *Et ether*
4,8-Dinitro-1-naphthol, *Et ether*
1,6-Dinitro-2-naphthol, *Et ether*
1,8-Dinitro-2-naphthol, *Et ether*
5,8-Dinitro-2-naphthol, *Et ether*
4-Hydroxy-6-nitroquinoline-3-carboxylic Acid, *Et ester*
4-Hydroxy-7-nitroquinoline-3-carboxylic Acid, *Et ester*
4-Hydroxy-8-nitroquinoline-3-carboxylic Acid, *Et ester*

$C_{12}H_{10}N_2O_5S$
4-Nitrodiphenylamine-2-sulphonic Acid
2-Nitrodiphenylamine-4-sulphonic Acid

$C_{12}H_{10}N_2O_6$
1,4-Dihydroxy-2,3-dinitronaphthalene, *Di-Me ether*
1,5-Dihydroxy-2,6-dinitronaphthalene, *Di-Me ether*
1,8-Dihydroxy-2,4-dinitronaphthalene, *Di-Me ether*
2,6-Dihydroxy-1,5-dinitronaphthalene, *Di-Me ether*
2,7-Dihydroxy-1,8-dinitronaphthalene, *Di-Me ether*

$C_{12}H_{10}N_2O_6S_2$
Azobenzene-3,3′-disulphonic Acid
Azobenzene-3,4′-disulphonic Acid
Azobenzene-4,4′-disulphonic Acid

$C_{12}H_{10}N_2O_7S_2$
Azoxybenzene-3,3′-disulphonic Acid

$C_{12}H_{10}N_4$
1,3-Diaminophenazine
2,3-Diaminophenazine
2,6-Diaminophenazine
3,6-Diaminophenazine

$C_{12}H_{10}N_4O_2$
4-Amino-4′-nitroazobenzene
4′-Amino-2-nitroazobenzene
4′-Amino-3-nitroazobenzene
2-Nitrodiazoaminobenzene
3-Nitrodiazoaminobenzene
4-Nitrodiazoaminobenzene

$C_{12}H_{10}N_4O_3$
2-Amino-4′-hydroxy-5-nitroazobenzene
5-Amino-2-hydroxy-4′-nitroazobenzene
3-Phenyluric Acid, 1-*Me*

$C_{12}H_{10}N_4O_4$
4,4′-Diamino-2,2′-dinitrobiphenyl
4,4′-Diamino-2,3′-dinitrobiphenyl
4,4′-Diamino-3,3′-dinitrobiphenyl
4,4′-Diamino-3,5′-dinitrobiphenyl
2,2′-Dinitrohydrazobenzene
2,4-Dinitrohydrazobenzene
4,4′-Dinitrohydrazobenzene

$C_{12}H_{10}O$
1-Acetonaphthone
2-Acetonaphthone
Diphenyl Ether
2-Hydroxybiphenyl
3-Hydroxybiphenyl
4-Hydroxybiphenyl
1-Hydroxyacenaphthene
5-Hydroxyacenaphthene
4-Methyl-1-naphthaldehyde
3-Methyl-2-naphthaldehyde†
1-Naphthylacetaldehyde
2-Naphthylacetaldehyde

$C_{12}H_{10}OS$
Diphenyl sulphoxide
p-Hydroxybenzenethiol, S-*Phenyl*

$C_{12}H_{10}OS_2$
5-(4-Hydroxy-1-butynyl)-2,2′-bithienyl†

$C_{12}H_{10}O_2$
2-Acetyl-1-naphthol
3-Acetyl-1-naphthol
4-Acetyl-1-naphthol
1-Acetyl-2-naphthol
3-Acetyl-2-naphthol
7-Acetyl-2-naphthol
8-Acetyl-2-naphthol
Azulene-1-carboxylic Acid, *Me ester*
Azulene-5-carboxylic Acid, *Me ester*
1,2-Dihydroxyacenaphthene
5,6-Dihydroxyacenaphthene
2,2′-Dihydroxybiphenyl
2,4′-Dihydroxybiphenyl
2,5-Dihydroxybiphenyl
3,3′-Dihydroxybiphenyl
4,4′-Dihydroxybiphenyl
3,6-Dimethyl-1,2-naphthoquinone
3,7-Dimethyl-1,2-naphthoquinone
6,7-Dimethyl-1,2-naphthoquinone
6,8-Dimethyl-1,2-naphthoquinone
2,3-Dimethyl-1,4-naphthoquinone
2,5-Dimethyl-1,4-naphthoquinone
2,6-Dimethyl-1,4-naphthoquinone
2,7-Dimethyl-1,4-naphthoquinone
2,8-Dimethyl-1,4-naphthoquinone
3,5-Dimethyl-1,4-naphthoquinone

$C_{12}H_{10}O_2$ *(continued)*
5,6-Dimethyl-1,4-naphthoquinone
5,7-Dimethyl-1,4-naphthoquinone
5,8-Dimethyl-1,4-naphthoquinone
6,7-Dimethyl-1,4-naphthoquinone
Drosphilin C, *Me ester* †
6-Ethyl-1,4-naphthaquinone †
2-Hydroxy-1-naphthaldehyde, *Me ether*
3-Hydroxy-1-naphthaldehyde, *Me ether*
4-Hydroxy-1-naphthaldehyde, *Me ether*
5-Hydroxy-1-naphthaldehyde, *Me ether*
1-Hydroxy-2-naphthaldehyde, *Me ether*
3-Hydroxy-2-naphthaldehyde, *Me ether*
6-Hydroxy-2-naphthaldehyde, *Me ether*
2-Hydroxyphenyl phenyl Ether
3-Hydroxyphenyl phenyl Ether
4-Hydroxyphenyl phenyl Ether
3-Methoxybenzocyclohepten-2-one †
2-Methylazulene-5-carboxylic Acid
2-Methylazulene-6-carboxylic Acid
2-Methyl-1-naphthoic Acid
4-Methyl-1-naphthoic Acid
6-Methyl-1-naphthoic Acid
7-Methyl-1-naphthoic Acid
8-Methyl-1-naphthoic Acid
1-Methyl-2-naphthoic Acid
4-Methyl-2-naphthoic Acid
6-Methyl-2-naphthoic Acid
4-Methyl-1,2-naphthoquinone, *Me ether*
2-Naphthoic Acid, *Me ester*
α-Naphthoxyacetaldehyde
β-Naphthoxyacetaldehyde
1-Naphthylacetic Acid
2-Naphthylacetic Acid
Odyssin
Podophyllomerol

$C_{12}H_{10}O_2S$
Di-*o*-hydroxyphenyl sulphide
Di-*m*-hydroxyphenyl sulphide
Di-*p*-hydroxyphenyl sulphide
Diphenyl sulphone

$C_{12}H_{10}O_2S_2$
Benzenethiolsulphonic Acid, *Phenyl ester*
Di-2-hydroxyphenyl disulphide
Di-3-hydroxyphenyl disulphide
Di-4-hydroxyphenyl disulphide
Thiolbenzenesulphonic Acid, *Phenyl ester*

$C_{12}H_{10}O_3$
3-Acetyl-4-methylcoumarin
3-Acetyl-5-methylcoumarin
3-Acetyl-6-methylcoumarin
3-Acetyl-7-methylcoumarin
3-Acetyl-8-methylcoumarin
2,2′-Dihydroxydiphenyl Ether
2,3-Dihydroxydiphenyl Ether
2,3′-Dihydroxydiphenyl Ether
2,4′-Dihydroxydiphenyl Ether
3,3′-Dihydroxydiphenyl Ether
3,4-Dihydroxydiphenyl Ether
3,4′-Dihydroxydiphenyl Ether
4,4′-Dihydroxydiphenyl Ether
1,4-Dihydroxy-2-naphthaldehyde, 4-*Me ether*
1,5-Dihydroxy-2-naphthaldehyde, 5-*Me ether*
5-Methyl-2-phenylfuran-3-carboxylic Acid
α-Naphthoxyacetic Acid
β-Naphthoxyacetic Acid
2,1-Naphthylglycollic Acid
2,2-Naphthylglycollic Acid
Pentenedial, *Benzoyl*
1,2,3,4-Tetrahydronaphthalene-1,2-dicarboxylic Acid, *Anhydride*
1,2,3,4-Tetrahydronaphthalene-2,3-dicarboxylic Acid, *Anhydride*

$C_{12}H_{10}O_3S$
Acenaphthene-3-sulphonic Acid
Benzenesulphonic Acid, *Phenyl ester*
Phenol, *Sulphite*

$C_{12}H_{10}O_3S_2$
Benzenesulphinic Acid, *Anhydride*

$C_{12}H_{10}O_4$
6-Acetoxymethyl-2-formylbenzofuran †
Coumarin-3-carboxylic Acid, *Et ester*
2,8-Decadiene-4,6-diynedioic Acid, *Di-Me ester*
2,4,6-Decatriynedioic Acid, *Di-Me ester*
1,4-Dihydroxy-2-naphthoic Acid, 4-*Me ether*
1,5-Dihydroxy-2-naphthoic Acid, 5-*Me ether*
1,5-Dihydroxy-2-naphthoic Acid, *Me ester*
3,4-Dihydroxy-2-naphthoic Acid, *Me ester*
2,3-Dihydroxy-1,4-naphthoquinone, *Di-Me ether*
1,3-Dioxoindane-2-carboxylic Acid, *Et ester*
Eleutherolic Acid
Isochavicic Acid
6-Methylcoumarin-4-acetic Acid
7-Methylcoumarin-4-acetic Acid
8-Methylcoumarin-4-acetic Acid
5-(3,4-Methylenedioxyphenyl)-2,4-pentadienoic Acid
4-Phenyl-1,3-butadiene-1,1-dicarboxylic Acid
Phthaloylacetic Acid, *Et ester*
Phyllomeronic Acid
Quinhydrone
2,2′,4,4′-Tetrahydroxybiphenyl
2,2′,4,5′-Tetrahydroxybiphenyl
2,2′,5,5′-Tetrahydroxybiphenyl
2,2′,6,6′-Tetrahydroxybiphenyl
2,3′,4,4′-Tetrahydroxybiphenyl
3,3′,4,4′-Tetrahydroxybiphenyl
3,3′,5,5′-Tetrahydroxybiphenyl

$C_{12}H_{10}O_4S$
Camphor-10(or 6-)-sulphonic Acid, *Et ester*
Di-(2,5-dihydroxyphenyl)sulphide
2,5-Dihydroxyphenyl phenyl sulphone
Di-*o*-hydroxyphenyl sulphone
Di-*m*-hydroxyphenyl sulphone
Di-*p*-hydroxyphenyl sulphone

$C_{12}H_{10}O_5$
4-Acetyl-6,8-dihydroxy-5-methyl-2-benzopyran-1-one †
Anhydrobrazilic Acid
2-Ethyl-3,5,8-trihydroxy-1,4-naphthaquinone
5-Hydroxy-2,7-dimethoxy-1,4-naphthaquinone †
Ostholic Acid
Phloroglucide
β-Sorigenin, *Mono-Me ether*
2,5,7-Trihydroxy-1,4-naphthoquinone, 5,7-*Di-Me ether*

$C_{12}H_{10}O_5S_2$
Benzenesulphonic Acid, *Anhydride*

$C_{12}H_{10}O_6$
5,8-Dihydroxy-2,7-dimethoxy-1,4-naphthoquinone†
2,2′,3,3′,4,4′-Hexahydroxybiphenyl
2,2′,4,4′,5,5′-Hexahydroxybiphenyl
2,2′,4,4′,6,6′-Hexahydroxybiphenyl
2,3,3′,4,4′,5′-Hexahydroxybiphenyl
3,3′,4,4′,5,5′-Hexahydroxybiphenyl
2,3,5,8-Tetrahydroxy-1,4-naphthoquinone, 5,8-*Di-Me ether*

$C_{12}H_{10}O_6S_2$
Acenaphthene-3,8-disulphonic Acid
Acenaphthene-5,6-disulphonic Acid

$C_{12}H_{10}O_7$
Echinochrome A
Spinochrome P

$C_{12}H_{10}O_8$
Benzene-1,2,3,4-tetracarboxylic Acid, 1,4-*Di-Me ester*
Benzene-1,2,4,5-tetracarboxylic Acid, 1,4-*Di-Me ester*

$C_{12}H_{10}S$
Diphenyl sulphide
4-Mercaptobiphenyl

$C_{12}H_{10}S_2$
Diphenyl disulphide

$C_{12}H_{10}Te_2$
Diphenyl ditelluride

$C_{12}H_{10}Zn$
Zinc diphenyl

$C_{12}H_{11}As$
Diphenylarsine

$C_{12}H_{11}AsO_2$
Diphenylarsinic Acid

$C_{12}H_{11}BO$
Diphenylboric Acid

$C_{12}H_{11}Br$
1-α-Bromoethylnaphthalene
1-β-Bromoethylnaphthalene
2-β-Bromoethylnaphthalene

$C_{12}H_{11}BrN_2$
2-Amino-5-bromodiphenylamine
4-Amino-4′-bromodiphenylamine

$C_{12}H_{11}BrO$
6-Bromo-2-hydroxy-1-methylnaphthalene, *Me ether*
4-Bromo-1-naphthol, *Et ether*
1-Bromo-2-naphthol, *Et ether*
3-Bromo-2-naphthol, *Et ether*
6-Bromo-2-naphthol, *Et ether*

$C_{12}H_{11}BrO_6$
6-Bromobenzene-1,2,4-tricarboxylic Acid, *Tri-Me ester*

$C_{12}H_{11}Cl$
1-β-Chloroethylnaphthalene

$C_{12}H_{11}ClN_2$
2-Amino-4-chlorodiphenylamine
2-Amino-4′-chlorodiphenylamine
2-Amino-5-chlorodiphenylamine
4-Amino-4′-chlorodiphenylamine
p-Chlorohydrazobenzene

$C_{12}H_{11}ClN_2O$
2-Chloroquinoline-4-carboxylic Acid, *Di-Me amide*

$C_{12}H_{11}ClO$
2-Chloroethyl-1-naphthyl Ether
2-Chloroethyl-2-naphthyl Ether
1-Chloro-2-naphthol, *Et ether*

$C_{12}H_{11}ClO_2$
α-Chlorocinnamic Acid, *Allyl ester*

$C_{12}H_{11}ClO_3S$
6-Chloronaphthalene-1-sulphonic Acid, *Et ester*
8-Chloronaphthalene-1-sulphonic Acid, *Et ester*
1-Chloronaphthalene-2-sulphonic Acid, *Et ester*
4-Chloronaphthalene-2-sulphonic Acid, *Et ester*
5-Chloronaphthalene-2-sulphonic Acid, *Et ester*
6-Chloronaphthalene-2-sulphonic Acid, *Et ester*
7-Chloronaphthalene-2-sulphonic Acid, *Et ester*
8-Chloronaphthalene-2-sulphonic Acid, *Et ester*
1-Naphthol-4-sulphonic Acid, *Et ether*, *Chloride*
2-Naphthol-1-sulphonic Acid, *Et ether*, *Chloride*
2-Naphthol-6-sulphonic Acid, *Et ether*, *Chloride*
2-Naphthol-7-sulphonic Acid, *Et ether*, *Chloride*
2-Naphthol-8-sulphonic Acid, *Et ether*, *Chloride*

$C_{12}H_{11}FO_3S$
4-Fluoronaphthalene-1-sulphonic Acid, *Et ester*
5-Fluoronaphthalene-1-sulphonic Acid, *Et ester*

$C_{12}H_{11}IN_2$
4,4′-Diamino-2-iodobiphenyl

$C_{12}H_{11}N$
1-Aminoacenaphthene
3-Aminoacenaphthene
4-Aminoacenaphthene
5-Aminoacenaphthene
o-Aminobiphenyl
m-Aminobiphenyl
p-Aminobiphenyl
2-Benzylpyridine
3-Benzylpyridine
4-Benzylpyridine
Diphenylamine
2,2′-Indenylpropionic Acid, *Nitrile*
2,3′-Indenylpropionic Acid, *Nitrile*
2-Methyl-4-phenylpyridine
2-Methyl-6-phenylpyridine

$C_{12}H_{11}NO$

3-Acetyl-2-methylquinoline
6-Acetyl-2-methylquinoline
2-Amino-2′-hydroxybiphenyl
2-Amino-4′-hydroxybiphenyl
2-Amino-5-hydroxybiphenyl
3-Amino-2-hydroxybiphenyl
4-Amino-3-hydroxybiphenyl
4′-Amino-2-hydroxybiphenyl
4-Amino-4′-hydroxybiphenyl
5-Amino-2-hydroxybiphenyl
N-Diphenylhydroxylamine
2-Hydroxyindene-3-carboxylic Acid, *Nitrile, Et ether*
4-Hydroxy-2-methylpyridine, *Phenyl ether*
2-Hydroxy-*N*-phenylaniline
3-Hydroxy-*N*-phenylaniline
4-Hydroxy-*N*-phenylaniline
4-Hydroxypyridine, *Benzyl ether*
2-Methylazulene-6-carboxylic Acid, *Amide*
2-Methyl-1-naphthoic Acid, *Amide*
4-Methyl-1-naphthoic Acid, *Amide*
7-Methyl-1-naphthoic Acid, *Amide*
1-Naphthylacetic Acid, *Amide*
2-Naphthylacetic Acid, *Amide*
o-Phenoxyaniline
m-Phenoxyaniline
p-Phenoxyaniline
α-2-Pyridylbenzyl Alcohol
α-4-Pyridylbenzyl Alcohol

$C_{12}H_{11}NO_2$

3-Amino-2-naphthoic Acid, *Me ester*
o-Carboxycinnamic Acid, *Et ester*
α-Cyanocinnamic Acid, *Et ester*
1,3-Diacetylindolizine
4,4′-Dihydroxydiphenylamine
2,3-Dimethyl-1-nitronaphthalene
2,6-Dimethyl-1-nitronaphthalene
3,7-Dimethyl-1-nitronaphthalene
3,8-Dimethylquinoline-2-carboxylic Acid
4,6-Dimethylquinoline-2-carboxylic Acid
6,8-Dimethylquinoline 2-carboxylic Acid
2,4-Dimethylquinoline-3-carboxylic Acid
2,3-Dimethylquinoline-4-carboxylic Acid
2,6-Dimethylquinoline-4-carboxylic Acid
2,8-Dimethylquinoline-4-carboxylic Acid
2,3-Dimethylquinoline-6-carboxylic Acid
2,8-Dimethylquinoline-6-carboxylic Acid
2,3-Dimethylquinoline-8-carboxylic Acid
2,4-Dimethylquinoline-8-carboxylic Acid
4-Hydroxy-1-naphthoic Acid, *Me ether*, *Amide*
3-Hydroxy-2-naphthoic Acid, *Me ether*, *Amide*
1-(2-Methoxycarbonylphenyl)pyrrole†
3-Methylfuran-2-carboxylic Acid, *Anilide*
2-Methylquinoline-3-carboxylic Acid, *Me ester*
2-Methylquinoline-4-carboxylic Acid, *Me ester*
3-Methylquinoline-4-carboxylic Acid, *Me ester*
α-Naphthoxyacetic Acid, *Amide*
β-Naphthoxyacetic Acid, *Amide*
2,1-Naphthylglycollic Acid, *Amide*
2,2-Naphthylglycollic Acid, *Amide*
2-Nitroso-1-naphthol, *Et ether*
1-Nitroso-2-naphthol, *Et ether*
1-Phenylpyrrole-2-carboxylic Acid, *Me ester*
2-Quinolineacetic Acid, *Me ester*
8-Quinolineacetic Acid, *Me ester*
Quinoline-2-carboxylic Acid, *Et ester*
Quinoline-3-carboxylic Acid, *Et ester*
Quinoline-4-carboxylic Acid, *Et ester*
Quinoline-5-carboxylic Acid, *Et ester*
Quinoline-6-carboxylic Acid, *Et ester*
Quinoline-8-carboxylic Acid, *Et ester*
2-Quinolinepropionic Acid
4-Quinolinepropionic Acid

$C_{12}H_{11}NO_2S$

Acenaphthene-3-sulphonic Acid, *Amide*
1,8-Naphthosultam, N-*Et*

$C_{12}H_{11}NO_3$

4-Amino-1-hydroxy-2-naphthoic Acid, *Me ether*
4-Amino-3-hydroxy-2-naphthoic Acid, *Me ester*
7-Amino-3-hydroxy-2-naphthoic Acid, *Me ether*
Benzoylmalonic Acid, *Et ester-nitrile*
2-Benzyl-4-hydroxymethylene-5-oxazolone, *Me ether*
1-Hydroxyisoquinoline-3-carboxylic Acid, N-*Et*
1-Hydroxyisoquinoline-4-carboxylic Acid, *Et ester*
2-Hydroxy-3-methylquinoline-4-carboxylic Acid, *Me ester*
3-Hydroxy-2-methylquinoline-4-carboxylic Acid, *Me ester*
2-Hydroxyquinoline-3-carboxylic Acid, *Et ether*
4-Hydroxyquinoline-3-carboxylic Acid, *Et ester*
2-Hydroxyquinoline-4-carboxylic Acid, *Et ester*
2-Hydroxyquinoline-4-carboxylic Acid, *Et ether*
6-Hydroxyquinoline-4-carboxylic Acid, *Et ester*
6-Hydroxyquinoline-4-carboxylic Acid, *Et ether*
8-Hydroxyquinoline-5-carboxylic Acid, *Et ester*
8-Hydroxyquinoline-6-carboxylic Acid, *Et ester*
2-3′-Indolylglyoxylic Acid, *Et ester*
6-Methoxyquinoline-4-carboxylic Acid, *Me ester*
2-Nitro-1-naphthol, *Et ether*
4-Nitro-1-naphthol, *Et ether*
1-Nitro-2-naphthol, *Et ether*
5-Nitro-2-naphthol, *Et ether*
6-Nitro-2-naphthol, *Et ether*
8-Nitro-2-naphthol, *Et ether*
6-*o*-Nitrophenyl-3,5-hexadien-2-one
6-*p*-Nitrophenyl-3,5-hexadien-2-one

$C_{12}H_{11}NO_4$

Casimiroin†
6,7-Dihydroxyquinoline-2-carboxylic Acid, *Di-Me ether*
2,4-Dihydroxyquinoline-3-carboxylic Acid, *Et ester*
2,4-Dihydroxyquinoline-6-carboxylic Acid, *Et ester*
2-Hydroxy-4,8-dimethoxyquinoline-3-aldehyde

$C_{12}H_{11}NO_4$ (*continued*)
Indole-2,3-dicarboxylic Acid, *Di-Me ester*
Indole-2,5-dicarboxylic Acid, 2-*Et ester*
Phthalimidoacetic Acid, *Et ester*
Tricarballylic Acid, αβ-*Anil*

$C_{12}H_{11}NO_5$
2-Hydroxy-4,8-dimethoxyquinoline-3-carboxylic Acid

$C_{12}H_{11}NO_5S$
4-Nitronaphthalene-1-sulphonic Acid, *Et ester*
5-Nitronaphthalene-1-sulphonic Acid, *Et ester*
8-Nitronaphthalene-1-sulphonic Acid, *Et ester*
4-Nitronaphthalene-2-sulphonic Acid, *Et ester*
5-Nitronaphthalene-2-sulphonic Acid, *Et ester*
8-Nitronaphthalene-2-sulphonic Acid, *Et ester*

$C_{12}H_{11}NO_6$
3-(4,5-Methylenedioxy-2-nitrophenyl)acrylic Acid, *Et ester*
o-Nitrobenzylidenemalonic Acid, *Di-Me ester*
m-Nitrobenzylidenemalonic Acid, *Di-Me ester*
p-Nitrobenzylidenemalonic Acid, *Di-Me ester*

$C_{12}H_{11}NO_6S$
2-Nitro-1-naphthol-4-sulphonic Acid, *Et ether*

$C_{12}H_{11}NS$
2-Aminodiphenyl sulphide
4-Aminodiphenyl sulphide

$C_{12}H_{11}N_3$
o-Aminoazobenzene
m-Aminoazobenzene
p-Aminoazobenzene
2,7-Diaminocarbazole
3,4(or 2,3)-Diaminocarbazole
3,6-Diaminocarbazole
Diazoaminobenzene

$C_{12}H_{11}N_3O$
o-Aminoazoxybenzene
p-Aminoazoxybenzene
2-Amino-4-hydroxyazobenzene
3-Amino-4′-hydroxyazobenzene
4-Amino-2-hydroxyazobenzene
4-Amino-4′-hydroxyazobenzene

$C_{12}H_{11}N_3O_2$
2-Amino-2′-nitrodiphenylamine
2-Amino-4-nitrodiphenylamine
2-Amino-4′-nitrodiphenylamine
2-Amino-6-nitrodiphenylamine
3-Amino-2′-nitrodiphenylamine
3-Amino-4′-nitrodiphenylamine
4-Amino-2-nitrodiphenylamine
4-Amino-2′-nitrodiphenylamine
4-Amino-4′-nitrodiphenylamine
Caerulomycin
1-Naphthylbiuret
2-Naphthylbiuret
2-Nitrobenzidine
3-Nitrobenzidine
m-Nitrohydrazobenzene
p-Nitrohydrazobenzene

$C_{12}H_{11}N_3O_2S$
Azobenzene-3-sulphonic Acid, *Amide*

$C_{12}H_{11}N_3O_3$
2-Amino-4′-hydroxy-4-nitrodiphenylamine

$C_{12}H_{11}N_3O_4$
2,4-Dinitro-1-naphthylamine, N-*Et*

$C_{12}H_{11}N_3O_4S$
4-Nitrodiphenylamine-2-sulphonic Acid, *Amide*
2-Nitrodiphenylamine-4-sulphonic Acid, *Amide*

$C_{12}H_{11}N_5O_2$
2,4-Diamino-3′-nitroazobenzene

$C_{12}H_{11}O_2P$
Diphenylphosphinic Acid

$C_{12}H_{11}O_3P$
Phenol, *Hydrogen phosphite*
Phenylphosphonic Acid, *Phenyl ester*

$C_{12}H_{11}O_4P$
Phenol, *Hydrogen phosphate*

$C_{12}H_{11}P$
Diphenylphosphine

$C_{12}H_{12}$
Agropyrene
9,10-Dihydro-9,10-ethanonaphthalene†
1,2-Dimethylazulene
1,3-Dimethylazulene
1,4-Dimethylazulene
1,5-Dimethylazulene
1,7-Dimethylazulene
2,6-Dimethylazulene
4,5-Dimethylazulene
4,6-Dimethylazulene
4,7-Dimethylazulene
4,8-Dimethylazulene
1,2-Dimethylnaphthalene
1,3-Dimethylnaphthalene
1,4-Dimethylnaphthalene
1,5-Dimethylnaphthalene
1,6-Dimethylnaphthalene
1,7-Dimethylnaphthalene
1,8-Dimethylnaphthalene
2,3-Dimethylnaphthalene
2,6-Dimethylnaphthalene
2,7-Dimethylnaphthalene
2-Ethylazulene
4-Ethylazulene
5-Ethylazulene
1-Ethylnaphthalene
2-Ethylnaphthalene

$C_{12}H_{12}BrNO_2$
Phthalimide, N-4-*Bromobutyl*

$C_{12}H_{12}Br_2N_2$
9,10-Dihydro-8α, 10α-diazoniaphenanthrene dibromide

$C_{12}H_{12}Br_2O_4$
4,5-Dibromophthalic Acid, *Di-Et ester*
2,5-Dibromoterephthalic Acid, *Di-Et ester*

$C_{12}H_{12}ClNO$
2-Chloro-6-hydroxy-4-methylquinoline, *Et ether*

$C_{12}H_{12}ClNO_2S$
5-Dimethylamino-1-naphthalenesulphonyl chloride†

$C_{12}H_{12}Cl_2O_4$
5,7-Dichloro-3,4-dihydro-8-hydroxy-6-methoxy-3-methylisocoumarin, *Me ether*†
3,5-Dichlorophthalic Acid, *Di-Et ester*
3,6-Dichlorophthalic Acid, *Di-Et ester*

$C_{12}H_{12}FeO_2$
Ferrocenecarboxylic Acid, *Me ester*

$C_{12}H_{12}HgN_2$
p-Mercuri-dianiline

$C_{12}H_{12}INO_2$
Phthalimide, N-4-*Iodobutyl*

$C_{12}H_{12}N_2$
o-Aminodiphenylamine
m-Aminodiphenylamine
p-Aminodiphenylamine
Benzidine
2-Benzylaminopyridine †
3,7-Diaminoacenaphthene
4,5-Diaminoacenaphthene
5,6-Diaminoacenaphthene
2,2′-Diaminobiphenyl
2,3′-Diaminobiphenyl
2,4′-Diaminobiphenyl
3,3′-Diaminobiphenyl
3,4-Diaminobiphenyl
3,3′-Dimethyl-2,2′-bipyridyl
4,4′-Dimethyl-2,2′-bipyridyl
6,6′-Dimethyl-2,2′-bipyridyl
2,6-Dimethyl-2′,4-bipyridyl
2,2′-Dimethyl-4,4′-bipyridyl
2,6-Dimethyl-4,4′-bipyridyl
3,3′-Dimethyl-4,4′-bipyridyl
1,1-Diphenylhydrazine
Hydrazobenzene
Withasomnine †

$C_{12}H_{12}N_2O$
2-Amino-4′-hydroxydiphenylamine
4-Amino-4′-hydroxydiphenylamine
2-Aminophenyl-4-aminophenyl Ether
2,4′-Diamino-5-hydroxybiphenyl
4,4′-Diamino-2-hydroxybiphenyl
4,4′-Diamino-3-hydroxybiphenyl
Di-(2-aminophenyl) Ether
Di-(4-aminophenyl) Ether
2,4-Dimethylquinoline-3-carboxylic Acid, *Amide*
Harmalol
3-Hydroxyhydrazobenzene
4-Hydroxyhydrazobenzene
N^1-Methyl-N^2-1-naphthylurea
N^1-Methyl-N^2-2-naphthylurea
2-Nitroso-1-naphthylamine, N-*Et*
4-Nitroso-1-naphthylamine, N-*Et*
1-Nitroso-2-naphthylamine, N-*Et*
2-Quinolinepropionic Acid, *Amide*
6,7,8,9-Tetrahydropyrido[2,1-*b*]quinazolin-11-one †

$C_{12}H_{12}N_2OS$
Di-(4-aminophenyl)sulphoxide

$C_{12}H_{12}N_2O_2$
Adrenodiamine
2-Aminoquinoline-3-carboxylic Acid, *Et ester*
2,2′-Diamino-5,5′-dihydroxybiphenyl
3,3′-Diamino-2,2′-dihydroxybiphenyl
3,3′-Diamino-4,4′-dihydroxybiphenyl
4,4′-Diamino-2,2′-dihydroxybiphenyl
4,4′-Diamino-3,3′-dihydroxybiphenyl
5,5′-Diamino-2,2′-dihydroxybiphenyl
1,3-Diamino-2-naphthoic Acid, *Me ester*
4-Formyl-2,3-dimethyl-1-phenyl-5-pyrazolone
2,2′-Hydrazodiphenol
5-Methyl-3-phenylpyrazole-1-carboxylic Acid, *Me ester*
1-Methyl-4-phenylpyrazole-3-carboxylic Acid, *Me ester*
5-Methyl-1-phenylpyrazole-3-carboxylic Acid, *Me ester*
5-Methyl-1-phenylpyrazole-4-carboxylic Acid, *Me ester*
1-Methyl-4-phenylpyrazole-5-carboxylic Acid, *Me ester*
3-Methyl-1-phenylpyrazole-5-carboxylic Acid, *Me ester*
2-Nitro-1-naphthylamine, N-*Et*
4-Nitro-1-naphthylamine, N-*Et*
1-Nitro-2-naphthylamine, N-*Et*
3-Phenylpyrazole-1-carboxylic Acid, *Et ester*
4-Phenylpyrazole-1-carboxylic Acid, *Et ester*
5-Phenylpyrazole-1-carboxylic Acid, *Et ester*
4-Phenylpyrazole-3-carboxylic Acid, *Et ester*
5(3)-Phenylpyrazole-3(5)-carboxylic Acid, *Et ester*
5-Phenylpyrazole-4-carboxylic Acid, *Et ester*
1-Phenylpyrazole-4-carboxylic Acid, *Et ester*

$C_{12}H_{12}N_2O_2S$
2-Aminophenyl 3-aminophenyl sulphone
2-Aminophenyl 4-aminophenyl sulphone
3-Aminophenyl 4-aminophenyl sulphone
2,4-Diaminophenyl phenyl sulphone
Di-(2-aminophenyl)sulphone
Di-(3-aminophenyl)sulphone
Di-(4-aminophenyl)sulphone

$C_{12}H_{12}N_2O_3$
7-Amino-8-nitro-2-naphthol, N,O-*Di-Me*
2-Aminoquinoline-3-carboxylic Acid, *Et ester*
4-Hydroxy-3-nitroquinoline, *Propyl ether*
6-Hydroxy-8-nitroquinoline, *Propyl ether*
Luminal
1-Phenyl-4-pyrazolone-3-carboxylic Acid, *Et ester*
1-Phenyl-5-pyrazolone-3-carboxylic Acid, *Et ester*
1-Phenyl-5-pyrazolone-3-carboxylic Acid, *Et ether*
1-Phenyl-5-pyrazolone-4-carboxylic Acid, *Et ester*

$C_{12}H_{12}N_2O_4S$
8-Nitronaphthalene-1-sulphonic Acid, *Di-Me amide*
8-Nitronaphthalene-1-sulphonic Acid, *Et-amide*

$C_{12}H_{12}N_2O_4S_2$
Acenaphthene-3,8-disulphonic Acid, *Diamide*
Acenaphthene-5,6-disulphonic Acid, *Diamide*

$C_{12}H_{12}N_2O_5S$
1-Nitro-2-naphthol-6-sulphonic Acid, *Et ether, Amide*

$C_{12}H_{12}N_2O_6S_2$
Benzidine-2,2′-disulphonic Acid
Benzidine-3,3′-disulphonic Acid

$C_{12}H_{12}N_2O_7$
2-(2,4-Dinitrophenyl)acetoacetic Acid, *Et ester* †

$C_{12}H_{12}N_2O_8$
4,6-Dinitroisophthalic Acid, *Di-Et ester*
3,4-Dinitrophthalic Acid, *Di-Et ester*
3,5-Dinitrophthalic Acid, *Di-Et ester*
4,5-Dinitrophthalic Acid, *Di-Et ester*
2,5-Dinitroterephthalic Acid, *Di-Et ester*

$C_{12}H_{12}N_2S$
2-Aminophenyl 4-aminophenyl sulphide
Di-(2-aminophenyl)sulphide
Di-(4-aminophenyl)sulphide
N^1-Methyl-N^2-1-naphthylthiourea
N^1-Methyl-N^1-2-naphthylthiourea
N^1-Methyl-N^2-2-naphthylthiourea

$C_{12}H_{12}N_2S_2$
Di-(2-aminophenyl)disulphide
Di-(3-aminophenyl)disulphide
Di-(4-aminophenyl)disulphide

$C_{12}H_{12}N_4$
2,2′-Diaminoazobenzene
2,4-Diaminoazobenzene
3,3′-Diaminoazobenzene
4,4′-Diaminoazobenzene

$C_{12}H_{12}N_4O$
2,2′-Diaminoazoxybenzene
3,3′-Diaminoazoxybenzene
4,4′-Diaminoazoxybenzene

$C_{12}H_{12}N_4O_2$
Lumichrome

$C_{12}H_{12}N_4O_4S_2$
Azobenzene-3,3′-disulphonic Acid, *Diamide*

$C_{12}H_{12}N_4O_5S_2$
Azoxybenzene-3,3′-disulphonic Acid, *Diamide*

$C_{12}H_{12}O$
2,3-Dimethyl-1-naphthol
2,4-Dimethyl-1-naphthol
2,6-Dimethyl-1-naphthol
2,7-Dimethyl-1-naphthol
3,4-Dimethyl-1-naphthol
3,7-Dimethyl-1-naphthol
4,6-Dimethyl-1-naphthol
4,7-Dimethyl-1-naphthol
5,7-Dimethyl-1-naphthol
5,8-Dimethyl-1-naphthol
6,7-Dimethyl-1-naphthol
1,3-Dimethyl-2-naphthol
1,4-Dimethyl-2-naphthol
1,5-Dimethyl-2-naphthol
1,6-Dimethyl-2-naphthol
3,4-Dimethyl-2-naphthol
3,6-Dimethyl-2-naphthol
3,7-Dimethyl-2-naphthol
3,8-Dimethyl-2-naphthol
4,8-Dimethyl-2-naphthol
5,7-(or 6,8) Dimethyl-2-naphthol
6,7-Dimethyl-2-naphthol
Ethyl 1-naphthyl Ether
Ethyl 2-naphthyl Ether
2-Hydroxyazulene, *Et ether*†
3-Isopropyl-2-cyclopenten-1-one
4-Methyl-1-naphthol, *Me ether*
1-Methyl-2-naphthol, *Me ether*
5-Methyl-2-naphthol, *Me ether*
6-Methyl-2-naphthol, *Me ether*
8-Methyl-2-naphthol, *Me ether*
2-Methyl-1-naphthylmethanol
4-Methyl-1-naphthylmethanol
1-α-Naphthylethanol
1-β-Naphthylethanol
2-α-Naphthylethanol
2-β-Naphthylethanol
6-Phenyl-3,5-hexadiene-2-one

$C_{12}H_{12}O_2$
2-Allyl-5-methylbenzofuran-3-one
3-Butylidene phthalide†
ββ-Diacetylstyrene
2,8-Decadiene-4,6-triyn-1-ol, *Ac*
1,2-Dihydroxynaphthalene, *Di-Me ether*
1,4-Dihydroxynaphthalene, *Di-Me ether*
1,4-Dihydroxynaphthalene, *Et ether*
1,5-Dihydroxynaphthalene, *Di-Me ether*
1,6-Dihydroxynaphthalene, *Di-Me ether*
1,8-Dihydroxynaphthalene, *Di-Me ether*
2,3-Dihydroxynaphthalene, *Di-Me ether*
2,6-Dihydroxynaphthalene, *Di-Me ether*
2,7-Dihydroxynaphthalene, *Di-Me ether*
2-Ethyl-6-methylchromone
3-Ethyl-2-methylchromone
2-Hydroxyethyl 1-naphthyl Ether
2-Hydroxyethyl 2-naphthyl Ether
2-(1-Hydroxy-2-indanyl)propionic Acid, *Lactone*
1-Hydroxymethyl-2-naphthol, 2-*Me ether*
Indene-2-carboxylic Acid, *Et ester*
Indene-3-carboxylic Acid, *Et ester*
2,2′-Indenylpropionic Acid
2,3′-Indenylpropionic Acid
3,1′(or 3′)-Indenylpropionic Acid
3-Isobutylidene phthalide
3-Methylindene-2-carboxylic Acid, *Me ester*
1-Methylindene-3-carboxylic Acid, *Me ester*
6-Phenyl-3,5-hexadienoic Acid
5-Phenyl-2,4-pentadienoic Acid, *Me ester*
Phyllomerol, *Me ether*

$C_{12}H_{12}O_3$
Croweacin
αα-Diacetylacetophenone
Flemingin
3-Hydroxyindene-2-carboxylic Acid, *Et ester*
2-Hydroxyindene-3-carboxylic Acid, *Et ester*
2-Hydroxy-1-naphthoic Acid, *Me ester*
2-Hydroxy-1-naphthoic Acid, *Me ether*
3-Hydroxy-1-naphthoic Acid, *Me ester*
3-Hydroxy-1-naphthoic Acid, *Me ether*
4-Hydroxy-1-naphthoic Acid, *Me ester*
4-Hydroxy-1-naphthoic Acid, *Me ether*
5-Hydroxy-1-naphthoic Acid, *Me ester*
5-Hydroxy-1-naphthoic Acid, *Me ether*
6-Hydroxy-1-naphthoic Acid, *Me ester*
6-Hydroxy-1-naphthoic Acid, *Me ether*
7-Hydroxy-1-naphthoic Acid, *Me ester*
7-Hydroxy-1-naphthoic Acid, *Me ether*
8-Hydroxy-1-naphthoic Acid, *Me ether*
1-Hydroxy-2-naphthoic Acid, *Me ester*
3-Hydroxy-2-naphthoic Acid, *Me ester*
3-Hydroxy-2-naphthoic Acid, *Me ether*
4-Hydroxy-2-naphthoic Acid, *Me ether*
5-Hydroxy-2-naphthoic Acid, *Me ether*
6-Hydroxy-2-naphthoic Acid, *Me ether*

$C_{12}H_{12}O_3$ (*continued*)
7-Hydroxy-2-naphthoic Acid, *Me ether*
8-Hydroxy-2-naphthoic Acid, *Me ester*
8-Hydroxy-2-naphthoic Acid, *Me ether*
2-Hydroxyphenylpropiolic Acid, *Me ether, Et ester*
3-Hydroxyphenylpropiolic Acid, *Me ether, Et ester*
3-Methylbenzofuran-2-carboxylic Acid, *Et ester*
6-Methylbenzofuran-2-carboxylic Acid, *Et ester*
2-Methyl-3-phenylglutaric Acid, *Anhydride*
Nemotinic Acid★, *Me ester*†
Odyssic Acid
4-Phenylbutane-1,2-dicarboxylic Acid, *Anhydride*
1,2,3-Triacetylbenzene†
1,2,4-Triacetylbenzene†
1,3,5-Triacetylbenzene

$C_{12}H_{12}O_3S$
Naphthalene-1-sulphonic Acid, *Et ester*
Naphthalene-2-sulphonic Acid, *Et ester*

$C_{12}H_{12}O_4$
Benzoylpyruvic Acid, *Et ester*
Benzylidenemalonic Acid, *Di-Me ester*
Benzylidenemalonic Acid, *Et ester*
o-Carboxycinnamic Acid, ω-*Mono-Me ester*
Daphnetin, 7-*Me*-8-*Et ether*
2-Decene-4,6-diynedioic Acid, *Di-Me ester*
6,7-Dihydroxy-4-methylcoumarin, *Di-Me ether*
Eugenitin
Isoeugenitin
Isotubaic Acid
Mesaconic Acid, α-*Benzyl ester*
3-Methylazelaic Acid, *Di-Me ester*
4-Methylazelaic Acid, *Di-Me ester*
5-Methylazelaic Acid, *Di-Me ester*
5-(3,4-Methylenedioxy)-3-pentenoic Acid
3-(3,4-Methylenedioxyphenyl)acrylic Acid, *Et ester*
5-(3,4-Methylenedioxyphenyl)-2-pentenoic Acid
1-Phenyl-1-butene-2,4-dicarboxylic Acid
Phthalide-3-acetic Acid, *Et ester*
1,2,3,4-Tetrahydronaphthalene-1,2-dicarboxylic Acid
1,2,3,4-Tetrahydronaphthalene-1,8-dicarboxylic Acid
1,2,3,4-Tetrahydronaphthalene-2,3-dicarboxylic Acid
Tetrahydro-5-oxo-2-phenylfuran-3-carboxylic Acid, *Me ester*
1,3,6,8-Tetrahydroxynaphthalene, 1,3-*Di-Me ether*†
1,3,6,8-Tetrahydroxynaphthalene, 3,6-*Di-Me ether*†
Tubaic Acid

$C_{12}H_{12}O_5$
Cyclopolide, *Me ether*
3,5-Dihydroxyphthalic Acid, *Di-Et ether, Anhydride*
Fraxetin, *Di-Me ether*
Fraxetin, *Et ether*
Furan-2,4-dicarboxylic Acid, *Di-allyl ester*
Gladiolic Acid, *Me ester*
1-Hydroxy-2-indanylmalonic Acid
Khellinone
Radicinin
Rosellinic Acid, *Mono-Me ether*†

$C_{12}H_{12}O_6$
Benzene-1,2,3-tricarboxylic Acid, *Tri-Me ester*
Benzene-1,2,4-tricarboxylic Acid, *Tri-Me ester*
Benzene-1,3,5-tricarboxylic Acid, *Tri-Me ester*
Brazilic Acid
3-Hydroxy-4-methoxycinnamic Acid, *Carbomethoxyl*
2-Phenyltricarballylic Acid

$C_{12}H_{12}O_6S_2$
Naphthalene-1,5-disulphonic Acid, *Di-Me ester*
Naphthalene-1,5-disulphonic Acid, *Mono-Et ester*

$C_{12}H_{12}O_7$
5Hydroxybenzene-1,2,3-tricarboxylic Acid, *Me ester, Mono-Et ester*
5-Hydroxytoluene-2,3,6-tricarboxylic Acid, *Me ether, Me ester*★†
3-Hydroxytoluene-2,4,6-tricarboxylic Acid, *Mono-Et ester*
3-Hydroxytoluene-2,4,6-tricarboxylic Acid, *Et ether*

$C_{12}H_{12}O_8S$
Thiophene-tetracarboxylic Acid, *Tetra-Me ester*

$C_{12}H_{12}O_9$
Furan-tetracarboxylic Acid, *Tetra-Me ester*

$C_{12}H_{12}S$
1-Naphthalenethiol, *Et ether*
2-Naphthalenethiol, *Et ether*

$C_{12}H_{12}S_2$
1,5-Naphthalenedithiol, *Di-Me ether*
1,8-Naphthalenedithiol, *Di-Me ether*

$C_{12}H_{13}BrO_4$
4-Bromoisophthalic Acid, *Di-Et ester*
4-Bromophthalic Acid, *Di-Et ester*
Ethoxyacetic Acid, p-*Bromophenacyl ester*

$C_{12}H_{13}ClN_4$
Pyrimethamine

$C_{12}H_{13}ClO$
3-*p*-Isopropylphenylacrylic Acid, *Chloride*
1-Phenylcyclopentane-1-carboxylic Acid, *Chloride*

$C_{12}H_{13}ClO_4$
5-Chloroisophthalic Acid, *Di-Et ester*
4-Chlorophthalic Acid, *Di-Et ester*

$C_{12}H_{13}IO_4$
2-Iodobenzene-1,3-diacetic Acid, *Di-Me ester*†
5-Iodoisophthalic Acid, *Di-Et ester*
3-Iodophthalic Acid, *Di-Et ester*
4-Iodophthalic Acid, *Di-Et ester*

$C_{12}H_{13}I_2NO_5$
3,5-Di-iodo-1-methyl-4-pyridone-2,6-dicarboxylic Acid, *Di-Et ester*

$C_{12}H_{13}N$
1-Amino-2,3-dimethylnaphthalene
1-Amino-2,6-dimethylnaphthalene
1-Amino-3,7-dimethylnaphthalene

$C_{12}H_{13}N$ (*continued*)

1-Amino-4,7-dimethylnaphthalene
2-Amino-1,4-dimethylnaphthalene
2-Amino-3,6-dimethylnaphthalene
2-Amino-3,7-dimethylnaphthalene
N-Dimethyl-1-naphthylamine
N-Dimethyl-2-naphthylamine
2-Ethyl-3-methylquinoline
2-Ethyl-4-methylquinoline
2-Ethyl-6-methylquinoline
3-Ethyl-2-methylquinoline
3-Ethyl-4-methylquinoline
3-Ethyl-6-methylquinoline
3-Ethyl-7-methylquinoline
4-Ethyl-2-methylquinoline
6-Ethyl-2-methylquinoline
6-Ethyl-8-methylquinoline
8-Ethyl-3-methylquinoline
N-Ethyl-1-naphthylamine
N-Ethyl-2-naphthylamine
2-Isopropylquinoline
3-Isopropylquinoline
4-Isopropylquinoline
7-Isopropylquinoline
1-α-Naphthylethylamine
1-β-Naphthylethylamine
2-α-Naphthylethylamine
2-β-Naphthylethylamine
1-Phenylcyclopentane-1-carboxylic Acid, *Nitrile*
1-Propylisoquinoline
3-Propylisoquinoline
2-Propylquinoline
4-Propylquinoline
6-Propylquinoline
7-Propylquinoline
8-Propylquinoline
Tetrahydrocarbazole
2,3,4-Trimethylquinoline
2,3,6-Trimethylquinoline
2,3,8-Trimethylquinoline
2,4,6-Trimethylquinoline
2,4,7-Trimethylquinoline
2,4,8-Trimethylquinoline
2,5,8-Trimethylquinoline
2,6,8-Trimethylquinoline
3,4,6-Trimethylquinoline
3,4,7-Trimethylquinoline
3,4,8-Trimethylquinoline
4,5,8-Trimethylquinoline
4,6,8-Trimethylquinoline
5,6,8-Trimethylquinoline

$C_{12}H_{13}NO$

2-Amino-1-naphthol, *Et ether*
4-Amino-1-naphthol, *Et ether*
8-Amino-1-naphthol *Et ether*
1-Amino-2-naphthol, *Et ether*
6-Amino-2-naphthol, *Et ether*
8-Amino-2-naphthol, *Et ether*
Benzylpenaldic Acid, *Me ester*
3-Ethoxyquinaldine
5-Ethoxyquinaldine
6-Ethoxyquinaldine
7-Ethoxyquinaldine
3-Hydroxy-2-methylquinoline, *Et ether*
4-Hydroxy-2-methylquinoline, *Et ether*
5-Hydroxy-2-methylquinoline, *Et ether*
6-Hydroxy-2-methylquinoline, *Et ether*
5-Phenyl-2,4-pentadienoic Acid, *Methylamide*

$C_{12}H_{13}NO_2$

ω-Camphanic Acid, *Nitrile*
3-*o*-Carboxyphenylpropionic Acid, 2-*Nitrile*, *Et ester*
2-Cyano-3-phenylpropionic Acid, *Et ester*
2,3-Dihydro-5-hydroxy-1,3-dimethyl-2-oxo-3-vinylindole
4,6-Dihydroxy-2-methylquinoline, *Di-Me ether*
6,7-Dihydroxy-2-methylquinoline, *Di-Me ether*
2,7-Dihydroxy-1-naphthylamine, 2,7-*Di-Me ether*
2-Indolylacetic Acid, *Et ester*
4-3′-Indolylbutyric Acid
3,3′-Indolylpropionic Acid, *Me ester*
Methsuximide
1-Methylindole-2-carboxylic Acid, *Et ester*
3-Methylindole-2-carboxylic Acid, *Et ester*
5-Methylindole-2-carboxylic Acid, *Et ester*
7-Methylindole-2-carboxylic Acid, *Et ester*
2-Methylindole-3-carboxylic Acid, *Et ester*
5-Methylindole-3-carboxylic Acid, *Et ester*
2-Methyl-3-phenylglutaric Acid, *Imide*
3-Methyl-3-phenylglutaric Acid, *Imide*
Phenylsuccinic Acid, β-*Et ester*, α-*Nitrile*
Phthalimide, N-*Butyl*

$C_{12}H_{13}NO_3$

1-Butane-1,3-dicarboxylic Acid, *Mono-anilide*
5-Hydroxyindole-3-acetic Acid, *Me ether*, *Me ester*†
3-Hydroxy-5-methylindole-2-carboxylic Acid, *Et ester*
Mesacon-α-anilic Acid, *Me ester*
Mesacon-β-anilic Acid, *Me ester*
Mesaconic Acid, α-p-*Toluidide*
2-Methylpropene-1,3-dicarboxylic Acid, *Mono-anilide*
3-Phenylmalic Acid, β-*Et ester*, α-*Nitrile*
Succinimide, N-p-*Hydroxyphenyl*, *Et ether*

$C_{12}H_{13}NO_3S$

1-Naphthol-4-sulphonic Acid, *Et ether*, *Amide*
2-Naphthol-6-sulphonic Acid, *Di-Me amide*
2-Naphthol-6-sulphonic Acid, *Et ether*, *Amide*
1-Naphthylamine-4-sulphonic Acid, N-*Di-Me*

$C_{12}H_{13}NO_4$

2-Methyl-3-*m*-nitrophenylacrylic Acid, *Et ester*
3-(4-Methyl-3-nitrophenyl)acrylic Acid, *Et ester*
3-*p*-Nitrophenylcrotonic Acid, *Et ester*
1-Phenylpyrrolidine-2,2′-dicarboxylic Acid
1-Phenylpyrrolidine-2,5-dicarboxylic Acid

$C_{12}H_{13}NO_5$

Elatinic Acid
p-Hydroxybenzylpenaldic Acid, *Me ester*
3-(2-Hydroxy-4-methyl-5-nitrophenyl)acrylic Acid, *Me ester*, *Me ether*
α-Hydroxy-2-nitrocinnamic Acid, *Me ether*, *Et ester*
α-Hydroxy-2-nitrocinnamic Acid, *Et ether*, *Me ester*
3-(2-Hydroxy-5-nitrophenyl)acrylic Acid, *Et ether*, *Me ester*

$C_{12}H_{13}NO_5$ (*continued*)
3-(2-Hydroxy-5-nitrophenyl)acrylic Acid, *Me ether, Et ester*
3-(4-Hydroxy-3-nitrophenyl)acrylic Acid, *Me ether, Et ester*
Methyl-lycoctinic Acid

$C_{12}H_{13}NO_6$
2,3-Dihydroxy-5-nitrocinnamic Acid, 2-*Me*-3-*Et ether*
p-Nitrobenzylmalonic Acid, *Di-Me ester*
5-Nitroisophthalic Acid, *Di-Et ester*
3-Nitrophthalic Acid, *Di-Et ester*
3-Nitrophthalic Acid, 2-*Butyl ester*
4-Nitrophthalic Acid, *Di-Et ester*
3-Nitrophthalic Acid, 1-*Isobutyl ester*
3-Nitrophthalic Acid, 2-*Isobutyl ester*
2-Nitroterephthalic Acid, *Di-Et ester*
2-Nitroterephthalic Acid, 1-*Butyl ester*
2-Nitroterephthalic Acid, 4-*Butyl ester*
Pyridine-2,3,4-tricarboxylic Acid, 3,4-*Di-Et Ester*

$C_{12}H_{13}NO_7$
3-Nitro-opianic Acid, *Et ester*

$C_{12}H_{13}NO_8$
3,4-Dimethoxy-6-nitrobenzene-1,2-dicarboxylic Acid, *Di-Me ester*

$C_{12}H_{13}N_3$
o-Aminohydrazobenzene
m-Aminohydrazobenzene
p-Aminohydrazobenzene
2,2′-Diaminodiphenylamine
2,3′-Diaminodiphenylamine
2,4-Diaminodiphenylamine
2,4′-Diaminodiphenylamine
2,6-Diaminodiphenylamine
3,3′-Diaminodiphenylamine
3,4′-Diaminodiphenylamine
4,4′-Diaminodiphenylamine
2,4,4′-Triaminobiphenyl
2,4′,5-Triaminobiphenyl
3,4,4′-Triaminobiphenyl

$C_{12}H_{13}N_3O_6$
1,2,3,4-Tetrahydro-6,6-dinitroquinoline-1-carboxylic Acid, *Et ester*

$C_{12}H_{13}N_3O_8$
2,4,6-Trinitrobenzoic Acid, *Pentyl ester*

$C_{12}H_{13}N_5O_2$
Prontosil

$C_{12}H_{13}N_5O_4$
Toyocamycin†

$C_{12}H_{14}$
1-Phenylcyclohexene
3-Phenylcyclohexene
4-Phenylcyclohexene
1-Phenyl-1,3-hexadiene
6-Phenyl-1,3-hexadiene
1-Phenyl-1,5-hexadiene
3-Phenyl-1,5-hexadiene
[4,4,2]Propella-3,8,11-triene†
Tetrahydroacenaphthene

$C_{12}H_{14}As_2Cl_2N_2O_2$
Salvarsan

$C_{12}H_{14}Br_2O_2$
4-Allyl-3,6-dibromoguaiacol, *Et ether*
2,3-Dibromo-3-phenylpropionic Acid, *Propyl ester*
3,6-Dibromo-4-propenylguaiacol, *Et ether*

$C_{12}H_{14}Br_2O_3$
3,5-Dibromo-2,4-dihydroxyacetophenone, *Di-Et ether*

$C_{12}H_{14}Cl_2N_2$
Paraquat

$C_{12}H_{14}Cl_2O_3$
2,4-Dichlorophenoxyacetic Acid, *Butyl ester*

$C_{12}H_{14}Cl_4O_2$
Tetrachlororesorcinol, *Dipropyl ether*

$C_{12}H_{14}NO_4PS$
O,O-Diethyl phthalimidophosphonothionate†

$C_{12}H_{14}N_2$
13,14-Diazatricyclo[6,4,1,1^{2,7}]tetradeca-3,5,9,11-tetraene†
Eleagnine
1-Ethyl-3-methyl-5-phenylpyrazole
1-Ethyl-5-methyl-3-phenylpyrazole
3-Ethyl-4-methyl-1-phenylpyrazole
3-Ethyl-4-methyl-5-phenylpyrazole
4-Ethyl-3-methyl-1-phenylpyrazole
1,8-Naphthylenediamine★, NN′-*Di-Me*†
2,3′-Pyridyl-pyrrolidine, N-*Allyl*
Tetrahydroharman†
6,7,8,9-Tetrahydro-11*H*-pyrido[2,1-*b*]quinazoline†

$C_{12}H_{14}N_2O$
Dehydrobufotenine
1-Ethyl-5-methyl-2-phenyl-3-pyrazolone
4-Ethyl-1-methyl-2-phenyl-3-pyrazolone
4-Ethyl-5-methyl-1-phenyl-3-pyrazolone
4-Ethyl-5-methyl-2-phenyl-3-pyrazolone
5-Ethyl-4-methyl-3-phenyl-3-pyrazolone

$C_{12}H_{14}N_2O_2$
Abrine
2,2′-Indazolylpropionic Acid, *Et ester*
3,2′-Indazolylpropionic Acid, *Et ester*
Mesaconic Acid, β-*Amide*, α-p-*Toluidide*
Methoin
β-Methyltryptophan†
2-Methyltryptophane
5-Methyltryptophane
Primidone
Tryptophane, *Me ester*
Tryptophane, α-N-*Me*

$C_{12}H_{14}N_2O_3$
5-Allyl-5-(cyclopent-2-enyl)barbituric Acid

$C_{12}H_{14}N_2O_4$
N-Benzoylglycylalanine
N-Benzoylglycyl-3-aminopropionic Acid
Methyl-lycoctinic Acid, β-*Amide*

$C_{12}H_{14}N_2O_5$
Dinex

$C_{12}H_{14}N_2O_5S_2$
2-Naphthol-1,6-disulphonic Acid, *Et ether, Diamide*

$C_{12}H_{14}N_2O_7$
2-Hydroxy-3,5-dinitrobenzoic Acid, 3-*Methylbutyl ester*

$C_{12}H_{14}N_4$
3,3′-Diaminohydrazobenzene
4,4′-Diaminohydrazobenzene
2,2′,4,4′-Tetra-aminobiphenyl
2,2′,5,5′-Tetra-aminobiphenyl
3,3′,4,4′-Tetra-aminobiphenyl

$C_{12}H_{14}N_4OS$
Thiochrome

$C_{12}H_{14}N_4O_2S$
Elkosin
Sulphamethazine

$C_{12}H_{13}N_4O_4S$
Sulphadimethoxine

$C_{12}H_{14}N_4O_4S_2$
Benzidine-2,2′-disulphonic Acid, *Diamide*
Thiostreptoic Acid†

$C_{12}H_{14}N_4O_6$
Sangivamycic Acid†
1,1′,3,3′-Tetramethyl-5,5′-bibarbituric Acid

$C_{12}H_{14}N_4O_8$
Amalic Acid

$C_{12}H_{14}O$
2-Benzylcyclopentanone†
2-Hexenophenone
4-Hexenophenone
4-Isopropyl-7-methylbenzofuran
7-Isopropyl-4-methylbenzofuran
2-Phenylcyclohexanone
3-Phenylcyclohexanone
4-Phenylcyclohexanone
4-Phenyl-3-hexen-2-one
6-Phenyl-3-hexen-2-one
4-Phenyl-4-hexen-2-one
6-Phenyl-4-hexen-2-one
6-Phenyl-5-hexen-2-one
4-Phenyl-5-hexen-2-one
1-Phenyl-1-hexen-3-one
5-Phenyl-4-hexen-3-one
4-Phenyl-5-hexen-3-one
3-Phenyl-2-propyn-1-ol, *Propyl ether*

$C_{12}H_{14}O_2$
2-Benzylacrylic Acid, *Et ester*
3-Butyl phthalide†
Cinnamic Acid, *Propyl ester*
5,6-Dihydro-2,7-dimethylnaphthalene-1,4-diol†
1,6-Dihydroxy-1-vinyltetralin†
3-(2,4-Dimethylphenyl)crotonic Acid
3-(2,5-Dimethylphenyl)crotonic Acid
α-Ethylcinnamic Acid, *Me ester*
β-Ethylcinnamic Acid, *Me ester*
4-Hydroxy-4-phenylhexanoic Acid, *γ-Lactone*
5-Hydroxy-3-phenylhexanoic Acid, *Lactone*
5-Hydroxy-5-phenylhexanoic Acid, *Lactone*
Indane-1-carboxylic Acid, *Et ester*
3-Isobutylidene-3α,4-dihydrophthalide
3-*p*-Isopropylphenylacrylic Acid
Ligustilide†
7-Methoxy-2,2-dimethylchromene†
1-Methylindane-2-carboxylic Acid, *Me ester*
2-Methyl-5-(3-methylbut-2-enyl)-1,4-benzoquinone†
2-Methyl-3-phenylacrylic Acid, *Et ester*
2-Methyl-3-*o*-tolylcrotonic Acid
2-Methyl-3-*p*-tolylcrotonic Acid
4-Phenyl-3-butenoic Acid, *Et ester*
2-Phenylcrotonic Acid, *Et ester*
3-Phenylcrotonic Acid, *Et ester*
1-Phenylcyclopropane-1-carboxylic Acid, *Et ester*
2-Phenylcyclopropane-1-carboxylic Acid, *Et ester*
1-Phenylcyclopentane-1-carboxylic Acid
2-Phenylcyclopentane-1-carboxylic Acid
3-Phenylcyclopentane-1-carboxylic Acid
3-*o*-Tolylacrylic Acid, *Et ester*
3-*p*-Tolylacrylic Acid, *Et ester*
3-(2,3,4-Trimethylphenyl)acrylic Acid
3-(2,4,5-Trimethylphenyl)acrylic Acid
3-(2,4,6-Trimethylphenyl)acrylic Acid

$C_{12}H_{14}O_2S_2$
Ditophal

$C_{12}H_{14}O_3$
Aceteugenol
2-Allyloxybenzoic Acid, *Et ester*
4-Allyloxybenzoic Acid, *Et ester*
Benzoylacetic Acid, *Propyl ester*
2-Benzoylbutyric Acid, *Me ester*
4-Benzoylbutyric Acid, *Me ester*
Benzoylformic Acid, *Isobutyl ester*
2-Benzoyl-3-methylpropionic Acid, *Me ester*
2-Benzoylpropionic Acid, *Et ester*
3-Benzoylpropionic Acid, *Et ester*
2-Benzylacetoacetic Acid, *Me ester*
3-Benzyl-levulinic Acid
5-Benzyl-levulinic Acid
2,4-Dimethylbenzoylformic Acid, *Et ester*
2,5-Dimethylbenzoylformic Acid, *Et ester*
α-Ethyl-*o*-hydroxy-*trans*-cinnamic Acid, *Me ether*
Eugenyl acetate
2-(1-Hydroxy-2-indanyl)propionic Acid
3-(2-Hydroxy-4-methylphenyl)acrylic Acid, *Et ester*
3-*o*-Hydroxyphenylcrotonic Acid, *Me ether, Me ester*
3-*m*-Hydroxyphenylcrotonic Acid, *Me ether, Me ester*
3-*p*-Hydroxyphenylcrotonic Acid, *Et ether*
3-*o*-Hydroxyphenyl-2-methylacrylic Acid, *Me ether, Me ester*
3-*o*-Hydroxyphenyl-2-methylacrylic Acid, *Et ether*
3-*m*-Hydroxyphenyl-2-methylacrylic Acid, *Et ether*
Levulinic Acid, *Benzyl ester*
3-*o*-Methoxyphenylacrylic Acid, *Et ester*
3-*p*-Methoxyphenylacrylic Acid, *Et ester*
2-Methoxy-4-propenylphenol, *Ac*
2-Methoxy-5-propenylphenol, *Ac*
5-Methylmellein, *Me ether*†
3-Oxo-2-phenylbutyric Acid, *Et ester*
3-Oxo-4-phenylbutyric Acid, *Et ester*
4-Oxo-2-phenylvaleric Acid, *Me ester*
2,4,6-Trimethylbenzoylformic Acid, *Me ester*

$C_{12}H_{14}O_4$

Apiol
Benzene-1,3-diacetic Acid, *Di-Me ester*
Benzene-1,4-diacetic Acid, *Di-Me ester*
Benzene-1,2-dipropionic Acid
Benzene-1,3-dipropionic Acid
Benzene-1,4-dipropionic Acid
O-Benzoyl-lactic Acid, *Et ester*
Benzylmalonic Acid, *Di-Me ester*
Caffeic Acid, *Me ester*
3-*m*-Carboxyphenylpropionic Acid, *Di-Me ester*
3-*p*-Carboxyphenylpropionic Acid, *Di-Me ester*
4,6-Diacetylresorcinol, *Di-Me ether*
4,6-Diacetylresorcinol, *Mono-Et ether*
o-Di-2,3-epoxypropoxybenzene
m-Di-2,3-epoxypropoxybenzene
p-Di-2,3-epoxypropoxybenzene
Diethyl phthalate
Dill-apiol
3,4-Dimethylphthalic Acid, *Di-Me ester*
2,5-Dimethylterephthalic Acid, *Di-Me ester*
2,6-Dimethylterephthalic Acid, 1-*Et ester*
2,6-Dimethylterephthalic Acid, 4-*Et ester*
3,4-Dihydro-8-hydroxy-6-methoxy-3-methylisocoumarin, *Me ether* †
3-Formyl-6-hydroxy-5-isopropyl-*o*-toluic Acid
4-Hydroxy-3-methoxycinnamic Acid, *Et ester*
4-Hydroxyphenylpyruvic Acid, *Et ester*
Isoapiol
Isophthalic Acid, *Di-Et ester*
p-Methoxybenzoylacetic Acid, *Et ester*
3-(3,4-Methylenedioxyphenyl)propionic Acid, *Et ester*
2-Methyl-2-phenylglutaric Acid
2-Methyl-3-phenylglutaric Acid
3-Methyl-3-phenylglutaric Acid
2-Phenyladipic Acid
3-Phenyladipic Acid
2-Phenylbutane-1,1-dicarboxylic Acid
4-Phenylbutane-1,1-dicarboxylic Acid
3-Phenylbutane-1,2-dicarboxylic Acid
4-Phenylbutane-1,2-dicarboxylic Acid
Phenylsuccinic Acid, *Di-Me ester*
Phenylsuccinic Acid, α-*Et ester*
Phenylsuccinic Acid, β-*Et ester*
Phthalic Acid, *Isobutyl ester*
Sclerolide †
Terephthalic Acid, *Di-Et ester*
Terephthalic Acid, *Mono-butyl ester*

$C_{12}H_{14}O_5$

Coriarin(?)
Curvulinic Acid, *Di-Me ether* †
Curvulinic Acid, *Et ester* †
Dihydrogladiolic Acid, *Me ester*
Dihydroradicin †
2,4-Dihydroxybenzoylformic Acid, *Di-Et ether*
3,4-Dihydroxybenzoylformic Acid, *Di-Me ether, Et ester*
2-Formyl-5,6-dimethoxybenzoic Acid, α-*Et ester*
3-(4-Hydroxy-3,5-dimethoxyphenyl)acrylic Acid, *Me ester*
3-(4-Hydroxy-3,5-dimethoxyphenyl)acrylic Acid, 4-*Me ether*
2-Hydroxyisophthalic Acid, *Di-Et ester*
4-Hydroxyisophthalic Acid, *Di-Et ester*
5-Hydroxyisophthalic Acid, *Di-Et ester*
Netoric Acid
3-Phenylmalic Acid, β-*Et ester*
Sclerotinin B †
3-(2,3,4-Trihydroxyphenyl)acrylic Acid, *Tri-Me ether*
3-(2,4,5-Trihydroxyphenyl)acrylic Acid, *Tri-Me ether*
3-(2,4,6-Trihydroxyphenyl)acrylic Acid, *Tri-Me ether*

$C_{12}H_{14}O_6$

Curvulic Acid, *Me ester* †
2,4-Dihydroxyisophthalic Acid, *Di-Et ester*
4,6-Dihydroxyisophthalic Acid, *Di-Me ether, Di-Me ester*
4,6-Dihydroxyisophthalic Acid, *Di-Et ether*
3,5-Dihydroxyphthalic Acid, *Di-Me ether, Di-Me ester*
3,6-Dihydroxyphthalic Acid, *Di-Et ester*
3,6-Dihydroxyphthalic Acid, *Di-Me ether, Di-Me ester*
4,5-Dihydroxyphthalic Acid, *Di-Et ester*
2,3-Dihydroxyterephthalic Acid, *Di-Et ester*
2,5-Dihydroxyterephthalic Acid, *Di-Et ester*
2,6-Dihydroxyterephthalic Acid, *Di-Et ester*
3,4-Dimethoxyphthalic Acid, *Di-Me ester*
3,4-Dimethoxyphthalic Acid, 1-*Et ester*
Metahemipinic Acid, *Di-Me ester*
Metahemipinic Acid, *Mono-Et ester*
Puberulic Acid, *Tetra-Me ether*
2,4,5-Trihydroxyphenylglyoxylic Acid, *Tri-O-Me ether, Me ester*

$C_{12}H_{14}O_7$

2-Carboxymethyl-4,5-dimethoxyphenoxyacetic Acid
Piscidic Acid, *Mono-Me ester* †
Piscidic Acid, p-O-*Me ether* †
2,4,6-Trihydroxyisophthalic Acid, *Di-Et ester*
3,4,5-Trihydroxyphthalic Acid, *Tri-Me ether, 1-Me ester*
2,3,5-Trihydroxyterephthalic Acid, *Di-Et ester*
Ustic Acid, 4-*Me ether*

$C_{12}H_{15}BrO$

2-Bromo-1,2,3,4-tetrahydro-1-naphthol, *Et ether*

$C_{12}H_{15}BrO_2$

o-Bromobenzoic Acid, d-*Amyl ester*
m-Bromobenzoic Acid, d-*Amyl ester*
p-Bromobenzoic Acid, d-*Amyl ester*
2-Bromo-4,6-dimethylbenzoic Acid, *Propyl ester*
α-Bromophenylacetic Acid, *Isobutyl ester*
2-Bromo-4-phenylbutyric Acid, *Et ester*

$C_{12}H_{15}Br_3O_3$

Tribromophloroglucinol, *Tri-Et ether*
Tribromopyrogallol, *Tri-Et ether*

$C_{12}H_{15}ClNO_4PS_2$

Phosalone †

$C_{12}H_{15}ClNO_5PS$

Phosalone-oxon †

$C_{12}H_{15}ClO$
2-Isopropyl-3-phenylpropionic Acid, *Chloride*
4-Phenylhexanoic Acid, *Chloride*
5-Phenylhexanoic Acid, *Chloride*
6-Phenylhexanoic Acid, *Chloride*

$C_{12}H_{15}ClO_2$
α-Chlorophenylacetic Acid, *Butyl ester*
p-Pentoxybenzoic Acid, *Chloride*

$C_{12}H_{15}ClO_3$
3-Chloro-6-hydroxy-*p*-toluic Acid, *Propyl ether, Me ester*
3-Chloro-6-hydroxy-*p*-toluic Acid, *Butyl ether*

$C_{12}H_{15}IO_4$
3-Iodo-4-methoxybenzoic Acid, 2-*Ethoxyethyl ester*

$C_{12}H_{15}N$
6,7-Benzomorphan
1,2-Dihydro-2,2,4-trimethylquinoline
1,2,3,4,4α,9α-Hexahydrocarbazole
1,2,3,4,6,10*b*-Hexahydropyrido[2,1-*a*]isoindole†
2-Isopropyl-3-methylindole
3-Isopropyl-2-methylindole
Julolidine
4-Methyl-2-phenylvaleric Acid, *Nitrile*
Pentamethylbenzoic Acid, *Nitrile*
2-Phenylhexanoic Acid, *Nitrile*
4-Phenylhexanoic Acid, *Nitrile*
6-Phenylhexanoic Acid, *Nitrile*

$C_{12}H_{15}NO$
4-Hydroxy-5-isopropyl-*o*-toluic Acid, *Me ether, Nitrile*
4-3′-Indolylbutanol
3-*p*-Isopropylphenylacrylic Acid, *Amide*
1-Phenylcyclopentane-1-carboxylic Acid, *Amide*
Phthalimidine, N-*Isobutyl*
Physostigmol, *Et ether*

$C_{12}H_{15}NO_2$
3-Anilinocrotonic Acid, *Et ester*
Bicyclo[2,2,1]hept-5-ene-2,3-dicarboxylic Acid, *Propylimide*
Dipropionanilide
3-Methylamino-3-phenylacrylic Acid, *Et ester*
o-Nitrophenylcyclohexane
p-Nitrophenylcyclohexane
1-Nitro-1-phenylcyclohexane
1-Phenylpiperidine-4-carboxylic Acid
Piperidine-*N*-carboxylic Acid, *Phenyl ester*
1,2,3,4-Tetrahydroisoquinoline-3-carboxylic Acid, *Et ester*

$C_{12}H_{15}NO_3$
Anhalonine
Hydrocotarnine
2-Methylglutaric Acid, *Monoanilide*
3-Methylglutaric Acid, *Monoanilide*
Methylmalonic Acid, *Mono-Et ester, Anilide*
Oxanilic Acid, 3-*Methylbutyl ester*
Phenaceturic Acid, *Et ester*
Phenylsuccinic Acid, α-*Et ester*, β-*Amide*
Succinanilic Acid, *Et ester*

$C_{12}H_{15}NO_4$
4-Acetyl-3-isovalerylpyridin-2,6(1*H*,3*H*)dione†
4-Aminoisophthalic Acid, *Di-Et ester*
4-Aminoisophthalic Acid, N-*Di-Me*, *Di-Me ester*
5-Aminoisophthalic Adid, *Di-Et ester*
4-Aminophthalic Acid, *Di-Et ester*
2-Benzylglutamic Acid
Carbethoxy-DL-alanine, *Phenyl ester*
Cotarnine
Isatoic Acid, *Di-Et ester*
4-Isopropyl-3-nitrobenzoic Acid, *Et ester*
p-Methoxybenzamidoacetic Acid, *Et ester*
6-Methylpyridine-3,4-dicarboxylic Acid, *Di-Et ester*
o-Nitrobenzoic Acid, (+)-2-*Methylbutyl ester*
m-Nitrobenzoic Acid, (+)-2-*Methylbutyl ester*
p-Nitrophenylacetic Acid, *Butyl ester*
2-*p*-Nitrophenylbutyric Acid, *Et ester*
N-Phenylglycine-*o*-carboxylic Acid, α-*Et ester*, β-*Me ester*
2-Phenyliminodiacetic Acid, *Di-Me ester*
N-Phenyliminodiacetic Acid, *Di-Me ester*
N-Phenyliminodiacetic Acid, *Mono-Et ester*

$C_{12}H_{15}NO_5$
2-Hydroxy-4-nitrobenzoic Acid, 3-*Methylbutyl ester*
4-Hydroxy-3-nitrobenzoic Acid, 3-*Methylbutyl ester*
Metahemipinic Acid, *Mono-Et amide*
Tartranilic Acid, *Et ester*

$C_{12}H_{15}NO_6$
1-(*o*-Carboxyphenylamino)-1-deoxy-D-ribulose†

$C_{12}H_{15}NO_7$
2-Nitrogallic Acid, *Tri-Me ether, Et ester*

$C_{12}H_{15}N_3$
5-Amino-3-methyl-1-phenylpyrazole, 5-N-*Et*

$C_{12}H_{15}N_3O_2$
1-Phenylurazole, 3,5-*Di-Et ether*

$C_{12}H_{15}N_3O_3$
Cyanuric Acid, O-*Tri-allyl ester*

$C_{12}H_{15}N_3O_6$
5-*tert*-Butyl-2,4,6-trinitro-*m*-xylene
Triazine-tricarboxylic Acid, *Tri-Et ester*

$C_{12}H_{15}N_3O_9$
1,3,5-Trihydroxy-2,4,6-trinitrobenzene, *Tri-Et ether*

$C_{12}H_{15}N_5O_5$
Sangivamycin†

$C_{12}H_{15}NS_2$
Piperidine-*N*-carbodithioic Acid, *Phenyl ester*

$C_{12}H_{15}O_4P$
Diallyl phenyl phosphate

$C_{12}H_{16}$
Butylcyclo-octatetraene
4-Methyl-1-phenyl-1-pentene
2-Methyl-3-phenyl-2-pentene
2-Methyl-5-phenyl-2-pentene
3-Methyl-2-phenyl-2-pentene
Phenylcyclohexane
6-Phenyl-1-hexene
1-Phenyl-2-hexene
2-Phenyl-2-hexene
1-Phenyl-3-hexene
2-Phenyl-3-hexene
3-Phenyl-3-hexene
1,2,3,4-Tetrahydro-1,1-dimethylnaphthalene

$C_{12}H_{16}$ (*continued*)
1,2,3,4-Tetrahydro-1,2-dimethylnaphthalene
1,2,3,4-Tetrahydro-1,4-dimethylnaphthalene
1,2,3,4-Tetrahydro-1,6-dimethylnaphthalene
1,2,3,4-Tetrahydro-1,7-dimethylnaphthalene
1,2,3,4-Tetrahydro-2,2-dimethylnaphthalene
1,2,3,4-Tetrahydro-2,3-dimethylnaphthalene
1,2,3,4-Tetrahydro-2,5-dimethylnaphthalene
1,2,3,4-Tetrahydro-2,6-dimethylnaphthalene
1,2,3,4-Tetrahydro-2,7-dimethylnaphthalene
1,2,3,4-Tetrahydro-5,7-dimethylnaphthalene
1,2,3,4-Tetrahydro-5,8-dimethylnaphthalene
1,2,3,4-Tetrahydro-6,7-dimethylnaphthalene

$C_{12}H_{16}Cl_2NO_2$
Neburon

$C_{12}H_{16}F_3N$
Fenfluramine†

$C_{12}H_{16}NO_9P$
1-(*o*-Carboxyphenylamino)-1-deoxy-D-ribulose 5-phosphate†

$C_{12}H_{16}N_2$
N,N-Dimethyltryptamine†

$C_{12}H_{16}N_2O$
Bufotenine
Caulophylline
N,N-Dimethyltryptamine, N-*oxide*†
5-Methoxy-3-ω-methylaminoethylindole†
Psilocin★†

$C_{12}H_{16}N_2O_2$
Benzene-1,2-dipropionic Acid, *Diamide*
Piperazine-*N*-carboxylic Acid, *Benzyl ester*

$C_{12}H_{16}N_2O_3$
Cyclobarbitone
Evipal

$C_{12}H_{16}N_2O_4$
Butane-1,1,4,4-tetracarboxylic Acid, 1,4-*Dinitrile*-1,4-*di-Et ester*
4,6-Diaminoisophthalic Acid, *Di-Et ester*
2,5-Diaminoterephthalic Acid, *Di-Et ester*
3,4-Methylenedioxy-6-nitroaniline, N-*Amyl*
o-Nitrophenylcarbamic Acid, *Amyl ester*

$C_{12}H_{16}N_2O_7$
1,2,3-Trihydroxy-4,5-dinitrobenzene, *Tri-Et ether*
1,3,5-Trihydroxy-2,4-dinitrobenzene, *Tri-Et ether*

$C_{12}H_{16}N_2S$
2-3′-Methylbutylaminobenzothiazole

$C_{12}H_{16}O$
Cyclohexyl phenyl Ether
Hexanophenone
Isobutyl *o*-tolyl Ketone
Isobutyl *m*-tolyl Ketone
Isobutyl *p*-tolyl Ketone
4-Isopropyl-2-methylacetophenone
5-Isopropyl-2-methylacetophenone
3-Methyl-1-phenylpentan-2-one
3-Methyl-3-phenylpentan-2-one
3-Methyl-4-phenylpentan-2-one
3-Methyl-5-phenylpentan-2-one
4-Methyl-1-phenyl-2-pentanone
4-Methyl-3-phenyl-2-pentanone
4-Methyl-4-phenyl-2-pentanone
2-Methyl-4-phenyl-3-pentanone
2-Methyl-5-phenyl-3-pentanone
2-Methyl-5-phenylvaleraldehyde
3-Methyl-5-phenylvaleraldehyde
4-Methylvalerophenone
1-Phenylcyclohexanol
2-Phenylcyclohexanol
3-Phenylcyclohexanol
4-Phenylcyclohexanol
3-Phenyl-2-hexanone
5-Phenyl-2-hexanone
6-Phenyl-2-hexanone
1-Phenyl-3-hexanone
2-Phenyl-3-hexanone
4-Phenyl-3-hexanone
6-Phenyl-3-hexanone
2-Phenylvinyl Alcohol, *Isobutyl ether*
5,6,7,8-Tetrahydro-3-methyl-1-naphthol, *Me ether*
5,6,7,8-Tetrahydro-1-naphthol, *Et ether*
5,6,7,8-Tetrahydro-2-naphthol, *Et ether*
p-Valeryltoluene

$C_{12}H_{16}O_2$
Benzoic Acid, 3-*Methylbutyl ester*
2-Benzyl-2-methylbutyric Acid
2-Benzylpropionic Acid, *Et ester*
5-*tert*-Butyl-2-hydroxybenzaldehyde, *Me ether*
2,4-Dimethylphenylacetic Acid, *Et ester*
2,5-Dimethylphenylacetic Acid, *Et ester*
3,5-Dimethylphenylacetic Acid, *Et ester*
2,2-Dimethyl-3-phenylpropionic Acid, *Me ester*
Eugenol, *Et ether*
Hexanoic Acid, *Phenyl ester*
p-Hydroxybutyrophenone, *Et ether*
4-Hydroxy-6-isopropyl-*m*-tolualdehyde, *Me ether*
4-Hydroxy-5-isopropyl-*o*-toluic Aldehyde, *Me ether*
2-Hydroxy-5-methylpropiophenone, *Et ether*
γ-Hydroxypropiophenone, *Propyl ether*
ω-Hydroxyvalerophenone, *Me ether*
m-Hydroxyvalerophenone, *Me ether*
p-Hydroxyvalerophenone, *Me ether*
2-Isobutyryl-*p*-cresol, *Me ether*
p-Isopropylbenzoic Acid, *Et ester*
2-Isopropyl-3-phenylpropionic Acid
3-*p*-Isopropylphenylpropionic Acid
5-Isopropyl-*o*-toluic Acid, *Me ester*
6-Isopropyl-*m*-toluic Acid, *Me ester*
Isosafroeugenol, *Me ether*
2-Methoxy-4-propenylphenol, 1-*Et ether*
2-Methoxy-5-propenylphenol, *Et ether*
4-(2-Methyl-2-butyl)benzoic Acid
2-Methyl-2-phenylbutyric Acid, *Me ester*
3-Methyl-3-phenylbutyric Acid, *Me ester*
3-Methyl-4-phenylbutyric Acid, *Me ester*
2-Methyl-2-phenylpropionic Acid, *Et ester*
3-Methyl-3-phenylvaleric Acid
4-Methyl-2-phenylvaleric Acid
4-Methyl-4-phenylvaleric Acid
4-Methyl-5-phenylvaleric Acid
5-Methyl-2-phenylvaleric Acid
1,2,3,4,4*a*,5,10,10*a*-Octahydrophenazine

$C_{12}H_{16}O_2$ (*continued*)
1,2,3,4,6,7,8,9-Octahydrophenazine
Pentamethylbenzoic Acid
Phenylacetic Acid, *Isobutyl ester*
3-Phenylbutyric Acid, *Et ester*
4-Phenylbutyric Acid, *Et ester*
2-Phenylhexanoic Acid
3-Phenylhexanoic Acid
4-Phenylhexanoic Acid
5-Phenylhexanoic Acid
6-Phenylhexanoic Acid
5-Phenylvaleric Acid, *Me ester*
o-Propylbenzoic Acid, *Et ester*
Resorcinol, *Mono-cyclohexyl ether*
2,3,4,5-Tetramethylbenzoic Acid, *Me ester*
2,3,5,6-Tetramethylbenzoic Acid, *Me ester*
2-*p*-Tolylpropionic Acid, *Et ester*
3-*p*-Tolylpropionic Acid, *Et ester*
2,4,6-Trimethylphenylacetic Acid, *Me ester*

$C_{12}H_{16}O_2S$
Phenylmercaptoacetic Acid, *Butyl ester*

$C_{12}H_{16}O_3$
Acetylquinol, 2,5-*Di-Et ether*
Anisic Acid, *Butyl ester*
Anisic Acid, *Isobutyl ester*
Asarone
3-Benzyl-4-hydroxyvaleric Acid
p-Butoxybenzoic Acid, *Me ester*
Calythrone
2,4-Dihydroxyacetophenone, *Di-Et ether*
2,4-Dihydroxyacetophenone, 4-*Butyl ether*
Elemicin
3-α-Furylacrylic Acid, *Pentyl ester*
3-Hydroxy-4-isopropylbenzoic Acid, *Et ester*
2-Hydroxy-2-(4-isopropylphenyl)acetic Acid, *Me ether*
2-Hydroxy-2-(4-isopropylphenyl)acetic Acid, *Me ester*
4-Hydroxy-5-isopropyl-*o*-toluic Acid, *Me ether*
4-Hydroxy-5-isopropyl-*o*-toluic Acid, *Me ester*
6-Hydroxy-5-isopropyl-*o*-toluic Acid, *Me ester*
2-Hydroxy-5-isopropyl-*m*-toluic Acid, *Me ester*
4-Hydroxy-6-isopropyl-*m*-toluic Acid, *Me ether*
5′-Hydroxyjasmonic Acid Lactone†
p-Hydroxymethyl-benzoic Acid, *Butyl ester*
p-Hydroxymethyl-benzoic Acid, *Et ether*, *Et ester*
2-Hydroxy-2-phenylbutyric Acid, *Et ester*
3-Hydroxy-3-phenylbutyric Acid, *Et ester*
3-Hydroxy-4-phenylbutyric Acid, *Me ether*, *Me ester*
4-Hydroxy-4-phenylbutyric Acid, *Et ester*
3-Hydroxy-3-phenylhexanoic Acid
4-Hydroxy-3-phenylhexanoic Acid
4-Hydroxy-4-phenylhexanoic Acid
5-Hydroxy-3-phenylhexanoic Acid
5-Hydroxy-5-phenylhexanoic Acid
4-*o*-Hydroxyphenylhexanoic Acid
4-*p*-Hydroxyphenylhexanoic Acid
3-*o*-Hydroxyphenylpropionic Acid, *Propyl ether*
3-Hydroxy-5-phenylvaleric Acid, *Me ester*
4-*p*-Hydroxyphenylvaleric Acid, *Me ether*
5-*p*-Hydroxyphenylvaleric Acid, *Me ether*
5-Hydroxy-*o*-toluic Acid, *Et ester*, *Et ether*
4-Hydroxy-*m*-toluic Acid, *Et ether*, *Et ester*
2-Hydroxy-2-(2,4,6-trimethylphenyl)acetic Acid, *Me ester*
Isoelemicin†
p-Isopropoxybenzoic Acid, *Et ester*
Mandelic Acid, *Butyl ester*
Mandelic Acid, *Isobutyl ester*
Mandelic Acid, *Et ether*, *Et ester*
3-Methylbutyl salicylate
2-Methyl-2-phenoxypropionic Acid, *Et ester*
4,4′-Methylvalerylresorcinol
m-Pentoxybenzoic Acid
p-Pentoxybenzoic Acid
2-Phenoxybutyric Acid, *Et ester*
4-Phenoxybutyric Acid, *Et ester*
Zingerone, *Me ether*

$C_{12}H_{16}O_4$
Aspidinol
Curvulol, *Di-Me ether*†
2,4-Dihydroxybenzoic Acid, *Di-Et ether*, *Me ester*
2,4-Dihydroxy-6-pentylbenzoic Acid
3,5-Dihydroxy-*o*-toluic Acid, n-*Butyl ester*
2,6-Dimethoxybenzoic Acid, *Propyl ester*
3,4-Dimethoxyphenylacetic Acid, *Et ester*
3-Hydroxy-4-methoxybenzoic Acid, *Et ester*, *Et ether*
2-Hydroxy-4-methoxy-3,6-dimethylbenzoic Acid, *Et ester*
2-Hydroxy-4-methoxy-3,6-dimethylbenzoic Acid, *Me ester*, *Me ether*
4-Hydroxy-2-methoxy-3,6-dimethylbenzoic Acid, *Et ester*
3,4-Methylenedioxybenzaldehyde, *Di-Et acetal*
4-Methyl-5-(1-methyl 2-formyloxypropyl)-resorcinol†
α,2,4-Trihydroxyacetophenone, α,4-*Di-Me ether*, *Et ether*
α,2,4-Trihydroxyacetophenone, α,4-*Di-Et ether*
2,4,6-Trihydroxyacetophenone, 2,4-*Di-Me*-6-*Et ether*
2,4,6-Trihydroxyacetophenone, 2,6-*Di-Me*-4-*Et ether*
2,4,6-Trihydroxyacetophenone, 2,4-*Di-Et ether*
2,4,6-Trihydroxybutyrophenone, 2,4-*Di-Me ether*
2,4,6-Trihydroxybutyrophenone, 2,6-*Di-Me ether*
2,4,6-Trihydroxyisovalerophenone, 2-*Me ether*†
2,4,6-Trihydroxyisovalerophenone, 4-*Me ether*†
2,4,5-Trihydroxypropiophenone, *Tri-Me ether*
3,4,5-Trihydroxypropiophenone, *Tri-Me ether*
Vanillic Acid, *Butyl ester*

$C_{12}H_{16}O_5$
2,4-Hydroxymandelic Acid, 2,4-*Di-Et ether*
2,5-Dimethylfuran-3,4-dicarboxylic Acid, *Di-Et ester*
Furan-2,5-dicarboxylic Acid, *Dipropyl ester*
Gallic Acid, 3,4-*Di-Me ether*, *Me ester*, *5-Et ether*
Gallic Acid, *Tri-Me ether*, *Et ester*
Methronic Acid, *Di-Et ester*
α-2,4,6-Tetrahydroxyacetophenone, *Tetra-Me ether*
α-3,4,5-Tetrahydroxyacetophenone, *Tetra-Me ether*

$C_{12}H_{16}O_5$ *(continued)*
2,3,4,6-Tetrahydroxyacetophenone, *Tetra-Me ether*
2,4,6-Trihydroxybenzoic Acid, 2,4,6-*Tri-Me ether*, *Et ester*
2,4,6-Trihydroxybenzoic Acid, 4-*Me*-2,6-*Di-Et ether*
2,4,5-Trimethoxybenzoic Acid, *Et ester*

$C_{12}H_{16}O_6$
2,5-Dioxocyclohexane-1,4-dicarboxylic Acid, *Di-Et ester*
2,5-Dioxocyclohexane-1,4-dicarboxylic Acid, *Isobutyl ester*
α,2,3,4,6-Pentahydroxyacetophenone, α-3,4,6-*Tetra-Me ether*
Phenol, α-D-Glucoside
2,3,4,6-Tetrahydroxybenzoic Acid, *Tetra-Me ether*, *Me ester*
Viridicatic Acid

$C_{12}H_{16}O_7$
Arbutin
Montagnetol

$C_{12}H_{16}O_8$
Phlorin

$C_{12}H_{17}AsO_2$
Benzenearsonous Acid, *Di-propyl ester*

$C_{12}H_{17}BrO_8$
2,3,4-Tri-*O*-acetyl-α-D-glucopyranosyl Bromide†

$C_{12}H_{17}HgNO_2$
Mercuri-*p*-diethylaminophenyl acetate

$C_{12}H_{17}IO$
5-Hydroxy-2-iodo-4-isopropyltoluene, *Et ether*

$C_{12}H_{17}N$
N-Benzylpiperidine
2-Benzylpiperidine
3-Benzylpiperidine
4-Benzylpiperidine
N-Cyclohexylaniline
1-Ethyl-1,2,3,4-tetrahydro-2-methylisoquinoline
1-Ethyl-1,2,3,4-tetrahydro-2-methylquinoline
2-Ethyl-1,2,3,4-tetrahydro-1-methylquinoline
2-Ethyl-1,2,3,4-tetrahydro-3-methylquinoline
2-Methyl-1-phenylpiperidine
2-Methyl-3-phenylpiperidine
2-Methyl-6-phenylpiperidine
2-Phenylcyclohexylamine
4-Phenylcyclohexylamine
Piperidine, N-*Benzyl*
1,2,3,4-Tetrahydroisoquinoline, N-*Propyl*
1,2,3,4-Tetrahydroisoquinoline, N-*Isopropyl*
5,6,7,8-Tetrahydro-1-naphthylamine, N-*Di-Me*
5,6,7,8-Tetrahydro-1-naphthylamine, N-*Et*
1,2,3,4-Tetrahydro-2-naphthylamine, N-*Di-Me*
1,2,3,4-Tetrahydro-2-naphthylamine, N-*Et*
5,6,7,8-Tetrahydro-2-naphthylamine, N-*Di-Me*
5,6,7,8-Tetrahydro-2-naphthylamine, N-*Et*
1,2,3,4-Tetrahydro-*N*-propylquinoline
1,2,3,4-Tetrahydro-2-propylquinoline
5,6,7,8-Tetrahydro-2-propylquinoline
1,2,3,4-Tetrahydro-1,2,2-trimethylquinoline
1,2,3,4-Tetrahydro-1,2,4-trimethylquinoline
1,2,3,4-Tetrahydro-1,2,6-trimethylquinoline
1,2,3,4-Tetrahydro-1,2,8-trimethylquinoline
1,2,3,4-Tetrahydro-1,6,8-trimethylquinoline
1,2,3,4-Tetrahydro-2,2,4-trimethylquinoline
1,2,3,4-Tetrahydro-2,4,6-trimethylquinoline
1,2,3,4-Tetrahydro-2,4,7-trimethylquinoline
1,2,3,4-Tetrahydro-2,4,8-trimethylquinoline
1,2,3,4-Tetrahydro-2,6,8-trimethylquinoline
1,2,3,4-Tetrahydro-5,6,8-trimethylquinoline
5,6,7,8-Tetrahydro-2,3,4-trimethylquinoline
5,6,7,8-Tetrahydro-2,4,6-trimethylquinoline
5,6,7,8-Tetrahydro-2,4,7-trimethylquinoline
5,6,7,8-Tetrahydro-2,4,8-trimethylquinoline
5,6,7,8-Tetrahydro-2,6,7-trimethylquinoline

$C_{12}H_{17}NO$
o-*n*-Butylaniline, N-*Ac*
p-*n*-Butylaniline, N-*Ac*
p-2-Butylaniline, N-*Ac*
o-*tert*-Butylaniline, N-*Ac*
m-*tert*-Butylaniline, N-*Ac*
p-*tert*-Butylaniline, N-*Ac*
N-Butylaniline, N-*Ac*
N-*tert*-Butylaniline, N-*Ac*
NN-Diethyl-*m*-toluamide†
p-Isobutylaniline, N-*Ac*
2-Isopropyl-3-phenylpropionic Acid, *Amide*
3-*p*-Isopropylphenylpropionic Acid, *Amide*
Pentamethylbenzoic Acid, *Amide*
2-Phenylhexanoic Acid, *Amide*
6-Phenylhexanoic Acid, *Amide*
1,2,3,4-Tetrahydro-6-methoxyquinoline, N-*Et*
2,3,4,6-Tetramethylaniline, N-*Ac*
2,3,5,6-Tetramethylaniline, N-*Ac*

$C_{12}H_{17}NOS$
Phenylthiocarbamic Acid, O-3-*Methylbutyl ester*
Phenylthiocarbamic Acid, S-3-*Methylbutyl ester*

$C_{12}H_{17}NO_2$
m-Aminobenzoic Acid, (+)-2-*Pentyl ester*
p-Aminobenzoic Acid, 1-*Pentyl ester*
p-Aminobenzoic Acid, (+)-2-*Pentyl ester*
4-*o*-Aminophenylbutyric Acid, *Et ester*
2-Amino-4-phenylbutyric Acid, *Et ester*
2-Anilinobutyric Acid, *Et ester*
Anthranilic Acid, 3-*Methylbutyl ester*
Bicyclo[2,2,2]octane-1,4-dicarboxylic Acid, *Mono-Et ester*, *Nitrile*
p-Butylaminobenzoic Acid, *Me ester*
α-Diethylamino-*o*-toluic Acid
α-Diethylamino-*p*-toluic Acid
Ethyl *N*-ethyl-*o*-toluidinoformate
4-Hydroxy-5-isopropyl-*o*-toluic Acid, *Me ether*, *Amide*
4-Hydroxy-6-isopropyl-*m*-toluic Acid, *Me ether*, *Amide*
3-Methylamino-3-phenylpropionic Acid, *Et ester*
Phenylcarbamic Acid, 1-*Methylbutyl ester*
N-Phenylglycine, N-*Et*, *Et ester*
N-Phenylsarcosine, *Propyl ester*
Pyridine-4-carboxylic Acid, n-*Hexyl ester*
Salsolidine
o-Tolylcarbamic Acid, *Butyl ester*
Valine, *Benzyl ester*

$C_{12}H_{17}NO_3$
Angoline†
Anhalidine
Anhalinine
Anhalonidine†
Calycotomine
6-Isopropyl-2-nitro-*m*-cresol, *Et ether*
6-Isopropyl-4-nitro-*m*-cresol, *Et ether*
Lodal
2-(3,4,5-Trimethoxyphenyl)cyclopropylamine†

$C_{12}H_{17}NO_4$
1,2-Dihydroxy-4-nitro-5-propylbenzene, 2-*Me ether*, 1-*Et ether*
4,5-Dimethylpyrrole-2,3-dicarboxylic Acid, *Di-Et ester*
3,5-Dimethylpyrrole-2,4-dicarboxylic Acid, *Di-Et ester*
2,5-Dimethylpyrrole-3,4-dicarboxylic Acid, *Di-Et ester*
Pyrrole-2,5-diacetic Acid, *Di-Et ester*
1,2,3-Trihydroxy-5-nitrobenzene, *Tri-Et ether*
1,2,4-Trihydroxy-5-nitrobenzene, *Tri-Et ether*
3,4,5-Trimethoxyphenylalanine

$C_{12}H_{17}NO_6$
Mannonic Acid, *Anilide*

$C_{12}H_{17}NS_2$
Phenyldithiocarbamic Acid, 3-*Methylbutyl ester*

$C_{12}H_{17}N_2O_4P$
Psilocybin★†

$C_{12}H_{17}N_3O_4$
Agaritine

$C_{12}H_{17}N_5O_4$
N^6-Dimethyladenosine

$C_{12}H_{18}$
tert-Butyl-2,6-dimethylbenzene†
5-*tert*-Butyl-*m*-xylene
Cogeijerene
1,5,9-Cyclododecatriene†
Geijerene★†
Hexamethylbenzene
Hexamethylbicyclo[2,2,0]hexa-2,5-diene†
Hexamethylprismane†
4-3′-Methylbutyltoluene
1-Phenylhexane
2-Phenylhexane
3-Phenylhexane
Tetracyclo[6,3,1,$0^{2,6}$,$0^{5,10}$]dodecane†
1,2,4-Triethylbenzene
1,3,5-Triethylbenzene
Tri-isopropylidenecyclopropane†

$(C_{12}H_{19}CaO_{14}S_2)_n$
Fucoidin

$C_{12}H_{18}Cl_2N_4OS$
Aneurin

$C_{12}H_{18}N_2O$
α-Diethylamino-*o*-toluic Acid, *Amide*
α-Diethylamino-*p*-toluic Acid, *Amide*
p-Nitrosoaniline, N-*Dipropyl*

$C_{12}H_{18}N_2O_2$
p-Dimethylamino-*N*-phenylalanine†
p-Nitrobenzylamine, N-*Isopentyl*
2-Nitro-*N*-dipropylaniline
4-Nitro-*N*-dipropylaniline

$C_{12}H_{18}N_2O_3$
4-Amino-6-isopropyl-2-nitro-*m*-cresol, *Et ether*
Nealbarbitone

$C_{12}H_{18}N_2O_3S$
Tolbutamide

$C_{12}H_{18}N_2O_4$
Furosine†
2-Methylenecycloheptene-1,3-diglycine†

$C_{12}H_{18}N_2O_5$
Bacilysin†
Hypoglycin B†

$C_{12}H_{18}N_2O_{12}$
Endecaphyllin D
Endecaphyllin E

$C_{12}H_{18}N_6$
Triallyltricyanamide

$C_{12}H_{18}O$
Benzyl 3-methylbutyl Ether
5-*tert*-Butyl-*o*-cresol, *Me ether*
4-*tert*-Butyl-*m*-cresol, *Me ether*
6-2′-Butyl-*m*-cresol, *Me ether*
6-*tert*-Butyl-*m*-cresol, *Me ether*
2-*tert*-Butyl-*p*-cresol, *Me ether*
o-Butylphenol, *Et ether*
p-*tert*-Butylphenol, *Et ether*
Carvacrol, *Et ether*
Hexamethyl-2,4-cyclohexadienone†
4-Isopropyl-*o*-cresol, *Et ether*
6-Isopropyl-*m*-cresol, *Et ether*
o-Isopropylphenol, *Isopropyl ether*
p-(2-Methyl-2-butyl)phenol, *Me ether*
p-(3-Methyl-1-butyl)phenol, *Me ether*
2-Methyl-1-phenyl-1-pentanol
2-Methyl-5-phenyl-1-pentanol
3-Methyl-1-phenyl-1-pentanol
3-Methyl-5-phenyl-1-pentanol
4-Methyl-1-phenyl-1-pentanol
4-Methyl-2-phenyl-1-pentanol
2-Methyl-5-phenyl-2-pentanol
3-Methyl-2-phenyl-2-pentanol
3-Methyl-4-phenyl-2-pentanol
4-Methyl-1-phenyl-1-pentanol
4-Methyl-1-phenyl-2-pentanol
4-Methyl-4-phenyl-2-pentanol
2-Methyl-3-phenyl-3-pentanol
2-Methyl-4-phenyl-3-pentanol
2-Methyl-5-phenyl-3-pentanol
Pentamethylphenol, *Me ether*
1-Phenyl-1-hexanol
2-Phenyl-1-hexanol
3-Phenyl-1-hexanol
4-Phenyl-1-hexanol
6-Phenyl-1-hexanol
2-Phenyl-2-hexanol
6-Phenyl-2-hexanol
1-Phenyl-3-hexanol
2-Phenyl-3-hexanol
3-Phenyl-3-hexanol
4-Phenyl-3-hexanol
1-Phenyl-2-propanol, *Propyl ether*
2,3,4,6-Tetramethylphenol, *Et ether*

$C_{12}H_{18}O$ (*continued*)
α,α,4-Trimethylbenzyl Alcohol, *Et ether*
2,4,6-Trimethylbenzyl Alcohol, *Et ether*

$C_{12}H_{18}OS$
p-Hydroxybenzenethiol, S-n-*Pentyl, Me ether*
p-Hydroxybenzenethiol, S-n-*Hexyl*

$C_{12}H_{18}O_2$
3-Acetylcamphor
Cnidilide†
Dicadic Acid, *Me ester*
α,β-Dihydroxyethylbenzene, *Di-Et ether*
3,6-Dihydroxy-*o*-xylene, *Di-Et ether*
4,6-Dihydroxy-*o*-xylene, *Di-Et ether*
4,6-Dihydroxy-*m*-xylene, *Di-Et ether*
2,5-Dihydroxy-*p*-xylene, *Di-Et ether*
Dodeca-2,4,6,8-tetraene-1,12-diol†
2-Hexylresorcinol
4-Hexylresorcinol
3-Hydroxymethylenecamphor, *Me ether*
Isophthalyl Alcohol, *Di-Et ether*
p-Isopropylbenzaldehyde, *Di-Me acetal*
2-Isopropyl-5-methylquinol, *Di-Me ether*
4,4′-Methylpentylresorcinol
Neocnidilide†
Norecsantalic Acid, *Me ester*
Phthalyl Alcohol, *Di-Et ether*
Resorcinol, *Dipropyl ether*
Resorcinol, *Monohexyl ether*
Sedanolic Acid, *Lactone*
Shonanic Acid, *Et ester*
Terephthalyl Alcohol, *Di-Et ether*
p-Toluic Aldehyde, *Di-Et acetal*
Tricyclenic Acid, *Et ester*

$C_{12}H_{18}O_2P_6S$
O,O-Dibutylphosphorothiolic *O,O*-diethylphosphorothiolic anhydrosulphide

$C_{12}H_{18}O_3$
Corchoritin
Cyclopentanecarboxylic Acid, *Anhydride*
Filicinic Acid, *Di-Et ether*
Furan-2-carboxylic Acid, *Heptyl ester*
Jasmonic Acid
Ligusticumic Acid, *Me ester*
9-Oxo-2-decenoic Acid, *Me ester*
Phloroglucinol, *Tri-Et ether*
Pyrogallol, *Tri-Et ether*
Sedanonic Acid
1,2,4-Trihydroxybenzene, 1,2,4-*Tri-Et ether*

$C_{12}H_{18}O_3S$
Toluene-*p*-sulphonic Acid, *Pentyl ester*

$C_{12}H_{18}O_4$
Bicyclo[2,2,2]octane-1,4-dicarboxylic Acid, *Mono-Et ester*
ω-Camphanic Acid, *Et ester*
1-Cyclohexene-1,2-dicarboxylic Acid, *Di-Et ester*
1-Cyclohexene-1,3-dicarboxylic Acid, *Di-Et ester*
2-Cyclohexene-1,2-dicarboxylic Acid, *Di-Et ester*
1,2,3,5-Tetrahydroxybenzene, 1,2,3-*Tri-Et ether*
1,2,4,5-Tetrahydroxybenzene, 1,4-*Dipropyl ether*

$C_{12}H_{18}O_5$
8-Hydroxy-2,7-dimethyldeca-2,4-dienedioic Acid†
α-Longinecic Acid, *Di-Me ester*
Seneciphyllic Acid, *Di-Me ester*†

$C_{12}H_{18}O_6$
Aconitic Acid, *Tri-Et ester*
Biacetyl, *Trimer*
Cyclopropane-1,2,3-tricarboxylic Acid, *Tri-Et ester*
2,3-Diacetylsuccinic Acid, *Di-Et ester*
Hexa(hydroxymethyl)benzene
Isaconitic Acid, *Tri-Et ester*
cis-2-Pentene-2,3,5-tricarboxylic Acid, *Di-Et ester*

$C_{12}H_{18}O_6S_2$
m-Xylene-4,6-disulphonic Acid, *Di-Et ester*

$C_{12}H_{18}O_7$
Fructuronic Acid, *Di-O-isopropylidene deriv.*
Oxalsuccinic Acid, *Tri-Et ester*

$C_{12}H_{18}O_8$
Butane-1,1,4,4-tetracarboxylic Acid, *Tetra-Me ester*
Butane-1,2,3,4-tetracarboxylic Acid, *Tetra-Me ester*
Butane-1,2,3,4-tetracarboxylic Acid, *Di-Et ester*
Hexane-1,3,4,6-tetracarboxylic Acid, *Di-Me ester*

$C_{12}H_{18}O_9$
3-Formylpropionic Acid, *Trimer*

$C_{12}H_{18}O_9S_3$
Benzene-1,3,5-trisulphonic Acid, *Tri-Et ester*

$C_{12}H_{19}BrO_4S$
3-Bromo-(+)-camphor-10-sulphonic Acid, *Et ester*

$C_{12}H_{19}ClN_4O_2P_7S$
Cocarboxylase

$C_{12}H_{19}N$
γ-Aminobutylbenzene, N-*Di-Me*
N-Benzyl-3-methylbutylamine
N-Butyl-*N*-ethylaniline
$\Delta^{4\alpha}$-Decahydrocarbazole
N-Di-isopropylaniline
2,4-Dimethylaniline, N-*Butyl*
2,6-Dimethylaniline, N-*Di-Et*
N-Dipropylaniline
N-Ethyl-*N*-isobutylaniline
N-3-Methylbutyl-*o*-toluidine
N-Methyl-*N*-2-methylbutylaniline
N-Methyl-*N*-3-methylbutylaniline
2,3,4,5,6-Pentamethylaniline, N-*Me*
3-Phenyl-1-propylamine, N-*Propyl*

$C_{12}H_{19}NO$
2-Amino-6-isopropyl-*m*-cresol, *Et ether*
4-Amino-6-isopropyl-*m*-cresol, *Et ether*
Dicadylamine, *Ac*
o-Diethylaminophenol, *Et ether*
m-Diethylaminophenol, *Et ether*

$C_{12}H_{19}NO_2$
α-Aminomethyl-*p*-hydroxybenzyl, N-*Butyl*
α-Aminomethyl-*p*-hydroxybenzyl, N-*Isobutyl*
α-Aminomethyl-*p*-hydroxybenzyl, N-sec-*Butyl*
α-Aminomethyl-*p*-hydroxybenzyl, N-tert-*Butyl*
α-Aminomethyl-*p*-hydroxybenzyl, N-*Di-Et*

$C_{12}H_{19}NO_2$ *(continued)*
2-Butylamino-1-*p*-hydroxyphenylethanol
3-Carboxy-2,2,3-trimethylcyclopentylacetonitrile, *Me ester*
1,2,2-Trimethylcyclopentane-1,3-dicarboxylic Acid, α-*Nitrile*, *Et ester*

$C_{12}H_{19}NO_2S$
Planchonelline†

$C_{12}H_{19}NO_2S_2$
Brugine†

$C_{12}H_{19}NO_2Si$
p-Trimethylsilylphenylalanine

$C_{12}H_{19}NO_3$
Bicyclo[2,2,2]octane-1,4-dicarboxylic Acid, *Mono-Et ester*, *Amide*
Macromerine†
(3,4,5-Trimethoxyphenyl)-2-propylamine

$C_{12}H_{19}NO_4$
3-Carboxymethyl-4-isopropenylpyrrolidine-2-carboxylic Acid, *Dimethyl ester*
3-Methylbutane-1,1,2-tricarboxylic Acid, 4-*Mono-nitrile*, *Di-Et ester*
2-Methylbutane-1,1,3-tricarboxylic Acid, 1-*Mono-nitrile*, *Di-Et ester*
2-Methylbutane-1,1,4-tricarboxylic Acid, 1-*Mono-nitrile*, *Di-Et ester*
2-Methylbutane-1,2,4-tricarboxylic Acid, 2-*Mono-nitrile*, *Di-Et ester*
2-Methylbutane-1,3,3-tricarboxylic Acid, 3-*Mono-nitrile*, *Di-Et ester*
3-Methylbutane-2,2,3-tricarboxylic Acid, 3-*Mono-nitrile*, *Di-Et ester*

$C_{12}H_{19}NO_6$
Fructuronic Acid, *Amide*

$C_{12}H_{19}N_3O_2$
4,6-Diamino-2-nitro-*m*-xylene, N,N′-*Di-Et*

$C_{12}H_{19}O_2PS_2$
O,O-Diethyl-*S*-α-methylbenzylphosphorothiolothionate

$C_{12}H_{20}$
Bicyclohexylidene†
[4,4,2]Propellane†

$C_{12}H_{20}ClN_3O$
Isochaksine, *Chloride*

$C_{12}H_{20}N_2$
α,α′-Diamino-*o*-xylene, N-*Tetra-Me*
Dodecanedioic Acid, *Dinitrile*
α-Matrinidine
p-Phenylenediamine, N,N′-*Dipropyl*
Tetraethylsuccinic Acid, *Dinitrile*

$C_{12}H_{20}N_2O$
Ammodendrine
5-(2-Butyl)-3-hydroxy-2-isobutylpyrazine
Flavacol
Isoammodendrine

$C_{12}H_{20}N_2O_2$
Aspergillic Acid
Granegillin
Neoaspergillic Acid†

$C_{12}H_{20}N_2O_3$
Hydroxyaspergillic Acid
Neohydroxyaspergillic Acid†

$C_{12}H_{20}N_2O_4$
Pulcherriminic Acid

$C_{12}H_{20}N_4O_3$
L-Isoleucyl-L-histidine

$C_{12}H_{20}O$
5,9-Dimethyl-4,8-decadien-2-one
3-*cis*-6-*cis*-8-*trans*-Dodecatrien-1-ol†
6-Isopropylidene-3-methyl-1-cyclohexen-1-ol, *Et ether*

$C_{12}H_{20}O_2$
Aleprylic Acid
Buchu-camphor, *Et ether*
1,2-Cyclododecanedione†
2,2′-Cyclopentenylpropionic Acid, *Isobutyl ester*
2,8-Dodecadienoic Acid
Dodec-2-en-5-olide†
Geranic Acid, *Et ester*
Isoborneol, *Ac*
2-Methylcyclohexaneacetic Acid, *Et ester*
3-Methylcyclohexaneacetic Acid, *Et ester*
2-Octynoic Acid, *Isobutyl ester*
2,6,6-Trimethyl-2-cyclohexene-1-carboxylic Acid, *Et ester*
2,2,6-Trimethyl-3-cyclohexene-1-carboxylic Acid, *Et ester*
2,6,6-Trimethyl-3-cyclohexene-1-carboxylic Acid, *Et ester*
3,4,6-Trimethyl-3-cyclohexene-1-carboxylic Acid, *Et ester*
2,2,3-Trimethyl-3-cyclopentenylacetic Acid, *Et ester*
2,3,3-Trimethyl-1-cyclopentenylacetic Acid, *Et ester*
9-Undecynoic Acid, *Me ester*
10-Undecynoic Acid, *Me ester*

$C_{12}H_{20}O_3$
3-Acetyl-2,2-dimethylcyclobutaneacetic Acid, *Et ester*
10-Hydroxy-3,7-dimethyldeca-2,6-dienoic Acid†
Sedanolic Acid
Tetraethylsuccinic Acid, *Anhydride*

$C_{12}H_{20}O_4$
α-Carboxy-γ-undecalactone†
Cyclohexane-1,2-dicarboxylic Acid, *Di-Et ester*
Cyclohexane-1,3-dicarboxylic Acid, *Di-Et ester*
Cyclohexane-1,4-dicarboxylic Acid, *Di-Et ester*
Decanedioic Acid, *Glycol ester*
2,2-Dimethylcyclobutane-1,3-dicarboxylic Acid, *Di-Et ester*
2-Dodecenedioic Acid
4-Ethyl-3-methylglutaconic Acid, *Di-Et ester*
Fumaric Acid, *Di-butyl ester*
Isobutylmalonic Acid, *Di-Et ester*
2-Methylcyclopentane-1,1-dicarboxylic Acid, *Di-Et ester*
3-Methylcyclopentane-1,1-dicarboxylic Acid, *Di-Et ester*
1-Methylcyclopentane-1,2-dicarboxylic Acid, *Di-Et ester*

$C_{12}H_{20}O_4$ (*continued*)
3-Methylcyclopentane-1,2-dicarboxylic Acid, *Di-Et ester*
1,2,2-Trimethylcyclopentane-1,3-dicarboxylic Acid, *Di-Me ester*
1,2,2-Trimethylcyclopentane-1,3-dicarboxylic Acid, α-*Et ester*
1,2,2-Trimethylcyclopentane-1,3-dicarboxylic Acid, β-*Et ester*

$C_{12}H_{20}O_4Ti$
Tetra-allyl titanate

$C_{12}H_{20}O_5$
3-Acetonylglutaric Acid, *Di-Et ester*†
Phoronic Acid, *Me ester*
Tetrahydro-2,6,6-trimethylpyran-2,5dicarboxylic Acid, *Di-Me ester*

$C_{12}H_{20}O_6$
2,3-Dimethylbutane-1,2,3-tricarboxylic Acid, *Tri-Me ester*
Hexane-1,2,6-tricarboxylic Acid, *Tri-Me ester*
Propionic Acid, *Glycerol tri-ester*
Tricarballylic Acid, *Tri-Et ester*

$C_{12}H_{20}O_7$
Isocitric Acid, *Tri-Et ester*
Triethyl citrate

$C_{12}H_{20}O_9$
Lactal
Maltal
ψ-Maltal

$C_{12}H_{20}O_{10}$
Agarobiose
Carrobiose†
Di-D-fructose Anhydride I†
Lactosan
Maltosan

$C_{12}H_{20}O_{12}$
Acaciabiuronic Acid
Maltobiouronic Acid†

$C_{12}H_{20}S_2$
2,2:4,4-Bispentamethylene-1,3-dithietan†

$C_{12}H_{21}ClO_2$
tert-Butyl-4-chloro-2-methylcyclohexanecarboxylate†
tert-Butyl-5-chloro-2-methylcyclohexanecarboxylate†
3-*p*-Menthoxyacetic Acid, *Chloride*

$C_{12}H_{21}Cl_3O_3$
3-Chlorobutyraldehyde, *Trimeride*

$C_{12}H_{21}N$
2,5-Di-*tert*-butylpyrrole†
11-Dodecenoic Acid, *Nitrile*

$C_{12}H_{21}NO$
2,2,6,6-Tetramethyl-4-piperidone, N-*Allyl*

$C_{12}H_{21}NO_2$
Butropine
Isoporoidine
Poroidine

$C_{12}H_{21}NO_3$
3-Carbamoyl-1,2,2-trimethylcyclopentanecarboxylic Acid, N-*Me, Me ester*
3-Carbamoyl-1,2,2-trimethylcyclopentanecarboxylic Acid, N-*Di-Me*
3-Carbamoyl-1,2,2-trimethylcyclopentanecarboxylic Acid, N-*Et*
3-Carbamoyl-2,2,3-trimethylcyclopentanecarboxylic Acid, *Et ester*
3-Carbamoyl-2,2,3-trimethylcyclopentanecarboxylic Acid, N-*Me, Me ester*
Fuchsisenecionine

$C_{12}H_{21}NO_4$
Tropinic Acid, *Di-Et ester*

$C_{12}H_{21}NO_6$
Tri-(carboxymethyl)amine, *Tri-Et ester*

$C_{12}H_{21}NO_8$
Streptobiosamine

$C_{12}H_{21}NO_9S_2$
Glucobrassicanapin

$C_{12}H_{21}NO_9S_3$
trans-4-Methylthio-3-butenylglucosinolate†

$C_{12}H_{21}NO_{10}$
N-Glycollyl-8-*O*-methylneuraminic Acid†
Melibionic Acid, *Nitrile*

$C_{12}H_{21}NO_{11}$
Chondrosine
Hyalbiuronic Acid

$C_{12}H_{21}N_2O_3PS$
Diazinon

$C_{12}H_{21}N_3O_3$
Cyanuric Acid, O-*Tri*-n-*propyl ester*
Cyanuric Acid, O-*Tri-isopropyl ester*
Hexahydro-1,3,5-tripropionyl-*s*-triazine†

$C_{12}H_{21}N_3O_8$
2-Acetamido-1β-(L-β-aspartamido)-1,2-dideoxy-D-glucose†

$C_{12}H_{21}N_3O_{12}P_2$
Cytidine diphosphate glycerol

$C_{12}H_{22}$
Bicyclohexyl
Cyclododecene†
2-Dodecyne

$C_{12}H_{22}N_2$
Quinanthradine

$C_{12}H_{22}N_2O_4$
2,2′-Azodi-2-methylpropionic Acid, *Di-Et ester*
Cobactin

$C_{12}H_{22}O$
Borneol, *Et ether*
Cyclododecanone
2-Dodecenal†
Geraniol, *Et ether*
Isoborneol, *Et ether*
2-Isopropenyl-5-methylcyclohexanol, *Et ether*
Linalool, *Et ether*
9-Methoxyundecyne

$C_{12}H_{22}O_2$
Citronellic Acid, *Et ester*
2-Dodecenoic Acid
5-Dodecenoic Acid
6-Dodecenoic Acid

$C_{12}H_{22}O_2$ (*continued*)
7-Dodecenoic Acid
9-Dodecenoic Acid
10-Dodecenoic Acid
11-Dodecenoic Acid
2-Isopropyl-5-methylcyclopentane-1-carboxylic Acid, *Et ester*
3-Isopropyl-1-methylcyclopentane-1-carboxylic Acid, *Et ester*
2-*p*-Menthylacetic Acid
1,2,2,3-Tetramethylcyclopentane-1-carboxylic Acid, *Et ester*
2,2,3,3-Tetramethylcyclopentane-1-carboxylic Acid, *Et ester*
2,2,3-Trimethylcyclopentylacetic Acid, *Et ester*
10-Undecenoic Acid, *Me ester*

$C_{12}H_{22}O_3$
2,2-Dimethylbutyric Acid, *Anhydride*
2-Ethylbutyric Acid, *Anhydride*
Hexanoic Acid, *Anhydride*
2-Hydroxy-1-methylcyclohexane-1-carboxylic Acid, *Isopropyl ester*, *Me ether*
Levulinic Acid, *Heptyl ester*
3-*p*-Menthoxyacetic Acid
4-Methylvaleric Acid, *Anhydride*
3-Oxodecanoic Acid, *Et ester*
8-Oxodecanoic Acid, *Et ester*
9-Oxodecanoic Acid, *Et ester*
9-Oxododecanoic Acid
4-Oxoundecanoic Acid, *Me ester*

$C_{12}H_{22}O_4$
Adipic Acid, *Dipropyl ester*
Adipic Acid, *Di-isopropyl ester*
Butylmethylmalonic Acid, *Di-Et ester*
(2-Butyl)succinic Acid, *Di-Et ester*
Decane-1,9-dicarboxylic Acid
Decanedioic Acid, *Di-Me ester*
Decanedioic Acid, *Et ester*
2,3-Diethylsuccinic Acid, *Di-Et ester*
2,2-Dimethyladipic Acid, *Di-Et ester*
2,4-Dimethyladipic Acid, *Di-Et ester*
Dodecanedioic Acid
2-Ethylbutane-1,1-dicarboxylic Acid, *Di-Et ester*
Hexane-3,3-dicarboxylic Acid, *Di-Et ester*
Homononactinic Acid, *Me ester*
2-Isopropylpropane-1,3-dicarboxylic Acid, *Di-Et ester*
(2-Methyl-1-butyl)malonic Acid, *Di-Et ester*
(2-Methyl-2-butyl)malonic Acid, *Di-Et ester*
6-Methylheptane-2,5-dicarboxylic Acid, *Di-Me ester*
4-Methylpentane-1,1-dicarboxylic Acid, *Di-Et ester*
4-Methylpentane-1,3-dicarboxylic Acid, *Di-Et ester*
4-Methylpentane-2,2-dicarboxylic Acid, *Di-Et ester*
2-Methylpimelic Acid, *Di-Et ester*
3-Methylpimelic Acid, *Di-Et ester*
4-Methylpimelic Acid, *Di-Et ester*
Octanedioic Acid, *Di-Et ester*
Oxalic Acid, *Di-pentyl ester*
2-Pentylmalonic Acid, *Di-Et ester*
3-Propylglutaric Acid, *Di-Et ester*
Succinic Acid, *Dibutyl ester*
Tetraethylsuccinic Acid
Tetramethylsuccinic Acid, *Di-Et ester*
2,2,3-Trimethylglutaric Acid, *Di-Et ester*
2,2,4-Trimethylglutaric Acid, *Di-Et ester*
2,3,3-Trimethylglutaric Acid, *Di-Et ester*
2,2,3-Trimethylpimelic Acid, *Di-Me ester*
Undecanedioic Acid, *Mono-Me ester*

$C_{12}H_{22}O_4Si$
Triallyl isopropyl orthosilicate

$C_{12}H_{22}O_5$
2-Ethyl-3-hydroxymethylglutaric Acid, *Di-Et ester*
3-Hydroxy-2,2-dimethylbutyric Acid, *Anhydride*
3-Hydroxy-2,2,3-trimethylglutaric Acid, *Di-Et ester*
Malic Acid, *Dibutyl ester*
Malic Acid, *Di-isobutyl ester*
Malic Acid, *Et ether*, *Dipropyl ester*
Olivomycose, 4-O-*Isobutyrylmethylglycoside*†

$C_{12}H_{22}O_6$
1,2-Diethoxysuccinic Acid, *Di-Et ester*
Tartaric Acid, *Dibutyl ester*
Tartaric Acid, *Di-isobutyl ester*

$C_{12}H_{22}O_7$
*Gluco*heptonic Acid, *γ-Lactone*, 1,2,4,5,6-*Penta-Me ether*

$C_{12}H_{22}O_9$
Digilanidobiose

$C_{12}H_{22}O_{10}$
Eryscenobiose†
Neohesperidose†
Rutinose
Scillabiose
Sophorabiose

$C_{12}H_{22}O_{10}S$
Thiodigalactoside†

$C_{12}H_{22}O_{11}$
Cellobiose
Cellobiulose
Galactinol
Galactobiose
3-*O*-β-D-Galactopyranosyl-D-galactopyranose
4-*O*-α-D-Galactopyranosyl-D-galactopyranose
4-*O*-β-D-Galactopyranosyl-D-galactopyranose
O-α-D-Galactopyranosyl-α-D-galactopyranoside
O-α-D-Galactopyranosyl-β-D-galactopyranoside
O-β-D-Galactopyranosyl-β-D-galactopyranoside
Gentiobiose
6-*O*-β-D-Galactofuranosyl-D-galactose†
O-β-D-Galactopyranosyl-(1 → 4)-D-galactose†
O-β-D-Glucopyranosyl-(1 → 6)-D-galactose†
Inulobiose
Isotrehalose
Kojibiose
Lactose
Lactulose
Laminaribiose
Leucrose
Lycobiose
Maltose
Mannobiose
1-*O*-α-D-Mannopyranosyl-*myo*inositol

$C_{12}H_{22}O_{11}$ (*continued*)
Melibiose
Neolactose
Sakebiose
Sophorose
Sucrose
Solabiose†
α,α-Trehalose
α,β-Trehalose
β,β-Trehalose
Turanose

$C_{12}H_{22}O_{12}$
Lactobionic Acid
Maltobionic Acid
Melibionic Acid

$C_{12}H_{23}BrO_2$
2-Bromodecanoic Acid, *Et ester*
2-Bromododecanoic Acid
11-Bromododecanoic Acid
12-Bromododecanoic Acid

$C_{12}H_{23}ClO$
Dodecanoic Acid, *Chloride*

$C_{12}H_{23}N$
4-(2-Aminoethyl)-1,5,5-trimethylcyclopentene, N-*Di-Me*
Dicyclohexylamine
Dodecanoic Acid, *Nitrile*

$C_{12}H_{23}NO$
Azacyclotridecan-2-one
1-Lupinine, *Et ether*
4-Lupinine, *Et ether*
Undecyl isocyanate†

$C_{12}H_{23}NOS_2$
(−)-Methylsulphinyldecyl isothiocyanate

$C_{12}H_{23}NO_2$
3-*p*-Menthoxyacetic Acid, *Amide*
(−)-3-Menthylglycine

$C_{12}H_{23}NO_4$
Iminodiacetic Acid, *Dibutyl ester*
3,3′-Iminodibutyric Acid, *Di-Et ester*
2,3′-Iminodipropionic Acid, *Di-Butyl ester*

$C_{12}H_{23}NO_9S_3$
Glucoerucin

$C_{12}H_{23}NO_{10}$
O-α-D-Glucopyranosyl(1→4)-2-amino-2-deoxy-α-D-glucopyranose†
Maltosimine

$C_{12}H_{23}N_3O_4$
Glycyl-leucyl-glycine, *Et ester*

$C_{12}H_{23}N_5O_7$
N-Guan-Streptolidyl Gulosaminide

$C_{12}H_{23}O_{14}P$
Trehalose-6-phosphate†

$C_{12}H_{24}$
Butylcyclo-octane
2-Butylcyclo-octane
Cyclododecane†

$C_{12}H_{24}BrN$
5-Azoniaspiro[4,8]tridecane, *Bromide*

$C_{12}H_{24}Br_2$
1,2-Dibromododecane
1,5-Dibromododecane
1,12-Dibromododecane
6,7-Dibromododecane

$C_{12}H_{24}Cl_2$
1,12-Dichlorododecane

$C_{12}H_{24}N_2$
4,4′-Diaminobicyclohexyl

$C_{12}H_{24}N_2O_2$
Dodecanedioic Acid, *Diamide*
Tetraethylsuccinic Acid, *Diamide*

$C_{12}H_{24}N_2O_4$
Carisoprodol
2,2′-Hydrazodi-(2-methylpropionic) Acid, *Di-Et ester*

$C_{12}H_{24}N_2O_4S$
Thiodiglycollic Acid, Bis-*dimethylaminoethyl ester*

$C_{12}H_{24}N_2O_5$
Cobactinin

$C_{12}H_{24}N_2O_7$
Kasuganobiosamine†

$C_{12}H_{24}N_4$
1,4,7,10-Tetra-azatricyclo[8,2,2,$2^{4,7}$]hexadecane

$C_{12}H_{24}O$
Cyclododecanol
Dodecanal
2-Dodecanone
3-Dodecanone
6-Dodecanone
Dodec-7-en-1-ol†
2-Isopropyl-5-methylcyclohexanol, *Et ether*
Lanolin Alcohol
3-Methyl-1,3′-methylbutylcyclohexanol
2-Methyl-3-undecanone

$C_{12}H_{24}O_2$
Decanoic Acid, *Et ester*
Dodecanoic Acid
2-Hexanol, *Hexanoyl*
Isovaleric Acid, *Heptyl ester*
Ketene, *Di-isopentyl acetal*
2-Methyldecanoic Acid, *Me ester*
3-Methyldecanoic Acid, *Me ester*
4-Methyldecanoic Acid, *Me ester*
8-Methyldecanoic Acid, *Me ester*
2-Methylnonanoic Acid, *Et ester*
3-Methylnonanoic Acid, *Et ester*
4-Methylnonanoic Acid, *Et ester*
5-Methylnonanoic Acid, *Et ester*
Nonanoic Acid, n-*Propyl ester*
Octanoic Acid, n-*Butyl ester*
2-Octanol, *Butyryl*
2-Octanol, *Isobutyryl*
Undecanoic Acid, *Me ester*

$C_{12}H_{24}O_3$
2-Hydroxydecanoic Acid, *Et ester*
2-Hydroxydodecanoic Acid
3-Hydroxydodecanoic Acid
4-Hydroxydodecanoic Acid
12-Hydroxydodecanoic Acid
2-Hydroxymethylnonanoic Acid, *Et ester*
11-Hydroxyundecanoic Acid, *Me ester*

$C_{12}H_{24}O_4$
Azelaialdehydic Acid, *Di-Me acetal*

$C_{12}H_{24}O_5$
Hydroxy-methylmalonic Acid, *Di-butyl ester*

$C_{12}H_{24}O_7$
*Gluco*heptose, 2,3,4,6,7-*Penta-Me ether*
Gluconic Acid, *Penta-Me ether*, *Me ester*

$C_{12}H_{24}O_{11}$
O-β-D-Galactofuranosyl(1 → 3)-D-mannitol†
1-*O*-β-D-Glucopyranosyl-D-mannitol
3-*O*-β-D-Glucopyranosyl-D-mannitol
Lactositol
D-Mannitol-1-β-D-glucoside
Melibiotol

$C_{12}H_{24}S_2$
1,8-Dithiacyclotetradecane

$C_{12}H_{25}Br$
1-Bromododecane

$C_{12}H_{25}BrO_2$
Bromoacetaldehyde, *Di-isoamyl acetal*

$C_{12}H_{25}Cl$
1-Chlorododecane

$C_{12}H_{25}ClO_2$
Chloroacetaldehyde, *Dipentyl acetal*

$C_{12}H_{25}ClO_2S$
1-Dodecanesulphonic Acid, *Chloride*

$C_{12}H_{25}I$
1-Iodododecane

$C_{12}H_{25}N$
Dodecamethyleneimine

$C_{12}H_{25}NO$
5-Azoniaspiro(4,8)tridecane(hydroxide)
Dodecanoic Acid, *Amide*

$C_{12}H_{25}NO_2$
Piperidine-3-aldehyde, *Et acetal*

$C_{12}H_{25}NO_3$
Desosamine, *Bu glycoside*
Pentylpenilloaldehyde, *Di-Et acetal*

$C_{12}H_{25}N_3O_7$
Paromamine†

$C_{12}H_{26}$
Dodecane

$C_{12}H_{26}N_2$
2,5-Di-2-butylpiperazine

$C_{12}H_{26}N_4O_6$
Neomycin A★†

$C_{12}H_{26}O$
1-Dodecanol
3-Dodecanol
6-Dodecanol
2-Methyl-2-undecanol
2-Methyl-3-undecanol
2-Methyl-5-undecanol
6-Methyl-6-undecanol
1-Octanol, n-*Butyl ether*
2-Octanol, n-*Butyl ether*
2,5,8-Trimethyl-5-nonanol

M

$C_{12}H_{26}O_2$
Acetaldehyde, *Dipentyl acetal*
Acetaldehyde, *Di*-3-*methylbutyl acetal*
1,2-Dodecanediol
1,9-Dodecanediol
1,10-Dodecanediol
1,12-Dodecanediol
2,3-Dodecanediol
2,11-Dodecanediol
4,9-Dodecanediol
6,7-Dodecanediol

$C_{12}H_{26}O_3$
Di(-2-hydroxyethyl) Ether, *Dibutyl ether*
Orthoacetic Acid, *Et di-butyl ester*
Orthoformic Acid, *Dipropyl*-3-*methylbutyl ester*

$C_{12}H_{26}O_3S$
1-Dodecanesulphonic Acid

$C_{12}H_{26}O_4S$
1-Dodecyl sulphate
2-Dodecyl sulphate

$C_{12}H_{26}O_6$
Mannitol, 1,2,3,4-*Tetra-Me ether*, 5-*Et ether*

$C_{12}H_{26}S$
1-Dodecanethiol

$C_{12}H_{27}AlO_3$
Aluminium butoxide
Aluminium 2-butoxide
Aluminium *tert*-butoxide
Aluminium isobutoxide

$C_{12}H_{27}BO_3$
Tributyl borate
Tri-2-butyl borate
Tri-*tert*-butyl borate
Tri-isobutyl borate

$C_{12}H_{27}N$
1-Aminododecane
Dihexylamine
Tributylamine
Tri-isobutylamine

$C_{12}H_{27}NO$
2-Di-(3-methylbutyl)aminoethanol

$C_{12}H_{27}O_3P$
Tributyl phosphite
Tri-*tert*-butyl phosphite

$C_{12}H_{27}O_3Sb$
Tributyl antimonate

$C_{12}H_{27}O_4P$
Tributyl phosphate
Tri-*tert*-butyl phosphate†

$C_{12}H_{27}P$
Tri-*tert*-butylphosphine†

$C_{12}H_{27}Tl$
Thallium tri-isobutyl

$C_{12}H_{28}N_2S_2$
4-Dimethylamino-1-butanethiol, *Disulphide*

$C_{12}H_{28}N_6$
Synthalin

$C_{12}H_{28}Pb$
Lead dibutyldiethyl
Lead tetra-isopropyl
Lead tetrapropyl

$C_{12}H_{28}S_2Si_2$
Tetrapropylcyclodisilthiane
$C_{12}H_{28}Sn$
Tin tetrapropyl
$C_{12}H_{29}N_3$
Solamine†
$C_{12}H_{30}Br_2N_2$
Hexamethonium, *Bromide*
$C_{12}H_{30}Cl_2N_2$
Hexamethonium, *Chloride*
$C_{12}H_{30}I_2N_2$
Hexamethonium, *Iodide*
$C_{12}H_{30}O_{13}P_4$
Tetraethyl pyrophosphate
$C_{12}H_{32}O_4Si_4$
2,4,6,8-Tetraethyl-2,4,6,8-tetramethylcyclotetrasiloxane
$C_{12}H_{36}Si_6$
Dodecamethylcyclohexasilane
$C_{12}O_9$
Mellitic Acid, *Trianhydride*

C_{13}

$C_{13}Cl_{11}$
Perchlorodiphenylmethyl†
$C_{13}F_8O$
Octafluorofluoren-9-one†
$C_{13}HBr_8O_2$
2,2′-Dibromobenzophenone
2,4-Dibromobenzophenone
2,4′-Dibromobenzophenone
2,6-Dibromobenzophenone
3,3′-Dibromobenzophenone
3,4-Dibromobenzophenone
3,4′-Dibromobenzophenone
3,5-Dibromobenzophenone
4,4′-Dibromobenzophenone
$C_{13}H_5N_3O_7$
2,4,7-Trinitrofluoroenone
$C_{13}H_5N_5O_9$
2,4,5,7-Tetranitroacridone
$C_{13}H_6Br_2O$
1,3-Dibromofluorenone
2,7-Dibromofluorenone
3,6-Dibromofluorenone
$C_{13}H_6Cl_2O$
1,6-Dichlorofluorenone
1,8-Dichlorofluorenone
2,5-Dichlorofluorenone
2,7-Dichlorofluorenone
3,6-Dichlorofluorenone
$C_{13}H_6Cl_4O$
2,2′,4,4′-Tetrachlorobenzophenone
2,2′,4′,5-Tetrachlorobenzophenone
2,2′,5,5′-Tetrachlorobenzophenone
3,3′,4,4′-Tetrachlorobenzophenone
$C_{13}H_6Cl_5NO_3$
Oxyclozanide†
$C_{13}H_6Cl_6O_2$
3,3′,5,5′,6,6′-Hexachloro-2,2′-dihydroxydiphenylmethane
$C_{13}H_6N_2O_5$
1,8-Dinitrofluorenone
2,4-Dinitrofluorenone
2,5-Dinitrofluorenone
2,7-Dinitrofluorenone
3,6-Dinitrofluorenone
4,5-Dinitrofluorenone
$C_{13}H_6N_2O_6$
1,5-Dinitroxanthone
1,6-Dinitroxanthone
1,7-Dinitroxanthone
2,4-Dinitroxanthone
2,5-Dinitroxanthone
2,6-Dinitroxanthone
2,7-Dinitroxanthone
3,5-Dinitroxanthone
$C_{13}H_6N_8O_{13}$
1,3-Di-(2,4,6-trinitrophenyl)urea
$C_{13}H_6O_3$
2,3-Dihydro-1*H*-phenalene-1,2,3-trione
$C_{13}H_6O_5$
Naphthalene-1,4,5-tricarboxylic Acid, *Anhydride*
$C_{13}H_7BrO$
1-Bromofluorenone
2-Bromofluorenone
3-Bromofluorenone
4-Bromofluorenone
$C_{13}H_7Br_2NO$
1,4-Dibromoacridone
3,7-Dibromoacridone
$C_{13}H_7Br_3O$
2,4,6-Tribromobenzophenone
$C_{13}H_7ClO$
1-Chlorofluorenone
2-Chlorofluorenone
3-Chlorofluorenone
$C_{13}H_7Cl_2NO$
1,4-Dichloroacridone
1,6-Dichloroacridone
2,3-Dichloroacridone
2,4-Dichloroacridone
$C_{13}H_7Cl_3O$
2,2′,5-Trichlorobenzophenone
2,4,4′-Trichlorobenzophenone
2,4,6-Trichlorobenzophenone
$C_{13}H_7NO$
Dibenzofuran-3-carboxylic Acid, *Nitrile*
$C_{13}H_7NO_2$
Benz[*g*]isoquinoline-5-10-dione
Benzo[*g*]quinoline-5,10-dione
$C_{13}H_7NO_3$
1-Nitrofluorenone
2-Nitrofluorenone
3-Nitrofluorenone
4-Nitrofluorenone

$C_{13}H_7NO_3S$
1-Nitrothioxanthen-9-one
2-Nitrothioxanthen-9-one
3-Nitrothioxanthen-9-one
4-Nitrothioxanthen-9-one

$C_{13}H_7NO_4$
2-Hydroxy-7-nitrofluorenone
1-Nitro-9-xanthenone
2-Nitro-9-xanthenone
3-Nitro-9-xanthenone
4-Nitro-9-xanthenone

$C_{13}H_7N_3$
Phenazine-1-carboxylic Acid, *Nitrile*
Phenazine-2-carboxylic Acid, *Nitrile*
4,5,9-Triazapyrene†

$C_{13}H_7N_3O$
4,5,9-Triazapyrene, *Oxide*†

$C_{13}H_7N_3O_4$
2,4-Dinitroacridine

$C_{13}H_7N_3O_5$
1,3-Dinitroacridanone
2,4-Dinitroacridanone
2,7-Dinitroacridanone
2,8-Dinitroacridanone
3,5-Dinitroacridanone
3,6-Dinitroacridanone
3,7-Dinitroacridanone
3,8-Dinitroacridanone
4,7-Dinitroacridanone
4,8-Dinitroacridanone

$C_{13}H_7N_3O_8$
2,4,6-Trinitrobenzoic Acid, *Phenyl ester*

$C_{13}H_7N_5O_8$
1,3,6,8-Tetranitrocarbazole, N-*Me*

$C_{13}H_7N_7O_{12}$
Di-2,4,6-trinitrophenylamine, N-*Me*

$C_{13}H_8$
1-Phenylhepta-1,3,5-triyne†

$C_{13}H_8BrN$
1-Bromoacridine†
2-Bromoacridine†
3-Bromoacridine†
4-Bromoacridine†
9-Bromoacridine
1-Bromophenanthridine†
2-Bromophenanthridine†
3-Bromophenanthridine†
4-Bromophenanthridine†
6-Bromophenanthridine†
7-Bromophenanthridine†
8-Bromophenanthridine†
9-Bromophenanthridine†
10-Bromophenanthridine†

$C_{13}H_8BrNO$
2-Bromoacridone
3-Bromoacridone

$C_{13}H_8BrNO_4$
3-Bromo-2-nitrophenol, *Benzoyl*

$C_{13}H_8Br_2$
2,7-Dibromofluorene
2,9-Dibromofluorene
3,9-Dibromofluorene
4,9-Dibromofluorene
9,9-Dibromofluorene

$C_{13}H_8Br_2O_3$
5,5′-Dibromo-2,2′-dihydroxybenzophenone
3,5-Dibromo-2-hydroxybenzoic Acid, *Phenyl ester*

$C_{13}H_8Br_3NO_3$
2,4,6-Tribromophenol, p-*Nitrobenzyl ether*

$C_{13}H_8ClN$
2-Chloroacridine
4-Chloroacridine
9-Chloroacridine

$C_{13}H_8ClNO$
1-Chloroacridone
2-Chloroacridone
3-Chloroacridone
4-Chloroacridone

$C_{13}H_8ClNO_3$
2-Chloro-5-nitrobenzophenone
4-Chloro-3-nitrobenzophenone
4-Chloro-4′-nitrobenzophenone
4′-Chloro-2-nitrobenzophenone
4′-Chloro-3-nitrobenzophenone
4′-Nitrobiphenyl-4-carboxylic Acid, *Chloride*

$C_{13}H_8ClNO_4$
4-(*p*-Nitrophenoxy)benzoic Acid, *Chloride*

$C_{13}H_8ClNO_5S$
4-Nitro-2-phenylsulphonylbenzoic Acid, *Chloride*

$C_{13}H_8ClNO_6S$
4-Nitro-2-sulphobenzoic Acid, 1-*Phenyl ester*, 2-*Chloride*

$C_{13}H_8ClN_3O_3$
2′-Nitroazobenzene-4-carboxylic Acid, *Chloride*
3′-Nitroazobenzene-4-carboxylic Acid, *Chloride*
4′-Nitroazobenzene-4-carboxylic Acid, *Chloride*

$C_{13}H_8ClN_3O_5$
2′,4′-Dinitrodiphenylamine-2-carboxylic Acid, *Chloride*

$C_{13}H_8Cl_2N_2O_3$
o-Nitrobenzoic Acid, 2,4-*Dichloroanilide*

$C_{13}H_8Cl_2O$
2,2′-Dichlorobenzophenone
2,3′-Dichlorobenzophenone
2,4-Dichlorobenzophenone
2,4′-Dichlorobenzophenone
2,5-Dichlorobenzophenone
3,3′-Dichlorobenzophenone
3,4-Dichlorobenzophenone
3,4′-Dichlorobenzophenone
3,5-Dichlorobenzophenone
4,4′-Dichlorobenzophenone

$C_{13}H_8Cl_2O_2$
2,5-Dichloro-4′-hydroxybenzophenone
2′,5-Dichloro-2-hydroxybenzophenone
3,2′-Dichloro-4-hydroxybenzophenone
3,4′-Dichloro-4-hydroxybenzophenone
3,5-Dichloro-2-hydroxybenzophenone
3,5-Dichloro-4-hydroxybenzophenone

$C_{13}H_8Cl_2O_3$
3,5-Dichloro-2-hydroxybenzoic Acid, *Phenyl ester*
$C_{13}H_8Cl_2O_5S_2$
Benzophenone-3,3′-disulphonic Acid, *Dichloride*
$C_{13}H_8IN$
9-Iodoacridine
$C_{13}H_8INO$
4′-Iodobiphenyl-4-isocyanate
$C_{13}H_8I_2O$
2,2′-Di-iodobenzophenone
3,5-Di-iodobenzophenone
4,4′-Di-iodobenzophenone
$C_{13}H_8I_2O_3$
2-Hydroxy-3,5-di-iodobenzoic Acid, *Phenyl ester*
$C_{13}H_8N_2$
Benz[*e*]indane-1,3-dione
$C_{13}H_8N_2O_2$
1-Nitroacridine
2-Nitroacridine
3-Nitroacridine
4-Nitroacridine
5-Nitrobiphenyl-2-carboxylic Acid, *Nitrile*
2-Nitrophenanthridine
7-Nitrophenanthridine
8-Nitrophenanthridine
Phenazine-1-carboxylic Acid
Phenazine-2-carboxylic Acid
$C_{13}H_8N_2O_3$
1-Amino-2-nitrofluorenone
1-Amino-4-nitrofluorenone
1-Amino-7-nitrofluorenone
2-Amino-3-nitrofluorenene
5-Amino-2-nitrofluorenone
7-Amino-2-nitrofluorenone
2-Hydroxyphenazine-1-carboxylic Acid†
1-Nitroacridone
2-Nitroacridone
3-Nitroacridone
4-Nitroacridone
3-Nitro-4-phenoxybenzoic Acid, *Nitrile*
$C_{13}H_8N_2O_4$
2,5-Dinitrofluorene
2,7-Dinitrofluorene
2,9-Dinitrofluorene
9,9-Dinitrofluorene
$C_{13}H_8N_2O_5$
2,2′-Dinitrobenzophenone
2,3′-Dinitrobenzophenone
2,4-Dinitrobenzophenone
2,4′-Dinitrobenzophenone
3,3′-Dinitrobenzophenone
3,4′-Dinitrobenzophenone
3,5-Dinitrobenzophenone
4,4′-Dinitrobenzophenone
$C_{13}H_8N_2O_6$
3,5-Dinitrobenzoic Acid, *Phenyl ester*
2′,4-Dinitrobiphenyl-2-carboxylic Acid
4,4′-Dinitrobiphenyl-2-carboxylic Acid
2-Hydroxy-3,5-dinitrobenzophenone
4-Hydroxy-3,3′-dinitrobenzophenone
4-Hydroxy-3,4′-dinitrobenzophenone
4-Hydroxy-3,5-dinitrobenzophenone
$C_{13}H_8N_2O_7$
2-Hydroxy-3,5-dinitrobenzoic Acid, *Phenyl ether*
$C_{13}H_8N_4O_4$
4,6-Dinitrodiphenylamine-2-carboxylic Acid, *Nitrile*
$C_{13}H_8N_4O_8$
2,2′,4,4′-Tetranitrodiphenylmethane
$C_{13}H_8O$
Fluorenone
1*H*-Phenalen-1-one
Trideca-2,10,12-triene-4,6,8-triyn-1-al†
$C_{13}H_8OS$
Thioxanthone
Xanthione
$C_{13}H_8O_2$
5,6-Benzocoumarin
6,7-Benzocoumarin
7,8-Benzocoumarin
Dibenzofuran-2-aldehyde
Dibenzo[*b,d*]pyran-6-one
2,3-Dihydro-1*H*-phenalene-1,3-dione
1-Hydroxyfluorenone
2-Hydroxyfluorenone
3-Hydroxyfluorenone
4-Hydroxyfluorenone
1*H*-Naphtho[2,1-*b*]pyran-4-one
4*H*-Naphtho[1,2-*b*]pyran-4-one
3,1-Naphthylpropiolic Acid
4-(Nona-6,8-dien-2,4-diynylidene)butenolide†
Xanthone
$C_{13}H_8O_3$
Dibenzofuran-1-carboxylic Acid
Dibenzofuran-2-carboxylic Acid
Dibenzofuran-3-carboxylic Acid
Dibenzofuran-4-carboxylic Acid
2′,4′-Dihydroxybiphenyl-2-carboxylic Acid (2,2′)-lactone†
1,4-Dihydroxyfluorenone
2,3-Dihydroxyfluorenone
2,7-Dihydroxyfluorenone
3,6-Dihydroxyfluorenone
3-Hydroxyacenaphthenequinone, *Me ether*
1-Hydroxyxanthone
2-Hydroxyxanthone
3-Hydroxyxanthone
4-Hydroxyxanthone
Urolithin B
$C_{13}H_8O_3S$
Thioxanthone-5,5-dioxide
$C_{13}H_8O_4$
1,3-Dihydroxyxanthone★†
1,4-Dihydroxyxanthone
1,5-Dihydroxyxanthone†
1,6-Dihydroxyxanthone†
1,7-Dihydroxyxanthone★†
1,8-Dihydroxyxanthone
2,3-Dihydroxyxanthone
3,4-Dihydroxyxanthone
3,6-Dihydroxyxanthone
2-Hydroxynaphthalene-1,8-dicarboxylic Acid, *Me ether*, *Anhydride*
3-Hydroxynaphthalene-1,8-dicarboxylic Acid, *Me ether*, *Anhydride*

$C_{13}H_8O_4$ (*continued*)
4-Hydroxynaphthalene-1,8-dicarboxylic Acid, *Me ether*, *Anhydride*
Isoeuxanthone
β-Isoeuxanthone
2′,4,4′-Trihydroxybiphenyl-2-carboxylic Acid †
Urolithin A

$C_{13}H_8O_5$
1,2,8-Trihydroxyxanthone †
1,3,5-Trihydroxyxanthone †
1,3,7-Trihydroxyxanthone
1,4,7-Trihydroxyxanthone
1,5,6-Trihydroxyxanthone †
2,3,4-Trihydroxyxanthone †
2,3,8-Trihydroxyxanthone
2,4,7-Trihydroxyxanthone

$C_{13}H_8O_6$
Naphthalene-1,2,5-tricarboxylic Acid
Naphthalene-1,3,8-tricarboxylic Acid
Naphthalene-1,4,5-tricarboxylic Acid
Oxypeucedanic Acid
1,3,5,6-Tetrahydroxyxanthone †
1,3,5,8-Tetrahydroxyxanthone †
1,3,6,7-Tetrahydroxyxanthone †
1,3,6,8-Tetrahydroxyxanthone †
1,3,7,8-Tetrahydroxyxanthone †
3,4,5,6-Tetrahydroxyxanthone

$C_{13}H_8O_7$
1,3,4,5,8-Pentahydroxyxanthone †
1,3,4,7,8-Pentahydroxyxanthone †

$C_{13}H_8O_8$
Brevifolincarboxylic Acid

$C_{13}H_9Br$
1-Bromocyclotrideca-1,2-diene-4,8,10-triene †
2-Bromofluorene
3-Bromofluorene
4-Bromofluorene
9-Bromofluorene

$C_{13}H_9BrHgO$
Mercuri-*o*-benzoylphenyl bromide

$C_{13}H_9BrNO_3$
2-Bromo-4-nitroaniline, N-*Benzoyl*
4-Bromo-2-nitroaniline, N-*Benzoyl*

$C_{13}H_9BrO$
2-Bromobenzophenone
3-Bromobenzophenone
4-Bromobenzophenone
2-Bromo-9-fluorenol
3-Bromo-9-fluorenol
4-Bromo-9-fluorenol

$C_{13}H_9BrO_2$
m-Bromobenzoic Acid, *Phenyl ester*
p-Bromobenzoic Acid, *Phenyl ester*
2′-Bromobiphenyl-2-carboxylic Acid
2′-Bromobiphenyl-4-carboxylic Acid
3-Bromobiphenyl-4-carboxylic Acid
3′-Bromobiphenyl-4-carboxylic Acid
4-Bromobiphenyl-2-carboxylic Acid
4-Bromobiphenyl-3-carboxylic Acid
4′-Bromobiphenyl-4-carboxylic Acid
5-Bromobiphenyl-2-carboxylic Acid
6-Bromobiphenyl-2-carboxylic Acid
2-Bromo-2′-hydroxybenzophenone
2-Bromo-4′-hydroxybenzophenone
3-Bromo-4-hydroxybenzophenone
3-Bromo-4′-hydroxybenzophenone
3′-Bromo-2-hydroxybenzophenone
4-Bromo-4′-hydroxybenzophenone

$C_{13}H_9BrO_3$
5-Bromo-2-hydroxybenzoic Acid, *Phenyl ester*

$C_{13}H_9Br_2N_2O_4$
α-(4-Bromophthalimido)glutarimide †

$C_{13}H_9Br_4N$
2,2′,4,4′-Tetrabromodiphenylamine, N-*Me*

$C_{13}H_9Cl$
2-Chlorofluorene
3-Chlorofluorene
9-Chlorofluorene

$C_{13}H_9ClHgO$
Mercuri-*o*-benzoylphenyl chloride

$C_{13}H_9ClO$
Biphenyl-4-carboxylic Acid, *Chloride*
2-Chlorobenzophenone
3-Chlorobenzophenone
4-Chlorobenzophenone
3-α-Naphthylacrylic Acid, *Chloride*

$C_{13}H_9ClO_2$
o-Chlorobenzoic Acid, *Phenyl ester*
m-Chlorobenzoic Acid, *Phenyl ester*
p-Chlorobenzoic Acid, *Phenyl ester*
6-Chlorobiphenyl-3-carboxylic Acid
4′-Chlorobiphenyl-3-carboxylic Acid
4′-Chlorobiphenyl-4-carboxylic Acid
2-Chloro-4′-hydroxybenzophenone
3-Chloro-2-hydroxybenzophenone
3-Chloro-4-hydroxybenzophenone
3-Chloro-4′-hydroxybenzophenone
4-Chloro-4′-hydroxybenzophenone
4′-Chloro-3-hydroxypenzophenone
5-Chloro-2-hydroxybenzophenone

$C_{13}H_9ClO_3$
5-Chloro-2-hydroxybenzoic Acid, *Phenyl ester*
5-Chloro-2-hydroxybenzoic Acid, *Phenyl ether*

$C_{13}H_9ClO_3S$
Benzophenone-2-sulphonic Acid, *Chloride*
o-Phenylsulphonylbenzoic Acid, *Chloride*
p-Phenylsulphonylbenzoic Acid, *Chloride*

$C_{13}H_9ClO_4S$
o-Sulphobenzoic Acid, 1-*Phenyl ester*, *Chloride*

$C_{13}H_9Cl_3O$
2,4,6-Trichloro-*m*-cresol, *Phenyl ether*

$C_{13}H_9Cl_4N$
2,2′4,4′-Tetrachlorodiphenylamine, N-*Me*

$C_{13}H_9FO$
2-Fluorobenzophenone
4-Fluorobenzophenone

$C_{13}H_9FO_3S$
3-Acetyl-5(4-fluorobenzylidene)-4-hydroxy-2-oxo-2,5-dihydrothiophene †

$C_{13}H_9IO$
2-Iodobenzophenone
4-Iodobenzophenone

$C_{13}H_9N$
Acridine
Benz[*g*]isoquinoline
Benzo[*f*]quinoline
Benzo[*g*]quinoline
Benzo[*h*]quinoline
Biphenyl-4-carboxylic Acid, *Nitrile*
Cyclohept[*b*]indole
Phenanthridine

$C_{13}H_9NO$
Acridone
1-Aminofluorenone
2-Aminofluorenone
3-Aminofluorenone
4-Aminofluorenone
Dibenz[*b,f*]oxepin
1-Hydroxyacridine
2-Hydroxyacridine
3-Hydroxyacridine
4-Hydroxyacridine
1-Hydroxybenzo[*f*]quinoline
3-Hydroxybenzo[*f*]quinoline
5-Hydroxybenzo[*f*]quinoline
7-Hydroxybenzo[*f*]quinoline
8-Hydroxybenzo[*f*]quinoline
9-Hydroxybenzo[*f*]quinoline
2-Hydroxybenzo[*g*]quinoline
4-Hydroxybenzo[*g*]quinoline
10-Hydroxybenzo[*g*]quinoline
2-Hydroxybenzo[*h*]quinoline
4-Hydroxybenzo[*h*]quinoline
8-Hydroxybenzo[*h*]quinoline
9-Hydroxybenzo[*h*]quinoline
3-Hydroxyphenanthridine
4-Hydroxyphenanthridine
Phenanthridone
N-Phenylanthranil

$C_{13}H_9NOS$
1-Aminothioxanthone
2-Aminothioxanthone
3-Aminothioxanthone
4-Aminothioxanthone
2-*o*-Hydroxyphenylbenzothiazole

$C_{13}H_9NO_2$
1-Aminoxanthone
2-Aminoxanthone
3-Aminoxanthone
4-Aminoxanthone
1,3-Dihydroxyacridine
1,4-Dihydroxyacridine
2,3-Dihydroxyacridine
2,5-Dihydroxyacridine
2,7-Dihydroxyacridine
3,4-Dihydroxyacridine
3,5-Dihydroxyacridine
3,6-Dihydroxyacridine
2-Hydroxyacridone
4-Hydroxyacridone
10-Hydroxyacridone
2-*o*-Hydroxyphenylbenzoxazole
Naphthalene-1,8-dicarboxylic Acid, N-*Me imide*
1-Naphthylglyoxylic Acid, *Amide*
2-Nitrofluorene
3-Nitrofluorene
4-Nitrofluorene
9-Nitrofluorene

$C_{13}H_9NO_3$
1,3-Dihydroxyacridone
1,4-Dihydroxyacridone
2,3-Dihydroxyacridone
2,5-Dihydroxyacridone
2,7-Dihydroxyacridone
3,4-Dihydroxyacridone
3,6-Dihydroxyacridone
4,5-Dihydroxyacridone
4-Formyl-2′-nitrobiphenyl
4-Formyl-4′-nitrobiphenyl
2-Hydroxycarbazole-1-carboxylic Acid
1-Hydroxycarbazole-2-carboxylic Acid
2-Hydroxycarbazole-3-carboxylic Acid
2-Nitrobenzophenone
3-Nitrobenzophenone
4-Nitrobenzophenone
2-Nitro-1-fluorenol
4-Nitro-1-fluorenol
3-Nitro-2-fluorenol
7-Nitro-2-fluorenol
2-Nitro-3-fluorenol
2-Nitro-9-fluorenol
3-Nitro-9-fluorenol

$C_{13}H_9NO_4$
2-Formyl-2′-nitrodiphenyl Ether
2-Formyl-4′-nitrodiphenyl Ether
4-Formyl-2′-nitrodiphenyl Ether
4-Formyl-4′-nitrodiphenyl Ether
2-Hydroxy-4′-nitrobenzophenone
2-Hydroxy-5-nitrobenzophenone
4-Hydroxy-2′-nitrobenzophenone
4-Hydroxy-3-nitrobenzophenone
4-Hydroxy-3′-nitrobenzophenone
4-Hydroxy-4′-nitrobenzophenone
Kokusagine
Maculine†
p-Nitrobenzoic Acid, *Phenyl ester*
2′-Nitrobiphenyl-2-carboxylic Acid
3-Nitrobiphenyl-2-carboxylic Acid
4-Nitrobiphenyl-2-carboxylic Acid
4′-Nitrobiphenyl-2-carboxylic Acid
5-Nitrobiphenyl-2-carboxylic Acid
6-Nitrobiphenyl-2-carboxylic Acid
2′-Nitrobiphenyl-3-carboxylic Acid
4′-Nitrobiphenyl-3-carboxylic Acid
6-Nitrobiphenyl-3-carboxylic Acid
2-Nitrobiphenyl-4-carboxylic Acid
2′-Nitrobiphenyl-4-carboxylic Acid
3-Nitrobiphenyl-4-carboxylic Acid
4′-Nitrobiphenyl-4-carboxylic Acid
4-Phenylpyridine-2,3-dicarboxylic Acid
6-Phenylpyridine-2,3-dicarboxylic Acid
6-Phenylpyridine-3,4-dicarboxylic Acid
4-Phenylpyridine-3,5-dicarboxylic Acid
Phthaloylmalonic Acid, *Et ester-nitrile*

$C_{13}H_9NO_4S$
2-*o*-Nitrophenylmercaptobenzoic Acid
2-*m*-Nitrophenylmercaptobenzoic Acid
2-*p*-Nitrophenylmercaptobenzoic Acid
4-Nitro-2-phenylmercaptobenzoic Acid

$C_{13}H_9NO_5$
2,4-Dihydroxy-3′-nitrobenzophenone

$C_{13}H_9NO_5$ (*continued*)
2,4-Dihydroxy-4′-nitrobenzophenone
2,5-Dihydroxy-4′-nitrobenzophenone
3-Nitro-4-phenoxybenzoic Acid
4-Nitro-2-phenoxybenzoic Acid
5-Nitro-2-phenoxybenzoic Acid
2-(*o*-Nitrophenoxy)benzoic Acid
2-(*m*-Nitrophenoxy)benzoic Acid
2-(*p*-Nitrophenoxy)benzoic Acid
3-(*m*-Nitrophenoxy)benzoic Acid
3-(*p*-Nitrophenoxy)benzoic Acid
4-(*o*-Nitrophenoxy)benzoic Acid
4-(*p*-Nitrophenoxy)benzoic Acid
Phenyl 3-nitrosalicylate
Phenyl 5-nitrosalicylate

$C_{13}H_9NO_5S$
2-*o*-Nitrophenylsulphinylbenzoic Acid
2-*m*-Nitrophenylsulphinylbenzoic Acid
2-*p*-Nitrophenylsulphinylbenzoic Acid

$C_{13}H_9NO_6S$
2-*o*-Nitrophenylsulphonylbenzoic Acid
2-*m*-Nitrophenylsulphonylbenzoic Acid
2-*p*-Nitrophenylsulphonylbenzoic Acid
3-Nitro-4-phenylsulphonylbenzoic Acid
4-Nitro-2-phenylsulphonylbenzoic Acid

$C_{13}H_9NS$
9-Acridinethiol
2-Phenylbenzothiazole

$C_{13}H_9N_3$
Azobenzene-4-carboxylic Acid, *Nitrile*

$C_{13}H_9N_3O$
3-Hydroxyazobenzene-4-carboxylic Acid, *Nitrile*
Oxychlororaphine
Phenazine-2-carboxylic Acid, *Amide*

$C_{13}H_9N_3O_2$
9-Amino-1-nitroacridine
9-Amino-2-nitroacridine
9-Amino-3-nitroacridine
9-Amino-4-nitroacridine

$C_{13}H_9N_3O_4$
5-Nitroazobenzene-2-carboxylic Acid
2-Nitroazobenzene-4-carboxylic Acid
2′-Nitroazobenzene-4-carboxylic Acid
3′-Nitroazobenzene-4-carboxylic Acid
4′-Nitroazobenzene-4-carboxylic Acid

$C_{13}H_9N_3O_5$
4-Hydroxy-2′-nitroazobenzene-3-carboxylic Acid
4-Hydroxy-3′-nitroazobenzene-3-carboxylic Acid
4-Hydroxy-4′-nitroazobenzene-3-carboxylic Acid
4′-Hydroxy-4-nitroazobenzene-3-carboxylic Acid
4-Hydroxy-5-nitroazobenzene-3-carboxylic Acid
o-Nitrobenzoic Acid, o-*Nitroanilide*

$C_{13}H_9N_3O_6$
2′,4′-Dinitrodiphenylamine-2-carboxylic Acid
2′,6-Dinitrodiphenylamine-2-carboxylic Acid
3′,4-Dinitrodiphenylamine-2-carboxylic Acid
3′,5-Dinitrodiphenylamine-2-carboxylic Acid
3′,5′-Dinitrodiphenylamine-2-carboxylic Acid
3′,6-Dinitrodiphenylamine-2-carboxylic Acid
4,5-Dinitrodiphenylamine-2-carboxylic Acid
4,6-Dinitrodiphenylamine-2-carboxylic Acid
2′,4′-Dinitrodiphenylamine-3-carboxylic Acid
2′,4′-Dinitrodiphenylamine-4-carboxylic Acid
2,6-Dinitrodiphenylamine-4-carboxylic Acid

$C_{13}H_9N_3O_7$
Picric Acid, *Benzyl ether*

$C_{13}H_9P$
Dibenzo[*b*,*d*]phosphorin
Dibenzo[*b*,*e*]phosphorin

$C_{13}H_{10}$
2*H*-Benz[*cd*]azulene†
Fluorene
1*H*-Phenalene
1-Phenylhept-5-ene-1,3-diyne†

$C_{13}H_{10}AsN$
Diphenylcyanoarsine

$C_{13}H_{10}BrNO$
o-Bromoaniline, N-*Benzoyl*
m-Bromoaniline, N-*Benzoyl*
p-Bromoaniline, N-*Benzoyl*

$C_{13}H_{10}BrNO_4$
8-Bromo-3-nitro-1-naphthoic Acid, *Et ester*
8-Bromo-5-nitro-1-naphthoic Acid, *Et ester*

$C_{13}H_{10}Br_2$
Di-2-bromophenylmethane
Di-4-bromophenylmethane

$C_{13}H_{10}Br_2O$
3,5-Dibromo-α-phenylbenzyl Alcohol
Di-4-bromophenylmethanol

$C_{13}H_{10}Br_2O_2$
5,8-Dibromo-2-naphthoic Acid, *Et ester*

$C_{13}H_{10}ClN$
Benzylidene-*o*-chloroaniline
Benzylidene-*m*-chloroaniline
Benzylidene-*p*-chloroaniline

$C_{13}H_{10}ClNO$
2-Amino-5-chlorobenzophenone
N-Chlorobenzanilide
Diphenylcarbamic Acid, *Chloride*
Diphenylcarbamoyl Chloride†

$C_{13}H_{10}ClNO_2$
2-Amino-5-chlorobenzoic Acid, N-*Phenyl*

$C_{13}H_{10}Cl_2$
α,α-Dichlorodiphenylmethane
3,5-Dichloro-4-methylbiphenyl
Di-4-chlorophenylmethane

$C_{13}H_{10}Cl_2N_2O$
1,3-Di-2-chlorophenylurea
1,3-Di-3-chlorophenylurea
1,3-Di-4-chlorophenylurea

$C_{13}H_{10}Cl_2N_2S$
1,3-Di-(2-chlorophenyl)thiourea
1,3-Di-(3-chlorophenyl)thiourea
1,3-Di-(4-chlorophenyl)thiourea

$C_{13}H_{10}Cl_2O$
α-2,6-Dichlorophenylbenzyl Alcohol
Di-4-chlorophenylmethanol

$C_{13}H_{10}Cl_2O_2$
5,8-Dichloro-1-naphthoic Acid, *Et ester*
4,5-Dichloro-2-naphthoic Acid, *Et ester*
5,8-Dichloro-2-naphthoic Acid, *Et ester*
Dichlorophen

$C_{13}H_{10}Cl_2S$
p-Chlorobenzyl *p*-chlorophenyl sulphide

$C_{13}H_{10}INO$
m-Iodoaniline, N-*Benzoyl*

$C_{13}H_{10}N$ (ion)
Benzo[*a*]quinolizinium †
Benzo[*b*]quinolizinium †
Benzo[*c*]quinolizinium †

$C_{13}H_{10}N_2$
1-Aminoacridine
2-Aminoacridine
3-Aminoacridine
4-Aminoacridine
9-Aminoacridine
Diphenylcarbodiimide
Diphenylcyanamide
1-Methylphenazine
2-Methylphenazine
1-Phenylbenzimidazole
2-Phenylbenzimidazole
1-Phenylindazole
2-Phenylindazole
3-Phenylindazole

$C_{13}H_{10}N_2O$
1-Aminoacridone
2-Aminoacridone
3-Aminoacridone
4-Aminoacridone
p-Benzeneazobenzaldehyde
1,2-Diaminofluorenone
1,4-Diaminofluorenone
1,7-Diaminofluorenone
2,3-Diaminofluorenone
2,5-Diaminofluorenone
2,7-Diaminofluorenone
3,6-Diaminofluorenone
2-Hydroxyphenazine, *Me ether*
Pyocyanine

$C_{13}H_{10}N_2O_2$
2-Amino-3-nitrofluorene
2-Amino-4-nitrofluorene
2-Amino-5-nitrofluorene
5-Amino-2-nitrofluorene
7-Amino-2-nitrofluorene
Azobenzene-2-carboxylic Acid
Azobenzene-3-carboxylic Acid
Azobenzene-4-carboxylic Acid
N-Benzylidene-*m*-nitroaniline
N-Benzylidene-*p*-nitroaniline
β-Carboline-1-carboxylic Acid, *Me ester*
o-Nitrobenzylideneaniline
m-Nitrobenzylideneaniline
p-Nitrobenzylideneaniline
3-Nitrocarbazole, N-*Me*
2-*o*-Nitrostyrylpyridine
2-*m*-Nitrostyrylpyridine
2-*p*-Nitrostyrylpyridine
4-*o*-Nitrostyrylpyridine
4-*m*-Nitrostyrylpyridine
4-*p*-Nitrostyrylpyridine
3-Nitro-4-styrylpyridine

$C_{13}H_{10}N_2O_3$
2-Amino-2′-nitrobenzophenone
2-Amino-3′-nitrobenzophenone
2-Amino-4-nitrobenzophenone
2-Amino-5-nitrobenzophenone
2-Amino-6-nitrobenzophenone
4-Amino-3-nitrobenzophenone
4-Amino-4′-nitrobenzophenone
4′-Amino-3-nitrobenzophenone
5-Amino-3-nitrobenzophenone
3-Hydroxyazobenzene-4-carboxylic Acid
4-Hydroxyazobenzene-3-carboxylic Acid
o-Nitrobenzoic Acid, *Anilide*
2-*o*-Nitrobenzylideneaminophenol
2-*m*-Nitrobenzylideneaminophenol
2-*p*-Nitrobenzylideneaminophenol
3-*o*-Nitrobenzylideneaminophenol
4-*o*-Nitrobenzylideneaminophenol
4-*m*-Nitrobenzylideneaminophenol
4-*p*-Nitrobenzylideneaminophenol

$C_{13}H_{10}N_2O_4$
Myxin †
3-Nitrodiphenylamine-2-carboxylic Acid
3′-Nitrodiphenylamine-2-carboxylic Acid
4-Nitrodiphenylamine-2-carboxylic Acid
5-Nitrodiphenylamine-2-carboxylic Acid
6-Nitrodiphenylamine-2-carboxylic Acid
2-Nitrodiphenylamine-4-carboxylic Acid
2-Nitro-α-(2-nitrophenyl)toluene
2-Nitro-α-(4-nitrophenyl)toluene
3-Nitro-α-(3-nitrophenyl)toluene
3-Nitro-α-(4-nitrophenyl)toluene
4-Nitro-α-(4-nitrophenyl)toluene
4-(*p*-Nitrophenoxy)benzoic Acid, *Amide*
α-Phthalimidoglutarimide

$C_{13}H_{10}N_2O_5$
2-(4-Hydroxyanilino)-5-nitrobenzoic Acid
4-(2-Hydroxyanilino)-3-nitrobenzoic Acid
5-(4-Hydroxyanilino)-2′-nitrobenzoic Acid
2-Hydroxy-3,5-dinitrophenyl, *Me ether*
2-Hydroxy-4′,5-dinitrophenyl, *Me ether*
2-(4-Hydroxy-2-nitroanilino)benzoic Acid

$C_{13}H_{10}N_2O_5S$
4-Nitro-2-phenylsulphonylbenzoic Acid, *Amide*

$C_{13}H_{10}N_2O_6$
2,2′-Dihydroxy-5,5′-dinitrobiphenyl, *Me ether*
3,6-Dinitro-1-naphthoic Acid, *Et ester*
4,5-Dinitro-1-naphthoic Acid, *Et ester*
5,8-Dinitro-1-naphthoic Acid, *Et ester*
4,5-Dinitro-2-naphthoic Acid, *Et ester*

$C_{13}H_{10}N_2S$
2-Aminobenzothiazole, N-*Phenyl*
2-Mercaptobenzimidazole, 1-N-*Phenyl*

$C_{13}H_{10}N_4O_5$
2′,4′-Dinitrodiphenylamine-2-carboxylic Acid, *Amide*
1-(2,4-Dinitrophenyl)-3-phenylurea
1-(3,5-Dinitrophenyl)-3-phenylurea
1,1-Di-(4-nitrophenyl)urea
1,3-Di-(2-nitrophenyl)urea
1,3-Di-(3-nitrophenyl)urea
1,3-Di-(4-nitrophenyl)urea
1-(2-Nitrophenyl)-3-(3-nitrophenyl)urea
1-(2-Nitrophenyl)-3-(4-nitrophenyl)urea
1-(3-Nitrophenyl)-3-(4-nitrophenyl)urea

$C_{13}H_{10}O$
Acenaphthene-3-aldehyde
Acenaphthene-5-aldehyde
Benz[*e*]indane-3-one
Benzophenone
Biphenyl-2-aldehyde
Biphenyl-4-aldehyde
Carlina oxide
6*H*-Dibenzo[*b*,*d*]pyran
2,5-Dihydro-2-[*cis*,*cis*-nona-6,8-dien-2,4-diynylidene]furan †
2,3-Dihydro-1*H*-phenalen-1-one
1-Fluorenol
2-Fluorenol
3-Fluorenol
9-Fluorenol
2-Methylnaphtho[1,2-*b*]furan
3-Methylnaphtho[1,2-*b*]furan
2-Methylnaphtho[2,1-*b*]furan
2-Methylnaphtho[2,3,*b*]furan
3-Methylnaphtho[2,1-*b*]furan
3-α-Naphthylacrolein
3-β-Naphthylacrolein
7-Phenylhept-2-ene-4,6-diyn-1-ol †
Xanthene

$C_{13}H_{10}OS$
Thiobenzoic Acid, S-*Phenyl ester*

$C_{13}H_{10}OS_2$
Dithiocarbonic Acid, *Diphenyl ester*

$C_{13}H_{10}OS_3$
α-Terthienylmethanol †

$C_{13}H_{10}O_2$
Acenaphthene-1-carboxylic Acid
Acenaphthene-3-carboxylic Acid
Acenaphthene-5-carboxylic Acid
Biphenyl-2-carboxylic Acid
Biphenyl-3-carboxylic Acid
Biphenyl-4-carboxylic Acid
Capillarin †
3-(2-Furyl)acrylophenone
p-Hydroxybenzaldehyde, *Phenyl ether*
2-Hydroxybenzophenone
3-Hydroxybenzophenone
4-Hydroxybenzophenone
Isomycomycin
Mycomycin
2-α-Naphthylacrylic Acid
3-α-Naphthylacrylic Acid
3-β-Naphthylacrylic Acid
Phenyl benzoate
Xanthydrol

$C_{13}H_{10}O_2S$
o-Mercaptobenzoic Acid, *Phenyl ester*
o-Phenylmercaptobenzoic Acid
p-Phenylmercaptobenzoic Acid
Thiocarbonic Acid, O,S-*Diphenyl ester*

$C_{13}H_{10}O_3$
1,5-Difuryl-1,4-pentadien-3-one
2,2′-Dihydroxybenzophenone
2,3-Dihydroxybenzophenone
2,3′-Dihydroxybenzophenone
2,4-Dihydroxybenzophenone
2,4′-Dihydroxybenzophenone
2,5-Dihydroxybenzophenone
2,6-Dihydroxybenzophenone
3,3′-Dihydroxybenzophenone
3,4-Dihydroxybenzophenone
3,4′-Dihydroxybenzophenone
3,5-Dihydroxybenzophenone
4,4′-Dihydroxybenzophenone
1,2-Dihydroxydibenzofuran, 2-*Me ether*
1,4-Dihydroxydibenzofuran, 4-*Me ether*
3,8-Dihydroxydibenzofuran, 3-*Me ether*
2,3-Dihydroxyxanthene
Diphenyl carbonate
2′-Hydroxybiphenyl-2-carboxylic Acid
3-Hydroxybiphenyl-2-carboxylic Acid
4′-Hydroxybiphenyl-2-carboxylic Acid
5-Hydroxybiphenyl-2-carboxylic Acid
6-Hydroxybiphenyl-2-carboxylic Acid
2-Hydroxybiphenyl-3-carboxylic Acid
4′-Hydroxybiphenyl-4-carboxylic Acid
5-Hydroxytoluquinone, *Phenyl ether*
1,8-Naphthaldehydic Acid, *Me ester*
1-Naphthylglyoxylic Acid, *Me ester*
β-Naphthylpyruvic Acid
o-Phenoxybenzoic Acid
m-Phenoxybenzoic Acid
p-Phenoxybenzoic Acid
Salol

$C_{13}H_{10}O_3S$
o-Phenylsulphinylbenzoic Acid
p-Phenylsulphinylbenzoic Acid

$C_{13}H_{10}O_4$
Allovisnagin
3,4-Dihydroxybenzoic Acid, *Phenyl ester*
1-Naphthylmalonic Acid
2-Naphthylmalonic Acid
Podophyllomeronic Acid
Rhamnin B
Santal
Stypandrone †
2,2′,6-Trihydroxybenzophenone
2,3,4-Trihydroxybenzophenone
2,3,4′-Trihydroxybenzophenone
2,4,4′-Trihydroxybenzophenone
2,4,5-Trihydroxybenzophenone
2,4,6-Trihydroxybenzophenone
3,3′,4-Trihydroxybenzophenone
3,4,4′-Trihydroxybenzophenone
3,4,5-Trihydroxybenzophenone
3,4′,5-Trihydroxybenzophenone

$C_{13}H_{10}O_4S$
Benzophenone-2-sulphonic Acid
o-Phenylsulphonylbenzoic Acid
p-Phenylsulphonylbenzoic Acid

$C_{13}H_{10}O_5$
Citromycin
Hispidin
1-Hydroxynaphthalene-2,4-dicarboxylic Acid, *Me ether*
2-Hydroxynaphthalene-1,8-dicarboxylic Acid, *Me ether*
3-Hydroxynaphthalene-1,8-dicarboxylic Acid, *Me ether*
4-Hydroxynaphthalene-1,8-dicarboxylic Acid, *Me ether*
Isopimpinellin
Khellol

$C_{13}H_{10}O_5$ (*continued*)
Pimpinellin †
α-Sorigenin
2,2′,3,3′-Tetrahydroxybenzophenone
2,2′,3,4-Tetrahydroxybenzophenone
2,2′,4,4′-Tetrahydroxybenzophenone
2,2′,4,6-Tetrahydroxybenzophenone
2,2′,4,6′-Tetrahydroxybenzophenone
2,2′,5,5′-Tetrahydroxybenzophenone
2,2′,5,6′-Tetrahydroxybenzophenone
2,2′,6,6′-Tetrahydroxybenzophenone
2,3,3′,4-Tetrahydroxybenzophenone
2,3,4,4′-Tetrahydroxybenzophenone
2,3,4,5-Tetrahydroxybenzophenone
2,3,4,6-Tetrahydroxybenzophenone
2,3′,4,4′-Tetrahydroxybenzophenone
2,3′,4,5-Tetrahydroxybenzophenone
2,3′,4,6-Tetrahydroxybenzophenone
2,3′,4′,5-Tetrahydroxybenzophenone
2,4,4′,5-Tetrahydroxybenzophenone
2,4,4′,6-Tetrahydroxybenzophenone
3,3′,4,4′-Tetrahydroxybenzophenone
3,3′,4,5′-Tetrahydroxybenzophenone

$C_{13}H_{10}O_5S$
o-Sulphobenzoic Acid, 1-*Phenyl ester*
o-Sulphobenzoic Acid, 2-*Phenyl ester*

$C_{13}H_{10}O_6$
Maclurin
2,2′,3,4,4′-Pentahydroxybenzophenone
2,2′,4,4′,6-Pentahydroxybenzophenone
2,2′,4,6,6′-Pentahydroxybenzophenone
2,3,3′,4,4′-Pentahydroxybenzophenone
2,3,3′,4,5′-Pentahydroxybenzophenone
2,3′,4,5′,6-Pentahydroxybenzophenone
2′,3,4,4′,5-Pentahydroxybenzophenone
3,3′,4,4′,5-Pentahydroxybenzophenone

$C_{13}H_{10}O_7$
2,2′,3,3′,4,4′-Hexahydroxybenzophenone
2,3,3′,4,4′,5′-Hexahydroxybenzophenone

$C_{13}H_{10}O_7S_2$
Benzophenone-3,3′-disulphonic Acid

$C_{13}H_{10}S$
2-(But-3-en-1-ynyl)-5-(pent-3-en-1-ynyl)thiophene †
2-(Hexa-3,5-diynyl)-5-prop-1-ynylthiophene †
2-Phenyl-5-propynylthiophene †
Thiobenzophenone
Thioxanthene

$C_{13}H_{11}BF_4$
Phenyltropylium fluoroborate

$C_{13}H_{11}Br$
α-Bromodiphenylmethane
2-Bromomethylbiphenyl
2-Bromo-2′-methylbiphenyl
2-Bromo-5-methylbiphenyl
3-Bromo-4-methylbiphenyl
4-Bromo-3-methylbiphenyl
4-Bromo-4′-methylbiphenyl

$C_{13}H_{11}BrO_2$
4-Bromo-1-naphthoic Acid, *Et ester*
5-Bromo-1-naphthoic Acid, *Et ester*
7-Bromo-1-naphthoic Acid, *Et ester*
8-Bromo-1-naphthoic Acid, *Et ester*

$C_{13}H_{11}Cl$
4-Chlorodiphenylmethane
2-Chloro-2′-methylbiphenyl
4-Chloro-4′-methylbiphenyl
4′-Chloro-2-methylbiphenyl

$C_{13}H_{11}ClN_2O_2$
4-Chloro-2-nitrodiphenylamine, N-*Me*

$C_{13}H_{11}ClO$
3-α-Naphthylpropionic Acid, *Chloride*

$C_{13}H_{11}ClO_2$
5-Chloro-1-naphthoic Acid, *Et ester*
8-Chloro-1-naphthoic Acid, *Et ester*
3-Chloro-2-naphthoic Acid, *Et ester*
5-Chloro-2-naphthoic Acid, *Et ester*
8-Chloro-2-naphthoic Acid, *Et ester*
Resorcinol, *Mono*-p-*chlorobenzyl ether*

$C_{13}H_{11}ClO_3$
4-Chloro-1-hydroxy-2-naphthoic Acid, *Et ester*

$C_{13}H_{11}ClO_7$
Caffeic Acid, *Dicarbomethoxyl*, *Chloride*

$C_{13}H_{11}IO_2$
7-Iodo-1-naphthoic Acid, *Et ester*
8-Iodo-1-naphthoic Acid, *Et ester*
3-Iodo-2-naphthoic Acid, *Et ester*
5-Iodo-2-naphthoic Acid, *Et ester*

$C_{13}H_{11}N$
Acridan
1-Aminofluorene
2-Aminofluorene
3-Aminofluorene
4-Aminofluorene
9-Aminofluorene
Benzylideneaniline
1-Methylcarbazole
2-Methylcarbazole
3-Methylcarbazole
N-Phenylbenzoazetine †
2-Styrylpyridine
4-Styrylpyridine

$C_{13}H_{11}NO$
o-Aminobenzophenone
m-Aminobenzophenone
p-Aminobenzophenone
3-Aminodibenzofuran, N-*Me*
Benzanilide
o-Benzylideneaminophenol
p-Benzylideneaminophenol
Biphenyl-2-carboxylic Acid, *Amide*
Biphenyl-4-carboxylic Acid, *Amide*
o-Hydroxybenzylideneaniline
p-Hydroxybenzylideneaniline

$C_{13}H_{11}NOS$
Phenylthiocarbamic Acid, O-*Phenyl ester*
Phenylthiocarbamic Acid, S-*Phenyl ester*

$C_{13}H_{11}NO_2$
4-Acetoacetylquinoline
2-Amino-4′-hydroxybenzophenone
4-Amino-2-hydroxybenzophenone
4-Amino-4′-hydroxybenzophenone
5-Amino-2-hydroxybenzophenone
Anthranilic Acid, *Phenyl ester*
N-Benzoyl-2-hydroxyaniline

$C_{13}H_{11}NO_2$ (*continued*)
N-Benzoyl-3-hydroxyaniline
N-Benzoyl-4-hydroxyaniline
2-Cyano-5-phenyl-2,4-pentadienoic Acid, *Me ester*
Diphenylamine-2-carboxylic Acid
Diphenylcarbamic Acid
2-Methyl-2′-nitrobiphenyl
2-Methyl-4-nitrobiphenyl
2-Methyl-4′-nitrobiphenyl
2-Methyl-6-nitrobiphenyl
3-Methyl-4-nitrobiphenyl
3-Methyl-6-nitrobiphenyl
4-Methyl-2-nitrobiphenyl
4-Methyl-3-nitrobiphenyl
4-Methyl-4′-nitrobiphenyl
2-Methyl-6-phenylpyridine-3-carboxylic Acid
2-Methyl-6-phenylpyridine-4-carboxylic Acid
3-[2-Methyl-5-quinolyl]acrylic Acid
3-[2-Methyl-6-quinolyl]acrylic Acid
3-[2-Methyl-7-quinolyl]acrylic Acid
α-*o*-Nitrophenyltoluene
α-*m*-Nitrophenyltoluene
α-*p*-Nitrophenyltoluene
α-Nitro-α-phenyltoluene
o-Phenoxybenzoic Acid, *Amide*
m-Phenoxybenzoic Acid, *Amide*

$C_{13}H_{11}NO_2S$
o-Nitrophenyl-*o*-tolyl sulphide
p-Nitrophenyl-*p*-tolyl sulphide
4-Nitro-2-tolyl phenyl sulphide
4-Nitro-3-tolyl phenyl sulphide
6-Nitro-3-tolyl phenyl sulphide

$C_{13}H_{11}NO_3$
5-Anilino-2-hydroxybenzoic Acid
4-Anilino-3-hydroxybenzoic Acid
3-Anilino-5-hydroxybenzoic Acid
Evolitrine
γ-Fagarine
2-(2-Hydroxyanilino)benzoic Acid
2-(3-Hydroxyanilino)benzoic Acid
2-(4-Hydroxyanilino)benzoic Acid
2-Hydroxy-2′-nitrobiphenyl, *Me ether*
2-Hydroxy-3-nitrobiphenyl, *Me ether*
2-Hydroxy-3′-nitrobiphenyl, *Me ether*
2-Hydroxy-4′-nitrobiphenyl, *Me ether*
2-Hydroxy-5-nitrobiphenyl, *Me ether*
4-Hydroxy-2-nitrobiphenyl, *Me ether*
4-Hydroxy-3-nitrobiphenyl, *Me ether*
4-Hydroxy-4′-nitrobiphenyl, *Me ether*
6-Methoxydictamnine†
2-Methyl-2′-nitrodiphenyl Ether
2-Methyl-3′-nitrodiphenyl Ether
2-Methyl-4′-nitrodiphenyl Ether
3-Methyl-2′-nitrodiphenyl Ether
3-Methyl-3′-nitrodiphenyl Ether
3-Methyl-4′-nitrodiphenyl Ether
3-Methyl-6-nitrodiphenyl Ether
4-Methyl-2′-nitrodiphenyl Ether
4-Methyl-3′-nitrodiphenyl Ether
4-Methyl-4′-nitrodiphenyl Ether
1-*N*-Naphthoylglycine
2-*N*-Naphthoylglycine
o-Nitrophenol, *Benzyl ether*
p-Nitrophenol, *Benzyl ether*

$C_{13}H_{11}NO_3S$
o-Phenylsulphonylbenzoic Acid, *Amide*
p-Phenylsulphonylbenzoic Acid, *Amide*

$C_{13}H_{11}NO_4$
Gallic Acid, *Anilide*
2-Hydroxy-4-nitrophenyl phenyl Ether, *Me ether*
2-Hydroxy-5-nitrophenyl phenyl Ether, *Me ether*
4-Hydroxy-3-nitrophenyl phenyl Ether, *Me ether*
Maltol, *Carbanilide*
3-Nitro-1-naphthoic Acid, *Et ester*
4-Nitro-1-naphthoic Acid, *Et ester*
5-Nitro-1-naphthoic Acid, *Et ester*
6-Nitro-1-naphthoic Acid, *Et ester*
8-Nitro-1-naphthoic Acid, *Et ester*
5-Nitro-2-naphthoic Acid, *Et ester*
6-Nitro-2-naphthoic Acid, *Et ester*
7-Nitro-2-naphthoic Acid, *Et ester*
8-Nitro-2-naphthoic Acid, *Et ester*
(1-Nitro-2-naphthoxy)acetone
2-Nitroquinol, *Benzyl ether*
Quinoline-2,3-dicarboxylic Acid, *Di-Me ester*
Quinoline-2,3-dicarboxylic Acid, *Et ester*
Quinoline-5,6-dicarboxylic Acid, *Di-Me ester*

$C_{13}H_{11}NO_4S$
o-Nitrophenyl *p*-tolyl sulphone
p-Nitrophenyl-*p*-tolyl sulphone
4-Nitro-2-tolyl phenyl sulphone
5-Nitro-2-tolyl phenyl sulphone
o-Sulphobenzoic Acid, 1-*Phenyl ester*, *Amide*
o-Sulphobenzoic Acid, 2-*Phenyl ester*, *Amide*

$C_{13}H_{11}NO_5$
1-Hydroxy-4-nitro-2-naphthoic Acid, *Et ester*
3-Hydroxy-4-nitro-2-naphthoic Acid, *Et ester*

$C_{13}H_{11}NO_5S$
6-Nitroacenaphthene-3-sulphonic Acid, *Me ester*
6-Nitroacenaphthene-4-sulphonic Acid, *Me ester*
2-Nitrobenzenesulphonic Acid, o-*Tolyl ester*
4-Nitrobenzenesulphonic Acid, p-*Tolyl ester*
4-Nitrotoluene-2-sulphonic Acid, *Phenyl ester*
2-Nitrotoluene-4-sulphonic Acid, *Phenyl ester*
Toluene-*o*-sulphonic Acid, o-*Nitrophenyl ester*

$C_{13}H_{11}NS$
6*H*-[2]-Benzothiapyranol[4,3-*b*]quinoline
Phenothiazine, N-*Me*

$C_{13}H_{11}NS_2$
Phenyldithiocarbamic Acid, *Phenyl ester*

$C_{13}H_{11}NSe$
Selenazine, *N-Me*

$C_{13}H_{11}N_3$
1,5-Diaminoacridine
1,6-Diaminoacridine
1,7-Diaminoacridine
1,9-Diaminoacridine
2,4-Diaminoacridine
2,5-Diaminoacridine
2,6-Diaminoacridine
2,7-Diaminoacridine
2,9-Diaminoacridine

$C_{13}H_{11}N_3$ (*continued*)
3,5-Diaminoacridine
3,6-Diaminoacridine
3,9-Diaminoacridine
4,5-Diaminoacridine
Isomacrorine†
Macrorine†

$C_{13}H_{11}N_3O$
Azobenzene-3-carboxylic Acid, *Amide*
Azobenzene-4-carboxylic Acid, *Amide*
1,5-Diaminoacridone
1,6-Diaminoacridone
1,7-Diaminoacridone
2,4-Diaminoacridone
2,5-Diaminoacridone
2,6-Diaminoacridone
2,7-Diaminoacridone
3,5-Diaminoacridone
3,6-Diaminoacridone
4,5-Diaminoacridone
Macrorungine†

$C_{13}H_{11}N_3O_2$
2-Amino-4-nitrodiphenylamine, 1-N-*Me*
4-Hydroxyazobenzene-3-carboxylic Acid, *Amide*
2-Methyl-2′-nitroazobenzene
2-Methyl-3-nitroazobenzene
2-Methyl-4-nitroazobenzene
2-Methyl-5-nitroazobenzene
2-Methyl-6-nitroazobenzene
4-Methyl-2-nitroazobenzene
4-Methyl-2′-nitroazobenzene
4-Methyl-3-nitroazobenzene
4-Methyl-4′-nitroazobenzene

$C_{13}H_{11}N_3O_2S$
o-Nitrothiocarbanilide
m-Nitrothiocarbanilide
p-Nitrothiocarbanilide

$C_{13}H_{11}N_3O_3$
2-Amino-5-nitrobenzoic Acid, *Anilide*
2-Amino-6-nitrobenzoic Acid, *Anilide*
2-Hydroxy-5-methyl-2′-nitroazobenzene
2-Hydroxy-5-methyl-3′-nitroazobenzene
2-Hydroxy-5-methyl-4′-nitroazobenzene
4-Hydroxy-2-methyl-4′-nitroazobenzene
4-Hydroxy-2-methyl-5-nitroazobenzene
4-Hydroxy-2′-methyl-3-nitroazobenzene
4-Hydroxy-3-methyl-2′-nitroazobenzene
4-Hydroxy-3-methyl-4′-nitroazobenzene
4-Hydroxy-3′-methyl'-3-nitroazobenzene
4-Hydroxy-4′-methyl-3-nitroazobenzene
4′-Hydroxy-2-methyl-5-nitroazobenzene
4′-Hydroxy-4-methyl-2-nitroazobenzene
4′-Hydroxy-4-methyl-3-nitroazobenzene
4-Hydroxy-3-nitroazobenzene, *Me ether*
4-Hydroxy-4′-nitroazobenzene, *Me ether*
2-Nitrocarbanilide
3-Nitrocarbanilide
4-Nitrocarbanilide
2-Nitrodiphenylamine-4-carboxylic Acid, *Amide*

$C_{13}H_{11}N_3O_4$
2,4-Dinitroaniline, N-*Benzyl*
3,4-Dihydroxy-2′-nitroazobenzene, 3-*Me ether*
3,4-Dihydroxy-3′-nitroazobenzene, 3-*Me ether*
3,4-Dihydroxy-4′-nitroazobenzene, 3-*Me ether*
N-2,4-Dinitrobenzyl-aniline
N-2,6-Dinitrobenzyl-aniline

$C_{13}H_{11}N_3O_5$
4-Amino-2,6-dinitrophenol, N-*Benzoyl*
2-Amino-4,6-dinitrophenol, N-*Benzoyl*
2-Hydroxy-4,6-dinitrodiphenylamine, *Me ether*
2′-Hydroxy-2,4-dinitrodiphenylamine, *Me ether*
4-Hydroxy-2,2′-dinitrodiphenylamine, *Me ether*
4′-Hydroxy-2,4-dinitrodiphenylamine, *Me ether*

$C_{13}H_{11}N_5O_2$
N-(6-Purinyl)-α-phenylglycine†

$C_{13}H_{11}N_5O_4$
1,3-Di-(3-nitrophenyl)guanidine
1,3-Di-(4-nitrophenyl)guanidine

$C_{13}H_{12}$
1-Allylnaphthalene
Benz[*e*]indane
2,3-Dihydro-1*H*-phenalene★†
Diphenylmethane
2-Methylbiphenyl
3-Methylbiphenyl
4-Methylbiphenyl
6-Methyl-6-phenylfulvene†
Trideca-2,4,6,8-tetrayne†

$C_{13}H_{12}BN$
10-Methyl-10,9-borazarophenanthrene†

$C_{13}H_{12}BrClO$
6-Bromo-1-chloro-2-naphthol, n-*Propyl ether*

$C_{13}H_{12}Br_2O$
1,6-Dibromo-2-naphthol, *Propyl ether*

$C_{13}H_{12}ClNO$
2-Amino-4-chlorophenol, m-*Tolyl ether*
2-Amino-4-chlorophenol, p-*Tolyl ether*

$C_{13}H_{12}Cl_2O_4$
Ethacrynic Acid†

$C_{13}H_{12}I_3NO_3$
3-Amino-2,4,6-tri-iodobenzoic Acid, N-*Adipoyl*

$C_{13}H_{12}N_2$
2-Aminoacridan
3-Aminoacridan
4-Aminoacridan
2-(2-Aminostyryl)pyridine
2-(3-Aminostyryl)pyridine
2-(4-Aminostyryl)pyridine
4-(2-Aminostyryl)pyridine
4-(4-Aminostyryl)pyridine
1,2-Diaminofluorene
1,9-Diaminofluroene
2,3-Diaminofluorene
2,5-Diaminofluorene
2,7-Diaminofluorene
2,9-Diaminofluorene
3,6-Diaminofluorene
Diphenylformamidine
2-Methylazobenzene
3-Methylazobenzene
4-Methylazobenzene
Naphth[1,2-*d*]imidazole, 1-N-*Et*
N-1-Naphthylalanine, *Nitrile*
N-Phenylbenzamidine†

$C_{13}H_{12}N_2$ (di-ion)
9-Methyl-8*a*,10*a*-diazoniaphenanthrene

$C_{13}H_{12}N_2O$
m-Aminobenzoic Acid, *Anilide*
Anthranilic Acid, *Anilide*
N-Benzyl-*p*-nitrosoaniline
Carbanilide
2,2′-Diaminobenzophenone
2,3′-Diaminobenzophenone
2,4-Diaminobenzophenone
2,4′-Diaminobenzophenone
3,3′-Diaminobenzophenone
3,4′-Diaminobenzophenone
1,1-Diphenylurea
Gyrilone
Harmine
2-Hydroxyazobenzene, *Me ether*
3-Hydroxyazobenzene, *Me ether*
4-Hydroxyazobenzene, *Me ether*
N-Hydroxydiphenylformamidine
2-Hydroxy-4-methylazobenzene
2′-Hydroxy-4-methylazobenzene
2-Hydroxy-5-methylazobenzene
4-Hydroxy-2-methylazobenzene
4-Hydroxy-3-methylazobenzene
4′-Hydroxy-2-methylazobenzene
4′-Hydroxy-3-methylazobenzene
4-Hydroxy-4′-methylazobenzene

$C_{13}H_{12}N_2OS$
1-*o*-Hydroxyphenyl-3-phenylthiourea
1-*m*-Hydroxyphenyl-3-phenylthiourea
1-*p*-Hydroxyphenyl-3-phenylthiourea

$C_{13}H_{12}N_2O_2$
2′-Aminodiphenylamine-2-carboxylic Acid
3′-Aminodiphenylamine-2-carboxylic Acid
4-Aminodiphenylamine-2-carboxylic Acid
4′-Aminodiphenylamine-2-carboxylic Acid
2-Aminodiphenylamine-4-carboxylic Acid
2-Amino-5-nitrodiphenylmethane
4-Amino-3-nitrodiphenylmethane
Gyrolone
3-Hydroxy-1,1-diphenylurea
1-*o*-Hydroxyphenyl-3-phenylurea
1-*p*-Hydroxyphenyl-3-phenylurea
N-Methyl-2-nitrodiphenylamine
2-Methyl-2′-nitrodiphenylamine
2-Methyl-4-nitrodiphenylamine
2-Methyl-4′-nitrodiphenylamine
3-Methyl-6-nitrodiphenylamine
4-Methyl-2′-nitrodiphenylamine
4-Methyl-4′-nitrodiphenylamine
o-Nitrobenzylaniline
m-Nitrobenzylaniline
p-Nitrobenzylaniline

$C_{13}H_{12}N_2O_2S$
1,3-Di-*m*-hydroxyphenylthiourea
1,3-Di-*p*-hydroxyphenylthiourea

$C_{13}H_{12}N_2O_3$
α-4-Aminophenyl-3-nitrobenzyl Alcohol
α-4-Aminophenyl-4-nitrobenzyl Alcohol
1,3-Di-*o*-hydroxyphenylurea
1,3-Di-*m*-hydroxyphenylurea
1,3-Di-*p*-hydroxyphenylurea
1-Hydroxynaphthalene-2,4-dicarboxylic Acid, *Diamide*
α-Hydroxy-*N*-*p*-nitrophenylbenzylamine
2-Hydroxy-*N*-*p*-nitrophenylbenzylamine
2-Hydroxy-*N*-*m*-nitrophenylbenzylamine
2-Hydroxy-*N*-*p*-nitrophenylbenzylamine
N-2-Hydroxyphenyl-2-nitroaniline, *Me ether*
N-4-Hydroxyphenyl-2-nitroaniline, *Me ether*
N-4-Nitrobenzyl-*p*-aminophenol

$C_{13}H_{12}N_2O_4$
2-Methyl-5-nitroquinoline-3-carboxylic Acid, *Et ester*
2-Methyl-8-nitroquinoline-3-carboxylic Acid, *Et ester*
4-Phenylpyrazole-1,3(5)-dicarboxylic Acid, 3(5)-*Et ester*
3-Phenylpyrazole-1,4-dicarboxylic Acid, 4-*Et ester*
1-Phenylpyrazole-3,4-dicarboxylic Acid, *Di-Me ester*
1-Phenylpyrazole-3,5-dicarboxylic Acid, *Di-Me ester*
1-Phenylpyrazole-4,5-dicarboxylic Acid, *Di-Me ester*

$C_{13}H_{12}N_2O_4S$
2-Nitrobenzenesulphonic Acid, N-*Me-anilide*
3-Nitrobenzenesulphonic Acid, *Benzylamide*
3-Nitrobenzenesulphonic Acid, o-*Toluidide*
4-Nitrobenzenesulphonic Acid, N-*Me-anilide*
4-Nitrotoluene-α-sulphonic Acid, *Anilide*
4-Nitrotoluene-2-sulphonic Acid, *Anilide*
2-Nitrotoluene-4-sulphonic Acid, *Anilide*

$C_{13}H_{12}N_2O_4S_2$
Dehydrogliotoxin†

$C_{13}H_{12}N_2O_5S$
Benzophenone-3,3′-disulphonic Acid, *Diamide*

$C_{13}H_{12}N_2S$
Thiocarbanilide

$C_{13}H_{12}N_4O$
Diphenylcarbazone

$C_{13}H_{12}N_4O_2$
4-Amino-2-methyl-2′-nitroazobenzene
4-Amino-2-methyl-3′-nitroazobenzene
4-Amino-2-methyl-4′-nitroazobenzene
4-Amino-3-methyl-2′-nitroazobenzene
4-Amino-3-methyl-3′-nitroazobenzene
4-Amino-3-methyl-4′-nitroazobenzene
4-Amino-4′-nitroazobenzene, N-*Me*
Lumilactoflavin

$C_{13}H_{12}N_4O_3$
2-Amino-4′-hydroxy-5-nitroazobenzene, *Me ether*

$C_{13}H_{12}N_4O_8S_2$
Methionic Acid, *Di*-p-*nitroanilide*

$C_{13}H_{12}N_4S$
Diphenylthiocarbazone

$C_{13}H_{12}O$
2-Benzylphenol
3-Benzylphenol
4-Benzylphenol
Benzyl phenyl Ether
o-Cresol, *Phenyl ether*

$C_{13}H_{12}O$ (*continued*)
- *m*-Cresol, *Phenyl ether*
- *p*-Cresol, *Phenyl ether*
- Diphenylmethanol
- 2-Hydroxybiphenyl, *Me ether*
- 4-Hydroxybiphenyl, *Me ether*
- 2-Hydroxymethyl-biphenyl
- 3-Hydroxymethyl-biphenyl
- 2-Hydroxy-5-methylbiphenyl
- 4-Hydroxy-4′-methylbiphenyl
- 1-α-Naphthyl-2-propen-1-ol
- 1-β-Naphthyl-2-propen-1-ol
- 3-α-Naphthyl-2-propen-1-ol
- 3-β-Naphthyl-2-propen-1-ol
- 2-α-Naphthylpropionaldehyde
- 2-β-Naphthylpropionaldehyde
- 1-Propionylnaphthalene
- 2-Propionylnaphthalene

$C_{13}H_{12}OS$
- *p*-Hydroxybenzenethiol, S-*Phenyl*, *Me ether*
- *p*-Hydroxybenzenethiol, S-p-*Tolyl*
- Phenyl *o*-tolyl sulphoxide
- Phenyl *m*-tolyl sulphoxide
- Phenyl *p*-tolyl sulphoxide

$C_{13}H_{12}O_2$
- 2-Acetyl-1-naphthol, *Me ether*
- 4-Acetyl-1-naphthol, *Me ether*
- 1-Acetyl-2-naphthol, *Me ether*
- 3-Acetyl-2-naphthol, *Me ether*
- 7-Acetyl-2-naphthol, *Me ether*
- Benzyl *p*-hydroxyphenyl Ether
- Dehydrotremetone
- 2,5-Dihydroxydiphenylmethane
- 4,8-Dimethylazulene-6-carboxylic Acid
- 2-Hydroxy-α-4-hydroxyphenyltoluene
- 3-Hydroxy-α-3-hydroxyphenyltoluene
- 4-Hydroxy-α-4-hydroxyphenyltoluene
- 2-Hydroxy-1-naphthaldehyde, *Et ether*
- 4-Hydroxy-1-naphthaldehyde, *Et ether*
- 2-Hydroxyphenyl phenyl Ether, *Me ether*
- 3-Hydroxyphenyl phenyl Ether, *Me ether*
- 4-Hydroxyphenyl phenyl Ether, *Me ether*
- 2-Methylazulene-5-carboxylic Acid, *Me ester*
- 2-Methylazulene-6-carboxylic Acid, *Me ester*
- 2-Methyl-1-naphthoic Acid, *Me ester*
- 4-Methyl-1-naphthoic Acid, *Me ester*
- 6-Methyl-1-naphthoic Acid, *Me ester*
- 4-Methyl-2-naphthoic Acid, *Me ester*
- 6-Methyl-2-naphthoic Acid, *Me ester*
- 1-Naphthoic Acid, *Et ester*
- 2-Naphthoic Acid, *Et ester*
- α-Naphthoxyacetone
- β-Naphthoxyacetone
- 1-Naphthylacetic Acid, *Me ester*
- 2-α-Naphthylpropionic Acid
- 2-β-Naphthylpropionic Acid
- 3-α-Naphthylpropionic Acid
- 3-β-Naphthylpropionic Acid
- 7-Phenylhepta-4,6-diyne-2,3-diol†
- 2-Propionyl-1-naphthol
- 4-Propionyl-1-naphthol
- 1-Propionyl-2-naphthol
- 6-Propionyl-2-naphthol
- 2-Propyl-1,4-naphthoquinone
- Resorcinol, *Monobenzyl ether*

$C_{13}H_{12}O_2S$
- Benzyl phenyl sulphone
- Phenyl *o*-tolyl sulphone
- Phenyl *m*-tolyl sulphone
- Phenyl *p*-tolyl sulphone

$C_{13}H_{12}O_2S_2$
- Thiolbenzenesulphonic Acid, p-*Tolyl ester*

$C_{13}H_{12}O_3$
- 2-Acetyl-1,8-dihydroxy-3-methylnaphthalene
- Allenolic Acid
- 4,8-Dihydroxy-1-naphthaldeyhde, *Di-Me ether*
- 1,4-Dihydroxy-2-naphthaldehyde, *Di-Me ether*
- Euparin
- 2-Hydroxy-1-naphthoic Acid, *Et ester*
- 2-Hydroxy-1-naphthoic Acid, *Me ether*, *Me ester*
- 2-Hydroxy-1-naphthoic Acid, *Et ether*
- 4-Hydroxy-1-naphthoic Acid, *Et ester*
- 5-Hydroxy-1-naphthoic Acid, *Et ester*
- 5-Hydroxy-1-naphthoic Acid, *Et ether*
- 6-Hydroxy-1-naphthoic Acid, *Et ester*
- 8-Hydroxy-1-naphthoic Acid, *Me ether*, *Me ester*
- 8-Hydroxy-1-naphthoic Acid, *Et ether*
- 1-Hydroxy-2-naphthoic Acid, *Et ester*
- 1-Hydroxy-2-naphthoic Acid, *Me ether*, *Me ester*
- 3-Hydroxy-2-naphthoic Acid, *Et ester*
- 3-Hydroxy-2-naphthoic Acid, *Me ether*, *Me ester*
- 4-Hydroxy-2-naphthoic Acid, *Et ester*
- 5-Hydroxy-2-naphthoic Acid, *Et ester*
- 6-Hydroxy-2-naphthoic Acid, *Et ester*
- 6-Hydroxy-2-naphthoic Acid, *Et ether*
- 7-Hydroxy-2-naphthoic Acid, *Et ester*
- 8-Hydroxy-2-naphthoic Acid, *Et ester*
- Methysticone
- 2,1-Naphthylglycollic Acid, *Me ester*
- 2,2-Naphthylglycollic Acid, *Me ester*
- 7-Phenylhepta-4,6-diyne-1,2,3-triol†

$C_{13}H_{12}O_3S$
- Acenaphthene-3-sulphonic Acid, *Me ester*
- Toluene-α-sulphonic Acid, *Phenyl ester*
- Toluene-*o*-sulphonic Acid, *Phenyl ester*
- Toluene-*p*-sulphonic Acid, *Phenyl ester*

$C_{13}H_{12}O_4$
- 3-Carboxylic-6-phenyl-3,5-hexadienoic Acid
- 2,7-Dihydroxy-1-naphthoic Acid, *Di-Me ether*
- 4,8-Dihydroxy-1-naphthoic Acid, *Di-Me ether*
- 1,4-Dihydroxy-2-naphthoic Acid, 4-*Et ether*
- 1,5-Dihydroxy-2-naphthoic Acid, *Di-Me ether*
- 3,6-Dihydroxy-2-naphthoic Acid, *Di-Me ether*
- 6-Hydroxy-4,5-dimethoxy-2-naphthaldehyde†
- 3-Methylchromone-2-carboxylic Acid, *Et ester*
- 4-Methylcoumarin-3-carboxylic Acid, *Et ester*
- 5-Methylcoumarin-3-carboxylic Acid, *Et ester*
- 6-Methylcoumarin-3-carboxylic Acid, *Et ester*
- 7-Methylcoumarin-3-carboxylic Acid, *Et ester*
- 8-Methylcoumarin-3-carboxylic Acid, *Et ester*
- 6-Methylcoumarin-4-carboxylic Acid, *Et ester*
- 7-Methylcoumarin-4-carboxylic Acid, *Et ester*
- 5-(3,4-Methylenedioxyphenyl)-2,4-pentadienoic Acid, *Me ester*
- Phyllomeronic Acid, *Me ester*
- Trideca-5,7,9,11-tetrayne-1,2,3,4-tetraol†

$C_{13}H_{12}O_4S$
4-Hydroxytoluene-3-sulphonic Acid, *Phenyl ester*

$C_{13}H_{12}O_5$
4-Acetyl-6,8-dihydroxy-5-methyl-2-benzopyran-1-one, 8-*Me ether*†
5-Hydroxy-2,7-dimethoxy-1,4-naphthaquinone, *Me ether*†
Ostholic Acid, *Me ester*
Phloroglucide, *Mono-Me ether*
6,7,8-Trihydroxy-1-naphthoic Acid, *Et ester*
2,5,7-Trihydroxy-1,4-naphthaquinone, *Tri-Me ether*

$C_{13}H_{12}O_6$
Di-2,4,6-trihydroxyphenylmethane
Di-3,4,5-trihydroxyphenylmethane
Spinochrome B, *Tri-Me ether*†
2,5,7,8-Tetrahydroxy-1,4-naphthaquinone, 5,7,8-*Tri-Me ether*†

$C_{13}H_{12}O_6S_2$
Methionic Acid, *Di-phenyl ester*

$C_{13}H_{12}O_8$
Phaselic Acid

$C_{13}H_{12}S$
Diphenylmethanethiol
4-Mercaptobiphenyl, S-*Me*
Phenyl *o*-tolyl sulphide
Phenyl *m*-tolyl sulphide
Phenyl *p*-tolyl sulphide
α-Toluenethiol, *Phenyl ether*

$C_{13}H_{13}As_2N_2NaO_4$
Neosalvarsan

$C_{13}H_{13}BrO$
6-Bromo-2-hydroxy-1-methylnaphthalene, *Et ether*
1-Bromo-2-naphthol, *Propyl ether*
6-Bromo-2-naphthol, *Propyl ether*

$C_{13}H_{13}ClO_4$
Sordidone *Di-Me ether*†

$C_{13}H_{13}N$
α-Aminodiphenylmethane
2-Aminodiphenylmethane
3-Aminodiphenylmethane
4-Aminodiphenylmethane
2-Amino-4-methylbiphenyl
2-Amino-4′-methylbiphenyl
2-Amino-6-methylbiphenyl
3-Amino-4-methylbiphenyl
3-Amino-5-methylbiphenyl
4-Amino-3-methylbiphenyl
4-Amino-4′-methylbiphenyl
Benzylaniline
N-Methyldiphenylamine
2-Methyldiphenylamine
3-Methyldiphenylamine
4-Methyldiphenylamine
2-Phenethylpyridine
3-Phenethylpyridine
4-Phenethylpyridine
1,2,3,4-Tetrahydroacridine

$C_{13}H_{13}NO$
o-Aminobenzyl Alcohol, *Phenyl ether*
m-Aminophenol, *Benzyl ether*
p-Aminophenol, *Benzyl ether*
α-*o*-Aminophenylbenzyl Alcohol
α-*m*-Aminophenylbenzyl Alcohol
α-*p*-Aminophenylbenzyl Alcohol
N-2-Hydroxybenzylaniline
N-3-Hydroxybenzylaniline
N-4-Hydroxybenzylaniline
3-Hydroxy-4′-methyldiphenylamine
3-Hydroxy-5-methyldiphenylamine
3′-Hydroxy-2-methyldiphenylamine
4′-Hydroxy-2-methyldiphenylamine
4′-Hydroxy-3-methyldiphenylamine
4′-Hydroxy-4-methyldiphenylamine
2-Hydroxy-4-methylpyridine, *Benzyl ether*
2-Hydroxy-*N*-phenylaniline, *Me ether*
4-Hydroxy-*N*-phenylaniline, *Me ether*
2-Methylaminophenyl phenyl Ether
3-α-Naphthylpropionic Acid, *Amide*
3-β-Naphthylpropionic Acid, *Amide*
1,2,3,4-Tetrahydroacridone

$C_{13}H_{13}NO_2$
4-Amino-1-naphthoic Acid, *Et ester*
5-Amino-1-naphthoic Acid, *Et ester*
3-Amino-2-naphthoic Acid, *Et ester*
2,3-Dimethylquinoline-4-carboxylic Acid, *Me ester*
4-Hydroxy-1-naphthoic Acid, *Et ether*, *Amide*
Lactic Acid, 1-*Naphthalide*
2-Methylquinoline-3-carboxylic Acid, *Et ester*
2-Methylquinoline-4-carboxylic Acid, *Et ester*
3-Methylquinoline-4-carboxylic Acid, *Et ester*
7-Methylquinoline-4-carboxylic Acid, *Et ester*
N-1-Naphthylalanine
N-2-Naphthylalanine
α-Naphthylurethane
β-Naphthylurethane
1-Phenylpyrrole-2-carboxylic Acid, *Et ester*
2-Phenylpyrrole-3-carboxylic Acid, *Et ester*
2-Quinolineacetic Acid, *Et ester*
3-Quinolineacetic Acid, *Et ester*
4-Quinolineacetic Acid, *Et ester*
6-Quinolineacetic Acid, *Et ester*
8-Quinolineacetic Acid, *Et ester*
1,2,3,4-Tetrahydrocarbazole-2-(or 4)-carboxylic Acid
1,2,3,4-Tetrahydrocarbazole-3-carboxylic Acid
1,2,3,4-Tetrahydrocarbazole-5-carboxylic Acid
1,2,3,4-Tetrahydrocarbazole-3-carboxylic Acid
1,2,3,4-Tetrahydrocarbazole-5-carboxylic Acid
1,2,3,4-Tetrahydrocarbazole-6-carboxylic Acid
1,2,3,4-Tetrahydrocarbazole-7-carboxylic Acid
1,2,3,4-Tetrahydrocarbazole-8-carboxylic Acid
1,2,3,4-Tetrahydrocarbazole-9-carboxylic Acid

$C_{13}H_{13}NO_3$
2-Benzyl-4-hydroxymethylene-5-oxazolone, *Me ether*
6-Hydroxyisoquinoline-3-carboxylic Acid, *Me ether*, *Et ester*
2-Hydroxy-4-methylquinoline-3-carboxylic Acid, *Et ester*
4-Hydroxy-2-methylquinoline-3-carboxylic Acid, *Et ester*

$C_{13}H_{13}NO_3$ (*continued*)
2-Hydroxy-3-methylquinoline-4-carboxylic Acid, *Et ester*
3-Hydroxy-2-methylquinoline-4-carboxylic Acid, *Et ether*
3-Hydroxy-2-methylquinoline-4-carboxylic Acid, *Me ester*, *Me ether*
2-Hydroxyquinoline-3-carboxylic Acid, *Propyl ether*
6-Methoxyquinoline-4-carboxylic Acid, *Et ester*
1-Nitro-2-naphthol, *Propyl ether*

$C_{13}H_{13}NO_4$
1-Methylindole-2,3-dicarboxylic Acid, *Mono-Et ester*
2-Phthalimidopropionic Acid, *Et ester*
3-Phthalimidopropionic Acid, *Et ester*
Tricarballylic Acid, αβ-*Anil*, *Me ester*

$C_{13}H_{13}N_2$ (ion)
Melinonine F

$C_{13}H_{13}N_3$
Agroflavine
1,1-Diphenylguanidine
1,3-Diphenylguanidine
4-Methylaminoazobenzene

$C_{13}H_{13}N_3O$
1,1-Diphenylsemicarbazide
1,4-Diphenylsemicarbazide
2,4-Diphenylsemicarbazide
4,4-Diphenylsemicarbazide

$C_{13}H_{13}N_3O_2$
2-Amino-α-4-aminophenyl-4-nitrotoluene
4-Amino-α-4-aminophenyl-2-nitrotoluene

$C_{13}H_{13}N_3S$
1,1-Diphenylthiosemicarbazide
1,4-Diphenylthiosemicarbazide
2,4-Diphenylthiosemicarbazide

$C_{13}H_{13}P$
2,6-Dimethyl-4-phenylphosphorin †

$C_{13}H_{14}$
Capillene
1-Ethyl-2-methylnaphthalene
1-Ethyl-4-methylnaphthalene
1-Ethyl-5-methylnaphthalene
1-Ethyl-6-methylnaphthalene
1-Ethyl-7-methylnaphthalene
2-Ethyl-1-methylnaphthalene
2-Ethyl-5-methylnaphthalene
2-Ethyl-6-methylnaphthalene
2-Ethyl-7-methylnaphthalene
2-Ethyl-8-methylnaphthalene
1-Isopropylazulene
2-Isopropylazulene
5-Isopropylazulene
6-Isopropylazulene
1-Isopropylnaphthalene
2-Isopropylnaphthalene
1-Propylnaphthalene
2-Propylnaphthalene
1,2,3-Trimethylnaphthalene
1,2,4-Trimethylnaphthalene
1,2,5-Trimethylnaphthalene
1,2,6-Trimethylnaphthalene
1,2,7-Trimethylnaphthalene
1,2,8-Trimethylnaphthalene
1,3,5-Trimethylnaphthalene
1,3,6-Trimethylnaphthalene
1,3,7-Trimethylnaphthalene
1,3,8-Trimethylnaphthalene
1,4,5-Trimethylnaphthalene
1,4,6-Trimethylnaphthalene
1,6,7-Trimethylnaphthalene
2,3,6-Trimethylnaphthalene

$C_{13}H_{14}N_2$
2-Amino-α-(4-aminophenyl) toluene
3-Amino-α-(4-aminophenyl) toluene
o-Aminobenzylamine, αN-*Phenyl*
m-Aminobenzylamine, αN-*Phenyl*
p-Aminobenzylamine, αN-*Phenyl*
2-Amino-4-methyldiphenylamine
2-Amino-5-methyldiphenylamine
2′-Amino-4-methyldiphenylamine
3-Amino-4-methyldiphenylamine
4-Amino-2-methyldiphenylamine
4-Amino-3-methyldiphenylamine
4-Amino-4′-methyldiphenylamine
4′-Amino-2-methyldiphenylamine
5-Amino-2-methyldiphenylamine
1-Benzyl-1-phenylhydrazine
1-Benzyl-2-phenylhydrazine
4,4′-Diaminodiphenylmethane †
Di-(2-aminophenyl)methane
Di-(3-aminophenyl)methane
Di-(4-aminophenyl)methane
Hydrazobenzene, N-*Me*
2-Methylbenzidine
3-Methylbenzidine
Methylenedianiline
2-Methylhydrazobenzene
3-Methylhydrazobenzene
4-Methylhydrazobenzene
4,5,6,7-Tetrahydro-1-phenylindazole
4,5,6,7-Tetrahydro-2-phenylindazole

$C_{13}H_{14}N_2O$
3-Amino-α-(4-aminophenyl) benzyl Alcohol
4-Amino-4-hydroxydiphenylamine, *Me ether*
4,4′-Diamino-2-hydroxybiphenyl, *Me ether*
Di-(3-aminophenyl)methanol
Di-(4-aminophenyl)methanol
Harmaline
Harmidine
1-Nitroso-2-naphthylamine, N-*Propyl*

$C_{13}H_{14}N_2O_2$
1,3-Diamino-2-naphthoic Acid, *Et ester*
1,4-Diamino-2-naphthoic Acid, *Et ester*
3-Methyl-5-phenylpyrazole-1-carboxylic Acid, *Et ester*
5-Methyl-3-phenylpyrazole-1-carboxylic Acid, *Et ester*
5-Methyl-1-phenylpyrazole-4-carboxylic Acid, *Et ester*
1-Methyl-4-phenylpyrazole-5-carboxylic Acid, *Et ester*

$C_{13}H_{14}N_2O_3$
4-Hydroxy-3-nitroquinoline, *Butyl ether*
Prominal

$C_{13}H_{14}N_2O_4S_2$
Gliotoxin
Methionic Acid, *Dianilide*

$C_{13}H_{14}N_4O$
N,N′-Di-(2-aminophenyl)urea
N,N′-Di-(3-aminophenyl)urea
N,N′-Di-(4-aminophenyl)urea
Diphenylcarbazide
2,2′,4,4′-Tetra-aminobenzophenone
3,3′,4,4′-Tetra-aminobenzophenone

$C_{13}H_{14}N_4O_4$
4-Amino-2-hydroxybenzoic Acid, *Isonicotinyl hydrazide ester*

$C_{13}H_{14}N_4S$
Diphenylthiocarbazide
Thiocarbazide, 1,5-*Diphenyl*

$C_{13}H_{14}O$
2,4-Dimethyl-1-naphthol, *Me ether*
1,4-Dimethyl-2-naphthol, *Me ether*
3,8-Dimethyl-2-naphthol, *Me ether*
4,8-Dimethyl-2-naphthol, *Me ether*
2-Isopropyl-1-naphthol
4-Isopropyl-1-naphthol
7-Isopropyl-1-naphthol
1-Methyl-2-naphthol, *Et ether*
6-Methyl-2-naphthol, *Et ether*
1-Naphthol, *Propyl ether*
2-Naphthol, *Propyl ether*
1-β-Naphthylpropanol
3-α-Naphthylpropanol
Tridec-2-ene-4,6,8-triyn-1-ol †
2,4,7-Trimethyl-1-naphthol
3,4,6-Trimethyl-1-naphthol
1,7,8-Trimethyl-2-naphthol
3,4,8-Trimethyl-2-naphthol
3,5,7-Trimethyl-2-naphthol

$C_{13}H_{14}OS$
9-(2-Thienyl)-non-6-en-8-yn-3-one †

$C_{13}H_{14}O_2$
1,4-Dihydro-1-naphthoic Acid, *Et ester*
3,4-Dihydro-1-naphthoic Acid, *Et ester*
1,2-Dihydro-2-naphthoic Acid, *Et ester*
1,4-Dihydro-2-naphthoic Acid, *Et ester*
3,4-Dihydro-2-naphthoic Acid, *Et ester*
1,4-Dihydroxy-2-methylnaphthalene, *Di-Me ether*
1,4-Dihydroxy-5-methylnaphthalene, *Di-Me ether*
1,6-Dihydroxy-4-methylnaphthalene, *Di-Me ether*
2,6-Dihydroxy-1-methylnaphthalene, *Di-Me ether*
2,7-Dihydroxy-1-methylnaphthalene, *Di-Me ether*
8-Isopropyl-5-methylcoumarin
3-Isovalidenephthalide
3-Methylindene-2-carboxylic Acid, *Et ester*
1-Methylindene-3-carboxylic Acid, *Et ester*
5-Phenyl-2,4-pentadienoic Acid, *Et ester*
Phyllomerol, *Di-Me ether*
Pyroguaiacin
Tremetone

$C_{13}H_{14}O_3$
Diaporthin
Hydroxytremetone
Macassar II †
Toxol
1,2,7-Trihydroxynaphthalene, *Tri-Me ether*
1,6,7-Trihydroxynaphthalene, *Tri-Me ether*

$C_{13}H_{14}O_3S$
6-Methylnaphthalene-2-sulphonic Acid, *Et ester*

$C_{13}H_{14}O_4$
Aesculetin, *Di-Et ether*
Anhydrosepedonin, *Di-Me ether* †
Daphnetin, 7,8-*Di-Et ether*
3,7-Dihydroxychromone, *Di-Et ether*
8-Hydroxy-6-methoxy-3-propylisocoumarin
Indane-1,2-dicarboxylic Acid, *Mono-Et ester*
p-Isopropylbenzylidenemalonic Acid
γ-Phenylallylsuccinic Acid †
3-Phenylcyclopropane-1,2-dicarboxylic Acid, *Di-Me ester*
Phenylitaconic Acid, *Di-Me ester*
Phenylitaconic Acid, α-*Et ester*
Phenylitaconic Acid, β-*Et ester*
3-Phenyl-2-pentenedioic Acid, *Mono-Et ether*
Sclerin †
Tetrahydro-5-oxo-2-phenylfuran-3-carboxylic Acid, *Et ester*
1,3,6,8-Tetrahydroxynaphthalene, 1,3,6-*Tri-Me ether* †
Tubaic Acid, *Me ester*
Tubaic Acid, *Me ether*
6,7,8-Trimethoxy-1-naphthol †
Undec-2-ene-4,6-diyne-1,11-dioic Acid, *Di-Me ester* †

$C_{13}H_{14}O_4S$
1-Naphthol-4-sulphonic Acid, *Et ether*, *Me ester*

$C_{13}H_{14}O_5$
Citrinin
Fraxetin, *Me-Et ether*
Gladiolic Acid, *Et ester*
3-Hydroxy-4-methoxycinnamic Acid, *Me ester*, *Ac*
Radicinin, *Me ether*
Rosellinic Acid, *Me ester*, *Me ether* †

$C_{13}H_{14}O_6$
3,4-Dihydro-6,8-dihydroxy-3,4,5-trimethylisocoumarin-7-carboxylic Acid
Toluene-2,4,6-tricarboxylic Acid, *Tri-Me ester*

$C_{13}H_{14}O_7$
4-Hydroxybenzene-1,2,3-tricarboxylic Acid, *Me ether*, *Tri-Me ester*
5-Hydroxybenzene-1,2,3-tricarboxylic Acid, *Me ether*, *Tri-Me ester*
3-Hydroxybenzene-1,2,4-tricarboxylic Acid, *Me Ether*, *Tri-Me ester*
5-Hydroxybenzene-1,2,4-tricarboxylic Acid, *Me Ether*, *Tri-Me ester*
6-Hydroxybenzene-1,2,4-tricarboxylic Acid, *Me Ether*, *Tri-Me ester*
2-Hydroxybenzene-1,3,5-tricarboxylic Acid, *Me ether*, *Tri-Me ester*
5-Hydroxytoluene-2,3,6-tricarboxylic Acid, *Tri-Me ester* †
5-Hydroxytoluene-2,3,6-tricarboxylic Acid, *Me ether*, *Di-Me ester* †

$C_{13}H_{15}BrO$
2-Bromo-1,2,3,4-tetrahydro-1-naphthol, *Allyl ether*

$C_{13}H_{15}ClO_2$
p-Cyclohexoxybenzoic Acid, *Chloride*

$C_{13}H_{15}N$
1-Butylisoquinoline
3-Butylisoquinoline
1-*tert*-Butylisoquinoline
2-Butylquinoline
2-(2-Butyl)quinoline
4-Butylquinoline
6-Butylquinoline
7-Butylquinoline
3,4-Diethylquinoline
2-Isobutylquinoline
3-Isobutylquinoline
N-Isopropyl-2-naphthylamine
1-Naphthylamine, N-*Propyl*
2-Naphthylamine, N-*Propyl*
2,3,4,6-Tetramethylquinoline
2,3,4,7-Tetramethylquinoline
2,4,5,7-Tetramethylquinoline
2,4,5,8-Tetramethylquinoline
2,4,6,8-Tetramethylquinoline
2,5,6,8-Tetramethylquinoline
3,4,6,8-Tetramethylquinoline

$C_{13}H_{15}NO$
1-Amino-2-naphthol, *Propyl ether*
1-Amino-2-naphthol, *Isopropyl ether*
6-Hydroxy-2,4-dimethylquinoline, *Et ether*
4-Hydroxy-2-methylquinoline, *Propyl ether*
5-Phenyl-2,4-pentadienoic Acid, *Ethylamide*

$C_{13}H_{15}NO_2$
Glutethimide
N-Hydroxy-2-fluorenylbenzamide†
4-3′-Indolylbutyric Acid, *Me ester*
Phthalimide, N-*Pentyl*
Securinine

$C_{13}H_{15}NO_3$
Cinnamoylglycine, *Et ester*
5-Hydroxyindole-3-acetic Acid, *Me ether*, *Et ester*†
Mesacon-α-anilic Acid, *Et ester*
Mesacon-β-anilic Acid, *Et ester*
Mesaconic Acid, α-*Me ester*, β-p-*Toluidide*
1-Phenyl-3-pyrrolidone-2-carboxylic Acid, *Et ester*
Succinimide, N-p-*Hydroxyphenyl*, *Propyl ether*

$C_{13}H_{15}NO_4$
Benzylpenaldic Acid, *Et ester*

$C_{13}H_{15}NO_5$
3-(2-Hydroxy-5-nitrophenyl)acrylic Acid, *Et ether*, *Et ester*
Phenylacetylglutamic Acid

$C_{13}H_{15}NO_6$
2,3-Dihydroxy-5-nitrocinnamic Acid, *Di-Et ether*
3-*o*-Nitrophenylglutaric Acid, *Di-Me ester*
3-*p*-Nitrophenylglutaric Acid, *Di-Me ester*
3-Nitrophthalic Acid, 1-2′-*Methylbutyl ester*
3-Nitrophthalic Acid, 2-2′-*Methylbutyl ester*
3-Nitrophthalic Acid, 1-3′-*Methylbutyl ester*
3-Nitrophthalic Acid, 2-3′-*Methylbutyl ester*
5-Nitrotoluene-α,2-dicarboxylic Acid, *Di-Et ester*

$C_{13}H_{15}NO_7$
4-Hydroxy-3-nitrophthalic Acid, *Me ether*, *Di-Et ester*
o-Nitrophenoxymalonic Acid, *Di-Et ester*
m-Nitrophenoxymalonic Acid, *Di-Et ester*
p-Nitrophenoxymalonic Acid, *Di-Et ester*

$C_{13}H_{15}N_2NaO_2S$
Kemithal

$C_{13}H_{16}$
1-Ethyl-3,4-dihydro-5-methylnaphthalene
1,1α,2,3,4,4α-Hexahydrofluorene

$C_{13}H_{16}Br_2O_2$
2,3-Dibromo-3-phenylpropionic Acid, *Isobutyl ester*

$C_{13}H_{16}Br_2O_5$
2,6-Dibromo-3,4,5-trihydroxybenzoic Acid, *Tri-Et ether*

$C_{13}H_{16}N_2$
1,8-Naphthylenediamine★, NN′-*Tri-Me*†
Tetrahydroharman, N_b-*Me*†
Tetrahydrozoline

$C_{13}H_{16}N_2O$
1,2,3,4-Tetrahydroharmine†
1,2,3,4-Tetrahydro-6-methoxy-1-methyl-2-carboline†

$C_{13}H_{16}N_2O_2$
Melatonin†
Rubreserine
Tryptophane, *Et ester*
Tryptophane, α-N-*Di-Me*

$C_{13}H_{16}N_2O_3$
2-Ethyl-3-methyl-5-oxo-1-phenyl-3-pyrazoline-4-carboxylic Acid
6-Hydroxymelatonin†

$C_{13}H_{16}N_2O_4$
N-Benzoylglycylalanine, *Me ester*
N-Benzoylglycyl-3-aminobutyric Acid
N-Benzoylglycyl-4-aminobutyric Acid
N-Benzoylglycylglycine, *Et ester*

$C_{13}H_{16}N_2S_2$
2-Benzothiazolesulphenamide, N-*Cyclohexyl*

$C_{13}H_{16}N_4$
2,2′,4,4′-Tetra-aminodiphenylmethane
3,3′,4,4′-Tetra-aminodiphenylmethane

$C_{13}H_{16}O$
Cyclohexyl phenyl Ketone
4,4-Dimethyl-1-phenyl-1-penten-3-one
2-Phenylcycloheptanone†
1-Phenyl-1-hepten-3-one

$C_{13}H_{16}O_2$
Cinnamic Acid, *Isobutyl ester*
α-Ethylcinnamic Acid, *Et ester*
4,5′-Indanylbutyric Acid
3-Isovalidene-3*a*,4-dihydrophthalide
1-Methylcyclopentane-1-carboxylic Acid, *Phenyl ester*
1-Methylindane-2-carboxylic Acid, *Et ester*
2-Methyl-3-phenylacrylic Acid, *Propyl ester*

$C_{13}H_{16}O_2$ (*continued*)
1-Phenylcyclohexanecarboxylic Acid
2-Phenylcyclohexanecarboxylic Acid
4-Phenylcyclohexanecarboxylic Acid
1,2,3,4-Tetrahydro-1-naphthoic Acid, *Et ester*
5,6,7,8-Tetrahydro-1-naphthoic Acid, *Et ester*

$C_{13}H_{16}O_3$
Ageratochromene
Benzoylacetic Acid, *Butyl ester*
Benzoylacetic Acid, *Isobutyl ester*
2-Benzoylbutyric Acid, *Et ester*
3-Benzoylbutyric Acid, *Et ester*
4-Benzoylbutyric Acid, *Et ester*
Benzoylformic Acid, (+)-*Pentyl ester*
2-Benzoyl-2-methylpropionic Acid, *Et ester*
2-Benzoylpropionic Acid, *Isopropyl ester*
2-Benzylacetoacetic Acid, *Et ester*
p-Cyclohexoxybenzoic Acid
Cyclohexyl-2,4-dihydroxyphenyl Ketone
α-Ethyl-*o*-hydroxy-*trans*-cinnamic Acid, *Me ether, Me ester*
3-α-Furylacrylic Acid, *Cyclohexyl ester*
3-*m*-Hydroxyphenyl-2-methylacrylic Acid, *Me ester*
Mandelic Acid, *Et ester, Allyl ether*
2-Methoxy-4-propenylphenol, *Propionyl*
4-Oxo-2-phenylvaleric Acid, *Et ester*
2,4,5-Trimethylbenzoylformic Acid, *Et ester*
2,4,6-Trimethylbenzoylformic Acid, *Et ester*

$C_{13}H_{16}O_4$
Benzylsuccinic Acid, *Di-Me ester*
2-Carboxyphenylacetic Acid, *Di-Et ester*
4-Carboxyphenylacetic Acid, *Di-Et ester*
4,6-Diacetylresorcinol, *Me-Et ether*
4,6-Diacetylresorcinol, *Monopropyl ether*
2,6-Dimethylterephthalic Acid, 1-*Propyl ester*
4-Hydroxy-3-methoxycinnamic Acid, *Propyl ester*
3-*p*-Hydroxyphenylcrotonic Acid, *Me ether, Et ester*
5-Methylisophthalic Acid, *Di-Et ester*
2-Methyl-2-phenylglutaric Acid, 5-*Me ester*
3-Phenylglutaric Acid, *Di-Me ester*
Phenylmalonic Acid, *Di-Et ester*
Umbellic Acid, *Et ester, Di-Me ether*
Umbellic Acid, *Di-Et ether*

$C_{13}H_{16}O_4S$
S-p-Carboxyphenylthioglycollic Acid, *Di-Et ester*

$C_{13}H_{16}O_5$
Curvulinic Acid, *Me ester, Di-Me ether*†
1,3-Diacetyl-2,4,5-trihydroxybenzene, *Tri-Me ether*
1,5-Diacetyl-2,3,4-trihydroxybenzene, *Tri-Me ether*
Eugenone
2-Formyl-5,6-dimethoxybenzoic Acid, ψ-*Propyl ester*
3-(4-Hydroxy-3,5-dimethoxyphenyl)acrylic Acid, *Et ester*
3-(4-Hydroxy-3,5-dimethoxyphenyl)acrylic Acid, 4-*Me ether, Me ester*
2-Hydroxy-4-methylisophthalic Acid, *Di-Et ester*
4-Hydroxy-5-methylisophthalic Acid, *Di-Et ester*
4-Hydroxy-6-methylisophthalic Acid, *Di-Et ester*
Netoric Acid, *Me ester*
Sepedonin, *Di-Me ether*†
3-(2,4,5-Trihydroxyphenyl)acrylic Acid, *Tri-Me ether, Me ester*

$C_{13}H_{16}O_6$
Curvulic Acid, *Et ester*†
Curvulic Acid, *Di-Me ether*†
Cyclopolic Acid, *Me ester*
3-Isopropyl-4,5-dimethoxyphthalic Acid
Quinic Acid, *Phenyl ester*

$C_{13}H_{16}O_6S$
p-Ethylsulphonylbenzaldehyde, *Diacetate*

$C_{13}H_{16}O_7$
2-Carboxy-4,5-dimethoxyphenoxyacetic Acid, β-*Di-Me ester*
2-Carboxy-4,5-dimethoxyphenoxyacetic Acid, α-*Et ester*
Furan-2,3,4-tricarboxylic Acid, *Tri-Et ester*
Furan-2,3,5-tricarboxylic Acid, *Tri-Et ester*
o-Glucosyloxybenzaldehyde
3-Hydroxy-4-oxo-4*H*-pyran-2,6-dicarboxylic Acid, *Dipropyl ester*
Piscidic Acid, *Di-Me ester*†
Piscidic Acid, *Mono-Et ester*†
Piscidic Acid, *Di-Me ether*†
Purpurogallin-carboxylic Acid, *Me ester*
2,4,6-Trihydroxyisophthalic Acid, *Tri-Me ether, Di-Me ester*
3,4,5-Trihydroxyphthalic Acid, *Tri-Me ether, Di-Me ester*
Vacciniin

$C_{13}H_{16}O_8$
Caffeic Acid, *Dicarbomethoxyl, Et ester*

$C_{13}H_{16}O_9$
Gentisic Acid 2-β-D-glucopyranose†
Gentisic Acid 5-β-D-glucopyranose†

$C_{13}H_{16}O_{10}$
3-Glucogallic Acid
4-Glucogallic Acid
5-Glucogallic Acid

$C_{13}H_{17}BrO_2$
2-Bromo-4,6-dimethylbenzoic Acid, *Isobutyl ester*

$C_{13}H_{17}ClO_3$
p-Butoxybenzoic Acid, 2-*Chloroethyl ester*
3-Chloro-6-hydroxy-*p*-toluic Acid, *Butyl ether, Me ester*
3-Chloro-6-hydroxy-*p*-toluic Acid, 3-*Methylbutyl ether*

$C_{13}H_{17}N$
1,2,3,4,10,11-Hexahydro-1-methylcarbazole
1,2,3,4,10,11-Hexahydro-2-methylcarbazole
1,2,3,4,10,11-Hexahydro-3-methylcarbazole
1,2,3,4,10,11-Hexahydro-6-methylcarbazole
1,2,3,4,10,11-Hexahydro-9-methylcarbazole
1,2,3,4,10,11-Hexahydro-11-methylcarbazole
1,2,3,4,4*a*,9,9*a*,10-Octahydroacridine
1,2,3,4,5,6,7,8-Octahydroacridine

$C_{13}H_{17}NO$
4-Methyl-2-hexenoic Acid, *Anilide*
4-Methyl-3-hexenoic Acid, *Anilide*
1-Phenylcyclohexanecarboxylic Acid, *Amide*
1,2,3,4-Tetrahydro-6-methoxyquinoline, N-*Allyl*

$C_{13}H_{17}NO_2$
Bicyclo[2,2,1]hept-5-ene-2,3-dicarboxylic Acid, *Butylimide*
Bicyclo[2,2,1]hept-5-ene-2,3-dicarboxylic Acid, *Isobutylimide*
Heptaphylline†
1-Phenylpiperidine-4-carboxylic Acid, *Me ester*

$C_{13}H_{17}NO_3$
2-Methyladipic Acid, *Monoanilide*
3-Methyladipic Acid, *Monoanilide*
2-Methylglutaric Acid, *Mono*-p-*toluidide*
3-Methylglutaric Acid, *Mono*-p-*toluidide*
Methylmalonic Acid, *Mono-Et ester*, p-*Toluidide*
Phenaceturic Acid, *Propyl ester*

$C_{13}H_{17}NO_4$
Anilinomalonic Acid, *Di-Et ester*
2,6-Dimethylpyridine-3,4-dicarboxylic Acid, *Di-Et ester*
2,6-Dimethylpyridine-3,5-dicarboxylic Acid, *Di-Et ester*
N-Phenylglycine-*o*-carboxylic Acid, *Di-Et ester*

$C_{13}H_{17}NO_5$
Benzylpenaldic Acid, *Di-Me acetal*
Tartranilic Acid, *Propyl ester*

$C_{13}H_{17}NO_7$
2-Nitrogallic Acid, *Tri-Et ether*

$C_{13}H_{17}NS_2$
Piperidine-*N*-carbodithioic Acid, p-*Tolyl ester*

$C_{13}H_{17}N_3O$
Pyramidone
Tryptophane, *Et-amide*

$C_{13}H_{17}N_3O_2$
Methyl 5(6)-butylbenzimidazole-2-ylcarbamate†

$C_{13}H_{17}N_3O_3$
N-Benzoylglycyl-3-aminobutyric Acid, *Amide*

$C_{13}H_{18}$
1-Ethyl-1,2,3,4-tetrahydro-1-methylnaphthalene
1-Ethyl-1,2,3,4-tetrahydro-6-methylnaphthalene
2-Ethyl-1,2,3,4-tetrahydro-5-methylnaphthalene
2-Ethyl-1,2,3,4-tetrahydro-8-methylnaphthalene
6-Ethyl-1,2,3,4-tetrahydro-3-methylnaphthalene
Ionene
4-Phenyl-3-heptene
1,2,3,4-Tetrahydro-1-isopropylnaphthalene
1,2,3,4-Tetrahydro-2-isopropylnaphthalene

$C_{13}H_{18}BrNO_3$
2(?)-Bromo-3′-nitrobenzophenone
2(?)-Bromo-4′-nitrobenzophenone
2-Bromo-5-nitrobenzophenone
4′-Bromo-2-nitrobenzophenone
4-Bromo-2-nitrobenzophenone
4-Bromo-4′-nitrobenzophenone
4′-Bromo-3-nitrobenzophenone

$C_{13}H_{18}Cl_2N_2O_2$
Melphalan

$C_{13}H_{18}N_2$
Leptocladine

$C_{13}H_{18}N_2O$
Bufotenidine
Bufotenine, *Me ether*
Eseroline
Lespedamine†
5-Methoxy-*N*,*N*-dimethyltryptamine†
Noresermethole

$C_{13}H_{18}N_2O_2$
3-Dimethylamino-1,5-(dimethoxy)indole†
Geneseroline
5-Methoxy-*N*,*N*-dimethyltryptamine, N-*Oxide*†

$C_{13}H_{18}N_2O_2S$
Probenecid, *Nitrile*

$C_{13}H_{18}N_2O_3$
2-Anilino-2-methylsuccinic Acid, α-*Amide*-β-*Et ester*
Heptabarbitone

$C_{13}H_{18}N_2O_4$
Benzylidenediurethane
1-*tert*-Butyl-3,4,5-trimethyl-2,6-dinitrobenzene
5-Octanoylaminopyridine-2,3,6(1*H*)-trione†

$C_{13}H_{18}N_2O_5$
Antimycic Acid, *Me-ester*, *Me ether*

$C_{13}H_{18}N_4O_6$
G-Compound

$C_{13}H_{18}N_4O_7$
9-β-D-Glucopyranosyltheophylline†

$C_{13}H_{18}O$
Hexyl phenyl Ketone
3-*p*-Isopropylphenyl-2-methylpropanal
5-Methyl-1-phenyl-2-hexanone
2-Phenylcyclohexanol, *Me ether*
3-Phenyl-2-heptanone
6-Phenyl-2-heptanone
2-Phenyl-3-heptanone
4-Phenyl-3-heptanone
3-Phenyl-4-heptanone
1-Phenylvinyl Alcohol, 3-*Methylbutyl ether*

$C_{13}H_{18}O_2$
4-Allylcatechol, *Di-Et ether*
Benzoic Acid, *Hexyl ester*
2-(Buta-1,3-dienyl)-3-hydroxy-4-(penta-1,3-dienyl)-tetrahydrofuran†
Cyclohexyl-2,4-dihydroxyphenylmethane
2,4-Dimethylphenylacetic Acid, n-*Propyl ester*
2,2-Dimethyl-3-phenylpropionic Acid, *Et ester*
γ-Hydroxypropiophenone, *Butyl ether*
4-Hydroxypropiophenone, *Isobutyl ether*
ω-Hydroxyvalerophenone, *Et ether*
p-Hydroxyvalerophenone, *Et ether*
p-Isopropylbenzoic Acid, *Propyl ester*
5-Isopropyl-*o*-toluic Acid, *Et ester*
6-Isopropyl-*m*-toluic Acid, *Et ester*
2-Methoxy-5-propenylphenol, *Propyl ether*
3-Methyl-1-phenyl-1-butanol, *Ac*
2-Methyl-2-phenylbutyric Acid, *Et ester*
2-Methyl-3-phenylbutyric Acid, *Et ester*
2-Methyl-4-phenylbutyric Acid, *Et ester*
3-Methyl-4-phenylbutyric Acid, *Et ester*

$C_{13}H_{18}O_2$ (*continued*)
2-Methyl-2-phenylpropionic Acid, *Propyl ester*
2-Methyl-3-*p*-tolylpropionic Acid, *Et ester*
Microl
Pentamethylbenzoic Acid, *Me ester*
3-Phenylvaleric Acid, *Et ester*
4-Phenylvaleric Acid, *Et ester*
5-Phenylvaleric Acid, *Et ester*
o-Toluic Acid, 2-*Methylbutyl ester*
m-Toluic Acid, 2-*Methylbutyl ester*
p-Toluic Acid, 2-*Methylbutyl ester*
p-Toluic Acid, 3-*Methylbutyl ester*
3-*p*-Tolylbutyric Acid, *Et ester*
4-*o*-Tolylbutyric Acid, *Et ester*
4-*p*-Tolylbutyric Acid, *Et ester*
2,4,6-Triethylbenzoic Acid
2,4,6-Trimethylphenylacetic Acid, *Et ester*

$C_{13}H_{18}O_3$
Anisic Acid, *Isoamyl ester*
2,4-Dihydroxypropiophenone, *Di-Et ether*
4-Heptanoylresorcinol
2,2,4,4,6,6-Hexamethylcyclohexane-1,3,5-trione†
5-Hexyl-2-hydroxybenzoic Acid
2-Hydroxy-2-(4-isopropylphenyl)acetic Acid, *Et ester*
4-Hydroxy-5-isopropyl-*o*-toluic Acid, *Et ether*
6-Hydroxy-5-isopropyl-*o*-toluic Acid, *Et ester*
4-Hydroxy-6-isopropyl-*m*-toluic Acid, *Et ether*
4-Hydroxy-2-phenylbutyric Acid, *Me ether, Et ester*
4-*o*-Hydroxyphenylhexanoic Acid, *Me ether*
4-*p*-Hydroxyphenylhexanoic Acid, *Me ether*
3-Hydroxy-3-phenylvaleric Acid, *Et ester*
3-Hydroxy-5-phenylvaleric Acid, *Et ester*
2-Hydroxy-2-(2,4,6-trimethylphenyl)acetic Acid, *Et ester*
Mandelic Acid, DL-*Pentyl ester*
Mandelic Acid, *Propyl ether, Et ester*
Methyl salicylate, 3-*Methylbutyl ether*
p-Pentoxybenzoic Acid, *Me ester*
Zingerone, *Et ether*

$C_{13}H_{18}O_4$
Agglomerone†
Baeckeol
Bicyclo[2,2,1]hept-5-ene-2,2-dicarboxylic Acid, *Di-Et ester*
Bicyclo[2,2,1]hept-5-ene-2,3-dicarboxylic Acid, *Di-Et ester*
Bicyclo[2,2,1]hept-5-ene-2,3-dicarboxylic Acid, *Mono-butyl ester*
Conglomerone
3,4-Dihydroxybenzoic Acid, *Di-Et ether, Et ester*
3,5-Dihydroxybenzoic Acid, *Di-Et ether, Et ester*
2,4-Dihydroxy-6-pentylbenzoic Acid, 4-*Me ether*
3,5-Dihydroxy-*o*-toluic Acid, *Pentyl ester*
2,3,4-Trihydroxybenzaldehyde, 2,3,4-*Tri-Et ether*
2,4,5-Trihydroxybenzaldehyde, 2,4,5-*Tri-Et ether*
3,4,5-Trihydroxybutyrophenone, *Tri-Me ether*

$C_{13}H_{18}O_5$
Gallic Acid, 3,4-*Di-Me ether, Et ester*, 5-*Et ether*
Gallic Acid, *Tri-Et ether*
2,3,4-Trihydroxybenzoic Acid, 2,3,4-*Tri-Et ether*
2,4,5-Trihydroxybenzoic Acid, 2,4,5-*Tri-Et ether*
2,4,6-Trihydroxybenzoic Acid, *Et ester*, 2,6-*Di-Et ether*

$C_{13}H_{18}O_6$
2-*O*-Benzylglucose†
2,3,4,5-Tetrahydroxybenzoic Acid, 2,5-*Di-Me ether, Di-Et ether*

$C_{13}H_{18}O_7$
Arbutin, *Me ether*
Guaiacol, D-*Glucoside*
Salicin

$C_{13}H_{18}O_8$
Salirepin†

$C_{13}H_{19}ClN_2O_4$
Glycyltyrosine, *Et ester hydrochloride*

$C_{13}H_{19}N$
2-Phenethylpiperidine
4-Phenethylpiperidine
1,2,3,4-Tetrahydroisoquinoline, N-*Butyl*
1,2,3,4-Tetrahydro-2-propylquinoline, N-*Me*

$C_{13}H_{19}NO$
Diethylpropion
1-Ethyl-1,2,3,4-tetrahydro-8-hydroxyquinoline, *Et ether*
2,4,6-Triethylbenzoic Acid, *Amide*

$C_{13}H_{19}NO_2$
2-Anilino-3-methylbutyric Acid, *Et ester*
Benzylideneaminoacetal
p-Butylaminobenzoic Acid, *Et ester*
Carnegine
Dioscorine
4-Hydroxy-6-isopropyl-*m*-toluic Acid, *Et ether, Amide*
4-*o*-Hydroxyphenylhexanoic Acid, *Me ether, Amide*
N-Phenylglycine, 3-*Methylbutyl ester*

$C_{13}H_{19}NO_3$
3,4-Dimethoxybenzoic Acid, *Amide*, N-*Di-Et*
Gigantine†
Pellotine
2-Phenoxypropionic Acid, *Di-methylamino-ethyl ester*
Retronecine, 7-*Angelyl ester*
1,2,3,4-Tetrahydro-6,7,8-trimethoxy-1-methyl-isoquinoline†

$C_{13}H_{19}NO_4S$
Probenecid

$C_{13}H_{19}NO_6$
Mannonic Acid, m-*Toluidide*

$C_{13}H_{19}NO_9$
Gynocardin

$C_{13}H_{19}N_3OS$
6-(2-Diethylaminoethoxy)-2-dimethylamino-benzothiazole

$C_{13}H_{19}N_4O_{12}P$
N-(5-Amino-1-β-D-ribofuranosylimidazole-4-carbonyl)-L-aspartic Acid 5′-phosphate

$C_{13}H_{20}$
tert-Butyl-2,4,6-trimethylbenzene †
Heptylbenzene
3-Phenylheptane
4-Phenylheptane

$C_{13}H_{20}N_2$
β-Matrinidine

$C_{13}H_{20}N_2O$
Conolline

$C_{13}H_{20}N_2O_2$
2-Isobutylaminoethyl *p*-aminobenzoate †
Novocaine

$C_{13}H_{20}N_2O_2S$
2-Pentenylpenillamine

$C_{13}H_{20}N_2O_3$
4-Amino-2-hydroxybenzoic Acid, *Diethylaminoethyl ester*

$C_{13}H_{20}N_2O_5S$
Ascarylose, *Toluenesulphonylhydrazone*
Carbon Dioxide Biotin †

$C_{13}H_{20}N_2O_6$
Actinobolin †

$C_{13}H_{20}N_4O_6$
Eisenine

$C_{13}H_{20}O$
2,4-Di-*tert*-butylcyclopentadienone †
α-Hydroxy-*o*-xylene, *3-Methylbutyl ether*
α-Ionone
β-Ionone
γ-Ionone
ψ-Ionone
6-Isopropyl-*m*-cresol, *Propyl ether*
p-(2-Methyl-2-butyl)phenol, *Et ether*
1-Phenyl-1-heptanol
3-Phenyl-1-heptanol
5-Phenyl-1-heptanol
7-Phenyl-1-heptanol
7-Phenyl-2-heptanol
2-Phenyl-3-heptanol
7-Phenyl-3-heptanol
3-Phenyl-4-heptanol
4-Phenyl-4-heptanol
6-Phenyl-1-hexanol, *Me ether*
1-Phenyl-2-propanol, *Butyl ether*

$C_{13}H_{20}OS$
p-Hydroxybenzenethiol, S-n-*Hexyl*, *Me ether*

$C_{13}H_{20}O_2$
Actinidol †
4-Heptylresorcinol
2-3′-Methylbutylresorcinol, *Di-Me ether*
2-Nonenoic Acid, *Et ester*
8-Nonenoic Acid, *Et ester*
5-Pentylresorcinol, *Di-Me ether*
1-Phenyl-2-propanone, *Di-Et ketal*
5-Phenylvaleraldehyde, *Di-Me acetal*
Resorcinol, *Monoheptyl ether*
Theaspirone †

$C_{13}H_{20}O_3$
4-(2,4-Dihydro-2,6,6-trimethylcyclohexylidene)-but-3-en-2-one †
Jasmonic Acid, *Me ester*
Orthobenzoic Acid, *Tri-Et ester*

$C_{13}H_{20}O_4$
Diallylmalonic Acid, *Di-Et ester*
3-Methyl-3-cyclohexene-1,1-dicarboxylic Acid, *Di-Et ester*
4-Methyl-3-cyclohexene-1,2-dicarboxylic Acid, *Di-Et ester*
6-Methyl-3-cyclohexene-1,2-dicarboxylic Acid, *Di-Et ester*
3-Methyl-4-cyclohexene-1,2-dicarboxylic Acid, *Di-Et ester*
4-Methyl-4-cyclohexene-1,2-dicarboxylic Acid, *Di-Et ester*

$C_{13}H_{20}O_6$
3-Methylcyclopropane-1,1,2-tricarboxylic Acid, *Tri-Et ester*

$C_{13}H_{20}O_7$
Varicinal A †

$C_{13}H_{20}O_8$
Methane-tetracarboxylic Acid, *Tetra-Et ester*
Methane-tetracetic Acid, *Tetra-Me ester*

$C_{13}H_{20}O_{11}$
Gynocardinic Acid

$C_{13}H_{21}N$
N-butyl-*N*-ethylbenzylamine
$\Delta^{4\alpha}$-Decahydrocarbazole, N-*Me*
2,6-Di-*tert*-butylpyridine †
N-Ethyl-*N*-3-methylbutylaniline
2,3,4,5,6-Pentamethylaniline, N-*Di-Me*
5-Phenyl-1-pentylamine, N-*Di-Me*
5-Phenyl-1-pentylamine, N-*Et*
3-Phenyl-1-propylamine, N-*Di-Et*

$C_{13}H_{21}NO_2$
3-Carboxy-2,2,3-trimethylcyclopentylacetonitrile, *Et ester*
Dihydrodioscorine
Dioscine †
Tigloidine

$C_{13}H_{21}NO_3$
N-*tert*-Butyl-β,4-dihydroxy-3-hydroxymethylphenethylamine †
3α,6β-Tropanediol, 3α-*Tigloyl ester*
3α,6β-Tropanediol, 6β-*Tigloyl ester*
Turneforcine

$C_{13}H_{21}NO_4$
Meteloidine

$C_{13}H_{21}N_3O$
Procaine Amide

$C_{13}H_{22}$
1,3-Cyclotridecadiene
Ionane
ψ-Ionane
[4,4,3]Propellane †

$C_{13}H_{22}N_2$
Brassylic Acid, *Dinitrile*

$C_{13}H_{22}N_2O$
Ammodendrine, N-*Me*

$C_{13}H_{22}N_2O_2S$
Pentylpenillamine

$C_{13}H_{22}N_2O_8$
Aspergillomarasmine B, *Tetra-me ester* †

$C_{13}H_{22}O$
Albolineol †
Dicyclohexyl Ketone
α-Ionol
β-Ionol
Linalool, *Allyl ether* †
Luparone †
Solanone †

$C_{13}H_{22}O_2$
Bicyclohexyl-1-carboxylic Acid
Bicyclohexyl-2-carboxylic Acid
Bicyclohexyl-4-carboxylic Acid
Isoborneol, *Propionyl*
2-Octynoic Acid, 3-*Methylbutyl ester*
9-Undecynoic Acid, *Et ester*
10-Undecynoic Acid, *Et ester*

$C_{13}H_{22}O_3$
10-Hydroxy-3,7-dimethyldeca-2,6-dienoic Acid, *Me ester* †
1-Isopropyl-3-methyl-2-cyclohexanone-1-carboxylic Acid, *Et ester*
1-Isopropyl-4-methyl-2-cyclohexanone-1-carboxylic Acid, *Et ester*
1-Isopropyl-5-methyl-2-cyclohexanone-1-carboxylic Acid, *Et ester*
Lactic Acid, (−)-*Bornyl ester*
Lanolic Acid

$C_{13}H_{22}O_4$
Allylisopropylmalonic Acid, *Di-Et ester*
3-Carboxy-2,2-dimethylcyclobutylacetic Acid, *Di-Et ester*
α-Carboxy-γ-dodecalactone †
3-Carboxy-2,2,3-trimethyl-1-cyclopentylacetic Acid, α-*Et ester*
Cyclohexylmalonic Acid, *Di-Et ester*
Decanedioic Acid, *Trimethylene glycol ester*
1,2-Dimethylcyclopentane-1,3-dicarboxylic Acid, *Di-Et ester*
2-Methylcyclohexane-1,1-dicarboxylic Acid, *Di-Et ester*

$C_{13}H_{22}O_5$
Acetonedicarboxylic Acid, *Di-isobutyl ester*
Phoronic Acid, *Di-Me ester*

$C_{13}H_{22}O_6$
Butane-1,1,2-tricarboxylic Acid, *Tri-Et ester*
Butane-1,1,3-tricarboxylic Acid, *Tri-Et ester*
Butane-1,1,4-tricarboxylic Acid, *Tri-Et ester*
Butane-1,2,2-tricarboxylic Acid, *Tri-Et ester*
Butane-1,2,4-tricarboxylic Acid, *Tri-Et ester*
Butane-1,3,3-tricarboxylic Acid, *Tri-Et ester*
Butane-2,2,3-tricarboxylic Acid, *Tri-Et ester*
Methane-triacetic Acid, *Tri-Et ester*
2-Methyltricarballylic Acid, *Tri-Et ester*
3-Methyltricarballylic Acid, *Tri-Et ester*

$C_{13}H_{22}O_7$
Homoisocitric Acid, *Tri-Et ester* †

$C_{13}H_{22}O_8$
Spinochrome E, *Tetra-Me ether*

$C_{13}H_{23}ClO$
Bicyclohexyl-1-carboxylic Acid, *Chloride*
2-Tridecenoic Acid, *Chloride*

$C_{13}H_{23}N$
2-*p*-Menthyl-β-propionic Acid, *Nitrile*
Perhydroacridine

$C_{13}H_{23}NO_2$
Valtropine

$C_{13}H_{23}NO_3$
3-Carbamoyl-1,2,2-trimethylcyclopentane-carboxylic Acid, N-*Propyl*
Valeroidine

$C_{13}H_{23}NO_5$
2-Pentenylpenaldic Acid, *Di-Et acetal*

$C_{13}H_{24}N_2O$
Anaferine
Anahygrine †
Cuskhygrine

$C_{13}H_{24}N_2O_3$
Hygroscopin A

$C_{13}H_{24}N_2O_3S$
Pentylpenilloic Acid

$C_{13}H_{24}N_2O_{11}$
Macrozamin

$C_{13}H_{24}N_4O_4$
Methane-tetracetic Acid, *Tetra-di-Me amide*

$C_{13}H_{24}N_4O_{11}P_2$
Cytidine diphosphate choline

$C_{13}H_{24}O$
Cyclotridecanone
2-Isopropyl-5-methylcyclohexanol, *Allyl ether*
Tetrahydroionone

$C_{13}H_{24}O_2$
11-Dodecenoic Acid, *Me ester*
2-*p*-Menthyl-β-propionic Acid
2-Tridecenoic Acid
11-Tridecenoic Acid
12-Tridecenoic Acid
9-Undecenoic Acid, *Et ester*
10-Undecenoic Acid, *Et ester*

$C_{13}H_{24}O_3$
2-Ethyl-3-oxohexanoic Acid, 3-*Methylbutyl ester*
Lactic Acid, (−)-*Menthyl ester*
3-*p*-Menthoxyacetic Acid, *Me ester*
2-Oxoundecanoic Acid, *Et ester*
4-Oxoundecanoic Acid, *Et ester*
9-Oxoundecanoic Acid, *Et ester*
10-Oxoundecanoic Acid, *Et ester*

$C_{13}H_{24}O_4$
Azelaic Acid, *Di-Et ester*
Brassylic Acid
3-Carboxymethyl-5-methylhexanoic Acid, *Di-Et ester*
Dipropylmalonic Acid, *Di-Et ester*
Dodecanedioic Acid, *Mono-Me ester*
3-Ethyl-4-methyladipic Acid, *Di-Et ester*
Heptane-1,1-dicarboxylic Acid, *Di-Et ester*
Heptane-1,5-dicarboxylic Acid, *Di-Et ester*
Heptane-2,6-dicarboxylic Acid, *Di-Et ester*
Heptane-3,3-dicarboxylic Acid, *Di-Et ester*

$C_{13}H_{24}O_4$ (*continued*)
Heptane-3,4-dicarboxylic Acid, *Di-Et ester*
2-Isopropylbutane-1,4-dicarboxylic Acid, *Di-Et ester*
3-Methyladipic Acid, *Dipropyl ester*
2-Methyldodecanedioic Acid
3-Methyldodecanedioic Acid
4-Methyldodecanedioic Acid
6-Methyldodecanedioic Acid
5-Methylhexane-1,1-dicarboxylic Acid, *Di-Et ester*
5-Methylhexane-1,4-dicarboxylic Acid, *Di-Et ester*
5-Methylhexane-2,2-dicarboxylic Acid, *Di-Et ester*
2-Methylhexane-3,3-dicarboxylic Acid, *Di-Et ester*
Undecanedioic Acid, *Di-Me ester*
Undecanedioic Acid, *Mono-Et ester*

$C_{13}H_{24}O_4Si$
Triallyl isobutyl orthosilicate

$C_{13}H_{24}O_5$
Isovaleric Acid, *Di-glyceryl ester*
Malic Acid, *Isopropyl, Di-isopropyl ester*
Methoxysuccinic Acid, *Dibutyl ester*

$C_{22}H_{24}O_9$
Homonataloin

$C_{13}H_{24}O_{10}$
Condurangobiose

$C_{13}H_{24}O_{11}$
Cellobiose, α-*Me cellobioside*
Cellobiose, β-*Me-cellobioside*
Gentiobiose, *Methyl glycoside*
Maltose, β-*Methylglycoside*
Melibiose, β-*Methyl ether*

$C_{13}H_{25}BrO_2$
2-Bromododecanoic Acid, *Me ester*

$C_{13}H_{25}ClO$
2-Methyldodecanoic Acid, *Chloride*

$C_{13}H_{25}N$
2-Methyldodecanoic Acid, *Nitrile*
Tridecanoic Acid, *Nitrile*

$C_{13}H_{25}NO$
1-Lupinine, *Propyl ether*
2-Tridecenoic Acid, *Amide*

$C_{13}H_{25}NO_2$
Bicyclohexyl-1-carboxylic Acid, *Amide*
Bicyclohexyl-4-carboxylic Acid, *Amide*

$C_{18}H_{25}NO_3$
Mydriasin

$C_{13}H_{25}NO_5$
Pentylpenaldic Acid, *Di-Et acetal*
Serratamic Acid

$C_{13}H_{25}N_5O_7$
N-Guan-Streptolidyl Gulosaminide★, N-*Me*†

$C_{13}H_{26}$
1-Tridecene

$C_{13}H_{26}Br_2$
1,12-Dibromotridecane
1,13-Dibromotridecane

$C_{13}H_{26}N_2O_2$
Brassylic Acid, *Diamide*

$C_{13}H_{26}N_2O_3$
Elaiomycin
D-Methionyl-D-methionine, *Isopropyl ester*

$C_{13}H_{26}O$
Cyclotridecanol
2-Isopropyl-5-methylcyclohexanol, *Propy ether*
Tridecanal
2-Tridecanone
3-Tridecanone
7-Tridecanone
Dodecanoic Acid, *Me ester*
Ficocerylic Acid
Isovaleric Acid, *Octyl ester*
5-Methyldecanoic Acid, *Et ester*
2-Methyldodecanoic Acid
3-Methyldodecanoic Acid
10-Methyldodecanoic Acid
Nonanoic Acid, n-*Butyl ester*
Nonanoic Acid, tert-*Butyl ester*
2-Octanol, *Valeryl*
Propionic Acid, *Decyl ester*
Tridecanoic Acid
Undecanoic Acid, *Et ester*

$C_{13}H_{26}O_3$
3-Ethyl-2-hydroxynonanoic Acid, *Et ester*
12-Hydroxydodecanoic Acid, *Me ester*
2-Hydroxyundecanoic Acid, *Et ester*
10-Hydroxyundecanoic Acid, *Et ester*
11-Hydroxyundecanoic Acid, *Me ester, Me ether*

$C_{13}H_{26}O_4$
Glycerol, 1-*Decanoyl*

$C_{13}H_{26}O_7$
*Gluco*heptose, 2,3,4,6-*Penta-Me ether, Methyl-glycoside*

$C_{13}H_{27}Br$
1-Bromotridecane

$C_{13}H_{27}Cl$
1-Chlorotridecane

$C_{13}H_{27}ClO$
2,4,6-Triethylbenzoic Acid, *Chloride*

$C_{13}H_{27}N$
Azacyclotetradecane

$C_{13}H_{27}NO$
2-Methyldodecanoic Acid, *Amide*
10-Methyldodecanoic Acid, *Amide*
Tridecanoic Acid, *Amide*

$C_{13}H_{27}NO_2$
2-Amino-2-methylbutyric Acid, *Octyl ester*
13-Aminotridecanoic Acid

$C_{13}H_{27}N_3O_3$
Fragin†

$C_{13}H_{28}$
2,4,6,8-Tetramethylnonane†
Tridecane

$C_{13}H_{28}N_2O$
1,3-Di-(4-methylpentyl)urea

$C_{13}H_{28}O$
2-Methyl-1-dodecanol
3-Methyl-3-dodecanol
2-Octanol, n-*Pentyl ether*
1-Tridecanol
2-Tridecanol
3-Tridecanol
7-Tridecanol

$C_{13}H_{28}O_3$
Orthoformic Acid, *Tributyl ester*

$C_{13}H_{28}O_4$
Orthocarbonic Acid, *Tetrapropyl ester*
Pentaerythritol, *Tetra-Et ether*

$C_{13}H_{29}N$
1-Aminotridecane

$C_{13}H_{29}N_3$
Dodine

$C_{13}H_{31}N_2OP$
Methylphosphonic Acid, *Di-dipropylamide*

$C_{13}H_{33}Br_2N_3$
Azamethonium bromide

$C_{13}H_{35}N_3O_2$
[(Methylimino)diethylene]*bis*[ethyldimethyl]-ammonium hydroxide

C_{14}

$C_{14}Cl_8O_2$
Octachloroanthraquinone

$C_{14}Cl_{10}$
Decachlorophenanthrene†

$C_{14}Cl_{12}$
Dodecachlorodihydrophenanthrene†

$C_{14}H_3Cl_7$
Heptachloroanthracene

$C_{14}H_3Cl_7O_2$
2-(2,5-Dichlorobenzoyl)-3,4,5,6-tetrachlorobenzoic Acid, *Chloride*

$C_{14}H_4Br_4O_2$
1,2,3,4-Tetrabromoanthraquinone
1,3,5,7-Tetrabromoanthraquinone
1,3,6,8-Tetrabromoanthraquinone

$C_{14}H_4Br_6O_3$
2-(2,5-Dibromobenzoyl)-3,4,5,6-tetrabromobenzoic Acid

$C_{14}H_4Cl_4O_2$
1,2,3,4-Tetrachloroanthraquinone
1,2,5,8-Tetrachloroanthraquinone
1,2,6,7-Tetrachloroanthraquinone
1,4,5,8-Tetrachloroanthraquinone
1,4,6,7-Tetrachloroanthraquinone
2,3,6,7-Tetrachloroanthraquinone

$C_{14}H_4Cl_6O_3$
2-(2,5-Dichlorobenzoyl)-3,4,5,6-tetrachlorobenzoic Acid

$C_{14}H_4N_4O_{12}$
1,5-Dihydroxy-2,4,6,8-tetranitroanthraquinone
1,8-Dihydroxy-2,4,5,7-tetranitroanthraquinone
2,6-Dihydroxy-1,3,5,7-tetranitroanthraquinone

$C_{14}H_4N_6O_{15}$
2,4,6-Trinitrobenzoic Acid, *Anhydride*

$C_{14}H_4O_6$
Naphthalene-1,2,4,5-tetracarboxylic Acid, *Dianhydride*
Naphthalene-1,4,5,8-tetracarboxylic Acid, *Dianhydride*

$C_{14}H_5Cl_2NO_4$
1,2-Dichloro-4-nitroanthraquinone
1,4-Dichloro-5-nitroanthraquinone
1,5-Dichloro-4-nitroanthraquinone
4,5-Dichloro-1-nitroanthraquinone
6,7-Dichloro-1-nitroanthraquinone

$C_{14}H_5Cl_3O_2$
1,2,3-Trichloroanthraquinone
1,2,4-Trichloroanthraquinone
1,2,5-Trichloroanthraquinone
1,2,6-Trichloroanthraquinone
1,2,7-Trichloroanthraquinone
1,3,5-Trichloroanthraquinone
1,3,6-Trichloroanthraquinone
1,3,7-Trichloroanthraquinone
1,3,8-Trichloroanthraquinone
1,4,5-Trichloroanthraquinone
1,4,6-Trichloroanthraquinone
2,3,5-Trichloroanthraquinone
2,3,6-Trichloroanthraquinone

$C_{14}H_6Br_2$
1,2-Dibromopyracylene†

$C_{14}H_6Br_2O_2$
1,2-Dibromoanthraquinone
1,3-Dibromoanthraquinone
1,4-Dibromoanthraquinone
1,5-Dibromoanthraquinone
1,6-Dibromoanthraquinone
1,7-Dibromoanthraquinone
1,8-Dibromoanthraquinone
2,3-Dibromoanthraquinone
2,6-Dibromoanthraquinone
2,7-Dibromoanthraquinone
2,7-Dibromophenanthraquinone
3,6-Dibromophenanthraquinone

$C_{14}H_6Br_2O_3$
4,4′-Dibromobiphenyl-2,2′-dicarboxylic Acid, *Anhydride*

$C_{14}H_6Br_2O_3$
1,3-Dibromo-2-hydroxyanthraquinone
2,4-Dibromo-1-hydroxyanthraquinone

$C_{14}H_6Br_2O_4$
1,2-Dibromo-3,4-dihydroxyanthraquinone
1,2-Dibromo-5,8-dihydroxyanthraquinone
1,3-Dibromo-2,4-dihydroxyanthraquinone
1,4-Dibromo-5,8-dihydroxyanthraquinone
1,5-Dibromo-4,8-dihydroxyanthraquinone
2,3-Dibromo-1,4-dihydroxyanthraquinone
2,6-Dibromo-1,5-dihydroxyanthraquinone
6,7-Dibromo-1,4-dihydroxyanthraquinone

$C_{14}H_6Br_4$
2,3,9,10-Tetrabromoanthracene
2,6,9,10-Tetrabromoanthracene

$C_{14}H_6Br_4N_2O_2$
1,5-Diamino-2,4,6,8-tetrabromoanthraquinone
1,8-Diamino-2,4,5,7-tetrabromoanthraquinone
2,6-Diamino-1,3,5,7-tetrabromoanthraquinone

$C_{14}H_6ClNO_4$
1-Chloro-2-nitroanthraquinone
1-Chloro-4-nitroanthraquinone
1-Chloro-5-nitroanthraquinone
1-Chloro-8-nitroanthraquinone
2-Chloro-1-nitroanthraquinone
6-Chloro-1-nitroanthraquinone
7-Chloro-1-nitroanthraquinone
7-Chloro-2-nitroanthraquinone
7-Nitrofluorenone-4-carboxylic Acid, *Chloride*

$C_{14}H_6ClNO_6S$
5-Nitroanthraquinone-1-sulphonic Acid, *Chloride*
8-Nitroanthraquinone-1-sulphonic Acid, *Chloride*
5-Nitroanthraquinone-2-sulphonic Acid, *Chloride*

$C_{14}H_6Cl_2N_2O_6$
4,4′-Dinitrobiphenyl-2,2′-dicarboxylic Acid, *Dichloride*
6,6′-Dinitrobiphenyl-2,2′-dicarboxylic Acid, *Dichloride*

$C_{14}H_6Cl_2O_2$
1,2-Dichloroanthraquinone
1,3-Dichloroanthraquinone
1,4-Dichloroanthraquinone
1,5-Dichloroanthraquinone
1,6-Dichloroanthraquinone
1,7-Dichloroanthraquinone
1,8-Dichloroanthraquinone
2,3-Dichloroanthraquinone
2,6-Dichloroanthraquinone
2,7-Dichloroanthraquinone

$C_{14}H_6Cl_2O_3$
3,3′-Dichlorobiphenyl-2,2′-dicarboxylic Acid
5,5′-Dichlorobiphenyl-2,2′-dicarboxylic Acid

$C_{14}H_6Cl_2O_4$
1,2-Dichloro-5,8-dihydroxyanthraquinone
1,3-Dichloro-2,4-dihydroxyanthraquinone
1,3-Dichloro-5,8-dihydroxyanthraquinone
1,4-Dichloro-5,8-dihydroxyanthraquinone
1,4-Dichloro-6,7-dihydroxyanthraquinone
1,5-Dichloro-4,8-dihydroxyanthraquinone
1,8-Dichloro-4,5-dihydroxyanthraquinone
5,8-Dichloro-1,2-dihydroxyanthraquinone
6,7-Dichloro-1,4-dihydroxyanthraquinone

$C_{14}H_6Cl_2O_4S$
8-Chloroanthraquinone-2-sulphonic Acid

$C_{14}H_6Cl_2O_6S_2$
Anthraquinone-1,5-disulphonic Acid, *Dichloride*
Anthraquinone-1,6-disulphonic Acid, *Dichloride*
Anthraquinone-1,7-disulphonic Acid, *Dichloride*
Anthraquinone-1,8-disulphonic Acid, *Dichloride*
Anthraquinone-2,6-disulphonic Acid, *Dichloride*
Anthraquinone-2,7-disulphonic Acid, *Dichloride*

$C_{14}H_6Cl_4$
1,2,3,4-Tetrachloranthracene
1,3,9,10-Tetrachloroanthracene
1,4,5,8-Tetrachloroanthracene
2,3,9,10-Tetrachloroanthracene

$C_{14}H_6Cl_4N_2O_2$
1,5-Diamino-2,4,6,8-tetrachloroanthraquinone
1,8-Diamino-2,4,5,7-tetrachloroanthraquinone

$C_{14}H_6Cl_4O_5$
Thiophan Acid†

$C_{14}H_6N_2O_4$
Naphthalene-1,4,5,8-tetracarboxylic Acid, *Di-imide*

$C_{14}H_6N_2O_6$
1,3-Dinitroanthraquinone
1,5-Dinitroanthraquinone
1,6-Dinitroanthraquinone
1,7-Dinitroanthraquinone
1,8-Dinitroanthraquinone
2,6-Dinitroanthraquinone
2,7-Dinitroanthraquinone
2,7-Dinitrophenanthraquinone
3,6-Dinitrophenanthraquinone
4,5-Dinitrophenanthraquinone

$C_{14}H_6N_2O_7$
5,5′-Dinitrobiphenyl-2,2′-dicarboxylic Acid, *Anhydride*
1-Hydroxy-2,4-dinitroanthraquinone
2-Hydroxy-1,3-dinitroanthraquinone

$C_{14}H_6N_2O_8$
1,5-Dihydroxy-4,8-dinitroanthraquinone
1,8-Dihydroxy-4,5-dinitroanthraquinone
2,3-Dihydroxy-1,4-dinitroanthraquinone
2,6-Dihydroxy-1,5-dinitroanthraquinone

$C_{14}H_6N_4O_8$
Phthalimide, N-*Picryl*

$C_{14}H_6N_4O_{11}$
2,4-Dinitrobenzoic Acid, *Anhydride*
3,5-Dinitrobenzoic Acid, *Anhydride*

$C_{14}H_6N_6O_{12}$
2,2′,4,4′,6,6′-Hexanitrostilbene

$C_{14}H_6O_2$
Pyracyloquinone†

$C_{14}H_6O_4$
1,4,9,10-Anthradiquinone
6,6′-Dihydroxybiphenyl-2,2′dicarboxylic Acid, *Dilactone*
Metellagic Acid

$C_{14}H_6O_6$
NasutinA★†

$C_{14}H_6O_7$
Resoflavin

$C_{14}H_6O_8$
Ellagic Acid

$C_{14}H_6O_9$
Flavellagic Acid

$C_{14}H_7BrO_2$
1-Bromoanthraquinone
2-Bromoanthraquinone
2-Bromo-3,4-phenanthraquinone
1-Bromo-9,10-phenanthraquinone
2-Bromo-9,10-phenanthraquinone
3-Bromo-9,10-phenanthraquinone
4-Bromo-9,10-phenanthraquinone

$C_{14}H_7BrO_3$
1-Bromo-2-hydroxyanthraquinone
1-Bromo-3-hydroxyanthraquinone
1-Bromo-4-hydroxyanthraquinone
2-Bromo-1-hydroxyanthraquinone
2-Bromo-3-hydroxyanthraquinone

$C_{14}H_7BrO_4$
1-Bromo-3,4-dihydroxyanthraquinone
1-Bromo-5,8-dihydroxyanthraquinone
2-Bromo-1,4-dihydroxyanthraquinone
3-Bromo-1,2-dihydroxyanthraquinone
6-Bromo-1,4-dihydroxyanthraquinone

$C_{14}H_7BrO_5$
1-Bromo-2,3,4-trihydroxyanthraquinone
2-Bromo-1,3,4-trihydroxyanthraquinone

$C_{14}H_7Br_2NO_2$
1-Amino-2,4-dibromoanthraquinone
2-Amino-1,3-dibromoanthraquinone
Phthalimide, N-2,4-*Dibromophenyl*

$C_{14}H_7Br_3$
2,9,10-Tribromoanthracene

$C_{14}H_7ClO_2$
1-Chloroanthraquinone
2-Chloroanthraquinone
Fluorenone-1-carboxylic Acid, *Chloride*
Fluorenone-2-carboxylic Acid, *Chloride*
Fluorenone-4-carboxylic Acid, *Chloride*

$C_{14}H_7ClO_3$
1-Chloro-2-hydroxyanthraquinone
1-Chloro-4-hydroxyanthraquinone
1-Chloro-5-hydroxyanthraquinone
2-Chloro-3-hydroxyanthraquinone
Xanthone-2-carboxylic Acid, *Chloride*
Xanthone-4-carboxylic Acid, *Chloride*

$C_{14}H_7ClO_4$
2-Chloro-1,4-dihydroxyanthraquinone
3-Chloro-1,2-dihydroxyanthraquinone
4-Chloro-1,2-dihydroxyanthraquinone
4-Chloro-1,5-dihydroxyanthraquinone
5-Chloro-1,4-dihydroxyanthraquinone
6-Chloro-1,4-dihydroxyanthraquinone

$C_{14}H_7ClO_4S$
Anthraquinone-1-sulphonic Acid, *Chloride*
Anthraquinone-2-sulphonic Acid, *Chloride*
9,10-Phenanthraquinone-2-sulphonic Acid, *Chloride*
9,10-Phenanthraquinone-3-sulphonic Acid, *Chloride*

$C_{14}H_7ClO_5$
1-Chloro-2,3,4-trihydroxyanthraquinone

$C_{14}H_7ClO_5S$
8-Chloroanthraquinone-1-sulphonic Acid
5-Chloroanthraquinone-2-sulphonic Acid
6-Chloroanthraquinone-2-sulphonic Acid
8-Chloroanthraquinone-2-sulphonic Acid
4-Hydroxyanthraquinone-1-sulphonic Acid, *Chloride*

$C_{14}H_7Cl_2NO_2$
1-Amino-2,3-dichloroanthraquinone
1-Amino-2,4-dichloroanthraquinone
1-Amino-4,8-dichloroanthraquinone
1-Amino-5,8-dichloroanthraquinone
1-Amino-6,7-dichloroanthraquinone
1-Amino-6,8-dichloroanthraquinone
2-Amino-1,3-dichloroanthraquinone
3,5-Dichlorophthalic Acid, *Anil*
3,6-Dichlorophthalic Acid, *Anil*
Phthalimide, N-2,4-*Dichlorophenyl*

$C_{14}H_7Cl_2NO_4$
4-Nitrobiphenic Acid, *Dichloride*
6-Nitrobiphenic Acid, *Dichloride*

$C_{14}H_7Cl_3O_5$
Arthothelin †

$C_{14}H_7FO_4$
2-Fluoro-1,4-dihydroxyanthraquinone

$C_{14}H_7F_7O_2$
Heptafluorobutyric Acid, α-*Naphthyl ester*

$C_{14}H_7IO_2$
1-Iodoanthraquinone
2-Iodoanthraquinone
2-Iodophenanthrenequinone

$C_{14}H_7NO_3$
1-Nitrosoanthraquinone

$C_{14}H_7NO_4$
1-Nitroanthraquinone
2-Nitroanthraquinone
2-Nitrophenanthraquinone
3-Nitrophenanthraquinone
4-Nitrophenanthraquinone

$C_{14}H_7NO_5$
1-Hydroxy-2-nitroanthraquinone
1-Hydroxy-3-nitroanthraquinone
1-Hydroxy-4-nitroanthraquinone
1-Hydroxy-5-nitroanthraquinone
1-Hydroxy-8-nitroanthraquinone
2-Hydroxy-1-nitroanthraquinone
2-Hydroxy-3-nitroanthraquinone
2-Hydroxy-5-nitroanthraquinone
7-Hydroxy-1-nitroanthraquinone
2-Hydroxy-5-nitrophenanthraquinone
4-Nitrobiphenic Acid, *Anhydride*
6-Nitrobiphenic Acid, *Anhydride*
2-Nitrofluorenone-1-carboxylic Acid
3-Nitrofluorenone-1-carboxylic Acid
6-Nitrofluorenone-4-carboxylic Acid
7-Nitrofluorenone-4-carboxylic Acid

$C_{14}H_7NO_6$
1,2-Dihydroxy-3-nitroanthraquinone
1,2-Dihydroxy-4-nitroanthraquinone
1,4-Dihydroxy-2-nitroanthraquinone
1,4-Dihydroxy-5-nitroanthraquinone
1,8-Dihydroxy-4-nitroanthraquinone
2,3-Dihydroxy-1-nitroanthraquinone

$C_{14}H_7NO_7$
1,2,3-Trihydroxy-4-nitroanthraquinone
1,2,6-Trihydroxy-3-nitroanthraquinone
1,2,6-Trihydroxy-4-nitroanthraquinone

$C_{14}H_7NO_7S$
5-Nitroanthraquinone-1-sulphonic Acid
8-Nitroanthraquinone-1-sulphonic Acid
5-Nitroanthraquinone-2-sulphonic Acid
7-Nitroanthraquinone-2-sulphonic Acid
8-Nitroanthraquinone-2-sulphonic Acid

$C_{14}H_7N_3O_6$
4,4′-Dinitrobiphenyl-2,2′-dicarboxylic Acid, *Mononitrile*

$C_{14}H_8$
Bicyclo[9,3,0]tetradeca-1,5,7,11,13-pentanene-3,9-diyne †
Pyracylene †
1,5,9-Tridehydro[14]annulene †

$C_{14}H_8BrNO_2$
1-Amino-2-bromoanthraquinone
1-Amino-3-bromoanthraquinone
1-Amino-4-bromoanthraquinone
2-Amino-1-bromoanthraquinone
2-Amino-3-bromoanthraquinone
Phthalimide, N-p-*Bromophenyl*

$C_{14}H_8BrNO_3$
1-Amino-2-bromo-4-hydroxyanthraquinone
2-Amino-3-bromo-1-hydroxyanthraquinone

$C_{14}H_8Br_2$
9,10-Dibromoanthracene
2,7-Dibromophenanthrene
3,6-Dibromophenanthrene
3,9-Dibromophenanthrene
9,10-Dibromophenanthrene

$C_{14}H_8Br_2N_2O_2$
1,5-Diamino-2,6-dibromoanthraquinone
1,5-Diamino-3,7-dibromoanthraquinone
1,8-Diamino-3,6-dibromoanthraquinone
2,6-Diamino-1,5-dibromoanthraquinone
2,6-Diamino-3,7-dibromoanthraquinone
2,7-Diamino-3,6-dibromoanthraquinone

$C_{14}H_8Br_2O_2$
4,4′-Dibromobenzil

$C_{14}H_8Br_2O_3$
m-Bromobenzoic Acid, *Anhydride*
p-Bromobenzoic Acid, *Anhydride*

$C_{14}H_8Br_2O_4$
4,4′-Dibromobiphenyl-2,2′-dicarboxylic Acid

$C_{14}H_8ClNO_2$
1-Amino-2-chloroanthraquinone
1-Amino-3-chloroanthraquinone
1-Amino-4-chloroanthraquinone
1-Amino-5-chloroanthraquinone
1-Amino-6-chloroanthraquinone
1-Amino-7-chloroanthraquinone
1-Amino-8-chloroanthraquinone
2-Amino-1-chloroanthraquinone
2-Amino-3-chloroanthraquinone
2-Amino-7-chloroanthraquinone
Phthalimide, N-p-*Chlorophenyl*

$C_{14}H_8ClNO_4$
4-*m*-Nitrobenzoylbenzoic Acid, *Chloride*
4-*p*-Nitrobenzoylbenzoic Acid, *Chloride*

$C_{14}H_8Cl_2$
1,4-Dichloroanthracene
1,5-Dichloroanthracene
1,6-Dichloroanthracene
1,7-Dichloroanthracene
1,8-Dichloroanthracene
1,9-Dichloroanthracene
2,3-Dichloroanthracene
9,10-Dichloroanthracene

$C_{14}H_8Cl_2N_2O_2$
Azobenzene-4,4′-dicarboxylic Acid, *Dichloride*
Azoxybenzene-3,3′-dicarboxylic Acid, *Dichloride*
1,4-Diamino-2,3-dichloroanthraquinone
1,5-Diamino-2,4-dichloroanthraquinone
1,5-Diamino-4,8-dichloroanthraquinone

$C_{14}H_8Cl_2O$
1,3-Dichloroanthrone
1,4-Dichloroanthrone
1,5-Dichloroanthrone
1,8-Dichloroanthrone
2,3-Dichloroanthrone
2,4-Dichloroanthrone
4,5-Dichloroanthrone
4,10-Dichloroanthrone
10,10-Dichloroanthrone

$C_{14}H_8Cl_2O_2$
2-Benzoyl-4-chlorobenzoic Acid, *Chloride*
2,2′-Dichlorobenzil
4,4′-Dichlorobenzil
Diphenic Acid, *Dichloride*

$C_{14}H_8Cl_2O_2S_2$
Diphenyl disulphide 2,2′-dicarboxylic Acid, *Dichloride*

$C_{14}H_8Cl_2O_3$
2-Benzoyl-3,6-dichlorobenzoic Acid
2-Benzoyl-4,5-dichlorobenzoic Acid
o-Chlorobenzoic Acid, *Anhydride*
m-Chlorobenzoic Acid, *Anhydride*
p-Chlorobenzoic Acid, *Anhydride*
2-(2,5-Dichlorobenzoyl)benzoic Acid
2,2′-Oxydibenzoic Acid, *Dichloride*

$C_{14}H_8Cl_2O_4$
3,3′-Dichlorobiphenyl-2,2′-dicarboxylic Acid
4,6-Dichlorobiphenyl-2,2′-dicarboxylic Acid
5,5′-Dichlorobiphenyl-2,2′-dicarboxylic Acid
6,6′-Dichlorobiphenyl-2,2′-dicarboxylic Acid

$C_{14}H_8Cl_2O_4S_2$
Anthracene-1,5-disulphonic Acid, *Dichloride*
Anthracene-1,8-disulphonic Acid, *Dichloride*

$C_{14}H_8Cl_4$
9,9,10,10-Tetrachloroanthracene †

$C_{14}H_8INO_2$
Phthalimide, N-p-*Iodophenyl*

$C_{14}H_8I_2O_3$
m-Iodobenzoic Acid, *Anhydride*
p-Iodobenzoic Acid, *Anhydride*

$C_{14}H_8I_4O_4$
4-(4′-Hydroxy-3′,5′-di-iodophenoxy)-3,5-di-iodophenylacetic Acid

$C_{14}H_8N_2$
Benzo[*f*]quinoline-1-carboxylic Acid, *Nitrile*
Benzo[*h*]quinoline-2-carboxylic Acid, *Nitrile*
Benzo[*h*]quinoline-4-carboxylic Acid, *Nitrile*
Biphenyl-2,4′-dicarboxylic Acid, *Dinitrile*
Biphenyl-3,4-dicarboxylic Acid, *Dinitrile*
Biphenyl-4,4′-dicarboxylic Acid, *Dinitrile*
4,9-Diazapyrene †
4,10-Diazapyrene †

$C_{14}H_8N_2O$
Canthin-6-one †

$C_{14}H_8N_2O_2$
Bisanthranil †

$C_{14}H_8N_2O_4$
1-Amino-2-nitroanthraquinone
1-Amino-3-nitroanthraquinone
1-Amino-4-nitroanthraquinone
1-Amino-5-nitroanthraquinone
1-Amino-7-nitroanthraquinone
1-Amino-8-nitroanthraquinone
2-Amino-1-nitroanthraquinone
2-Amino-3-nitroanthraquinone
6-Amino-1-nitroanthraquinone
2-Amino-5-nitrophenanthenequinone
9,10-Dinitroanthracene
7-Nitrofluorenone-4-carboxylic Acid, *Amide*
3-Nitrophthalanil
4-Nitrophthalanil
Phthalimide, N-o-*Nitrophenyl*

$C_{14}H_8N_2O_5$
Tramesanguin

$C_{14}H_8N_2O_6$
Cinnabarinic Acid
2,2′-Dinitrobenzil
2,3′-Dinitrobenzil
2,4-Dinitrobenzil
2,4′-Dinitrobenzil
3,3′-Dinitrobenzil
3,4′-Dinitrobenzil
4,4′-Dinitrobenzil

$C_{14}H_8N_2O_7$
o-Nitrobenzoic Acid, *Anhydride*
m-Nitrobenzoic Acid, *Anhydride*
p-Nitrobenzoic Acid, *Anhydride*

$C_{14}H_8N_2O_8$
4,4′-Dinitrobiphenyl-2,2′-dicarboxylic Acid
4,6′-Dinitrobiphenyl-2,2′-dicarboxylic Acid
5,5′-Dinitrobiphenyl-2,2′-dicarboxylic Acid
6,6′-Dinitrobiphenyl-2,2′-dicarboxylic Acid
Oxalic Acid, *Di-o-nitrophenyl ester*

$C_{14}H_8N_2S_4$
2-Mercaptobenzothiazole, 2,2′-Dithio*bis*benzthiazole

$C_{14}H_8N_4$
Dipyridol[2,3-*f*,2′,3′-*h*]quinoxaline
Dipyridol[2,3-*f*,3′,2′-*h*]quinoxaline
Quinoxalo[2,3-*b*]quinoxaline

$C_{14}H_8N_4O$
Azoxybenzene-2,2′-dicarboxylic Acid, *Dinitrile*

$C_{14}H_8N_4O_8$
2,2′,4,4′-Tetranitrostilbene
2,2′,4,6′-Tetranitrostilbene
2,2′,6,6′-Tetranitrostilbene
2,3′,4,6′-Tetranitrostilbene
2,4,4′,6-Tetranitrostilbene

$C_{14}H_8O_2$
1,2-Anthraquinone
1,4-Anthraquinone
9,10-Anthraquinone
Morphenol
1,2-Phenanthraquinone
1,4-Phenanthraquinone
3,4-Phenanthraquinone
9,10-Phenanthraquinone

$C_{14}H_8O_2S$
1-Anthraquinonethiol
2-Anthraquinonethiol

$C_{14}H_8O_3$
Acenaphthene-5,6-dicarboxylic Acid, *Anhydride*
Biphenyl-2,3-dicarboxylic Acid, *Anhydride*
Biphenyl-3,4-dicarboxylic Acid, *Anhydride*
Diphenic Acid, *Anhydride*
Fluorenone-1-carboxylic Acid
Fluorenone-2-carboxylic Acid
Fluorenone-3-carboxylic Acid
Fluorenone-4-carboxylic Acid
2-Hydroxy-1,4-anthraquinone
1-Hydroxy-9,10-anthraquinone
2-Hydroxy-9,10-anthraquinone
3-Hydroxy-1,4-phenanthraquinone
2-Hydroxy-9,10-phenanthraquinone
3-Hydroxy-9,10-phenanthraquinone
4-Hydroxy-9,10-phenanthraquinone

$C_{14}H_8O_4$
Alizarin
1,3-Dihydroxyanthraquinone
1,4-Dihydroxyanthraquinone
1,5-Dihydroxyanthraquinone
1,6-Dihydroxyanthraquinone
1,7-Dihydroxyanthraquinone
1,8-Dihydroxyanthraquinone
2,3-Dihydroxyanthraquinone
2,6-Dihydroxyanthraquinone
2,7-Dihydroxyanthraquinone
3,7-Dihydroxy-2,6-anthraquinone †
5,6-Dihydroxy-1,2-anthraquinone †
1,2-Dihydroxyphenanthraquinone
1,4-Dihydroxyphenanthraquinone
2,3-Dihydroxyphenanthraquinone
2,4-Dihydroxyphenanthraquinone
2,6-Dihydroxyphenanthraquinone
2,7-Dihydroxyphenanthraquinone
3,4-Dihydroxyphenanthraquinone
3,6-Dihydroxyphenanthraquinone
4,5-Dihydroxyphenanthraquinone
α-Disalicylide
Xanthone-2-carboxylic Acid
Xanthone-4-carboxylic Acid

$C_{14}H_8O_5$
4-Hydroxy-2,3-methylenedioxyxanthone †
Lambertellin †
Naphthalene-1,4,5-tricarboxylic Acid, *Anhydride, Me ester*
1,2,3-Trihydroxyanthraquinone
1,2,4-Trihydroxyanthraquinone
1,2,5-Trihydroxyanthraquinone
1,2,6-Trihydroxyanthraquinone
1,2,7-Trihydroxyanthraquinone
1,2,8-Trihydroxyanthraquinone
1,3,5-Trihydroxyanthraquinone
1,3,7-Trihydroxyanthraquinone
1,3,8-Trihydroxyanthraquinone
1,4,5-Trihydroxyanthraquinone
1,4,6-Trihydroxyanthraquinone
1,2,4-Trihydroxyphenanthraquinone
1,3,4-Trihydroxyphenanthraquinone
2,3,4-Trihydroxyphenanthraquinone

$C_{14}H_8O_5S$
Anthraquinone-1-sulphonic Acid
Anthraquinone-2-sulphonic Acid
9,10-Phenanthraquinone-2-sulphonic Acid
9,10-Phenanthraquinone-3-sulphonic Acid
1,2,3,4-Tetrahydroxyanthraquinone
1,2,3,5-(or 1,2,3,7-)Tetrahydroxyanthra-quinone
1,2,3,7-(or 1,2,3,5-)Tetrahydroxyanthra-quinone
1,2,4,6-Tetrahydroxyanthraquinone
1,2,4,7-Tetrahydroxyanthraquinone
1,2,5,6-Tetrahydroxyanthraquinone
1,2,5,8-Tetrahydroxyanthraquinone
1,2,6,7-Tetrahydroxyanthraquinone
1,2,7,8-Tetrahydroxyanthraquinone
1,3,5,7-Tetrahydroxyanthraquinone
1,3,6,8-Tetrahydroxyanthraquinone★†
1,4,5,8-Tetrahydroxyanthraquinone
1,4,6,7-Tetrahydroxyanthraquinone
2,3,6,7-Tetrahydroxyanthraquinone

$C_{14}H_8O_6S$
4-Hydroxyanthraquinone-1-sulphonic Acid
5-Hydroxyanthraquinone-1-sulphonic Acid
8-Hydroxyanthraquinone-1-sulphonic Acid
3-Hydroxyanthraquinone-2-sulphonic Acid
5-Hydroxyanthraquinone-2-sulphonic Acid
6-Hydroxyanthraquinone-2-sulphonic Acid
7-Hydroxyanthraquinone-2-sulphonic Acid
8-Hydroxyanthraquinone-2-sulphonic Acid

$C_{14}H_8O_7$
1,2,3,5,7-Pentahydroxyanthraquinone
1,2,4,5,7-Pentahydroxyanthraquinone†
1,2,4,5,8-Pentahydroxyanthraquinone

$C_{14}H_8O_8$
1,2,3,5,6,7-Hexahydroxyanthraquinone
1,2,4,5,6,8-Hexahydroxyanthraquinone
1,2,4,5,7,8-Hexahydroxyanthranquinone
Naphthalene-1,2,4,5-tetracarboxylic Acid
Naphthalene-1,4,5,8-tetracarboxylic Acid

$C_{14}H_8O_8S_2$
Anthraquinone-1,5-disulphonic Acid
Anthraquinone-1,6-disulphonic Acid
Anthraquinone-1,7-disulphonic Acid
Anthraquinone-1,8-disulphonic Acid
Anthraquinone-2,6-disulphonic Acid
Anthraquinone-2,7-disulphonic Acid

$C_{14}H_8S_2$
Thianaphtheno[3,2-*b*]thianaphthene
Thianaphtheno[4,5-*e*]thianaphthene
Thianaphtheno[5,4-*e*]thianaphthene

$C_{14}H_9Br$
9-Bromoanthracene
1-Bromophenanthrene
2-Bromophenanthrene
3-Bromophenanthrene
9-Bromophenanthrene

$C_{14}H_9BrO_3$
2-Benzoyl-4-bromobenzoic Acid
2-Benzoyl-3(or 6)-bromobenzoic Acid
o-4-Bromobenzoylbenzoic Acid

$C_{14}H_9BrO_4$
4-Bromodiphenic Acid
5-Bromodiphenic Acid

$C_{14}H_9Cl$
1-Chloroanthracene
2-Chloroanthracene
9-Chloroanthracene
1-Chlorophenanthrene
2-Chlorophenanthrene
3-Chlorophenanthrene
9-Chlorophenanthrene

$C_{14}H_9ClO$
Fluorene-9-carboxylic Acid, *Chloride*

$C_{14}H_9ClO_2$
o-Benzoylbenzoic Acid, *Chloride*

$C_{14}H_9ClO_2S$
Phenanthrene-2-sulphonic Acid, *Chloride*
Phenanthrene-3-sulphonic Acid, *Chloride*
Phenanthrene-9-sulphonic Acid, *Chloride*

$C_{14}H_9ClO_3$
2-Benzoyl-4-chlorobenzoic Acid
4-Benzoyl-2-chlorobenzoic Acid
o-2-Chlorobenzoylbenzoic Acid
o-3-Chlorobenzoylbenzoic Acid
o-4-Chlorobenzoylbenzoic Acid

$C_{14}H_9ClO_3S$
Anthracene-1-sulphonic Acid, *Chloride*
Anthracene-2-sulphonic Acid, *Chloride*

$C_{14}H_9ClO_4$
4-Chlorodiphenic Acid
O-Salicyloylsalicylic Acid, *Chloride*

$C_{14}H_9Cl_3F_2$
1,1,1-Trichloro-2,2-*di*-(*p*-fluorophenyl)ethane

$C_{14}H_9Cl_5$
1,1-Di-4-chlorophenyl-2,2,2-trichloroethane

$C_{14}H_9N$
Fluorene-2-carboxylic Acid, *Nitrile*
Fluorene-9-carboxylic Acid, *Nitrile*

$C_{14}H_9NO$
Acridine-9-aldehyde
Dibenzo[*b,f*]azepin-2-one†
2-Dibenzofuranylacetic Acid, *Nitrile*

$C_{14}H_9NO_2$
Acridine-2-carboxylic Acid
Acridine-4-carboxylic Acid
Acridine-9-carboxylic Acid
1-Aminoanthraquinone
2-Aminoanthraquinone
2-Aminophenanthrenequinone
3-Aminophenanthrenequinone
4-Aminophenanthrenequinone
Benzo[*f*]quinoline-1-carboxylic Acid
Benzo[*f*]quinoline-1-carboxylic Acid
Benzo[*f*]quinoline-3-carboxylic Acid
Benzo[*f*]quinoline-5-carboxylic Acid
Benzo[*g*]quinoline-4-carboxylic Acid
Benzo[*h*]quinoline-2-carboxylic Acid
Benzo[*h*]quinoline-4-carboxylic Acid
Benzoylanthranil
Diphenic Acid, *Mononitrile*
Dubamine†
Fluorenone-1-carboxylic Acid, *Amide*
Fluorenone-1-carboxylic Acid, *Amide*
9-Nitroanthracene

$C_{14}H_9NO_2$ (*continued*)
1-Nitrophenanthrene
2-Nitrophenanthrene
3-Nitrophenanthrene
4-Nitrophenanthrene
9-Nitrophenanthrene
Phenanthridine-6-carboxylic Acid
N-Phenylisatin
Phthalanil

$C_{14}H_9NO_3$
1-Amino-2-hydroxyanthraquinone
1-Amino-4-hydroxyanthraquinone
1-Amino-5-hydroxyanthraquinone
1-Amino-8-hydroxyanthraquinone
2-Amino-1-hydroxyanthraquinone
3-Amino-1-hydroxyanthraquinone
3-Hydroxy-9-nitrophenanthrene
2-Methyl-6-phenylpyridine-3,4-dicarboxylic Acid, *Anhydride*
ms-Nitroanthranol
ms-Nitroanthrone
Xanthone-2-carboxylic Acid, *Amide*
Xanthone-4-carboxylic Acid, *Amide*

$C_{14}H_9NO_4$
1-Amino-2,3-dihydroxyanthraquinone
1-Amino-2,7-dihydroxyanthraquinone
1-Amino-4,5-dihydroxyanthraquinone
2-Amino-1,3-dihydroxyanthraquinone
2-Amino-1,4-dihydroxyanthraquinone
3-Amino-1,2-dihydroxyanthraquinone
3-Amino-1,8-dihydroxyanthraquinone
4-Amino-1,2-dihydroxyanthraquinone
4-Amino-1,3-dihydroxyanthraquinone
5-Amino-1,4-dihydroxyanthraquinone
2-Hydroxy-7-nitrofluorenone, *Me ether*
2-Nitrobenzil
3-Nitrobenzil
4-Nitrobenzil
2-Nitrofluorene-9-carboxylic Acid
2-Nitrophenanthrahydroquinone
3-Nitrophenanthrahydroquinone
4-Nitrophenanthrahydroquinone

$C_{14}H_9NO_4S$
Anthraquinone-2-sulphonic Acid, *Amide*

$C_{14}H_9NO_5$
2-Benzoyl-3-nitrobenzoic Acid
2-Benzoyl-4-nitrobenzoic Acid
2-Benzoyl-5-nitrobenzoic Acid
3-Benzoyl-6-nitrobenzoic Acid
9-Hydroxy-2-nitrofluorene-9-carboxylic Acid
9-Hydroxy-3-nitrofluorene-9-carboxylic Acid
9-Hydroxy-4-nitrofluorene-9-carboxylic Acid
Narciprimine†
2-*o*-Nitrobenzoylbenzoic Acid
4-*o*-Nitrobenzoylbenzoic Acid
4-*m*-Nitrobenzoylbenzoic Acid
4-*p*-Nitrobenzoylbenzoic Acid

$C_{14}H_9NO_6$
4-Nitrobiphenic Acid
5-Nitrobiphenic Acid
6-Nitrobiphenic Acid

$C_{14}H_9NS$
10*H*-Thieno[3,2-α]carbazole

$C_{14}H_9N_3$
Benzimidazo[2,1-α]phthalazine
2-Phenylindazole-3-carboxylic Acid, *Nitrile*

$C_{14}H_9N_3O_4$
1,4-Diamino-2-nitroanthraquinone
1,4-Diamino-5-nitroanthraquinone
Indazole-2-carboxylic Acid, o-*Nitrophenyl ester*

$C_{14}H_9N_3O_6$
2,2′,4-Trinitrostilbene
2,3′,4-Trinitrostilbene
2,4,4′-Trinitrostilbene

$C_{14}H_9N_3O_9$
Vanillin, *Picryl ether*

$C_{14}H_9N_5$
Diazoaminobenzene-2,2′-dicarboxylic Acid, *Dinitrile*

$C_{14}H_9N_5O_8$
1,3,6,8-Tetranitrocarbazole, N-*Et*

$C_{14}H_9N_7O_{12}$
Di-2,4,6-trinitrophenylamine, N-*Et*

$C_{14}H_{10}$
Anthracene
Benz[*a*]azulene
Benz[*e*]azulene
Benz[*f*]azulene★
1,8-Bisdehydro[14]annulene†
Cyclohepta[*de*]naphthylene
Phenanthrene
Tolane

$C_{14}H_{10}Br_2$
1,1-Dibromo-2,2-diphenylethylene

$C_{14}H_{10}Br_2$
α,β-Dibromostilbene
2,2′-Dibromostilbene
2,4′-Dibromostilbene
3,3′-Dibromostilbene
4,4′-Dibromostilbene

$C_{14}H_{10}Br_2O$
4-Bromo-α-(4-bromophenyl)acetophenone
αα-Dibromo-α-phenylacetophenone

$C_{14}H_{10}Br_2O_2$
3,3′-Dibromobenzoin

$C_{14}H_{10}Br_2O_3$
5,5′-Dibromo-2,2′-dihydroxybenzophenone, 2-*Me ether*

$C_{14}H_{10}ClN$
9-Chloro-2-methylacridine
9-Chloro-3-methylacridine

$C_{14}H_{10}ClNO$
1-Chloro-4-methylacridone

$C_{14}H_{10}ClNO_4$
2-Chloro-3-nitrobenzoic Acid, *Phenyl ester*

$C_{14}H_{10}Cl_2$
1,2-Di-*p*-chlorophenylethylene
α,β-Dichlorostilbene

$C_{14}H_{10}Cl_2O$
α-Chlorodiphenylacetic Acid, *Chloride*
2,2′-Dichloro-α,α′-epoxybibenzyl†
α,α-Dichloro-α-phenylacetophenone

$C_{14}H_{10}Cl_2O_2$
2,2′-Dichlorobenzoin
3,3′-Dichlorobenzoin
4,4′-Dichlorobenzoin
3,5-Dichloro-4-hydroxybenzophenone

$C_{14}H_{10}Cl_2O_4$
Mollisin★†

$C_{14}H_{10}Cl_4$
1,1-Dichloro-2,2-di-(4-chlorophenyl)ethane

$C_{14}H_{10}N_2$
Dibenzo[*b,f*][1,5]diazocine†
2-Phenylquinazoline
4-Phenylquinazoline
2-Phenylquinoxaline

$C_{14}H_{10}N_2$ (dication)
Pyrazino(1,2,3,4-*l,m,n*]-1,10-phenanthrolium†

$C_{14}H_{10}N_2O$
Benzo[*f*]quinoline-1-carboxylic Acid, *Amide*
Benzo[*f*]quinoline-3-carboxylic Acid, *Amide*
Benzo[*f*]quinoline-5-carboxylic Acid, *Amide*
Benzoylanthranilic Acid, *Nitrile*
α-Diazo-α-phenylacetophenone
3,5-Diphenyl-1,2,4-oxadiazole
3,4-Diphenyl-1,2,5-oxadiazole
2,5-Diphenyl-1,3,4-oxadiazole
Isatin α-anilide

$C_{14}H_{10}N_2O_2$
Anthranoylanthranilic Acid, *Anhydride*
1,2-Diaminoanthraquinone
1,3-Diaminoanthraquinone
1,4-Diaminoanthraquinone
1,5-Diaminoanthraquinone
1,6-Diaminoanthraquinone
1,7-Diaminoanthraquinone
1,8-Diaminoanthraquinone
2,3-Diaminoanthraquinone
2,6-Diaminoanthraquinone
2,7-Diaminoanthraquinone
2,7-Diaminophenanthraquinone
4,5-Diaminophenanthraquinone
4,4′-Diformylazobenzene
Diphenylfuroxan
Halfordinol†
2-Methyl-4-nitroacridine
9-Methyl-2-nitroacridine
6-Methyl-2-nitrophenanthridine
6-Methyl-8-nitrophenanthridine
2-Methyl-6-phenylpyridine-3,4-dicarboxylic Acid, *Imide*
3-*o*-Nitrophenylindole
3-Nitro-2-phenylindole
Phenazine-1-carboxylic Acid, *Me ester*
Phenazine-2-carboxylic Acid, *Me ester*
3-Phenylindazole-2-carboxylic Acid
2-Phenylindazole-3-carboxylic Acid
Phenylphthaloylhydrazine

$C_{14}H_{10}N_2O_3$
1,3-Diamino-4-hydroxyanthraquinone
2,2′-Diformylazoxybenzene
3,3′-Diformylazoxybenzene
4,4′-Diformylazoxybenzene
2-Hydroxyphenazine-1-carboxylic Acid, *Me ester*†
2-Hydroxyphenazine-1-carboxylic Acid, *Me ether*†
2-Methyl-4-nitroacridone
3-Methyl-6-nitroacridone
4-Methyl-1-nitroacridone
5-Methyl-4-nitroacridone
10-Methyl-2-nitroacridone
10-Methyl-4-nitroacridone
10-Methyl-5-nitroacridone
2-Nitroacridone, N-*Me*
4-Nitroacridone, N-*Me*

$C_{14}H_{10}N_2O_3S$
4,6-Diphenyl-1,2,3,5-oxathiadiazine 2,2-Dioxide†

$C_{14}H_{10}N_2O_4$
Azobenzene-2,2′-dicarboxylic Acid
Azobenzene-2,3′-dicarboxylic Acid
Azobenzene-3,3′-dicarboxylic Acid
Azobenzene-4,4′-dicarboxylic Acid
1,4-Diamino-2,3-dihydroxyanthraquinone
1,4-Diamino-5,8-dihydroxyanthraquinone
1,5-Diamino-2,6-dihydroxyanthraquinone
1,5-Diamino-4,8-dihydroxyanthraquinone
1,8-Diamino-4,5-dihydroxyanthraquinone
2,6-Diamino-1,5-dihydroxyanthraquinone
α,α′-Dinitrostilbene
α,2′-Dinitrostilbene
α,3-Dinitrostilbene
α,4′-Dinitrostilbene
2,2′-Dinitrostilbene
2,4-Dinitrostilbene
2,6-Dinitrostilbene
3,4′-Dinitrostilbene
4,4′-Dinitrostilbene
4-*m*-Nitrobenzoylbenzoic Acid, *Amide*

$C_{14}H_{10}N_2O_5$
Azoxybenzene-2,2′-dicarboxylic Acid
Azoxybenzene-3,3′-dicarboxylic Acid
Azoxybenzene-4,4′-dicarboxylic Acid
Cinnabarine
3-Nitrophthalic Acid, *Anilide*
4-Nitrophthalic Acid, *Anilide*

$C_{14}H_{10}N_2O_6$
3,5-Dinitrobenzoic Acid, o-*Tolyl ester*
2-Hydroxy-3,5-dinitrobenzophenone, *Me ether*
4-Hydroxy-3,3′-dinitrobenzophenone, *Me ether*
4-Hydroxy-3,4′-dinitrobenzophenone, *Me ether*
4-Hydroxy-3,5-dinitrobenzophenone, *Me ether*

$C_{14}H_{10}N_2O_6S_2$
Anthraquinone-1,5-disulphonic Acid, *Diamide*
Anthraquinone-1,8-disulphonic Acid, *Diamide*

$C_{14}H_{10}N_2O_7$
5-Hydroxy-2-nitrobenzoic Acid, p-*Nitrobenzyl ester*
Vanillin, 2,4-*Dinitrophenyl ether*

$C_{14}H_{10}N_2O_8$
4,5-Dinitronaphthalene-1,8-dicarboxylic Acid, *Di-Me ester*

$C_{14}H_{10}N_2S$
4,5-Diphenyl-1,2,3-thiadiazole
3,5-Diphenyl-1,2,4-thiadiazole
2,5-Diphenyl-1,3,4-thiadiazole

$C_{14}H_{10}N_4$
Fluoflavine

$C_{14}H_{10}N_4O_3$
6-Nitroindazole-1-carboxylic Acid, *Anilide*

$C_{14}H_{10}N_4O_6$
6,6′-Dinitrobiphenyl-2,2′-dicarboxylic Acid, *Diamide*

$C_{14}H_{10}N_4O_8$
6-Amino-4-nitro-*m*-toluic Acid, N-2,4-*Dinitrophenyl*

$C_{14}H_{10}O$
1-Anthrol
2-Anthrol
9-Anthrol
Anthrone
Diphenylketene
1-Hydroxyphenanthrene
2-Hydroxyphenanthrene
3-Hydroxyphenanthrene
4-Hydroxyphenanthrene
9-Hydroxyphenanthrene
1-Methylfluorenone
2-Methylfluorenone
3-Methylfluorenone
4-Methylfluorenone
2-Phenylbenzo[*b*]furan
3-Phenylbenzo[*b*]furan

$C_{14}H_{10}O_2$
Benzil
4-Benzoylbenzaldehyde
2,3-Dihydro-1*H*-phenalene-1,3-dione, 1-*Me enol-ether*
1,2-Dihydroxyanthracene
1,5-Dihydroxyanthracene
1,8-Dihydroxyanthracene
1,9-Dihydroxyanthracene
2,3-Dihydroxyanthracene
2,6-Dihydroxyanthracene
2,7-Dihydroxyanthracene
9,10-Dihydroxyanthracene
1,2-Dihydroxyphenanthrene
1,4-Dihydroxyphenanthrene
1,9-Dihydroxyphenanthrene
2,3-Dihydroxyphenanthrene
2,5-Dihydroxyphenanthrene
2,6-Dihydroxyphenanthrene
2,7-Dihydroxyphenanthrene
2,8-Dihydroxyphenanthrene
3,6-Dihydroxyphenanthrene
3,8-Dihydroxyphenanthrene
Fluorene-1-carboxylic Acid
Fluorene-2-carboxylic Acid
Fluorene-3-carboxylic Acid
Fluorene-4-carboxylic Acid
Fluorene-9-carboxylic Acid
1-Hydroxy-9-anthrone
2-Hydroxy-9-anthrone
3-Hydroxy-9-anthrone
4-Hydroxy-9-anthrone
1-Hydroxyfluorenone, *Me ether*
2-Hydroxyfluorenone, *Me ether*
3-Hydroxyfluorenone, *Me ether*
2-Methylnaphtho[1,2-*b*]pyran-4-one
2-Methyl-4-oxobenzo[*g*]chromene
3-Methyl-2-oxobenzo[*f*]chromene
4-Methyl-2-oxobenzo[*f*]chromene
4-Methyl-2-oxobenzo[*g*]chromene
2-Methyl-9-xanthenone
3-Methyl-9-xanthenone
4-Methyl-9-xanthenone
Morphol
Oxanthranol
9,10-Phenanthraquinol

$C_{14}H_{10}O_2S$
Dibenzoyl sulphide

$C_{14}H_{10}O_2S_2$
Dibenzoyl disulphide

$C_{14}H_{10}O_2S_4$
Dibenzoyl tetrasulphide

$C_{14}H_{10}O_3$
Benzoic Anhydride
o-Benzoylbenzoic Acid
m-Benzoylbenzoic Acid
p-Benzoylbenzoic Acid
Dibenzofuran-1-carboxylic Acid, *Me ester*
Dibenzofuran-3-carboxylic Acid, *Me ester*
Dibenzofuran-4-carboxylic Acid, *Me ester*
2-Dibenzofuranylacetic Acid
3,4-Dihydroxy-9-anthrone
2′,4′-Dihydroxybiphenyl-2-carboxylic Acid (2,2′)-lactone, *Me ether*†
Disalicylaldehyde
2-Ethylnaphtho[2,3-*b*]furan-4,9-dione†
Frutescinol-lactone†
2-Hydroxyfluorene-1-carboxylic Acid
2-Hydroxyfluorene-3-carboxylic Acid
9-Hydrofluorene-4-carboxylic Acid
9-Hydroxyfluorene-9-carboxylic Acid
1-(2-Hydroxyphenyl)-2-phenylglyoxal
1-(4-Hydroxyphenyl)-2-phenylglyoxal
1-Hydroxyxanthone, *Me ether*
2-Hydroxyxanthone, *Me ether*
3-Hydroxyxanthone, *Me ether*
4-Hydroxyxanthone, *Me ether*
3-Hydroxyacenaphthenequinone, *Et ether*
Maturinone†
3-Methyl-2*H*-naphtho[2,3-*b*]pyran-5,10-dione†
Oroselone
1,8,9-Trihydroxyanthracene
1,3,4-Trihydroxyphenanthrene
1,5,6-Trihydroxyphenanthrene
3,4,5-Trihydroxyphenanthrene
3,4,6-Trihydroxyphenanthrene
Xanthene-9-carboxylic Acid

$C_{14}H_{10}O_3S$
Anthracene-1-sulphonic Acid
Anthracene-2-sulphonic Acid
Phenanthrene-1-sulphonic Acid
Phenanthrene-2-sulphonic Acid
Phenanthrene-3-sulphonic Acid
Phenanthrene-9-sulphonic Acid

$C_{14}H_{10}O_4$
Acenaphthene-3,8-dicarboxylic Acid
Acenaphthene-5,6-dicarboxylic Acid
2-Benzoyl-4-hydroxybenzoic Acid
3-Benzoyl-4-hydroxybenzoic Acid
5-Benzoyl-2-hydroxybenzoic Acid

$C_{14}H_{10}O_4$ (*continued*)
Benzoyl peroxide
Biphenyl-2,3-dicarboxylic Acid
Biphenyl-2,3′-dicarboxylic Acid
Biphenyl-2,4-dicarboxylic Acid
Biphenyl-2,4′-dicarboxylic Acid
Biphenyl-2,5-dicarboxylic Acid
Biphenyl-2,6-dicarboxylic Acid
Biphenyl-3,3′-dicarboxylic Acid
Biphenyl-3,4-dicarboxylic Acid
Biphenyl-3,4′-dicarboxylic Acid
Biphenyl-3,5-dicarboxylic Acid
Biphenyl-4,4′-dicarboxylic Acid
3,6-Dihydroxy-1,2-acenaphthenequinone, *Di-Me ether*
3,8-Dihydroxy-1,2-acenaphthenequinone, *Di-Me ether*
1,3-Dihydroxyxanthone, 3-*Me ether* (incorrectly given as 2-*Me ether*)
1,5-Dihydroxyxanthone, 5-*Me ether* †
1,6-Dihydroxyxanthone, 1-*Me ether* †
1,6-Dihydroxyxanthone, 6-*Me ether* †
1,7-Dihydroxyxanthone, 1-*Me ether* (incorrectly given as 6-*Me ether*)
1,7-Dihydroxyxanthone, 7-*Me ether* (incorrectly given as 3-*Me ether*)
Diphenic Acid
2-(2-Hydroxybenzoyl)benzoic Acid
2-(3-Hydroxybenzoyl)benzoic Acid
2-(4-Hydroxybenzoyl)benzoic Acid
3-Hydroxy-2-methyl-2*H*-naphtho[2,3-*b*]pyran-5,10-dione †
Isoeuxanthone, 6-*Me ether*
Leuco-quinizarin
Maturone †
1,8-Naphthaldehydic Acid, *Ac*
Oxalic Acid, *Diphenyl ester*
Phthalic Acid, *Phenyl ester*
1,2,5,10-Tetrahydroxyanthracene
1,2,6,10-Tetrahydroxyanthracene
1,2,7,10-Tetrahydroxyanthracene
1,2,9,10-Tetrahydroxyanthracene
1,3,9,10-Tetrahydroxyanthracene
1,4,9,10-Tetrahydroxyanthracene
2,3,6,7-Tetrahydroxyanthracene
2,3,9,10-Tetrahydroxyanthracene
2,6,9,10-Tetrahydroxyanthracene
2,7,9,10-Tetrahydroxyanthracene
1,4,5,6-Tetrahydroxyphenanthrene
2,3,5,6-Tetrahydroxyphenanthrene
2′,4,4′-Trihydroxybiphenyl-2-carboxylic Acid, 4-*Me ether* †

$C_{14}H_{10}O_4S_2$
Diphenyldisulphide 2,2′-dicarboxylic Acid
Diphenyldisulphide 3,3′-dicarboxylic Acid

$C_{14}H_{10}O_5$
Alternariol
2′,4′-Dihydroxybenzophenone-2-carboxylic Acid
2′,5′-Dihydroxybenzophenone-2-carboxylic Acid
2′,6-Dihydroxybenzophenone-2-carboxylic Acid
3′,4′-Dihydroxybenzophenone-2-carboxylic Acid
4,4′-Dihydroxybenzophenone-2-carboxylic Acid
4,5-Dihydroxybenzophenone-2-carboxylic Acid
4′,5-Dihydroxybenzophenone-2-carboxylic Acid
4′,6-Dihydroxybenzophenone-2-carboxylic Acid
5,6-Dihydroxybenzophenone-2-carboxylic Acid
2′,4′-Dihydroxybenzophenone-4-carboxylic Acid
1,7-Dihydroxy-3-methoxyxanthone
1,5-Dihydroxy-3-methoxyxanthone †
1,5-Dihydroxy-6-methoxyxanthone †
1,6-Dihydroxy-5-methoxyxanthone †
2,8-Dihydroxy-1-methoxyxanthone †
3,4-Dihydroxy-2-methoxyxanthone †
4-Hydroxybiphenyl-2,2′-dicarboxylic Acid
3-Hydroxybiphenyl-4,4′-dicarboxylic Acid
Nor-rubrofusarin
2,2′-Oxydibenzoic Acid
Ptaeroxylone †
Ravenelin
O-Salicyloylsalicylic Acid
2,2′,4-Trihydroxybenzil
2,3,4-Trihydroxybenzil
2,4,4′-Trihydroxybenzil
2,4,6-Trihydroxybenzil
1,3,6-Trihydroxy-8-methylxanthone †
1,3,7-Trihydroxyxanthone, 7-*Me ether*
2,3,8-Trihydroxyxanthone, 8-*Me ether*

$C_{14}H_{10}O_6$
Altertenuol
Bellidifolin †
4,4′-Dihydroxybiphenyl-2,2′dicarboxylic Acid
5,5′-Dihydroxybiphenyl-2,2′-dicarboxylic Acid
6,6′-Dihydroxybiphenyl-2,2′-dicarboxylic Acid
2′,3-Dihydroxybiphenyl-2,6-dicarboxylic Acid
4,4′,Dihydroxybiphenyl-3,3′-dicarboxylic Acid
6,6′-Dihydroxybiphenyl-3,3′-dicarboxylic Acid
3,3′-Dihydroxybiphenyl-4,4′-dicarboxylic Acid
Isobellifolin †
Isohalfordin †
Isophenicin
Oxypeucedanic Acid, *Me ester*
Phoenicine
1,3,5,6-Tetrahydroxyxanthone, 1-*Me ether* †
1,3,6,7-Tetrahydroxyxanthone, 6-*Me ether* †
1,3,7,8-Tetrahydroxyxanthone, 7-*Me ether* †
5,8,10-Trihydroxy-2-methyl-2*H*-naphtho[2,3-*b*]-pyran-6,9-dione †

$C_{14}H_{10}O_6S_2$
Anthracene-1,5-disulphonic Acid
Anthracene-1,8-disulphonic Acid

$C_{14}H_{10}O_7$
Citromycetin
1,3,4,5,8-Pentahydroxyxanthone, 3-O-*Me ether* †

$C_{14}H_{10}O_8$
Iso-oösporein

$C_{14}H_{10}O_9$
Digallic Acid

$C_{14}H_{10}S$
- 2-Anthracenethiol
- 9-Anthracenethiol
- 2-Phenanthrenethiol
- 3-Phenanthrenethiol

$C_{14}H_{10}S_4$
- Dithiobenzoic Acid, *Disulphide*

$C_{14}H_{11}Br$
- 1-Bromo-1,2-diphenylethylene
- 1-Bromo-2,2-diphenylethylene
- 1-*p*-Bromophenyl-1-phenylethylene
- 1-*p*-Bromophenyl-2-phenylethylene

$C_{14}H_{11}BrO$
- α-Bromo-α-phenylacetophenone

$C_{14}H_{11}BrO_2$
- 4-Bromobiphenyl-3-carboxylic Acid, *Me ester*
- 4′-Bromobiphenyl-4-carboxylic Acid, *Me ester*
- 5-Bromobiphenyl-2-carboxylic Acid
- 3-Bromo-4-hydroxybenzophenone, *Me ether*
- 4-Bromo-4′-hydroxybenzophenone, *Me ether*

$C_{14}H_{11}BrO_3$
- *p*-Bromophenoxyacetic Acid, *Phenyl ester*

$C_{14}H_{11}Cl$
- 1-Chloro-1,2-diphenylethylene
- 1-*o*-Chlorophenyl-2-phenylethylene
- 1-*m*-Chlorophenyl-2-phenylethylene
- 1-*p*-Chlorophenyl-2-phenylethylene

$C_{14}H_{11}ClN_2O_4S$
- Chlorthalidone

$C_{14}H_{11}ClO$
- 4-Chloromethylbenzophenone
- 2-Chloro-4′-methylbenzophenone
- 2-Chloro-5-methylbenzophenone
- 3-Chloro-4-methylbenzophenone
- 3-Chloro-4′-methylbenzophenone
- 4-Chloro-4′-methylbenzophenone
- 4′-Chloro-2-methylbenzophenone
- 5-Chloro-2-methylbenzophenone
- α-Chloro-α-phenylacetophenone
- 2-Chloro-α-phenylacetophenone
- 3-Chloro-α-phenylacetophenone
- 4-Chloro-α-phenylacetophenone
- α-2-Chlorophenylacetophenone
- α-3-Chlorophenylacetophenone
- α-4-Chlorophenylacetophenone
- Diphenylacetic Acid, *Chloride*

$C_{14}H_{11}ClO_2$
- α-Chlorodiphenylacetic Acid
- 2-Chloro-4′-hydroxybenzophenone, *Me ether*
- 3-Chloro-2-hydroxybenzophenone, *Me ether*
- 3-Chloro-4-hydroxybenzophenone, *Me ether*
- 4-Chloro-4′-hydroxybenzophenone, *Me ether*
- 4′-Chloro-3-hydroxybenzophenone, *Me ether*
- 5-Chloro-2-hydroxybenzophenone, *Me ether*

$C_{14}H_{11}ClO_3$
- 3-Chloro-6-hydroxy-*p*-toluic Acid, *Phenyl ester*

$C_{14}H_{11}ClO_4S$
- *o*-Sulphobenzoic Acid, 1-o-*Tolyl ester*

$C_{14}H_{11}N$
- 1-Aminoanthracene
- 2-Aminoanthracene
- 9-Aminoanthracene
- 1-Aminophenanthrene
- 2-Aminophenanthrene
- 3-Aminophenanthrene
- 4-Aminophenanthrene
- 9-Aminophenanthrene
- Diphenylacetic Acid, *Nitrile*
- Diphenylmethane-2-carboxylic Acid, *Nitrile*
- Diphenylmethane-4-carboxylic Acid, *Nitrile*
- 1-Methylacridine
- 2-Methylacridine
- 3-Methylacridine
- 4-Methylacridine
- 9-Methylacridine
- 2-Methylbenzo[*f*]quinoline
- 4-Methylbenzo[*f*]quinoline
- 2-Methylbenzo[*g*]quinoline
- 4-Methylbenzo[*g*]quinoline
- 2-Methylbenzo[*h*]quinoline
- 4-Methylbenzo[*h*]quinoline
- 2-Methylphenanthridine
- 3-Methylphenanthridine
- 4-Methylphenanthridine
- 6-Methylphenanthridine
- 8-Methylphenanthridine
- 1-Phenylindole
- 2-Phenylindole
- 3-Phenylindole
- 6-Phenylindole

$C_{14}H_{11}NO$
- Acridone, *N*-Me
- 1-Amino-2-hydroxyphenanthrene
- 2-Amino-9-hydroxyphenanthrene
- 2-Amino-10-hydroxyphenanthrene
- 4-Amino-1-hydroxyphenanthrene
- 4-Amino-3-hydroxyphenanthrene
- 9-Amino-3-hydroxyphenanthrene
- 10-Amino-9-hydroxyphenanthrene
- Fluorene-9-carboxylic Acid, *Amide*
- 4-Hydroxyacridine, *Me ether*
- 2-Hydroxy-1-anthramine
- 1-Hydroxy-2-anthramine
- *o*-Hydroxymethyl-benzoic Acid, *Phenyl ether*, *Nitrile*
- 4-Hydroxyphenanthridine, *Me ether*
- 1-Methylacridone
- 2-Methylacridone
- 3-Methylacridone
- 4-Methylacridone
- 10-Methylacridone
- 1-Phenyloxindole
- 3-Phenyloxindole
- 7-Phenyloxindole
- *N*-Phenylphthalimidine
- 3-Phenylphthalimidine

$C_{14}H_{11}NO_2$
- Benzil monoxime
- *o*-Benzoylbenzoic Acid, *Amide*
- Dibenzamide
- 2-Hydroxyacridone, *Me ether*
- 4-Hydroxyacridone, *Me ether*
- 10-Hydroxyacridone, *Me ether*
- 9-Hydroxyfluorene-4-carboxylic Acid, *Amide*
- 3,4-Methylenedioxybenzaldehyde, *Anil*
- Murrayanine†
- Naphthalene-1,8-dicarboxylic Acid, N-*Et imide*

$C_{14}H_{11}NO_2$ (*continued*)
2-Nitrostilbene
3-Nitrostilbene
4-Nitrostilbene
N-Phenyl-3-hydroxyphthalimidine†
Succinimide, N-1-*Naphthyl*
Succinimide, N-2-*Naphthyl*

$C_{14}H_{11}NO_2S$
Anthracene-1-sulphonic Acid, *Amide*
Anthracene-2-sulphonic Acid, *Amide*
Phenanthrene-2-sulphonic Acid, *Amide*
Phenanthrene-3-sulphonic Acid, *Amide*
Phenanthrene-9-sulphonic Acid, *Amide*

$C_{14}H_{11}NO_3$
4-Acetyl-2′-nitrobiphenyl
4-Acetyl-3′-nitrobiphenyl
4-Acetyl-4′-nitrobiphenyl
4-Amino-2-benzoylbenzoic Acid
5-Amino-2-benzoylbenzoic Acid
2-(3-Aminobenzoyl)benzoic Acid
4-(3-Aminobenzoyl)benzoic Acid
4-(4-Aminobenzoyl)benzoic Acid
Benzoylanthranilic Acid
Dibenzhydroxamic Acid
1,3-Dihydroxyacridone, *Mono-Me ether*
Diphenic Acid, *Mono-amide*
2-Hydroxy-2′-nitrostilbene
2-Hydroxy-3′-nitrostilbene
2-Hydroxy-4′-nitrostilbene
2-Hydroxy-β-nitrostilbene
3-Hydroxy-4′-nitrostilbene
4-Hydroxy-2-nitrostilbene
4-Hydroxy-2′-nitrostilbene
4-Hydroxy-3′-nitrostilbene
4-Hydroxy-4′-nitrostilbene
4-Hydroxy-β-nitrostilbene
α-Hydroxy-4′-nitrostilbene
α-Hydroxy-β-nitrostilbene
3-Methyl-4-nitrobenzophenone
4-Methyl-2-nitrobenzophenone
4-Methyl-2′-nitrobenzophenone
4-Methyl-3-nitrobenzophenone
4-Methyl-3′-nitrobenzophenone
4-Methyl-4′-nitrobenzophenone
α-Nitro-α-phenylacetophenone
α-*o*-Nitrophenylacetophenone
α-*p*-Nitrophenylacetophenone
Phthalanilic Acid
Resorufin, *Et ether*

$C_{14}H_{11}NO_4$
4-Acetyl-4′-nitrodiphenyl Ether
2-Benzamido-3-(2-furyl)acrylic Acid
Disalicylamide
2-Hydroxy-4-methyl-5-nitrobenzophenone
2-Hydroxy-5-methyl-4′-nitrobenzophenone
2-Hydroxy-4′-nitrobenzophenone, *Me ether*
4-Hydroxy-3-nitrobenzophenone, *Me ether*
4-Hydroxy-3′-nitrobenzophenone, *Me ether*
4-Hydroxy-4′-nitrobenzophenone, *Me ether*
α-Hydroxy-α-*p*-nitrophenylacetophenone
2′-Methyl-6-nitrobiphenyl-2-carboxylic Acid
2′-Methyl-6′-nitrobiphenyl-2-carboxylic Acid
6-Methyl-2′-nitrobiphenyl-2-carboxylic Acid
2-Methyl-6-phenylpyridine-3,4-dicarboxylic Acid
p-Nitrobenzoic Acid, *Benzyl ester*
3-Nitrobiphenyl-2-carboxylic Acid, *Me ester*
4-Nitrobiphenyl-2-carboxylic Acid, *Me ester*
4′-Nitrobiphenyl-2-carboxylic Acid, *Me ester*
4′-Nitrobiphenyl-3-carboxylic Acid, *Me ester*
4′-Nitrobiphenyl-4-carboxylic Acid, *Me ester*
Resazurin, *Et ether*
O-Salicyloylsalicylic Acid, *Amide*

$C_{14}H_{11}NO_4S$
2-*o*-Nitrophenylmercaptobenzoic Acid, *Me ester*
2-*m*-Nitrophenylmercaptobenzoic Acid, *Me ester*
2-*p*-Nitrophenylmercaptobenzoic Acid, *Me ester*

$C_{14}H_{11}NO_5$
Flindersiamine
4-Hydroxy-3-nitrobenzoic Acid, 3-*Methylbutyl ester*, *Benzyl ester*
2-(*p*-Nitrophenoxy)benzoic Acid, *Me ester*
4-(*o*-Nitrophenoxy)benzoic Acid, *Me ester*
4-(*p*-Nitrophenoxy)benzoic Acid, *Me ester*

$C_{14}H_{11}NO_5S$
2-*o*-Nitrophenylsulphinylbenzoic Acid, *Me ester*
2-*m*-Nitrophenylsulphinylbenzoic Acid, *Me ester*
2-*p*-Nitrophenylsulphinylbenzoic Acid, *Me ester*

$C_{14}H_{11}NO_6S$
2-*o*-Nitrophenylsulphonylbenzoic Acid, *Me ester*
2-*p*-Nitrophenylsulphonylbenzoic Acid, *Me ester*

$C_{14}H_{11}NS$
9-Acridinethiol, *S*-Me

$C_{14}H_{11}N_3$
1,4-Diphenyl-1,2,3-triazole
1,5-Diphenyl-1,2,3-triazole
2,4-Diphenyl-1,2,3-triazole
4,5-Diphenyl-1,2,3-triazole
1,5-Diphenyl-1,2,4-triazole
2,3-Diphenyl-1,2,4-triazole
2,5-Diphenyl-1,2,4-triazole
3,4-Diphenyl-1,2,4-triazole
3,5-Diphenyl-1,2,4-triazole

$C_{14}H_{11}N_3O$
Benzo[*f*]quinoline-1-carboxylic Acid, *Hydrazide*
Benzo[*f*]quinoline-3-carboxylic Acid, *Hydrazide*
Benzo[*f*]quinoline-5-carboxylic Acid, *Hydrazide*
2-Phenylindazole-3-carboxylic Acid, *Amide*

$C_{14}H_{11}N_3O_2$
Aeruginosin A†
3,5-Dioxo-1,4-diphenyl-1,2,4-triazolidine
4-Nitroindazole, 1-o-*Tolyl*
4-Nitroindazole, 1-p-*Tolyl*
4-Nitroindazole, 1-*Benzyl*
4-Nitroindazole, 2-*Benzyl*
6-Nitroindazole, N-*Benzyl*
1,2,3-Triaminoanthraquinone
1,2,4-Triaminoanthraquinone
1,4,5-Triaminoanthraquinone

$C_{14}H_{11}N_3O_3$
Azobenzene-2,2′-dicarboxylic Acid, *Monoamide*

$C_{14}H_{11}N_3O_4$
Diazoaminobenzene-2,2′-dicarboxylic Acid
Diazoaminobenzene-3,3′-dicarboxylic Acid
Diazoaminobenzene-4,4′-dicarboxylic Acid

$C_{14}H_{11}N_3O_5$
4-Hydroxy-3′-nitroazobenzene-3-carboxylic Acid, *Me ester*
4-Hydroxy-4′-nitroazobenzene-3-carboxylic Acid, *Me ester*
4-Hydroxy-5-nitroazobenzene-3-carboxylic Acid, *Me ester*

$C_{14}H_{11}N_3O_5S$
Aeruginosin B†

$C_{14}H_{11}N_3O_6$
2′,4′-Dinitrodiphenylamine-3-carboxylic Acid, *Me ester*
2′,4′-Dinitrodiphenylamine-4-carboxylic Acid, *Me ester*

$C_{14}H_{12}$
9,10-Dihydroanthracene
6,7-Dihydrocyclohepta[*d,e*]naphthalene†
9,10-Dihydrophenanthrene
1,1-Diphenylethylene
Heptafulvalene†
1-Methylfluorene
2-Methylfluorene
3-Methylfluorene
4-Methylfluorene
9-Methylfluorene
Monodehydro[14]annulene†
Phenylcyclo-octatetraene
Pyracene†
Stilbene

$C_{14}H_{12}As_2$
5,10-Ethanoarsanthrene

$C_{14}H_{12}BN_2$
1,2,3,6-Tetrahydro-4,5,7,8-dibenzo[2,1,3]boradiazapentalene

$C_{14}H_{12}Br_2$
2,2′-Dibromobibenzyl
4,4′-Dibromobibenzyl
1,2-Dibromo-1,2-diphenylethane

$C_{14}H_{12}ClNO$
α-Chlorodiphenylacetic Acid, *Amide*

$C_{14}H_{12}Cl_2$
1,2-Dichloro-1,2-diphenylethane

$C_{14}H_{12}Cl_2O$
1,1-Di-(*p*-chlorophenyl)ethanol

$C_{14}H_{12}F_{14}O_4$
Heptafluorobutyric Acid, *Hexamethylene ester*

$C_{14}H_{12}N_2$
4-Aminoacridine, *N*-Me
1-(3-Aminophenyl)-2-(4-aminophenyl)acetylene
α-Anilinophenylacetic Acid, *Nitrile*
Benzylideneazine
1,4-Diaminoanthracene
9,10-Diaminoanthracene
9,10-Diaminophenanthrene
Di-(2-aminophenyl)acetylene
Di-(4-aminophenyl)acetylene
3,4-Dihydro-3-phenylquinazoline

$C_{14}H_{12}N_2$ (di-ion)
5,6-Dihydropyrazino[1,2,3,4-*lmn*][1,10]phenanthrolinium Dication

$C_{14}H_{12}N_2O$
2-Hydroxyphenazine, *Et ether*

$C_{14}H_{12}N_2O_2$
1-Amino-1,2-diphenyl-2-nitroethylene
1-(2-Amino-4-nitrophenyl)-2-phenylethylene
1-(4-Amino-2-nitrophenyl)-2-phenylethylene
1-(2-Aminophenyl)-2-(2-nitrophenyl)ethylene
1-(4-Aminophenyl)-2-(4-nitrophenyl)ethylene
Azobenzene-2-carboxylic Acid, *Me ester*
Azobenzene-4-carboxylic Acid, *Me ester*
Benzil dioxime
Benzoylanthranilic Acid, *Amide*
m-Benzylidenehydrazinobenzoic Acid
p-Benzylidenehydrazinobenzoic Acid
Biphenyl-2,3′-dicarboxylic Acid, *Diamide*
1,2-Dibenzoylhydrazine
Diphenic Acid, *Diamide*
2′-Methylazobenzene-2-carboxylic Acid
4′-Methylazobenzene-2-carboxylic Acid
6-Methylazobenzene-2-carboxylic Acid
2′-Methylazobenzene-3-carboxylic Acid
3′-Methylazobenzene-3-carboxylic Acid
4′-Methylazobenzene-3-carboxylic Acid
2-*o*-Nitrobenzylideneaminotoluene
2-*m*-Nitrobenzylideneaminotoluene
3-*o*-Nitrobenzylideneaminotoluene
4-*o*-Nitrobenzylideneaminotoluene
4-*m*-Nitrobenzylideneaminotoluene
4-*p*-Nitrobenzylideneaminotoluene
3-Nitrocarbazole, N-*Et*
Oxanilide
Salazine

$C_{14}H_{12}N_2O_2S_2$
Diphenyl disulphide 2,2′-dicarboxylic Acid, *Diamide*

$C_{14}H_{12}N_2O_3$
Anthranoylanthranilic Acid
2-Benzamido-3-(2-furyl)acrylic Acid, *Amide*
4-Hydroxyazobenzene-3-carboxylic Acid, *Me ether*
4-Hydroxyazobenzene-3-carboxylic Acid, *Me ester*
2-*N*-Methylamino-5-nitrosobenzoic Acid, *Phenyl ester*
2-Methyl-6-phenylpyridine-3,4-dicarboxylic Acid, 4-*Amide*
o-Nitrobenzoic Acid, N-*Me-anilide*
2-*o*-Nitrobenzylideneaminoanisole
2-*p*-Nitrobenzylideneaminoanisole
3-*o*-Nitrobenzylideneaminoanisole
4-*o*-Nitrobenzylideneaminoanisole
4-*m*-Nitrobenzylideneaminoanisole
4-*p*-Nitrobenzylideneaminoanisole
p-Nitrophenylacetic Acid, *Anilide*
2,2′-Oxydibenzoic Acid, *Diamide*
Pyrogalline

$C_{14}H_{12}N_2O_4$
2-Amino-5-nitrobenzoic Acid, N-o-*Tolyl*
2-Amino-5-nitrobenzoic Acid, N-p-*Tolyl*

$C_{14}H_{12}N_2O_4$ (*continued*)
2,2′-Dimethyl-4,4′-dinitrobiphenyl
2,2′-Dimethyl-5,5′-dinitrobiphenyl
2,2′-Dimethyl-6,6′-dinitrobiphenyl
3,3′-Dimethyl-2,2′-dinitrobiphenyl
3,3′-Dimethyl-4,4′-dinitrobiphenyl
4,4′-Dimethyl-2 2′-dinitrobiphenyl
4,4′-Dimethyl-3,3′-dinitrobiphenyl
5,5′-Dimethyl-2,2′-dinitrobiphenyl
α,β-Dinitrobibenzyl
α,4-Dinitrobibenzyl
2,2′-Dinitrobibenzyl
2,4′-Dinitrobibenzyl
4,4′-Dinitrobibenzyl
2,2′-Hydrazodibenzoic Acid
2,3′-Hydrazodibenzoic Acid
3,3′-Hydrazodibenzoic Acid
4,4′-Hydrazodibenzoic Acid
N-Methyl-2′-nitrodiphenylamine-2-carboxylic Acid
2′-Methyl-4-nitrodiphenylamine-2-carboxylic Acid
2′-Methyl-5′-nitrodiphenylamine-2-carboxylic Acid
3′-Methyl-4-nitrodiphenylamine-2-carboxylic Acid
4-Methyl-6-nitrodiphenylamine-2-carboxylic Acid
4′-Methyl-4-nitrodiphenylamine-2-carboxylic Acid
5-Methyl-4-nitrodiphenylamine-2-carboxylic Acid
2′-Methyl-2-nitrodiphenylamine-4-carboxylic Acid
4′-Methyl-2-nitrodiphenylamine-4-carboxylic Acid
N-Methylphenylcarbamic Acid, o-*Nitrophenyl ester*
Myxin, 1-*Me ether*†
5-Nitrodiphenylamine-2-carboxylic Acid, *Me ester*
2-Nitrodiphenylamine-4-carboxylic Acid, *Me ester*
o-Nitrophenoxyacetic Acid, *Anilide*
m-Nitrophenoxyacetic Acid, *Anilide*
p-Nitrophenoxyacetic Acid, *Anilide*

$C_{14}H_{12}N_2O_4S_2$
Anthracene-1,5-disulphonic Acid, *Diamide*
Anthracene-1,8-disulphonic Acid, *Diamide*

$C_{14}H_{12}N_2O_5$
2-(4-Hydroxy-2-nitroanilino)benzoic Acid, *Me ether*

$C_{14}H_{12}N_2O_6$
2,2′-Dihydroxy-3,5′-dinitrobiphenyl, *Di-Me ether*
2,2′-Dihydroxy-5,5′-dinitrobiphenyl, *Di-Me ether*
2,2′-Dihydroxy-5,5′-dinitrobiphenyl, *Et ether*
4,4′-Dihydroxy-3,3′-dinitrobiphenyl, *Di-Me ether*

$C_{14}H_{12}N_2S$
2-*o*-Aminophenyl-4-methylbenzothiazole
2-*p*-Aminophenyl-6-methylbenzothiazole
2-Mercaptobenzimidazole, 1-N-*Benzyl*
2-Mercaptobenzimidazole, S-*Benzyl*

$C_{14}H_{12}N_2S_2$
2-Benzothiazolesulphenamide, N-*Benzyl*

$C_{14}H_{12}N_4$
5-Amino-1-phenyl-1,2,3-triazole, 5-N-*Phenyl*

$C_{14}H_{12}N_4O_2$
Azobenzene-2,2′-dicarboxylic Acid, *Diamide*
Azoformic Acid, *Dianilide*
Isonitrosoacetylaminoazobenzene
1,4,5,8-Tetra-aminoanthraquinone

$C_{14}H_{12}N_4O_3$
10-Formylmethyl-7,8-dimethylisoalloxazine†

$C_{14}H_{12}N_4O_6$
Indigoidin, *Di-Ac*†

$C_{14}H_{12}N_6O_3$
Pteroic Acid

$C_{14}H_{12}O$
2-Acetylbiphenyl
3-Acetylbiphenyl
4-Acetylbiphenyl
Deoxybenzoin
2,3-Dihydro-2-phenylbenzo[*b*]furan
2,3-Dihydro-3-phenylbenzo[*b*]furan
3,4-Dihydro-1(2*H*)-phenanthrone
2,3-Dihydro-4(1*H*)-phenanthrone
Diphenylacetaldehyde
2,2-Diphenyloxiran
2,3-Diphenyloxiran
1-Fluorenol, *Me ether*
2-Fluorenol, *Me ether*
3-Fluorenol, *Me ether*
9-Fluorenol, *Me ether*
2-Hydroxystilbene
3-Hydroxystilbene
4-Hydroxystilbene
2-Methyl-9-fluorenol
3-Methyl-9-fluorenol
4-Methyl-9-fluorenol
9-Methyl-9-fluorenol
3-Methylxanthene
Phenyl *o*-tolyl Ketone
Phenyl *m*-tolyl Ketone
Phenyl *p*-tolyl Ketone
1-Phenylvinyl Alcohol, *Phenyl ether*
2-Phenylvinyl Alcohol, *Phenyl ether*
α-Stilbenol

$C_{14}H_{12}OSe$
Selenobenzoic Acid, p-*Tolyl ester*

$C_{14}H_{12}O_2$
Acenaphthene-3-carboxylic Acid, *Me ester*
Benzoin
Benzoic Acid, *Benzyl ester*
Benzyl benzoate
Biphenyl-2-carboxylic Acid, *Me ester*
Biphenyl-3-carboxylic Acid *Me ester*
Biphenyl-4-carboxylic Acid *Me ester*
2,3-Dihydrophenalene-2-carboxylic Acid†
9,10-Dihydrophenanthrene-9,10-diol†
α,β-Dihydroxystilbene
2,2′-Dihydroxystilbene
3,5-Dihydroxystilbene
4,4′-Dihydroxystilbene
Diphenylacetic Acid
Diphenylglycollaldehyde

$C_{14}H_{12}O_2$ (*continued*)
Diphenylmethane-2-carboxylic Acid
Diphenylmethane-3-carboxylic Acid
Diphenylmethane-4-carboxylic Acid
4-Hydroxyacetophenone, *Phenyl ether*
α-Hydroxyacetophenone, *Phenyl ether*
p-Hydroxybenzaldehyde, *Benzyl ether*
2-Hydroxybenzophenone, *Me ether*
3-Hydroxybenzophenone, *Me ether*
4-Hydroxybenzophenone, *Me ether*
2-Hydroxy-4-methylbenzophenone
2-Hydroxy-5-methylbenzophenone
2′-Hydroxy-4-methylbenzophenone
4-Hydroxy-2-methylbenzophenone
4′-Hydroxy-2-methylbenzophenone
4-Hydroxy-3-methylbenzophenone
4′-Hydroxy-4-methylbenzophenone
4-Hydroxymethyl-benzophenone
2-Hydroxy-α-phenylacetophenone
4-Hydroxy-α-phenylacetophenone
α-3-Hydroxyphenylacetophenone
α-4-Hydroxyphenylacetophenone
Marmesin, *Anhydro*
2′-Methylbiphenyl-2-carboxylic Acid
4′-Methylbiphenyl-2-carboxylic Acid
3′-Methylbiphenyl-3-carboxylic Acid
2′-Methylbiphenyl-4-carboxylic Acid
2-Methylbiphenyl-4-carboxylic Acid
3-Methylbiphenyl-4-carboxylic Acid
4′-Methylbiphenyl-4-carboxylic Acid
Mycomycin, *Me ester*
Phenylacetic Acid, *Phenyl ester*
p-Phenylphenacyl Alcohol
Salicylaldehyde, *Benzyl ether*
1,2,3,4-Tetrahydroanthraquinone
1,4,4α,9α-Tetrahydroanthraquinone
o-Toluic Acid, *Phenyl ester*
m-Toluic Acid, *Phenyl ester*
p-Toluic Acid, *Phenyl ester*

$C_{14}H_{12}O_2S$
Thiobenzilic Acid

$C_{14}H_{12}O_2S_2$
Dibenzo[1,3,6,8]dioxadithiecin

$C_{14}H_{12}O_2Se$
Selenobenzoic Acid, p-*Methoxyphenyl ester*

$C_{14}H_{12}O_3$
Anisic Acid, *Phenyl ester*
Benzilic Acid
3-Benzylsalicylic Acid
5-Benzylsalicylic Acid
Desmethylfrutescin
2,4-Dihydroxybenzophenone, 4-*Me ether*
2,5-Dihydroxybenzophenone, 5-*Me ether*
3,4-Dihydroxybenzophenone, *Me ether*
4,4′-Dihydroxybenzophenone, *Me ether*
1,2-Dihydroxydibenzofuran, *Di-Me ether*
1,4-Dihydroxydibenzofuran, *Di-Me ether*
2,4-Dihydroxy-2′-methylbenzophenone
2,4-Dihydroxy-3-methylbenzophenone
2,4-Dihydroxy-4′-methylbenzophenone
2,4-Dihydroxy-6-methylbenzophenone
2,4′-Dihydroxy-5-methylbenzophenone
2,6-Dihydroxy-2′-methylbenzophenone
2,6-Dihydroxy-4′-methylbenzophenone
3,4-Dihydroxy-2′-methylbenzophenone
3,4′-Dihydroxy-4-methylbenzophenone
3-α-Furylacrylic Acid, *Benzyl ester*
6-Hydroxybiphenyl-2-carboxylic Acid, *Me ester*
4′-Hydroxybiphenyl-4-carboxylic Acid, *Me ester*
o-Hydroxymethyl-benzoic Acid, *Phenyl ether*
α-2-Hydroxyphenoxyacetophenone
α-4-Hydroxyphenoxyacetophenone
2-Hydroxy-*m*-toluic Acid, *Phenyl ester*
6-Hydroxy-*m*-toluic Acid, *Phenyl ester*
2-Hydroxy-*p*-toluic Acid *Phenyl ester*
Mandelic Acid, *Phenyl ether*
o-Methoxybenzoic Acid, *Phenyl ester*
m-Methoxybenzoic Acid, *Phenyl ester*
Methyl salicylate, *Phenyl ether*
3-α-Naphthoylpropionic Acid
3-β-Naphthoylpropionic Acid
1-Naphthylglyoxylic Acid, *Et ester*
Phenoxyacetic Acid, *Phenyl ester*
o-Phenoxybenzoic Acid, *Me ester*
m-Phenoxybenzoic Acid, *Me ester*
p-Phenoxybenzoic Acid, *Me ester*
Salicylic Acid, *Benzyl ester*
Salicylic Acid, o-*Tolyl ester*
Salicylic Acid, m-*Tolyl ester*
Salicylic Acid, p-*Tolyl ester*
Salicylic Acid, *Benzyl ether*
Salol, *Me ether*
Seselin
3,4′,5-Trihydroxystilbene
4,5′,8-Trimethylpsoralen
Xanthyletin

$C_{14}H_{12}O_4$
Cotoin
2,5-Dihydroxybenzoic Acid, 5-*Me*-2-*Phenyl ether*
6,7-Dihydro-8,8-dimethyl-2*H*,8*H*-benzo-[1,2-*b*:5,4-*b*′]dipyran-2,6-dione†
2,4-Dihydroxy-6-methoxybenzophenone
Diphenyloxyacetic Acid
Discophoridin
Eleutherol
Fulvoplumierin
Guaiacol, *Phenyl ester*
Naphthalene-1,2-dicarboxylic Acid, *Di-Me ester*
Naphthalene-1,4-dicarboxylic Acid, *Di-Me ester*
Naphthalene-1,5-dicarboxylic Acid, *Di-Me ester*
Naphthalene-1,6-dicarboxylic Acid, *Di-Me ester*
Naphthalene-1,7-dicarboxylic Acid, *Di-Me ester*
Naphthalene-1,8-dicarboxylic Acid, *Di-Me ester*
Naphthalene-2,3-dicarboxylic Acid, *Di-Me ester*
Naphthalene-2,6-dicarboxylic Acid, *Di-Me* ester
Naphthalene-2,7-dicarboxylic Acid, *Di-Me ester*
2α-Naphthylsuccinic Acid
Obliquin†

$C_{14}H_{12}O_4$ (*continued*)
Oroselol †
Piceatannol
Podophyllomeronic Acid, *Me ester*
Stypandrone, *Me ether* †
Tetradec-9-ene-2,4,6-triynedioic Acid
3,3′,4,5′-Tetrahydroxystilbene †
α,2,4-Trihydroxyacetophenone, *α-Phenyl ether*
2,3,4-Trihydroxybenzophenone, 3-(or-4)-*Me ether*
2,4,4′-Trihydroxybenzophenone, 4′-*Me ether*
2,4,5-Trihydroxybenzophenone, 5-*Me ether*
2,4,6-Trihydroxy-3-methylbenzophenone
Yangonalactone

$C_{14}H_{12}O_4S$
o-Phenylsulphonylbenzoic Acid, *Me ester*

$C_{14}H_{12}O_5$
Clausenin †
5,7-Dimethoxy-9-oxocyclopenteno[*c*]coumarin †
1-Hydroxynaphthalene-2,4-dicarboxylic Acid, *Di-Me ester*
Khellin
Purpurogenone
2,3′,4,4′-Tetrahydroxybenzophenone, 3′-*Me ether*

$C_{14}H_{12}O_6$
Halfordin †
Isohalfordin
Maclurin, 3′-*Me ether*

$C_{14}H_{12}O_7$
4-Carboxy-2-oxo-3-phenyl-3-heptenedioic Acid

$C_{14}H_{12}O_8$
Fulvic Acid
Pleomycin
Xanthophanic Acid †

$C_{14}H_{12}O_9$
Clavorubin

$C_{14}H_{12}S_2$
2,7-Dimethylthianthrene †
Dithiobenzoic Acid, *Benzyl ester*

$C_{14}H_{13}Br$
1-*p*-Bromophenyl-2-phenylethane

$C_{14}H_{13}BrO$
3-Bromo-5-hydroxybiphenyl, *Et ether*

$C_{14}H_{13}BrO_2$
Bromoacetaldehyde, *Di-phenyl acetal*
2-Bromobutyric Acid, 1-*Naphthol ester*

$C_{14}H_{13}ClN_2O_2$
4-Chloro-2-nitrodiphenylamine, N-*Et*

$C_{14}H_{13}ClO$
2-Chloro-1,1-diphenylethanol
2-Methyl-3-α-naphthylpropionic Acid, *Chloride*
2-Methyl-3-β-naphthylpropionic Acid, *Chloride*

$C_{14}H_{13}Cl_2N$
2-Chloro-*N*-3-chlorobenzylbenzylamine
2-Chloro-*N*-4-chlorobenzylbenzylamine
3-Chloro-*N*-4-chlorobenzylbenzylamine
Di-(2-chlorobenzyl)amine
Di-(3-chlorobenzyl)amine
Di-(4-chlorobenzyl)amine

$C_{14}H_{13}Cl_2NO_3$
Pyoluteorin★,NOO-*Tri-Me* †

$C_{14}H_{13}N$
2-Aminostilbene
3-Aminostilbene
N-Benzylidene-*o*-toluidine
N-Benzylidene-*m*-toluidine
N-Benzylidene-*p*-toluidine
6,7-Dihydro-5*H*-dibenz[*c*,*e*]azepine
2,3-Dihydro-2-phenylindole
2,3-Dihydro-7-phenylindole
1,3-Dimethylcarbazole
1,4-Dimethylcarbazole
1,8-Dimethylcarbazole
2,3-Dimethylcarbazole
2,4-Dimethylcarbazole
2,6-Dimethylcarbazole
2,7-Dimethylcarbazole
3,6-Dimethylcarbazole
2-Ethylcarbazole
3-Ethylcarbazole
9-Ethylcarbazole
2-Methylacridan
3-Methylacridan
4-Methylacridan
9-Methylacridan
3′-Methyl-2-styrylpyridine
3′-Methyl-4-styrylpyridine
4-Methyl-2-styrylpyridine
4′-Methyl-2-styrylpyridine
4′-Methyl-4-styrylpyridine
6-Methyl-2-styrylpyridine
2-α-Naphthylbutyric Acid, *Nitrile*

$C_{14}H_{13}NO$
3-Aminodibenzofuran, N-*Di-Me*
2-Amino-2′-methylbenzophenone
2-Amino-3′-methylbenzophenone
2-Amino-4-methylbenzophenone
2-Amino-4′-methylbenzophenone
2-Amino-5-methylbenzophenone
3-Amino-4-methylbenzophenone
3-Amino-4′-methylbenzophenone
4-Amino-3-methylbenzophenone
4-Amino-4′-methylbenzophenone
5-Amino-2-methylbenzophenone
α-Amino-α-phenylacetophenone
4-Amino-α-phenylacetophenone
α-Anilinoacetophenone
Benzylbenzamide
o-Benzylideneaminobenzyl Alcohol
p-Benzylideneaminobenzyl Alcohol
p-Benzylideneaminophenol, *Me ether*
Diphenylacetic Acid, *Amide*
Diphenylamine, N-*Ac*
Diphenylmethane-2-carboxylic Acid, *Amide*
Glycozoline †
o-Hydroxybenzylideneaniline, *Me ether*
o-Hydroxybenzylidene-*o*-toluidine
p-Hydroxybenzylidene-*o*-toluidine
o-Hydroxybenzylidene-*m*-toluidine
m-Hydroxybenzylidene-*m*-toluidine
p-Hydroxybenzylidene-*m*-toluidine
o-Hydroxybenzylidene-*p*-toluidine
m-Hydroxybenzylidene-*p*-toluidine
p-Hydroxybenzylidene-*p*-toluidine
2-Hydroxycarbazole, *Et ether*
2-Methylaminobenzophenone

$C_{14}H_{13}NO$ (*continued*)
4-Methylaminobenzophenone
2′-Methylbiphenyl-2-carboxylic Acid, *Amide*

$C_{14}H_{13}NO_2$
2-Amino-4′-hydroxybenzophenone, *Me ether*
α-Anilinophenylacetic Acid
Anthranilic Acid, *Benzyl ester*
Benzilic Acid, *Amide*
N-Benzoyl-2-hydroxyaniline, *Me ether*
N-Benzoyl-4-hydroxyaniline, *Me ether*
2-Cyano-5-phenyl-2,4-pentadienoic Acid, *Et ester*
2,3-Dihydroxycarbazole, *Di-Me ether*
2,2′-Dimethyl-6-nitrobiphenyl
4,4′-Dimethyl-2-nitrobiphenyl
4,4′-Dimethyl-3-nitrobiphenyl
Diphenylcarbamic Acid, *Me ester*
N-Diphenylglycine
Flindersine
Mandelamide, *Phenyl ether*
DL-Mandelanilide
o-Methoxybenzoic Acid, *Anilide*
o-Methylaminobenzoic Acid, *Phenyl ester*
2′-Methyldiphenylamine-2-carboxylic Acid
3′-Methyldiphenylamine-2-carboxylic Acid
4-Methyldiphenylamine-2-carboxylic Acid
4′-Methyldiphenylamine-2-carboxylic Acid
N-Methylphenylcarbamic Acid, *Phenyl ester*
o-Nitrophenyl-*p*-tolylmethane
N-Phenylglycine, *Phenyl ester*
2-Quinolineacrylic Acid, *Et ester*
4-Quinolinecarboxylic Acid, *Et ester*

$C_{14}H_{13}NO_3$
Diphenoxyacetic Acid, *Amide*
2-(2-Hydroxyanilino)benzoic Acid, *Me ether*
2-(3-Hydroxyanilino)benzoic Acid, *Me ether*
2-(4-Hydroxyanilino)benzoic Acid, *Me ether*
2-Hydroxy-5-nitrobiphenyl, *Et ether*
2-Nitro-1,1-diphenylethanol
2-Nitro-1,2-diphenylethanol
2-Nitro-4-tolyl 4-tolyl Ether
3-Nitro-4-tolyl 4-tolyl Ether
4-Nitro-2-tolyl 2-tolyl Ether
4-Nitro-3-tolyl 3-tolyl Ether
N-(*trans*-Non-2-ene-4,6,8-triynol)-L-valine

$C_{14}H_{13}NO_4$
2,5-Dihydroxy-3′-nitrobiphenyl, *Di-Me ether*
2,5-Dihydroxy-4-nitrobiphenyl, *Di-Me ether*
2,5-Dihydroxy-4′-nitrobiphenyl, *Di-Me ether*
Kokusaginine
Maculosidine†
2-Methylquinoline-3,4-dicarboxylic Acid, *Di-Me ester*
5-Nitro-1-naphthoic Acid, *Isopropyl ester*
5-Nitro-2-naphthoic Acid, *Isopropyl ester*
Skimmianine

$C_{14}H_{13}NO_4S$
o-Sulphobenzoic Acid, 1-o-*Tolyl ester*, *Amide*

$C_{14}H_{13}NO_5$
(1-Nitro-2-naphthoxy)acetic Acid, *Et ester*

$C_{14}H_{13}NO_5S$
2,5-Dimethyl-3-nitrobenzenesulphonic Acid, *Phenyl ester*
2,5-Dimethyl-4-nitrobenzenesulphonic Acid, *Phenyl ester*
2,5-Dimethyl-6-nitrobenzenesulphonic Acid, *Phenyl ester*
4′-Nitrobiphenyl-4-sulphonic Acid, *Et ester*

$C_{14}H_{13}NO_7$
Dhurrin
Narciclasine†

$C_{14}H_{13}NS$
Phenothiazine, N-*Et*

$C_{14}H_{13}NS_2$
Phenyldithiocarbamic Acid, *Benzyl ester*

$C_{14}H_{13}N_2O_7PS$
O-Ethyl-*O*,*O*-di-*p*-nitrophenylphosphorothionate

$C_{14}H_{13}N_3O_2$
2,4-Dimethyl-4′-nitroazobenzene
2,4′-Dimethyl-6-nitroazobenzene
3,4-Dimethyl-4′-nitroazobenzene
4,4′-Dimethyl-2-nitroazobenzene
1,1-Diphenylbiuret
1,3-Diphenylbiuret
1,5-Diphenylbiuret

$C_{14}H_{13}N_3O_3$
2-Hydroxy-3,5-dimethyl-4′-nitroazobenzene
4-Hydroxy-2,3-dimethyl-4′-nitroazobenzene
4-Hydroxy-2,5-dimethyl-4′-nitroazobenzene
4-Hydroxy-2,6-dimethyl-4′-nitroazobenzene
4-Hydroxy-3,5-dimethyl-4′-nitroazobenzene
4-Hydroxy-2-methyl-4′-nitroazobenzene, *Me ether*
4′-Hydroxy-2-methyl-5-nitroazobenzene, *Me ether*

$C_{14}H_{13}N_3O_4$
3,4-Dihydroxy-2′-nitroazobenzene, *Di-Me ether*

$C_{14}H_{13}N_3O_5$
4-Amino-2,6-dinitrophenol, N-*Me*, N-*Benzyl*

$C_{14}H_{13}N_3S$
2-Mercapto-1-(2-4′-pyridylethyl)benzimidazole†

$C_{14}H_{13}N_5O_2$
N-(6-Purinyl)-β-phenylalanine†

$C_{14}H_{13}N_7O_2$
Aminopteroic Acid

$C_{14}H_{14}$
[14]Annulene†
Bibenzyl
2,2′-Dimethylbiphenyl
2,3-Dimethylbiphenyl
2,3′-Dimethylbiphenyl
2,4-Dimethylbiphenyl
2,4′-Dimethylbiphenyl
2,6-Dimethylbiphenyl
3,3′-Dimethylbiphenyl
3,4-Dimethylbiphenyl
3,4′-Dimethylbiphenyl
3,5-Dimethylbiphenyl
4,4′-Dimethylbiphenyl
1,1-Diphenylethane
2-Ethylbiphenyl
3-Ethylbiphenyl
4-Ethylbiphenyl

$C_{14}H_{14}$ (*continued*)
1-Isopropenyl-4-methylnaphthalene
3-Isopropenyl-1-methylnaphthalene
7-Isopropenyl-1-methylnaphthalene
Phenyl-*m*-tolylmethane
Phenyl-*p*-tolylmethane
1,2,3,4-Tetrahydroanthracene
1,2,3,4-Tetrahydrophenanthrene

$C_{14}H_{14}As_2$
o-Arsenotoluene
m-Arsenotoluene
p-Arsenotoluene

$C_{14}H_{14}BrN$
2-Amino-3′-bromobiphenyl, N-*Di-Me*

$C_{14}H_{14}Br_2O_2$
2,6-Dibromo-1,5-dihydroxynaphthalene, *Di-Et ether*

$C_{14}H_{14}Br_4$
2,3,4,5-Tetrabromo-1-(2,3-dipropylcyclopropenylidene)cyclopentadiene†

$C_{14}H_{14}ClO_3P$
Dibenzyl phosphorochloridate

$C_{14}H_{14}Cl_4$
2,3,4,5-Tetrachloro-1-(2,3-dipropylcyclopropenylidene)cyclopentadiene†

$C_{14}H_{14}FeO_4$
1,1′-Ferrocenedicarboxylic Acid, *Di-Me ester*

$C_{14}H_{14}Hg$
Mercury dibenzyl
Mercury di-*o*-tolyl
Mercury di-*m*-tolyl
Mercury di-*p*-tolyl

$C_{14}H_{14}I_4O_4$
Tetraiodoterephthalic Acid, *Dipropyl ester*

$C_{14}H_{14}NO_4PS$
O-Ethyl-*O*-*p*-nitrophenylphenylphosphonothionate†

$C_{14}H_{14}N_2$
Benzylidenemethylphenylhydrazine
2,2′-Diaminostilbene
2,4-Diaminostilbene
2,4′-Diaminostilbene
4,4′-Diaminostilbene
2,2′-Dimethylazobenzene
2,3′-Dimethylazobenzene
2,4-Dimethylazobenzene
2,4′-Dimethylazobenzene
3,3′-Dimethylazobenzene
3,4′-Dimethylazobenzene
3,5-Dimethylazobenzene
4,4′-Dimethylazobenzene
N-Methyldiphenylformamidine
Naphazoline
p-Tolamidine, β-N-*Phenyl*

$C_{14}H_{14}N_2O$
Anthranilic Acid, o-*Toluidide*
Anthranilic Acid, m-*Toluidide*
Anthranilic Acid, p-*Toluidide*
1-Benzyl-3-phenylurea
2,2′-Dimethylazoxybenzene
3,3′-Dimethylazoxybenzene
4,4′-Dimethylazoxybenzene
Hydrazobenzene, N-*Ac*
2-Hydroxyazobenzene, *Et ether*
3-Hydroxyazobenzene, *Et ether*
4-Hydroxyazobenzene, *Et ether*
2-Hydroxy-3,5-dimethylazobenzene
2-Hydroxy-4,4′-dimethylazobenzene
2-Hydroxy-4′,5-dimethylazobenzene
4-Hydroxy-2,2′-dimethylazobenzene
4-Hydroxy-2′,3-dimethylazobenzene
4-Hydroxy-2,3-dimethylazobenzene
4-Hydroxy-2,3′-dimethylazobenzene
4-Hydroxy-2,4′-dimethylazobenzene
4′-Hydroxy-2,4-dimethylazobenzene
4-Hydroxy-2,5-dimethylazobenzene
4-Hydroxy-2,6-dimethylazobenzene
4-Hydroxy-3,3′-dimethylazobenzene
4-Hydroxy-3,4′-dimethylazobenzene
4-Hydroxy-3,5-dimethylazobenzene
5-Hydroxy-2,4-dimethylazobenzene
4′-Hydroxy-2-methylazobenzene, *Me ether*
4-Hydroxy-4′-methylazobenzene, *Me ether*
N-Methylcarbanilide
N-Phenyl-*N′*-*o*-tolylurea
N-Phenyl-*N′*-*m*-tolylurea
N-Phenyl-*N′*-*p*-tolylurea

$C_{14}H_{14}N_2OS$
1-*o*-Hydroxyphenyl-3-phenylthiourea, *Me ether*
1-*p*-Hydroxyphenyl-3-phenylthiourea, *Me ether*

$C_{14}H_{14}N_2O_2$
2-Amino-4,4′-dimethyl-2′-nitrobiphenyl
2-Amino-6,6′-dimethyl-2′-nitrobiphenyl
4-Amino-2,2′-dimethyl-4′-nitrobiphenyl
Di-(4-aminophenyl)acetic Acid
2,4-Dicyano-2-phenylbutyric Acid, *Et ester*†
2,2′-Dihydroxyazobenzene, *Di-Me ether*
2,4-Dihydroxyazobenzene, *Di-Me ether*
3,3′-Dihydroxyazobenzene, *Di-Me ether*
3,4-Dihydroxyazobenzene, *Di-Me ether*
4,4′-Dihydroxyazobenzene, *Di-Me ether*
4,4′-Dimethyl-2-nitrodiphenylamine
o-Nitrobenzylaniline, N-*Me*
m-Nitrobenzylaniline, N-*Me*
N-2-Nitrobenzyl-*o*-toluidine
N-3-Nitrobenzyl-*o*-toluidine
N-4-Nitrobenzyl-*o*-toluidine
N-3-Nitrobenzyl-*m*-toluidine
N-2-Nitrobenzyl-*p*-toluidine
N-3-Nitrobenzyl-*p*-toluidine
N-4-Nitrobenzyl-*p*-toluidine
2-β-Phenylhydrazinoacetic Acid, *Phenyl ester*

$C_{14}H_{14}N_2O_3$
α-4-Aminophenyl-4-nitrobenzyl Alcohol, N-*Me*
2,2′-Dihydroxyazoxybenzene, *Di-Me ether*
5-Hydroxy-2-nitro-*N*-phenylaniline, *Et ether*
N-2-Nitrobenzyl-*o*-anisidine
N-4-Nitrobenzyl-*o*-anisidine
N-2-Nitrobenzyl-*p*-anisidine

$C_{14}H_{14}N_2O_4$
4-Phenylpyrazole-1,3(5)-dicarboxylic Acid, 1-*Me*-3(5)-*Et ester*
3(5)-Phenylpyrazole-1,5(3)-dicarboxylic Acid, 1-*Me*-5(3)-*Et ester*
4-Phenylpyrazole-3,5-dicarboxylic Acid, *Me-Et ester*

$C_{14}H_{14}N_2O_4S$
5-Nitrotoluene-2-sulphonic Acid, o-*Toluidide*
2-Nitrotoluene-4-sulphonic Acid, o-*Toluidide*

$C_{14}H_{14}N_2S$
N-Phenyl-*N′*-*o*-tolylthiourea
N-Phenyl-*N′*-*m*-tolylthiourea
N-Phenyl-*N′*-*p*-tolylthiourea

$C_{14}H_{14}N_4O_2$
2,2′-Hydrazodibenzoic Acid, *Diamide*
Lumilactoflavin, N-*Me*

$C_{14}H_{14}N_4O_3$
3-Phenyluric Acid, 1,7,9-*Tri-Me*

$C_{14}H_{14}N_4O_4$
4,4′-Diamino-3,3′-dimethyl-5,5′-dinitrobiphenyl
4,4′-Diamino-5,5′-dimethyl-2,2′-dinitrobiphenyl

$C_{14}H_{14}O$
2-Benzylphenol, *Me ether*
4-Benzylphenol, *Me ether*
1-Butyronaphthone
2-Butyronaphthone
Dibenzyl Ether
1,1-Diphenylethanol
1,2-Diphenylethanol
2,2-Diphenylethanol
Di-*o*-tolyl Ether
Di-*m*-tolyl Ether
Di-*p*-tolyl Ether
2-Hydroxybiphenyl, *Et ether*
3-Hydroxybiphenyl, *Et ether*
3-Hydroxymethylbiphenyl, *Me ether*
1-*o*-Hydroxyphenyl-2-phenylethane
1-*p*-Hydroxyphenyl-2-phenylethane
1-Isobutyronaphthone
2-Isobutyronaphthone
Pentacyclo[7,5,0,$0^{2,7}$,$0^{5,13}$,$0^{6,12}$]tetradeca-3,10-dien-8-one†
2-Phenylethanol, *Phenyl ether*
Phenyl-*o*-tolylmethanol
Phenyl-*m*-tolylmethanol
Phenyl-*p*-tolylmethanol
1,2,3,4-Tetrahydroanthranol

$C_{14}H_{14}OS$
Dibenzyl sulphoxide
Di-*o*-tolyl sulphoxide
Di-*m*-tolyl sulphoxide
Di-*p*-tolyl sulphoxide
p-Hydroxybenzenethiol, S-p-*Tolyl*, *Me ether*
o-Tolyl *m*-tolyl sulphoxide
o-Tolyl *p*-tolyl sulphoxide
m-Tolyl *p*-tolyl sulphoxide

$C_{14}H_{14}O_2$
2-Acetyl-1-naphthol, *Et ether*
4-Acetyl-1-naphthol, *Et ether*
Butyric Acid, β-*Naphthyl ester*
2-Butyryl-1-naphthol
4-Butyryl-1-naphthol
2,2′-Dihydroxybiphenyl, *Di-Me ether*
2,4′-Dihydroxybiphenyl, *Di-Me ether*
3,3′-Dihydroxybiphenyl, *Di-Me ether*
4,4′-Dihydroxybiphenyl, *Di-Me ether*
4,8-Dimethylazulene-6-carboxylic Acid, *Me ester*
1,2-Diphenyl-1,2-ethanediol
Ethylene Glycol, *Diphenyl ether*
Guaiacol, *Benzyl ether*
2(2,4-Hexadiynylidene)-1,6-di-oxaspiro-[4,5]-dec-3-ene
Hexahydroanthraquinone
3-Hydroxy-2-(*trans*-non-1-ene-3,5,7-triynyl)-tetrahydropyran†
2-Methyl-1-naphthoic Acid, *Et ester*
4-Methyl-1-naphthoic Acid, *Et ester*
6-Methyl-1-naphthoic Acid, *Et ester*
1-Methyl-2-naphthoic Acid, *Et ester*
2-Methyl-3-α-naphthylpropionic Acid
2-Methyl-3-β-naphthylpropionic Acid
1-Naphthylacetic Acid, *Et ester*
2-Naphthylacetic Acid, *Et ester*
2-α-Naphthylbutyric Acid
3-α-Naphthylbutyric Acid
4-α-Naphthylburytic Acid
3-β-Naphthylbutyric Acid
4-β-Naphthylbutyric Acid
3-α-Naphthylpropionic Acid, *Me ester*
2-Propionyl-1-naphthol, *Me ether*
4-Propionyl-1-naphthol, *Me ether*
6-Propionyl-2-naphthol, *Me ether*
Resorcinol, *Mono-phenylethyl ether*
1,2,3,4-Tetrahydroanthrahydroquinone
Xanthorrhoein†

$C_{14}H_{14}O_2S$
Dibenzyl sulphone
Di-*o*-hydroxyphenyl sulphide, *Di-Me ether*
Di-*p*-hydroxyphenyl sulphide, *Di-Me ether*
Di-*o*-tolyl sulphone
Di-*m*-tolyl sulphone
Di-*p*-tolyl sulphone
Phenyl 2,4-xylyl sulphone
Toluene-*p*-sulphinic Acid, *Benzyl ester*
o-Tolyl *m*-tolyl sulphone
o-Tolyl *p*-tolyl sulphone
m-Tolyl *p*-tolyl sulphone

$C_{14}H_{14}O_2S_2$
Di-2-hydroxyphenyl disulphide, *Di-Me ether*
Di-4-hydroxyphenyl disulphide, *Di-Me ether*

$C_{14}H_{14}O_3$
Aucuparin
2,2′-Dihydroxydiphenyl Ether, *Di-Me ether*
2,3-Dihydroxydiphenyl Ether, *Di-Me ether*
2,3′-Dihydroxydiphenyl Ether, *Di-Me ether*
2,4′-Dihydroxydiphenyl Ether, *Di-Me ether*
3,3′-Dihydroxydiphenyl Ether, *Di-Me ether*
3,4-Dihydroxydiphenyl Ether, *Di-Me ether*
3,4′-Dihydroxydiphenyl Ether, *Di-Me ether*
4,4′-Dihydroxydiphenyl Ether, *Di-Me ether*
Euparin, *Me ether*
4-Hydroxy-1-naphthoic Acid, *Propyl ester*
4-Hydroxy-1-naphthoic Acid, *Propyl ether*
5-Hydroxy-1-naphthoic Acid, *Propyl ether*
8-Hydroxy-1-naphthoic Acid, *Propyl ether*
1-Hydroxy-2-naphthoic Acid, *Me ether*, *Et ester*
6-Hydroxy-2-naphthoic Acid, *Et ester*, *Me ether*
6-Hydroxy-2-naphthoic Acid, *Propyl ether*
Kawaic Acid
Kawain
5-Methyl-2-phenylfuran-3-carboxylic Acid, *Et ester*

$C_{14}H_{14}O_3$ (*continued*)
α-Naphthoxyacetic Acid, *Et ester*
β-Naphthoxyacetic Acid, *Et ester*
2,1-Naphthylglycollic Acid, *Et ester*
2,2-Naphthylglycollic Acid, *Et ester*
Osthenol

$C_{14}H_{14}O_3S$
Acenaphthene-3-sulphonic Acid, *Et ester*
Toluene-*o*-sulphonic Acid, o-*Tolyl ester*
Toluene-*o*-sulphonic Acid, m-*Tolyl ester*
Toluene-*o*-sulphonic Acid, p-*Tolyl ester*
Toluene-*p*-sulphonic Acid, *Benzyl ester*

$C_{14}H_{14}O_3S_2$
Toluene-*p*-sulphinic Acid, *Anhydride*

$C_{14}H_{14}O_4$
Coriariadilactone
Corticrocin
Dihydro-oroselol†
Eleutherolic Acid, *Et ester*
Lomatin†
Marmesin
2-Methyl-1,4-benzoquinone, *Toluhydroquinone*
6-Methylcoumarin-4-acetic Acid, *Et ester*
7-Methylcoumarin-4-acetic Acid, *Et ester*
8-Methylcoumarin-4-acetic Acid, *Et ester*
5-(3,4-Methylenedioxyphenyl)-2,4-pentadienoic Acid, *Et ester*
Obliquetol†
4-Phenyl-1,3-butadiene-1,1-dicarboxylic Acid, *Di-Me ester*
Phyllomeronic Acid, *Di-Me ester*
Prenyletin†
4,5,6-Trimethoxy-2-naphthaldehyde†

$C_{14}H_{14}O_4S$
Di-*o*-hydroxyphenyl sulphone, *Di-Me ether*
Di-*p*-hydroxyphenyl sulphone, *Di-Me ether*

$C_{14}H_{14}O_5$
4-Acetyl-6,8-dihydroxy-5-methyl-2-benzopyran-1-one, *Di-Me ether*†
Itaconitin†
Kellactone
Obliquol†
Purpurogallin, *Tri-Me ether*
6,7,8-Trihydroxy-1-naphthoic Acid, *Tri-Me ether*

$C_{14}H_{14}O_5S$
Toluene-*p*-sulphonic Acid, *Anhydride*

$C_{14}H_{14}O_6$
Decevinic Acid
Leucophenicin
Spinochrome B, *Tetra-Me ether*†
2,5,7,8-Tetrahydroxy-1,4-naphthaquinone, *Tetra-Me ether*†

$C_{14}H_{14}O_6S_2$
Acenaphthene-5,6-disulphonic Acid, *Di-Me ester*

$C_{14}H_{14}O_8$
Aphloiol†
Benzene-1,2,3,4-tricarboxylic Acid, *Tetra-Me ester*
Benzene-1,2,3,5-tricarboxylic Acid, *Tetra-Me ester*
Benzene-1,2,4,5-tricarboxylic Acid, *Tetra-Me ester*

$C_{14}H_{14}O_{10}$
3,6-Dihydroxybenzene-1,2,4,5-tetracarboxylic Acid, *Tetra-Me ether*

$C_{14}H_{14}O_{11}$
Chebulic Acid

$C_{14}H_{14}S$
Dibenzyl sulphide
Di-*o*-tolyl sulphide
Di-*m*-tolyl sulphide
Di-*p*-tolyl sulphide
o-Tolyl *m*-tolyl sulphide
o-Tolyl *p*-tolyl sulphide
m-Tolyl *p*-tolyl sulphide

$C_{14}H_{14}S_2$
Dibenzyl disulphide★†
Di-*o*-tolyl disulphide
Di-*m*-tolyl disulphide
Di-*p*-tolyl disulphide

$C_{14}H_{14}S_3$
Dibenzyl trisulphide

$C_{14}H_{14}S_4$
Dibenzyl tetrasulphide

$C_{14}H_{15}ClN_2O$
2-Chloroquinoline-4-carboxylic Acid, *Di-Et amide*

$C_{14}H_{15}Cl_2N_3$
Acriflavine

$C_{14}H_{15}N$
2-Aminobibenzyl
3-Aminobibenzyl
4-Aminobibenzyl
2-Amino-2′,6-dimethylbiphenyl
2-Amino-4,4′-dimethylbiphenyl
2-Amino-4′,5-dimethylbiphenyl
3-Amino-4,4′-dimethylbiphenyl
3-Amino-4,5-dimethylbiphenyl
4-Amino-2′,3-dimethylbiphenyl
4-Amino-3,3′-dimethylbiphenyl
4-Amino-3,4′-dimethylbiphenyl
4-Amino-3,5-dimethylbiphenyl
5-Amino-2,3-dimethylbiphenyl
2-Amino-1,7-dimethylbiphenyl
α-Aminodiphenylmethane, N-*Me*
N-Benzyl-*o*-toluidine
N-Benzyl-*m*-toluidine
N-Benzyl-*p*-toluidine
Dibenzylamine
1,1-Diphenylethylamine
1,2-Diphenylethylamine
2,2-Diphenylethylamine
Di-2-tolylamine
Di-3-tolylamine
Di-4-tolylamine
N-Ethyldiphenylamine
N-Methyldiphenylmethylamine
N-Methylbenzylaniline
p-α-Methylbenzylaniline
N-*o*-Tolyl-*m*-tolylamine
N-*o*-Tolyl-*p*-tolylamine
N-*m*-Tolyl-*p*-tolylamine

$C_{14}H_{15}NO$
2-Amino-1,2-diphenylethanol
2-Amino-2,2-diphenylethanol
N-Dibenzylhydroxylamine
N-4-Hydroxybenzylaniline, *Me ether*
3-Hydroxy-4′-methyldiphenylamine, *Me ether*
3′-Hydroxy-2-methyldiphenylamine, *Me ether*
4-Hydroxy-*N*-phenylaniline, *Et ether*
4-α-Naphthylbutyric Acid, *Amide*

$C_{14}H_{15}NO_2$
4,4′-Dihydroxydiphenylamine, *Di-Me ether*
Khaplofoline†
2-Methylquinoline-3-carboxylic Acid, *Propyl ester*
1,2,3,4-Tetrahydrocarbazole-5-carboxylic Acid, *Me ester*
1,2,3,4-Tetrahydrocarbazole-6-carboxylic Acid, *Me ester*
1,2,3,4-Tetrahydrocarbazole-7-carboxylic Acid, *Me ester*
1,2,3,4-Tetrahydrocarbazole-8-carboxylic Acid, *Me ester*
1,2,3,4-Tetrahydrocarbazole-9-carboxylic Acid, *Me ester*

$C_{14}H_{15}NO_3$
2-(4-Hydroxyanilino)benzoic Acid, *Et ether*
2-Hydroxyquinoline-3-carboxylic Acid, *Butyl ether*
4-Hydroxyquinoline-3-carboxylic Acid, *Butyl ether*
2-Hydroxyquinoline-4-carboxylic Acid, *Et ether, Et ester*

$C_{14}H_{15}NO_4$
Indole-1,3-dicarboxylic Acid, *Di-Et ester*
Indole-2,5-dicarboxylic Acid, *Di-Et ester*
Indole-2,6-dicarboxylic Acid, *Di-Et ester*
Tricarballylic Acid, αβ-*Anil, Et ester*

$C_{14}H_{15}NO_6$
o-Nitrobenzylidenemalonic Acid, *Di-Et ester*
m-Nitrobenzylidenemalonic Acid, *Di-Et ester*
p-Nitrobenzylidenemalonic Acid, *Di-Et ester*

$C_{14}H_{15}NO_7$
Isatan B†

$C_{14}H_{15}N_3$
2-Amino-4′,5-dimethylazobenzene
4-Amino-2,2′-dimethylazobenzene
4-Amino-2,3-dimethylazobenzene
4-Amino-2,3′-dimethylazobenzene
4-Amino-2,4′-dimethylazobenzene
4-Amino-2′,3-dimethylazobenzene
4-Amino-2,5-dimethylazobenzene
4-Amino-2,6-dimethylazobenzene
4-Amino-3,3′-dimethylazobenzene
4-Amino-3,4′dimethylazobenzene
4-Amino-3,5-dimethylazobenzene
p-Aminohydrazobenzene, N-*Ac*
4-Dimethylaminoazobenzene
2,2′-Dimethyldiazoaminobenzene
2,4′-Dimethyldiazoaminobenzene
2′,3-Dimethyldiazoaminobenzene
3,3′-Dimethyldiazoaminobenzene
3,4′-Dimethyldiazoaminobenzene
4,4′-Dimethyldiazoaminobenzene
Phenyl-*o*-tolylguanidine
Phenyl-*p*-tolylguanidine

$C_{14}H_{15}N_3O$
4-Amino-4′-hydroxyazobenzene, N-*Di-Me*
Histamine★, N$^{\alpha}$-*Cinnamoyl*†

$C_{14}H_{15}N_3O_2$
4-Amino-2-nitrodiphenylamine, 1-N-*Et*
4,4′-Diamino-3′,5-dimethyl-2-nitrobiphenyl
2,3-Diamino-5-nitro-*p*-xylene
Indolmycin

$C_{14}H_{15}O_2P$
Diphenylphosphinic Acid, *Et ester*

$C_{14}H_{15}O_3P$
Dibenzyl hydrogen phosphite
Dibenzylphosphonate

$C_{14}H_{15}O_4P$
Dibenzyl hydrogen phosphate

$C_{14}H_{16}$
1-Butylnaphthalene
2-Butylnaphthalene
1-(2-Butyl)naphthalene
2-(2-Butyl)naphthalene
1-*tert*-Butylnaphthalene
2-*tert*-Butylnaphthalene
Chamazulene★ (given in error as $C_{15}H_{18}$)
Hexahydroanthracene
β-Hexahydroanthracene
γ-Hexahydroanthracene
Hexahydrophenanthrene
1-Isopropyl-4-methylnaphthalene
1-Isopropyl-7-methylnaphthalene
3-Isopropyl-1-methylnaphthalene
7-Isopropyl-1-methylnaphthalene
1,2,3,4-Tetramethylnaphthalene
1,2,4,7-Tetramethylnaphthalene
1,2,4,8-Tetramethylnaphthalene
1,2,5,6-Tetramethylnaphthalene
1,2,5,7-Tetramethylnaphthalene
1,2,5,8-Tetramethylnaphthalene
1,2,6,8-Tetramethylnaphthalene
1,3,6,8-Tetramethylnaphthalene
1,4,5,7-Tetramethylnaphthalene
1,4,5,8-Tetramethylnaphthalene
1,4,6,7-Tetramethylnaphthalene
2,3,6,7-Tetramethylnaphthalene
1,2,3,5-Tetramethylnaphthalene†
1,2,3-8-Tetramethylnaphthalene†

$C_{14}H_{16}HgN_2$
p-Mercuri-di-methylaniline

$C_{14}H_{16}NOP$
NN-Dimethyldiphenylphosphinamide†

$C_{14}H_{16}N_2$
2-Amino-4,4′-dimethyldiphenylamine
2-Amino-4′,5-dimethyldiphenylamine
3-Amino-4,4′-dimethyldiphenylamine
4-Amino-2,5-dimethyldiphenylamine
4-Amino-2′,3-dimethyldiphenylamine
4-Amino-3,4′-dimethyldiphenylamine
2,2′-Diaminobibenzyl
3,4′-Diaminobibenzyl
4,4′-Diaminobibenzyl
2,2′-Diamino-4,4′-dimethylbiphenyl

$C_{14}H_{16}N_2$ (*continued*)
2,2-Diamino-6,6′-dimethylbiphenyl
4,4′-Diamino-2,2′-dimethylbiphenyl
4,4′-Diamino-2,3′-dimethylbiphenyl
4,4′-Diamino-2,6-dimethylbiphenyl
4,4′-Diamino-3,3′-dimethylbiphenyl
5,5′-Diamino-2,2′-dimethylbiphenyl
2,4-Diaminotoluene, 2-N-*Benzyl*
1,1-Dianilinoethane
1,2-Dianilinoethane
1,1-Dibenzylhydrazine
1,2-Dibenzylhydrazine
2,2′-Dimethylhydrazobenzene
2,4-Dimethylhydrazobenzene
3,3′-Dimethylhydrazobenzene
3,4′-Dimethylhydrazobenzene
3,5-Dimethylhydrazobenzene
4,4′-Dimethylhydrazobenzene
1,2-Diphenylethylenediamine
1,1-Di-*p*-tolylhydrazine
Hydrazobenzene, NN′-*Di-Me*

$C_{14}H_{16}N_2O$
2-Amino-4′-hydroxydiphenylamine, *Et ether*
4-Amino-4′-hydroxydiphenylamine, *Et ether*
2,4′-Diamino-5-hydroxybiphenyl, *Et ether*
4,4′-Diamino-3-hydroxybiphenyl, *Et ether*
4-Dimethylamino-3′-hydroxydiphenylamine
4-Dimethylamino-4′-hydroxydiphenylamine
3-Hydroxyhydrazobenzene, *Et ether*
4-Hydroxyhydrazobenzene, *Et ether*
4-Nitroso-1-naphthylamine, N-*Di-Et*
1-Nitroso-2-naphthylamine, N-*Butyl*

$C_{14}H_{16}N_2O_2$
2,2′-Diamino-4,4′-dimethoxybiphenyl
4,4′-Diamino-3,3′-dimethoxybiphenyl
2,2′-Di(hydroxymethyl)hydrazobenzene
3,3′-Di(hydroxymethyl)hydrazobenzene
2,2′-Dimethoxyhydrazobenzene
Ethylene Glycol, *Di-p-aminophenyl ether*
Etimidate†

$C_{14}H_{16}N_2O_2S$
Di-(4-aminophenyl) sulphone, N,N′-*Di-Me*

$C_{14}H_{16}N_2O_3$
6-Hydroxy-8-nitroquinoline, 3-*Methylbutyl ether*

$C_{14}H_{16}N_2O_4S$
8-Nitronaphthalene-1-sulphonic Acid, *Di-Et-amide*

$C_{14}H_{16}N_2O_6$
Indicaxanthin†

$C_{14}H_{16}N_2S$
Di-2-aminobenzyl sulphide
Di-4-aminobenzyl sulphide
Di-(2-amino-5-methylphenyl)sulphide
Di-(4-amino-3-methylphenyl)sulphide

$C_{14}H_{16}N_2S_2$
Di-2-aminobenzyl disulphide
Di-4-aminobenzyl disulphide

$C_{14}H_{16}N_4O$
2,2′-Diamino-5,5′-dimethylazoxybenzene
3,3′-Diamino-2,2′-dimethylazoxybenzene
3,3′-Diamino-4,4′-dimethylazoxybenzene
4,4′-Diamino-2,2′-dimethylazoxybenzene
4,4′-Diamino-3,3′-dimethylazoxybenzene
5,5′-Diamino-2,2′-dimethylazoxybenzene

$C_{14}H_{16}N_6$
Diamidrazone

$C_{14}H_{16}O$
Dehydrochamaecynenal†
2-Naphthol, *Isobutyl ether*
Tetradeca-4,6,12-triene-8,10-diyn-1-ol†
3,4,6-Trimethyl-1-naphthol, *Me ether*
1,7,8-Trimethyl-2-naphthol, *Me ether*

$C_{14}H_{16}O_2$
1,5-Dihydroxynaphthalene, *Di-Et ether*
1,6-Dihydroxynaphthalene, *Di-Et ether*
1,7-Dihydroxynaphthalene, *Di-Et ether*
2,3-Dihydroxynaphthalene, *Di-Et ether*
2,6-Dihydroxynaphthalene, *Di-Et ether*
2,7-Dihydroxynaphthalene, *Di-Et ether*
Hexahydroanthrahydroquinone
3,1′(or 3′)-Indenylpropionic Acid, *Et ester*
1,2,3,4,5,6,7,8-Octahydroanthraquinone
1,4,4*a*,5,8,8*a*,9*a*,10*a*-Octahydroanthraquinone
1,2,3,4,5,6,7,8-Octahydrophenanthraquinone
5-Phenyl-2,4-pentadienoic Acid, *Isopropyl ester*
3-Phenyl-2-propynoic Acid, *Pentyl ester*
Pyroguaiacin, 2-*Me ether*

$C_{14}H_{16}O_3$
Fraxinellone
Macassar III†
Mexicanin E
Sorbicillin

$C_{14}H_{16}O_4$
Benzylidenemalonic Acid, *Di-Et ester*
3,3′-Diamino-2,2′-dimethylazobenzene
3,3′-Diamino-4,4′-dimethylazobenzene
4,4′-Diamino-3,3′-dimethylazobenzene
5,5′-Diamino-2,2′-dimethylazobenzene
Evodionol
Olivetonide
Phthalic Acid, *Cyclohexyl ester*
1,2,3,4-Tetrahydronaphthalene-1,8-dicarboxylic Acid, *Di-Me ester*
1,4,5,8-Tetrahydroxynaphthalene, *Tetra-Me ether*
6,7,8-Trimethoxy-1-naphthol, *Me ether*†
Tubaic Acid, *Me ether*, *Me ester*

$C_{14}H_{16}O_4S$
1-Naphthol-4-sulphonic Acid, *Et ether*, *Et ester*

$C_{14}H_{16}O_5$
Fraxetin, *Di-Et ether*
Phloroglucide, *Mono-Et ether*

$C_{14}H_{16}O_6$
2,5-Dioxocyclohexane-1,4-dicarboxylic Acid, *Diallyl ester*

$C_{14}H_{16}O_7$
5-Hydroxybenzene-1,2,3-tricarboxylic Acid, *Mono-Et-Di-Me ester*
5-Hydroxytoluene-2,3,4-tricarboxylic Acid, *Me ether*, *Tri-Me ester*†
5-Hydroxytoluene-2,3,6-tricarboxylic Acid, *Me ether*, *Tri-Me ester*†
3-Hydroxytoluene-2,4,6-tricarboxylic Acid, *Di-Et ester*

$C_{14}H_{16}O_7$ (*continued*)
3-Hydroxytoluene-2,4,6-tricarboxylic Acid, *Et ether*, *Mono-Et ester*

$C_{14}H_{16}O_9$
Bergenin

$C_{14}H_{17}BrO_3$
3,3-Dimethylbutyric Acid, p-*Bromophenacyl ester*

$C_{14}H_{17}ClN_2O_3$
5-Chloro-2-cyclohexyl-1-oxo-6-sulphamoyliso-indoline

$C_{14}H_{17}ClN_2O_3S$
Clorexolone †

$C_{14}H_{17}ClO$
α-Phenylcyclohexaneacetic Acid, *Chloride*

$C_{14}H_{17}N$
N-Diethyl-1-naphthylamine
N-Diethyl-2-naphthylamine
1-Ethyl-1,2,3,4-tetrahydrocarbazole
6-Ethyl-1,2,3,4-tetrahydrocarbazole
9-Ethyl-1,2,3,4-tetrahydrocarbazole
2-Pentylquinoline
1,2,3,4-Tetrahydro-1,8-dimethylcarbazole
1,2,3,4-Tetrahydro-1,9-dimethylcarbazole
1,2,3,4-Tetrahydro-2,4-dimethylcarbazole
1,2,3,4-Tetrahydro-2,6-dimethylcarbazole
1,2,3,4-Tetrahydro-3,6-dimethylcarbazole
1,2,3,4-Tetrahydro-3,9-dimethylcarbazole
1,2,3,4-Tetrahydro-5,8-dimethylcarbazole
1,2,3,4-Tetrahydro-6,7-dimethylcarbazole
1,2,3,4-Tetrahydro-6,8-dimethylcarbazole
1,2,3,4-Tetrahydro-6,9-dimethylcarbazole
1,2,3,4-Tetrahydro-7,8-dimethylcarbazole
2,3,4,11-Tetrahydro-1,11-dimethylcarbazole
2,3,4,11-Tetrahydro-9,11-dimethylcarbazole

$C_{14}H_{17}NO$
1-Amino-2-naphthol, *Isobutyl ether*
4-Hydroxy-2-methylquinoline, *Butyl ether*

$C_{14}H_{17}NO_2$
4-Phenylbutane-1,1-dicarboxylic Acid, *Et ester-nitrile*

$C_{14}H_{17}NO_3$
Fagaramide
Galantidine
Mesaconic Acid, α-*Et ester*, β-p-*Toluidide*

$C_{14}H_{17}NO_4$
Peyoglutam †
1-Phenylpyrrolidine-2,5-dicarboxylic Acid, *Di-Me ester*

$C_{14}H_{17}NO_6$
Anthranilinodiacetic Acid, *Tri-Me ester*
Indican
p-Nitrobenzylmalonic Acid, *Di-Et ester*
2-Nitroterephthalic Acid, *Dipropyl ester*
Prulaurasine
Prunasine
Pyridine-2,3,4-tricarboxylic Acid, *Tri-Et ester*
Sambunigrin

$C_{14}H_{17}NO_7$
Taxiphyllin †
Zierin

$C_{14}H_{17}NO_8$
Blepharin

$C_{14}H_{17}NO_9$
2,4-Dihydroxy-1,4-benzoxazolin-3-one, D-*Glucoside*

$C_{14}H_{17}N_3$
2,4′,5-Triaminobiphenyl, 5-N-*Di-Me*

$C_{14}H_{17}N_3O_4S_2$
Elektyl

$C_{14}H_{17}O_5PS$
O,O-Diethyl 4-methylumbelliferone thio-phosphate †

$C_{14}H_{18}$
1,2,3,4,5,6,7,8-Octahydroanthracene
1,2,3,4,4*a*,9,10,10*a*-Octahydrophenanthrene
1,2,3,4,5,6,7,8-Octahydrophenanthrene

$C_{14}H_{18}ClN_3S$
Chlorothen

$C_{14}H_{18}KNO_9S_2$
Glucotropaeolin

$C_{14}H_{18}N_2$
13,14-Diazatricyclo[6,4,1,$1^{2,7}$]tetradeca-3,5,9,11-tetraene, N,N′-*Di-Me* †
1,8-Naphthylenediamine★, NN′-*Tetra-Me* †

$C_{14}H_{18}N_2O_2$
Hypaphorine
Tryptophane, *Isopropyl ester*

$C_{14}H_{18}N_2O_3$
5,6-Dimethyl-1-(2′-deoxy-α-D-ribofuranosyl)-benzimidazole †
5,6-Dimethyl-1-(2′-deoxy-β-D-ribofuranosyl)-benzimidazole †
Physovenine †

$C_{14}H_{18}N_2O_4$
N-Benzoylglycylalanine, *Et ester*
N-Benzoylglycl-3-aminobutyric Acid, *Me ester*
Ribazole

$C_{14}H_{18}N_2O_5$
4-Acetyl-5-(2′-methyl-2′-propyl)-2,6-dinitro-*m*-xylene
γ-L-Glutamyl-L-phenylalanine

$C_{14}H_{18}N_2O_6$
γ-L-Glutamyl-γ-tyrosine

$C_{14}H_{18}N_2O_7$
Dinobuton †

$C_{14}H_{18}N_4$
3,3′-Diamino-4,4′-dimethylhydrazobenzene
5,5′-Diamino-2,2′-dimethylhydrazobenzene

$C_{14}H_{18}N_5O_{11}P$
Adenylosuccinic Acid

$C_{14}H_{18}N_6O$
Neotropine

$C_{14}H_{18}O$
2-Benzylideneheptanal
Chamaecynone †
Dehydrochamaecynenol †
Isochamaecynone †
Khusilal †
1,2,3,4,5,6,7,8-Octahydroanthranol

$C_{14}H_{18}O_2$
Cinnamic Acid, *Pentyl ester*
3-(2,4-Dimethylphenyl)crotonic Acid, *Et ester*
3-(2,5-Dimethylphenyl)crotonic Acid, *Et ester*
4-Hydroxyisochamaecynone†
3-*p*-Isopropylphenylacrylic Acid, *Et ester*
5-Methyl-2-pentylbenzofuran-3-one
2-Methyl-3-*o*-tolylcrotonic Acid, *Et ester*
2-Methyl-3-*p*-tolylcrotonic Acid, *Et ester*
1,2,3,4,5,6,7,8-Octahydroanthraquinol
α-Phenylcyclohexaneacetic Acid
2-Phenylcyclohexaneacetic Acid
4-Phenylcyclohexanecarboxylic Acid, *Me ester*
1-Phenylcyclopentane-1-carboxylic Acid, *Et ester*
3-(2,4,6-Trimethylphenyl)acrylic Acid, *Et ester*

$C_{14}H_{18}O_3$
2-Benzoylpropionic Acid, tert-*Butyl ester*
α-Hydroxy-α-phenylcyclohexaneacetic Acid

$C_{14}H_{18}O_4$
Benzene-1,2-diacetic Acid, *Di-Et ester*
Benzene-1,3-diacetic Acid, *Di-Et ester*
Benzene-1,4-diacetic Acid, *Di-Et ester*
Benzene-1,3-dipropionic Acid, *Di-Me ester*
Benzene-1,4-dipropionic Acid, *Di-Me ester*
Benzylmalonic Acid, *Di-Et ester*
3,3-Dimethylglutaric Acid, *Monobenzyl ester*
Biphenyl-3,3′-dicarboxylic Acid, *Di-Et ester*
Biphenyl-3,4-dicarboxylic Acid, *Di-Et ester*
Biphenyl-4,4′-dicarboxylic Acid, *Di-Et ester*
3-*o*-Carboxyphenylpropionic Acid, *Di-Et ester*
4,6-Diacetylresorcinol, *Di-Et ether*
4,6-Diacetylresorcinol, *Me-Propyl ether*
4,6-Diacetylresorcinol, *Monobutyl ether*
3,5-Dimethylphthalic Acid, *Di-Et ester*
2,6-Dimethylterephthalic Acid, *Di-Et ester*
6-Hydroxy-5-isopropyl-*o*-toluic Acid, *Acetonyl ester*
Methylphenylmalonic Acid, *Di-Et ester*
4-Phenylbutane-1,1-dicarboxylic Acid, *Di-Me ester*
Phenylsuccinic Acid, *Di-Et ester*
Terephthalic Acid, *Dipropyl ester*
Terephthalic Acid, *Di-isopropyl ester*

$C_{14}H_{18}O_5$
Cinchodectonic Acid
2,4-Dihydroxy-6,2′-oxoheptylbenzoic Acid
2-Phenylmalic Acid, *Di-Et ester*
3-(2,3,4-Trihydroxyphenyl)acrylic Acid, 4-*Me*-2,3-*Di-Et ether*

$C_{14}H_{18}O_6$
4,6-Dihydroxyisophthalic Acid, *Di-Me ether, Di-Et ester*
3,6-Dihydroxyphthalic Acid, *Di-Me ether, Di-Et ester*
2,5-Dihydroxyterephthalic Acid, *Di-Me ether, Di-Et ester*
3,4-Dimethoxyphthalic Acid, *Di-Et ester*
Metahemipinic Acid, *Di-Et ester*

$C_{14}H_{18}O_7$
2-Carboxymethyl-4,5-dimethoxyphenoxyacetic Acid, *Di-Me ester*
Ipecacuanhic Acid
Piceoside
Piscidic Acid, *Di-Me ester*, p-O-*Me ether*†
3,4,6-Triethoxy-2-hydroxybenzoylformic Acid

$C_{14}H_{18}O_8$
6-*O*-Acetylarbutin†
Gaultherin
Glucovanillin
Prulaurasinic Acid
Prunasinic Acid
Sambunigrinic Acid

$C_{14}H_{18}O_9$
Gentisic Acid 2-β-D-glucopyranose, *Me ester*†
Gentisic Acid 5-β-D-glucopyranose, *Me ester*†
Glucovanillic Acid

$C_{14}H_{18}O_{10}$
4-Glucogallic Acid, *Me ether*

$C_{14}H_{19}BrO_2$
2-Bromo-4,6-dimethylbenzoic Acid, *Pentyl ester*

$C_{14}H_{19}BrO_9$
Acetobromoglucose

$C_{14}H_{19}ClO_3$
3-Chloro-6-hydroxy-*p*-toluic Acid, 3-*Methylbutyl ether, Me ester*

$C_{14}H_{19}Cl_2NO_2$
Chlorambucil

$C_{14}H_{19}N$
9-Ethyl-1,2,3,4,4α,9α-hexahydrocarbazole
3-Hexylindole
2-Methyl-3,3′-methylbutylindole

$C_{14}H_{19}NO$
4-Methyl-2-hexenoic Acid, p-*Toluidide*

$C_{14}H_{19}NO_2$
Methylphenidate
Piperoxan

$C_{14}H_{19}NO_3$
3-Ethyl-3-methylglutaric Acid, *Monoanilide*
Pleurospermine

$C_{14}H_{19}NO_4$
Anisomycin
2-Phenyliminodiacetic Acid, *Di-Et ester*
N-Phenyliminodiacetic Acid, *Di-Et ester*

$C_{14}H_{19}NO_5$
Dicrotaline
4-Methoxy-3-nitrobenzoic Acid, *Hexyl ester*
Tartranilic Acid, *Isobutyl ester*

$C_{14}H_{19}NO_6$
2-Nitro-3-phenylglutaric Acid, *Di-Et ester*

$C_{14}H_{19}NO_{10}S_2$
Glucosinalbin†

$C_{14}H_{19}N_2O_2$(?)
Etheserolene

$C_{14}H_{19}N_3O_2$
Isoeserine

$C_{14}H_{19}N_3S$
Thenyldiamine

$C_{14}H_{20}$
Diamantane†
Irene

$C_{14}H_{20}ClNO_8$
2-Acetamido-3,4,6-tri-*O*-acetyl-2-deoxy-α-D-glucopyranosyl Chloride†

$C_{14}H_{20}Cl_6N_2$
Chlorisondamine chloride

$C_{14}H_{20}N_2O$
Bufotenine, *Et ether*
Noreserethole
Velbsimine

$C_{14}H_{20}N_2O_3$
Feruloylputrescine †

$C_{14}H_{20}N_2O_4S$
2-Pentenylpenicillinic Acid
2-Pentenylpenillic Acid

$C_{14}H_{20}N_2O_5$
2-6-Dinitro-4-(1-propylpentyl)phenol) †
4-(1-Ethylhexyl)-2,6-dinitrophenol †
6-(1-Ethylhexyl)-2,4-dinitrophenol †
6-(1-Methylheptyl)-2,4-dinitrophenol †

$C_{14}H_{20}N_2O_7$
Ydiginic Acid †

$C_{14}H_{20}N_2O_{10}$
5-Ribosyluridine †

$C_{14}H_{20}N_6O_5$
Blasticifin S

$C_{14}H_{20}N_6O_5S$
S-Adenosyl-L-homocysteine

$C_{14}H_{20}O$
1-Benzyl-2-methylcyclohexanol
1-Benzyl-3-methylcyclohexanol
1-Benzyl-4-methylcyclohexanol
2-Benzyl-4-methylcyclohexanol
2-Benzyl-5-methylcyclohexanol
2-Benzyl-6-methylcyclohexanol
Chamaecynenol †
Khusitone †
Mayurone †
2-Phenylcyclohexanol, *Et ether*

$C_{14}H_{20}O_2$
2-Benzyl-2-methylbutyric Acid, *Et ester*
2,4-Dimethylphenylacetic Acid, n-*Butyl ester*
3-Hydroxymethylenecamphor, *Et ether*
p-Isopropylbenzoic Acid, n-*Butyl ester*
2-Isopropyl-3-phenylpropionic Acid, *Et ester*
2-Methyl-2-phenylpropionic Acid, *Isobutyl ester*
Octanoic Acid, *Phenyl ester*
3-Phenylhexanoic Acid, *Et ester*
5-Phenylhexanoic Acid, *Et ester*
6-Phenylhexanoic Acid, *Et ester*
5-Phenyloctanoic Acid
8-Phenyloctanoic Acid
2,4,6-Triethylbenzoic Acid, *Me ester*
Zanthoxylol †

$C_{14}H_{20}O_3$
2,4-Dihydroxyacetophenone, *Dipropyl ether*
4-Hydroxy-5-isopropyl-*o*-toluic Acid, *Me ether, Et ester*
3-Hydroxy-2-phenyloctanoic Acid
3-Hydroxy-3-phenyloctanoic Acid
8-*m*-Hydroxyphenyloctanoic Acid
4-*p*-Hydroxyphenylvaleric Acid, *Me ether, Et ester*
2-Hydroxy-2-(2,4,6-trimethylphenyl)acetic Acid, *Isopropyl ester*

$C_{14}H_{20}O_4$
Cohumulinic Acid A★†
Cohumulinic Acid B †
Cohumulinic Acid C †
Curvulol, *Di-Et ether* †
2,4-Dihydroxy-6-pentylbenzoic Acid, *Et ester*
Flavesone †
Frequentin
Tasmanone
α,2,4-Trihydroxyacetophenone, *Tri-Et ether*
2,4,6-Trihydroxyacetophenone, *Tri-Et ether*

$C_{14}H_{20}O_5$
α,2,4,6-Tetrahydroxyacetophenone, α-4,6- *Tri-Et ether*

$C_{14}H_{20}O_6$
Colletodiol †
2,5-Dioxocyclohexane-1,4-dicarboxylic Acid, *Dipropyl ester*

$C_{14}H_{20}O_7$
Salidroside †

$C_{14}H_{20}O_8$
4-Glucosyloxy-α-hydroxy-3-methoxytoluene
Loroglossin

$C_{14}H_{20}O_9$
Unedoside †

$C_{14}H_{21}AsO_2$
Benzenearsonous Acid, *Di-butyl ester*

$C_{14}H_{21}N$
1,2,3,4-Tetrahydroisoquinoline, N-3-*Methylbutyl*

$C_{14}H_{21}NO$
Fabianine
4-Methylhexane-3-carboxylic Acid, *Anilide*
5-Methylhexane-3-carboxylic Acid, *Anilide*
8-Phenyloctanoic Acid, *Amide*
Sedamine †

$C_{14}H_{21}NO_2$
α-Diethylamino-*p*-toluic Acid, *Et ester*
N-Phenylsarcosine, 3-*Methylbutyl ester*
Stovaine

$C_{14}H_{21}NO_3$
p-Propoxybenzoic Acid, *Dimethylaminoethyl ester*

$C_{14}H_{21}NO_4$
Fermicidin

$C_{14}H_{21}NO_6$
Mannonic Acid, β-*Phenylethylamide*

$C_{14}H_{21}N_2O_3$ (ion)
o-(*N*-Benzoylglycyl)choline

$C_{14}H_{21}N_3O_3S$
Metahexamide

$C_{14}H_{21}N_3O_5$
Leonurine★†

$C_{14}H_{21}N_3O_6S$
[D-4-Amino-4-carboxybutyl]penicillinic Acid
[L-4-Amino-4-carboxybutyl]penicillinic Acid

$C_{14}H_{21}N_3O_{10}S$
S-(αβ-Dicarboxyethyl)glutathione

$C_{14}H_{21}N_4O_{12}P$
N-(5-Amino-1-β-D-ribofuranosylimidazole-4-carbonyl)-L-*threo*-β-methylaspartic Acid 5′-Phosphate †

$C_{14}H_{21}OP$
2,2,3,4,4-Pentamethyl-1-phenylphosphetan 1-oxide †

$C_{14}H_{21}P$
2,2,3,4,4-Pentamethyl-1-phenylphosphetan †

$C_{14}H_{22}$
$\Delta^{4a(b)}$-Dodecahydrophenanthrene †
n-Octylbenzene

$C_{14}H_{22}ClN_3O_2$
Metoclopramide †

$C_{14}H_{22}N_2O$
Angustifoline
Lignocaine
p-Nitrosoaniline, N-*Dibutyl*

$C_{14}H_{22}N_2O_4S$
Pentylpenicillenic Acid
Pentylpenicillinic Acid
Pentylpenillic Acid

$C_{14}H_{22}N_2O_5S$
2-Pentenylpenicilloic Acid

$C_{14}H_{22}N_2S_4$
Cassipourine †

$C_{14}H_{22}O$
Apoaromadendrone
α-Aromadendrone
Irone
ψ-Irone
Isolongifolic Aldehyde
6-Isopropyl-*m*-cresol, *Butyl ether*
1-Phenyl-2-propanol, *Pentyl ether*
2,4,5-Trimethylphenol, 3-*Methylbutyl ether*

$C_{14}H_{22}O_2$
p-Isopropylbenzaldehyde, *Di-Et acetal*
2-*n*-Octylresorcinol
4-*n*-Octylresorcinol
5-*n*-Octylresorcinol
Resorcinol, *Mono-octyl ether*

$C_{14}H_{22}O_3$
9-Oxo-2-decenoic Acid, *Isopropyl ester*

$C_{14}H_{22}O_3S$
p-Octylbenzenesulphonic Acid

$C_{14}H_{22}O_4$
Bicyclo[2,2,2]octane-1,4-dicarboxylic Acid, *Di-Et ester*
Isoborneol, *Acid succinyl*
Oxalic Acid, *Dicyclohexyl ester*
Palitantin
Succinic Acid, *Bornyl ester*
1,2,4,5-Tetrahydroxybenzene, *Tetra-Et ether*

$C_{14}H_{22}O_5$
8-Hydroxy-2,7-dimethyldeca-2,4-dienedioic Acid, *Di-Me ester* †

$C_{14}H_{22}O_6$
cis-2-Pentene-2,3,5-tricarboxylic Acid, *Tri-Et ester*

$C_{14}H_{22}O_8$
Butane-1,2,3,4-tetracarboxylic Acid, *Di-propyl ester*
Ethane-tetracarboxylic Acid, *Tetra-Et ester*

$C_{14}H_{23}ClO_6$
Caryophyllene Chlorohydrin †

$C_{14}H_{23}N$
N-Dibutylaniline

$C_{14}H_{23}NO$
Affinin
Apoaromadendrone, *Oxime*

$C_{14}H_{23}NO_2$
α-Aminomethyl-*p*-hydroxybenzyl, N-*Di-iso-propyl*
Bicyclo[2,2,1]hept-5-ene-2-carboxylic Acid, 2-*Diethylaminoethyl ester*

$C_{14}H_{23}N_3O_8S$
γ-L-Glutamyl-*S*[2-carboxy-1-propyl]cysteinylglycine

$C_{14}H_{23}N_5O_7S$
3-(2-Aminoethyl)-5-hydroxyindole, Neutral salt with creatinine sulphate

$C_{14}H_{23}O_2PS_2$
S-α-Methylbenzyl-*O,O*-dipropylphosphorothiolothionate

$C_{14}H_{24}$
1,3-Cyclotetradecadiene
Dispiro[5,1,5,1]tetradecane
Perhydroanthracene
Perhydrophenanthrene
[4,4,4]Propellane †

$C_{14}H_{24}NO$
1-Camphenyl *tert*-butylnitroxide

$C_{14}H_{24}NO_{10}S$
Gluconorcappasalin

$C_{14}H_{24}NO_{10}S_2$
Gluccocapangulin

$C_{14}H_{24}N_2$
p-Phenylenediamine, N,N′-*Tetra-Et*

$C_{14}H_{24}N_2O_5S$
Pentylpenicilloic Acid

$C_{14}H_{24}N_2O_7$
Actinospectacin

$C_{14}H_{24}O$
Aromadendrol
α-Ionol, *Me ether*
Isolongifolol

$C_{14}H_{24}O_2$
Alepric Acid
Isoborbeol, *Butyryl*
10-Undecynoic Acid, *Propyl ester*

$C_{14}H_{24}O_4$
Camphenic Acid, *Di-Et ester*
α-Carboxy-γ-tridecalactone
Maleic Acid, *Di-2-methylbutyl ester*
6-Methyl-2-heptene-2,3-dicarboxylic Acid, *Di-Et ester*
6-Methyl-1-heptene-4,7-dicarboxylic Acid, *Di-Et ester*
Succinic Acid, *Menthyl ester*

$C_{14}H_{24}O_4$ *(continued)*
1,2,2-Trimethylcyclopentane-1,3-dicarboxylic Acid, *Di-Et ester*

$C_{14}H_{24}O_5$
Azalomycin B
Oxalacetic Acid, *Di-3-methylbutyl ester*
Tetrahydro-2,6,6-trimethylpyran-2,5-dicarboxylic Acid, *Di-Et ester*

$C_{14}H_{24}O_6$
3-Methylbutane-1,1,2-tricarboxylic Acid, *Tri-Et ester*
2-Methylbutane-1,1,3-tricarboxylic Acid, *Tri-Et ester*
2-Methylbutane-1,1,4-tricarboxylic Acid, *Tri-Et ester*
3-Methylbutane-1,2,3-tricarboxylic Acid, *Tri-Et ester*
2-Methylbutane-1,2,4-tricarboxylic Acid, *Tri-Et ester*
3-Methylbutane-1,2,4-tricarboxylic Acid, *Tri-Et ester*
2-Methylbutane-1,3,3-tricarboxylic Acid, *Tri-Et ester*
Pentane-1,3,5-tricarboxylic Acid, *Tri-Et ester*†

$C_{14}H_{24}O_7$
Triethyl citrate, *Et ether*

$C_{14}H_{25}N$
Apoaromadendrylamine

$C_{14}H_{25}NO$
2,4-Decadienoic Acid, N-*Isobutylamide*
2,6-Decadienoic Acid, N-*Isobutylamide*
4,6-Decadienoic Acid, N-*Isobutylamide*

$C_{14}H_{25}NO_3$
3-Carbamoyl-1,2,2-trimethylcyclopentanecarboxylic Acid, N-*Di-Et*

$C_{14}H_{25}NO_4$
Strigosine†

$C_{14}H_{25}N_3O_9$
Kasugamycin†

$C_{14}H_{26}$
1,4-Di-*tert*-butylcyclohexene†

$C_{14}H_{26}N_2$
Leontamine

$C_{14}H_{26}N_2O_3S$
Pentylpenilloic Acid, *Me ester*

$C_{14}H_{26}N_2O_4$
Piperazine-*N,N'*-dicarboxylic Acid, *Dibutyl ester*

$C_{14}H_{26}O$
Cyclotetradecanone

$C_{14}H_{26}O_2$
11-Dodecenoic Acid, *Et ester*
2-*p*-Menthylacetic Acid, *Et ester*
12-Methyl-10-oxotridecanoic Acid
2-Tetradecenoic Acid
4-Tetradecenoic Acid
5-Tetradecenoic Acid
8-Tetradecenoic Acid
9-Tetradecenoic Acid
12-Tridecenoic Acid, *Me ester*

$C_{14}H_{26}O_3$
Heptanoic Acid, *Anhydride*
3-Oxotetradecanoic Acid
4-Oxotetradecanoic Acid
6-Oxotetradecanoic Acid★†
13-Oxotetradecanoic Acid
Pinol Glycol, *Di-Et ether*

$C_{14}H_{26}O_4$
Adipic Acid, *Dibutyl ester*
Adipic Acid, *Di-2-butyl ester*
Brassylic Acid, *Me ester*
Decanedioic Acid, *Di-Et ester*
Dodecanedioic Acid, *Di-Me ester*
3-Methylazelaic Acid, *Di-Et ester*
2-Methylheptane-1,1-dicarboxylic Acid, *Di-Et ester*
6-Methylheptane-2,5-dicarboxylic Acid, *Di-Et ester*
2-Methylheptane-4,4-dicarboxylic Acid, *Di-Et ester*
2-Methyltridecanedioic Acid
3-Methyltridecanedioic Acid
4-Methyltridecanedioic Acid
12-Methyltridecanedioic Acid
Oxalic Acid, *Di-hexyl ester*
Succinic Acid, *Di-pentyl ester*
Tetradecanedioic Acid

$C_{14}H_{26}O_5$
Malic Acid, *Di-(+)-2-methylbutyl ester*
Malic Acid, *Et ether*, *Dibutyl ester*

$C_{14}H_{26}O_6$
Tartaric Acid, *Di-(+)-2-methylbutyl ester*
Tartaric Acid, *Di-(±)-2-methylbutyl ester*
Tartaric Acid, *Di-3-methylbutyl ester*

$C_{14}H_{26}O_{11}$
Maltose, β-*Ethylglycoside*
Carrobiose, *Di-Me Acetal*†

$C_{14}H_{27}BrO_2$
2-Bromododecanoic Acid, *Et ester*
12-Bromododecanoic Acid, *Et ester*

$C_{14}H_{27}ClO$
Tetradecanoic Acid, *Chloride*

$C_{14}H_{27}N$
Tetradecanoic Acid, *Nitrile*

$C_{14}H_{27}NO$
2-Hydroxytetradecanoic Acid, *Nitrile*

$C_{14}H_{27}NO_2$
(−)-3-Menthylglycine, *Et ester*

$C_{14}H_{27}NO_3$
Carpamic Acid
ψ-Carpamic Acid

$C_{14}H_{27}NO_5$
Amicetamine†

$C_{14}H_{28}$
1-Tetradecene

$C_{14}H_{28}Br_2$
1,2-Dibromotetradecane
1,14-Dibromotetradecane

$C_{14}H_{28}N_2O_4S_2$
Tetramethylene-*S,S'*-dicystine, *Di-Et ester*

$C_{14}H_{28}N_2O_4S$
Thiodiglycollic Acid, Bis-*diethylaminoethyl ester*

$C_{14}H_{28}O$
Cyclotetradecanol
2-Methyl-3-tridecanone
Tetradecanal
2-Tetradecanone
3-Tetradecanone

$C_{14}H_{28}O_2$
Dodecanoic Acid, *Et ester*
10-Methyldodecanoic Acid, *Me ester*
Nonanoic Acid, *Pentyl ester*
2-Octanol, *Hexanoyl*
Tetradecanoic Acid
2,6,10-Trimethylundecanoic Acid†

$C_{14}H_{28}O_3$
Dodecanoic Acid, 2-*Hydroxyethyl ester*
2-Hydroxydodecanoic Acid, *Et ester*
2-Hydroxytetradecanoic Acid
3-Hydroxytetradecanoic Acid
6-Hydroxytetradecanoic Acid†
10-Hydroxytetradecanoic Acid
11-Hydroxytetradecanoic Acid

$C_{14}H_{28}O_4$
9,10-Dihydroxytetradecanoic Acid†
Ipurolic Acid

$C_{14}H_{28}O_5$
Braziliolic Acid†

$C_{14}H_{28}S_2$
1,9-Dithiacyclohexadecane

$C_{14}H_{29}Br$
12-Methyltridecyl bromide

$C_{14}H_{29}I$
1-Iodotetradecane

$C_{14}H_{29}N$
Azacyclopentadecane

$C_{14}H_{29}NO$
Decanoic Acid, N-*Isobutylamide*
Tetradecanoic Acid, *Amide*

$C_{14}H_{29}NO_2$
2-Aminotetradecanoic Acid
2-Hydroxytetradecanoic Acid, *Amide*

$C_{14}H_{30}$
Tetradecane

$C_{14}H_{30}Br_2N_2O_4$
Suxamethonium bromide

$C_{14}H_{30}Cl_2N_2O_4$
Suxamethonium chloride

$C_{14}H_{30}O$
12-Methyl-1-tridecanol
2-Methyl-3-tridecanol
2-Methyl-4-tridecanol
4-Methyl-4-tridecanol
1-Tetradecanol
3-Tetradecanol
2-Octanol, n-*Hexyl ether*

$C_{14}H_{30}O_2$
1,10-Decanediol, *Di-Et ether*
1,12-Dodecanediol, *Di-Me ether*

$C_{14}H_{30}O_3$
Orthoacetic Acid, *Tri-isobutyl ester*
Orthoformic Acid, *Propyl-di-3-methylbutyl ester*

$C_{14}H_{30}O_3S$
Tetradecane-1-sulphonic Acid

$C_{14}H_{31}N$
1-Aminotetradecane
Diheptylamine

C_{15}

$C_{15}H_6ClNO_2$
2-Chloroanthraquinone-1-carboxylic Acid, *Nitrile*
4-Chloroanthraquinone-1-carboxylic Acid, *Nitrile*
1-Chloroanthraquinone-2-carboxylic Acid, *Nitrile*
3-Chloroanthraquinone-2-carboxylic Acid, *Nitrile*
6-Chloroanthraquinone-2-carboxylic Acid, *Nitrile*

$C_{15}H_6Cl_2O_3$
6-Chloroanthraquinone-1-carboxylic Acid, *Chloride*
1-Chloroanthraquinone-2-carboxylic Acid, *Chloride*

$C_{15}H_6N_2O_4$
5-Nitroanthraquinone-1-carboxylic Acid, *Nitrile*

$C_{15}H_6N_4O_{13}$
1,8-Dihydroxy-3-hydroxymethyl-2,4,5-tetranitroanthraquinone

$C_{15}H_6O_4$
Fluorenone-1,2-dicarboxylic Acid, *Anhydride*
Fluorenone-2,3-dicarboxylic Acid, *Anhydride*
Pyrenic Acid, *Anhydride*

$C_{15}H_7ClO_3$
Anthraquinone-2-carboxylic Acid, *Chloride*

$C_{15}H_7ClO_4$
2-Chloroanthraquinone-1-carboxylic Acid
3-Chloroanthraquinone-1-carboxylic Acid
4-Chloroanthraquinone-1-carboxylic Acid
5-Chloroanthraquinone-1-carboxylic Acid
6-Chloroanthraquinone-1-carboxylic Acid
7-Chloroanthraquinone-1-carboxylic Acid
8-Chloroanthraquinone-1-carboxylic Acid
1-Chloroanthraquinone-2-carboxylic Acid
3-Chloroanthraquinone-2-carboxylic Acid
4-Chloroanthraquinone-2-carboxylic Acid
6-Chloroanthraquinone-2-carboxylic Acid
4-Hydroxyanthraquinone-2-carboxylic Acid, *Chloride*
5-Hydroxyanthraquinone-2-carboxylic Acid, *Chloride*

$C_{15}H_7Cl_2F_3N_2O_2$
5,6-Dichloro-1-phenoxycarbonyl-2-trifluoromethylbenzimidazole†

$C_{15}H_7NO_2$
- Anthraquinone-1-carboxylic Acid, *Nitrile*
- 9,10-Phenanthraquinone-2-carboxylic Acid, *Nitrile*
- 9,10-Phenanthraquinone-3-carboxylic Acid, *Nitrile*

$C_{15}H_7NO_6$
- 3-Nitroanthraquinone-1-carboxylic Acid
- 5-Nitroanthraquinone-1-carboxylic Acid
- 6-Nitroanthraquinone-1-carboxylic Acid
- 1-Nitroanthraquinone-2-carboxylic Acid
- 5-Nitroanthraquinone-2-carboxylic Acid

$C_{15}H_8Br_2O_2$
- 2-Dibromomethyl-anthraquinone

$C_{15}H_8Br_2O_3$
- 1,3-Dibromo-2-hydroxyanthraquinone, *Me ether*
- 2,4-Dibromo-1-hydroxyanthraquinone, *Me ether*

$C_{15}H_8ClNO_3$
- 1-Chloroanthraquinone-2-carboxylic Acid, *Amide*

$C_{15}H_8Cl_2O_3$
- Benzophenone-2,4′-dicarboxylic Acid, *Dichloride*
- Benzophenone-4,4′-dicarboxylic Acid, *Dichloride*

$C_{15}H_8Cl_2O_5$
- 2,4-Dichloro-1,3,8-trihydroxy-6-methyl-anthraquinone†

$C_{15}H_8N_2O$
- Benzophenone-4,4′-dicarboxylic Acid, *Dinitrile*

$C_{15}H_8N_2O_2$
- 2-Aminoanthraquinone-1-carboxylic Acid, *Nitrile*
- 1-Aminoanthraquinone-2-carboxylic Acid, *Nitrile*
- 1-Aminoanthraquinone-5-carboxyic Acid, *Nitrile*

$C_{15}H_8N_2O_5$
- 1-Nitroanthraquinone-2-carboxylic Acid, *Amide*
- 5-Nitroanthraquinone-2-carboxylic Acid, *Amide*

$C_{15}H_8O_3$
- Anthraquinone-1-aldehyde
- Anthraquinone-2-aldehyde

$C_{15}H_8O_4$
- Anthraquinone-1-carboxylic Acid
- Anthraquinone-2-carboxylic Acid
- Benzophenone-2,2′-dicarboxylic Acid, *Anhydride*
- 9,10-Phenanthraquinone-2-carboxylic Acid
- 9,10-Phenanthraquinone-3-carboxylic Acid

$C_{15}H_8O_5$
- Coumestrol
- Fluorenone-1,2-dicarboxylic Acid
- Fluorenone-1,5-dicarboxylic Acid
- Fluorenone-1,7-dicarboxylic Acid
- Fluorenone-2,3-dicarboxylic Acid
- Fluorenone-2,7-dicarboxylic Acid
- Fluorenone-4,5-dicarboxylic Acid
- 4-Hydroxyanthraquinone-1-carboxylic Acid
- 1-Hydroxyanthraquinone-2-carboxylic Acid
- 3-Hydroxyanthraquinone-2-carboxylic Acid
- 4-Hydroxyanthraquinone-2-carboxylic Acid
- 5-Hydroxyanthraquinone-2-carboxylic Acid
- Pyrenic Acid

$C_{15}H_8O_6$
- 5,8-Dihydroxyanthraquinone-1-carboxylic Acid
- 1,3-Dihydroxyanthraquinone-2-carboxylic Acid
- 1,4-Dihydroxyanthraquinone-2-carboxylic Acid
- 5,8-Dihydroxyanthraquinone-2-carboxylic Acid
- Lucernol†
- Rhein
- 7,11,12-Trihydroxycoumestan†

$C_{15}H_8O_7$
- Boletol
- Emodic Acid
- 1,3,4-Trihydroxyanthraquinone-2-carboxylic Acid

$C_{15}H_8O_8$
- Gossypitone
- Isoquercetone

$C_{15}H_9BrO_2$
- 2-Bromomethylanthraquinone
- 1-Bromo-3-methylanthraquinone
- 2-Bromo-3-methylanthraquinone

$C_{15}H_9BrO_3$
- 1-Bromo-2-hydroxyanthraquinone, *Me ether*
- 1-Bromo-4-hydroxyanthraquinone, *Me ether*

$C_{15}H_9BrO_3S$
- 1-*p*-Brompbenzenesulphonyloxymethylbicyclo-[2,2,2]octane†

$C_{15}H_9ClO$
- Phenanthrene-2-carboxylic Acid, *Chloride*
- Phenanthrene-3-carboxylic Acid, *Chloride*
- Phenanthrene-9-carboxylic Acid, *Chloride*

$C_{15}H_9ClO_2$
- 1-Chloro-2-methylanthraquinone
- 1-Chloro-3-methylanthraquinone
- 1-Chloro-4-methylanthraquinone
- 2-Chloro-1-methylanthraquinone
- 2-Chloro-3-methylanthraquinone
- 3-Chloro-1-methylanthraquinone

$C_{15}H_9ClO_3$
- Benzil-*o*-carboxylic Acid, *Chloride*
- 1-Chloro-2-hydroxyanthraquinone, *Me ether*
- 1-Chloro-4-hydroxyanthraquinone, *Me ether*

$C_{15}H_9ClO_5$
- 2-Chloro-1,3,8-trihydroxy-6-methylanthra-quinone†

$C_{15}H_9FO_2$
- 1-Fluoro-3-methylanthraquinone

$C_{15}H_9N$
- 1-Azafluoranthene
- 3-Azafluoranthene
- 2-Azapyrene†
- 4-Azapyrene
- Phenanthrene-1-carboxylic Acid, *Nitrile*
- Phenanthrene-2-carboxylic Acid, *Nitrile*
- Phenanthrene-3-carboxylic Acid, *Nitrile*
- Phenanthrene-9-carboxylic Acid, *Nitrile*

$C_{15}H_9NO$
Benzofuro[2,3-*f*]quinoline
Benzofuro[3,2-*g*]quinoline
Phenanthranil

$C_{15}H_9NO_3$
Anthraquinone-1-carboxylic Acid, *Amide*
Anthraquinone-2-carboxylic Acid, *Amide*
Benzophenone-2,2′-dicarboxylic Acid, *Imide*
9,10-Phenanthraquinone-3-carboxylic Acid, *Amide*

$C_{15}H_9NO_4$
2-Aminoanthraquinone-1-carboxylic Acid
4-Aminoanthraquinone-1-carboxylic Acid
5-Aminoanthraquinone-1-carboxylic Acid
6-Aminoanthraquinone-1-carboxylic Acid
1-Aminoanthraquinone-2-carboxylic Acid
3-Aminoanthraquinone-2-carboxylic Acid
5-Aminoanthraquinone-2-carboxylic Acid
2-Methyl-1-nitroanthraquinone

$C_{15}H_9NO_5$
1-Hydroxy-3-methyl-2-nitroanthraquinone
1-Hydroxy-3-methyl-4-nitroanthraquinone
1-Hydroxy-4-methyl-3-nitroanthraquinone
1-Hydroxy-4-methyl-6-nitroanthraquinone
2-Hydroxy-1-methyl-3-nitroanthraquinone
2-Hydroxy-3-methyl-1-nitroanthraquinone
2-Hydroxy-4-methyl-1-nitroanthraquinone
1-Hydroxy-4-nitroanthraquinone, *Me ether*
2-Hydroxy-1-nitroanthraquinone, *Me ether*
2-Hydroxy-5-nitroanthraquinone, *Me ether*
7-Hydroxy-1-nitroanthraquinone, *Me ether*
7-Nitrofluorenone-4-carboxylic Acid, *Me ester*

$C_{15}H_9NO_6$
1,2-Dihydroxy-4-nitroanthraquinone, 2-*Me ether*

$C_{15}H_9NS$
[1]Benzothiopyreno[4,3-*b*]indole†

$C_{10}H_9N_5O$
Kinetin

$C_{15}H_{10}$
4*H*-Cyclopenta[*def*]phenanthrene

$C_{15}H_{10}BrNO_2$
1-Amino-4-bromo-2-methylanthraquinone
2-Amino-1-bromo-3-methylanthraquinone
1-Bromo-2-methylaminoanthraquinone
1-Bromo-4-methylaminoanthraquinone
2-Bromo-1-methylaminoanthraquinone
2-Bromo-3-methylaminoanthraquinone

$C_{15}H_{10}ClN$
2-Chloro-3-phenylquinoline
3-Chloro-2-phenylquinoline
4-Chloro-2-phenylquinoline
4-Chloro-3-phenylquinoline
4-Chloro-6-phenylquinoline
4-Chloro-7-phenylquinoline
6-Chloro-2-phenylquinoline
8-Chloro-2-phenylquinoline

$C_{15}H_{10}ClNO_3$
3-*o*-Nitrophenyl-2-phenylacrylic Acid, *Chloride*

$C_{15}H_{10}ClNO_4$
2-(3′-Nitro-4′-toluyl)benzoic Acid, *Chloride*

$C_{15}H_{10}ClNO_5$
3-Nitrophthalic Acid, 2-*Benzyl ester, Chloride*

$C_{15}H_{10}Cl_2O_2$
Diphenylmalonic Acid, *Dichloride*
Diphenylmethane-2,4′-dicarboxylic Acid, *Dichloride*

$C_{15}H_{10}Cl_2O_4$
6,6′-Dichlorobiphenyl-2,2′-dicarboxylic Acid, *Mono-Me ester*

$C_{15}H_{10}N_2$
Benzimidazo[2,1-*a*]isoquinoline
Benzimidazo[1,2-*a*]quinoline
Quindoline
Quinindoline

$C_{15}H_{10}N_2O$
Norisotuboflavin†

$C_{15}H_{10}N_2O_2$
5-Methoxycanthin-6-one†
2-*p*-Nitrophenyl-3-phenylacrylic Acid, *Nitrile*
3-*o*-Nitrophenyl-2-phenylacrylic Acid, *Nitrile*
3-*m*-Nitrophenyl-2-phenylacrylic Acid, *Nitrile*
3-*p*-Nitrophenyl-2-phenylacrylic Acid, *Nitrile*
2-*o*-Nitrophenylquinoline
2-*m*-Nitrophenylquinoline
2-*p*-Nitrophenylquinoline
3-Nitro-4-phenylquinoline
3-Nitro-4-phenylquinoline
8-Nitro-6-phenylquinoline
4-Nitrostilbene-2-carboxylic Acid, *Nitrile*
4-Nitrostilbene-3-carboxylic Acid, *Nitrile*
2-Nitrostilbene-4-carboxylic Acid, *Nitrile*

$C_{15}H_{10}N_2O_3$
1-Aminoanthraquinone-2-carboxylic Acid, *Amide*
2-Aminoanthraquinone-1-carboxylic Acid, *Amide*
1,3-Diphenylimidazolidine-2,4,5-trione
1-Hydroxy-3-*m*-nitrophenylisoquinoline
1-Hydroxy-4-nitro-3-phenylisoquinoline
6-Hydroxy-8-nitroquinoline, *Phenyl ether*
4′-Hydroxy-2-nitrostilbene-4-carboxylic Acid, *Nitrile*
4′-Hydroxy-4-nitrostilbene-2-carboxylic Acid *Nitrile*

$C_{15}H_{10}N_2O_4$
1-Amino-4-nitroanthraquinone, N-*Me*
1-Amino-5-nitroanthraquinone, N-*Me*
1-Amino-8-nitroanthraquinone, N-*Me*
2-Amino-1-nitroanthraquinone, N-*Me*

$C_{15}H_{10}N_2O_5$
2,2′-Dinitrochalcone
2,3′-Dinitrochalcone
2,4-Dinitrochalcone
2,4′-Dinitrochalcone
3,3′-Dinitrochalcone
3′,4-Dinitrochalcone
Tramesanguin, *Me ester*

$C_{15}H_{10}O$
9-Anthraldehyde
Dibenzo[*a,c*]cycloheptatrien-5-one
Dibenzo[*a,d*]cycloheptatrien-5-one
Diphenylcyclopropenone†

$C_{15}H_{10}O$ (*continued*)
Methylene-anthrone
Phenanthrene-1-aldehyde
Phenanthrene-2-aldehyde
Phenanthrene-3-aldehyde
Phenanthrene-9-aldehyde

$C_{15}H_{10}OS$
2-Phenyl-4*H*-benzo[*b*]thiin-4-one

$C_{15}H_{10}O_2$
Anthracene-1-carboxylic Acid
Anthracene-2-carboxylic Acid
Anthracene-9-carboxylic Acid
Benz[*a*]azulene-7-carboxylic Acid
2-Benzofuranyl phenyl Ketone
Benzylidenephthalide
Dibenzotropolone
Flavone
2-Hydroxyanthracene-1-aldehyde
Isoflavone
1-Methylanthraquinone
2-Methylanthraquinone
1-Methylphenanthraquinone
2-Methylphenanthraquinone
3-Methylphenanthraquinone
4-Methylphenanthraquinone
Morphenol, *Me ether*
Phenanthrene-1-carboxylic Acid
Phenanthrene-2-carboxylic Acid
Phenanthrene-3-carboxylic Acid
Phenanthrene-4-carboxylic Acid
Phenanthrene-9-carboxylic Acid
3-Phenyl-2*H*-chromen-2-one
4-Phenyl-2*H*-chromen-2-one
6-Phenyl-2*H*-chromen-2-one
8-Phenylchromone
2-Phenyl-1,3-indanedione
3-Phenyl-1,2-indanedione

$C_{15}H_{10}O_2S$
1-Anthraquinonethiol, S-*Me ether*
2-Anthraquinonethiol, S-*Me ether*

$C_{15}H_{10}O_3$
Benzofuran-2-carboxylic Acid, *Phenyl ester*
Flavonol
Fluorenone-1-carboxylic Acid, *Me ester*
Fluorenone-2-carboxylic Acid, *Me ester*
Fluorenone-4-carboxylic Acid, *Me ester*
2-Hydroxyanthracene-1-carboxylic Acid
10-Hydroxyanthracene-2-carboxylic Acid
1-Hydroxyanthracene-2-carboxylic Acid
3-Hydroxyanthracene-2-carboxylic Acid
9-Hydroxyanthracene-2-carboxylic Acid
1-Hydroxy-9,10-anthraquinone, *Me ether*
2-Hydroxy-9,10-anthraquinone, *Me ether*
2′-Hydroxyflavone
3′-Hydroxyflavone
4′-Hydroxyflavone
5-Hydroxyflavone
6-Hydroxyflavone
7-Hydroxyflavone
8-Hydroxyflavone
5-Hydroxyisoflavone
7-Hydroxyisoflavone
1-Hydroxy-2-methylanthraquinone
1-Hydroxy-3-methylanthraquinone
1-Hydroxy-4-methylanthraquinone
1-Hydroxy-6-methylanthraquinone
2-Hydroxy-1-methylanthraquinone
2-Hydroxy-3-methylanthraquinone
2-Hydroxy-6-methylanthraquinone
3-Hydroxy-1-methylanthraquinone
3-Hydroxy-1,4-phenanthraquinone, *Me ether*
2-Hydroxy-9,10-phenanthraquinone, *Me ether*
3-Hydroxy-9,10-phenanthraquinone, *Me ether*
4-Hydroxyphenanthrene-1-carboxylic Acid
3-Hydroxyphenanthrene-2-carboxylic Acid
2-Hydroxyphenanthrene-3-carboxylic Acid
4-Hydroxyphenanthrene-3-carboxylic Acid
2-Hydroxyphenanthrene-9-carboxylic Acid
4-Hydroxyphenanthrene-9-carboxylic Acid
6-Hydroxyphenanthrene-9-carboxylic Acid
8-Hydroxyphenanthrene-9-carboxylic Acid
10-Hydroxyphenanthrene-9-carboxylic Acid

$C_{15}H_{10}O_4$
Alizarin, 1-*Me ether*
Alizarin, 2-*Me ether*
Benzil-*o*-carboxylic Acid
Chrysin
Chrysophanic Acid
Daidzein
Deuticulatol
1,8-Dihydroxyanthraquinone, *Mono-Me ether*
2,3-Dihydroxyanthraquinone, *Mono-Me ether*
1,3-Dihydroxyanthraquinone, 1-*Me ether*
1,3-Dihydroxyanthraquinone, 3-*Me ether*
2′,4′-Dihydroxyflavone
2′,6′-Dihydroxyflavone
2′,7′-Dihydroxyflavone
3′,4′-Dihydroxyflavone
3′,6-Dihydroxyflavone
3′,7-Dihydroxyflavone
4′,5-Dihydroxyflavone
4′,6-Dihydroxyflavone
4′,7-Dihydroxyflavone
5,6-Dihydroxyflavone
5,8-Dihydroxyflavone
6,7-Dihydroxyflavone
6,8-Dihydroxyflavone
7,8-Dihydroxyflavone
1,2-Dihydroxy-3-methylanthraquinone
1,2-Dihydroxy-4-methylanthraquinone
1,2-Dihydroxy-6-methylanthraquinone
1,2-Dihydroxy-7-methylanthraquinone
1,3-Dihydroxy-2-methylanthraquinone
1,3-Dihydroxy-4-methylanthraquinone
1,3-Dihydroxy-5-methylanthraquinone
1,3-Dihydroxy-6-methylanthraquinone
1,3-Dihydroxy-7-methylanthraquinone
1,3-Dihydroxy-8-methylanthraquinone
1,4-Dihydroxy-2-methylanthraquinone
1,4-Dihydroxy-5-methylanthraquinone
1,4-Dihydroxy-6-methylanthraquinone
1,5-Dihydroxy-2-methylanthraquinone
1,5-Dihydroxy-3-methylanthraquinone
1,6-Dihydroxy-2-methylanthraquinone
1,6-Dihydroxy-3-methylanthraquinone
1,7-Dihydroxy-3-methylanthraquinone
1,8-Dihydroxy-2-methylanthraquinone
2,3-Dihydroxy-6-methylanthraquinone
Fluorene-9,9-dicarboxylic Acid
5-Hydroxy-7-methoxyphthalide, *Me ether*†

$C_{15}H_{10}O_4$ (*continued*)
7-Hydroxy-5-methoxyphthalide, *Me ether*†
Nordalbergin†
Xanthone-2-carboxylic Acid, *Me ester*
Xanthone-4-carboxylic Acid, *Me ester*

$C_{15}H_{10}O_5$
Aloe-emodin
Apigenin
Baicalein
Benzophenone-2,2′ dicarboxylic Acid
Benzophenone-2,4′dicarboxylic Acid
Benzophenone-3,4′-dicarboxylic Acid
Benzophenone-4,4′-dicarboxylic Acid
2-Benzoylisophthalic Acid
4-Benzoylisophthalic Acid
3-Benzoylphthalic Acid
4-Benzoylphthalic Acid
Benzoylterephthalic Acid
Carajuretin
Cladofulvin†
Funiculosin
Galangin
Genistein
4-Hydroxy-2,3-methylenedioxyxanthone, *Me ether*†
Islandicin
Isocarajuretin
Isogenistein
Lambertellin, *Me ether*†
Lucidin (Anthraquinone)
Norwogonin
Pulcheremodin
Resokaempferol
Strepsilin
1,2,3-Trihydroxyanthraquinone, 1-*Me ether*
1,2,3-Trihydroxyanthraquinone, 2-*Me ether*
1,2,3-Trihydroxyanthraquinone, 3-*Me ether*
1,2,4-Trihydroxyanthraquinone, 2-*Me ether*
1,2,5-Trihydroxyanthraquinone, 2-*Me ether*
1,2,7-Trihydroxyanthraquinone, 1-*Me ether*
1,2,7-Trihydroxyanthraquinone, 2-*Me ether*
1,2,8-Trihydroxyanthraquinone, 2-*Me ether*
2′,3,7-Trihydroxyflavone
2′,3′,6′-Trihydroxyflavone
2′,5,6-Trihydroxyflavone
3,3′,4′-Trihydroxyflavone
3,3′,7-Trihydroxyflavone
3,4′,5-Trihydroxyflavone
3,5,6-Trihydroxyflavone
3,5,8-Trihydroxyflavone
3,6,7-Trihydroxyflavone
3,7,8-Trihydroxyflavone
3′,4′,5-Trihydroxyflavone
3′,4′,5′-Trihydroxyflavone
3′,4′,7-Trihydroxyflavone
3′,4′,8-Trihydroxyflavone
3′,5,7-Trihydroxyflavone
4′,5,6-Trihydroxyflavone
4′,6,7-Trihydroxyflavone
4′,6,8-Trihydroxyflavone
4′,7,8-Trihydroxyflavone
6,7,8-Trihydroxyflavone
4′,6,7-Trihydroxyisoflavone
1,2,3-Trihydroxy-5-methylanthraquinone
1,2,3-Trihydroxy-6-methylanthraquinone
1,2,3-Trihydroxy-7-methylanthraquinone
1,2,3-Trihydroxy-8-methylanthraquinone
1,2,5-Trihydroxy-6-methylanthraquinone
1,2,5-Trihydroxy-8-methylanthraquinone
1,2,6-Trihydroxy-7-methylanthraquinone
1,2,6-Trihydroxy-8-methylanthraquinone
1,2,7-Trihydroxy-6-methylanthraquinone
1,2,8-Trihydroxy-6-methylanthraquinone
1,2,8-Trihydroxy-7-methylanthraquinone
1,3,6-Trihydroxy-8-methylanthraquinone†
1,3,8-Trihydroxy-6-methylanthraquinone
1,4,5-Trihydroxy-6-methylanthraquinone
1,4,5-Trihydroxy-7-methylanthraquinone
1,4,5-Trihydroxy-8-methylanthraquinone
1,4,6-Trihydroxy-5-methylanthraquinone
1,4,6-Trihydroxy-7-methylanthraquinone
1,4,6-Trihydroxy-8-methylanthraquinone

$C_{15}H_{10}O_5S$
Anthraquinone-2-sulphonic Acid, *Me ester*
9,10-Phenanthraquinone-2-sulphonic Acid, *Me ester*
9,10-Phenanthraquinone-3-sulphonic Acid, *Me ester*

$C_{15}H_{10}O_6$
Aphloiol
Aureusidin
Biphenyl-2,2′,3-tricarboxylic Acid
Biphenyl-2,3′,4-tricarboxylic Acid
Catenarin
Citreorosein
Coelulatin (incorrectly given as $C_{15}H_{40}O_6$)
Cynodontin
Datiscetin
Fisetin
Kaempferol
Luteolin
Orobol
Pratoletin
Resomorin
Scutellarein
1,3,6,8-Tetrahydroxyanthraquinone, 3-*Me ether*†
3′,4′,5′,6-Tetrahydroxyaurone†
3,′4′,6,7-Tetrahydroxyaurone†
2′,3,4′,6-Tetrahydroxyflavone
2′,3,7,8-Tetrahydroxyflavone
2′,4′,5,7-Tetrahydroxyflavone★†
2′,4′,6′,7-Tetrahydroxyflavone
3,3′,4′,6-Tetrahydroxyflavone
3,3′,7,8-Tetrahydroxyflavone
3,4′,5,6-Tetrahydroxyflavone
3,4′,6,7-Tetrahydroxyflavone
3,4′,7,8-Tetrahydroxyflavone★†
3,5,6,7-Tetrahydroxyflavone
3,5,7,8-Tetrahydroxyflavone
3,6,7,8-Tetrahydroxyflavone
3′,4′,5′,6-Tetrahydroxyflavone
3′,4′,5′,7-Tetrahydroxyflavone
3′,4′,7,8-Tetrahydroxyflavone
4′,5,7,8-Tetrahydroxyflavone
4′,6,7,8-Tetrahydroxyflavone
5,6,7,8-Tetrahydroxyflavone★†
1,2,4,6-Tetrahydroxy-8-methylanthraquinone
1,2,6,8-Tetrahydroxy-3-methylanthraquinone
1,3,5,8-Tetrahydroxy-2-methylanthraquinone

$C_{15}H_{10}O_6$ (*continued*)
1,3,5,6-Tetrahydroxy-8-methylanthraquinone
1,5,6,7-Tetrahydroxy-2-methylanthraquinone
Versicolorin

$C_{15}H_{10}O_6S$
5-Hydroxyanthraquinone-1-sulphonic Acid, *Me ether*
5-Hydroxyanthraquinone-2-sulphonic Acid, *Me ether*
8-Hydroxyanthraquinone-2-sulphonic Acid, *Me ether*

$C_{15}H_{10}O_7$
Herbacetin
1,2,4,5,7-Pentahydroxyanthraquinone, 2-*Me ether*†
3′,4,4′,5′,6-Pentahydroxyaurone†
3′,4′,5,5′,6-Pentahydroxyaurone†
3′,4′,5′,6,7-Pentahydroxyaurone†
2′,3,5,5′,7-Pentahydroxyflavone
2′,3,5,7,8-Pentahydroxyflavone
2′,4′,5,7,8-Pentahydroxyflavone
3,3′,4′,5,8-Pentahydroxyflavone
3,3′,4′,6,7-Pentahydroxyflavone
3,3′,4′,7,8-Pentahydroxyflavone
3′,4′,5,6,7-Pentahydroxyflavone
3,4′,5,6,7-Pentahydroxyflavone★†
3,4′,6,7,8-Pentahydroxyflavone
3,5,6,7,8-Pentahydroxyflavone
3′,4′,5,7,8-Pentahydroxyflavone
3′,4′,5′,7,8-Pentahydroxyflavone
4′,5,6,7,8-Pentahydroxyflavone★†
1,3,4,5,6-Pentahydroxy-2-methylanthraquinone
Quercetin
Robinetin
Tricetin
Tritisporin
Morin

$C_{15}H_{10}O_8$
Asperthecin
Gossypetin
3′,4′,5,6,7,8-Hexahydroxyflavone†
Irigenol
Myricetin
Quercetagetin

$C_{15}H_{10}O_9$
Hibiscetin

$C_{15}H_{11}BrO_2$
Dibenzoylbromomethane

$C_{15}H_{11}ClN_2$
4-Amino-2-chloroquinoline, 4-N-*Phenyl*

$C_{15}H_{11}ClN_2O_2$
Oxazepam†

$C_{15}H_{11}ClO_2$
4-*p*-Toluylbenzoic Acid, *Chloride*

$C_{15}H_{11}ClO_3$
2-Benzoyl-4-chlorobenzoic Acid, *Me ester*
o-4-Chlorobenzoylbenzoic Acid, *Me ester*

$C_{15}H_{11}ClO_4$
Apigeninidin
Galanginidin chloride

$C_{15}H_{11}ClO_5$
Fisetinidin chloride
Luteolinidin chloride
Pelargonidin chloride

$C_{15}H_{11}ClO_6$
Cyanidin chloride
Morindin chloride
3,3′,4′,5′,7-Pentahydroxyflavylium chloride

$C_{15}H_{11}FO_3$
2-(2-Fluoro-*p*-toluyl)benzoic Acid

$C_{15}H_{11}I_4NO_4$
Thyroxine

$C_{15}H_{11}N$
Acenaphtheno[5,4-*b*]pyridine
2,3-Diphenylacrylic Acid, *Nitrile*
3,3-Diphenylacrylic Acid, *Nitrile*
1-Phenylisoquinoline
3-Phenylisoquinoline
4-Phenylisoquinoline
2-Phenylquinoline
3-Phenylquinoline
4-Phenylquinoline
5-Phenylquinoline
6-Phenylquinoline
8-Phenylquinoline

$C_{15}H_{11}NO$
Anthracene-1-carboxylic Acid, *Amide*
Anthracene-2-carboxylic Acid, *Amide*
9-Anthraldehyde, *Oxime*
a-Benzoylphenylacetic Acid, *Nitrile*
2,4-Diphenyloxazole
2,5-Diphenyloxazole
4,5-Diphenyloxazole
2-Hydroxy-α-phenylcinnamic Acid, *Nitrile*
3-Hydroxy-α-phenylcinnamic Acid, *Nitrile*
4-Hydroxy-α-phenylcinnamic Acid, *Nitrile*
1-Hydroxy-3-phenylisoquinoline
2-*o*-Hydroxyphenylquinoline
2-*m*-Hydroxyphenylquinoline
2-*p*-Hydroxyphenylquinoline
4-*o*-Hydroxyphenylquinoline
4-*m*-Hydroxyphenylquinoline
4-*p*-Hydroxyphenylquinoline
2-Hydroxy-3-phenylquinoline
2-Hydroxy-4-phenylquinoline
3-Hydroxy-2-phenylquinoline
4-Hydroxy-2-phenylquinoline
4-Hydroxy-3-phenylquinoline
4-Hydroxy-6-phenylquinoline
4-Hydroxy-7-phenylquinoline
4-Hydroxy-8-phenylquinoline
6-Hydroxy-2-phenylquinoline
6-Hydroxy-4-phenylquinoline
7-Hydroxy-2-phenylquinoline
8-Hydroxy-2-phenylquinoline
Phenanthrene-1-carboxylic Acid, *Amide*
Phenanthrene-2-carboxylic Acid, *Amide*
Phenanthrene-3-carboxylic Acid, *Amide*
Phenanthrene-9-carboxylic Acid, *Amide*
N-Phenylisocarbostyril

$C_{15}H_{11}NOS$
2-Oxo-3,4-diphenylthiazoline

$C_{15}H_{11}NO_2$
Acridine-9-carboxylic Acid, *Me ester*
1-Amino-2-methylanthraquinone
3-Amino-2-methylanthraquinone
Benzo[*f*]quinoline-1-carboxylic Acid, *Me ester*
Benzo[*f*]quinoline-3-carboxylic Acid, *Me ester*
Benzo[*f*]quinoline-5-carboxylic Acid, *Me ester*
4,6-Dihydroxy-2-phenylquinoline
6,7-Dihydroxy-2-phenylquinoline
1-Methylaminoanthraquinone
2-Methylaminoanthraquinone
1-Phenylindole-2-carboxylic Acid
3-Phenylindole-2-carboxylic Acid
6-Phenylindole-3-carboxylic Acid
Phthalimide, N-*Benzyl*
Phthalimide, N-o-*Tolyl*
Pyrenic Acid, *Imide*
Viridicatin

$C_{15}H_{11}NO_3$
1-Amino-4-hydroxyanthraquinone, N-*Me*
α-Nitrochalcone
β-Nitrochalcone
2-Nitrochalcone
2′-Nitrochalcone
3-Nitrochalcone
3′-Nitrochalcone
4-Nitrochalcone
4′-Nitrochalcone
Viridicatol†

$C_{15}H_{11}NO_4$
4-Amino-1,3-dihydroxyanthraquinone, 3-*Me ether*
Benzoyl-2-nitrobenzoylmethane
Benzoyl-3-nitrobenzoylmethane
Benzoyl-4-nitrobenzoylmethane
Evoxanthidine
2-*o*-Nitrophenyl-3-phenylacrylic Acid
2-*p*-Nitrophenyl-3-phenylacrylic Acid
3-*o*-Nitrophenyl-2-phenylacrylic Acid
3-*m*-Nitrophenyl-2-phenylacrylic Acid
3-*p*-Nitrophenyl-2-phenylacrylic Acid
4-Nitrostilbene-2-carboxylic Acid
4-Nitrostilbene-3-carboxylic Acid
2-Nitrostilbene-4-carboxylic Acid
Phthalonic Acid, *Anilide*

$C_{15}H_{11}NO_5$
Actinomycinol
2-Benzoyl-3-nitrobenzoic Acid, *Me ester*
2-Benzoyl-4-nitrobenzoic Acid, *Me ester*
2-Benzoyl-5-nitrobenzoic Acid, *Me ester*
2-Benzoyl-6-nitrobenzoic Acid, *Me ester*
Bostrycoidin†
2,4-Dihydroxy-5-nitrochalcone
3,4-Dihydroxy-2-nitrochalcone
3,4-Dihydroxy-3′-nitrochalcone
4,4′-Dihydroxy-3-nitrochalcone
4,5-Dihydroxy-2-nitrochalcone
3-Hydroxy-2-nitro-α-phenylcinnamic Acid
3-Hydroxy-4-nitro-α-phenylcinnamic Acid
5-Hydroxy-2-nitro-α-phenylcinnamic Acid
2-Hydroxy-α-(4-nitrophenyl)cinnamic Acid
4-Hydroxy-α-(4-nitrophenyl)cinnamic Acid
2′-Hydroxy-4-nitrostilbene-2-carboxylic Acid
2′-Hydroxy-2-nitrostilbene-4-carboxylic Acid
3′-Hydroxy-2-nitrostilbene-4-carboxylic Acid
4′-Hydroxy-2-nitrostilbene-4-carboxylic Acid
4′-Hydroxy-4-nitrostilbene-2-carboxylic Acid
4′-Hydroxy-4-nitrostilbene-3-carboxylic Acid
α-(2-Hydroxyphenyl)-2-nitrocinnamic Acid
α-(4-Hydroxyphenyl)-2-nitrocinnamic Acid
2-*o*-Nitrobenzoylbenzoic Acid, *Me ester*
2-(3′-Nitro-4′-toluyl)benzoic Acid
2-Nitro-6-*p*-toluybenzoic Acid
3-Nitro-2-*p*-toluylbenzoic Acid
4-Nitro-2-*p*-toluylbenzoic Acid
5-Nitro-2-*p*-toluylbenzoic Acid

$C_{15}H_{11}NO_6$
3-Nitrophthalic Acid, 1-*Benzyl ester*
3-Nitrophthalic Acid, 2-*Benzyl ester*
4-Nitrophthalic Acid, 1-*Benzyl ester*

$C_{15}H_{11}NS$
2,4-Diphenylthiazole
2,5-Diphenylthiazole
4,5-Diphenylthiazole

$C_{15}H_{11}N_3$
2,4-Diphenyl-1,3,5-triazine
2,2′,6′,2″, Terpyridine

$C_{15}H_{11}N_3O_2$
1-(4-Aminophenyl)-2-(4-carboxy-2-nitrophenyl)-ethylene, *Nitrile*

$C_{15}H_{12}$
5*H*-Dibenzo[*a,d*]cycloheptatriene
1,3-Diphenylallene†
1-Methylanthracene
2-Methylanthracene
9-Methylanthracene
10-Methylbenz[*a*]azulene
1-Methylphenanthrene
2-Methylphenanthrene
3-Methylphenanthrene
4-Methylphenanthrene
9-Methylphenanthrene

$C_{15}H_{12}Br_2O_2$
2,3-Dibromo-3-phenylpropionic Acid, *Phenyl ester*

$C_{15}H_{12}Br_2O_3$
5,5′-Dibromo-2,2′-dihydroxybenzophenone, 2,2′-*Di-Me ether*
5,5′-Dibromo-2,2′-dihydroxybenzophenone, 2-*Et ether*

$C_{15}H_{12}Cl_2O_4$
Mollisin, *Me ester*†

$C_{15}H_{12}NO_7$
Corymbiferin

$C_{15}H_{12}N_2$
2-*m*-Aminophenylquinoline
2-*p*-Aminophenylquinoline
4-*m*-Aminophenylquinoline
4-*p*-Aminophenylquinoline
2-Amino-3-phenylquinoline
3-Amino-2-phenylquinoline
3-Amino-4-phenylquinoline
4-Amino-2-phenylquinoline
4-Amino-8-phenylquinoline
6-Amino-2-phenylquinoline
6-Amino-4-phenylquinoline
7-Amino-2-phenylquinoline
8-Amino-2-phenylquinoline

$C_{15}H_{12}N_2$ (*continued*)
8-Amino-4-phenylquinoline
8-Amino-5-phenylquinoline
2-Aminoquinoline, N-*Phenyl*
4-Aminoquinoline, N-*Phenyl*
2-Aminostilbene-4-carboxylic Acid, *Nitrile*
2-Anilinoquinoline
2,5-Diphenylimidazole
4,5-Diphenylimidazole
1,3-Diphenylpyrazole
1,4-Diphenylpyrazole
1,5-Diphenylpyrazole
3,4-Diphenylpyrazole
3,5-Diphenylpyrazole
4,5-Diphenylpyrazole

$C_{15}H_{12}N_2O$
Glycosminine
3-Oxo-1,2-diphenyl-4-pyrazoline
3-Oxo-1,5-diphenyl-4-pyrazoline
3-Oxo-2,4-diphenyl-4-pyrazoline
3-Oxo-2,5-diphenyl-4-pyrazoline

$C_{15}H_{12}N_2OS$
Benzothiazole-2-acetic Acid, *Anilide*
2-Thiohydantoin, 1,3-N-*Diphenyl*

$C_{15}H_{12}N_2O_2$
1-Amino-4-methylaminoanthraquinone
Cyanomycin
1,3-Diamino-2-methylanthraquinone
1,3-Diamino-6-methylanthraquinone
1,4-Diamino-2-methylanthraquinone
1,4-Diamino-5-methylanthraquinone
1,5-Diamino-2-methylanthraquinone
1,8-Diamino-2-methylanthraquinone
1,3-Diphenylhydantoin
1,5-Diphenylhydantoin
3,5-Diphenylhydantoin
5,5-Diphenylhydantoin
5-Hydroxyisatin, *Anil*
3-Phenylindazole-2-carboxylic Acid, *Me ester*
2-Phenylindazole-3-carboxylic Acid, *Me ester*

$C_{15}H_{12}N_2O_3$
Benzophenone-2,4′-dicarboxylic Acid, *Diamide*
Benzophenone-4,4′-dicarboxylic Acid, *Diamide*
1,1-Dibenzoylurea
1,3-Dibenzoylurea
Furfuramide
2-Hydroxyphenazine-1-carboxylic Acid, *Me ester, Me ether*†
3-*o*-Nitrophenyl-2-phenylacrylic Acid, *Amide*

$C_{15}H_{12}N_2O_4$
1-(4-Aminophenyl)-2-(2-carboxy-4-nitrophenyl)-ethylene
1-(4-Aminophenyl)-2-(4-carboxy-2-nitrophenyl)-ethylene
2-(3′-Nitro-4′-toluyl)benzoic Acid, *Amide*

$C_{15}H_{12}N_2O_7$
2-Hydroxy-3,5-dinitrobenzoic Acid, *Et ester, Et ester*

$C_{15}H_{12}N_6O_4$
Rhizopterin

$C_{15}H_{12}O$
1-Anthrol, *Me ether*
2-Anthrol, *Me ether*
Chalcone
1,3-Dimethylfluorenone
1,4-Dimethylfluorenone
1,7-Dimethylfluorenone
2,3-Dimethylfluorenone
2,7-Dimethylfluorenone
3,4-Dimethylfluorenone
3,6-Dimethylfluorenone
2,3-Diphenylacrolein
3,3-Diphenylacrolein
1-Hydroxyphenanthrene, *Me ether*
2-Hydroxyphenanthrene, *Me ether*
3-Hydroxyphenanthrene, *Me ether*
4-Hydroxyphenanthrene, *Me ether*
9-Hydroxyphenanthrene, *Me ether*
1-Methylanthrone
2-Methylanthrone
3-Methylanthrone
4-Methylanthrone
10-Methylanthrone
1-Phenanthrenemethanol
2-Phenanthrenemethanol
3-Phenanthrenemethanol
4-Phenanthrenemethanol
9-Phenanthrenemethanol
2-Phenyl-1-indanone
3-Phenyl-1-indanone
4-Phenyl-1-indanone

$C_{15}H_{12}OS$
2,3-Dihydro-2-phenyl-4-*H*-benzo[*b*]thiin-4-one

$C_{15}H_{12}O_2$
Acenaphthene-5-acrylic Acid
2-Biphenylacrylic Acid
4-Biphenylacrylic Acid
Cinnamic Acid, *Phenyl ester*
Dibenzoylmethane
2,3-Dihydro-1*H*-phenalene-1,3-dione, 1-*Et enol-ether*
9,10-Dihydroxyanthracene, *Mono-Me ether*
1,10-Dihydroxy-2-methylanthracene
3,10-Dihydroxy-1-methylanthracene
1,9-Dihydroxyphenanthrene, 9-*Me ether*
3,6-Dihydroxyphenanthrene, 3-*Me ether*
2,4-Dimethylxanthone
2,7-Dimethylxanthone
3,6-Dimethylxanthone
4,5-Dimethylxanthone
2,3-Diphenylacrylic Acid
3,3-Diphenylacrylic Acid
1,3-Diphenyl-2,3-propanedione
Flavanone
Fluorene-2-carboxylic Acid, *Me ester*
Fluorene-4-carboxylic Acid, *Me ester*
Fluorene-9-carboxylic Acid, *Me ester*
1-Hydroxy-9-anthrone, *Me ether*
4-Hydroxy-9-anthrone, *Me ether*
2-Hydroxychalcone
2′-Hydroxychalcone
3-Hydroxychalcone
3′-Hydroxychalcone
4-Hydroxychalcone
4′-Hydroxychalcone
1-Hydroxyfluorenone, *Et ether*
1-Hydroxy-2-methylanthrone
1-Hydroxy-3-methylanthrone

$C_{15}H_{12}O_2$ (*continued*)
1-Hydroxy-4-methylanthrone
2-Hydroxy-3-methylanthrone
4-Hydroxy-1-methylanthrone
4-Hydroxy-2-methylanthrone
4-Hydroxy-3-methylanthrone
9-Methylanthracene, *Photo-oxide*
4-Methylbenzil
3,4-Methylenedioxystilbene
2-Methylfluorene-9-carboxylic Acid
3-Methylfluorene-9-carboxylic Acid
9-Methylfluorene-9-carboxylic Acid
Morphol, *Mono-Me ether*
Oxanthranol, *Me ether*
Pterocarpan †

$C_{15}H_{12}O_3$
Anisindione
o-Benzoylbenzoic Acid, *Me ester*
m-Benzoylbenzoic Acid, *Me ester*
p-Benzoylbenzoic Acid, *Me ester*
α-Benzoyl-phenylacetic Acid
p-Benzoyl-phenylacetic Acid
2-Carboxy-α-phenylacetophenone
Dehydro-α-lapachone †
Dehydro-β-lapachone †
2,3-Dihydro-2-isopropenylnaphtho[2,3-*b*]-furan-4,9-dione †
6*a*,12*a*-Dihydrorotoxen-12(6*H*)-one †
2,2′-Dihydroxychalcone
2,4′-Dihydroxychalcone
2′,3-Dihydroxychalcone
2′,4-Dihydroxychalcone
2′,4′-Dihydroxychalcone
2′,5′-Dihydroxychalcone
2′,6′-Dihydroxychalcone
3,4-Dihydroxychalcone
3,4′-Dihydroxychalcone
3′,4-Dihydroxychalcone
3′,4′-Dihydroxychalcone
4,4′-Dihydroxychalcone
1,4-Dihydroxyfluorenone, *Di-Me ether*
2,3-Dihydroxyfluorenone, *Di-Me ether*
1,8-Dihydroxy-3-methyl-9-anthrone
Freelingyne †
4′-Hydroxyflavanone
7-Hydroxyflavanone
9-Hydroxyfluorene-9-carboxylic Acid, *Me ester*
9-Hydroxyfluorene-9-carboxylic Acid, *Me ether*
2-Hydroxy-α-phenylcinnamic Acid
3-Hydroxy-α-phenylcinnamic Acid
4-Hydroxy-α-phenylcinnamic Acid
α-2-Hydroxyphenylcinnamic Acid
α-4-Hydroxyphenylcinnamic Acid
1-(2-Hydroxyphenyl)-2-phenylglyoxal, *Me ether*
1-(4-Hydroxyphenyl)-2-phenylglyoxal, *Me ether*
Mansonone F †
α-Phenylacetophenone-2-carboxylic Acid
2-*o*-Toluylbenzoic Acid
2-*m*-Toluylbenzoic Acid
2-*p*-Toluylbenzoic Acid
4-*p*-Toluylbenzoic Acid
Xanthene-9-carboxylic Acid, *Me ester*

$C_{15}H_{12}O_3S$
Anthracene-2-sulphonic Acid, *Me ester*
Phenanthrene-1-sulphonic Acid, *Me ester*
Phenanthrene-2-sulphonic Acid, *Me ester*
Phenanthrene-3-sulphonic Acid, *Me ester*
Phenanthrene-9-sulphonic Acid, *Me ester*

$C_{15}H_{12}O_4$
O-Acetylsalicylic Acid, *Phenyl ester*
5-Benzoyl-2-hydroxybenzoic Acid, *Me ester*
5-Benzoyl-2-hydroxybenzoic Acid, *Me ether*
Dehydroptaeroxylin †
1,3-Dihydroxyanthrone, *Di-Me ether*
1,5-Dihydroxyxanthone, *Di-Me ether* †
1,6-Dihydroxyxanthone, *Di-Me ether* †
1,7-Dihydroxyxanthone, *Di-Me ether*
3,4-Dihydroxyxanthone, *Di-Me ether*
Diphenic Acid, *Me ester*
Diphenylmalonic Acid
Diphenylmethane-2,2′-dicarboxylic Acid
Diphenylmethane-2,4-dicarboxylic Acid
Diphenylmethane-2,4′-dicarboxylic Acid
Diphenylmethane-3,3′-dicarboxylic Acid
Diphenylmethane-4,4′-dicarboxylic Acid
Eleutherinol
Emodinanthranol
Frutescinone
Hydrangenol †
2-(2-Hydroxybenzoyl)benzoic Acid, *Me ether*
2-(4-Hydroxybenzoyl)benzoic Acid, *Me ester*
2-(4-Hydroxybenzoyl)benzoic Acid, *Me ether*
β-Isoeuxanthone, *Di-Me ether*
Liquiritigenin
Lucidone †
Malonic Acid, *Di-phenyl ester*
Methylene dibenzoate
Phthalic Acid, *Benzyl ester*
1,2,5,10-Tetrahydroxyanthracene, *Mono-Me ether*
2′,4,4′-Trihydroxybiphenyl-2-carboxylic Acid (2,2′)-Lactone, *Di-Me ether* †

$C_{15}H_{12}O_5$
4-Acetoxy-8,9-epoxy-7-[hexa-,2,4-diynylidene]-1,6-dioxaspiro[4,4]non-2-ene †
Alternariol, *Mono-Me ether*
ψ-Baptigenetin
Butein
Butin
2′,4′-Dihydroxybenzophenone-2-carboxylic Acid, 4′-*Me ether*
4,4′-Dihydroxybenzophenone-2-carboxylic Acid, 4′-*Me ether*
4,5-Dihydroxybenzophenone-2-carboxylic Acid, 4-*Me ether*
4′,5-Dihydroxybenzophenone-2-carboxylic Acid, 4′-*Me ether*
4′,6-Dihydroxybenzophenone-2-carboxylic Acid, 4′-*Me ether*
2′-4′-Dihydroxybenzophenone-2-carboxylic Acid, *Me ester*
3′,4′-Dihydroxybenzophenone-2-carboxylic Acid, *Me ester*
1,6-Dihydroxy-5-methoxyxanthone, 6-*Me ether* †
2,8-Dihydroxy-1-methoxyxanthone, 2-*Me ether* †

$C_{15}H_{12}O_5$ (*continued*)
Flavoskyrin
Fonsecin
Guaiacol, *Di-guaiacol carbonate*
3-Hydroxybiphenyl-4,4′-dicarboxylic Acid, 4-*Me ester*
4-Hydroxy-2,3-dimethoxyxanthone †
5-Hydroxy-1,3-dimethoxyxanthone †
8-Hydroxy-1,2-dimethoxyxanthone †
Hydroxyterephthalic Acid, *Benzyl ether*
Naringenin
1-Phenyl-3-(2,4,6-trihydroxyphenyl)propane-1,3-dione †
Pinobanksin
Rubrofusarin
O-Salicyloylsalicylic Acid, *Me ester*
2,2′,3,3′-Tetrahydroxychalcone
2,2′,3′,4′-Tetrahydroxychalcone
2,2′,4,4′-Tetrahydroxychalcone
2,2′,4,5′-Tetrahydroxychalcone
2,2′,4′,5′-Tetrahydroxychalcone
2,2′,4′,6′-Tetrahydroxychalcone
2′,3,3′,4-Tetrahydroxychalcone
2′,3,3′,4′-Tetrahydroxychalcone
2,3,4,4′-Tetrahydroxychalcone
2′,3′,4,4′,-Tetrahydroxychalcone
2′,3,4,5′-Tetrahydroxychalcone
2′,3′,4′,6′-Tetrahydroxychalcone
2′,4,4′,5′-Tetrahydroxychalcone
2,4,4′,6-Tetrahydroxychalcone
2′,4,4′,6′-Tetrahydroxychalcone
3,3′,4,4′-Tetrahydroxychalcone
3,3′,4′,5-Tetrahydroxychalcone
1,3,7-Trihydroxyxanthone, 3,7-*Di-Me ether*
2,3,8-Trihydroxyxanthone, 2,3-*Di-Me ether*
1,3,5-Trihydroxyxanthone, 3,5-*Di-Me ether* †
1,5,6-Trihydroxyxanthone, 5,6-*Di-Me ether* †
2,2′,4-Trihydroxybenzil, 2′-*Me ether*
2,4,4′-Trihydroxybenzil, 4′-*Me ether*
2′,5,7-Trihydroxyflavanone
2′,7,8-Trihydroxyflavanone
3′,5,7-Trihydroxyflavanone
4′,7,8-Trihydroxyflavanone †

$C_{15}H_{12}O_6$
Aromadendrin
Bellidifolin, 5-*Me ether* †
Carthamidin
Cyanomaclurin
Dehydroaltenusin
Eriodictyol
Fustin
Isocarthamidin
Maesopsin
Micromelin †
2′,3,3′,4,4′-Pentahydroxychalcone †
2′,3,4,4′,5-Pentahydroxychalcone
Swerchirin †
Swertinin †
3,4′,5,7-Tetrahydroxyflavanone
3′,4′,5,7-Tetrahydroxyflavanone
3′,4′,5′,7-Tetrahydroxyflavanone
3,4′,7,8-Tetrahydroxyflavanone †
3′,4′,7,8-Tetrahydroxyflavanone †

$C_{15}H_{12}O_7$
Alphitonin
3,3′,4′,5,7-Pentahydroxyflavanone
3,3′,4′,5′,7-Pentahydroxyflavanone
3,3′,4′,7,8-Pentahydroxyflavanone †
1,3,4,7,8-Pentahydroxyxanthone, 4,7-*Di-Me ether* †

$C_{15}H_{12}O_8$
Ampelopsin
Isogalloflavin, *Tri-Me ether*

$C_{15}H_{12}O_9$
Digallic Acid, *Me ester*

$C_{15}H_{12}S$
3-Phenanthrenethiol, S-*Me*

$C_{15}H_{13}BrN_2$
Bromomalondialdehyde, *Anilino-anil*

$C_{15}H_{13}BrO_2$
3′-Bromodiphenyl-4-carboxylic Acid
2-Bromo-4′-hydroxybenzophenone, *Et ether*
3-Bromo-4-hydroxybenzophenone, *Et ether*
3-Bromo-4′-hydroxybenzophenone, *Et ether*
4-Bromo-4′-hydroxybenzophenone, *Et ether*

$C_{15}H_{13}ClO$
2,2-Diphenylpropionic Acid, *Chloride*
2,3-Diphenylpropionic Acid, *Chloride*

$C_{15}H_{13}I_2NO_4$
β-[4-(*p*-Hydroxyphenoxy)-3,5-di-iodophenyl]-alanine

$C_{15}H_{13}N$
1-Benzylindole
2-Benzylindole
3-Benzylindole
3,4-Dihydro-1-phenylisoquinoline
2,3-Dimethylacridine
2,4-Dimethylacridine
2,7-Dimethylacridine
2,9-Dimethylacridine
3,9-Dimethylacridine
4,5-Dimethylacridine
1,4-Dimethylphenanthridine
2,4-Dimethylphenanthridine
3,6-Dimethylphenanthridine
6,9-Dimethylphenanthridine
2,2-Diphenylpropionic Acid, *Nitrile*
2,3-Diphenylpropionic Acid, *Nitrile*
3,3-Diphenylpropionic Acid, *Nitrile*
9*H*-9,10-Ethanoacridine
9-Methylaminoanthracene
9-Phenanthrenemethylamine
2-Phenylindole, N-*Me*
3-Phenylindole, N-*Me*

$C_{15}H_{13}NO$
Acridone, *N*-Et
2-Aminochalcone
3-Aminochalcone
4-Aminochalcone
2′-Aminochalcone
β-Aminochalcone
1,10-Dimethylacridone
2,3-Dimethylacridone
2,4-Dimethylacridone
2,10-Dimethylacridone
3,10-Dimethylacridone
4,10-Dimethylacridone

$C_{15}H_{13}NO$ (*continued*)
2,3-Diphenylacrylic Acid, *Amide*
2-Hydroxyacridine, *Et ether*
4-Hydroxyacridine, *Et ether*
Phthalimidine, N-*Benzyl*

$C_{15}H_{13}NO_2$
2-Aminostilbene-4-carboxylic Acid
α-Benzoylphenylacetic Acid, *Amide*
Bicyclo[2,2,1]hept-5-ene-2,3-dicarboxylic Acid, *Anil*
4-Hydroxyacridone, *Et ether*
2-Methyl-3′-nitrostilbene
4-Methyl-2′-nitrostilbene
4-Methyl-3′-nitrostilbene
4-Methyl-4′-nitrostilbene
Murrayanine, N-*Me*†
1′-Nitro-1-phenyl-2-*o*-tolylethylene
1-Nitro-1-phenyl-2-*m*-tolylethylene
1-Nitro-1-phenyl-2-*p*-tolylethylene
1-Nitro-2-phenyl-1-*o*-tolylethylene
1-Nitro-2-phenyl-1-*m*-tolylethylene
1-Nitro-2-phenyl-1-*p*-tolylethylene
α-Phenylacetophenone-2-carboxylic Acid, *Amide*
Phthalimidine, N-o-*Hydroxybenzyl*
Phthalimidine, N-p-*Hydroxybenzyl*
2-*p*-Toluylbenzoic Acid, *Amide*
4-*p*-Toluylbenzoic Acid, *Amide*

$C_{15}H_{13}NO_3$
Benzoylanthranilic Acid, *Me ester*
1,3-Dihydroxyacridone, *Di-Me ether*
1,4-Dihydroxyacridone, *Di-Me ether*
2,3-Dihydroxyacridone, *Di-Me ether*
3,4-Dihydroxyacridone, *Di-Me ether*
3,6-Dihydroxyacridone, *Di-Me ether*
4,5-Dihydroxyacridone, *Di-Me ether*
2,4-Dimethyl-3-nitrobenzophenone
2,4-Dimethyl-3′-nitrobenzophenone
2,4-Dimethyl-5-nitrobenzophenone
2,5-Dimethyl-3′-nitrobenzophenone
3,4-Dimethyl-3′-nitrobenzophenone
Hippuric Acid, *Phenyl ester*
2-Hydroxy-2′-nitrostilbene, *Me ether*
2-Hydroxy-3′-nitrostilbene, *Me ether*
2-Hydroxy-4′-nitrostilbene, *Me ether*
2-Hydroxy-β-nitrostilbene, *Me ether*
3-Hydroxy-4′-nitrostilbene, *Me ether*
4-Hydroxy-2-nitrostilbene, *Me ether*
4-Hydroxy-2′-nitrostilbene, *Me ether*
4-Hydroxy-3′-nitrostilbene, *Me ether*
4-Hydroxy-4′-nitrostilbene, *Me ether*
4-Hydroxy-β-nitrostilbene, *Me ether*
α-Hydroxy-β-nitrostilbene, *Me ether*
Oxanilic Acid, m-*Tolyl ester*
Oxanilic Acid, p-*Tolyl ester*
N-Phenacyl-*o*-aminobenzoic Acid
N-Phenacyl-*m*-aminobenzoic Acid
N-Phenacyl-*p*-aminobenzoic Acid
2-Phenylhippuric Acid

$C_{15}H_{13}NO_4$
p-Acetamidophenyl salicylate
O-Benzamidoacetylresorcinol
2-Benzamido-3-(2-furyl)acrylic Acid, *Me ester*
2-Hydroxy-5-methyl-4′-nitrobenzophenone, *Me ether*
4-Hydroxy-2-methyl-*a*-*p*-nitrophenylacetophenone
4-Hydroxy-2′-nitrobenzophenone, *Et ether*
4-Hydroxy-3′-nitrobenzophenone, *Et ether*
4-Hydroxy-4′-nitrobenzophenone, *Et ether*
2′-Nitrobiphenyl-3-carboxylic Acid, *Et ester*
4′-Nitrobiphenyl-4-carboxylic Acid, *Et ester*
p-Nitrophenylacetic Acid, *Benzyl ester*
Orixidine
N-Phenylglycine-*o*-carboxylic Acid, N-*Phenyl*
6-Phenylpyridine-3,4-dicarboxylic Acid, *Di-Me ester*
o-Toluic Acid, p-*Nitrobenzyl ester*
m-Toluic Acid, p-*Nitrobenzyl ester*
p-Toluic Acid, p-*Nitrobenzyl ester*
Xanthoxoline

$C_{15}H_{13}NO_4S$
2-*o*-Nitrophenylmercaptobenzoic Acid, *Et ester*
2-*p*-Nitrophenylmercaptobenzoic Acid, *Et ester*

$C_{15}H_{13}NO_5$
Anisic Acid, 4-*Nitrobenzyl ester*
2,4-Dihydroxy-3′-nitrobenzophenone, *Di-Me ether*
2,4-Dihydroxy-4′-nitrobenzophenone, *Di-Me ether*
2,5-Dihydroxy-4′-nitrobenzophenone, *Di-Me ether*
2-Hydroxy-*m*-toluic Acid, p-*Nitrobenzyl ester*
6-Hydroxy-*m*-toluic Acid, p-*Nitrobenzyl ester*
2-Hydroxy-*p*-toluic Acid, p-*Nitrobenzyl ester*
3-Nitro-4-phenoxybenzoic Acid, *Et ester*
4-(*p*-Nitrophenoxy)benzoic Acid, *Et ester*
Vanillin, p-*Nitrobenzyl ether*

$C_{15}H_{13}NO_5S$
2-*o*-Nitrophenylsulphinylbenzoic Acid, *Et ester*
2-*p*-Nitrophenylsulphinylbenzoic Acid, *Et ester*

$C_{15}H_{13}NO_6$
Vanillic Acid, p-*Nitrobenzyl ester*

$C_{15}H_{13}NO_6S$
2-*p*-Nitrophenylsulphonylbenzoic Acid, *Et ester*

$C_{15}H_{13}NS$
9-Acridinethiol, *S*-Et

$C_{15}H_{13}N_3O_4$
Diazoaminobenzene-2,2′-dicarboxylic Acid, *Me ester*
2-Nitroazobenzene-4-carboxylic Acid, *Et ester*
2′-Nitroazobenzene-4-carboxylic Acid, *Et ester*
3′-Nitroazobenzene-4-carboxylic Acid, *Et ester*
4′-Nitroazobenzene-4-carboxylic Acid, *Et ester*
Nitromalonic Acid, *Dianilide*

$C_{15}H_{13}N_3O_5$
4-Hydroxy-4′-nitroazobenzene-3-carboxylic Acid, *Et ester*
4-Hydroxy-5-nitroazobenzene-3-carboxylic Acid, *Et ester*

$C_{15}H_{13}N_3O_6$
2′,4′-Dinitrodiphenylamine-2-carboxylic Acid, *Et ester*
2′,4′-Dinitrodiphenylamine-3-carboxylic Acid, *Et ester*
2′,4′-Dinitrodiphenylamine-4-carboxylic Acid, *Et ester*
2,6-Dinitrodiphenylamine-4-carboxylic Acid, *Et ester*

$C_{15}H_{14}$
9,10-Dihydro-2-methylanthracene
9,10-Dihydro-9-methylanthracene
1,2-Dimethylfluorene
1,3-Dimethylfluorene
1,4-Dimethylfluorene
2,7-Dimethylfluorene
2,3-Dimethylfluorene
2,5-Dimethylfluorene
2,6-Dimethylfluorene
2,7-Dimethylfluorene
3,4-Dimethylfluorene
3,6-Dimethylfluorene
3,7-Dimethylfluorene
4,6-Dimethylfluorene
9,9-Dimethylfluorene
1,2-Diphenylcyclopropane†
1,1-Diphenylpropene
1,2-Diphenylpropene
1,3-Diphenylpropene
3,3-Diphenylpropene
3-Methylstilbene
4-Methylstilbene
1-Phenylindane
2-Phenylindane

$C_{15}H_{14}Br_2O$
Di-4-bromophenylmethanol, *Et ether*

$C_{15}H_{14}ClNO_2$
Diphenylcarbamic Acid, 2-*Chloroethyl ester*

$C_{15}H_{14}ClN_3O$
4-(7-Chloro-4-quinolylamino)-α-amino-*o*-cresol†

$C_{15}H_{14}FNO_4$
2-Fluorothyronine
5′-Fluorothyronine

$C_{15}H_{14}N_2$
4-Aminoacridine, *N*-Et
3-Amino-2,7-dimethylacridine
9-Amino-3,4-dimethylacridine
9-Amino-4,5-dimethylacridine
Dibenzylcyanamide
1,3-Diphenylpyrazoline
1,5-Diphenylpyrazoline
3,5-Diphenylpyrazoline

$C_{15}H_{14}N_2O$
3-Amino-3,4-dihydro-2-quinolone, N-*Phenyl*

$C_{15}H_{14}N_2O$
Phthalimidine, N-o-*Aminobenzyl*
Phthalimidine, N-p-*Aminobenzyl*

$C_{15}H_{14}N_2O_2$
Azobenzene-2-carboxylic Acid, *Et ester*
Azobenzene-4-carboxylic Acid, *Et ester*
Benzil dioxime, O-*Me*
Diphenylmethane-2,4′-dicarboxylic Acid, *Diamide*
Malonanilide

$C_{15}H_{14}N_2O_3$
4-Amino-3-nitrobenzophenone, N-*Di-Me*
4-Amino-3-nitrobenzophenone, N-*Et*
4′-Amino-3-nitrobenzophenone, N-*Di-Me*
Anthranoylanthranilic Acid, *Me ester*
4-Hydroxyazobenzene-3-carboxylic Acid, *Et ester*
4-*o*-Nitrobenzylideneaminophenetole
4-*p*-Nitrobenzylideneaminophenetole
Pyrogalline, *Me ether*

$C_{15}H_{14}N_2O_4$
2′-Methyl-5′-nitrodiphenylamine-2-carboxylic Acid, *Me ester*
5-Methyl-4-nitrodiphenylamine-2-carboxylic Acid, *Me ester*
4-Nitrodiphenylamine-2-carboxylic Acid, *Et ester*
2-Nitrodiphenylamine-4-carboxylic Acid, *Et ester*

$C_{15}H_{14}N_4O_2S$
Sulphaphenazole

$C_{15}H_{14}N_4O_6S_2$
Althiomycin

$C_{15}H_{14}N_6O_3$
N^{10}-Methylpteroic Acid

$C_{15}H_{14}O$
ω-Benzylacetophenone
4-Cinnamylphenol
2,2′-Dimethylbenzophenone
2,3′-Dimethylbenzophenone
2,4-Dimethylbenzophenone
2,4′-Dimethylbenzenophenone
2,5-Dimethylbenzophenone
3,3′-Dimethylbenzophenone
3,4-Dimethylbenzophenone
3,4′-Dimethylbenzophenone
4,4′-Dimethylbenzophenone
2,2-Dimethylnaphtho[1,2-*b*]pyran†
1,1-Diphenylacetone
1,3-Diphenylacetone
Flavan
2-Hydroxystilbene, *Me ether*
3-Hydroxystilbene, *Me ether*
4-Hydroxystilbene, *Me ether*
Lactaroviolin
Linderazulene
α-Methyl-α-phenylacetophenone
2-Methyl-α-phenylacetophenone
3-Methyl-α-phenylacetophenone
4-Methyl-α-phenylacetophenone
4-Phenylchroman
α-*o*-Tolylacetophenone
α-*m*-Tolylacetophenone
α-*p*-Tolylacetophenone

$C_{15}H_{14}O_2$
Benzoic Acid, *Phenethyl ester*
Benzoin, *Me ether*
Biphenyl-2-carboxylic Acid, *Et ester*
Biphenyl-4-carboxylic Acid, *Et ester*
3,5-Dihydroxystilbene, *Me ether*

$C_{15}H_{14}O_2$ (*continued*)
2-(γ,γ-Dimethylallyl)-1,4-naphthoquinone
Diphenylacetic Acid, *Me ester*
2,2-Diphenylpropionic Acid
2,3-Diphenylpropionic Acid
3,3-Diphenylpropionic Acid
Flavanol
Flavan-4β-ol†
2-Hydroxybenzophenone, *Et ether*
4-Hydroxybenzophenone, *Et ether*
2-Hydroxy-3,5-dimethylbenzophenone
2-Hydroxy-4′,5-dimethylbenzophenone
3-Hydroxy-4,6-dimethylbenzophenone
4-Hydroxy-2,5-dimethylbenzophenone
4-Hydroxy-3,5-dimethylbenzophenone
α-Hydroxy-4-methylacetophenone, *Phenyl ether*
1-Hydroxy-3-phenylacetone, *Phenyl ether*
4-Hydroxy-α-phenylacetophenone, *Me ether*
α-4-Hydroxyphenylacetophenone, *Me ether*
β-(*p*-Hydroxyphenyl)-propiophenone
γ-(*o*-Hydroxyphenyl)-propiophenone
γ-(*p*-Hydroxyphenyl)-propiophenone
β-Hydroxy-β-phenylpropiophenone
β-Hydroxy-γ-phenylpropiophenone
γ-Hydroxy-β-phenylpropiophenone
p-Hydroxy-β-phenylpropiophenone
p-Hydroxy-γ-phenylpropiophenone
β-Hydroxypropiophenone, *Phenyl ether*
γ-Hydroxypropiophenone, *Phenyl ether*
4-Hydroxypropiophenone, *Phenyl ether*
2-Methylbiphenyl-4-carboxylic Acid, *Me ester*
3-Methylbiphenyl-4-carboxylic Acid, *Me ester*
Phenylacetic Acid, *Benzyl ester*
Phenyl-*o*-toluylmethanol
Phenyl-*p*-toluylmethanol
Resorcinol, *Me-α-phenylvinyl ether*
o-Toluic Acid, *Benzyl ester*
m-Toluic Acid, *Benzyl ester*
p-Toluic Acid, *Benzyl ester*

$C_{15}H_{14}O_2S$
α-Mercaptophenylacetic Acid, S-*Benzyl*†
o-Phenylmercaptobenzoic Acid, *Et ester*

$C_{15}H_{14}O_2Se$
Selenobenzoic Acid, p-*Ethoxyphenyl ester*

$C_{15}H_{14}O_3$
Allodunnione
Anisic Acid, *Benzyl ester*
Benzilic Acid, *Me ester*
Benzilic Acid, *Me ether*
5-Benzylsalicylic Acid, *Me ester*
5-Benzylsalicylic Acid, *Me ether*
α,4-Dihydroxyacetophenone, *Phenyl ether*
2,2′-Dihydroxybenzophenone, *Di-Me ether*
2,3-Dihydroxybenzophenone, *Di-Me ether*
2,4-Dihydroxybenzophenone, *Di-Me ether*
2,4′-Dihydroxybenzophenone, *Di-Me ether*
2,5-Dihydroxybenzophenone, *Di-Me ether*
3,4-Dihydroxybenzophenone, *Di-Me ether*
3,4′-Dihydroxybenzophenone, *Di-Me ether*
4,4′-Dihydroxybenzophenone, *Di-Me ether*
4,4′-Dihydroxybenzophenone, *Et ether*
4′,7-Dihydroxyisoflavan
2,4-Dihydroxy-3-methylbenzophenone, 4-*Me ether*
2,4′-Dihydroxy-5-methylbenzophenone, 6-*Me ether*
2,4′-Dihydroxy-5-methylbenzophenone, 4′-*Me ether*
4,4′-Di[hydroxymethyl]benzophenone
Dunnione
Flavan-3,4-diol†
Frutescine
Hircinol†
3-Hydroxybiphenyl-2-carboxylic Acid, *Et ester*
6-Hydroxybiphenyl-2-carboxylic Acid, *Et ester*
2-Hydroxy-2,3-diphenylpropionic Acid
2-Hydroxy-3,3-diphenylpropionic Acid
o-Hydroxymethyl-benzoic Acid, *Phenyl ether, Me ester*
p-Hydroxymethyl-benzoic Acid, *Benzyl ester*
α-2-Hydroxyphenoxyacetophenone, *Me ether*
α-4-Hydroxyphenoxyacetophenone, *Me ether*
2-Hydroxy-3-phenylpropionic Acid, *Phenyl ether*
3-Hydroxy-3-phenylpropionic Acid, *Phenyl ether*
Isolapachol
Lapachol
α-Lapachone
β-Lapachone
Mandelic Acid, *Benzyl ester*
Mansonone D†
Mansonone E†
Methyl salicylate, *Benzyl ether*
3-α-Naphthoylpropionic Acid, *Me ester*
3-β-Naphthoylpropionic Acid, *Me ester*
o-Phenoxybenzoic Acid, *Et ester*
2-Phenoxypropionic Acid, *Phenyl ester*
Vanillin, *Benzyl ether*
Xanthorrhoeol†

$C_{15}H_{14}O_4$
Alloptaeroxylin†
Alloxanthoxyletin
Braylin
Catalpalactone†
Cotoin, *Me ether*
Dehydrogeijerin†
3,4-Dicarbomethoxybicyclo[3,2,2]nonatriene†
1,6-Diphenyl-1,3,4,6-hexanetetrone
Eleutherol, *Me ether*
Luvangetin
Mansonone H†
Methyl 3-[5-(hept-*cis*-4-en-2-ynoyl)-2-furyl]-*trans*-acrylate†
1-Naphthylmalonic Acid, *Di-Me ester*
Peucedanine
Phebalosin†
Ptaeroxylin†
Rhapontigenin
Santal, *Di-Me ether*
Tetradec-9-ene-2,4,6-triynedioic Acid, *Me ester*
2,3,4-Trihydroxybenzophenone, 3,4-*Di-Me ether*
2,4,5-Trihydroxybenzophenone, 4,5-*Di-Me ether*
2,4,6-Trihydroxybenzophenone, 2,4-*Di-Me ether*
2,4,6-Trihydroxybenzophenone, 2,6-*Di-Me ether*

$C_{15}H_{14}O_4$ (*continued*)
3′,4′,7-Trihydroxyflavan
Xanthoxyletin
Yangonin

$C_{15}H_{14}O_4S$
o-Phenylsulphonylbenzoic Acid, *Et ester*

$C_{15}H_{14}O_5$
(−)-*epi*-Afzelechin
Clausenin, *Me ether* †
Karenin †
Methysticic Acid
Methysticin
Methysticone, *γ-Carbomethoxyl*
Phloretin
Ptaerochromenol †
Ptaeroxylinol †
Scleroin †
α-Sorigenin, *Di-Me ether*
2,3,4,5-Tetrahydroxybenzophenone, 3,4-*Di-Me ether*
3,3′,4′,7-Tetrahydroxyflavan †
3,4,4′,7-Tetrahydroxyflavan★ †
3′,4′,5′,7-Tetrahydroxyflavan
Umtatin †

$C_{15}H_{14}O_6$
Altenusin
Brevifolin, *Tri-Me ether*
Catechin
5,8-Dihydroxy-6-methoxy-2-methyl-3,2′-oxopropyl-1,4-naphthaquinone
Mikanolide †
2,2′,4,6,6′-Pentahydroxybenzophenone, 2′,6′-*Di-Me ether*
3,3′,4,4′,7-Pentahydroxyflavan
3,3′,4′,5′,7-Pentahydroxyflavan
3,4,4′,7,8-Pentahydroxyflavan
Phthaloylmalonic Acid, *Di-Et ester*
Ptaeroglycol †

$C_{15}H_{14}O_7$
Canescin †
Fusarubin
Gallocatechin
epi-Gallocatechin
3,3′,4,4′,5′,7-Hexahydroxyflavan
3,3′,4,4′,7,8-Hexahydroxyflavan
Ptaerocyclin †
Spinochrome C, *Tri-Me ether* †

$C_{15}H_{14}O_8$
Altenuic Acid I
Altenuic Acid II
Altenuic Acid III
3,3′,4,4′,5,5′,7-Heptahydroxyflavan †

$C_{15}H_{14}S$
3,5,8-Trimethylazuleno[6,5-*b*]thiophen †

$C_{15}H_{15}Ce$
Tricyclopentadienyl Cerium

$C_{15}H_{15}Cl_2N$
Di-(4-chlorobenzyl)amine, N-*Me*

$C_{15}H_{15}Gd$
Tricyclopentadienyl Gadolinium

$C_{15}H_{15}La$
Tricyclopentadienyl Lanthanum

O

$C_{15}H_{15}N$
N-Benzylidene-2,4-xylidine
N-Benzylidene-2,5-xylidine
N-Benzylidine-2,6-xylidine
2,4-Dimethyl-6-styrylpyridine
2-Methyl-6-(4-methylstyryl)pyridine
4-Methyl-2-(4-methylstyryl)pyridine
1,2,3,4-Tetrahydro-1-phenylisoquinoline
1,2,3,4-Tetrahydro-2-phenylisoquinoline
1,2,3,4-Tetrahydro-3-phenylisoquinoline
1,2,3,4-Tetrahydro-2-phenylquinoline
1,2,3,4-Tetrahydro-3-phenylquinoline
1,2,3,4-Tetrahydro-4-phenylquinoline
1,2,3,4-Tetrahydro-6-phenylquinoline
5,6,7,8-Tetrahydro-2-phenylquinoline
5,6,7,8-Tetrahydro-3-phenylquinoline
1,3,6-Trimethylcarbazole
1,3,7-Trimethylcarbazole
1,4,8-Trimethylcarbazole
1,4,9-Trimethylcarbazole
2,4,6-Trimethylcarbazole

$C_{15}H_{15}NO$
α-Benzylaminoacetophenone
p-Benzylideneaminophenol, *Et ether*
3-Dimethylaminobenzophenone
4-Dimethylaminobenzophenone
2,2-Diphenylpropionic Acid, *Amide*
2,3-Diphenylpropionic Acid, *Amide*
3,3-Diphenylpropionic Acid, *Amide*
N-Ethylaniline, N-*Benzoyl*
o-Ethylaniline, N-*Benzoyl*
p-Ethylaniline, N-*Benzoyl*
p-Hydroxybenzylidene-*m*-toluidine, *Me ether*

$C_{15}H_{15}NO_2$
2-Anilinoethanol, O-*Benzoyl*
Apo-β-erythroidine
N-Benzoyl-3-hydroxyaniline, *Et ether*
N-Benzoyl-4-hydroxyaniline, *Et ether*
p-Dimethylaminobenzoic Acid, *Phenyl ester*
Diphenylamine-2-carboxylic Acid, *Et ester*
Diphenylcarbamic Acid, *Et ester*
Mandelic Acid, o-*Toluidide*
Mandelic Acid, p-*Toluidide*
N-Methylphenylcarbamic Acid, p-*Tolyl ester*
2-Methyl-6-phenylpyridine-3-carboxylic Acid, *Et ester*
2-Methyl-6-phenylpyridine-4-carboxylic Acid, *Et ester*
N-Phenylglycine, p-*Tolyl ester*

$C_{15}H_{15}NO_2S$
2-(*p*-Toluenesulphonyl)dihydroisoindole †

$C_{15}H_{15}NO_3$
Ismine
2-Nitro-1,2-diphenylethanol, *Me ether*
Palasonin, N-*Phenylimide* †

$C_{15}H_{15}NO_4$
Quinoline-2,3-dicarboxylic Acid, *Di-Et ester*

$C_{15}H_{15}NO_4S$
Benzo[*h*]quinoline, *Methosulphate*

$C_{15}H_{15}NO_5$
Acronycidine

$C_{15}H_{15}NO_5S$
2,5-Dimethyl-6-nitrobenzenesulphonic Acid, o-*Tolyl ester*

$C_{15}H_{15}NO_6$
3,4-Dimethoxybenzoic Acid, p-*Nitrobenzyl ester*
$C_{15}H_{15}NO_7$
Narciclasine, *Me ether* †
$C_{15}H_{15}N_3$
3,6-Diamino-2,7-dimethylacridine
$C_{15}H_{15}N_3O$
3,9-Diamino-7-ethoxyacridine
$C_{15}H_{15}N_3O_3$
4-Hydroxy-2,3-dimethyl-4′-nitroazobenzene, *Me ether*
4-Hydroxy-2,5-dimethyl-4′-nitroazobenzene, *Me ether*
4-Hydroxy-2,6-dimethyl-4′-nitroazobenzene, *Me ether*
2-Hydroxy-5-methyl-3′-nitroazobenzene, *Et ether*
4-Hydroxy-2′-methyl-3-nitroazobenzene, *Et ether*
4-Hydroxy-3′-methyl-3-nitroazobenzene, *Et ether*
4-Hydroxy-4′-methyl-3-nitroazobenzene, *Et ether*
4′-Hydroxy-2-methyl-5-nitroazobenzene, *Et ether*
$C_{15}H_{15}N_3O_3S_2$
Amiphenazole, *Benzenesulphonate*
$C_{15}H_{15}N_3O_5$
4-Amino-2,6-dinitrophenol, N-*Benzoyl*, N-*Et*
$C_{15}H_{15}N_7O_2$
Amino-methylpteroic Acid
$C_{15}H_{15}Nd$
Tricyclopentadienyl Neodymium
$C_{15}H_{15}Pr$
Tricyclopentadienyl Praseodymium
$C_{15}H_{15}Sc$
Tricyclopentadienyl Scandium
$C_{15}H_{15}Sm$
Tricyclopentadienyl Samarium
$C_{15}H_{15}Y$
Tricyclopentadienyl Yttrium
$C_{15}H_{16}$
1,1-Diphenylpropane
1,2-Diphenylpropane
1,3-Diphenylpropane
2,2-Diphenylpropane
Di-*p*-tolylmethane
Lactarazulene
1-Phenyl-1-*p*-tolylethane
1-Phenyl-2-*p*-tolyethane
2-Phenyl-2-*p*-tolylethane
$C_{15}H_{16}ClN_3O_4S_3$
Benzthiazide
$C_{15}H_{16}N_2O$
N-Dibenzylurea
NN′-Dibenzylurea
1,3-Di-*o*-tolylurea
1,3-Di-*m*-tolylurea
1,3-Di-*p*-tolylurea
4-Hydroxyazobenzene, *Propyl ester*
2′-Hydroxy-4-methylazobenzene, *Et ether*
2-Hydroxy-5-methylazobenzene, *Et ether*
4-Hydroxy-2-methylazobenzene, *Et ether*
4-Hydroxy-3-methylazobenzene, *Et ether*
4′-Hydroxy-2-methylazobenzene, *Et ether*
4′-Hydroxy-3-methylazobenzene, *Et ether*
4-Hydroxy-4′-methylazobenzene, *Et ether*
2-Hydroxy-3,4′,5-trimethylazobenzene
2′-Hydroxy-2,4,5′-trimethylazobenzene
2-Hydroxy-3′,4′,5-trimethylazobenzene
4-Hydroxy-2′,4′,5′-trimethylazobenzene
o-Tolyl-*m*-tolylurea
o-Tolyl-*p*-tolylurea
$C_{15}H_{16}N_2O_2$
Albonoursin †
2-Aminodiphenylamine-4-carboxylic Acid, *Et ester*
1-*o*-Hydroxyphenyl-3-phenylurea, *Et ether*
o-Nitrobenzylaniline, N-*Et*
m-Nitrobenzylaniline, N-*Et*
p-Nitrobenzylaniline, N-*Et*
$C_{15}H_{16}N_2O_2S$
1,3-Di-*p*-hydroxyphenylthiourea, *Di-Me ether*
$C_{15}H_{16}N_2O_3$
α-4-Aminophenyl-3-nitrobenzyl Alcohol, N-*Di-Me*
α-4-Aminophenyl-4-nitrobenzyl Alcohol, N-*Di-Me*
α-4-Aminophenyl-4-nitrobenzyl Alcohol, N-*Et*
1,3-Di-*o*-hydroxyphenylurea, *Di-Me ether*
1,3-Di-*p*-hydroxyphenylurea, *Di-Me ether*
N-2-Nitrobenzyl-*p*-phenetidine
$C_{15}H_{16}N_2O_3S$
3,4-Dimethyl-*N*-sulphanilylbenzamide
$C_{15}H_{16}N_2O_4$
3-Phenylpyrazole-1,4-dicarboxylic Acid, *Di-Et ester*
4-Phenylpyrazole-3,5-dicarboxylic Acid, *Di-Et ester*
$C_{15}H_{16}N_2O_5S_2$
Gliotoxin, *Ac*
7-Methoxymitosene †
$C_{15}H_{16}N_2S$
N-Dibenzylthiourea
NN′-Dibenzylthiourea
1,3-Di-*o*-tolylthiourea
1,3-Di-*m*-tolylthiourea
1,3-Di-*p*-tolylthiourea
2-Isobutylaminonaphtho[1,2-*d*]thiazole
$C_{15}H_{16}N_4O_6$
Araboflavin
$C_{15}H_{16}O$
4-Benzylphenol, *Et ether*
2,2-Diphenylethanol, *Me ether*
1,1-Diphenyl-1-propanol
1,2-Diphenyl-1-propanol
1,3-Diphenyl-1-propanol
2,2-Diphenyl-1-propanol
2,3-Diphenyl-1-propanol
1,2-Diphenyl-2-propanol
1,3-Diphenyl-2-propanol
Di-*o*-tolylmethanol

$C_{15}H_{16}O$ (*continued*)
Di-*p*-tolylmethanol
3,4-Dihydro-2,2-dimethylnaphtho[1,2-*b*]pyran †
3-Hydroxymethylbiphenyl, *Et ether*
1-*o*-Hydroxyphenyl-2-phenylethane, *Me ether*
1-*p*-Hydroxyphenyl-2-phenylethane, *Me ether*
1-Naphthyl isopentenyl Ether †
Phenyl-*p*-tolylmethanol, *Me ether*
Pyrocurzerenone †
1,2,3,4-Tetrahydroanthranol, *Me ether*
o-Tolyl-*p*-tolylmethanol

$C_{15}H_{16}O_2$
2-Butyryl-1-naphthol, *Me ether*
4-Butyryl-1-naphthol, *Me ether*
3,3′-Dihydroxy-6-methylbibenzyl †
2,2-Di-*p*-hydroxyphenylpropane
4,8-Dimethylazulene-6-carboxylic Acid, *Et ester*
2-(3,8-Dimethyl-5-azulenyl)propionic Acid
1,2-Diphenyl-1,2-ethanediol, *Me ether*
2-Hydroxy-α-4-hydroxyphenyltoluene
4-Hydroxy-α-4-hydroxyphenyltoluene
Mansonone C †
2-Methyl-4,1′-naphthylbutyric Acid
4-Methyl-4,1′-naphthylbutyric Acid
2-Methyl-4,2′-naphthylbutyric Acid
3-[6-Methyl-1-naphthyl]-butyric Acid
4-[5-Methyl-1-naphthyl]-butyric Acid
4-[6-Methyl-1-naphthyl]-butyric Acid
4-α-Naphthylbutyric Acid, *Me ester*
3-α-Naphthylpropionic Acid, *Et ester*
3-β-Naphthylpropionic Acid, *Et ester*
Nordihydrolapachenole †
3-Phenyl-1,2-propanediol, 1-*Phenyl ether*
1,3-Propanediol, *Diphenyl ether*
2-Propionyl-1-naphthol, *Et ether*
Resorcinol, *Mono-phenylpropyl ether*

$C_{15}H_{16}O_3$
Cacalone †
Desmotropo-ψ-santonin
3,3′-Dihydroxydiphenyl Ether, 3-*Me*, 3′-*Et ether*
3-(1,1-Dimethylallyl)herniarin †
Glycerol, 1,3-*Diphenyl ether*
4-Hydroxy-1-naphthoic Acid, *Butyl ester*
4-Hydroxy-1-naphthoic Acid, *Butyl ether*
5-Hydroxy-1-naphthoic Acid, *Butyl ether*
8-Hydroxy-1-naphthoic Acid, *Butyl ether*
3-Hydroxy-2-naphthoic Acid, *Et ether*, *Et ester*
6-Hydroxy-2-naphthoic Acid, *Butyl ether*
Isolinderalactone †
Kawaic Acid, *Me ester*
Linderalactone †
Mansonone G †
Osthol
Suberosin
Virginolide †
Yomogin †

$C_{15}H_{16}O_4$
Allopeucenin
Aurapten
Auraptenol †
Cedrelopsin †
Fuscin
Geijerin
Glaupalol †
Griesenin †
Heteropeucenin †
Indene-2,3-dicarboxylic Acid, *Di-Et ester*
Isabelin †
Isoaurapten
Linderane †
Methoxyaucuparin
7-Methoxy-8-(2-formyl-2-methylpropyl)-coumarin †
3-Methylchromone-2-carboxylic Acid, n-*Butyl ester*
Nieshoutin †
Neolinderane †
Obliquetin †
Peucenin
Phyllomeronic Acid, *Di-Me ether*, *Me ester*
Prenyletin, 6-O-*Me ether* †
Pseudoneolinderane †

$C_{15}H_{16}O_5$
Ascochitine †
Linderadine †
Nieshoutol †
Purpurogallin, *Tetra-Me ether*
Vernolepin †
Vernomenin †
Visamminol

$C_{15}H_{16}O_6$
Coriaric Acid
Dihydromikanolide †
Elephantol †
Picrotoxinin

$C_{15}H_{16}O_6S_2$
Methionic Acid, *Di-o-tolylester*
Propane-1,1-disulphonic Acid, *Diphenyl ester*
Propane-2,2-disulphonic Acid, *Diphenyl ester*

$C_{15}H_{16}O_7$
Spinochrome P, *Tri-Me ether*

$C_{15}H_{16}O_8$
Caffeic Acid, *Et ester*
4-Hydroxycoumarin, *Glucoside*
Leucodrin
Skimmin

$C_{15}H_{16}O_9$
Aesculin
Cichoriin
Daphnin

$C_{15}H_{17}HgNO_2S$
N-Ethylmercuri-*p*-toluenesulphonamide

$C_{15}H_{17}N$
α-Aminodiphenylmethane, N-*Et*
N-Benzyl-*N*-ethylaniline
N-Benzyl-1-phenylethylamine
N-Benzyl-2-phenylethylamine
2,5-Dimethylaniline, N-*Benzyl*
1,3-Diphenylpropylamine
2,2-Diphenylpropylamine
2,3-Diphenylpropylamine
3,3-Diphenylpropylamine
N-Methyldibenzylamine
N-Methyl-4,4′-ditolylamine
1-Phenyl-1-propylamine, N-*Phenyl*

$C_{15}H_{17}NO$
4-Amino-3,5-xylenol, N-*Benzyl*
2-Anilinoethanol, p-*Tolyl ether*
N-4-Hydroxybenzylaniline, *Et ether*
3-Hydroxy-4′-methyldiphenylamine, *Et ether*
3′-Hydroxy-2-methyldiphenylamine, *Et ether*
4′-Hydroxy-2-methyldiphenylamine, *Et ether*

$C_{15}H_{17}NO_2$
α-Aminomethyl-*p*-hydroxybenzyl Alcohol, N-*Benzyl*
3-Dimethylallyl-4-methoxy-2-quinoline †
1,2-Dimethylcyclopentane-1,3-dicarboxylic Acid, *Anil*
Ifflaiamine
Khaplofoline, N-*Me* †
N-1-Naphthylalanine, *Et ester*
N-2-Naphthylalanine, *Et ester*
1,2,3,4-Tetrahydrocarbazole-6-carboxylic Acid, *Et ester*
1,2,3,4-Tetrahydrocarbazole-7-carboxylic Acid, *Et ester*
1,2,3,4-Tetrahydrocarbazole-8-carboxylic Acid, *Et ester*
1,2,3,4-Tetrahydrocarbazole-9-carboxylic Acid, *Et ester*

$C_{15}H_{17}NO_3$
1-Acetoxymethyl-2-propyl-4-quinolone †
Platydesmine †
Ribalinine †

$C_{15}H_{17}NO_4$
Actiphenol
Dubinidine †
1-Methylindole-2,3-dicarboxylic Acid, *Di-Et ester*
3-Methylindole-2,5-dicarboxylic Acid, *Di-Et ester*
Ribalindine †

$C_{15}H_{17}NO_5$
Penitrinic Acid
N-(3,4,5-Trimethoxyphenylethyl)maleimide †

$C_{15}H_{17}NO_6$
Adrenolutin, *Tri-Ac*

$C_{15}H_{17}N_3$
Dibenzylguanidine
1,3-Di-*o*-tolylguanidine
1,3-Di-*m*-tolylguanidine
1,3-Di-*p*-tolylguanidine

$C_{15}H_{17}N_3O$
4-Amino-4′-hydroxyazobenzene, N-*Di-Me*, *Me ether*

$C_{15}H_{17}N_5O_4$
Cytimidine

$C_{15}H_{17}O_2P$
Diphenylphosphinic Acid, *Isopropyl ester*

$C_{15}H_{18}$
Cadalene
Se-Chamazulene
Eucazulene
7-Isopropyl-1,4-dimethylazulene
7-Isopropyl-2,4-dimethylazulene
8-Isopropyl-2,4-dimethylazulene
1-Isopropyl-4,6-dimethylnaphthalene
3-Isopropyl-1,5-dimethylnaphthalene
6-Isopropyl-1,4-dimethylnaphthalene
7-Isopropyl-1,2-dimethylnaphthalene
7-Isopropyl-1,3-dimethylnaphthalene
7-Isopropyl-1,6-dimethylnaphthalene
Vetivazulene

$C_{15}H_{18}BrN$
Leucotrope, *Bromide*

$C_{15}H_{18}BrNO_2$
9-Benzoyl-3*a*-bromo-2β-hydroxy-9-azabicyclo-[3,3,1]nonane †

$C_{15}H_{18}ClN$
Leucotrope, *Chloride*

$C_{15}H_{18}IN$
Leucotrope, *Iodide*

$C_{15}H_{18}N_2$
2-Amino-5-methylphenyl-(4-amino-3-methyl-phenyl)methane
Di-(2-amino-5-methylphenyl)methane
Di-(3-amino-4-methylphenyl)methane
Di-(4-amino-2-methylphenyl)methane
Di-(4-amino-3-methylphenyl)methane
Di-(5-amino-2-methylphenyl)methane
2,4-Diaminotoluene, 4-N-*Et*-4-*Phenyl*
1,2,3,4,6 7,12,12*b*-Octahydroindole[2,3-*a*]quino-lizine †

$C_{15}H_{18}N_2O$
1-Nitroso-2-naphthylamine, N-3′-*Methylbutyl*
Selagine

$C_{15}H_{18}N_2O_2$
Propoxate †

$C_{15}H_{18}N_2O_2S$
Benzylpenillamine

$C_{15}H_{18}N_2O_3S$
p-Hydroxybenzylpenillamine

$C_{15}H_{18}N_4O_5$
Mitomycin C

$C_{15}H_{18}O$
1-Benzoyl-2-cyclohexylethylene
2-Hydroxycadalene †
5-Hydroxycadalene †
7-Hydroxycadalene
Lindestrene †
1-Naphthol, *Isopentyl ether*
2-Naphthol, *Isopentyl ether*
1-Naphthyl isopentyl Ether †

$C_{15}H_{18}O_2$
Cacalol †
Coriarialactone
Curzerenone †
epi-Curzerenone †
2,2′-Cyclopentenylpropionic Acid, *Benzyl ester*
Dehydrococtus Lactone †
Furanoligularenone †
2-(1-Hydroxy-2-indanyl)hexanoic Acid, *Lactone*
Hyposantonin
Ligularenolide †
Linderene
Lindestrenolide †
Procerin
2,3,3-Trimethyl-1-cyclopentene 1 carboxylic Acid, *Phenyl ester*
Warburgiadione †

$C_{15}H_{18}O_3$
Achillin†
Ambrosin
Aromaticin†
Aromatin†
Desacetoxymatricarin
Desmotroposantonin
Dihydroisolinderalactone†
epi-Dihydroisolinderalactone†
Estafiatin
Hydroxylindestrenolide†
Ligustrin†
Parvifloral†
Perezinone†
Pinnatifidin
Santonin
β-Santonin
Serticenic Acid†
Tuberiferine†
Xanthatin
Zaluzanin C†

$C_{15}H_{18}O_4$
Arglanine†
Artemisin
Dihydrogriesenin†
Evodionol, *Me ether*
Fomannosin†
Helenalin
Hydroxyachillin
α-Hydroxysantonin
β-Hydroxysantonin
Indane-2,2-dicarboxylic Acid, *Di-Et ester*
Jacquinelin†
Marasmic Acid★†
Mexicanin A
Mexicanin D
Mexicanin H†
Mexicanin I†
Olivetonide, 6-*Me ether*
Olivetonide, 8-*Me ether*
Parthenin
3-Phenylcyclopropane-1,2-dicarboxylic Acid, *Di-Et ester*
Phenylitaconic Acid, *Di-Et ester*
3-Phenyl-2-pentenedioic Acid, *Di-Et ester*
4-Phenyl-2-pentenedioic Acid, *Di-Et ester*
Psilostachyin B†

$C_{15}H_{18}O_5$
Auraptenic Acid
Co-anisation
Coriamyrtin
Isocoriamyrtin
Paniculide C†
Pulvilloric Acid★†

$C_{15}H_{18}O_6$
Benzene-1,2,3-tricarboxylic Acid, *Tri-Et ester*
Benzene-1,2,4-tricarboxylic Acid, *Tri-Et ester*
Benzene-1,3,5-tricarboxylic Acid, *Tri-Et ester*
Tutin
ψ-Tutin

$C_{15}H_{18}O_7$
2-Hydroxybenzene-1,3,5-tricarboxylic Acid, *Tri-Et ester*
Hyenanchin★†
Isohyenanchin
Mellitoxin†
Picrotin
Picrotoxic Acid
α-Picrotoxinic Acid
β-Picrotoxinic Acid

$C_{15}H_{18}O_8$
Evodionic Acid
Melilotoside

$C_{15}H_{19}BrO$
Aplysin†
Laurinterol†

$C_{15}H_{19}BrO_2$
Aplysinol†

$C_{15}H_{19}BrO_3$
4-Methylhexanoic Acid, p-*Bromophenacyl ester*
Santonic Acid, *Bromide*

$C_{15}H_{19}ClO_3$
Santonic Acid, *Chloride*

$C_{15}H_{19}IO_3$
Santonic Acid, *Iodide*

$C_{15}H_{19}NO$
2-1′-Cyclohexenylpropionic Acid, *Anilide*
2-Cyclohexylidenepropionic Acid, *Anilide*
Leucotrope
4-Methoxy-2-pentylquinoline
5-Phenyl-2,4-pentadienoic Acid, *Diethylamide*
Pronethalol†
Undeca-2,4-diene-8,10-diynoic Acid *N*-isobutylamide†

$C_{15}H_{19}NO_2$
Tropacocaine

$C_{15}H_{19}NO_3$
Cocculolidine†
Tropacocaine, N-*Oxide*

$C_{15}H_{19}NO_4$
Mescalotam†

$C_{15}H_{19}NO_5$
N-(3,4,5-Trimethoxyphenylethyl)succinimide†

$C_{15}H_{19}NO_6$
N-(3,4,5-Trimethoxyphenylethyl)malimide†

$C_{15}H_{19}NO_{10}$
2,4-Dihydroxy-7-methoxy-1,4-benzoxazolin-3-one

$C_{15}H_{20}$
Laurene†

$C_{15}H_{20}Br_2O_2$
Isolaureatin†
Laureatin†

$C_{15}H_{20}N_2O$
Anagyrine
Rhombifolin
Sophoramine
Thermopsine

$C_{15}H_{20}N_2O_2$
Baptifoline
Uretropine

$C_{15}H_{20}N_2O_3S$
Benzylpenilloic Acid

$C_{15}H_{20}N_2O_4$
N-Benzoylglycyl-3-aminobutyric Acid, *Et ester*
N-Benzoylglycyl-4-aminobutyric Acid, *Et ester*

$C_{15}H_{20}N_2O_4S$
p-Hydroxybenzylpenilloic Acid

$C_{15}H_{20}N_2S$
Methaphenilene

$C_{15}H_{20}N_4$
2,2′,4,4′-Tetra-aminodiphenylmethane, 4-N-*Di-Me*

$C_{15}H_{20}N_6O_3$
Monilin

$C_{15}H_{20}O$
Atractylon†
α-Cuparenone†
β-Cuparenone†
Debromoaplysin†
Debromolaurinterol†
Furanodiene†
Isofuranogermacrene†
Isogermafurene†
Melalinol
Nuciferal†
ar-Turmerone

$C_{15}H_{20}O_2$
Alantolactone
Aristolactone
Bilobanone†
Furanoeremophilone
Isoalantolactone
Isogermafurenolide†
Mansonone A†
Nootkatin
6-Oxofuranoeremophilane†
1-Phenylcyclohexanecarboxylic Acid, *Et ester*
Torreyal†

$C_{15}H_{20}O_3$
Arborescine
Artabsin
Balchanolide†
2-Benzoylbutyric Acid, *Butyl ester*
Bisabolangelone†
Carabrone†
Damsin★†
Douglanine†
Euryopsonol★†
Finitin
2-(1-Hydroxy-2-indanyl)hexanoic Acid
Hydroxyisogermafurenolide†
Illudin M
Ivalin
Isotelekin†
Mansonone B†
Parthenolide
Perezone
α-Pipitzol†
β-Pipitzol†
Pseudoivalin†
Santamarine†
Tamaulipin B†
Telekin†
Santonene
Winterin

$C_{15}H_{20}O_4$
Abscisic Acid†
Amaralin†
Bicyclo[2,2,1]hept-5-ene-2,3-dicarboxylic Acid, *Di-allyl ester*
Chamissonin†
Coronopilin
Costunolide†
Crotocol†
Deacetylconfertiflorin†
4,6-Diacetylresorcinol, *Et-Propyl ether*
4,6-Diacetylresorcinol, *Me-Butyl ether*
Dihydrofomannosin†
Geigerin
Helicobasidin†
Hirsutic Acid C†
Hulupinic Acid†
Illudin S
Isosantonic Acid
Lampterol
Mexicanin C
Paniculide A†
Peruvin†
Peruvinin†
3-Phenylglutaric Acid, *Di-Et ester*
Psilostachyin C†
Pulchellin C†
Santonic Acid
ψ-Santonin
Tanacetin
Vermeerin†
Vulgarin
Zaluzanin A†

$C_{15}H_{20}O_5$
5-Acetyl-2,4-dihydroxybenzoic Acid, *Et ester, Di-Et ether*
2,4-Dihydroxy-6,2′-oxoheptylbenzoic Acid, *Me ester*
2,4-Dihydroxy-6,2′-oxoheptylbenzoic Acid, 2-*Me ether*
2-Formyl-5,6-dimethoxybenzoic Acid, ψ-tert-*Pentyl ester*
Icaritol II
Jalaric Acid A
Paniculide B†
Phaeseic Acid†
Psilostachyin†
3-(2,3,4-Trihydroxyphenyl)acrylic Acid, *Tri-Et ether*

$C_{15}H_{20}O_6$
Coriatin
Shellolic Acid
Tricarballylic Acid, *Triallyl ester*

$C_{15}H_{20}O_7$
3-Hydroxy-4-oxo-4*H*-pyran-2,6-dicarboxylic Acid, *Di-isobutyl ester*
Isodihydrohyenanchin†
Neoanisatin†
Piscidic Acid, *Di-Me ester, Di-Me ether*†
Piscidic Acid, *Di-Et ester*†
2,4,6-Trihydroxyisophthalic Acid, *Tri-Me ether, Di-Et ester*

$C_{15}H_{20}O_8$
Androsin
Anisatin★†
Anisatinic Acid
2-Hydroxy-4-methoxyacetophenone, *Glucoside*
4-Hydroxy-2-methoxyacetophenone, *Glucoside*
α-Picrotinic Acid
β-Picrotinic Acid
Sambunigrinic Acid, *Me ester*

$C_{15}H_{20}O_9$
Glucovanillic Acid, *Me ester*

$C_{15}H_{21}BrO$
2-Bromo-1,2,3,4-tetrahydro-1-naphthol, *Amyl ether*

$C_{15}H_{21}N$
Epiguaipyridine†
3-Heptylindole
Patchoulipyridine†

$C_{15}H_{21}NO$
2,6-Dimethylcyclohexanecarboxylic Acid, *Anilide*
3,4-Dimethylcyclohexanecarboxylic Acid, *Anilide*
Metazocine

$C_{15}H_{21}NO_2$
4-Benzoyloxy-2,2,6-trimethylpiperidine
β-Eucaine
Ketobemidone
Pethidine

$C_{15}H_{21}NO_3S_2$
Benzylpenaldic Acid, *Di-Et mercaptal*

$C_{15}H_{21}NO_4$
Dehydroactidione†

$C_{15}H_{21}NO_5$
Benzylpenaldic Acid, *Di-Et acetal*
Tartranilic Acid, 3-*Methylbutyl ester*

$C_{15}H_{21}NO_6$
p-Hydroxybenzylpenaldic Acid, *Di-Et acetal*

$C_{15}H_{21}NO_{10}S_2$
Glucoaubrietin†
Glucobarbarin

$C_{15}H_{21}N_3O$
Primaquine
Tryptophane, *Di-Et amide*

$C_{15}H_{21}N_3O_2$
Eserine

$C_{15}H_{21}N_3O_3$
Geneserine

$C_{15}H_{21}N_3O_{15}$
Endecaphyllin A
Endecaphyllin B
Endecaphyllin C
Karakin

$C_{15}H_{21}N_5O_4$
*N*⁶-(γ,γ-Dimethylallyl)adenosine†

$C_{15}H_{21}N_7O_6$
Alazopeptin

$C_{15}H_{22}$
Bicyclo[9,2,2]pentadeca-11,13,14-triene
Calamene
Copadiene†
Cuparene†
α-Curcumene
ar-Himachalene†
Leddiene
Nootkatene†

$C_{15}H_{22}N_2O$
Aphyllidine★†
Aponucidine
Monspessulanine†

$C_{15}H_{22}N_2O_2$
Argyrolobine†
Geneserethol

$C_{15}H_{22}N_2O_4$
Benzylpenaldic Acid, *Amide*
Leucyltyrosine

$C_{15}H_{22}N_4O_2S_2$
Allithiamine

$C_{15}H_{22}O$
Aristolone
Atlantone
Chamigrenal†
α-Costal†
β-Costal†
α-Cuparenol†
β-Cuparenol†
γ-Cuparenol†
Cyclocolorenone
1-*epi*-Cyclocolorenone†
α-Cyperone
6-*epi*-α-Cyperone
β-Cyperone
Cyperotundone†
Dendrolasin
Elemenal†
Eremophilone
allo-Eremophilone†
Furanoeremophilane
Germacrone†
α-Isocuparenol†
Mustakone†
Nardostachone†
Nootkatone
Nuciferol†
Occidol
Patchoulenone†
Rotundone†
α-Santalal†
α-Sinensal†
β-Sinensal†
Turmerone
Valerenal†
α-Vetivone†
β-Vetivone†
Zerumbone
Zierone

$C_{15}H_{22}O_2$
Apoaromadendrone, Hydroxymethylene comp.
Benzoic Acid, *Octyl ester*
14,8-Cedranolide†
Cinnamolide†
Confertifolin†

$C_{15}H_{22}O_2$ (*continued*)
(α)-Costic Acid†
Dihydrocostunolide†
2,2-Dimethyl-3-phenylpropionic Acid, *Butyl ester*
Drimenin†
Eremophilenolide
2-Ethylhexanoic Acid, *Benzyl ester*
Fukinanolide†
Furopelargone A†
Furopelargone B†
Helminthosporal
Hinokiic Acid
8α-Hydroxy-7α(*H*)-eremophila-1,11-dien-9-one†
8α-Hydroxy-7α(*H*)-eremophila-1(10),11-dien-9-one†
2-Hydroxyeremophilone
p-Isopropylbenzoic Acid, n-*Pentyl ester*
Isodrimenin†
Isovalencenic Acid†
Neotorreyol†
2-Octanol, *Benzoyl*
p-Octylbenzoic Acid
Petasalbin
Polygodial
Saussurea Lactone†
Tadeonal
Valerenic Acid†
Zizanoic Acid†
5-*epi*-Zizanoic Acid†

$C_{15}H_{22}O_3$
Adhumulinic Acid
Alantolic Acid
Albopetasol†
Balchanin†
Dacrinol†
Dihydroperezone
Dihydropseudoivalin†
Futronolide
Hydronootkatinol†
3-Hydroxyeremophilenolide†
6β-Hydroxyeremophilenolide†
9-Hydroxynonanoic Acid, *Phenyl ether*
8-*m*-Hydroxyphenyloctanoic Acid, *Me ether*
Ipomeamarone
Millefolide
Nardosinone†
Ngaione
Nootkatinol†
2-Octanol, *Salicylol*
Roridin C
Santonan
Trichodermol†
Valdiviolide

$C_{15}H_{22}O_4$
Ambrosiol†
Avenaciolide†
Brefeldin B†
Calendin
Cumanin†
Euryopsol†
Fuegin
Fukinolidiol†
Geigerinin
Humulinic Acid★†
Humulinic Acid C†
Humulinic Acid D†
Iresin
Isoiresin†
Leptospermone
Mibulactone
Microcephalin†
Verrucarol
Zinniol†

$C_{15}H_{22}O_5$
Gallic Acid, *Tri-Et ether*, *Et ester*
Laksholic Acid†
epi-Laksholic Acid†
2,4,5-Trimethoxybenzoic Acid, 3-*Methylbutyl ester*

$C_{15}H_{22}O_9$
Aucubine

$C_{15}H_{22}O_{10}$
Catalpol† (incorrectly given as Catapol)

$C_{15}H_{23}N$
Longifolic Acid, *Nitrile*

$C_{15}H_{23}NO$
2,5-Dimethylhexanoic Acid, p-*Toluidide*
Nuphenine†

$C_{15}H_{23}NO_2$
Alantolic Acid, *Amide*
Castoramin
Lophocerine†
2-Octanol, (−)-o-*Aminobenzoyl*
2-Octanol, m-*Aminobenzoyl*
2-Octanol, p-*Aminobenzoyl*
N-Phenylglycine, N-*Et*, *Isoamyl ester*
Tolpronine

$C_{15}H_{23}NO_3$
p-Butoxybenzoic Acid, *Dimethylaminoethyl ester*

$C_{15}H_{23}NO_4$
Actidione
Naramycin B

$C_{15}H_{23}NO_4S$
p-Sulphobenzoic Acid, 4-*Dibutylamide*

$C_{15}H_{23}NO_5$
Streptovitacin A
Streptovitacin B
Streptovitacin C
Streptovitacin D

$C_{15}H_{23}N_3O$
Glochidicine†

$C_{15}H_{23}N_3OS$
Diamthazole

$C_{15}H_{23}N_5O_{14}P_2$
Adenosinediphosphate ribose

$C_{15}H_{24}$
Albicaulene
Alloaromadendrene†
Aristolene†
Aromadendrene
Bergamotene

$C_{15}H_{24}$ (*continued*)
β-Bergmotene
Bisabolene
β-Bisabolene†
α-Bourbonene†
β-Bourbonene
ε-Bulgarene†
α-Bulnesene
β-Bulnesene
Canadene
9,11-Cadinadiene
Cadinene
β-Cadinene
γ-Cadinene
γ_1-Cadinene
γ_2-Cadinene
δ-Cadinene
ε-Cadinene
ω-Cadinene†
Calarene★† (incorrectly given as Calerene†)
β-Caryophyllene
γ-Caryophyllene
Cedrene
β-Cedrene†
α-Chamigrene†
β-Chamigrene†
α-Chigadmarene
Clovene
ψ-Clovene
pseudo-Clovene A†
Copaene
α-Cubebene†
β-Cubebene†
Cuprenene†
β-Curcumene
γ-Curcumene
Cyclosativene†
Cyperene†
Daucene
α-Elemene
β-Elemene
δ-Elemene
4-Ethyl-4-phenylheptane
Eremophila-3,11-diene†
Eremophilene†
α-Farnesene†
β-Farnesene†
α-Ferulene†
Germacratriene†
β-Gorgonene†
3,7-Guaiadiene†
α-Guaiene
β-Guaiene
γ-Guaiene
ε-Guaiene
α-Gurjunene
β-Gurjunene
α-Himachalene
β-Himachalene
Humulene
β-Humulene
Isobisabolene
Isocadinene
Isoclovene
Isolongifolene†
Lansene
Ledene
Longicyclene
Longifolene
α-Longipinene†
γ-Maaliene†
ε-Muurolene†
Neoclovene†
Orthodonene
β-Patchoulene
Patschulene
α-Santalene
β-Santalene
γ-Santalene
Sativene†
Selina-3,7(11)-diene†
Selina-4(14),7(11)-diene†
α-Selinene
β-Selinene
Sesquicamphene
Sesquicarene†
Sesquicitronellene
β-Sesquiphellandrene†
Seychellene†
Thujopsene
Tricyclovetivene†
1,2,4-Tri-isopropylbenzene
1,3,5-Tri-isopropylbenzene
3,7,11-Trimethyl-1,3,6,10-dodecatetraene
Valencene†
Ylangene
Zingiberene

$C_{15}H_{24}IN_2O_3$
Caryophyllene iodonitrosite†

$C_{15}H_{24}N_2O$
Allomatrine
Aphylline
Leontine†
Lupanine
Matrine
Nonalupine
Oxysparteine
Sophocarpine

$C_{15}H_{24}N_2O_2$
13-Hydroxylupanine
17-Hydroxylupanine
Jamaidine
Virgiline

$C_{15}H_{24}N_2O_2S$
4-Methylhexanoic Acid, S-*Benzylthiouronium salt*

$C_{15}H_{24}N_2O_4S$
Pentylpenicillenic Acid, *Me ester*

$C_{15}H_{24}N_2O_{17}P$
Uridinediphosphateglucose

$C_{15}H_{24}N_4O_2S_2$
Alinamine

$C_{15}H_{24}O$
Acorenone†
Acorenone B†
α-Agarofuran†
β-Agarofuran†

$C_{15}H_{24}O$ (*continued*)
- Araucariol †
- α-Betulenol
- β-Betulenol
- α-Biotol
- β-Biotol
- Calacone
- Calarenol † (incorrectly given as Calerenol)
- Campherenone †
- Canarone †
- Caryophyllenol
- 8,14-Cedranoxide †
- α-Costal †
- Cyperenol †
- 2,6-Di-*tert*-butyl-*p*-cresol †
- 2,6-Di-*tert*-butyl-4-methylphenol
- δ-Elemenol †
- *epi*-δ-Elemenol †
- Epoxyguaiene †
- 7α(*H*)-Eudesm-4(14)-en-9-one †
- Humulene epoxide I †
- Humulene epoxide II †
- Humulenol I †
- Humulenol II †
- 6-Isopropyl-*m*-cresol, (+)-2-*Methylbutyl ether*
- Isoshyobunone †
- Khusinol
- Khusol
- Lanceol
- Lansol
- Longihomocamphenylone †
- Luparenol
- Occidentalol †
- 2-Octanol, *Benzyl ether*
- Partheniol
- Patchoulenol †
- α-Santalol
- β-Santalol
- Shyobunone †
- *epi*-Shyobunone †
- Spathulenol
- Tricyclovetivenol †
- 3,7,11-Trimethyl-2,6,10-dodecatrienal
- β-Vetivone
- Zingiberone

$C_{15}H_{24}O_2$
- Acorone
- Carissone
- Cyperolone †
- 2,6-Di-*tert*-butyl-4-hydroxymethylphenol †
- 1,2-Dihydro-2-hydroxyeremophilone
- Helminthosporol †
- 9α-Hydroxycadin-3-en-2-one †
- Isolongifolic Acid
- 4-Isopropyl-*m*-toluic Aldehyde, *Di-Et acetal*
- Khusinoloxide †
- Longifolic Acid
- Occidiol †
- Partheniol, *Epoxide*
- Resorcinol, *Mono-nonyl ether*
- Sironin †
- 3,7,11-Trimethyl-2,6,10(or 11)-dodecatrienoic Acid

$C_{15}H_{24}O_3$
- Cedrolic Acid
- 5-Hydroxymethyltetradeca-2,4,6-trienoic Acid
- Ilicic Acid †
- Illudol †
- Ngaiol
- Todomatuic Acid

$C_{15}H_{24}O_4$
- Clovenic Acid
- ψ-Clovenic Acid
- Pulchellin A †
- Pyrethic Acid, tert-*Bu ester* †
- Santonanic Acid

$C_{15}H_{24}O_6$
- Aconitic Acid, *Tripropyl ester*

$C_{15}H_{24}O_{10}$
- Harpagide †

$C_{15}H_{24}O_{11}$
- Procumbid †

$C_{15}H_{25}BrO$
- Humulene bromohydrin †

$C_{15}H_{25}ClN_2O_2$
- Decicaine

$C_{15}H_{25}N$
- 3-Phenyl-1-propylamine, N-*Dipropyl*

$C_{15}H_{25}NO$
- Longifolic Acid, *Amide*

$C_{15}H_{25}NO_2$
- α-Aminomethyl-*p*-hydroxylbenzyl Alcohol, N-*Di-isopropyl*, *Me ether*

$C_{15}H_{25}NO_4$
- Supinine

$C_{15}H_{25}NO_5$
- Echinatine
- Indicine
- Intermedine †
- Lycopsamine †

$C_{15}H_{25}NO_5$
- Rinderine †

$C_{15}H_{25}N_3O_2$
- Histamine★, N^{α}-(4-*Oxodecanoyl*) †

$C_{15}H_{26}$
- Ferulene

$C_{15}H_{26}Br_2$
- Bulgarene Dihydrobromide †
- Cadinene Dihydrobromide †

$C_{15}H_{26}Cl_2$
- Cadinene Dihydrochloride †
- Muurolene Dihydrochloride †

$C_{15}H_{26}NO_{10}S_2$ (ion)
- Glucocappasalin

$C_{15}H_{26}N_2$
- Allomatridine
- Isosparteine
- α-Isosparteine
- Matridine
- 3-Methyltetradecanedioic Acid, *Dinitrile*
- Sparteine

$C_{15}H_{26}N_2O$
Retamine
Sophoridine
Sparteine★, 16-N-*Oxide*†

$C_{15}H_{26}N_2O_5S$
Pentylpenicilloic Acid, *α-Me ester*

$C_{15}H_{26}O$
Agarol†
Agarospirol†
Allohimachalol†
Anymol
Bicyclovetivenol★†
Bisabolol
Bulnesol
Calamone
α-Cadinol
δ-Cadinol
T-Cadinol†
Campherenol†
epi-Campherenol†
Carotol
α-Caryophyllene Alcohol†
Cedrol
Copaborneol†
Cryptomeriol
Cubenol†
epi-Cubenol†
Cyclofarnesol
Cyclopentadec-2-en-1-one
Cymbopol
α-Elemol
Eremoligenol†
α-Eudesmol
β-Eudesmol
γ-Eudesmol
Farnesol
Galbanol
Galipol
Globulol
Guaiol
Guaioxide†
Hedycaryol†
Himachalol†
Hinesol†
Humbertiol†
Intermedeol
α-Ionol, *Et ether*
Isoclovene Alcohol
Junenol†
Juniperol†
Kessane
Ledol
Liguloxide†
Maaliol†
T-Muurolol†
Nardol†
Neointermedeol†
Nerolidol
Palustrol★†
Patchouli Alcohol
Selin-11-en-4α-ol†
Trichothecane†
Valeranone
Valerianol†
α-Verbesinol
β-Verbesinol
Veticadinol
Viridiflorol
Widdrol
Zingiberol

$C_{15}H_{26}O_2$
Bicyclohexyl-1-carboxylic Acid, *Et ester*
Calameone
Culmorin
Daucol
Eudesm-4(15)-ene-2α,11-diol†
Isoborneol, *Valeryl*
Kanokonol†
Kessyl Alcohol
Kessyl Ketone
Liguloxidol†
2-Methylcrotonic Acid, (+)-*Citronellyl ester*
Opiodiol†
Oplopanone†

$C_{15}H_{26}O_3$
Isocaucalol†
Kessyl Glycol
Lactic Acid, *Et ether*, 1-*Bornyl ester*
Shiromodiol†

$C_{15}H_{26}O_4$
3-Carboxy-2,2,3-trimethyl-1-cyclopentylacetic Acid, *Di-Et ester*
3-Methylcyclohexane-1,1-diacetic Acid, *Di-Et ester*
4-Methylcyclohexane-1,1-diacetic Acid, *Di-Et ester*
2-Propene-1,2-dicarboxylic Acid, *Di-2-methylbutyl*

$C_{15}H_{26}O_4Si$
Triallyl cyclohexyl orthosilicate

$C_{15}H_{26}O_5$
Heptanoylsuccinic Acid, *Di-Et ester*†
Phoronic Acid, *Di-Et ester*

$C_{15}H_{26}O_6$
3-Carboxymethyl-2-methylbutane-2,4-dicarboxylic Acid, *Tri-Et ester*
2,3-Dimethylbutane-1,2,3-tricarboxylic Acid, *Tri-Et ester*
Hexane-1,1,6-tricarboxylic Acid, *Tri-Et ester*
Hexane-1,2,2-tricarboxylic Acid, *Tri-Et ester*
Hexane-1,2,6-tricarboxylic Acid, *Tri-Et ester*
Hexane-1,3,3-tricarboxylic Acid, *Tri-Et ester*
Hexane-1,4,4-tricarboxylic Acid, *Tri-Et ester*
Hexane-2,4,4-tricarboxylic Acid, *Tri-Et ester*
Tributyrin
Tri-isobutyrin

$C_{15}H_{27}ClO_3$
Brassylic Acid, *Et ester-chloride*

$C_{15}H_{27}NO_4$
Lindelofine
Trachelantamine
Viridiflorine

$C_{15}H_{27}NO_5$
Macrotomine†
Trachelantine

$C_{15}H_{27}N_3O_3$
Cyanuric Acid, O-*Tri*-n-*butyl ester*
Cyanuric Acid, O-*Tri-isobutyl ester*
Cyanuric Acid, O-*Tri*-sec-*butyl ester*

$C_{15}H_{27}N_3O_4$
L-Valyl-L-valyl-D-valine

$C_{15}H_{28}$
Agarospirane†
Fukinane†
Isoguaiazulene

$C_{15}H_{28}N_2O_3$
Hygroscopin B

$C_{15}H_{28}N_2O_3S$
Heptylpenilloic Acid

$C_{15}H_{28}O$
Cyclopentadecanone

$C_{15}H_{28}O_2$
Caparrapidiol†
Cryptomeridiol
2-Hydroxycyclopentadecanone
4-Hydroxypentadecanoic Acid, *Lactone*
15-Hydroxypentadecanoic Acid, *Lactone*
2-*p*-Menthyl-β-propionic Acid, *Et ester*
4-Tetradecenoic Acid, *Me ester*
9-Tetradecenoic Acid, *Me ester*
12-Tridecenoic Acid, *Et ester*
Valeric Acid, *Menthyl ester*

$C_{15}H_{28}O_3$
Lactic Acid, *Et ether (−)-Menthyl ester*
3-*p*-Menthoxyacetic Acid, *Propyl ester*
2-Oxopentadecanoic Acid
3-Oxopentadecanoic Acid
11-Oxopentadecanoic Acid
3-Oxotetradecanoic Acid, *Me ester*

$C_{15}H_{28}O_4$
Brassylic Acid, *Di-Me ester*
3-Methyladipic Acid, *Di-isobutyl ester*
2-Methyldodecanedioic Acid, *Di-Me ester*
3-Methyldodecanedioic Acid, *Di-Me ester*
6-Methyldodecanedioic Acid, *Di-Me ester*
Methylsuccinic Acid, *Di-2-methylbutyl ester*
3-Methyltetradecanedioic Acid
5-Methyltetradecanedioic Acid
Nonane-1,1-dicarboxylic Acid, *Di-Et ester*
Undecanedioic Acid, *Di-Et ester*

$C_{15}H_{28}O_{13}$
O-α-D-Galactopyranosyl-(1 → 2)-*O*-α-D-glucopyranosyl-(1 → 1)-D-glycerol†
Xylotriose

$C_{15}H_{29}BrO_2$
2-Bromododecanoic Acid, *Propyl ester*

$C_{15}H_{29}ClO$
Pentadecanoic Acid, *Chloride*

$C_{15}H_{29}N$
1-Azacyclohexadecane

$C_{15}H_{29}NO$
2-Hydroxypentadecanoic Acid, *Nitrile*

$C_{15}H_{29}NO_3$
Brassylic Acid, *Et ester-amide*

$C_{15}H_{30}$
Cyclopentadecane
1-(1,5-Dimethylhexyl)-4-methylcyclohexane
1-Ethyl-2,4-di-isopropyl-1-methylcyclohexane
Germacrane†
1-Pentadecene
7-Pentadecene

$C_{15}H_{30}Br_2$
1,14-Dibromopentadecane
1,15-Dibromopentadecane

$C_{15}H_{30}N_2$
3-(3-Dimethylaminopropyl)-1,8,8-trimethyl-3-azabicyclo[3,2,1]octane†

$C_{15}H_{30}N_2O_9$
Marcomycin

$C_{15}H_{30}O$
Cyclopentadecanol
8-Methyl-6-tetradecanone
Pentadecanal
2-Pentadecanone
3-Pentadecanone
8-Pentadecanone
13-Pentadecen-1-ol
14-Pentadecen-1-ol
1-Pentadecen-3-ol

$C_{15}H_{30}O_2$
2-Octanol, *Heptanoyl*
Pentadecanoic Acid
Tetradecanoic Acid, *Me ester*
Tridecanoic Acid, *Et ester*
3,7,11-Trimethyldodecanoic Acid †
2,6,10-Trimethylundecanoic Acid, *Me ester*†

$C_{15}H_{30}O_3$
Caparrapitriol†
2,2-Dimethylpropanal, *Trimer*
10-Hydroxydecanoic Acid, *Pentyl ester*
2-Hydroxypentadecanoic Acid
3-Hydroxypentadecanoic Acid
4-Hydroxypentadecanoic Acid
11-Hydroxypentadecanoic Acid
15-Hydroxypentadecanoic Acid
2-Hydroxytetradecanoic Acid, *Me ester*
6-Hydroxytetradecanoic Acid, *Me ester*†

$C_{15}H_{30}O_4$
9,10-Dihydroxytetradecanoic Acid, *Me ester*†
α-Monolaurin
β-Monolaurin

$C_{15}H_{30}O_8$
O-α-D-Galactopyranosyl-(1 → 2)-glycerol, *Hexa-Me ether*†

$C_{15}H_{31}N$
Cyclopentadecylamine

$C_{15}H_{31}NO$
Pentadecanoic Acid, *Amide*

$C_{15}H_{31}NO_2$
2-Aminotetradecanoic Acid, *Me ester*
13-Aminotridecanoic Acid, *Et ester*
2-Hydroxypentadecanoic Acid, *Amide*

$C_{15}H_{32}$
Pentadecane
2,6,10-Trimethyldodecane

$C_{15}H_{32}O$
1-Octanol, n-*Heptyl ether*
2-Octanol, n-*Heptyl ether*
Pentadecanol
3-Pentadecanol
8-Pentadecanol

$C_{15}H_{32}O_2$
1,5-Pentanediol, *Di-3-methylbutyl ether*

$C_{15}H_{33}BO_3$
Tri-(3-methylbutyl)borate
Tripentyl borate

$C_{15}H_{33}N$
1-Aminopentadecane
Tri-(2-methylbutyl)amine
Tri-(3-methylbutyl)amine
Tripentylamine

$C_{15}H_{33}O_4P$
Phosphoric Acid, *Tripentyl ester*

$C_{15}H_{40}O_5Si_5$
2,4,6,8,10-Pentaethyl-2,4,6,8,10-pentamethyl-cyclopentasiloxane

C_{16}

$C_{16}H_6Cl_2O_4$
Anthraquinone-1,4-dicarboxylic Acid, *Dichloride*
Anthraquinone-1,5-dicarboxylic Acid, *Dichloride*
Anthraquinone-2,6-dicarboxylic Acid, *Dichloride*
Anthraquinone-2,7-dicarboxylic Acid, *Dichloride*

$C_{16}H_6N_2O_2$
Anthraquinone-1,5-dicarboxylic Acid, *Dinitrile*
Anthraquinone-1,8-dicarboxylic Acid, *Dinitrile*

$C_{16}H_6O_5$
Anthraquinone-1,2-dicarboxylic Acid, *Anhydride*
Anthraquinone-1,5-dicarboxylic Acid, *Anhydride*
Anthraquinone-2,3-dicarboxylic Acid, *Anhydride*

$C_{16}H_8Br_2N_2O_2$
6,6′-Dibromoindigotin †

$C_{16}H_8Cl_2O_4$
Benzil-2,2′-dicarboxylic Acid, *Dichloride*

$C_{16}H_8Cl_4$
1,2,3,4-Tetrachlorobenzo[*g*]sesquifulvalene †

$C_{16}H_8O_2$
Aceanthrenequinone
Fluoranthene-2,3-dione
Pyrene-1,6-quinone
Pyrene-1,8-quinone

$C_{16}H_8O_2S_2$
Thioindigo

$C_{16}H_8O_3$
Anthracene-1,9-dicarboxylic Acid, *Anhydride*
Anthra[1,9-*bc*]pyran-2,7-quinone
Phenanthrene-1,2-dicarboxylic Acid, *Anhydride*
Phenanthrene-3,4-dicarboxylic Acid, *Anhydride*
Phenanthrene-4,5-dicarboxylic Acid, *Anhydride*
Phenanthrene-9,10-dicarboxylic Acid, *Anhydride*

$C_{16}H_8O_3S_2$
Thioindigo, S-*Oxide*

$C_{16}H_8O_4$
Diphthalyl

$C_{16}H_8O_5$
Benzil-2,2′-dicarboxylic Acid, *Anhydride*

$C_{16}H_8O_6$
Anthraquinone-1,2-dicarboxylic Acid
Anthraquinone-1,3-dicarboxylic Acid
Anthraquinone-1,4-dicarboxylic Acid
Anthraquinone-1,5-dicarboxylic Acid
Anthraquinone-1,6-dicarboxylic Acid
Anthraquinone-1,7-dicarboxylic Acid
Anthraquinone-1,8-dicarboxylic Acid
Anthraquinone-2,3-dicarboxylic Acid
Anthraquinone-2,6-dicarboxylic Acid
Anthraquinone-2,7-dicarboxylic Acid
Medicagol †

$C_{16}H_9Cl$
1-Chloropyrene

$C_{16}H_9ClO_3$
2-Methylanthraquinone-1-carboxylic Acid, *Chloride*

$C_{16}H_9ClO_4$
4-Chloroanthraquinone-1-carboxylic Acid, *Me ester*
5-Chloroanthraquinone-1-carboxylic Acid, *Me ester*
6-Chloroanthraquinone-1-carboxylic Acid, *Me ester*
7-Chloroanthraquinone-1-carboxylic Acid, *Me ester*
8-Chloroanthraquinone-1-carboxylic Acid, *Me ester*
1-Chloroanthraquinone-2-carboxylic Acid, *Me ester*

$C_{16}H_9N$
N-Benzyl-α-ethylbenzylamine

$C_{16}H_9NO_2$
Anthracene-1,9-dicarboxylic Acid, *Imide*
Pyridanthrone

$C_{16}H_9NO_5$
Anthraquinone-2,3-dicarboxylic Acid, *Mono-amide*

$C_{16}H_9NO_6$
Aristolochic Acid II
1-Nitroanthraquinone-2-carboxylic Acid, *Me ester*

$C_{16}H_9NO_8$
Aristolochic Acid C

$C_{16}H_9N_5O_8$
2,4,5,7-Tetranitro-1-naphthylamine, N-*Phenyl*
2,4,5,8-Tetranitro-1-naphthylamine, N-*Phenyl*

$C_{16}H_{10}$
Cyclohept[*fg*]acenaphthylene †
Dicyclopenta[*ef,kl*]heptalene
1,4-Diphenylbutadi-yne
Fluoranthene
Pyrene
1,3,9-Trihydro[16]annulene †

$C_{16}H_{10}ClNO$
- 4-Phenylquinoline-2-carboxylic Acid, *Chloride*
- 2-Phenylquinoline-4-carboxylic Acid, *Chloride*

$C_{16}H_{10}Cl_2O_2$
- Stilbene-3,3′-dicarboxylic Acid, *Dichloride*

$C_{16}H_{10}Cl_2O_4$
- 1,4-Dichloro-5,5-dihydroxyanthraquinone, *Di-Me ether*
- 6,7-Dichloro-1,4-dihydroxyanthraquinone, *Di-Me ether*

$C_{16}H_{10}Cl_2O_7$
- Erdin

$C_{16}H_{10}Cl_4O_5$
- Diploicin

$C_{16}H_{10}N_2$
- Benzo[*f*]-4,7-phenanthroline
- Benzo[*f*]-1,7-phenanthroline
- Benzo[*a*]phenazine
- Benzo[*b*]phenazine
- Dibenzo[*b,h*][1,6]naphthyridine
- Dibenzo[*c,h*][2,6]naphthyridine
- Dibenzo[*f,h*]quinoxaline
- Diphenylmaleic Acid, *Dinitrile*
- 2-Phenylquinoline-3-carboxylic Acid, *Nitrile*
- 2-Phenylquinoline-4-carboxylic Acid, *Nitrile*
- Quino[6,5-*f*]quinoline
- Quino[7,8-*f*]quinoline
- Quino[7,8-*h*]quinoline
- Quino[8,7-*f*]quinoline
- Quino[8,7-*h*]quinoline

$C_{16}H_{10}N_2O$
- Benzo[*a*]phenazine, 7-*Oxide*

$C_{16}H_{10}N_2O_2$
- Indigotin
- Indirubin
- β-Isoindigo
- 2,3-Methylenedioxyquinodoline
- Quindoline-10-carboxylic Acid

$C_{16}H_{10}N_2O_4$
- Anthraquinone-2,6-dicarboxylic Acid, *Diamide*
- 1,3-Diphenylalloxan
- 4-Nitrostilbene-2,2′-dicarboxylic Acid, 2-*Nitrile*
- 2-Nitrostilbene-2′,4-dicarboxylic Acid, 4-*Nitrile*

$C_{16}H_{10}N_2O_5$
- 2-Naphthol, 2,4-*Dinitrophenyl ether*

$C_{16}H_{10}N_2O_7$
- 2-Hydroxy-1,3-dinitroanthraquinone, *Et ether*

$C_{16}H_{10}N_2O_8$
- 1,5-Dihydroxy-4,8-dinitroanthraquinone, *Di-Me ether*

$C_{16}H_{10}N_4O_6$
- *N*-2,4,6-Trinitrophenyl-1-naphthylamine
- *N*-2,4,6-Trinitrophenyl-2-naphthylamine

$C_{16}H_{10}O$
- Benzo[*b*]naphtho[2,1-*d*]furan
- Benzo[*kl*]xanthene
- Brazan

$C_{16}H_{10}O_2$
- 1,3-Dihydroxypyrene
- 1,6-Dihydroxypyrene
- 1,8-Dihydroxypyrene
- 4,9-Dihydroxypyrene

$C_{16}H_{10}O_2S$
- 1-Anthraquinonethiol, S-*Vinyl ether*
- 2-Anthraquinonethiol, S-*Vinyl ether*

$C_{16}H_{10}O_2S_2$
- Thioindigo White

$C_{16}H_{10}O_3$
- Diphenylmaleic Acid, *Anhydride*
- 5-Formyl-4-phenanthroic Acid†

$C_{16}H_{10}O_4$
- Anthracene-1,5-dicarboxylic Acid
- Anthracene-1,9-dicarboxylic Acid
- Anthraquinone-1-carboxylic Acid, *Me ester*
- Phenanthrene-1,2-dicarboxylic Acid
- Phenanthrene-1,7-dicarboxylic Acid
- Phenanthrene-3,4-dicarboxylic Acid
- Phenanthrene-3,9-dicarboxylic Acid
- Phenanthrene-3,10-dicarboxylic Acid
- Phenanthrene-4,5-dicarboxylic Acid
- Phenanthrene-9,10-dicarboxylic Acid
- 2-Methylanthraquinone-1-carboxylic Acid
- 3-Methylanthraquinone-1-carboxylic Acid
- 4-Methylanthraquinone-1-carboxylic Acid
- 6-Methylanthraquinone-1-carboxylic Acid
- 7-Methylanthraquinone-1-carboxylic Acid
- 6-Methylanthraquinone-2-carboxylic Acid
- 3′,4′-Methylenedioxyflavone

$C_{16}H_{10}O_5$
- ψ-Baptigenin
- Damnacanthal
- Fluorenone-1,2-dicarboxylic Acid, *Mono-Me ester*
- 2-Formylbenzoic Acid, *Anhydride*
- 5-Hydroxyanthraquinone-2-carboxylic Acid, *Me ether*
- 3′,4′-Methylenedioxyflavonol
- Pyrenic Acid, α-*Me ester*
- Pyrenic Acid, β-*Me ester*

$C_{16}H_{10}O_6$
- Benzil-2,2′-dicarboxylic Acid
- Fallacinal
- Juzunal
- Mopanin†
- Nasutin A, *Di-Me ether*†
- Piperil
- Rhein, *Me ester*
- Sativol†
- Trifoliol†

$C_{16}H_{10}O_7$
- Clavoxanthin
- Emodic Acid, *Me ester*
- Emodic Acid, 7-*Me ether*
- Endocrocin
- Laccaic Acid D†
- Naphthalene-1,2,4,5-tetracarboxylic Acid, *Di-Me ester*
- Parietinic Acid
- Porphyrilic Acid
- Variolaric Acid
- Wedelolactone

$C_{16}H_{10}O_8$
Biphenyl-2,2′,3,3′-tetracarboxylic Acid
Biphenyl-2,2′,3,4′-tetracarboxylic Acid
Biphenyl-2,2′,4,4′-tetracarboxylic Acid
Biphenyl-2,2′,5,5′-tetracarboxylic Acid
Biphenyl-2,2′,6,6′-tetracarboxylic Acid
Biphenyl-2,3,5,6-tetracarboxylic Acid
Biphenyl-3,3′,4,4′-tetracarboxylic Acid
Biphenyl-3,3′,5,5′-tetracarboxylic Acid
Ceroalbolinic Acid†
Kermesic Acid†
3,5,6,7-Tetrahydroxy-1-methylanthraquinone-2-carboxylic Acid†

$C_{16}H_{10}O_9$
Distemonanthin

$C_{16}H_{10}S$
Phenanthro[3,2-*b*]thiophene

$C_{16}H_{11}BrO_4$
6-Bromo-1,4-dihydroxyanthraquinone, *Di-Me ether*

$C_{16}H_{11}ClO_5$
Fragilin†

$C_{16}H_{11}FO_2$
1-Ethyl-4-fluoroanthraquinone
2-Ethyl-3-fluoroanthraquinone

$C_{16}H_{11}IO_2$
1-Iodo-2,4-dimethylanthraquinone

$C_{16}H_{11}N$
1-Aminopyrene
2-Aminopyrene
4-Aminopyrene
5*H*-Benzo[*b*]carbazole
7*H*-Benzo[*c*]carbazole
11*H*-Benzo[*a*]carbazole
Dibenzo[*e,g*]indolizine
11*H*-Indeno[1,2-*b*]quinoline
2-Phenanthreneacetic Acid, *Nitrile*
3-Phenanthreneacetic Acid, *Nitrile*
9-Phenanthreneacetic Acid, *Nitrile*

$C_{16}H_{11}NO$
7*H*-Benzo[*c*]phenoxazine
12*H*-Benzo[*a*]phenoxazine
12*H*-Benzo[*b*]phenoxazine
2-Benzoylquinoline
3-Benzoylquinoline
4-Benzoylquinoline
6-Benzoylquinoline
8-Benzoylquinoline
Chalcone-4-carboxylic Acid, *Nitrile*
Chalcone-β-carboxylic Acid, *Nitrile*

$C_{16}H_{11}NO_2$
Dibenzoylacetic Acid, *Nitrile*
1,1-Diphenylethylene-2,2-dicarboxylic Acid, *Mononitrile*
3-Phenylquinoline-2-carboxylic Acid
4-Phenylquinoline-2-carboxylic Acid
2-Phenylquinoline-3-carboxylic Acid
2-Phenylquinoline-4-carboxylic Acid
3-Phenylquinoline-4-carboxylic Acid
2-Phenylquinoline-5-carboxylic Acid
2-Phenylquinoline-6-carboxylic Acid
2-Phenylquinoline-8-carboxylic Acid

$C_{16}H_{11}NO_3$
1-Hydroxyisoquinoline-3-carboxylic Acid, N-*Phenyl*
2-Hydroxy-3-phenylquinoline-4-carboxylic Acid
2-Naphthol, o-*Nitrophenyl ether*
Oxycinchophen
Phenanthrene-4,5-dicarboxylic Acid, 5-*Mono-Amide*

$C_{16}H_{11}NO_4$
1-Aminoanthraquinone-2-carboxylic Acid, *Me ester*
2,4-Dimethyl-1-nitroanthraquinone
1-Methylaminoanthraquinone-2-carboxylic Acid

$C_{16}H_{11}NO_5$
2-Hydroxy-1-methyl-3-nitroanthraquinone, *Me ether*
2-Hydroxy-3-methyl-1-nitroanthraquinone, *Me ether*
2-Hydroxy-1-nitroanthraquinone, *Et ether*

$C_{16}H_{11}NO_5S$
8-Nitronaphthalene-1-sulphonic Acid, *Phenyl ester*

$C_{16}H_{11}NO_6$
1,2-Dihydroxy-4-nitroanthraquinone, *Di-Me ether*
1,8-Dihydroxy-4-nitroanthraquinone, *Di-Me ether*
4-Nitrostilbene-2,2′-dicarboxylic Acid
2-Nitrostilbene-2′,4-dicarboxylic Acid

$C_{16}H_{11}NS$
7*H*-Benzo[*c*]phenothiazine
12*H*-Benzo[*a*]phenothiazine

$C_{16}H_{11}N_3O$
Quindoline-10-carboxylic Acid, *Amide*

$C_{16}H_{11}N_3O_3$
Imasatin
Isamic Acid★† (*see also* Sixth Supplement)
1,2-Naphthoquinone, 1-o-*Nitrophenylhydrazone*
1,2-Naphthoquinone, 1-m-*Nitrophenylhydrazone*
1,2-Naphthoquinone, 1-p-*Nitrophenylhydrazone*
1,2-Naphthoquinone, 2-o-*Nitrophenylhydrazone*
1,2-Naphthoquinone, 2-p-*Nitrophenylhydrazone*
1,4-Naphthoquinone, o-*Nitrophenylhydrazone*
1,4-Naphthoquinone, m-*Nitrophenylhydrazone*
1,4-Naphthoquinone, p-*Nitrophenylhydrazone*

$C_{16}H_{12}$
Aceanthrene
Benzo[*g*]sesquifulvalene
1,3-Bisdehydro[16]annulene†
1,9-Bisdehydro[16]annulene†
Cyclohept [*fg*]acenaphthene
Cyclo-octa[*a,e*]dibenzene
1,2,3,4-Dibenzo-cyclo-octatetraene†
1,2-Diphenyltricyclo[1,1,0,0^{2,4}]butane†
1-Phenylazulene
2-Phenylazulene
4-Phenylazulene
5-Phenylazulene
6-Phenylazulene
1-Phenylnaphthalene
2-Phenylnaphthalene

$C_{16}H_{12}BN$
5,6-Borazarobenz[*a*]anthracene†
6,5-Borazarobenz[*a*]anthracene†

$C_{16}H_{12}BrNO_2$
1-Amino-4-bromoanthraquinone, N-*Di-Me*

$C_{16}H_{12}Br_2O_4$
4,4′-Dibromobiphenyl-2,2′-dicarboxylic Acid, *Di-Me ester*

$C_{16}H_{12}ClNO_2$
1-Amino-4-chloroanthraquinone, N-*Di-Me*

$C_{16}H_{12}Cl_2O_2$
Bibenzyl-4,4′-dicarboxylic Acid, *Dichloride*

$C_{16}H_{12}Cl_2O_3$
2-Benzoyl-3,6-dichlorobenzoic Acid, *Et ester*

$C_{16}H_{12}Cl_2O_4$
6,6′-Dichlorobiphenyl-2,2′-dicarboxylic Acid, *Mono-Et ester*

$C_{16}H_{12}N_2$
1-Benzeneazonaphthalene
2-Benzeneazonaphthalene
Bibenzyl-2,2′-dicarboxylic Acid, *Dinitrile*
Bibenzyl-4,4′-dicarboxylic Acid, *Dinitrile*
Cryptolepine
1,4-Diaminopyrene
1,6-Diaminopyrene
2,3-Diphenylpyrazine
2,5-Diphenylpyrazine
2,6-Diphenylpyrazine
3,4-Diphenylpyridazine
3,5-Diphenylpyridazine
3,6-Diphenylpyridazine
7-Methylquindoline

$C_{16}H_{12}N_2O$
1-Benzeneazo-2-naphthol
2-Benzeneazo-1-naphthol
4-Benzeneazo-1-naphthol
Indileucin
Isotuboflavin†
4-Nitroso-1-naphthylamine, N-*Phenyl*
2-Phenylquinoline-4-carboxylic Acid, *Amide*
3-Phenylquinoline-4-carboxylic Acid, *Amide*
2-Phenylquinoline-8-carboxylic Acid, *Amide*
Tuboflavin

$C_{16}H_{12}N_2O_2$
Indigo White
2-*o*-Nitrophenyl-3-*p*-tolylacrylic Acid, *Nitrile*

$C_{16}H_{12}N_2O_3$
1-Hydroxy-4-nitro-3-phenylisoquinoline, *Me ether*
2′-Hydroxy-4-nitrostilbene-2-carboxylic Acid, *Me ether*, *Nitrile*
2′-Hydroxy-2-nitrostilbene-4-carboxylic Acid, *Me ether*, *Nitrile*
3′-Hydroxy-2-nitrostilbene-4-carboxylic Acid, *Me ether*, *Nitrile*
4′-Hydroxy-2-nitrostilbene-4-carboxylic Acid, *Me ether*, *Nitrile*
4′-Hydroxy-4-nitrostilbene-2-carboxylic Acid, *Me ether*, *Nitrile*
α-(4-Hydroxyphenyl)-2-nitrocinnamic Acid, *Nitrile*
Isatan†

$C_{16}H_{12}N_2O_4$
1-Amino-5-nitroanthraquinone, N-*Et*
1-Amino-8-nitroanthraquinone, N-*Di-Me*
1-Amino-8-nitroanthraquinone, N-*Et*
Isatide

$C_{16}H_{12}N_2O_4S$
2-Nitrobenzenesulphonic Acid, β-*Naphthalide*

$C_{16}H_{12}N_2O_7$
p-Nitrophenylacetic Acid, *Anhydride*

$C_{16}H_{12}N_2O_8$
4,4′-Dinitrobiphenyl-2,2′-dicarboxylic Acid, *Di-Me ester*
5,5′-Dinitrobiphenyl-2,2′-dicarboxylic Acid, *Di-Me ester*
6,6′-Dinitrobiphenyl-2,2′-dicarboxylic Acid, *Di-Me ester*

$C_{16}H_{12}N_2O_8$
Oxalic Acid, *Di-*p*-nitrobenzyl ester*
Succinic Acid, *Di-*o*-nitrophenyl ester*
Succinic Acid, *Di-*m*-nitrophenyl ester*
Succinic Acid, *Di-*p*-nitrophenyl ester*

$C_{16}H_{12}N_2S$
Benzeneazo-2-naphthol sulphide

$C_{16}H_{12}N_4O_2$
Isamic Acid, *Amide*★†

$C_{16}H_{12}O$
9-Acetylanthracene†
1-Acetylphenanthrene
2-Acetylphenanthrene
3-Acetylphenanthrene
4-Acetylphenanthrene
9-Acetylphenanthrene
2,4-Diphenylfuran
2,5-Diphenylfuran
3,4-Diphenylfuran
1-Naphthol, *Phenyl ether*
2-Naphthol, *Phenyl ether*
1-Naphthyl phenyl Ether
2-Naphthyl phenyl Ether
Octaleno[3,4-*c*]furan†
2-Phenyl-1-naphthol
3-Phenyl-1-naphthol
4-Phenyl-1-naphthol
6-Phenyl-2-naphthol
7-Phenyl-2-naphthol

$C_{16}H_{12}O_2$
Anthracene-1-carboxylic Acid, *Me ester*
Anthracene-9-carboxylic Acid, *Me ester*
1,2-Dimethylanthraquinone
1,3-Dimethylanthraquinone
1,4-Dimethylanthraquinone
1,5-Dimethylanthraquinone
2,3-Dimethylanthraquinone
2,6-Dimethylanthraquinone
2,7-Dimethylanthraquinone
1,2-Dimethylphenanthraquinone
1,3-Dimethylphenanthraquinone
1,4-Dimethylphenanthraquinone
1,6-Dimethylphenanthraquinone
1,7-Dimethylphenanthraquinone
1,8-Dimethylphenanthraquinone
2,3-Dimethylphenanthraquinone

$C_{16}H_{12}O_2$ (*continued*)
- 2,4-Dimethylphenanthraquinone
- 2,5-Dimethylphenanthraquinone
- 2,6-Dimethylphenanthraquinone
- 2,7-Dimethylphenanthraquinone
- 3,4-Dimethylphenanthraquinone
- 3,6-Dimethylphenanthraquinone
- 4,5-Dimethylphenanthraquinone
- 1,4-Diphenyl-2-butene-1,4-dione
- 2-Methylanthracene-1-carboxylic Acid
- 4-Methylanthracene-1-carboxylic Acid
- 6(or 7)Methylanthracene-1-carboxylic Acid
- 6-Methylanthracene-2-carboxylic Acid
- 2-Methylanthracene-9(or 10)-carboxylic Acid
- 10-Methylanthracene-9-carboxylic Acid
- 5-Methylflavone
- 6-Methylflavone
- 7-Methylflavone
- 8-Methylflavone
- Morphenol, *Et ether*
- 2-Phenanthreneacetic Acid
- 3-Phenanthreneacetic Acid
- 4-Phenanthreneacetic Acid
- 9-Phenanthreneacetic Acid
- Phenanthrene-1-carboxylic Acid, *Me ester*
- Phenanthrene-2-carboxylic Acid, *Me ester*
- Phenanthrene-3-carboxylic Acid, *Me ester*

$C_{16}H_{12}O_2S$
- 1-Anthraquinonethiol, S-*Et ether*
- 2-Anthraquinonethiol, S-*Et ether*
- 1-Naphthyl phenyl sulphone
- 2-Naphthyl phenyl sulphone

$C_{16}H_{12}O_3$
- Chalcone-4-carboxylic Acid
- Chalcone-4′-carboxylic Acid
- Chalcone-α-carboxylic Acid
- Chalcone-β-carboxylic Acid
- 2,2-Diphenylsuccinic Acid, *Anhydride*
- 2,3-Diphenylsuccinic Acid, *Anhydride*
- Fluorenone-1-carboxylic Acid, *Et ester*
- Fluorenone-2-carboxylic Acid, *Et ester*
- Fluorenone-4-carboxylic Acid, *Et ester*
- 2-Hydroxy-9,10-anthraquinone, *Et ether*
- 1-Hydroxy-3,4-dimethylanthraquinone
- 4-Hydroxy-1,3-dimethylanthraquinone
- 5-Hydroxy-2,3-dimethylanthraquinone
- 7-Hydroxyisoflavone, *Me ether*
- 1-Hydroxy-2-methylanthraquinone, *Me ether*
- 1-Hydroxy-3-methylanthraquinone, *Me ether*
- 1-Hydroxy-4-methylanthraquinone, *Me ether*
- 2-Hydroxy-1-methylanthraquinone, *Me ether*
- 2-Hydroxy-6-methylanthraquinone, *Me ether*
- 3-Hydroxy-1-methylanthraquinone, *Me ether*
- 2-Hydroxy-9,10-phenanthraquinone, *Et ether*
- 3-Hydroxy-9,10-phenanthraquinone, *Et ether*
- 4-Hydroxyphenanthrene-1-carboxylic Acid, *Me ether*
- 3-Hydroxyphenanthrene-2-carboxylic Acid, *Me ester*
- 2-Hydroxyphenanthrene-3-carboxylic Acid, *Me ester*
- 4-Hydroxyphenanthrene-3-carboxylic Acid, *Me ether*
- 2-Hydroxyphenanthrene-9-carboxylic Acid, *Me ether*
- 4-Hydroxyphenanthrene-9-carboxylic Acid, *Me ether*
- 6-Hydroxyphenanthrene-9-carboxylic Acid, *Me ether*
- 8-Hydroxyphenanthrene-9-carboxylic Acid, *Me ether*
- 10-Hydroxyphenanthrene-9-carboxylic Acid, *Me ester*
- 10-Hydroxyphenanthrene-9-carboxylic Acid, *Me ether*
- 3,4-Methylenedioxychalcone
- 6-Methylflavonol
- 7-Methylflavonol
- 8-Methylflavonol
- Omephine

$C_{16}H_{12}O_3S$
- Naphthalene-1-sulphonic Acid, *Phenyl ester*
- Naphthalene-2-sulphonic Acid, *Phenyl ester*
- 1-Phenylnaphthalene-4-sulphonic Acid

$C_{16}H_{12}O_4$
- Alizarin, *Di-Me ether*
- Alizarin, 2-*Et ether*
- Benzil-*o*-carboxylic Acid, *Me ester*
- Chrysin, 7-*Me ether*
- Dalbergin
- Dibenzoylacetic Acid
- Digitolutein
- 1,3-Dihydroxyanthraquinone, *Di-Me ether*
- 1,4-Dihydroxyanthraquinone, *Di-Me ether*
- 1,4-Dihydroxyanthraquinone, *Et ether*
- 1,5-Dihydroxyanthraquinone, *Di-Me ether*
- 1,6-Dihydroxyanthraquinone, *Di-Me ether*
- 1,7-Dihydroxyanthraquinone, *Di-Me ether*
- 1,8-Dihydroxyanthraquinone, *Di-Me ether*
- 2,3-Dihydroxyanthraquinone, *Di-Me ether*
- 2,6-Dihydroxyanthraquinone, *Di-Me ether*
- 2,7-Dihydroxyanthraquinone, *Di-Me ether*
- 1,2-Dihydroxy-6,7-dimethylanthraquinone
- 1,4-Dihydroxy-2,3-dimethylanthraquinone
- 1,4-Dihydroxy-5,8-dimethylanthraquinone
- 1,4-Dihydroxy-6,7-dimethylanthraquinone
- 1,5-Dihydroxy-2,6-dimethylanthraquinone
- 1,5-Dihydroxy-3,7-dimethylanthraquinone
- 1,7-Dihydroxy-2,6-dimethylanthraquinone
- 1,7-Dihydroxy-3,5-dimethylanthraquinone
- 1,8-Dihydroxy-2,7-dimethylanthraquinone
- 2,6-Dihydroxy-1,5-dimethylanthraquinone
- 2,6-Dihydroxy-3,7-dimethylanthraquinone
- 2,7-Dihydroxy-1,5-dimethylanthraquinone
- 5,8-Dihydroxy-1,2-dimethylanthraquinone
- 5,8-Dihydroxy-1,3-dimethylanthraquinone
- 4′,5-Dihydroxyflavone, 4′-*Me ether*
- 5,6-Dihydroxyflavone, 5-*Me ether*
- 5,8-Dihydroxyflavone, 8-*Me ether*
- 7,8-Dihydroxyflavone, 7-*Me ether*
- 1,3-Dihydroxy-2-methylanthraquinone, 1-*Me ether*
- 2,3-Dihydroxyphenanthraquinone, *Di-Me ether*
- 4,5-Dihydroxyphenanthraquinone, *Di-Me ether*
- 1,1-Diphenylethylene-2-dicarboxylic Acid
- Diphenylfumaric Acid
- Diphenylmaleic Acid
- Formo-ononetin
- Fumaric Acid, *Di-Phenyl ester*
- Isoformo-ononetin

$C_{16}H_{12}O_4$ (*continued*)
- Maleic Acid, *Diphenyl ester*
- 3′,4′-Methylenedioxyflavanone
- Pratol
- Stilbene-2,2′-dicarboxylic Acid
- Stilbene-2,3-dicarboxylic Acid
- Stilbene-2,4′-dicarboxylic Acid
- Stilbene-3,3′-dicarboxylic Acid
- Stilbene-4,4′-dicarboxylic Acid
- Xanthone-2-carboxylic Acid, *Et ester*
- Xanthone-4-carboxylic Acid, *Et ester*

$C_{16}H_{12}O_4S$
- 2-Naphthol-6-sulphonic Acid, *Phenyl ester*

$C_{16}H_{12}O_5$
- Acacetin
- Biochanin A
- Brazilein
- Carajurone
- Damnacanthol
- Demethylpterocarpin
- Echioidinin †
- Galangin, 3-*Me ether*
- Genistein, 5-*Me ether*
- Genkwanin
- 2′-Hydroxyflavone, *Me ether*
- 3′-Hydroxyflavone, *Me ether*
- 4′-Hydroxyflavone, *Me ether*
- 5-Hydroxyflavone, *Me ether*
- 6-Hydroxyflavone, *Me ether*
- 7-Hydroxyflavone, *Me ether*
- 8-Hydroxyflavone, *Me ether*
- 2-(6-Hydroxy-2-methoxy-3,4-methylene-biophenyl)benzofuran
- Isowogonin
- Izalpinin
- Lucidin, 3-*Me ether* (Anthraquinone)
- Maackiain
- Macrosporin
- 2-Methylgenistein
- Oroxylin-A
- Physcione
- Questin †
- 1,2,3-Trihydroxyanthraquinone, 1,2-*Di-Me ether*
- 1,2,3-Trihydroxyanthraquinone, 1,3-*Di-Me ether*
- 1,2,3-Trihydroxyanthraquinone, 2,3-*Di-Me ether*
- 1,2,3-Trihydroxyanthraquinone, 2-*Et ether*
- 1,2,3-Trihydroxyanthraquinone, 3-*Et ether*
- 1,2,4-Trihydroxyanthraquinone, 2,4-*Di-Me ether*
- 1,2,5-Trihydroxyanthraquinone, 1,2-*Di-Me ether*
- 1,2,6-Trihydroxyanthraquinone, 2,6-*Di-Me ether*
- 1,2,7-Trihydroxyanthraquinone, 1,7-*Di-Me ether*
- 1,2,7-Trihydroxyanthraquinone, 2,7-*Di-Me ether*
- 1,2,8-Trihydroxyanthraquinone, 2,8-*Di-Me ether*
- 2′,3,7-Trihydroxyflavone, 7-*Me ether*
- 3,3′,7-Trihydroxyflavone, 7-*Me ether*
- 3′,5,7-Trihydroxyflavone, 7-*Me ether*
- 1,2,5-Trihydroxy-6-methylanthraquinone, *Mono-Me ether*
- 1,2,8-Trihydroxy-6-methylanthraquinone, 1-*Me ether*
- 1,3,8-Trihydroxy-6-methylanthraquinone, 1-*Me ether*
- 1,3,8-Trihydroxy-6-methylanthraquinone, 3-*Me ether*
- 1,4,6-Trihydroxy-5-methylanthraquinone, 6-*Me ether*
- Wogonin

$C_{16}H_{12}O_5S$
- Anthraquinone-2-sulphonic Acid, *Et ester*

$C_{16}H_{12}O_6$
- Artocarpetin †
- Carviolin
- Chrysoeriol
- Datiscetin, 7-*Me ether*
- Detmoglaucin †
- Diosmetin
- Ferreirin
- Haematein
- Isoshehkangenin
- Kaempferide
- Luteolin, 3′-*Me ether*
- Luteolin, 7-*Me ether*
- Pedicinin
- Piperoin
- Pratensein
- Questinol †
- Rengasin
- Rhamnocitrin
- Sophorol
- Tectorigenin
- Teloschistin
- 1,2,5,6-Tetrahydroxyanthraquinone, 1,2-*Di-Me ether*
- 1,3,5,7-Tetrahydroxyanthraquinone, *Di-Me ether*
- 1,3,6,8-Tetrahydroxyanthraquinone, 1,3-*Di-Me ether* †
- 1,3,6,8-Tetrahydroxyanthraquinone, 3,6-*Di-Me ether* †
- 3′,4′,7-Trihydroxy-3-methoxyflavone †
- 3,5,7-Trihydroxy-2′-methoxyflavone †
- 4′,5,7-Trihydroxy-6-methoxyflavone (incorrectly given as $C_{15}H_{12}O_6$) ★ †
- 1,2,6,8-Tetrahydroxy-3-methylanthraquinone ★, 6-*Me ether* †
- Xanthorin †

$C_{16}H_{12}O_7$
- Dermocybin
- Isorhamnetin
- Pannaric Acid
- Quercetin, 5-*Me ether*
- Rhamnetin
- Tambuletin
- 3′,4′,5,7-Tetrahydroxy-6-methoxyflavone †
- Vogeletin †

$C_{16}H_{12}O_8$
- Balmacamgenin
- Laccainic Acid
- Mearnsetin †

$C_{16}H_{12}O_8$ *(continued)*
- 2-Methylirigenol
- Patuletin

$C_{16}H_{12}S$
- 2,4-Diphenylthiophene
- 2,5-Diphenylthiophene
- 3,4-Diphenylthiophene
- 1-Naphthyl phenyl sulphide
- 2-Naphthyl phenyl sulphide

$C_{16}H_{13}B$
- 3-Phenyl[3]benzoborepin †

$C_{16}H_{13}BrO_3$
- *o*-Toluic Acid, p-*Bromophenacyl ester*
- *m*-Toluic Acid, p-*Bromophenacyl ester*
- *p*-Toluic Acid, p-*Bromophenacyl ester*

$C_{16}H_{13}ClN_2O$
- Diazepam †

$C_{16}H_{13}ClO_2$
- 3′,4′-Dimethylbenzophenone-2-carboxylic Acid, *Chloride*
- Diphenacyl chloride

$C_{16}H_{13}ClO_3$
- *o*-4-Chlorobenzoylbenzoic Acid, *Et ester*

$C_{16}H_{13}ClO_4$
- Apigeninidin, 4′-*Me ether*
- Galanginidin chloride, 3-*Me ether*

$C_{16}H_{13}ClO_5$
- Luteolinidin chloride, 3′-*Me ether*
- Pelargonidin chloride, 5-*Me ether*

$C_{16}H_{13}ClO_6$
- Morindin chloride, 3-*Me ether*
- Peonidin chloride
- Rhamnetidin chloride

$C_{16}H_{13}ClO_7$
- Myrtillidin chloride
- Petunidin (chloride)

$C_{16}H_{13}FO_3$
- *o*-[2-Ethyl-5-fluorobenzoyl]-benzoic Acid
- *o*-[4-Ethyl-3-fluorobenzoyl]-benzoic Acid

$C_{16}H_{13}IO_2$
- 3-*o*-Iodophenylacrylic Acid, m-*Tolyl ester*
- 3-*m*-Iodophenylacrylic Acid, m-*Tolyl ester*
- 3-*p*-Iodophenylacrylic Acid, m-*Tolyl ester*

$C_{16}H_{13}I_4NO_4$
- Thyroxine, *Me ester*

$C_{16}H_{13}N$
- 1-Benzylisoquinoline
- 3-Benzylisoquinoline
- 4-Benzylisoquinoline
- 2-Benzylquinoline
- 4-Benzylquinoline
- 6-Benzylquinoline
- 1,2-Diphenylpyrrole
- 2,4-Diphenylpyrrole
- 2,5-Diphenylpyrrole
- 1-Methyl-3-phenylisoquinoline
- 3-Methyl-4-phenylisoquinoline
- 8-Methyl-3-phenylisoquinoline
- 2-Methyl-3-phenylquinoline
- 2-Methyl-4-phenylquinoline
- 3-Methyl-2-phenylquinoline
- 4-Methyl-2-phenylquinoline
- 5-Methyl-2-phenylquinoline
- 6-Methyl-2-phenylquinoline
- 7-Methyl-2-phenylquinoline
- 8-Methyl-2-phenylquinoline
- *N*-Phenyl-1-naphthylamine
- *N*-Phenyl-2-naphthylamine
- 2-Phenyl-1-naphthylamine
- 4-Phenyl-1-naphthylamine
- 6-Phenyl-1-naphthylamine
- 7-Phenyl-1-naphthylamine
- 6-Phenyl-2-naphthylamine
- 2-*o*-Tolylquinoline
- 2-*m*-Tolylquinoline
- 2-*p*-Tolylquinoline

$C_{16}H_{13}NO$
- 3-Benzoyl-2-phenylpropionic Acid, *Nitrile*
- 4-Hydroxy-2-methyl-3-phenylquinoline
- 6-Hydroxy-2-methyl-4-phenylquinoline
- 7-Hydroxy-2-methyl-3-phenylquinoline
- 7-Hydroxy-2-methyl-4-phenylquinoline
- 4-Hydroxy-α-phenylcinnamic Acid, *Me ether, Nitrile*
- 4-*o*-Hydroxyphenyl-2-methylquinoline
- 4-*m*-Hydroxyphenyl-2-methylquinoline
- 4-*p*-Hydroxyphenyl-2-methylquinoline
- 4-*p*-Hydroxyphenylquinoline, *Me ether*
- 4-Hydroxy-2-phenylquinoline, *Me ether*
- 4-Hydroxy-3-phenylquinoline, *Me ether*
- 6-Hydroxy-2-phenylquinoline, *Me ether*
- 6-Hydroxy-4-phenylquinoline, *Me ether*
- 7-Hydroxy-2-phenylquinoline, *Me ether*
- 8-Hydroxy-2-phenylquinoline, *Me ether*
- 9-Phenanthreneacetic Acid, *Amide*

$C_{16}H_{13}NO_2$
- Arboricine
- Benzo[*f*]quinoline-1-carboxylic Acid, *Et ester*
- 1-Dimethylaminoanthraquinone
- 2-Dimethylaminoanthraquinone
- 2,2-Diphenylsuccinic Acid, *Imide*
- 2,3-Diphenylsuccinic Acid, *Imide*
- Phenanthridine-6-carboxylic Acid, *Et ester*
- Viridicatin, 3-O-*Me-ether*

$C_{16}H_{13}NO_3$
- 1-Amino-2-hydroxyanthraquinone, *Et ether*
- 1-Amino-4-hydroxyanthraquinone, N-*Di-Me*
- Viridicatol, 3′-O-*Me* †

$C_{16}H_{13}NO_3S$
- 8-Anilinonaphthalene-1-sulphonic Acid
- 1-Naphthylamine-2-sulphonic Acid, N-*Phenyl*
- 1-Naphthylamine-5-sulphonic Acid, N-*Phenyl*
- 1-Naphthylamine-6-sulphonic Acid, N-*Phenyl*
- 1-Naphthylamine-7-sulphonic Acid, N-*Phenyl*
- 1-Naphthylamine-8-sulphonic Acid, N-*Phenyl*
- 2-Naphthylamine-5-sulphonic Acid, N-*Phenyl*
- 2-Naphthylamine-6-sulphonic Acid, N-*Phenyl*
- 2-Naphthylamine-7-sulphonic Acid, N-*Phenyl*
- 2-Naphthylamine-8-sulphonic Acid, N-*Phenyl*
- Quinoline-8-sulphonic Acid, *Benzyl ester*

$C_{16}H_{13}NO_4$
- 1-Amino-2,3-dihydroxyanthraquinone, *Di-Me ether*
- 3-Amino-1,8-dihydroxyanthraquinone, *Di-Me ether*

$C_{16}H_{13}NO_4$ (*continued*)
4-Amino-1,2-dihydroxyanthraquinone, *Di-Me ether*
4-Amino-1,3-dihydroxyanthraquinone, *Di-Me ether*
5-Amino-1,4-dihydroxyanthraquinone, *Di-Me ether*
Evoxanthine
2-[5-Methyl-2-nitrophenyl]-3-phenylacrylic Acid
3-[3-Methyl-2-nitrophenyl]-2-phenylacrylic Acid
3-[5-Methyl-2-nitrophenyl]-2-phenylacrylic Acid
2-Nitrophenanthrahydroquinone, *Di-Me ether*
2-*p*-Nitrophenyl-3-phenylacrylic Acid, *Me ester*
3-*o*-Nitrophenyl-2-phenylacrylic Acid, *Me ester*
3-*m*-Nitrophenyl-2-phenylacrylic Acid, *Me ester*
3-*p*-Nitrophenyl-2-phenylacrylic Acid, *Me ester*
2-*o*-Nitrophenyl-3-*m*-tolylacrylic Acid
2-*o*-Nitrophenyl-3-*p*-tolylacrylic Acid
3-*o*-Nitrophenyl-2-*m*-tolylacrylic Acid
2-Nitrostilbene-4-carboxylic Acid, *Me ester*
Papveroline

$C_{16}H_{13}NO_4S$
4-Anilino-5-hydroxynaphthalene-1-sulphonic Acid
6-Anilino-4-hydroxynaphthalene-2-sulphonic Acid
7-Anilino-4-hydroxynaphthalene-2-sulphonic Acid

$C_{16}H_{13}NO_5$
2-Benzoyl-3-nitrobenzoic Acid, *Et ester*
2-Benzoyl-5-nitrobenzoic Acid, *Et ester*
3-Hydroxy-2-nitro-α-phenylcinnamic Acid, *Me ether*
5-Hydroxy-2-nitro-α-phenylcinnamic Acid, *Me ether*
2-Hydroxy-α-(4-nitrophenyl)cinnamic Acid, *Me ether*
4-Hydroxy-α-(4-nitrophenyl)cinnamic Acid, *Me ether*
2′-Hydroxy-4-nitrostilbene-2-carboxylic Acid, *Me ether*
2′-Hydroxy-2-nitrostilbene-4-carboxylic Acid, *Me ether*
3′-Hydroxy-2-nitrostilbene-4-carboxylic Acid, *Me ether*
4′-Hydroxy-2-nitrostilbene-4-carboxylic Acid, *Me ether*
4′-Hydroxy-4-nitrostilbene-2-carboxylic Acid, *Me ether*
4′-Hydroxy-4-nitrostilbene-3-carboxylic Acid, *Me ether*
α-(2-Hydroxyphenyl)-2-nitrocinnamic Acid, *Me ether*
α-(4-Hydroxyphenyl)-2-nitrocinnamic Acid, *Me ether*
Narciprimine, *Di-Me ether*†
2-*o*-Nitrobenzoylbenzoic Acid, *Et ester*
Xanthevodine

$C_{16}H_{13}NO_6$
4-Nitrobiphenic Acid, *Di-Me ester*
6-Nitrobiphenic Acid, *Di-Me ester*

$C_{16}H_{13}NO_7$
Papaveric Acid

$C_{16}H_{13}N_3$
4-Benzeneazo-1-naphthylamine

$C_{16}H_{13}N_3O_4$
Biochanin C

$C_{16}H_{13}N_7O_{12}$
Di-(3,5-dimethyl-2,4,6-trinitrophenyl)amine
Tetramethyldipicrylamine†

$C_{16}H_{14}$
1,3-Dimethylanthracene
1,4-Dimethylanthracene
1,5-Dimethylanthracene
2,3-Dimethylanthracene
2,6-Dimethylanthracene
2,7-Dimethylanthracene
2,9-Dimethylanthracene
3,9-Dimethylanthracene
9,10-Dimethylanthracene
1,2-Dimethylphenanthrene
1,3-Dimethylphenanthrene
1,4-Dimethylphenanthrene
1,5-Dimethylphenanthrene
1,6-Dimethylphenanthrene
1,7-Dimethylphenanthrene
1,8-Dimethylphenanthrene
1,9-Dimethylphenanthrene
2,3-Dimethylphenanthrene
2,4-Dimethylphenanthrene
2,5-Dimethylphenanthrene
2,6-Dimethylphenanthrene
2,7-Dimethylphenanthrene
2,9-Dimethylphenanthrene
3,4-Dimethylphenanthrene
3,5-Dimethylphenanthrene
3,6-Dimethylphenanthrene
3,9-Dimethylphenanthrene
3,10-Dimethylphenanthrene
4,5-Dimethylphenanthrene
9,10-Dimethylphenanthrene
1,4-Diphenyl-1,3-butadiene
1,2-Diphenylcyclobutene†
(Diphenylmethylene)cyclopropane†
1-Ethylphenanthrene
2-Ethylphenanthrene
3-Ethylphenanthrene
9-Ethylphenanthrene
3,4,8,9-Tetrahydropyrene

$C_{16}H_{14}ClN_3O$
α-Amino-4(7′-chloro-4′-quinolylamino)-*o*-cresol
Chlordiazepoxide

$C_{16}H_{14}Cl_2O_3$
Chlorobenzilate

$C_{16}H_{14}Cl_2O_6$
Estin

$C_{16}H_{14}Fe$
Phenylferrocene

$C_{16}H_{14}N_2$
2-*p*-Aminophenyl-4-methylquinoline
2-*p*-Aminophenylquinoline, N-*Me*
Indole, *Dimer*
1,2-Naphthylenediamine, 1-N-*Phenyl*

$C_{16}H_{14}N_2$ (*continued*)
1,2-Naphthylenediamine, 2-N-*Phenyl*
1,4-Naphthylenediamine, N-*Phenyl*
1,8-Naphthylenediamine, 1-N-*Phenyl*
N[1]-1-Naphthyl-*N*[2]-phenylhydrazine
m-Phenylenediamine, N-1-*Naphthyl*
m-Phenylenediamine, N-2-*Naphthyl*
4-Phenyl-1,2-naphthylenediamine
2-Phenyl-1,3-naphthylenediamine
2-Phenyl-1,4-naphthylenediamine

$C_{16}H_{14}N_2O$
Arborine
2-(1-Imidazolin-2-yl)benzophenone†
Methaqualone

$C_{16}H_{14}N_2O_2$
Anilinosuccinanil
2,3-Dioxo-1,4-diphenylpiperazine
2,5-Dioxo-1,4-diphenylpiperazine
2,5-Dioxo-3,6-diphenylpiperazine
2,6-Dioxo-1,4-diphenylpiperazine
Maleanilide
3-Phenylindazole-2-carboxylic Acid, *Et ester*

$C_{16}H_{14}N_2O_4$
1-(4-Aminophenyl)-2-(4-carboxy-2-nitrophenyl) ethylene, *Me ester*
Azobenzene-2,2′-dicarboxylic Acid, *Di-Me ester*
Azobenzene-3,3′-dicarboxylic Acid, *Di-Me ester*
Nybomycin

$C_{16}H_{14}N_2O_5$
Azoxybenzene-2,2′-dicarboxylic Acid, *Di-Me ester*
Azoxybenzene-3,3′-dicarboxylic Acid, *Di-Me ester*

$C_{16}H_{14}N_2O_7$
3,4-Dihydroxy-6-nitrophthalic Acid, 4-*Me ether* 1-*Anilide*
3,4-Dihydroxy-6-nitrophthalic Acid, 4-*Me ether* 2-*Anilide*

$C_{16}H_{14}N_2O_8$
4,5-Dinitronaphthalene-1,8-dicarboxylic Acid, *Di-Et ester*

$C_{16}H_{14}N_6O_4$
Rhizopterin, *Me ether*

$C_{16}H_{14}O$
1-Anthrol, *Et ether*
2-Anthrol, *Et ether*
Benzyl Styryl Ketone
2-Hydroxyphenanthrene, *Et ether*
3-Hydroxyphenanthrene, *Et ether*
2-Methylanthrone, *Me ether*
α-Methylchalcone
2-Methylchalcone
2′-Methylchalcone
3-Methylchalcone
3′-Methylchalcone
4-Methylchalcone
4′-Methylchalcone

$C_{16}H_{14}O_2$
Benzyl cinnamate
α-Benzylcinnamic Acid
β-Benzylcinnamic Acid
Cinnamic Acid, *Benzyl ester*
2,2′-Diacetylbiphenyl
4,4′-Diacetylbiphenyl
1,5-Dihydroxyanthracene, *Di-Me ether*
1,8-Dihydroxyanthracene, *Di-Me ether*
2,3-Dihydroxyanthracene, *Di-Me ether*
2,6-Dihydroxyanthracene, *Di-Me ether*
2,7-Dihydroxyanthracene, *Di-Me ether*
9,10-Dihydroxyanthracene, *Di-Me ether*
1,2-Dihydroxyphenanthrene, *Di-Me ether*
1,9-Dihydroxyphenanthrene, *Di-Me ether*
2,3-Dihydroxyphenanthrene, *Di-Me ether*
2,5-Dihydroxyphenanthrene, *Di-Me ether*
2,6-Dihydroxyphenanthrene, *Di-Me ether*
2,7-Dihydroxyphenanthrene, *Di-Me ether*
3,6-Dihydroxyphenanthrene, *Di-Me ether*
3,8-Dihydroxyphenanthrene, *Di-Me ether*
2,2′-Dimethylbenzil
3,3′-Dimethylbenzil
4,4′-Dimethylbenzil
2,3-Diphenylacrylic Acid, *Me ester*
3,3-Diphenylacrylic Acid, *Me ester*
1,4-Diphenyl-1,4-butanedione
Fluorene-1-carboxylic Acid, *Et ester*
Fluorene-9-carboxylic Acid, *Et ester*
2-Hydroxychalcone, *Me ether*
2′-Hydroxychalcone, *Me ether*
3-Hydroxychalcone, *Me ether*
3′-Hydroxychalcone, *Me ether*
4-Hydroxychalcone, *Me ether*
4′-Hydroxychalcone, *Me ether*
1-Hydroxy-3-phenylacetone, *Benzyl ether*
4-Methoxychalcone
6-Methylflavanone
7-Methylflavanone
2-Methylfluorene-9-carboxylic Acid, *Me ester*
9-Methylfluorene-9-carboxylic Acid, *Me ester*
Morphol, *Di-Me ether*
Peltogynan†
3-Phenylindane-1-carboxylic Acid
2-Phenylindane-2-carboxylic Acid

$C_{16}H_{14}O_2S_2$
Dithio-oxalic Acid, *Di-*p*-tolyl ester*

$C_{16}H_{14}O_3$
o-Benzoylbenzoic Acid, *Et ester*
p-Benzoylbenzoic Acid, *Et ester*
α-Benzoyl-phenylacetic Acid, *Me ester*
p-Benzoylphenylacetic Acid, *Me ester*
2-Benzoyl-3-phenylpropionic Acid
3-Benzoyl-2-phenylpropionic Acid
3-Benzoyl-3-phenylpropionic Acid
Chalepensin†
Dalbergione
3,4-Dihydroxy-9-anthrone, *Di-Me ether*
2,2′-Dihydroxychalcone, 2-*Me ether*
2,4′-Dihydroxychalcone, 4′-*Me ether*
2′,3-Dihydroxychalcone, 3-*Me ether*
2′,4-Dihydroxychalcone, 4-*Me ether*
2′,4′-Dihydroxychalcone, 4′-*Me ether*
2′,5′-Dihydroxychalcone, 5′-*Me ether*
4,4′-Dihydroxychalcone, 4-*Me ether*
4,4′-Dihydroxychalcone, 4′-*Me ether*
2′,4′-Dimethylbenzophenone-2-carboxylic Acid
2,6-Dimethylbenzophenone-4-carboxylic Acid
3′,4′-Dimethylbenzophenone-2-carboxylic Acid

$C_{16}H_{14}O_3$ (*continued*)
4,6-Dimethylbenzophenone-2-carboxylic Acid
6-Hydroxy-2,4-dimethoxyphenanthrene
7-Hydroxy-2,4-dimethoxyphenanthrene
2-Hydroxy-2,3-diphenyl-3-butenoic Acid
2-Hydroxy-3,4-diphenyl-3-butenoic Acid
2-Hydroxy-3,4-diphenylbutyrolactone
4′-Hydroxyflavanone, *Me ether*
9-Hydroxyfluorene-9-carboxylic Acid, *Et ester*
9-Hydroxyfluorene-9-carboxylic Acid, *Me ether*, *Me ester*
9-Hydroxyfluorene-9-carboxylic Acid, *Et ether*
2-Hydroxy-α-phenylcinnamic Acid, *Me ether*
3-Hydroxy-α-phenylcinnamic Acid, *Me ether*
4-Hydroxy-α-phenylcinnamic Acid, *Me ether*
4-Hydroxy-α-phenylcinnamic Acid, *Me ester*
α-2-Hydroxyphenylcinnamic Acid, *Me ether*
α-4-Hydroxyphenylcinnamic Acid, *Me ether*
Maturinin†
4-Methoxydalbergione† (Dalbergione★)
Obtusaquinone†
Phenylacetic Acid, *Anhydride*
Styracol
Thebaol
o-Toluic Acid, *Anhydride*
m-Toluic Acid, *Anhydride*
p-Toluic Acid, *Anhydride*
2-*p*-Toluylbenzoic Acid, *Me ester*
4-*p*-Toluylbenzoic Acid, *Me ester*
1,5,6-Trihydroxyphenanthrene, 1,5-*Di-Me ether*
1,5,6-Trihydroxyphenanthrene, 5,6-*Di-Me ether*
3,4,6-Trihydroxyphenanthrene, 3,6-*Di-Me ether*

$C_{16}H_{14}O_3S$
Anthracene-2-sulphonic Acid, *Et ester*
Phenanthrene-2-sulphonic Acid, *Et ester*
Phenanthrene-3-sulphonic Acid, *Et ester*
Phenanthrene-9-sulphonic Acid, *Et ester*

$C_{16}H_{14}O_4$
O-Acetylsalicylic Acid, *Benzyl ester*
Alloimperatorin
Alpinetin
5-Benzoyl-2-hydroxybenzoic Acid, *Et ester*
5-Benzoyl-2-hydroxybenzoic Acid, *Et ether*
Bibenzyl-2,2′-dicarboxylic Acid
Bibenzyl-4,4′-dicarboxylic Acid
Biphenyl-2,3-dicarboxylic Acid, *Di-Me ester*
Biphenyl-2,3′-dicarboxylic Acid, *Di-Me ester*
Biphenyl-3,3′-dicarboxylic Acid, *Di-Me ester*
Biphenyl-3,4-dicarboxylic Acid, *Di-Me ester*
Biphenyl-3,4′-dicarboxylic Acid, *Di-Me ester*
Biphenyl-4,4′-dicarboxylic Acid, *Di-Me ester*
Cryptostrobin
Di-4-methoxyphenylglyoxal
Diphenic Acid, *Di-Me ester*
Diphenic Acid, *Et ester*
2,2-Diphenylsuccinic Acid
2,3-Diphenylsuccinic Acid
Emodinanthranol, *Mono-Me ether*
Hydrangenol, *Mono-Me ether*†
2-(2-Hydroxybenzoyl)benzoic Acid, *Et ester*
2-(3-Hydroxybenzoyl)benzoic Acid, *Et ester*
2-(4-Hydroxybenzoyl)benzoic Acid, *Et ether*
4′-Hydroxy-4-methoxydalbergione†
7-Hydroxy-4′-methoxypterocarpan (3-Hydroxy-9-methoxypterocarpan)†
2-Hydroxy-*m*-toluic Acid, *Phenacyl ester*
2-Hydroxy-*p*-toluic Acid, *Phenacyl ester*
Imperatorin
Isoimperatorin
Mandelic Acid, *Phenacyl ester*
Maturin†
Methyl-lucidone†
Oxalic Acid, *Di-*o*-tolyl ester*
Oxalic Acid, *Di-*m*-tolyl ester*
Oxalic Acid, *Di-*p*-tolyl ester*
Pinostrobin†
Succinic Acid, *Diphenyl ester*
2,3,5,6-Tetrahydroxyphenanthrene, 2,6-*Di-Me ether*
2′,4,4′-Trihydroxybiphenyl-2-carboxylic Acid (2,2′)-lactone, 4-*Me ether*†

$C_{16}H_{14}O_4S_2$
Diphenyl disulphide 2,2′-dicarboxylic Acid, *Di-Me ester*

$C_{16}H_{14}O_5$
3-Acetyl-4,7-dimethylnaphthalene-1,6-dicarboxylic Acid†
Alpinone
Anisic Acid, *Anhydride*
Apoalpinone
Asperxanthone
ψ-Baptigenetin, *Me ether*
Brazilin
Butein, 4′-*Me ether*
Citronetin
Decussatin
Diglycollic Acid, *Di-Phenyl ester*
2′,4′-Dihydroxybenzophenone-2-carboxylic Acid, *Et ester*
2′,4′-Dihydroxybenzophenone-2-carboxylic Acid, *Di-Me ether*
2′,4′-Dihydroxybenzophenone-2-carboxylic Acid, 4′-*Et ether*
2′,5′-Dihydroxybenzophenone-2-carboxylic Acid, *Di-Me ether*
3′,4′-Dihydroxybenzophenone-2-carboxylic Acid, *Di-Me ether*
4,5-Dihydroxybenzophenone-2-carboxylic Acid, *Di-Me ether*
5,6-Dihydroxybenzophenone-2-carboxylic Acid, *Di-Me ether*
2,8-Dihydroxy-1-methoxyxanthone, *Di-Me ether* (1,2,8-Trimethoxyxanthone)†
3,4-Dihydroxy-2-methoxyxanthone, *Di-Me ether* (2,3,4-Trimethoxyxanthone)†
Fonsecin, *Mono-Me ether*
Heraclenin†
3-Hydroxybiphenyl-4,4′-dicarboxylic Acid, *Di-Me ester*
4-Hydroxy-2,3-dimethoxyxanthone, *Me ether*†
5-Hydroxy-1,3-dimethoxyxanthone, *Me ether*†
8-Hydroxy-1,2-dimethoxyxanthone, *Me ether*†
Iso-oxypeucedanine
Isosakuranetin

$C_{16}H_{14}O_5$ *(continued)*
Lichexanthone
Linderone
Noralpinone
Nor-rubrofusarin, *Di-Me ether*
2,2′-Oxydibenzoic Acid, *Di-Me ester*
Oxypeucedanine
Padmakasein
Phenoxyacetic Acid, *Anhydride*
Phyllodulcin★ (Phyllodulcinol)†
Pterofuran†
Ravenelin, *Di-Me ether*
Rubrofusarin, *Me ether*
Sakuranetin
O-Salicyloylsalicylic Acid, *Et ester*
2′,5,7-Trihydroxyflavanone, 5(7)-*Me ether*
3′,5,7-Trihydroxyflavanone, 3′-*Me ether*
3′,5,7-Trihydroxyflavanone, 5(7)-*Me ether*
1,2,8-Trihydroxyxanthone, *Tri-Me ether*†
1,3,5-Trihydroxyxanthone, *Tri-Me ether*†
2,3,4-Trihydroxyxanthone, *Tri-Me ether*†
2,3,8-Trihydroxyxanthone, *Tri-Me ether*

$C_{16}H_{14}O_6$
Aromadendrin, 7-*Me ether*
Artocarpanone†
Bellidifolin, *Di-Me ether*†
5,5′-Dihydroxybiphenyl-2,2′-dicarboxylic Acid, *Di-Me ether*
6,6′-Dihydroxybiphenyl-2,2′-dicarboxylic Acid, *Di-Me ether*
6,6′-Dihydroxybiphenyl-2,2′-dicarboxylic Acid, *Di-Me ester*
2′,3-Dihydroxybiphenyl-2,6-dicarboxylic Acid, *Di-Me ether*
6,6′-Dihydroxybiphenyl-3,3′-dicarboxylic Acid, *Di-Me ester*
6,6′-Dihydroxybiphenyl-3,3′-dicarboxylic Acid, *Di-Me ether*
3,3′-Dihydroxybiphenyl-4,4′-dicarboxylic Acid, *Di-Me ester*
3,3′-Dihydroxybiphenyl-4,4′-dicarboxylic Acid, *Di-Me ether*
Guaiacol, *Di-guaiacol oxalate*
Haematoxylin
Hesperetin
Homoeriodictyol
Isoprotocotoin
Mopanol†
Naphthalene-1,2,5-tricarboxylic Acid, *Tri-Me ester*
Peltogynol
Protocotoin
Tartaric Acid, *Diphenyl ester*
1,3,5,6-Tetrahydroxyxanthone, 3,5,6-*Tri-Me ether*†
1,3,6,7-Tetrahydroxyxanthone, 3,6,7-*Tri-Me ether*

$C_{16}H_{14}O_7$
Deodarin†
4′-Hydroxydalbergione
Lecanoric Acid
Lithospermic Acid†

$C_{16}H_{14}O_8$
Diploschistesic Acid
Galloflavin, *Tetra-Me ether*
Isogalloflavin, *Tri-Me ether, Me ester*

$C_{16}H_{14}O_{12}$
Mellitic Acid, *Tetra-Me ester*

$C_{16}H_{15}ClO$
Dibenzylacetic Acid, *Chloride*
2,4-Diphenylbutyric Acid, *Chloride*

$C_{16}H_{15}ClO_2$
α-Chlorodiphenylacetic Acid, *Et ester*

$C_{16}H_{15}ClO_6$
Griseofulvic Acid

$C_{16}H_{15}Cl_3O_2$
1,1,1-Trichloro-2,2-di-(*p*-methoxyphenyl)ethane

$C_{16}H_{15}N$
2-Benzylindole, N-*Me*
Dibenzylacetic Acid, *Nitrile*
2,2-Diphenylbutyric Acid, *Nitrile*
2,3-Diphenylbutyric Acid, *Nitrile*
2,4-Diphenylbutyric Acid, *Nitrile*
2-Methyl-2,3-diphenylpropionic Acid, *Nitrile*
2-Phenylindole, N-*Et*

$C_{16}H_{15}NO$
Dibenzylglycollic Acid, *Nitrile*
3-Phenylcrotonic Acid, *Anilide*

$C_{16}H_{15}NO_2$
Apocrysopine
3-Benzoyl-2-phenylpropionic Acid, *Amide*
Bicyclo[2,2,1]hept-5-ene-2,3-dicarboxylic Acid, *Benzylimide*
2′,4′-Dimethylbenzophenone-2-carboxylic Acid, *Amide*
4-Hydroxy-α-phenylcinnamic Acid, *Me ether, Amide*
αα′-Iminodiacetophenone
18-Nor-13-azaequilenin†

$C_{16}H_{15}NO_3$
Benzoylanthranilic Acid, *Et ester*
2,5-Dihydroxyacridone, 9-*Me*, 3-*Et ether*
2,2-Diphenylsuccinic Acid, α-*Amide*
Hippuric Acid, *Benzyl ester*
Hypnoacetin
Phenaceturic Acid, *Phenyl ester*

$C_{16}H_{15}NO_4$
Arborinine
2-Benzamido-3-(2-furyl)acrylic Acid, *Et ester*
Fumarinine
4-Hydroxy-2-methyl-α-*p*-nitrophenylacetophenone, *Me ether*
2-Methyl-6-phenylpyridine-3,4-dicarboxylic Acid, 3-*Et ester*
N-Phenylglycine-*o*-carboxylic Acid, N-*Benzyl*

$C_{16}H_{15}NO_5$
2-Hydroxyphenylacetic Acid, *Me ether, Et ester*
3-Hydroxyphenylacetic Acid, *Me ether, Et ester*

$C_{16}H_{15}NO_6$
4-Hydroxy-3-methoxy-2-nitrophenylacetic Acid, *Benzyl ether*
4-Hydroxy-5-methoxy-2-nitrophenylacetic Acid, *Benzyl ether*
4-Nitronaphthalic Acid, *Di-Et ester*

$C_{16}H_{15}N_3$
5-Amino-3-methyl-1-phenylpyrazole, 5-N-*Phenyl*

$C_{16}H_{15}N_3O_4$
Diazoaminobenzene-3,3′-dicarboxylic Acid, *Di-Me ester*

$C_{16}H_{16}$
[16]Annulene †
α,β-Dimethylstilbene
α,4′-Dimethylstilbene
2,2′-Dimethylstilbene
3,3′-Dimethylstilbene
4,4′-Dimethylstilbene
1,1-Diphenyl-1-butene
1,2-Diphenyl-1-butene
1,3-Diphenyl-1-butene
1,4-Diphenyl-1-butene
2,4-Diphenyl-1-butene
1,4-Diphenyl-2-butene
2,4-Diphenyl-2-butene
Di-*p*-xylylene
Equinene †
[2,2]Metaparacyclophane †
[2,2]Paracyclophane †
1,2,3,4-Tetrahydro-1-phenylnaphthalene
1,2,3,4-Tetrahydro-2-phenylnaphthalene

$C_{16}H_{16}Cl_2O_2$
1,1-Dichloro-2,2-di-(4-methoxyphenyl)ethane

$C_{16}H_{16}N_2O_2$
Azobenzene-2-carboxylic Acid, *Propyl ester*
Azobenzene-4-carboxylic Acid, *Propyl ester*
Benzil dioxime, O,O-*Di-Me*
p-Benzylidenehydrazinobenzoic Acid, *Et ester*
1,2-Di-*o*-toluylhydrazine
1,2-Di-*m*-toluylhydrazine
1,2-Di-*p*-toluylhydrazine
Isolysergic Acid
Lysergic Acid
6-Methylergol-8-ene-8-carboxylic Acid †
Methylmalonic Acid, *Dianilide*
Salazine, *Di-Me ether*
Succinanilide

$C_{16}H_{16}N_2O_3$
Anthranoylanthranilic Acid, *Et ester*
Malanilide

$C_{16}H_{16}N_2O_4$
2,3-Dianilinosuccinic Acid
1,2-Di-*p*-methoxybenzoylhydrazine
4-Methyl-6-nitrodiphenylamine-2-carboxylic Acid, *Et ester*
2′-Methyl-2-nitrodiphenylamine-4-carboxylic Acid, *Et ester*
4′-Methyl-2-nitrodiphenylamine-4-carboxylic Acid, *Et ester*
Tartranilic Acid, *Anilide*

$C_{16}H_{16}N_2O_6$
2,2′-Dihydroxy-5,5′-dinitrobiphenyl, *Di-Et ether*
4,4′-Dihydroxy-3,3′-dinitrobiphenyl *Di-Et ether*

$C_{16}H_{16}N_2S$
2-(4-Amino-3-methylphenyl)-4,6-dimethyl-benzothiazole

$C_{16}H_{16}N_6O$
Hydroxystilbamidine

$C_{16}H_{16}O$
2,4-Dimethyl-α-phenylacetophenone
2,5-Dimethyl-α-phenylacetophenone
3,4-Dimethyl-α-phenylacetophenone
α-(2,4-Dimethylphenyl)acetophenone
1,4-Diphenyl-2-butanone
4-Ethyl-2-methylbenzophenone
4-Ethyl-4′-methylbenzophenone
9-Methyl-9-fluorenol, *Et ether*
4-Methyl-*a,p*-tolylacetophenone
2,2′,6-Trimethylbenzophenone
2,3′,4-Trimethylbenzophenone
2,4,4′-Trimethylbenzophenone
2,4,5-Trimethylbenzophenone
2,4,6-Trimethylbenzophenone
2,4′,5-Trimethylbenzophenone

$C_{16}H_{16}O_2$
Benzoin, *Et ether*
4-Cinnamyl-3-methoxyphenol †
Dibenzylacetic Acid
α,β-Dihydroxystilbene, *Di-Me ether*
2,2′-Dihydroxystilbene, *Di-Me ether*
3,5-Dihydroxystilbene, *Di-Me ether*
4,4′-Dihydroxystilbene, *Di-Me ether*
2,2′-Dimethylbenzoin
4,4′-Dimethylbenzoin
2,4-Dimethylphenylacetic Acid, *Phenyl ester*
Diphenylacetic Acid, *Et ester*
2,2-Diphenylbutyric Acid
2,3-Diphenylbutyric Acid
2,4-Diphenylbutyric Acid
3,3-Diphenylbutyric Acid
3,4-Diphenylbutyric Acid
4,4-Diphenylbutyric Acid
2,3-Diphenyl-1,4-dioxan
2,5-Diphenyl-1,4-dioxan
2,2-Diphenylpropionic Acid, *Me ester*
2,3-Diphenylpropionic Acid, *Me ester*
3,3-Diphenylpropionic Acid, *Me ester*
Di-*p*-tolylacetic Acid
3-Hydroxy-4,6-dimethylbenzophenone, *Me ether*
4-Hydroxy-2,5-dimethylbenzophenone, *Me ether*
4-Hydroxy-3,5-dimethylbenzophenone, *Me ether*
2-Hydroxy-5-methylbenzophenone, *Et ether*
β-(*p*-Hydroxyphenyl)propiophenone, *Me ether*
γ-(*o*-Hydroxyphenyl)propiophenone, *Me ether*
γ-(*p*-Hydroxyphenyl)propiophenone, *Me ether*
γ-Hydroxy-β-phenylpropiophenone, *Me ether*
p-Hydroxy-β-phenylpropiophenone, *Me ether*
p-Hydroxy-γ-phenylpropiophenone, *Me ether*
Lapachenole †
Menaquinone-1 †
2-Methyl-2,3-diphenylpropionic Acid
2-Methyl-3,3-diphenylpropionic Acid
o-Tolylacetic Acid, *Benzyl ester*

$C_{16}H_{16}O_3$
Acetylxanthorrhoein †
Benzilic Acid, *Et ester*
Benzilic Acid, *Et ether*
Dibenzylglycollic Acid

$C_{16}H_{16}O_3$ *(continued)*
α-4-Dihydroxyacetophenone, *Phenyl ether*
2,4-Dihydroxy-4′-methylbenzophenone, *Di-Me ether*
2,4′-Dihydroxy-5-methylbenzophenone, *Di-Me ether*
2,6-Dihydroxy-2′-methylbenzophenone, *Di-Me ether*
2,6-Dihydroxy-4′-methylbenzophenone, *Di-Me ether*
3,4-Dihydroxy-2′-methylbenzophenone, *Di-Me ether*
3,4′-Dihydroxy-4-methylbenzophenone, *Di-Me ether*
Hydroxy di-*o*-tolylacetic Acid
Hydroxy di-*m*-tolylacetic Acid
Hydroxy di-*p*-tolyacetic Acid
7-Hydroxy-5-methoxy-flavan
α-2-Hydroxyphenoxyacetophenone, *Et ether*
Lapachol★, *Me ether*†
3-β-Naphthoylpropionic Acid, *Et ester*
Obtusafuran†
Orchinol
2-Phenoxybutyric Acid, *Phenyl ester*

$C_{16}H_{16}O_4$
Alloeleutherin
Alloisoeleutherin
Angolensin
Cotoin, *Di-Me ether*
2′,7-Dihydroxy-4′-methoxyisoflavan†
Diphenoxyacetic Acid, *Et ester*
Eleutherin
Hexadecaheptaenedioic Acid
α-Hydroxy-4-methoxy-α-(4-methoxyphenyl)-acetophenone
Isoeleutherin
Mansonone H, *Me ether*†
Naphthalene-1,5-dicarboxylic Acid, *Di-Et ester*
Naphthalene-1,8-dicarboxylic Acid, *Di-Et ester*
Naphthalene-2,7-dicarboxylic Acid, *Di-Et ester*
α,2,4-Trihydroxyacetophenone, α-*Phenyl ether*, 2,4-*Di-Me ether*
2,3,4-Trihydroxybenzophenone, *Tri-Me ether*
2,4,4′-Trihydroxybenzophenone, *Tri-Me ether*
2,4,5-Trihydroxybenzophenone, 2,4,5-*Tri-Me ether*
3,3′,4-Trihydroxybenzophenone, *Tri-Me ether*
3,4,4′-Trihydroxybenzophenone, *Tri-Me ether*
3,4,5-Trihydroxybenzophenone, *Tri-Me ether*
3,4′,5-Trihydroxybenzophenone, *Tri-Me ether*
2,4,6-Trihydroxy-3-methylbenzophenone, 2,4-*Di-Me ether*
Warburgin†

$C_{16}H_{16}O_5$
Alkannin
Asebogenin
Di-*p*-methoxyphenylglycollic Acid
Gallic Acid, *Tri-Me ether*, *Phenyl ester*
1-Hydroxynaphthalene-2,4-dicarboxylic Acid, *Di-Et ester*
Methysticic Acid, *Me ester*
Shikonin
2,3,4,6-Tetrahydroxybenzophenone, *Tri-Me ether*
2,4,5-Trimethoxybenzoic Acid, *Phenyl ester*
Ventilagone

$C_{16}H_{16}O_6$
Coerulignone
Heraclenol†
Maclurin, 2,4,6-*Tri-Me ether*
2′,3,4,4′,5-Pentahydroxybenzophenone, 3,4,5-*Tri-Me ether*

$C_{16}H_{16}O_7$
Fusarubin, *Mono-Me ether*
3,3′,5,5′,7-Pentahydroxy-4′-methoxy-*epi*-catechin†

$C_{16}H_{16}O_8$
Altersolanol A†
Bostrycin†
3-*O*-Caffeoylshikimic Acid†
Xanthophanic Acid, *Di-Me ester*†

$C_{16}H_{16}O_{10}$
Benzenepentacarboxylic Acid, *Penta-Me ester*

$C_{16}H_{17}N$
1-Isopropyl-4-methylcarbazole
1,2,3,4-Tetrahydro-2-phenylquinoline, N-*Me*
1,2,3,4-Tetrahydro-4-phenylquinoline, N-*Me*
1,2,3,4-Tetrahydro-6-phenylquinoline, N-*Me*
1,2,7,8-Tetramethylcarbazole†
1,3,6,8-Tetramethylcarbazole†
1,2,5,7-Tetramethyl-1*H*-cyclohepta[*d*,*e*]-1-pyridine

$C_{16}H_{17}NO$
3-Aminodibenzofuran, N-*Di-Et*
Dibenzylacetic Acid, *Amide*
2,3-Diphenylbutyric Acid, *Amide*
2,4-Diphenylbutyric Acid, *Amide*
3,4-Diphenylbutyric Acid, *Amide*
Lysergic Acid, *Amide*
2-Methyl-3,3-diphenylpropionic Acid, *Amide*

$C_{16}H_{17}NO_2$
α-Anilinophenylacetic Acid, *Et ester*
N-Benzoyl-L-phenylalaninol†
Dibenzylglycollic Acid, *Amide*
N-Diphenylglycine, *Et ester*
Piperidine-*N*-carboxylic Acid, 2-*Naphthyl ester*

$C_{16}H_{17}NO_2S$
S-Benzyl-L-cysteine†

$C_{16}H_{17}NO_2Si$
2-(Methylphenylsila)-3-phenyl-4-methyl-oxazolid-5-one†

$C_{16}H_{17}NO_3$
Amygdophenine
N-Benzoyl-2-hydroxy-2-(4-methoxyphenyl)-ethylamine†
Caranine
Crinine
epi-Crinine
α-Demethoxyhaemanthamine
β-Demethoxyhaemanthamine
Diphenylcarbamic Acid, 2-*Methoxyethyl ester*
Haemultine
2-Nitro-1,1-diphenylethanol, *Et ether*
Normorphine
Powellamine
Tembamide†
Tyrosine, *Benzyl ester*

$C_{16}H_{17}NO_4$
Coranicine
Daphnarcine
2,5-Dihydroxy-4-nitrobiphenyl, *Di-Et ether*
Flexinine
Lunine †
Lycorine
2-Methylquinoline-3,4-dicarboxylic Acid, *Di-Et ester*
Pancracine †
Poetaricine

$C_{16}H_{17}NO_5$
Dubinine †
Sincamidine †

$C_{16}H_{17}N_3O_4$
Anthramycin †

$C_{16}H_{17}N_3O_4S_2$
Chetomin

$C_{16}H_{17}N_3O_5$
7-Methoxy-1,2-(*N*-methylaziridino)mitosene †

$C_{16}H_{18}$
1-Benzyl-2,4,6-trimethylbenzene
1-Cyclohexylnaphthalene
2-Cyclohexylnaphthalene
1,1-Diphenylbutane
1,2-Diphenylbutane
1,3-Diphenylbutane
1,4-Diphenylbutane
2,2-Diphenylbutane
2,3-Diphenylbutane
1,2-Di-*o*-tolylethane
1,2-Di-*m*-tolylethane
1,2-Di-*p*-tolylethane
Phenyl-1,3,5-trimethylphenylmethane
2,2′,4,4′-Tetramethylbiphenyl
2,2′,4,6′-Tetramethylbiphenyl
2,2′,5,5′-Tetramethylbiphenyl
2,2′,6,6′-Tetramethylbiphenyl
3,3′,4,4′-Tetramethylbiphenyl

$C_{16}H_{18}Hg$
Di-[2-phenyl]mercury †

$C_{16}H_{18}INO_2$
Apo-β-erythroidine, *Methiodide*

$C_{16}H_{18}N_2$
Agroclavine
2,2′-Diethylazobenzene
4,4′-Diethylazobenzene
1,4-Diphenylpiperazine
2,3-Diphenylpiperazine
2,2′,3,3′-Tetramethylazobenzene
2,2′,4,4′-Tetramethylazobenzene
2,2′,5,5′-Tetramethylazobenzene
2,2′,6,6′-Tetramethylazobenzene
2,3′,4,5′-Tetramethylazobenzene
2,4,4′,5-Tetramethylazobenzene
3,3′,4,4′-Tetramethylazobenzene
3,3′,5,5′-Tetramethylazobenzene

$C_{16}H_{18}N_2O$
De-*N*-methyldasycarpidone †
Elymoclavine
2-Hydroxy-4,4′-dimethylazobenzene, *Et ether*
2-Hydroxy-4′,5-dimethylazobenzene, *Et ether*
4-Hydroxy-2,2′-dimethylazobenzene, *Et ether*
4-Hydroxy-2′,3′-dimethylazobenzene, *Et ether*
4-Hydroxy-2,3′-dimethylazobenzene, *Et ether*
4-Hydroxy-2,4′-dimethylazobenzene, *Et ether*
4′-Hydroxy-2,4-dimethylazobenzene, *Et ether*
4-Hydroxy-3,3′-dimethylazobenzene, *Et ether*
4-Hydroxy-3,4′-dimethylazobenzene, *Et ether*
4-Hydroxy-2,2′,4′,6-tetramethylazobenzene
4-Hydroxy-2′,4′,5′-trimethylazobenzene, *Me ether*
Isosetoclavine
Setoclavine★ †

$C_{16}H_{18}N_2O_2$
2,2′-Dihydroxyazobenzene, *Di-Et ether*
2,4-Dihydroazobenzene, *Di-Et ether*
2,4′-Dihydroxyazobenzene, *Di-Et ether*
3,3′-Dihydroxyazobenzene, *Di-Et ether*
3,4′-Dihydroxyazobenzene, *Di-Et ether*
4,4′-Dihydroxyazobenzene, *Di-Et ether*
Isopenniclavine
Penniclavine★ †

$C_{16}H_{18}N_2O_3$
2,2′-Dihydroxyazoxybenzene, *Di-Et ether*
3,3′-Dihydroxyazoxybenzene, *Di Et ether*
4,4′-Dihydroxyazoxybenzene, *Di-Et ether*
Pilosine
Spinamycin †

$C_{16}H_{18}N_2O_4S$
Benzylpenicillenic Acid
Benzylpenicillinic Acid
Benzylpenillic Acid
Benzylpenillonic Acid

$C_{16}H_{18}N_2O_5S$
p-Hydroxybenzylpenicillinic Acid
p-Hydroxybenzylpenillic Acid
Phenoxymethylpenicillinic Acid
5,6-*trans*-Phenoxymethylpenicillinic Acid †

$C_{16}H_{18}N_2O_5S_2$
7-(Thiophene-2-acetamido)cephalosporanic Acid

$C_{16}H_{18}N_2O_6S_2$
Azobenzene-3,3′-disulphonic Acid, *Di-Et ester*

$C_{16}H_{18}N_2S$
Anergen
2-3′-Methylbutylaminonaphtho[1,2-*d*]thiazole

$C_{16}H_{18}N_4O$
Isolysergic Acid, *Hydrazide*

$C_{16}H_{18}N_4O_2$
Nialamide

$C_{16}H_{18}N_4O_5$
Erythroflavin †
Threoflavin †

$C_{16}H_{18}O$
6-Isopropyl-*m*-cresol, *Phenyl ether*
2,4,5-Trimethylphenol, *Benzyl ether*

$C_{16}H_{18}O_2$
2-Butyryl-1-naphthol, *Et ether*
Dihydrolapachenole †
2,2′-Dihydroxybiphenyl, *Di-Et ether*
4,4′-Dihydroxybiphenyl, *Di-Et ether*

$C_{16}H_{18}O_2$ *(continued)*
2-(3,8-Dimethyl-5-azulenyl)propionic Acid, *Me ester*
1,2-Diphenyl-1,2-ethanediol, *Di-Me ether*
12-Diphenyl-1,2-ethanediol, *Mono-Et ether*
1-Naphthoic Acid, (+)-2-*Methylbutyl ester*
2-Naphthoic Acid, (+)-2-*Methylbutyl ester*
4-α-Naphthylbutyric Acid, *Et ester*
4-β-Naphthylbutyric Acid, *Et ester*
6-α-Naphthylhexanoic Acid

$C_{16}H_{18}O_2S$
Di-*p*-hydroxyphenyl sulphide, *Di-Et ether*
Thiodiglycol, *Diphenyl ether*

$C_{16}H_{18}O_2S_2$
Di-2-hydroxyphenyl disulphide, *Di-Et ether*
Di-4-hydroxyphenyl disulphide, *Di-Et ether*

$C_{16}H_{18}O_3$
3,3′-Dihydroxydiphenyl Ether, *Di-Et ether*
Diphenylglycollaldehyde, *Di-Me acetal*
4-Hydroxy-1-naphthoic Acid, *Pentyl ester*
Mansonone G, *Me ether*†

$C_{16}H_{18}O_4$
Adipic Acid, *Di-phenyl ester*
Coumurrayin†
Glaupalol, *Me ether*†
Heteropeucenin, 7-*Me ether*†
Methoxyaucuparin, *Me ether*
Peucenin, 7-*Me ether*
4-Phenyl-1,3-butadiene-1,1-dicarboxylic Acid, *Di-Et ester*
2,2′,4,4′-Tetrahydroxybiphenyl, *Tetra-Me ether*
2,2′,5,5′-Tetrahydroxybiphenyl, *Tetra-Me ether*
2,2′,6,6′-Tetrahydroxybiphenyl, *Tetra-Me ether*
2,3′,4,4′-Tetrahydroxybiphenyl, *Tetra-Me ether*
3,3′,4,4′-Tetrahydroxybiphenyl, *Tetra-Me ether*
3,3′,5,5′-Tetrahydroxybiphenyl, *Tetra-Me ether*
Toddaculine

$C_{16}H_{18}O_4S$
Di-*p*-hydroxyphenyl sulphone, *Di-Et ether*

$C_{16}H_{18}O_5$
Aculeatin
αβ-Dehydrocurvularin†
5,7-Dimethoxy-8-(3-methyl-2-oxobutyl)-coumarin†
Globuxanthone†
8-(2-Hydroxy-1-methoxy-3-methylbut-3-enyl)-7-methoxycoumarin†
Hyposalazinol
Sibiricin†
Visamminol, *Me ether*

$C_{16}H_{18}O_6$
3,3′,4,4′,5,5′-Hexahydroxybiphenyl, 3,5,3′,5′-*Tetra-Me ether*

$C_{16}H_{18}O_7$
1-*O*-Cinnamoylquinic Acid†
3-*O*-Cinnamoylquinic Acid†
4-*O*-Cinnamoylquinic Acid†
5-*O*-Cinnamoylquinic Acid†
Nemotinic Acid★, 4-D-*Xyloside*†
Purpurogallin-carboxylic Acid, *Tetra-Me ether*

$C_{16}H_{18}O_8$
1-*O*-*p*-Coumaroylquinic Acid†
3-*O*-*p*-Coumaroylquinic Acid
4-*O*-*p*-Coumaroylquinic Acid†
5-*O*-*p*-Coumaroylquinic Acid†
Leucodrin, *Mono-Me ether*

$C_{16}H_{18}O_9$
Aesculin, *Me ether*
1-*O*-Caffeoylquinic Acid†
4-*O*-Caffeoylquinic Acid†
5-*O*-Caffeoylquinic Acid†
Chlorogenic Acid
Fabiatrin

$C_{16}H_{18}O_{10}$
Fraxin

$C_{16}H_{19}BrN_2$
Brompheniramine

$C_{16}H_{19}BrO_5$
Octanedioic Acid, p-*Bromophenacyl ester*

$C_{16}H_{19}ClN_2$
Chloroprophenpyridamine

$C_{16}H_{19}ClN_2O$
Carbinoxamine

$C_{16}H_{19}N$
α-Aminodiphenylmethane, N-*Isopropyl*
Di-2-methylbenzylamine
Di-3-methylbenzylamine
Di-4-methylbenzylamine
Di-1-phenylethylamine
Di-2-phenylethylamine
N-Ethyldibenzylamine
2,2′,4,4′-Tetramethyldiphenylamine
3,3′,4,4′-Tetramethyldiphenylamine

$C_{16}H_{19}NO_2$
α-Aminomethyl-*p*-hydroxybenzyl Alcohol, N-*Benzyl, Me ether*
4,4′-Dihydroxydiphenylamine, *Di-Et ether*
Elaeocarpine†
Isoelaeocarpine†
Isonitrosocamphor, *Phenyl ether*
2-Methylanthraquinone-1-carboxylic Acid, *Nitrile*
Phenylephrine★, 3-*Benzyl ether*†
1,2,2-Trimethylcyclopentane-1,3-dicarboxylic Acid, α-*Nitrile, Phenyl ester*

$C_{16}H_{19}NO_3$
Demethylgalanthamine†
Erythroidine
Lunacrine
Norpluviine†
Piperlonguminine†

$C_{16}H_{19}NO_4$
Balfourodine†
Dihydrolycorine†
Isobalfourodine†
2-Pentenylpenaldic Acid, *Benzyl ester*
3-Phthalimidopropionic Acid, 3-*Methylbutyl ester*
Ribaldinine, *Mono-Me ether*†
Zephyranthine†

$C_{16}H_{19}NO_5$
Macranthin †
Peepuloidin †
Peyonine †
$C_{16}H_{19}N_2O_9S_2$
Glucobrassicin
$C_{16}H_{19}N_3$
4-Amino-2,3′-dimethylazobenzene, N-*Di-Me*
4-Amino-2,4′-dimethylazobenzene, N-*Di-Me*
2-Amino-2′,3,4′,5-tetramethylazobenzene
2-Amino-3′,4,4′,5-tetramethylazobenzene
4-Amino-2,2′,3,3′-tetramethylazobenzene
4-Amino-2,2′,5,5′-tetramethylazobenzene
4-Amino-2,3′,5′,6-tetramethylazobenzene
4-Amino-2′,3,4′-6-tetramethylazobenzene
4-Amino-2′,3,5,6′-tetramethylazobenzene
$C_{16}H_{19}N_3O_3$
Febrifugine
Isofebrifugine
$C_{16}H_{19}N_3O_4S$
Ampicillin
$C_{16}H_{19}N_3O_6$
Mitiromycin A
Mitomycin A
Mitomycin B
$C_{16}H_{19}N_3S$
Isothipendyl
Prothipendyl
$C_{16}H_{19}N_5O_2$
2,4-Diamino-3′-nitroazobenzene, *Tetra-Me*
$C_{16}H_{19}N_7$
Cypridina-Etioluciferin †
$C_{16}H_{19}O_2P$
Diphenylphosphinic Acid, *Isobutyl ester*
$C_{16}H_{20}$
4-Isopropyl-1,2,7-trimethylnaphthalene †
2-Methylcadalene †
3-Methylcadalene †
5-Methylcadalene †
7-Methylcadalene †
8-Methylcadalene †
$C_{16}H_{20}ClN_3$
2-[*p*-Chlorobenzyl-(2-dimethylaminoethyl)-amino]pyridine
$C_{16}H_{20}HgN_2$
p-Mercuri-di-ethylaniline
p-Mercuri-di-dimethylaniline
$C_{16}H_{20}NO_4$ (ion)
Ribalinium †
$C_{16}H_{20}N_2$
Agroclavine, *Dihydro*
Costaclavine★ †
2,2′-Di-dimethylaminobiphenyl
3,3′-Di-dimethylaminobiphenyl
4,4′-Di-methylaminobiphenyl
N,N′-Diethylbenzidine
Dihydroagroclavine
1,2-Di-*o*-toluidinoethane
1,2-Di-*m*-toluidinoethane
1,2-Di-*p*-toluidinoethane
2,2′,4,4′,6,6′-Hexamethyl-3,3′-dipyridyl
Hydrazobenzene, NN′-*Di-Et*
Pheniramine
Pyroclavine
2,2′,3,3′-Tetramethylhydrazobenzene
2,2′,4,4′-Tetramethylhydrazobenzene
2,2′,5,5′-Tetramethylhydrazobenzene
3,3′,4,4′-Tetramethylhydrazobenzene
3,3′,5,5′-Tetramethylhydrazobenzene
$C_{16}H_{20}N_2O$
Chanoclavine★ †
Dihydroelymoclavine
$C_{16}H_{20}N_2O_2$
β-(2,3-Dehydro-3,3-dimethylpyrrolo[1,2-*g*]-indol-9-yl)-alanine †
4,4′-Diamino-3,3′-diethoxybiphenyl
2,2′-Diethoxyhydrazobenzene
3,3′-Diethoxyhydrazobenzene
4,4′-Diethoxyhydrazobenzene
4-Dimethylallyltryptophan †
$C_{16}H_{20}N_2O_2S$
Di-(4-aminophenyl)sulphone, N-N′-*Tetra-Me*
$C_{16}H_{20}N_2O_4$
Dethiobenzylpenicillin
$C_{16}H_{20}N_2O_4S_2$
Pyrithioxin †
$C_{16}H_{20}N_2O_5S$
Benzylpenicilloic Acid
$C_{16}H_{20}N_2S_2$
Di-(4-dimethylaminophenyl)disulphide
$C_{16}H_{20}N_4O_5$
Porfiromycin
$C_{16}H_{20}O_2$
Cacalol, *Me ether* †
Isolinderoxide †
Linderoxide †
Procerin, *Me ether*
$C_{16}H_{20}O_3$
Desmotropo-ψ-santonin, *Me ether*
Isosericenine †
Sericenic Acid, *Me ester* †
$C_{16}H_{20}O_4$
Deoxysericealactone †
Marasmic Acid, *Me ester*★ †
Olivetonide, *Di-Me ether*
1,2,3,4-Tetrahydronaphthalene-1,8-dicarboxylic Acid, *Di-Et ester*
Tocopheronic Acid, *Lactone*
$C_{16}H_{20}O_4S$
Camphor-10(or 6-)-sulphonic Acid, *Phenyl ester*
$C_{16}H_{20}O_5$
Curvularin
Octanedioic Acid, *Phenacyl ester*
Sericealactone †
$C_{16}H_{20}O_6$
Mexoticin †
Pyrenophorin †
Toddalolactone
Toluene-2,4,6-tricarboxylic Acid, *Tri-Et ester*

$C_{16}H_{20}O_7$
3-Hydroxytoluene-2,4,6-tricarboxylic Acid, *Tri-Et ester*
Picrotoxic Acid, *Me ester*
α-Picrotoxinic Acid, *Me ester*
β-Picrotoxinic Acid, *Me ester*

$C_{16}H_{20}O_8S$
Thiophene-tetracarboxylic Acid, *Tetra-Et ester*

$C_{16}H_{20}O_9$
Bergenin, *Di-Me ether*
Furan-tetracarboxylic Acid, *Tetra-Et ester*
Gentiopicrin

$C_{21}H_{21}N$
Erythrinane
Morphinan
1-Naphthylamine, N-*Dipropyl*

$C_{16}H_{21}NO$
2-1′-Cyclohexenylpropionic Acid, p-*Toluidine*
2-Cyclohexylidenepropionic Acid, p-*Toluidide*
2-Heptyl-4-hydroxyquinoline

$C_{16}H_{21}NO_2$
Elaeocarpilinine†
2-Heptyl-4-hydroxyquinoline
Isoelaeocarpiline†
Phthalimide, N-*Octyl*
Propanolol†

$C_{16}H_{21}NO_3$
Annotine★†
Annotinine
Demethyldihydrogalanthamine†
Homatropine★†
Isolaeocarpicine†
Noratropine
Norhyoscyamine

$C_{16}H_{21}NO_4$
Convolvine
Edulinine†
1-Phenylpyrrolidine-2,5-dicarboxylic Acid, *Di-Et ester*
Phyllalbine†

$C_{16}H_{21}NS_2$
Diethylthiambutene

$C_{16}H_{21}N_3$
Tripelennamine

$C_{16}H_{21}N_3O_3$
Dichroine

$C_{16}H_{21}N_3O_8S$
Cephalosporin C

$C_{16}H_{22}N_2$
2,3-Di-*tert*-butylquinoxaline†
Lycodine

$C_{16}H_{22}N_2O_4$
Inproquone

$C_{16}H_{22}N_2O_5$
O-Acetyltyrosylvaline†
Leucylglycine, N-*Carbobenzyloxy*

$C_{16}H_{22}N_4$
2,2′,4,4′-Tetra-aminobiphenyl, 2,2′-N-*Tetra-Me*
3,3′,4,4′-Tetra-aminobiphenyl, 4,4′-N-*Tetra-Me*
2,2′,4,4′-Tetra-aminodiphenylmethane, 4-N-*Me*-4′-N-*Di-Me*

$C_{16}H_{22}N_4O$
Thonzylamine

$C_{16}H_{22}N_4O_3$
Eseramine†

$C_{16}H_{22}O_2$
Nootkatin, *Me ether*
1,2,2,3-Tetramethylcyclopentane-1-carboxylic Acid, *Phenyl ester*

$C_{16}H_{22}O_3$
Perezone, *Me ether*

$C_{16}H_{22}O_4$
Benzene-1,2-dipropionic Acid, *Di-Et ester*
4,6-Diacetylresorcinol, *Dipropyl ether*
4,6-Diacetylresorcinol, *Et-Butyl ether*
Decanedioic Acid, *Phenyl ester*
Hirsutic Acid C, *Me ester*†
Isosantonic Acid, *Me ester*
2-Phenyladipic Acid, *Di-Et ester*
3-Phenyladipic Acid, *Di-Et ester*
2-Phenylbutane-1,1-dicarboxylic Acid, *Di-Et ester*
4-Phenylbutane-1,1-dicarboxylic Acid, *Di-Et ester*
Phthalic Acid★, *Di-butyl ester*†
Santonic Acid, *Me ester*
Terephthalic Acid, *Dibutyl ester*
Terephthalic Acid, *Di-isobutyl ester*
Terephthalic Acid, *Di*-tert-*butyl ester*

$C_{16}H_{22}O_5$
2,4-Dihydroxy-6,2′-oxoheptylbenzoic Acid, 2-*Me ether*, *Me ester*
2,4-Dihydroxy-6,2′-oxoheptylbenzoic Acid, *Di-Me ether*
Phaseic Acid, *Me ester*†
Tocopheronic Acid

$C_{16}H_{22}O_6$
Penicillic Acid, *Dimedone deriv.*

$C_{16}H_{22}O_7$
2-Carboxymethyl-4,5-dimethoxyphenoxyacetic Acid, *Di-Et ester*
Isodihydrohyenanchin, 2-O-*Me ether*†

$C_{16}H_{22}O_8$
Anisatinic Acid, *Me ester*
Coniferin
3-Hydroxy-3-(3,4-dihydroxy-4-methylpentanoyl)-5-isovalerylcyclopentane-1,2,4-trione†
α-Picrotinic Acid, *Me ester*
β-Picrotinic Acid, *Me ester*

$C_{16}H_{22}O_9$
Sweroside†
2,4,6-Trihydroxyisobutyrophenone, 2-β-Glucopyranoside†

$C_{16}H_{22}O_{10}$
Swertiamarin

$C_{16}H_{22}O_{11}$
Monotropein†
Scandoside†

$C_{16}H_{23}I$
1-Iodohexadecane

$C_{16}H_{23}NO$
Fawcettidine†

$C_{16}H_{23}NO$ (*continued*)
8-Nonenoic Acid, p-*Toluidide*
2,2,6,6-Tetramethyl-4-piperidone, N-*Benzyl*

$C_{16}H_{23}NO_2$
Acrifoline
Ethoheptazine
Piperocaine
Serratidine †

$C_{16}H_{23}NO_3$
Decanedioic Acid, *Monoanilide*

$C_{16}H_{23}NO_4$
Annotine
Streptimidone

$C_{16}H_{23}NO_5$
Benzylpenaldic Acid, *Me ester*
Crispatine
Fulvine

$C_{16}H_{23}NO_6$
Monocrotaline

$C_{18}H_{23}NO_7$
Grantianine

$C_{16}H_{23}N_3O_{15}$
Endecaphyllin B_1

$C_{16}H_{24}$
1,5,9,13-Cyclohexadecatetraene
1,4-Decamethylenebenzene

$C_{16}H_{24}N_2O$
De-*N*-methyl-α-obscurine †

$C_{16}H_{24}N_2O_2$
Hydroxy-de-*N*-methyl-α-obscurine †
Molindone †
Tetraethylsuccinic Acid, *Di-Et ester*

$C_{16}H_{24}N_5O_{16}P_2$
Guanosine diphosphate mannose

$C_{16}H_{24}O_2$
(α)-Costic Acid, *Me ester* †
2,2-Dimethyl-3-phenylpropionic Acid, 3-*Methylbutyl ester*
Hexadeca-2,6,8,12-tetraenoic Acid †
Hinokiic Acid, *Me ester*
Isovalencenic Acid, *Me ester* †
2-Octanol, *Phenylacetal*
2-Phenyldecanoic Acid
3-Phenyldecanoic Acid
6-Phenyldecanoic Acid
7-Phenyldecanoic Acid
10-Phenyldecanoic Acid
5-Phenyloctanoic Acid, *Et ester*
Valerenic Acid, *Me ester* †
Zizanoic Acid, *Me ester* †
5-*epi*-Zizanoic Acid, *Me ester* †

$C_{16}H_{24}O_3$
Alantolic Acid, *Me ester*
Dehydrojuvabione †
2,4-Dihydroxyacetophenone, *Dibutyl ether*

$C_{16}H_{24}O_4$
Brefeldin A †
Humulinic Acid D, *Me ether* †
Prehumulinic Acid
Torquatone

$C_{16}H_{24}O_6$
2,5-Dioxocyclohexane-1,4-dicarboxylic Acid, *Di-isobutyl ester*
Icaritol
α,2,3,4,6-Pentahydroxyacetophenone, α-2,3,4-*Tetra-Et ether*

$C_{16}H_{24}O_7$
Betuloside
Rhododendrin

$C_{16}H_{24}O_{10}$
Catalpol, 6-*Me ether* †
Cephalaroside
Idomethylitol, *Penta-Ac*
Loganic Acid †

$C_{16}H_{25}NO$
Echinacein
Lycopodine
Neoherculin

$C_{16}H_{25}NO_2$
Annofoline
Clavatine
Clavolonine
Dendrobine †
Fawcettimine
Flabelliformine
Gravitole
6-α-Hydroxylycopodine
Lophocerine, *Me ether* †
Lycodoline
Lycofoline

$C_{16}H_{25}NO_3$
p-Pentoxybenzoic Acid, *Dimethylaminoethyl ester*
Serratinine †

$C_{16}H_{25}NO_6$
Sinapin

$C_{16}H_{25}NO_9$
Bacancosin

$C_{16}H_{25}N_7O_8$
Gougerotin †

$C_{16}H_{26}ClNO_2$
Falicain

$C_{16}H_{26}N_2O$
Cernuine †

$C_{16}H_{26}N_2O_2$
Lamprolobine †
Lyrocernuine †

$C_{16}H_{26}N_2O_4S$
Heptylpenicillinic Acid
Heptylpenillic Acid

$C_{16}H_{26}O_2$
Longifolic Acid, *Me ester*
Luparol
5-*n*-Octylresorcinol, *Di-Me ether*
3,7,11-Trimethyl-2,6,10(or 11)-dodecatrienoic Acid, *Me ester*
Resorcinol, *Di-3-methylbutyl ether*

$C_{16}H_{26}O_3$
Cedrolic Acid, *Me ether*
5-Hydroxymethyltetradeca-2,4,6-trienoic Acid
Ilicic Acid, *Me ester* †

$C_{16}H_{26}O_3$ (*continued*)
Juvabione†
Ngaiol, *Me ether*

$C_{16}H_{26}O_4$
Fumagillol

$C_{16}H_{26}O_7$
Minioluteic Acid
Picrocrocin

$C_{16}H_{26}O_8$
Butane-1,1,2,2-tetracarboxylic Acid, *Tetra-Et ester*
Butane-1,1,2,4-tetracarboxylic Acid, *Tetra-Et ester*
Butane-1,1,3,3-tetracarboxylic Acid, *Tetra-Et ester*
Butane-1,1,3,4-tetracarboxylic Acid, *Tetra-Et ester*
Butane-1,1,4,4-tetracarboxylic Acid, *Tetra-Et ester*
Butane-1,2,2,4-tetracarboxylic Acid, *Tetra-Et ester*
Butane-1,2,3,3-tetracarboxylic Acid, *Tetra-Et ester*
Butane-1,2,3,4-tetracarboxylic Acid, *Tetra-Et ester*
Butane-2,2,3,3-tetracarboxylic Acid, *Tetra-Et ester*

$C_{16}H_{26}O_{10}$
Lamiol†

$C_{16}H_{27}ClN_2O_2$
Larocaine

$C_{16}H_{27}N$
Hydnocarpic Acid, *Nitrile*
5-Phenyl-1-pentylamine, N-*Pentyl*

$C_{16}H_{27}NO_5$
Heliotrine

$C_{16}H_{27}O_3PS_2$
O,O-Dibutyl-*S*-α-methylbenzylphosphorothiolothionate

$C_{16}H_{28}$
Dispiro[5,1,5,3]hexadecane†
Dispiro[5,2,5,2]hexadecane†

$C_{16}H_{28}N_2$
α,α′-Diamino-*o*-xylene, N-*Tetra-Et*
Isoparteine, N-*Me*

$C_{16}H_{28}N_2O_5S$
Heptylpenicilloic Acid
Pentylpenicilloic Acid, *Di-Me ester*
Pentylpenicilloic Acid, *α-Et ester*

$C_{16}H_{28}N_4O_4S$
Biocytin

$C_{16}H_{28}O$
Drimenol

$C_{16}H_{28}O_2$
Ambrettolide
1,9-Cyclohexadecanedione
Grifolin
Hydnocarpic Acid
16-Hydroxy-5-hexadecenoic Acid, *Lactone*
16-Hydroxy-6-hexadecenoic Acid, *Lactone*
16-Hydroxy-9-hexadecenoic Acid, *Lactone*

$C_{16}H_{28}O_3$
Hexadecanedioic Acid, *Anhydride*

$C_{16}H_{28}O_6$
Heptane-1,1,7-tricarboxylic Acid, *Tri-Et ester*
Heptane-1,2,2-tricarboxylic Acid, *Tri-Et ester*
Heptane-1,3,3-tricarboxylic Acid, *Tri-Et ester*
Heptane-1,3,7-tricarboxylic Acid, *Tri-Et ester*
Heptane-1,5,5-tricarboxylic Acid, *Tri-Et ester*
Heptane-2,4,4-tricarboxylic Acid, *Tri-Et ester*

$C_{16}H_{29}N$
2,3,5-Tri-*tert*-butylpyrrole†

$C_{16}H_{29}NO$
Hydnocarpic Acid, *Amide*

$C_{16}H_{30}$
1-Hexadecyne
Muscene

$C_{16}H_{30}Br_2O_2$
2,3-Dibromohexadecanoic Acid

$C_{16}H_{30}Cl_2O_2$
2,2-Dichlorohexadecanoic Acid

$C_{16}H_{30}O$
Cyclohexadecanone
Drimanol
10,12-Hexadecadien-1-ol
Muscone
2-Methylcyclopentadecanone
4-Methylcyclopentadecanone
5-Methylcyclopentadecanone
7-Methylcyclopentadecanone
10-Propyltrideca-*trans*-5,9-dien-1-ol†

$C_{16}H_{30}O_2$
2-Hexadecenoic Acid
3-Hexadecenoic Acid†
9-Hexadecenoic Acid
10-Hexadecenoic Acid
14-Methyl-10-oxopentadecanoic Acid
2-Tetradecenoic Acid, *Et ester*
4-Tetradecenoic Acid, *Et ester*

$C_{16}H_{30}O_3$
Ambrettolic Acid
16-Hydroxy-5-hexadecenoic Acid
16-Hydroxy-6-hexadecenoic Acid
12-Hydroxy-9-hexadecenoic Acid†
16-Hydroxy-9-hexadecenoic Acid
Octanoic Acid, *Anhydride*
2-Oxohexadecanoic Acid
3-Oxohexadecanoic Acid
4-Oxohexadecanoic Acid
5-Oxohexadecanoic Acid
7-Oxohexadecanoic Acid
8-Oxohexadecanoic Acid
9-Oxohexadecanoic Acid
10-Oxohexadecanoic Acid
11-Oxohexadecanoic Acid
3-Oxotetradecanoic Acid, *Et ester*
13-Oxotetradecanoic Acid, *Et ester*

$C_{16}H_{30}O_4$
Dodecandioic Acid, *Di-Et ester*
1,12-Dodecanediol, *Di-Ac*
Hexadecanedioic Acid
2-Methyltridecanedioic Acid, *Di-Me ester*
3-Methyltridecanedioic Acid, *Di-Me ester*

$C_{16}H_{30}O_4$ *(continued)*
- Oxalic Acid, *Di-heptyl ester*
- Tetradecanedioic Acid, *Di-Me ester*
- Tetraethylsuccinic Acid, *Di-Et ester*

$C_{16}H_{30}O_9S_2$
- Carrobiose, *Di-Et Mercaptal*†

$C_{16}H_{30}O_{15}$
- 6,8-Di-*C*-glucosylapigenin†

$C_{16}H_{31}BrO_2$
- 2-Bromododecanoic Acid, *Butyl ester*
- 2-Bromohexadecanoic Acid
- 3-Bromohexadecanoic Acid
- 16-Bromohexadecanoic Acid

$C_{16}H_{31}ClO$
- Hexadecanoic Acid, *Chloride*

$C_{16}H_{31}FO_2$
- 16-Fluorohexadecanoic Acid†

$C_{16}H_{31}IO_2$
- 2-Iodohexadecanoic Acid
- 3-Iodohexadecanoic Acid
- 16-Iodohexadecanoic Acid

$C_{16}H_{31}N$
- Hexadecanoic Acid, *Nitrile*

$C_{16}H_{31}NO$
- 2-Hydroxyhexadecanoic Acid, *Nitrile*
- 12-Methyltridecanoic Acid, *Amide*

$C_{16}H_{32}$
- Cyclohexadecane
- 1-Hexadecene

$C_{16}H_{32}Br_2$
- 1,2-Dibromohexadecane
- 1,16-Dibromohexadecane

$C_{16}H_{32}INO$
- 2-Iodohexadecanoic Acid, *Amide*

$C_{16}H_{32}O$
- Cyclohexadecanol
- Hexadecanal
- 2-Hexadecanone
- 3-Hexadecanone
- 9-Methyl-7-pentadecanone
- Muscol

$C_{16}H_{32}O_2$
- Hexadecanoic Acid
- 14-Methylpentadecanoic Acid
- 12-Methyltridecanoic Acid, *Et Ester*
- Nonanoic Acid, *Heptyl ester*
- 2-Octanol, *Octanoyl*
- Pentadecanoic Acid, *Me ester*
- Tetradecanoic Acid, *Et ester*
- Tridecanoic Acid, *Propyl ester*
- 3,7,11-Trimethyldodecanoic Acid, *Me ester*†

$C_{16}H_{32}O_3$
- 2-Hydroxyhexadecanoic Acid
- 3-Hydroxyhexadecanoic Acid
- 4-Hydroxyhexadecanoic Acid
- 8-Hydroxyhexadecanoic Acid†
- 11-Hydroxyhexadecanoic Acid
- 14-Hydroxyhexadecanoic Acid
- 16-Hydroxyhexadecanoic Acid
- 2-Hydroxypentadecanoic Acid, *Me ether*
- 11-Hydroxypentadecanoic Acid, *Me ester*
- 3-Hydroxytetradecanoic Acid, *Et ester*
- Lanopalminic Acid

$C_{16}H_{32}O_4$
- 2,3-Dihydroxyhexadecanoic Acid
- 3,12-Dihydroxyhexadecanoic Acid
- 9,10-Dihydroxyhexadecanoic Acid

$C_{16}H_{32}O_5$
- Aleuritic Acid

$C_{16}H_{32}S_2$
- 1,10-Dithiacyclo-octadecane

$C_{16}H_{33}Br$
- 1-Bromohexadecane

$C_{16}H_{33}BrHg$
- Mercuri-cetyl bromide

$C_{16}H_{33}Cl$
- 1-Chlorohexadecane

$C_{16}H_{33}ClHg$
- Mercuri-cetyl chloride

$C_{16}H_{33}HgI$
- Mercuri-cetyl iodide

$C_{16}H_{33}N$
- Azacycloheptadecane

$C_{16}H_{33}NO$
- Hexadecanoic Acid, *Amide*
- 14-Methylpentadecanoic Acid, *Amide*

$C_{16}H_{33}NO_2$
- 2-Aminohexadecanoic Acid
- 2-Aminotetradecanoic Acid, *Et ester*
- Dodecanoic Acid, 2-*Dimethylaminoethyl ester*
- 2-Hydroxyhexadecanoic Acid, *Amide*

$C_{16}H_{34}$
- 5-Ethyl-5-propylundecane†
- Hexadecane

$C_{16}H_{34}Br_2N_2O_4$
- Suxethonium bromide

$C_{16}H_{34}HgO$
- Mercuri-cetyl hydroxide

$C_{16}H_{34}I_2N_2O_4$
- Celocurin

$C_{16}H_{34}N_2$
- 1,10-Diazacyclo-hexadecane

$C_{16}H_{34}N_4$ (di-ion)
- 5,7,7,12,14-14-Hexamethyl-1,4,8,11-tetra-azacyclotetradeca-4,11-diene†

$C_{16}H_{34}O$
- 1-Hexadecanol
- 2-Hexadecanol
- 3-Hexadecanol
- 4-Hexadecanol
- 5-Hexadecanol
- 6-Hexadecanol
- 7-Hexadecanol
- 8-Hexadecanol
- 2-Methyl-1-pentadecanol
- 3-Methyl-1-pentadecanol
- 6-Methyl-6-pentadecanol
- 9-Methyl-7-pentadecanol
- 8-Methyl-8-pentadecanol
- 2-Octanol, n-*Octyl ether*

$C_{16}H_{34}O_3$
Orthoformic Acid, *Tri-3-methylbutyl ester*

$C_{16}H_{34}O_6$
Sorbitol, *Penta-Et ether*

$C_{16}H_{34}S$
1-Hexadecanethiol

$C_{16}H_{35}N$
1-Aminohexadecane

$C_{16}H_{35}O_4P$
Phosphoric Acid, *Monohexadecyl ester*

$C_{16}H_{36}B_2O$
Dibutylborinic Acid, *Anhydride*

$C_{16}H_{36}GeO$
Tetrabutyl germanate
Tetra-2-butyl germanate
Tetra-isobutyl germanate

$C_{16}H_{36}O_4Si$
Tetrabutyl silicate
Tetra-2-butyl silicate
Tetra-*tert*-butyl silicate
Tetra-isobutyl silicate

$C_{16}H_{36}O_4Sn$
Tetrabutyl stannate

$C_{16}H_{36}O_4Ti$
Tetrabutyl titanate
Tetra-2-butyl tinanate
Tetra-*tert*-butyl titanate
Tetra-isobutyl titanate

$C_{16}H_{36}O_7P_2$
Tetrabutyl pyrophosphate

$C_{16}H_{36}Pb$
Lead tetrabutyl
Lead tetra-isobutyl

$C_{16}H_{36}Sn$
Tin tetrabutyl
Tin tetraisobutyl

$C_{16}H_{38}N_2$ (di-ion)
Decamethonium

$C_{16}H_{40}Al_2MgO_8$
Aluminium magnesium ethoxide

$C_{16}H_{40}B_2N_2O_2$
1,1,4,4-Tetrabutylbon-bon

$C_{16}H_{40}O_{12}Si_4$
2,2,4,4,6,6,8,8-Octaethoxycyclotetrasiloxane

C_{17}

$C_{17}H_7ClO_2$
1-Chlorophenanthraquinone
2-Chlorophenanthraquinone
3-Chlorophenanthraquinone
4-Chlorophenanthraquinone

$C_{17}H_8Br_2O$
2,3-Dibromobenz[*de*]anthracen-7-one
3,9-Dibromobenz[*de*]anthracen-7-one

P

$C_{17}H_8O_8$
Anthraquinone-1,2,3-tricarboxylic Acid
Anthraquinone-1,2,4-tricarboxylic Acid
Anthraquinone-1,2,5-tricarboxylic Acid
Anthraquinone-1,2,6-tricarboxylic Acid

$C_{17}H_8O_8$
Anthraquinone-1,2,7-tricarboxylic Acid
Anthraquinone-1,3,6-tricarboxylic Acid

$C_{17}H_9BrO$
3-Bromobenzanthrone

$C_{17}H_9ClO$
2-Chloro-7*H*-benz[*de*]anthracen-7-one
3-Chloro-7*H*-benz[*de*]anthracen-7-one
4-Chloro-7*H*-benz[*de*]anthracen-7-one
5-Chloro-7*H*-benz[*de*]antgracen-7-one
6-Chloro-7*H*-benz[*de*]anthracen-7-one
7-Chloro-7*H*-benz[*de*]anthracen-7-one
8-Chloro-7*H*-benz[*de*]anthracen-7-one
9-Chloro-7*H*-benz[*de*]anthracen-7-one
10-Chloro-7*H*benz[*de*]anthracen-7-one

$C_{17}H_9NO_2$
α-Anthraquinolinequinone
β-Anthraquinolinequinone
γ-Anthraquinolinequinone

$C_{17}H_9NO_3$
Liriodenine
2-Nitrobenzanthrone
3-Nitrobenzanthrone
5-Nitrobenzanthrone
Spermatheridine†

$C_{17}H_9NO_4$
4-Nitrobenzophenenone, *Di-Et ketal*

$C_{17}H_{10}Br_2O_3$
3,5-Dibromo-2-hydroxybenzoic Acid, *α-Naphthyl ester*

$C_{17}H_{10}N_2O_2$
2-Phenylquinoline-3,4-dicarboxylic Acid, 3-*Nitrile*

$C_{17}H_{10}N_2O_7$
2-Hydroxy-3,5-dinitrobenzoic Acid, *α-Naphthyl ester*
2-Hydroxy-3,5-dinitrobenzoic Acid, *β-Naphthyl ester*

$C_{17}H_{10}N_4$
3*H*-Benz[*a*]imidazo[4,5-*c*]phenazine
Dibenzo[*f, h*]pyrazolo[3,4-*b*]quinoxaline†
Pyrrolo[1,2-*a*:4,5-*b'*]diquinoxaline†

$C_{17}H_{10}O$
7*H*-Benz[*de*]anthracene-7-one
Benzo[*a*]fluoren-11-one
7*H*-Benzo[*c*]fluoren-7-one
11*H*-Benzo[*b*]fluoren-11-one
2*H*-Dibenz[*c,d,h*]azulen-2-one†

$C_{17}H_{10}O_2$
1-Hydroxybenzanthrone
2-Hydroxybenzanthrone
3-Hydroxybenzanthrone
4-Hydroxybenzanthrone
5-Hydroxybenzanthrone
6-Hydroxybenzanthrone

$C_{17}H_{10}O_2S$
2,3-Diphenylvinylene Sulphone†

$C_{17}H_{10}O_3$
1,2-Dihydroxybenz[*de*]anthracen-7-one
2,3-Dihydroxybenz[*de*]anthracen-7-one
4,5-Dihydroxybenz[*de*]anthracen-7-one
5,6-Dihydroxybenz[*de*]anthracen-7-one
6,11-Dihydroxybenz[*de*]anthracen-7-one
8,9-Dihydroxybenz[*de*]anthracen-7-one
10,11-Dihydroxybenz[*de*]anthracen-7-one

$C_{17}H_{10}O_4$
Karanjonol

$C_{17}H_{10}O_6$
Anthracene-1,2,4-tricarboxylic Acid
Anthracene-1,5,9-tricarboxylic Acid

$C_{17}H_{11}ClO_3$
4-Chloro-1-hydroxy-2-naphthoic Acid, *Phenyl ester*

$C_{17}H_{11}ClO_4$
1-Chloroanthraquinone-2-carboxylic Acid, *Et ester*
3-Chloroanthraquinone-2-carboxylic Acid, *Et ester*

$C_{17}H_{11}ClO_5$
Rhein, *Di-Me ether*, *Chloride*

$C_{17}H_{11}N$
α-Anthraquinoline
β-Anthraquinoline
Benz[*a*]acridine
Benz[*b*]acridine
Benz[*c*]acridine
Benzo[6,7]cyclohept[1,2-*b*]indole
Benzo[*a*]phenanthridine
Benzo[*b*]phenanthridine
Benzo[*c*]phenanthridine
Benzo[*i*]phenanthridine
Benzo[*k*]phenanthridine
p-1-Naphthylbenzoic Acid, *Nitrile*
4-Phenyl-1-naphthoic Acid, *Nitrile*
5-Phenyl-1-naphthoic Acid, *Nitrile*

$C_{17}H_{11}NO$
Benz[*a*]acridone
Benz[*b*]acridone
Benz[*c*]acridone
Benzo[*a*]phenanthridone
Benzo[*b*]phenanthridone
Benzo[*c*]phenanthridone
Benzo[*j*]phenanthridone
Benzo[*k*]phenanthridone
N-Phenylnaphthastyril
N-Phenylnaphtho[2,3-*b*]azetinone †

$C_{17}H_{11}NO_2$
Pyridanthrone

$C_{17}H_{11}NO_3$
1-Naphthyl-2-nitrophenyl Ketone
1-Naphthyl-3-nitrophenyl Ketone
1-Naphthyl-4-nitrophenyl Ketone

$C_{17}H_{11}NO_4$
p-Nitrobenzoic Acid, 2-*Naphthyl ester*
4-Phenylquinoline-2,3-dicarboxylic Acid
3-Phenylquinoline-2,4-dicarboxylic Acid
2-Phenylquinoline-3,4-dicarboxylic Acid
2-Phenylquinoline-4,6-dicarboxylic Acid
2-Phenylquinoline-4,7-dicarboxylic Acid
2-Phenylquinoline-4,8-dicarboxylic Acid

$C_{17}H_{11}NO_6$
1-Nitroanthraquinone-2-carboxylic Acid, *Et ester*

$C_{17}H_{11}NO_7$
Aristolochic Acid

$C_{17}H_{12}$
11*H*-Benzo[*a*]fluorene
11*H*-Benzo[*b*]fluorene
7*H*-Benzo[*c*]fluorene
17*H*-Cyclopenta[*a*]phenanthrene
4-Methylpentaleno[6,6*a*,1,2-*def*]heptalene †
1-Methylpyrene
2-Methylpyrene
4-Methylpyrene

$C_{17}H_{12}BrO$
5-Bromo-1-naphthoic Acid, *Anilide*
7-Bromo-1-naphthoic Acid, *Anilide*
8-Bromo-1-naphthoic Acid, *Anilide*

$C_{17}H_{12}ClNO$
6-Methyl-2-phenylquinoline-4-carboxylic Acid, *Chloride*
2-*p*-Tolylquinoline-4-carboxylic Acid, *Chloride*

$C_{17}H_{12}Cl_2O_7$
Geodin

$C_{17}H_{12}Cl_2O_8$
Geodoxin

$C_{17}H_{12}Cl_4O_5$
Thiophan Acid, *Tri-Me ether* †

$C_{17}H_{12}N_2O_2$
N-*o*-Nitrobenzylidene-1-naphthylamine
N-*m*-Nitrobenzylidene-1-naphthylamine
N-*o*-Nitrobenzylidene-2-naphthylamine
N-*m*-Nitrobenzylidene-2-naphthylamine
N-*p*-Nitrobenzylidene-2-naphthylamine
2-*o*-Nitrostyrylquinoline
2-*m*-Nitrostylrylquinoline
2-*p*-Nitrostyrylquinoline
4-*o*-Nitrostyrylquinoline
4-*m*-Nitrostyrylquinoline
4-*p*-Nitrostyrylquinoline
5-Nitro-2-styrylquinoline
6-Nitro-2-styrylquinoline
8-Nitro-2-styrylquinoline
Quindoline-10-carboxylic Acid, *Me ester*

$C_{17}H_{12}N_2O_4$
4-Amino-3-nitrobenzoic Acid, N-α-*Naphthyl*
4-Amino-3-nitrobenzoic Acid, N-β-*Naphthyl*

$C_{17}H_{12}O$
11*H*-Benzo[*a*]fluoran-11-ol
11*H*-Benzo[*b*]fluoran-11-ol
1-Benzoylnaphthalene
2-Benzoylnaphthalene
1,5-Diphenyltricyclo[2,1,0,$0^{2,5}$]pentan-3-one †
7-Hydroxybenzo[*a*]fluorene
2-Oxo-4,5-diphenyltricyclo[1,1,1,$0^{4,5}$]pentane †
4,6-Diphenyl-2-pyranone
2,6-Diphenyl-4-pyranone
3,5-Diphenyl-4-pyranone
o-1-Naphthylbenzoic Acid
o-2-Naphthylbenzoic Acid
p-1-Naphthylbenzoic Acid
2-Phenyl-1-naphthoic Acid

$C_{17}H_{12}O$ (*continued*)
4-Phenyl-1-naphthoic Acid
5-Phenyl-1-naphthoic Acid
1-Phenyl-2-naphthoic Acid

$C_{17}H_{12}O_2S$
2-Anthraquinonethiol, S-*Allyl ether*

$C_{17}H_{12}O_3$
4,4-Diphenylitaconic Acid, *Anhydride*
8-Hydroxy-1-naphthoic Acid, *Phenyl ether*

$C_{17}H_{12}O_4$
Anthraquinone-1-carboxylic Acid, *Et ester*
2-Methylanthraquinone-1-carboxylic Acid, *Me ester*

$C_{17}H_{12}O_5$
Coumestrol, *Di-Me ether*
Fluorenone-1,2-dicarboxylic Acid, *Di-Me ester*
Fluorenone-1,5-dicarboxylic Acid, *Di-Me ester*
Fluorenone-1,7-dicarboxylic Acid, *Di-Me ester*
Fluorenone-2,3-dicarboxylic Acid, *Di-Me ester*
Fluorenone-2,7-dicarboxylic Acid, *Di-Me ester*
Fluorenone-4,5-dicarboxylic Acid, *Di-Me ester*
3′,4′-Methylenedioxyflavonol, *Me ether*

$C_{17}H_{12}O_6$
Aflatoxin B
Aflatoxin B_1†
7-Hydroxy-11,12-dimethoxycoumestan†
Irisolone
Rhein, *Et ester*
Rhein, *Di-Me ether*
Trifoliol, *Mono-Me ether*†

$C_{17}H_{12}O_7$
Aflatoxin G
Aflatoxin G_1†
Aflatoxin M_1†
Emodic Acid, *Et ester*
Gemmatein
Laccaic Acid, *Me ester*†
Resoflavin, *Tri-Me ether*

$C_{17}H_{12}O_8$
Ceroalbolinic Acid, *Me ester*†
Kermesic Acid, *Me ester*†
3,5,6,7-Tetrahydroxy-1-methylanthraquinone-2-carboxylic Acid, *Me ester*†

$C_{17}H_{12}O_9$
Equisporol
Rhodocadonic Acid
3,3′,4-Tri-*O*-methylflavellagic Acid†

$C_{17}H_{13}N$
1-Azabicyclo[3,2,1]octane
N-Benzylidene-1-naphthylamine
N-Benzylidene-2-naphthylamine
2,4-Diphenylpyridine
2,6-Diphenylpyridine
3,5-Diphenylpyridine
3-α-Pyridylbiphenyl
3-β-Pyridylbiphenyl
3-γ-Pyridylbiphenyl
4-α-Pyridylbiphenyl
4-β-Pyridylbiphenyl
4-γ-Pyridylbiphenyl
2-Styrylquinoline
4-Styrylquinoline

$C_{17}H_{13}NO$
o-Hydroxybenzylidene-1-naphthylamine
p-Hydroxybenzylidene-1-naphthylamine
o-Hydroxybenzylidene-2-naphthylamine
p-Hydroxybenzylidene-2-naphthylamine
Phenanthranil, *Et ether*

$C_{17}H_{13}NO_2$
3-Methyl-2-phenylquinoline-4-carboxylic Acid
6-Methyl-2-phenylquinoline-4-carboxylic Acid
7-Methyl-2-phenylquinoline-4-carboxylic Acid
8-Methyl-2-phenylquinoline-4-carboxylic Acid
2-α-Naphthylaminobenzoic Acid
2-β-Naphthylaminobenzoic Acid
3-Phenylquinoline-2-carboxylic Acid, *Me ester*
2-Phenylquinoline-3-carboxylic Acid, *Me ester*
2-Phenylquinoline-4-carboxylic Acid, *Me ester*
3-Phenylquinoline-4-carboxylic Acid, *Me ester*
Phthalimide, N-*Cinnamyl*
2-*p*-Tolylquinoline-4-carboxylic Acid

$C_{17}H_{13}NO_3$
Graveoline
Graveolinine
Isatophan
1-Naphthol, p-*Nitrobenzyl ether*
2-Naphthol, p-*Nitrobenzyl ether*

$C_{17}H_{13}NO_4$
1-Aminoanthraquinone-2-carboxylic Acid, *Et ester*
2-Phthalimidopropionic Acid, *Phenyl ester*

$C_{17}H_{13}NO_5$
Rhein, *Di-Me ether*, *Amide*

$C_{17}H_{13}NS$
7*H*-Benzo[*c*]phenothiazine, N-*Me*

$C_{17}H_{13}N_3O_3$
1,2-Naphthoquinone, 1-o-*Nitrophenylhydrazone*, *Me ether*
1,2-Naphthoquinone,1-m-*Nitrophenylhydrazone*, *Me ether*
1,2-Naphthoquinone, 1-p-*Nitrophenylhydrazone*, *Me ether*
1,4-Naphthoquinone, p-*Nitrophenylhydrazone*, *Me ether*

$C_{17}H_{14}$
1-Benzylnaphthalene
2-Benzylnaphthalene
1,2-Cyclopentenophenanthrene
1,4-Diphenylcyclopentadiene†

$C_{17}H_{14}Cl_2O_5$
Vicanicin

$C_{17}H_{14}Cl_2O_7$
Methyl 3,5-dichlorolecanorate†

$C_{17}H_{14}I_2O_3$
Benziodarone

$C_{17}H_{14}N_2$
Dibenzylmalonic Acid, *Dinitrile*
Ellipticine
Flavopereirine
Olivacine

$C_{17}H_{14}N_2O$
1-Benzeneazo-2-naphthol, *Me ether*
2-Benzeneazo-1-naphthol, *Me ether*
4-Benzeneazo-1-naphthol, *Me ether*

$C_{17}H_{14}N_2O$ (*continued*)
Indileucin, *Me ether*
3-Methyl-2-phenylquinoline-4-carboxylic Acid, *Amide*
8-Methyl-2-phenylquinoline-4-carboxylic Acid, *Amide*
3-Methylquinoline-4-carboxylic Acid, *Anilide*
Olivacine, *N*-oxide†
2-*o*-Tolueneazo-1-naphthol
2-*m*-Tolueneazo-1-naphthol
2-*p*-Tolueneazo-1-naphthol
4-*o*-Tolueneazo-1-naphthol
4-*p*-Tolueneazo-1-naphthol
1-*o*-Tolueneazo-2-naphthol
1-*m*-Tolueneazo-2-naphthol
1-*p*-Tolueneazo-2-naphthol
2-*p*-Tolylquinoline-4-carboxylic Acid, *Amide*

$C_{17}H_{14}N_2OS$
p-Methoxybenzenediazo-2-naphthyl sulphide

$C_{17}H_{14}N_2O_2$
N-*o*-Nitrobenzyl-1-naphthylamine
N-*m*-Nitrobenzyl-1-naphthylamine
N-*p*-Nitrobenzyl-1-naphthylamine
N-*o*-Nitrobenzyl-2-naphthylamine
N-*m*-Nitrobenzyl-2-naphthylamine
N-*p*-Nitrobenzyl-2-naphthylamine

$C_{17}H_{14}N_2O_3$
Cyclopenin
Piricularin (or $C_{18}H_{14}N_2O_3$)

$C_{17}H_{14}N_2O_4$
Cyclopenol†

$C_{17}H_{14}N_2O_6$
Griseolutein A

$C_{17}H_{14}O$
4-Benzyl-1-naphthol
1-Benzyl-2-naphthol
Benzyl-1-naphthyl Ether
1,5-Diphenyl-2,4-pentadien-1-one
1,5-Diphenyl-1,4-pentadien-3-one
1,5-Diphenyltricyclo[2,1,0,0^{2,5}]pentan-3-ol†
2-Naphthol, *Benzyl ether*
6-Phenyl-2-naphthol, *Me ether*
7-Phenyl-2-naphthol, *Me ether*

$C_{17}H_{14}O_2$
Anthracene-2-carboxylic Acid, *Et ester*
Benz[*a*]azulene-7-carboxylic Acid, *Et ester*
4-Benzoyl-3-phenylbutyric Acid, *Lactone*
2-Phenanthreneacetic Acid, *Me ester*
3-Phenanthreneacetic Acid, *Me ester*
9-Phenanthreneacetic Acid, *Me ester*
Phenanthrene-2-carboxylic Acid, *Et ester*
Phenanthrene-3-carboxylic Acid, *Et ester*
Phenanthrene-9-carboxylic Acid, *Et ester*
1-Phenanthrenepropionic Acid
2-Phenanthrenepropionic Acid
3-Phenanthrenepropionic Acid
9-Phenanthrenepropionic Acid
1-Propylphenanthraquinone
1,2,4-Trimethylanthraquinone
1,3,6-Trimethylanthraquinone★†
1,3,7-Trimethylanthraquinone★†
1,4,5-Trimethylanthraquinone
1,4,6-Trimethylanthraquinone
2,3,6-Trimethylanthraquinone

$C_{17}H_{14}O_3$
Chalcone-4-carboxylic Acid, *Me ester*
1,1-Dibenzoylacetone
1,5-Di-*o*-hydroxyphenyl-1,4-pentadien-3-one
1,5-Di-*p*-hydroxyphenyl-1,4-pentadien-3-one
2,2-Diphenylglutaric Acid, *Anhydride*
2,3-Diphenylglutaric Acid, *Anhydride*
2,3-Diphenylpropane-1,2-dicarboxylic Acid, *Anhydride*
Drachorhodin
4-Hydroxy-1,3-dimethylanthraquinone, *Me ether*
3′-Hydroxyflavone, *Et ether*
4′-Hydroxyflavone, *Et ether*
6-Hydroxyflavone, *Et ether*
7-Hydroxyflavone, *Et ether*
4-Hydroxyphenanthrene-1-carboxylic Acid, *Me ether*, *Me ester*
2-Hydroxyphenanthrene-3-carboxylic Acid, *Me ether*, *Me ester*
10-Hydroxyphenanthrene-9-carboxylic Acid, *Me ether*, *Me ester*
6-Hydroxyphenanthrene-9-carboxylic Acid, *Et ether*
Morphol, 3-*Me ether*-4-*Ac*
Thebenol

$C_{17}H_{14}O_3S$
Toluene-*p*-sulphonic Acid, 1-*Naphthyl ester*

$C_{17}H_{14}O_4$
Alizarin, 1-*Et*, 2-*Me ether*
Benzil-*o*-carboxylic Acid, *Et ester*
Chrysin, *Mono-Et ether*
Chrysophanic Acid, *Di-Me ether*
Dalbergin, *Me ether*
1,7-Dihydroxy-2,6-dimethylanthraquinone, 7-*Me ether*
2′,4′-Dihydroxyflavone, *Di-Me ether*
2′,7-Dihydroxyflavone, *Di-Me ether*
3′,4′-Dihydroxyflavone, *Di-Me ether*
3′,7-Dihydroxyflavone, *Di-Me ether*
3′,7-Dihydroxyflavone, 7-*Et ether*
4′,7-Dihydroxyflavone, *Di-Me ether*
5,6-Dihydroxyflavone, *Di-Me ether*
5,8-Dihydroxyflavone, *Di-Me ether*
6,7-Dihydroxyflavone, *Di-Me ether*
6,8-Dihydroxyflavone, *Di-Me ether*
7,8-Dihydroxyflavone, *Di-Me ether*
1,2-Dihydroxy-4-methylanthraquinone, *Di-Me ether*
1,3-Dihydroxy-2-methylanthraquinone, *Di-Me ether*
1,3-Dihydroxy-4-methylanthraquinone, *Di-Me ether*
1,5-Dihydroxy-2-methylanthraquinone, *Di-Me ether*
1,6-Dihydroxy-2-methylanthraquinone, *Di-Me ether*
3,9-Dimethoxypterocarpen†
4,4-Diphenylitaconic Acid
Fluorene-9,9-dicarboxylic Acid, *Di-Me ester*
Mesaconic Acid, *Di-phenyl ester*
Nordalbergin, *Di-Me ether*†

$C_{17}H_{14}O_5$
Apigenin, 5,4′-*Di-Me ether*

$C_{17}H_{14}O_5$ (*continued*)
Apigenin, 7,4′-*Di-Me ether*
Baicalein, 6,7-*Di-Me ether*
Benzophenone-2,2′-dicarboxylic Acid, *Di-Me ester*
Benzophenone-2,4′-dicarboxylic Acid, *Di-Me ester*
Benzophenone-4,4′-dicarboxylic Acid, *Di-Me ester*
4-Benzoylisophthalic Acid, *Di-Me ester*
Benzoylterephthalic Acid, *Di-Me ester*
Carajurin
Cladofulvin, *Di-Me ether*†
Echioidinin, 2′-*Me ether*†
Eucomin†
Galangin, 5,7-*Di-Me ether*
Genistein, 5,4′-*Di-Me ether*
Genkwanin★, 4′-*Me ether*†
Genkwanin, 5-*Me ether*
Glyceraldehyde, *Dibenzoyl*
7-Hydroxy-4′,6-dimethoxyflavone†
Kuhlmannin†
Lanigerin
Maackiain, *Mono*-O-*Me*
2-Methylgenistein, 4′-*Me ether*
Oroxylin-A, 7-*Me ether*
Pterocarpine
Questin, 7-*Me ether*†
Resokaempferol, 3,4′-*Di-Me ether*
Strepsilin, *Di-Me ether*
1,2,3-Trihydroxyanthraquinone, *Tri-Me ether*
1,2,5-Trihydroxyanthraquinone, 1,2,5-*Tri-Me ether*
1,2,6-Trihydroxyanthraquinone, 1,2,6-*Tri-Me ether*
1,2,7-Trihydroxyanthraquinone, *Tri-Me ether*
1,2,8-Trihydroxyanthraquinone, 1,2,8-*Tri-Me ether*
1,3,7-Trihydroxyanthraquinone, *Tri-Me ether*
1,3,8-Trihydroxyanthraquinone, *Tri-Me ether*
3,3′,7-Trihydroxyflavone, 3,3′-*Di-Me ether*
3,5,8-Trihydroxyflavone, 5,8-*Di-Me ether*
3′,4′,5-Trihydroxyflavone, 3′,4′-*Di-Me ether*
3′,4′,7-Dihydroxyflavone, 3′,4′-*Di-Me ether*
3′,4′,8-Trihydroxyflavone, 3′,4′-*Di-Me ether*
3′,5,7-Trihydroxyflavone, 5,7-*Di-Me ether*
4′,6,7-Trihydroxyflavone, *Di-Me ether*
1,4,5-Trihydroxy-6-methylanthraquinone, 1,4-*Di-Me ether*
1,4,5-Trihydroxy-8-methylanthraquinone, 1,4-*Di-Me ether*
1,2,5-Trihydroxy-6-methylanthraquinone, *Di-Me ether*
1,2,5-Trihydroxy-8-methylanthraquinone, 1,2-*Di-Me ether*
1,2,7-Trihydroxy-6-methylanthraquinone, 1,2-*Di-Me ether*
1,2,8-Trihydroxy-6-methylanthraquinone, 1,2-*Di-Me ether*
1,2,8-Trihydroxy-7-methylanthraquinone, 1,2-*Di-Me ether*
1,3,8-Trihydroxy-6-methylanthraquinone, 1,3-*Di-Me ether*
1,3,6-Trihydroxy-8-methylanthraquinone, 3,6-*Di-Me ether*†
1,3,8-Trihydroxy-6-methylanthraquinone, 1,8-*Di-Me ether*
Wogonin, *Mono-Me ether*

$C_{17}H_{14}O_6$
Aflatoxin B_2†
Daticetin, 3,2′-*Di-Me ether*
Dermoglaucin, *Me ether*†
3,5-Dihydroxy-4′,7-dimethoxyflavone†
4′,5-Dihydroxy-3,7-dimethoxyflavone†
4′,5-Dihydroxy-6,7-dimethoxyisoflavone†
3-Hydroxy-4-methoxy-8,9-methylenedioxypterocarpan†
Involutin†
Isorhodoptilometrin†
Leptosidin
Luteolin, 4′,7-*Di-Me ether*
Melannein†
Pectolinarigenin
Pisatin
Questinol, 7-*Me ether*†
Rhodoptilometrin†
Scutellarein, 4′-6-*Di-Me ether*
Scutellarein, 4′,7-*Di-Me ether*
1,2,5,6-Tetrahydroxyanthraquinone, 1,2,6-*Tri-Me ether*
1,2,6,7-Tetrahydroxyanthraquinone, 1,2,6-*Tri-Me ether*
1,3,5,7-Tetrahydroxyanthraquinone, *Tri-Me ether*
2′,4′,5,7-Tetrahydroxyflavone, 2′,4′-*Di-Me ether*
3,5,6,7-Tetrahydroxyflavone, 3,7-*Di-Me ether*
1,3,5,6-Tetrahydroxy-8-methylanthraquinone, *Di-Me ether*
1,3,6,8-Tetrahydroxyanthraquinone, 1,3,6-*Tri-Me ether*†

$C_{17}H_{14}O_7$
Aflatoxin B_{2a}†
Aflatoxin G_2†
Aflatoxin M_2†
Bisdechlorogeodin
Caryatin†
Demethoxysudachitin†
Morin, *Di-Me ether*
2′,3,5,7,8-Pentahydroxyflavone, 2′,3-*Di-Me ether*
3′,4′,5,6,7-Pentahydroxyflavone, 3′,6-*Di-Me ether*
1,3,4,5,6-Pentahydroxy-2-methylanthraquinone, 4,5-*Di-Me ether*
Podospicatin†
Prudomestin†
Rhamnazin
Tricin
3′,5,7- Trihydroxy-4′,6-dimethoxyflavone†
4′,5,7-Trihydroxy-3,8-dimethoxyflavone†

$C_{17}H_{14}O_8$
Aflatoxin G_{2a}†
Asperthecin, *Di-Me ether*
Limocitrin
Spinacetin†
Syringetin
3′,4′,5,7-Tetrahydroxy-3,6-dimethoxyflavone†
3′,4′,5,7-Tetrahydroxy-3,8-dimethoxyflavone†

$C_{17}H_{14}S$
1-Naphthalenethiol, o-*Tolyl ether*
1-Naphthalenethiol, m-*Tolyl ether*
1-Naphthalenethiol, p-*Tolyl ether*
2-Naphthalenethiol, o-*Tolyl ether*
2-Naphthalenethiol, m-*Tolyl ether*
2-Naphthalenethiol, p-*Tolyl ether*

$C_{17}H_{15}ClO$
2,2-Diphenyl-4-pentenoic Acid, *Chloride*

$C_{17}H_{15}ClO_3$
Dibenzylmalonic Acid, *Chloride*

$C_{17}H_{15}ClO_5$
Pelargonidin chloride, 3,4′-*Di-Me ether*

$C_{17}H_{15}ClSi$
Methyl-α-naphthylphenylsilyl chloride†

$C_{17}H_{15}Cl_3N_2O$
7-Chloro-5-(2,4-dichlorophenyl)-4,5-dihydro-1,4-dimethyl-3*H*-1,4-benzodiazepin-2-one†

$C_{17}H_{15}I_4NO_4$
Thyroxine, *Me ester*, *Me ether*
Thyroxine, *Et ester*

$C_{17}H_{15}N$
N-Benzyl-1-naphthylamine
N-Benzyl-2-naphthylamine
4-Benzyl-1-naphthylamine
6-Benzyl-2-naphthylamine
2-Phenethylquinoline
4-Phenethylquinoline
N-Phenyl-2-naphthylamine, N-*Me*

$C_{17}H_{15}NO$
6-Hydroxy-2-methyl-4-phenylquinoline, *Me ether*
2-Hydroxy-α-phenylcinnamic Acid, *Et ether*
1-Hydroxy-3-phenylisoquinoline, *Et ether*
4-*o*-Hydroxyphenylquinoline, *Et ether*
6-Hydroxy-2-phenylquinoline, *Et ether*

$C_{17}H_{15}NO_2$
Anonaine†
Dibenzylmalonic Acid, *Nitrile*
4,6-Dihydroxy-2-phenylquinoline, *Di-Me ether*
2,3-Diphenylglutaric Acid, 1-*Nitrile*
Diphenylmalonic Acid, *Et ester-nitrile*
Edulein†
3-Phenylindole-2-carboxylic Acid, *Et ester*
Viridication, *Di-Me ether*

$C_{17}H_{15}NO_3$
Anolobine
Mesacon-α-anilic Acid, *Phenyl ester*
Mesacon-β-anilic Acid, *Phenyl ester*
Norushinsunine†
Viridicatol, 3′-O,N-*Di-Me*†
Vitricin

$C_{17}H_{15}NO_4$
2-*p*-Nitrophenyl-3-phenylacrylic Acid, *Et ester*
3-*o*-Nitrophenyl-2-phenylacrylic Acid, *Et ester*
3-*o*-Nitrophenyl-2-*m*-tolylacrylic Acid, *Me ester*
2-Nitrostilbene-4-carboxylic Acid, *Et ester*

$C_{17}H_{15}NO_5$
3,4-Dihydroxy-2-nitrochalcone, *Di-Me ether*
4,4′-Dihydroxy-3-nitrochalcone, *Di-Me ether*
4,5-Dihydroxy-2-nitrochalcone, *Di-Me ether*
4′-Hydroxy-2-nitrostilbene-4-carboxylic Acid, *Me ether*, *Me ester*
α-(4-Hydroxyphenyl)-2-nitrocinnamic Acid, *Et ether*
Melicopidine†
Melicopine†
2-(3′-Nitro-4′-toluyl)benzoic Acid, *Et ester*

$C_{17}H_{15}NO_7$
Papaveric Acid, 3-*Me ester*
Papaveric Acid, 4-*Me ester*

$C_{17}H_{15}N_3O$
5-Methyl-1-phenylpyrazole-3-carboxylic Acid, *Anilide*
5-Methyl-1-phenylpyrazole-4-carboxylic Acid, *Anilide*

$C_{17}H_{16}$
1-Ethyl-2-methylphenanthrene
1-Ethyl-4-methylphenanthrene
1-Ethyl-7-methylphenanthrene
1-Ethyl-8-methylphenanthrene
2-Ethyl-1-methylphenanthrene
2-Ethyl-8-methylphenanthrene
2-Ethyl-10-methylphenanthrene
3-Ethyl-5-methylphenanthrene
3-Ethyl-6-methylphenanthrene
6-Ethyl-1-methylphenanthrene
7-Ethyl-1-methylphenanthrene
9-Ethyl-1-methylphenanthrene
9-Ethyl-10-methylphenanthrene
1-Isopropylphenanthrene
2-Isopropylphenanthrene
9-Isopropylphenanthrene
1-Propylphenanthrene
2-Propylphenanthrene
9-Propylphenanthrene
1,2,4-Trimethylanthracene★†
1,2,10-Trimethylanthracene
1,3,6-Trimethylanthracene★†
1,3,7-Trimethylanthracene†
1,3,10-Trimethylanthracene
1,4,9-Trimethylanthracene
1,9,10-Trimethylanthracene
2,3,6-Trimethylanthracene
2,3,9-Trimethylanthracene
2,7,9-Trimethylanthracene
2,9,10-Trimethylanthracene
1,2,3-Trimethylphenanthrene
1,2,4-Trimethylphenanthrene
1,2,6-Trimethylphenanthrene
1,2,7-Trimethylphenanthrene
1,2,8-Trimethylphenanthrene
1,3,4-Trimethylphenanthrene
1,3,7-Trimethylphenanthrene
1,3,8-Trimethylphenanthrene
1,4,5-Trimethylphenanthrene
1,4,7-Trimethylphenanthrene
1,4,10-Trimethylphenanthrene
1,6,7-Trimethylphenanthrene
2,6,9-Trimethylphenanthrene

$C_{17}H_{16}Br_2O_3$
5,5′-Dibromo-2,2′-dihydroxybenzophenone, 2,2′-*Di-Et ether*

$C_{17}H_{16}N_2$
2-*p*-Aminophenylquinoline, N-*Et*
1,2-Dihydroellipticine

$C_{17}H_{16}N_2O$
Dibenzylmalonic Acid, *Nitrile*, *Amide*

$C_{17}H_{16}N_2O_2$
Mesaconic Acid, *Dianilide*

$C_{17}H_{16}N_2O_4$
1-(4-Aminophenyl)-2-(2-carboxy-4-nitrophenyl)-ethylene, N-*Di-Me*
1-(4-Aminophenyl)-2-(4-carboxy-2-nitrophenyl)-ethylene, *Et ester*
1-(4-Aminophenyl)-2-(4-carboxy-2-nitrophenyl)-ethylene, N-*Di Me*

$C_{17}H_{16}N_2O_5$
1,3-Di-*p*-methoxybenzoylurea

$C_{17}H_{16}N_2O_6$
Griseolutein B

$C_{17}H_{16}N_2O_7$
3,4-Dihydroxy-6-nitrophthalic Acid, 4-*Me ether*, 2-*Me ester* 1-*Anilide*

$C_{17}H_{16}O_2$
4-Biphenylacrylic Acid, *Et ester*
2,3-Diphenylacrylic Acid, *Et ester*
3,3-Diphenylacrylic Acid, *Et ester*
1,1-Diphenyl-1-butene-2-carboxylic Acid
1,4-Diphenyl-1-butene-2-carboxylic Acid
1,4-Diphenyl-1-butene-3-carboxylic Acid
1,5-Diphenyl-1,5-pentanedione
2,5-Diphenyl-2-pentenoic Acid
4,5-Diphenyl-2-pentenoic Acid
2,5-Diphenyl-3-pentenoic Acid
2,2-Diphenyl-4-pentenoic Acid
3,5-Diphenyl-4-pentenoic Acid
4,5-Diphenyl-4-pentenoic Acid
5,5-Diphenyl-4-pentenoic Acid
Hinokiresinol†
3-Hydroxychalcone, *Et ether*
4-Hydroxychalcone, *Et ether*
4′-Hydroxychalcone, *Et ether*
3-Phenylindane-1-carboxylic Acid, *Me ester*

$C_{17}H_{16}O_3$
2-Benzoyl-2-benzylpropionic Acid
3-Benzoyl-2-benzylpropionic Acid
3-Benzoyl-2-methyl-3-phenylpropionic Acid
α-Benzoylphenylacetic Acid, *Et ester*
2-Benzoyl-4-phenylbutyric Acid
3-Benzoyl-2-phenylbutyric Acid
3-Benzoyl-4-phenylbutyric Acid
4-Benzoyl-2-phenylbutyric Acid
4-Benzoyl-3-phenylbutyric Acid
4-Benzoyl-4-phenylbutyric Acid
2-Benzoyl-3-phenylpropionic Acid, *Me ester*
3-Benzoyl-2-phenylpropionic Acid, *Me ester*
3-Benzoyl-3-phenylpropionic Acid, *Me ester*
2,2′-Dihydroxychalcone, 2-*Et ether*
2′,4′-Dihydroxychalcone, 4′-*Et ether*
2′,5′-Dihydroxychalcone, 5′-*Et ether*
2-Hydroxy-3,4-diphenyl-3-butenoic Acid, *Me ester*
2-Hydroxy-1,2-diphenyl-1,4-pentanedione
9-Hydroxyfluorene-9-carboxylic Acid, *Me ether*, *Et ester*
9-Hydroxyfluorene-9-carboxylic Acid, *Me ester*
3-Hydroxy-α-phenylcinnamic Acid, *Et ester*
2-*p*-Toluylbenzoic Acid, *Et ester*
1,5,6-Trihydroxyphenanthrene, *Tri-Me ether*
3,4,5-Trihydroxyphenanthrene, *Tri-Me ether*
3,4,6-Trihydroxyphenanthrene, *Tri-Me ether*

$C_{17}H_{16}O_4$
Alpinetin, *Me ether*
Dibenzylmalonic Acid
1,6-Dihydroxyxanthone, *Di-Et ether*†
1,7-Dihydroxyxanthone, *Di-Et ether*
3,4-Dimethoxydalbergione†
4,4′-Dimethoxydalbergione†
Diphenic Acid, *Propyl ester*
2,2-Diphenylglutaric Acid
2,3-Diphenylglutaric Acid
2,4-Diphenylglutaric Acid
3,3-Diphenylglutaric Acid
Diphenylmalonic Acid, *Di-Me ester*
Diphenylmethane-2,2′-dicarboxylic Acid, *Di-Me ester*
Diphenylmethane-2,4′-dicarboxylic Acid, *Di-Me ester*
Diphenylmethane-4,4′-dicarboxylic Acid, *Di-Me ester*
1,3-Diphenylpropane-1,2-dicarboxylic Acid
2,3-Diphenylpropane-1,2-dicarboxylic Acid
3,3-Diphenylpropane-1,2-dicarboxylic Acid
2,2-Diphenylsuccinic Acid, α-*Me ester*
Eleutherinol, *Di-Me ether*
Homopterocarpin
Hydrangenol, *Di-Me ether*†
2′-Hydroxy-4′,6′-dimethoxychalcone†
Isoeuxanthone, *Di-Et ether*
Kuhlmannene†
Pinostrobin, *Me ether*†
1,2,6,10-Tetrahydroxyanthracene, *Tri-Me ether*
1,2,7,10-Tetrahydroxyanthracene, 1,2,7-*Tri-Me ether*

$C_{17}H_{16}O_5$
Alternariol, *Tri-Me ether*
Apoalpinone, 4-*Mono-Me ether*
ψ-Baptigenetin, *Di-Me ether*
ψ-Baptigenetin, *Et ether*
Butein, 2′,4′-*Di-Me ether*
Citronetin, *Mono-Me ether*
Damnidin
Decussatin, *Me ether*
2′,4′-Dihydroxybenzophenone-2-carboxylic Acid, *Propyl ester*
2′,4′-Dihydroxychalcone, *Di-Me ether*
2′,5′-Dihydroxychalcone, *Di-Me ether*
2′,6′-Dihydroxychalcone, *Di-Me ether*
3,4-Dihydroxychalcone, *Di-Me ether*
3,4′-Dihydroxychalcone, *Di-Me ether*
3′,4-Dihydroxychalcone, *Di-Me ether*
3′,4′-Dihydroxychalcone, *Di-Me ether*
4,4′-Dihydroxychalcone, *Di-Me ether*
Dracoic Acid
Farrerol†
Fonsecin, *Di-Me ether*
2-Formyl-5,6-dimethoxylbenzoic Acid, α-*Benzyl ester*
Glycerol, 1,2-*Dibenzoyl*
3-Hydroxy-4,9-dimethoxypterocarpan†
Isosakuranetin, *Me ether*
Methyl-linderone

$C_{17}H_{16}O_5$ (*continued*)
Padmakasein, *Mono-Me ether*
Phellopterin †
1-Phenyl-3-(2,4,6-trihydroxyphenyl)propane-1,3-dione, 4,6-*Di-Me ether*
Phyllodulcinol, 3′-*Me ether* †
Pinobanksin, 5,7-*Di-Me ether*
Ravenelin, *Tri-Me ether*
Rubrofusarin, *Di-Me ether*
2′,3′,4,4′-Tetrahydroxychalcone, 3′,4′-*Di-Me ether*
2′,3,4,5′-Tetrahydroxychalcone, 3,5′-*Di-Me ether*
2′,4,4′,6′-Tetrahydroxychalcone, 4′,6′-*Di-Me ether*
3′,5,7-Trihydroxyflavanone, 3′-*Me ether*, 5(7)-*Me ether*
1,3,6-Trihydroxy-8-methylxanthone, *Tri-Me ether* †

$C_{17}H_{16}O_6$
Aromadendrin, 4′,7-*Di-Me ether*
Aromadendrin, 5,7-*Di-Me ether*
Byak-angelicol
4′,5-Dihydroxy-3′,7-dimethoxyflavanone †
5,7-Dihydroxy-2′,4′-dimethoxyisoflavanone
Eucomol †
Hesperetin*, 7-*Me ether* †
Mucroquinone †
Protocotoin, *Me ether*
Swerchirin, *Di-Me ether* †
1,3,5,6-Tetrahydroxyxanthone, *Tetra-Me ether* †
1,3,5,8-Tetrahydroxyxanthone, *Tetra-Me ether* †
1,3,6,8-Tetrahydroxyxanthone, *Tetra-Me ether* †
1,3,7,8-Tetrahydroxyxanthone, *Tetra-Me ether* †

$C_{17}H_{16}O_7$
Evernic Acid
Lecanoric Acid, *Me ester*
Lithospermic Acid, *Me ester* †
1,3,4,7,8-Pentahydroxyxanthone, 1,3,4,7-*Tetra-Me ether* †
Sulochrin

$C_{17}H_{16}O_8$
Asterric Acid
Brevifolincarboxylic Acid, *Tri-Me ether*, *Me ester*

$C_{17}H_{16}O_{12}$
Mellitic Acid, *Penta-Me ester*

$C_{17}H_{16}Si$
Methyl-α-naphthylphenylsilane †

$C_{17}H_{17}BrO_6$
7-Bromo-7-dechlorogriseofulvin

$C_{17}H_{17}ClO_6$
Griseofulvin

$C_{17}H_{17}FO_6$
7-Dechloro-7-fluorogriseofulvin †
7-Fluoro-7-dechlorogriseofulvin

$C_{17}H_{17}N$
6-Allyl-6,7-dihydro-5*H*-dibenz[*c*,*e*]azepine
Aporphine
2,3-Diphenylvaleric Acid, *Nitrile*
2,5-Diphenylvaleric Acid, *Nitrile*

$C_{17}H_{17}NO$
2,2-Diphenyl-4-pentenoic Acid, *Amide*

$C_{17}H_{17}NO_2$
Apomorphine
6-Azaequilenin †
Caaverin †
3,5-Dihydroxyacridine, *Di-Et ether*
4-Hydroxy-3-aza-1,3,5(10),6,8-oestrapentaen-17-one
18-Nor-13-azaequilenin, *Me ether* †

$C_{17}H_{17}NO_3$
Acrophylline †
Crotonosine †
Crotsparine †
Dibenzylmalonic Acid, *Amide*
2,3-Diphenylglutaric Acid, *Monoamide*
Phenaceturic Acid, *Benzyl ester*
2-Phenylhippuric Acid, *Et ester*
Sparsiflorine †

$C_{17}H_{17}NO_4$
Biflorine II
Biflorone

$C_{17}H_{17}NO_5$
Benzoyl-4-nitrobenzoylmethane, *Di-Me acetal*
Hippeastrine †
Tubispacin

$C_{17}H_{17}NO_6$
Norneronine †

$C_{17}H_{17}N_3O_3$
4-Hydroxy-2,6-dimethyl-4′-nitroazobenzene, *Allyl ether*

$C_{17}H_{17}N_3O_4$
2′-Nitroazobenzene-4-carboxylic Acid, n-*Butyl ester*
3′-Nitroazobenzene-4-carboxylic Acid, n-*Butyl ester*
4′-Nitroazobenzene-4-carboxylic Acid, n-*Butyl ester*
Nitromalonic Acid, *Di*-N-*methylanilide*

$C_{17}H_{18}$
1-Ethyl-3,4-dihydro-2-methylphenanthrene
[2,3]Paracyclophane †
1,2,7,8-Tetramethylfluorene †
1,4,5,8-Tetramethylfluorene †

$C_{17}H_{18}N_2$
Dehydro-de-*N*-methyluleine †
Tröger's Base

$C_{17}H_{18}N_2O$
3-(2-Aminoethyl)-4-hydroxyindole, 4-*Benzyl ether* †
3-(2-Aminoethyl)-5-hydroxyindole, *Benzyl ether*

$C_{17}H_{18}N_2O_2$
Dibenzylmalonic Acid, *Diamide*
Isolysergic Acid, *Me ester*
Lysergic Acid, *Me ester*
Malonanilide, N,N′-*Di-Me*

$C_{17}H_{18}N_2O_3$
Malanilide, *Me ether*

$C_{17}H_{18}N_2O_6S$
Carbenicillin †

$C_{17}H_{18}N_8O_3$
Aminopteroylalanine

$C_{17}H_{18}O$
1,5-Diphenyl-3-pentanone
Gibberone
2,2′,4,4′-Tetramethylbenzophenone
2,2′,4,6-Tetramethylbenzophenone
2,2′,5,5′-Tetramethylbenzophenone
2,2′,6,6′-Tetramethylbenzophenone
2,3,4,6-Tetramethylbenzophenone
2,3,5,6-Tetramethylbenzophenone
2,4,4′,5-Tetramethylbenzophenone
3,3′,4,4′,-Tetramethylbenzophenone
3,3′,5,5′-Tetramethylbenzophenone

$C_{17}H_{18}O_2$
Dibenzylacetic Acid, *Me ester*
2,4-Dimethylphenylacetic Acid, *Benzyl ester*
2,2-Diphenylbutyric Acid, *Me ester*
3,3-Diphenylbutyric Acid, *Me ester*
4,4-Diphenylbutyric Acid, *Me ester*
2,3-Diphenylpropionic Acid, *Et ester*
3,3-Diphenylpropionic Acid, *Et ester*
2,2-Diphenylvaleric Acid
2,3-Diphenylvaleric Acid
2,5-Diphenylvaleric Acid
3,4-Diphenylvaleric Acid
3,5-Diphenylvaleric Acid
4,4-Diphenylvaleric Acid
4,5-Diphenylvaleric Acid
Di-*p*-tolylacetic Acid, *Me ester*
3-Hydroxy-4,6-dimethylbenzophenone, *Et ether*
β-Hydroxyvalerophenone, *Phenyl ether*
6-Isopropyl-*m*-toluic Acid, *Phenyl ester*
2-Methoxy-4-propenylphenol, 1-*Benzyl ether*
2-Methyl-3,3-diphenylpropionic Acid, *Me ester*

$C_{17}H_{18}O_3$
Benzilic Acid, *Propyl ester*
4-Cinnamyl-2,3-dimethoxyphenol†
4-Cinnamyl-2,5-dimethoxyphenol†
5-Cinnamyl-2,4-dimethoxyphenol†
Dibenzylglycollic Acid, *Me ester*
2,2′-Dihydroxybenzophenone, *Di-Et ether*
4,4′-Dihydroxybenzophenone, *Di-Et ether*
4′,7-Dihydroxyisoflavan, *Di-Me ether*
2-Hydroxy-3,3-diphenylpropionic Acid, *Et ester*
Hydroxy di-*p*-tolylacetic Acid, *Me ester*
3,4′,5-Trihydroxystilbene, *Tri-Me ether*

$C_{17}H_{18}O_4$
Agatharesinol†
Angolensin, 4-O-*Me*
α-Hydroxy-4-methoxy-α-(4-methoxyphenyl)-acetophenone, *Me ether*
1-(2-Hydroxyphenyl)-3-(4-hydroxy-2,3-dimethoxyphenyl)propene†
Latifolin★†
1-Naphthylmalonic Acid, *Di-Et ester*
2-Naphthylmalonic Acid, *Di-Et ester*
Sugiresinol†
2,4,6-Trihydroxy-3-methylbenzophenone, *Tri-Me ether*

$C_{17}H_{18}O_5$
3′,7-Dihydroxy-2′,4′-dimethoxyisoflavan†
Di-*p*-methoxyphenylglycollic Acid, *Me ester*
Giloinin
Litsealactone†
Sequirin B†
Sequirin C†
Shikonin, *Me ether*
2,2′,3,3′-Tetrahydroxybenzophenone, *Tetra-Me ether*
2,2′,4,4′-Tetrahydroxybenzophenone, *Tetra-Me ether*
2,2′,5,5′-Tetrahydroxybenzophenone, *Tetra-Me ether*
2,2′,6,6′-Tetrahydroxybenzophenone, *Tetra-Me ether*
2,3,4,6-Tetrahydroxybenzophenone, *Tetra-Me ether*
2,3′,4,4′-Tetrahydroxybenzophenone, *Tetra-Me ether*
2,3′,4,5′-Tetrahydroxybenzophenone, *Tetra-Me ether*
2,3′,4′,5-Tetrahydroxybenzophenone, *Tetra-Me ether*
2,4,4′,5-Tetrahydroxybenzophenone, *Tetra-Me ether*
2,4,4′,6-Tetrahydroxybenzophenone, *Tetra-Me ether*
3,3′,4,4′-Tetrahydroxybenzophenone, *Tetra-Me ether*
3,3′,4,5′-Tetrahydroxybenzophenone, *Tetra-Me ether*
3,4,4′,7-Tetrahydroxyflavan★, 4′,7-*Di-Me ether*†
Ventilagone, *Mono-Me ether*
Zeylanine†

$C_{17}H_{18}O_6$
7-Dechlorogriseofulvin
5-(2′,3′-Dihydroxy-3′-methylbutyl)-8-methoxypsoralen†
Litseaculane†
Mycochromenic Acid†
Zeylanane†
Zeylanicine†

$C_{17}H_{18}O_7$
Byak-angelicin
Zeylanidine†

$C_{17}H_{19}BrO_4$
4-Methyl-3-oxo-1-cyclohexaneacetic Acid, p-*Bromophenacyl ester*

$C_{17}H_{19}ClN_2S$
Chlorpromazine

$C_{17}H_{19}N$
2,6-Diphenylpiperidine
3,4-Diphenylpiperidine
3,5-Diphenylpiperidine

$C_{17}H_{19}NO$
Chlophedianol
2,2-Diphenylvaleric Acid, *Amide*
4,5-Diphenylvaleric Acid, *Amide*
2-*p*-Tolypropionic Acid, p-*Toluidide*

$C_{17}H_{19}NO_2$
Dihydro-6-azaequilenin †
Diphenylcarbamic Acid, *Butyl ester*
Ethyl di-*p*-tolylcarbamate

$C_{17}H_{19}NO_3$
Buphanisine
epi-Buphanisine
Coclaurine
Dihydromorphinone †
Diphenylcarbamic Acid, 2-*Ethoxyethyl ester*
Erysonine
Erysopine
Isococlaurine
α-Isomorphine
β-Isomorphine
γ-Isomorphine
6-Isopropyl-*m*-cresol, p-*Nitrobenzyl ether*
Jacularine †
Morphine
Narwedine †
Norcodeine
Piperine

$C_{17}H_{19}NO_4$
Acrophyllidine †
Buphanamine †
Coccinine
Crinamine
9-Demethylhomolycorine
Falcatine
Haemanthamine
Montanine †
Morphine *N*-oxide
Oxymorphone

$C_{17}H_{19}NO_5$
Clivonine
Crinalbine
Crinamidine
Evellerine †
Haemanthidine
6-Hydroxycrinamine †
Lunidonine †
Piplartine †

$C_{17}H_{19}N_3$
Antazoline
3,6-Di-(dimethylamino)-acridine

$C_{17}H_{19}N_3O$
Phentolamine

$C_{17}H_{19}N_3O_3$
4-Hydroxy-2,6-dimethyl-4′-nitroazobenzene, *Isopropyl ether*

$C_{17}H_{19}N_3O_4$
Anthramycin, *Me ether* †

$C_{17}H_{19}N_3O_5$
4-Amino-2,6-dinitrophenol, N-*Benzoyl, Et ether*

$C_{17}H_{20}BrNO$
Bromodiphenylhydramine

$C_{17}H_{20}I_2O_3$
2-Hydroxy-3,5-di-iodobenzoic Acid, *Bornyl ester*

$C_{17}H_{20}NO_4$ (ion)
O-Methyl-luninium Cation †

$C_{17}H_{20}N_2$
De-*N*-methyluleine †
Ruban

$C_{17}H_{20}N_2O$
Dasycarpidone †
3-*epi*-Dasycarpidone †
m-Dimethylaminophenyl *p*-dimethylaminophenyl Ketone
Di-*o*-dimethylaminophenyl Ketone
Di-*m*-dimethylaminophenyl Ketone
Di-*p*-dimethylaminophenyl Ketone
2-Hydroxy-3,4′,5-trimethylazobenzene, *Et ether*
2′-Hydroxy-2,4,5′-trimethylazobenzene, *Et ether*
Ruban-5-ol
Ruban-9-ol
2,2′,3,3′-Tetramethylcarbanilide
2,2′,4,4′-Tetramethylcarbanilide
2,2′,5,5′-Tetramethylcarbanilide
3,3′,4,4′-Tetramethylcarbanilide

$C_{17}H_{20}N_2O_2S$
1,3-Di-*p*-hydroxyphenylthiourea, *Di-Et ether*

$C_{17}H_{20}N_2O_3$
α-4-Aminophenyl-3-nitrobenzyl Alcohol, N-*Di-Et*
α-4-Aminophenyl-4-nitrobenzyl Alcohol, N-*Di-Et*
1,3-Di-*p*-hydroxyphenylurea
m-Dimethylaminophenol, *Carbonate*

$C_{17}H_{20}N_2O_5S$
5,6-*trans*-Phenoxymethylpenicillinic Acid, *Me ester* †

$C_{17}H_{20}N_2S$
Isophenergan
Promazine
Promethazine

$C_{17}H_{20}N_4O_2$
4,4′-Trimethylenedioxydibenzamidine

$C_{17}H_{20}N_4O_6$
Lactoflavine
Lyxoflavin

$C_{17}H_{20}O$
1,5-Diphenyl-3-pentanol
1,1-Diphenyl-1-propanol, *Et ether*
Enanthetone
Heptadeca-2,8,10,16-tetraene-4,6-diyn-1-ol †
Hexyl-2-naphthyl Ketone
6-Isopropyl-*m*-cresol, *Benzyl ether*
5-Phenyl-1-pentanol, *Phenyl ether*

$C_{17}H_{20}O_2$
2,2-Di-*p*-hydroxyphenylpropane, *Di-Me ether*
1,5-Diphenylpentane-1,3-diol
2-Hydroxy-α-4-hydroxyphenyltoluene, *Di-Et ether*
4-Hydroxy-α-4-hydroxyphenyltoluene, *Di-Et ether*
Naphthotocopherol

$C_{17}H_{20}O_3$
Glycerol, 1,3-*Di-*o*-tolyl ether*
Glycerol, 1,3-*Dibenzyl ether*
Inophyllic Acid
Linderene, *Ac*

$C_{17}H_{20}O_4$
Heteropeucenin, *Di-Me ether*†

$C_{17}H_{20}O_5$
Balduilin
Bigelovin†
Guaiacol, 1,3-*Diglyceryl ether*
Linifolin A
Linifolin B
Phloroglucide, *Penta-Me ether*

$C_{17}H_{20}O_6$
Mycophenolic Acid

$C_{17}H_{20}O_8$
Psilotin†

$C_{17}H_{20}O_9$
1-*O*-Feruloylquinic Acid†
3-*O*-Feruloylquinic Acid
4-*O*-Feruloylquinic Acid†
5-*O*-Feruloylquinic Acid†

$C_{17}H_{21}N$
N-Benzyl-*Na*-dimethylphenethylamine

$C_{17}H_{21}NO$
Dicadic Acid, *Anilide*
Diphenhydramine
Phenyltoloxamine†

$C_{17}H_{21}NO_2$
Apoatropine
6-Azoestrone†
8-Azoestrone†
Cocculine
Desomorphine

$C_{17}H_{21}NO_3$
Chlidanthine
Dihydromorphine†
Galanthamine
Mesembrinine†
Pluviine†
Rotundine

$C_{17}H_{21}NO_4$
α-Cocaine
β-Cocaine
ψ-Cocaine
Goleptin
Habranthine†
Hyoscine
Isoborneol, p-*Nitrobenzoyl*

$C_{17}H_{21}NO_5$
Buphanitine★†
Lunidine†
Nerbowdine
Parkacin
Poetaminine

$C_{17}H_{21}NO_6$
Orixine

$C_{17}H_{21}N_2O_{10}S_2$
Neoglucobrassicin

$C_{17}H_{21}N_3$
Auramine
Auramine G
Elaeocarpidine†

$C_{17}H_{21}N_3O_2$
4-Amino-α-4-aminophenyl-2-nitrotoluene, 4,4′-N-*Tetra-Me*

$C_{17}H_{21}N_5O_4S$
6-[D(−)-α-Guanidinophenylacetamido]-penicillanic Acid†
6-(*p*-Guanidinophenylacetamido)-penicillanic Acid†

$C_{17}H_{22}NO_3$ (ion)
Lunasine†

$C_{17}H_{22}NO_4$ (ion)
O-Methylbalfourodinium†
Ribalinium, O-*Me ether*†

$C_{17}H_{22}N_2$
Antergan
Di-(*m*-dimethylaminophenyl)methane
Di-(*p*-dimethylaminophenyl)methane
Rubatoxan

$C_{17}H_{22}N_2O$
Dasycarpidol†
m-Dimethylaminophenyl-*p*-dimethylamino-phenyl-methanol
Di-*m*-dimethylaminophenylmethanol
Di-*p*-dimethylaminophenylmethanol
Doxylamine

$C_{17}H_{22}N_2O_3$
6-Hydroxy-8-nitroquinoline, *Octyl ether*

$C_{17}H_{22}N_2O_4$
Dethiobenzylpenicillin, *Me ester*

$C_{17}H_{22}N_6$ (di-ion)
Antrycide

$C_{17}H_{22}O$
Cicutol
Enanthetol
Falcarinone†

$C_{17}H_{22}O_2$
Cicutoxin
Enanthotoxin
5-Geranyl-2-methyl-1,4-benzoquinone†
Isoborneol, *Benzoyl*

$C_{17}H_{22}O_3$
2,6-Dimethylcyclohexanecarboxylic Acid, *Phenacyl ester*
Doisynolic Acid
6-Oxa-oestrone†
6-Oxa-8α-oestrone†
Podocarpic Acid

$C_{17}H_{22}O_4$
3-Carboxy-2,2,3-trimethyl-1-cyclopentylacetic Acid, β-*Phenyl ester*
p-Isopropylbenzylidenemalonic Acid, *Di-Et ester*

$C_{17}H_{22}O_5$
Calocephalin
Chrysanthin
Confertiflorin†
Gaillardin†
Globicin
Isogaillardin†
Isotenulin
Pulchellin B†

$C_{17}H_{22}O_5$ (*continued*)
Pyrethrosin†
Tenulin
Xanthinin
Xanthumin†
Zaluzanin A, 3-*Ac*†

$C_{17}H_{22}O_6$
Gaillardilin†

$C_{17}H_{22}O_7$
Picrotoxic Acid, *Et ester*
α-Picrotoxinic Acid, *Et ester*
β-Picrotoxinic Acid, *Et ester*

$C_{17}H_{22}O_9$
Bergenin, *Tri-Me ether*†

$C_{17}H_{23}BrO_3$
Laurencin†

$C_{17}H_{23}IO_2$
o-Iodobenzoic Acid, (—)-*Menthyl ester*
m-Iodobenzoic Acid, (—)-*Menthyl ester*
p-Iodobenzoic Acid, (—)-*Menthyl ester*

$C_{17}H_{23}NO$
2-Allyl-2′-hydroxy-5,9-dimethyl-6,7-benzomorphan†
Dextrorphan

$C_{17}H_{23}NO_2$
4-Azaoestra-1,3,5(10)trien-3,17β-diol

$C_{17}H_{23}NO_3$
Atropine
Hyoscyamine
Littorine†
Lycoramine★†
Mesembrine★†
3-Methylcyclohexane-1,1-diacetic Acid, *Monoanilide*
4-Methylcyclohexane-1,1-diacetic Acid, *Monoanilide*
Methylhomatropine
Oreoline†
2-Phenylpropionic Acid, *Tropine ester*

$C_{17}H_{23}NO_4$
Convolvamine
Galanthine
Genatropine
6-Hydroxyhyoscyamine
Lunacridine
Morphine, *Methoxymethyl ether*
o-Nitrobenzoic Acid, (—)-*Menthyl ester*
m-Nitrobenzoic Acid, (—)-*Menthyl ester*
p-Nitrobenzoic Acid, (—)-*Menthyl ester*
Porphyroxine

$C_{17}H_{23}NO_5$
Balfourolone†
Nilgirine†

$C_{17}H_{23}NO_9S_2$
Glucobenzosisymbrin

$C_{17}H_{23}N_3O$
Pyrilamine

$C_{17}H_{23}N_7O_8$
Duazomycin B†

$C_{17}H_{24}N_2O$
Dimethisoquin
β-Obscurine

$C_{17}H_{24}N_2O_2$
2,3-Diaza-17β-hydroxy-4-methyloestra-3,5(10)-dien-1-one
Phenglutarimide

$C_{12}H_{24}N_2O_3$
Leucyl-leucine

$C_{17}H_{24}N_4O_7$
5-Azoniaspiro(4,7)dodecane (hydroxide), *Picrate*

$C_{17}H_{24}O$
Falcarinol†

$C_{17}H_{24}O_2$
A-Nor-19-nortestosterone†
Podocarpinol

$C_{17}H_{24}O_3$
Cyclandelate
Shogaol

$C_{17}H_{24}O_4$
Acetylbalchanolide
4,6-Diacetylresorcinol, *Propyl-Butyl ether*
Hulupinic Acid, *Di-Me ether*†
Isosantonic Acid, *Et ester*
Santonic Acid, *Et ester*
Trichodermin†

$C_{17}H_{24}O_5$
Cinnamodial†
Cinnamosmolide†
Gafrinin
Hysterin†
Mexicanin B
Pulchellin D†
Stevin†
3-(2,4,5-Trihydroxyphenyl)acrylic Acid, *Tri-Et ether*, *Et ester*

$C_{17}H_{24}O_6$
Flexuosin A†
Shellolic Acid, *Di-Me ester*

$C_{17}H_{24}O_8$
α-Picrotonic Acid, *Et ester*

$C_{17}H_{24}O_9$
Syringin

$C_{17}H_{24}O_{10}$
Secologanin†
Verbenalin

$C_{17}H_{24}O_{11}$
Asperuloside
Monotropein, *Me ester*†
Scandoside, *Me ester*†

$C_{17}H_{24}O_{12}$
Piptoside†

$C_{17}H_{25}NO_2$
Alphameprodine
Anthranilic Acid, (—)-*Menthyl ester*
Betaprodine
Calabatin†
Lettocine

$C_{17}H_{25}NO_3$
Annopodine†
Cyclopentolate
Decanedioic Acid monoanilide, *Me ester*

$C_{17}H_{25}NO_3$ (*continued*)
Eucatropine
Lyconnotine†
Undecanedioic Acid, *Monoanilide*
Variotin†

$C_{17}H_{25}NO_5$
Benzylpenaldic Acid, *Et ester*

$C_{17}H_{25}NO_6$
3-[2-(3,5-Dimethyl-5-acetoxy-2-oxocyclohexyl)-2-hydroxyethyl]glutarimide

$C_{17}H_{25}N_3O_3$
Calabacin†

$C_{17}H_{25}N_5O_{13}$
Polyoxin B†

$C_{17}H_{25}N_5O_{15}P_2$
Adenosine 5′-diphosphate glucose†

$C_{17}H_{26}ClNO_2$
Alphaprodine

$C_{17}H_{26}N_2O$
N-(1-Methyl-2-piperidinoethyl)propionanilide†
α-Obscurine
Phenampromide
Sauroxine†

$C_{17}H_{26}N_2O_2$
2,3-Diaza-17β-hydroxy-4-methyloestr-3-en-1-one

$C_{17}H_{26}N_8O_5$
Blasticidin S★†

$C_{17}H_{26}O$
2-Benzoyl-6-isopropyl-3-methylcyclohexanol

$C_{17}H_{26}O_2$
Decanoic Acid, *Benzyl ester*
Heptadeca-10,16-dien-8-ynoic Acid†
2,4,6-Triethylbenzoic Acid, *Butyl ester*

$C_{17}H_{26}O_3$
Alantolic Acid, *Et ester*
7-Hydroxyheptadeca-10,16-dien-8-ynoic Acid†

$C_{17}H_{26}O_3S$
Toluene-*o*-sulphonic Acid, *Menthyl ester*
Toluene-*p*-sulphonic Acid, *Menthyl ester*

$C_{17}H_{26}O_4$
5-Acetoxymethyltetradeca-*trans*-2-*trans*-4-*trans*-6-trienoic Acid
Bicyclo[2,2,1]hept-5-ene-2,3-dicarboxylic Acid, *Di-butyl ester*
Embelin
Gingerol

$C_{17}H_{26}O_7$
Arbutin, *Penta-Me ether*

$C_{17}H_{26}O_{10}$
Loganin

$C_{17}H_{27}N$
5-Phenyl-1-pentylamine, N-*Cyclohexyl*

$C_{17}H_{27}NO_2$
Clavatoxine
Lophocerine, *Et ether*†
Stenine†

$C_{17}H_{27}NO_3$
p-Butoxybenzoic Acid, *Diethylaminoethyl ester*
Nobiline†

$C_{17}H_{27}NO_3S$
Penthienate

$C_{17}H_{28}$
Heptadecatetraene

$C_{17}H_{28}O$
Ficoceryl Alcohol
Inusterol B

$C_{17}H_{28}O_2$
Gurjuresene
Heptadec-10-en-8-ynoic Acid†
3,7,11-Trimethyl-2,6,10(or 11)-dodecatrienoic Acid, *Et ester*

$C_{17}H_{28}O_3$
7-Hydroxyheptadec-10-en-8-ynoic Acid†

$C_{17}H_{28}O_4$
Acarenoic Acid†
Clovenic Acid, *Di-Me ester*
ψ-Clovenic Acid, *Di-Me ester*
Nephrosterinic Acid

$C_{17}H_{28}O_6$
Gitalin
Spiculisporic Acid

$C_{17}H_{28}O_8$
Methane-tetracarboxylic Acid, *Tetra-propyl ester*
Methane-tetracarboxylic Acid, *Tetra-isopropyl ester*
Methane-tetracetic Acid, *Tetra-Et ester*

$C_{17}H_{29}N$
2,4,6-Tri-*tert*-butylpyridine†

$C_{17}H_{29}O$ (ion)
Tri-*tert*-butylpyrylium†

$C_{17}H_{29}P$
2,4,6-Tri-*tert*-butylphosphorin†

$C_{17}H_{30}N_2$
4-Methylhexadecanedioic Acid, *Dinitrile*

$C_{17}H_{30}N_2O_5S$
Pentylpenicilloic Acid, α-*Et* β-*Me ester*

$C_{17}H_{30}O$
Civetone
Ladaniol

$C_{17}H_{30}O_2$
Bicyclohexyl-1-carboxylic Acid, *Bu ester*
Geranic Acid, n-*Heptyl ester*
Hydnocarpic Acid, *Me ester*

$C_{17}H_{30}O_4$
Acaronic Acid†
Civetic Acid
Cyclohexylmalonic Acid, *Di-*tert-*butyl ester*
Nephrosteranic Acid

$C_{17}H_{30}O_5$
9-Oxoheptadecanedioic Acid

$C_{17}H_{30}O_6$
6-Methylheptane-1,1,7-tricarboxylic Acid, *Tri-Et ester*
6-Methylheptane-1,2,2-tricarboxylic Acid, *Tri-Et ester*

$C_{17}H_{30}O_{14}$
Glucofrangulin

$C_{17}H_{31}NO_2$
Pentadecane-1,1-dicarboxylic Acid, *Mono-nitrile*

$C_{17}H_{31}N_3O_2$
Palustrine

$C_{17}H_{32}BrNO_2$
N-2-(1-Cyclohexenylacetoxy)ethyl-*N*-di-isopropyl methylammonium bromide

$C_{17}H_{32}N_2O_6S$
Antibiotic U-11,973†
Antibiotic U-21,699†

$C_{17}H_{32}O$
Givetol
Cycloheptadecanone

$C_{17}H_{32}O_2$
11-Cyclohexylundecanoic Acid†
2-Heptadecenoic Acid
9-Heptadecenoic Acid
9-Hexadecenoic Acid, *Me ester*

$C_{17}H_{32}O_3$
12-Hydroxy-9-hexadecenoic Acid, *Me ester*†
9-Oxoheptadecanoic Acid
2-Oxohexadecanoic Acid, *Me ester*
3-Oxohexadecanoic Acid, *Me ester*
3-Oxopentadecanoic Acid, *Et ester*

$C_{17}H_{32}O_4$
Brassylic Acid, *Di-Et ester*
Hexadecanedioic Acid, *Mono-Me ester*
2-Methyldodecanedioic Acid, *Di-Et ester*
3-Methyldodecanedioic Acid, *Di-Et ester*
4-Methyldodecanedioic Acid, *Di-Et ester*
2-Methylhexadecanedioic Acid
3-Methylhexadecanedioic Acid
4-Methylhexadecanedioic Acid
5-Methylhexadecanedioic Acid
8-Methylhexadecanedioic Acid
Pentadecane-1,1-dicarbosylic Acid
Roccellic Acid

$C_{17}H_{32}O_{16}$
6,8-Di-*C*-glucosyldiosmetin†

$C_{17}H_{33}BrO_2$
2-Bromododecanoic Acid, *Isoamyl ester*

$C_{17}H_{33}N$
1-Amino-12-(2-cyclopentenyl)dodecane
Heptadecanoic Acid, *Nitrile*

$C_{17}H_{33}NO_2$
9-Oxoheptadecanoic Acid, *Amide*

$C_{17}H_{33}NO_3$
Pentadecane-1,1-dicarboxylic Acid, *Mono-amide*

$C_{17}H_{33}NO_{10}S_3$
Glucoarabin

$C_{17}H_{34}$
1-Heptadecene
8-Heptadecene

$C_{17}H_{34}Br_2$
1,17-Dibromoheptadecane

$C_{17}H_{34}O$
Cycloheptadecanol
Heptadecanal
2-Heptadecanone
9-Heptadecanone

$C_{17}H_{34}O_2$
Heptadecanoic Acid
Hexadecanoic Acid, *Me ester*
2-Methylhexadecanoic Acid
14-Methylhexadecanoic Acid
2-Octanol, *Nonanoyl*
Pentadecanoic Acid, *Et ester*
Tetradecanoic Acid, *Isopropyl ester*

$C_{17}H_{34}O_3$
2-Hydroxyhexadecanoic Acid, *Me ester*
11-Hydroxyhexadecanoic Acid, *Me ester*
16-Hydroxyhexadecanoic Acid, *Me ester*
3-Hydroxypentadecanoic Acid, *Et ester*

$C_{17}H_{34}O_4$
3,12-Dihydroxyhexadecanoic Acid, *Me ester*
α-Monomyristin
β-Monomyristin

$C_{17}H_{34}O_5$
Aleuritic Acid, *Me ester*

$C_{17}H_{35}Br$
1-Bromoheptadecane★†

$C_{17}H_{35}Cl$
1-Chloroheptadecane

$C_{17}H_{35}I$
1-Iodoheptadecane

$C_{17}H_{35}N$
1-Azacyclo-octadecane
12-Cyclopentyldodecylamine

$C_{17}H_{35}NO$
Heptadecanoic Acid, *Amide*

$C_{17}H_{35}NO_2$
1-Hexadecanol, *Urethane*

$C_{17}H_{35}N_3O_2$
DL-2-Guanidinohexadecanoic Acid

$C_{17}H_{36}$
Heptadecane
2,6,10-Trimethyltetradecane†

$C_{17}H_{36}ClNO_2$
Dodecanoic Acid, 2-*Dimethylaminoethyl ester, Methochloride*

$C_{17}H_{36}INO_2$
Dodecanoic Acid, 2-*Dimethylaminoethyl ester, Methiodide*

$C_{17}H_{36}O$
1-Heptadecanol
2-Heptadecanol
3-Heptadecanol
4-Heptadecanol
5-Heptadecanol
6-Heptadecanol
7-Heptadecanol
8-Heptadecanol
9-Heptadecanol
2-Octanol, n-*Nonyl ether*

$C_{17}H_{36}O_2$
Heptadecane-1,14-diol†

$C_{17}H_{36}O_3$
Sukesyl Alcohol

$C_{17}H_{36}O_4$
Orthocarbonic Acid, *Tetra-isobutyl ester*

$C_{17}H_{37}N$
1-Aminoheptadecane

C_{18}

$C_{18}H_6$
1,3,7,9,13,15-Hexadehydro[18]annulene†

$C_{18}H_6O_6$
Phenanthrene-1,8,9,10-tetracarboxylic Acid, *Di-anhydride*

$C_{18}H_8BrN_3O_4$
4-Bromo-2,5-dinitrodiphenylamine
4-Bromo-2,6-dinitrodiphenylamine
4′-Bromo-2,4-dinitrodiphenylamine
5-Bromo-2,4-dinitrodiphenylamine

$C_{18}H_8O_6$
Erosnine†

$C_{18}H_8O_8$
Thelephoric Acid★†

$C_{18}H_8O_{10}$
Anthraquinone-1,2,5,6-tetracarboxylic Acid
Anthraquinone-1,2,6,7-tetracarboxylic Acid
Anthraquinone-1,2,7,8-tetracarboxylic Acid
Anthraquinone-1,3,5,7-tetracarboxylic Acid
Anthraquinone-1,4,5,8-tetracarboxylic Acid

$C_{18}H_8O_{14}$
1,5-Dihydroxy-*S*-indacene-2,3,4,6,7,8-hexacarboxylic Acid†

$C_{18}H_9NO_4$
1-Nitrobenz[α]anthracene-7,12-dione
4-Nitrobenz[α]anthracene-7,12-dione

$C_{18}H_9N_5O_8$
1,3,6,8-Tetranitrocarbazole, N-*Phenyl*

$C_{18}H_{10}$
Benzo[*ghi*]fluoranthene
Tetradehydro[18]annulene†

$C_{18}H_{10}Cl_6O_4$
2,4-Dichloro-2,4-di-(2,4-dichlorophenyl)-cyclobutane-1,3-dicarboxylic Acid

$C_{18}H_{10}N_2$
Acenaphtheno[1,2-*b*]quinoxaline

$C_{18}H_{10}N_2O$
10-Phthaloperinone

$C_{18}H_{10}N_2O_2$
p-Dioxino[2,3-*h*, 5,6-*h′*]diquinoline
p-Dioxino[*ff′*]diquinoline

$C_{18}H_{10}N_2S$
Pyrido[3,2-*f*-]thianaphtheno[3,2-*b*]quinoline†

$C_{18}H_{10}N_4$
Dipyrido[2,3-*a*, 2′,3′-*c*]phenazine
Dipyrido[2,3-*a*, 3′,2′-*c*]phenazine
Dipyrido[3,2-*a*:3′,2′-*h*]phenazine†
Quinoxalo[2,3-*a*]phenazine

$C_{18}H_{10}N_4O_{10}$
Resorcinol, *Di*-2,4-*dinitrophenyl ether*

$C_{18}H_{10}O_2$
1,2-Benz[*a*]anthracene-7,12-dione
5,6-Chrysenedione
Naphthacenequinone
7,12-Pleiadenedione

$C_{18}H_{10}O_3$
7,12-Dihydro-7,12-dioxopleiaden-1-ol
7-Oxo-7*H*-benz[*de*]anthracene-2-carboxylic Acid
7-Oxo-7*H*-benz[*de*]anthracene-3-carboxylic Acid
7-Oxo-7*H*-benz[*de*]anthracene-4-carboxylic Acid
7-Oxo-7*H*-benz[*de*]anthracene-11-carboxylic Acid

$C_{18}H_{10}O_4$
1,11-Dihydroxytetracene-5,12-quinone†
Pulvinic Acid Lactone

$C_{18}H_{10}O_5$
Calycin
Pongaglabrone

$C_{18}H_{10}O_8$
Phenanthrene-1,8,9,10-tetracarboxylic Acid
Variegatic Acid, *Lactone*†

$C_{18}H_{11}ClO_2$
8-Benzoyl-1-naphthoic Acid, *Chloride*

$C_{18}H_{11}NO_2$
Naphthalene-1,8-dicarboxylic Acid, N-*Phenyl imide*
3-Nitro-1,7,13-tridehydro[18]annulene†
7-Oxo-7*H*-benz[*de*]anthracene-2-carboxylic Acid, *Amide*
Phthalimide, N-1-*Naphthyl*
Phthalimide, N-2-*Naphthyl*
Quinophthalone

$C_{18}H_{11}NO_3$
Pulvinic Acid, *Nitrile*

$C_{18}H_{11}NO_6$
1-Nitroanthraquinone-2-carboxylic Acid, *Allyl ester*

$C_{18}H_{12}$
Benz[*a*]anthracene
Benz[*c*]phenanthrene
Chrysene
Dibenz[*a,e*]azulene
sym-Dibenzfulvalene†
Naphthacene
Pleiadene†
1,7,13-Tridehydro[18]annulene†
Triphenylene

$C_{18}H_{12}Br_2O_3$
α-Bromocinnamic Acid, *Anhydride*

$C_{18}H_{12}N_2$
2,2′-Biquinolyl
2,3′-Biquinolyl
2,5′-Biquinolyl
2,6′-Biquinolyl
2,7′-Biquinolyl
3,3′-Biquinolyl
3,4′-Biquinolyl
4,4′-Biquinolyl
4,6′-Biquinolyl
5,5′-Biquinolyl
6,6′-Biquinolyl

$C_{18}H_{12}N_2$ (*continued*)
6,8′-Biquinolyl
7,7′-Biquinolyl
8,8′-Biquinolyl

$C_{18}H_{12}N_2O_2$
Safranol
Xanthocillin X

$C_{18}H_{12}N_2O_6$
4,6-Dinitroresorcinol, *Diphenyl ether*

$C_{18}H_{12}N_3O_{10}P$
Tri-*p*-nitrophenyl phosphate†

$C_{18}H_{12}N_4$
2,2′-Azoquinoline
4,4′-Azoquinoline
5,5′-Azoquinoline
6,6′-Azoquinoline
7,7′-Azoquinoline
Homofluorindine

$C_{18}H_{12}O$
1-Hydroxybenz[*a*]anthracene
2-Hydroxybenz[*a*]anthracene
4-Hydroxybenz[*a*]anthracene
7-Hydroxybenz[*a*]anthracene
8-Hydroxybenz[*a*]anthracene
12-Hydroxybenz[*a*]anthracene
2-Methylbenzanthrone
4-Methylbenzanthrone
8-Methylbenzanthrone
9-Methylbenzanthrone
10-Methylbenzanthrone

$C_{18}H_{12}OS_2$
[18]Annulene 1(4)-oxide 7(10),13(16)-disulphide†

$C_{18}H_{12}O_2$
5,6-Chrysenediol
2,5-Diphenyl-1,4-benzoquinone
2,6-Diphenyl-1,4-benzoquinone
1-Hydroxybenzanthrone, *Me ether*
4-Hydroxybenzanthrone, *Me ether*
α-Truxone

$C_{18}H_{12}O_2S$
[18]Annulene 1(4), 7(10)-dioxide 13(16)-sulphide†

$C_{18}H_{12}O_2S_2$
2,4-Diphenacylidene-1,3-dithietan†

$C_{18}H_{12}O_3$
[18]Annulene Trioxide†
1-Benzoyl-2-naphthoic Acid
2-Benzoyl-1-naphthoic Acid
3-Benzoyl-2-naphthoic Acid
4-Benzoyl-1-naphthoic Acid
6-Benzoyl-2-naphthoic Acid
8-Benzoyl-1-naphthoic Acid

$C_{18}H_{12}O_3$
10,11-Dihydroxybenz[*de*]anthracen-7-one, 10-*Me ether*
o-1-Naphthoylbenzoic Acid
o-2-Naphthoylbenzoic Acid
Tanshinone I

$C_{18}H_{12}O_4$
Karanjin
6-Methoxy-4-oxo-2-phenylfuro[2,3-*h*]-1-benzopyran
2-Phenylnaphthalene-1,2′-dicarboxylic Acid
1-Phenylnaphthalene-2,3-dicarboxylic Acid
Pinnatin
Polyporic Acid
Valucrisporin

$C_{18}H_{12}O_5$
Pulvinic Acid

$C_{18}H_{12}O_6$
Anthraquinone-1,2-dicarboxylic Acid, *Di-Me ester*
Anthraquinone-1,5-dicarboxylic Acid, *Di-Me ester*
Anthraquinone-2,3-dicarboxylic Acid, *Di-Me ester*
Atromentin
Sterigmatocystin
Tetrahydroxy-*p*-benzoquinone, *Diphenyl ether*

$C_{18}H_{12}O_7$
Atromentinic Acid
Leucomelone

$C_{18}H_{12}O_9$
Norstictic Acid
Variegatic Acid†

$C_{18}H_{12}O_{10}$
Salazinic Acid

$C_{18}H_{12}S$
2,2′-Naphthylbenzo[*b*]thiophen

$C_{18}H_{12}S_3$
[18]Annulene Trisulphide†

$C_{18}H_{13}ClN_2O_8$
Chlorofumaric Acid, *Di-*p*-nitrobenzyl ester*

$C_{18}H_{13}Cl_2NO_3$
Pyoluteorin★, N-*Benzyl*†

$C_{18}H_{13}N$
N-Phenylcarbazole
1-Phenylcarbazole

$C_{18}H_{13}NO_2$
2,6-Diphenylpyridine-4-carboxylic Acid
Isotetrophane
o-1-Naphthoylbenzoic Acid, *Amide*
Tetrophane

$C_{18}H_{13}NO_3$
Lysicamine†
2-Phenylnaphthalene-1,2′-dicarboxylic Acid, 1-*Amide*
2-Phenylnaphthalene-1,2′-dicarboxylic Acid, 2′-*Amide*

$C_{18}H_{13}NO_4$
Atherospermidine†
Moschatoline†
p-Nitrophenylacetic Acid, 1-*Naphthyl ester*
Pulvinic Acid, *Amide*

$C_{18}H_{13}NO_6$
1-Nitroanthraquinone-2-carboxylic Acid, *Propyl ester*

$C_{18}H_{13}NO_7$
Aristolochic Acid, *Me ester*

$C_{18}H_{13}NS$
Phenothiazine, N-*Phenyl*

$C_{18}H_{13}NS_2$
1,4-Epimino[18]annulene 7(10),13(16)-disulphide †

$C_{18}H_{13}N_3$
Di-2-quinolylamine

$C_{18}H_{13}N_3O$
Rutaecarpine

$C_8H_{13}N_3O_5$
5-Hydroxy-2,4-dinitrodiphenylamine, *Phenyl ether*

$C_{18}H_{14}$
Dicyclo-octatetreno[1,2:4,5]benzene †
1,2-Diphenylbenzene
1,3-Diphenylbenzene
1,4-Diphenylbenzene
Diphenylfulvene
1-Methylbenzo[*a*]fluorene †
3-Methylbenzo[*a*]fluorene †
4-Methylbenzo[*a*]fluorene †
10-Methylbenzo[*a*]fluorene †

$C_{18}H_{14}Cl_2O_2$
Neotruxinic Acid, *Dichloride*
α-Truxillic Acid, *Dichloride*
γ-Truxillic Acid, *Dichloride*
ε-Truxillic Acid, *Dichloride*
epi-Truxillic Acid, *Dichloride*
β-Truxinic Acid, *Dichloride*
δ-Truxinic Acid, *Dichloride*
ζ-Truxinic Acid, *Dichloride*

$C_{18}H_{14}Cl_2O_7$
Gangaleoidin
Geodin, *Mono-Me ether*

$C_{18}H_{14}N_2$
α-Truxillic Acid, *Dinitrile*
γ-Truxillic Acid, *Dinitrile*
ε-Truxillic Acid, *Dinitrile*
epi-Truxillic Acid, *Dinitrile*

$C_{18}H_{14}N_2O$
4-Hydroxyazobenzene, *Phenyl ether*
Tetrophane, *Amide*

$C_{18}H_{14}N_2O_2$
Tetrophane, *Amide*

$C_{18}H_{14}N_2O_2$
4,5-Dianilino-1,2-benzoquinone
2,3-Dianilino-1,4-benzoquinone
2,5-Dianilino-1,4-benzoquinone
3,6-Dibenzylidene-2,6-dioxopiperazine †
Flavocarpine
o-Nitrophenyldiphenylamine
m-Nitrophenyldiphenylamine
p-Nitrophenyldiphenylamine

$C_{18}H_{14}N_2O_3$
Piricularin (or $C_{17}H_{14}N_2O_3$)

$C_{18}N_{14}N_2O_4$
4-Nitrostilbene-2,2′-dicarboxylic Acid, 2-*Nitrile, Et ester*

$C_{18}H_{14}O$
Benzyl 2-naphthyl Ketone

$C_{18}H_{14}O$
1,6-Diphenyltricyclo[3,1,0,0^{2,6}]hexan-3-one †

$C_{18}H_{14}O_2$
1,3-Dihydroxypyrene, *Di-Me ether*
1,6-Dihydroxypyrene, *Di-Me ether*
o-1-Naphthylbenzoic Acid, *Me ester*
o-2-Naphthylbenzoic Acid, *Me ester*
2-Phenyl-1-naphthoic Acid, *Me ester*
Resorcinol, *Diphenyl ether*

$C_{18}H_{14}O_3$
Cinnamic Acid, *Anhydride*
3,4-Dihydroxydiphenyl Ether, 3-*Phenyl ether*
8-Hydroxy-1-naphthoic Acid, *Benzyl ether*
2-(*p*-Methoxyphenyl)-6-phenyl-4*H*-pyran-4-one †
γ-Truxillic Acid, *Anhydride*
η-Truxillic Acid, *Anhydride*
β-Truxinic Acid, *Anhydride*
ζ-Truxinic Acid, *Anhydride*

$C_{18}H_{14}O_3S_2$
β-Lipoic Acid

$C_{18}H_{14}O_4$
Anthracene-1,9-dicarboxylic Acid, *Di-Me ester*
Dibenzylidenesuccinic Acid
2-Methylanthraquinone-1-carboxylic Acid, *Et ester*
Phenanthrene-1,2-dicarboxylic Acid, *Di-Me ester*
Phenanthrene-1,7-dicarboxylic Acid, *Di-Me ester*
Phenanthrene-3,4-dicarboxylic Acid, *Di-Me ester*
Phenanthrene-3,9-Dicarboxylic Acid, *Di-Me ester*
Phenanthrene-9,10-dicarboxylic Acid, *Di-Me ester*
Pongamol

$C_{18}H_{14}O_5$
ψ-Baptigenin, *Et ether*
Dalbergin, *Ac*
6-Deoxyjacareubin †
Digitolutein, *Ac*
Osajaxanthone★ †

$C_{18}H_{14}O_6$
Benzil-2,2′-dicarboxylic Acid, *Di-Me ester*
7-Hydroxy-11,12-dimethoxycoumestan, *Me ether* †
Irisolone, *Mono-Me ether*
Jacareubin †
Lucernol, *Tri-Me ether* †
Rhein, *Propyl ester*
Sativol, *Di-Me ether* †
Tlatlancuayin
Trifoliol, *Di-Me ether* †
7,11,12-Trihydroxycoumestan, *Tri-Me ether* †

$C_{18}H_{14}O_7$
Bostrycoidin
Emodic Acid, *Tri-Me ether*
Ptilometric Acid †
Rhodocomatulin †
Variolaric Acid, *Di-Me ether*

$C_{18}H_{14}O_8$
Biphenyl-2,2′,3,3′-tetracarboxylic Acid, 2,2′-*Di-Me ester*
Hyposalazinic Acid
Lucidin (flavone)†
Psoromic Acid
Rubrocomatulin†

$C_{18}H_{14}O_9$
Distemonanthin, 3,10-*Di-Me ether*
Flavellagic Acid, 4,5,4′,5′-*Tetra-Me ether*

$C_{18}H_{14}O_{10}$
Barbatolic Acid

$C_{18}H_{15}Al$
Triphenylaluminium†

$C_{18}H_{15}As$
Triphenylarsine

$C_{18}H_{15}BO_3$
Phenol, *Borate*

$C_{18}H_{15}ClO_5$
Fragilin, *Di-Me ether*†

$C_{18}H_{15}ClO_6$
Nalgiolaxin
Pannarin

$C_{18}H_{15}Cr$
Triphenyl chromium

$C_{18}H_{15}IN_2$
Benzo[*a*]phenazine, *Mono-ethiodide*

$C_{18}H_{15}IO_4$
Iodosobenzene, *Dibenzoate*

$C_{18}H_{15}N$
2-Methyl-4,6-diphenylpyridine
3-Methyl-2,6-diphenylpyridine
4-Methyl-2,6-diphenylpyridine
3-Methyl-2-styrylquinoline
4-Methyl-2-styrylquinoline
4′-Methyl-2-styrylquinoline
6-Methyl-2-styrylquinoline
8-Methyl-2-styrylquinoline
Triphenylamine

$C_{18}H_{15}NO$
p-Hydroxybenzylidene-1-naphthylamine, *Me ether*
p-Hydroxybenzylidene-2-naphthylamine, *Me ether*

$C_{18}H_{15}NO_2$
3-Methyl-2-phenylquinoline-4-carboxylic Acid, *Me ester*
6-Methyl-2-phenylquinoline-4-carboxylic Acid, *Me ester*
8-Methyl-2-phenylquinoline-4-carboxylic Acid, *Me ester*
2-Methylquinoline-3-carboxylic Acid, *Benzyl ester*
Murrayacine†
2-β-Naphthylaminobenzoic Acid, *Me ester*
2-Phenylquinoline-4-carboxylic Acid, *Et ester*
2-*p*-Tolyquinoline-4-carboxylic Acid, *Me ester*
γ-Truxillic Acid, *Imide*
ε-Truxillic Acid, *Imide*
η-Truxillic Acid, *Imide*
β-Truxinic Acid, *Imide*

$C_{18}H_{15}NO_3$
De-*N*-methylnoracronycine†

$C_{18}H_{15}NO_4$
Lunamarine
Ovigerine†

$C_{18}H_{15}NO_6$
2-Nitrostilbene-2′,4-dicarboxylic Acid, *Di-Me ester*

$C_{18}H_{15}N_3$
4-Anilinoazobenzene

$C_{18}H_{15}N_3O_3$
1,2-Naphthoquinone, 1-o-*Nitrophenylhydrazone, Et ether*
1,2-Naphthoquinone, 1-m-*Nitrophenylhydrazone, Et ether*
1,2-Naphthoquinone, 1-*p*-*Nitrophenylhydrazone, Et ether*
1,4-Naphthoquinone, o-*Nitrophenylhydrazone, Et ether*

$C_{18}H_{15}N_3O_4S$
4-Nitrodiphenylamine-2-sulphonic Acid, *Anilide*
2-Nitrodiphenylamine-4-sulphonic Acid, *Anilide*

$C_{18}H_{15}OP$
Triphenylphosphine oxide†

$C_{18}H_{15}O_2P$
Diphenylphosphinic Acid, *Phenyl ester*
Phenol, *Phosphite*
Phenylphosphonic Acid, *Diphenyl ester*

$C_{18}H_{15}O_3PS$
O,O,O-Triphenylphosphorothionate

$C_{18}H_{15}O_4P$
Triphenyl phosphate

$C_{18}H_{15}P$
Triphenylphosphine

$C_{18}H_{15}Sb$
Triphenylstibine

$C_{18}H_{15}Tl$
Thallium triphenyl

$C_{18}H_{16}$
15,16-Dihydro-15,16-*trans*-dimethylpyrene†
7-Methyl-2-*p*-tolylnaphthalene
Truxane

$C_{18}H_{16}Br_2O_4$
4,4′-Dibromobiphenyl-2,2′-dicarboxylic Acid, *Di-Et ester*

$C_{18}H_{16}Cl_2O_5$
Vicanicin

$C_{18}H_{16}Cl_2O_8$
Nordin

$C_{18}H_{16}N_2$
N-o-Aminophenyldiphenylamine
N-m-Aminophenyldiphenylamine
N-p-Aminophenyldiphenylamine
N,N′-Diphenyl-*o*-phenylenediamine
N,N′-Diphenyl-*m*-phenylenediamine
N,N′-Diphenyl-*p*-phenylenediamine

$C_{18}H_{16}N_2O$
1-Benzeneazo-2-naphthol, *Et ether*
2-Benzeneazo-1-naphthol, *Et ether*
4-Benzeneazo-1-naphthol, *Et ether*

$C_{18}H_{16}N_2O$ (*continued*)
9-Methoxyellipticine †
2-*o*-Tolueneazo-1-naphthol, *Me ether*
2-*m*-Tolueneazo-1-naphthol, *Me ether*
2-*p*-Tolueneazo-1-naphthol, *Me ether*
4-*o*-Tolueneazo-1-naphthol, *Me ether*
4-*p*-Tolueneazo-1-naphthol, *Me ether*
1-*o*-Tolueneazo-2-naphthol, *Me ether*
1-*m*-Tolueneazo-2-naphthol, *Me ether*
1-*p*-Tolueneazo-2-naphthol, *Me ether*

$C_{18}H_{16}N_2O_2$
3-Benzyl-6-benzylidenepiperazine-2,5-dione †

$C_{18}H_{16}N_2O_4$
1-Amino-8-nitroanthraquinone, N-*Di-Et*
Isatide, N,N′-*Di-Me*

$C_{18}H_{16}N_2O_5$
Flazine

$C_{18}H_{16}N_2O_8$
Betanidin †
6,6′-Dinitrobiphenyl-2,2′-dicarboxylic Acid, *Di-Et ester*
Isobetanidin †

$C_{18}H_{16}N_2S$
1-Benzyl-3-(1-naphthyl)thiourea

$C_{18}H_{16}N_4O_6S$
3-Carboxy-2-quinoxalinylpenicillinic Acid †

$C_{18}H_{16}O$
4-Benzyl-1-naphthol, *Me ether*

$C_{18}H_{16}OS$
1-Methyl-3,5-diphenylthiabenzene-1-oxide †

$C_{18}H_{16}O_2$
Cinnamyl cinnamate
3-Ethyl-6-methylflavone
7-Isopropyl-1-methylphenanthraquinone
3-Methoxygona-1,3,5(10),8,11,13-hexaen-17-one †
2-Methylanthracene-1-carboxylic Acid, *Et ester*
2-Phenanthreneacetic Acid, *Et ester*
1-Phenanthrenepropionic Acid, *Me ester*
2-Phenanthrenepropionic Acid, *Me ester*
3-Phenanthrenepropionic Acid, *Me ester*
9-Phenanthrenepropionic Acid, *Me ester*
1,2,3,4-Tetramethylanthraquinone
1,2,5,6-Tetramethylanthraquinone
1,3,5,7-Tetramethylanthraquinone
1,3,6,7-Tetramethylanthraquinone
1,3,6,8-Tetramethylanthraquinone †
1,4,5,8-Tetramethylanthraquinone †
1,4,6,7-Tetramethylanthraquinone
2,3,6,7-Tetramethylanthraquinone

$C_{18}H_{16}O_3$
Chalcone-4-carboxylic Acid, *Et ester*
Chalcone-β-carboxylic Acid, *Et ester*
2-Methyl-2-phenoxypropionic Acid, *Phenyl ester*
Oestra-1,5(10),6,8-tetraene-3,4,17-trione
Thebenol, *Me ether*

$C_{18}H_{16}O_4$
Alizarin, *Di-Et ether*
Daidzein, 4′-*Et*-7-*Me-ether*
Dibenzoylacetic Acid, *Et ester*
1,4-Dihydroxyanthraquinone, *Di-Et ether*
1,5-Dihydroxyanthraquinone, *Di-Et ether*
2,6-Dihydroxy-1,5-dimethylanthraquinone, *Di-Me ether*
2,4-Diphenyl-1-butene-1,3-dicarboxylic Acid
1,4-Diphenyl-2-butene-1,4-dicarboxylic Acid
2,3-Diphenyl-2-butene-1,4-dicarboxylic Acid
1,1-Diphenyl-1-butene-2,3-dicarboxylic Acid
1,4-Diphenyl-1-butene-2,3-dicarboxylic Acid
1,2-Diphenyl-1-butene-3,4-dicarboxylic Acid
1,3-Diphenyl-1-butene-4,4-dicarboxylic Acid
Diphenylfumaric Acid, *Di-Me ester*
Diphenylmaleic Acid, *Di-Me ester*
Fumaric Acid, *Di-benzyl ester*
α-Isatropic Acid
β-Isotropic Acid
1-(*p*-Methoxyphenyl)-5-phenylpentane-1,2,5-trione †
Neotruxinic Acid
Stilbene-2,2′-dicarboxylic Acid, *Di-Me ester*
Stilbene-2,3-dicarboxylic Acid, *Di-Me ester*
Stilbene-4,4′-dicarboxylic Acid, *Di-Me ester*
α-Truxillic Acid
γ-Truxillic Acid
ε-Truxillic Acid
η-Truxillic Acid
epi-Truxillic Acid
β-Truxinic Acid
δ-Truxinic Acid
ξ-Truxinic Acid

$C_{18}H_{16}O_5$
Aloe-emodin, *Tri-Me ether*
Apigenin, *Tri-Me ether*
Baicalein, *Tri-Me ether*
Cladofulvin, *Tri-Me ether* †
2-(3,3-Dimethylallyl)-1,3,5-trihydroxyxanthone †
2-(3,3-Dimethylallyl)-1,3,7-trihydroxyxanthone †
Echioidinin, *Di-Me ether* †
Eucomin, 7-*Me ether* †
Galangin, 3,5,7-*Tri-Me ether*
Genistein, *Tri-Me ether*
7-Hydroxy-4′,6-dimethoxyisoflavone, *Me ether* †
5-Hydroxy-4′,7-dimethoxy-6-methylflavone †
3-Hydroxy-4-methoxycinnamic Acid, *Me ester, Benzoyl*
Izalpinin
Kuhlmannin, *Me ether* †
Lapodin
Lucidin, *Tri-Me ether* (Anthraquinone)
Macrosporin, *Mono-Et ether*
Mbarraxanthone †
2-Methylgenistein, 4′,7-*Di-Me ether*
Norwogonin, *Tri-Me ether*
Oroxylin-A, 5,7-*Di-Me ether*
Padmakasein, *Di-Me ether*
Questin, *Di-Me ether* †
Resokaempferol, *Tri-Me ether*
1,2,6-Trihydroxyanthraquinone, 2,6-*Di-Et ether*
2′,5,6-Trihydroxyflavone, *Tri-Me ether*
3,4′,5-Trihydroxyflavone, *Tri-Me ether*
3,5,6-Trihydroxyflavone, *Tri-Me ether*

$C_{18}H_{16}O_5$ (*continued*)

3,5,8-Trihydroxyflavone, 3,5,8-*Tri-Me ether*
3′,4′,5-Trihydroxyflavone, *Tri-Me ether*
3′,4′,5′-Trihydroxyflavone, *Tri-Me ether*
3′,4′,8-Trihydroxyflavone, *Tri-Me ether*
4′,5,6-Trihydroxyflavone, *Tri-Me ether*
4′,5,6-Trihydroxyflavone, *Tri-Me ether*
4′,6,7-Trihydroxyflavone, *Tri-Me ether*
4′,6,8-Trihydroxyflavone, *Tri-Me ether*
4′,7,8-Trihydroxyflavone, *Tri-Me ether*
6,7,8-Trihydroxyflavone, *Tri-Me ether*
4′,6,7-Trihydroxyisoflavone, *Tri-Me ether*
1,2,5-Trihydroxy-6-methylanthraquinone, *Tri-Me ether*
1,2,6-Trihydroxy-6-methylanthraquinone, *Tri-Me ether*
1,2,6-Trihydroxy-8-methylanthraquinone, *Tri-Me ether*
1,2,8-Trihydroxy-6-methylanthraquinone, *Tri-Me ether*
1,2,8-Trihydroxy-7-methylanthraquinone, *Tri-Me ether*★†
1,3,6-Trihydroxy-8-methylanthraquinone, *Tri-Me ether*†
1,3,8-Trihydroxy-6-methylanthraquinone, *Tri-Me ether*
1,4,5-Trihydroxy-6-methylanthraquinone, *Tri-Me ether*
1,4,5-Trihydroxy-8-methylanthraquinone, *Tri-Me ether*
Wogonin, *Di-Me ether*

$C_{18}H_{16}O_6$

Datiscetin, 3,7,2′-*Tri-Me ether*
Datiscetin, 5,7,2′-*Tri-Me ether*
Decussatin, *Ac*
4′,5-Dihydroxy-3,7-dimethoxyflavone, 4′-*Me ether*†
4′,5-Dihydroxy-6,7-dimethoxyisoflavone, 4′-*Me ether*†
3,4-Dimethoxy-8,9-methylenedioxypterocarpan†
2-(3,3-Dimethylallyl)-1,3,5,6-tetrahydroxanthone†
Elliptol
Ferreirin, 2′,7-*Di-Me ether*
Fisetin, 3,3′,4′-*Tri-Me ether*
5-Hydroxy-2′,6,7-trimethoxyflavone†
5-Hydroxy-3,4′,7-trimethoxyflavone†
Isoshehkangenin, *Di-Me ether*
Kaempferol, *Tri-Me ether*
Leptosidin, *Mono-Me ether*
Luteolin, 3′,4′,7-*Tri-Me ether*
Nalgiovensin
Pectolinarigenin, 7-*Me ether*
Pratoletin, *Tri-Me ether*
Resomorin, 2′,4′,7-*Tri-Me ether*
Rhodoptilometrin, 6-*Me ether*†
Scutellarein, 4′,5,7-*Tri-Me ether*
Scutellarein, 4′,6,7-*Tri-Me ether*
Sophorol, *Di-Me ether*
Symphoxanthone†
Tectorigenin, *Di-Me ether*
1,2,6,7-Tetrahydroxyanthraquinone, *Tetra-Me ether*
1,3,5,7-Tetrahydroxyanthraquinone, *Tetra-Me ether*
1,3,6,8-Tetrahydroxyanthraquinone, *Tetra-Me ether*†
1,4,5,8-Tetrahydroxyanthraquinone, *Tetra-Me ether*
2,3,6,7-Tetrahydroxyanthraquinone, *Tetra-Me ether*
2′,3,4′,6-Tetrahydroxyflavone, 2′,4′,6-*Tri-Me ether*
2′,3,7,8-Tetrahydroxyflavone, 2′,7,8-*Tri-Me ether*
3,3′,4′,6-Tetrahydroxyflavone, 3′,4′,6-*Tri-Me ether*
3,3′,7,8-Tetrahydroxyflavone, 3′,7,8-*Tri-Me ether*
3,4′,5,8-Tetrahydroxyflavone, 4′,5,8-*Trimethyl ether*†
3,4′,7,8-Tetrahydroxyflavone, 4′,7,8-*Tri-Me ether*
3′,4′,5′,6-Tetrahydroxyflavone, 3′,4′,5′-*Tri-Me ether*
4′,5,7,8-Tetrahydroxyflavone, 4′,5,8-*Tri-Me ether*
Ugazanthone†
Zapotinin

$C_{18}H_{16}O_7$

Amarbelin
Ayanin
Bisdechlorogeodin, *Me ether*†
4′,5-Dihydroxy-3,7,8-trimethoxyflavone†
5,7-Dihydroxy-3,4′,8-trimethoxyflavone†
5,7-Dimethoxy-4′,6,8-trimethoxyflavone†
Isousnic Acid†
Morin, *Tri-Me ether*
Penduletin
4′,5,6,7,8-Pentahydroxyflavone, *Tri-Me ether*†
1,3,4,5,6-Pentahydroxy-2-methylanthraquinone, *Tri-Me ether*
Quercetin, 3,3′,4′-*Tri-Me ether*
Tambulin
Tricetin, 3′,4′,5′-*Tri-Me ether*
Trypacidin†
Usnic Acid
Wightin†
Xanthomicrol

$C_{18}H_{16}O_8$

Acerosin†
Centaureidin†
1,2,3,5,6,7-Hexahydroxyanthraquinone, *Tetra-Me ether*
Irigenin
Jaceidin†
Limocitrin, 5-*Me ether*
Rosmarinic Acid
Sudachitin†
3′,4′,5-Trihydroxy-3,6,7-trimethoxyflavone†
3′,4′,5-Trihydroxy-3,7,8-trimethoxyflavone†
3′,5,5′-Trihydroxy-3,4′,7-trimethoxyflavone†
3′,5,7-Trihydroxy-3,4′,8-trimethoxyflavone†
4′,5,7-Trihydroxy-3,3′,8-trimethoxyflavone†

$C_{18}H_{16}O_9$
Isolimocitrol†
Limocitrol†

$C_{18}H_{17}ClO_2S$
7-Isopropyl-1-methylphenanthrene-2-sulphonic Acid, *Chloride*
2-Isopropyl-8-methylphenanthrene-3-sulphonic Acid, *Chloride*

$C_{18}H_{17}ClO_4$
Apigeninidin, *Tri-Me ether*
Galanginidin, *Tri-Me ether*

$C_{18}H_{17}ClO_6$
Morindin chloride, 2′,3,4′-*Tri-Me ether*
Radicicol†

$C_{18}H_{17}ClO_7$
Hirsutidin chloride

$C_{18}H_{17}N$
N-Phenyl-2-naphthylamine, *N-Et*

$C_{18}H_{17}NO$
2-Anilinoethanol, 2-*Naphthyl ether*
Girinimbine†
7-Hydroxy-2-methyl-4-phenylquinoline, *Et ether*

$C_{18}H_{17}NO_2$
Nitro[18]annulene†
Roemerine

$C_{18}H_{17}NO_3$
10-Hydroxy-1,2-methylenedioxyaporphine†
Mecambrine†
Mecambroline†
Mesaconic Acid, α-*Phenyl ester*, β-p-*Toluidide*
Neotruxinic Acid, 1-*Amide*
Neotruxinic Acid, 2-*Amide*
Pukateine
α-Truxillic Acid, 1-*Amide*
γ-Truxillic Acid, 3-*Amide*
ε-Truxillic Acid, 3-*Amide*
η-Truxillic Acid, 1-*Amide*
epi-Truxillic Acid, 3-*Amide*
β-Truxinic Acid, 2-*Amide*
δ-Truxinic Acid, 2-*Amide*
ζ-Truxinic Acid, 1-*Amide*
ζ-Truxinic Acid, 2-*Amide*
Ushinsunine†
Xylopine

$C_{18}H_{17}NO_4$
Acronidine
Actinodaphnine
Benzylpenaldic Acid, *Benzyl ester*
Cularine†
1-Hydroxy-2-methoxy-9,10-methylenedioxynor-aporphine†
Laurepukin
Nandigerine†
2-*o*-Nitrophenyl-3-*p*-tolylacrylic Acid, *Et ester*

$C_{18}H_{17}NO_5$
Fontaphilline†
2′-Hydroxy-2-nitrostilbene-4-carboxylic Acid, *Me ether*, *Et ester*
4′-Hydroxy-2-nitrostilbene-4-carboxylic Acid, *Me ether*, *Et ester*
Jonquillin†

$C_{18}H_{17}NO_7$
Papaveric Acid, *Di-Me ester*
Papaveric Acid, 3-*Et ester*
Papaveric Acid, 4-*Et ester*

$C_{18}H_{17}N_5O_6S_2$
Acetylalthiomycin†

$C_{18}H_{17}N_7O_6$
Pteroylaspartic Acid

$C_{18}H_{18}$
[18]-Annulene†
5,5*a*,6,11,11*a*,12-Hexahydrochrysene
Hexavinylbenzene†
1-Isopropyl-7-methylphenanthrene
3-Isopropyl-1-methylphenanthrene
4-Isopropyl-1-methylphenanthrene
6-Isopropyl-1-methylphenanthrene
7-Isopropyl-1-methylphenanthrene
7-Isopropyl-1-methylphenanthrene
9-Isopropyl-1-methylphenanthrene
Scianthrene
3,5,8,10-Tetramethylaceheptylene†
1,4,5,8-Tetramethylanthracene†
1,2,3,4-Tetramethylanthracene
1,2,5,6-Tetramethylanthracene
1,2,7,8-Tetramethylanthracene
1,2,9,10-Tetramethylanthracene
1,3,5,7-Tetramethylanthracene
1,3,6,7-Tetramethylanthracene
1,3,6,8-Tetramethylanthracene†
1,3,9,10-Tetramethylanthracene
2,3,6,7-Tetramethylanthracene★†
2,3,9,10-Tetramethylanthracene
1,2,3,4-Tetramethylphenanthrene†

$C_{18}H_{18}N_2$
2-Phenyl-1,3-naphthylenediamine, 1,3-N-*Di-Me*

$C_{18}H_{18}N_2O_2$
3,6-Dibenzyl-2,5-dioxopiperazine
Maleic Acid, *Di*-p-*toluidide*
Neotruxinic Acid, *Diamide*
epi-Truxillic Acid, *Diamide*
Tryptophane, *Benzyl ester*

$C_{18}H_{18}N_2O_3$
5-Hydroxytryptophan, 5-*Benzyl ether*†
7-Hydroxytryptophan, 7-*Benzyl ether*†

$C_{18}H_{18}N_2O_4$
1-(4-Aminophenyl)-2-(2-carboxy-4-nitrophenyl)-ethylene, N-*Di-Me*, *Me ester*
Azobenzene-2,2′-dicarboxylic Acid, *Di-Et ester*
Azobenzene-3,3′-dicarboxylic Acid, *Di-Et ester*
Piperazine-*N*,*N*′-dicarboxylic Acid, *Diphenyl ester*

$C_{18}H_{18}N_2O_5$
Azoxybenzene-2,2′-dicarboxylic Acid, *Di-Et ester*
Azoxybenzene-3,3′-dicarboxylic Acid, *Di-Et ester*

$C_{18}H_{18}N_4$
1,2,4,5-Tetra-aminobenzene, 2,4-N-*Diphenyl*

$C_{18}H_{18}N_8O_5$
Aminopteroylaspartic Acid

$C_{18}H_{18}O$
Oestra-1,3,5(10),6,8-pentaen-17-one
2-Retenol
6-Retenol

$C_{18}H_{18}O_2$
1,5-Dihydroxyanthracene, *Di-Et ether*
1,8-Dihydroxyanthracene, *Di-Et ether*
2,6-Dihydroxyanthracene, *Di-Et ether*
2,7-Dihydroxyanthracene, *Di-Et ether*
9,10-Dihydroxyanthracene, *Di-Et ether*
1,6-Diphenyl-1,5-hexadiene-3,4-diol
2,2-Diphenyl-4-pentenoic Acid, *Me ester*
3,5-Diphenyl-4-pentenoic Acid, *Me ester*
5,5-Diphenyl-4-pentenoic Acid, *Me ester*
3-Hydroxyoestra-1,3,5(10),6,8-pentaene-15-one
Isoequilenin
3-Phenylindane-1-carboxylic Acid, *Et ester*

$C_{18}H_{18}O_2S$
7-Isopropyl-1-methylphenanthrene-2-sulphonic Acid
2-Isopropyl-8-methylphenanthrene-3-sulphonic Acid

$C_{18}H_{18}O_3$
3-Benzoyl-2-benzylpropionic Acid, *Me ester*
3-Benzoyl-2-phenylbutyric Acid, *Me ester*
4-Benzoyl-3-phenylbutyric Acid, *Me ester*
4-Benzoyl-4-phenylbutyric Acid, *Me ester*
2-Benzoyl-3-phenylpropionic Acid, *Et ester*
3-Benzoyl-2-phenylpropionic Acid, *Et ester*
Bibenzyl-2,2′-dicarboxylic Acid, *Di-Me ester*
Bibenzyl-4,4′-dicarboxylic Acid, *Di-Me ester*
3,4-Dihydroxy-9-anthrone, *Di-Et ether*
3-Phenylpropionic Acid, *Anhydride*
1,5,6-Trihydroxyphenanthrene, 1-*Et ether*

$C_{18}H_{18}O_4$
2,2′-Diethoxybenzil
3,3′-Diethoxybenzil
4,4′-Diethoxybenzil
Diphenic Acid, *Di-Et ester*
2,2-Diphenylglutaric Acid, 1-*Me ester*
2,2-Diphenylglutaric Acid, 5-*Me ester*
2,2-Diphenylsuccinic Acid, *Di-Me ester*
2,2-Diphenylsuccinic Acid, α-*Et ester*
2,3-Diphenylsuccinic Acid, *Di-Me ester*
Propylmalonic Acid, *Diphenyl ester*
Succinic Acid, *Di-*o-*tolyl ester*
Succinic Acid, *Di-*p-*tolyl ester*
Succinic Acid, *Di-*m-*tolyl ester*
Succinic Acid, *Dibenzyl ester*
2,3,6,7-Tetrahydroxyanthracene, *Tetra-Me ether*
1,4,5,6-Tetrahydroxyphenanthrene, *Tetra-Me ether*
2,3,5,6-Tetrahydroxyphenanthrene, *Tetra-Me ether*
Thermophillin

$C_{18}H_{18}O_4S_2$
Diphenyl disulphide 2,2′-dicarboxylic Acid, *Di-Et ester*

$C_{18}H_{18}O_5$
3-Acetyl-4,7-dimethylnaphthalene-1,6-dicarboxylic Acid, *Di-Me ester*†
Agrimolide†
Apoalpinone, 2,4-*Di-Me ether*
Butein, 2′,3,4-*Tri-Me ether*
Butin, *Tri-Me ether*
Citronetin, *Di-Me ether*
Dracoic Acid, *Me ester*
2′-Hydroxy-4,4′,6-trimethoxychalcone†
Isosakuranetin, *Di-Me ether*
Isosakuranetin, *Et ether*
Matteucinol
2,2′-Oxydibenzoic Acid, *Di-Et ester*
Phyllodulcin, *Di-Me ether*
Phyllodulcinol, *Di-Me ether*†
Piceatannol†
Pinobanksin, 3,5,7-*Tri-Me ether*
Pterofuran, *Di-Me ether*†
2,2′,3′,4′-Tetrahydroxychalcone, 2,3′,4′-*Tri-Me ether*
2,2′,4,4′,-Tetrahydroxychalcone, 2,4,4′-*Tri-Me ether*
2,2′,4,4′-Tetrahydroxychalcone, 2′,4,4′-*Tri-Me ether*
2,2′,4,5′-Tetrahydroxychalcone, 2,4,5′-*Tri-Me ether*
2,2′,4′,5′-Tetrahydroxychalcone, 2′,4′,5′-*Tri-Me ether*
2,2′,4′,6′-Tetrahydroxychalcone, 2,4′,6′-*Tri-Me ether*
2′,3,3′,4-Tetrahydroxychalcone, *Tri-Me ether*
2′,3,3′,4′-Tetrahydroxychalcone, 3,3′,4′-*Tri-Me ether*
2′,3′,4,4′-Tetrahydroxychalcone, 3′,4,4′-*Tri-Me ether*
2′,4,4′,5′-Tetrahydroxychalcone, 4,4′,5′-*Tri-Me ether*
2′,4,4′,6′-Tetrahydroxychalcone, 4,4′,6′-*Tri-Me ether*
2′,4,4′,6′-Tetrahydroxychalcone, 2′,4′,6′-*Tri-Me ether*
2′,5,7-Trihydroxyflavone, *Tri-Me ether*
3,4,9-Trimethoxypterocarpan†

$C_{18}H_{18}O_6$
Aromadendrin, 5,7,4′-*Tri-Me ether*
Cyanomaclurin★, *Tri-Me ether*†
Eriodictyol, 7,3′,4′-*Tri-Me ether*
Eucomol, 7-*Me ether*†
Fustin, 3′,4′,7-*Tri-Me ether*
Homoeriodictyol, 4′,7-*Di-Me ether*
2-Hydroxyethyl phenyl Ether
Isopedicin
Pedicin
Samaderine A†
Tartaric Acid, *Dibenzyl ester*

$C_{18}H_{18}O_7$
2,3-Dimethoxybenzoic Acid, *Anhydride*
2,4-Dimethoxybenzoic Acid, *Anhydride*
3,4-Dimethoxybenzoic Acid, *Anhydride*
3,5-Dimethoxybenzoic Acid, *Anhydride*
Evernic Acid, *Me ester*
Frenolicin†
Mikanin†
Obtusatic Acid
Sulochrin★, *Mono-Me ether*†

$C_{18}H_{18}O_8$
Crotepoxide†
Leprapric Acid†

$C_{18}H_{18}O_{12}$
Mellitic Acid, *Hexa-Me ester*

$C_{18}H_{19}F_3N_2S$
Fluopromazine

$C_{18}H_{19}NO$
1-*tert*-Butyl-3,3-diphenylaziridinone†
1-Methyl-2,6-diphenylpiperid-4-one†

$C_{18}H_{19}NO_2$
Apocodeine
6-Azaequilenin, *Me ether*†
Floribundine
2-Hydroxy-1-methoxyaporphine†
(—)-*N*-Nornuciferine

$C_{18}H_{19}NO_2S$
2-Isopropyl-8-methylphenanthrene-3-sulphonic Acid, *Amide*

$C_{18}H_{19}NO_3$
Aegeline†
Bervulvine†
Bibenzyl-2,2′-dicarboxylic Acid, *Et ester amide*
Codeinone
Crotsparine, N-*Me*†
1,10-Dihydroxy-2-methoxyaporphine†
Erythraline
Gindaricine
Glaziovine†
Morphothebaine
Northebaine†
Stepharine†
Thebenine
Tuduranine
Vulracin

$C_{18}H_{19}NO_4$
Flavinine†
Hernovine†
Laurelliptine★†
Laurolitsine†
Lindcarpine†
2-Methyl-6-phenylpyridine-3,4-dicarboxylic Acid, *Di-Et ester*
Norsinoacutine†
N-Phenylglycine-*o*-carboxylic Acid, *Di-Me ester*

$C_{18}H_{19}NO_5$
Choisyine†
Evodine
Evoxoidine
Macronin†
Melicopicine†
Poetaminine†

$C_{18}H_{19}NO_6$
Candimine†
Neronine

$C_{18}H_{19}N_3O_4$
Diazoaminobenzene-3,3′-dicarboxylic Acid, *Di-Et ester*

$C_{18}H_{20}$
1,2-Diphenylcyclohexane
1,3-Diphenylcyclohexane
1,4-Diphenylcyclohexane
Methronol
1,2,3,4,7,8,9,10-Octahydrochrysene
[2,4]Paracyclophane†

$C_{18}H_{20}BrNO_2$
Apomorphine, *Methobromide*

$C_{18}H_{20}ClN_3O_5S_2$
Sporidesmin B

$C_{18}H_{20}ClN_3O_6S_3$
Sporidesmin C†
Sporidesmin E†

$C_{18}H_{20}F_{14}O_4$
Heptafluorobutyric Acid, *Decamethylene ester*

$C_{18}H_{20}N_2$
Apparicine†
Guatambuine
N-Methyltetrahydroellipticine
Tabernoschizine

$C_{18}H_{20}N_2O$
1-Methyl-2,6-diphenylpiperid-4-one oxime†

$C_{18}H_{20}N_2O_2$
Methylmalonic Acid, *Di*-p-*toluidide*
Salazine, *Di-Et ether*

$C_{18}H_{20}N_2O_3$
Malic Acid, *Di*-o-*toluidide*

$C_{18}H_{20}N_2O_4$
4,4′-Hydrazodibenzoic Acid, *Di-Et ester*
Quitenol

$C_{18}H_{20}N_2O_8$
β-[3,4-Dihydroxyphenyl]alanine, 5,5′-*Dimer*

$C_{18}H_{20}N_4O_2$
1,4,5,8-Tetra-aminoanthraquinone, 1,4,5,8-N-*Tetra-Me*

$C_{18}H_{20}O$
3,4-Diphenylvaleric Acid, *Me ester*
4,5-Diphenylvaleric Acid, *Me ester*
2,2′,4,5′,6-Pentamethylbenzophenone
2,2′,4,6,6′-Pentamethylbenzophenone
2,3,4,5,6-Pentamethylbenzophenone
2,3′,4,4′,5-Pentamethylbenzophenone
2,3′,4,5′,6-Pentamethylbenzophenone

$C_{18}H_{20}O_2$
5-*tert*-Butyl-2-hydroxybenzaldehyde, *Benzyl ether*
Dibenzylacetic Acid, *Et ester*
Diethylstilboestrol
4,4′-Dihydroxystilbene, *Di-Et ether*
2,4-Diphenylbutyric Acid★, *Et ester*†
Di-*p*-tolyacetic Acid, *Et ester*
3-Hydroxyoestra-1,3,5(10),7-tetraen-17-one

$C_{18}H_{20}O_3$
Dibenzylglycollic Acid, *Et ester*
3,4-Di[-*p*-hydroxyphenyl]-2-hexanone
Gibberic Acid†
Hydroxy-di-*o*-tolylacetic Acid, *Et ester*

$C_{18}H_{20}O_4$
Angolensin, 4-O-*Et*
3-(4-Hydroxy-2,3-dimethoxyphenyl)-1-(2-methoxyphenyl)propene†
α-Hydroxy-4-methoxy-α-(4-methoxyphenyl)-acetophenone, *Et ether*

$C_{18}H_{20}O_4$ *(continued)*
- Piceatannol, *Tetra*-O-*Me*
- 3,3′,4,5′-Tetrahydroxystilbene, *Tetra-Me ether* †
- 4′,5,7-Trimethoxyflavan †

$C_{18}H_{20}O_5$
- 3,4-Dimethoxy-α-(3,4-dimethoxyphenyl)-acetophenone
- 1-(2-Hydroxyphenyl)-3-(5-hydroxy-2,3,4-trimethoxyphenyl)propene †
- Kuhlmanniquinol †
- Phloretin, 2′,4,4′-*Tri-Me ether*
- 3,3′,4′,7-Tetrahydroxyflavan, *Tri-Me ether* †

$C_{18}H_{20}O_6$
- Byssochlamic Acid
- 3′,7-Dihydroxy-2′,4′,8-trimethoxyisoflavan †
- Glaucanic Acid
- Maclurin, *Penta-Me ether*
- Nagilactone D †
- 2,2′,4,4′,6-Pentahydroxybenzophenone, *Penta-Me ether*
- 2,3′,4,5′,6-Pentahydroxybenzophenone, *Penta-Me ether*
- 3,3′,4,4′,5-Pentahydroxybenzophenone, *Penta-Me ether*
- 3,3′,4,4′,7-Pentahydroxyflavan, 3′,4′,7-*Tri-Me ether*
- 3,4,4′,7,8-Pentahydroxyflavan★, 4′,7,8-*Tri-Me ether* †

$C_{18}H_{20}O_7$
- Glauconic Acid

$C_{18}H_{20}O_8$
- Inumakilactone A †
- Xanthophanic Acid, *Di-Et ester* †

$C_{18}H_{21}ClN_2$
- 1-(*p*-Chloro-α-phenylbenzyl)-4-methylpiperazine

$C_{18}H_{21}N$
- 6-Isopropyl-1,2,8-trimethyl-1-*H*-cyclohepta[*d,e*]-1-pyrindine

$C_{18}H_{21}NO$
- α,α-Diphenyl-2-piperidinemethanol
- α,α-Diphenyl-3-piperidinemethanol
- α,α-Diphenyl-4-piperidinemethanol

$C_{18}H_{21}NO_2$
- Phthalimide, N-α-*Bornyl*

$C_{18}H_{21}NO_3$
- Allo-ψ-codeine
- Burmannine
- Codeine
- Dihydrocodeinone
- Erysocine
- Erysodine
- Erysovine
- Erysovine
- Erythramine
- Isocodeine
- Linearisine †
- Metopon
- Neopine
- (−)-*N*-Norarmepavine
- Oripavine
- Sendaverine †
- Thebainone A
- β-Thebainone A
- Thebainone-B
- Thebainone C

$C_{18}H_{21}NO_4$
- Buphanidrin
- Cephalotaxine †
- 8,14-Dihydronorsalutaridine †
- Erythratine
- Haplophyllidine
- Manthine †
- Narcipoetine★ †
- Oxycodone

$C_{18}H_{21}NO_5$
- Ambellin
- Crispanin
- Criucelline
- Criwelline †
- Krigenamine
- Pretazettine †
- Tazettine

$C_{18}H_{21}NO_6$
- Evolatine
- Evoxine
- Krigeine
- Macranthin, O-*Ac* †

$C_{18}H_{21}N_3O_2$
- Lysergic Acid, α-Hydroxyethylamide

$C_{18}H_{21}N_3O_6$
- 10-Deazariboflavin †

$C_{18}H_{21}N_3O_7S$
- Ascarylose, pp′-*Nitrobiphenylsulphonylhydrazone*

$C_{18}H_{22}$
- 2,2′,4,4′,5,5′-Hexamethylbiphenyl
- 2,2′,4,4′,6,6′-Hexamethylbiphenyl
- 3,3′,4,4′,5,5′-Hexamethylbiphenyl

$C_{18}H_{22}ClNO$
- Chlorphenoxamine
- Dibenzyline

$C_{18}H_{22}ClNO_6$
- Acutumidine †

$C_{18}H_{22}ClN_3O_6S_2$
- Sporidesmin

$C_{18}H_{22}NO$ (ion)
- *N*-(*p*-Hydroxyphenethyl)actinidine †

$C_{18}H_{22}NO_8$ (ion)
- Xanthofagarine

$C_{18}H_{22}N_2$
- Aspidospermatidine †
- Azocumene †
- Condyfoline †
- Cyclizine
- 4 4′-Di-isopropylazobenzene
- 2,2′,4,4′,5,5′-Hexamethylazobenzene
- Piperazine, N,N-*Dibenzyl*
- Piperazine, N,N-*Di*-o-*tolyl*
- Piperazine, N,N-*Di*-m-*tolyl*
- Piperazine, N,N-*Di*-p-*tolyl*
- Propanal, *Anil, Dimer*

$C_{18}H_{22}N_2$ *(continued)*
Tubifoline†
Uleine
3-*epi*-Uleine†

$C_{18}H_{22}N_2O_2$
Holocain

$C_{18}H_{22}N_2S$
10-(2′-Diethylaminoethyl)phenothiazine
Trimeprazine

$C_{18}H_{22}N_4O_5$
Galactose formazan
Glucose formazan

$C_{18}H_{22}O_2$
Dihydrodiethylstilboestrol
Isoestrone
Lumiestrone
Oestrone
16-Oxo-octadeca-9,17-diene-12,14-diyn-1-al†

$C_{18}H_{22}O_3$
Diphenylglycollaldehyde, *Di-Et acetal*
6α-Hydroxyoestrone†
6β-Hydroxyoestrone†
18-Hydroxyoestrone†
Methallenoestril

$C_{18}H_{22}O_4$
Isoborneol, *Hydrogen phthalate*

$C_{18}H_{22}O_5$
Marrianolic Acid
Zearalenone†

$C_{18}H_{22}O_6$
2,2′,3,3′,4,4′-Hexahydroxybiphenyl, *Hexa-Me ether*
2,2′,3,3′,5,5′-Hexahydroxybiphenyl, *Hexa-Me ether*
2,2′,4,4′,6,6′-Hexahydroxybiphenyl, *Hexa-Me ether*
2,3,3′,4,4′,5′-Hexahydroxybiphenyl, *Hexa-Me ether*
3,3′,4,4′,5,5′-Hexahydroxybiphenyl, *Hexa-Me ether*

$C_{18}H_{22}O_8$
Benzene-1,2,4,5-tetracarboxylic Acid, *Tetra-Et ester*
Loroglossigenin

$C_{18}H_{22}O_{10}$
3,6-Dihydroxybenzene-1,2,4,5-tetracarboxylic Acid, *Tetra-Et ester*

$C_{18}H_{22}O_{10}S$
Paederoside†

$C_{18}H_{22}O_{11}$
Asperuloside★†

$C_{18}H_{23}BrO_3$
9-Decenoic Acid, p-*Bromophenacyl ester*

$C_{18}H_{23}N$
Di-1-phenylpropylamine
Di-3-phenylpropylamine

$C_{18}H_{23}NO$
4-Hydroxy-2-non-1-enylquinoline†
Orphenadrine

$C_{18}H_{23}NO_2$
6-Azaoestrone, *Me ether*†
8-Azaoestrone, *Me ether*†
3-Carboxy-2,2,3-trimethylcyclopentylacetonitrile, *Benzyl ester*
Cocculidine
Oestrone, *Oxime*

$C_{18}H_{23}NO_3$
Dihydrocodeine†
Isoxsuprine
Neolitsinine†
Norpluviine, *Et ether*†
Pulchelline†
Thebainol

$C_{18}H_{23}NO_4$
Amaryllisine†
Lycorenine

$C_{18}H_{23}NO_5$
Coruscine
Galanthamidine
Narcissidine†
Spartioidine

$C_{18}H_{23}NO_6$
Crispine
Jacozine†
3-Nitrophthalic Acid, 2-(−)-*Menthyl ester*
Riddelliine

$C_{18}H_{23}N_5O_6S$
6-(4-Guanidino-2,6-dimethoxybenzamido)-penicillanic Acid†

$C_{18}H_{24}$
Octamethylnaphthalene†
Triamantane†

$C_{18}H_{24}BrNO_3S$
Bretylium tosylate

$C_{18}H_{24}ClN_5O_3$
Akrinor†

$C_{18}H_{24}ClN_5O_5$
Akrinor†

$C_{18}H_{24}N_2$
Tubifolidine†

$C_{18}H_{24}N_2O$
1,13-Dihydro-13-hydroxyuleine†
Di-*p*-dimethylaminophenylmethanol, *Me ether*

$C_{18}H_{24}N_2O_3$
Julocrotine

$C_{18}H_{24}N_4O_{18}$
Endecaphyllin X

$C_{18}H_{24}O$
Bakuchiol†

$C_{18}H_{24}O_2$
9-Heptadecenoic Acid, *Me ester*
Nimbiol
17α-Oestradiol
17β-Oestradiol

$C_{18}H_{24}O_2S$
3,6-Di-*tert*-butylnaphthalene-1-sulphonic Acid†
3,7-Di-*tert*-butylnaphthalene-1-sulphonic Acid†

$C_{18}H_{24}O_3$
Doisynolic Acid, *Me ester*
Oestriol
16-*epi*-Oestriol
6-Oxa-oestrone, *Me ether*†
6-Oxa-8α-oestrone, *Me ether*†
Podocarpic Acid, *Me ester*
Podocarpic Acid, *Me ether*

$C_{18}H_{24}O_4$
Oestra-1,3,5(10)-triene-3,15α,16α,17β-tetrol†

$C_{18}H_{24}O_5$
Streptolic Acid†

$C_{18}H_{24}O_6$
2-Phenyltricarballylic Acid, *Tri-Et ester*

$C_{18}H_{24}O_7$
3-Hydroxytoluene-2,4,6-tricarboxylic Acid, *Et ether*, *Tri-Et ester*
Renifolin†

$C_{18}H_{24}O_8$
Hyptolide†

$C_{18}H_{24}O_{11}S$
Paederosidic Acid†

$C_{18}H_{25}NO$
Anacyclin
Dextromethorphan
4-Hydroxy-2-nonylquinoline†

$C_{18}H_{25}NO_2$
Allylprodine
Isolobinine
Lobinine
Lobinone

$C_{18}H_{25}NO_3$
Oreoline, O-*Me ether*†
Oreoline, N-*Me deriv.*†

$C_{20}H_{25}NO_4$
Porphyroxine, *Me ether*

$C_{18}H_{25}NO_5$
Aquaticine
Integerrimine
Renardine
Senecionine†
Seneciphylline

$C_{18}H_{25}NO_6$
Anacrotine†
Indican, *Tetra-Me ether*
Jacobine
Madurensine†
Mucronatinine†
Retrorsine†
Usaramine†

$C_{18}H_{25}NO_7$
Isatidine
Retrorsine *N*-Oxide†

$C_{18}H_{26}$
2,6-Di-*tert*-butylnaphthalene†
2,7-Di-*tert*-butylnaphthalene†

$C_{18}H_{26}ClNO_6$
Chlorodeoxysceleratine†
Jaconine

$C_{18}H_{26}ClN_3$
Chloroquine

$C_{18}H_{26}ClN_3O$
7-Chloro-4-[4-(*N*-ethyl-*N*-2-hydroxyethyl-amino)-1-methylbutylamino]quinoline

$C_{18}H_{26}N_2O_5$
Leucylglycine, N-*Carbobenzyloxy*, *Et ester*

$C_{18}H_{26}N_{10}O_3$
Netropsin

$C_{18}H_{26}O_2$
Erythrogenic Acid
13-Octadecen-9,11-diynoic Acid
o-Toluic Acid, *Menthyl ester*
m-Toluic Acid, *Menthyl ester*
p-Toluic Acid, *Menthyl ester*

$C_{18}H_{26}O_3$
Isanolic Acid
Shogaol, *Me ether*
2,3,3-Trimethyl-1-cyclopentene-1-carboxylic Acid, *Anhydride*
Trisporic B Acid†

$C_{18}H_{26}O_4$
Benzylmalonic Acid, *Di-butyl ester*
4,6-Diacetylresorcinol, *Dibutyl ether*
Santonic Acid, *Propyl ester*
Trisporic C Acid†

$C_{18}H_{27}NO_2$
Caramiphen
Lobinanidine

$C_{18}H_{27}NO_3$
Capsaicin
Lycofoline, 7-*Mono-Ac*

$C_{18}H_{27}NO_4$
Dodecanoic Acid, 9-*Nitrophenyl ester*
3α,6β-Tropanediol, *Di-tigloyl ester*

$C_{18}H_{27}NO_5$
Platyphylline
Teloidine, 3,6-*Ditigloyl ester*

$C_{18}H_{27}NO_6$
Hydrophylline†
Rosmarinine

$C_{18}H_{27}NO_7$
Axillarin†
Jacoline†
Sceleratine

$C_{18}H_{27}NO_8$
Senecifoline

$C_{18}H_{27}N_3O$
8-(5-Isopropylaminopentylamino)-6-methoxy-quinoline

$C_{18}H_{27}N_3O_6$
Triazine-tricarboxylic Acid, *Tri-isobutyl ester*

$C_{18}H_{28}$
Bicyclo[12,2,2]octadeca-14,16,18-triene

$C_{18}H_{28}ClNO$
Chlorphenoyolan†

$C_{18}H_{28}N_2$
2-Ethyl-3-[2-(3-ethylpiperidino)ethyl]indole†

$C_{18}H_{28}N_2O$
Flabellidine†
Flabelline

$C_{18}H_{28}N_2O_2$
Jaborandine (or $C_{18}H_{30}N_2O_2$)
Serratinidine†

$C_{18}H_{28}N_2O_3$
Lycocernuine, O-*Ac*†

$C_{18}H_{28}O$
Dodecanophenone

$C_{18}H_{28}O_2$
Dodecanoic Acid, *Phenyl ester*
Heptadeca-10,16-dien-8-ynoic Acid, *Me ester*†
Octadeca-11,13-dien-9-ynoic Acid
Octadeca-11,17-dien-9-ynoic Acid†
Octadeca-9,14-dien-12-ynoic Acid†
3,9,12,15-Octadecatetraenoic Acid†
6,9,12,15-Octadecatetraenoic Acid†
9,11,13,15-Octadecatetraenoic Acid
6,9,12,15-Octadecatetraenoic Acid
10-Phenyldecanoic Acid, *Et ester*

$C_{18}H_{28}O_3$
α-Couepic Acid
β-Couepic Acid
7-Hydroxyheptadeca-10,16-dien-8-ynoic Acid, *Me ester*†
8-Hydroxyoctadeca-11,17-dien-9-ynoic Acid†

$C_{18}H_{28}O_3S$
1,2-Methylenedioxy-4-octylsulphinylpropylbenzene

$C_{18}H_{28}O_4$
5-Acetoxymethyltetradeca-*trans*-2-*trans*-4-*trans*-6-trienoic Acid, *Me ester*
Gingerol, *Me ether*

$C_{18}H_{28}O_7$
Salicin, *Penta-Me ether*

$C_{18}H_{28}O_{11}$
Lamioside†

$C_{18}H_{29}NO_2$
o-(*N*-Di-isopropylaminoethoxy)butyrophenone
Lobinol
N-Phenylglycine, 3-*Methylbutyl ester*, N-*Isoamyl*
Thelepogidine

$C_{18}H_{29}NO_3$
Amprotropine
Fawcettiine
α-Lofoline
Lycoclavine
p-Pentoxybenzoic Acid, *Diethylaminoethyl ester*
p-Propoxybenzoic Acid, *Dipropylaminoethyl ester*

$C_{18}H_{29}NO_4$
Lycofawcine†

$C_{18}H_{30}$
Dodecylbenzene
Hexadecahydrotriphenylene†
Hexa-ethylbenzene
1,3,5-Tri-*tert*-butylbenzene†

$C_{18}H_{30}BrNO_3S$
Penthienate, *Monodral*

$C_{18}H_{30}Br_6O_2$
9,10,12,13,15,16-Hexabromo-octadecanoic Acid

$C_{18}H_{30}FeN_3O_6$
Pulcherrimin

$C_{18}H_{30}N_2$
Dodecahydrobenzo[*ij,i'j'*]diquinolizine

$C_{18}H_{30}N_2O_2$
Butacaine
Butyn
Jaborandine (or $C_{18}H_{28}N_2O_2$)

$C_{18}H_{30}N_2O_6$
Kafiroic Acid

$C_{18}H_{30}O_2$
Gorlic Acid
Heptadec-10-en-8-ynoic Acid, *Me ester*†
α-Kamlolenic Acid
β-Kamlolenic Acid
11-(2-Methylcyclohex-2-enyl)undec-9-enoic Acid†
5,9,12-Octadecatrienoic Acid
6,9,12-Octadecatrienoic Acid
6,10,14-Octadecatrienoic Acid
8,10,12-Octadecatrienoic Acid
9,11,13-Octadecatrienoic Acid
9,12,15-Octadecatrienoic Acid
10,12,14-Octadecatrienoic Acid
11-Octadecen-9-ynoic Acid
Octadec-17-en-9-ynoic Acid†
9-Octadecen-12-ynoic Acid†

$C_{18}H_{30}O_3$
Coronaric Acid
15,16-Epoxyoctadeca-9,12-dienoic Acid†
8-[2-(5-Hexylfuryl)]-octanoic Acid†
9-Hydroxyoctadec-10-ene-12-ynoic Acid†
Juvenile Hormone†

$C_{18}H_{30}O_3S$
o-Dodecylbenzenesulphonic Acid
m-Dodecylbenzenesulphonic Acid
p-Dodecylbenzenesulphonic Acid

$C_{18}H_{30}O_8$
Hexane-1,1,3,3-tetracarboxylic Acid, *Tetra-Et ester*
Hexane-1,1,6,6-tetracarboxylic Acid, *Tetra-Et ester*
Hexane-2,2,5,5-tetracarboxylic Acid, *Tetra-Et ester*

$C_{18}H_{31}ClO$
Chaulmoogric Acid, *Chloride*
9,12-Octadecadienoic Acid, *Chloride*
9-Octadecynoic Acid, *Chloride*

$C_{18}H_{31}N$
Chaulmoogric Acid, *Nitrile*

$C_{18}H_{31}NO$
6-Aza-5-ζ-androstan-17β-ol
Gorlic Acid, *Amide*
11-Octadecen-9-ynoic Acid, *Amide*
Solanidine-*S*

$C_{18}H_{32}$
1,3-Cyclo-octadecadiene
$C_{18}H_{32}Br_4O_2$
6,6,7,7-Tetrabromo-octadecanoic Acid
9,9,10,10-Tetrabromo-octadecanoic Acid
9,10,12,13-Tetrabromo-octadecanoic Acid
$C_{18}H_{32}N_4O_6$
Cytosamine
$C_{18}H_{32}O_2$
Chaulmoogric Acid
1,10-Cyclo-octadecanedione
Hydnocarpic Acid, *Et ester*
Malvalic Acid†
5,6-Octadecadienoic Acid†
6,8-Octadecadienoic Acid
8,10-Octadecadienoic Acid
8,11-Octadecadienoic Acid
9,11-Octadecadienoic Acid
5,12-Octadecadienoic Acid
9,12-Octadecadienoic Acid
10,12-Octadecadienoic Acid
10,13-Octadecadienoic Acid
11,14-Octadecadienoic Acid
6-Octadecynoic Acid
9-Octadecynoic Acid
11-Octadecynoic Acid
Propylure†
Vernolic Acid
$C_{18}H_{32}O_3$
9-Hydroxy-10,12-octadecadienoic Acid†
13-Hydroxy-9,11-octadecadienoic Acid†
12-Hydroxy-9,15-octadecadienoic Acid
12-Hydroxy-9-octadecynoic Acid
Pedicellic Acid, *Anhydride*†
$C_{18}H_{32}O_4$
6,7-Dioxo-octadecanoic Acid
9,10-Dioxo-octadecanoic Acid
$C_{18}H_{32}O_4Si$
Diallyl dicyclohexyl orthosilicate
$C_{18}H_{32}O_6$
Gentianose
Tri-isovalerin
$C_{18}H_{32}O_{14}$
O-α-D-Mannopyranosyl (1 → 6)-*O*-α-D-manno-pyranosyl (1 → 1)-1-*R*-*myo*-inositol
Rhamninose
$C_{18}H_{32}O_{15}$
Solatriose†
$C_{18}H_{32}O_{16}$
Cellotriose
Dectrantriose
4,6-Di-*O*-(α-D-glucopyranosyl)-D-gluco-pyranose†
Gentiotriose
Kestose
Lactulosucrose†
Lycotriose I
α-Maltosylfructose
Maltotriose
Melezitose
Panose
Raffinose

$C_{18}H_{33}ClO$
6-Octadecenoic Acid, *Chloride*
9-Octadecenoic Acid, *Chloride*
$C_{18}H_{33}FO_2$
18-Fluoro-9-octadecenoic Acid
$C_{18}H_{33}N$
9-Octadecenoic Acid, *Nitrile*
$C_{18}H_{33}NO$
Chaulmoogric Acid, *Amide*
9,12-Octadecadienoic Acid, *Amide*
Tetradeca-2,4-dienoic Acid, *N*-isobutylamide†
$C_{18}H_{33}N_3O_3$
Cyanuric Acid, O-*Tri*-n-*pentyl ester*
$C_{18}H_{34}$
1-Octadecyne
$C_{18}H_{34}Br_2O_2$
2,3-Dibromo-octadecanoic Acid
3,4-Dibromo-octadecanoic Acid
4,5-Dibromo-octadecanoic Acid
9,10-Dibromo-octadecanoic Acid
$C_{18}H_{34}Cl_2O_2$
2,2-Dichloro-octadecanoic Acid
6,7-Dichloro-octadecanoic Acid
9,10-Dichloro-octadecanoic Acid
$C_{18}H_{34}N_2O_6S$
Antibiotic U-20,943†
Lincomycin†
$C_{18}H_{34}O$
Chaulmoogryl Alcohol
Cyclo-octadecanone
2-Methylcycloheptadecanone
9,12-Octadecadien-1-ol
9-Octadecenal
$C_{18}H_{34}O_2$
Cycloheptadecane-carboxylic Acid
11-Cyclohexylundecanoic Acid, *Me ester*†
Glenocardic Acid
Hexadecanoic Acid, *Vinyl ester*
2-Hexadecenoic Acid, *Et ester*
9-Hexadecenoic Acid, *Et ester*
4-Hydroxyoctadecanoic Acid, *Lactone*
2-Methyl-16-heptadecenoic Acid
2-Octadecenoic Acid
3-Octadecenoic Acid
4-Octadecenoic Acid
5-Octadecenoic Acid
6-Octadecenoic Acid
7-Octadecenoic Acid
8-Octadecenoic Acid
9-Octadecenoic Acid
10-Octadecenoic Acid
11-Octadecenoic Acid
12-Octadecenoic Acid
15-Octadecenoic Acid
16-Octadecenoic Acid
17-Octadecenoic Acid
$C_{18}H_{34}O_3$
9,10-Epoxyoctadecanoic Acid†
5-Hydroxy-2-octadecenoic Acid
10-Hydroxy-8-octadecenoic Acid
8-Hydroxy-9-octadecenoic Acid
12-Hydroxy-9-octadecenoic Acid

$C_{18}H_{34}O_3$ (*continued*)
- 17-Hydroxy-9-octadecenoic Acid
- 18-Hydroxy-9-octadecenoic Acid
- 9-Hydroxy-12-octadecenoic Acid
- 2-Methyl-4-oxoheptadecanoic Acid
- 16-Methyl-10-oxoheptadecanoic Acid
- 9-Oxoheptadecanoic Acid, *Me ester*
- 3-Oxohexadecanoic Acid, *Et ester*
- 3-Oxo-octadecanoic Acid
- 4-Oxo-octadecanoic Acid
- 5-Oxo-octadecanoic Acid
- 6-Oxo-octadecanoic Acid
- 7-Oxo-octadecanoic Acid
- 9-Oxo-octadecanoic Acid
- 10-Oxo-octadecanoic Acid
- 11-Oxo-octadecanoic Acid
- 12-Oxo-octadecanoic Acid

$C_{18}H_{34}O_4$
- Decanedioic Acid, *Dibutyl ester*
- 12,13-Dihydroxy-9-octadecenoic Acid
- 9,10-Dihydroxy-12-octadecenoic Acid
- Hexadecanedioic Acid, *Di-Me ester*
- Hexadecanedioic Acid, *Mono-Et ester*
- Oxalic Acid, *Di-nonyl ester*
- Pedicellic Acid†
- Succinic Acid, *Diheptyl ester*
- Tetradecanedioic Acid, *Di-Et ester*

$C_{18}H_{34}O_6$
- Phloionic Acid

$C_{18}H_{34}O_{11}$
- Cellobiose, *Hexa-Me ether*

$C_{18}H_{34}O_{16}$
- Clavicepsin
- D-Mannitol-1,6-(β-D-glucoside)

$C_{18}H_{35}BrO_2$
- 2-Bromohexadecanoic Acid, *Et ester*
- 2-Bromo-octadecanoic Acid
- 18-Bromo-octadecanoic Acid
- Hexadecanoic Acid, 2-*Bromoethyl ester*

$C_{18}H_{35}ClO$
- Octadecanoic Acid, *Chloride*

$C_{18}H_{35}ClO_2$
- Hexadecanoic Acid, 2-*Chloroethyl ester*

$C_{18}H_{35}FO_2$
- 18-Fluoro-octadecanoic Acid

$C_{18}H_{35}IO_2$
- Hexadecanoic Acid, 2-*Iodoethyl ester*
- 2-Iodo-octadecanoic Acid
- 3-Iodo-octadecanoic Acid
- 4-Iodo-octadecanoic Acid
- 10-Iodo-octadecanoic Acid
- 11-Iodo-octadecanoic Acid

$C_{18}H_{35}N$
- Octadecanonitrile

$C_{18}H_{35}NO$
- 2-Hydroxyoctadecanoic Acid, *Nitrile*
- 2-Octadecenoic Acid, *Amide*
- 6-Octadecenoic Acid, *Amide*
- 9-Octadecenoic Acid, *Amide*
- 10-Octadecenoic Acid, *Amide*

$C_{18}H_{35}NO_2$
- Cassine†
- 12-Hydroxy-9-octadecenoic Acid, *Amide*

$C_{18}H_{35}NO_2$
- 9-Oxo-octadecanoic Acid, *Amide*
- 10-Oxo-octadecanoic Acid, *Amide*

$C_{18}H_{35}NO_3$
- *N*-Hexadecanoylglycine

$C_{18}H_{35}NO_{10}S_3$
- Glucocamelinin

$C_{18}H_{36}$
- 2-Methyl-2-heptadecene
- 1-Octadecene

$C_{18}H_{36}Br_2$
- 1,2-Dibromo-octadecane
- 1,18-Dibromo-octadecane

$C_{18}H_{36}INO$
- 2-Iodo-octadecanoic Acid, *Amide*

$C_{18}H_{36}N_2O_2$
- *N*-Hexadecanoylglycine, *Amide*

$C_{18}H_{36}N_2O_3$
- Allophanic Acid, *Cetyl ester*
- 1-Hexadecanol, *Allophanate*

$C_{18}H_{36}N_2S_2$
- Bis-2,2,6,6-tetramethyl-1-piperidyl disulphide†

$C_{18}H_{36}N_4O_{10}$
- Gentiomycin A†

$C_{18}H_{36}N_4O_{11}$
- Kanamycin
- Kanamycin C†

$C_{18}H_{36}N_{10}O_5$
- L-Alanyl-L-arginyl-L-alanyl-L-arginine
- L-Arginyl-L-alanyl-L-arginyl-L-alanine
- L-Arginyl-D-alanyl-L-arginyl-D-alanine

$C_{18}H_{36}O$
- Cyclo-octadecanol
- Octadecanal
- 2-Octadecanone
- 3-Octadecanone
- 9-Octadecen-1-ol
- Octadec-11-en-1-ol†

$C_{18}H_{36}O_2$
- Hexadecanoic Acid, *Et ester*
- Hexadecyl acetate
- Heptadecanoic Acid, *Me ester*
- 2-Methylheptadecanoic Acid
- 16-Methylheptadecanoic Acid
- Octadecanoic Acid
- Tetradecanoic Acid, *Butyl ester*

$C_{18}H_{36}O_2S$
- 2-Mercapto-octadecanoic Acid

$C_{18}H_{36}O_3$
- Hexadecanoic Acid, *Ethylene glycol mono-ester*
- 2-Hydroxyhexadecanoic Acid, *Et ester*
- 2-Hydroxyhexadecanoic Acid, *Et ether*
- 11-Hydroxyhexadecanoic Acid, *Et ester*
- 2-Hydroxyoctadecanoic Acid
- 3-Hydroxyoctadecanoic Acid
- 4-Hydroxyoctadecanoic Acid

$C_{18}H_{36}O_3$ *(continued)*
5-Hydroxyoctadecanoic Acid
6-Hydroxyoctadecanoic Acid
7-Hydroxyoctadecanoic Acid
8-Hydroxyoctadecanoic Acid
9-Hydroxyoctadecanoic Acid
10-Hydroxyoctadecanoic Acid
11-Hydroxyoctadecanoic Acid
12-Hydroxyoctadecanoic Acid
13-Hydroxyoctadecanoic Acid
14-Hydroxyoctadecanoic Acid
15-Hydroxyoctadecanoic Acid
16-Hydroxyoctadecanoic Acid
17-Hydroxyoctadecanoic Acid
18-Hydroxyoctadecanoic Acid
2-Hydroxypentadecanoic Acid, *Me ether, Et ester*
Trichocarpinic Acid

$C_{18}H_{36}O_4$
3,12-Dihydroxyhexadecanoic Acid, *Et ester*
2,3-Dihydroxyoctadecanoic Acid
5,6-Dihydroxyoctadecanoic Acid
6,7-Dihydroxyoctadecanoic Acid
7,8-Dihydroxyoctadecanoic Acid
8,9-Dihydroxyoctadecanoic Acid
9,10-Dihydroxyoctadecanoic Acid
9,12-Dihydroxyoctadecanoic Acid
10,11-Dihydroxyoctadecanoic Acid
11,12-Dihydroxyoctadecanoic Acid
12,13-Dihydroxyoctadecanoic Acid
15,16-Dihydroxyoctadecanoic Acid†

$C_{18}H_{36}O_5$
Phloionolic Acid
9,10,18-Trihydroxyoctadecanoic Acid†

$C_{18}H_{36}O_6$
9,10,12,13-Tetrahydroxyoctadecanoic Acid

$C_{18}H_{36}O_8$
8,9,11,12,14,15-Hexahydroxyoctadecanoic Acid

$C_{18}H_{37}Br$
1-Bromo-octadecane

$C_{18}H_{37}Cl$
1-Chloro-octadecane

$C_{18}H_{37}I$
1-Iodo-octadecane

$C_{18}H_{37}N$
Octadecamethyleneimine

$C_{18}H_{37}NO$
Octadecanoamide

$C_{18}H_{37}NO_2$
2-Amino-octadecanoic Acid
12-Amino-octadecanoic Acid
Carnavaline†
Dodecanoic Acid, 2-*Diethylaminoethyl ester*
2-Hydroxyoctadecanoic Acid, *Amide*
Sphingosine

$C_{18}H_{37}N_5O_{10}$
Kanamycin B†

$C_{18}H_{38}$
2-Methylheptadecane
9-Methylheptadecane
Octadecane

$C_{18}H_{38}O$
1-Octadecanol
3-Octadecanol

$C_{18}H_{38}O_2$
1,2-Octadecanediol
1,18-Octadecanediol
1,8-Octanediol, *Di-pentyl ether*

$C_{18}H_{38}O_4S$
Octadecylsulphuric Acid†

$C_{18}H_{39}N$
1-Amino-octadecane

$C_{18}H_{39}NO_3$
Cerebrin base

$C_{18}H_{40}Br_2N_4O_4$
Carbolonium bromide

$C_{18}H_{40}N_2$
1,2-Diamino-octadecane
1,18-Diamino-octadecane
7,12-Diamino-octadecane

$C_{18}H_{48}O_6Si_6$
1,3,5,7,9,11-Hexa-ethyl-1,3,5,7,9,11-hexamethylcyclohexasiloxane

$C_{18}H_{54}Fe_2O_6Si_6$
Tris-trimethylsiloxyiron

C_{19}

$C_{19}HF_{15}$
Tris(pentafluorophenyl)methane†

$C_{19}HF_{15}O$
Tris(pentafluorophenyl)methanol†

$C_{19}H_{10}O_4$
Sesquixanthydrol†

$C_{19}H_{10}N_6O_{12}$
Tri-(2,4-dinitrophenyl)methane

$C_{19}H_{11}$
Benzo[*cd*]pyrenium (ion)

$C_{19}H_{11}N$
Benz[*a*]anthracene-7-carboxylic Acid, *Nitrile*

$C_{19}H_{11}N_3O_{11}$
4-Nitro-2-sulphobenzoic Acid, *Di-*o*-nitrophenyl ester*

$C_{19}H_{12}$
Benz[*bc*]aceanthrylene

$C_{19}H_{12}N_2$
Benzimidazo[1,2-*f*]phenanthridine

$C_{19}H_{12}N_2O$
Di-2-quinolyl Ketone
Di-6-quinolyl Ketone

$C_{19}H_{12}O$
Benzylideneacenaphthenone

$C_{19}H_{12}O_2$
Benz[*a*]anthracene-4-carboxylic Acid
Benz[*a*]anthracene-7-carboxylic Acid
5,6-Benzoflavone
6,7-Benzoflavone

$C_{19}H_{12}O_2$ *(continued)*
7,8-Benzoflavone
1-Methylbenz[*a*]anthracene-7,12-dione
2-Methylbenz[*a*]anthracene-7,12-dione
3-Methylbenz[*a*]anthracene-7,12-dione
4-Methylbenz[*a*]anthracene-7,12-dione
6-Methylbenz[*a*]anthracene-7,12-dione
8-Methylbenz[*a*]anthracene-7,12-dione
9-Methylbenz[*a*]anthracene-7,12-dione
10-Methylbenz[*a*]anthracene-7,12-dione
11-Methylbenz[*a*]anthracene-7,12-dione

$C_{19}H_{12}O_3$
5,6-Benzoflavonol
7,8-Benzoflavonol
7,12-Dihydro-7,12-dioxopleiaden-1-ol, *Me ether*
7-Oxo-7*H*-benz[*de*]anthracene-11-carboxylic Acid, *Me ester*
Resorcin-benzein

$C_{19}H_{12}O_4$
Phlebiarubrone
Tetrangulol †

$C_{19}H_{12}O_6$
Dehydroneotenone †
Dolineone
Gamatin
3,3′-Methylene*bis*-4-hydroxycoumarin
Pachyrrhizin
Pongapin

$C_{19}H_{12}O_7$
Coumestrol, *Di-O-Ac*
Daphnoretin

$C_{19}H_{13}BrOS_4$
3-Benzoyl-5-*p*-bromophenyl-2-methylthio-6*a*-thiathiophthen †

$C_{19}H_{13}N$
6-Azacholanthrene
Azatriptyrene †
2-Phenylacridine
9-Phenylacridine
6-Phenylphenanthridine

$C_{19}H_{13}NO$
Acridone, N-*Phenyl*
N-Phenylacridone

$C_{19}H_{13}NO_4$
4-Hydroxy-3′-nitrobenzophenone, *Phenyl ether*
4-Hydroxy-4′-nitrobenzophenone, *Phenyl ether*
Pyridine-2,4-dicarboxylic Acid, *Di-phenyl ester*
Pyridine-2,5-dicarboxylic Acid, *Di-phenyl ester*
Pyridine-2,6-dicarboxylic Acid, *Di-phenyl ester*

$C_{19}H_{13}NO_5$
1,2-Dimethoxy-9,10-methylenedioxy-7-oxodibenzo[*de,g*]quinoline †

$C_{19}H_{13}NO_6$
1,3,10,11,12-Pentahydroxynaphthacene-2-carboximide

$C_{19}H_{13}NO_7S$
4-Nitro-2-sulphobenzoic Acid, *Diphenyl ester*

$C_{19}H_{13}NS$
9-Acridinethiol, N-*Phenyl*

$C_{19}H_{13}N_3O_5$
4-Hydroxy-4′-nitroazobenzene-3-carboxylic Acid, *Phenyl ester*

$C_{19}H_{13}N_3O_6$
Tri-(4-nitrophenyl)methane

$C_{19}H_{13}N_3O_7$
Tri-(4-nitrophenyl)methanol

$C_{19}H_{13}P$
10-Phenyldibenzo[*b,e*]phosphorin †

$C_{19}H_{14}$
1-Methylbenz[α]anthracene
2-Methylbenz[α]anthracene
3-Methylbenz[α]anthracene
4-Methylbenz[α]anthracene
5-Methylbenz[α]anthracene
6-Methylbenz[α]anthracene
7-Methylbenz[α]anthracene
8-Methylbenz[α]anthracene
9-Methylbenz[α]anthracene
10-Methylbenz[α]anthracene
11-Methylbenz[α]anthracene
12-Methylbenz[α]anthracene
7-Methylbenzo[*c*]phenanthrene
8-Methylbenzo[*c*]phenanthrene
9-Methylbenzo[*c*]phenanthrene
10-Methylbenzo[*c*]phenanthrene
11-Methylbenzo[*c*]phenanthrene
12-Methylbenzo[*c*]phenanthrene
1-Methylchrysene
2-Methylchrysene
3-Methylchrysene
4-Methylchrysene
5-Methylchrysene
6-Methylchrysene
9-Phenylfluorene

$C_{19}H_{14}ClN_2O_7$
4-Aminodedimethylamino-anhydrodemethylchlorotetracycline †

$C_{19}H_{14}N_2$
4-*p*-Aminophenylacridine
2-Amino-9-phenylacridine
9-Amino-2-phenylacridine
9-Amino-4-phenylacridine
1,2-Diphenylbenzimidazole
2,5-(or 2,6)-Diphenylbenzimidazole
Di-2-quinolylmethane
Di-6-quinolylmethane

$C_{19}H_{14}N_2O$
1-Benzoyl-2-phenylpyrazole[1,2-*a*]pyrazole †

$C_{19}H_{14}N_2O_2$
Safranol, *Me ether*

$C_{19}H_{14}N_2O_3$
2-Amino-5-nitrobenzophenone, N-*Phenyl*
4-Amino-3-nitrobenzophenone, N-*Phenyl*
Gelsedine
4-Hydroxyazobenzene-3-carboxylic Acid, *Phenyl ester*

$C_{19}H_{14}N_3$
1,3-Diphenyl-1,2,4-benzotriazinyl radical †

$C_{19}H_{14}N_4O_3$
5-Nitrobenzene-2-carboxylic Acid, *Anilide*

$C_{19}H_{14}O$
Fuchsone
1-Hydroxybenz[*a*]anthracene, *Me ether*
4-Hydroxybenz[*a*]anthracene, *Me ether*

$C_{19}H_{14}O$ (*continued*)
8-Hydroxybenz[*a*]anthracene, *Me ether*
12-Hydroxybenz[*a*]anthracene, *Me ether*
2-Phenylbenzophenone
3-Phenylbenzophenone
4-Phenylbenzophenone
9-Phenylxanthene

$C_{19}H_{14}O_3$
2,3-Dihydroxybenz[*de*]anthracen-7-one, *Di-Me ether*
4,5-Dihydroxybenz[*de*]anthracen-7-one, *Di-Me ether*
10,11-Dihydroxybenz[*de*]anthracen-7-one, *Di-Me ether*
o-1-Naphthoylbenzoic Acid, *Me ester*
o-Phenoxybenzoic Acid, *Phenyl ester*
p-Phenoxybenzoic Acid, *Phenyl ester*

$C_{19}H_{14}O_4$
3′,4′:7,8-Dimethylenedioxyisoflavone
Ochromycinone†
2-Phenylnaphthalene-1,2′-dicarboxylic Acid, 1-*Me ester*
2-Phenylnaphthalene-1,2′-dicarboxylic Acid, 2′-*Me ester*
1-Phenylnaphthalene-2,3-dicarboxylic Acid, 2-*Me ester*

$C_{19}H_{14}O_5$
Tetrangomycin†
Vulpinic Acid

$C_{19}H_{14}O_5S$
o-Sulphobenzoic Acid, *Diphenyl ester*

$C_{19}H_{14}O_6$
Neotenone
Sterigmatocystin, O-*Me ether*
Tectoleafquinone†

$C_{19}H_{14}O_6S$
2-Hydroxy-5-sulphobenzoic Acid, *Di-phenyl ester*

$C_{19}H_{14}O_7$
Aspertoxin†
Fukugenetin
5-Methoxysterigmatocystin★†

$C_{19}H_{14}O_8$
Meliternatin

$C_{19}H_{14}O_9$
Stictic Acid
Variegatic Acid, *Me ester*†

$C_{19}H_{14}O_{10}$
Glaucophanic Acid†
Salazinic Acid, α-*Me ether*
Salazinic Acid, β-*Me ether*

$C_{19}H_{15}Br$
α-Bromotriphenylmethane

$C_{19}H_{15}Cl$
α-Chlorotriphenylmethane

$C_{19}H_{15}ClN_4$
2,3,5-Triphenyltetrazolium chloride

$C_{19}H_{15}Cl_3O_5$
Nornidulin

$C_{19}H_{15}I$
α-Iodotriphenylmethane

$C_{19}H_{15}N$
9,10-Dihydro-9-phenylacridine
9,10-Dihydro-10-phenylacridine

$C_{19}H_{15}NO$
Benz[*b*]acridone, N-*Et*
N-Diphenylbenzamide

$C_{19}H_{15}NO_2$
4-Amino-4′-hydroxybenzophenone, *Phenyl ether*
o-Diphenylaminobenzoic Acid
o-Nitrophenyldiphenylmethane
m-Nitrophenyldiphenylmethane
p-Nitrophenyldiphenylmethane

$C_{19}H_{15}NO_3$
m-Nitrophenyldiphenylmethanol
p-Nitrophenyldiphenylmethanol

$C_{19}H_{15}NO_4$
αα-Di-(*p*-hydroxyphenyl)-2-nitrotoluene
αα-Di-(*p*-hydroxyphenyl)-3-nitrotoluene
αα-Di-(*p*-hydroxyphenyl)-4-nitrotoluene
Eupaverine
Galipoidine
4-Hydroxy-3-nitrophenyldiphenylmethanol
3-Phenylquinoline-2,4-dicarboxylic Acid, *Di-Me ester*

$C_{19}H_{15}NO_5$
Atheroline†
Coptisine

$C_{19}H_{15}NO_6$
Nicoumalone
1-Nitroanthraquinone-2-carboxylic Acid, *Butyl ester*
1-Nitroanthraquinone-2-carboxylic Acid, 2-*Butyl ester*

$C_{19}H_{15}NO_9$
4-Hydroxytetracycloxide†

$C_{19}H_{15}NS$
Phenothiazine, N-*Benzyl*

$C_{19}H_{15}N_3$
1,3-Diphenyl-1,2,4-benzotriazine†

$C_{19}H_{15}N_3O_2$
4-Hydroxyazobenzene-3-carboxylic Acid, *Anilide*

$C_{19}H_{15}N_3O_3$
4-Nitrodiphenylamine-2-carboxylic Acid, *Anilide*
2-Nitrodiphenylamine-4-carboxylic Acid, *Anilide*

$C_{19}H_{15}N_3O_5S$
4-Nitro-2-sulphobenzoic Acid, *Dianilide*

$C_{19}H_{16}$
4-Benzylbiphenyl
Triphenylmethane

$C_{19}H_{16}ClNO_2$
3-Methyl-2-phenylquinoline-4-carboxylic Acid, 2-*Chloroethyl ester*
6-Methyl-2-phenylquinoline-4-carboxylic Acid, 2-*Chloroethyl ester*
8-Methyl-2-phenylquinoline-4-carboxylic Acid, 2-*Chloroethyl ester*
2-*p*-Tolylquinoline-4-carboxylic Acid, 2-*Chloroethyl ester*

$C_{19}H_{16}Cl_2O_5$
Dechloronornidulin

$C_{19}H_{16}NO_4$ (ion)
Thalifendine†

$C_{19}H_{16}NO_5$ (ion)
Thalidastine†

$C_{19}H_{16}N_2$
Sempervirine
Tetrabyrine
Yobyrine

$C_{19}H_{16}N_2O$
4-Hydroxyazobenzene, *Benzyl ether*

$C_{19}H_{16}N_2O_2$
4-Amino-3-nitrophenyldiphenylmethane

$C_{19}H_{16}N_2O_4$
4-Amino-3-nitrobenzoic Acid, N-α-*Naphthyl, Et ester*
4-Amino-3-nitrobenzoic Acid, N-β-*Naphthyl, Et ester*

$C_{19}H_{16}O$
α,α-Diphenyl-*o*-cresol
α,α-Diphenyl-*m*-cresol
α,α-Diphenyl-*p*-cresol
1-Naphthyl phenylethyl Ketone
2-Naphthyl phenylethyl Ketone
Triphenylmethanol

$C_{19}H_{16}O_2$
3-Carbethoxymethylene-1,2-diphenylcyclopropene†
2,5-Dihydroxy-α,α-diphenyltoluene
3,4-Dihydroxy-α,α-diphenyltoluene
2,2′-Dihydroxytriphenylmethane
4,4′-Dihydroxytriphenylmethane
2,6-Dimethyl-3,5-diphenyl-4*H*-pyran-4-one†
2-Hydroxyphenyldiphenylmethanol
3-Hydroxyphenyldiphenylmethanol
4-Hydroxyphenyldiphenylmethanol
α-2-Hydroxyphenyl-α-4-hydroxyphenyltoluene

$C_{19}H_{16}O_3$
3,4-Dihydroxydiphenyl Ether, 4-*Me*, 3-*Phenyl ether*
Leucaurine
Orthoformic Acid, *Triphenyl ester*

$C_{19}H_{16}O_3S$
6-Benzoyl-3-hydroxy-1-methyl-5-phenylthiabenzene 1-oxide†

$C_{19}H_{16}O_4$
3-(2-Acetyl-1-phenylethyl)-4-hydroxycoumarin†
1,7-Diphenyl-1,3,5,7-tetraoxoheptane†
7-Methylcoumarin-4-acetic Acid, m-*Tolyl ester*

$C_{19}H_{16}O_5$
6-Deoxyjacareubin, *Mono-Me ether*†
Fluorenone-1,7-dicarboxylic Acid, *Di-Et ester*
Fluorenone-2,3-dicarboxylic Acid, *Di-Et ester*
Osajaxanthone★, *Mono-Me ether*†
Tanshinonic Acid†

$C_{19}H_{16}O_6$
Aureusidin, *Tetra-Me ether*
Celebixanthone
Mopanin, *Tri-Me ether*†
Rhein, *Di-Me ether*, *Et ester*

Q

Rhein, *Isobutyl ester*
Viridin

$C_{19}H_{16}O_7$
Alternariol, *Di-Ac*
Emodic Acid, *Isobutyl ester*
Kanugin
Milldurone†
Ptilometric Acid, *Me ester*†
Rhodocomatulin, 6-*Me ether*†
β_1-Rhodomycinone†
4,5,7-Trimethoxy-3-(3,4-methylenedioxyphenyl)-coumarin†
Wedelolactone, *Tri-Me ether*

$C_{19}H_{16}O_8$
Ceroalbolinic Acid, *Tri-Me ether*†
Isoquercetone, *Tetra-Me ether*
Melisimplin
α_1-Rhodomycinone†
Rubrocomatulin, 2-*Me ether*†

$C_{19}H_{16}O_9$
Flavellagic Acid, 3,4,5,4′,5′-*Penta-Me ether*
3,3′,4-Tri-*O*-methylflavellagic Acid, *Di-Me ether*†

$C_{19}H_{16}O_{10}$
Barbatolic Acid, *Me ester*
Euxanthic Acid
Haemathamnolic Acid†
Isoeuxanthic Acid

$C_{19}H_{16}O_{11}$
Thamnolic Acid

$C_{19}H_{17}ClO_8$
Chloroatranorin

$C_{19}H_{17}Cl_2N_3O_5S$
3-(2,6-Dichlorophenyl)-5-methyl-4-isoxazolylpenicillin†

$C_{19}H_{17}N$
α-Aminodiphenylmethane, N-*Phenyl*
o-Aminophenyldiphenylmethane
m-Aminophenyldiphenylmethane
p-Aminophenyldiphenylmethane
α-Aminotriphenylmethane
N-Benzyldiphenylamine
N-Diphenyl-*m*-toluidine
7-Isopropyl-1-methylphenanthrene-3-carboxylic Acid, *Nitrile*

$C_{19}H_{17}NO$
o-Aminophenyldiphenylmethanol
m-Aminophenyldiphenylmethanol
p-Aminophenyldiphenylmethanol

$C_{19}H_{17}NO_2$
3-Methyl-2-phenylquinoline-4-carboxylic Acid, *Et ester*
6-Methyl-2-phenylquinoline-4-carboxylic Acid, *Et ester*
8-Methyl-2-phenylquinoline-4-carboxylic Acid, *Et ester*
2-Phenylquinoline-4-carboxylic Acid, *Propyl ester*
1-Piperidinoanthraquinone

$C_{19}H_{17}NO_3$
Cusparine
De-*N*-methylnoracronycine, O-*Me ether*†

$C_{19}H_{17}NO_3$ (*continued*)
4-(4′-Hydroxybenzylidene)-1-phenethyl-2,3-dioxopyrrolidine†
Isocusparine
Noracronycine†

$C_{19}H_{17}NO_4$
Eschscholtzine†
Neolitsine†
Ovigerine, N-*Me*†
Tetrahydrocoptisine

$C_{19}H_{17}NO_5$
Cassythidine†

$C_{19}H_{17}NO_8$
Phomazarin†

$C_{19}H_{17}N_3$
α-Triphenylguanidine
β-Triphenylguanidine

$C_{19}H_{17}N_3O$
Evodiamine

$C_{19}H_{17}N_3O_2$
αα-Di-(4-aminophenyl)-2-nitrotoluene
αα-Di-(4-aminophenyl)-3-nitrotoluene
αα-Di-(4-aminophenyl)-4-nitrotoluene
Rhetsinine

$C_{19}H_{17}N_3O_3S_2$
Cephaloridine†

$C_{19}H_{18}ClN_3O_5S$
3-*o*-Chlorophenyl-5-methyl-4-isoxazolylpenicillin

$C_{19}H_{18}Cl_2O_5$
Vicanicin, *Et ether*

$C_{19}H_{18}MgN_8O_5$
Aminopterin, *Mg salt*

$C_{19}H_{18}N_2$
α-3,4-Diaminophenyldiphenylmethane
αα-Di-(4-aminophenyl)toluene
3,5-Diaminotoluene, 3,5-N-*Diphenyl*
Flavocoryline

$C_{19}H_{18}N_2O$
1-Benzeneazo-2-naphthol, *Propyl ether*
1-Benzeneazo-2-naphthol, *Isopropyl ether*
αα-Di-(4-aminophenyl)benzyl Alcohol
2-*o*-Tolueneazo-1-naphthol, *Et ether*
2-*m*-Tolueneazo-1-naphthol, *Et ether*
2-*p*-Tolueneazo-1-naphthol, *Et ether*
4-*o*-Tolueneazo-1-naphthol, *Et ether*
4-*p*-Tolueneazo-1-naphthol, *Et ether*
1-*o*-Tolueneazo-2-naphthol, *Et ether*
1-*m*-Tolueneazo-2-naphthol, *Et ether*
1-*p*-Tolueneazo-2-naphthol, *Et ether*

$C_{19}H_{18}N_2O_3$
Halfordinone†

$C_{19}H_{18}N_2O_4$
Cyclopenol, *Di-Me*†

$C_{19}H_{18}O_2$
7-Isopropyl-1-methylphenanthrene-3-carboxylic Acid
9-Isopropyl-1-methylphenanthrene-3-carboxylic Acid
2-Isopropyl-8-methylphenanthrene-4-carboxylic Acid
5-Isopropyl-8-methylphenanthrene-9-carboxylic Acid
Robinetin, 3,3′,4′,5′-*Tetra-Me ether*

$C_{19}H_{18}O_3$
1,5-Di-*o*-hydroxyphenyl-1,4-pentadien-3-one, *Di-Me ether*
1,5-Di-*p*-methoxyphenyl-1,4-pentadien-3-one
Tanshinone II
Tanshinone IIA†
Thebenol, *Et ether*

$C_{19}H_{18}O_4$
2′,6′-Dihydroxyflavone, *Di-Et ether*
2′,7-Dihydroxyflavone, *Di-Et ether*
3′,6-Dihydroxyflavone, *Di-Et ether*
3′,7-Dihydroxyflavone, *Di-Et ether*
4′,6-Dihydroxyflavone, *Di-Et ether*
4,4-Diphenylitaconic Acid, 1-*Et ester*
4,4-Diphenylitaconic Acid, 3-*Et ester*
Fluorene-9,9-dicarboxylic Acid, *Di-Et ester*
Hydroxytanshinone†
Mesaconic Acid, *Dibenzyl ester*
Neotruxinic Acid, 1-*Me ester*
Neotruxinic Acid, 2-*Me ester*
Tanshinone IIB†
α-Truxillic Acid, *Mono-Me ester*
γ-Truxillic Acid, *Mono-Me ester*
ε-Truxillic Acid, *Mono-Me ester*
η-Truxillic Acid, *Mono-Me ester*
epi-Truxillic Acid, 1-*Mono-Me ester*
epi-Truxillic Acid, 3-*Mono-Me ester*
β-Truxinic Acid, *Mono-Me ester*
ζ-Truxinic Acid, 1-*Mono-Me ester*
ζ-Truxinic Acid, 2-*Mono-Me ester*

$C_{19}H_{18}O_5$
Apigenin, 7,4′-*Di-Et ether*
Benzophenone-2,2′-dicarboxylic Acid, *Di-Et ester*
4-Benzoylisophthalic Acid, *Di-Et ester*
Benzoylterephthalic Acid, *Di-Et ester*
Egonol
Eucalyptin
Eucomin, *Di-Me ether*†
Geiparvarin†
Genistein, 7,4′-*Di-Et ether*
2-Methylgenistein, *Tri-Me ether*
Norwogonin, 5-*Et*-7,8-*Di-Me ether*

$C_{19}H_{18}O_6$
Artocarpetin, *Tri-Me ether*†
Atrovenetin
4′,5-Dihydroxy-6,7-dimethoxyisoflavone, *Di-Me ether*†
Ferreirin, *Tri-Me ether*
Fisetin, *Tetra-Me ether*
Kaempferol, *Tetra-Me ether*
Leptosidin, *Di-Me ether*
Melannein, *Di-Me ether*†
Munduserone†
Pectolinarigenin, *Di-Me ether*
Pratoletin, *Tetra-Me ether*
Questinol, *Tri-Me ether*†
Rengasin, *Tri-Me ether*
Scriblitifolic Acid†
Scutellarein, *Tetra-Me ether*
2,3,6,7-Tetrahydroxyanthraquinone, 2,3,6-*Tri-Me*-7-*Et ether*

$C_{19}H_{18}O_6$ *(continued)*
3′,4′,6,7-Tetrahydroxyaurone, *Tetra-Me ether*†
2′,4′,5,7-Tetrahydroxyflavone, *Tetra-Me ether*†
3,4′,5,6-Tetrahydroxyflavone, *Tetra-Me ether*
3,4′,5,8-Tetrahydroxyflavone★, *Tetra-Me ether*†
3,4′,6,7-Tetrahydroxyflavone, *Tetra-Me ether*
3,5,7,8-Tetrahydroxyflavone, *Tetra-Me ether*
3,6,7,8-Tetrahydroxyflavone, *Tetra-Me ether*
3′,4′,5′,7-Tetrahydroxyflavone, *Tetra-Me ether*
3′,4′,7,8-Tetrahydroxyflavone, *Tetra-Me ether*
4′,5,7,8-Tetrahydroxyflavone, *Tetra-Me ether*
4′,6,7,8-Tetrahydroxyflavone, *Tetra-Me ether*
5,6,7,8-Tetrahydroxyflavone, *Tetra-Me ether*★†
1,2,6,8-Tetrahydroxy-3-methylanthraquinone, *Tetra*-O-*Me*
1,3,5,6-Tetrahydroxy-8-methylanthraquinone, *Tetra-Me ether*
1,5,6,7-Tetrahydroxy-2-methylanthraquinone, *Tetra-Me ether*
2′,5,6,7-Tetramethoxyflavone†
3′,4′,7-Trihydroxy-3-methoxyflavone, *Tri-Me ether*†
3,5,7-Trihydroxy-2′-methoxyflavone, *Tri-Me ether*†
4′,5,7-Trihydroxy-6-methoxyflavone, *Tri-Me ether*†
Zapotin

$C_{19}H_{18}O_7$
Flindulatin
5-Hydroxy-3′,4′,5′,7-tetramethoxyflavone†
5-Hydroxy-3,4′,6,7-tetramethoxyflavone†
5-Hydroxy-4′,6,7,8-tetramethoxyflavone†
Mikanin, 3-*Me ether*†
Morin, 2′,3,4′,7-*Tetra-Me ether*
Norherqucinone
1,2,4,5,7-Pentahydroxyanthraquinone, *Penta-Me ether*†
3,3′,4′,6,7-Pentahydroxyflavone, 3′,4′,6,7-*Tetra-Me ether*
3,3′,4′,6,7-Pentahydroxyflavone, 3,3′,4′,7-*Tetra-Me ether*
3,4′,5,6,7-Pentahydroxyflavone, 3,4′,6,7-*Tetra-Me ether*
4′,5,6,7,8-Pentahydroxyflavone, 4′,6,7,8-*Tetra-Me ether*†
Podospicatin, 2′,7-*Di-Me ether*†
Quercetin, 3,3′,4′,5-*Tetra-Me ether*
Quercetin, 3,3′,4′,7-*Tetra-Me ether*
Quercetin, 3′,4′,5,7-*Tetra-Me ether*
Usnic Acid, *Me ether*
Wightin, *Mono-Me ether*†
Xanthomicrol, *Mono-Me ether*

$C_{19}H_{18}O_8$
Atranorin
Baecomycesic Acid
Beomycesic Acid
Calycopterin
Casticin†
Caviunin
3′,5-Dihydroxy-3,4′,5′,7-tetramethoxyflavone†
3′,5-Dihydroxy-3,4′,7,8-tetramethoxyflavone†
1,2,3,5,6,7-Hexahydroxyanthraquinone, *Penta-Me ether*
Hymenoxin†
Myricetin, 3,3′,4′,5′-*Tetra-Me ether*
Quercetagetin, *Tetra-Me ether*

$C_{19}H_{12}O_9$
Leprarin
Scaposin†
Squamatic Acid
3′,5,5′-Trihydroxy3,4′,6,7-tetramethoxyflavone†

$C_{19}H_{18}O_{10}$
Hypothammolic Acid

$C_{19}H_{18}O_{11}$
Mangiferin

$C_{19}H_{19}ClO_5$
Luteolinidin chloride, *Tetra-Me ether*

$C_{19}H_{19}N$
Phenindamine

$C_{19}H_{19}NO$
Apocinchene
7-Isopropyl-1-methylphenanthrene-3-carboxylic Acid, *Amide*

$C_{19}H_{19}NO_2$
Dibenzylmalonic Acid, *Nitrile*, *Et ester*
Koenimbin†

$C_{19}H_{19}NO_3$
Galipoline
Isolaureline
Laureline
Pukateine, *Me ether*
Stephanine

$C_{19}H_{19}NO_4$
Actinodaphnine, *Me ether*
Amurensine†
Amurine†
Bulbocapnine
Caryachine†
Cheilanthifoline
Cularicine, O-*Me ether*†
1,2-Dimethoxy-9,10-methylenedioxynoraporphine†
Domesticine★†
9-Hydroxy-10-methoxy-1,2-methylenedioxyaporphine†
Isobulbocapnine
Nandigerine, N-*Me*†
Phanostenine†

$C_{19}H_{19}NO_5$
3,4-Dihydroxy-3′-nitrochalcone, *Di Et ether*
Hernandine†
9-Hydroxy-3,10-dimethoxy-1,2-methylenedioxynoraporphine†

$C_{19}H_{19}N_3$
2,4′,4″-Triaminotriphenylmethane
3,4′,4″-Triaminotriphenylmethane
4,4′,4″-Triaminotriphenylmethane

$C_{19}H_{19}N_3O$
Tri-(*p*-aminophenyl)methanol

$C_{19}H_{19}N_3O_2$
Sessiflorine†

$C_{19}H_{19}N_7O_6$
Pteroylglutamic Acid

$C_{19}H_{20}$
1-Ethyl-7-isopropylphenanthrene

$C_{19}H_{20}ClN_3$
Clemizole

$C_{19}H_{20}INO_3$
Michepressine, *Iodide* †

$C_{19}H_{20}INO_4$
Cryptowoline

$C_{19}H_{20}NO_3$ (ion)
Michepressine †

$C_{19}H_{20}N_2$
Mebhydrolin
2-Phenyl-1,3-naphthylenediamine, N,N-*Tri-Me*

$C_{19}H_{20}N_2O$
Norfluorocurarine †

$C_{19}H_{20}N_2O_2$
Mesaconic Adid, *Di-*p-*Toluidide*
Phenylbutazone

$C_{19}H_{20}N_2O_4$
1-(4-Aminophenyl)-2-(2-carboxy-4-nitrophenyl)-ethylene, N-*Di-Me, Et ester*
Halfordine †
Methane-triacetic Acid, *Dianilide*
Ornithuric Acid

$C_{19}H_{20}N_4$
α,4,4′,4″-Tetra-aminotriphenylmethane

$C_{19}H_{20}N_8O_5$
Aminopterin

$C_{19}H_{20}O_2$
1,1-Diphenyl-1-butene-2-carboxylic Acid, *Et ester*
Hinokiresinol, *Di-Me ether* †
D-Homoequilenin
Isoequilenin, *Me ether*
2,3,3-Trimethyl-1-cyclopentene-1-carboxylic Acid, *α-Naphthyl ester*
2,3,3-Trimethyl-1-cyclopentene-1-carboxylic Acid, *β-Naphthyl ester*

$C_{19}H_{20}O_3$
2-Benzoyl-2-benzylpropionic Acid, *Et ester*
2-Benzoyl-4-phenylbutyric Acid, *Et ester*
4-Benzoyl-3-phenylbutyric Acid, *Et ester*
4-Benzoyl-4-phenylbutyric Acid, *Et ester*
Columbianadin †
Cryptotanshinone †
2′,4′-Dihydroxychalcone, *Di-Et ether*
2′,5′-Dihydroxychalcone, *Di-Et ether*
2′,6′-Dihydroxychalcone, *Di-Et ether*
Mundulea Lactone †

$C_{19}H_{20}O_3S$
7-Isopropyl-1-methylphenanthrene-2-sulphonic Acid, *Me ester*
2-Isopropyl-8-methylphenanthrene-3-sulphonic Acid, *Me ester*

$C_{19}H_{20}O_4$
Deoxybruceol †
Diethylmalonic Acid, *Diphenyl ester*
Diphenylmalonic Acid, *Di-Et ester*
2,3-Diphenylpropane-1,2-dicarboxylic Acid, *Di-Me ester*

$C_{19}H_{20}O_5$
Agrimolide, *Mono-O-Me ether* †
Brazilin, *Tri-Me ether*
Bruceol †
Clausenidin †
Decursin †
Dihydro-oroselol, *Tigoyl ester* †
Samaderol
Selinidin †
2,2′,3,3′-Tetrahydroxychalcone, *Tetra-Me ether*
2,2′,4,4′-Tetrahydroxychalcone, *Tetra-Me ether*
2,2′,4,5′-Tetrahydroxychalcone, *Tetra-Me ether*
2,2′,4′,6′-Tetrahydroxychalcone, *Tetra-Me ether*
2′,3′,4,4′-Tetrahydroxychalcone, *Tetra-Me ether*
2′,3′,4′,6′-Tetrahydroxychalcone, *Tetra-Me ether*
2′,4,4′,5′-Tetrahydroxychalcone, *Tetra-Me ether*
2′,4,4′,6′-Tetrahydroxychalcone, *Tetra-Me ether*
3,3′,4,4′-Tetrahydroxychalcone, *Tetra-Me ether*
3,3′,4′,5-Tetrahydroxychalcone, *Tetra-Me ether*

$C_{19}H_{20}O_6$
Artocarpanone †
Coleon B
Columbianadinoxide †
5,7-Dihydroxy-2′,4′-dimethoxyisoflavanone, *Di-Me ether*
Eucomol, 5,7-*Di-Me ether* †
Maesopsin, *Tetra-Me ether*
Mopanol, *Tri-Me ether* †
Peltogynol, *Tri-Me ether*
3′,4′,5,7-Tetrahydroxyflavanone, *Tetra-Me ether*
3′,4′,5′,7-Tetrahydroxyflavanone, *Tetra-Me ether*
3′,4′,7,8-Tetrahydroxyflavanone, *Tetra-Me ether* †

$C_{19}H_{20}O_7$
Barbatic Acid
Elephantopin †
Evernic Acid, *Et ester*
Frenolicin, *Me ester* †
3,3′,4′,5′,7-Pentahydroxyflavanone, *Tetra-Me*
3,3′,4′,7,8-Pentahydroxyflavanone, 3′,4′,7,8-*Tetramethyl ether* †
Sulochrin, *Di-Me ether*

$C_{19}H_{20}O_9$
Cervicarcin †
Digallic Acid, *Penta-Me ether*

$C_{19}H_{20}O_{10}$
Khellinin

$C_{19}H_{20}O_{11}$
2-*O*-Galloylarbutin †
3-*O*-Galloylarbutin †
6-*O*-Galloylarbutin †
p-Galloyloxyphenyl-β-D-glucoside †

$C_{19}H_{21}N$
1,2,4,5,7,8-Hexamethylacridine

$C_{19}H_{21}NO_2$
Apomorphine, *Di-Me ether*
1-Dimethylaminoethyl-3-hydroxy-4-methoxy-phenanthrene †
Nuciferine

$C_{19}H_{21}NO_3$
Isothebaine
Pronuciferine †
Schelhammeridine †
Thebaine
Tuduranine, *Me ether*
Tuduranine, N-*Me*

$C_{19}H_{21}NO_4$
Bisnorargemonine †
Boldine
Bracteine
Caseamine †
Coreximine
Corytuberine
Cularidine †
Cularimine
2,11-Dihydroxy-1,10-dimethoxyaporphine †
Flavinantine †
Floripavine
Hernovine, N-*Me* †
Hernovine, 10-O-*Me ether* †
Isoboldine †
Isocorytuberine †
Laurotetanine
Lindcarpine, N-*Me* †
Munitagine †
Norisocorydine †
Nudaurine †
Orientalinone †
Salutaridine †
Scoulerine †
Sinoacutine †
Suaveoline †
1,2,3,4-Tetrahydro-6,7-dimethoxy-2-methyl-1-(3,4-methylenedioxyphenyl)isoquinoline †
Wilsonirine †

$C_{19}H_{21}NO_7S$
Erysothiopine

$C_{19}H_{21}NS$
6,11-Dihydro-11-(3-dimethylaminopropylidene)dibenzo[*b*,*e*]thiepin

$C_{19}H_{21}N_5O_6S$
7-(*p*-Guanidinophenylacetamido)cephlosporanic Acid †

$C_{19}H_{22}$
1,3,4,6,7,9-Hexamethylphenalene

$C_{19}H_{22}N_2$
Alstyrine
Ammocalline †
Eburnamenine

$C_{19}H_{22}N_2O$
Apocinchonine
Cinchonidine
Cinchonine
Cinchotoxine
Eburnamonine
Homocinchonidine
Isocinchonine
Isocinchotoxine
Tombozine
Yohimbone

$C_{19}H_{22}N_2OS$
Acepromazine

$C_{19}H_{22}N_2O_2$
N-2-Aminoethyl-*N*-benzyl-*m*-methoxycinnamamide †
Apoquinidine
Apoquinine
Caracurine VII
Cupreine
Epicupreine
Isoapoquinidine
Isocinchonine
2-Methyladipic Acid, *Dianilide*
3-Methyladipic Acid, *Dianilide*
Peraksine †
Sarpagine

$C_{19}H_{22}N_2O_3$
Quininal

$C_{19}H_{22}N_2O_3S$
2-(2-Dimethylaminoethoxy)ethyl phenothiazine-10-carboxylate

$C_{19}H_{22}N_2O_4$
Quitenine

$C_{19}H_{22}N_2S$
Pecazine

$C_{19}H_{22}N_4$
Corrin †

$C_{19}H_{22}O_2$
4,5-Diphenylvaleric Acid, *Et ester*

$C_{19}H_{22}O_3$
Auroglaucin
Benzilic Acid, *Isopentyl ester*
7-Geranyloxycoumarin †
Gibberic Acid, *Me ester* †
Gravelliferone †
Ostruthin

$C_{19}H_{22}O_4$
Chalepin †
Diphenic Acid, *Dipropyl ester*
Franklinone
Grandiflorone †
Heliettin †
3α-Hydroxy-1β-methylgibb-4-ene-1α,10β-dicarboxylic Acid (1 → 3α)-Lactone †
Latifolin, *Di*-O-*Me*
Oduratin †
Sugiresinol, *Di-Me ether* †

$C_{19}H_{22}O_5$
Gibberellin A_5
Gibberellin A_7
Gibberellin A_{11} †
3,3′,4′,7-Tetrahydroxyflavan, *Tetra-Me ether* †
3′,4′,5′,7-Tetrahydroxyflavan, *Tetra-Me ether*

$C_{19}H_{22}O_6$
Catechin, 3′,4′,5,7-*Tetra-Me ether*
Gibberellenic Acid
Gibberellic Acid
Gibberellin A_6★ †
Gibberellin A_{22} †
Vaginidin †

$C_{19}H_{22}O_7$
Gaillardipinnatin †
Gibberellin A_{21} †
3,3′,4,4′,5′,7-Hexahydroxyflavan, 3′-4′-5′-*Tetra-O-Me ether*
3,3′,4,4′,5′,7-Hexahydroxyflavan, *Tetra-Me ether*
3,3′,4,4′,7,8-Hexahydroxyflavan, 3′,4′,7,8-*Tetra-Me ether*
Nagilactone C †
Samaderine B †
Vernolide †

$C_{19}H_{23}ClN_2S$
3-Chloro-10-(3-diethylaminopropyl)phenothiazine

$C_{19}H_{23}NO$
Diphenylpyraline

$C_{19}H_{23}NO_3$
Amuronine †
Armepavine
Magnocurarine †
Mevalonic Acid, *Benzhydrylamide*
Morphine, 3,6-*Di-Me ether*
Morphine, 3-*Et ether*
Schelhammericine †
Thebainone A, *Me ether*

$C_{19}H_{23}NO_4$
Cepharamine †
Cinnamoylcocaine
Coclandine
Dihydro-orientalinone †
8,14-Dihydrosalutaridine †
Isosinomenine †
Morphine, N-*oxide, Di-Me ether*
Orientaline †
Reticuline †
Salutaridinol †
Schelhammerine †
Sinomenine

$C_{19}H_{23}NO_5$
Albomaculine †
Metaphanine †

$C_{19}H_{23}N_2O_2$ (ion)
Calebassinine

$C_{19}H_{23}N_3O$
Benzydamine †
Norprodigiosin †

$C_{19}H_{23}N_3O_2$
Ergometrine
Ergometrinine

$C_{19}H_{23}N_3O_4S$
Hetacillin †
6-*epi*-Hetacillin †

$C_{19}H_{24}ClNO_6$
Acutumine †

$C_{19}H_{24}N_2$
Alloyohimban
Aspidofractinine †
4-*N*-Benzanilino-1-methylpiperidine
Cleavamine
1,2-Dehydroaspidospermine
Ibogamine
epi-Ibogamine †
Imipramine
Tuboxenin †
Yohimbane †

$C_{19}H_{24}N_2O$
Adenocarpine
Antirhine †
Aspidoalbidine †
Cinchonamine
epi-Cinchonamine
Cinchotine
Conoflorin †
γ-Dimethylamino-α,α-diphenylvaleramide
Eburnamine
Fendleridine †
Hydrocinchonidine
Hydrocuprean
Isoeburnamine
Rhazinine
Yohimbol
epi-Yohimbol

$C_{19}H_{24}N_2OS$
Escorpal †
Methotrimeprazine

$C_{19}H_{24}N_2O_2$
Deoxyaspidodispermine †
Hunterburnine
Hydrocupreidine
Isoniquidine
Isoquinamine
Niquidine
Niquine
Quinamine
Strychnosplendine †

$C_{19}H_{24}N_2O_2S$
1,3-Di(4-propoxyphenyl)thiourea

$C_{19}H_{24}N_2O_3$
Aspidodispermine †

$C_{19}H_{24}N_2S$
Ethopropazine

$C_{19}H_{24}N_4O_2$
4,4′-Pentamethylenedioxydibenzamidine

$C_{19}H_{24}N_7O_{12}P$
Adenosine 5′-uridine 5′-phosphate

$C_{19}H_{24}O_2$
2,2-Di-*p*-hydroxyphenylpropane, *Di-Et ether*
3-Hydroxy-D-homo-oestra-1,3,5(10)-trien-17α-one
Lumiestrone, *Me ether*
Oestrone, *Me ether*

$C_{19}H_{24}O_3$
Adrenosterone
p-Geranyloxycinnamic Acid †
6α-Hydroxyoestrone, 6α-*Me ether* †
6β-Hydroxyoestrone, 6β-*Me ether* †

$C_{19}H_{24}O_4$
Adenostylone †
Crotonin †
Gibberellin A_9
Isoadenostylone †

$C_{19}H_{24}O_5$
Crotocin†
7-(6,7-Dihydroxy-3,7-dimethyloct-2-enyl)oxy-coumarin
Gibberellin A_4
Gibberellin A_{20}†
Hyposalazinol, *Tri-Me ether*
Marmin†
Marrianolic Acid, 2-*Me* ester
Trichothecin
Zearalenone, 2-*Mono-Me ether*†
Zearalenone, 4-*Mono-Me ether*†

$C_{19}H_{24}O_6$
Gibberellin A_1
Gibberellin A_6
Gibberellin A_{16}†
Mycophenolic Acid★, *Et ester*†
Nagilactone A†

$C_{19}H_{24}O_7$
Gibberellin A_8
Nagilactone B†
Samaderine C†

$C_{19}H_{24}O_8$
Leucodrin, *Tetra-Me ether*

$C_{19}H_{24}O_9$
4β,15-Diacetoxy-3α,7α-dihydroxyscirp-9-en-8-one†

$C_{19}H_{24}O_{10}$
Anisatin, *Diacetate*

$C_{19}H_{25}NO$
Levallorphan

$C_{19}H_{25}NO_2$
Buphenine

$C_{19}H_{25}NO_3$
Amuroline†
Belladine
Delatine

$C_{19}H_{25}NO_4$
Amaryllisine, *Me ether*†
Erythratidine

$C_{19}H_{25}NO_5$
Nerinine

$C_{19}H_{25}NO_7$
Retusamine

$C_{19}H_{25}NO_{10}$
Vicianin

$C_{19}H_{25}N_2O_4$ (?)
Kouminidine

$C_{19}H_{26}N_2$
Aspidospermidine†
Quebrachamine

$C_{19}H_{26}N_2O$
Astrophylline†
Dihydrocorynantheol†
Geissoschizoline
Pereirine
Rhazidine★†
Velhanamine†
Vinrosamine†

$C_{19}H_{26}N_2O_2$
Dihydroniquidine
Rauqofinine

$C_{19}H_{26}O$
3,5-Androstadien-17-one
4,16-Androstadien-3-one
Bakuchiol, *Me ether*†

$C_{19}H_{26}O_2$
13,14-*seco*-4-*cis*-13(17)-Androstadiene-3,14-dione†
4-Androstene-3,17-dione
5-Androstene-3,17-dione
5α-Androst-1-ene-3,17-dione
5α-Androst-2-ene-6,17-dione
Podocarpic Acid, *Et ester*

$C_{19}H_{26}O_3$
Allethrin
5α-Androstane-3,11,17-trione
3-Hydroxy-5α-androst-3-ene-2,17-dione

$C_{19}H_{26}O_4$
Cohulupone

$C_{19}H_{26}O_4S$
S-Petasitolide A★†
S-Petasitolide B†

$C_{19}H_{26}O_5$
Gibberellin A_{10}†
Rubrosterone†

$C_{19}H_{26}O_6$
Gibberellin A_2
Melicopol†

$C_{19}H_{26}O_7$
Diacetoxyscirpenol†

$C_{19}H_{26}O_9$
Bergenin, *Penta-Me ether*†

$C_{19}H_{26}O_{12}$
Violutin

$C_{19}H_{27}NO$
Pentazocine†

$C_{19}H_{27}NO_3$
Protoemetine†
Tetrabenazine

$C_{19}H_{27}NO_4$
α-Eucaine

$C_{19}H_{27}NO_6$
Senkirkine† (Renardine★)

$C_{19}H_{27}NO_7$
Otosenine

$C_{19}H_{27}N_2O$ (ion)
Rhazidine Salt†

$C_{19}H_{28}KNO_{13}S_2$
Glucochlearine, *Tetra-Ac*

$C_{19}H_{28}O$
4-Androsten-17-one
5-Androsten-17-one
5α-Androst-2(or 3)-en-17-one
5α-Androst-16-en-3-one
3,5-Cycloandrostan-6-one

$C_{19}H_{28}O_2$
- 5,7-Androstadien-3β,17β-diol
- 5α-Androstane-3,17-dione
- 5α-Androstane-6,17-dione
- Hinokiol
- 3α-Hydroxyandrost-5-en-17-one
- 3β-Hydroxyandrost-5-en-17-one
- 3α-Hydroxy-5α-androst-9(11)-en-17-one
- 3β-Hydroxy-5α-androst-9(11)-en-17-one
- 3α-Hydroxy-5α-androst-11(?)-en-17-one
- 17α-Hydroxyandrost-4-en-3-one
- 17β-Hydroxyandrost-4-en-3-one
- 17β-Hydroxyandrost-5-en-3-one
- 17β-Hydroxy-9β,10α-androst-4-en-3-one †
- 6β-Hydroxy-3,5-cycloandrostan-17-one
- Pipataline †

$C_{19}H_{28}O_3$
- Flavoglaucin
- Shogaol, *Et ether*
- Trisporic B Acid, *Me ester* †

$C_{19}H_{28}O_4$
- Atractyligenin †
- Santonic Acid, *Isobutyl ester*
- Trisporic C Acid, *Me ester* †

$C_{19}H_{28}O_5$
- Movragenic Acid

$C_{19}H_{28}O_6$
- Arctiopicrin

$C_{19}H_{29}IO_2$
- Iophendylate

$C_{19}H_{29}NO$
- 3-Amino-5*a*-androst-16-ene
- 2-*p*-Menthylacetic Acid, p-*Toluidide*
- Procyclidine

$C_{19}H_{29}NO_2$
- Cycloneosamandione
- Samandarone

$C_{19}H_{29}NO_3$
- Capsaicin, *Me ether*
- Dihydroprotoemetine †

$C_{19}H_{29}NO_4$
- Ankorine †
- Dendrine †

$C_{19}H_{29}NO_5$
- Benzylpenaldic Acid, *Di-butyl acetal*

$C_{19}H_{29}NO_6$
- Crosemperine †

$C_{19}H_{29}NO_8$
- Onetine †

$C_{19}H_{29}N_3O$
- Pamaquin

$C_{19}H_{30}$
- 3,5-Cycloandrostane

$C_{19}H_{30}O$
- 5α-Androstan-3-one
- 5α-Androstan-15-one †
- 5α-Androstan-16-one †
- 5α-Androstan-17-one
- 4-Androsten-17-ol
- 5-Androsten-3β-ol
- 5-Androsten-17-ol
- 5α-Androst-16-en-3α-ol
- 5α-Androst-16-en-3β-ol
- 3,5-Cycloandrostane-6α-ol
- 3,5-Cycloandrostane-6β-ol
- 18-Nor-isopimaradien-4α-ol †

$C_{19}H_{30}O_2$
- 4-Androstene-3α,17β-diol
- 4-Androstene-3β,17β-diol
- 5-Androstene-3α,17β-diol
- 5-Androstene-3β,17α-diol
- 5-Androstene-3β,17β-diol
- Androsterone
- *epi*-Androsterone
- Colensenone
- 3,5-Cycloandrostane-6α,17β-diol
- 3,5-Cycloandrostane-6β,17β-diol
- Dodecanoic Acid, p-*Tolyl ester*
- 3α-Hydroxy-5β-androstan-17-one
- 3β-Hydroxy-5β-androstan-17-one
- 17α-Hydroxy-5α-androstan-3-one
- 17β-Hydroxy-5α-androstan-3-one
- Isonoragathic Acid
- Noragathic Acid
- Octadeca-11,17-dien-9-ynoic Acid, *Me ester* †
- Octadeca-9,14-dien-12-ynoic Acid, *Me ester* †
- 3,9,12,15-Octadecatetraenoic Acid, *Me ester* †
- Sterculynic Acid †

$C_{19}H_{30}O_3$
- 5-Androstene-3β,16α,17α-triol
- 5-Androstene-3β,16α,17β-triol
- 3α,11β-Dihydroxy-5α-androstan-17-one
- 3β,11β-Dihydroxy-5α-androstan-17-one
- 8-Hydroxyoctadeca-11,17-dien-9-ynoic Acid, *Me ester* †
- 2-Oxo-3-oxamanoyl oxide †

$C_{19}H_{30}O_4$
- 2//3-5α-Androstane-2,3-dioic Acid
- 3//4-5α-Androstane-3,4-dioic Acid
- Rapanone

$C_{19}H_{30}O_5$
- Caucalol Diacetate †

$C_{19}H_{31}NO_2$
- Samandarin

$C_{19}H_{31}NO_3$
- 3α,17α-Dihydroxy-17α-aza-D-homo-5α-androstan-17-one

$C_{19}H_{31}NO_5$
- Isostemonidine
- Stemonidine

$C_{19}H_{32}$
- 5α-Androstane
- 2,4,6-Tri-*tert*-butyltoluene †

$C_{19}H_{32}Br_6O_2$
- 9,10,12,13,15,16-Hexabromo-octadecanoic Acid, *Me ester*

$C_{19}H_{32}O$
- 5α-Androstan-3α-ol
- 5α-Androstan-3β-ol
- 5α-Androstan-15α-ol †
- 5α-Androstan-15β-ol †
- 5α-Androstan-17α-ol

$C_{19}H_{32}O$ (*continued*)
5α-Androstan-17β-ol
$C_{19}H_{32}O_2$
5α-Androstane-3α,17α-diol
5α-Androstane-3α,17β-diol
5α-Androstane-3β,17α-diol
5α-Androstane-3β,17β-diol
5β-Androstane-3α,17β-diol
α-Kamlolenic Acid, *Me ester*
β-Kamlolenic Acid, *Me ester*
6,10,14-Octadecatrienoic Acid, *Me ester*
9,11,13-Octadecatrienoic Acid, *Me ester*
9,12,15-Octadecatrienoic Acid, *Me ester*
10,12,14-Octadecatrienoic Acid, *Me ester*
11-Octadecen-9-ynoic Acid, *Me ester*
Octadec-17-en-9-ynoic Acid, *Me ester* †
5-Tridecylresorcinol †
$C_{19}H_{32}O_3$
5α-Androstane-3β,16α,17α-triol
5α-Androstane-3β,16β,17α-triol
5α-Androstane-3β,16β,17β-triol
8-[2-(5-Hexylfuryl)]-octanoic Acid, *Me ester* †
9-Hydroxyoctadec-10-ene-12-ynoic Acid, *Me ester* †
$C_{19}H_{32}O_4$
Lichesteric Acid
Protolichesteric Acid
$C_{19}H_{32}O_8$
Heptane-1,3,3,7-tetracarboxylic Acid, *Tetra-Et ester*
Heptane-1,4,4,7-tetracarboxylic Acid, *Tetra-Et ester*
Heptane-2,2,6,6-tetracarboxylic Acid, *Tetra-Et ester*
Heptane-3,3,5,5-tetracarboxylic Acid, *Tetra-Et ester*
$C_{19}H_{33}N$
3-Aza-*A*-homo-5β-androstane
4-Aza-*A*-homo-5α-androstane
4-Aza-*A*-homo-5β-androstane
Neosaman
$C_{19}H_{34}$
Fichtelite
$C_{19}H_{34}Br_4O_2$
9,10,12,13-Tetrabromo-octadecanoic Acid, *Me ester*
$C_{19}H_{34}N_8O_8$
Streptothricin
$C_{19}H_{34}O_2$
Chaulmoogric Acid, *Me ester*
9-Hydroxy-10,12-octadecadienoic Acid, *Me ester* †
13-Hydroxy-9,11-octadecadienoic Acid, *Me ester* †
Malvalic Acid, *Me ester* †
5,6-Octadecadienoic Acid, *Me ester* †
6,8-Octadecadienoic Acid, *Me ester*
9,11-Octadecadienoic Acid, *Me ester*
9,12-Octadecadienoic Acid, *Me ester*
10,12-Octadecadienoic Acid, *Me ester*
Sterculic Acid †
$C_{19}H_{34}O_3$
12-Hydroxy-9,16-octadecadienoic Acid, *Me ester*
12-Hydroxy-9-octadecynoic Acid, *Me ester*
2-Hydroxysterculic Acid †
$C_{19}H_{34}O_4$
Nephromopsic Acid
$C_{19}H_{34}O_5$
9-Oxoheptadecanedioic Acid, *Di-Me ester*
$C_{19}H_{34}O_{14}$
Strophanthotriose
$C_{19}H_{35}NO_2$
Dicyclomine
$C_{19}H_{35}N_3O_5$
Actinonin
$C_{19}H_{36}N_2O_6S$
Antibiotic U-11,921 †
$C_{19}H_{36}O$
Cyclononadecanone
$C_{19}H_{36}O_2$
Cycloheptadecylacetic Acid
Gynocardic Acid, *Me ester*
Hexadecanoic Acid, *Allyl ester*
Lactobacillic Acid
2-Methyl-16-heptadecenoic Acid, *Me ester*
2-Octadecenoic Acid, *Me ester*
3-Octadecenoic Acid, *Me ester*
6-Octadecenoic Acid, *Me ester*
9-Octadecenoic Acid, *Me ester*
11-Octadecenoic Acid, *Me ester*
16-Octadecenoic Acid, *Me ester*
$C_{19}H_{36}O_3$
9,10-Epoxyoctadecanoic Acid, *Me ester* †
12-Hydroxy-9-octadecenoic Acid, *Me ester*
17-Hydroxy-9-octadecenoic Acid, *Me ester*
18-Hydroxy-9-octadecenoic Acid, *Me ester*
9-Oxoheptadecanoic Acid, *Et ester*
10-Oxononadecanoic Acid
3-Oxo-octadecanoic Acid, *Me ester*
5-Oxo-octadecanoic Acid, *Me ester*
Pyruvic Acid, *Hexadecyl ester*
$C_{19}H_{36}O_4$
Hexadecylmalonic Acid
2-Methylhexadecanedioic Acid, *Di-Me ester*
3-Methylhexadecanedioic Acid, *Di-Me ester*
4-Methylhexadecanedioic Acid, *Di-Me ester*
5-Methylhexadecanedioic Acid, *Di-Me ester*
8-Methylhexadecanedioic Acid, *Di-Me ester*
Roccellic Acid, *Di-Me ester*
$C_{19}H_{36}O_6$
9-Heptadecenoic Acid, *Et ester*
$C_{19}H_{36}O_{10}$
Rutinose, *Hepta-Me ether*
$C_{19}H_{36}O_{11}$
Cellobiose, *Hepta-Me ether*
Turanose, *Hepta-Me ether*
$C_{19}H_{37}IO_2$
3-Iodo-octadecanoic Acid, *Me ester*
$C_{19}H_{37}NO_2$
Bicyclohexyl-1-carboxylic Acid, *Diethylaminoethyl ester*

$C_{19}H_{37}NO_3$
N-Hexadecanoylglycine, *Me ester*
Hexadecylmalonic Acid, *Monamide*

$(C_{19}H_{37}NO_7)_n$
Primycin

$C_{19}H_{38}$
1-Nonadecene
Zamene

$C_{19}H_{38}O$
Cyclononadecanol
2-Nonadecanone
4-Nonadecanone
10-Nonadecanone

$C_{19}H_{38}O_2$
Dodecanoic Acid, *Heptyl ester*
Heptadecanoic Acid, *Et ester*
Hexadecanoic Acid, *Propyl ester*
2-Methyloctadecanoic Acid
3-Methyloctadecanoic Acid
4-Methyloctadecanoic Acid
5-Methyloctadecanoic Acid
6-Methyloctadecanoic Acid
7-Methyloctadecanoic Acid
8-Methyloctadecanoic Acid
9-Methyloctadecanoic Acid
10-Methyloctadecanoic Acid
11-Methyloctadecanoic Acid
12-Methyloctadecanoic Acid
13-Methyloctadecanoic Acid
14-Methyloctadecanoic Acid
15-Methyloctadecanoic Acid
16-Methyloctadecanoic Acid
17-Methyloctadecanoic Acid
Nonadecanoic Acid
Octadecanoic Acid, *Me ester*
2-Octanol, *Undecanoyl*
2,6,10,14-Tetramethylpentadecanoic Acid†

$C_{19}H_{38}O_3$
1-*O*-Hexadec-1-enylglycerol†
2-Hydroxyoctadecanoic Acid, *Me ester*
2-Hydroxyoctadecanoic Acid, *Me ether*
3-Hydroxyoctadecanoic Acid, *Me ester*
4-Hydroxyoctadecanoic Acid, *Me ester*
5-Hydroxyoctadecanoic Acid, *Me ester*
6-Hydroxyoctadecanoic Acid, *Me ester*
7-Hydroxyoctadecanoic Acid, *Me ester*
8-Hydroxyoctadecanoic Acid, *Me ester*
9-Hydroxyoctadecanoic Acid, *Me ester*
10-Hydroxyoctadecanoic Acid, *Me ester*
11-Hydroxyoctadecanoic Acid, *Me ester*
12-Hydroxyoctadecanoic Acid, *Me ester*
13-Hydroxyoctadecanoic Acid, *Me ester*
14-Hydroxyoctadecanoic Acid, *Me ester*
15-Hydroxyoctadecanoic Acid, *Me ester*
16-Hydroxyoctadecanoic Acid, *Me ester*
17-Hydroxyoctadecanoic Acid, *Me ester*
18-Hydroxyoctadecanoic Acid, *Me ester*

$C_{19}H_{38}O_4$
9,10-Dihydroxyoctadecanoic Acid, *Me ester*
12,13-Dihydroxyoctadecanoic Acid, *Me ester*
15,16-Dihydroxyoctadecanoic Acid, *Me ester*†
α-Monopalmitin
β-Monopalmitin

$C_{19}H_{38}O_5$
Phloionolic Acid, *Me ester*
9,10,18-Trihydroxyoctadecanoic Acid, *Me ester*†

$C_{19}H_{38}O_6$
9,10,12,13-Tetrahydroxyoctadecanoic Acid, *Me ester*

$C_{19}H_{39}N$
Nonadecamethyleneimine

$C_{19}H_{39}NO$
2-Methyloctadecanoic Acid, *Amide*
3-Methyloctadecanoic Acid, *Amide*
4-Methyloctadecanoic Acid, *Amide*
5-Methyloctadecanoic Acid, *Amide*
6-Methyloctadecanoic Acid, *Amide*
7-Methyloctadecanoic Acid, *Amide*
8-Methyloctadecanoic Acid, *Amide*
9-Methyloctadecanoic Acid, *Amide*
10-Methyloctadecanoic Acid, *Amide*
11-Methyloctadecanoic Acid, *Amide*
12-Methyloctadecanoic Acid, *Amide*
13-Methyloctadecanoic Acid, *Amide*
14-Methyloctadecanoic Acid, *Amide*
15-Methyloctadecanoic Acid, *Amide*
16-Methyloctadecanoic Acid, *Amide*
17-Methyloctadecanoic Acid, *Amide*
2,6,10,14-Tetramethylpentadecanoic Acid, *Amide*†

$C_{19}H_{39}NO_2$
2-Amino-octadecanoic Acid, *Me ester*
Sphingosine, *Me ether*

$C_{19}H_{40}$
Nonadecane
Pristane

$C_{19}H_{40}O$
1-Nonadecanol
1-Octadecanol, *Me ether*

$C_{19}H_{40}O_3$
Chimyl Alcohol

C_{20}

$C_{20}H_6Cl_6$
Hexachloroperylene

$C_{20}H_{10}$
Dibenzo[*ghi*,*mno*]fluoranthene†

$C_{20}H_{10}Cl_4$
2,3,4,5-Tetrachloro-1-(2,3-diphenylcycloprop-enylidene)cyclopentadiene†

$C_{20}H_{10}I_2O_5$
4,5-Di-iodofluorescein

$C_{20}H_{10}I_4O_4$
3′,5′,3″,5″-Tetraiodophenolphthalein

$C_{20}H_{10}O_2$
1,12-Perylenequinone
3,9-Perylenequinone
3,10-Perylenequinone
peri-Xanthenoxanthene

$C_{20}H_{10}O_4$
4,9-Dihydroxyperylene-3,10-dione

$C_{20}H_{10}O_6$
Leprapric Acid

$C_{20}H_{10}O_{11}$
Flavogallone

$C_{20}H_{11}ClO_2$
14-Hydroxy-5,14-dihydronaphtho[3,2,1-*kl*]xanthen-5-one, *Chloride*

$C_{20}H_{11}NO$
13*H*-Phenanthro[10,1-*fg*]quinolin-13-one

$C_{20}H_{12}$
Acepleiadylene†
Benz[*e*]acephenanthrylene
Benzo[*j*]fluoranthrene
Benzo[*k*]fluoranthrene
Benzo[*a*]pyrene
Benzo[*e*]pyrene
Cyclobuta[1,2-*a*, 3,4-*c′*]dinaphthalene
Cyclobuta[2,3-*a*, 2,3-*c*]dinaphthalene
Dibenzo[*a,c*]biphenylene†
Dibenzo[*a,g*]biphenylene†
Dibenzo[*a,h*]biphenylene†
Dibenzo[*a,i*]biphenylene†
Dibenzo[*b,h*]biphenylene†
Perylene

$C_{20}H_{12}ClNO_2$
1-Amino-5-chloroanthraquinone, N-*Phenyl*
2-Amino-7-chloroanthraquinone, N-*Phenyl*

$C_{20}H_{12}N_2$
Benzo[*a*]pyridol[3,2-*h*]acridine
Benzo[*a*]pyridol[4,3-*h*]acridine
Dibenzo[*b,j*]-4,7-phenanthroline
Dibenzo[*a,c*]phenazine
Dibenzo[*a,h*]phenazine
Dibenzo[*a,i*]phenazine
Dibenzo[*a,j*]phenazine

$C_{20}H_{12}N_2O_3$
Quinoline-4-carboxylic Acid, *Anhydride*
2,3,4-Trihydroxyphenanthraquinone, *Phenazine*

$C_{20}H_{12}N_2O_4$
1-Amino-8-nitroanthraquinone, N-*Phenyl*

$C_{20}H_{12}N_2O_7$
Phthalimidoacetic Acid, *Anhydride*

$C_{20}H_{12}O$
1-Cholanthrenone
Dinaphthol[1,2-*b*, 1′,2′-*d*]furan
Dinaphthol[1,2-*b*, 2′,1′-*d*]furan
Dinaphthol[2,1-*b*, 1,2-*d*]furan
Dinaphthol[2,1-*b*, 2,3-*d*]furan
3-Hydroxybenzo[*a*]pyrene
5-Hydroxybenzo[*a*]pyrene
6-Hydroxybenzo[*a*]pyrene
7-Hydroxybenzo[*a*]pyrene
8-Hydroxybenzo[*a*]pyrene†
9-Hydroxybenzo[*a*]pyrene†
6-Hydroxyperylene
Naphtho[3,2,1-*kl*]xanthene

$C_{20}H_{12}O_2$
3,10-Dihydroxyperylene
1-Phenylanthraquinone
2-Phenylanthraquinone

$C_{20}H_{12}O_2S$
1-Anthraquinonethiol, S-*Phenyl ether*

$C_{20}H_{12}O_3$
Fluoran
1-Hydroxy-9,10-anthraquinone, *Phenyl ether*
2-Hydroxy-9,10-anthraquinone, *Phenyl ether*
14-Hydroxy-5,14-dihydronaphtho[3,2,1-*kl*]xanthen-5-one

$C_{20}H_{12}O_5$
Fluorescein

$C_{20}H_{12}O_6$
Helioxanthin†
Taiwanin C†

$C_{20}H_{12}O_6S$
5-Hydroxyanthraquinone-1-sulphonic Acid, *Phenyl ether*
5-Hydroxyanthraquinone-2-sulphonic Acid, *Phenyl ether*

$C_{20}H_{12}O_7$
Elliptonone
Gallein
Mycochrysone
Taiwanin E†

$C_{20}H_{12}O_8$
Di-(4-hydroxy-3-coumarinyl)acetic Acid

$C_{20}H_{13}N$
11*H*-Benz[*h*]indeno[1,2-*c*]quinoline
13*H*-Benz[*f*]indeno[1,2-*c*]quinoline
Dibenzo[*a,c*]carbazole
Dibenzo[*a,g*]carbazole
Dibenzo[*a,i*]carbazole
Dibenzo[*b,g*]carbazole
Dibenzo[*b,h*]carbazole†
Dibenzo[*c,g*]carbazole
Naphth-[1,8-*bc*]-acridine
Naphtho[2,3-*a*]carbazole
Naphtho[2,3-*b*]carbazole
Naphtho[2,3-*c*]carbazole

$C_{20}H_{13}NO$
Dibenzo[*a,c*]phenoxazine
Dibenzo[*c,h*]phenoxazine

$C_{20}H_{13}NO_2$
1-Anilinoanthraquinone
2-Anilinoanthraquinone

$C_{20}H_{13}NO_3$
1-Amino-4-hydroxyanthraquinone, N-*Phenyl*

$C_{20}H_{13}NO_5$
Oxyavicine†

$C_{20}H_{13}NS$
Dibenzo[*a,j*]phenothiazine
Dibenzo[*b,i*]phenothiazine
Dibenzo[*c,h*]phenothiazine

$C_{20}H_{13}N_3O_3$
Violacein

$C_{20}H_{13}N_3O_7$
Xanthomattin

$C_{20}H_{14}$
Acepleiadene †
1,1′-Binaphthyl
1,2′-Binaphthyl
2,2′-Binaphthyl
Cholanthrene
10,12-Dihydroindeno[2,1,*b*]fluorene
5,8-Dihydroindeno[2,1-*c*]fluorene
Naphthacenaphthene
1-Phenylanthracene
2-Phenylanthracene
9-Phenylanthracene
9-Phenylphenanthrene †
Triptycene †

$C_{20}H_{14}Cl_6O_4$
2,4-Dichloro-2,4-di-(2,4-dichlorophenyl)cyclobutane-1,3-dicarboxylic Acid, *Di-Me ester*

$C_{20}H_{14}Hg$
Mercury di-1-naphthyl
Mercury di-2-naphthyl

$C_{20}H_{14}I_6N_2O_6$
3-Amino-2,4,6-tri-iodobenzoic Acid, N-*Adipoyl*, 3-*Carboxy*-2,4,6-*tri-iodoanilide*

$C_{20}H_{14}NO_4$ (ion)
Avicine †

$C_{20}H_{14}N_2$
1,1′-Azonaphthalene
1,2′-Azonaphthalene
2,2′-Azonaphthalene
2,3-Diphenylquinoxaline
2,4-Diphenylquinazoline

$C_{20}H_{14}N_2O$
1,1′-Azoxynaphthalene
2,2′-Azoxynaphthalene
2-Hydroxy-1,1′-azonaphthalene
2-Hydroxy-1,2′-azonaphthalene
1-Hydroxy-1′,4-azonaphthalene
1-Hydroxy-2,2′-azonaphthalene
2,4′-Dihydroxy-1,1′-azonaphthalene
2,7-Dihydroxy-1,2′-azonaphthalene

$C_{20}H_{14}O$
Benz[*a*]anthracene-7-acetaldehyde
5,6:7,8-Dibenzocyclodeca[1,2-*c*]furan †
1,1′-Dinaphthyl Ether
1,2′-Dinaphthyl Ether
2,2′-Dinaphthyl Ether
2,5-Diphenyl-3,4-isobenzofuran †
9-Hydroxyphenanthrene, *Phenyl ether*

$C_{20}H_{14}OS$
1,1′-Dinaphthyl sulphoxide
2,2′-Dinaphthyl sulphoxide

$C_{20}H_{14}O_2$
Benz[*a*]anthracene-1-acetic Acid
Benz[*a*]anthracene-7-acetic Acid
Benz[*a*]anthracene-8-acetic Acid
1,2-Dibenzoylbenzene
1,3-Dibenzoylbenzene
1,4-Dibenzoylbenzene
2,2′-Dihydroxy-1,1′-binaphthyl
4,4′-Dihydroxy-1,1′-binaphthyl
1,1′-Dihydroxy-2,2′-binaphthyl
1-Hydroxyfluorenone, *Benzyl ether*
2-Hydroxyphenyldiphenylacetic Acid, *Lactone*
2-Phenylbenzil
4-Phenylbenzil
Phthalophenone

$C_{20}H_{14}O_2S$
1,1′-Dinaphthyl sulphone
1,2′-Dinaphthyl sulphone
2,2′-Dinaphthyl sulphone

$C_{20}H_{14}O_3$
o-(5-Acenaphthoyl)benzoic Acid)
7,12-Dihydro-7,12-dioxopleiaden-1-ol, *Et ether*
Florantyrone
Resorcin-benzein, 6-*Me ether*

$C_{20}H_{14}O_4$
Haemocorin Aglycone †
2-(4-Hydroxybenzoyl)benzoic Acid, *Phenyl ether*
Isophthalic Acid, *Di-phenyl ester*
Phenolphthalein I
Phenolphthalein III
Phthalic Acid, *Diphenyl ester*
Terephthalic Acid, *Diphenyl ester*
2,2′,6,6′-Tetrahydroxy-1,1′-binaphthyl
3,3′,4,4′-Tetrahydroxy-1,1′-binaphthyl
4,4′,5,5′-Tetrahydroxy-1,1′-binaphthyl
1,1′,4,4′-Tetrahydroxy-2,2′-binaphthyl
Tetrangulol, *Mono-Me ether* †

$C_{20}H_{14}O_5$
Fluorescin

$C_{20}H_{14}O_6$
Taiwanin A †

$C_{20}H_{14}O_7$
Gallin
Pachyrrhizone †

$C_{20}H_{14}O_8$
Anthraquinone-1,2,3-tricarboxylic Acid, *Tri-Me ester*
Anthraquinone-1,2,5-tricarboxylic Acid, *Tri-Me ester*
Anthraquinone-1,2,6-tricarboxylic Acid, *Tri-Me ester*
Anthraquinone-1,2,7-tricarboxylic Acid, *Tri-Me ester*

$C_{20}H_{14}S$
1,1′-Dinaphthyl sulphide
1,2′-Dinaphthyl sulphide
2,2′-Dinaphthyl sulphide

$C_{20}H_{15}Cl$
Chlorotriphenylethylene

$C_{20}H_{15}ClO$
p-Diphenylmethylbenzoic Acid, *Chloride*
Triphenylacetic Acid, *Chloride*

$C_{20}H_{15}N$
1,1′-Dinaphthylamine
1,2′-Dinaphthylamine
2,2′-Dinaphthylamine
1,3-Diphenylisoindole †
o-Diphenylmethylbenzoic Acid, *Nitrile*
p-Diphenylmethylbenzoic Acid, *Nitrile*
Triphenylacetic Acid, *Nitrile*

$C_{20}H_{15}NO$
Benzil-anil

$C_{20}H_{15}NO_2$
Dibenzanilide

$C_{20}H_{15}NO_4$
Disalicylanilide

$C_{20}H_{15}NO_5$
Sanguinarine

$C_{20}H_{15}NO_7$
4-Hydroxy-6-methylpretetramid†

$C_{20}H_{15}NO_7S$
4-Nitro-6-sulpho-*m*-toluic Acid, *Diphenyl ester*

$C_{20}H_{15}N_3$
2-Amino-1,1′-azonaphthalene
4-Amino-1,1′-azonaphthalene
2-Amino-1,2′-azonaphthalene
4-Amino-1,2′-azonaphthalene
6-Amino-2,3-diphenylquinoxaline
1,1′-Diazoaminonaphthalene
2,2′-Diazoaminonaphthalene

$C_{20}H_{15}N_3O$
7′-Amino-2-hydroxy-1,2′-azonaphthalene

$C_{20}H_{15}N_3O_3$
Euxylophoricine B†

$C_{20}H_{15}N_3O_4$
3-Nitrophthalic Acid, *Dianilide*

$C_{20}H_{15}OP$
Triphenylphosphoranylideneketene†

$C_{20}H_{16}$
9,10-Dihydro-2-phenylanthracene
9,10-Dihydro-9-phenylanthracene
1,7-Dimethylbenz[*a*]anthracene
1,8-Dimethylbenz[*a*]anthracene
1,12-Dimethylbenz[*a*]anthracene
2,9-Dimethylbenz[*a*]anthracene
2,10-Dimethylbenz[*a*]anthracene
3,9-Dimethylbenz[*a*]anthracene
3,10-Dimethylbenz[*a*]anthracene
4,7-Dimethylbenz[*a*]anthracene
5,12-Dimethylbenz[*a*]anthracene
6,7-Dimethylbenz[*a*]anthracene
6,12-Dimethylbenz[*a*]anthracene
7,8-Dimethylbenz[*a*]anthracene
7,10-Dimethylbenz[*a*]anthracene
7,11-Dimethylbenz[*a*]anthracene
7,12-Dimethylbenz[*a*]anthracene
8,9-Dimethylbenz[*a*]anthracene
8,10-Dimethylbenz[*a*]anthracene
8,12-Dimethylbenz[*a*]anthracene
9,10-Dimethylbenz[*a*]anthracene
9,11-Dimethylbenz[*a*]anthracene
1,2-Dimethylchrysene†
1,4-Dimethylchrysene
1,7-Dimethylchrysene†
2,3-Dimethylchrysene
2,8-Dimethylchrysene
2,12-Dimethylchrysene†
3,4-Dimethylchrysene†
3,5-Dimethylchrysene†
4,5-Dimethylchrysene
6,12-Dimethylchrysene
1,2-Diphenylbenzocyclobutene
7-Ethylbenz[*a*]anthracene
1,1,2-Triphenylethylene

$C_{20}H_{16}ClNO_7$
Berberilic Anhydride, *Chloride*

$C_{20}H_{16}N_2$
2,2′-Diamino-1,1′-binaphthyl
3,3′-Diamino-1,1′-binaphthyl
4,4′-Diamino-1,1′-binaphthyl
1,1′-Diamino-2,2′-binaphthyl
3,3′-Diamino-2,2′-binaphthyl
Dibenzylidene-*o*-phenylenediamine
Dibenzylidene-*m*-phenylenediamine
Dibenzylidene-*p*-phenylenediamine
4,4′-Dimethyl-2,2′-biquinolyl
6,6′-Dimethyl-2,2′-biquinolyl
7,7′-Dimethyl-2,2′-biquinolyl
8,8′-Dimethyl-2,2′-biquinolyl
2,2′-Dimethyl-3,3′-biquinolyl
2,2′-Dimethyl-6,6′-biquinolyl
8,8′-Dimethyl-6,6′-biquinolyl
5,5′-Dimethyl-8,8′-biquinolyl
7,7′-Dimethyl-8,8′-biquinolyl
1,1′-Hydrazodinaphthalene
2,2′-Hydrazodinaphthalene

$C_{20}H_{16}N_2O_2$
3,3-Di-(4-aminophenyl)phthalide
Phthalanilide
Safranol, *Et ether*
Xanthocillin X, *Di-Me ether*

$C_{20}H_{16}N_2O_4$
2-Amino-5-nitrobenzophenone, N-o-*Methoxyphenyl*
Camptothecin†

$C_{20}H_{16}N_2S_2$
Dithio-2,2′-bis-1-naphthylamine
Dithio-4,4′-bis-1-naphthylamine
Dithio-5,5′-bis-1-naphthylamine
Dithio-6,6′-bis-1-naphthylamine
Dithio-8,8′-bis-1-naphthylamine

$C_{20}H_{16}N_4$
Nitron

$C_{20}H_{16}O$
αα-Diphenylacetophenone

$C_{20}H_{16}O_2$
o-Diphenylmethylbenzoic Acid
m-Diphenylmethylbenzoic Acid
p-Diphenylmethylbenzoic Acid
α-Phenylbenzoin
Triphenylacetic Acid
Tropolone, *Diphenylmethyl ether*

$C_{20}H_{16}O_3$
8-Benzoyl-1-naphthoic Acid, *Et ester*
m-(Hydroxydiphenylmethyl)benzoic Acid
p-(Hydroxydiphenylmethyl)benzoic Acid
2-Hydroxyphenyldiphenylacetic Acid
4-Hydroxyphenyldiphenylacetic Acid
2-(4-Hydroxyphenylphenylmethyl)benzoic Acid
o-1-Naphthoylbenzoic Acid, *Et ester*

$C_{20}H_{16}O_4$
1,3-Cyclohexadiene-1,4-dicarboxylic Acid, *Diphenyl ester*
2,5-Cyclohexadiene-1,4-dicarboxylic Acid, *Diphenyl ester*
Phenolphthalin

$C_{20}H_{16}O_4$ (*continued*)
- 2-Phenylnaphthalene-1,2′-dicarboxylic Acid, *Di-Me ester*
- 1-Phenylnaphthalene-2,3-dicarboxylic Acid, *Di-Me ester*
- 1-Phenylnaphthalene-2,3-dicarboxylic Acid, 2-*Et ester*
- Polyporic Acid, *Di-Me ether*
- Volucrisporin, *Di-Me ether*

$C_{20}H_{16}O_5$
- Isopsoralidin
- Phenolphthalein II
- Psoralidin
- Pulvinic Acid, *Et ester*
- Vulpinic Acid, *Me ether*

$C_{20}H_{16}O_6$
- Anthraquinone-1,5-dicarboxylic Acid, *Di-Et ester*
- Atromentin, 4′,4″-*Di-Me ether*
- Elliptone
- Erosone
- Isopinastric Acid†
- Leprapric Acid
- Pinastric Acid
- Savinin
- Tetrahydroxy-*p*-benzoquinone, *Diphenyl ether, Di-Me ether*
- Viridin★†
- β-Viridin★†

$C_{20}H_{16}O_7$
- Atromentinic Acid, *Et ester*
- Averufin†
- Aversin
- Carviolacin
- Elliptolone
- Malaccol

$C_{20}H_{16}O_8$
- Lycopersin (?)

$C_{20}H_{17}BrO_2$
- 5-Bromoresorcinol, *Di-benzyl ether*

$C_{20}H_{17}Cl_3O_5$
- Nidulin

$C_{20}H_{17}NO$
- α-Anilino-α-phenylacetophenone
- *p*-Benzylideneaminophenol, *Benzyl ether*
- Triphenylacetic Acid, *Amide*

$C_{20}H_{17}NO_2$
- Benzilic Acid, *Anilide*
- 6-Methyl-2-phenylquinoline-4-carboxylic Acid, *Allyl ester*
- Tetrophane, *Et ester*

$C_{20}H_{17}NO_3$
- *O*-Benzoyl-lactic Acid, α-*Naphthalide*
- *O*-Benzoyl-lactic Acid, β-*Naphthalide*

$C_{20}H_{17}NO_4$
- 2-Nitroquinol, *Dibenzyl ether*

$C_{20}H_{17}NO_5$
- Berlambine
- Oxyberberine
- 1,2,9,10-Tetramethoxy-7-oxodibenzo[*de,g*]-quinoline†

$C_{20}H_{17}NO_6$
- Bicuculline
- 1-Nitroanthraquinone-2-carboxylic Acid, *Pentyl ester*

$C_{20}H_{17}NO_7$
- Berberal

$C_{20}H_{17}NO_8$
- Berberilic Anhydride

$C_{20}H_{17}N_3O_3$
- Euxylophoricine A†

$C_{20}H_{17}N_3O_5$
- 4-Amino-2,6-dinitrophenol, N-*Dibenzyl*

$C_{20}H_{17}N_3O_5S$
- 6-Nitro-4-sulpho-*m*-toluic Acid, *Dianilide*

$C_{20}H_{18}$
- 1,2-Dibenzylbenzene
- 1,3-Dibenzylbenzene
- 1,4-Dibenzylbenzene
- Diphenyl-*o*-tolylmethane
- Diphenyl-*m*-tolylmethane
- Diphenyl-*p*-tolylmethane
- 1,1,1-Triphenylethane
- 1,1,2-Triphenylethane

$C_{20}H_{18}BrN_3$
- Dimidium bromide

$C_{20}H_{18}ClNO_6$
- Ochratoxin A†

$C_{20}H_{18}ClN_3$
- Dimidium chloride

$C_{20}H_{18}Fe_2$
- Biferrocenyl

$C_{20}H_{18}NO_4$ (ion)
- Berberine★†

$C_{20}H_{18}N_2O_2$
- *N*-Dibenzyl-2-nitroaniline
- *N*-Dibenzyl-3-nitroaniline
- *N*-Dibenzyl-4-nitroaniline

$C_{20}H_{18}N_2O_4$
- Perloline

$C_{20}H_{18}N_2O_6S_2$
- Apoaranotin†

$C_{20}H_{18}N_2O_7$
- Berberilic Anhydride, *Amide*

$C_{20}H_{18}O$
- α,α-Diphenyl-*o*-cresol, *Me ether*
- α,α-Diphenyl-*m*-cresol, *Me ether*
- α,α-Diphenyl-*p*-cresol, *Me ether*
- Diphenyl-*o*-tolylmethanol
- Diphenyl-*m*-tolylmethanol
- Diphenyl-*p*-tolylmethanol
- Phenyl-*p*-tolylmethanol, *Phenyl ether*
- 1,1,2-Triphenylethanol
- 1,2,2-Triphenylethanol
- 2,2,2-Triphenylethanol
- Triphenylmethanol, *Me ether*

$C_{20}H_{18}O_2$
- 2-Hydroxyphenyldiphenylmethanol, *Me ether*
- 3-Hydroxyphenyldiphenylmethanol, *Me ether*
- 4-Hydroxyphenyldiphenylmethanol, *Me ether*
- Resorcinol, *Dibenzyl ether*

$C_{20}H_{18}O_3$
4-Phenyl-3-butenoic Acid, *Anhydride*
Phloroglucinol, *Dibenzyl ether*

$C_{20}H_{18}O_4$
Anthracene-1,5-dicarboxylic Acid, *Di-Et ester*
Cyclocoumarol
Dibenzylidenesuccinic Acid, *Di-Me ester*
Phaseollin†
Phenanthrene-1,2-dicarboxylic Acid, *Di-Et ester*
Phenanthrene-3,10-dicarboxylic Acid, *Di-Et ester*

$C_{20}H_{18}O_5$
Artocarpesin†
Madagascin†
Osajaxanthone★, *Di-Me ether*†
Tanshinonic Acid *Me ester*†

$C_{20}H_{18}O_6$
Asarinin
Averythrin†
Benzil-2,2′-dicarboxylic Acid, *Di-Et ester*
γ-Citromycinone†
Hinokinin†
Isohinokinin
Jacareubin, *Di-Me ether*†
Nepseudin
Nor-β-anhydroicaritin
Sesamin

$C_{20}H_{18}O_7$
α-Citromycinone†
Laccaic Acid D, *Me ester, Tri-Me ether*†
Norsolorinic Acid†
Rhodocomatulin, 3,6-*Di-Me ether*†
Rhodocomatulin, 6,8-*Di-Me ether*†
γ-Rhodomycinone

$C_{20}H_{18}O_8$
Biphenyl-2,3,5,6-tetracarboxylic Acid, *Tetra-Me ester*
Biphenyl-3,3′,4,4′-tetracarboxylic Acid, *Tetra-Me ester*
Biphenyl-3,3′,5,5′-tetracarboxylic Acid, *Tetra-Me ester*
Ceroalbolinic Acid, *Tetra-Me ether*†
Ceroalbolinic Acid, *Me ester*, 2,6,7-*Tri-Me ether*†
Elliptic Acid
γ-Isorhodomycinone†
Lucidin, *Di-Me ether* (flavone)†
Melisimplexin
Meliternin
α-Rhodomycinone†
α_2-Rhodomycinone†
β-Rhodomycinone
Tartaric Acid, *Diphenacyl ester*
3,5,6,7-Tetrahydroxy-1-methylanthraquinone-2-carboxylic Acid, *Me ester*, 3,6,7-*Tri-Me ether*†

$C_{20}H_{18}O_9$
Distemonanthin, *Tetra-Me ether*
2-Formyl-4,5-dimethoxybenzoic Acid, *Anhydride*
2-Formyl-5,6-dimethoxybenzoic Acid, *Anhydride*
α-Isorhodomycinone†
β-Isorhodomycinone†

$C_{20}H_{18}O_{10}$
Euxanthic Acid, *Me ester*
Haemathamnolic Acid, *Me ester*†
Juglanin†

$C_{20}H_{18}O_{11}$
Avicularin†
Guaijaverin†
Polystachoside†
Quercetin★, 3-β-D-*Xyloside*†
Reynoutrin†

$C_{20}H_{18}O_{14}$
Lapathinic Acid

$C_{20}H_{19}N$
o-Aminophenyldiphenylmethane, N-*Me*
α-Aminotriphenylmethane, N-*Me*
N-Methyl-*o*-diphenylmethylaniline
N-Methyltriphenylmethylamine
N-Phenyldibenzylamine

$C_{20}H_{19}NO_2$
3-Methyl-2-phenylquinoline-4-carboxylic Acid, *Propyl ester*
6-Methyl-2-phenylquinoline-4-carboxylic Acid, *Propyl ester*
2-Phenylquinoline-4-carboxylic Acid, *Butyl ester*
2-Phenylquinoline-4-carboxylic Acid, *Isobutyl ester*
2-*p*-Tolylquinoline-4-carboxylic Acid, *Propyl ester*
γ-Truxillic Acid, *Et-imide*
ε-Truxillic Acid, *Et imide*

$C_{20}H_{19}NO_3$
Acronycine

$C_{20}H_{19}NO_4$
Annuloline
Dehydrodicentrine†
Ficine†
4-(4′-Hydroxy-3′-methoxybenzylidene)-2,3-dioxo-1-phenethylpyrrolidine†
Isoficine†

$C_{20}H_{19}NO_5$
Berberine
Chelidonine
Papaveraldine
Protopine

$C_{20}H_{19}NO_6$
Cornigereine†
Isorhoeagenine†
Ochratoxin B†
Ochrobirine
Papaverrubine A†
Papaverrubine E†
Rhoeagenine†

$C_{20}H_{19}NO_8$
Phomazarin, *Me ester*†

$C_{20}H_{19}NO_9$
Berberilic Acid

$C_{20}H_{19}N_3$
N^1,N^2-Diphenyl-N^3-*o*-tolylguanidine
N^1,N^2-Diphenyl-N^3-*m*-tolylguanidine
N^1,N^2-Diphenyl-N^3-*p*-tolylguanidine

$C_{20}H_{19}N_3O_3$
Anisessine †
Anisotine †

$C_{20}H_{20}$
trans-15,16-Diethyldihydropyrene †

$C_{20}H_{20}NO_4$ (ion)
Jatrorrhizine

$C_{20}H_{20}N_2$
α,α′-Diamino-*o*-xylene, N,N′-*Diphenyl*
N-Dibenzyl-*o*-phenylenediamine
NN′-Dibenzyl-*o*-phenylenediamine
N-Dibenzyl-*m*-phenylenediamine
N-Dibenzyl-*p*-phenylenediamine
N,N′-Di-*p*-tolyl-*m*-phenylenediamine
Di-*o*-tolyl-*p*-phenylenediamine
Di-*p*-tolyl-*p*-phenylenediamine
o-Phenylenediamine, N-*Dibenzyl*
m-Phenylenediamine, N-*Dibenzyl*

$C_{20}H_{20}N_2O$
αα-Di-(4-aminophenyl)benzyl Alcohol, *Me ether*

$C_{20}H_{20}N_2O_2$
10,22-Dioxokopsane †

$C_{20}H_{20}N_2O_3$
2-Phenyl-3-*n*-propylisoxazolidine-4,5-*cis*-dicarboxylic Acid *N*-Phenylimide †
Picrorocellin, *Anhydropicrorocellin*
Schizozygine

$C_{20}H_{20}N_2O_4$
Caffaeoschizin (?)

$C_{20}H_{20}N_2O_6$
Butane-1,2,3,4-tetracarboxylic Acid, *Dianilide*

$C_{20}H_{20}O$
Acetyl[18]annulene †

$C_{20}H_{20}O_2$
9-Isopropyl-1-methylphenanthrene-3-carboxylic Acid, *Me ester*

$C_{20}H_{20}O_3$
Biflorine I
2-Isovaleryl-4-(3-methylbut-2-eneylidine)cyclopentane-1,3-dione †
Thebenol, *Propyl ether*

$C_{20}H_{20}O_3S$
7-Isopropyl-1-methylphenanthrene-2-sulphonic Acid, *Et ester*
2-Isopropyl-8-methylphenanthrene-3-sulphonic Acid, *Et ester*

$C_{20}H_{20}O_4$
Avicennin
Bavachin †
Chrysin, *Mono-isopentyl ether*
Di-isosafrol
2,3-Diphenyl-2-butene-1,4-dicarboxylic Acid, *Di-Me ester*
1,3-Diphenyl-1-butene-4,4-dicarboxylic Acid, *Di-Me ester*
1,1-Diphenylethylene-2,2-dicarboxylic Acid, *Di-Et ester*
Diphenylfumaric Acid, *Di-Et ester*
Diphenylmaleic Acid, *Di-Et ester*
Durlettone †
α-Isatropic Acid, *Mono-Et ester*
β-Isatropic Acid, *Mono-Et ester*
Isobavachalcone †
Isobavachin †
Madagascin Anthrone †
Neotruxinic Acid, *Di-Me ester*
Neotruxinic Acid, 1-*Et ester*
Neotruxinic Acid, 2-*Et ester*
Otobain
Stilbene-2,2′-dicarboxylic Acid, *Di-Et ester*
Stilbene-2,4′-dicarboxylic Acid, *Di-Et ester*
Stilbene-3,3′-dicarboxylic Acid, *Di-Et ester*
Stilbene-4,4′-dicarboxylic Acid, *Di-Et ester*
α-Truxillic Acid, *Di-Me ester*
α-Truxillic Acid, *Mono-Et ester*
γ-Truxillic Acid, *Di-Me ester*
γ-Truxillic Acid, *Mono-Et ester*
ε-Truxillic Acid, *Di-Me ester*
η-Truxillic Acid, *Di-Me ester*
epi-Truxillic Acid, *Di-Me ester*
β-Truxinic Acid, *Di-Me ester*
β-Truxinic Acid, *Mono-Et ester*
δ-Truxinic Acid, *Di-Me ester*
ζ-Truxinic Acid, *Di-Me ester*
ζ-Truxinic Acid, *Mono-Et ester*

$C_{20}H_{20}O_5$
2-(3,3-Dimethylallyl)-1,3,5-trihydroxyxanthone, 3,5-*Di-Me ether* †
Futoenone †
Galbacin
Hydroxyotobain †
Mbarraxanthone, 3,7-*Di-Me ether* †
Selinone †
Utahin †

$C_{20}H_{20}O_6$
Angolensin, *Di-O-Ac*
Cubebin★ †
Haematein, *Tetra-Me ether*
Involutin, *Tri-Me ether* †
Isorhodoptilometrin, *Tri-Me ether* †
Kaempferide, 3,7-*Di-Et ether*
Leptosidin, *Tri-Me ether*
Luteolin, 3′,5,7-*Tri-Me-4*′-*Et ether*
Nalgiovensin★, *Di-Me ether* †
Phellamuretin
Rhamnin A
Rhodoptilometrin, 1,6,8-*Tri-Me ether* †
Scriblitifolic Acid, *Me ether* †
Scriblitifolic Acid, *Me ester* †
2,3,6,7-Tetrahydroxyanthraquinone, 2,6-*Di-Me*-3,7-*di-Et ether*
Tsugaresinol

$C_{20}H_{20}O_7$
Averantin †
Demethoxysudachitin, *Tri-Me ether* †
Fibraurin †
Herbacetin, *Penta-Me ether*
5-Hydroxy-3,4′,6,7-tetramethoxyflavone, *Me ether* †

$C_{20}H_{20}O_7$ (*continued*)
Mikanin, *Di-Me ether*†
Morin, *Penta-Me ether*
Norherqueinone, *Mono-Me ether*
Penduletin, *Di-Me ether*
1,3,4,5,6-Pentahydroxy-2-methylanthraquinone, 1,3,4,5,6-*Penta-Me ether*
Pinoresinolide
Podospicatin, *Tri-Me ether*†
Prudomestin, *Tri-Me ether*†
Quercetin, *Penta-Me ether*
Robinetin, *Penta-Me ether*
Tangeritin
Tricetin, *Penta-Me ether*
Tricin, *Tri-Me ether*
Vogeletin, *Tetra-Me ether*†
Wightin, *Di-Me ether*†
Xanthomicrol, *Di-Me ether*

$C_{20}H_{20}O_8$
Artemetin
Asperthecin, *Penta-Me ether*
Centaureidin, 3,5-*Di-O-Me ether*†
Combretol†
Gentiogenin, *Dimer*
Gossypetin, 3,3′,4′,5,8-*Penta-Me ether*
1,2,3,5,6,7-Hexahydroxyanthraquinone, *Hexa-Me ether*
1,2,3,5,6,7-Hexahydroxyanthraquinone, *Tri-Et ether*
3′,4′,5,6,7,8-Hexahydroxyflavone, 3′,4′,6,7,8-*Penta-Me ether*†
6-Hydroxyfibraurin†
5-Hydroxy-3,3′,4′,7,8-pentamethoxyflavone†
Irigenin, 5,5′-*Di-Me ether*
Limocitrin, 3,3′,7-*Tri-Me ether*
Myricetin, 3′,4′,5′,7-*Penta-Me ether*
Serpyllin†

$C_{20}H_{20}O_9$
Erianthin
ε-Isorhodomycinone

$C_{20}H_{20}O_{14}$
Hamameli-tannin

$C_{20}H_{21}ClO_6$
Morindin chloride, *Penta-Me ether*
Radicicol, *Di-Me ether*

$C_{20}H_{21}NO$
Apocinchene, *Me ether*

$C_{20}H_{21}NO_2$
C-Xanthocurine

$C_{20}H_{21}NO_3$
Galipine
Isogalipine

$C_{20}H_{21}NO_4$
Amurensine, O-*Me ether*†
Bulbocapnine, *Me ether*
Canadine
Crebanine
Cularicine, O-*Et ether*†
Dicentrine
Eschscholtzidine†
Evoprenine†
Laurepukin, *Di-Me ether*
Nandigerine, N,O-*Di-Me*†
Nantenine†
Papaverine
Sinactine

$C_{20}H_{21}NO_5$
Columbamine
Fumaritine
Hunnemanine†
9-Hydroxy-3,10-dimethoxy-1,2-methylenedioxy-noraporphine, N-*Me*†
9-Hydroxy-3,10-dimethoxy-1,2-methylenedioxy-noraporphine, O-*Me ether*†
Ocokryptine†
Ophiocarpine
13-*epi*-Ophiocarpine
Papaverinol

$C_{20}H_{21}NO_6$
Papaverrubine C†
Papaverrubine D†

$C_{20}H_{21}N_2O$ (ion)
Fedamazine
Melinonine H (or $C_{20}H_{23}N_2O$)

$C_{20}H_{21}N_3O$
Tri-(*p*-aminophenyl)methanol, *Me ether*

$C_{20}H_{21}N_3O_2$
Deoxyaniflorine†

$C_{20}H_{21}N_3O_3$
Aniflorine†

$C_{20}H_{22}ClN$
Pyrrobutamine

$C_{20}H_{22}N_2$
2-Phenyl-1,3-naphthylenediamine, N,N-*Tetra-Me*

$C_{20}H_{22}N_2O$
Kopsanone†
Koumine
Quinene

$C_{20}H_{22}N_2O_2$
Akuammicine
Condylocarpine†
Gelsemine
Isoschizogalin
epi-Kopsanol-10-lactam†
Pericyclivine†
Pleiocarpamine†
Quininone
Schizogalin
Strictamine†
Thalloperazine

$C_{20}H_{22}N_2O_3$
Echitamidine
Lochrovicine†
Mossambine
Perivine★†
Picrinine†

$C_{20}H_{22}N_2O_4$
Lysuric Acid
Ornithuric Acid, *Me ester*
Perividine†
Picrorocellin

$C_{20}H_{22}N_2O_4$ (*continued*)
Piperazine-*N*,*N'*-dicarboxylic Acid, *Dibenzyl ester*
α-Schizozygol
β-Schizozygol

$C_{20}H_{22}N_8O_5$
Amethopterin

$C_{20}H_{22}O$
Asperyellone †

$C_{20}H_{22}O_2$
4,4′-Di-isopropylbenzil
D-Homoequilenin, O-*Me ether*

$C_{20}H_{22}O_3$
2,2-Dimethylbutyric Acid, p-*Phenylphenacyl ester*
2,4-Dimethylphenylacetic Acid, *Anhydride*

$C_{20}H_{22}O_4$
Bibenzyl-2,2′-dicarboxylic Acid, *Di-Et ester*
Bibenzyl-4,4′-dicarboxylic Acid, *Di-Et ester*
Cervicarcin, *Me ether* †
2,2-Diphenylsuccinic Acid, *Di-Et ester*
2,3-Diphenylsuccinic Acid, *Di-Et ester*
Otobaphenol †

$C_{20}H_{22}O_5$
3-Acetyl-4,7-dimethylnaphthalene-1,6-dicarboxylic Acid, *Di-Et ester* †
Agrimolide, *Di-O-Me ether* †
Brazilin, *Tetra-Me ether*
Collinin
2-(3,4-Dimethoxyphenyl)-5-(3-hydroxypropyl)-7-methoxybenzofuran †
Farrerol, *Tri-Me ether* †
Pleurotin
Sorbifolin †

$C_{20}H_{22}O_6$
Coleon A
Columbin
Haematoxylin, *Tetra-Me ether*
Isocolumbin
Matai-resinol
Miroestrol †
Pedicellin
Peltogynol, *Tetra-Me ether*
Pinoresinol

$C_{20}H_{22}O_7$
Barbatic Acid, *Me ester*
Chasmanthin †

$C_{20}H_{22}O_7$
Deodarin, *Tetra-Me ether* †
Diffractaic Acid
Elephantin †
Jateorin †
8-*epi*-Jateorin †
Lecanoric Acid, *Tri-Me ether*
Palmarin
3,3′,4′,7,8-Pentamethoxyflavanone †

$C_{20}H_{22}O_8$
Populin

$C_{20}H_{22}O_9$
Digallic Acid, *Penta-Me ether*, *Me ester*
Olivin †
Salireposide †
Trichocarpin †

$C_{20}H_{22}O_{10}$
Plicatic Acid †
Polydine †

$C_{20}H_{22}O_{11}$
Lecanoric Acid, *Erythritol ester*

$C_{20}H_{23}ClN_2O_4$
Chlorprophenpyridamine, *Maleate*

$C_{20}H_{23}NO_2$
1-*N*,*N*-Dimethylaminoethyl-3,4-dimethoxyphenanthrene †
Macranthin, *Di-O-Ac* †

$C_{20}H_{23}NO_3$
Morphothebaine, *Di-Me ether*
Thebenine, N,1-*Di-Me*
Thebenine,1-*Et ether*
Tuduranine, *Et ether*

$C_{20}H_{23}NO_4$
Acedicon
Caseadine †
Corydine
Corypalmine
Cularine
Flavinantine, *Me ether* †
Flavinine, N,O-*Di-Me* †
Glaucentrine
Hernovine, *Di-O-Me ether* †
Hernovine, N,10-*Di-Me* †
1-Hydroxy-2,9,10-trimethoxyaporphine †
10-Hydroxy-1,2,9-trimethoxyaporphine †
Isocordine
Isocorydine
Kreysiginone †
Laurotetanine, *Me ether*
Norargemonine †
Pavine
Romneine †
Sinoacutine, O-*Me* †
Tetrahydrocolumbanine
Thalicmidine

$C_{20}H_{23}NO_5$
Capaurimine †
Ethyl 6,7-di-(cyclopropylmethoxy)-4-hydroxyquinoline-3-carboxylate †

$C_{20}H_{23}NO_7S$
Erysothiovine

$(C_{20}H_{23}N_2)_n$ (ion)
C-Isodihydrotoxiferine I

$C_{20}H_{23}N_2O$ (ion)
C-Alkaloid P
C-Curarine III
Hemidihydrotoxiferine
Melinonine E
Melinonine H (or $C_{20}H_{21}N_2O$)

$C_{20}H_{23}N_2O_2$ (ion)
C-Fluorocurine

$C_{20}H_{23}N_3O_2$
Pheneserine

$C_{20}H_{23}N_7O_7$
Folinic Acid SF

$C_{20}H_{24}$
2,2′,4,4′,5,5′-Hexamethylstilbene
2′,4-Octamethylenebiphenyl†

$C_{20}H_{24}ClNO_2$
1-(4′-Chlorophenethyl)-1,2,3,4-tetrahydro-6,7-dimethoxy-2-methylisoquinoline†

$C_{20}H_{24}ClNO_4$
Laurifoline, *Chloride*†

$C_{20}H_{24}ClN_3S$
Prochlorperazine

$C_{20}H_{24}INO_4$
Cryptaustoline
Magnoflorine, *Iodide*†

$C_{20}H_{24}NO_4$ (ion)
Cyclanoline†
Laurifoline†
Magnoflorine†

$C_{20}H_{24}N_2O$
Affinisine†
Kopsanol†
epi-Kopsanol†
Maviensine
Taberpsychine†
Tetraphyllicine

$C_{20}H_{24}N_2OS$
Lucanthone
Propiomazine†

$C_{20}H_{24}N_2O_2$
N-Acetyl-11-hydroxyaspidospermatidine†
Ajmalidine
Alloquinidine
Apoquinidine, *Me ether*
Apoquinine, *Me ether*
Calebassine A
Cavincine†
19,20-Dihydroakuammicine†
α-Isoquinidine
β-Isoquinidine
γ-Isoquinidine
Lochnerine
11-Methoxyeburnamonine†
2-Methylpimelic Acid, *Dianilide*
3-Methylpimelic Acid, *Dianilide*
4-Methylpimelic Acid, *Dianilide*
Quinicine
Quinidine
epi-Quinidine
Quinine
epi-Quinine
Quinotinone
Tubotaiwin

$C_{20}H_{24}N_2O_2S$
Hycanthone†

$C_{20}H_{24}N_2O_3$
Corynanthic Acid
Corynanthidic Acid
Isoyohimbic Acid
Kopsidine
Lochneridine†
Vincamidine
Yohimbic Acid
α-Yohimbic Acid
β-Yohimbic Acid

$C_{20}H_{24}N_2O_4$
Compactinervine†
Quitenine, *Et ester*

$C_{20}H_{24}N_2O_4S_2$
Cystine, *Di-Benzyl ester*

$C_{20}H_{24}N_4O_2$
2,3-Dianilinosuccinic Acid, *Di-Et ester*

$C_{20}H_{24}N_{10}O_{21}P_4$
P^1,P^4-Diguanosine 5′-tetraphosphate†

$C_{20}H_{24}O_2$
Diethylstilboestrol, *Di-Me ether*
4,4′-Di-isopropylbenzoin
17-Ethynyloestra-1,3,5(10)-triene-3,17-diol
Isoanethole

$C_{20}H_{24}O_3$
4,4′-Di-isopropylbenzilic Acid
Gravelliferone, *Me ether*†
Ostruthin, *Me ether*

$C_{20}H_{24}O_4$
Angolensin, 2,4-*Di*-O-*Et*
Brayleyanin
Crocetin
Di-isochavibetol
Guaiaretic Acid
Naphthalene-1,8-dicarboxylic Acid, *Dibutyl ester*
Ovatodioide
Ovatodiolide†
Sciadin
Sciadinone

$C_{20}H_{24}O_5$
Eriostemoic Acid
Eriostoic Acid
Gibberellin A_5, *Me ester*
Gibberellin A_7, *Me ester*
Gibberellin A_{11}, *Me ester*†
Sequirin B, *Tri-Me ether*†
Sequirin C, *Tri-Me ether*†

$C_{20}H_{24}O_6$
Catechin, *Penta-Me ether*
α-Conidendryl Alcohol
β-Conidendryl Alcohol
Gibberellenic Acid, 2-*Me ether*
Gibberellic Acid, *Me ester*
Gibberellin A_{21}, *Me ester*†
Glaucanic Acid, *Di-Me ester*
Grayanotoxin III
Lariciresinol

$C_{20}H_{24}O_7$
Ailanthone†
Angelol†
Cyclo-olivil
Euparotin†
Eupatundin†
Gallocatechin, *Penta-Me ether*
epi-Gallocatechin, *Penta-Me ether*
Glauconic Acid, *Di-Me ester*
3,3′,4,4′,7,8-Hexahydroxyflavan, 3,3′,4′,7,8(or 3′,4,4′,7,8)-*Penta-Me ether*
Olivil
3,3′,5,5′,7-Pentahydroxy-4′-methoxy-*epi*-catechin, *Tetra-Me ether*†
3,3′,4′,7,8-Pentamethoxyflavan-4-ol†
Phloraspyron†

$C_{20}H_{24}O_8$
Capenicin
Eupatoroxin†
10-*epi*-Eupatoroxin†
3,3′,4,4′,5,5′,7-Heptahydroxyflavan, 3′,4′,5,5′,7-*Penta-Me ether*†
Vellein

$C_{20}H_{24}O_9$
Dihydro-oroselol, *Glucoside*†
Ginkgolide A†

$C_{20}H_{24}O_{10}$
Ginkolide B†
Ginkolide M†

$C_{20}H_{24}O_{11}$
Ginkolide C†

$C_{20}H_{25}ClO_7$
Eupachlorin†

$C_{20}H_{25}ClO_8$
Eupachloroxin†

$C_{20}H_{25}NO$
Isoamidone

$C_{20}H_{25}NO_2$
Cuspareine
Spiradine A†

$C_{20}H_{25}NO_3$
Armepavine, *Me ether*
Dimenoxadole

$C_{20}H_{25}NO_3$
Hemirepanduline†

$C_{20}H_{25}NO_4$
Codamine†
Corpaverine
Dihydrokreysiginone†
2,5-Dihydroxy-4-nitrobiphenyl, *Dibutyl ether*
Eccaine
Laudanidine
Laudanine
ψ-Laudanine★†
Thalifendlerine†

$C_{20}H_{25}NO_5$
Prometaphanine†

$C_{20}H_{25}N_2IO$
Macusine B, *Iodide*★†

$C_{20}H_{25}N_2O$ (ion)
Macusine B★†
Mavacurine

$C_{20}H_{25}N_2O_2$ (ion)
Calebassine F
ψ-Fluorocurine
Hemitoxiferine I

$C_{20}H_{25}N_3O$
Lysergide†
Prodigiosine

$C_{20}H_{25}N_3O_2$
Methylergometrine

$C_{20}H_{25}N_3O_3$
2,3-Dehydro-*O*-(2-pyrrolylcarbonyl)virgiline†

$C_{20}H_{25}N_3O_4S$
Hetacillin, *Me ester*†
6-*epi*-Hetacillin, *Me ester*†

$C_{20}H_{26}ClNO_2$
Adiphenine

$C_{20}H_{26}ClNO_4$
Tembetarine, *Chloride*†

$C_{20}H_{26}ClN_3O_6S_2$
Sporidesmin D†

$C_{20}H_{26}NO_3$ (ion)
Colletine†
Petaline†

$C_{20}H_{26}NO_4$ (ion)
Tembetarine†

$C_{20}H_{26}N_2O$
Astrocasine†
Ibogaine
17-Methoxyaspidofractinine†
Tabernanthine†

$C_{20}H_{26}N_2O_2$
Ajmaline
Aspidofiline
Beninine★†
Burnamicine†
19,20-Dehydro-10-methoxydihydro-corynantheol†
Hydroquinidine
Iboluteine★†
Iboxygaine
Quinoticine
Sandwicine
Vincaminol†

$C_{20}H_{26}N_2O_3$
Ajmalinine
Aspidodispermine, *Me ether*†
Lanceine (or $C_{24}H_{36}N_2O_4$)†

$C_{20}H_{26}N_2O_4$
Gelsemicine
Melinonine L

$C_{20}H_{26}N_2O_6$
Nitrosporin

$C_{20}H_{26}O$
Retinene$_2$

$C_{20}H_{26}O_2$
Δ^5-Dehydrosugiol†
1,2-Dihydroxy-1,2-di-(2,4,6-trimethylphenyl)-ethane
Eremolactone
Iso-eremolactone†
Norethisterone
Norethynodrel†
Octofollin
Promethestrol

$C_{20}H_{26}O_3$
9-Dehydroroyleanone
Fuerstiaquinone
p-Geranyloxycinnamic Acid, *Me ester*†
Hautriwaic Acid, *Lactone*†
Kahweol
Taxodione†
Xanthoperol

$C_{20}H_{26}O_4$
Carnosol†
Fujenal

$C_{20}H_{26}O_4$ (*continued*)
p-Geranylferulic Acid†
Gibberellin A_9, *Me ester*
Gibberellin A_{15}†
Phthalic Acid, *Di-cyclohexyl ester*
Picrosalvin
3,3′,5,5′-Tetrahydroxybiphenyl, *Tetra-Et ether*

$C_{20}H_{26}O_5$
Fujenoic Acid
Gibberellin A_4, *Me ester*
Gibberellin A_{20}, *Me ester*†
Gibberellin A_{24}†
Isodocarpin†
Marrianolic Acid, *Di-Me ester*
Marrubiin-3-one†
Olearin†
Thurberilin†
Zearalenone, 2,4-*Di-Me ether*†

$C_{20}H_{26}O_6$
Eupatoriopicrin
Fastigilin A†
Fastigilin B†
Fastigilin C†
Gibberellin A_1, *Me ester*
Gibberellin A_6, *Me ester*
Gibberellin A_{16}, *Me ester*†
Gibberellin A_{19}†
Heliangine†
seco-Isolariciresinol
Nodosin†
epi-Nodosin†

$C_{20}H_{26}O_7$
Chaparrinone†
Cnicin†
Gibberellin A_8, *Me ester*
Gibberellin A_{13}†
Gibberellin A_{17}†
Gibberellin A_{23}†
2′,3,5,5′,7-Pentahydroxyflavone, *Penta-Me ether*
2′,3,5,7,8-Pentahydroxyflavone, *Penta-Me ether*
2′,4′,5,7,8-Pentahydroxyflavone, *Penta-Me ether*
3,3′,4′,5,8-Pentahydroxyflavone, *Penta-Me ether*
3,3′,4′,6,7-Pentahydroxyflavone, *Penta-Me ether*
3,3′,4′,7,8-Pentahydroxyflavone, *Penta-Me ether*
3′,4′,5,6,7-Pentahydroxyflavone, *Penta-Me ether*
3,4′,6,7,8-Pentahydroxyflavone, *Penta-Me ether*
3,5,6,7,8-Pentahydroxyflavone, *Penta-Me ether*
3′,4′,5,7,8-Pentahydroxyflavone, *Penta-Me ether*
4′,5,6,7,8-Pentahydroxyflavone, *Penta-Me ether*
Visnagan (or $C_{20}H_{28}O_7$)

$C_{20}H_{26}O_8$
Bruceine G†
Glaucarubolone†

$C_{20}H_{27}BrN_2O$
Ambutonium bromide

$C_{20}H_{27}BrO_3$
9α-Bromo-17β-hydroxy-17α-methylandrost-4-ene-3,11-dione†

$C_{20}H_{27}ClN_2O_2$
Hunterburnine, β-*Methchloride*

$C_{20}H_{27}IN_2O$
Ambutonium iodide

$C_{20}H_{27}IN_2O_2$
Hunterburnine, α-*Methiodide*

$C_{20}H_{27}N$
Di-(4-isopropylbenzyl)amine

$C_{20}H_{27}NO_2$
Galbulimina alkaloid G.B.16†
Kobusine★†
Spiradine B†

$C_{20}H_{27}NO_3$
Calpurnine
Hetisine
Pseudokobusine†

$C_{20}H_{27}NO_5$
Buquinolate†
Intensatin†
Phalaenopsin†

$C_{20}H_{27}NO_7$
Latifoline

$C_{20}H_{27}NO_{11}$
Amygdalin

$C_{20}H_{27}N_2O$ (ion)
C-Alkaloid O
Melinonine B

$C_{20}H_{27}N_2O_3$ (ion)
C-Alkaloid Y

$C_{20}H_{27}N_3O_3$
O-(2-Pyrrolylcarbonyl)virgiline†

$C_{20}H_{28}$
Anhydrovitamin A

$C_{20}H_{28}ClNO_8P_2$
Pyrrobutamine, *Diphosphate*

$C_{20}H_{28}HgN_2$
p-Mercuri-di-diethylanilinc

$C_{20}H_{28}N_2$
Vallesamidine†

$C_{20}H_{28}N_2O$
Astrophylline, N-*Me deriv.*†
Deacetylaspidospermine†

$C_{20}H_{28}N_2O_2$
Affinine†
Cryptocurine (or $C_{20}H_{30}N_2O_2$)
10-Methoxydihydrocorynantheol†

$C_{20}H_{28}N_2O_3$
Oxyphencyclimine
1,2,2-Trimethylcyclopentane-1,3-dicarboxylic Acid, α-*Nitrile*, *Anhydride*
1,2,2-Trimethylcyclopentane-1,3-dicarboxylic Acid, β-*Nitrile*, *Anhydride*

$C_{20}H_{28}N_2O_5$
- 3-Nitro-(+)-camphor, *Anhydride*

$C_{20}H_{28}N_2S_2$
- Di-(4-diethylaminophenyl)disulphide

$C_{20}H_{28}O$
- Retinene$_1$
- Vitamin A$_2$

$C_{20}H_{28}O_2$
- Dehydroabietic Acid
- 4-*epi*-Dehydroabietic Acid†
- Hinokione
- 17β-Hydroxy-17α-methylandrosta-1,4-dien-3-one†
- 3-Hydroxyretinal†
- Kaura-9(11),16-dien-19-oic Acid†
- Kaurenolide†
- 5,6-Monoepoxyvitamin A Aldehyde†
- 5,8-Monoepoxyvitamin A Aldehyde†
- 7-Oxototarol
- 16-Oxototarol
- Retinoic Acid†
- Sugiol
- Totarolone

$C_{20}H_{28}O_3$
- Cafestol
- Cinerin I
- Daniellic Acid†
- Hardwickiic Acid†
- 13-Hydroxy-14-isopropylpodocarpa-8,11,13-trien-16-oic Acid
- 7-Hydroxykaurenolide
- Illuric Acid
- Isocafestol
- Lambertianic Acid†
- 6α-Methyl-5α-androstane-3,7,17-trione†
- 11-Oxaprogesterone†
- Petasalbin, *Angelate ester*
- Polyalthic Acid
- Rosenonolactone
- Rosonolactone
- Royleanone
- Solidagenone†
- Taxodone†
- Vinhaticoic Acid
- Vouacapenic Acid

$C_{20}H_{28}O_4$
- Adhulupone†
- Araucarenolone†
- Callicarpone†
- 4-Deoxycohumulone†
- 7,18-Dihydroxykaurenolide
- 11α,21-Dihydroxy-4-pregnen-3,20-dione
- Flexuosin B†
- Furanopetasin
- Gibberellin A$_{12}$†
- Hautriaic Acid†
- Horminone†
- Hulupone
- 16-Hydroxyeperu-8(20),13-diene-15,18-dioic Acid (15 → 16) Lactone†
- 6β-Hydroxyrosenolactone†
- Marrubiin
- Petasitin†
- Petasitolide A
- Petasitolide B
- Salvin†
- Taxoquinone†

$C_{20}H_{28}O_5$
- Cascarillin A†
- Cohumulone
- Gibberellin A$_{10}$, *Me ester*†
- Gibberellin A$_{14}$†
- Pododacric Acid†

$C_{20}H_{28}O_6$
- Amarolide†
- β-Caesalpin†
- Gibberellin A$_2$, *Me ester*
- Gibberellin A$_{18}$†
- Klaineanone†
- Oridonin†
- Phorbol†
- Poncidin†

$C_{20}H_{28}O_7$
- Chaparrin
- Visnagan (or $C_{20}H_{26}O_7$)

$C_{20}H_{28}O_8$
- Glaucarubol

$C_{20}H_{28}O_{12}$
- Neolloydosin†

$C_{20}H_{28}O_{13}$
- Primeverin
- Primulaverin

$C_{20}H_{29}ClN_2O_2$
- Ochrosandwine, *Chloride*†
- Melinonine B, *Chloride*

$C_{20}H_{29}FO_3$
- Fluoxymesterone

$C_{20}H_{29}IO_3$
- Dodecanoic Acid, p-*Iodophenacyl ester*

$C_{20}H_{29}NO$
- 4-Hydroxy-2-*n*-undecylquinoline†

$C_{20}H_{29}NO_2$
- 17-Azaprogesterone†
- Galbulimina alkaloid G.B.13†
- 4-Hydroxy-2-*n*-undecylquinoline, N-*Oxide*†

$C_{20}H_{29}NO_3$
- Talatisine

$C_{20}H_{29}NO_4$
- Lycofoline, *Di-Ac*

$C_{20}H_{29}NO_8$
- Isorhodomycin A

$C_{20}H_{29}N_2O_2$ (ion)
- Ochrosandwine†
- Melinonine B

$C_{20}H_{29}N_3O_2$
- Percaine

$C_{20}H_{30}$
- Axerophthene

$C_{20}H_{30}N_2O_2$
- Cryptocurine (or $C_{20}H_{28}N_2O_2$)

$C_{20}H_{30}N_4O_7$
Grisamine (or $C_{28}H_{38}N_6O_{10}$)

$C_{20}H_{30}O$
Cryptopinone
Ferruginol
Isodextropimarinal
Isopimara-7,15-dien-18-al†
Sandaracopimaradien-3-one†
Sempervirol†
Stachenone
Totarol
Visnagin
Vitamin A_1★†

$C_{20}H_{30}O_2$
Abieta-8,12-dienoic Acid†
Abieta-8,13-dienoic Acid†
Abieta-12,14-dienoic Acid†
Abietic Acid
Communic Acid
9-Deoxorosenonolactone
Hinokiol, *Me ether*
7β-Hydroxytotarol
16-Hydroxytotarol
Isopimaric Acid†
$\Delta^{8(9)}$-Isopimaric Acid†
Kaurenic Acid
Kaur-16-en-19-oic Acid†
Laricinoleic Acid
14α-Methyltestosterone
17-Methyltestosterone
5,6-Monoepoxyvitamin A†
5,8-Monoepoxyvitamin A†
Norethandrolone
Ozic Acid
Photolevopimaric Acid†
Pimaric Acid
Pimara-8(14),15-dien-19-oic Acid†
Sandaracopimaric Acid
Sapietic Acid
Trachylobanic Acid†
enantio-Trachylobanic Acid†
Vouacapenol

$C_{20}H_{30}O_2S$
"Carvone Hydrosulphide"

$C_{20}H_{30}O_3$
Agathalic Acid†
Dodecanoic Acid, *Phenacyl ester*
3-Hydroxycyclokauranic Acid
15α-Hydroxykaur-16-en-19-oic Acid†
15β-Hydroxykaur-16-en-19-oic Acid†
2-Hydroxy-4-phenylbutryic Acid, (—)-*Menthyl ester*
6α-Hydroxysandaracopimaric Acid†
12β-Hydroxysandaracopimaric Acid†
17-Hydroxy-3,4-secobeyer-4(18),15-dien-3-oic Acid†
Isorosenolic Acid†
Iso-steviol
α-Levantenolide
β-Levantenolide
Psiadiol†
Rosololactone
Steviol
1,2,2-Trimethyl-3-cyclohexene-1-carboxylic Acid, *Anhydride*

$C_{20}H_{30}O_4$
Agathic Acid
Araucarolone†
Atractyligenin, *Me ester*
Cassaic Acid
Eunicin†
19-Hydroxy-3-oxo-16α-kauran-17-oic Acid†
16α-Kaurane-17,19-dioic Acid†
Kolavic Acid†
Vaginatin†

$C_{20}H_{30}O_5$
Andrographolide
Prostaglandin E_3★†
7,16,18-Trihydroxykaurenolide

$C_{20}H_{30}O_6$
δ-Caesalpin†

$C_{20}H_{31}NO$
Benzhexol
2-Benzylpropionic Acid, (—)-*Menthylamide*
Thelepogine

$C_{20}H_{31}NO_2$
Himgaline†

$C_{20}H_{31}NO_4$
α-Lofoline★, O-*Ac*†

$C_{20}H_{31}NO_5$
Benzylpenaldic Acid, *Me ester*

$C_{20}H_{31}NO_6$
Echiumine†

$C_{20}H_{31}NO_7$
Heliosupine†

$C_{20}H_{31}NO_8$
Heliosupine, N-*Oxide*†

$C_{20}H_{31}N_3$
Ormojanine†

$C_{20}H_{32}$
Atisirene†
Biformene
α-Camphorene
γ-Camphorene
Cembrene
Devadarene†
β-Difenchene
Dipinene
Dolabradiene†
Hibaene†
Iso-atisirene†
α-Isocamphorene
Isohibaene†
Isokaurene
Isomanoene
Isophyllocladene
Isopimara-7,15-diene†
Kaurene
Phyllocladene
Pimara-8(14),15-diene†
Rimuene
Sandaracopimaradiene†
Sclarene†
Stach-15-ene★†

$C_{20}H_{32}ClNO$
Rimuene, *Nitroso-chloride*

$C_{20}H_{32}NO$
Hexadeca-2,6,8,12-tetraenoic Acid, *Isobutylamide* †

$(C_{20}H_{32}N_2O_9)_n$
Mycospocidin

$C_{20}H_{32}O$
Elliotinol †
Geranylgeranial †
Isopimara-7,15-dien-3β-ol †
Isopimara-7,15-dien-18-ol †
10-Isopropyl-3,7,13-trimethyltetradeca-2,6,11,13-tetraen-1-al †
Kaur-16-en-19-ol †
Ozol †
Sandaracopimaradien-3β-ol †
Stach-15-ene epoxide †
Stach-15-en-17-ol †
Stach-15-en-19-ol †

$C_{20}H_{32}O_2$
Abietinol
5β-Androstane-17β-carboxylic Acid
Arachidonic Acid
Copalic Acid★ †
Dihydroiso-$\Delta^{13,14}$-abietic Acid †
Eicosa-5,11,14,17-tetraenoic Acid †
Eperua-7,13-dien-15-oic Acid †
17β-Hydroxy-4,4-dimethyl-19-nor-5α-androstan-3-one †
17β-Hydroxy-4,4-dimethyl-19-nor-5β-androstan-3-one †
16α-Hydroxy-(−)-kauran-19-al †
15-Hydroxylabda-8(20),13-dien-19-al †
17β-Hydroxy-17α-methyl-5α-androstan-3-one †
Isonoragathic Acid, *Me ester*
Kaur-15-ene-17,19-diol †
Kaur-16-ene-3α,19-diol †
Kolavenic Acid †
Labda-8(20),13-dien-15-oic Acid †
enantio-Labda-8(20),13-dien-15-oic Acid †
Noragathic Acid, *Me ester*
Oblongifoliol †
2-Oxomanoyl oxide †
3-Oxomanoyl oxide †
Rubenol
Sandaracopimaradiene-3β,18-diol †
Sandaracopimaradiene-3β,19-diol †
Stach-15-ene-3α,17-diol †
Stach-15-ene-3α,19-diol †
Stach-15-ene-17,19-diol †
Stach-15-en-19-ol epoxide †
Sterculynic Acid, *Me ester* †
Tetradecanoic Acid, *Phenyl ester*
Torulosal †
Virescenol B †

$C_{20}H_{32}O_3$
Abbeokutone †
Araucarol †
Beyerol †
8β,13β-Epoxyeperu-14-en-18-oic Acid †
Grindelic Acid
3α-Hydroxy-5α-androstane-17β-carboxylic Acid
16α-Hydroxy-(−)-kauran-19-oic Acid †
19-Hydroxy-16α-kauran-17-oic Acid †
17-Hydroxy-16α-kauran-19-oic Acid †
18-Hydroxy-2-oxomanoyl oxide †
α_1-Levantanolide
α_2-Levantanolide
Sandaracopimara-8(14),16-diene-2α,18,19-triol †
Sandaracopimara-8(14),16-diene-3β,18,19-triol †
Torulosic Acid †
Virescenol A †

$C_{20}H_{32}O_4$
1α,19-Dihydroxy-16α-kauran-17-oic Acid †
16,17-Dihydroxy-16β-kauran-19-oic Acid †
Eperu-8(20)-en-15,18-dioic Acid †
enantio-Labd-8(20)-ene-15,18-dioic Acid †
Marrubenol †
Oxygrindelic Acid
Pinifolic Acid
Rapanone, *Mono-Me ether*
Sandaracopimara-8(14),16-diene-2α,3β,18,19-tetrol †
Taxa-4(16),11-diene-5α,9α,10β,13α-tetraol †

$C_{20}H_{32}O_5$
2,3-Dicarboxy-2,3-secomanoyl oxide †
Grayanotoxin II
Prostaglandin E_2

$C_{20}H_{32}O_6$
Andrographolic Acid

$C_{20}H_{33}Cl$
Rimuene, *Hydrochloride*

$C_{20}H_{33}NO_2$
Alvanidine
Samandarin, N-*Me*

$C_{20}H_{33}NO_5$
Lindelofamine

$C_{20}H_{33}N_3$
Ormosajine †
Ormosinine †
Panamine †

$C_{20}H_{34}$
1,19-Eicosadiyne

$C_{20}H_{34}Br_6O_2$
9,10,12,13,15,16-Hexabromo-octadecanoic Acid, *Et ester*

$C_{20}H_{34}O$
Abienol †
Geranylgeraniol †
Geranyl-linalool †
8β-Hydroxysandaracopimar-15-ene †
Isoborneol, *Isobornyl ether*
Kolavelool †
Kolavenol †
Labda-8(20),13-dien-15-ol †
Manool
Manoyl oxide
13-*epi*-Manool †

$C_{20}H_{34}O$ (*continued*)
13-*epi*-Manoyl oxide†
Phyllocladan-16α-ol
Plathyterpol†
Thunbergol†
Verticillol†

$C_{20}H_{34}O_2$
2,2′-Bi-(1,3,3-trimethylbicyclo[1,2,2]heptan-2-ol)
Cassanic Acid
Cativic Acid
4,8,13-Duvatriene-1,3-diol
3,8,13-Duvatriene-1,5-diol
Eperu-7-en-15-oic Acid†
8β,13β-Epoxyeperu-14-en-18-ol†
Eperuic Acid
Erythroxydiol X†
Erythroxydiol Y†
Erythroxydiol Z†
15-Hydroxylabd-8(20)-en-19-al†
18-Hydroxy-13-*epi*-manool†
2α-Hydroxymanoyl oxide†
Incensole†
Isocassanic Acid
16α-Kaurane-17,19-diol†
Labda-8(20),13-diene-15,19-diol†
Labda-8(20),14-diene-13α,19-diol†
enantio-Labd-8(20)-en-15-oic Acid†
Larixol†
11-(2-Methylcyclohex-2-enyl)undec-9-enoic Acid, *Et ester*†
9,11,13-Octadecatrienoic Acid, *Et ester*
9,12,15-Octadecatrienoic Acid, *Et ester*
11-Octadecen-9-ynoic Acid, *Et ester*
Tetradecylquinol
Torulosol†

$C_{20}H_{34}O_3$
Corymbol†
Darutigenol
Erythroxytriol Q†
Hydroxydevadarool†
15-Hydroxyeperu-8(20)-en-18-oic Acid†
3β-Hydroxylabd-8(20)-en-15-oic Acid†
enantio-8β-Hydroxylabd-13-en-15-oic Acid†
12-Hydroxylabd-8(20)-en-19-oic Acid†
15-Hydroxylabd-8(20)-en-19-oic Acid†
enantio-18-Hydroxylabd-8(20)-en-15-oic Acid†
Isodarutigenol B
Isodarutigenol C
16α-Kaurane-3α,17,19-triol†
16β-Kaurane-16α,17,19-triol†
1,2,2,3-Tetramethylcyclopentane-1-carboxylic Acid, *Anhydride*

$C_{20}H_{34}O_4$
15,16-Dihydroxyeperu-8(20)-en-18-oic Acid†
Lichesteric Acid, *Me ester*

$C_{20}H_{34}O_5$
6,8-Dihydroxy-11-isopropyl-4,8-dimethyl-14-oxopentadeca-4,9-dienoic Acid†
Prostaglandin E_1★†
8-*epi*-Prostaglandin E_1†
11-*epi*-Prostaglandin E_1†
15-*epi*-Prostaglandin E_1†
11,15-*epi*-Prostaglandin E_1†

$(C_{20}H_{34}O_8)_x$
Dioscin

$C_{20}H_{34}O_{12}$
Xylotetraose

$C_{20}H_{35}BrO_2$
Aplysin-20†

$C_{20}H_{35}N_3$
Dasycarpine
Ormosanine†
Piptanthine†

$C_{20}H_{36}$
Abietane†
enantio-Abietane†
Taxane†

$C_{20}H_{36}Br_4O_2$
9,10,12,13-Tetrabromo-octadecanoic Acid, *Et ester*

$C_{20}H_{36}N_2O$
Kurchicine

$C_{20}H_{36}N_2O_{12}$
Hexamethonium, *Tartrate*

$C_{20}H_{36}O_2$
Chaulmoogric Acid, *Et ester*
Cycloeicosane-1,11-dione
11,14-Eicosadienoic Acid
Eperu-13-ene-8β,15-diol†
Labd-8(20)-ene-3β,15-diol†
enantio-Labd-8(20)-ene-15,18-diol†
Labd-8(20)-ene-15,19-diol†
9,11-Octadecadienoic Acid, *Et ester*
5,12-Octadecadienoic Acid, *Et ester*
9,12-Octadecadienoic Acid, *Et ester*
9-Octadecynoic Acid, *Et ester*
9-Octadecenoic Acid, *Vinyl ester*
2,6-Phytadienoic Acid
Sclareol

$C_{20}H_{36}O_3$
Erythroxytriol P†
11-Hydroxyeicosa-12,14-dienoic Acid†
enantio-8β-Hydroxylabdan-15-oic Acid†
12-Hydroxy-9-octadecynoic Acid, *Et ester*
2-Hydroxysterculic Acid, *Me ester*†
Labdanolic Acid
enantio-Labdanolic Acid†
enantio-13-*epi*-Labdanolic Acid†

$C_{20}H_{36}O_4$
6β,8β-Dihydroxy-*enantio*-labdan-15-oic Acid†
7α,8β-Dihydroxy-*enantio*-labdan-15-oic Acid†

$C_{20}H_{36}O_4Ti$
Tetracyclopentyl titanate

$C_{20}H_{36}O_5$
Dihydroprostaglandin E_1†
10-Oxoeicosanedioic Acid
Prostaglandin $F_{1\alpha}$
Prostaglandin $F_{1\beta}$

$C_{20}H_{38}$
1,19-Eicosadiene
1-Eicosyne

$C_{20}H_{38}$ (*continued*)
Neophytadiene
1,3-Phytadiene†
2,4-Phytadiene†

$C_{20}H_{38}N_2O_3$
Holarrhine

$C_{20}H_{38}O_2$
2,19-Eicosanedione
4,5-Eicosanedione
7,14-Eicosanedione
9-Eicosenoic Acid
11-Eicosenoic Acid
14-Eicosenoic Acid
Eperuane-8β,15-diol†
Isogadoleic Acid
Labdane-8α,15-diol
enantio-13-*epi*-Labdane-8β,15-diol†
Menthopinacol
2-Octadecenoic Acid, *Et ester*
9-Octadecenoic Acid, *Et ester*
10-Octadecenoic Acid, *Et ester*
12-Octadecenoic Acid, *Et ester*

$C_{20}H_{38}O_3$
Decanoic Acid, *Anhydride*
Eperuane-8β,15,18-triol†
9,10-Epoxyoctadecanoic Acid, *Et ester*†
14-Hydroxy-*cis*-11-eicosenoic Acid
12-Hydroxy-9-octadecenoic Acid, *Et ester*
Labdane-8α,15,19α-triol
3-Oxo-octadecanoic Acid, *Et ester*
6-Oxo-octadecanoic Acid, *Et ester*
10-Oxo-octadecanoic Acid, *Et ester*

$C_{20}H_{38}O_4$
Decanedioic Acid, *Di-3-menthylbutyl ester*
Eicosanedioic Acid
Hexadecanedioic Acid, *Di-Et ester*
Pedicellic Acid, *Di-Me ester*†

$C_{20}H_{38}O_6$
Phloionic Acid, *Di-Me ester*

$C_{20}H_{38}O_{11}$
Maltose, β-*Methylglycoside, Hepta-Me ether*
Melibiose, β-*Methyl ether, Hepta-Me ether*
Sucrose, *Octa-Me ether*
α,α-Trehalose, *Octa-Me ether*
β,β-Trehalose, *Octa-Me ether*

$C_{20}H_{38}O_{12}$
Melibionic Acid, *Octa-Me ether*

$C_{20}H_{39}BrO_2$
2-Bromo-octadecanoic Acid, *Et ester*

$C_{20}H_{39}ClO_2$
Octadecanoic Acid, 2-*Chloroethyl ester*

$C_{20}H_{39}IO_2$
Octadecanoic Acid, 2-*Iodoethyl ester*

$C_{20}H_{39}N$
Eicosanoic Acid, *Nitrile*

$C_{20}H_{39}NO$
9-Eicosenoic Acid, *Amide*

$C_{20}H_{39}NO_3$
N-Hexadecanoylglycine, *Et ester*

$C_{20}H_{40}$
1-Eicosene
Thunbergane†

$C_{20}H_{40}O$
Cycloeicosanol (incorrectly given as Cyclo-eicosadecanol)
2-Eicosanone
3-Eicosanone
7-Eicosanone
Phytol

$C_{20}H_{40}O_2$
Eicosanoic Acid
Hexadecanoic Acid, *Butyl ester*
2-Methyloctadecanoic Acid, *Me ester*
3-Methyloctadecanoic Acid, *Me ester*
4-Methyloctadecanoic Acid, *Me ester*
5-Methyloctadecanoic Acid, *Me ester*
14-Methyloctadecanoic Acid, *Me ester*
16-Methyloctadecanoic Acid, *Me ester*
17-Methyloctadecanoic Acid, *Me ester*
Nonadecanoic Acid, *Me ester*★†
Octadecanoic Acid, *Et ester*
2-Octanol, *Dodecanoyl*
2,6,10,14-Tetramethylpentadecanoic Acid, *Me ester*†
3,7,11,15-Tetramethylhexadecanoic Acid†

$C_{20}H_{40}O_3$
2-Hydroxyeicosanoic Acid
3-Hydroxyeicosanoic Acid
2-Hydroxyoctadecanoic Acid, *Et ester*
3-Hydroxyoctadecanoic Acid, *Et ester*
7-Hydroxyoctadecanoic Acid, *Et ester*
11-Hydroxyoctadecanoic Acid, *Et ester*
13-Hydroxyoctadecanoic Acid, *Et ester*

$C_{20}H_{40}O_4$
9,10-Dihydroxyoctadecanoic Acid, *Et ester*
12,13-Dihydroxyoctadecanoic Acid, *Et ester*
15,16-Dihydroxyoctadecanoic Acid, *Et ester*†

$C_{20}H_{41}I$
1-Iodoeicosane

$C_{20}H_{41}N$
Eicosamethyleneimine

$C_{20}H_{41}NO$
Eicosanoic Acid, *Amide*
3,7,11,15-Tetramethylhexadecanoic Acid, *Amide*†

$C_{20}H_{41}NO_2$
2-Amino-octadecanoic Acid, *Et ester*
Sphingosine, *Di-Me ether*

$C_{20}H_{42}$
Crocetane
Eicosane
Phytane

$C_{20}H_{42}N_2$
1,12-Diazacycloeicosane

$C_{20}H_{42}O$
1-Eicosanol
3-Eicosanol

$C_{20}H_{42}O_4S$
Didecyl sulphate

$C_{20}H_{43}N$
1-Aminoeicosane

$C_{20}H_{43}NO_3$
Cerebrin (C_{20}-phytosphingosine)

$C_{20}H_{44}N_2$
Necrosamine

$C_{20}H_{44}Pb$
Lead tetra-3-methylbutyl
Lead tetra-pentyl
Lead tetra-*tert*-pentyl

$C_{20}H_{44}Sn$
Tin tetra-2-methylbutyl
Tin tetra-3-methylbutyl
Tin tetrapentyl

$C_{20}H_{46}N_8O_4S$
Guanethidine

$C_{20}H_{50}O_5Si_5$
1,1,3,3,5,5,7,7,9,9-Decaethylcyclopentasiloxane

C_{21}

$C_{21}H_8O_{12}$
Flavogallol

$C_{21}H_{10}O_{13}$
Flavogallonic Acid
Valoneaic Acid, *Dilactone*

$C_{21}H_{12}Br_2O_3$
4-Bromo-1-naphthol, *Carbonate*
1-Bromo-2-naphthol, *Carbonate*

$C_{21}H_{12}Cl_2O_3$
4-Chloro-1-naphthol, *Carbonate*

$C_{21}H_{12}N_4$
1*H*-Dibenz[*a,h*]imidazole[*c*]phenazine
11*H*-Dibenzo[*a,c*]pyrazolo[3,4-*i*]phenazine
Tricycloquinazoline

$C_{21}H_{12}N_4O_7$
1-Azafluoranthene, *Picrate*

$C_{21}H_{12}O$
Ceranthrone
Naphtho[2,1-*a*]fluoren-11-one
Naphtho[1,2-*b*]fluoren-12-one

$C_{21}H_{12}O_2$
Dibenzo[*a,c*]xanthen-14-one
Dibenzo[*a,j*]xanthen-14-one
Dibenzo[*b,i*]xanthen-14-one
Dibenzo[*c,h*]xanthen-14-one
Perylene-3-carboxylic Acid

$C_{21}H_{12}O_3$
1-Benzoylanthraquinone

$C_{21}H_{12}O_5$
1-Hydroxyanthraquinone-2-carboxylic Acid, *Phenyl ether*

$C_{21}H_{12}O_6$
Rhein, *Phenyl ester*
Trisalicylide

$C_{21}H_{13}N$
Benzo[*h*]naphtho[1,2-*f*]quinoline
Benzo[*h*]naphtho[2,1-*f*]quinoline†
Dibenz[*a,c*]acridine
Dibenz[*a,h*]acridine
Dibenz[*a,i*]acridine
Dibenz[*a,j*]acridine
Dibenz[*b,h*]acridine
Dibenz[*c,h*]acridine
Naphth[1,2-*c*]acridine
Naphth[2,1-*c*]acridine
Phenanthro[9,10-*f*]quinoline†
Phenanthro[9,10-*h*]quinoline†

$C_{21}H_{13}NO$
Dibenz[*a,i*]acridine
Dibenz[*a,j*]acridine

$C_{21}H_{13}NO_3$
Anthraquinone-1-carboxylic Acid, *Anilide*

$C_{21}H_{13}NO_4$
1-Aminoanthraquinone-2-carboxylic Acid, *Phenyl ester*

$C_{21}H_{14}$
13*H*-Dibenzo[*a,c*]fluorene
13*H*-Dibenzo[*a,g*]fluorene
13*H*-Dibenzo[*a,h*]fluorene
13*H*-Dibenzo[*a,i*]fluorene
12*H*-Dibenzo[*b,h*]fluorene
7*H*-Dibenzo[*c,g*]fluorene
2-Methylbenzo[*a*]pyrene
3-Methylbenzo[*a*]pyrene
4-Methylbenzo[*a*]pyrene
5-Methylbenzo[*a*]pyrene
6-Methylbenzo[*a*]pyrene
8-Methylbenzo[*a*]pyrene
9-Methylbenzo[*a*]pyrene
10-Methylbenzo[*a*]pyrene
Naphtho[2,1-*a*]fluorene
Naphtho[1,2-*b*]fluorene
Naphtho[2,3-*b*]fluorene

$C_{21}H_{14}N_2O_4$
1-Amino-5-nitroanthraquinone, N-p-*Tolyl*
1-Amino-7-nitroanthraquinone, N-p-*Tolyl*
1-Amino-8-nitroanthraquinone, N-p-*Tolyl*

$C_{21}H_{14}O$
Anthryl phenyl Ketone
1,1′-Dinaphthyl Ketone
1,2′-Dinaphthyl Ketone
2,2′-Dinaphthyl Ketone
3-Hydroxybenzo[*a*]pyrene, *Me ether*
6-Hydroxybenzo[*a*]pyrene, *Me ether*
7-Hydroxybenzo[*a*]pyrene, *Me ether*
6-Hydroxyperylene, *Me ether*
Picylene-carbinol

$C_{21}H_{14}O_2$
Anthracene-1-carboxylic Acid, *Phenyl ester*

$C_{21}H_{14}O_2S$
1-Anthraquinonethiol, S-o-*Tolyl ether*
1-Anthraquinonethiol, S-p-*Tolyl ether*
1-Anthraquinonethiol, S-*Benzyl ether*

$C_{21}H_{14}O_3$
1-Hydroxy-9,10-anthraquinone, p-*Tolyl ether*
1-Hydroxy-2-methylanthraquinone, *Phenyl ether*

$C_{21}H_{14}O_4$
2,3-Dibenzoylbenzoic Acid
2,6-Dibenzoylbenzoic Acid

$C_{21}H_{14}O_5$
Fluorescein, *Me ether*
Fluorescein, *Me ester*

$C_{21}H_{14}O_7$
Gallein, *Me ester*
η_1-Pyrromycinone†
$C_{21}H_{14}O_8$
Di-(4-hydroxy-3-coumarinyl)acetic Acid, *Me ester*
$C_{21}H_{14}O_{13}$
m-Trigallic Acid†
$C_{21}H_{14}O_{15}$
Valoneaic Acid
$C_{21}H_{15}Br$
Triphenylcyclopropenyl bromide
$C_{21}H_{15}N$
α,β-Diphenylcinnamic Acid, *Nitrile*
2,3-Diphenylquinoline
2,4-Diphenylquinoline
$C_{21}H_{15}NO$
2,4,5-Triphenyloxazole
$C_{21}H_{15}NO_2$
1-Amino-2-methylanthraquinone, N-*Phenyl*
$C_{21}H_{15}NO_3$
Tribenzamide
$C_{21}H_{15}NO_4$
3-*p*-Nitrophenyl-2-phenylacrylic Acid, *Phenyl ester*
$C_{21}H_{15}N_3O_3$
6-Methylpyridine-2,4,5-tricarboxylic Acid, *Anilide-phenylimide*
$C_{21}H_{16}$
1,1′-Dinaphthylmethane
1,2′-Dinaphthylmethane
2,2′-Dinaphthylmethane
1,1-Diphenylindene
1,2-Diphenylindene
1,3-Diphenylindene
2,3-Diphenylindene
1,2-Diphenylnaphthalene
1,3-Diphenylnaphthalene
1,4-Diphenylnaphthalene
1,5-Diphenylnaphthalene
1,6-Diphenylnaphthalene
2,3-Diphenylnaphthalene
2,6-Diphenylnaphthalene
1-Methylcholanthrene
3-Methylcholanthrene
4-Methylcholanthrene
5-Methylcholanthrene
7-Methylcholanthrene
$C_{21}H_{16}N_2$
Lophine
1,3,4-Triphenylpyrazole
1,3,5-Triphenylpyrazole
1,4,5-Triphenylpyrazole
3,4,5-Triphenylpyrazole
$C_{21}H_{16}N_2O$
1,1-Di-(2-naphthyl)urea
1,3-Di-(1-naphthyl)urea
1,3-Di-(2-naphthyl)urea
$C_{21}H_{16}N_2O_3$
Oxogambirtannine†
Phthalonic Acid, *Dianilide*
$C_{21}H_{16}N_2S$
1,3-Di-(1-naphthyl)thiourea
1,3-Di-(2-naphthyl)thiourea
$C_{21}H_{16}O$
1,1′-Dinaphthylmethanol
1,2′-Dinaphthylmethanol
2,2′-Dinaphthylmethanol
2,3-Diphenylacrylophenone
3-Hydroxyphenanthrene, *Benzyl ether*
$C_{21}H_{16}O_2$
2,2′-Di-(2-hydroxy-1-naphthyl)methane
2,3-Diphenylacrylic Acid, *Phenyl ester*
α,β-Diphenylcinnamic Acid
4′-Hydroxychalcone, *Phenyl ether*
5-Isopropylbenz[*a*]anthraquinone
8-Isopropylbenz[*a*]anthraquinone
9-Isopropylbenz[*a*]anthraquinone
10-Isopropylbenz[*a*]anthraquinone
2-Isopropylnaphthacene-5,12-dione
$C_{21}H_{16}O_3$
o-(5-Acenaphthoyl)benzoic Acid, *Me ester*
o-Benzoylbenzoic Acid, *Benzyl ester*
$C_{21}H_{16}O_4$
Haemocorin Aglycone, *Mono-Me ether A*†
Haemocorin Aglycone, *Mono-Me ether B*†
Phenolphthalein I, *Me ether*
Phenolphthalein I, *Me ester*
Phenolphthalein II, *Me ether*
Phenolphthalein II, *Me ester*
Phenolphthalein III, *Me ether*
Phenolphthalein III, *Me ester*
Tetrangulol, *Di-Me ether*†
$C_{21}H_{16}O_6$
Justicidin B†
$C_{21}H_{16}O_7$
Diphyllin†
$C_{21}H_{17}ClO$
2,2,3-Triphenylpropionic Acid, *Chloride*
2,3,3-Triphenylpropionic Acid, *Chloride*
3,3,3-Triphenylpropionic Acid, *Chloride*
$C_{21}H_{17}N$
1,3-Diphenylisoindole, N-*Me*†
2,2,3-Triphenylpropionic Acid, *Nitrile*
2,3,3-Triphenylpropionic Acid, *Nitrile*
3,3,3-Triphenylpropionic Acid, *Nitrile*
$C_{21}H_{17}NO$
α,β-Diphenylcinnamic Acid, *Amide*
$C_{21}H_{17}NO_3$
2-Phenylhippuric Acid, *Phenyl ester*
$C_{21}H_{17}NO_4$
2-Styrylquinoline, *Maleate*
$C_{21}H_{17}NO_5$
Oxynitidine†
$C_{21}H_{17}NO_7S$
4-Nitro-2-sulphobenzoic Acid, *Di-*p*-tolyl ester*
$C_{21}H_{17}N_2O_2$ (ion)
Ourouparine†
$C_{21}H_{17}N_3$
1,3-Di-(1-naphthyl)guanidine
1,3-Di-(2-naphthyl)guanidine

$C_{21}H_{18}$
5-Isopropylbenz[*a*]anthracene
7-Isopropylbenz[*a*]anthracene
8-Isopropylbenz[*a*]anthracene
9-Isopropylbenz[*a*]anthracene
10-Isopropylbenz[*a*]anthracene
11-Isopropylbenz[*a*]anthracene
2-Isopropylnaphthacene

$C_{21}H_{18}NO_4$ (ion)
Nitidine†

$C_{21}H_{18}N_2$
Hydrobenzamide†
Isomarine
2,4,5-Triphenylimidazoline

$C_{21}H_{18}N_2O_2$
Gambirtannine†
Malonamide, N,N'-*Dibenzyl*

$C_{21}H_{18}N_8O_{12}$
Flavipin, 2,4-*Dinitrophenylhydrazone*

$C_{21}H_{18}O_2$
m-Cresolbenzein
o-Diphenylmethylbenzoic Acid, *Me ester*
α-Phenylbenzoin, *Me ether*
Triphenylacetic Acid, *Me ester*
2,2,3-Triphenylpropionic Acid
2,3,3-Triphenylpropionic Acid
3,3,3-Triphenylpropionic Acid

$C_{21}H_{18}O_3$
Benzilic Acid, *Benzyl ester*
p-(Hydroxydiphenylmethyl)benzoic Acid, *Me ester*
2-Hydroxyphenyldiphenylacetic Acid, *Me ether*
4-Hydroxyphenyldiphenylacetic Acid, *Me ether*

$C_{21}H_{18}O_5$
Pulvinic Acid, *Propyl ester*
Vulpinic Acid, *Et ether*

$C_{21}H_{18}O_6$
Collinusin†
Isopinastric Acid, *Me ether*†
Leprapic Acid★, *Me ether*†

$C_{21}H_{18}O_7$
Lisetin†
Mitorubrin†
Narcindone

$C_{21}H_{18}O_8$
Daunomycinone†
Mitorubrinol†

$C_{21}H_{18}O_9$
Stictic Acid, *Di-Me ether*

$C_{21}H_{18}O_{10}$
Chrysin, 7-*Glucuronide*
Glaucophanic Acid, *Di-Me ester*†

$C_{21}H_{18}O_{11}$
Baicalin

$C_{21}H_{18}O_{13}$
Quercetin, *Glucuronide*

$C_{21}H_{19}Cl_3O_5$
Nornidulin, *Di-Me ether*

$C_{21}H_{19}NO$
2,3,5-Triphenylisoxazolidine†
2,2,3-Triphenylpropionic Acid, *Amide*
2,3,3-Triphenylpropionic Acid, *Amide*
3,3,3-Triphenylpropionic Acid, *Amide*

$C_{21}H_{19}NO_4$
Dihydrochelerythrine
αα-Di-(*p*-hydroxyphenyl)-2-nitrotoluene, *Di-Me ether*
αα-Di-(*p*-hydroxyphenyl)-3-nitrotoluene, *Di-Me ether*
αα-Di-(*p*-hydroxyphenyl)-4-nitrotoluene, *Di-Me ether*
Methylanhydrochelidonine

$C_{21}H_{19}NO_5$
Chelerythrine

$C_{21}H_{19}NO_6$
1-Nitroanthraquinone-2-carboxylic Acid, *Hexyl ester*

$C_{21}H_{19}NO_8$
Berberilic Anhydride, *Me ester*

$C_{21}H_{19}N_3O_3$
Euxylophorine†

$C_{21}H_{19}N_3O_5$
4-Amino-2,6-dinitrophenol, N-*Dibenzyl*, *Me ether*

$C_{21}H_{20}$
Phenyldi-*o*-tolylmethane
Phenyldi-*p*-tolylmethane

$C_{21}H_{20}ClNO_6$
Ochratoxin A, *Me ester*†

$C_{21}H_{20}N_2O_2$
Dihydrogambirtannine†

$C_{21}H_{20}N_2O_3$
Alstonine

$C_{21}H_{20}N_4O_5$
Calliastin (?)

$C_{21}H_{20}O$
α,α-Diphenyl-*o*-cresol, *Et ether*
α,α-Diphenyl-*p*-cresol, *Et ether*
Triphenylmethanol, *Et ether*

$C_{21}H_{20}O_2$
2,5-Dihydroxy-α,α-diphenyltoluene, *Di-Me ether*
3,4-Dihydroxy-α,α-diphenyltoluene, *Di-Me ether*
4,4'-Dihydroxytriphenylmethane, *Di-Me ether*
4-Hydroxyphenyldiphenylmethanol, *Di-Me ether*

$C_{21}H_{20}O_4$
8-Cinnamoyl-5,7-dihydroxy-2,2,6-trimethylchrom-3-ene†

$C_{21}H_{20}O_5$
Piloquinone

$C_{21}H_{20}O_6$
β-Anhydroicaritin
Curcurmin
Jacareubin, *Tri-Me ether*†

$C_{21}H_{20}O_7$
7-Dehydroxy-4-demethylpodophyllotoxin
4'-Demethyl-deoxypodophyllotoxin†
Piscerythrone†

$C_{21}H_{20}O_7$ *(continued)*
Piscidone†
Rhodocomatulin, 1,3,6-*Tri-Me ether*†
Rhodocomatulin, 3,6,8-*Tri-Me ether*†
Sesangolin★†
Solorinic Acid

$C_{21}H_{20}O_8$
Ceroalbolinic Acid, *Me ester, Tetra-Me ether*†
Elliptic Acid, *Me ester*
Hyposalazinic Acid, *Di-Me ether, Me ester*
Kermesic Acid, *Me ester, Tetra-Me ether*†
Narceonic Acid
α-Peltatin
3,5,6,7-Tetrahydroxy-1-methylanthraquinone-2-carboxylic Acid, *Me ester, Tetra-Me ether*†

$C_{21}H_{20}O_9$
Bayin
Daidzin
1,3-Dihydroxy-2-methylanthraquinone, 3-β-*Glucoside*
Frangulin
Frangulin A†
Frangulin B†
Melibentin†
Rhodocadonic Acid, *Tetra-Me ether*
Toringin

$C_{21}H_{20}O_{10}$
Afzelin
Aloe-emodin, 8-*Mono*-β-D-*glucoside*★†
Cosmosiin†
Euxanthic Acid, *Et ester*
Genistin
Isovitexin
Puerarin†
Sophoricoside
Vitexin

$C_{21}H_{20}O_{11}$
Astragalin
Aureusin
Cernuoside
Cosmetin
Glucoluteolin
Iso-orientin†
Isorhamnetin, 3-β-D-*Xyloside*†
Orientin
epi-Orientin†
Populnin
Quercitrin

$C_{21}H_{20}O_{12}$
Bractein
Herbacitrin
Hyperin
Isoquercitrin
Myricitrin
Quercetin, ?-*Glucoside*
Quercimeritrin

$C_{21}H_{20}O_{13}$
Gossypetin, 7-*Glucoside*
Gossytrin
Quercetagitrin

$C_{21}H_{20}O_{14}$
Cannabiscitrin

$C_{21}H_{21}ClN_2O_8$
7-Chloro-6-demethyltetracycline

$C_{21}H_{21}ClO_{10}$
Callistephin chloride
Pelargonenin chloride

$C_{21}H_{21}ClO_{11}$
Chrysanthemin
Cyanenin chloride
Idaein chloride

$C_{21}H_{21}ClO_{12}$
Empetrin

$C_{21}H_{21}N$
p-Aminophenyldiphenylmethane, N-*Di-Me*
α-Aminotriphenylmethane, N-*Di-Me*
α-Aminotriphenylmethane, N-*Et*
Cyproheptadine
Tribenzylamine

$C_{21}H_{21}NO$
o-Aminophenyldiphenylmethanol, N-*Di-Me*
m-Aminophenyldiphenylmethanol, N-*Di-Me*
p-Aminophenyldiphenylmethanol, N-*Di-Me*

$C_{21}H_{21}NO_2$
6-Methyl-2-phenylquinoline-4-carboxylic Acid, *Butyl ester*
6-Methyl-2-phenylquinoline-4-carboxylic Acid, *Isobutyl ester*
2-Phenylquinoline-4-carboxylic Acid, 3-*Methylbutyl ester*

$C_{21}H_{21}NO_4$
Ochotensine

$C_{21}H_{21}NO_5$
Corycavamine
Corycavine

$C_{21}H_{21}NO_6$
Adlumine
Corlumine
Cornigerine†
Coulteropine†
α-(−)-Hydrastine
β-(−)-Hydrastine
Hydrastine *a*
Hydrastine *b*
Isorhoeadine†
Rhoeadine

$C_{21}H_{21}NO_7$
Narcotoline

$C_{21}H_{21}N_3$
Hexahydro-1,3,5-triphenyltriazine

$C_{21}H_{21}N_3O_2$
αα-Di-(4-aminophenyl)-2-nitrotoluene, 4,4′-N-*Di-Me*
αα-Di-(4-aminophenyl)-3-nitrotoluene, 4,4′-N-*Di-Me*
αα-Di-(4-aminophenyl)-4-nitrotoluene, 4,4′-N-*Di-Me*

$C_{21}H_{21}O_3P$
Tribenzyl phosphite

$C_{21}H_{21}O_4P$
Phosphoric Acid, *Tribenzyl ester*
Tri-*o*-tolyl phosphate
Tri-*m*-tolyl phosphate
Tri-*p*-tolyl phosphate

$C_{21}H_{21}Pb$
Lead tri-*m*-tolyl

$C_{21}H_{22}ClN_2O_8$
7-Chloro-5α(11α)-dehydrotetracycline

$C_{21}H_{22}NO_4$ (ion)
Isocryptopine

$C_{21}H_{22}N_2$
αα-Di-(2-amino-5-methylphenyl)toluene
αα-Di-(4-amino-3-methylphenyl)toluene
2,5-Diaminotoluene, 2,5-N-*Di-*p*-tolyl*

$C_{21}H_{22}N_2O$
3-Methyl-2-phenylquinoline-4-carboxylic Acid, *Diethylamide*
6-Methyl-2-phenylquinoline-4-carboxylic Acid, *Diethylamide*

$C_{21}H_{22}N_2O_2$
1-Demethyl-21-deoxyajamaline-1,19-diene 17-*O*-acetate†
10,22-Dioxokopsane, N_a-*Me deriv.*†
Isostrychnine I
Isostrychnine II
Strychnine

$C_{21}H_{22}N_2O_3$
4-Hydroxystrychnine†
Perakine
Polyneuridine Aldehyde†
Serpentine
ψ-Strychnine
Vallesiachotamine†
Vomilenin

$C_{21}H_{22}N_2O_4$
Periformyline†
Perimivine†

$C_{21}H_{22}N_2O_5$
Fumaramine

$C_{21}H_{22}N_2O_7$
6-Demethyl-6-deoxytetracycline

$C_{21}H_{22}N_2O_8$
6-Demethyltetracycline

$C_{21}H_{22}O_2$
Trachylobanic Acid, *Me ester*†

$C_{21}H_{22}O_3$
1,5-Di-*o*-hydroxyphenyl-1,4-pentadien-3-one, *Di-Et ether*

$C_{21}H_{22}O_4$
Archangelin†
Bavalchalcone†
Bavachinin†
Bergamottin
4,4-Diphenylitaconic Acid, *Di-Et ester*

$C_{21}H_{22}O_5$
Apigenin, *Tri-Et ether*
Biochanin A, 5,7-*Di-Et ether*
5-[(3,6-Dimethyl-6-formyl-2-heptenyl)-oxy]-psoralen
Globuxanthone, *Tri-Me ether*†
Isoxanthohumol
Rubropunctatin
Selinone, 7-*Me ether*†
Xanthohumol

$C_{21}H_{22}O_6$
Derritol
4′,5-Dihydroxy-6,7-dimethoxyisoflavone, *Di-Et ether*†
2-(3,3-Dimethylallyl)-1,3,5,6-tetrahydroxy-xanthone, 3,5,6-*Tri-Me ether*†
Involutin, *Tetra-Me ether*†
Isoderritol
Luteolin, 3′,4′,7-*Tri-Et ether*
Melannein, *Di-Et ether*†
Phellamuretin, *Me ether*
Rhodoptilometrin, *Tetra-Me ether*†
Scriblitifolic Acid, *Et ester*†
Symphoxanthone, *Tri-Me ether*†
Ugaxanthone, 3,5,6-*Tri-Me ether*†

$C_{21}H_{22}O_7$
Edulitin†
Icaritin
Ostruthol
Prudomestin, 3,7-*Di-Et ether*†
Pteryxin†
Quercetin, 3′,4′,7-*Tri-Et ether*
Samidin†

$C_{21}H_{22}O_8$
Atranorin, *Di-Me ether*
Calycopterin, *Di-Me ether*
Centaureidin, 3′,5,7-*Tri-O-Me ether*† (Quercetagetin, *Hexa-Me ether*)
Gossypetin, *Hexa-Me ether*
5-Hydroxy-3,3′,4′,7,8-pentamethoxyflavone, *Me ether*†
Irigenin, *Tri-Me ether*
Limocitrin, *Tetra-Me ether*
Nobiletin
Quercetagetin, *Hexa-Me ether*
Serpyllin, *Me ether*†

$C_{21}H_{22}O_9$
Barbaloin
Limocitrol, 3,4′,7-*Tri-Me ether*†
Liquiritin
Squamatic Acid, *Di-Me ester*
3′,5,5′-Trihydroxy-3,4′,6,7-tetramethoxy-flavone, 3′,5′-*Di-Me ether*†

$C_{21}H_{22}O_{10}$
2-*O*-Caffeoylarbutin†
Coreopsin
Digicitrine
Floribundoside★†
Glucobutein
Heliochrysin A†
Hemiphloin†
Isohemiphloin†

$C_{21}H_{22}O_{11}$
Astilbin
Carthamin
6-*C*-Glucosyldiosmetin†
8-*C*-Glucosyldiosmetin†
Isocarthamine
Keyakinin
Mangiferin, *Di-Me ether*
Saponaretin

$C_{21}H_{22}O_{12}$
Homo-orientin

$C_{21}H_{23}ClN_2O_9$
6-Demethyl-7-chlorotetracycline
$C_{21}H_{23}ClO_5$
Sclerotiorin
7-*epi*-Sclerotiorin
$C_{21}H_{23}NO$
Apocinchene, *Et ether*
$C_{21}H_{23}NO_2$
Norlobelanine
$C_{21}H_{23}NO_3$
Northebaine, N-*Allyl*†
$C_{21}H_{23}NO_4$
Burmannaline
Dehydroglaucine†
$C_{21}H_{23}NO_5$
Cryptocavine
Cryptopine†
Heroin
α-Homochelidonine
β-Homochelidonine
γ-Homochelidonine
Ocokryptine, *Me ether*†
Ocopodine†
Ocoteine
Palmatine
$C_{21}H_{23}NO_6$
Allocolchiceine
Colchiceine
N-Deacetyl-*N*-formylcolchicine†
2-Demethylcolchicine†
3-Demethylcolchicine
Oreogenine†
Papaverrubine B†
Papaverrubine C, O-*Me ether*†
Papaverrubine F†
$C_{21}H_{24}F_3N_3S$
Trifluoperazine†
$C_{21}H_{24}NO_4$ (ion)
Takatonine†
$C_{21}H_{24}N_2O$
Isostrychnidine
$C_{21}H_{24}N_2O_2$
Apoyohimbine†
Caracurine IV
Catharanthine
α-Isoyohimbine
Tabersonine
Venalstonine†
Vindolinine
$C_{21}H_{24}N_2O_3$
Akuammidine
Akuammigine
Alangine A (or $C_{21}H_{26}N_2O_3$)
Alangine B (or $C_{21}H_{26}N_2O_3$)
Diaboline
Geissoschizine
Gelseverine (or $C_{21}H_{26}N_2O_3$)
Isoajmalicine
Isoschizogamin
Kopsinilam
Lochnericine
Mayumbine
Minoricine
Minovincine
Polyneuridine
Raubasine
Rauniticine†
Schizogamin
Tetrahydroalstonine
Venalstonidine†
Vincoridine
Vobasine
$C_{21}H_{24}N_2O_4$
Burnamine†
Dichotamine†
Isomitraphylline†
Isopteropodine
Lysuric Acid, *Me ester*
Mitraphylline
Ornithuric Acid, *Et ester*
Pteropodine†
Uncarine A†
Uncarine B†
Uncarine C†
Uncarine D†
Uncarine E†
Uncarine F†
Vincaminine
Vincareine
Vincoline†
Voacarpine†
$C_{21}H_{24}N_2O_5$
Voamonine†
$C_{21}H_{24}N_8O_5$
Adenopterin
Aminopterin, 4-N-*Di-Me*
$C_{21}H_{24}O_2$
1-Naphthoic Acid, *Bornyl ester*
$C_{21}H_{24}O_4$
Azelaic Acid, *Diphenyl ester*
Dibenzylmalonic Acid, *Di-Et ester*
2,3-Diphenylglutaric Acid, *Di-Et ester*
2,4-Diphenylglutaric Acid, *Di-Et ester*
3,3-Diphenylglutaric Acid, *Di-Et ester*
2,3-Diphenylpropane-1,2-dicarboxylic Acid, *Di-Et ester*
Galcatin
8-Geranyloxypsoralen†
Isogalcatin†
$C_{21}H_{24}O_5$
Calopiptin
$C_{21}H_{24}O_6$
(−)-Arctigenin
Forsythigenol
Gallacetonin
Haematoxylin, *Penta-Me ether*
Miroestrol, 3-*Me ether*†
Tinophyllone†
4′,5,7-Trihydroxy-6-methoxyflavone, *Tri-Et ether*
$C_{21}H_{24}O_7$
Barbatic Acid, *Et ester*
Diffractaic Acid, *Me ester*

$C_{21}H_{24}O_7$ (*continued*)
Dihydrosamidin †
Divaricatic Acid
8-*epi*-Jateorin, *Me ether* †
Lithospermic Acid, *Penta*-O-*Me* †
Obtusatic Acid, *Di-Me ether*
Suksdorfin †
Visnadin †

$C_{21}H_{24}O_8$
Albaspidin★ †

$C_{21}H_{24}O_9$
Chromomycinone †
Rhapontin

$C_{21}H_{24}O_{10}$
Phloridzin
Plicatic Acid, *Me ester* †

$C_{21}H_{24}O_{11}$
Aspalathin †

$C_{21}H_{25}NO$
Benztropine

$C_{21}H_{25}NO_2$
1-Dimethylaminoethyl-3-hydroxy-4-methoxyphenanthrene, *Et ether* †
Guaianine (?)

$C_{21}H_{25}NO_3$
1-*N*,*N*-Dimethylaminoethyl-2,3,4-trimethoxyphenanthrene †
Pipenzolute
Thebenine, N-*Me*, 1-*Et ether*
Thebenine, 1-*Propyl ether*

$C_{21}H_{25}NO_4$
Caseadine, *Me ether* †
Caseamine, *Di-Me ether* †
Coreximine, *Di-Me ether*
Corybulbine
Glaucine
Isocorybulbine
Munitagine, *Di-Me ether* †
Norcoralydine
Pavine, N-*Me*
Tetrahydropalmatine
Thalmidine

$C_{21}H_{25}NO_5$
Androcymbine †
Capaurine †
Colchamine
Demecolcine
Floramultine †
Multifloramine †
Oconovine †
Preocoteine †
Clivatin

$C_{21}H_{25}N_3OS$
Macusine B, *Thiocyanate*★ †

$C_{21}H_{25}O_9$
Brucenol (?)

$C_{21}H_{26}$
[3,6]Paracyclophane †

$C_{21}H_{26}BrNO_3$
Methantheline

$C_{21}H_{26}ClNO_4$
N-Methylisocorydinium, *Chloride* †

R

$C_{21}H_{26}ClN_3OS$
Perphenazine

$C_{21}H_{26}F_2O_6$
6α-9α-Difluoro-11β,16α,17α,21-tetrahydroxypregna-1,4-diene-3,20-dione

$C_{21}H_{26}INO_4$
N-Methylisocorydinium, *Iodide* †
Xanthoplanine, *Iodide* †

$C_{21}H_{26}NO_4$ (ion)
Cocsarmine †
Corydine, N-*Me* †
1-Hydroxy-2,9,10-trimethoxy-*N*-methylaporphinium †
N-Methylisocorydinium★ †
Xanthoplanine †

$C_{21}H_{26}N_2O$
Semperflorine

$C_{21}H_{26}N_2O_2$
1-Acetylaspidoalbidine †
Apoquinine, *Et ether*
Aspidospermatine †
Coronaridine
Cupreine, *Mono-Et ether*
Kopsinine
N-Formyl-17-methoxyaspidofractinine †
Retuline★ †
Spermostrychnine
Vincadifformine
Vincamajoreine★ †

$C_{21}H_{26}N_2O_3$
1-Acetyl-17-hydroxyaspidoalbidine †
Alangine A (or $C_{21}H_{24}N_2O_3$)
Alangine B (or $C_{21}H_{24}N_2O_3$)
Alloyohimbine
Corynanthine
14,19-Dihydro-11-methoxycondylocarpine †
Dregamine †
Gelseverine (or $C_{21}H_{24}N_2O_3$)
Haplocidine
Heyneanine †
Isositsirikine †
Isovincamine
Kopsinine, N_b-*Oxide* †
Minovincinine
Pelirine †
Rauhimbine
Schizophyllin
Stemmadenine
Sitsirikine †
Tabernaemontanine
Vincamine
14-*epi*-Vincamine †
Yohimbene
Yohimbine
α-Yohimbine
β-Yohimbine

$C_{21}H_{26}N_2O_4$
Aspidodasycarpine †
20-Hydroxyvincamine †
Isostrychnic Acid
Powerine †
Splendoline †

$C_{21}H_{26}N_2S_2$
Thioridazine
$C_{21}H_{26}O_2$
Cannabinol
1-Naphthoic Acid, (−)-*Menthyl ester*
2-Naphthoic Acid, (±)-*Menthyl ester*
$C_{21}H_{26}O_3$
20(*S*)-Hydroxy-3-oxo-4-pregnene-18-carboxylic Acid (18 → 20)-Lactone †
$C_{21}H_{26}O_4$
Crocetin, *Mono-Me ester*
Furethrolone, *Chrysanthemum-monocarboxylyl*
Latifolin, *Di-O-Et*
$C_{21}H_{26}O_5$
Mammea B/BC †
Monascin
Odoratin, *Ethylene ketal* †
Prednisone
2,2′,5,6′-Tetrahydroxybenzophenone, *Tetra-Et ether*
$C_{21}H_{26}O_7$
Arctigenic Acid
Gibberellin A_{21}, *Di-Me ester* †
Olivil, *Me ether*
Phloropyron
$C_{21}H_{26}O_{10}$
Bruceolide †
$C_{21}H_{26}O_{11}$
Anisatin, *Triacetate*
$C_{21}H_{26}O_{12}$
Plumieride
$C_{21}H_{26}O_{13}$
Leucoglycodrin †
$C_{21}H_{27}ClN_2O_2$
Hydroxyzine
$C_{21}H_{27}FO_5$
9α-Fluoro-17,21-dihydroxypregn-4-ene-3,11,20-trione
$C_{21}H_{27}FO_6$
9α-Fluoro-11β,16α,17α,21-tetrahydroxypregna-1,4-diene-3,20-dione †
$C_{21}H_{27}NO$
Amidone
Apoaromadendrone, *Anilinomethylene comp.*
Isoamidone II
(−)-Isomethadone
Malarboreine †
$C_{21}H_{27}NO_2$
Hasubanonine †
Norlobelanidine
$C_{21}H_{27}NO_4$
Laudanosine
Protostephanine★ †
Thalifendlerine, O-*Me ether* †
$C_{21}H_{27}NO_4S$
Diphemanilmethyl sulphate
$C_{21}H_{27}NO_5$
Kreysiginine †
$C_{21}H_{27}NO_6$
Floramultinine †

$C_{21}H_{27}N_2O_2$ (ion)
Macroalhine †
$C_{21}H_{27}N_3O_6$
Casimiroedine †
$C_{21}H_{27}N_7O$
Cypridina-Oxyluciferin †
$C_{21}H_{27}N_7O_{14}P_2$
Coenzyme I
$C_{21}H_{27}Na_2O_8P$
Prednisolone, 21-*Disodiumphosphate*
$C_{21}H_{28}INO_3$
Magnocurarine, *Di-Me ether iodide* †
$C_{21}H_{28}KNO_{15}S_2$
Glucorapiferin, *Penta-Ac K salt*
$C_{21}H_{28}N_2O$
N(α)-Acetyl-7-ethyl-5-de-ethylaspidospermine †
Demethoxyaspidospermine †
Di-(4-diethylaminophenyl) Ketone
$C_{21}H_{28}N_2O_2$
Demethylaspidospermine †
16,17-Dimethoxyaspidofractinine †
Optochine
Vallesine
Vincadine
$C_{21}H_{28}N_2O_3$
O-Demethylaspidocarpine
Dihydrositsirikine †
Limapodin
Lochneram
Raugalline
Spegazzinine
$C_{21}H_{28}N_2O_4$
Amsonine
Spegazzidnidine
$C_{21}H_{28}N_7O_{17}P_3$
Coenzyme II
$C_{21}H_{28}O$
2-Octanol, *Diphenylmethyl ether*
$C_{21}H_{28}O_2$
2,2-Di-*p*-hydroxyphenylpropane, *Di-*n*-propyl ether*
13β-Ethyl-17-hydroxy-18,19-dinor-17α-pregn-4-en-20-yn-3-one †
17α-Hydroxy-4-pregnen-20-yn-3-one
$C_{21}H_{28}O_3$
1,3-Dioxoferruginyl methyl Ether †
1,3-Dioxototaryl methyl Ether †
12β-Hydroxypregna-4,6-diene-3,20-dione †
Pyrethrin I
Xanthoperol, *Me ether*
$C_{21}H_{28}O_4$
Anhydrohirundigenin †
Diginigenin
16α,20(*S*)-Dihydroxy-3-oxo-4-pregnene-18-carboxylic Acid (→ 20)-Lactone
p-Geranylferulic Acid, *Me ester*
Gibberellin A_{15}, *Me ester* †
Holadysone †
21-Hydroxypregn-4-ene-3,11,20-trione

$C_{21}H_{28}O_5$
- Aldosterone
- Cinerin II†
- Cortisone
- Digifologenin
- 14α-Digipronogenin
- Fujenoic Acid, *Me ester*
- Prednisolone

$C_{21}H_{28}O_6S$
- *S*-Fukinolide†

$C_{21}H_{28}O_8$
- Scabiolide†

$C_{21}H_{28}O_9$
- Grandidentatin

$C_{21}H_{29}ClN_2O_{13}$
- Betanin

$C_{21}H_{29}ClO_5$
- 12α-Chlorohydrocortisone

$C_{21}H_{29}NO$
- Alphamethadol
- Apoaromadendrylamine, *Benzoyl*
- Betamethadol
- Biperiden
- Isomethadol†

$C_{21}H_{29}NO_5$
- Floripavidine

$C_{21}H_{29}NO_8$
- Florosenine†

$C_{21}H_{29}N_2O_2$ (ion)
- *C*-Fluorocurinine

$C_{21}H_{29}N_3O_2$
- 4-Amino-α-4-aminophenyl-2-nitrotoluene, 4,4′-N-*Tetra-Et*

$C_{21}H_{30}NS_2$
- Captodiamine

$C_{21}H_{30}N_2$
- Di-(4-diethylaminophenyl)methane

$C_{21}H_{30}N_2O$
- Di-(4-diethylaminophenyl)methanol

$C_{21}H_{30}O$
- 3-(8′,11′,14′-Pentadecatrienyl)phenol

$C_{21}H_{30}O_2$
- Cannabichromene†
- Cannabicyclol†
- Cannabidiol
- Dehydroabietic Acid, *Me ester*
- 20α-Hydroxypregna-4,6-dien-3-one†
- 7-Oxototarol, *Me ether*
- 16-Oxototarol, *Me ether*
- Progesterone
- 9(10 → 19)*abeo*-$\Delta^{9(11)}$-5α,10α-Pregnene-3,20-dione†
- 9(10 → 19)*abeo*-$\Delta^{9(11)}$-5α,10β-Pregnene-3,20-dione†
- Retinoic Acid, *Me ester*†
- Sugiol, *Me ether*
- Δ^8-Tetrahydrocannabinol
- Δ^9-Tetrahydrocannabinol★†

$C_{21}H_{30}O_3$
- Daniellic Acid, *Me ester*†
- Deoxycorticosterone
- Hardwickiic Acid, *Me ester*†
- 13-Hydroxy-14-isopropylpodocarpa-8,11,13-trien-16-oic Acid, *Me ester*
- Jasmolin I†
- Lambertianic Acid, *Me ester*†
- Royleanone, *Me ether*
- Vinhaticoic Acid, *Me ester*
- Vouacapenic Acid, *Me ester*

$C_{21}H_{30}O_4$
- Corticosterone
- 4-Deoxyhumulone†
- 17α,21-Dihydroxypregn-4-en-3,20-dione
- 19,21-Dihydroxypregn-4-ene-3,20-dione†
- Euonymol
- Horminone, *Me ether*†
- 16-Hydroxyeperu-8(20),13-diene-15,18-dioic Acid (15 → 16)-Lactone, *Me ester*†
- 2,3,4-Trimethoxyoesta-1,3,5(10)-triene-17β-ol†

$C_{21}H_{30}O_5$
- Adhumulone
- Cortisol
- Desacetyldigacetigenin
- Hirundigenin†
- Humulone
- Isoadhumulone
- Isohumulone A†
- Isohumulone B†
- Purprogenin
- 2β,3β,14α-Trihydroxy-5β-pregn-7-ene-6,20-dione†
- Withaferin

$C_{21}H_{30}O_6$
- Humulinone

$C_{21}H_{30}O_7$
- Isodotricin†

$C_{21}H_{30}O_8$
- Alternaric Acid

$C_{21}H_{30}O_{11}$
- Gein

$C_{21}H_{31}NO$
- Malarborine†

$C_{21}H_{31}NO_2$
- Himbadine

$C_{21}H_{31}NO_3$
- 12*a*-Aza-*C*-homo-5α-pregnane-3,12,20-trione
- Galbulimina alkaloid G.B.17†
- Heterophyllidine†
- Samandaridine

$C_{21}H_{31}NO_4$
- Heterophylline†

$C_{21}H_{31}NO_9$
- Floridanine†

$C_{21}H_{32}N_2$
- Concurchine
- Irehline

$C_{21}H_{32}N_2O_2$
- Kurchenine

$C_{21}H_{32}O$
- Agaurol β
- Allyloestrenol

$C_{21}H_{32}O$ (*continued*)
Ferruginol, *Me ether*
3-(8′,11′-Pentadecadienyl)phenol
Totarol, *Me ether*

$C_{21}H_{32}O_2$
Abieta-8,13-dienoic Acid, *Me ester*†
Bhilawanol
Cannabigerol†
Cardol
Cyclopregnol
20α-Hydroxypregn-4-en-3-one†
20β-Hydroxypregn-4-en-3-one†
3β-Hydroxy-5-pregnen-20-one
Kaur-16-en-19-oic Acid, *Me ester*†
Ozic Acid, *Me ester*†
Pimaric Acid, *Me ester*
5β-Pregnane-3,20-dione
Photolevopimaric Acid, *Me ester*†
Pimara-8(14),15-dien-19-oic Acid, *Me ester*†
Sandaracopimaric Acid, *Me ester*
Sapietic Acid, *Me ester*

$C_{21}H_{32}O_3$
15β-Hydroxykaur-16-en-19-oic Acid, *Me ester*†
21-Hydroxypregnane-3,20-dione
6α-Hydroxysandaracopimaric Acid, *Me ester*†
17-Hydroxy-3,4-secobeyer-4(18),15-dien-3-oic Acid, *Me ester*†
Steviol, *Me ester*

$C_{21}H_{32}O_4$
Anhydrodigipurpurogenin
Digipurpurogenin I†
Diginpurpurogenin II†
Isopimaric Acid, *Me ester*†
$\Delta^{8(9)}$-Isopimaric Acid, *Me ester*†
16α-Kaurane-17,19-dioic Acid, 17-*Mono-Me ester*†
16α-Kaurane-17,19-dioic Acid, 19-*Mono-Me ester*†
Purpnigenin
17α,20β,21-Trihydroxy-4-pregnen-3-one

$C_{21}H_{32}O_5$
ρ-Isohumulone†
3β,17α,21-Trihydroxy-5α-pregnane-11,20-dione

$C_{21}H_{32}O_8$
ψ-Anisatin
5-(2,3-Dihydroxy-3-methylbutyl)-4-(3,4-epoxy-4-methylpentanoyl)-3,4-dihydroxy-2-iso-valerylcyclopent-2-en-1-one†

$C_{21}H_{33}N$
Aconitane

$C_{21}H_{33}NO$
Holamine
Holaphyllamine
Norlatifolin

$C_{21}H_{33}NO_2$
Himandravine
Himbeline

$C_{21}H_{33}NO_7$
Lasiocarpin

$C_{21}H_{33}NO_{14}S_3$
Glucoalyssin

$C_{21}H_{34}O$
Dibornyl Ketone
Mangiferene
3-(8′-Pentadecenyl)phenol

$C_{21}H_{34}O_2$
Bilobol
Dihydroiso-$\Delta^{13,14}$-abietic Acid, *Me ester*†
1,3-Dihydroxy-5-(10′-pentadecenyl)benzene
Eperua-7,13-dien-15-oic Acid, *Me ester*†
3α-Hydroxy-5β-pregnan-20-one
Labda-8(20),13-dien-15-oic Acid, *Me ester*†
enantio-Labda-8(20),13-dien-15-oic Acid, *Me ester*†
Kolavenic Acid, *Me ester*†
Renghol
Tetradecanoic Acid, p-*Tolyl ester*

$C_{21}H_{34}O_3$
2//3-5α-Androstane-2,3-dioic Acid, *Di-Me ester*
3α,20α-Dihydroxy-5-β-pregnane-11-one
16α-Hydroxy-(−)-kauran-19-oic Acid, *Me ester*†
19-Hydroxy-16α-kauran-17-oic Acid, *Me ester*†

$C_{21}H_{34}O_4$
1α,19-Dihydroxy-16α-kauran-17-oic Acid, *Me ester*†
16,17-Dihydroxy-16β-kauran-19-oic Acid, *Me ester*†
Methyl sciadopate†
Oxygrindelic Acid, *Me ester*
Tetrahydrocorticosterone†
3α,11β,21-Trihydroxy-5α-pregnan-20-one
3β,11β,21-Trihydroxy-5α-pregnan-20-one
3β,17α,21-Trihydroxy-5α-pregnan-20-one

$C_{21}H_{34}O_5$
Cortolone
β-Cortolone
3α,11β,17α,21-Tetrahydroxy-5α-pregnan-20-one
3β,11β,17α,21-Tetrahydroxy-5α-pregnan-20-one

$C_{21}H_{35}NO$
Funtumine
Holafebrine

$C_{21}H_{35}N_3$
Jamine†

$C_{21}H_{36}$
5β-Pregnane
5α,17α-Pregnane
Trispiro[5,2,2,5,2,2]heneicosane†

$C_{21}H_{36}Br_4O_2$
9,10,12,13-Tetrabromo-octadecanoic Acid, *Allyl ester*

$C_{21}H_{36}Br_6O_2$
9,10,12,13,15,16-Hexabromo-octadecanoic Acid, *Propyl ester*

$C_{21}H_{36}N_2$
Irehdiamine A

$C_{21}H_{36}N_2O$
Holarrhidine
Holarrhimine

$C_{21}H_{36}N_2O_5S$
Hexocyclium methyl sulphate
$C_{21}H_{36}N_7O_{16}P_3S$
Coenzyme A
$C_{21}H_{36}N_7O_{16}P_3Se$
Selenocoenzyme A†
$C_{21}H_{36}N_7O_{17}P_3$
Oxy-Coenzyme A†
$C_{21}H_{36}O$
3-Pentadecylphenol
5β-Pregnan-3α-ol
Pregnan-3β-ol
Pregnan-7α-ol
Pregnan-12α-ol
$C_{21}H_{36}O_2$
Cativic Acid, *Me ester*
Chaulmoogric Acid, *Allyl ester*
1,2-Dihydroxy-3-pentadecylbenzene
1,3-Dihydroxy-5-pentadecylbenzene
Eperu-7-en-15-oic Acid, *Me ester*†
Eperuic Acid, *Me ester*
Isocassanic Acid, *Me ester*
enantio-Labd-8(20)-en-15-oic Acid, *Me ester*†
5β-Pregnane-3α,20α-diol
5β-Pregnane-3α,20β-diol
5β-Pregnane-3β-20α-diol
5β-Pregnane-3β-20β-diol
Tetradecylquinol, *Mono-Me ether*
5-Tridecylresorcinol, *Di-Me ether*†
$C_{21}H_{36}O_3$
3β-Hydroxylabd-8(20)-en-15-oic Acid, *Me ester*†
enantio-8β-Hydroxylabd-13-en-15-oic Acid, *Me ester*†
12-Hydroxylabd-8(20)-en-19-oic Acid, *Me ester*
15-Hydroxylabd-8(20)-en-19-oic Acid, *Me ester*†
enantio-18-Hydroxylabd-8(20)-en-15-oic Acid, *Me ester*†
5α-Pregnane-3β,17α,20α-triol
5α-Pregnane-3β,17α,20β-triol
$C_{21}H_{36}O_4$
15,16-Dihydroxyeperu-8(20)-en-18-oic Acid, *Me ester*†
α-Monolinolenin
5α-Pregnane-3β,17α,20β,21-tetrol
$C_{21}H_{36}O_5$
Cortol
β-Cortol
6,8-Dihydroxy-11-isopropyl-4,8-dimethyl-14-oxopentadeca-4,9-dienoic Acid, *Me ester*†
5α-Pregnane-3β,11β,17α,20β,21-pentol
$C_{21}H_{36}O_6$
Turbicorytin
$C_{21}H_{36}O_8$
Methane-tetracarboxylic Acid, *Tetrabutyl ester*
Methane-tetracetic Acid, *Tetrapropyl ester*
$C_{21}H_{37}NO$
Funtumidine
Funtuphyllamine A†
$C_{21}H_{37}N_3$
Piptanthine, N-*Me*†
$C_{21}H_{38}BrN$
Hexadecylpyridinium Bromide
$C_{21}H_{38}Br_4O_2$
9,10,12,13-Tetrabromo-octadecanoic Acid, *Propyl ester*
9,10,12,13-Tetrabromo-octadecanoic Acid, *Isopropyl ester*
$C_{21}H_{38}ClN$
Hexadecylpyridinium Chloride
$C_{21}H_{38}IN$
Hexadecylpyridinium Iodide
$C_{21}H_{38}O_2$
9-Octadecenoic Acid, *Allyl ester*
$C_{21}H_{38}O_3$
enantio-8β-Hydroxylabdan-15-oic Acid, *Me ester*†
enantio-Labdanolic Acid, *Me ester*†
enantio-13-*epi*-Labdanolic Acid, *Me ester*†
$C_{21}H_{38}O_4$
6β,8β-Dihydroxy-*enantio*-labdan-15-oic Acid, *Me ester*†
7α,8β-Dihydroxy-*enantio*-labdan-15-oic Acid, *Me ester*†
α-Monolinolein
9-Octadecynoic Acid, 1-*Glycerol ester*
$C_{21}H_{38}O_6$
Hexanoic Acid, *Glycerol ester*
Rangiformic Acid
$C_{21}H_{38}O_7$
Caperatic Acid
$C_{21}H_{39}NO_2$
Nonadecane-1,1-dicarboxylic Acid, *Mononitrile*
$C_{21}H_{39}N_5O_{14}$
Bluensomycin
$C_{21}H_{39}N_7O_{12}$
Streptomycin
$C_{21}H_{39}N_7O_{13}$
Hydroxystreptomycin
$C_{21}H_{40}N_5O_8P$
Azathymidine, 5′-*Phosphate*
$C_{21}H_{40}O_2$
11-Eicosenoic Acid, *Me ester*
2-Isopropyl-5-methylcyclohexanol, *Methylene ether*
Isovaleric Acid, *Cetyl ester*
$C_{21}H_{40}O_2$
9-Octadecenoic Acid, *Propyl ester*
9-Octadecenoic Acid, *Isopropyl ester*
$C_{21}H_{40}O_3$
12-Hydroxy-9-octadecenoic Acid, *Propyl ester*
$C_{21}H_{40}O_4$
1,21-Henicosanedioic Acid
Hexadecylmalonic Acid, *Di-Me ester*
3-Methylhexadecanedioic Acid, *Di-Et ester*
4-Methylhexadecanedioic Acid, *Di-Et ester*
5-Methylhexadecanedioic Acid, *Di-Et ester*
2-Monoelaidin
α-Mono-olein
β-Mono-olein
Nonadecane-1,1-dicarboxylic Acid
Pentadecane-1,1-dicarboxylic Acid, *Di-Et ester*

$C_{21}H_{40}O_{12}$
Maltobionic Acid, *Me ester, Octa-Me ether*
$C_{21}H_{41}NO_3$
Nonadecane-1,1-dicarboxylic Acid, *Amide*
$C_{21}H_{41}N_7O_{12}$
Dihydrostreptomycin
$C_{21}H_{42}$
9-Heneicosene
$C_{21}H_{42}O$
2-Heneicosanone
4-Heneicosanone
7-Heneicosanone
8-Heneicosanone
9-Heneicosanone
10-Heneicosanone
11-Heneicosanone
$C_{21}H_{42}O_2$
Eicosanoic Acid, *Me ester*
Hexadecanoic Acid, *Pentyl ester*
Nonadecanoic Acid, *Et ester*★†
Octadecanoic Acid, *Propyl ester*
3,7,11,15-Tetramethylhexadecanoic Acid, *Me ester*†
$C_{21}H_{42}O_3$
2-Hydroxyeicosanoic Acid, *Me ester*
13-Hydroxyoctadecanoic Acid, *Propyl ester*
Selachyl Alcohol
$C_{21}H_{42}O_4$
α-Monostearin
β-Monostearin
$C_{21}H_{44}$
Heneicosane
2,6,10,14-Tetramethylheptadecane†
$C_{21}H_{44}O_2$
3,6-Heneicosanediol
$C_{21}H_{44}O_3$
Batyl Alcohol
1-Octadecanol, β-*Glyceryl ether*
$C_{21}H_{45}N_3$
Hexetidine

C_{22}

$C_{22}H_{10}Cl_2O_2$
Perylene-3,9-dicarboxylic Acid, *Dichloride*
Perylene-3,10-dicarboxylic Acid, *Dichloride*
$C_{22}H_{10}N_2$
Perylene-3,9-dicarboxylic Acid, *Dinitrile*
Perylene-3,10-dicarboxylic Acid, *Dinitrile*
$C_{22}H_{10}O_2$
Anthanthrone
$C_{22}H_{12}$
Benzo[*ghi*]perylene
Dibenzo[*cd,jk*]pyrene
$C_{22}H_{12}NS$
12*H*-Benzo[*g*]thianaphtheno[2,3-*a*]carbazole
14*H*-Benzo[*a*]thianaphtheno[3,2,-*i*]carbazole
$C_{22}H_{12}N_4$
9,10,11,16-Tetrazatribenz[*a,c,h*]anthracene
$C_{22}H_{12}O_2$
Dibenz[*a,c*]anthracene-9,14-dione
Dibenz[*a,h*]anthracene-7,14-dione
Dibenz[*a,j*]anthracene-7,14-dione
6,13-Pentacenedione
Picenequinone
$C_{22}H_{12}O_4$
Perylene-3,9-dicarboxylic Acid
Perylene-3,10-dicarboxylic Acid
$C_{22}H_{12}O_{13}$
Flavogallonic Acid, *Me ester*
Valoneaic Acid, *Dilactone, Me ester*
$C_{22}H_{12}S_2$
Benzo[*d*]naphtho[1,2-*d*]benzo[1,2-*b*:4,3-*b*′]dithiophene†
$C_{22}H_{13}BrO_4$
2-Methylanthraquinone-1-carboxylic Acid, p-*Bromophenyl ester*
$C_{22}H_{13}ClO_4$
1-Chloroanthraquinone-2-carboxylic Acid, *Benzyl ester*
$C_{22}H_{13}NO_2$
7-Nitrodibenz[*a,h*]anthracene
$C_{22}H_{13}N_5O_5S_4$
Micrococcinic Acid†
$C_{22}H_{14}$
Benzo[*b*]chrysene
Benzo[*c*]chrysene
Benzo[*g*]chrysene
Dibenz[*a,c*]anthracene
Dibenz[*a,n*]anthracene
Dibenz[*a,j*]anthracene
Dibenzo[*b,g*]phenanthrene
Dibenzo[*c,g*]phenanthrene
Pentacene
Pentaphene
1-Phenylpyrene†
2-Phenylpyrene†
4-Phenylpyrene†
Picene
Tribenz[*a,e,h*]azulene
3,4:5,6:9,10-Tribenzobicyclo[6,2,0]decapentaene†
$C_{22}H_{14}N_2$
2,3-Diphenylquinoline-4-carboxylic Acid, *Nitrile*
$C_{22}H_{14}O_2$
Di-1-naphthylglyoxal
Di-2-naphthylglyoxal
3-Hydroxybenzo[*a*]pyrene, *Ac*
7-Hydroxybenzo[*a*]pyrene, *Ac*
Perylene-3-carboxylic Acid, *Me ester*
$C_{22}H_{14}O_4$
2-Methylanthraquinone-1-carboxylic Acid, *Phenyl ester*
α-Naphthoyl peroxide
β-Naphthoyl peroxide
$C_{22}H_{14}O_6$
2,2′-Dihydroxy-1,1′-binaphthyl-3,3′-dicarboxylic Acid†

$C_{22}H_{14}O_6$ (*continued*)
Diospyrin †
Elliptinone †
Isodiospyrin †
Rhein, *Benzyl ester*

$C_{22}H_{15}ClO$
Di-1-naphthylacetic Acid, *Chloride*

$C_{22}H_{15}NO_2$
2,3-Diphenylquinoline-4-carboxylic Acid
2,8-Diphenylquinoline-4-carboxylic Acid

$C_{22}H_{15}NO_4$
1-Aminoanthraquinone-2-carboxylic Acid, *Benzyl ester*

$C_{22}H_{16}$
7,12-Dihydro-8,9-benzopleiadene †
1,1-Di-(1-naphthyl)ethylene
1,2-Di-(1-naphthyl)ethylene
1,2-Di-(2-naphthyl)ethylene
1,7-Diphenylnaphthalene †
1,8-Diphenylnaphthalene
1,2 : 3,4 : 7,8-Tribenz[10]annulene †

$C_{22}H_{16}N_2O$
2,3-Diphenylquinoline-4-carboxylic Acid, *Amide*

$C_{22}H_{16}N_2O_3$
2,3-Bis-salicylideneaminobenzofuran †

$C_{22}H_{16}O$
2-Methyl-1,1′-dinaphthyl Ketone
2-Methyl-1,2′-dinaphthyl Ketone
4-Methyl-1,2′-dinaphthyl Ketone

$C_{22}H_{16}O_2$
Di-1-naphthylacetic Acid
Di-2-naphthylacetic Acid
α-Hydroxy-α-2-naphthylacetonaphthone

$C_{22}H_{16}O_3$
Tribenzoylmethane

$C_{22}H_{16}O_4$
Haemocorin Aglycone, *Di-Me ether A* †
Haemocorin Aglycone, *Di-Me ether B* †

$C_{22}H_{16}O_5$
Fluorescein, *Et ether*
Fluorescein, *Me ester*, *Me ether*
Fluorescein, *Et ester*

$C_{22}H_{16}O_6$
6α,12α-Dehydromilletone
Resistomycin★ †

$C_{22}H_{16}O_7$
η-Pyrromycinone

$C_{22}H_{16}O_8$
Di-(4-hydroxy-3-coumarinyl)acetic Acid, *Et ester*
Ethyl *bis*coumacetate
Garcinin
Thelephoric Acid, *Tetra-Me ether* †

$C_{22}H_{16}O_9$
Di-(4-hydroxy-3-coumarinyl)acetic Acid, 2-*Hydroxyethyl ester*

$C_{22}H_{16}O_{10}$
Anthraquinone-1,2,5,6-tetracarboxylic Acid, *Tetra-Me ester*
Anthraquinone-1,2,6,7-tetracarboxylic Acid, *Tetra-Me ester*
Anthraquinone-1,2,7,8-tetracarboxylic Acid, *Tetra-Me ester*

$C_{22}H_{16}O_{12}$
Fumarprotocetraric Acid

$C_{22}H_{16}O_{13}$
m-Trigallic Acid, *Me ester* †

$C_{22}N_{17}N$
N-Diphenyl-1-naphthylamine

$C_{22}H_{17}NO_4$
3-*o*-Nitrophenyl-2-phenylacrylic Acid, o-*Tolyl ester*
3-*m*-Nitrophenyl-2-phenylacrylic Acid, o-*Tolyl ester*

$C_{22}H_{18}$
2,2′-Dimethyl-1,1′-binaphthyl †

$C_{22}H_{18}Cl_6O_4$
2,4-Dichloro-2,4-di-(2,4-dichlorophenyl)cycylbutane-1,3-dicarboxylidc Acid, *Di-Et ester*

$C_{22}H_{18}N_2$
1,4-Dianilinonaphthalene
1,6-Dianilinonaphthalene
2,3-Dianilinonaphthalene
2,6-Dianilinonaphthalene
2,7-Dianilinonaphthalene

$C_{22}H_{18}N_2O_2$
1-Methylamino-4-*p*-toluidinoanthraquinone
1-Methylamino-5-*p*-toluidinoanthraquinone

$C_{22}H_{18}N_2O_3$
Alstoniline

$C_{22}H_{18}N_2O_4$
Alstoniline oxide

$C_{22}H_{18}O$
1,1′-Dinaphthylmethanol, *Me ether*

$C_{22}H_{18}O_2$
2,2′-Dihydroxy-1,1′-binaphthyl, *Di-Me ether*
4,4′-Dihydroxy-1,1′-binaphthyl, *Di-Me ether*
1,1′-Dihydroxy-2,2′-binaphthyl, *Di-Me ether*
2,3-Diphenylacrylic Acid, o-*Tolyl ester*
α,β-Diphenylcinnamic Acid, *Me ester*
9,10-Phenanthraquinol, *Mono*-o-*xylyl ether*

$C_{22}H_{18}O_3$
o-(5-Acenaphthoyl)benzoic Acid, *Et ester*
5-Phenyl-2,4-pentadienoic Acid, *Anhydride*

$C_{22}H_{18}O_4$
Diospyrol †
Phenolphthalein I, *Di-Me ether*
Phenolphthalein II, *Di-Me ether*
Phenolphthalein III, *Di-Me ether*
Phthalic Acid, *Dibenzyl ester*

$C_{22}H_{18}O_5$
Fluorescin, *Di-Me ether*
Fluorescin, *Et ester*

$C_{22}H_{18}O_6$
Croceomycin
Durmillone †
Isomilletone †
Jamaicin †
Millettone †

$C_{22}H_{18}O_7$
Justicidin A†
Millettosin†
1-Phenylcyclopropane-1,2-dicarboxylic Acid, *Anhydride*

$C_{22}H_{18}O_8$
Phenanthrene-1,8,9,10-tetracarboxylic Acid, *Tetra-Me ester*

$C_{22}H_{18}O_{11}$
2,2′,4,5′-Tetra-carbomethoxybenzoic Anhydride†

$C_{22}H_{18}O_{12}$
Chicoric Acid

$C_{22}H_{19}NO_4$
Aminoterephthalic Acid, *Dibenzyl ester*
Bisacodyl

$C_{22}H_{19}NO_6$
3,4-Dihydroxy-2-nitrophenylacetic Acid, 3,4-*Dibenzyl ether*

$C_{22}H_{19}NO_9$
Berberilic Anhydride, *Ac*

$C_{22}H_{20}N_2$
N,*N*′-Di-(1-naphthyl)ethylenediamine
N,*N*′-Di-(2-naphthyl)ethylenediamine

$C_{22}H_{20}N_2O_7$
Adifoline†

$C_{22}H_{20}N_2O_{11}$
2-*p*-Nitrophenylglutaric Acid, *Anhydride*
3-*o*-Nitrophenylglutaric Acid, *Anhydride*
3-*m*-Nitrophenylglutaric Acid, *Anhydride*
3-*p*-Nitrophenylglutaric Acid, *Anhydride*

$C_{22}H_{20}O_2$
Octahydrodibenz[*a*,*c*]anthraquinone
a-Phenylbenzoin, *Et ether*
Triphenylacetic Acid, *Et ester*
2,2,3-Triphenylpropionic Acid, *Me ester*
2,3,3-Triphenylpropionic Acid, *Me ester*
3,3,3-Triphenylpropionic Acid, *Me ester*

$C_{22}H_{20}O_3$
2-Hydroxyphenyldiphenylacetic Acid, *Me ether*, *Me ester*
2-Hydroxyphenyldiphenylacetic Acid, *Et ether*

$C_{22}H_{20}O_4$
Phenolphthalin, *Et ester*
Phenolphthalin, *Di-Me ether*
1-Phenylnaphthalene-2,3-dicarboxylic Acid, *Di-Et ester*
Polyporic Acid, *Di-Et ether*

$C_{22}H_{20}O_5$
Psoralidin, *Di-Me ether*

$C_{22}H_{20}O_6$
Atromentin, *Tetra-Me ether*
Robustic Acid†
Tetrahydroxy-*p*-benzoquinone, *Diphenyl ether*, *Di-Et ether*

$C_{22}H_{20}O_7$
α-Apopicropodophyllin
β-Apopicropodophyllin
γ-Apopicropodophyllin
Chrysomycin (?)
7-Deoxyaklavinone†

$C_{22}H_{20}O_8$
Aklavinone
ζ-Pyrromycinone
ζ-Rhodomycinone†

$C_{22}H_{20}O_9$
Chromocycline
ε-Pyrromycinone
δ-Rhodomycinone
ε-Rhodomycinone

$C_{22}H_{20}O_{10}$
Granatacin
ε-Isorhodomycinone★†

$C_{21}H_{20}O_{11}$
Oroboside
Thamnolic Acid, *Di-Me ester*

$C_{22}H_{20}O_{13}$
Carminic Acid
Ellagic Acid, 3,3′-*Di-Me ether*

$C_{22}H_{21}Cl$
α-Chlorotri-*o*-tolylmethane
α-Chlorotri-*p*-tolylmethane

$C_{22}H_{21}NO_5$
Angoline†

$C_{22}H_{21}N_3O_5S$
6-Nitro-4-sulpho-*m*-toluic Acid, *Di-*o*-toluidide*
6-Nitro-4-sulpho-*m*-toluic Acid, *Di-*m*-toluidide*
6-Nitro-4-sulpho-*m*-toluic Acid, *Di-*p*-toluidide*

$C_{22}H_{22}$
Tri-*o*-tolymethane
Tri-*p*-tolylmethane

$C_{22}H_{22}ClNO_6$
Ochratoxin A, *Me ester*, *Me ether*†
Ochratoxin A, *Et ester*†

$C_{22}H_{22}NO_6$ (ion)
Alborine†

$C_{22}H_{22}N_2O_4$
N_a-Carbomethoxy-10,22-dioxokopsane†

$C_{22}H_{22}O$
Triphenylmethanol, *Propyl ether*
Triphenylmethanol, *Isopropyl ether*
Tri-*o*-tolylmethanol
Tri-*p*-tolylmethanol

$C_{22}H_{22}O_2$
Terephthalyl Alcohol, *Dibenzyl ether*

$C_{22}H_{22}O_3$
Leucaurine, *Tri-Me ether*
Orthoformic Acid, *Tri-*o*-tolyl ester*

$C_{22}H_{22}O_4$
Dibenzylidenesuccinic Acid, *Di-Et ester*

$C_{22}H_{22}O_6$
Diethyl dibenzoylsuccinate
Nor-β-anhydroicaritin, *Di-Me ether*

$C_{22}H_{22}O_7$
Amorphigenin
Deoxypicropodophyllin
Deoxypodophyllotoxin
Ptilometric Acid, *Me ester*, *Tri-Me ether*†
Rhodocomatulin, 1,3,6,8-*Tetra-Me ether*†

$C_{22}H_{22}O_8$
Elliptic Acid, *Et ester*
Lucidin, *Di-Et ether* (flavone)†
Narceonic Acid, *Me ester*
β-Peltatin
Picropodophyllin
1-*epi*-Picropodophyllin
Podophyllotoxin
1-*epi*-Podophyllotoxin
Toringin, *Me ether*

$C_{22}H_{22}O_{10}$
Echioidin†
Glucogenkwanin
Isocytisoside†
4′-*O*-Methylvitexin†
Sissotrin†
Swertisin†
Tilianin†
Trifolirhizin

$C_{22}H_{22}O_{11}$
Azalein
Isorhamnetin★, 3-β-D-*Rhamnoside*†
Isoshehkanin
Keyakinin†
Leptosin
Orientin, 5-*Me ether*†
Scoparin
Swertiajaponin†

$C_{22}H_{22}O_{12}$
Brassicin†
Cacticin
Isorhamnetin★, 3-β-D-*Glucoside*†

$C_{22}H_{22}O_{13}$
Belmacamdin

$C_{22}H_{23}BrN_2O_8$
Bromotetracycline

$C_{22}H_{23}ClN_2O_8$
Aureomycin

$C_{22}H_{23}ClN_2O_9$
7-Chloro-5-hydroxytetracycline†

$C_{22}H_{23}ClO_{12}$
Myrtillin chloride

$C_{22}H_{23}IN_2$
Erythroapocyanine

$C_{22}H_{23}NO$
m-Aminophenyldiphenylmethanol, N-*Tri-Me*

$C_{22}H_{23}NO_4$
Methyl benzoquate†
Ochotensimine†

$C_{22}H_{23}NO_5$
Heteratisine

$C_{22}H_{23}NO_6$
Aureothin
Methylhydrastine
Ochratoxin B, *Me ester*, *Me ether*†

$C_{22}H_{23}NO_7$
β-Gnoscopine
Isonarcotine
α-Narcotine
β-Narcotine

$C_{22}H_{23}NO_9$
Berberilic Acid, *Di-Me ester*

$C_{22}H_{24}ClN_4O_4$
Azacrin

$C_{22}H_{24}NO_6$ (?)
Mycolutein

$C_{22}H_{24}N_2$
α,α′-Diamino-*o*-xylene, N,N′-*Di-Me*

$C_{22}H_{24}N_2O_3$
α-Colubrin
β-Colubrin
Dehydrovoachalotine†
Icajine†

$C_{22}H_{24}N_2O_4$
Akuammiline
Alstonidine
Fruticosamine
Fruticosine
Kopsine
Voacafricine (or $C_{22}H_{26}N_2O_4$)
Vomicine

$C_{22}H_{24}N_2O_8$
Tetracycline

$C_{22}H_{24}N_2O_9$
Terramycin

$C_{22}H_{24}O_4$
α-Isatropic Acid, *Di-Et ester*
Neotruxinic Acid, *Di-Et ester*
α-Truxillic Acid, *Di-Et ester*
γ-Truxillic Acid, *Di-Et ester*
ε-Truxillic Acid, *Di-Et ester*
β-Truxinic Acid, *Di-Et ester*
ζ-Truxinic Acid, *Di-Et ester*

$C_{22}H_{24}O_5$
Tomentolide B†

$C_{22}H_{24}O_6$
4′,5-Dihydroxy-3′,7-dimethoxyflavanone, 4′-[1-(3-*Methylbutenyl*)]-*ether*†
Phellamuretin, *Di-Me ether*
Rengasin, *Tri-Et ether*
Scriblitifolic Acid, *Et ester*, *Me ether*†
Symphoxanthone, *Tetra-Me ether*†
3,5,7-Trihydroxy-2′-methoxyflavone, *Tri-Et ether*
Tsugaresinol, *Di-Me ether*

$C_{22}H_{24}O_7$
5,7-Dihydroxy-3,4′,8-trimethoxyflavone, *Di-Et ether*†
Pinoresinolide, *Di*-O-*Me ether*

$C_{22}H_{24}O_8$
Atranorin, *Tri-Me ether*
Combretol, 5-*Et ether*†
1,2,3,5,6,7-Hexahydroxyanthraquinone, *Tetra-Et ether*
2-Methylirigenol, *Hexa-Me ether*

$C_{22}H_{24}O_9$
3,3′,4′,5,5′,6,7-Heptamethoxyflavone†
Limocitrol, *Tetra-Me ether*†
Podophyllic Acid
epi-Podophyllic Acid
Thomasic Acid†

$C_{22}H_{24}O_{10}$
Sakuranin
$C_{22}H_{24}O_{11}$
Mangiferin★, *Tri-Me ether*†
$C_{22}H_{24}O_{16}$
Digicitrine, 3′-*Me ether*
$C_{22}H_{25}NO_2$
Lobelanine
$C_{22}H_{25}NO_5$
Corycavidine
Hunnemanine, *Et ether*†
$C_{22}H_{25}NO_6$
Allocolchicine
Colchicine
Cordrastine I
Cordrastine II
Glaudine†
Gloriosine
β-Lumicolchicine
γ-Lumicolchicine
Mecambridine†
Oreodine†
$C_{22}H_{25}NO_7$
Methylhydrasteine
Oxycolchicine†
$C_{22}H_{25}NO_8$
Nornarceine
$C_{22}H_{25}N_2OS$ (ion)
Trimethaphan
$C_{22}H_{26}NO_4$ (?)
Fagarine III
$C_{22}H_{26}N_2O_2$
Pleiocarpinine
$C_{22}H_{26}N_2O_3$
ψ-Akuammigine
Aspidofractine
Ochropamine†
Pleiocarpinilam
Powerchrine†
Vincamajine
Voachalotine
$C_{22}H_{26}N_2O_4$
Akuammine
Aricine
Cimicine†
Corymine★†
Corynoxeine
Gambirine
Holstiine (or $C_{23}H_{28}NO_4$)
Isocorymine†
Isoraunitidine
Isorespinine
Lochnerinine
Lochrovidine†
16-Methoxyminovincine
Minoriceine
Mitrajavanine†
Mitraversine
Picraphylline†
Raunitidine
Reserpinine†
Raumitorine†
Tetraphylline
Vincadiffine†
Virosine
Voacafricine (or $C_{22}H_{24}N_2O_4$)
Voacafrine
Voacryptine
$C_{22}H_{26}N_2O_5$
Henningsoline†
Vincinine
Vinosidine†
$C_{22}H_{26}N_2O_6$
Methylhydrasteine, *Amide*
$C_{22}H_{26}N_4$
Calycanthine
Chimonanthine
Hodgkinsine
$C_{22}H_{26}O_4$
Bakalactone
Cannabinolic Acid †
Otobaphenol, *Et ether*†
$C_{22}H_{26}O_5$
Didymic Acid
$C_{22}H_{26}O_6$
Diaeudesmin†
Eudesmin†
epi-Eudesmin
Matai-resinol, *Di-Me ether*
Pinoresinol, *Di-Me ether*
$C_{22}H_{26}O_7$
Diffractaic Acid, *Et ester*
Divaricatic Acid, *Me ester*
Gmelinol
Isogmelinol
Neogmelinol†
$C_{22}H_{26}O_8$
Desaspidin AB†
Flavaspidic Acid AB†
Lirioresinol A†
Lirioresinol B†
Lirioresinol C†
Picropoline†
Sekikaic Acid
$C_{22}H_{26}O_{10}$
p-Asebotin
Glycosmin
$C_{22}H_{26}O_{12}$
Catalposide
$C_{22}H_{27}ClN_2O_3$
Melinonine A, *Chloride*
$C_{22}H_{27}NO$
Phenazocine
$C_{22}H_{27}NO_2$
Lobeline
$C_{22}H_{27}NO_3$
Dioxaphetyl butyrate
Tuduranine, N-*Et*-O-*Et*
$C_{22}H_{27}NO_4$
Corydaline

$C_{22}H_{27}NO_5$
Capaurimine, *Di-Me ether* †
Capaurine, *Me ether* †
Kreysigine †
Muramine †
Oconovine, *Me ether* †
Preocoteine, *Me ether* †

$C_{22}H_{27}NO_6$
Alpinigenine †
Papaverrubine G †

$C_{22}H_{27}N_2O_3$ (ion)
Macusine A
Macusine C †
Melinonine A

$C_{22}H_{27}N_3O_3$
(+)-Lysergyl-L-valine methyl ester

$C_{22}H_{27}N_3O_7S$
Griseoviridin

$C_{22}H_{27}N_5O_4$
Distamycin A †

$C_{22}H_{28}$
2′,4-Decamethylenebiphenyl †
Nonacyclo[11,7,1,$1^{2,18}$,$0^{3,16}$,$0^{4,13}$,$0^{5,10}$,-$0^{6,14}$,$0^{7,11}$,$0^{15,20}$]-docosane †

$C_{22}H_{28}BrNO_3$
Benziloyloxy-1,1-diethylpyrrolidium bromide

$C_{22}H_{28}N_2O_2$
Anileridine
Cupreine, *Monopropyl ether*
Minovine

$C_{22}H_{28}N_2O_3$
1-Acetyl-17-hydroxy-aspidoalbidine, O-*Me ether* †
Corynantheidine
Corynantheine
N-Formyl-16,17-dimethoxyaspidofractinine †
Haplocine
Hirsutine †
Isovoacangine
Isoyohimbic Acid, *Et ester*
Pleiocarpinine★, N_b-*Oxide* †
Strychnospermine
Vincorine
Voacangine
Yohimbic Acid, *Et ester*

$C_{22}H_{28}N_2O_4$
Aspidolimidine
Catharosine †
Corynoxine★ †
Echitamine
Gambirine †
Isorhyncophylline
Isovenenatine †
Isovoacristine †
16-Methoxyminovincinine †
11-Methoxyvincamine
Mitrinermine
Rhyncophylline
Rupicoline †
Venenatine †
Voacristine
Voaluteine †

$C_{22}H_{28}N_2O_5$
Mitragynol
Montanine †
Mycelianamide
Reserpic Acid
Rotundifoline
Speciofoline †
Tetraphyllinine †
Venoxidine †

$C_{22}H_{28}O_2$
Cannabinol, *Me ether*

$C_{22}H_{28}O_4$
Crocetin, *Di-Me ester*
Galbulin
Guaiaretic Acid, (±)-*Di-Me ether*

$C_{22}H_{28}O_5$
Galgravin
Mammea B/BA †
Mammea B/BB †
Mammein
16α-Methylprednisone
16β-Methylprednisone
Pyrethrin II
Veraguensin

$C_{22}H_{28}O_6$
Apetalic Acid †
Dimethyl sciadinonate
Lariciresinol, *Di-Me ether*
Picrasmin

$C_{22}H_{28}O_7$
Glauconic Acid, *Di-Et ester*
Isodonal †
Olivil, *Di-Me ether*
Olivil, *Et ether*
Trichodonin †

$C_{22}H_{28}O_8$
Lyoniresinol †

$C_{22}H_{29}ClN_4O_4$
Azacrin

$C_{22}H_{29}FO_5$
9α-Fluoro-11β-17,21-trihydroxy-16α-methyl-pregna-1,4-diene-3,20-dione
9α-Fluoro-11β-17α,21-trihydroxy-16β-methyl-pregna-1,4-diene-3,20-dione
9α-Fluoro-11α-17,21-trihydroxypregn-4-ene-3,20-dione
12α-Fluoro-11β-17α,21-trihydroxypregn-4-en-3,20-dione

$C_{22}H_{29}N$
Lobelan

$C_{22}H_{29}NO_2$
8,10-Diphenyl-lobelidiol
Lobelanidine
Spiradine D †

$C_{22}H_{29}NO_4S$
Benztropine, *Methanesulphonate*

$C_{22}H_{29}N_3O_4$
Reserpic Acid, *Amide*

$C_{22}H_{29}N_3S_2$
Thiethylperazine †
$C_{22}H_{29}N_7O_5$
Puromycin
$C_{22}H_{30}Cl_2N_{10}$
Chlorhexidine
$C_{22}H_{30}N_2$
Lysuric Acid, *Et ester*
$C_{22}H_{30}N_2O$
1-Acetyl-16-methylaspidospermidine †
Demethoxypalosine
$C_{22}H_{30}N_2O_2$
Aspidospermine
O-Demethylpalosine †
Vincaminoreine
Vincaminorine
$C_{22}H_{30}N_2O_3$
Aspidocarpine
Limaspermine
Spegazzinine, O-*Me ether*
Vincaminoridine
$C_{22}H_{30}N_2O_4$
3′-Methoxylimapodin
$C_{22}H_{30}N_8O_{16}$
Alanyl-(3′-[5′-adenyl])-5′-uridylate
$C_{22}H_{30}O_2$
1,2-Dihydroxy-1,2-di-(2,3,4,6-tetramethylphenyl)ethane
Siccanochromene A †
$C_{22}H_{30}O_3$
17α-Hydroxy-6-methylpregna-4,6-diene-3,20-dione †
2-Hydroxy-6-(8,11,14-pentadecatrienyl)-benzoic Acid
Siccanin †
Siccanochromene B †
$C_{22}H_{30}O_4$
Cannabidiol Acid
Carnosol, *Di-Me ether* †
Picrosalvin, *Di-O-Me ether*
Tauranin †
$C_{22}H_{30}O_5$
Acetoxyroyleanone
Gibberellin A_{24}, *Di-Me ester* †
Jasmolin II †
16β-Methylcortisone
6α-Methylprednisolone
16β-Methylprednisolone
$C_{22}H_{30}O_6$
6-Acetoxymethyl-5-(2,3′-furylethyl)-3,4,5,6,7,8,9,10-octahydro-9-hydroxymethyl-5-methyl-l-naphthoic Acid †
Fukinolide †
Gibberellin A_{19}, *Di-Me ester* †
Laserolide †
epi-Nodosin, O-*Et* †
$C_{22}H_{30}O_7$
Gibberellin A_{23}, *Di-Me ester* †
$C_{22}H_{30}O_8$
Valepotriate †

$C_{22}H_{30}O_{14}$
ψ-Maltal, *Penta-Ac*
$C_{22}H_{31}BrO_2$
17α-Bromo-6α-methylprogesterone
$C_{22}H_{31}NO_3$
Songorine †
Spiradine G †
$C_{22}H_{31}NO_5$
Oxotuberostemonine
$C_{22}H_{32}$
2,6-Dihexylnaphthalene †
$C_{22}H_{32}O_2$
Dehydroabietic Acid, *Et ester*
4,7,10,13,16,19-Docosahexaenoic Acid
4,8,12,15,18,21-Docosahexaenoic Acid
Grifolin †
6-α-Methylprogesterone
14α-Methylprogesterone
$C_{22}H_{32}O_3$
17α-Hydroxy-6α-methylpregn-4-ene-3,20-dione †
2-Hydroxy-6-(8,11-pentadecadienyl)benzoic Acid
Vinhaticoic Acid, *Et ester*
$C_{22}H_{32}O_3S$
3,7-Dihexylnaphthalene-1-sulphonic Acid †
$C_{22}H_{32}O_4$
Cannabigerolic Acid
Gibberellin A_{12}, *Di-Me ester* †
Xylopic Acid
$C_{22}H_{32}O_5$
Gibberellin A_{14}, *Di-Me ester* †
16β-Methylhydrocortisone
Prehumulone
Torilin †
$C_{22}H_{32}O_6$
Genin (from *Digitalis* spp.)
$C_{22}H_{32}O_7$
Cascarillin †
$C_{22}H_{32}O_8$
Alternaric Acid, *Me ester*
$C_{22}H_{32}O_8$
5,6-Dihydrovalepotriate †
$C_{22}H_{33}NO$
Latifolinine
$C_{22}H_{33}NO_2$
Atisine
Cuauchichicine
Denudatine
Garryfoline
Garryine
Geralbine
Himgravine
Paravallarine
$C_{22}H_{33}NO_3$
Ajaconine
Atidine
Cyclomethycaine
Galbulimina alkaloid G.B.18 †
Himgrine

$C_{22}H_{33}NO_3$ *(continued)*
Isonapelline†
Luciculine★†
Napelline
Paravallaridine
Xylopic Acid, *Amide*

$C_{22}H_{33}NO_4$
Heterophyllisine†
Stemonine
Tuberostemonine
Tuberostemonine-A

$C_{22}H_{34}BrNO_2$
N-2-α-1′-Cyclopentenylphenylacetoxyethyl-*N*-di-isopropylmethylammonium bromide

$C_{22}H_{34}N_2$
Conessidine

$C_{22}H_{34}O_2$
Clupanodonic Acid
Pimaric Acid, *Et ester*
Sapietic Acid, *Et ester*

$C_{22}H_{34}O_3$
5-Androstene-3β,16α,17α-triol, *Acetone condensation product*
2-Hydroxy-6-(8-pentadecenyl)benzoic Acid

$C_{22}H_{34}O_4$
enantio-18-Acetoxylabd-8(20),13-dien-15-oic Acid†
Agathic Acid, *Di-Me ester*
Ancepsenolide†
Isoborneol, *Oxalyl*
16α-Kaurane-17,19-dioic Acid, *Di-Me ester*†
Kolavic Acid, *Di-Me ester*†
Solidagonic Acid†

$C_{22}H_{34}O_5$
Acovenosigenin A
Pleuromutilin

$C_{22}H_{34}O_6$
Diginatigenin
Drevogenin D

$C_{22}H_{34}O_{12}$
Piptoside, *Penta-*O*-Me deriv.*†

$C_{22}H_{35}NO$
Holaphylline
Latifoline (5-Conanen-3β-ol)

$C_{22}H_{35}NO_2$
Funtuline†
Himbacine

$C_{22}H_{35}NO_3$
Galbulimina alkaloid G.B.15†

$C_{22}H_{35}NO_4$
Hexadecanoic Acid, o-*Nitrophenyl ester*

$C_{22}H_{35}NO_7$
N-De-ethyldelcosine

$C_{22}H_{36}N_6O_{12}S_2$
Bis-γ-L-glutamyl-L-cystinyl-bis-β-alanine†

$C_{22}H_{36}O$
Palmitophenone
3-(8′-Pentadecenyl)phenol, *Me ester*

$C_{24}H_{36}O_2$
Chaulmoogric Acid, *Phenyl ester*
7,10,13,16-Docosatetraenoic Acid
Hexadecanoic Acid, *Phenyl ester*
2-Hydroxyhexadecanophenone
4-Hydroxyhexadecanophenone
8-Phenylhexadecanoic Acid
16-Phenylhexadecanoic Acid
Renghol, *Me ether*

$C_{22}H_{36}O_3$
Grindelic Acid, *Et ester*
2-Hydroxy-6-pentadecylbenzoic Acid

$C_{22}H_{36}O_4$
Eperu-8(20)-en-15,18-dioic Acid, *Di-Me ester*†
enantio-Labd-8(20)-ene-15,18-dioic Acid, *Di-Me ester*†
Pinifolic Acid, *Di-Me ester*

$C_{22}H_{36}O_5$
2,3-Dicarboxy-2,3-secomanoyl oxide, *Di-Me ester*†

$C_{22}H_{36}O_6$
Oligomycin B

$C_{22}H_{36}O_{13}$
6-*O*-Oleuropeoylsucrose†

$C_{22}H_{37}NO$
Funtumufrine B
Irehamine†

$C_{22}H_{38}$
1,2,4,5-Tetra-*tert*-butylbenzene†

$C_{22}H_{38}Br_6O_2$
9,10,12,13,15,16-Hexabromo-octadecanoic Acid, n-*Butyl ester*

$C_{22}H_{38}N_2$
Irehdiamine B

$C_{22}H_{38}N_2O_6$
Angolide†

$C_{22}H_{38}O$
3-Pentadecylphenol, *Me ether*

$C_{22}H_{38}O_2$
Cativic Acid, *Et ester*
1,2-Dihydroxy-3-pentadecylbenzene, *Me ether*
Hexadecylquinol
2-Methyl-5-pentadecylresorcinol†
Tetradecylquinol, *Di-Me ether*

$C_{22}H_{38}O_3$
enantio-18-Hydroxylabd-8(20)-en-15-oic Acid, *Me ester*, 18-*Me ether*†
10-Undecenoic Acid, *Anhydride*

$C_{22}H_{38}O_6$
Ungulinic Acid

$C_{22}H_{39}ClO$
13-Docosynoic Acid, *Chloride*

$C_{22}H_{39}NO$
Funtuphyllamine B†

$C_{22}H_{39}N_5O_4$
Cerevioccidin

$C_{22}H_{40}BrNO$
Domiphen

$C_{22}H_{40}N_2$
Docosanedioic Acid, *Dinitrile*†

$C_{22}H_{40}O_2$
Cyclododocosane-1,12-dione
13-Docosynoic Acid
5,13-Docosadienoic Acid
Isoborneol, *Dodecanoyl*

$C_{22}H_{40}O_4$
13,14-Dioxodocosanoic Acid

$C_{22}H_{40}O_5$
11-Oxodocosanedioic Acid
10-Oxoeicosanedioic Acid, *Di-Me ester*

$C_{22}H_{40}O_7$
Agaric Acid

$C_{22}H_{41}ClO$
Brassidic Acid, *Chloride*

$C_{22}H_{41}NO$
13-Docosynoic Acid, *Amide*

$C_{22}H_{41}NO_2$
(−)-3-Menthylglycine, (−)-*Menthyl ester*

$C_{22}H_{42}O_2$
Brassidic Acid
11-Docosenoic Acid
9-Eicosenoic Acid, *Et ester*
Erucic Acid
9-Octadecenoic Acid, *Butyl ester*
9-Octadecenoic Acid, *Isobutyl ester*
9-Octadecenoic Acid, tert-*Butyl ester*

$C_{22}H_{42}O_3$
12-Hydroxy-9-octadecenoic Acid, *Butyl ester*
4-Oxodocosanoic Acid
10-Oxodocosanoic Acid
14-Oxodocosanoic Acid

$C_{22}H_{42}O_4$
Docosanedioic Acid†
Eicosanedioic Acid, *Di-Me ester*

$C_{22}H_{43}ClO$
Docosanoic Acid, *Chloride*

$C_{22}H_{43}IO_2$
13-(or 14)-Iodocosanoic Acid

$C_{22}H_{43}NO$
Brassidic Acid, *Amide*

$C_{22}H_{44}N_2O_2$
Glyodin

$C_{22}H_{44}O$
13-Docosen-1-ol

$C_{22}H_{44}O_2$
1,22-Docosanediol†
Docosanoic Acid
Eicosanoic Acid, *Et ester*
2-Hexanol, *Hexadecanoyl*
Octadecanoic Acid, *Isobutyl ester*
2-Octanol, *Tetradecanoyl*

$C_{22}H_{44}O_3$
2-Hydroxydocosanoic Acid
3-Hydroxydocosanoic Acid
22-Hydroxydocosanoic Acid
2-Hydroxyeicosanoic Acid, *Et ester*
2-Hydroxyeicosanoic Acid, *Et ether*
13-Hydroxytetradecanoic Acid, *Butyl ester*

$C_{22}H_{44}O_6$
Ventosic Acid

$C_{22}H_{45}Br$
1-Bromodocosane

$C_{22}H_{45}I$
1-Iododocosane

$C_{22}H_{45}NO$
Docosanoic Acid, *Amide*

$C_{22}H_{45}NO_2$
Sphingosine, *Di-Et ether*

$C_{22}H_{46}$
Docosane

$C_{22}H_{46}N_4O_2$
Pithecolobine

$C_{22}H_{46}O$
Bridelyl Alcohol
1-Docosanol

$C_{22}H_{47}N$
Diundecylamine

C_{23}

$C_{23}H_{13}N_3$
Dibenzo[*f*,*h*]quino[3,4-*b*]quinoxaline

$C_{23}H_{14}$
13*H*-Acenaphtheno[1,8-*ab*]phenanthrene
11*H*-Indeno[*b*,*f*,*g*]pyrene†

$C_{23}H_{14}O$
1-Phenylbenzanthrone
2-Phenylbenzanthrone
3-Phenylbenzanthrone
6-Phenylbenzanthrone
10-Phenylbenzanthrone

$C_{23}H_{14}O_{13}$
Flavogallonic Acid, *Et ester*

$C_{23}H_{16}$
2-Methyldibenz[*a*,*h*]anthracene
3-Methyldibenz[*a*,*h*]anthracene
6-Methyldibenz[*a*,*h*]anthracene
7-Methyldibenz[*a*,*h*]anthracene
10-Methyldibenz[*a*,*h*]anthracene

$C_{23}H_{16}O$
2,3,4-Triphenylcyclopentadienone†
2,3,5-Triphenylcyclopentadienone†

$C_{23}H_{16}O_2$
Perylene-3-carboxylic Acid, *Et ester*

$C_{23}H_{16}O_2S$
3,8-Diphenyl-2*H*-naphth[2,3-*b*]thiete-1,1-dioxide†

$C_{23}H_{16}O_3$
Diphenadione

$C_{23}H_{16}O_5$
Dalbergin, *Benzoyl*

$C_{23}H_{17}N$
2,3,6-Triphenylpyridine
2,4,5-Triphenylpyridine
2,4,6-Triphenylpyridine

$C_{23}H_{17}NO$
Dibenz[*a*,*i*]acridone, N-*Et*

$C_{23}H_{17}P$
2,4,6-Triphenylphosphorin†

$C_{23}H_{18}$
1,2,3-Triphenylcyclopentadiene†
1,2,4-Triphenylcyclopentadiene†

$C_{23}H_{18}N_2O$
4-Benzeneazo-1-naphthol, *Benzyl ether*
1-(2-Naphthyl)-1,3-diphenylurea

$C_{23}H_{18}O$
1-Naphthyldiphenylmethanol
2-Naphthyldiphenylmethanol

$C_{23}H_{18}O_5$
Dalatinone

$C_{23}H_{18}O_8$
Di-(4-Hydroxy-3-coumarinyl)acetic Acid, *Propyl ester*

$C_{23}H_{18}O_{11}$
Leucocyanidin Gallate†

$C_{23}H_{19}NO_4$
N-Phenacyl-*o*-aminobenzoic Acid, *Phenacyl ester*
N-Phenacyl-*p*-aminobenzoic Acid, *Phenacyl ester*

$C_{23}H_{20}N_2$
Lophine

$C_{23}H_{20}N_2O_3S$
Sulphinpyrazone

$C_{23}H_{20}O$
1,1′-Dinaphthylmethanol, *Et ether*

$C_{23}H_{20}O_2$
1,3-Dibenzoyl-2-phenylpropane
2,2′-Di-(2-hydroxy-1-naphthyl)methane, *Di-Me ether*
α,β-Diphenylcinnamic Acid, *Et ester*

$C_{23}H_{20}O_5$
Fluorescin, *Di-Me ether*, *Me ester*

$C_{23}H_{20}O_7$
Ichthynone

$C_{23}H_{20}O_8$
Anthraquinone-1,2,4-tricarboxylic Acid, *Tri-Et ester*

$C_{23}H_{20}O_{10}$
Aromadendrin, *Tetra-Ac*

$C_{23}H_{21}ClO_3$
Chlorotrianisene

$C_{23}H_{21}NO_2$
5,6-Dihydroxyskatole, 5,6-*Dibenzyl ether*†

$C_{23}H_{21}NO_6$
4-Hydroxy-3-methoxy-2-nitrophenylacetic Acid, *Benzyl ether*, *Benzyl ester*

$C_{23}H_{22}O_2$
Dibenzylacetic Acid, *Benzyl ester*
Triphenylacetic Acid, *Propyl ester*
Triphenylacetic Acid, *Isopropyl ester*
2,3,3-Triphenylpropionic Acid, *Et ester*
3,3,3-Triphenylpropionic Acid, *Et ester*

$C_{23}H_{22}O_3$
4-Hydroxy-3,4,5-triphenylvaleric Acid

$C_{23}H_{22}O_6$
Deguelin
Isodeguelin
Macluraxanthone
Myriconol
Robustic Acid, *Me ether*†
Rotenone

$C_{23}H_{22}O_7$
Amorphigenin★†
Fukugenetin, *Tetra-Me ether*
Isotephrosin
Sumatrol
Tephrosin
α-Toxicarol
β-Toxicarol
Toxicarol Isoflavone†

$C_{23}H_{22}O_{10}$
Glaucophanic Acid, *Di-Et ester*†

$C_{23}H_{22}O_{14}$
Equisporoside

$C_{23}H_{23}IN_2S_2$
Dithiazanine

$C_{23}H_{23}NO$
2-Benzylpropionic Acid, *Benzylanilide*
N-Tritylmorpholine†

$C_{23}H_{23}NO_8$
Chelocardin†

$C_{23}H_{24}ClNO_6$
Ochratoxin A, *Et ester*, *Me ether*†

$C_{23}H_{24}N_2O_6$
10,11-Dioxopleiocarpine†

$C_{23}H_{24}O$
Triphenylmethanol, *Butyl ether*

$C_{23}H_{24}O_2$
4-Hydroxyphenyldiphenylmethanol, *Di-Et ether*

$C_{23}H_{24}O_5$
Rotiorin

$C_{23}H_{24}O_6$
Alvaxanthone†
β-Anhydroicaritin, *Di-Me ether*
Averythrin, *Tri-Me ether*†
β-Mangostin
Physodylic Acid

$C_{23}H_{24}O_8$
Hydroxytoxicarol
Hyposalazinic Acid, *Penta-Me*
Isoderrisic Acid
Narceonic Acid, *Et ester*
α-Peltatin, *Di-Me ether*
Rubrocomatulin, 1,2,4,5,7-*Penta-Me ether*†
Wortmannin†

$C_{23}H_{24}O_9$
Bayin, *Di-Me ether*†

$C_{23}H_{24}O_{10}$
Cosmosiin, *Di-Me ether*†
Haemathamnolic Acid, *Me ester*, *Tri-Me ether*†
Kutkin
Nataloin
Puerarin, *Di-Me ether*†

$C_{23}H_{24}O_{12}$
Alboside†
$C_{23}H_{25}ClO_{12}$
Malvenin chloride
Oenin chloride
$C_{23}H_{25}NO$
Mahanimbin†
$C_{23}H_{25}NO_4$
Tylophorinine
$C_{23}H_{25}NO_7$
Narcindonine
$C_{23}H_{25}NO_8$
2-Acetyl-2-decarboxamidotetracycline†
$C_{23}H_{25}NO_9$
2-Acetyl-2-decarboxamido-oxytetracycline
$C_{23}H_{25}N_3O_2$
αα-Di-(4-aminophenyl)-2-nitrotoluene, 4,4′-N-*Di-Et*
αα-Di-(4-aminophenyl)-3-nitrotoluene, 4,4′-N-*Di-Et*
αα-Di-(4-aminophenyl)-4-nitrotoluene, 4,4′-N-*Di-Et*
$C_{23}H_{26}N_2$
αα-Di-(4-amino-2,5-dimethylphenyl)toluene
3,4′-Di-dimethylaminotriphenylmethane
4,4′-Di-dimethylaminotriphenylmethane
$C_{23}H_{26}N_2O_3$
Vincolidine†
$C_{23}H_{26}N_2O_4$
Brucine
Henningsamine†
Neblinine†
Vomicine, *Me ether*
$C_{23}H_{26}N_2O_5$
Fumaridine
Picraline
$C_{23}H_{26}N_2O_6$
Kopsaporine†
$C_{23}H_{26}N_2O_9$
Ethylmethyloxytetracycline
$C_{23}H_{26}O_3$
2,6-Dimethylcyclohexanecarboxylic Acid, p-*Phenylphenacyl ester*
$C_{23}H_{26}O_4$
Bavalchalchone, *Di-Me ether*†
$C_{23}H_{26}O_5$
Monascorubrin
Xanthohumol, *Di-Me ether*
$C_{23}H_{26}O_6$
Luteolin, *Tetra-Et ether*
$C_{23}H_{26}O_7$
Caryatin, *Tri-Et ether*†
Demethoxysudachitin, *Tri-Et ether*†
Morin, 2′,3,4′,7-*Tetra-Et ether*
Norberqueinone, *Tetra-Me ether*
Quercetin, 3,3′,4′,7-*Tetra-Et ether*
4′,5,7-Trihydroxy-3,8-dimethoxyflavone, *Tri-Et ether*†
$C_{23}H_{26}O_8$
Calycopterin, *Di-Et ether*
$C_{23}H_{26}O_9$
Squamatic Acid, *Me ether*
Thomasic Acid, *Me ester*†
$C_{23}H_{26}O_{11}$
Mangiferin★, *Tetra-Me ether*†
$C_{23}H_{26}O_{13}$
β-Sorinin
$C_{23}H_{26}O_{16}$
Digicitrine, *Di-Me ester*
$C_{23}H_{27}NO_4$
Thevinone†
$C_{23}H_{27}NO_5$
Octaverine
Talatizidine†
$C_{23}H_{27}NO_6$
Allocolchiceine, *Et ester*
Allocolchicine, N-*Me*
$C_{23}H_{27}NO_7$
Methylhydrasteine, *Me ester*
$C_{23}H_{27}NO_8$
Narceine
$C_{23}H_{27}N_3O_7$
Takawo base I
$C_{23}H_{28}ClNO_2S$
Thiopropazate
$C_{23}H_{28}FeO_4$
Retinal Iron Tricarbonyl†
$C_{23}H_{28}N_2O_3$
Majoridine†
$C_{23}H_{28}N_2O_4$
Echitovenine†
Ochropine†
Paynantheine†
Pleiocarpine
Refractine
Vellosine
$C_{23}H_{28}N_2O_5$
Cimicidine
Isoreserpiline
10-Oxo-cylindrocarpidine†
Pleiocarpamine, N_b-*Oxide*†
Rauvanine
Reserpiline
$C_{23}H_{28}N_2O_6$
Carapanaubine
Isoreserpiline-ψ-indoxyl
Rauvoxine†
Rauvoxinine†
$C_{23}H_{28}N_2O_7$
Narceine, *Amide*
$C_{23}H_{28}N_4$
Calycanthidine
$C_{23}H_{28}O_4$
Cannabinolic Acid, *Me ester*†
$C_{23}H_{28}O_5$
Piceatannol, *Penta-Me ether*†
$C_{23}H_{28}O_7$
Diffractaic Acid, *Propyl ester*
Flavogenin

$C_{23}H_{28}O_7$ *(continued)*
Imbricaric Acid
Sphaerophorin

$C_{23}H_{28}O_8$
Albaspidin PP★†
Desaspidin PB†
Flavaspidic Acid PB†
Phloraspin
Ramalinolic Acid

$C_{23}H_{28}O_{10}$
Plicatic Acid, *Tri-Me ether*†

$C_{23}H_{28}O_{11}$
Bruceine B†
Paeoniflorin†

$C_{23}H_{28}O_{12}$
Catalposide

$C_{23}H_{28}O_{15}$
Dryophantin

$C_{23}H_{29}ClO_4$
Chloromadinone acetate†

$C_{23}H_{29}NO$
Dipipanone

$C_{23}H_{29}NO_2$
Phenadoxone

$C_{23}H_{29}NO_3$
Benzethidine

$C_{23}H_{29}NO_4$
Scoulerine, *Di-Et ether*†
Thevinol†

$C_{23}H_{29}NO_5$
Capaurine, *Et ether*†
Floramultine, *Di-*O-*Me ether*†
6β-Isovaleryloxy-3α-tigloyloxytropan-7β-ol†

$C_{23}H_{29}NO_6$
Alpinine†

$C_{23}H_{29}NO_{12}$
Hygromycin A

$C_{23}H_{29}N_3O_2S_2$
Thiothixene†

$C_{23}H_{29}N_3O_7$
Xanthomycin A (or $C_{23}H_{31}N_3O_7$)

$C_{23}H_{30}ClN_3O$
Atebrin

$C_{23}H_{30}N_2O_2$
Undecanedioic Acid, *Dianilide*

$C_{23}H_{30}N_2O_3$
Haplocine★, O-*Me ether*†
Lochrovine†
Yohimbic Acid, *Propyl ester*

$C_{23}H_{30}N_2O_4$
Conopharyngine
Cylindrocarpidine
Fendlerine†
Mitraciliatine†
Mitragynine
3-(2-Morpholinoethyl)morphine†
Poweramine†
Speciociliatine†
Speciogynine

$C_{23}H_{30}N_2O_5$
N-Acetyl-*N*-depropionylaspidoalbine†
Deacetylvindoline†
Herbaceine†
Jollyanine†
Reserpic Acid, *Me ester*
Seredine†

$C_{23}H_{30}N_2O_6$
Herbaline†

$C_{23}H_{30}N_2O_8$
Kopsiflorine

$C_{23}H_{30}N_2O_9$
Phyllomycin

$C_{23}H_{30}O_3$
Canariengenin A
3,5-Dianhydroperiplogenin

$C_{23}H_{30}O_4$
Anhydroperiplogenone

$C_{23}H_{30}O_5$
Apocynamarin
Ferruol A†

$C_{23}H_{30}O_6$
Apetalic Acid, *Me ester*†
Citreoviridin†
Decogenin

$C_{23}H_{30}O_7$
Isostrophanthidic Acid
Olivil, *Me-Et ether*

$C_{23}H_{30}O_7$
Sarverogenin†

$C_{23}H_{30}O_8$
Isostrophanthonic Acid

$C_{23}H_{31}NO_4$
Reticuline, O,O-*Di-Et ether*†

$C_{23}H_{31}NO_{12}$
Homomycin

$C_{23}H_{31}N_3O_7$
Xanthomycin A (or $C_{23}H_{29}N_3O_7$)

$C_{23}H_{32}BrNO_2$
2-Diphenylacetoxyethyldi-isopropylmethyl-ammonium Bromide

$C_{23}H_{32}N_2O_2$
Palosine

$C_{23}H_{32}N_2O_3$
Aspidolimine
Pyrifolidine

$C_{23}H_{32}N_2O_4$
3′-Methoxylimaspermine
Spegazzidnidine, *Di-Me ether*

$C_{23}H_{32}N_6O_{14}$
Polyoxin A†

$C_{23}H_{32}O_2$
Dehydroabietic Acid, *Allyl ester*
2,2-Di-*p*-hydroxyphenylpropane, *Di-*n*-butyl ether*
17β-Hydroxy-6α,21-dimethylpregn-4-en-20-yn-3-one
Plastoquinone-3†

$C_{23}H_{32}O_4$
Adynerigenin
Canarigenin †
Neriantogenin
Tanghiferigenin
Glaucorigenin
Glaucotoxigenin
Isocorchortoxin
Isostrophanthidin
17-Isostrophanthidin
Quilengenin
3β,5,14-Trihydroxy-19-oxo-5β-card-20(22)-enolide

$C_{23}H_{32}O_5$
Corotoxigenin
Tanghinigenin (or $C_{23}H_{34}O_5$)
3-*epi*-Tanghinigenin (or $C_{23}H_{34}O_5$)

$C_{23}H_{32}O_6$
Alloglaucotoxigenin
Calotropagenin
Caudogenin

$C_{23}H_{32}O_7$
Antiarigenin
Averufin, *Tri-O-Me* †
Gibberellin A_{13}, *Tri-Me ester* †
Gibberellin A_{17}, *Tri-Me ester* †
Nigrescigenin †
Strophadogenin

$C_{23}H_{32}O_8$
Isostrophanthic Acid

$C_{23}H_{32}O_{10}$
Giloin

$C_{23}H_{33}IN_2O$
Isopropamide

$C_{23}H_{33}NO$
Evocarpine †

$C_{23}H_{33}NO_2$
Veatchine

$C_{23}H_{33}NO_3$
Yuzurimine B †

$C_{23}H_{33}NO_4$
Macrodaphniphyllamine †

$C_{23}H_{33}NO_5$
Galbulimina alkaloid GB.8 †

$C_{23}H_{33}NO_9$
Genatropine, *Glucoside*

$C_{23}H_{33}NO_{10}$
Floricaline †

$C_{23}H_{34}KNO_{14}S_2$
Glucocappasalin, *Tetra-Ac K salt*

$C_{23}H_{34}O$
Agaurol B

$C_{23}H_{34}O_2$
4,8,12,15,18,21-Docosahexaenoic Acid, *Me ester*

$C_{23}H_{34}O_4$
Cannabigerolic Acid, *Me ester* †
Chiogralactone †
3β,14-Dihydroxy-5β-card-20(22)-enolide
Isouzarigenin
17-Isouzarigenin
Salvin, *Di-O-Me, Me ester* †
Xylopic Acid, *Me ester*
Xysmalogenin

$C_{23}H_{34}O_5$
Coroglaucigenin
17-Isoperiplogenin
Periplogenin
Tanghinigenin (or $C_{23}H_{32}O_5$)
3-*epi*-Tanghinigenin (or $C_{23}H_{32}O_5$)
3β,11α,14β-Trihydroxy-5β-card-20(22)-enolide
3β,11β,14β-Trihydroxy-5β-card-20(22)-enolide
3α,12,14β-Trihydroxy-5β-card-20(22)-enolide
3β,14β,16β-Trihydroxy-5β-card-20-enolide

$C_{23}H_{34}O_6$
Anhydroafrogenin
Antiogenin

$C_{23}H_{36}N_4O_{10}S_2$
4,4′-Pentamethylenedioxydibenzamidine

$C_{23}H_{36}N_{12}O_8$
Viomycin †

$C_{23}H_{36}O_2$
Clupanodonic Acid, *Me ester*
3-Heptadecadienylcatechol
Laccol

$C_{32}H_{36}O_3$
2-Hydroxy-6-(8-pentadecenyl)benzoic Acid, *Me ether*

$C_{23}H_{36}O_4$
enantio-18-Acetoxylabd-8(20),13-dien-15-oic Acid, *Me ester* †
Calabarol
Solidagonic Acid, *Me ester* †

$C_{23}H_{36}O_6$
Digacetigenin

$C_{23}H_{37}NO_2$
Raddeamine

$C_{23}H_{37}NO_4$
Hexadecanoic Acid, p-*Nitrobenzyl ester*

$C_{23}H_{37}NO_5$
Delavaconine
Isotalatizidine †

$C_{23}H_{37}NO_6$
Lappaconine

$C_{23}H_{38}N_2$
Conessimine
Isoconessimine
Kurchine
Norconessine

$C_{23}H_{38}N_2O$
12β-Hydroxyconessimine †

$C_{23}H_{38}O_2$
Benzoic Acid, *Cetyl ester*
Bilobol, *Di-Me ether*
Heptadecanoic Acid, *Phenyl ester*
3-Heptadecenylcatechol
Hexadecanoic Acid, p-*Tolyl ester*
Hexadecanoic Acid, *Benzyl ester*
4-Hydroxyhexadecanophenone, *Me ether*

$C_{23}H_{38}O_2$ *(continued)*
Norallocholanic Acid
23-Nor-5β-cholanoic Acid
Renghol, *Di-Me ether*

$C_{23}H_{38}O_4$
Grindelol (?)

$C_{23}H_{38}O_{10}$
Arasaponin B

$C_{23}H_{39}NO$
Funtumufrine C
Irehine†

$C_{23}H_{39}NO_3$
Alginine

$C_{23}H_{40}O_2$
1,2-Dihydroxy-3-pentadecylbenzene
Guaiacol, *Cetyl ether*
1-Heptadecyl-2,3-dihydroxybenzene
4-Heptadecyl-1,2-dihydroxybenzene

$C_{23}H_{40}O_6$
Corymbositin†
Turbicorytin†

$C_{23}H_{40}O_{20}$
Lycotetraose

$C_{23}H_{41}N$
N-Heptadecylaniline

$C_{23}H_{41}NO$
2-Azacholan-24-ol
3-Azacholan-24-ol
Funtuphyllamine C†

$C_{23}H_{41}NO_2$
Terminaline†

$C_{23}H_{42}N_2$
Chonemorphine
Epipachysamine F†

$C_{23}H_{42}N_2O_{12}$
Pentolium Tartrate

$C_{23}H_{43}NO_2$
Heneicosane-1,1-dicarboxylic Acid, *Mono-nitrile*

$C_{23}H_{44}$
11,15,19-Trimethyl-1,11-eicosadiene

$C_{23}H_{44}O_2$
Brassidic Acid, *Me ester*
13-Docosynoic Acid, *Me ester*

$C_{23}H_{42}O_4$
13,14-Dioxodocosanoic Acid, *Me ester*

$C_{23}H_{44}O_2$
9-Octadecenoic Acid, 3-*Methylbutyl ester*
9-Octadecenoic Acid, 2-*Methyl*-2-*butyl ester*

$C_{23}H_{44}O_3$
4-Oxodocosanoic Acid, *Me ester*
14-Oxodocosanoic Acid, *Me ester*

$C_{23}H_{44}O_4$
Docosanedioic Acid, *Mono-Me ester*†
Heneicosane-1,1-dicarboxylic Acid
1,21-Henicosanedioic Acid, *Di-Me ester*
Hexadecylmalonic Acid, *Di-Me ester*
Tricosanedioic Acid

$C_{23}H_{45}N$
Tricosanoic Acid, *Nitrile*

$C_{23}H_{45}N_5O_{14}$
Paromomycin★(Paromomycin I†)
Paromomycin II†

$C_{23}H_{46}ClNO_4$
Pahutoxin†

$C_{23}H_{46}N_6O_{12}$
Neomycin B
Neomycin C

$C_{23}H_{46}O$
12-Tricosanone

$C_{23}H_{46}O_2$
Docosanoic Acid, *Me ester*
Hexadecanoic Acid, *Heptyl ester*
21-Methyldocosanoic Acid
Octadecanoic Acid, 3-*Methylbutyl ester*
Tricosanoic Acid

$C_{23}H_{46}O_3$
2-Hydroxydocosanoic Acid, *Me ester*
22-Hydroxydocosanoic Acid, *Me ester*

$C_{23}H_{48}$
Tricosane

$C_{23}H_{48}O$
21-Methyldocosanol
1-Tricosanol
12-Tricosanol

C_{24}

$C_{24}H_{10}O_2$
1,2-Coronenequinone

$C_{24}H_{12}$
Dibenzo[*b,mno*]fluoranthene
Coronene
1,2:5,6:9,10-Tribenzocyclododeca-1,5,9-triene-3,7,11-triyne†

$C_{24}H_{12}O_2$
1,1′-Biacenaphthylidene-2,2′-dione

$C_{24}H_{12}O_3$
Truxenequinone

$C_{24}H_{12}O_8$
Perylene-3,4,9,10-tetracarboxylic Acid

$C_{24}H_{12}S_3$
Naphtho[1,2-*d*]benzo[*b″*]thieno[4,5-*d′*]benzo-[1,2-*b*,4,3-*b′*]dithiophene†

$C_{24}H_{14}$
Benzo[*b*]perylene
Dibenzo[*a,e*]fluoranthene†
Dibenzo[*de,mn*]naphthacene
Dibenzo[*de,qr*]naphthacene
Dibenzo[*a,e*]pyrene
Dibenzo[*a,h*]pyrene
Dibenzo[*a,i*]pyrene★†
Dibenzo[*e,l*]pyrene
Naphtho[1,2-*a*]pyrene
Naphtho[2,1-*a*]pyrene
Naphtho[3,2-*a*]pyrene

$C_{24}H_{14}N_2$
Chrysophenazine
Di-indole[3,2,1-*de*:3′,2′,1′-*kl*]phenazine
Tribenzo[*a*,*c*,*h*]phenazine

$C_{24}H_{14}N_2O_2S_2$
Indophenin

$C_{24}H_{14}O_3$
1-Hydroxy-9,10-anthraquinone, 1-*Naphthyl ether*

$C_{24}H_{15}N$
Anthraceno[2,3-*b*]carbazole†
Anthraceno[2,3-*c*]carbazole†
13*H*-Benzo[*g*]naphtho[2,3-*a*]carbazole
15*H*-Benzo[*a*]naphtho[2,3-*i*]carbazole

$C_{24}H_{15}N_3O_9$
1,3,5-Trihydroxy-2,4,6-trinitrobenzene, *Triphenyl ether*

$C_{24}H_{16}$
Benzo[5,6]cyclohepta[1,2,3-*cd*]fluoranthene
9*H*-Benzo[*b*]naphtho[1,2-*g*]fluorene
9*H*-Benzo[*a*]naphtho[1,2-*g*]fluorene
1,2,3,6,7,8-Di(1′,8′-naphtho)[10]annulene†
1,2-Diphenylnaphtho[*b*]cyclobutadiene†
Heptacyclene

$C_{24}H_{16}N_2$
1,1′-Bicarbazyl
3,3′-Bicarbazyl
3,9′-Bicarbazyl
9,9′-Dicarbazyl

$C_{24}H_{16}N_2O_4$
1-Amino-5-nitroanthraquinone, N-*a-Naphthyl*

$C_{24}H_{16}O_4$
Naphthalene-1,5-dicarboxylic Acid, *Diphenyl ester*
Naphthalene-2,7-dicarboxylic Acid, *Diphenyl ester*
2,8,14,20-Tetraoxa-penta-cyclo[19,3,1,1^{3,7},1^{9,13},1^{15,19}]-octacosa-1(25),3,5,7(28),9,11,13(27),15,17,19(26),21,23-dodecaene†

$C_{24}H_{16}O_8$
Fukugetin

$C_{24}H_{16}O_{12}$
Laccaic Acid B†

$C_{24}H_{17}NO$
Benz[*b*]acridone, N-*Benzyl*

$C_{24}H_{17}N_5O_5S_4$
Micrococcinic Acid, *Di-Me ester*†

$C_{24}H_{18}$
1,1′-Biacenaphthene
2,9-Dimethylpicene
2,11-Dimethylpicene
4,9-Dimethylpicene
5,8-Dimethylpicene
2,2′-Diphenylbiphenyl
2,3′-Diphenylbiphenyl
2,4′-Diphenylbiphenyl
3,3′-Diphenylbiphenyl
3,4′-Diphenylbiphenyl
4,4′-Diphenylbiphenyl
Tribenzo[12]annulene†
1,2,3-Triphenylbenzene
1,2,4-Triphenylbenzene
1,3,5-Triphenylbenzene
Truxene

$C_{24}H_{18}N_2$
o-Azobiphenyl
p-Azobiphenyl

$C_{24}H_{18}N_2O$
o-Azoxybiphenyl
p-Azoxybiphenyl

$C_{24}H_{18}O_3$
3,3′-Dihydroxydiphenyl Ether, *Di-phenyl ether*
4,4′-Dihydroxydiphenyl Ether, *Di-phenyl ether*
Phloroglucinol, *Triphenyl ether*

$C_{24}H_{18}O_4$
Succinic Acid, *Di-1-naphthyl ester*
Succinic Acid, *Di-2-naphthyl ester*

$C_{24}H_{18}O_6$
2,2′-Dihydroxy-1,1′-binaphthyl-3,3′-dicarboxylic Acid, *Di-Me ester*†
Diospyrin, *Di-Me ether*†
Elliptinone, *Di-Me ether*†
Isodiospyrin, *Di-Me ether*†

$C_{24}H_{19}NO_2$
2,3-Diphenylquinoline-4-carboxylic Acid, *Et ester*
2,8-Diphenylquinoline-8-carboxylic Acid, *Et ester*
γ-Truxillic Acid, *Phenylimide*
ε-Truxillic Acid, *Phenylimide*
β-Truxinic Acid, *Phenylimide*
ζ-Truxinic Acid, *Phenylimide*

$C_{24}H_{19}P_3$
1,2,3-Triphenyl-1,2,3-triphosphaindane†

$C_{24}H_{20}$
Dibenzoequinene†

$C_{24}H_{20}As_2$
Phenylcacodyl

$C_{24}H_{20}N_2$
4,4′-Diamino-2,2′-diphenylbiphenyl
4,4′-Diamino-3,3′-diphenylbiphenyl
4,4′-Dianilinobiphenyl
2,2′-Hydrazodibiphenyl
4,4′-Hydrazodibiphenyl
Tetraphenylhydrazine

$C_{24}H_{20}N_2O_6S_2$
Benzidine-2,2′-disulphonic Acid, (+)-*Diphenyl ester*

$C_{24}H_{20}N_2O_8$
Benzylmalonic Acid, *Di*-p-*nitrobenzyl Ester*

$C_{24}H_{20}O_3$
Cortisalin

$C_{24}H_{20}O_4Si$
Phenol, *Orthosilicate*

$C_{24}H_{20}O_5$
Fluorescein, *Et ester*, *Et ether*

$C_{24}H_{20}O_6$
Draconol
Glycerol, *Tribenzoyl*

$C_{24}H_{20}O_7$
Gallein, *Me ester*, *Tri-Me ether*
Tricin, 4′-*Benzyl ether*

$C_{24}H_{20}O_8$
Di-(4-hydroxy-3-coumarinyl)acetic Acid, *Et Ester*, *Di-Me ether*
Di-(4-hydroxy-3-coumarinyl)acetic Acid, *Butyl ester*
Di-(4-hydroxy-3-coumarinyl)acetic Acid, *Isobutyl ester*
Syringetin, 4′-*Benzyl ether*

$C_{24}H_{20}O_{10}$
Gyrophoric Acid
Hiascininic Acid

$C_{24}H_{20}O_{14}$
1,5-Dihydroxy-*S*-indacene-2,3,4,6,7,8-hexacarboxylic Acid, *Hexa-Me ester*†

$C_{24}H_{20}P_4$
Phosphobenzene

$C_{24}H_{20}Pb$
Lead tetraphenyl

$C_{24}H_{20}Si$
Tetraphenylsilane

$C_{24}H_{20}Sn$
Tin tetraphenyl

$C_{24}H_{21}N$
N-Dibenzyl-1-naphthylamine
N-Dibenzyl-2-naphthylamine

$C_{24}H_{21}NO_2$
Anonaine, N-*Benzyl*†

$C_{24}H_{21}NO_3$
α-Truxillic Acid, 3-*Anilide*
γ-Truxillic Acid, 3-*Anilide*
ε-Truxillic Acid, 3-*Anilide*
η-Truxillic Acid, 1-*Anilide*
β-Truxinic Acid, 2-*Anilide*
δ-Truxinic Acid, 2-*Anilide*
ζ-Truxinic Acid, 1-*Anilide*
ζ-Truxinic Acid, 2-*Anilide*

$C_{24}H_{21}N_3$
Indole, *Trimer*

$C_{24}H_{22}N_2$
N,N′-Di-*o*-tolyl-1,4-naphthylenediamine
N,N′-Di-*p*-tolyl-1,6-naphthylenediamine
N,N′-Di-*o*-tolyl-2,6-naphthylenediamine
N,N′-Di-*p*-tolyl-2,6-naphthylenediamine
N,N′-Di-*o*-tolyl-2,7-naphthylenediamine
N,N′-Di-*p*-tolyl-2,7-naphthylenediamine

$C_{24}H_{22}O_2$
2,2′-Dihydroxy-1,1′-binaphthyl, *Di-Et ether*
4,4′-Dihydroxy-1,1′-binaphthyl, *Di-Et ether*

$C_{24}H_{22}O_2S$
Thiodiglycol, *Di*-1-*naphthyl ether*
Thiodiglycol, *Di*-2-*naphthyl ether*

$C_{24}H_{22}O_4$
Diospyrol, *Di-Me ether*†
Phenolphthalein I, *Di-Et ether*
Phenolphthalein I, *Et ester*, *Et ether*
Phenolphthalein II, *Di-Et ether*
Phenolphthalein II, *Et ester*, *Et ether*
Phenolphthalein III, *Di-Et ether*
Phenolphthalein III, *Et ester*, *Et ether*
3,3′,4,4′-Tetrahydroxy-1,1′-binaphthyl, *Tetra-Me ether*

$C_{24}H_{22}O_5$
Fluorescin, *Di-Et ether*

$C_{24}H_{22}O_6$
Croceomycin, *Di-Me*
Phoenicine, *Cyclopentadiene add. comp.*

$C_{24}H_{22}O_{11}$
Actinorhodine

$C_{24}H_{23}N$
2,2′-Di-(α-naphthylethyl)amine
2,2′-Di-(β-naphthylethyl)amine

$C_{24}H_{23}NO_9$
Streptovarone†

$C_{24}H_{24}$
[24]Annulene†

$C_{24}H_{24}B_3N_3O_3$
Tribenzotalarene

$C_{24}H_{24}O_2$
Triphenylacetic Acid, *Butyl ester*
Triphenylacetic Acid, 2-*Butyl ester*
Triphenylacetic Acid, *Isobutyl ester*
2,3,3-Triphenylpropionic Acid, *Propyl ester*
2,3,3-Triphenylpropionic Acid, *Isopropyl ester*

$C_{24}H_{24}O_3$
2-Hydroxyphenyldiphenylacetic Acid, *Et ether*, *Et ester*

$C_{24}H_{24}O_5$
Mesuol★†

$C_{24}H_{24}O_7$
Fukugenetin, *Penta-Me ether*
Lisetin, *Tri-Me ether*†
Toxicarol Isoflavone, *Me ether*†

$C_{24}H_{24}O_9$
Variegatic Acid, *Me ester*, *Penta-Me ether*†

$C_{24}H_{26}Br_2N_2$
Hedaquinium, *Bromide*

$C_{24}H_{26}Cl_2N_2$
Hedaquinium, *Chloride*

$C_{24}H_{26}N_2O_{13}$
Betanin†

$C_{24}H_{26}O$
Tri-*p*-tolylmethanol, *Et ether*

$C_{24}H_{26}O_6$
Averythrin, *Tetra-Me ether*†
α-Mangostin★†
Phellamuretin, *Di-Me ether*, *Ac*

$C_{24}H_{26}O_7$
Archangelicin†
Norsolorinic Acid, *Tetra-Me ether*†
Siphulin†
Solorinic Acid★, *Tri-Me ether*†

$C_{24}H_{26}O_8$
Biphenyl-3,3′,5,5′-tetracarboxylic Acid, *Tetra-Et ester*

$C_{24}H_{26}O_9$
Bayin, *Tri-Me ether* †
$C_{24}H_{26}O_{10}$
Genistin, *Tri-Me ether*
Swertisin, *Di-O-Me* †
$C_{24}H_{26}O_{11}$
Avicularin, *Tetra-Me ether* †
Guaijaverin, *Tetra-Me ether* †
Polystachoside, *Tetra-Me ether* †
Reynoutrin, 3′,4′,5,7-*Tetra-Me ether* †
Scoparin, *Di-Me ether*
$C_{24}H_{26}O_{12}$
Pendulin
$C_{24}H_{26}O_{13}$
Centaureine †
Iridin
Jacein †
$C_{24}H_{26}O_{14}$
Isolimocitrol, 3β-D-*Glucosyl* †
Limocitrol, 3-β-*Glucosyl* †
$C_{24}H_{27}NO_3$
Cryptopleurine
Isocryptopleurine
$C_{24}H_{27}NO_4$
Tylocrebrine
Tylophorine
$C_{24}H_{28}ClN_5O_3$
Dimenhydrinate
$C_{24}H_{28}NO_5$
Condensamine (?)
$C_{24}H_{28}N_2O_4$
Rubradinine
Vincamedine
$C_{24}H_{28}N_2O_5$
Lochnerivine †
11-Methoxyhenningsamine †
Novacine †
Obscurinervidine †
$C_{24}H_{28}N_2O_5S$
Propiomazine, *Hydrogen maleate* †
$C_{24}H_{28}N_2O_6$
21,22α-Epoxy-*N*-methyl-*sec*-pseudobrucine †
$C_{24}H_{28}N_2O_7$
Kopsamine
Kopsingine †
$C_{24}H_{28}N_2O_8$
Decanedioic Acid, *Di*-p-*nitrobenzyl ester*
$C_{24}H_{28}N_4O_4$
Ergosecalinine
$C_{24}H_{28}O_4$
Chlorophorin
Isonorbixin
Norbixin
$C_{24}H_{28}O_6$
Didymic Acid, *Ac*
1,2,2-Trimethylcyclopentane-1,3-dicarboxylic Acid, *Di-guaiacol ester*
Tsugaresinol, *Di-Et ether*
$C_{24}H_{28}O_7$
Siphulin
Vogeletin, *Tetra-Et ether* †
$C_{24}H_{28}O_8$
Evogin
Gossypolic Acid
Isopeulustrin †
Jaceidin, 4′,5,7-*Tri-Et ether* †
Peulustrin †
Sudachitin, *Tri-Et ether* †
3′,4′,5-Trihydroxy-3,6,7-trimethoxyflavone, *Tri-Et ether* †
3′,4′,5-Trihydroxy-3,7,8-trimethoxyflavone, *Tri-Et ether* †
3′,5,7-Trihydroxy-3,4′,8-trimethoxyflavone, *Tri-Et ether* †
4′,5,7-Trihydroxy-3,3′,8-trimethoxyflavone, *Tri-Et ether* †
$C_{24}H_{28}O_9$
Thomasic Acid, 4′,7-*Di-Me ether* †
$C_{24}H_{28}O_{14}$
α-Sorinin
$C_{24}H_{29}NO_4$
Septicine
$C_{24}H_{29}NO_6$
Allocolchiceine, *Propyl ester*
$C_{24}H_{29}NO_7$
Methylhydrasteine, *Et ester*
$C_{24}H_{29}NO_8$
Narceine, *Me ester*
$C_{24}H_{30}$
Hexaisopropenylbenzene †
$C_{24}H_{30}ClNO_4$
Perparin
$C_{24}H_{30}N_2O_3$
Methyl-ψ-brucidine
$C_{24}H_{30}N_2O_5$
Dihydro-obscurinervidine †
Elliptamine †
Poweridine †
Rindline †
Vindolidine
$C_{24}H_{30}N_2O_6$
21-Oxoaspidoalbine †
$C_{24}H_{30}N_2O_8$
Kopsilongine
$C_{24}H_{30}N_4$
Folicanthine
$C_{24}H_{30}O_2$
Partheniol, *Cinnamoyl ester*
$C_{24}H_{30}O_3$
Bufa-3,5,20,22-tetraenolide
Bufotalein
Ferulenol †
Umbelliprenin
$C_{24}H_{30}O_4$
Ammoresinol
Decanedioic Acid, *Dibenzyl ester*
Farnesiferol A †

$C_{24}H_{30}O_4$ (*continued*)
Farnesiferol B†
Farnesiferol C†

$C_{24}H_{30}O_5$
Didymic Acid, *Me ether*, *Me ester*
Resibufagin†

$C_{24}H_{30}O_6$
Bufotalinin
Isocafestol, *Maleic anhydride adduct*

$C_{24}H_{30}O_7$
Anziaic Acid
Athamantin†
Diffractaic Acid, *Butyl ester*
Hypophyllanthin†

$C_{24}H_{30}O_8$
Desaspidin BB†
Desaspidin i-BB†
Homosekikaic Acid
Lirioresinol A, *Di-Me ether*†
Lirioresinol B, *Di-Me ether*†
Lirioresinol C, *Di-Me ether*†
Margaspidin†
Ortho-desaspidin†
Phloroaspidinol†

$C_{24}H_{30}O_9$
Kilimandscharogenin B

$C_{24}H_{30}O_{10}$
Phloridzin, *Tri-Me ether*
Plicatic Acid, *Me ester*, *Tri-Me ether*†

$C_{24}H_{30}O_{11}$
Harpagoside†

$C_{24}H_{30}O_{12}$
Mellitic Acid, *Hexa-Et ester*

$C_{24}H_{31}NO_3$
Tururanine, N-*Di-Et*, O-*Et*

$C_{24}H_{31}NO_6$
Erythroskyrine

$C_{24}H_{32}$
4,4′-Dodecamethylenebiphenyl†

$C_{24}H_{32}ClFO_7$
3-(2′-Chloroethoxy)-9α-fluoro-6-formyl-11β,16α,17α,21-tetrahydroxypregna-3,5-dien-20-one†

$C_{24}H_{32}N_2O_2$
Cupreine, *Monoisopentyl ether*

$C_{24}H_{32}N_2O_3$
Yohimbic Acid, *Butyl ester*

$C_{24}H_{32}N_2O_5$
Aspidoalbine†
Reserpic Acid, *Et ester*

$C_{24}H_{32}O_4$
Artebufogenin A
Artebufogenin B
Di-isochavibetol, *Di-Et ether*
Guaiaretic Acid, *Di-Et ether*
3β-Hydroxy-14,15β-epoxybufa-20,22-dienolide
17α-Hydroxy-6-methylpregna-4,6-diene-3,20-dione, *Ac*†
Scillarenin†

$C_{24}H_{32}O_4S$
Spironolactone

$C_{24}H_{32}O_5$
Asperugin†
Bufagin
Mammea C/BB†
Marinobufagin†

$C_{24}H_{32}O_6$
Apetalic Acid, *Me ether*†
Arenobufagin†
Hellebrigenin†
Lariciresinol, *Di-Et ether*
Salvin, *Di-Ac*†

$C_{24}H_{32}O_7$
Isostrophanthidic Acid, *Me ester*
Olivil, *Di-Et ether*
Schizandrin

$C_{24}H_{32}O_8$
α-Caesalpin†
Lyoniresinol, *Di-Me ether*†

$C_{24}H_{32}O_9$
3,17β-Dihydroxyoestra-1,3,5(10)-trien-16α-yl-β-D-glucopyranosiduronic Acid†

$C_{24}H_{32}O_{12}$
Isochollepidanic Acid (or $C_{24}H_{34}O_{12}$)

$C_{24}H_{32}O_{15}$
Maltal, *Hexa-Ac*

$C_{24}H_{33}NO_4$
Batrachotoxin†

$C_{24}H_{33}NO_5$
Galbulimina alkaloid G.B.14†

$C_{24}H_{33}NO_6$
Galbulimina alkaloid G.B.11†

$C_{24}H_{33}NO_7$
Galbulimina alkaloid G.B.5†

$C_{24}H_{33}N_3O_4$
Lunariamine

$C_{24}H_{33}N_7O_5$
Puromycin★, N,N-*Di-Me*†

$C_{24}H_{34}N_2O_3$
4,4′-Dihydroxyazoxybenzene★, *Dihexyl ether*†

$C_{24}H_{34}O_4$
Bufalin
Canarigenin, 3-*Me ether*
17α-Hydroxy-6α-methylpregn-4-en-3,20-dione, *Ac*†
Isobufalin†

$C_{24}H_{34}O_5$
Dehydrocholic Acid
Kilimandscharogenin A
Telocirobufogin
3β,11α,14-Trihydroxy-5β-bufa-20,22-dienolide

$C_{24}H_{34}O_6$
Cinobufotal idin
Gitaloxigenin
Phyllanthin†

$C_{24}H_{34}O_7$
ε-Caesalpin†
Clerodin
Trichokaurin†

$C_{24}H_{34}O_8$
Isostrophanthic Acid, *Mono-Me ester*
$C_{24}H_{34}O_9$
Bovogenin E
4β,15-Diacetoxy-8α-(3-methylbutyryloxy)scirp-9-en-3α-ol†
$C_{24}H_{34}O_{10}$
Acetoxyvalepotriate†
$C_{24}H_{34}O_{12}$
Isochollepidanic Acid (or $C_{24}H_{32}O_{12}$)
$C_{24}H_{35}NO$
Cyclosuffrobuxinine†
$C_{24}H_{35}NO_4$
Lucidusculine†
$C_{24}H_{35}NO_5$
Batrachotoxinin A†
Ethyl 6-decyloxy-7-ethoxy-4-hydroxyquinoline-3-carboxylate†
$C_{24}H_{35}NO_8$
Demethanolaconinone†
$C_{24}H_{36}N_2O_4$
Lanceine (or $C_{20}H_{26}N_2O_3$)†
$C_{24}H_{36}N_2O_6$
ψ-Morphine
$C_{24}H_{36}N_2O_9S$
Celesticetin
$C_{24}H_{36-40}N_2O_9S$
Celesticetin I
$C_{24}H_{36}O_3$
Lochnerol
Pelandjauic Acid
$C_{24}H_{36}O_6$
Benzene-1,3,5-tricarboxylic Acid, *Tri-3-methylbutyl ester*
$C_{24}H_{36}O_8$
Glaucarubol★, *Tetra-O-Me ether*†
$C_{24}H_{36}O_9$
Gigantin
$C_{24}H_{36}O_{10}$
β-Isocholoidanic Acid
$C_{24}H_{37}BrO_3$
Hexadecanoic Acid, p-*Bromophenacyl ester*
$C_{24}H_{37}ClO_3$
Hexadecanoic Acid, p-*Chlorophenacyl ester*
$C_{24}H_{37}IO_3$
Hexadecanoic Acid, p-*Iodophenacyl ester*
$C_{24}H_{37}NO_2$
Cyclomicrobuxinine†
$C_{24}H_{37}NO_3$
Veratrobasine
$C_{24}H_{37}NO_5$
Hexadecanoic Acid, p-*Nitrophenacyl ester*
Talatisidine
$C_{24}H_{38}N_2O$
Holarrhenine
$C_{24}H_{38}O_2$
9-Octadecenoic Acid, *Phenyl ester*
$C_{24}H_{38}O_3$
Hexadecanoic Acid, *Phenacyl ester*
6-Hydroxy-3,5-cyclocholan-24-oic Acid
2-Hydroxy-6-(8-pentadecenyl)benzoic Acid, *Me ether, Me ester*
11-Lithocholenic Acid
$C_{24}N_{38}O_4$
3α,12α-Dihydroxy-5β-chol-8(14)-enoic Acid
3α-Hydroxy-6-oxo-5α-cholanoic Acid†
Isoborneol, *Succinyl*
Succinic Acid, *Di(—)-Bornyl ester*
$C_{24}H_{38}O_6$
Lithobilianic Acid
$C_{24}H_{38}O_8$
Grayanotoxin III, *Di-Ac*
9(11)-Lithocholenic Acid
$C_{24}H_{38}O_{10}$
Megacidin
$C_{24}H_{39}NO_2$
Raddeanine
$C_{24}H_{39}NO_4$
Cassaine
Octadecanoic Acid, o-*Nitrophenyl ester*
$C_{24}H_{39}NO_5$
Talatisamine
$C_{24}H_{39}NO_6$
De(oxymethylene)lycoctonine†
Neoline
$C_{24}H_{39}NO_7$
Delcosine
Delkine
Delphonine
$C_{24}H_{39}N_5O_8$
Desmosine†
Isodesmosine†
$C_{24}H_{40}N_2$
Conessine
$C_{24}H_{40}N_2O$
α-Hydroxyconessine†
Malouphyllamine
$C_{24}H_{40}O$
Cholan-24-al†
Kutkisterol
Octadecanophenone
$C_{24}H_{40}O_2$
Cholanic Acid
5α-Cholanic Acid
4-Hydroxyhexadecanophenone, *Et ether*
Norallocholanic Acid, *Me ester*
23-Nor-5β-cholanoic Acid, *Me ester*
Octadecanoic Acid, *Phenyl ester*
8-Phenylhexadecanoic Acid, *Et ester*
2-Phenyloctadecanoic Acid
9-Phenyloctadecanoic Acid
10-Phenyloctadecanoic Acid
18-Phenyloctadecanoic Acid
$C_{24}H_{40}O_3$
Lithocholic Acid
β-Lithocholic Acid

$C_{24}H_{40}O_4$
3α,6α-Dihydroxy-5β-cholan-24-oic Acid
3α,7α-Dihydroxy-5α-cholan-24-oic Acid†
3α,7α-Dihydroxy-5β-cholan-24-oic Acid
3α,7β-Dihydroxy-5β-cholanoic Acid
3α,12α-Dihydroxy-5α-cholan-24-oic Acid†
3α,12α-Dihydroxy-5β-cholan-24-oic Acid

$C_{24}H_{40}O_5$
Isocholic Acid
3α,6α,7α-Trihydroxy-5β-cholanoic Acid†
3α,7α,12α-Trihydroxy-5β-cholan-24-oic Acid

$C_{24}H_{40}O_6$
Oligomycin A

$C_{24}H_{41}NO$
Cholanic Acid, *Amide*
Octadecanoanilide
2-Phenyloctadecanoic Acid, *Amide*

$C_{24}H_{41}NO_3$
3α,12α-Dihydroxy-5β-cholan-24-oic Acid, *Amide*

$C_{24}H_{41}NO_4$
Cassaidine

$C_{24}H_{41}NO_6$
Delsonine
Isodelsonine

$C_{24}H_{41}O$ (?)
Chondristerol

$C_{24}H_{42}$
Octadecylbenzene

$C_{24}H_{42}N_2O$
Cyclobuxamine†

$C_{24}H_{42}N_2O_4$
Azimine†

$C_{24}H_{42}N_2O_6$
Enniatin C

$C_{24}H_{42}N_2O_7$
Sambucinin

$C_{24}H_{42}O_2$
Cativic Acid, *Butyl ester*
Octadecylquinol
Tetradecylquinol, *Di-Et ether*

$C_{24}H_{42}O_4$
Succinic Acid, *Di(−)-menthyl ester*

$C_{24}H_{42}O_{21}$
Cellotetraose
Gentiotetraose
Maltotetraose
Scorodose
Stachyose

$C_{24}H_{42}O_{22}$
Maltotetraoinic Acid

$C_{24}H_{43}NO$
7-Aza-*B*-homocholan-24-ol

$C_{24}H_{43}NO_9$
Pseudopederin†

$C_{24}H_{44}INO$
13-(or 14)-Iodocosanoic Acid, *Amide*

$C_{24}H_{44}N_2$
Pachysamine A†

$C_{24}H_{44}N_2O$
Pachysandrine C†

$C_{24}H_{44}N_2O_2$
Kurcholessin†

$C_{24}H_{44}O_2$
13-Docosynoic Acid, *Et ester*

$C_{24}H_{44}O_5$
11-Oxodocosanedioic Acid, *Di-Me ester*

$C_{24}H_{44}O_{21}$
Maltotetraitol

$C_{24}H_{44}Sn$
Tin tetracyclohexyl

$C_{24}H_{46}O_2$
Brassidic Acid, *Et ester*
Nervonic Acid
15-Tetracosenoic Acid

$C_{24}H_{46}O_3$
Dodecanoic Acid, *Anhydride*
14-Oxodocosanoic Acid, *Et ester*

$C_{24}H_{46}O_4$
Docosane-11,12-dicarboxylic Acid†
Docosanedioic Acid, *Di-Me ester*†
Eicosanedioic Acid, *Di-Et ester*
Tetracosanedioic Acid
Tricosanedioic Acid, *Me ester*

$C_{24}H_{47}IO_2$
13-(or 14)-Iodocosanoic Acid, *Et ester*

$C_{24}H_{47}N$
Tetracosanoic Acid, *Nitrile*

$C_{24}H_{47}NO$
Nervonic Acid, *Amide*
15-Tetracosenoic Acid, *Amide*

$C_{24}H_{48}$
2-Methyl-2-tricosene

$C_{24}H_{48}N_4O_9S_2$
Homopantethine

$C_{24}H_{48}O_2$
2,4-Dimethyldocosanoic Acid†
Docosanoic Acid, *Et ester*
Dodecanoic Acid, *Dodecyl ester*
Hexadecanoic Acid, *Octyl ester*
2-Hexanol, *Octadecanoyl*
2-Octanol, *Hexadecanoyl*
Tetracosanoic Acid
Tricosanoic Acid, *Me ester*

$C_{24}H_{48}O_3$
2-Hydroxydocosanoic Acid, *Et ester*
2-Hydroxydocosanoic Acid, *Et ether*
2-Hydroxyeicosanoic Acid, *Et ether*, *Et ester*

$C_{24}H_{49}I$
1-Iodotetracosane

$C_{24}H_{50}$
Bixane
Lignocerane
2-Methyltricosane
12-Methyltricosane
Tetracosane

$C_{24}H_{50}N_2O$
Buxeridine†

$C_{24}H_{50}O$
Lignoceryl Alcohol
Loranthyl Alcohol
2-Methyl-2-tricosanol
12-Methyl-1-tricosanol
1-Tetracosanol
$C_{24}H_{50}O_2$
1,24-Tetracosanediol †
$C_{24}H_{50}O_4S$
Didodecyl sulphate
$C_{24}H_{51}N$
Didodecylamine
$C_{24}H_{52}N_8O_2$
Eulicin
$C_{24}H_{56}B_2N_2O_2$
1,1,3,4,4,6-Hexabutylbon-bon

C_{25}

$C_{25}H_{14}O_5$
1-Hydroxyanthraquinone-2-carboxylic Acid, 2-*Naphthyl ether*
$C_{25}H_{15}N$
Benzo[*a*]naphtho[1,2-*j*]acridine †
Benzo[*c*]naphtho[1,2-*j*]acridine †
Tribenz[*a,c,h*]acridine †
Tribenz[*a,c,j*]acridine †
$C_{25}H_{16}$
8*H*-Benzo[*fg*]pentacene
Tribenzo[*a,c,h*]fluorene
Tribenzo[*a,c,i*]fluorene
$C_{25}H_{16}O_2$
3,6-Diphenylxanthone
$C_{25}H_{16}O_9$
Muscarufin
$C_{25}H_{17}NO_4$
3-*p*-Nitrophenyl-2-phenylacrylic Acid, 1-*Naphthyl ester*
$C_{25}H_{18}O$
5,5′-Diacenaphthyl Ketone
9,9-Diphenylxanthene
$C_{25}H_{20}$
5,5′-Diacenaphthylmethane
Tetraphenylmethane
2,9,10-Trimethylpicene
$C_{25}H_{20}O$
Triphenylmethanol, *Phenyl ether*
$C_{25}H_{20}OS$
3-Benzyl-2,6-diphenyl-2*H*-thiopyran-5-carboxaldehyde †
$C_{25}H_{20}O_2$
2-Hydroxyphenyldiphenylmethanol, *Phenyl ether*
$C_{25}H_{20}O_6$
Caffeic Acid, *Me ester*
$C_{25}H_{20}O_{11}$
Collinomycin (?)

$C_{25}H_{21}N$
α-Aminotriphenylmethane, N-*Phenyl*
$C_{25}H_{21}NO$
o-Aminophenyldiphenylmethanol, N-*Phenyl*
$C_{25}H_{21}N_3$
1,1,3,3-Tetraphenylguanidine
1,1,3,4-Tetraphenylguanidine
$C_{25}H_{22}N_4O_8$
Streptonigrin
$C_{25}H_{22}O$
1-Naphthyldiphenylmethanol, *Et ether*
$C_{25}H_{22}O_5$
Inophyllolide★ †
Tomentolide A †
$C_{25}H_{22}O_6$
Cyclomulberrochromene †
$C_{25}H_{22}O_7$
5,7-Dihydroxy-3,4′,8-trimethoxyflavone, 7-*Benzyl ether* †
$C_{25}H_{22}O_{10}$
Silybin †
Umbilicaric Acid
$C_{25}H_{22}O_{11}$
Carajuretin, *Penta*-O-*Ac*
Hiascininic Acid, *Me ester*
$C_{25}H_{22}O_{12}$
Daphnorin †
$C_{25}H_{24}$
1,2,3,4-Tetrahydro-2,2,9-trimethylpicene
$C_{25}H_{24}O$
5-Methyl-1,3,5-triphenyl-2-cyclohexen-1-ol (incorrectly given as $C_{32}H_{28}O_2$)
$C_{25}H_{24}O_5$
Iso-osajin
Mammeigin †
Osajin
Scandenone †
Sericetin †
$C_{25}H_{24}O_6$
Calphyllic Acid
Cyclomulberrin †
Isopomiferin
Mulberrochromene †
Pomiferin
$C_{25}H_{24}O_7$
Gallin, *Tetra-Me ether Me ester*
$C_{25}H_{24}O_{12}$
Cynarine
3,4-Dicaffeoylquinic Acid †
3,5-Dicaffeoylquinic Acid †
4,5-Dicaffeoylquinic Acid †
$C_{25}H_{25}NO_6$
Epanorin
$C_{25}H_{26}N_2O_7S_2$
Bisdethiodi(methylthio)acetylapoaranotin †
$C_{25}H_{26}N_2O_8S_2$
Bisdethiodi(methylthio)acetylaranotin †
$C_{25}H_{26}O_2$
Triphenylacetic Acid, *Phenyl ester*

$C_{25}H_{26}O_5$
5,7-Dihydroxy-8-isopentenyl-6-isovaleryl-4-phenylcoumarin
Mammea A/AB†
Mammea A/BA†
Mammea A/BB†

$C_{25}H_{26}O_6$
Mulberrin†

$C_{25}H_{26}O_7$
Fukugenetin, *Hexa-Me ether*

$C_{25}H_{26}O_{13}$
Ruberythric Acid

$C_{25}H_{27}ClN_2$
Meclozine

$C_{25}H_{27}N$
Cevanthridine

$C_{25}H_{27}NO_5$
Lyfoline†
Verticillatine

$C_{25}H_{28}$
Tri-(2,5-dimethylphenyl)methane

$C_{25}H_{28}O_6$
α-Mangostin, *Me ether*★†

$C_{25}H_{28}O_7$
Piscerythrone, *Tetra-Me ether*†
Piscidone, *Tetra-Me ether*†
Siphulin, *Me ester*†

$C_{25}H_{28}O_8$
Isoderrisic Acid, *Et ester*
Lobaric Acid

$C_{25}H_{28}O_9$
Bayin, *Di-Et ether*†

$C_{25}H_{28}O_{11}$
epi-Orientin, *Tetra-Me ether*†
Scoparin, *Tri-Me ether*

$C_{25}H_{28}O_{14}$
Gentioside

$C_{25}H_{29}NO_5$
Decodine

$C_{25}H_{29}N_3O_2$
1,6-Dimethyl-8β-carbobenzyloxyaminomethyl-10α-ergoline†

$C_{25}H_{30}N_2O_5$
Obscurinervine†

$C_{25}H_{30}N_2O_8$
Chandrine

$C_{25}H_{30}O_4$
Bixin
Isobixin
Isonorbixin, *Me ester*
Tingenone

$C_{25}H_{30}O_6$
α-Mangostin, *Di-Me ether*★†

$C_{25}H_{30}O_7$
Picrolichenic Acid
Quercetin, *Penta-Et ether*

$C_{25}H_{30}O_8$
Flavaspidic Acid
Flavogenin, *Ac*
Glomelliferic Acid
Limocitrin, *Tetra-Et ether*
3′,4′,5,7-Tetrahydroxy-3,6-dimethoxyflavone, *Tetra-Et ether*†
3′,4′,5,7-Tetrahydroxy-3,8-dimethoxyflavone, *Tetra-Et ether*†

$C_{25}H_{30}O_9$
Thomasic Acid, *Me ester*, 4′,7-*Di-Me ether*†
3′,5,5′-Trihydroxy-3,4′,6,7-tetramethoxyflavone, *Tri-Et ether*†

$C_{25}H_{30}O_{12}$
Homo-orientin

$C_{25}H_{31}NO_3$
Isodihydrohomocryptopleurine

$C_{25}H_{31}NO_8$
Narceine, *Et ester*

$C_{25}H_{31}N_3O_4$
Lunaridine
Lunarine

$C_{25}H_{31}N_3O_6$
PA114A (or $C_{35}H_{42}N_4O_9$)

$C_{25}H_{32}N_2O_2$
Dextromoramide

$C_{25}H_{32}N_2O_4$
Norcephaeline

$C_{25}H_{32}N_2O_5$
Dihydro-obscurinervine†

$C_{25}H_{32}N_2O_6$
21-Oxoaspidoalbine, 17-O-*Me ether*†
Vindolicine†
Vindoline

$C_{25}H_{32}N_4$
α,4,4′,4″-Tetra-aminotriphenylmethane, 4,4′,4″-N-*Hexa-Me*

$C_{25}H_{32}O_5$
Anhydroabogenin

$C_{25}H_{32}O_7$
Diffractaic Acid, *Amyl ester*
Perlatolic Acid

$C_{25}H_{32}O_8$
Albaspidin
Albaspidin BB★†
Aspidin
Boninic Acid
Desaspidin VB†
Homosekikaic Adid, *Me ester*
α-Kosin
β-Kosin
Para-aspidin

$C_{25}H_{32}O_{11}$
Aspalathin, 3,4,4′,6-*Tetra-Me ether*†

$C_{25}H_{32}O_{13}$
Oleuropein

$C_{25}H_{33}Cl_2N_5O_8$
Islanditoxin

$C_{25}H_{33}NO_4$
19-Propylorvinol†
Severine†

$C_{25}H_{33}N_3O_4$
Numismine
$C_{25}H_{34}N_2O_5$
Aspidoalbine, *Me ether*†
$C_{25}H_{34}O$
β-Apo-4-carotenal
β-Apo-12′-carotenal†
$C_{25}H_{34}O_4$
17α-Hydroxy-6-methylpregna-4,6-dien-3,20-dione, *Propionyl*†
$C_{25}H_{34}O_6$
Kablicin†
Regularobufagin
Salvin, *Di-Ac, Me ester*†
$C_{25}H_{34}O_7$
Gibberellic Acid, 2-*Butoxyethyl ester*
$C_{25}H_{34}O_8$
Boninic Acid, *Me ester*
Isostrophanthonic Acid, *Di-Me ester*
$C_{25}H_{34}O_{12}$
Gibberellin A_8★, 3-O-β-D-*Glucopyranosyl*†
$C_{25}H_{35}NO$
6-Aza-*N* benzyl-17β-hydroxy-5ξ–androstane
$C_{25}H_{35}NO_5$
Yuzurimine A†
$C_{25}H_{35}NO_6$
Galbulimina alkaloid G.B.9†
$C_{25}H_{35}NO_9$
Ryanodin
$C_{25}H_{35}N_3O$
Undecylprodigiosin†
$C_{25}H_{35}N_5O_7$
Plicacetin
$C_{25}H_{36}O_2$
2,2-Di-*p*-hydroxyphenylpropane, *Di-pentyl ether*
$C_{25}H_{36}O_4$
Colupulone
17α-Hydroxy-6α-methylpregn-4-ene-3,20-dione, *Propionyl*†
Isobufalin, *Me ester*†
Ophiobolin A†
Ophiobolin D†
$C_{25}H_{36}O_6$
Abogenin
$C_{25}H_{36}O_8$
Isostrophanthic Acid, *Di-Me ester*
$C_{25}H_{36}O_{10}$
Glaucarubin
$C_{25}H_{37}NO$
Cyclomicrobuxeine†
Cyclosuffrobuxinine, N-*Me*†
$C_{25}H_{37}NO_4$
Piericidin A†
$C_{25}H_{37}NO_6$
Dehydrodelpheline
$C_{25}H_{38}O_2$
Ceroplasteric Acid†
Chaulmoogric Acid, *Benzyl ester*
$C_{25}H_{38}O_2$
Gascardic Acid†
$C_{25}H_{38}O_3$
Ophiobolin C†
Pelandjauic Acid, *Me ester*
$C_{25}H_{38}O_4$
Githagonolic Acid
Ophiobolin B†
$C_{25}H_{38}O_6$
Couminginic Acid
$C_{25}H_{38}O_7$
Lasperitine†
$C_{25}H_{38}O_8$
5α-Androstan-(3α → 1β-oside)-D-glucopyranouronic Acid-17-one
5β-Androstan-(3α → 1β-oside)-D-glucopyranouronic Acid-17-one
$C_{25}H_{39}NO$
Buxenone†
Cyclobuxomicreine†
Cyclobuxophyllinine†
Cyclobuxosuffrine†
$C_{25}H_{39}NO_2$
Cyclomicrobuxine†
$C_{25}H_{39}NO_5$
Cassamine
$C_{25}H_{39}NO_6$
Codelphine†
Delpheline
Erythrophamine
Erythrophleguine†
$C_{25}H_{39}NO_7$
Amaromycin
Eldelidine
$C_{25}H_{40}N_2O$
Isoconessimine, *Ac*
$C_{25}H_{40}N_7O_{19}PS$
Methylmalonylcoenzyme A†
$C_{25}H_{40}O$
Catuabol
Ceroplastol I†
Ceroplastol II†
$C_{25}H_{40}O_2$
Laccol, *Di-Me ether*
9-Octadecenoic Acid, m-*Tolyl ester*
$C_{25}H_{40}O_3$
9(11)-Lithocholenic Acid, *Me ester*
11-Lithocholenic Acid, *Me ester*
$C_{25}H_{40}O_4$
3α,12α-Dihydroxy-5β-chol-8(14)-enoic Acid, *Me ester*
3α-Hydroxy-6-oxo-5α-cholanoic Acid, *Me ester*†
$C_{25}H_{40}O_{10}$
Lepranthin
$C_{25}H_{41}NO_6$
Chasmanine†

$C_{25}H_{41}NO_7$
Browniine
Delphamine
Delsoline
Lycoctonine

$C_{25}H_{41}NO_8$
ψ-Aconine

$C_{25}H_{41}NO_9$
Aconine
Japaconine A
Japaconine B

$C_{25}H_{42}N_2$
Norbuxamine †

$C_{25}H_{42}N_2O$
Cyclobuxine

$C_{25}H_{42}N_2O_2$
Malouphylline

$C_{25}H_{42}O$
Didymocarpenol
Geranylnerolidol †
Ophiobola-7,18-dien-20α-ol †

$C_{25}H_{42}O_2$
Cholanic Acid, *Me ester*
5α-Cholanic Acid, *Me ester*
23-Nor-5β-cholanoic Acid, *Et ester*
Octadecanoic Acid, p-*Tolyl ester*
Octadecanoic Acid, *Benzyl ester*

$C_{25}H_{42}O_3$
Lithocholic Acid, *Me ester*
3α,7α,12α-Trihydroxyhomo-5α-chol-24-ene
3α,7α,12α-Trihydroxyhomo-5β-chol-24-ene

$C_{25}H_{42}O_4$
3α,6α-Dihydroxy-5β-cholan-24-oic Acid, *Me ester*
3α,7α-Dihydroxy-5β-cholan-24-oic Acid, *Me ester*
3α,12α-Dihydroxy-5β-cholan-24-oic Acid, *Me ester*

$C_{25}H_{42}O_5$
3α,6α,7α-Trihydroxy-5β-cholanoic Acid, *Me ester* †
3α,7α,12α-Trihydroxy-5β-cholan-24-oic Acid, *Me ester*

$C_{25}H_{42}O_{21}$
Xylopentaose

$C_{25}H_{43}NO$
N-Heptadecylaniline, N-*Ac*

$C_{25}H_{43}NO_7$
Argomycin
Methymycin
Neomethymycin
Pederone †

$C_{25}H_{44}N_5O_{16}$ (?)
Anthelmycin †

$C_{25}H_{44}N_2$
Kurchessine †

$C_{25}H_{44}N_2O$
20,25-Diazocholesterol

$C_{25}H_{44}N_2O_7$
Avenacein

$C_{25}H_{44}O_2$
Cativic Acid, 3-*Methylbutyl ester*
1-Heptadecyl-2,3-dihydroxybenzene, *Di-Me ether*
4-Heptadecyl-1,2-dihydroxybenzene, *Di-Me ether*
5-Nonadecylresorcinol †

$C_{25}H_{44}O_8$
Methane-tetracarboxylic Acid, *Tetra-pentyl ester*

$C_{25}H_{45}FeN_6O_8$
Ferrioxamine B
Ferrioxamine C

$C_{25}H_{45}NO_9$
Pederin †

$C_{25}H_{45}NO_{10}$
PA133B

$C_{25}H_{46}$
Ophiobolane †

$C_{25}H_{46}N_2$
Pachysamine A, N-*Me* †
Epipachysamine C †

$C_{25}H_{46}N_2O$
20,25-Diazacholestanol

$C_{25}H_{46}N_{10}O_9$
Streptolin

$C_{25}H_{46}O_7$
Agaric Acid, *Tri-Me ester*

$C_{25}H_{47}N_5O_{15}$
Hydroxymycin

$C_{25}H_{48}O_2$
Nervonic Acid, *Me ester*
15-Tetracosenoic Acid, *Me ester*

$C_{25}H_{48}O_3$
12-Hydroxy-9-octadecenoic Acid, *Heptyl ester*

$C_{25}H_{48}O_4$
Tricosanedioic Acid, *Di-Me ester*
Tricosanedioic Acid, *Et ester*

$C_{25}H_{48}SiO_4$
Hexadecyl triallyl orthosilicate

$C_{25}H_{49}N$
Pentacosanoic Acid, *Nitrile*

$C_{25}H_{50}O$
Pentacosanal
9-Pentacosanone

$C_{25}H_{50}O_2$
Hyenic Acid
21-Methyldocosanoic Acid, *Et ester*
23-Methyltetracosanoic Acid
Pentacosanoic Acid
Tetracosanoic Acid, *Me ester*
Tricosanoic Acid, *Et ester*
3,12,15-Trimethyldocosanoic Acid †

$C_{25}H_{51}I$
1-Iodo-23-methyltetracosane

$C_{25}H_{52}$
2-Methyltetracosane
Pentacosane

$C_{25}H_{52}O$
23-Methyltetracosanol
1-Pentacosanol

C_{26}

$C_{26}H_{12}N_2$
Acenaphthazine

$C_{26}H_{14}$
Rubicene

$C_{26}H_{14}N_2$
8,16-Diazabenzo[*b*,*k*]perylene

$C_{26}H_{14}O_2$
6,15-Hexacenedione

$C_{26}H_{16}$
Anthra[1,2-*b*]phenanthrene
Benzo[*a*]pentacene
Benzo[*c*]picene
1,2:7,8-Dibenzochrysene †
Dibenzo[*a*,*c*]naphthacene
Dibenzo[*a*,*j*]naphthacene
Dibenzo[*a*,*l*]naphthacene
Dibenzo[*a*,*c*]triphenylene
Difluorenylene
1,8-(Diphenylethynyl)naphthalene †
Fulminene †
Hexacene
Hexahelicene †
Hexaphene
Naphtho[2′,1′:1,2]tetracene †
7-Phenylbenzo[*k*]fluoranthene †

$C_{26}H_{16}O_4$
1,3-Dihydroxyanthraquinone, *Di-phenyl ether*

$C_{26}H_{16}O_5$
Xylerythrin †

$C_{26}H_{16}O_9$
Neoclauxin †

$C_{26}H_{17}NO_3$
1-Amino-8-hydroxyanthraquinone, N-*Phenyl, Phenyl ether*

$C_{26}H_{18}$
9,9′-Bifluoroenyl
9,10-Diphenylphenanthrene
1-1′-Naphthyl-8-phenylnaphthalene †

$C_{26}H_{18}N_2O_2$
1,4-Dianilinoanthraquinone
1,5-Dianilinoanthraquinone

$C_{26}H_{18}O$
9-Fluorenol, *Fluorenyl ether*
2,4,6-Triphenylcyclopenta[*b*]pyran †

$C_{26}H_{18}O_7S_2$
o-Phenylsulphonylbenzoic Acid, *Anhydride*

$C_{26}H_{18}O_8$
Dianellinone †

$C_{26}H_{18}O_{12}$
γ-Rubromycin †
γ-*iso*-Rubromycin †

$C_{26}H_{19}NO_{12}$
Laccaic Acid A_1 †

$C_{26}H_{20}$
Tetraphenylethylene
1,9,17-Tridehydro[26]annulene †

$C_{26}H_{20}N_2$
N,*N*′-Di-1-naphthyl-*m*-phenylenediamine
N,*N*′-Di-1-naphthyl-*p*-phenylenediamine
N,*N*′-Di-2-naphthyl-*m*-phenylenediamine
N,*N*′-Di-2-naphthyl-*p*-phenylenediamine

$C_{26}H_{20}N_2O_3$
Diphenylcarbamic Acid, *Anhydride*

$C_{26}H_{20}O$
Benzpinacolin

$C_{26}H_{20}O_4$
Perylene-3,9-dicarboxylic Acid, *Di-Et ester*
Perylene-3,10-dicarboxylic Acid, *Di-Et ester*

$C_{26}H_{21}NO_4$
2,5-Dihydroxy-4-nitrobiphenyl, *Dibenzyl ether*

$C_{26}H_{22}$
1,1,1,3-Tetraphenylethane
1,1,2,2-Tetraphenylethane

$C_{26}H_{22}N_2O_4$
Piperazine-*N*,*N*′-dicarboxylic Acid, *Di*-1-*naphthyl ester*

$C_{26}H_{22}N_2O_7$
Berberilic Anhydride, *Anilide*

$C_{26}H_{22}O$
1,1,2,2-Tetraphenylethanol
1,2,2,2-Tetraphenylethanol
Triphenylmethanol, *Benzyl ether*

$C_{26}H_{22}O_2$
Benzpinacol

$C_{26}H_{23}N$
α-Aminotriphenylmethane, N-*Benzyl*

$C_{26}H_{23}O_2PS$
p-Tolyl Triphenylphosphoranylidenemethyl Sulphone †

$C_{26}H_{24}O_5$
Apetalolide †
Calophyllolide
Munetone

$C_{26}H_{24}O_{10}$
Tenuiorin

$C_{26}H_{24}O_{16}$
Spinochrome E, leuco-*Octa-Ac*

$C_{26}H_{25}N$
Veranthridine

$C_{26}H_{26}N_2O_6$
Limocrocin

$C_{26}H_{26}O_2$
2,2′-Dihydroxy-1,1′-binaphthyl, *Di-isopropyl ether*

$C_{26}H_{26}O_4$
Diospyrol, *Tetra-Me ether* †

$C_{26}H_{26}O_5$
Fluorescin, *Di-Et ether*, *Et ester*
Osajin, *Me ether*
Scandinone †

$C_{26}H_{26}O_6$
Cycloartocarpin†
Lonchocarpic Acid†
Mundulone
Scandenin†

$C_{26}H_{26}O_{16}$
Galiosin

$C_{26}H_{28}$
1-Isopropylnaphthalene, *Dimeride*

$C_{26}H_{28}Br_2O_6$
3-Methylazelaic Acid, *Di-*p*-bromophenacyl ester*

$C_{26}H_{28}N_2O$
Cinchotoxine, N-*Benzyl*

$C_{26}H_{28}O_4$
Decanedioic Acid, *Diphenacyl ester*

$C_{26}H_{28}O_5$
Mesuol★, *Di-Me ether*†

$C_{26}H_{28}O_6$
Artocarpin
Citrolin

$C_{26}H_{28}O_8$
Helianthoidin†

$C_{26}H_{28}O_9$
Evodol

$C_{26}H_{28}O_{13}$
1,3-Dihydroxy-2-methylanthraquinone, *Primeveroside*

$C_{26}H_{28}O_{14}$
Apiin
Vitexin, 2″-O-β-D-*Xylosyl*†

$C_{26}H_{28}O_{15}$
Caesioside†

$C_{26}H_{29}ClO_{15}$
Ilicyanin chloride
Lycoricyanin chloride

$C_{26}H_{29}NO$
1-*p*-(2-Dimethylaminoethoxyphenyl)-1,2-diphenylbut-1-ene†

$C_{26}H_{29}NO_5$
Lythrine†
Nesodine†
Sinicuichine†
Vertine★†

$C_{26}H_{29}NO_6$
Heimine†

$C_{26}H_{30}ClO_{14}$
Fritillaricyanin

$C_{26}H_{30}O_5$
Cedrelone
7-Deacetoxy-7-oxogedunin†

$C_{26}H_{30}O_7$
Obacunone★†

$C_{26}H_{30}O_8$
Isophysodic Acid
Limonin
Lobaric Acid, *Me ester*
Physodic Acid
Zapoterin†

$C_{26}H_{30}O_9$
Rutaevin

$C_{26}H_{30}O_{11}$
Rubratoxin B†

$C_{26}H_{30}O_{12}$
Amurensin
Quercimeritrin, *Penta-Me ether*

$C_{26}H_{31}NO_5$
Decaline
Decamine
Decinine
Indicamine†
Vertaline

$C_{26}H_{31}NO_6$
Lythridine†

$C_{26}H_{32}N_2O_4$
Echitovenidine†

$C_{26}H_{32}O_2$
Dehydroabietic Acid, *Phenyl ester*
1-Phenyl-1-hepten-3-one, *Dimeride*

$C_{26}H_{32}O_4$
Isobixin, *Me ester*
Isonorbixin, *Di-Me ester*

$C_{26}H_{32}O_6$
Ageratochrome, *Dimer*†
α-Mangostin, *Tri-Me ether*★†

$C_{26}H_{32}O_7$
Angolensic Acid†

$C_{26}H_{32}O_8$
1α,2α-Epoxyscillirosidine†
Gossypolic Acid, *Di-Me ether*
Gossypolic Acid, *Di-Me ester*
1,2,3,5,6,7-Hexahydroxyanthraquinone, *Hexa-Et ether*
6-Hydroxyangolensic Acid†
Olivetoric Acid

$C_{26}H_{32}O_9$
Evodinone
Inchangin†

$C_{26}H_{32}O_{10}$
Samaderoside B†

$C_{26}H_{32}O_{11}$
Rubratoxin A†

$C_{26}H_{32}O_{12}$
Phellamurin

$C_{26}H_{33}BrO_4S$
17αβ-*p*-Bromobenzenesulphonyloxy-17α-methyl-19-nor-9β,10α-*D*-homoandrost-4-ene-3-one†

$C_{26}H_{33}NO_4$
N-Cyclopropylmethyl-19-methylnororvinol†

$C_{26}H_{33}NO_6$
Erythroskyrine★†

$C_{26}H_{33}NO_{12}$
Luteomycin

$C_{26}H_{34}O_4$
Trichilenone†

$C_{26}H_{34}O_6$
Cinobufagin

$C_{26}H_{34}O_7$
Cinobufotalin
Fumagillin
Scillirosidine†
Sphaerophorin, *Di-Me ether*, *Me ester*
$C_{26}H_{34}O_8$
Desaspidin CB†
$C_{26}H_{34}O_9$
3,17β-Dihydroxyoestra-1,3,5(10)-trien-16α-yl-β-D-glucopyranosiduronic Acid, *Me ester*†
$C_{26}H_{34}O_{10}$
Plicatic Acid, *Tri-Et ether*†
$C_{26}H_{34}O_{11}$
Aspalathin, 2′,3,4,4′,6′-*Penta-Me ether*†
Bruceine A†
$C_{26}H_{35}NO_4$
Lythranidine†
19-Propylthevinol†
$C_{26}H_{34}NO_8$
Galbulimina alkaloid G.B.3†
$C_{26}H_{36}$
4,4′-Tetradecamethylenebiphenyl†
$C_{26}H_{36}N_2O_9$
Antimycin A_{2a}
Antimycin A_{2b} (?)
Antimycin A_3
$C_{26}H_{36}O_4$
Di-isochavibetol, *Di-propyl ether*
17α-Hydroxy-6-methylpregna-4,6-dien-3,20-dione, *Butyryl*†
$C_{26}H_{36}O_6$
Bufotalin
$C_{26}H_{36}O_7$
Olivil, *Dipropyl ether*
$C_{26}H_{36}O_8$
Taxinine-A†
Taxinine-K†
$C_{26}H_{36}O_9$
3,17β-Dihydroxyoestra-1,3,5(10)-trien-16α-yl-β-D-glucopyranosiduronic Acid, *Me ester*, 3-*Me ether*†
$C_{26}H_{36}O_{12}$
Foliamenthin†
Menthiafolin†
$C_{26}H_{37}NO_3$
Isojervine
$C_{26}H_{37}NO_6$
Dehydrocholylglycine
$C_{26}H_{37}NO_8$
Malaxin†
$C_{26}H_{37}N_5O_6$
6-Octanoylamino-3-hydroxy-2-azabenzo-1,4-quinone-4-(5-octanoylamino-2,6-hydroxy-pyridyl-3-imide)†
$C_{26}H_{28}N_4O_4$
Ceanothine C†
$C_{26}H_{38}O$
Neoergosterol
epi-Neoergosterol
2-Palmitonaphthone
2-Tetraprenylphenol†
$C_{26}H_{38}O_2$
Chaulmoogric Acid, *Cinnamyl ester*
$C_{26}H_{38}O_4$
Adlupulone†
Lupulone
Ophiobolin D, *Me ester*†
$C_{26}H_{38}O_5$
Havanensin†
Neohavanensin†
$C_{26}H_{38}O_7$
Acovenosigenin A, *Di-Ac*
Ecballium Acid
Luteoleersin
$C_{26}H_{38}O_{12}$
Dihydrofoliamenthin†
$C_{26}H_{39}NO$
Buxpsiine†
$C_{26}H_{39}NO_4$
Piericidin A, 10-*Me ether*†
$C_{26}H_{40}BrO_2$
9-Octadecenoic Acid, p-*Bromophenacyl ester*
$C_{26}H_{40}ClO_2$
9-Octadecenoic Acid, p-*Chlorophenacyl ester*
$C_{26}H_{40}O_2$
Gascardic Acid, *Me ester*†
$C_{26}H_{40}O_4$
Githagonolic Acid, *Me ester*
$C_{26}H_{40}O_5$
Convallamaretin
$C_{26}H_{40}O_7$
Alboleersin
$C_{26}H_{40}O_8$
Neoandrographolide†
Oligomycin B, *Di-Ac*
$C_{26}H_{40}O_{12}$
Neolloydosin, *Hexa-Me ether*†
$C_{26}H_{41}NO$
Buxenone, N-*Me*†
Cyclobuxophyllinine, 3-N-*Me*†
Cyclobuxoviridine†
$C_{26}H_{41}NO_3$
Peiminine
$C_{26}H_{41}NO_7$
Elatidine
$C_{26}H_{42}O_2$
5α-Androstan-17α-ol, *Hexahydrobenzoyl*
$C_{26}H_{43}N$
6-Azacholesta-3,5-diene
$C_{26}H_{43}NO_2$
Cyclomikuranine†
$C_{26}H_{43}NO_3$
Alvanine
Glycocholanic Acid
Peimine
$C_{26}H_{43}NO_4$
N-3α-Hydroxy-5β-cholan-24-oylglycine

$C_{26}H_{43}NO_5$
N-3α,7α-Dihydroxy-5β-cholan-24-oylglycine
N-3α,7β-Dihydroxy-5β-cholan-24-oylglycine
3α,6α-Dihydroxy-5β-cholan-24-oylglycine
3α-12α-Dihydroxy-5β-cholan-24-oylglycine

$C_{26}H_{43}NO_6$
Homochasmanine†
N-3α,7α,12α-Trihydroxy-5β-cholan-24-oylglycine

$C_{26}H_{43}O_8$ (?)
Rammacin

$C_{26}H_{44}N_2$
Buxamine E†
NN′,*NN′*-Di-decamethylene-*p*-phenylenediamine

$C_{26}H_{44}N_2O$
Buxaminol E†
20α-Methylacetamido-3α-dimethylaminopregn-5-ene†
20α-Methylacetamido-3β-dimethylaminopregn-5-ene†

$C_{26}H_{44}N_2O_2$
Baleabuxidiene F†

$C_{26}H_{44}N_2O_3$
Baleabuxidine F†

$C_{26}H_{44}N_2O_7$
Fructigenin

$C_{26}H_{44}O_2$
Cholanic Acid, *Et ester*
5α-Cholanic Acid, *Et ester*
Eicosanoic Acid, *Phenyl ester*
Helisterol
Tocol

$C_{26}H_{44}O_3$
Lithocholic Acid, *Et ester*

$C_{26}H_{44}O_4$
3α,12α-Dihydroxy-5β-cholan-24-oic Acid, *Et ester*

$C_{26}H_{44}O_5$
3α,6α,7α-Trihydroxy-5β-cholanoic Acid, *Et ester*†
3α,7α,12α-Trihydroxy-5β-cholan-24-oic Acid, *Et ester*

$C_{26}H_{44}O_8$
Darutoside

$C_{26}H_{45}N$
4-Aza-4-cholestene
6-Aza-5-cholestene

$C_{26}H_{45}NO$
6-Aza-5-cholesten-3β-ol

$C_{26}H_{45}NO_6S$
N-3α,7α-Dihydroxy-5β-cholan-24-oyltaurine

$C_{26}H_{45}NO_7S$
N-4α,7α,12α-Trihydroxy-5β-cholan-24-oyltaurine

$C_{26}H_{46}N_2$
Cycloprotobuxine D†

$C_{26}N_{46}N_2O$
Cyclovirobuxine D†
S
20α-Methylacetamido-3α-dimethylamino-5α-pregnane†
20α-Methylacetamido-3β-dimethylamino-5α-pregnane†

$C_{26}H_{46}N_2O_2$
Dihydrocyclomicrophylline F†

$C_{26}H_{46}N_2O_3$
Baleabuxaline F†

$C_{26}H_{46}N_2O_4$
Azcarpine†
Azimine, *Bis*-N-*Me*†

$C_{26}H_{46}N_2O_7$
Lateritiin-II

$C_{26}H_{46}N_2O_8$
Serratamolide

$C_{26}H_{46}O$
Mochyl Alcohol

$C_{26}H_{46}O_2$
Octadecylquinol, *Di-Me ether*

$C_{26}H_{46}O_5$
3α,7α,12α,24ζ,26-Pentahydroxy*bis*homocholane
Ranol†

$C_{26}H_{47}N$
3-Aza-5α-cholestane
3-Aza-5β-cholestane
4-Aza-5α-cholestane
4-Aza-5β-cholestane
6-Azacholestane

$C_{26}H_{47}NO_2$
6-Azacholestan-3β-ol

$C_{26}H_{48}N_2O_6$
Baccatine A

$C_{26}H_{48}O_2$
Chaulmoogric Acid, n-*Octyl ester*

$C_{26}H_{50}O_3$
17-Oxohexacos-20-enoic Acid†

$C_{26}H_{50}O_4$
Docosane-11,12-dicarboxylic Acid, *Di-Me ester*†
Docosanedioic Acid, *Di-Et ester*†
Hexacosanedioic Acid
Tetracosanedioic Acid, *Di-Me ester*

$C_{26}H_{51}N$
Hexacosanoic Acid, *Nitrile*

$C_{26}H_{52}O$
9-Hexacosanone

$C_{26}H_{52}O_2$
Hexacosanoic Acid
Hexadecanoic Acid, *Decyl ester*
2-Octanol, *Octadecanoyl*
Pentacosanoic Acid, *Me ester*
Tetracosanoic Acid, *Et ester*
3,12,15-Trimethyldocosanoic Acid, *Me ester*†

$C_{26}H_{52}O_3$
Corchorolic Acid

$C_{26}H_{53}I$
1-Iodohexacosane

$C_{26}H_{54}$
Hexacosane
Isohexacosane

$C_{26}H_{54}O$
1-Hexacosanol
$C_{26}H_{54}O_2$
1,26-Hexacosanediol †

C_{27}

$C_{27}H_{17}NO_7S$
4-Nitro-2-sulphobenzoic Acid, *Di-β-naphthyl ester*
$C_{27}H_{18}O_3$
1,3,5-Tribenzoylbenzene
$C_{27}H_{18}O_5$
Xylerythrin, *Me ether* †
$C_{27}H_{20}O$
1,1′-Dinaphthylmethanol, *Phenyl ether*
$C_{27}H_{20}O_{11}$
Deacetylduclauxin †
$C_{27}H_{20}O_{12}$
α-Rubromycin †
β-Rubromycin †
$C_{27}H_{21}$ (ion)
1,3,5-Tris-(cycloheptatrienylium)benzene Cation †
$C_{27}H_{21}NO_{12}$
Laccaic Acid A_1, *Me ester* †
$C_{27}H_{21}N_3O_9$
1,3,5-Trihydroxy-2,4,6-trinitrobenzene, *Tribenzyl ether*
$C_{27}H_{22}$
1,1,2,3-Tetraphenylpropene
1,1,3,3-Tetraphenylpropene
1,2,3,3-Tetraphenylpropene
$C_{27}H_{22}O_2$
Triphenylacetic Acid, *Benzyl ester*
2,3,3-Triphenylpropionic Acid, *Phenyl ester*
$C_{27}H_{22}O_{13}$
Valoneaic Acid, *Dilactone*, *Hexa-Me ether*
$C_{27}H_{22}O_{14}$
Asperthecin, *Hexa-O-Ac*
$C_{27}H_{22}O_{18}$
Corilagin
$C_{27}H_{23}BrO_8$
2,3,6-Tri-*O*-benzoyl-α-D-glucopyransoyl Bromide †
$C_{27}H_{23}NO_{12}$
Valoneaic Acid, *Dilactone*, *Amide*
$C_{27}H_{24}$
1,1,1,3-Tetraphenylpropane
1,1,2,3-Tetraphenylpropane
1,1,3,3-Tetraphenylpropane
$C_{27}H_{24}O$
1,1,2,3-Tetraphenylpropanol
1,1,2,3-Tetraphenyl-2-propanol
1,1,3,3-Tetraphenylpropanol
1,1,3,3-Tetraphenyl-2-propanol
1,2,2,3-Tetraphenylpropanol
1,2,3,3-Tetraphenylpropanol
$C_{27}H_{24}O_3$
Phloroglucinol, *Tribenzyl ether*
$C_{27}H_{24}O_{13}$
m-Trigallic Acid, *Cyclohexyl ester* †
$C_{27}H_{26}O_8$
Fukugetin, *Penta-Me ether*
$C_{27}H_{26}O_9$
Glycerol, *Tri-phenoxyacetyl*
$C_{27}H_{28}BrO_8$
Gibberellic Acid, *p-Bromophenacyl ester*
$C_{27}H_{28}N_2O_4$
1-[2-[*p*-[*a*-(*p*-Methoxyphenyl)-β-nitrostyryl]-phenoxy]-ethyl]pyrrolidine †
$C_{27}H_{28}N_8O_{10}S_3$
Althiomycin★ †
$C_{27}H_{28}O_5$
Osajin, *Di-Me ether*
$C_{27}H_{28}O_6$
Isopomiferin, *Di-Me ether*
Lonchocarpic Acid, *Mono-Me ether* †
Pomiferin, *Di-Me ether*
$C_{27}H_{28}O_{15}$
Violanin
$C_{27}H_{29}NO_{10}$
Daunomycin †
Yohimbic Acid, *Benzyl ester*
$C_{27}H_{30}O_6$
Cyclotriveratrylene †
$C_{27}H_{30}O_7$
Fukugenetin, *Tetra-Et ether*
Nimbolide †
$C_{27}H_{30}O_{12}$
4′-Demethyl-deoxypodophyllotoxin-β-D-glucoside †
$C_{27}H_{30}O_{14}$
Apigenin, 7-β-*Rutinoside*★ †
Kaempferitrin
Lanceolarin †
Lespedin
Morindin
β-Morindin
Multiflorin
Rhoifolin
Sphaerobioside †
$C_{27}H_{30}O_{15}$
Morindonin
Peumoside †
Saponarin★ †
Violanthin †
$C_{27}H_{30}O_{16}$
Equisetrin
Lucenin-1 †
Rutin
Sophoraflavanoloside
$C_{27}H_{30}O_{17}$
Meratin
Quercetin★, 3,7-*Di-β-D-glucoside* †

$C_{27}H_{30}O_{19}$
Hibiscitrin

$C_{27}H_{31}ClO_{15}$
Cosmocyanin
Keracyanin
Monardin chloride
Pelargonin chloride

$C_{27}H_{31}ClO_{16}$
Cyanin chloride
Mecocyanin chloride

$C_{27}H_{31}ClO_{17}$
Delphin chloride

$C_{27}H_{31}NO_4$
Isohypognavine†

$C_{27}H_{31}NO_5$
Lythrine, *Me ether*†
Vertine, *Me ether*†

$C_{27}H_{31}NO_6$
Ignavine

$C_{27}H_{32}N_4O_8$
Pyridomycin★†

$C_{27}H_{32}O$
2-Octanol, *Triphenylmethyl ether*

$C_{27}H_{32}O_7$
Andirobin†
Carapin†
Mexicanolide†
Siphulin, *Me ester*, *Di-Me ether*†

$C_{27}H_{32}O_8$
Isophysodic Acid, *Me ester*
Physodic Acid, *Me ester*

$C_{27}H_{32}O_9$
Verrucarin B

$C_{27}H_{32}O_{10}$
Echioidin, *Penta-Me ether*†

$C_{27}H_{32}O_{14}$
Naringin
Narirutin†

$C_{27}H_{32}O_{15}$
Butrin
Isobutyrin

$C_{27}H_{32}O_{17}$
Lutonarin

$C_{27}H_{33}NO_{11}$
Colchicoside

$C_{27}H_{34}N_2$
Di-(4-diethylaminophenyl)phenylmethane
Lobinaline★†

$C_{27}H_{34}N_2O_4$
Demethylpsychotrine†

$C_{27}H_{34}N_2O_6$
Lycaconitine

$C_{27}H_{34}N_2O_9$
Isovincoside†
Strictosidine†
Vincoside†

$C_{27}H_{34}O_7$
Angolensic Acid, *Me ester*†
6-Deoxydetigloylswietenine†
6-Deoxyswietenolide†

$C_{27}H_{34}O_8$
Gossypetin, *Hexa-Et ether*
6-Hydroxyangolensic Acid, *Me ester*†
Olivetoric Acid, *Me ester*
Quercetagetin, *Hexa-Et ether*
Swietenolide†
Veprisone†

$C_{27}H_{34}O_9$
Verrucarin A

$C_{27}H_{34}O_{11}$
Arctiin
Forsythin

$C_{27}H_{35}NO_{12}$
Ipecoside

$C_{27}H_{36}O_7$
Fumagillin, *Me ester*

$C_{27}H_{36}O_8$
Albaspidin★†
Ramalinolic Acid, *Tri-Me ether*, *Me ester*

$C_{27}H_{36}O_9$
Simarolide†

$C_{27}H_{36}O_{12}$
Lyoniresinol, 2-β-*Xyloside*†

$C_{27}H_{37}NO_7$
Galbulimina alkaloid G.B.10†
Yuzurimine†

$C_{27}H_{38}N_2O_4$
Isoptin†

$C_{27}H_{38}O_2$
Luvigenin

$C_{27}H_{38}O_4$
Azafrin

$C_{27}H_{38}O_7$
Abogenin, *Ac*

$C_{27}H_{39}NO$
Verarine†

$C_{27}H_{39}NO_2$
Veratramine

$C_{27}H_{39}NO_3$
Jervine
Isojervine★†

$C_{27}H_{39}NO_7$
Macrodaphnine†

$C_{27}H_{40}$
Anthracholestatetraene

$C_{27}H_{40}ClO_{16}$
Exfoliatin

$C_{27}H_{40}O$
Balsaminasterol

$C_{27}H_{40}O_2$
δ-Tocotrienol†

$C_{27}H_{40}O_4$
Botogenin
β-Chlorogenone
3β-Hydroxy-20α,22α,25D-spirost-5-en-12-one
Neoruscogenin

$C_{27}H_{40}O_5$
Kammogenin

$C_{27}H_{40}O_7$
Ecballium Acid, *Me ester*
Drevogenin A
Drevogenin B
2,3-*seco*-12-Oxo-5α-spirostan-2,3-dioic Acid

$C_{27}H_{41}NO_2$
11-Deoxojervine†
Veralobine†

$C_{27}H_{41}NO_3$
Veratrobasine†

$C_{27}H_{41}NO_8$
Deltaline

$C_{27}H_{42}ClNO_2$
Benzethonium chloride

$C_{27}H_{42}FeN_9O_{12}$
Ferrichrome

$C_{27}H_{42}O$
Cholesta-5,7,22-trien-3β-ol†

$C_{27}H_{42}O_2$
Bourjotone†
4-Hydroxycholesta-4,6-dien-3-one

$C_{27}H_{42}O_3$
Diosgenin
Yamogenin

$C_{27}H_{42}O_4$
Albizziagenin
*Ana*hecogenin
Andogenin
Convallamarogenin
Cryptogenin
ψ-Hecogenin
Hispidogenin†
3β-Hydroxy-5α-spirostan-12-one
3β-Hydroxy-5β,20α,22α,25L-spirostan-12-one
3β-Hydroxy-5β-spirostan-12-one
Lilagenin
Nuatigenin†
Ruscogenin
5α,22α-Spirost-25(27)-ene-2α,3β-diol†
Yuccagenin

$C_{27}H_{42}O_5$
Manogenin
Mexogenin

$C_{27}H_{42}O_6$
Cacogenin
Genin
2,3-*seco*-5α,22α-Spirostan-2,3-dioic Acid

$C_{27}H_{43}N$
3,5-Cyclosolanidane
N-Heptadecyl-1-naphthylamine
N-Heptadecyl-2-naphthylamine

$C_{27}H_{43}NO$
Solanidine-T
Veralinine†
Verazine†

$C_{27}H_{43}NO_2$
Alloisorubijervine
3,16-Bisdehydrodihydrotomatidine A
3,16-Bisdehydrodihydrotomatidine B
Isorubijervine
Leptinidine†
Rubijervine
Solasodine
Tomatid-5-en-3β-ol
Veralkamine†

$C_{27}H_{43}NO_3$
Imperialine
Verticinone†

$C_{27}H_{43}NO_7$
Zygadenine

$C_{27}H_{43}NO_8$
Cevagenine
Cevine
Germine
Isogermine

$C_{27}H_{44}$
2,4-Cholestadiene
2,5-Cholestadiene
2,6-Cholestadiene
2,7-Cholestadiene
3,5-Cholestadiene
4,6-Cholestadiene

$C_{27}H_{44}N_2O_3$
Baleabuxoxazine C†

$C_{27}H_{44}O$
Actiniasterol
Anasterol (?)
Carneagenin
5,7-Cholestadien-3α-ol
5,7-Cholestadien-3β-ol
Cholesta-5,22-dien-3β-ol†
1-Cholesten-3-one
4-Cholesten-3-one
5-Cholesten-3-one
Cholest-7-en-3-one†
3,5-Cyclocholestan-6-one
Reineckiagenin
Tachysterol$_3$
Vitamin D_3
Zymosterol

$C_{27}H_{44}O_2$
25-Hydroxycholecalciferol†

$C_{27}H_{44}O_3$
Anatigogenin
5α-Furost-20(22)-en-3α,26-diol
5α-Furost-20(22)-en-3β,26-diol
Isotigogenin
Neotigogenin
epi-Sarsasapogenin
ψ-Sarsasapogenin
Smilagenin
5α,22α-Spirostan-3α-ol
5α,22α-Spirostan-3β-ol
5β,22β-Spirostan-3β-ol

$C_{27}H_{44}O_4$
Chlorogenin
ψ-Chlorogenin
epi-Chlorogenin
Cholegenin
Digalogenin

$C_{27}H_{44}O_4$ *(continued)*
3β-Hydroxy-6,7-*seco*cholestane-6,7-dioic Acid, 6 → 3 *Lactone*
25-Isocholegenin
20-Isogitogenin
Isorhodeasapogenin
Rhodeasapogenin
Samogenin
5α,22α-Spirostane-2α,3β-diol
5α,22α-Spirostane-3β,12α-diol
5α,22α-Spirostane-3β,12β-diol
(25*S*)-5α,22α-Spirostane-3β,6α-diol†
5β,25D-Spirostane-2β,3α-diol†
Texogenin

$C_{27}H_{44}O_5$
Agavogenin
Bufonic Acid I†
Bufonic Acid II†
Ficusogenin†
Isocarneagenin
Isoreineckiagenin
Metagenin
Nologenin
Paniculogenin†
5α,22α-Spirostan-2,3,15-triol
Tokorigenin
2β,3α,4β-Trihydroxy-5β,25D-spirostan†
2β,3β,4β-Trihydroxy-5β,25D-spirostan†

$C_{27}H_{44}O_6$
2-Deoxycrustecdysone†
Ecdysone
Kitigenin
Lithobilianic Acid, *Tri-Me ester*
Ponasterone A†
Ponasterone B†

$C_{27}H_{44}O_7$
Crustecdysone†
Inokosterone†
Ponasterone C†
Pterosterone†
Shidasterone†

$C_{27}H_{44}O_8$
5β,20ξ-Dihydroxyecdysone†
20ξ,26-Dihydroxyecdysone†
5β-Hydroxycrustecdysone†

$C_{27}H_{45}Br$
3β-Bromo-5(6)-cholestene

$C_{27}H_{45}Cl$
3β-Chloro-5(6)-cholestene

$C_{27}H_{45}I$
3β-Iodo-5-cholestene

$C_{27}H_{45}NO$
Demissidine

$C_{27}H_{45}NO_2$
(22*R*,25*S*)-3β-Amino-5α-spirostane†
16-Dehydrodihydrotomatidine B
Soladulcidine
Tomatidine

$C_{27}H_{45}NO_3$
15α-Hydroxysoladulcidine†
15α-Hydroxytomatidine†
6-Nitrocholesterol
Paniculidine†
Verticine†

$C_{27}H_{45}NO_6$
Neoline★, *Tri-Me ether*†

$C_{27}H_{45}NO_7$
Delphatine

$C_{27}H_{46}$
2-Cholestene
4-Cholestene
5-Cholestene
3,5-Cyclocholestane
Inagostene

$C_{27}H_{46}Br_2$
2α,3β-Dibromo-5α-cholestane†
2β,3α-Dibromocholestane†

$C_{27}H_{46}N_2O$
Cyclokoreanine B†
Cyclovirobuxeine B†

$C_{27}H_{46}N_2O_2$
Buxazidine B†
Cyclobuxoxazine†
Cyclomicrophylline B†
Cyclomicrophylline C†
Solanocapsine

$C_{27}H_{46}O$
3-Cholestanone
2-Cholesten-7α-ol
4-Cholesten-3α-ol
4-Cholesten-3β-ol
6-Cholesten-3β-ol
7-Cholesten-3β-ol
Cholesterol
epi-Cholesterol
Coprostanone
3,5-Cyclocholestan-6α-ol
3,5-Cyclocholestan-6β-ol
Erythrosterol
Inagosterol
Satisterol (?)

$C_{27}H_{46}O_2$
5α-Cholanic Acid, *Propyl ester*
Cholest-5-ene-3β,7α-diol†
Cholest-5-ene-3β,7β-diol†
Cholest-5-ene-3β,26-diol†
5-Methyltocol
Peniocerol
δ-Tocopherol
η-Tocopherol

$C_{27}H_{46}O_4$
3α,7α-Dihydroxy-5β-cholestanoic Acid †

$C_{27}H_{46}O_5$
Chiograsterol A†
3β-Hydroxy-6,7-*seco*cholestane-6,7-dioic Acid
Scymnol
3α,7α,12α-Trihydroxy-5β-cholestanoic Acid†

$C_{27}H_{46}O_{11}$
Turbicoryn

$C_{27}H_{47}FeN_6O_9$
Ferrioxamine B, N-*Ac*

$C_{27}H_{47}FeN_6O_{10}$
Ferrioxamine G
$C_{27}H_{47}N$
6ξ-Amino-3,5-cyclocholestane
$C_{27}H_{47}NO_2$
Dihydrotomatidine A
Dihydrotomatidine B
$C_{27}H_{47}N_{12}O_{13}P$
L-Arginyl-L-alanyl-L-arginyl-L-alanyluridine-5′-phosphate
$C_{27}H_{48}$
Cholestane
Coprostane
$C_{27}H_{48}N_2$
Buxocyclamine A†
Cycloprotobuxine C†
$C_{27}H_{48}N_2O$
Cyclobuxamine, N-*Isopropyl*†
Cyclobuxamine, N,N,N′-*Tri-Me ether*†
$C_{27}H_{48}O$
3α-Cholestanol
3β-Cholestanol
Coprostan-3α-ol
Coprostan-3β-ol
Inagostanol
$C_{27}H_{48}O_2$
5-Heneicosylresorcinol†
$C_{27}H_{48}O_3$
Cholestane-3β,6,7-triol
$C_{27}H_{48}O_4$
Myxinol†
Trillogenin
$C_{27}H_{48}O_5$
Chimaerol
Chiograsterol B†
Cyprinol†
3α,7α,12α,26,27-Pentahydroxycoprostane
3α,7α,12α,25ξ,26-Pentahydroxycoprostane
$C_{27}H_{49}N$
3-Aza-*A*-homo-5α-cholestane
3-Aza-*A*-homo-5β-cholestane
6-Aza-*B*-homocholestane
$C_{27}H_{49}NO$
7*a*-Aza-*B*-homocholestan-3β-ol
$C_{27}H_{49}NO_2$
3β-Hydroxy-15-aza-D-homocholestan-16-one
3β-Hydroxy-16-aza-D-homocholestan-17-one
$C_{27}H_{49}NO_3$
15-Aza-D-homocholestanol
16-Aza-D-homocholestanol
$C_{27}H_{49}N_7O_{17}$
Mannosidostreptomycin
$C_{27}H_{50}N_2$
3,6-Diaza-bishomo-A_1B_1-5α-cholestane-4,7-dione
$C_{27}H_{50}O_2$
Mycolipodienic Acid†
$C_{27}H_{50}O_6$
Octanoic Acid, *Glycerol ester*
$C_{27}H_{52}Cl_2N_2$
Malouetine, *Di-chloride*★†
$C_{27}H_{52}N_2$ (di-ion)
Malouetine★†
$C_{27}H_{52}O_2$
Mycolipenic Acid-I
C_{27}-Phthienoic Acid
$C_{27}H_{52}O_3$
17-Oxohexacos-20-enoic Acid, *Me ester*†
$C_{27}H_{52}O_4$
Tricosanedioic Acid, *Di-Et ester*
$C_{27}H_{52}O_5$
Glycerol, 1,3-*Dilauroyl*
$C_{27}H_{54}O$
14-Heptacosanone
$C_{27}H_{54}O_2$
Heptacosanoic Acid
Hexacosanoic Acid, *Me ester*
23-Methyltetracosanoic Acid, *Et ester*
Pentacosanoic Acid, *Et ester*
$C_{27}H_{56}$
Heptacosane
$C_{27}H_{56}O$
1-Heptacosanol†
14-Heptacosanol

C_{28}

$C_{28}H_{12}O_2$
Phenanthro[1,10,9,8-*fghij*]perylene-7,14-dione
$C_{28}H_{14}$
Naphtho[1,7,8-*efg*]anthanthrene
Phenanthro[1,10,9,8-*fghij*]perylene
$C_{28}H_{14}N_2$
Flavanthrine
$C_{28}H_{14}N_2O_4$
7,16-Dihydrodinaphtho[2,3-*b*:2′,3′-*i*]phenazine-5,9,14,18,-diquinone†
Indanthrone
$C_{28}H_{14}O_2$
Dibenzo[*a,j*]perylene-8,16-dione
Helianthrone
$C_{28}H_{14}O_4$
1,1′-Bianthraquinonyl
2,2′-Bianthraquinonyl
$C_{28}H_{15}NO_4$
1,1′-Iminodianthraquinone
1,2′-Iminodianthraquinone
2,2′-Iminodianthraquinone
$C_{28}H_{16}$
Benzo[*a*]naphtho[1,2-*j*]fluoranthene†
Benzo[*a*]naphtho[2,3-*h*]pyrene
Dibenzo[*fg,qr*]pentacene
Dibenzo[*h,rst*]pentaphene
13*H*-Dibenzo[*a,cd*]perylene
Dibenzo[*a,f*]perylene
Dibenzo[*a,j*]perylene

$C_{28}H_{16}$ (*continued*)
Dibenzo[*a,n*]perylene
Dibenzo[*a,o*]perylene
Dibenzo[*b,n*]perylene
Dibenzo[*b,pqr*]perylene
Dibenzo[*cd,lm*]perylene
Naphtho[1,2-*l*]benzo[*a*]pyrene†
Naphtho[2,1-*l*]benzo[*a*]pyrene†
Naphtho[2,3-*l*]benzo[*a*]pyrene†
Naphtho[1,2-*a*]perylene†
Naphtho[2′,3′:2,3]perylene†
Phenanthreno[9′,10′:1,2]pyrene†
Tribenzo[*a,i,l*]pyrene

$C_{28}H_{16}N_2$
Anthrazine
Tetrabenzo[*a,c,h,j*]phenazine

$C_{28}H_{16}O_2$
9,9′-Bianthrylidene-10,10′-quinone
Dibenzo[*a,o*]perylene-7,16-diol
Dibenzo[*a,j*]perylene-8,16-diol

$C_{28}H_{17}N$
Dinaptho[2′,3′:2,3; 2″,3″:6,7]carbazole

$C_{28}H_{17}NO$
18*H*-Tetrabenzo[*a,c,h,j*]phenoxazine

$C_{28}H_{18}$
2,2′-Bianthryl
9,9′-Bianthryl
9,9′-Dehydrodianthracene†
1,2-Diphenylphenanthro[*l*]cyclobutadiene
Isodianthranyl

$C_{28}H_{18}N_2$
6,11-Diphenyldibenzo[*b,f*][1,4]diazocine†

$C_{28}H_{18}O_2$
Bianthranol

$C_{28}H_{18}O_3$
Fluorene-9-carboxylic Acid, *Anhydride*

$C_{28}H_{18}O_4$
o,o′,α-Naphtholphthalein
Phthalic Acid, *Di-1-naphthyl ester*

$C_{28}H_{18}O_5$
o-Benzoylbenzoic Acid, *Anhydride*

$C_{28}H_{18}O_{11}$
Xenoclauxin†

$C_{28}H_{20}$
1,2-Diphenylphenanthro[*l*]cyclobutene†
Tetrabenzo[*a,c,g,i*]cyclododecahexene
9,10,11,16-Tetrahydro-9,10[9′,10′]anthracenoanthracene
Tetrahydrocyclobutadiphenanthrene

$C_{28}H_{20}Cl_2O_3$
α-Chlorodiphenylacetic Acid, *Anhydride*

$C_{28}H_{20}N_2$
2,3,5,6-Tetraphenylpyrazine

$C_{28}H_{20}N_2O_4$
Isatide, N,N′-*Diphenyl*

$C_{28}H_{20}O$
Lepidene

$C_{28}H_{20}O_4$
5,8-Dihydroxy-1,2-dimethylanthraquinone, *Diphenyl ether*
5,8-Dihydroxy-1,3-dimethylanthraquinone, *Diphenyl ether*

$C_{28}H_{20}O_5$
Xylerythrin, *Di-Me ether*†

$C_{28}H_{20}S$
Tetraphenylthiophene

$C_{28}H_{21}N$
1,2,3,4-Tetraphenylpyrrole
1,2,3,5-Tetraphenylpyrrole
2,3,4,5-Tetraphenylpyrrole

$C_{28}H_{22}N_2O_2$
Syphilobin A†

$C_{28}H_{22}O_2$
1,2,3,4-Tetraphenyl-1,4-butanedione

$C_{28}H_{22}O_3$
Diphenylacetic Acid, *Anhydride*

$C_{28}H_{22}O_8$
Dianellinone, *Di-Me ether*†

$C_{28}H_{22}O_{10}$
Cephalochromin†

$C_{28}H_{23}NO_4$
3,3,3-Triphenylpropionic Acid, p-*Nitrobenzyl ester*

$C_{28}H_{23}NO_6$
Rhizocarpic Acid

$C_{28}H_{24}$
Isodistilbene

$C_{28}H_{24}N_2O_2$
Salazine, *Dibenzyl ether*

$C_{28}H_{24}O_2$
2,2,3-Triphenylpropionic Acid, *Benzyl ester*
2,3,3-Triphenylpropionic Acid, *Benzyl ester*

$C_{28}H_{24}O_4$
3,4,5-Trihydroxybenzaldehyde, *Tribenzyl ether*

$C_{28}H_{24}O_5$
Gallic Acid, *Tribenzyl ether*

$C_{28}H_{24}O_{13}$
Valoneaic Acid, *Dilactone, Hexa-Me ether, Me ester*

$C_{28}H_{26}N_2O_3$
Syphilobin F†

$C_{28}H_{28}N_4$
Bis-[*N,N*′-ethylenebenzidine]

$C_{28}H_{28}O_4Si$
Tetrabenzyl silicate

$C_{28}H_{28}O_7$
Gallein, *Et ester tri-Et ether*

$C_{28}H_{28}O_7P_2$
Tetrabenzyl pyrophosphate

$C_{28}H_{28}O_{10}$
Silybin, *Tri-Me ether*†

$C_{28}H_{28}Pb$
Lead tetra-*o*-tolyl
Lead tetra-*p*-tolyl

$C_{28}H_{28}Sn$
Tin tetrabenzyl
Tin tetra-*o*-tolyl
Tin-tetra-*p*-tolyl

$C_{28}H_{29}NO$
3,4-Dihydro-6-methoxy-2-phenyl-1-(4′-β-pyrrolidylethoxyphenyl)naphthalene

$C_{28}H_{30}N_4O$
N-{*N*-[*N*-(*p*-Aminobenzyl)-*p*-aminobenzyl]-*p*-aminobenzyl}-*p*-aminobenzyl Alcohol

$C_{28}H_{30}O_4$
Thymolphthalein

$C_{28}H_{30}O_6$
Apogossypol
Cycloartocarpin, *Di-Me ether* †
Lonchocarpic Acid, *Di-Me ether* †
Pomiferin, *Tri-Me ether*
Scandenin, *Di*-O-*Me ether* †

$C_{28}H_{30}O_9$
Dictamnolide (or $C_{28}H_{34}O_9$)

$C_{28}H_{30}O_{12}$
Granaticin B †

$C_{28}H_{30}O_{14}$
ψ-Baptisin

$C_{28}H_{31}NO_6$
Speciosine †

$C_{28}H_{32}N_2O_{12}$
Cordifoline †

$C_{28}H_{32}O_4$
Ophiobalin

$C_{28}H_{32}O_7$
Anthothecol

$C_{28}H_{32}O_9$
Alectoronic Acid

$C_{28}H_{32}O_{10}$
Dictamnolic Acid (or $C_{28}H_{34}O_{10}$)

$C_{28}H_{32}O_{14}$
Acaciin
Fortunellin †
Lespedin, *Mono-Me ether*
Linarin

$C_{28}H_{32}O_{15}$
4′-*O*-Methylvitexin, 7-O-β-D-*Glucosyl* †

$C_{28}H_{32}O_{16}$
Boldoside †
Brassidine †
Narcissin †

$C_{28}H_{32}O_{17}$
Isorhamnetin★, 3-β-D-*Cellobioside* †

$C_{28}H_{33}ClN_2$
Buclizine

$C_{28}H_{33}ClO_{16}$
Peonin chloride

$C_{28}H_{33}ClO_{17}$
Muscadinin chloride
Petunin (chloride)

$C_{28}H_{34}O_4$
β-Truxinic Acid, *Mono*-(+)-*menthyl ester*
β-Truxinic Acid, *Mono*-(−)-*menthyl ester*

$C_{28}H_{34}O_5$
Azadiradione †

$C_{28}H_{34}O_6$
Epoxyazadiradione †
Nimbinin †

$C_{28}H_{34}O_7$
Gedunin
Siphulin, *Me ester*, *Tri-Me ether* †

$C_{28}H_{34}O_8$
Physodic Acid, *Me ester*, *Me ether*
Uliginosin B †

$C_{28}H_{34}O_9$
Dictamnolide (or $C_{28}H_{30}O_9$)

$C_{28}H_{34}O_{10}$
Dictamnolic Acid (or $C_{28}H_{32}O_{10}$)
Vitexin, *Hepta-Me ether*

$C_{28}H_{34}O_{13}$
Deoxypodophyllotoxin 1-β-D-glucopyranoside ester †

$C_{28}H_{34}O_{14}$
Didymin †
Poncirin

$C_{28}H_{34}O_{15}$
Butrin, *Mono-Me ether*
Hesperidin

$C_{28}H_{34}O_{16}$
Neohesperidin †

$C_{28}H_{34}O_{17}$
Lutonarin, 3′-*Me ether*

$C_{28}H_{35}N_3O_3$
Demethyltubulosine †

$C_{28}H_{35}N_3O_7$
Ostreogrycin A †

$C_{28}H_{36}Br_2O_4$
Tetradecanedioic Acid, *Di*-p-*bromophenacyl ester*

$C_{28}H_{36}N_2O_4$
Ipecamine
Psychotrine
Tecleanine †

$C_{28}H_{36}N_2O_5$
Alangicine †
Mitraspecine

$C_{28}H_{36}N_3O_8$ (?)
Staphylomycin M_1

$C_{28}H_{36}O_4$
Azadirone †

$C_{28}H_{36}O_6$
Helleborin

$C_{28}H_{36}O_7$
Dihydrogedunin

$C_{28}H_{36}O_8$
Confluentin Acid
7-Deacetoxy-3-deacetyl-7-oxokhivorin †
Uliginosin A †

$C_{28}H_{36}O_{12}$
Amurensin, 5,4′-*Di-Me ether*
Bruceine C †
Phellamurin, *Di-Me ether*

$C_{28}H_{37}NO_5$
Lythranine †

$C_{28}H_{37}NO_8$
Galbulimina alkaloid G.B.12†

$C_{28}H_{37}N_3O_7$
Ostreogrycin G†

$C_{28}H_{38}N_2O$
Lobinaline, *Hydrate*★†

$C_{28}H_{38}N_2O_4$
Cephaeline

$C_{28}H_{38}N_2O_{12}$
*Arabo*keturonic Acid, *Brucine salt dihydrate*

$C_{28}H_{38}N_6O_{10}$
Grisamine (or $C_{20}H_{30}N_4O_7$)

$C_{28}H_{38}N_8O_4$
Hordatine A†

$C_{28}H_{38}O_5$
Jaborosalactone A†
Jaborosalactone B†
Meldenin†

$C_{28}H_{38}O_6$
27-Deoxy-14-hydroxywithaferin A†
Withaferin A†

$C_{28}H_{39}BrO_9$
14-Bromo-2,5,9,10-tetra-acetyltaxinol†

$C_{28}H_{39}ClO_5$
Jaborosalactone C†
Jaborosalactone E†

$C_{28}H_{39}NO_2$
Paspalin†

$C_{28}H_{40}$
2,6-Dioctylnaphthalene†

$C_{28}H_{40}NO_2$
2,3′,5′,6-Tetra-*tert*-butylindophenoxyl†

$C_{28}H_{40}N_2O_2$
Bialamicol

$C_{28}H_{40}N_2O_9$
Antimycin-A

$C_{28}H_{40}N_6O_9$
Bamicetin

$C_{28}H_{40}O$
3β-Hydroxy-24,25-dimethylcholesta-5,7,22-trien-3β-ol
24(*S*)-Methylcholesta-4,6,8(14),22-tetraen-3-one†

$C_{28}H_{40}O_3S$
3,7-Di-octylnaphthalene-1-sulphonic Acid†

$C_{28}H_{40}O_4$
Azafrin, *Me ester*

$C_{28}H_{40}O_6$
JaborosalactoneD†

$C_{28}H_{40}O_7$
Diginin

$C_{28}H_{40}O_8$
Digifolein
Digitalonin
Lanafolein

$C_{28}H_{40}O_9$
14α-Digipronin

$C_{28}H_{41}N_3O_2$
Teleocidin B†

$C_{28}H_{42}O$
Isoergosterone
24(*S*)-Methylcholesta-5,7,14,22-tetraen-3β-ol†
24-Methylcholesta-5,7,22-trien-3-one

$C_{28}H_{42}O_2$
ε-Tocopherol
β-Tocotrienol†
γ-Tocotrienol†

$C_{28}H_{42}O_4$
Terephthalic Acid, *Di-menthyl ester*

$C_{28}H_{42}O_6$
Adigenin†

$C_{28}H_{42}O_7$
Acetylcolletotrichin†

$C_{28}H_{42}O_9$
Eunicellin†

$C_{28}H_{42}O_{10}$
Convallatoxin
Grayanotoxin III, *Tetra-Ac*

$C_{28}H_{43}NO_5$
Ivorine†

$C_{28}H_{43}NO_6$
Borrelidin

$C_{28}H_{43}N_3O_2$
Dihydroteleocidin B†

$C_{28}H_{44}FeN_9O_{13}$
Ferricrocin†

$C_{28}H_{44}NO_8$ (ion)
Kuramerine†

$C_{28}H_{44}N_4O_4$
Adouetin Y†
Ceanothamine A† (Frangulanine†)
Ceanothamine B† (Adouetin X†)
Franganine†
Frangulanine†

$C_{28}H_{44}O$
Calciferol
Ergosterol
Lumisterol
24(*S*)-Methylcholesta-7,22-diene-3-one†
24-Methyl-5α-cholesta-6,8(14),22-trien-3β-ol
24-Methyl-5β-cholesta-7,9(11),22-trien-3β-ol
24-Methyl-5β-cholesta-7,14,22-trien-3β-ol
24-Methyl-5β-cholesta-8(9),14,22-trien-3β-ol
Pyrocalciferol
Suprasterol II
Tachysterol

$C_{28}H_{44}O_3$
Ergosterol peroxide†

$C_{28}H_{44}O_4$
Bethogenin

$C_{28}H_{44}O_5$
Githagoic Acid
3α-7α12α-Trihydroxy-5β-cholestanoic Acid, *Me ester*†

$C_{28}H_{44}O_8$
Oligomycin A

$C_{28}H_{44}O_9$
Laxiflorin†
$C_{28}H_{45}NO_6$
Coumingidine
$C_{28}H_{46}O$
Ascosterol
Brassicasterol
Crinosterol†
Dihydrotachysterol
24-Methylcholesta-5,7-dien-3-ol
24-Methyl-5α-cholesta-7,22-dien-3β-ol
24-Methyl-5α-cholesta-8(14),22-dien-3β-ol
4α-Methylcholest-8(9)-en-3-one†
24-Methylenecholest-5-en-3β-ol
24-Methylene-5α-cholest-8(9)-en-3β-ol
24-Methylenecholest-7-en-3β-ol
Ostreasterol
Roridin B
Vitamin D_4
$C_{28}H_{46}O_2$
Brassidic Acid, *Phenyl ester*
$C_{28}H_{46}O_3$
Cerevisterol
$C_{28}H_{46}O_4$
5β,25D-Spirostane-2β,3α-diol, 2-*Me ether*†
5β,25D-Spirostane-2β,3α-diol, 3-*Me ether*
$C_{28}H_{46}O_6$
Oligomycin C
$C_{28}H_{46}O_7$
Makisterone A†
Makisterone B†
$C_{28}H_{47}NO_2$
3β-(β,β-Dimethylacryloxy)-20α-dimethylamino-5α-pregnane†
$C_{28}H_{47}NO_7$
Narbomycin
$C_{28}H_{47}NO_8$
Picromycin★†
$C_{28}H_{48}N_2O$
Cyclokoreanine B, N-*Me*†
Cyclovirobuxeine A†
3α-(β,β-Dimethylacryl)amino-20α-dimethylamino-5α-pregnane†
$C_{28}H_{48}N_2O$
Epipachysamine E†
$C_{28}H_{48}N_2O_2$
Buxazine†
Cyclobuxoxazine, N-*Me*†
Cyclomicrophylline A†
$C_{28}H_{48}O$
Campesterol
Docosanophenone
Gilonisterol
Lophenol†
Lumistanone
4α-Methylcholest-8(9)-en-3α-ol†
4α-Methylcholest-8(9)-en-3β-ol†
24-Methylcholest-7-en-3β-ol
24-Methylcholest-8-en-3β-ol
24-Methylcholest-8(14)-en-3β-ol
24-Methylcholest-14-en-3β-ol
Neospongosterol
Pollinasterol†
$C_{28}H_{48}O_2$
5α-Cholanic Acid, *Butyl ester*
Macdougallin
β-Tocopherol
γ-Tocopherol
ζ_2-Tocopherol
$C_{28}H_{48}O_4$
17α-Hydroxy-6α-methylpregn-4-ene-3,20-dione, *Hexanoyl*†
Polygonaquinone†
$C_{28}H_{48}O_{18}$
Gossypin
$C_{28}H_{49}NO$
Docosanoic Acid, *Anilide*
$C_{28}H_{49}NO_2$
6-Azacholestan-3β-ol, O-*Ac*
$C_{28}H_{50}$
24-Methyl-5α-cholestane
$C_{28}H_{50}N_2$
Cycloprotobuxine A†
$C_{28}H_{50}N_2O$
Cyclovirobuxine A†
$C_{28}H_{50}N_2O_2$
Dihydrocyclomicrophylline A†
$C_{28}H_{50}N_2O_4$
Carpaine★†
ψ-Carpaine†
$C_{28}H_{50}O$
Lumistanol
24-Methylcholestan-3β-ol
$C_{28}H_{50}O_2$
Octadecylquinol, *Di-Et ether*
$C_{28}H_{52}O_2$
Mycolipodienic Acid, *Me ester*†
9-Octadecenoic Acid, *Menthyl ester*
$C_{28}H_{52}O_7$
Agaric Acid, *Tri-Et ester*
$C_{28}H_{54}O_2$
Centipeda II
Octadecanoic Acid, (—)-*Menthyl ester*
$C_{28}H_{54}O_4$
Tetracosanedioic Acid, *Di-Et ester*
$C_{28}H_{55}ClO$
Montanic Acid, *Chloride*
$C_{28}H_{56}O_2$
Dodecanoic Acid, *Cetyl ester*
Heptacosanoic Acid, *Me ester*
Hexacosanoic Acid, *Et ester*
Hexadecanoic Acid, *Dodecyl ester*
Montanic Acid
Octacosanoic Acid
$C_{28}H_{57}I$
1-Iodo-octacosane
$C_{28}H_{57}NO$
Montanic Acid, *Amide*

$C_{28}H_{57}N_3O$
Solapalmitenine†
$C_{28}H_{58}$
Octacosane
$C_{28}H_{58}O$
1-Octacosanol
Thea-alcohol B (or $C_{30}H_{62}O$)
$C_{28}H_{58}O_2$
1,28-Octacosanediol†
$C_{28}H_{59}N_3O$
Solapalmitine†

C_{29}

$C_{29}H_{13}NO_2$
Pyranthridone
$C_{29}H_{14}O_5$
2,2′-Dianthraquinonyl Ketone
$C_{29}H_{17}N$
Pyranthridine
Tetrabenz[*a*,*c*,*h*,*j*]acridine†
$C_{29}H_{20}O$
2,3,4,5-Tetraphenylcyclopentadien-1-one†
$C_{29}H_{20}O_{10}$
Aspelein
$C_{29}H_{20}O_{11}$
Aspelacin
$C_{29}H_{21}N$
2,3,4,6-Tetraphenylpyridine
2,3,5,6-Tetraphenylpyridine
$C_{29}H_{21}NO$
3,4,5,6-Tetraphenyl-2(1*H*)-pyridone†
$C_{29}H_{22}$
1,2,3,4-Tetraphenylcyclopentadiene†
$C_{29}H_{22}O_{11}$
Duclauxin†
$C_{29}H_{22}O_{12}$
Cryptoclauxin†
$C_{29}H_{23}N$
2,3,4,5-Tetraphenylpyrrole, N-*Me*
$C_{29}H_{24}O_8$
Fortoin
$C_{29}H_{24}O_{12}$
Theaflavin†
$C_{29}H_{26}N_4O_5$
Inosine, *Triphenylmethyl ether*
$C_{29}H_{28}N_2O_3$
Syphilobin F, O-*Me ether*†
$C_{29}H_{29}NO_9$
Prestreptovarone†
$C_{29}H_{30}O_{10}$
Gyrophoric Acid, *Tetra-Me ether*
$C_{29}H_{30}O_{12}$
Xanthoaphin *py*

$C_{29}H_{30}O_{13}$
Amarogentin†
m-Trigallic Acid, *Me ester*, *Hepta-Me ether*†
$C_{29}H_{30}O_{14}$
Amaroswerin†
$C_{29}H_{30}O_{15}$
Valoneaic Acid, *Octa-Me ether*
$C_{29}H_{31}NO_7$
Hernandaline†
$C_{29}H_{32}N_2O_7$
Poetamine
$C_{29}H_{33}N_9O_{12}$
Pteroyldi-γ-glutamylglutamic Acid
$C_{29}H_{34}O_{10}$
Spathelin†
$C_{29}H_{34}O_{14}$
Lespedin, *Di-Me ether*
$C_{29}H_{34}O_{15}$
Pectolinarin
$C_{29}H_{35}ClO_{17}$
Malvin chloride
$C_{29}H_{35}NO_5$
Cytochalasin A†
$C_{29}H_{36}N_2O_4$
Emetamine
$C_{29}H_{36}N_4O_4$
Ceanothine B†
$C_{29}H_{36}N_6O_6S$
L-Tryptophanyl-L-methionyl-L-aspartyl-L-phenylalanine amide†
$C_{29}H_{36}O_6$
O-Cinnamoyltaxicin II†
β-Gambogic Acid
$C_{29}H_{36}O_7$
O-Cinnamoyltaxicin I†
$C_{29}H_{36}O_8$
Fissinolide†
$C_{29}H_{36}O_9$
Microphyllinic Acid
$C_{29}H_{36}O_{15}$
Butrin, *Mono-Et ether*
$C_{29}H_{36}O_{19}$
Malvone
$C_{29}H_{37}NO_5$
Cytochalasin B†
Lythramine†
$C_{29}H_{37}N_3O_2$
Deoxytubulosine†
$C_{29}H_{37}N_3O_3$
Alangirmarckine†
Isotubulosine†
Tubulosine
$C_{29}H_{38}N_2O_4$
Psychotrine, *Me ether*
$C_{29}H_{38}O_4$
Beyerol, 17-*Cinnamyl ester*†
Celastrol

$C_{29}H_{38}O_8$
Olivetoric Acid, 3′,6-*Di-Me ether, Me ester*

$C_{29}H_{38}O_9$
Uscharidin

$C_{29}H_{38}O_{16}$
Azadirachtin†

$C_{29}H_{39}N_3O_2$
Echinulin

$C_{29}H_{40}N_2O_4$
Emetine
Isoemetine

$C_{29}H_{40}N_2O_8$
Narceine, *Diethylaminoethyl ester*

$C_{29}H_{40}N_8O_5$
Hordatine B†

$C_{29}H_{40}O_5$
β-Guttilactone

$C_{29}H_{40}O_7$
Bongkrekic Acid

$C_{29}H_{40}O_8$
Albaspidin★ CC†

$C_{29}H_{40}O_9$
Calactin
Calotropin
Roridin A (or $C_{29}H_{42}O_9$)

$C_{29}H_{40}O_{10}$
Calotoxin

$C_{29}H_{40}O_{12}$
Lyoniresinol, 2-β-*Xyloside, Di-Me ether*†

$C_{29}H_{41}NO_{10}$
11β,17-Dihydroxy-3,20-dioxo-1,4-pregnadien-21-yl 2-acetamido-2-deoxy-β-D-glucopyranoside

$C_{29}H_{41}O_2$
2,6-Di-*tert*-butyl-α-(3,5-di-*tert*-butyl-4-oxo-2,5-cyclohexadiene-1-ylidenyl-*p*-tolyoxy

$C_{29}H_{42}N_6O_9$
Amicetin

$C_{29}H_{42}O_4$
Ceanothenic Acid
Eupteleogenin†

$C_{29}H_{42}O_5$
Antheridiol†

$C_{29}H_{42}O_6$
Githagic Acid

$C_{29}H_{42}O_8$
Boistroside

$C_{29}H_{42}O_9$
Afroside B
Corchoroside A
Erysimine
Neriantin
Roridin A (or $C_{29}H_{40}O_9$)
Vallarotoxin

$C_{29}H_{42}O_{10}$
Adonitoxin†
Mansonin C

$C_{29}H_{42}O_{11}$
Antiarin
Courmontoside A (or $C_{30}H_{44}O_{11}$)
6′-Hydroxyconvallatoxin
Sarmentoside A†
Tholloside†

$C_{29}H_{44}O_2$
Di-(3,5-di-*tert*-butyl-4-hydroxyphenyl) methane†
ζ₁-Tocopherol
α-Tocotrienol†

$C_{29}H_{44}O_3$
Hedragonic Acid

$C_{29}H_{44}O_4$
3,20-Dioxo-30-norlupan-28-oic Acid
Githagenin

$C_{29}H_{44}O_5$
Ceanothic Acid

$C_{29}H_{44}O_6$
Polygalic Acid†

$C_{29}H_{44}O_7$
Capitasterone†
2,3-*seco*-12-Oxo-5α-spirostan-2,3-dioic Acid, *Di-Me ester*

$C_{29}H_{44}O_8$
Ascleposide
Corchoroside B
Cyasterone†
Digiproside
Evomonoside

$C_{29}H_{44}O_9$
Courmontoside B
Courmontoside C
Rhodexin A
Rhodexin B

$C_{29}H_{44}O_{10}$
Antioside (incorrectly given as $C_{20}H_{45}O_{10}$)
Gitorin

$C_{29}H_{44}O_{12}$
Acolongifloroside K
Ouabain

$C_{29}H_{45}NO_6$
Borrelidin, *Me ester*†

$C_{29}H_{45}NO_8$
Zygacine

$C_{29}H_{45}NO_9$
Cevacine

$C_{29}H_{45}N_3O$
Epipachysamine B†

$C_{29}H_{46}FeN_9O_{14}$
Ferrichrysin†

$C_{29}H_{46}N_2O_2$
Holafrine†

$C_{29}H_{46}O$
Corbisterol
Ergosterol, *Me ether*

$C_{29}H_{46}O$ (*continued*)
Fucostadienone
4α-Methylzymosterol†
28-Nor-olean-12,17-dien-3β-ol†
Stigmasta-3,5-dien-7-one

$C_{29}H_{46}O_2$
Albigenin
Phytonivein

$C_{29}H_{46}O_4$
Platanic Acid

$C_{29}H_{46}O_6$
2,3-*seco*-5α,22α-Spirostan-2,3-dioic Acid, *Di-Me ester*

$C_{29}H_{47}NO_6$
Coumingine

$C_{29}H_{47}N_5O_7$
Destruxin A

$C_{29}H_{48}N_2$
1′,4′,5′,6′-Tetrahydropyriminido[*a*-4,3]-4-aza-5-cholestene†

$C_{29}H_{48}N_2O_2$
Pachysantermine A†
Pachystermine A†

$C_{29}H_{48}O$
Adiantone†
Δ^5-Avenasterol
Δ^7-Avenasterol
Allostigmasterol
Clerosterol†
7-Dehydrositosterol
4,4-Dimethylzymosterol†
24-Ethylidenecholest-5-en-3β-ol
24-Methylenelophenol†
4α-Methyl-24-methylene-24,25-dihydrozymosterol†
Oleanol
Poriferasterol
Sargasterol
β-Sitostenone
α_1-Sitosterol
Spinasterol
Stigmasterol
Stigmasta-7-24(28)-dien-3β-ol† (Δ^7-Avenasterol★)
Stigmasta-8(14),22-dien-3β-ol†

$C_{29}H_{48}O_2$
Adipedatol†
21β-Hydroxy-30-norhopan-22-one†
Ketohakonanol†
Saringosterol†

$C_{29}H_{48}O_3$
3β-Acetoxy-5,10-secocholest-1(10)-en-5-one†

$C_{29}H_{48}O_4$
5β,25D-Spirostane-2β,3α-diol, *Di-Me ether*†

$C_{29}H_{48}O_7$
Amarasterone A†
Amarasterone B†
Makisterone C†
Makisterone D†

$C_{29}H_{50}N_2O$
Pachysamine A, N-ββ-*Dimethylacrylyl*†

$C_{29}H_{50}N_2O_2$
Pachysandrine C, O-ββ-*Dimethylacrylyl ester*†
Pachystermine B†

$C_{29}H_{50}O$
Allositosterol
Clionasterol
Enotherastanone
Enotherasterol
31-Norcycloartanol†
β-Sitosterol
γ-Sitosterol (or $C_{28}H_{48}O_7$)
δ-Sitosterol
ε-Sitosterol
Δ^{22}-Stigmasten-3β-ol

$C_{29}H_{50}O_2$
Melletol
α-Tocopherol

$C_{29}H_{50}O_3$
9-Hydroxy-α-tocopherone†
α-Tocopherolquinone†

$C_{29}H_{50}O_5$
24-Ethyl-3α,7α,12α-trihydroxycoprostanic Acid†
3α,7α,12α-Trihydroxy-5β-cholestanoic Acid, *Et ester*†

$C_{29}H_{50}O_{11}$
Corymbosin†
Turbicoryn†

$C_{29}H_{52}$
γ-Sitostane (or $C_{28}H_{50}$)
Stigmastane

$C_{29}H_{52}O$
Corbistanol
Enotherastanol
α_1-Sitostanol
γ-Sitostanol (or $C_{28}H_{50}O$)
Stigmastanol

$C_{29}H_{52}O_3$
α-Tocopherolhydroquinone†

$C_{29}H_{54}O_6$
Melezitose, *Hendeca-Me ether*

$C_{29}H_{54}O_{16}$
Raffinose, *Undeca-Me ether*

$C_{29}H_{56}O$
Mycol

$C_{29}H_{58}O$
Ginnone

$C_{29}H_{58}O_2$
Montanic Acid, *Me ester*
Nonacosanoic Acid

$C_{29}H_{59}I$
Montanyl iodide

$C_{29}H_{60}$
Nonacosane

$C_{29}H_{60}O$
Ginnol
Montanyl Alcohol

C_{30}

$C_{30}H_{12}Cl_2O_7$
4-Chloroanthraquinone-1-carboxylic Acid, *Anhydride*

$C_{30}H_{14}O_2$
Acedianthrone
Isopyranthrone
Pyranthrone

$C_{30}H_{14}O_7$
Anthraquinone-1-carboxylic Acid, *Anhydride*

$C_{30}H_{14}O_8$
Hypericin

$C_{30}H_{16}$
Anthra[1,2,3,4-*ghi*]perylene †
Benzo[*fg*]naphtho[2,1,8,7-*stuv*]pentacene †
Chalkacene
Pyreno[1′,2′:1,2]pyrene †
Rhodacene
Terrylene
Tribenzo[*b,ghi,k*]perylene

$C_{30}H_{16}O_4$
Anthraflavone

$C_{30}H_{18}$
Benzo[*a*]hexacene
Benzo[*c*]naphtho[1,2-*a*]naphthacene †
Dibenzo[*a,l*]pentacene
Dibenzo[*b,n*]picene
Heptahelicene †
Heptaphene
Tetrabenz[*a,c,h,j*]anthracene

$C_{30}H_{18}Cl_4O_8$
Flavo-obscurin B †

$C_{30}H_{18}O_9$
Cassiamine †

$C_{30}H_{18}O_{10}$
Amentoflavone †
Cupressuflavone †
Hinokiflavone †
Iridoskyrine
Skyrine

$C_{30}H_{18}O_{11}$
Oxyskyrine

$C_{30}H_{18}O_{12}$
Aurofusarin★ †

$C_{30}H_{19}Cl_3O_8$
Flavo-obscurin A †

$C_{30}H_{20}$
2,3:6,7:2′,3′:6′,6′-Tetrabenzoheptafulvalene †

$C_{30}H_{20}FeN_2O_8$
Ferroverdin

$C_{30}H_{20}N_2O_9$
2-(3′-Nitro-4′-toluyl)benzoic Acid, *Anhydride*

$C_{30}H_{20}N_4$
2,3,6,7-Tetraphenyl-1,4,5,8-tetra-azafulvalene †

$C_{30}H_{20}O_6$
Tetrahydroxy-*p*-benzoquinone, *Tetraphenyl ether*

$C_{30}H_{20}O_{11}$
Morelloflavone †

$C_{30}H_{21}As$
Phenyl-*bis*-phenylenearsene †

$C_{30}H_{21}P$
Phenyl-*bis*-phenylenephosphoran †

$C_{30}H_{22}$
5,5*a*,6,11,11*a*,12-Hexahydro-5,12:6,11-di-*o*-benzonaphthacene †
p-Pentaphenyl

$C_{30}H_{22}N_2$
Tetrabenzo[*b,fg,k,op*][1,4]diazocyclo-octadecine †

$C_{30}H_{22}O_2$
Bianthranol, 10,10′-*Di-Me ether*

$C_{30}H_{22}O_3$
3,3-Diphenylacrylic Acid, *Anhydride*

$C_{30}H_{22}O_6$
Ararobinol

$C_{30}H_{22}O_8$
Erythroapin *fb*
Erythroaphin-*tt* †
Penicilliopsin

$C_{30}H_{22}O_{10}$
Rhodoaphin *be*★ †
Rugulosin
4′,5,7-Trihydroxy-3-(4′,5,6-trihydroxy-8-flavanonyl)-flavanone †

$C_{30}H_{22}O_{11}$
4′,5,7-Trihydroxy-3-(3,4′,5,7-tetrahydroxy-8-flavanonyl)flavanone †

$C_{30}H_{22}O_{12}$
Luteoskyrine
Rubroskyrine
Xanthomegnin

$C_{30}H_{23}As$
Triphenyl-biphenylenearsene †

$C_{30}H_{23}P$
Triphenylbiphenylenephosphoran †

$C_{30}H_{24}N_2$
m-Phenylenediamine, N,N′-*Tetraphenyl*

$C_{30}H_{24}N_4$
Azophenine

$C_{30}H_{24}O_4$
Dehydrotectol †

$C_{30}H_{24}O_8$
Erythroaphin *sl*

$C_{30}H_{24}O_9$
Chrysoaphin-*fb*★ †
Chrysoaphin-*sl*★ †

$C_{30}H_{24}O_{10}$
Elsinochrome A

$C_{30}H_{24}O_{12}$
Procyanidino-(−)-epicatechin †

$C_{30}H_{25}As$
Pentaphenylarsene†

$C_{30}H_{25}N$
2,3,4,5-Tetraphenylpyrrole, N-*Et*

$C_{30}H_{25}O_5P$
Pentaphenoxyphosphorane†

$C_{30}H_{25}P$
Pentaphenylphosphoran†

$C_{30}H_{25}P_5$
Phosphobenzene A†

$C_{30}H_{26}$
2,2′-Dimethyl-1,1′-bianthryl†

$C_{30}H_{26}N_4$
1,2,4,5-Tetra-aminobenzene, 1,2,4,5-N-*Tetraphenyl*

$C_{30}H_{26}O_3$
2,3-Diphenylpropionic Acid, *Anhydride*

$C_{30}H_{26}O_4$
Tectol†

$C_{30}H_{26}O_7$
Kaempferol, 7-p-*Coumaryl*-3-D-*glucoside*

$C_{30}H_{26}O_{10}$
Flavomannin†
Xanthoaphin-*fb*★†
Xanthoaphin-*sl*★†

$C_{30}H_{26}O_{11}$
Bileucofisetinidin†
Leucofisetinidin-(+)-catechin†

$C_{30}H_{26}O_{12}$
Leucorobinetinidin-(+)-catechin†
Proanthocyanidin (Cola nut)†

$C_{30}H_{26}O_{13}$
Leucorobinetinidin-(+)-gallocatechin†

$C_{30}H_{26}O_{14}$
Ergoflavine

$C_{30}H_{28}N_4$
$\Delta^{2,2'}$-Bis(1,3-diphenylimidazolidine)†

$C_{30}H_{28}O_6$
Thamnosin†

$C_{30}H_{28}O_8$
Rottlerin

$C_{30}H_{28}O_{10}$
Cercosporin

$C_{30}H_{28}O_{11}$
Rhododactynaphin-*jc*-2†
Xanthodactynaphin-*jc*-2†

$C_{30}H_{28}O_{12}$
Rhododactynaphin-*jc*-1†
Xanthodactynaphin-*jc*-1†

$C_{30}H_{30}$
[30]Annulene†

$C_{30}H_{30}N_4O_4$
Deuteroporphyrin III
Deuteroporphyrin IX
Deuteroporphyrin XIII

$C_{30}H_{30}O_4$
Tetrahydrotectol†

$C_{30}H_{30}O_8$
Gossypol

$C_{30}H_{30}O_{11}$
Xanthoaphin *sm*

$C_{30}H_{32}O_{12}$
4′,5,7-Trihydroxy-3-(3,3′,4′,5,7-pentahydroxy-8-flavanonyl)flavanone†

$C_{30}H_{32}O_{26}$
Maltopentaose

$C_{30}H_{34}N_2O_4$
Cylindrocarpine

$C_{30}H_{34}N_2O_6$
Rauvomitine

$C_{30}H_{34}N_4O_4$
Fradicin

$C_{30}H_{34}O_{10}$
Deacetylhirtin†

$C_{30}H_{34}O_{13}$
Picrotoxin

$C_{30}H_{36}N_2O_8$
Pithomycolide†

$C_{30}H_{36}N_4O_{13}$
Maltotriose, *Phenylflavazole deriv.*

$C_{30}H_{36}O_4$
Harongin Anthrone†
Harunganin

$C_{30}H_{36}O_9$
11β-Acetoxygedunin†
Anchusin
Nimbin

$C_{30}H_{36}O_{13}$
Cacao "Leucocyanidin 1"
Ruberythric Acid, *Penta-Me ether*

$C_{30}H_{37}ClO_{17}$
Hirsutin

$C_{30}H_{37}NO_6$
Cytochalasin C†
Cytochalasin D†
Himandrine

$C_{30}H_{37}NO_7$
Himandridine★†

$C_{30}H_{37}NO_{11}$
Aklavine

$C_{30}H_{37}N_5O_5$
Ergosine
Ergosinine

$C_{30}H_{37}O_{17}$
Hirsutin

$C_{30}H_{38}O_8$
Isophysodic Acid, *Me ester*, *Tri-Me ether*

$C_{30}H_{38}O_9$
7-Deacetoxy-7-oxokhivorin†
Microphyllinic Acid, *Me ester*

$C_{30}H_{38}O_{11}$
Scoparin, *Octa-Me ether*

$C_{30}H_{38}O_{15}$
Matteucinin†

$C_{30}H_{39}NO_{10}$
Galbulimina alkaloid G.B.2†
$C_{30}H_{39}N_3O_3$
Isotubulosine, *Me ether*†
Tubulosine, *Me ether*†
$C_{30}H_{39}N_5O_5$
Dihydroergosine†
$C_{30}H_{40}Cl_2N_4$
Dequalinium chloride
$C_{30}H_{40}N_2OS$
Thiobidesoxynupharidine†
$C_{30}H_{40}N_2O_2S$
Pseudothiobinupharidine†
$C_{30}H_{40}N_4O_4$
Ceanothine A
$C_{30}H_{40}O$
β-Apo-2-carotenal
β-Apo-8′-carotenal†
$C_{30}H_{40}O_2$
α-Citraurin
β-Citraurin
$C_{30}H_{40}O_4$
Celastrol, *Me ester*
$C_{30}H_{40}O_5$
Arjunolic Acid
$C_{30}H_{40}O_6$
Absinthin
Anabsinthine
$C_{30}H_{40}O_8$
Ardisiaquinone A†
Ardisiaquinone B†
Grandifolione
$C_{30}H_{40}O_9$
3-Deacetylkhivorin†
$C_{30}H_{40}O_{10}$
Scilliglaucoside†
$C_{30}H_{41}NO_6$
Delavaconitine
$C_{30}H_{42}N_2O_2S$
Allothiobinupharidine†
Neothiobinupharidine†
Thiobinupharidine†
$C_{30}H_{42}N_2O_{15}S$
Sinalbin†
$C_{30}H_{42}O_8$
Cymarin
Proscillaridin A†
$C_{30}H_{42}O_9$
Decoside
Scillarenin β-D-glucoside†
$C_{30}H_{44}K_2O_{16}S_2$
Atractyloside†
$C_{30}H_{44}N_2O_4$
"Isopilocereine"
$C_{30}H_{44}O_4$
17-Deoxy-22,25-epoxyholothurinogenin†
$C_{30}H_{44}O_5$
22,25-Epoxyholothurinogenin†
Holothurigenin
Liquoric Acid†
$C_{30}H_{44}O_6$
Cyclosenegenin†
$C_{30}H_{44}O_7$
Adynerin
Cucurbitacin I
$C_{30}H_{44}O_8$
Acofrioside L
Acolongifloroside E
Cucurbitacin J
Cucurbitacin K★†
$C_{30}H_{44}O_9$
Acolongifloroside H
Amboside
Caudoside
Desaroside
3,17β-Dihydroxyoestra-1,3,5(10)-trien-16-α-yl-β-D-glucopyranosiduronic Acid, *Me ester, Penta-Me ether*†
Gossweiloside
Lanadoxin
Quilengoside (or $C_{30}H_{46}O_9$)
$C_{30}H_{44}O_{10}$
Deacetylcerbertatin†
Deacetylcerbertin†
Vernadigin
$C_{30}H_{44}O_{11}$
Courmontoside A (or $C_{29}H_{42}O_{11}$)
$C_{30}H_{45}ClO_6$
Senegenin†
$C_{30}H_{46}N_2O$
Epipachysamine D†
$C_{30}H_{46}N_2O_2$
Epipachysandrine A†
$C_{30}H_{46}N_6O_5$
Cyclosemigramicidin S†
$C_{30}H_{46}O$
Glochidone †
$C_{30}H_{46}O_2$
Arnidenedione
Faradione
21-Hydroxylanosta-7,9(11),24-trien-3-one
Onocera-7,14(27)-diene-3,21-dione†
$C_{30}H_{46}O_3$
Agaurilol
Anthocleistin†
Betulonic Acid
Crataegus α-sapogenin
11,12-Dehydroursolic Acid Lactone†
Ebelin Lactone★†
Elemadienonic Acid
δ-Elemolic Acid
α-Elemonic Acid
Flindissol
Friedelane-3,*x*,*y*-trione
Hedragonic Acid, *Me ester*
Ifflaionic Acid
Machaerinic Acid, *Lactone*†
Masticadienonic Acid
Micromeric Acid†

$C_{30}H_{46}O_3$ (*continued*)
Pinicolic Acid A
Sanguisorbigenin
Taraxol
Tomentosolic Acid
Vanguerolic Acid

$C_{30}H_{46}O_4$
Cyclamigenin B†
Glycyrrhetic Acid
18α-Glycyrrhetic Acid
Gypsogenin
Hederagonic Acid
4-*epi*-Hederagonic Acid
Holothurinogenin†
2-Hydroxymachaerinic Acid Lactone†
19α-Hydroxy-3-oxours-12-en-28-oic Acid†
Lansic Acid†
Machaeric Acid
Melianone†
Momordic Acid†
Proceragenin A†

$C_{30}H_{46}O_5$
Bassic Acid
Cincholic Acid†
2ξ,20β-Dihydroxy-3-oxours-12-en-28-oic Acid†
Griseogenin†
Melaleucic Acid†
Quillaic Acid
Quinovaic Acid
Spregulagenic Acid†
2α,3β,23-Trihydroxyurs-11-en-13β,28-olide†

$C_{30}H_{46}O_6$
Castanogenin
22-Deoxocucurbitacin D†
22-Deoxoisocucurbitacin D†
Medicagenic Acid
2,3-*seco*-Olean-12-ene-2,3,28-trioic Acid
2α,3β,23-Trihydroxy-11-oxours-12-en-28-oic Acid†

$C_{30}H_{46}O_7$
Beaumontoside†
Cucurbitacin D
Cucurbitacin F
Cucurbitacin L
Pre-senegenin†
Wallichoside

$C_{30}H_{46}O_8$
Beauwalloside†
Divaricoside
Honkelin
17-Isoperiplocymarin
Neriifolin
Periplocymarin
Sarmentocymarin

$C_{30}H_{46}O_9$
Acolongifloroside G
Acovenoside A
Acovenoside B
Arriagoside
Cymarol
Emicymarin
17-Isomicymarin
Quilengoside (or $C_{30}H_{44}O_9$)
Sarnovide★†
11-*epi*-Sarnovide★†
Strospeside

$C_{30}H_{46}O_{10}$
Acofrioside M
Acolongifloroside J

$C_{30}H_{46}O_{12}$
Isochollepidanic Acid (or $C_{30}H_{48}O_{12}$)

$C_{30}H_{46}O_{20}$
Callengoside†

$C_{30}H_{47}NO_3$
Codaphniphylline†

$C_{30}H_{47}NO_9$
Delphelatine

$C_{30}H_{48}$
2,6-Didecylnaphthalene†
neo-Hopa-11,13(18)-diene†

$C_{30}H_{48}FeN_6O_9$
Ferrioxamine E
Ferrioxamine F

$C_{30}H_{48}N_2O_2$
Holarrhetine†

$C_{30}H_{48}O$
Agnosterol
Alnusenone
α-Amyrone
β-Amyrone
δ-Amyrone
Cycloartenone
Dammadienone
Filicenal†
Hopenone I†
Hopenone II†
Isocarborinol†
Lanostenone
Lupenone
Moretenone†
Taraxerone

$C_{30}H_{48}O_2$
Aegiceradiol†
Arnidenolone
Canophyllal†
Davallic Acid†
Echicerin
Friedelane-2,3-dione†
Friedelane-3,7-dione†
Friedelane-3,*x*-dione
Friedelane-3,*y*-dione★ (*y* = 25†)
3β-Hydroxycycloart-25-en-24-one
3β-Hydroxylanosta-8,24-dien-7-one†
3β-Hydroxy-21-oxo-Δ^{14}-serratene†
α-Isoamyrenonol
β-Isoamyrenonol
Kulinone†
Mudarol
Mycosterol
Neoilexonol
Nyctanthic Acid
Oleanolic Aldehyde†
Roburic Acid
Soya-sapogenol C

$C_{30}H_{48}O_3$
Aegicerin†
Arnidenaldiol
Betulinic Acid
Boswellic Acid
Bourjotinolone B†
Bryogenin†
Clerodolone†
Commic Acid B
Damarenonic Acid
Elemadienolic Acid
epi-Elemadienolic Acid
Elemenonic Acid
Gledigenin
Gummosogenin
x[eq]-Hydroxyfriedelane-3,*y*-dione
y-Hydroxyfriedelane-3,*x*-dione
3β-Hydroxylanosta-8,24-dien-21-oic Acid★†
3β-Hydroxylanost-8-ene-7,11-dione†
Isoelemenonic Acid
Mangiferolic Acid†
Morolic Acid
Olean-11,13(18)-diene-3,23,28-triol
Oleanolic Acid
Saikogenin B†
Saikogenin C†
Saikogenin E†
Schinol
Soya-sapogenol E†
Ursolic Acid

$C_{30}H_{48}O_3S$
3,7-Didecylnaphthalene-1-sulphonic Acid†

$C_{30}H_{48}O_4$
Alisol B†
Alphitolic Acid
Barringtogenol D
Bourjotinolone A†
Cochalic Acid
Commic Acid C
Commic Acid D
Crategolic Acid
Cyclamiretin
Cyclamiretin A†
3β,15α-Dihydroxylanosta-8,24-dien-21-oic Acid†
2α,3β-Dihydroxyurs-12-en-28-oic Acid†
3β,19α-Dihydroxyurs-12-en-28-oic Acid†
3β,20β-Dihydroxyurs-12-en-28-oic Acid†
Echinocystic Acid
Guaijavolic Acid
Hederagenin
Helisterol, *Di-Ac*
Machaerinic Acid†
Melianol†
Mesembryanthemoidigenic Acid†
Odoratone†
Platanic Acid, *Me ester*
Priverogenin A†
Saikogenin A†
Saikogenin D†
Saikogenin F†
Siaresinolic Acid
Sumaresinolic Acid★†

$C_{30}H_{48}O_5$
Acacic Acid†
Arjunolic Acid (erratum)†
Asiatic Acid
Camelliagenin B†
Cimigenol†
Commic Acid E
Entagenic Acid
Escigenin★†
Githagoic Acid, *Di-Me ester*
Isoescigenin★†
Jacquinic Acid†
Melianodiol†
Rotundic Acid†

$C_{30}H_{48}O_6$
Caccigenin†
Camelliagenin D†
Polygalacic Acid
Terminolic Acid
Theasapogenol E†
Tomentosic Acid

$C_{30}H_{48}O_7$
Platicodigenin†

$C_{30}H_{48}O_9$
Flavofungin

$C_{30}H_{48}O_{12}$
Isochollepidanic Acid (or $C_{30}H_{46}O_{12}$)

$C_{30}H_{49}ClO_3$
Bourjotinolone C†

$C_{30}H_{49}NO_3$
Hederagenin, *Amide*

$C_{30}H_{50}$
Adianene†
Diploptene†
Fernene
Fern-7-ene†
Fern-8-ene†
Filicene†
neo-Hop-12-ene†
Hopene I†
Hopene II†
Lanostadiene
γ-Lanostadiene
α-Lupene
γ-Lupene
Δ^2-Lupene
Lupene I
Serratene†
Serrat-13-ene†
Squalene
Taraxerene

$C_{30}H_{50}N_2O_2$
Baleabuxine†
Pachysantermine A, 3-N-*Me*†

$C_{30}H_{50}N_2O_3$
Baleabuxidiene F, N-*Isobutyryl*†

$C_{30}H_{50}N_2O_4$
Baleabuxidine F, N-*Isobutyryl*†

$C_{30}H_{50}N_2O_{10}$
Azalomycin F

$C_{30}H_{50}O$
Adiantoxide †
Allolupeol
α-Amyrin
β-Amyrin
δ-Amyrin
epi-β-Amyrin
Bauerenol
Butyrospermol
Clerodone †
Cycloartenol
Dammadienol
Epoxyglutinane †
17β,21β-Epoxyhopane †
2,3-Epoxysqualene †
24-Ethylidenecholest-5-en-3β-ol, *Me ether*
Euphol
β-Euphol
Fernenol †
Ferreol (or $C_{30}H_{52}O$)
Friedelan-2-one †
Friedelan-7-one †
Friedelan-*y*-al★ (*y* = 25 †)
Friedelane-*x*-one
Friedelin
Germanicol
Glut-5-en-3α-ol †
Glut-5-en-3β-ol †
Hop-17(21)-en-3β-ol †
Ilexol
Inusterol A (or $C_{30}H_{52}O$)
Isobauerenol †
Isoelemenal
Isoursenol †
Lanosterol
Lupanone
Lupeol
epi-Lupeol †
Multiflorenol
Moretenol †
3-*epi*-Moretenol †
Motiol †
Neomotiol †
Parkeol †
Resiniferol
Shionone †
Simiarenol †
Sitosterol α_2
Taraxasterol
ψ-Taraxasterol
Taraxerol
Tirucallol
α-Tritisterol
β-Tritisterol
Walsurenol †

$C_{30}H_{50}O_2$
Adianene-2β,3β-diol †
Adipedatol, *Me ether* †
Aglaiol †
Arnidenediol †
Betulin
Brein
Canophyllol †
Cerin
Cycloart-23-ene-3β,25-diol
Cycloart-25-ene-3β,24-diol
Elemenic Acid
Erythrodiol
Faradiol★ †
Fern-9(11)-ene-2α,3β-diol †
Germanidiol †
epi-Germanidiol †
Giganteol
Glochidiol †
Heterobetulin
Hydroxydammarenone I
Hydroxydammarenone II
x-Hydroxyfriedelan-3-one
y-Hydroxyfriedelan-3-one★ (*y* = 25 †)
3α-Hydroxyfriedelan-7-one †
3β-Hydroxyfriedelan-7-one †
Hydroxyhopanone
3β-Hydroxylanost-8-en-7-one †
4β-Hydroxymethylfusida-17(20)[*Z*], 24-dien-3β-ol †
Isoarnidinediol
Lanosta-8,24-diene-3β,23ξ-diol †
Lup-20-ene-3β,16β-diol †
Maniladiol
24-Methylcholest-7-en-3β-ol, *Ac*
24-Methylcholest-8(14)-en-3β-ol, *Ac*
24-Methylcholest-14-en-3β-ol, *Ac*
Motodiol †
Onocerol
Olean-12-ene-3β,11α-diol †
β-Onocerin †
γ-Onocerin †
Serrat-14-ene-3α,21β-diol †
Serrat-14-ene-3β,21α-diol †
Serrat-14-ene-3β,21β-diol †
epi-ψ-Taraxastanonol †
Zeorinone †

$C_{30}H_{50}O_3$
Apetulactone †
Canophyllic Acid †
Dammarenolic Acid
12β,20*S*-Dihydroxydammar-24-en-3-one †
Drybalanone †
Elemenolic Acid
γ-Elemolic Acid
Genin (from Holothurin)
3β-Hydroxylanostane-7,11-dione †
Longispinogenin †
Lycoclavanol †
Malabaricol †
Ocotillone I †
Ocotillone II †
Olean-12-ene-3β,22α,28-triol †
6-Oxoleucotylin †
Primulagenin A †
Protoprimulagenin A †
Serrat-14-ene-3β,21α,24-triol †
Soya-sapogenol B
Soya-sapogenol D

$C_{30}H_{50}O_4$
Camelliagenin A †
16-Deoxybarringtogenol C †

$C_{30}H_{50}O_4$ (*continued*)
Dodecanoic Acid, *Quinol ester*
Epoxymalabaricol†
Leucotylic Acid†
Odoratol†
epi-Odoratol†
Priverogenin B†
Pyxinic Acid†
Soya-sapogenol A
Spergulagenin A†

$C_{30}H_{50}O_5$
Alisol A†
epi-Alisol A†
A_1-Barrigenol†
Barringtogenol C†
Camelliagenin C†
Jegosapogenol
Melianotriol†

$C_{30}H_{50}O_6$
R_1-Barrigenol
Dammarolic Acid
Gymnemagenin †
3β,7β,15α,16β,27,28-Hexahydroxyolean-12-ene
Protoescigenin†
Tanginol†
Theasapogenol
Theasapogenol A†

$C_{30}H_{50}O_{25}$
Adenophorin
Xylohexaose

$C_{30}H_{51}N_5O_7$
Destruxin B*†

$C_{30}H_{52}$
Cycloartane
Elemene
Friedelane†
Gammacerane†
Hopane†
21α(*H*)-Hopane†
Isoelemene
Lanostene
Lupane
Ursane†

$C_{30}H_{52}N_2O_4$
Baleabuxaline F, N-*Isobutyryl*†

$C_{30}H_{52}O$
Ambrein
β-Amyranol
Cycloartanol†
Ferreol (or $C_{30}H_{50}O$)
Friedelinol
epi-Friedelinol
Hopan-22-ol†
Inusterol A (or $C_{30}H_{50}O$)
18-Iso-β-amyranol
Lupanol
epi-Lupanol
Neriifoliol†
Tetrahymanol

$C_{30}H_{52}O_2$
Arnidanediol
Canaric Acid†
Ceruchdiol†
Dammarenediol I
Dammarenediol II
Dustanin†
Hopane-3β,22-diol†
Hopane-7β,22-diol†
Hopane-15α,22-diol†
Lupane-3β,20-diol†
Pachysandiol A†
Pachysandiol B†
ψ-Taraxastane-3β,20ξ-diol†
20-*epi*-ψ-Taraxastane-3β,20ξ-diol†
Zeorin

$C_{30}H_{52}O_3$
Betulafolienetriol†
Folientriol
Kanjalagin
Leucotylin
Lupane-3β,20-28-triol
Malabaricanediol†
Ocotillol I†
Ocotillol II†
Panaxadiol
Physarosterol
Protopanaxadiol
20-*epi*-Protopanaxadiol†
Tohogenol†

$C_{30}H_{52}O_4$
Folientetrol
Panaxatriol†
Polygonaquinone, *Di-Me ether*†
Tohogeninol†

$C_{30}H_{52}O_9$
Cucurbitacin G

$C_{30}H_{52}O_{10}$
Arasaponin A

$C_{30}H_{52}O_{16}$
Cellopentaose

$C_{30}H_{53}NO_{11}$
Benzonatate

$C_{30}H_{54}$
Dammarane
Lanostane†
Onocerane†

$C_{30}H_{54}O$
β-Equistanol

$C_{30}H_{54}O_4$
Decanedioic Acid, *Dimenthyl ester*

$C_{30}H_{56}$
Ambrane†

$C_{30}H_{56}O_2$
1,16-Cyclotriacontanedione

$C_{30}H_{56}O_5$
15-Oxotricosanedioic Acid

$C_{30}H_{58}O_3$
Dodecyl 5-oxostearate
Lanoceric Acid, *Anhydride*

$C_{30}H_{58}O_4$
Triacontanedioic Acid

$C_{30}H_{60}$
Melene
$C_{30}H_{60}I_3N_3O_3$
Gallamine
$C_{30}H_{60}O_2$
Hexadecanoic Acid, *Tetradecyl ester*
Montanic Acid, *Et ester*
Nonacosanoic Acid, *Me ester*
Octacosanoic Acid, *Et ester*
Triacontanoic Acid
$C_{30}H_{60}O_4$
Lanoceric Acid
$C_{30}H_{62}$
Anthemene
28-Methylnonacosane
Triacontane
$C_{30}H_{62}O$
Gossypyl Alcohol
Thea-alcohol B (or $C_{28}H_{58}O$)
Thea-alcohol C (or $C_{32}H_{66}O$)
1-Triacontanol

C_{31}

$C_{31}H_{20}O_{10}$
Bilobetin
Cryptomerin A†
Isocryptomerin †
Neocryptomerin †
Podocarpusflavone A†
Sotetsuflavone†
$C_{31}H_{26}O_5$
Tabebuin†
$C_{31}H_{28}O_{12}$
Ouratea proanthocyanidin A†
$C_{31}H_{28}O_{14}$
Ergochrome AC(2,2′)†
Ergochrome BC(2,2′)†
Ergochrysin
Ergochrysin B†
Ergoxanthin†
$C_{31}H_{30}O_3$
Polyalthic Acid, *Me ester*
$C_{31}H_{30}O_{14}$
Ergochrome CD†
$C_{31}H_{35}NO_4$
Paspalicin†
$C_{31}H_{36}N_2O_{11}$
Novobiocin
$C_{31}H_{37}NO_8$
Galbulimina alkaloid G.B.4†
$C_{31}H_{38}O_{10}$
6-*epi*-Detigloylswietenine diacetate†
$C_{31}H_{39}N_3O_5$
Roemeridine
$C_{31}H_{39}N_3O_9$
Mikamycin A
$C_{31}H_{39}N_5O_4$
Americine†
$C_{31}H_{39}N_5O_5$
Ergocornine
Ergocorninine
$C_{31}H_{40}O_8$
Khayasin†
$C_{31}H_{40}O_{11}$
Bovogenin A
$C_{31}H_{40}O_{12}$
Bovocyanotoxin
$C_{31}H_{41}NO_8S$
Uscharin
$C_{31}H_{41}NO_{12}$
Oxonitine
$C_{31}H_{42}N_2O_5$
Annotoxine
$C_{31}H_{42}N_4O_4$
Adouetin Y′†
Frangufoline†
Integerrenine†
$C_{31}H_{42}O$
Sintaxanthin†
$C_{31}H_{42}O_{12}$
Bovoside B (or $C_{31}H_{44}O_{12}$)
Bovoxanthiotoxin
$C_{31}H_{43}NO_8S$
Voruscharin
$C_{31}H_{44}N_4$
α,4,4′,4″-Tetra-aminotriphenylmethane, 4,4′,4″-N-*Hexa-Et*
$C_{31}H_{44}N_4O_5$
Pandamine†
16β-Acetoxy-3-oxofusida-1,17(20)(*Z*),24-trien-21-oic Acid†
Lantadene A
$C_{31}H_{44}O_9$
Bovoside A
$C_{31}H_{44}O_{10}$
Bovoside D
Boweiatoxin A
$C_{31}H_{44}O_{11}$
Adonitoxin, 3-O-*Ac*†
Bovoside C
$C_{31}H_{44}O_{12}$
Bovoside B (or $C_{31}H_{42}O_{12}$)
$C_{31}H_{45}NO_8$
Auriculine†
$C_{31}H_{46}N_2O$
Spiropachysine†
$C_{31}H_{46}N_2O_8$
Anthraniloyl-lycoctonine
$C_{31}H_{46}O_2$
Vitamin K_1
$C_{31}H_{46}O_3$
Dehydroeburiconic Acid†
Lithocholic Acid, *Benzyl ester*

$C_{31}H_{46}O_4$
Polyporenic Acid C
$C_{31}H_{46}O_5$
Liquoric Acid, *Me ester* †
$C_{31}H_{46}O_{11}$
Ambostnoside
Mansonin A
Mansonin B
$C_{36}H_{48}N_2O_{12}$
Rhodomycin A
$C_{31}H_{48}O_3$
Abieslactone †
Ambonic Acid †
Anthocleistin, *Me ester* †
Dehydroeburicoic Acid †
Elemadienoic Acid, *Me ester*
Masticadienonic Acid, *Me ester*
Micromeric Acid, *Me ester* †
Sanguisorbigenin, *Me ester*
Tomentosolic Acid, *Me ester*
Vanguerolic Acid, *Me ester*
$C_{31}H_{48}O_4$
Dehydrotumulosic Acid
Glycyrrhetic Acid, *Me ester*
18α-Glycyrrhetic Acid, *Me ester*
Gypsogenin, *Me ester*
19α-Hydroxy-3-oxours-12-en-28-oic Acid, *Me ester* †
Machaeric Acid, *Me ester* †
Momordic Acid, *Me ester* †
$C_{31}H_{48}O_5$
2ξ,20β-Dihydroxy-3-oxours-12-en-28-oic Acid, *Me ester* †
Quillaic Acid, *Me ester*
Serjanic Acid †
$C_{31}H_{48}O_6$
Fusidic Acid
Polygalic Acid, *Di-Me ester* †
Polygalic Acid, 17-*Mono-Et ester* †
2α,3β,23-Trihydroxy-11-oxours-12-en-28-oic Acid, *Me ester* †
$C_{31}H_{48}O_7$
Phytolaccagenin †
$C_{36}H_{48}O_{16}$
Rubellin
$C_{31}H_{49}NO_3$
Daphnimacropine †
$C_{31}H_{50}O_2$
Δ^7-Avenasterol, *Ac*
Davallic Acid, *Me ester*
3β-Hydroxy-21-oxo-Δ^{14}-serratene, *Me ether* †
3β-Methoxyserrat-14-en-21-one †
Nyctanthic Acid, *Me ester*
$C_{31}H_{50}O_3$
Ambolic Acid †
Boswellic Acid, *Me ester*
Commic Acid B, *Me ester*
Dammarenonic Acid, *Me ester*
Eburicoic Acid
epi-Eburicoic Acid (?)
epi-Elemadienolic Acid, *Me ester*
3β-Hydroxylanosta-8,24-dien-21-oic Acid, *Me ester* †
Isoelemenonic Acid, *Me ester*
Mangiferolic Acid, *Me ester*
Morolic Acid, *Me ester*
Oleanolic Acid, *Me ester*
Ursolic Acid, *Me ester*
$C_{31}H_{50}O_4$
Alphitolic Acid, *Me ester*
Commic Acid A
Commic Acid C, *Me ester*
Commic Acid D, *Me ester*
3β,15α-Dihydroxylanosta-8,24-dien-21-oic Acid, *Me ester* †
2α,3β-Dihydroxyurs-12-en-28-oic Acid, *Me ester* †
3β,19α-Dihydroxyurs-12-en-28-oic Acid, *Me ester* †
3β,20β-Dihydroxyurs-12-en-28-oic Acid, *Me ester* †
Echinocystic Acid, *Me ester*
Hederagenin, *Me ester*
Machaerinic Acid, *Me ester* †
Mesembryanthemoidigenic Acid, *Me ester* †
Polyporenic Acid A
Polyporenic Acid B
Quercinic Acid †
Siaresinolic Acid, *Me ester*
Sulphurenic Acid †
Sumaresinolic Acid, *Me ester* ★ †
Tumulosic Acid
$C_{31}H_{50}O_5$
Arjunolic Acid, *Me ester*
Asiatic Acid, *Me ester*
Crataegus β-sapogenin (?)
Entagenic Acid, *Me ester*
Jacquinic Acid, *Me ester* †
Rotundic Acid, *Me ester* †
$C_{31}H_{50}O_6$
Caccigenin, *Me ester* †
Polygalacic Acid, *Me ester*
Terminolic Acid, *Me ester*
Tomentosic Acid, *Me ester*
$C_{31}H_{50}O_7$
Platicodigenin, *Me ester* †
$C_{31}H_{52}N_2O_3$
Pachysandrine B †
$C_{31}H_{52}O$
Arundoin †
Bauerenol, *Me ether* †
Agaurol D (or $C_{32}H_{54}O$)
Cyclindrin †
Cycloeucalenol
Cyclolaudenol †
Euphorbol
24-Methylenecycloartanol †
24-Methylenedihydrolanosterol †
Miliacin †
Multiflorenol★, *Me ether* †
Sawamilletin †
Stigmasterol, *Et ether*
$C_{31}H_{52}O_2$
Adipedatol, *Et ether* †
Elemenic Acid, *Me ester*

$C_{31}H_{52}O_2$ *(continued)*
Euonysterol
Serrat-14-ene-3α,21β-diol, 3-*Me ether*†
Serrat-14-ene-3β,21α-diol, 3-*Me ether*†
Serrat-14-ene-3β,21β-diol, 3-*Me ether*†

$C_{31}H_{52}O_3$
Canophyllic Acid, *Me ester*†
Dammarenolic Acid, *Me ester*

$C_{31}H_{52}O_4$
Leucotylic Acid, *Me ester*†
Pyxinic Acid, *Me ester*†

$C_{31}H_{53}N_7O_{15}$
N[α]-[2-(2-Acetamido-3-*O*-D-glucose)-D-propionyl-L-alanyl-D-γ-glutamyl]-L-lysyl-D-alanyl-D-alanine†

$C_{31}H_{53}N_{13}O$
Veragenine (or $C_{31}H_{55}N_{13}O$)

$C_{31}H_{54}O_2$
Canaric Acid, *Me ester*†
Carnaubadiol†

$C_{31}H_{55}N_{13}O$
Veragenine (or $C_{31}H_{53}N_{13}O$)

$C_{31}H_{56}O_3$
α-Tocopherolhydroquinone, *Di-Me ether*†

$C_{31}H_{57}NO_2$
Tetrahexylammonium benzoate†

$C_{31}H_{60}O$
Hentriacont-24-en-16-one†

$C_{31}H_{60}O_3$
8-(9-)-Hydroxyhentriacontane-14,16-dione†

$C_{31}H_{60}O_5$
Glycerol, 1,3-*Dimyristoyl*

$C_{31}H_{62}$
Hentriacontene

$C_{31}H_{62}O$
16-Hentriacontanone

$C_{31}H_{62}O_2$
Hexadecanoic Acid, *Pentadecyl ester*
Mycoceranic Acid
Nonacosanoic Acid, *Et ester*
Triacontanoic Acid, *Me ester*

$C_{31}H_{64}$
Hentriacontane

$C_{31}H_{64}O$
1-Hentriacontanol
6-Hentriacontanol
16-Hentriacontanol†

C_{32}

$C_{32}H_{16}$
Dinaphtho[*bcd*,*lmn*]perylene
Dinaphtho[2,1,8-*cde*, 2′,1′,8′-*lmn*]perylene
Periflanthene
1,2:5,6:9,10:13,14-Tetrabenzocyclohexadeca-1,5,9,13-tetraene-3,7,11,15-tetrayne†

$C_{32}H_{16}O_2$
Ixone

$C_{32}H_{18}N_8$
Phthalocyanine

$C_{32}H_{18}O_7$
2-Methylanthraquinone-1-carboxylic Acid, *Anhydride*

$C_{32}H_{20}N_2O_3$
2-Phenylquinoline-4-carboxylic Acid, *Anhydride*

$C_{32}H_{20}O_2$
Ixenediol

$C_{32}H_{20}O_3$
Tetraphenylphthalic Acid, *Anhydride*†

$C_{32}H_{22}O$
2,4,5,6-Tetraphenylcyclopenta[*b*]pyran†

$C_{32}H_{22}O_4$
Phenolphthalein I, *Diphenyl ether*
Phenolphthalein II, *Diphenyl ether*
Phenolphthalein III, *Diphenyl ether*
Tetraphenylphthalic Acid†

$C_{32}H_{22}O_6$
Fusaroskyrin

$C_{32}H_{22}O_{10}$
Chamaecyparin†
Cupressuflavone, *Di-Me ether*†
Ginkgetin
Isoginkgetin†
Podocarpus flavone B†
Skyrine, 2,2′-*Di-Me ether*

$C_{32}H_{22}O_{12}$
Aurofusarin★, *Di-Me ether*†

$C_{32}H_{24}$
Tetrabenzo[*a*,*e*,*i*,*m*]cyclohexadecaoctaene

$C_{32}H_{24}N_4O_{15}$
Pinoresinolide, bis-[2,4-*Dinitrophenyl*]*ether*

$C_{32}H_{24}O_5$
Dracorubin

$C_{32}H_{24}O_{10}$
Xylindein

$C_{32}H_{26}$
Pentaphenylethane

$C_{32}H_{26}O$
Albodypnopinacolone

$C_{32}H_{26}O_8$
Lucidin, *Di-Benzyl ether* (flavone)†

$C_{32}H_{26}O_{10}$
Aurasperone A†

$C_{32}H_{28}N_2$
α,α′-Diamino-*o*-xylene, N-*Tetraphenyl*
α,α′-Diamino-*p*-xylene, N-*Tetraphenyl*

$C_{32}H_{28}O_7$
Xanthorrhone†

$C_{32}H_{28}O_8$
14-Hydroxyxanthorrhone†

$C_{32}H_{28}O_{14}$
Protoleucomelone

$C_{32}H_{30}O_5$
Hydroxy di-*p*-tolylacetic Acid, *Anhydride*

$C_{32}H_{30}O_{12}$
Aurasperone B†

$C_{32}H_{30}O_{14}$
Chrysergonic Acid
Ergochrome AB(4,4′)†
Secalonic Acid
Secalonic Acid B†
Secalonic Acid C†

$C_{32}H_{32}$
Styrene, *Tetrastyrene*

$C_{32}H_{32}N_4O_4$
Pemptoporphyrin†

$C_{32}H_{32}O_8$
Rottlerin, *Di-Me Ether*

$C_{32}H_{32}O_{14}$
Chartreusin
Protoactinorhodin (?)

$C_{32}H_{32}O_{15}$
Ergochrome AD†
Ergochrome BD†

$C_{32}H_{34}N_4O_4$
Deuteroporphyrin III, *Di-Me ester*
Deuteroporphyrin IX, *Di-Me ester*

$C_{32}H_{34}O_4$
Tetrahydrotectol, *Di-Me ether*†

$C_{32}H_{34}O_{14}$
Haemocorin†

$C_{32}H_{34}O_{16}$
Ergochrome DD†

$C_{32}H_{36}ClFeN_4$
Aetioporphyrin II, *Fe salt*
Aetioporphyrin III, *Fe salt*

$C_{32}H_{36}CoN_4$
Aetioporphyrin I, *Co salt*

$C_{32}H_{36}CuN_4$
Aetioporphyrin II, *Cu salt*
Aetioporphyrin III, *Cu-salt*
Aetioporphyrin IV, *Cu salt*

$C_{32}H_{36}MgN_4$
Aetioporphyrin I, *Mg salt*
Aetioporphyrin II, *Mg salt*
Aetioporphyrin III, *Mg salt*

$C_{32}H_{36}N_2O_8$
Deserpideine†

$C_{32}H_{36}N_4$
Deoxophylloerythroetioporphyrin†

$C_{32}H_{36}N_4Ni$
Aetioporphyrin I, *Ni salt*

$C_{32}H_{36}N_4Zn$
Aetioporphyrin I, *Zn salt*

$C_{32}H_{36}O_{11}$
Hirtin†

$C_{32}H_{36}O_{15}$
Valoneaic Acid, *Octa-Me ether*, *Tri-Me ester*

$C_{32}H_{38}N_2O_8$
Deserpidine

$C_{32}H_{38}N_4$
Aetioporphyrin I
Aetioporphyrin II
Aetioporphyrin III
Aetioporphyrin IV

$C_{32}H_{38}O_{11}$
6α,11β-Diacetoxygedunin†

$C_{32}H_{39}NO_8$
Galbulimina alkaloid G.B.6†
Galbulimina alkaloid G.B.7†

$C_{32}H_{40}N_2O_5S_2$
Trimethaphan, *Camphorsulphonate*

$C_{32}H_{40}N_2O_6$
Narcissamine†

$C_{32}H_{40}O_9$
Alectoronic Acid, *Me ester-tri-Me ether*
Swietenine†

$C_{32}H_{41}NO_3$
Himandreline

$C_{32}H_{41}N_5O_5$
Ergocryptine
β-Ergocryptine†
Ergocryptinine
Ergomolline
Ergomollinine

$C_{32}H_{42}N_4O_6$
Urobilin (or $C_{33}H_{40}N_4O_6$)

$C_{32}H_{42}O$
β-Apo-10′-carotenal†

$C_{32}H_{42}O_7$
Kondurangogenin A

$C_{32}H_{42}O_8$
Acrovestone†

$C_{32}H_{42}O_9$
Khayanthone†

$C_{32}H_{42}O_{10}$
Khivorin†
Nyasin†

$C_{32}H_{43}NO_2$
Buxanine†

$C_{32}H_{44}N_2O_8$
Lycaconine

$C_{32}H_{44}N_2O_9$
Lappaconitine
Streptolydigin†

$C_{32}H_{44}O_2$
9-Octadecenoic Acid, p-*Phenylphenacyl ester*

$C_{32}H_{44}O_7$
Kondurangogenin C†

$C_{32}H_{44}O_8$
Cucurbitacin E

$C_{32}H_{44}O_{12}$
Scilliroside

$C_{32}H_{45}NO_{10}$
Japbenzaconine A
Japbenzaconine B

$C_{32}H_{46}O_3$
11,12-Dehydroursolic Acid Lactone, *Ac*†

$C_{32}H_{46}O_5$
16β-Acetoxy-3-oxofusida-1,17(20)(*Z*),24-trien-21-oic Acid, *Me ester*

$C_{32}H_{46}O_6$
Inophylloidic Acid†

$C_{32}H_{46}O_8$
Adynerin, 4'-*Mono-Ac*
Cucurbitacin B
2-*epi*-Cucurbitacin B

$C_{32}H_{46}O_9$
Cucurbitacin A
Tanghiferin (or $C_{32}H_{48}O_9$)

$C_{32}H_{46}O_{10}$
Cerbertin†
Tanghinin (or $C_{32}H_{48}O_{10}$)

$C_{32}H_{46}O_{11}$
Cerbertatin†

$C_{32}H_{47}N_5O_6$
Zizyphinin

$C_{32}H_{48}NO_8$ (ion)
Kumokirine†

$C_{32}H_{48}O_3$
Dehydroeburiconic Acid, *Me ester*†

$C_{32}H_{48}O_4$
Agaurilol, *Ac*
Polyporenic Acid C, *Me ester*

$C_{32}H_{48}O_6$
Grandifoliolenone†

$C_{32}H_{48}O_8$
Cucurbitacin C

$C_{32}H_{48}O_9$
Abomonoside
Cryptograndoside A
Oleandrin
Tanghiferin (or $C_{32}H_{46}O_9$)
Veneniferin

$C_{32}H_{48}O_{10}$
Tanghinin (or $C_{32}H_{46}O_{10}$)

$C_{32}H_{49}NO_4$
Daphmacrine†

$C_{32}H_{49}NO_5$
Daphniphylline†

$C_{32}H_{49}NO_9$
Cevadine

$C_{32}H_{50}O$
Docosanonaphthone

$C_{32}H_{50}O_3$
Ambonic Acid, *Me ester*†
Dehydroeburicoic Acid, *Me ester*†
Mudarol, *Ac*

$C_{32}H_{50}O_4$
Eburicoic Acid, *Ac*
Echinodol†
Lansic Acid, *Di-Me ester*†

$C_{32}H_{50}O_5$
Aphanamixin†
Cincholic Acid, *Di-Me ester*
Melaleucic Acid, *Di-Me ester*†
Quinovaic Acid, *Di-Me ester*
Serjanic Acid, *Me ester*†
Spergulagenic Acid, *Di-Me ester*†
Turraenthin†

$C_{32}H_{50}O_7$
Pre-senegenin, *Di-Me ester*†

$C_{32}H_{51}NO_4$
Daphmacropodine†

$C_{32}H_{51}NO_9$
Protoveratridine

$C_{32}H_{52}Br_2N_4O_4$
Demecarium bromide

$C_{32}H_{52}Br_2O_3$
3β-Acetoxy-7α,11α-dibromo-8α,9α-epoxylanostane†

$C_{32}H_{52}O_2$
Centipeda I
Dammadienol, *Ac*

$C_{32}H_{52}O_3$
Ambolic Acid, *Me ester*†
Eburicoic Acid, *Me ester*
Gledigenin
Ursolic Acid, *Et ester*

$C_{32}H_{52}O_4$
Hederagenin, *Et ester*
Polyporenic Acid A★, *Me ester*†
Quercinic Acid, *Me ester*†
Siaresinolic Acid, *Et ester*
Sulphurenic Acid, *Me ester*
Sumaresinolic Acid, *Et ester*★†
Tumulosic Acid, *Me ester*

$C_{32}H_{52}O_5$
Arjunolic Acid, *Et ester*
Jacquinic Acid, *Et ester*†

$C_{32}H_{52}O_6$
Terminolic Acid, *Et ester*

$C_{32}H_{52}O_8$
Yononin†

$C_{32}H_{52}O_{12}$
Canarienglucoside A

$C_{32}H_{54}NO_9$
Albomycetin (?)

$C_{32}H_{54}O_2$
Lupanol, *Ac*
epi-Lupanol, *Ac*
Serrat-14-ene-3β,21α-diol, 3,21-*Di-Me ether*†

$C_{32}H_{54}O_3$
Dammarenediol I, *Ac*
Dammarenediol II, *Ac*

$C_{32}H_{54}O_4$
Commic Acid A, *Me ester*

$C_{32}H_{56}N_4O_8$
Sporidesmolide III†

$C_{32}H_{62}O_3$
Corynomycolenic Acid
Hexadecanoic Acid, *Anhydride*

$C_{32}H_{64}O_2$
Dotriacontanoic Acid
Hexadecanoic Acid, *Hexadecyl (cetyl) ester*
Hexadecyl hexadecanoate
Triacontanoic Acid, *Et ester*

$C_{32}H_{64}O_3$
Corynomycolic Acid

$C_{32}H_{66}$
Dotriacontane

$C_{32}H_{66}O$
Dotriacontanol
Thea-alcohol A (or $C_{34}H_{70}O$)
Thea-alcohol C (or $C_{30}H_{62}O$)

$C_{32}H_{66}O_2$
Dotriacontanoic Acid, *Me ester*

C_{33}

$C_{33}H_{23}N_3O_3$
5-(α-Hydroxy-α-2-pyridylbenzyl)-7-(α-2-pyridylbenzylidene)-5-norbornene-2,3-dicarboxamide†

$C_{33}H_{24}O_{10}$
Hinokiflavone, *Tri-Me ether*†
Kayaflavone†
Sciadopitysin†

$C_{33}H_{26}O_6$
Brevifolin, *Tri-benzyl ether*

$C_{33}H_{28}O_8$
Methane-tetracetic Acid, *Tetraphenyl ester*

$C_{33}H_{30}N_8O_{14}$
Ractinomycin A
Ractinomycin B

$C_{33}H_{32}N_4O_5$
Chlorocruoroporphyrin†

$C_{33}H_{33}NO_{12}$
Laccaic Acid A_1, *Di-Me ester*, *penta-Me ether*†

$C_{33}H_{34}N_4MgO_4$
Glaucophyllin

$C_{33}H_{34}N_4O_6$
Biliverdin
Pterobilin

$C_{33}H_{34}O_{14}$
Chartreusin, *Mono-Me ether*†

$C_{33}H_{35}NO_4$
Orientaline, O,O′-*Dibenzyl ether*†
Reticuline, O,O′-*Dibenzyl ether*†

$C_{33}H_{35}N_5O_5$
Ergotamine
Ergotaminine

$C_{33}H_{36}N_4O_4$
Glaucoporphyrin

$C_{33}H_{36}N_4O_6$
Bilirubin

$C_{33}H_{36}O_6$
Tri-*o*-thymotide†

$C_{33}H_{36}O_7$
Isomorellin†
Morellin

$C_{33}H_{36}O_8$
Isomorellic Acid†
Morellic Acid†

$C_{33}H_{38}N_2O_9$
Raujemidine★†

$C_{33}H_{38}N_4O_4$
Integerressine†

$C_{33}H_{38}N_4O_6$
Glaucobilin III*a*
Glaucobilin IX*a*
Glaucobilin XIII*a*
Phycocyanobilin†
Phycoerythrobilin†

$C_{33}H_{38}N_6$
Hodgkinsine†

$C_{33}H_{38}O_6$
Deoxymorellin†

$C_{33}H_{38}O_7$
Dihydroisomorellin†

$C_{33}H_{38}O_8$
α-Guttiferin

$C_{33}H_{39}NO_9$
Galbulimina alkaloid G.B.1†

$C_{33}H_{40}N_2O_9$
Reserpine★†

$C_{33}H_{40}N_2O_{10}$
Reserpine★, N-*Oxide*†

$C_{33}H_{40}N_4O_6$
Mesobilirubin IC*a*
Phycocyanobilin†
Urobilin (or $C_{32}H_{42}N_4O_6$)

$C_{33}H_{40}O_{19}$
Robinin

$C_{33}H_{42}O_{16}$
Icatiin

$C_{33}H_{42}O_{18}$
Casanthranol

$C_{33}H_{44}O$
Citranaxanthin†

$C_{33}H_{44}O_2$
Reticulataxanthin★†

$C_{33}H_{44}O_8$
Helvolic Acid

$C_{33}H_{45}NO_8$
Taxine B†

$C_{33}H_{45}NO_9$
Delphinine

$C_{33}H_{45}NO_{10}$
Hypaconitine

$C_{33}H_{45}NO_{11}$
Mesaconitine

$C_{33}H_{46}N_4O_6$
Stercobilin

$C_{33}H_{46}O_2$
8′,9′-Dihydro-8′-hydroxycitranaxanthin†

$C_{33}H_{47}NO_{13}$
Pimaricin★†

$C_{33}H_{48}N_2O_2$
Buxatin†

$C_{33}H_{48}N_2O_3$
Baleabuxidiene F, N-*Benzoyl*†
Buxarine†
Buxidine†

$C_{33}H_{48}N_2O_4$
Baleabuxidine F, N-*Benzoyl*†
$C_{33}H_{48}N_4O_6$
Stercobilinogen†
$C_{33}H_{48}O_8$
Polygalic Acid, *Di-Ac*†
$C_{33}H_{49}NO_7$
Veratrosine
$C_{33}H_{49}NO_9$
Consolidine
$C_{33}H_{49}N_5O_6$
Zizyphin
$C_{33}H_{50}N_2O_3$
Pachysandrine A†
$C_{33}H_{50}O_8$
Cephalosporin P_1★†
Cephalosporin P_2
Cephalosporin P_3
Cephalosporin P_4
Tokorigenin, *Tri-Ac*
$C_{33}H_{50}O_9$
Kitigenin, *Ac*
$C_{33}H_{52}O_4$
Eburicoic Acid, *Ac, Me ester*
$C_{33}H_{52}O_8$
Methane-tetracetic Acid, *Tetracyclohexyl ester*
Trillin
$C_{33}H_{54}O_6$
Phellochryseine†
$C_{33}H_{54}O_{11}$
Chainin†
Ponasteroside A†
$C_{33}H_{56}O$
Stigmasterol, *Butyl ether*
$C_{33}H_{57}NO_8$
Jurubine†
$C_{33}H_{57}N_3O_9$
Enniatin B
$C_{33}H_{58}N_4O_8$
Sporidesmolide I
$C_{33}H_{58}N_8O_9$
Evolidine
$C_{33}H_{59}N_5O_7$
Isariin
$C_{33}H_{60}NO_{14}$
Desertomycin (or $C_{33}H_{62}NO_{14}$)
$C_{33}H_{62}NO_{14}$
Desertomycin (or $C_{33}H_{60}NO_{14}$)
$C_{33}H_{62}O_6$
Decanoic Acid, *Glycerol Tri-Ester*
$C_{33}H_{64}N_4O_5$
Spathulatine
$C_{33}H_{66}O_2$
Psyllostearic Acid
$C_{33}H_{66}O_3$
Phthiodiolone A†
$C_{33}H_{67}N_7O_{11}P_2$
Azathymidine, 3′,5′-Diphosphate
$C_{33}H_{68}O$
Psyllostearyl Alcohol
$C_{33}H_{68}O_3$
Phthiocerol B†

C_{34}

$C_{34}H_{16}O_2$
Dibenzanthrone
Isodibenzanthrone
$C_{34}H_{18}$
Benzo[*a*]benzonaphtheno[*cd*]perylene
2,3 : 7,8-Dibenzo-1,12-(*o*-phenylene)-perylene†
1,12-*o*-Phenylene-2,3 : 10,11-dibenzoperylene†
Tetrabenzo[*de,hi,cp,st*]pentacene
Tetrabenzo[*a,cd,f,lm*]perylene
$C_{34}H_{18}N_4$
Dibenz[*c,c′*]benzo[1,2-*a*, 4,5-*a′*]diphenazine
$C_{34}H_{18}O_2$
3,3′-Bibenzanthronyl
4,4′-Bibenzanthronyl
$C_{34}H_{20}$
7,8-Benzoheptaphene†
Benzo[*a*]phenanthro[9,10-*q*]naphthacene
Octahelicene†
Tetrapheno[6′,5′ : 5,6]tetraphene†
$C_{34}H_{22}O_2$
2,2′-Di-1-naphthoylbiphenyl
$C_{34}H_{24}$
1,2,3,4-Tetraphenylnaphthalene†
1,4,5,8-Tetraphenylnaphthalene†
$C_{34}H_{26}O_4$
Tetraphenylphthalic Acid, *Di-Me ester*†
$C_{34}H_{26}O_{10}$
Cupressuflavone, *Tetra-Me ether*†
5,5″-Dihydroxy-4′,4‴,7,7″-tetramethoxy-3‴,8-biflavan†
$C_{34}H_{28}$
3α,4,7,7α-Tetrahydro-1,2,6,7-tetraphenyl-4,7-methanoindene†
$C_{34}H_{28}O_2$
1,5-Diphenyl-2,4-pentadien-1-one, *Dimeride*
$C_{34}H_{28}O_{10}$
Glyceraldehyde, *Dibenzoyl*
Xylindein, *Di-Me ether*
$C_{34}H_{30}O_{10}$
Aurasperone A, *Di-Me ether*†
$C_{34}H_{32}ClFeN_4O_4$
Haemin
$C_{34}H_{32}N_2O_6$
Micranthine
$C_{34}H_{34}N_4O_4$
Protoporphyrin
$C_{34}H_{34}O_{14}$
Secalonic Acid B, *Di-Me ether*†
Secalonic Acid C, *Di-Me ether*†

$C_{34}H_{35}ClO_5$
2,3,4,6-Tetrabenzylgalactose, *Chloride* †

$C_{34}H_{36}N_4O_4$
Bonelline
Pemtoporphyrin, *Di-Me ester* †

$C_{34}H_{36}O_4$
Phenolphthalein I, *Dibenzyl ether*
Phenolphthalein II, *Dibenzyl ether*
Phenolphthalein III, *Dibenzyl ether*

$C_{34}H_{36}O_6$
2,3,4,6-Tetrabenzylgalactose †

$C_{34}H_{38}N_4O_4$
Mesoporphyrin I
Mesoporphyrin II
Mesoporphyrin III
Mesoporphyrin IV
Mesoporphyrin V
Mesoporphyrin VI
Mesoporphyrin VIII
Mesoporphyrin IX
Mesoporphyrin X †
Mesoporphyrin XI
Mesoporphyrin XII
Mesoporphyrin XIII
Mesoporphyrin XIV

$C_{34}H_{38}N_4O_6$
Haematoporphyrin

$C_{34}H_{38}O_7$
Isomorellin, *Me ether* †

$C_{34}H_{38}O_{18}$
Catalposide, *Hexa-Ac*

$C_{34}H_{40}N_2O_9$
Rescidine

$C_{34}H_{40}O_6$
Deoxymorellin, *Me ether* †

$C_{34}H_{40}O_{12}$
Filixic Acid PBP

$C_{34}H_{40}O_{16}$
Amorphin★ †

$C_{34}H_{41}NO_6$
β-Guttilactone, *Pyridine complex*

$C_{34}H_{42}O_{20}$
Xanthorhamnin

$C_{34}H_{42}O_{22}$
Brassicoside †

$C_{34}H_{44}O_2$
Tangeraxanthin

$C_{34}H_{44}O_9$
Salannin †

$C_{34}H_{44}O_{12}$
11β-Acetoxykhivorin †

$C_{34}H_{44}O_{21}$
Diosmin

$C_{34}H_{46}N_2O_2$
Herpestine

$C_{34}H_{46}N_2O_9$
Ajacine

$C_{34}H_{46}O_{18}$
Liriodendrin †

$C_{34}H_{47}NO_9$
Chasmaconitine †

$C_{34}H_{47}NO_{11}$
Aconitine
Japaconitine A
Japaconitine A_1
Japaconitine B

$C_{34}H_{48}O_3$
9-Eicosenoic Acid, p-*Phenylphenacyl ester*

$C_{34}H_{48}O_9$
Fabacein †

$C_{34}H_{48}O_{10}$
Acolongifloroside E, *Di-Ac*

$C_{34}H_{50}N_2O_3$
Buxarine, N-*Me* †
Cyclomicrosine †

$C_{34}H_{50}O_{10}$
Abomonoside, *Mono-Ac*

$C_{34}H_{50}O_{11}$
Acofrioside L, *Di-Ac*

$C_{34}H_{50}O_{15}$
Cheirotoxin (or $C_{35}H_{52}O_{15}$)

$C_{34}H_{52}O_6$
Acacic Acid, *Di-Ac lactone* †

$C_{34}H_{52}O_7$
Carboxyacetylquercinic Acid †

$C_{34}H_{53}NO_{10}$
Germidine
Isogermidine

$C_{34}H_{53}NO_{13}$
Leucomycin B_3

$C_{34}H_{53}NO_{14}$
Tetrine B

$C_{34}H_{54}O_5$
Quinovaic Acid, *Di-Et ester*

$C_{34}H_{56}$
2,6-Didodecylnaphthalene †

$C_{34}H_{56}O_3S$
3,7-Didodecylnaphthalene-1-sulphonic Acid †

$C_{34}H_{56}O_{16}$
Jalapin

$C_{34}H_{58}O_{14}$
17-(6′,6″-Di-*O*-acetyl-β-sophorosyloxy)-octadecanoic Acid 4″-Lactone †

$C_{34}H_{58}O_{29}$
Tamarindose

$C_{34}H_{60}N_4O_8$
Sporidesmolide II †
Sporidesmolide IV †

$C_{34}H_{64}O_2$
n-Tritriacontan-16,18-dione

$C_{34}H_{66}O_2$
β-Corinnic Acid

$C_{34}H_{66}O_4$
Hexadecanoic Acid, *Ethylene glycol di-ester*
Triacontanedioic Acid, *Di-Et ester*

$C_{34}H_{68}NO_8P$
Ditetradecanoyl-L-α-glycerylphosphoryl-L-2-amino-1-propanol†

$C_{34}H_{68}O_2$
Coccenyl Alcohol
Dotriacontanoic Acid, *Et ester*
Dotriacontanol, *Ac*
Gheddic Acid
Hexadecanoic Acid, *Octadecyl ester*
Hexadecyl octadecanoate
Octadecanoic Acid, *Hexadecyl ester*

$C_{34}H_{68}O_4$
Ascaroside A

$C_{34}H_{70}$
Phthiocerane (or $C_{35}H_{72}$)
Tetratriacontane

$C_{34}H_{70}O$
Incarnatyl Alcohol

$C_{34}H_{70}O$
Thea-alcohol A (or $C_{32}H_{66}O$)

$C_{34}H_{70}O_3$
Phthiocerol A†

C_{35}

$C_{35}H_{28}MgN_4O_5$
Chlorophyll c†

$C_{35}H_{28}N_2O_3$
Imabenzil

$C_{35}H_{28}O_2$
Benzamarone

$C_{35}H_{28}O_9$
Cassiamine, *Penta-Me ether*

$C_{35}H_{28}O_{10}$
Hinokiflavone, *Penta-Me ether*†

$C_{35}H_{30}MgN_4O_5$
Chlorophyll c†

$C_{35}H_{30}N_4O_5$
Protochlorophyll

$C_{36}H_{30}O_{10}$
Cupressuflavone, *Hexa-Me ether*†

$C_{35}H_{32}O_9$
Bayin, *Di-Benzyl ether*†

$C_{35}H_{32}O_{10}$
Rugulosin, *Penta-Me ether*

$C_{35}H_{34}O_{18}$
Cernuoside, *Hepta-Ac*

$C_{35}H_{36}N_2O_6$
Atherospermoline†
Chondrofoline
Daphnoline

$C_{35}H_{36}N_4O_4$
Protoporphyrin, *Mono-Me ester*

$C_{35}H_{36}N_4O_5$
Chlorocruoroporphyrin, *Di-Me ester*†

$C_{36}H_{38}N_2O_6$
Chondocurine
Curine

$C_{35}H_{38}N_4O_6$
Biliverdin, *Di-Me ester*
Pterobilin, *Di-Me ester*

$C_{36}H_{38}N_4O_8$
Coproporphyrin I
Coproporphyrin II
Coproporphyrin III
Coproporphyrin IV

$C_{35}H_{38}O_6$
2,3,4,6-Tetrabenzylgalactose, *Me galactoside*†

$C_{35}H_{38}O_8$
Rottlerin, *Penta-Me ether*

$C_{36}H_{38}O_{11}$
Bileucofisetinidin, *Hexa-Me ether*†

$C_{35}H_{39}N_5O_4$
Integerrine†

$C_{35}H_{39}N_5O_5$
Ergocristine
Ergocristinine

$C_{35}H_{40}N_4O_6$
Bilirubin, *Di-Me ester*

$C_{35}H_{40}N_4O_6Zn$
Glaucobilin IX*a*, *Zn deriv.*

$C_{36}H_{40}O_8$
Cyclotetraveratrylene†

$C_{35}H_{41}NO_{10}$
Himbosine★†
Vakognavine†

$C_{35}H_{42}N_2O_9$
Rescinnamine

$C_{35}H_{42}N_4O_6$
Glaucobilin III*a*, *Di-Me ester*
Glaucobilin IX*a*, *Di-Me ester*
Glaucobilin XIII*a*, *Di-Me ester*
Phycocyanobilin, *Di-Me ester*†
Phycoerythrobilin, *Di-Me ester*†

$C_{35}H_{42}N_4O_9$
PA114A (or $C_{25}H_{31}N_3O_6$)

$C_{35}H_{42}O_9$
Taxinine†

$C_{35}H_{42}O_{12}$
Filixic Acid PBB
Trisdesaspidin†
Trisflavaspidic Acid†

$C_{35}H_{44}N_4O_6$
Mesobilirubin IX*a*, *Di-Me ester*
Mesobilirubin IX*a*, *Di-Me ether*

$C_{35}H_{46}N_8O_{10}S$
Phalloidin

$C_{35}H_{46}O_2$
2-Demethylvitamin $K_2(25)$†
Neurosporaxanthin†

$C_{35}H_{46}O_{20}$
Echinacoside

$C_{35}H_{48}N_8O_{10}S$
Phalloin

$C_{35}H_{48}O_{14}$
Courmontoside A, *Acetate* (or $C_{36}H_{50}O_{14}$)
$C_{35}H_{49}NO_{12}$
Jesaconitine★†
$C_{35}H_{52}N_2O_3$
Cyclomicrophyllidine A†
$C_{35}H_{52}N_2O_4$
Buxandrine†
$C_{35}H_{52}O_5$
Lantadene B
$C_{35}H_{52}O_6$
Ichterogenin
$C_{35}H_{52}O_8$
Vilangin
$C_{35}H_{52}O_{10}$
Gomphoside
$C_{35}H_{52}O_{14}$
Erysimoside★†
Mansonin E
$C_{35}H_{52}O_{15}$
Cheirotoxin (or $C_{34}H_{50}O_{15}$)
Eryscenoside†
$C_{35}H_{53}NO_{13}$
Tetrine A
$C_{35}H_{53}NO_{14}$
PA166
$C_{35}H_{54}N_2O_3$
Dihydrocyclomicrophyllidine A†
$C_{35}H_{54}O_8$
Cocarcinogen B_2†
$C_{35}H_{54}O_9$
Adigoside †
$C_{35}H_{54}O_{13}$
Cheiroside A
Isocheiroside A
Isocheiroside H
$C_{35}H_{54}O_{14}$
Erysimosol†
Glucocymarol
Glucodifigucoside
Glucogitofucoside
Mansonin D
Rhodexin C
Xysmalorin
$C_{35}H_{55}N_3O_7$
Mycobactic Acid
$C_{35}H_{56}O_3$
Mudarol, *Isovaleric ester*
$C_{35}H_{56}O_5$
Rehmannic Acid
$C_{35}H_{56}O_6$
Barringtogenol B†
$C_{35}H_{56}O_6$
Jegosapogenin★†
$C_{35}H_{56}O_{14}$
Chalcomycin★†
$C_{35}H_{58}O_6$
Spinasterol, *Glucoside*
$C_{35}H_{58}O_{11}$
Filipin★†
$C_{35}H_{58}O_{12}$
Fungichromin
Lagosin†
$C_{35}H_{59}NO_{13}$
Leucomycin, B_1
$C_{35}H_{61}NO_{12}$
Oleandomycin
$C_{35}H_{65}NO_8$
Ciguatoxin†
$C_{35}H_{66}O$
9,26-Pentatriacontadien-18-one
$C_{35}H_{68}O_2$
α-Corinnic Acid
Diphtheric Acid
$C_{35}H_{68}O_5$
α,α′-Dipalmitin
α,β-Dipalmitin
$C_{35}H_{70}O$
18-Pentatriacontanone
$C_{35}H_{70}O_2$
Octadecanoic Acid, *Heptadecyl ester*
$C_{35}H_{70}O_3$
Phthiodiolone A†
$C_{35}H_{72}$
Pentatriacontane
Phthiocerane (or $C_{34}H_{70}$)
$C_{35}H_{72}O_3$
Phthiocerol
Phthiocerol B†

C_{36}

$C_{36}H_{16}$
5,10,15,20-Tetrazabisbenz[5,6]indeno[1,2,3-*cd*, 1′,2′,2′-*lm*]perylene
$C_{36}H_{18}$
Benzo[1,2-*a*′, 3,4-*a*′, 5,6-*a*′′]triacenaphthylene
1,12:2,3:6,7:8,9-Tetrabenzanthanthrene†
1,2:3,4:5,6-Tribenzocoronene†
$C_{36}H_{20}$
Dinaphtho[2′,3′:2,3][2′′,3′′:8,9]perylene†
$C_{36}H_{23}N_3$
Tricarbazyl†
$C_{36}H_{24}$
Hexa-*m*-phenylene†
3,4,5,6-Tetraphenyl-1,2-benzopentalene†
$C_{36}H_{26}$
3′,2′′-Diphenyl-*p*-quaterphenyl†
o-Sexiphenyl†
m-Sexiphenyl†
p-Sexiphenyl★†
$1^1,1^2:4^2,1^3:2^3,1^4:2^4,1^5:4^5,1^6$-Sexiphenyl†
$1^1,1^2:4^2,1^3:3^3,1^4:3^4,1^5:4^5,1^6$-Sexiphenyl†
$1^1,1^2:4^2,1^3:4^3,1^4:3^4,1^5:4^5,1^6$-Sexiphenyl†

$C_{36}H_{30}O_4$
Tetraphenylphthalic Acid, *Di-Et ester*†

$C_{36}H_{30}O_{10}$
5,5″-Dihydroxy-4,4‴,7,7″-tetramethoxy-3‴,8-biflavan, *Di-Me ether*†
Isoginkgetin, *Tetra-Me ether*†
Kayaflavone, *Tri-Me ether*†
Sciadopitysin, *Tri-Me ether*†
Skyrine, *Hexa-Me ether*
Sotetsuflavone, *Penta-Me ether*†

$C_{36}H_{30}P_6$
Phosphobenzene B

$C_{36}H_{30}Sn_2$
Hexaphenylditin

$C_{36}H_{34}N_2O_4$
Insularine

$C_{36}H_{34}N_2O_6$
Menisarine
Stebisimine†

$C_{36}H_{34}O_{10}$
Rugulosin, *Hexa-Me ether*

$C_{36}H_{36}N_2O_6$
Hydromenisarine
Thalmethine†

$C_{36}H_{37}ClO_{18}$
Shisonin†

$C_{36}H_{38}N_2O_6$
Aromoline
Daphnandrine
Demerarine†
Hayatin†
Isochondodendrin
Obamegine†
Sepeerine†

$C_{36}H_{38}N_4O_4$
Protoporphyrin, *Di-Me ester*

$C_{36}H_{38}O_{12}$
Theaflavin, *Hepta-Me ether*

$C_{36}H_{38}O_{16}$
Aphinin-*sm*†
Protoaphin *sl*†

$C_{36}H_{38}O_{17}$
Protodactynaphin-*jc*-1†

$C_{36}H_{40}N_2O_6$
Magnoline

$C_{36}H_{40}N_2O_7$
Magnolamine

$C_{36}H_{40}N_4O_6$
2-Devinyl-2-hydroxyethylphaeophodrbi *a*†

$C_{36}H_{40}O_{18}$
Corilagin, *Nona-Me ether*

$C_{36}H_{42}N_4O_4$
Mesoporphyrin I, *Di-Me ester*
Mesoporphyrin II, *Di-Me ester*
Mesoporphyrin III, *Di-Me ester*
Mesoporphyrin IV, *Di-Me ester*
Mesoporphyrin V, *Di-Me ester*
Mesoporphyrin VI, *Di-Me ester*
Mesoporphyrin VIII, *Di-Me ester*
Mesoporphyrin IX, *Di-Me ester*
Mesoporphyrin X, *Di-Me ester*†
Mesoporphyrin XI, *Di-Me ester*
Mesoporphyrin XII, *Di-Me ester*
Mesoporphyrin XIII, *Di-Me ester*
Mesoporphyrin XIV, *Di-Me ester*

$C_{36}H_{42}N_4O_6$
Haematoporphyrin, *Di-Me ester*

$C_{36}H_{42}O_8$
Gossypol★, *Hexa-Me ether*†

$C_{36}H_{44}O$
Stigmasterol, *Benzyl ether*

$C_{36}H_{44}O_{12}$
Filixic Acid
Triaspidin†

$C_{36}H_{44}O_{19}$
Protoaphin *fb*

$C_{36}H_{46}N_4$
Octaethylporphyrin†

$C_{36}H_{46}O_6$
Brasiliensic Acid†

$C_{36}H_{48}O_2$
Neurosporaxanthin, *Me ester*†

$C_{36}H_{49}NO_9$
Chasmanthinine†

$C_{36}H_{50}N_8O_{12}S$
Phallacidine

$C_{36}H_{50}O_{12}$
Acolongifloroside H, *Tri-Ac*

$C_{36}H_{50}O_{14}$
Courmontoside A, *Acetate* (or $C_{35}H_{48}O_{14}$)

$C_{36}H_{50}O_{15}$
Ambostnoside, *Ac*

$C_{36}H_{51}NO_{11}$
Bikhaconite†

$C_{36}H_{51}NO_{12}$
ψ-Aconitine

$C_{36}H_{52}O_{12}$
Acolongifloroside G, *Tri-Ac*
Acovenoside A, *Tri-Ac*
Acovenoside B, *Tri-Ac*
Cryptograndoside C, *Acetate*
Scillarenin di-L-rhamnoside†

$C_{36}H_{52}O_{13}$
Scillaren-A

$C_{36}H_{52}O_{14}$
Glucoscillipheoside

$C_{36}H_{52}O_{15}$
Hellebrin

$C_{36}H_{53}N$
Indolo[3′,2′-20,21]friedelene

$C_{36}H_{53}NO_{12}$
Nervosine†

$C_{36}H_{53}NO_{14}$
Leucensomycin†

$C_{36}H_{54}O_4$
Carpesterol

$C_{36}H_{54}O_8$
Arjunolic Acid, *Tri-acetate*

$C_{36}H_{56}O_3$
Agaurol C
$C_{36}H_{56}O_7$
Carboxyacetylquercinic Acid, *Di-Me ester*
$C_{36}H_{56}O_8$
Cocarcinogen A_1†
$C_{36}H_{56}O_9$
Quinovin
$C_{36}H_{56}O_{12}$
Echubioside
Fusicoccin A†
$C_{36}H_{56}O_{13}$
Graciloside
Periplocin
$C_{36}H_{56}O_{14}$
Digitalin
Emicin
$C_{36}H_{56}O_{15}$
Kisantoside†
$C_{36}H_{57}NO_{14}$
Etruscomycin
$C_{36}H_{58}O_5$
Rehmannic Acid, *Me ester*
$C_{36}H_{58}O_8$
γ-Ceasalpin†
Oleanoloyl-β-D-glucoside†
$C_{36}H_{58}O_{10}$
Longisporin
Mycotocin A†
$C_{36}H_{60}O_{14}$
Aldgamycin C†
$C_{36}H_{60}O_{30}$
Campanulin
Cyclohexa-amylose
$C_{36}H_{62}O_3$
9,12-Octadecadienoic Acid, *Anhydride*
$C_{36}H_{62}O_{11}$
Monensic Acid†
$C_{36}H_{62}O_{26}$
Verbascose
$C_{36}H_{62}O_{31}$
Ajugose
Cellohexaose
Lycopose
Maltohexaose
$C_{36}H_{63}N_3O_9$
Enniatin A
$C_{36}H_{64}O_2$
Chaulmoogric Acid, *Chaulmoogryl ester*
$C_{36}H_{65}NO_{13}$
Erythromycin C
$C_{36}H_{66}N_6O_{10}$
Viscosin
$C_{36}H_{66}O_3$
9-Octadecenoic Acid, *Anhydride*
$C_{36}H_{66}O_{21}$
Stachyose, *Tetradeca-Me ether*
$C_{36}H_{70}O_3$
Octadecanoic Acid, *Anhydride*
$C_{36}H_{70}O_4$
Succinic Acid, *Dihexadecyl ester*
$C_{36}H_{70}O_{11}$
Sucrose, *Octapropyl ether*
$C_{36}H_{74}$
Hexatriacontane
$C_{36}H_{74}O_3$
Phthiocerol A†
$C_{36}H_{75}N$
Dioctadecylamine

C_{37}

$C_{37}H_{34}O_{11}$
Morelloflavone, *Hepta-Me ether*†
$C_{37}H_{36}N_2O_7$
Repanduline★†
$C_{37}H_{38}N_2O_6$
Cepharanthine†
Cissampareine†
Ocotosine†
epi-Stephanine★†
Thalmethine, O-*Me ether*†
$C_{37}H_{38}O_{12}$
Procyanidino-(−)-epicatechin, *Hepta-Me ether*†
$C_{37}H_{40}N_2O_6$
Berbamine
Curine, 4″-*Me ether*†
Fangchinoline
Hayatidin†
Hayatin, O-*Me ether*†
Lauberine†
Limacine†
Limacusine†
2-Nortetrandrine†
Ocodemerine†
Ocotine†
Otocamine†
Oxyacanthine
Thalicherine†
Thalmine★†
$C_{37}H_{40}N_2O_{10}$
Methyl-lycaconitine
$C_{37}H_{40}N_4O_7$
Haplophytine★†
$C_{37}H_{40}O_{11}$
Leucofisetinidin(+)-catechin, *Hepta-Me ether*
$C_{37}H_{42}N_2O_6$
Cuspidaline†
Dirosine†
Isoliensinine†
Liensinine†
Norrodiasine†

$C_{37}H_{44}O_{11}$
2α,7β,9α,10β-Tetra-acetoxy-5α-cinnamoyl-oxytaxa-4(16),11-dien-13-one †

$C_{37}H_{45}NO_{12}$
Rifamycin S †

$C_{37}H_{46}O_{10}$
2α,9α,10β,13α-Tetra-acetoxy-5α-cinnamoyl-oxytaxa-4(16),11-diene †

$C_{37}H_{46}O_{12}$
Protokosin

$C_{37}H_{47}NO_{12}$
Rifamycin SV †

$C_{37}H_{48}N_4O_6$
Mesobilirubin XI*a*, *Di-Me ester*

$C_{37}H_{48}O_{20}$
Xanthorhamnin, *Tri-Me ether*

$C_{37}H_{49}NO_{10}$
Taxine I †

$C_{37}H_{49}N_7O_9S$
N-tert-Butoxycarbonyl-β-alanyl-L-tryptophyl-L-methionyl-L-aspartyl-L-phenylalanine amide

$C_{37}H_{52}O_{13}$
Courmontoside B, *Acetate*
Courmontoside C, *Acetate*

$C_{37}H_{56}O_{15}$
Glucoverodoxin

$C_{37}H_{56}O_{18}$
Helleborein

$C_{37}H_{57}N_3O_{10}$
Taxinine

$C_{37}H_{58}O_8$
Cocarcinogen B_1 †
Protoescigenin, 21-*Tigloyl*-22-*Ac* †

$C_{37}H_{58}O_{15}$
Aldgamycin E †

$C_{37}H_{59}NO_{11}$
Germerine

$C_{37}H_{59}NO_{12}$
Germbudine
Neogerbudine †

$C_{37}H_{60}O_{10}$
Gratioligenin
Mycotocin B †

$C_{37}H_{61}NO_{12}$
Veralbidine

$C_{37}H_{61}NO_{14}$
Leucomycin A_9 †
PA 153

$C_{37}H_{64}O_{11}$
Monensic Acid, *Me ester* †

$C_{37}H_{66}O_6$
2-*trans,trans*-Sorbo-1,3-dimyristin †

$C_{37}H_{67}NO_{12}$
Erythromycin B

$C_{37}H_{67}NO_{13}$
Erythromycin

$C_{37}H_{74}O_5$
Ascaroside B

C_{38}

$C_{38}H_{16}O_2$
Dianthra[8,9,1,2-*cdefg*, 8′,9′,1′,2′-*nopqr*]-pentacene-9,18-dione

$C_{38}H_{18}O_6$
Sesquixanthydryl †

$C_{38}H_{20}$
Benzo[*hi*]naphtho[2,1,8,7-*wxyz*]heptacene †

$C_{38}H_{22}$
7,8-Naphthoheptaphene
Nonahelicene †

$C_{38}H_{24}$
Hexabenzo-octalene

$C_{38}H_{24}O_2$
1,4,5,8-Tetraphenylanthraquinone †

$C_{38}H_{28}Cl_{12}Sb_2$
Biphenyl-4,4′-bis(diphenylmethyl)hexa-chloroantimonate †

$C_{38}H_{30}$
Hexaphenylethane

$C_{38}H_{30}O_2$
Benzpinacol, *Diphenyl ether*

$C_{38}H_{34}O_4$
Tetraphenylphthalic Acid, *Di-Propyl ester* †

$C_{38}H_{38}N_4O_2$
Caracurine II

$C_{38}H_{40}N_2O_2$ (di-ion)
C-Curarine II

$C_{38}H_{40}N_2O_6$
Cissampareine, O-*Me ether* †

$C_{38}H_{40}N_2O_7$
Repandinine
Thalsimine †

$C_{38}H_{40}N_4O_2$
Caracurine V

$C_{38}H_{40}O_{12}$
Procyanidino-(−)-epicatechin, *Octa-Me ether* †

$C_{38}H_{41}ClO_{17}$
Ensatin Chloride

$C_{38}H_{42}N_2O_6$
Berbamine, *Me ether*
Curine, *Di-Me ether* †
Cycleanine †
Isotetrandrine †
Lauberine, *Me ether* †
Melanthiodine †
Menisine
O-Methylisothalicherine †
Ocotine, N-*Me* †
Ocotine, O-*Me* †
Phaeanthine
Rodiasine †
Tetrandine
Thalicherine, O-*Me ether* †

$C_{38}H_{42}N_2O_7$
Thalfoetidine †
Thalidezine †

$C_{38}H_{42}O_{11}$
Gibberellic Acid, *Anhydride*
$C_{38}H_{42}O_{12}$
Leucorobinetinidin-(+)-catechin, *Octa-Me ether*†
Ouratea proanthocyanidin A, *Hepta-Me ether*†
Proanthocyanidin (Cola nut), 3′,3‴4′,4‴,5′,5″,7′,7″- *Octa-Me ether*†
$C_{38}H_{44}Cl_2N_2O_6$
Chondocurarine chloride
Tubocurarine chloride
$C_{38}H_{44}I_2N_2O_6$
Chondocurarine iodide
$C_{38}H_{44}N_2O_6$
Dauricine
$C_{38}H_{44}N_4O$
Geissolosimine
$C_{38}H_{44}O_4$
Violerythrin†
$C_{38}H_{44}O_8$
α-Gambogic Acid
$C_{38}H_{46}N_2O_7$
Thalictrinine
$C_{38}H_{46}N_2O_{12}$
Anisatinic Acid, *Brucine salt*
$C_{38}H_{46}N_4O_4$
Mesoporphyrin I, *Di-Et ester*
Mesoporphyrin IX, *Di-Et ester*
$C_{38}H_{50}N_2O_{10}$
Elatine
$C_{38}H_{50}O_{14}$
Pseudrelone-A_1†
$C_{38}H_{57-61}N_7O_{7-8}S$
Bottromycin
$C_{38}H_{58}O_{14}$
Abobioside
Cryptograndoside B
Cryptograndoside C
$C_{38}H_{59}NO_{16}$
Leucomycin B_4
$C_{38}H_{60}O_8$
Cocarcinogen A_3†
Stevioside
$C_{38}H_{61}NO_{12}$
Acumycin
$C_{38}H_{62}O_2$
Agaurol A
$C_{38}H_{63}NO_{13}$
Rimocidin†
$C_{38}H_{63}NO_{14}$
Leucomycin A_7†
PA108
$C_{38}H_{65}NO_{15}$
PA148
$C_{38}H_{65}NO_{16}$
Leucomycin B_2
$C_{38}H_{74}O_4$
Octadecanoic Acid, *Ethylene glycol di-ester*

C_{39}

$C_{39}H_{32}$
1,1,1,3,3,3-Hexaphenylpropane
$C_{39}H_{36}O_{10}$
Isocryptomerin, *Tetra-Et ether*†
Kayaflavone, *Tri-Et ether*†
$C_{39}H_{42}N_2O_6$
Cissampareine, O-*Et ether*†
$C_{39}H_{42}O_{12}$
Procyanidino-(−)-epicatechin, *Nona-Me ether*†
$C_{39}H_{44}N_2O_6$
Berbamine, *Et ether*
Fangchinoline, *Et ether*
Hayatidin, O-*Et ether*†
Lauberine, *Et ether*†
$C_{39}H_{44}N_2O_7$
Hernandezine
Thalidasine†
$C_{39}H_{44}O_{13}$
Leucorobinetinidin (+)-gallocatechin, *Nona-Me ether*†
$C_{39}H_{46}N_2O_6$
Cuspidaline, *Di*-O-*Me ether*†
Dauricine★, O-*Me ether*†
Isoliensinine, *Di*-O-*Me ether*†
Liensinine, *Di-Me ether*†
Magnoline, *Tri-Me ether*
$C_{39}H_{47}NO_{14}$
Rifamycin O†
$C_{39}H_{47}N_5O_5$
Scutianine†
$C_{39}H_{48}N_4O_2$
Hunteriamine
$C_{39}H_{48}O_{12}$
2α,7β,9α,10β,13α-Penta-acetoxy-5α-cinnamoyl-oxytaxa-4(16),11-diene†
$C_{39}H_{49}NO_{14}$
Rifamycin B†
$C_{39}H_{49}NO_{17}$
Nogalamycin†
$C_{39}H_{52}N_{10}O_{14}S$
α-Amantine
β-Amantine
$C_{39}H_{52}O_{14}$
Pseudrelone-A_1, *Me ether*†
$C_{39}H_{57}NO_{11}$
Germinitrine
$C_{39}H_{58}O_4$
Coenzyme Q_6
$C_{39}H_{59}NO_4$
Germanitrine
$C_{39}H_{60}O_{14}$
Purpronin
$C_{39}H_{62}O_{12}$
Gracillin (?)
Hispidin†

$C_{39}H_{62}O_{13}$
Purpnin
Trillarin

$C_{39}H_{63}NO_{11}$
γ_1-Solamarine†
γ_2-Solamarine†

$C_{39}H_{63}NO_{12}$
δ-Solamarine†

$C_{39}H_{64}O_5$
1,3-Dilinolenin

$C_{39}H_{64}O_{14}$
Digipurpurin

$C_{39}H_{65}NO_{14}$
Leucomycin A_5†

$C_{39}H_{67}N_5O_{11}$
Esperine
Esperine X

$C_{39}H_{68}O_5$
1,3-Dilinolein
9-Octadecynoic Acid, 1,2-*Glycerol di-ester*
9-Octadecynoic Acid, 1,3-*Glycerol di-ester*

$C_{39}H_{69}O_{11}$
Nigericin

$C_{39}H_{72}O_5$
1,3-Dielaidin
α,β-Diolein

$C_{39}H_{74}O_6$
Trilaurin

$C_{39}H_{76}O_4$
Octadecanoic Acid, *Propylene glycol di-ester*

$C_{39}H_{76}O_5$
α,α′-Distearin
α,β′-Distearin

$C_{39}H_{80}$
Nonatriacontane

C_{40}

$C_{40}H_{16}O_4$
Tetranaphthocyclo-octatetraene Tetra-oxide†

$C_{40}H_{18}N_2O_2$
Cyananthrene

$C_{40}H_{26}$
Bitriptoyl†

$C_{40}H_{30}$
9,10-Dimethyl-1,4,5,8-tetraphenylanthracene†

$C_{40}H_{30}O_3$
Triphenylacetic Acid, *Anhydride*

$C_{40}H_{37}NO_7$
Bephenium hydroxynaphthoate

$C_{40}H_{38}N_4O_{16}$
Uroporphyrin I
Uroporphyrin III

$C_{40}H_{38}O_4$
Tetraphenylphthalic Acid, *Di-Butyl ester*†

$C_{40}H_{38}O_{10}$
Hinokiflavone, *Penta-Et ether*†
Isoginkgetin, *Tetra-Et ether*†

$C_{40}H_{44}N_4O$ (di-ion)
C-Curarine I
Ultracurine A
Ultracurine B

$C_{40}H_{44}N_4O_2$ (di-ion)
C-Alkaloid G
C-Alkaloid E (or $C_{40}H_{46}N_4O_2$)

$C_{40}H_{46}N_2O_6$
Obamegine, *Di-Et ether*†

$C_{40}H_{46}N_2O_7$
Thalidezine, *Et ether*†

$C_{40}H_{46}N_2O_8$
Thalmelatine†

$C_{40}H_{46}N_4$ (di-ion)
C-Dihydrotoxiferine

$C_{40}H_{46}N_4$ (di-ion)
C-Alkaloid H

$C_{40}H_{46}N_4O_2$ (di-ion)
C-Alkaloid E (or $C_{40}H_{44}N_4O_2$)
Toxiferine I
Toxiferine II
Toxiferine III

$C_{40}H_{48}$
Leprotene†

$C_{40}H_{48}Cl_2N_2O_6$
Tubocurarine chloride, *Di-Me ether*

$C_{40}H_{48}N_4O_2$ (di-ion)
Calebassine
C-Alkaloid D

$C_{40}H_{48}N_4O_3$
Geissospermine

$C_{40}H_{48}N_4O_3$ (di-ion)
C-Alkaloid F

$C_{40}H_{48}N_4O_4$ (di-ion)
C-Alkaloid A

$C_{40}H_{48}O_4$
Astacene

$C_{40}H_{48}O_8$
Gambogic Acid★, *Me ester*, *Me ether*†

$C_{40}H_{50}N_4O_4$
Villalstonine

$C_{40}N_{50}O_2$
Rhodoxanthin

$C_{40}H_{51}NO_{14}$
Streptovaricin C†

$C_{40}H_{51}NO_{15}$
Streptovaricin G†

$C_{40}H_{52}$
Chlorobactene†
Dehydrolycopene†
β-Isorenieratene†

$C_{40}H_{52}O_2$
Pectenoxanthin†
Torularhodin†

$C_{40}H_{52}O_3$
Pectenolone†

$C_{40}H_{52}O_4$
Astaxanthin
$C_{40}H_{54}$
Isocarotene
Leprotene
$C_{40}H_{54}O$
Celaxanthin
OH-Chlorobactene †
Crocoxanthin †
Echinenone
4-Keto-γ-carotene †
Myoxanthin
$C_{40}H_{54}O_2$
Canthaxanthin
2-Demethylvitamin K_2(30) †
Diatoxanthin †
Eschscholtzxanthin
3-Hydroxyechinenone †
Monadoxanthin †
$C_{40}H_{54}O_3$
Diadinoxanthin †
Flexixanthin †
$C_{40}H_{54}O_6$
Macrophyllic Acid
$C_{40}H_{54}O_{13}$
Pseudrelone-A_2 †
$C_{40}H_{55}NO_{13}$
Streptovaricin E †
$C_{40}H_{56}$
α-Carotene
β-Carotene
γ-Carotene
δ-Carotene
ε-Carotene †
Lycopene
$C_{40}H_{56}N_3O_6$
Elatrine
$C_{40}H_{56}O$
Citroxanthin
Cryptoxanthin
3,4-Dehydrorhodopsin †
Flavochrome
Gazaniaxanthin
Lycoxanthin
Rubixanthin
$C_{40}H_{56}O_2$
Cryptoflavin
Deoxyflexixanthin †
1′2′-Dihydro-1′-hydroxy-4-keto-γ-carotene †
Kryptocapsin
Lutein
Lycophyll
Saproxanthin †
Zeaxanthin
$C_{40}H_{56}O_3$
Antheraxanthin
Chrysanthemaxanthin
Flavoxanthin
Sarcinaxanthin (?)
$C_{40}H_{56}O_4$
Auroxanthin
Neoxanthin †
Taraxanthin
Trollichrome
Violaxanthin
$C_{40}H_{56}O_5$
Trolliflavine
Trolliflor
$C_{40}H_{56}O_7$
Myxoxanthophyll (?)
$C_{40}H_{58}$
Dihydro-β-carotene
Neurosporene
$C_{40}H_{58}O$
1′,2′-Dihydro-1′-hydroxy-γ-carotene †
Rhodopin
$C_{40}H_{58}O_2$
Podototarin
$C_{40}H_{58}O_3$
Capsanthin
$C_{40}H_{58}O_4$
Trollixanthin
$C_{40}H_{60}$
ζ-Carotene †
$C_{40}H_{60}N_4O_{10}$
Bufotoxin
$C_{40}H_{60}N_{12}O_{13}S_2$
[Ser[3]-Ile[8]]-Oxytocin †
$C_{40}H_{60}O$
Chloroxanthin †
$C_{40}H_{60}O_2$
Kitol★ †
1,2,1′,2′-Tetrahydro-1,1′-dihydroxylycopene †
$C_{40}H_{60}O_4$
Capsorubin
Mangiferic Acid
Maytenone
$C_{40}H_{61}FeN_{10}O_{20}S$
Grisein
$C_{40}H_{61}N_{11}O_{11}S_2$
[Gly[4]]-Oxytocin †
$C_{40}H_{62}$
Phytofluene★ †
$C_{40}H_{62}N_{10}O_{12}S_2$
Deamino-Serine[4]-Isoleucine[8]-oxytocin †
$C_{40}H_{64}$
Phytoene★ †
$C_{40}H_{64}O_{12}$
Nonactin
$C_{40}H_{64}O_{13}$
Lustericin
$C_{40}H_{66}$
Lycopersene
$C_{40}H_{66}N_6$
Ormojine †
$C_{40}H_{67}NO_{14}$
Leucomycin A_1 †
$C_{40}H_{67}NO_{16}$
Carbomycin

$C_{40}H_{68}$
Phytofluene
$C_{40}H_{68}N_4O_{12}$
Amidomycin
$C_{40}H_{68}O_{11}$
Polyetherin A†
$C_{40}H_{70}O_4$
Phthalic Acid, *Dihexadecyl ester*
$C_{40}H_{70}O_{14}$
Digitoxin
$C_{40}H_{76}NO_8P$
Dihexadec-9-enoyl-α-lecithin
$C_{40}H_{78}$
Perhydrocarotene
$C_{40}H_{78}O_3$
Eicosanoic Acid, *Anydride*
$C_{40}H_{82}O$
Kutkiol

C_{41}

$C_{41}H_{30}S$
Diphenylsulphonium tetraphenylcyclopentadienylide†
$C_{41}H_{30}Se$
Diphenylselenonium tetraphenylcyclopentadienylide†
$C_{41}H_{32}O_{28}$
Chebulagic Acid★†
Neochebulagic Acid
$C_{41}H_{34}O_{28}$
Chebulinic Acid★†
Neochebulinic Acid★†
$C_{41}H_{40}O_{10}$
Sotetsuflavone, *Penta-Et ether*†
$C_{41}H_{46}N_2O_8$
Dehydrothalicarpine†
$C_{41}H_{47}NO_{17}$
Wilforzine
$C_{41}H_{47}NO_{19}$
Wilforgine
$C_{41}H_{47}NO_{20}$
Wilfortrine
$C_{41}H_{48}N_2O_8$
Thalicarpine★†
$C_{41}H_{48}N_4O_4$
Villalstonine†
$C_{41}H_{50}N_4O_2$
Pleiomutine†
$C_{41}H_{50}N_4O_3$
Macralstonidine
$C_{41}H_{50}N_4O_5$
Conodurine
$C_{41}H_{50}N_4O_6$
Conoduramine

$C_{41}H_{52}O_{17}$
Utilin†
$C_{41}H_{54}N_3O_9$
Leurosivine†
$C_{41}H_{54}O_2$
Torularhodin, *Me ester*†
$C_{41}H_{56}O_2$
Vitamin K_2(30)
$C_{41}H_{56}O_{18}$
Acolongifloroside K, *Hexa-Ac*
$C_{41}H_{57}O_4P$
24-Ethylidenecholest-5-en-3β-ol, *Diphenylphosphate*
$C_{41}H_{58}FeN_9O_{20}$
Ferrichrome A
$C_{41}H_{58}O$
Anhydrorhodovibrin†
$C_{41}H_{58}O_2$
Hydroxyspirilloxanthin
Lutein★, 3-*Me ether*†
Lutein★, 3′-*Me ether*†
Spheroidenone†
OH-Spirilloxanthin†
Zeaxanthin, *Mono-Me ether*
$C_{41}H_{60}N_{12}O_{13}S_2$
[Ser[4]-Gln[8]]-Oxytocin†
$C_{41}H_{60}O$
Spheroidene†
$C_{41}H_{60}O_2$
Rhodovibrin†
$C_{41}H_{61}NO_{13}$
Escholerine
$C_{41}H_{63}NO_{14}$
Germitetrine
Protoveratrine A
$C_{41}H_{63}NO_{15}$
Protoveratrine B
$C_{41}H_{63}N_{11}O_{11}S_2$
[β-Ala[4]]-Oxytocin†
$C_{41}H_{63}N_{11}O_{12}S_2$
Serine[4]-Isoleucine[8]-oxytocin★†
$C_{41}H_{64}FeN_9O_{17}$
Ferrirhodin†
$C_{41}H_{64}FeN_9O_{17}$
Ferrirubin†
$C_{41}H_{64}O_{11}$
Hederin (?)
Vanguerin
$C_{41}H_{64}O_{14}$
Digoxin
Gitoroside
Gitoxin
$C_{41}H_{64}O_{15}$
Diginatin
$C_{41}H_{65}FeN_{10}O_{14}$
Ferrimycin A_1†
$C_{41}H_{66}O_{12}$
Monactin

$C_{41}H_{68}O_{10}$
Flavomycoine †
$C_{41}H_{74}NO_8P$
Dilinoleoyl-L-α-cephalin †
$C_{41}H_{78}NO_8P$
Dioleoyl-L-α-cephalin †
$C_{41}H_{80}O_4$
Fomentaric Acid †
$C_{41}H_{82}NO_8P$
Distearoyl-L-α-cephalin †
$C_{41}H_{84}$
3,5,23-Trimethyloctatriacontane †

C_{42}

$C_{42}H_{18}$
Hexabenzo[*bc'*,*ef'*,*hi'*,*kl'*,*no'*,*pd'*]coronene
1:12,2:3,4:5,6:7,8:9,10:11-Hexaperi-benzocoronene †
$C_{42}H_{24}$
7,8-Anthraheptaphene †
3,6':3',6'':3'',6-Triphenanthrylene †
$C_{42}H_{28}$
Rubrene
$C_{42}H_{30}$
Hexaphenylbenzene †
p-Septiphenyl
$C_{42}H_{30}O_8$
Ellagorubin
$C_{42}H_{36}N_2O_{10}$
Fagopyrin
$C_{42}H_{38}BClO_6$
Rosocyanin
$C_{42}H_{38}O_{20}$
Sennoside A †
Sennoside B †
$C_{42}H_{42}O_4$
Tetraphenylphthalic Acid, *Di-Pentyl ester* †
$C_{42}H_{42}O_{10}$
Cupressuflavone, *Hexa-Et ether* †
$C_{42}H_{44}O_{19}$
Amurensin, *Acetate*
$C_{42}H_{45}N_5O_5$
Adouetin Z †
$C_{42}H_{47}ClO_{23}$
Nasunin †
$C_{42}H_{48}N_4O_6$
Vobtusine
$C_{42}H_{50}N_2O_9$
Adiantifoline †
$C_{42}H_{50}N_4O_5$
Gabunine †
$C_{42}H_{52}IN_2O_9$
Turumiquirensine, *Iodide* †
$C_{42}H_{52}N_4O_4$
Secamine †
$C_{42}H_{53}NO_{15}$
Streptovaricin B †
$C_{42}H_{53}NO_{16}$
Streptovaricin A †
$C_{42}H_{54}N_2O_{11}$
Turumiquirensine †
$C_{42}H_{54}N_4O_4$
Dihydrosecamine †
$C_{42}H_{55}NO_{15}$
Streptovaricin D †
$C_{42}H_{56}N_2O_2$
Δ^7-Avenasterol, *Azoyl*
$C_{42}H_{56}N_4O_4$
Tetrahydrosecamine †
$C_{42}H_{56}O_2$
2,2'-Dioxospirilloxanthin †
$C_{42}H_{58}O_6$
Fucoxanthin★ †
Isofucoxanthin †
Macrophyllic Acid, *Di-Me ester*
Macrophyllic Acid, *Di-O-Me*
$C_{42}H_{60}O_2$
Spirilloxanthin
Zeaxanthin, *Di-Me ether*
$C_{42}H_{60}O_{11}$
Heudelottin †
$C_{42}H_{62}O_2$
Podototarin
$C_{42}H_{62}O_{16}$
Glycyrrhinic Acid
$C_{42}H_{62}O_{18}$
Glucoscillaren A
Liriodendrin, *Octa-Me ether* †
$C_{42}H_{63}NO_{14}$
Germitetrine B
$C_{42}H_{64}O_{15}$
Gitaloxin
$C_{42}H_{65}N_{11}O_{11}S_2$
4-Decarboxamido-oxytocin †
5-Decarboxamido-oxytocin †
$C_{42}H_{66}O_{11}$
Hederin, *Me ester*
$C_{42}H_{66}O_{17}$
Echujin
$C_{42}H_{66}O_{18}$
Thevetin
$C_{42}H_{66}O_{19}$
Acovenoside C
Gitostin
Neogitostin
$C_{42}H_{67}NO_{13}$
Germitrine
$C_{42}H_{67}NO_{15}$
Carbomycin B
$C_{42}H_{67}NO_{16}$
Carbomycin★ †

$C_{42}H_{68}O_{12}$
Dinactin
$C_{42}H_{68}O_{13}$
Saikosaponin *a*†
Saikosaponin *d*†
$C_{42}H_{68}O_{14}$
Gratioside
$C_{42}H_{69}NO_{15}$
Leucomycin A_3†
$C_{42}H_{70}O_{35}$
Cyclohepta-amylose
$C_{42}H_{72}O_{14}$
Ginsenoside$_{Rg\text{-}1}$†
$C_{42}H_{72}O_{16}$
Lankamycin†
$C_{42}H_{72}O_{36}$
Celloheptaose
$C_{42}H_{78}NO_{10}P$
DL-α-(Dioleoyl)-phosphatidyl-DL-serine†
$C_{42}H_{83}NO_4$
Ergocerebrin†
$C_{42}H_{83}O_{10}P$
Diphosphatidylglycerol†
$C_{42}H_{84}NO_8P$
Di-octadecanoyl-L-α-glycerylphosphoryl-L-2-amino-1-propanol†
$C_{42}H_{84}O_2$
Hexadecanoic Acid, *Ceryl ester*

C_{43}

$C_{43}H_{43}N_3O_{12}$
Clivimine★†
$C_{43}H_{48}N_4O_6$
Deoxyvobtusine-lactone†
$C_{43}H_{48}N_4O_7$
Vobtusine-lactone†
$C_{43}H_{49}NO_{18}$
Wilforine
$C_{43}H_{49}NO_{19}$
Wilfordine
$C_{43}H_{49}N_7O_{10}$
Staphylomycin S
$C_{43}H_{50}N_4O_6$
Vobtusine†
$C_{43}H_{52}N_4O_5$
Voacamine
$C_{43}H_{52}N_4O_6$
Voacorine
$C_{43}H_{52}N_8O_{11}$
Doricin†
$C_{43}H_{60}O_{16}$
Canarienglucoside A
$C_{43}H_{64}N_{12}O_{12}S_2$
[Ile[8]]-Oxytocin†
$C_{43}H_{64}O_{16}$
1-Digacetinin
$C_{43}H_{65}N_{11}O_{11}S_2$
Desamino-desoxyoxytocin
$C_{43}H_{65}N_{11}O_{12}SSe$
1-Deamino-1-hemi-L-selenocystineoxytocin†
1-Deamino-6-hemi-L-selenocystineoxytocin†
$C_{43}H_{65}N_{11}O_{12}S_2$
Desamino-oxytocin
$C_{43}H_{65}N_{11}O_{12}Se_2$
[1-Deamino-1,6-selenocystine]oxytocin†
$C_{43}H_{65}N_{11}O_{13}S_2$
4-Deamido-oxytocin†
$C_{43}H_{65}N_{15}O_{12}S_2$
Arginine vasotocin
$C_{43}H_{66}N_{12}O_{11}S_2$
Desoxy-oxytocin
$C_{43}H_{66}N_{12}O_{12}SSe$
1-Hemi-L-selenocystineoxytocin†
6-Hemi-L-selenocystineoxytocin†
$C_{43}H_{66}N_{12}O_{12}S_2$
Oxytocin
$C_{43}H_{66}N_{14}O_{13}S_2$
8-L-Citrullineoxytocin
$C_{43}H_{66}O_{15}$
Digorid
$C_{43}H_{70}O$
Bombiprenone†
$C_{43}H_{70}O_{12}$
Trinactin
$C_{43}H_{71}NO_{12}$
Venturicidin
$C_{43}H_{71}NO_{15}$
Polyanine†
$C_{43}H_{74}N_2O_{14}$
Foromacidin A★†
$C_{43}H_{82}O_6$
α-Palmito-α′β-dilaurin
β-Palmito-αα′-dilaurin
$C_{43}H_{84}O_4$
Fomentaric Acid†
$C_{43}H_{84}O_8$
Ascaroside C
$C_{43}H_{86}O$
22-Tritetracontanone
$C_{43}H_{87}NO_2$
Caulerpicin†

C_{44}

$C_{44}H_{28}$
Verdene†
$C_{44}H_{30}$
Hexaphenylcyclopropenylidenecyclopentadiene†

$C_{44}H_{34}O_{18}$
Fusaroskyrin, *Hexa-Ac*
$C_{44}H_{42}O_{20}$
Sennoside A, *Di-Me ester* †
Sennoside B, *Di-Me ester* †
$C_{44}H_{44}N_{10}O_{14}$
Aetioporphyrin I, *Picrate*
$C_{44}H_{45}ClO_{23}$
Monardaein chloride
$C_{44}H_{52}N_4O_{10}$
Vincarodine †
$C_{44}H_{52}O$
α-Lumicolchicine
$C_{44}H_{54}N_4O_5$
Macralstonine
$C_{44}H_{55}NO_9$
Flavucidin
$C_{44}H_{56}N_4O_8$
Deacetylvinblastine †
$C_{44}H_{56}O_{18}$
Bussein †
$C_{44}H_{62}N_8O_{10}$
Etamycin
$C_{44}H_{64}O_{24}$
α-Crocin
$C_{44}H_{66}O_4$
Coenzyme Q_7
$C_{44}H_{66}O_8$
1,26-Hexacosanediol dicaffeate †
$C_{44}H_{68}N_{12}O_{12}S_2$
[1-(*N*-Methyl-hemi-L-cystine)]-Oxytocin †
$C_{44}H_{78}O_2$
Carcinolipin †
$C_{44}H_{86}NO_{10}P$
Phosphatidylcarnitine

C_{45}

$C_{45}H_{30}CoN_3O_{12}$
Ferroverdin, Cobalt analogue †
$C_{45}H_{33}Cl_{18}Sb_3$
1,3,5-Tris-(diphenylhexachloroantimony-methyl)benzene †
$C_{45}H_{54}N_8O_{10}$
Ostreogrycin B †
$C_{45}H_{56}N_4O_6$
Vocacamidine
$C_{45}H_{58}N_8O_{11}$
Mikamycin B (?)
$C_{45}H_{58}O_{18}$
Bussein, *Me ether* †
$C_{45}H_{60}CoN_4O_{14}$
Cobyrinic Acid †
$C_{45}H_{62}O_2$
2-Demethylvitamin $K_2(35)$ †
$C_{45}H_{65}N_3O_6$
Piloceredine
Pilocereine
$C_{45}H_{66}CoN_{10}O_8$
Cobyric Acid †
$C_{45}H_{66}O_2$
Chlorobiumquinone
$C_{45}H_{67}N_{13}O_{11}$
Alanyl[4]-isoleucyl[5]-hypertensin
$C_{45}H_{68}O_8$
1-Caffeoyl-26-feruloylhexacosanediol †
$C_{45}H_{73}NO_{15}$
Leptinine I †
β-Solamarine †
$C_{45}H_{73}NO_{16}$
Leptinine II †
α-Solamarine †
Solasonine
$C_{45}H_{76}N_2O_{15}$
Foromacidin B★ †
$C_{45}H_{76}O_2$
9,12-Octadecadienoic Acid, *Cholesteryl ester*
$C_{45}H_{79}NO_{17}$
Tylosin †
$C_{45}H_{80}O_{28}$
Convolvulinic Acid
$C_{45}H_{84}O_8$
Methane-tetracarboxylic Acid, *Tetra-decyl ester*
$C_{45}H_{85}N_{13}O_{10}$
Colistin
$C_{45}H_{86}O_6$
Trimyristin

C_{46}

$C_{46}H_{52}N_4O_9$
Catharine †
$C_{46}H_{52}N_4O_{10}$
Catharicine †
$C_{46}H_{56}N_4O_9$
Leurosine★ †
$C_{46}H_{56}N_4O_{10}$
Carosine †
Pleurosine †
Vincristine † (Leurocristine★)
$C_{46}H_{56}N_4O_{12}$
Neoleurocristine †
$C_{46}H_{58}N_4O_8$
Deoxyvinblastine A †
$C_{46}H_{58}N_4O_9$
Leurosidine★ †
Vinblastine †
$C_{46}H_{64}N_{12}O_{12}S_2$
1-Desamino-8-lysine vasopressin
$C_{46}H_{64}N_{14}O_{13}S_2$
8-L-Citrullinevasopressin

$C_{46}H_{64}O_2$
Vitamin K_2(35)
$C_{46}H_{65}N_{13}O_{11}S_2$
[2-Phenylalanine-8-lysine]vasopressin †
$C_{46}H_{65}N_{15}O_{12}S_2$
Arginine vasopressin
$C_{46}H_{67}N_3O_6$
Pilocereine, *Me ether*
$C_{46}H_{70}N_{12}O_{12}S_2$
Oxytocin, *Acetone deriv.* †
$C_{46}H_{70}O_8$
1,26-Hexacosanediol diferulate †
1,28-Octacosanediol dicaffeate (*see under* 1,26-Hexacosanediol dicaffeate †)
$C_{46}H_{70}O_{17}$
Eupteleoside A †
$C_{46}H_{73}NO_{20}$
Amphotericin B
$C_{46}H_{78}N_2O_{15}$
Foromacidin C★ †
$C_{46}H_{81}NO_{17}$
Leucomycin A_1
$C_{46}H_{92}O_2$
Hexadecanoic Acid, *Melissyl ester*
Hexatetracontanoic Acid

C_{47}

$C_{47}H_{35}As$
Triphenylarsonium tetraphenylcyclopentadienolide †
$C_{47}H_{35}Sb$
Triphenylstibonium tetraphenylcyclopentadienolide †
$C_{47}H_{56}O_{16}$
Methylene-bis-norflavaspidic Acid †
$C_{47}H_{65}N_{13}O_{12}$
Vasopressin
$C_{47}H_{66}O_{22}$
Olivomycin D †
$C_{47}H_{69}N_{15}O_{11}$
7-Glycine-bradykinin
$C_{47}H_{12}O_8$
1-Caffeoyl-28-feruloyl octacosanediol (*see under* 1-Caffeoyl-28-feruloyl hexacosanediol †)
$C_{47}H_{74}N_2O_{14}$
Perimycin (or $C_{47}H_{76}N_2O_{14}$) †
$C_{47}H_{74}O_{17}$
Digoxoside †
Neodigoxoside †
$C_{47}H_{75}NO_{16}$
Leptine I †
$C_{47}H_{75}NO_{18}$
Nystatin★ †
$C_{47}H_{75}N_5O_{10}$
Mycobactin
$C_{47}H_{76}N_2O_{14}$
Perimycin (or $C_{47}H_{74}N_2O_{14}$) †
$C_{47}H_{90}O_6$
α-Palmito-α′β-dimyristin
β-Palmito-αα′-dimyristin
$C_{47}H_{94}O_3$
Bruceolic Acid
$C_{47}H_{97}N_2O_7P$
Sphingomyelin

C_{48}

$C_{48}H_{24}$
1:2,3:4,5:6,7:8,9:10,11:12-Hexabenzocoronene †
$C_{48}H_{28}$
2,3,6,7-Tetrahydrotetra-acenaphthyleno[1,2-*a*, 1′,2′-*c*, 1″,2″-*e*, 1‴,2‴-*g*]cyclo-octatetraene
$C_{48}H_{54}N_4O_{16}$
Uroprophyrin I, *Octa-Me ester*
Uroprophyrin III, *Octa-Me ester*
$C_{48}H_{60}O_{10}$
Glutinosin
$C_{48}H_{62}N_4O_{11}$
Neoleurosidine †
$C_{48}H_{64}O_{21}$
Chromocyclomycin †
$C_{48}H_{68}O_{19}$
Abobioside, *Penta-Ac*
$C_{48}H_{68}O_{22}$
Chromomycin A_4 †
$C_{48}H_{74}O_8$
1,28-Octacosanediol diferulate (*see under* 1,26-Hexacosanediol diferulate †)
$C_{48}H_{76}O_{17}$
Saikosaponin *c* †
$C_{48}H_{78}$
Finene
$C_{48}H_{80}O_{40}$
Cyclo-octaamylose
Lobelinin
$C_{48}H_{82}O_2$
9-Octadecenoic Acid, *Stigmasteryl ester*
$C_{48}H_{91}NO_8$
Nervone
$C_{48}H_{93}NO_8$
Cerasine
Kerasin
$C_{48}H_{93}NO_9$
Cerebron★ †
$C_{48}H_{94}O_3$
Tetracosanoic Acid, *Anhydride*

C_{49}

$C_{49}H_{36}O_8$
Methane-tetracetic Acid, *Tetra-1-naphthyl ester*
$C_{49}H_{54}O_{27}$
Saponarin, Sub-*Ac*
$C_{49}H_{62}ClFeN_4O_6$
Cytohaemin †
$C_{49}H_{69}N_{13}O_{12}$
[5-Valyl]Angiotensin II †
$C_{49}H_{70}N_{14}O_{11}$
Antiotensin Amide †
$C_{49}H_{71}N_{15}O_{10}$
6-Glycine-bradykinin
$C_{49}H_{74}O_4$
Coenzyme Q_8
$C_{49}H_{74}O_{16}$
Gymnemic Acid A_1 †
$C_{49}H_{76}O_{19}$
Lanatoside A
$C_{49}H_{76}O_{20}$
Lanadigin (?)
Lanatoside B
Lanatoside C★ †
$C_{49}H_{76}O_{21}$
Lanatoside D★ †

C_{50}

$C_{50}H_{56}N_6O_{12}$
Amicetin, *Tribenzoyl*
$C_{50}H_{60}N_{12}O_{12}S_2$
Echinomycin
$C_{50}H_{62}N_{12}O_{12}S_2$
Triostin A †
$C_{50}H_{71}N_{13}O_{12}$
[5-Isoleucyl]Angiotensin II †
$C_{50}H_{72}N_{14}O_{11}$
[5-Isoleucyl]Angiotensin II, *Amide* †
$C_{50}H_{72}O_2$
Sarcinaxanthin †
$C_{50}H_{73}N_{15}O_{11}$
Bradykinin
$C_{50}H_{76}O_4$
Bacterioruberin †
$C_{50}H_{76}O_{21}$
Lanatoside E
$C_{50}H_{80}O_{23}$
Acrospirin
$C_{50}H_{82}O$
Spadicol
$C_{50}H_{82}O_{23}$
F-Gitonin †
$C_{50}H_{83}NO_{20}$
Demissine
$C_{50}H_{83}NO_{21}$
Tomatine
$C_{50}H_{89}N_7O_{11}$
Peptidolipin NA †
$C_{50}H_{100}O_5$
Didymocarpol

C_{51}

$C_{51}H_{34}N_6Na_6O_{24}S_6$
Bayer 205
$C_{51}H_{56}O_{28}$
Saponarin, *Ac*
$C_{51}H_{62}N_{12}O_{12}S_2$
Quinomycin D †
$C_{51}H_{72}O_2$
Vitamin K_2(40)
$C_{51}H_{78}O$
2-Nonaprenylphenol †
$C_{51}H_{82}O_{23}$
Convallamaroside★ †
$C_{51}H_{84}O_{22}$
Parillin
$C_{51}H_{86}O_{25}$
Gitonin
$C_{51}H_{98}O_6$
Tripalmitin

C_{52}

$C_{52}H_{54}N_2O_6$
Melanthioidine, OO-*Dibenzyl ether* †
$C_{52}H_{63}N_9O_{12}$
PA114B
$C_{52}H_{64}N_{12}O_{12}S_2$
Quinomycin B †
Quinomycin B_0 †
$C_{52}H_{66}N_{12}O_{12}S_2$
Triostin B †
$C_{52}H_{76}O_{24}$
Aureolic Acid †
$C_{52}H_{84}O_{25}$
Theaprosapogenol A
$C_{52}H_{92}O_{14}$
Ginsenoside$_{Rg-1}$, *Deca-Me ether* †
$C_{52}H_{93}N_7O_{11}$
[Val[6]]Peptidolipin NA †
$C_{52}H_{102}O_3$
Hexacosanoic Acid, *Anhydride*
$C_{52}H_{104}O_4$
Corynin

C_{53}

$C_{53}H_{66}N_{12}O_{12}S_2$
Quinomycin E†

$C_{53}H_{68}N_{12}O_{12}S_2$
Triostin B_0†

$C_{53}H_{72}O_8$
Amitenone†

$C_{53}H_{80}O_2$
Plastoquinone-9†

$C_{53}H_{80}O_3$
6-Methoxy-3-methyl-2-nonaprenylbenzoquinone†

$C_{53}H_{82}O_2$
Plastochromanol-8†

$C_{53}H_{100}O_6$
Oleodipalmitin

$C_{53}H_{103}NO$
Kerasin, *Penta-Me ether*

C_{54}

$C_{54}H_{32}$
Leucacene

$C_{54}H_{68}N_{12}O_{12}S_2$
Quinomycin C†

$C_{54}H_{70}N_{12}O_{12}S_2$
Triostin C†

$C_{54}H_{74}O_{36}$
Maltotetraitol

$C_{54}H_{75}N_9O_{11}$
Ilamycin†

$C_{54}H_{77}N_9O_9$
Ilamycin B_1†

$C_{54}H_{77}N_9O_{10}$
Ilamycin B_2†

$C_{54}H_{78}O_3$
2,4,6-Tri(3′,5′-di-*tert*-butyl-4′-hydroxybenzyl)-mesitylene†

$C_{54}H_{82}N_2O_{18}$
PA150

$C_{54}H_{82}O_4$
Coenzyme Q_9

$C_{54}H_{84}O_{23}$
Escin†

$C_{54}H_{85}N_{13}O_{15}S$
Eledoisin

$C_{54}H_{86}O_{17}$
Eupteleoside A, *Octa-Me ether*†

$C_{54}H_{88}O_{23}$
Asiaticoside

$C_{54}H_{90}N_6O_{18}$
Valinomycin

$C_{54}H_{96}O_{27}$
Convolvulin

$C_{54}H_{103}NO_{12}$
Cytolipin H†

C_{55}

$C_{55}H_{74}N_4O_6Mg$
Bacteriochlorophyll *a*
Bacteriochlorophyll *b*

$C_{55}H_{80}O_{25}$
Jegosaponin

$C_{55}H_{83}NO_{21}$
Avenacin†

$C_{55}H_{83}N_{17}O_{11}$
7-Glycine-kallidin

$C_{55}H_{90}O$
Undecaprenol†

$C_{55}H_{90}O_{22}$
Phytolaccatoxin†

$C_{55}H_{92}O$
Bactoprenol†

$C_{55}H_{94}O_{28}$
Gitin

$C_{55}H_{104}O_6$
Oleopalmitostearin

$C_{55}H_{106}O_6$
α-Palmito-α′β-distearin
β-Palmito-αα′-distearin

$C_{55}H_{110}O$
Montanone

$C_{55}H_{112}O$
Montanol

C_{56}

$C_{56}H_{40}$
Octaphenylcyclo-octatetraene†

$C_{56}H_{42}O_{12}$
Hopeaphenol†

$C_{56}H_{70}N_4O_{16}$
Uroporphyrin I, *Octa-Et ester*

$C_{56}H_{72}N_8O_8$
Fungisporin

$C_{56}H_{80}O_2$
Vitamin $K_2(45)$

$C_{56}H_{80}O_{26}$
Olivomycin B†

$C_{56}H_{82}O_2$
Vitamin $K_9(H)$†

$C_{56}H_{82}O_{25}$
Olivomycin C†

$C_{56}H_{85}N_{17}O_{12}$
Kallidin
$C_{56}H_{86}O_4$
2-Decaprenylphenol†
$C_{56}H_{86}O_2$
Ergopinacol
$C_{56}H_{92}O_4$
Di-α-tocopherone†
$C_{56}H_{92}O_{27}$
Tigonin
$C_{56}H_{96}N_{12}O_{13}$
Polypeptin
$C_{56}H_{98}N_{16}O_{13}$
Polymyxin B_1
$C_{56}H_{102}O_{31}$
Ajugose, *Eicosa-Me ether*
Lycopose, *Eicosa-Me ether*

C_{57}

$C_{57}H_{82}O_{26}$
Chromomycin A_3†
$C_{57}H_{84}O_{25}$
Aburamycin C†
$C_{57}H_{86}N_{12}O_{16}$
Actinomycin F_8
$C_{57}H_{88}O_2$
2-Decaprenyl-6-methoxyphenol†
$C_{57}H_{90}O_{26}$
Theasaponin
$C_{57}H_{92}O_6$
Trilinolenin
$C_{57}H_{96}O_{28}$
Sarsaparilloside†
$C_{57}H_{98}O_6$
9-Octadecynoic Acid, *Glycerol tri-ester*
Trilinolein
$C_{57}H_{98}O_9$
Trivernolin
$C_{57}H_{104}O_6$
Triolein
$C_{57}H_{108}O_6$
Oleodistearin
$C_{57}H_{110}O_6$
Tristearin

C_{58}

$C_{58}H_{44}$
Tetra-(ββ-diphenylvinyl)ethylene†
$C_{58}H_{80}O_4$
Bisgalvinoxyl†
$C_{58}H_{82}O_{27}$
Acovenoside C, *Octa-Ac*
$C_{58}H_{84}N_{14}O_{16}S$
Physalaemin†
$C_{58}H_{84}O_{26}$
Olivomycin A†
$C_{58}H_{88}N_{12}O_{16}$
Actinomycin F_1
$C_{58}H_{89}NO_3$
Rhodoquinone-10★†
$C_{58}H_{94}O_{27}$
Cyclamin†
$C_{58}H_{102}N_4O_6$
Datemycin

C_{59}

$C_{59}H_{77}N_{13}O_{19}$
Telomycin★†
$C_{59}H_{86}O_{26}$
Chromomycin A_2†
$C_{59}H_{90}N_{12}O_{16}$
Actinomycin F_3
$C_{59}H_{90}O_4$
Coenzyme Q_{10}
$C_{59}H_{92}O_4$
Coenzyme Q_{10}(*H*-10)
$C_{59}H_{94}O_4$
Tetrahydrocoenzyme Q_{10}
$C_{59}H_{94}O_{27}$
Cryptoescin

C_{60}

$C_{60}H_{64}N_2O_{14}$
Gossypurpurin
$C_{60}H_{82}N_{12}O_{16}$
Actinomycin (Ser-Val-Pro-Sar-Meval)†
1,8-Didemethylacetinomycin C_1†
$C_{60}H_{84}N_{12}O_{16}$
Actinomycin F_9
$C_{60}H_{90}N_{12}O_{16}$
Actinomycin F_2
$C_{60}H_{92}N_{12}O_{10}$
Gramicidin S
Retrogramicidin S†
$C_{60}H_{98}O$
Dodecaprenol†
$C_{60}H_{100}O_{50}$
Kiransin
$C_{60}H_{122}$
Hexacontane

$C_{60}H_{128}N_{20}O_{32}$
Racemomycin B
$C_{61}H_{87}N_{17}O_{14}$
[5-Valyl]Angiotensin †
$C_{61}H_{88}N_{18}O_{13}$
[5-Valyl]Angiotensin, *Amide* †
$C_{61}H_{88}O_2$
Vitamin K_2(50)
$C_{61}H_{90}N_{12}O_{17}$
Actinomycin $X_{0\beta}$
$C_{61}H_{92}N_{12}O_{16}$
Actinomycin F_4
$C_{61}H_{94}N_{18}O_{13}S$
Methionyl-lysyl-bradykinin †
$C_{62}H_{86}N_{12}O_{16}$
Actinomycin C_1(D)★ †
$C_{62}H_{86}N_{12}O_{17}$
Actinomycin X_2
$C_{62}H_{88}CoN_{14}O_{16}P$
Vitamin B_{12c}★ †
$C_{62}H_{89}CoN_{13}O_{15}P$
Vitamin B_{12b}★ †
$C_{62}H_{89}N_{17}O_{14}$
[5-Isoleucyl]Angiotensin I †
$C_{62}H_{90}CoN_{13}O_{15}P$
Vitamin B_{12a} †
$C_{62}H_{122}O_3$ (Mixture)
α′-Smegma-mycolic Acid †
$C_{62}H_{122}O_4$
Oxalic Acid, *Dimyricyl ester*
$C_{63}H_{87}CoN_{13}O_{15}P$
Vitamin B_{12} Monocarboxylic Acid (E_2) †
$C_{63}H_{88}CoN_{14}O_{14}P$
Vitamin B_{12}★ †
$C_{63}H_{88}N_{12}O_{16}$
Actinomycin C_2★ †
Actinomycin C_{2a}★ †
$C_{63}H_{93}CoN_{13}O_{14}P$
Methylcobamide †
$C_{63}H_{124}O_3$ (Mixture)
α′-Smegma-mycolic Acid, *Me ester* †
$C_{64}H_{78}N_{10}O_{10}$
Antamanide †
$C_{64}H_{90}N_{12}O_{16}$
Actinomycin C_3
$C_{64}H_{90}N_{12}O_{17}$
Actinomycin $X_{0\delta}$
$C_{64}H_{96}N_{12}O_{16}$
Actinomycin E_1
$C_{65}H_{85}N_{13}O_{30}$
Mycobacillin
$C_{65}H_{93}N_{17}O_{17}S$
Phyllokinin †
$C_{65}H_{98}N_{12}O_{16}$
Actinomycin E_2
$C_{65}H_{111}NO_{22}$
Leucomycin A_2
$C_{66}H_{86}N_{13}O_{13}$
Tyrocidine A
$C_{66}H_{99}N_{16}O_{17}S$
Bacitracin F
$C_{66}H_{103}N_{17}O_{16}S$
Bacitracin A
$C_{66}H_{112}O_{56}$
Hordeacin
$C_{67}H_{40}$
Tris-(7*H*-dibenzo[*c*,*g*]fluorenylidenemethyl)-methane †
$C_{67}H_{122}N_3O_{26}$
Ganglioside G_2
$C_{68}H_{88}N_{14}O_{13}$
Tyrocidine B
$C_{70}H_{125}N_9O_{15}$
Fortuitine★ †
$C_{72}H_{100}CoN_{18}O_{17}P$
5′-Deoxyadenosylcobalamin †
$C_{72}H_{116}O_4$
Helenien
Physalien
$C_{72}H_{120}O_4$
1,2,1′,2′-Tetrahydro-1,1′-dihydroxylycopene, *Di-hexadecanoate* †
$C_{72}H_{120}O_{60}$
Platycodonin
$C_{72}H_{125}N_9O_{15}$
Fortuitine★ †
$C_{75}H_{108}N_{22}O_{20}S$
α-Melanotropin †
$C_{75}H_{148}O_4$
Mycollic Acid★, Strains D.T., P.N., and C.-II †
$C_{76}H_{102}O_{51}$
Lycopose, *Eicosa-Ac*
$C_{76}H_{150}O_4$
Mycolic Acid★, Strains D.T., P.N., and C.-II, *Me ester* †
$C_{78}H_{152}O_3$
Mycollic Acid, Strain Test α †
$C_{79}H_{154}O_3$
α-Smegma-mycolic Acid †
$C_{80}H_{126}O_{44}$
Gyposide †
$C_{80}H_{156}O_3$
α-Smegma-mycolic Acid, *Me ester* †
$C_{82}H_{166}$
Do-octacontane
$C_{84}H_{156}O_{10}$
Mycoside B
$C_{90}H_{154}$
Enneaphyllin
$C_{90}H_{170}O_4$
Mycollic Acid★, Strains D.T., P.N., and C.-III †

$C_{94}H_{190}$
Tetranonacontane†
$C_{96}H_{142}N_{28}O_{27}S$
β-Melanotropin, *Bovine*†
$C_{98}H_{144}N_{28}O_{28}S$
β-Melanotropin, *Porcine*†
$C_{98}H_{144}N_{30}O_{28}S$
β-Melanotropin, *Monkey*†
$C_{99}H_{139}N_{19}O_{18}$
Gramicidin A★—Valine Gramicidin A†
$C_{99}H_{139}N_{19}O_{19}$
Gramicidin B—Valine Gramicidin B†
$C_{100}H_{141}N_{19}O_{18}$
Gramicidin A★—Isoleucine Gramicidin A†
$C_{100}H_{141}N_{19}O_{19}$
Gramicidin B—Isoleucine Gramicidin B†
$C_{100}H_{164}O$
Dolichol
$C_{103}H_{172}O_{44}$
Gyposide, *Me ether*†
$C_{126}H_{210}O_{105}$
Alopecurin P
$C_{132}H_{222}O_{111}$
Kritesin
$C_{138}H_{232}O_{116}$
Avenarin F
$C_{142}H_{222}N_{42}O_{31}$
[1-D-Seryl-4-L-norleucyl-24-L-valinamide]β-corticotropin (1 → 25)†
$C_{186}H_{366}O_{17}$
Cord Factor
$H_2N_3O_3P$
Phosphorazidic Acid†
H_8N_4Ge
Germanetetramine